Illustrated 2009 Building Code Handbook

Illustrated 2009 Building Code Handbook

Terry L. Patterson

McGraw Hill

New York Chicago San Francisco
Lisbon London Madrid Mexico City
Milan New Delhi San Juan
Seoul Singapore Sydney Toronto

The McGraw-Hill Companies

Cataloging-in-Publication Data is on file with the Library of Congress

Illustrated 2009 Building Code Handbook

1 2 3 4 5 6 7 8 9 0 DOC/DOC 0 1 4 3 2 1 0 9

ISBN 978-0-07-160618-9
MHID 0-07-160618-1

The pages within this book were printed on acid-free paper.

Sponsoring Editor
Joy Bramble Oehlkers

Acquisitions Coordinator
Michael Mulcahy

Editorial Supervisor
David E. Fogarty

Proofreader
Barbara Danziger

Production Supervisor
Richard C. Ruzycka

Art Director, Cover
Jeff Weeks

To Cary Herndon Wolfard
and Stephen L. Wolfard

ABOUT THE AUTHOR

Terry L. Patterson is a licensed architect and the W. Edwin Bryan, Jr. Professor of Architecture at the University of Oklahoma. He has more than 40 years of professional and teaching experience, and is the author of a dozen books on subjects including Frank Lloyd Wright, the philosophy of building materials, the teaching and practice of architectural technology, and building codes.

Contents

List of Tables

Chapter 21: Masonry

Chapter 26: Plastic

Chapter 29: Plumbing Systems

Chapter 32: Encroachments into the Public Right-of-Way

Preface

Every effort has been made to provide accurate clarifications of the code sections selected. To this end, I attended public hearings and have examined hundreds of proposals for changes that were subsequently approved or disapproved since the first drafts of the code. Proposal reasoning and comments of the technical committees were studied for additional insight to intentions. I attended annual code seminars to better understand code language in general and specific code sections. Every section of the handbook was traced back to its origin in a model code or a change proposal to verify accuracy and intent. Commentaries for earlier editions of the *International Building Code* were studied. Many cited standards such as those for accessibility regulations and fire codes were examined. Every cited reference was examined for content and accuracy. Through this research it has become apparent that in spite of the best intentions of the code sponsors, there will be differences of opinion regarding interpretation. The individual who proposed the original version of a section is not the person providing the official interpretation at the job site. Original intent is easily obscured in the several stages of review, modification, and approval that occur between the first proposal to the final interpretation at the local level. Consequently, the meaning imparted by the actual phrasing in the code dominates original intent.

Many actual building projects are used in the handbook to illustrate how real buildings comply with code requirements. This has been done for two reasons. It is intended to bring a sense of reality to students studying the code who otherwise would find it remote from their life experience. It also provides an opportunity to explore code intent as related to complex building circumstances. Such an explanation is not possible using only imaginary examples having simple rectangular shapes that neatly fit into conditions described in the code.

Terry L. Patterson
Norman, Oklahoma

Acknowledgments

A large work such as this handbook cannot be the result of a single person's efforts. Many people provided important assistance. First, I thank my wife, Jennie M. Patterson, for her significant and lengthy effort on this project. Jennie produced virtually all of the graphics for the examples from architectural firms, all the tables and supporting calculations, the index, and the table of contents. She edited the whole work and developed the raw manuscript into a camera-ready format. She kept the computers running with the necessary maintenance and software management. This handbook would not be possible without Jennie's competent and timely contributions.

I am indebted to David Pendley, AICP, CBO, for his indispensable help to me in understanding the *BOCA National Building Code*, the *Uniform Building Code*, and the *International Building Code*. David is the Chief Building Official for the City of North Richland Hills, Texas. He presently provides code interpretations, technical assistance, and building code administration to architects, engineers, builders, inspectors, and plan examiners, using the ICC International Building Code within the North Richland Hills jurisdiction. He graduated from the University of Nebraska School of Engineering and Technology in Lincoln, Nebraska, with a B.S. degree in Construction Management. David's experience includes that of serving as Chief Plans Examiner for the City of Fort Worth, Texas, and serving as the Building Official for Beatrice, Nebraska, and Norman, Oklahoma. David is an ICC Certified Building Official, an ICC Certified Building Code Official, an ICC Building Plans Examiner, an ICC Building Inspector, an ICC Plumbing Inspector, an ICC Mechanical Inspector, an ICC Commercial Energy Plans Examiner, an ICC Residential Electrical Inspector, a Texas Licensed Plumbing Inspector, and is a Texas E.I.T. David is also certified as a planner by the American Institute of Certified Planners (AICP). David's extensive and competent assistance with the model codes in no way implies responsibility on his part for my interpretations of the *International Building Code* in this handbook. Any errors or misunderstandings are entirely my responsibility as the sole author of the handbook.

Many thanks go to my graduate assistants for their important help with previous editions on which this edition is based. Srdan Kalajdzic was especially helpful in identifying examples from architectural working drawings that illustrate code requirements. He also produced a large number of the generic details for the handbook. Srdan performed his tasks with his usual high level of dependability and competence. Many thanks to Rene Spineto, who produced a large number of generic details and some graphics for the architectural project examples. Thanks to Arvind Vishnu Ram for his high-quality graphic production of numerous generic details. Thanks to Dana A. Templeton for her administrative help and graphic work on the architectural project examples. Thanks to Kevin Zhou for his help with the production of generic details.

Many thanks to Brett W. Johnson for his administrative assistance on the second edition. Thanks goes to A. Morris for his research into the referenced standards in the previous edition of the IBC to identify correct titles and to summarize their contents. Also thanks to Yi Zhao for her graphic work and research into the contact information for the organizations whose standards are referenced in the code. I am also grateful for the continued support for my publishing efforts provided by Charles Graham, Dean of the College of Architecture, and Nick Harm, Director of the Division of Architecture at the University of Oklahoma. I am indebted to the benefactor of the W. Edwin Bryan, Jr., Professorship of Architecture for the financial support provided to this project.

Thanks go to Samuel Ray Moore, Architect of Oklahoma City, and to John C. Womack, AIA, of the School of Architecture, Oklahoma State University, for their help in launching the first edition of this project. And special thanks to Joy Bramble Oehlkers, Senior Editor at McGraw-Hill, for her advice and help in seeing the fourth edition of this handbook through to a successful conclusion and for her continued support of my work.

Thanks are due the following architectural firms who generously permitted me to select examples from their work to illustrate methods of compliance with the code: Alt Breeding Schwarz Architects, LLC, Annapolis, Maryland; Ankrom Moisan Associated Architects, Portland, Oregon; C. Allen Mullins, Architect, Wilkes-Barre, Pennsylvania; Cromwell Architects Engineers, Little Rock, Arkansas; David Woodhouse Architects, LLC, Chicago, Illinois; Gossen Livingston Associates, Inc., Wichita, Kansas; HKS, Inc., Dallas, Texas; HKT Architects, Inc., Somerville, Massachusetts; The Hollis and Miller Group, Inc., Lee's Summit, Missouri; Overland Partners, Inc., San Antonio, Texas; PBK Architects, Inc., Houston, Texas; Perkins Eastman, Architects, New York, New York; Phillips Metsch Sweeney Moore, Architects/Marc A. Phillips, Project Architects, Santa Barbara, California; Spencer Godfrey Architects, Round Rock, Texas; Stephen Wen + Associates, Pasadena, California; Ed Vogt, Architect, Bloomington, Minnesota; Watkins Hamilton Ross Architects, Inc., Houston, Texas; and Wilson Darnell Mann, P.A., Wichita, Kansas.

I thank also the following building officials and other professionals for their helpful responses to my questions for the first edition of the handbook regarding various sections of model codes in their jurisdictions: Gene Abbot, Lakeville, Minnesota; Thomas Anderson, Hopkins, Minnesota; Dick Bower, Soldotna, Alaska; Christopher Caruso, Clinton, Pennsylvania; Kenneth Elsberry, Dallas, Georgia; John Graber, Milwaukee, Wisconsin; Robert Hegner, Northfield, Minnesota; Frank P. Hodge, Jr., Hilton Head Island, South Carolina; Charles M. Huss, R.C.I., Kotzebue, Alaska; Douglas Lalim, Williston, North Dakota; Clayton Larson, Coon Rapids, Minnesota; Duane Lasley, Duluth, Minnesota; John J. Mayo, Lac Du Flambeau, Wisconsin; Rick Murray, Fulton, Missouri; Aslam Rana, Dothan, Alabama; Donald Ranes, Casper, Wyoming; Jerry Ratzlaff, Devil's Lake, North Dakota; James Rich, Hermantown, Minnesota; Bruce J. Spiewak, AIA, West Haven, Connecticut; J. P. Swanson, Monticello, Minnesota; and Murray Ward, Grand Rapids, Minnesota. Their input continues to benefit subsequent editions of the handbook.

Introduction

Purpose

This handbook clarifies the sections of the 2009 *International Building Code* that are the most useful to designers, detailers, estimators, and students. It is not directed to specifiers or engineers. It is not intended to be a substitute for the code, but an aid to understanding it.

The 2009 *International Building Code* is owned by the International Code Council, Inc., of Washington, DC. This handbook is neither sponsored nor approved by this agency, which has no relationship to this project.

Code language

In their analysis of proposed change 1005.1-1 to the "First Draft" of the *International Building Code,* the Means of Egress Technical Subcommittee rejected the language of the proposal as being "commentary, not code text." This single statement succinctly summarizes the problem with codes for many people who must comply with them.

"Code text" is the language of building codes, a pseudo-legal kind of language intended to minimize variations in interpretation and withstand legal challenges. As in legal documents, the penalty for this special style is clarity to people who are not specialists in the language. The difficult language might be justified if interpretations among users and officials were consistent. This is not the case as evident in the various interpretations by code officials published on code discussion websites over the years. Code questions posted on such sites often generate conflicting responses from code officials and other knowledgeable parties.

Building codes have other readability problems. Sentences are often long and convoluted. Some items in the first part of a sentence affect some items in the second and third parts of the sentence but all items are not necessarily affected by all other items. Sorting out the relationships between words is complicated by the fact that some phrases affect previous or subsequent sentences and some do not. Too much substantive content is joined by too few words of clarification. Another problem in reading a code is letting expectations affect interpretation. The logic on which the code is based is not always accessible to the user and does not always reflect the experience of the professional. Statistics, tests, tradition, and other data and trends in life safety on which codes are based may not be available to the average user. In most cases, taking the literal meaning of code statements is more effective than is applying common sense. Since this approach is not 100 percent reliable, however, doubt makes the mental discipline required for understanding even more challenging.

Handbook language

The language of this handbook accommodates the needs of design and production professionals and students. It is one of illustrations, tables, and lists. Common phrasing is substituted for legalistic wording. Lengthy and convoluted code sentences are broken down into line items. Quick and easy readability is the goal.

Format

Drawings and diagrams illustrate numerous requirements. Actual building projects as well as generic examples are included. Tables are provided, many of which are based on mathematical equations that would otherwise require computation by the user. Large code tables are broken down into smaller tables and reformatted to reduce the number of variables that must be reconciled. Footnotes are integrated into the body of each table or the body of the text, which eliminates the fine print that is difficult to read and easily overlooked. Exceptions are integrated into the body of the basic requirements. This eliminates reversals of requirements where exceptions supersede the main text.

Several common-sense shortcuts were taken in the handbook to facilitate readability. First, the handbook refers to the *International Building Code* simply as the code or the IBC. The code consistently modifies references to residential occupancies as follows: "R-3 as applicable in Section 101.2." This indicates that the *International Residential Code* governs 1- and 2-family dwellings and townhouses ≤ 3 stories. By use of this phrase, the code is indicating to which residential occupancies it applies. It is sufficient to understand that the code does not address residences governed by the *International Residential Code*. Consequently, the reference to 101.2 is omitted throughout the handbook. The code refers to sprinklers as being automatic. Since it is understood that all sprinklers are to be automatic, the handbook omits this term.

The code often refers to "buildings and structures" so as not to exclude constructions such as stadiums, which may not be considered buildings. The handbook usually refers only to "buildings," which must be understood to include all the structures that the code governs. The handbook utilizes mathematical and other symbols instead of words to the greatest extent possible so as to provide visual relief to the text. For example, the symbols ≥ and ≤ are substituted, where readability is enhanced, for the terms "minimum" and "maximum." The code reports frequently that certain cases must comply with the code. Such comments are omitted as it must be understood that every entry of the code requires compliance.

The shortcuts and plain language used by this handbook lack the legal precision of the code. The code attempts to provide regulations that cannot be circumvented. This handbook makes selected regulations more accessible to designers, detailers, and estimators. Consequently, common sense must be applied to the guidance provided.

The need to refer to other pages in order to grasp the concept of a code requirement is minimized. Numerical references to other code sections are eliminated from the main text. Descriptions of such referenced data, the data itself, or the subject of the referenced data is substituted. This provides a more easily read text without the disruption of numbers that add no readily apparent meaning to the paragraph. The cited section number along with its name are listed below the body of the requirement text in italics. Comments on the citation are added where necessary for clarification. The reader may turn to the cited section if desired. The following example illustrates the contrast in formats:

Code entry:

"**407.2 Corridors.** Corridors in occupancies in Group I-2 shall be continuous to the exits and separated from other areas in accordance with Section 407.3 except spaces conforming to Sections 407.2.1 through 407.2.4."

Handbook clarification:

407.2 Corridors

- The enclosure of Occupancy I-2 corridors is governed as follows:
  - Each corridor must be continuous to an exit.
  - Corridors may be open to the spaces indicated below where design and construction meet minimum requirements for fire safety:

    Waiting areas.

    Nurses' stations.

    Mental health treatment areas.

    Gift shops.
  - Otherwise, corridors must be separated from other spaces for purposes of smoke protection.

Note: The following are cited as sources of requirements for space opening to a corridor:

407.2.1, "Waiting and similar areas," which addresses waiting rooms.

407.2.2, "Nurses' stations."

407.2.3, "Mental health treatment areas."

407.2.4, "Gift shops."

407.3, "Corridor walls," for walls required to separate corridors from other spaces.

Focus for design

The handbook focuses on code sections affecting design decisions at the schematic stage and design development phases such as in Chapters 3, 4, 5, 10, 11, 12, 30, and 32. Designers are provided with a clarification of requirements affecting floor plan configuration and building massing. Required heights, widths, lengths, clearances, and distances are among the data clarified. These sections are of particular interest to students, as much studio work is schematic in nature.

Focus for detailing

The handbook focuses on code sections affecting detailing decisions in the working drawing phase such as in Chapters 6, 7, 8, 9, 14, 15, 18, 19, 21, 22, 23, 24, 25, 26, and 31. Detailers are provided with a clarification of requirements affecting material choices and detail configuration. Clarification of these sections also helps the designer make spatially related decisions based on probable relative cost of the options as driven by fire protection requirements. These sections are of particular interest to students since they narrow the choices for material selection and detail composition.

Focus for cost estimating

The handbook's focus on code sections affecting detailing also helps estimators prepare construction bids. Where architectural working drawings require that the builder "meet current code requirements," this handbook can provide options for code compliance where certain detailing is vague or missing in the project drawings.

Sections de-emphasized

Material that is solely specification oriented is generally omitted from the handbook. That is, requirements referring to only specifications, tests, procedures, administration, other codes and standards, and paragraphs not related to space planning or detailing are not addressed. Chapters 1, 17, 33, and parts of other chapters are this type. These subjects are typically the responsibility of professionals who are familiar with code language. Specification type data is included in the handbook only where it is mixed with design and detailing information.

Requirements that are engineering oriented are generally omitted. This refers mainly to Chapter 16. Engineers and architects with responsibilities in Chapter 16 typically have the experience to respond directly to code language. Certain loading requirements from Chapter 16 are included in the handbook, as they may be useful to students and production personnel who need to approximate member sizes for detailing purposes.

Within chapters addressed by the handbook, certain paragraphs are omitted that are administrative in nature and contain no technical content. For example, sections are distributed throughout the code that establish the applicability of subsequent sections. For these to be useful, they must be referred to periodically as subsequent sections are studied. This requires turning pages, which interrupts concentration. In lieu of these scope-type paragraphs, the applicability of each section is reported in the handbook within the section itself, where such is not self-evident. Other sections are also omitted where they do not contribute to the needs of designers and detailers. Some of these are scattered and some are grouped. These various omissions result in occasional gaps in section numbering. When a numbered paragraph is selected for clarification, however, every item under the number is addressed.

In order to keep handbook chapter numbering continuous and consistent with the code, a few "place holder" pages are inserted to identify de-emphasized chapters that lack significant material of interest to designers, detailers, and estimators. Chapters 1, 2, 13, 17, 20, 27, 28, and 33 have such pages. For most of these, material of minor interest is included. For example, Chapter 17 deals with testing, a subject not featured in the handbook. Sections on performance labels for materials and assemblies were included on the Chapter 17 "place holder" page. Such label information is of general interest to the detailer and is of more value than would be an empty page. Code Chapters 13 and 28 merely refer to other codes with no further information, so the "place holder" pages for these chapters are correspondingly brief.

Code errors

This handbook is based on the first printing of the 2009 *International Building Code,* which contains errors. Included are common typographical errors or obsolete section reference numbers. Where these were discovered corrected data was included in the handbook without comment. It is anticipated that errors in the code will be corrected by errata and/or in a future supplement issued by the International Code Council.

Punctuation and capitalization in section titles, table references, and standards titles are listed exactly as shown in the code. These do not always match the format recommended by the manual of style used for this handbook. Consequently, there is some inconsistency in the handbook in this regard.

Illustrated 2009 Building Code Handbook

1

Scope and Administration

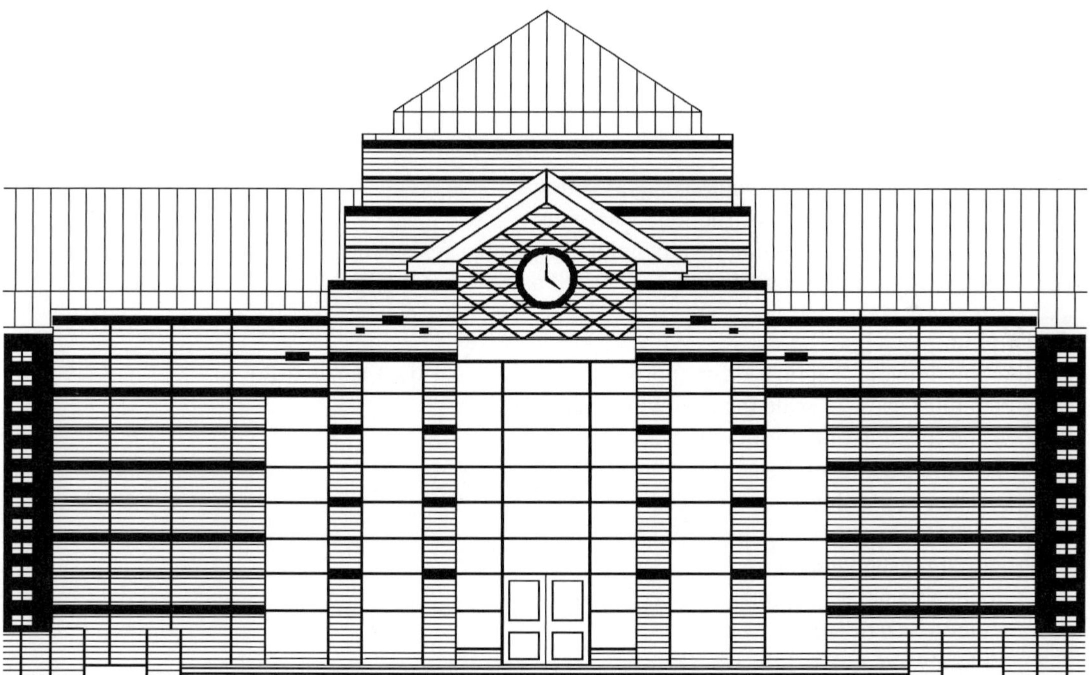

High School 6, Cypress-Fairbanks Independent School District.
Harris County, Texas. *(partial elevation)*
PBK Architects, Inc. Houston, Texas.

107 Submittal Documents

107.2.3 Means of egress

- Construction documents must show all parts of the means of egress as follows:
  - Location.
  - Construction.
  - Size.
- For occupancies other than R-2 and R-3, the following is required:
  - Construction documents must show the number of occupants:
    - On every floor.
    - In all rooms and spaces.

107.2.4 Exterior wall envelope

- Construction documents must describe the exterior wall envelope as follows:
  - Information must be adequate to verify code compliance, including:
    - Wall intersections with dissimilar materials.
    - Wall intersections with the roof.
    - Wall intersections with eaves.
    - Wall intersections with parapets.
    - Means of drainage.
    - Water-resistive membrane.
    - Details around openings.
    - Flashing.
    - Corners.
    - End details.
    - Control joints.
- Construction documents must include data as follows:
  - Manufacturers' installation instructions verifying:
    - That penetration and opening details maintain weather resistance of the exterior wall envelope.
  - Description of the exterior wall system as tested and the test method.

107.2.5 Site plan

- Construction drawings submitted for approval must include a site plan as follows:
  - Site plan must show to scale the following information:
    - Size and location of the following:
      - New and existing construction.
      - Distances to lot lines.
    - Established street grades.
    - Proposed finished grades:
      - Applicable flood-related aspects as follows:
        - Flood hazard areas.
        - Floodways.
        - Design flood elevations.
  - Site plan must be prepared in accordance with a boundary line survey.
- Where demolition will occur, the site plan must show the following:
  - Construction to be demolished and to remain.
- The building official may waive the site plan requirement as follows:
  - For alteration or repair or where otherwise warranted.

2

Definitions

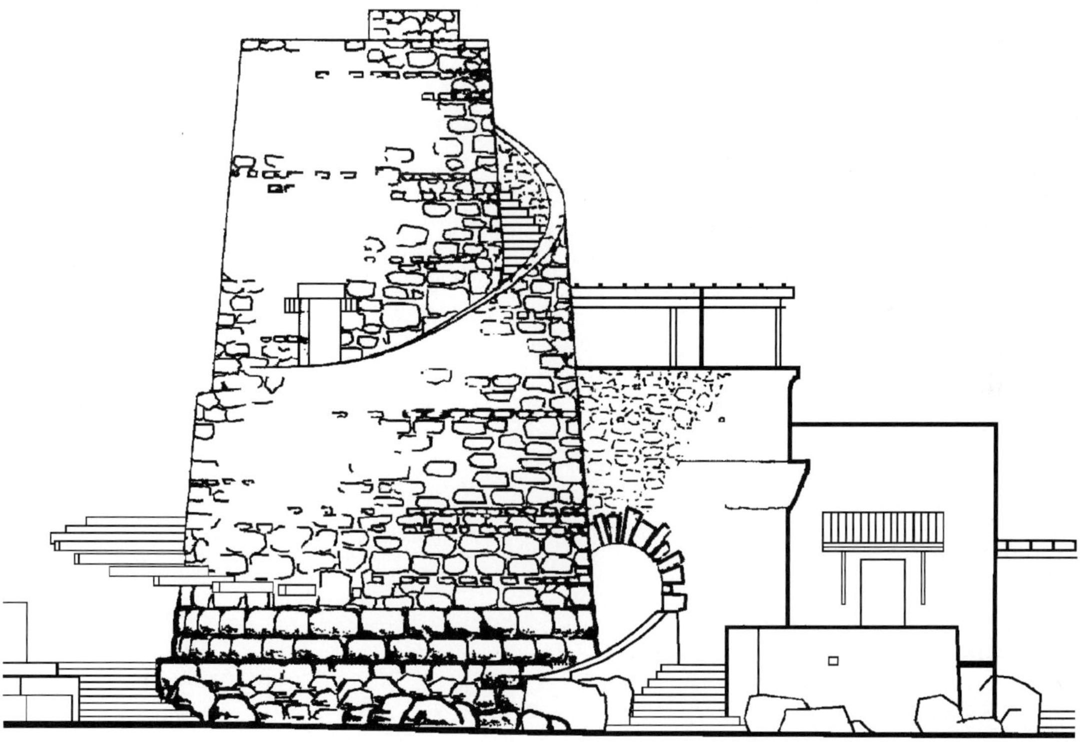

Lady Bird Johnson Wildflower Center. Austin, Texas. *(partial elevation)*
Overland Partners, Inc. San Antonio, Texas.

202 Definitions

- **Court**
  - An uncovered space open to the sky.
  - Enclosed on ≥ 3 sides by any of the following:
    - Exterior building walls.
    - Other enclosing elements.
- **Dwelling**
  - A building containing either of the following:
    - 1 dwelling unit.
    - 2 dwelling units.
  - To be occupied for living purposes by any of the following means:
    - Owned.
    - Rented.
    - Leased.
    - Let.
    - Hired out.
- **Grade floor opening**
  - One of the following:
    - Window.
    - Other opening.
  - Sill height is ≤ 44" from adjacent grade as follows:
    - Above finished grade.
    - Below finished grade.
- **Habitable space**
  - A space in a building for the following:
    - Sleeping.
    - Eating.
    - Cooking.
  - Does not include the following:
    - Bathrooms.
    - Toilet rooms.
    - Closets.
    - Halls.
    - Storage spaces.
    - Utility spaces.
    - Similar spaces.
- **Occupiable space**
  - A room or enclosed space.
  - Designed for human occupancy.
  - Where people congregate for the following:
    - Amusement.
    - Education.
    - Similar purposes.
    - Labor.
  - Has means of egress.
  - Has lighting.
  - Has ventilation.

202 Definitions

- **Primary structural frame**
  - Columns.
  - Structural members connected to columns as follows:
    - Girders.
    - Beams.
    - Trusses.
    - Spandrels.
  - Floor construction connected to columns.
  - Roof construction connected to columns.
  - Bracing necessary to vertical stability of the primary frame as follows:
    - In response to gravity loads on the primary frame.
    - Where bracing carries gravity loads.
    - Where bracing does not carry gravity loads.
- **Reflective plastic core insulation**
  - Roll insulation
  - < 0.5" thick
  - ≥ 1 exterior surface with an emittance ≤ 0.1.
  - Core material with the following cells or voids.
- **Secondary members**
  - Structural members not connected to columns.
  - Floor construction not connected to columns.
  - Bracing not part of the primary structural frame.
- **Site-fabricated stretch system**
  - Fabricated on the site of final installation.
  - Serves any or all of the following purposes:
    - Acoustical modification of the environment.
    - Tackable surface.
    - Aesthetic.
  - Includes the following elements:
    - Frame of one of the following to support fabric:
      - Plastic.
      - Wood.
      - Metal.
      - Other material.
    - Infill material with properties serving the application.
    - Surface material as follows:
      - Stretched tight and secured to the frame by one of the following:
        - Tension.
        - Mechanical fasteners.
      - One of the following materials:
        - Textile.
        - Fabric.
        - Vinyl.

Note: 802.1, "General," is cited as governing this definition which is summarized above.

202 Definitions

- **Skylight, unit**
  - Factory-assembled glazed fenestration unit.
  - Contains one panel of glazing as follows:
    Transmits natural light.
  - Penetrates the roof assembly as follows:
    Provides a weather barrier.
- **Skylights and sloped glazing**
  - Any of the following:
    Glass.
    Transparent glazing material.
    Translucent glazing material.
  - Installed at a slope ≥ 15° from vertical.
  - The following glazing is included:
    In skylights.
    In solariums.
    In sun spaces.
    In roofs.
    In sloped walls.
- **Sleeping unit**
  - A space where people sleep.
  - Can include permanent living facilities.
  - Can include permanent eating facilities.
  - Can include only one of the following:
    Sanitation facilities.
    Kitchen facilities.
  - Not part of a dwelling unit.
- **Story**
  - The segment of a building between the following levels:
    Upper surface of a floor.
    Upper surface of the floor or roof directly above.
  - For floors other than the top floor, a story is measured in one of the following ways:
    From top to top of successive tiers of beams.
    From top to top of successive tiers of finished floor surfaces.
  - For the top floor, a story is measured as follows:
    From top of finished floor to top of ceiling joists where there is a ceiling.
    From top of finished floor to top of roof rafters where there is no ceiling.

  Note: The following are cited as relating to this definition:
  "Basement" definition.
  "Mezzanine" definition.
  502.1, "Definitions."
- **Story above grade plane**
  - A story with any of the following conditions:
    The finished floor is all above the grade plane.
    The floor above is > 6' above grade plane.
    The floor above is > 12' above any point on the finished grade around the perimeter.

3

Use and Occupancy Classification

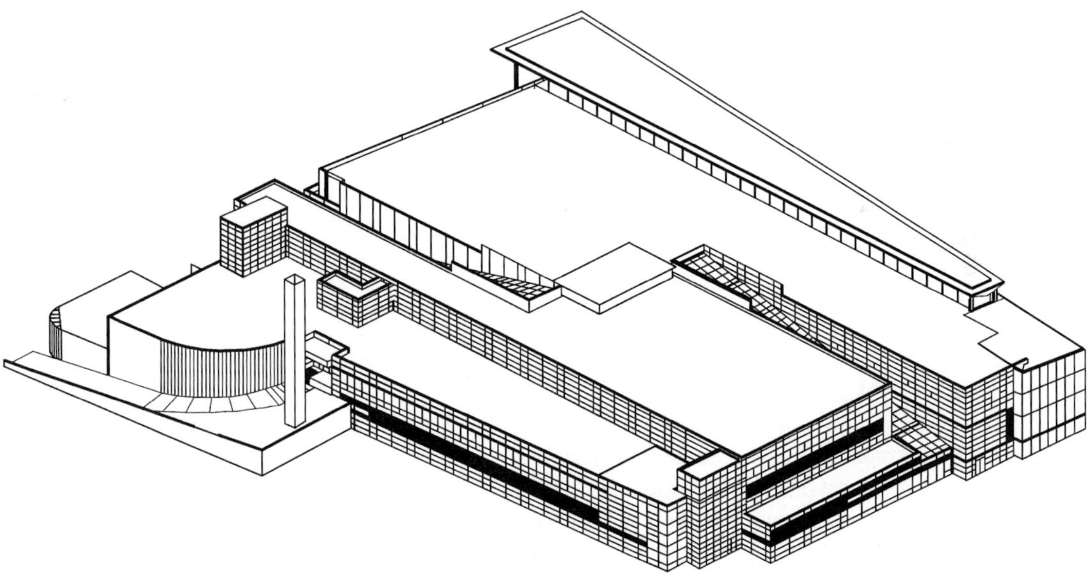

University of Connecticut New Downtown Campus at Stamford, Connecticut.
Perkins Eastman Architects, P.C. New York, New York.

302 Classification

302.1 General

- Buildings or parts of buildings are classified as one or more of the following occupancy categories:
  - A—Assembly:
    - A-1.
    - A-2.
    - A-3.
    - A-4.
    - A-5.
  - B—Business.
  - E—Educational.
  - F—Factory and Industrial:
    - F-1.
    - F-2.
  - H—High Hazard:
    - H-1.
    - H-2.
    - H-3.
    - H-4.
    - H-5.
  - I—Institutional:
    - I-1.
    - I-2.
    - I-3.
    - I-4.
  - M—Mercantile.
  - R—Residential:
    - R-1.
    - R-2.
    - R-3.
    - R-4.
  - S—Storage:
    - S-1.
    - S-2.
  - U—Utility and Miscellaneous.
- Buildings not listed in one of the occupancy categories by the code are to be assigned to the category with the most similar fire and other hazards.
- A space with changeable occupancies must comply with the rules of all occupancies involved.
- Buildings with ≥ 1 occupancy must comply with mixed occupancy requirements.

Note: Section 508, "Mixed Use and Occupancy," is cited as governing mixed occupancies.

303 Assembly Group A

303.1 Assembly Group A *(part 1 of 2)*

- Buildings or parts of buildings where people gather for activities such as the following:
  - Civic.
  - Religious functions.
  - Waiting for transportation.
  - Social.
  - Recreation.
  - Consumption of food or drink.
- A building is classified as Occupancy B where both of the following apply:
  - People assemble there.
  - Occupant load is < 50.
- The following space is classified as the occupancy noted below:
  - Space with all the following characteristics:
    - Where people assemble.
    - Occupant load is < 50.
    - Accessory to another occupancy.
  - Occupancy designation is either of the following:
    - Occupancy B.
    - Same occupancy as the space to which it is accessory.
- The following space is classified as the occupancy noted below:
  - Space with all of the following characteristics:
    - Where people assemble.
    - Area is < 750 sf.
    - Accessory to another occupancy.
  - Occupancy designation is either of the following:
    - Occupancy B.
    - Same occupancy as the space to which it is accessory.
- Both of the following apply to assembly spaces accessory to E occupancies :
  - They are regulated by E occupancy requirements.
  - They must meet accessibility requirements for assembly occupancies.

  Note: Chapter 11, "Accessibility," is cited as having accessibility requirements for assembly spaces.

- The following are not separate occupancies from the religious facility served:
  - Accessory religious education spaces with occupant loads < 100.
  - Religious auditoriums with occupant loads < 100.

- Assembly spaces are divided into the following designations:
  - A-1:
    - For the production and viewing of any of the following:
      - Motion pictures.
      - Performing arts.
    - Includes the following building types among others:

Movie theaters	Television studios with audience seating
Radio studios with audience seating	Theaters
Symphony and concert halls	Typically with fixed seating

303 Assembly Group A

303.1 Assembly Group A *(part 2 of 2)*

- A-2:
  - For the consumption of food and/or drink.
  - Includes the following building types among others:
    - Banquet halls.
    - Nightclubs.
    - Restaurants.
    - Taverns and bars.
- A-3:
  - Spaces for the following functions:
    - Amusement.
    - Gatherings not assigned to other assembly categories.
    - Recreation.
    - Worship.
  - Includes the following building types among others:
    - Amusement arcades
    - Art galleries
    - Bowling alleys
    - Community halls
    - Courtrooms
    - Dance halls, no food or drink
    - Exhibition halls
    - Funeral parlors
    - Gymnasiums, no spectator seating
    - Indoor swimming pools, no spectator seating
    - Indoor tennis courts, no spectator seating
    - Lecture halls
    - Libraries
    - Museums
    - Passenger waiting areas
    - Places of religious worship
    - Pool and billiard parlors
- A-4:
  - Spaces for viewing indoor sporting activities as follows:
    - With spectator seating.
  - Includes the following building types among others:
    - Arenas.
    - Skating rinks.
    - Swimming pools.
    - Tennis courts.
- A-5:
  - Spaces for the following functions:
    - Participating in outdoor activities.
    - Viewing outdoor activities.
  - Includes the following building types among others:
    - Amusement park structures.
    - Bleachers.
    - Grandstands.
    - Stadiums.

304 Business Group B

304.1 Business Group B

- Includes buildings or parts of buildings used for the following:
  - Offices.
  - Professional transactions.
  - Service transactions.
  - Storage of records.
  - Storage of accounts.
- Includes the following building types among others:
  - Airport traffic control towers.
  - Ambulatory health care facilities.
  - Animal hospital.
  - Animal kennel.
  - Animal pound.
  - Architect's office.
  - Attorney's office.
  - Bank.
  - Barber shop.
  - Beauty shop.
  - Car wash.
  - Civic administration.
  - Clinic, outpatient.
  - Dentist.
  - Dry cleaning pickup/drop off.
  - Dry cleaning, self-service.
  - Educational functions above 12th grade.
  - Electronic data processing.
  - Engineer's office.
  - Laboratories, testing and research.
  - Laundry pickup/drop off.
  - Laundry, self-service.
  - Motor vehicle showroom.
  - Physician's office.
  - Print shop.
  - Post office.
  - Professional services.
  - Radio station.
  - TV station.
  - Telephone exchange.
  - Training outside a school or academic program.

304 Business Group B

Case study: Fig. 304.1. Because people gather in the conference room of the architect's office, it must be determined whether or not the space is to be designated as a business occupancy like the office in general or as an assembly occupancy. This is done by computing the number of occupants based on the use of the space, which is assembly in nature. According to IBC Table 1004.1.1, 15 sf per occupant are assigned for an assembly use having tables and chairs. This yields an occupant load of 19 for the room. 303.1 indicates that an accessory gathering space with < 50 occupants is considered to be same occupancy as that served. The conference room is designated, therefore, as Occupancy B and must comply with means of egress requirements for a business.

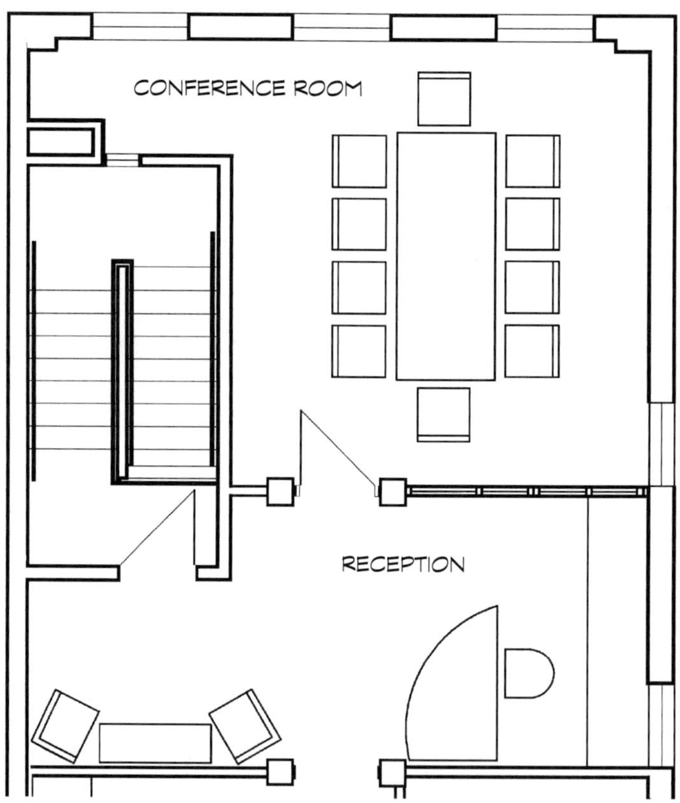

Fig. 304.1. Partial floor plan. Alterations to 209 Main Street, Annapolis, Maryland. Alt Breeding Schwarz Architects, LLC. Annapolis, Maryland.

305 Educational Group E

305.1 Educational Group E

- Includes buildings or parts of buildings used as follows:
  - Buildings with both the following characteristics:
    - Serving 12th grade and lower.
    - Serving $\geq$ 6 people at one time.
- Does not include the following spaces which are designated as Occupancy A-3:
  - Where all the following apply:
    - The space has one of the following uses:
      - Religious education room.
      - Religious auditorium.
    - The space is accessory to a place of religious worship.
    - The space has < 100 occupants.

Note: 303.1, "Assembly Group A," is cited as defining accessory spaces.

305.2 Day care

- Occupancy E includes buildings or parts of buildings serving the following purposes with all of the characteristics listed below:
  - Purposes:
    - Educational.
    - Supervisory.
    - Personal care.
  - Characteristics of services:
    - Serving > 5 children.
    - Serving children > 2 1/2 years old.

306 Factory Group F

306.1 Factory Industrial Group F

- Includes buildings or parts of buildings used for any of the following functions, which are not classified as Occupancy H or S:
  - Assembling.
  - Disassembling.
  - Fabricating.
  - Finishing.
  - Manufacturing.
  - Packaging.
  - Repair.
  - Processing operations.

306.2 Factory Industrial F-1 Moderate-hazard Occupancy

- Includes factory and industrial functions not classified as F-2: Low Hazard including the following:

  Agricultural machinery
  Aircraft manufacturing (no repair)
  Appliances
  Athletic equipment
  Automobiles
  Bakeries
  Beverages > 16% alcohol
  Bicycles
  Boats
  Brooms
  Brushes
  Cameras
  Canvas
  Carpets
  Carpet cleaning
  Clothing
  Construction machinery
  Detergents
  Disinfectants
  Door millwork
  Dry cleaning
  Dyeing
  Electric generation plants
  Electronics
  Engine rebuilding
  Engines
  Food processing
  Fabric similar to canvas
  Furniture
  Hemp products
  Jute products
  Laundries
  Leather products
  Machinery
  Metals
  Motion picture filming (no spectators)
  Motor vehicles
  Musical instruments
  Optical goods
  Paper mills
  Paper products
  Photographic equipment
  Photographic film
  Plastic products
  Printing
  Publishing
  Recreational vehicles
  Refuse incineration
  Rug cleaning
  Rugs
  Sash millwork
  Shoes
  Soaps
  Television filming (no spectators)
  Textiles
  Tobacco
  Trailers
  Upholstering
  Wood; distillation
  Woodworking (cabinet)

306 Factory Group F

306.3 Factory Industrial F-2 Low-hazard Occupancy

- Includes factory and industrial functions as follows:
  - Use of noncombustible materials in the following:
    - Manufacturing.
    - Fabrication.
  - Does not cause a significant fire hazard in the following:
    - Processing.
    - Finishing.
    - Packing.
  - Includes product types and processes similar to the following:
    - Beverages ≤ 16% alcohol.
    - Brick and masonry.
    - Ceramic products.
    - Foundries.
    - Glass products.
    - Gypsum.
    - Ice.
    - Metal products fabrication.
    - Metal product assembly.

307 High-Hazard Group H

307.1 High-hazard Group H *(part 1 of 3)*

- Includes buildings or building areas containing materials as follows:
  - That constitute either of the following hazards:
    - Health hazard.
    - Physical hazard.
  - That are involved in any of the following:
    - Manufacturing.
    - Processing.
    - Generation of materials.
    - Storage.
  - That are in quantities > than those permitted in control areas that are not in a high-hazard occupancy.

  Note: The following tables are cited as defining quantities of hazardous materials permitted in control areas that are not in a high-hazard occupancy:
  IBC Table 307.1(1), "Maximum Allowable Quantity per Control Area of Hazardous Materials Posing a Physical Hazard."
  IBC Table 307.1(2), "Maximum Allowable Quantity per Control Area of Hazardous Material Posing a Health Hazard."
  Section 414, "Hazardous Materials," is cited as governing the construction and location of these control areas.

- Includes the following occupancies:
  - H-1, H-2, H-3, H-4, H-5.

  Note: The following sources are cited as governing hazardous materials:
  This section.
  Section 415, "Groups H-1, H-2, H-3, H-4 and H-5."
  International Fire Code.

- The following are not classified as H but as the occupancy they are most similar to:
  - Buildings housing the application of flammable finishes.

  Note: Section 416, "Application of Flammable Finishes," is cited as governing these buildings.
  The International Fire Code *is cited as governing these buildings.*

  - Occupancy M buildings for wholesale or retail functions as follows:
    - Storage and sales of the following:
      - Flammable or combustible liquids.

  Note: The International Fire Code *is cited as governing these buildings.*

  - Closed piping containing the following:
    - Liquids or gases as follows:
      - Flammable or combustible.
      - For operating machinery or equipment.
  - Cleaning operations using liquid solvents with the following property:
    - Flash point ≥ 200° F.
  - Liquor establishments with no bulk storage as follows:
    - Distributors and stores.
  - Refrigeration systems.

307 High-Hazard Group H

307.1 High-hazard Group H *(part 2 of 3)*

- ○ Cleaning operations using combustible liquid solvents with all the following conditions:
  - Flash point ≥ 140° F.
  - Contained in closed systems.
  - In equipment listed by an approved testing agency.
  - In areas separated from the rest of the building by one or both of the following:
    - 1-hr fire barriers and/or 1-hr horizontal assemblies.

Note: Section 707, "Fire Barriers," is cited as governing these elements.
Section 712, "Horizontal Assemblies," is cited as governing these elements.

- ○ Agricultural materials used on the property where stored.
- ○ Stationary batteries meeting all the following conditions:
  - Where used for any of the following:
    - Facility emergency power.
    - Uninterruptible power supply.
    - Telecommunication facilities.
  - Where batteries comply with all the following:
    - Batteries must have safety venting caps.
    - Ventilation is provided.

Note: The International Mechanical Code *is cited as governing battery ventilation requirements.*

- ○ Buildings without corrosive materials as follows:
  - The following materials may be included in these buildings:
    - Retail displays in original packaging as follows:
      - Personal or household products.
    - Storage of common building materials.
- ○ Buildings used for aerosol storage as follows:
  - Such buildings are classified as S-1.

Note: The International Fire Code *is cited as governing these buildings.*

- ○ Buildings for display and storage of the following materials in the conditions listed below:
  - Materials:
    - Nonflammable solids and liquids.
    - Noncombustible liquids.
  - Conditions:
    - Quantities are ≤ limits for control areas in M or S occupancies.

Note: 414.2.5, "Hazardous material in Group M display and storage areas and in Group S storage areas," is cited as governing these materials.

- • Buildings storing materials as follows:
  - ○ Industrial explosives in the following occupancies:
    - B, F, M, S.
  - ○ The following materials in the M and R-3 occupancies:
    - Black powder Small arms primers Smokeless propellant

Note: The International Fire Code *is cited as governing the explosive materials.*

307 High-Hazard Group H

307.1 High-hazard Group H *(part 3 of 3)*

- Hazardous materials stored on roofs or canopies are classified outdoor storage.
- Hazardous materials used on roofs or canopies are classified as outdoor use.

Note: The International Fire Code *is cited as governing the above storage or use.*

307.3 High-hazard Group H-1 *(part 1 of 2)*

- Buildings containing materials with risk of explosion are classified as H-1 as follows:
  - Division 1.1:
    Risk of near simultaneous explosion of all material.
  - Division 1.2:
    No risk of near simultaneous explosion of all material.
    Risk of projectiles upon explosion.
  - Division 1.3:
    No risk of near simultaneous explosion of all material.
    Risk of limited explosion.
    Risk of limited projectiles upon explosion.
    Risk of fire.
    Does not include the following materials which are classified as H-2:
    Materials with either of the following conditions:
    Confinement of the substance prevents a near simultaneous explosion of all material due to fire.
    Configuration of the substance prevents a near simultaneous explosion of all material due to fire.
  - Division 1.4:
    Risk of limited explosion.
    No risk of significant projectiles upon explosion.
    Fire does not result in near simultaneous explosion of all material.
    Does not include the following materials which are classified as H-3:
    Materials not regulated as explosives by the Bureau of Alcohol, Tobacco, Firearms and Explosives.
    Materials used in processes that do not produce either of the following:
    Explosion of other products.
    Explosive-like combustion of other products.
  - Division 1.5:
    Potential for near simultaneous explosion of all material.
    Minimal risk for near simultaneous explosion of all material.
    Explosives have very low sensitivity.
  - Division 1.6:
    No risk of near simultaneous explosion of all material.
    Virtually no risk of accidental explosion.
    Explosives have extremely low sensitivity.
  - Organic peroxides, unclassified detonatable:
    Derivative of hydrogen peroxide.
    High risk of explosion.
  - Detonatable pyrophoric materials:
    Substances which can spontaneously combust at low temperatures.

307 High-Hazard Group H

307.3 High-hazard Group H-1 *(part 2 of 2)*

- Oxidizers, Class 4:
    - Substances which yield oxygen in a fire.
    - Risk of spontaneous combustion.
    - Risk of explosion with any of the following:
        - Contamination.
        - Thermal shock.
        - Physical shock.
- Unstable (reactive) materials of the two classes listed below:
    - Class 3 detonatable:
        - Not an explosive.
        - Risk of explosive reaction when subjected to the following:
            - Strong initiating source.
            - Thermal or physical shock at high temperatures and pressures.
    - Class 4:
        - Not an explosive.
        - Risk of explosive when subjected to the following:
            - Thermal or physical shock at normal temperatures and pressures.

307.4 High-hazard Group H-2 *(part 1 of 2)*

- Buildings containing materials subject to the following risks:
    - Accelerated burning.
    - Extremely rapid oxidation:
        - Potential for explosion if contained.
- Two types of substances and their storage methods classified as H-2:
    - Liquids:
        - Flammable:
            - Class I: Flash point < 100° F.
        - Combustible:
            - Class II: Flash point ≥ 100° F, < 140° F.
            - Class III-A: Flash point ≥ 140° F, < 200° F.
    - Oxidizers, Class 3:
        - Substances which yield oxygen in a fire.
        - Causes high increase in burning rate of other materials.
        - Undergoes vigorous decomposition when subjected to the following:
            - Contamination.
            - Heat.
    - Storage methods:
        - In open systems.
        - Closed systems > 15 psi.
- Selected examples of other H-2 substances:
    - Combustible dust.
    - Flammable cryogenic liquids:
        - Liquids with boiling point < 150° F.
    - Flammable gases:
        - Compressed gas that burns in air.

307 High-Hazard Group H

307.4 High-hazard Group H-2 *(part 2 of 2)*

- Organic peroxides, Class I:
  - Derivative of hydrogen peroxide.
  - Burns very rapidly.
- Pyrophoric materials:
  - Risk of spontaneous combustion at low temperatures.
- Class 3 water-reactive materials:
  - Reacts explosively to water without confinement or heat.
- Class 3, nondetonable unstable (reactive) materials:
  - Not an explosive.
  - Vigorous chemical reaction when subjected to the following:
    - Strong initiating source.
    - Thermal or physical shock at high temperatures and pressures.

307.5 High-hazard Group H-3 *(part 1 of 2)*

- Buildings containing materials with any of the following characteristics:
  - Readily support combustion.
  - Risk of explosion.
  - Risk of extremely rapid oxidation:
    - Potential for explosion if contained.
- Categories of H-3 substances include but are not limited to the following:
  - 2 types of liquids contained as indicated below:
    - Liquids:
      - Flammable:
        - Class I: Flash point < 100° F.
      - Combustible:
        - Class II: Flash point ≥ 100° F and < 140° F.
        - Class III-A: Flash point ≥ 140° F and < 200° F.
    - Contained as follows:
      - In closed systems.
      - @ < 15 psi.
  - Combustible fibers.
  - Oxidizing cryogenic fluids:
    - Liquids with boiling point < 150° F.
  - 2 classes of oxidizers:
    - Substances which yield oxygen in a fire.
    - Class 2:
      - Can cause moderate increase in burning rate of other materials.
      - Can cause spontaneous combustion of other materials.
    - Class 3:
      - Where contained in closed systems < 15 psi.
      - Causes high increase in burning rate of other materials.
      - Undergoes vigorous decomposition when subjected to the following:
        - Contamination.
        - Heat.

307 High-Hazard Group H

307.5 High-hazard Group H-3 *(part 2 of 2)*

- Consumer fireworks:
  - Division 1.4G substances as follows:
    - No risk of near simultaneous explosion of all materials due to the following:
      - Detonation of individual products.
      - Fire.
    - No significant projectile risk.
    - Can include products not considered explosives by the following:
      - Bureau of Alcohol, Tobacco, Firearms and Explosives.
- Flammable solids:
  - Blasting products are not included.
  - Explosives are not included.
  - Solids that readily ignite.
  - Solids that burn rapidly, vigorously, and persistently.
- 2 classes of organic peroxides:
  - Derivatives of hydrogen peroxide.
  - Class II:
    - Burns rapidly.
    - Severe risk of vigorous chemical reaction due to the following:
      - Heat.
      - Friction.
      - Shock.
  - Class III:
    - Burns rapidly.
    - Moderate risk of vigorous chemical reaction due to the following:
      - Heat.
      - Friction.
      - Shock.
- Oxidizing gases:
  - Sometimes yield oxygen.
  - Enhance burning of other materials.
- Tire storage with all the following characteristics:
  - $\geq$ 10,000 tires.
  - Passenger vehicle tires:
    - Average size.
    - Tire weight = about 25 lbs.
- Class 2 unstable (reactive) materials:
  - Not an explosive.
  - Subject to any of the following:
    - Rapid chemical reaction in the following conditions:
      - Normal temperatures and pressures.
    - Vigorous chemical reaction in the following conditions:
      - High temperatures and pressures.
- Class 2 water-reactive materials:
  - Contact with water can yield the following:
    - A potentially explosive substance.

307 High-Hazard Group H

307.6 High-hazard Group H-4

- Buildings containing health hazards among the following:
  - Corrosive substances:
    - Chemicals that cause any of the following upon contact with living tissue:
      - Visible destruction.
      - Irreversible alterations.
  - Toxic substances:
    - Lethal chemicals.
  - Highly toxic substances:
    - Chemicals that are lethal in smaller dose than are toxic substances.

307.7 High-hazard Group H-5 structures

- Includes buildings with either of the following processes using substances listed below:
  - Processes:
    - Semiconductor fabrication.
    - Semiconductor research and development.
  - Substances:
    - Hazardous production materials (HPM) as follows:
      - Where HPM are present in quantities greater than certain code limits.
      - Examples include any of the following among others:
        - Flammable substances:
          - Substances that ignite easily and burn readily.
        - Combustible liquids:
          - Liquids that burn but do not ignite as easily as flammable liquids.
        - Corrosive liquids:
          - Liquids that damage living tissue on contact.
          - Damage to nonliving materials is not required for a liquid to be corrosive.
        - Oxidizing substances:
          - Substances that enhance burning by the release of oxygen.
        - Organic peroxides:
          - Substances that have a risk of fire or explosion.
        - Toxic materials:
          - Substances that are poisonous to humans.

Note: The following are cited as sources of limits above which HPM result in an H-5 classification:

IBC Table 307.1(1), "Maximum Allowable Quantity per Control Area of Hazardous Materials Posing a Physical Hazard."

IBC Table 307.1(2), "Maximum Allowable Quantity per Control Area of Hazardous Materials Posing a Health Hazard."

415.8, "Group H-5," is cited as the source of requirements for H-5 structures.

308 Institutional Group I

308.1 Institutional Group I

- Includes buildings or parts of buildings where occupants are physically limited as follows:
  - Occupants live in a supervised environment.
  - Occupants receive the following due to health or age:
    Medical treatment.
    Other care.
    Other treatment.
  - Occupants are housed for penal or correctional purposes as follows:
    Occupants are detained.
    Liberty of occupants is restricted.

308.2 Group I-1

- Includes buildings or parts of buildings wherein people live in a residential environment as follows:
  - >16 occupants are housed for reasons related to the following:
    Age.
    Mental disability.
    Other reasons.
  - Occupants can respond to an emergency without assistance.
  - Personal care services are provided as follows:
    Occupants are supervised.
    Services are provided 24 hrs/day.
  - The following facility types among others in this category:
    Residential board and care facilities.
    Assisted living facilities.
    Halfway houses.
    Group homes.
    Congregate care facilities.
    Social rehabilitation facilities.
    Alcohol and drug centers.
    Convalescent facilities.
- The following facilities that are otherwise similar to those above are not included:
  - Facilities with ≤ 5 people are governed by one of the following:
    Occupancy R-3 requirements.
    International Residential Code requirements.
  - Facilities with ≥ 6 people and ≤ 16 people are classified as Occupancy R-4.

308 Institutional Group I

308.3 Group I-2

- Includes buildings or parts of buildings used for medical-related purposes as follows:
  - Functions:
    - Medical.
    - Surgical.
    - Psychiatric.
    - Nursing.
    - Custodial care.
  - Characteristics:
    - Clients are not capable of self-preservation.
  - Facility types include the following:
    - Child care facilities.
    - Detoxification facilities.
    - Hospitals.
    - Mental hospitals.
    - Nursing homes.

308.3.1 Definitions

- **Child care facilities**
  - Care is provided 24 hrs/day.
  - > 5 children are served.
  - Children are $\leq 2\frac{1}{2}$ years old.
- **Detoxification facilities**
  - Serves substance abusers.
  - Care is provided 24 hrs/day.
  - Patients are one of the following types:
    - Incapable of self-preservation.
    - Harmful to themselves or others.
- **Hospitals and mental hospitals**
  - A building or part of a building used as follows:
    - Provides inpatients incapable of self-preservation with the following types of care:
      - Medical.
      - Psychiatric.
      - Obstetrical.
      - Surgical.
- **Nursing homes**
  - Long-term care facility.
  - Care is provided 24 hrs/day.
  - > 5 people are served.
  - Any patients are not capable of self-preservation.
  - Includes the following:
    - Intermediate care facilities.
    - Skilled nursing facilities.

308 Institutional Group I

308.4 Group I-3

- Includes buildings wherein occupants are restrained for security purposes as follows:
  - > 5 restrained occupants.
  - Occupants are not capable of self-preservation due to security measures.
  - Facility types among others include the following:

Correctional centers	Jails	Prisons
Detention centers	Prerelease centers	Reformatories

  - Occupancy I-3 is subdivided into 5 conditions as follows:
    Degree of restraint is increased with each higher condition number.

Note: The following are cited as having requirements applicable to this section.
408.1, "General."
308.4.1, "Condition 1."
308.4.2, "Condition 2."
308.4.3, "Condition 3."
308.4.4, "Condition 4."
308.4.5, "Condition 5."

308.4.1 Condition 1

- This section addresses the least restrained condition of Occupancy I-3.
- Detainees may move freely without restraint to the exterior from the following areas:
  - Sleeping areas.
  - Other spaces occupied by detainees.
- Such buildings may be constructed as Occupancy R.

308.4.2 Condition 2

- This section addresses the Occupancy I-3 condition of restraint next higher above condition 1.
- Detainees may move freely between smoke compartments.
- Locked exits prevent free egress to the exterior.

308.4.3 Condition 3

- This section addresses the Occupancy I-3 condition of restraint next higher above condition 2.
- Detainees may move freely within a smoke compartment such as the following:
  - A residential unit containing the following:
    Individual sleeping units.
    Group activity spaces.
- Egress between smoke compartments is controlled by remote-controlled locks.

308.4.4 Condition 4

- This section addresses the Occupancy I-3 condition of restraint at the next higher level above that of condition 3.
- Detainee movement between spaces is restricted by remote-controlled locks as follows:
  - From the following within a smoke compartment:
    Sleeping units.
    Activity spaces.
    Other spaces occupied by detainees.
  - Between smoke compartments.

308 Institutional Group I

308.4.5 Condition 5

- This section addresses the most restrained condition of Occupancy I-3.
- Movement between occupied spaces is restricted by manual-release locks as follows:
  - From the following within a smoke compartment:
    - Sleeping units.
    - Activity spaces.
    - Other spaces occupied by detainees.
  - Between smoke compartments.

308.5 Group I-4, day care facilities

- Care during religious functions at places of worship is not governed by this section.
- Buildings wherein persons receive custodial care as follows:
  - Occupants served may be of any age.
  - Care is < 24 hrs/day.
  - Care is by individuals other than the following:
    - Parents.
    - Guardians.
    - Relatives by blood.
    - Relatives by marriage.
    - Relatives by adoption.
  - Care is at a location other than the home of person receiving care.
- Similar facilities caring for ≤ 5 persons are addressed in one of the following ways:
  - Designated as R-3.
  - Governed by the *International Residential Code*.

308.5.1 Adult care facility

- A facility providing supervision and personal care for adults as follows:
  - Care is provided < 24 hrs/day.
  - Care is provided for > 5 adults.
  - Adults are unrelated.
- Such facilities are classified as follows:
  - I-4 where occupants require help to respond to an emergency.
  - R-3 where occupants do not require help to respond to an emergency.

308.5.2 Child care facility

- Day care service is classified as Occupancy I-4 where the following apply:
  - Care is provided < 24 hrs/day.
  - Care is provided for > 5 children.
  - Children cared for are ≤ 2½ years old.
- Day care service is classified as Occupancy E where all the following apply:
  - Care is provided < 24 hrs/day.
  - Care is provided for > 5 children.
  - Care is provided for ≤ 100 children.
  - Children cared for are ≤ 2½ years old.
  - Child care rooms are on level of exit discharge that serves the rooms.
  - Each child care room has an exit door directly to the outside.

309 Mercantile Group M

309.1 Mercantile Group M

- Includes buildings or parts of buildings used for the display and sale of merchandise as follows:
  - Involves stocks of the following accessed by the public:
    - Goods.
    - Wares.
    - Merchandise.
    - Incidental items.
  - Includes the following functions:
    - Department stores.
    - Drug stores.
    - Markets.
    - Motor vehicle service stations.
    - Retail stores.
    - Wholesale stores.
    - Sales rooms.

309.2 Quantity of hazardous materials

- This section addresses hazardous materials stored or displayed in an Occupancy M control area.
- The total amount of the following hazardous materials must be within permitted limits:
  - Nonflammable solids.
  - Nonflammable liquids.
  - Noncombustible liquids.

Note: IBC Table 414.2.5(1), "Maximum Allowable Quantity per Indoor and Outdoor Control Area in Group M and S Occupancies: Nonflammable Solids and Nonflammable and Noncombustible Liquids," is cited as listing quantity limitations.

310 Residential Group R

310.1 Residential Group R *(part 1 of 2)*

- Includes buildings or parts of buildings as follows:
  - For sleeping.
  - Not classified as Occupancy I.

  Note: International Residential Code *is cited as governing facilities similar to those above that are not governed by this section.*

- Residential spaces are divided into the following designations:
  - R-1:
    - Residents are primarily transient.
    - Includes the following building types:
      - Hotels.
      - Motels.
      - Boarding houses.
    - Congregate living facilites with all of the following characteristics may meet R-3 requirements:
      - Sleeping units are present.
      - ≤ 10 occupants.
      - Residents share any of the following:
        - Bathrooms.
        - Kitchens.
  - R-2:
    - Residents are primarily permanent.
    - Buildings with any of the following:
      - Sleeping units.
      - > 2 dwelling units.
    - Includes the following nontransient building types:

Apartment houses	Fraternities	Motels
Boarding houses	Hotels	Sororities
Convents	Live/work units	Timeshare vacation properties
Dormitories	Monasteries	

    - Living facilities with all of the following characteristics may meet R-3 requirements:
      - Sleeping units are present.
      - ≤ 16 occupants.
      - Residents share any of the following:
        - Bathrooms.
        - Kitchens.

310 Residential Group R

310.1 Residential Group R *(part 2 of 2)*

- R-3:
    - Residents are primarily permanent.
    - Buildings not classified as any of the following:
        - R-1, R-2, R-4, I.
    - Includes the following:
        - Buildings with ≤ 2 dwelling units.
        - Adult and child care facilities as follows:
            - Care is provided for ≤ 5 persons.
            - Care is provided < 24 hrs/day.
        - Living facilities with all the following characteristcs:
            - Sleeping units are present.
            - ≤ 16 occupants.
            - Residents share any of the following:
                - Bathrooms.
                - Kitchens.
    - The following adult and child care facilities may be governed by the *International Residential Code*:
        - Located in a 1-family home as follows:
            - Care is provided for ≤ 5 persons.
            - Care is provided < 24 hrs/day.
- R-4:
    - Residential care/assisted living facilities as follows:
        - Number of residents served is > 5 and ≤ 16 as follows:
            - Staff is not included in the count.
    - Must comply with one of the following:
        - R-3 construction requirements as follows:
            - Unless superseded by other requirements in this code.
        - *International Residential Code* as follows:
            - The building must be sprinklered.

Note: 903.2.8, "Group R," is cited as governing the sprinkler system.

311 Storage Group S

311.1 Storage Group S

- Includes buildings or parts of buildings used for storage that are not classified as Occupancy H.

311.2 Moderate-hazard storage, Group S-1

- Includes storage not classified as S-2.
- Includes the following types of storage:

  Aerosols, Levels 2 and 3
  Aircraft hangar (storage and repair)
  Bags, cloth
  Bags, burlap
  Bags, paper
  Bamboos
  Baskets
  Belting, canvas
  Belting, leather
  Boat storage, indoor
  Books
  Boots
  Buttons
  Buttons, cloth-covered
  Buttons, pearl
  Buttons, bone
  Candles, wax
  Cardboard
  Cardboard boxes
  Cigars
  Cigarettes
  Clothing, wool
  Combs, not celluloid
  Cordage
  Flooring, resilient
  Furniture
  Furs
  Glue size
  Glues
  Grains
  Horns, not celluloid
  Leather
  Linoleum
  Lumber
  Mattresses
  Mucilage
  Paper, rolls
  Paper, packs
  Paste
  Photo engravings
  Rattan
  Shoes
  Silks
  Snuff
  Soaps
  Sugar
  Tires, bulk storage
  Tobacco
  Upholstery

- Includes motor vehicle repair garages containing limited hazardous materials.

  Note: The following are cited as governing repair garages:
  IBC Table 307.1(1), "Maximum Allowable Quantity per Control Area of Hazardous Materials Posing a Physical Hazard."
  406.6, "Repair garages."

311 Storage Group S

311.3 Low-hazard storage, Group S-2

- Includes the storage of noncombustible materials as follows:
  - Packaging allowed for stored materials includes the following:
    - On wood pallets.
    - In paper cartons with single-thickness divisions.
    - In paper wrappings.
    - In paper cartons without divisions.
  - Characteristics of all stored products:
    - A negligible amount of the following plastic trim materials is allowed:
      - Knobs.
      - Handles.
      - Film wrapping.
  - Items allowed to be stored, among others, are as follows:

Asbestos	Inert pigments
Cement in bags	Ivory
Chalk	Meats
Crayons	Metal cabinets
Dry cell batteries	Metal parts
Dryers, clothes	Metals
Electrical coils	Mirrors
Electrical motors	Porcelain
Empty cans	Pottery
Food products	Soapstones
Frozen foods	Stoves
Glass	Talc
Glass bottles, empty	Washers, clothes
Gypsum board	

    - Beverages ≤ 16% alcohol in the following containers:
      - Ceramic.
      - Glass.
      - Metal.
    - Dairy products in nonwaxed coated paper containers.
    - Distribution transformers that are not oil-filled.
    - Foods in noncombustible containers.
    - Fresh fruits in nonplastic containers.
    - Fresh vegetables in nonplastic containers.
    - Glass bottles containing noncombustible liquids.
    - Metal desks with plastic tops and trim.
    - Noncombustible liquids in glass bottles.
    - Oil-filled distribution transformers.
  - Facilities as follows:
    - Parking garages, enclosed.
    - Parking garages, open.

312 Utility and Miscellaneous Group U

312.1 General

- Occupancy U includes the following types of buildings and structures:
  - Accessory buildings.
  - Miscellaneous buildings not classified in another occupancy.
  - Building types and structures include the following:

    Agricultural buildings.
    Barns.
    Carports.
    Fences > 6' high.
    Greenhouses.
    Livestock shelters.
    Private garages.
    Sheds.
    Stables.
    Tanks.
    Towers.
  - Aircraft hangars are included as follows:

    Where accessory to a 1- or 2-family residence.

  Note: 412.5, "Residential aircraft hangars," is cited as governing these structures.

  - Grain silos are included as follows:

    Where accessory to a building in Occupancy R.

4

Special Detailed Requirements Based on Use and Occupancy

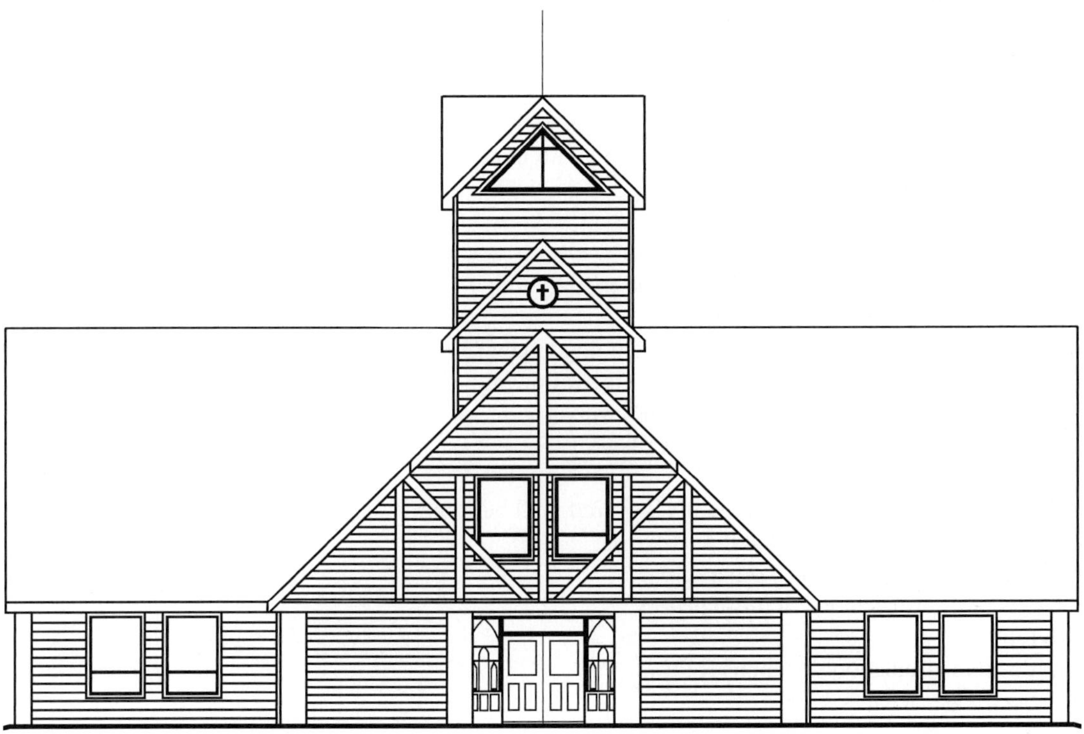

Glad Tidings Assembly of God Church. Naticoke, Pennsylvania.
Mullins and Weida, Architect and Associate. Bear Creek, Pennsylvania.

402 Covered Mall and Open Mall Buildings

402.1 Scope

- This section does not apply to foyers and lobbies in the following occupancies:
  - B, R-1, R-2.
- Otherwise, this section applies to covered mall buildings where both of the following apply:
  - Building is ≤ 3 floor levels at every location.
  - Building is ≤ 3 stories above the grade plane.
- Covered mall buildings are required to meet one of the following sets of requirements:
  - Applicable sections of the code in total excluding this section.
  - This section plus applicable sections of the code other than those addressed by this section.

402.2 Definitions *(part 1 of 2)*

- **Anchor building**
  - Located on the exterior perimeter of the mall building.
  - Is in an occupancy other than H.
  - Has direct access to a covered mall building.
  - Has required means of egress separate from the mall.
- **Covered mall building**
  - A single building housing tenants similar to the following types:
    - Retail stores and offices.
    - Eating and drinking establishments.
    - Entertainment and amusement facilities.
    - Passenger transportation terminals.
    - Similar functions.
  - ≥ 2 tenants have a main entrance into ≥ 1 mall areas.
  - Anchor buildings are not part of the covered mall building.
  - Open malls are included in this category.
- **Food court**
  - A public seating area in the mall area.
  - Serves adjacent tenants that provide food preparation.
- **Gross leasable area**
  - Total floor area for tenant occupancy including tenant storage.
  - Area of tenant occupancy is measured to the following:
    - To the outside of an individual tenant's walls.
    - To the centerline of shared tenant walls.
- **Mall**
  - Covered common pedestrian area in a covered mall building.
  - Provides access for ≥ 2 tenants.
  - Has ≤ 3 levels open to each other.
  - Open malls are included in this category.
- **Open mall**
  - Pedestrian circulation system as follows:
    - Serves ≥ 2 tenants.
    - Has no roof.
    - ≤ 3 levels.
    - Circulation above grade may include the following:
      - Exterior balconies with access to exits discharging at grade.

402 Covered Mall and Open Mall Buildings

402.2 Definitions *(part 2 of 2)*

- **Open mall building**
  - Several structures as follows:
    - Housing ≥ 2 tenants such as follows:
      - Retail stores.
      - Drinking and dining establishments.
      - Entertainment and amusement facilities.
      - Offices.
      - Similar uses.
  - ≥ 2 tenants have a main entrance as follows:
    - To ≥ 1 open mall.
  - Anchor buildings are not part of the open mall building.

402.4 Means of egress

- Means of egress is required for the following:
  - Each tenant space.
  - The covered mall building.
- Means of egress must comply with this section and other applicable sections of the code.
- Means of egress requirements of this section govern in the following case:
  - Where they conflict with requirements of other code sections.

402.4.1 Determination of occupant load

- Occupant load allowed in a tenant space is governed by the code.
- Means of egress requirements for a tenant space is based on the permitted occupant load of the space.

402.4.1.1 Occupant formula

- This section addresses occupant load for means of egress in a covered mall building.
- The following tenant areas are not included in the gross leasable area:
  - Attached tenants with means of egress systems entirely separate from that of the mall building.
  - Anchor buildings.
- The square feet required per occupant is calculated by the equation below as follows:
  - Gross leasable area does not include anchor buildings:

Sf required per person = (0.00007 × gross leasable area in sf) + 25

Note: Occupant load = gross leasable area ÷ sf required per person.

402.4.1.2 OLF range

- This section addresses the range permitted for square feet per occupant as calculated for various sizes of covered mall buildings.
- The sf required per person in covered mall buildings must be ≤ 50.
- 30 sf per occupant may be used where the sf required per person is calculated to be < 30.

402.4.1.3 Anchor buildings

- The occupant load of a covered mall building does not include the following:
  - The occupant load of anchor buildings opening into the mall.

402 Covered Mall and Open Mall Buildings

402.4.1.4 Food courts

- The occupant load of a food court is determined by requirements elsewhere in the code.

  Note: Section 1004, "Occupant Load," is cited as the source of requirements.

- The occupant load of the food court is added to the occupant load of the covered mall building as derived from gross leasable area.

402.4.2 Number of means of egress

- This section addresses tenant spaces in covered mall buildings.
- ≥ 2 means of egress are required in either of the following cases:
  - Travel distance for other than employees from any point in a tenant space to the mall > 75'.
  - Tenant space has an occupant load ≥ 50.

402.4.3 Arrangements of means of egress

- This section addresses assembly spaces in covered mall buildings.
- Assembly spaces with occupant loads ≥ 500 are governed as follows:
  - Entrance to assembly space must be immediately adjacent to a main entrance of the mall.
  - ≥ 1/2 the required means of egress for the assembly space must open directly to outside the mall building.

402.4.3.1 Anchor building means of egress

- The required means of egress for an anchor building is governed as follows:
  - It must be separate from the means of egress for the mall.
- Means of egress requirements for the mall do not include the following:
  - The occupant load of anchor buildings that open into the mall.
- Mall means of egress may not pass through anchor buildings.
- A mall terminating at an anchor building is a dead end in the following circumstance:
  - Where at the termination there is no means of egress independent from the anchor building.

402.4.4 Distance to exits

- This section addresses travel distance in a covered mall building.
- Travel distance must be ≤ 200' in the following cases:
  - From any point in a tenant space to the mall.
  - From any point in a tenant space to an exit.
  - From any point in a mall to an exit.

402.4.5 Access to exits

- Dead ends are permitted in malls in the following case:
  - Where dead-end length is ≤ 2 × its width at the narrowest point.
- Otherwise, where > 1 exit is required in a mall, the following applies:
  - At all mall locations, travel in different directions to separate exits must be possible.
- The width of the following routes from a mall must be ≥ 66":
  - Exit passageway.
  - Exit access corridor.

402 Covered Mall and Open Mall Buildings

402.4.5.1 Exit passageways

- This section addresses exit passageways that are a secondary means of egress from a tenant space.
- Doors from a tenant space to such passageways are governed as follows:
  - The doors must be 1-hr fire doors.
  - One of the following types of closing systems is required:
    The doors must be self-closing.
    The doors must be automatically closed by smoke detection.

Note: 715.4.8.3, "Smoke-activated doors," is cited as governing these doors.

402.4.6 Service areas fronting on exit passageways

- The following service areas may open directly into exit passageways where separated as indicated below:
  - Service areas:

Building service areas	Mechanical rooms
Electrical rooms	Service elevators

  - Separation of service areas from the exit passageway is required as follows:
    By fire barriers or horizontal assemblies with a fire-resistance rating ≥ 1-hr.
    Opening protectives in fire barriers are to be rated ≥ 1 hr.

Note: The following are cited as governing the separating assemblies:
Section 707, "Fire Barriers."
Section 712, "Horizontal Assemblies."

402.5 Mall width

- Malls are considered corridors for purposes of required egress.
- Requirements for mall widths and clearances specified in this section supersede those listed elsewhere in the code.

Note: 1005.1, "Minimum required egress width," is cited as the section having requirements that are superseded by requirements in this section.

402.5.1 Minimum width

- Mall width must accommodate the occupant load served.
- Mall width must be ≥ 20'.
- A clear exit width ≥ 10' wide by ≥ 8' high is required in a mall between any projection of a tenant space and any of the following:
  - Kiosk.
  - Vending machine.
  - Bench.
  - Display opening.
  - Food court.
  - Other obstruction to means of egress travel.

402.5.2 Minimum width open mall

- Open malls require widths ≥ 20' for the following openings:
  - Mall floor openings above grade.
  - Mall roof openings.

402 Covered Mall and Open Mall Buildings

402.6 Types of construction

- The area of covered mall buildings including anchor buildings is governed as follows:
  - Area is not limited where the buildings have all of the following characteristics:
    - The buildings are one of the following construction types:
      - Type I, Type II, Type III, Type IV.
    - The following buildings are surrounded on all sides by permanent open space ≥ 60':
      - Covered mall building.
      - Attached anchor buildings.
      - Parking garage.
    - The anchor buildings are ≤ 3 stories above the grade plane.

Note: The following are cited as governing height and area of anchor buildings > 3 stories:
Section 503, "General Building Height and Area Limitations."
Section 504, "Building Height."
Section 506, "Building Area Modifications."
The following are cited as governing the construction type of parking garages:
406.3, "Open parking garages."
406.4, "Enclosed parking garages."

402.7 Fire-resistance-rated separation

- A separation with a fire-resistance rating is not required between the following spaces:
  - Between a tenant space and the mall.
  - Between a food court and the mall.
  - Between a food court and adjacent tenant spaces.

402.7.1 Attached garage *(part 1 of 2)*

- Where a parking garage is located > 10' away from the following buildings, the requirements listed below apply:
  - Buildings:
    - Covered mall building.
    - Anchor building.
  - Requirements:
    - The walls separating the two structures must meet fire-resistance requirements.
    - The following elements connecting to the garage must meet code requirements:
      - Pedestrian walkways.
      - Tunnels.

Note: The following are cited as the source of applicable requirements:
IBC Table 602, "Fire-Resistance Rating Requirements for Exterior Walls Based on Fire Separation Distance."
3104, "Pedestrian Walkways and Tunnels."

402 Covered Mall and Open Mall Buildings

402.7.1 Attached garage *(part 2 of 2)*

- A covered mall building and attached parking garage are separate buildings as follows:
  - Where the following garage types are separated from the mall building as indicated:

    Garage types:

    Garages for vehicles carrying ≤ 9 passengers each.

    Open garages.

    Separation:

    Fire barrier with a fire-resistance rating ≥ 2 hrs.

Note: *The following are cited as governing the separating assemblies:*
Section 707, "Fire Barriers."
Section 712, "Horizontal Assemblies."

402.7.2 Tenant separations

- No separation wall is required between a tenant space and the mall.
- Fire partitions are required between tenant spaces.

Note: *Section 709, "Fire Partitions," is cited as the source of applicable requirements.*

402.7.3 Anchor building separation

- Anchor buildings with all the following characteristics are governed as indicated below:
  - Characteristics:

    ≤ 3 stories above the grade plane.

    Is classified as an occupancy that is permitted for tenants of the building.
  - Requirements:

    They must be separated from the covered mall building as follows:

    By fire barriers with the following a 2-hr fire-resistive rating:

Note: *Section 707, "Fire Barriers," is cited as governing these barriers.*

- Other anchor buildings must be separated from the covered mall building as follows:
  - By fire walls.

Note: *Section 706, "Fire Walls," is cited as governing these fire walls.*

402.7.3.1 Openings between anchor building and mall

- This section does not apply to the following openings:
  - Between Occupancy R-1 sleeping units and the mall.
- Openings between anchor buildings and the mall are governed as follows:
  - Openings are not required to be protected where anchor buildings are the following construction types:

    Type I, Type II.

402 Covered Mall and Open Mall Buildings

402.8 Interior finish

- The following mall and mall exit surfaces must have the characteristics shown below:
  - Surfaces:
    Walls.
    Ceilings.
  - Characteristics:
    Flame spread index ≥ Class B.
    Smoke developed index ≥ Class B.

Note: Chapter 8,"Interior Finishes," is cited as governing wall and ceiling finishes. Section 804, "Interior Floor Finish," is cited as governing floor finishes.

402.9 Automatic sprinkler system

- Sprinklers are not required in open parking garages.

Note: 406.2, "Parking garages," is cited as describing qualifying garages.

- Otherwise, covered mall buildings and connected buildings must be sprinklered as follows:
  - Prior to occupation by any tenant, the sprinkler system must be complete and operable throughout the covered mall building.
  - Empty tenant spaces must be protected with operable sprinklers as follows:
    Where approved alternative systems are not provided.
  - Mall sprinklers must be independent from sprinklers for tenant spaces and anchor buildings.
  - Sprinklers of the same system serving more than one tenant are governed as follows:
    They must be controlled independently.

Note: 903.3.1.1, "NFPA13 sprinkler systems," is cited as governing the sprinklers.

402.9.1 Standpipe system

- A standpipe system is required in covered mall buildings.

Note: 905.3.3, "Covered mall buildings," is cited as applicable to this requirement.

402.10 Smoke control

- Smoke control is not required in atriums that connect only 2 floors in covered mall buildings.
- Smoke control is required for atriums connecting > 2 stories in covered mall buildings.

Note: Section 404.5, "Smoke control," is cited as governing smoke control systems for atriums.

402.11 Kiosks *(part 1 of 2)*

- This section governs kiosks and like constructions in a mall as follows:
  - Temporary.
  - Permanent.
- Kiosks must be constructed of one of the following:
  - Noncombustible materials.
  - Fire-retardant-treated wood.

Note: 2303.2, "Fire-retardant-treated wood," is cited as governing this material.

402 Covered Mall and Open Mall Buildings

402.11 Kiosks *(part 2 of 2)*

- ◦ Foam plastics with a heat release rate ≤ the following:
  - 100 kW.
  - 105 Btu/h.

Note: UL 1975, "Fire Tests for Foam Plastics Used for Decorative Purposes," is cited as the required test for heat release using exhibit booth protocol.

- ◦ Aluminum composite material (ACM) with both the following characteristics:
  - Flame spread index ≤ 25.
  - Smoke-developed index ≤ 450.

Note: One of the following cited tests is required to be used for ACM in the maximum thickness to be used in the kiosk:
ASTM E 84, "Standard Test Method for Surface Burning Characteristics of Building Materials."
UL 723, "Standard for Test for Surface Burning Characteristics of Building Materials."

- Kiosks and like structures must have both of the following approved systems:
  - Fire detection.
  - Fire suppression.
- ≥ 20' separation is required between the following and other structures in a mall:
  - Kiosks.
  - Groups of kiosks.
- The following must be ≤ 300 sf in area:
  - Kiosks.
  - Groups of kiosks.

402.13 Security grilles and doors

- Security grilles or doors of the following types in a required means of egress have requirements as indicated below:
  - Types:
    - Horizontal sliding.
    - Vertical.
  - Requirements:
    - They must remain fully open during occupancy by the public.
    - They may not be closed while the served space has either of the following conditions:
      - Where ≥ 10 persons occupy a space with 1 exit.
      - Where ≥ 50 persons occupy a space with > 1 exit.
    - They must be operable when the served space is occupied as follows:
      - From inside the space.
      - Without special knowledge or effort.
- Where ≥ 2 exits are required, the following applies:
  - Security grilles or doors are limited to ≤ 1/2 the exits.

402.14 Standby power

- Standby power is required in mall buildings > 50,000 sf as follows:
  - Systems must be able to operate emergency voice/alarm systems.

402 Covered Mall and Open Mall Buildings

402.15 Emergency voice/alarm communication system

- Emergency voice/alarm systems are required in covered mall buildings > 50,000 sf of total floor area.
- Mall emergency voice/alarm systems must be accessible to the fire department as follows:
  - Where such systems are required.
  - Where such systems are not required but are provided.

Note: 907.5.2.2, "Emergency voice/alarm communication systems," is cited as governing the locations and characteristics of such systems.

402.16 Plastic signs

- Plastic signs are restricted in size and detail in covered mall buildings as follows:
  - Where on the front of any tenant space facing the mall.

Note: The following are cited as governing plastic signs:
402.16.1, "Area."
402.16.2, "Height and width."
402.16.3, "Location."
402.16.4, "Plastics other than foam plastics."
402.16.4.1, "Encasement."
402.16.5, "Foam plastics."
402.16.5.1, "Density."
402.16.5.2, "Thickness."

402.16.1 Area

- Plastic signs are limited to ≤ 20% of the tenant façade facing the mall.

402.16.2 Height and width

- The size of plastic signs is limited as follows:
  - Vertical dimension must be ≤ 36" for horizontal signs.
  - Vertical dimension must be ≤ 96" for vertical signs.
  - Horizontal dimension must be ≤ 36" for vertical signs.

402.16.3 Location

- Plastic signs must be ≥ 18" from adjacent tenants:
  - Measured to the center of the common wall between tenants.

402 Covered Mall and Open Mall Buildings

402.16.4 Plastics other than foam plastics

- This section does not apply to foam plastics used in signs.
- Other plastics are to be light-transmitting plastics complying with one of the following:
  - Specifications listed elsewhere in the code.

    Note: The following sections are cited as alternatives for governing plastics other than foam plastics:
    2606.4, "Specifications."
    803.1.2.1, "Acceptance criteria for NFPA 286," is an alternative when tests are in accordance with NFPA 286, "Standard Method of Fire Test for Evaluating Contribution of Wall and Ceiling Interior Finish to Room Fire Growth."

  - Specifications as follows:
    Self-ignition temperature ≥ 650° F.
    Flame-spread index ≤ 75.
    Smoke-developed index ≤ 450.

    Note: The following are cited as describing tests for verifying the properties above:
    ASTM D 1929, "Standard Test Method for Determining Ignition Temperature of Plastics."
    ASTM E 84, "Standard Test Method for Surface Burning Characteristics of Building Materials."

402.16.4.1 Encasement

- The backs and edges of plastic signs in the mall must be enclosed with metal.

402.16.5 Foam plastics

- Foam plastics used in signs are governed as follows:
  - The sign must have a heat release rate ≤ 150 kW.

    Note: UL 1975, "Fire Tests for Foamed Plastics Used for Decorative Purposes," is cited as describing the test verifying heat release rate.

  - The foam plastics must have the properties specified in this section series.
- Foam plastic signs meeting plastic-sign requirements need not have the following:
  - Flame-spread index ≤ 75.
  - Smoke-developed index ≤ 450.

    Note: 402.16, "Plastic signs," is cited as governing in lieu of the specified indexes.
    2603.3, "Surface-burning characteristics," is cited as the source of flame spread and smoke-developed indexes which do not apply in this case.

402.16.5.1 Density

- The density of foam plastic must be ≥ 20 lbs/cu ft.

402.16.5.2 Thickness

- The thickness of foam plastic signs must be ≤ 1/2".

402.17 Fire department access to equipment

- Areas housing the following controls are to be identified for use by the fire department:
  - Controls for air-conditioning systems.
  - Controls for automatic fire-extinguishing systems.
  - Controls for other detection, suppression, or control elements.

403 High-Rise Buildings

403.1 Applicability

- Buildings not governed by this section include the following:
  - Airport traffic control towers.
  - Open parking garages.
  - Occupancy A-5 buildings.
  - Occupancy F-2 buildings that require large heights to accommodate equipment such as the following:
    Craneways.
    Rolling mills.
    Structural metal fabrication.
    Production and distribution of power.
  - Buildings with the following occupancies:
    H-1, H-2, H-3.

Note: The following are cited as governing the buildings above:
303.1, "Assembly Group A."
406.3, "Open parking garages."
412.3, "Airport traffic control towers."
503.1.1, "Special industrial occupancies," for buildings with large equipment.
Section 415, "Groups H-1, H-2, H-3, H-4 and H-5."

- Otherwise, high-rise buildings are governed by this section:

Note: 403.2, "Construction," through 403.6, "Elevators," are cited as governing high-rise buildings.

403.2.1 Reduction in fire-resistance rating

- Reduction in requirements for fire-resistance ratings is permitted as follows:
  - Where sprinkler control valves are equipped as follows for specified conditions:
    Valves have supervisory initiating devices for each floor.
    Valves have water-flow initiating devices for each floor.

Note: The following are cited as indicating the rating reductions permitted:
403.2.1.1, "Type of construction."
403.2.1.2, "Shaft enclosures."

403.2.1.1 Type of construction

- Using certain sprinkler control valves permits reductions in required fire-resistance ratings as follows:
  - Required Type IA ratings may be reduced to those required in Type IB as follows:
    Allowed in buildings ≤ 420' high.
    Not allowed for columns supporting floors.
  - Required Type IB ratings may be reduced to those required in Type IIA as follows:
    Allowed in other than F-1, M, and S-1 occupancies.

Note: The following are cited as applicable to the reductions described above:
IBC Table 601, "Fire-Resistance Rating Requirements for Building Elements," lists required fire-resistance ratings.

  - Height and area limitations of the construction with reduced fire-resistance ratings are as follows:
    The same as the construction with the original fire-resistance ratings.

403 High-Rise Buildings

403.2.1.2 Shaft enclosures

- This section does not apply to exit enclosures.
- This section does not apply to elevator hoistway enclosures.
- Where sprinkler control valves are equipped as follows, required fire-resistance ratings may be reduced as indicated below:
  - Control valves:
    Valves have supervisory initiating devices for each floor.
    Valves have water-flow initiating devices for each floor.
  - The fire-resistance rating required for fire barriers enclosing vertical shafts may be reduced as follows:
    Rating reduced to ≥ 1 hr where all the following conditions apply:
    Where buildings are ≤ 420' high.
    Where sprinklers are provided in the shaft at the top.
    Where sprinklers are provided in the shaft at every other floor.

403.2.4 Sprayed fire-resistant materials (SFRM)

- The bond strength of SFRM is required as follows:
  - 430 psf for buildings of the following heights:
    > 75' and ≤ 420' above the lowest level accessible by fire department vehicles.
  - 1,000 psf for buildings of the following heights:
    > 420' above the lowest level accessible by fire department vehicles.

Note: IBC Table 403.2.4, "Minimum Bond Strength," is cited as governing bond strength of SFRM and is summarized above.

- The required bond strength applies to all SFRM installed throughout the building of the height specified.

403.3 Automatic sprinkler system *(part 1 of 2)*

- Sprinklers are not required in buildings used only to house telecommunications equipment as follows, with the conditions listed below:
  - Equipment:
    Telecommunications equipment.
    Associated electrical power distribution equipment.
    Batteries.
    Standby engines.
  - Conditions:
    Automatic fire detection system is required in equipment spaces.
    Equipment spaces must be isolated by one or both of the following as applicable:
    Fire barriers with a fire-resistance rating ≥ 1 hr.
    Horizontal assemblies with a fire-resistance rating ≥ 2 hrs.

Note: 907.2, "Where required—new buildings and structures," is cited as governing the fire detection system.
The following are cited as governing the separating assemblies:
Section 707, "Fire Barriers."
Section 712, "Horizontal Assemblies."

- Sprinklers are not required in open parking garages.

Note: 406.3, "Open parking garages," is cited as governing these facilities.

403 High-Rise Buildings

403.3 Automatic sprinkler system *(part 2 of 2)*

- Other high-rise buildings are required to be sprinklered.

Note: 903.3.1.1, "NFPA 13 sprinkler systems," is cited as governing the sprinklers.

- Secondary water supply systems for sprinklers are required for new construction as follows:
  - For high-rise buildings in Seismic Design Categories C, D, E, or F.

Note: 903.3.5.2, "Secondary water supply," is cited as the source of these requirements.

403.3.1 Number of sprinkler risers and system design

- This section addresses sprinklers in buildings > 420' high.
- ≥ 2 risers are required for each sprinkler zone as follows:
  - Each riser is to supply sprinklers on alternate floors.
  - The following applies where > 2 risers supply a sprinkler zone:
    Sprinklers on adjacent floors are to be supplied by different risers.

403.3.1.1 Riser location

- Sprinkler risers are to be located in exit enclosures as follows:
  - The stairway enclosures are to be located remotely from each other.

Note: 1015.2, "Exit or exit access doorway arrangement," is cited as governing remoteness of exits.

403.3.2 Water supply to required fire pumps

- Two fire-pump supply-pipe connections to the same water main are allowed in the following case:
  - Where an interruption of flow can be limited to only one connection while the other continues to flow.
- Otherwise required fire-pump supply pipe connections must be made to the following:
  - ≥ 2 water mains in different streets as follows:
    Connections to be made with separate piping.
    Piping to be sized adequately for flow and pressure required by the fire pumps.

403.4.1 Smoke detection

- Smoke detection is required in high-rise buildings.

Note: 907.2.13.1, "Automatic smoke detection," is cited as governing smoke detection.

403.4.2 Fire alarm systems

- A fire alarm system is required in high-rise buildings.

Note: 907.2.13, "High-rise buildings," is cited as governing the fire alarm systems required by this section.

403.4.3 Emergency voice/alarm communication system

- An emergency voice/alarm communication system is required in high-rise buildings.

Note: 907.5.2.2, "Emergency voice/alarm communication system," is cited as the source of requirements for such systems.

403 High-Rise Buildings

403.4.4 Emergency responder radio coverage

- The installation of an emergency responder radio communications system is required.

Note: The International Fire Code Section 510, "Emergency Responder Radio Coverage" is cited as governing this technology.

403.4.5 Fire command

- A fire command center is required in high-rise buildings as follows:
  - Location to be approved by the fire department.

Note: Section 911, "Fire Command Center," is cited as governing such a facility.

403.4.6 Smoke removal

- This section addresses removal of smoke and products of combustion after a fire is extinguished.
- A natural or mechanical ventilation system is required as follows:
  - In R-1 occupancies the following applies to guest rooms or suites with an exterior wall:
    Ventilation requirements can be met by 2 sf of ventilating opening.
  - Fixed windows are permitted to serve as post-fire ventilation in the following case:
    Where firefighters can penetrate the glazing.
  - Otherwise, the following system is required where natural ventilation is used:
    Windows or panels to the exterior are required as follows:
      Must be easily identifiable.
      Must be manually operable.
      Must be distributed along the perimater of each floor as follows:
        At a spacing ≤ 50'.
      Must have an opening area ≥ 40 sf/50' of perimeter.
  - A mechanical ventilatoin system must achieve ≥ 1 air change every 15 minutes as follows:
    Removed air must be exhausted directly to the exterior.
- Other approved systems are acceptable as follows:
  Where the same results as described above can be achieved.

403.4.7.1 Special requirements for standby power

- Where a generator inside a building provides standby power, the following applies:
  - The generator system must be located in its own room as follows:
    The room must be enclosed with the following as applicable:
      2-hr fire barriers.
      2-hr horizontal assemblies.

Note: The following are cited as governing the required enclosure:
Section 707, "Fire Barriers."
Section 712, "Horizontal Assemblies."

- The fire command center must provide the following:
  - System supervision.
  - Manual-start capability.
  - Transfer features.

403 High-Rise Buildings

403.4.7.2 Standby power loads

- Standby power loads include the following:
  - Fire command centers as follows:
    - Power.
    - Light.

  Note: 403.4.5, "Fire command," is cited as governing a fire command center.

  - Electric fire pumps.
  - Ventilation for smokeproof enclosures.
  - Automatic fire detection systems for smokeproof enclosures.
  - Elevators.

  Note: The following are cited as addressing standby power for elevators:
  1007.4, "Elevators."
  Section 3003, "Emergency Operations."
  Section 3007, "Fire Service Access Elevator."
  Section 3008, "Occupant Evacuation Elevators."

403.4.8.1 Emergency power loads

- Emergency power loads include the following:
  - Exit signs.
  - Lighting for means of egress.

  Note: Chapter 10, "Means of Egress," is cited as governing exit signs and egress lighting.

  - Lighting in elevator cars.
  - Emergency voice/alarm communication systems.
  - Automatic fire detection systems.
  - Fire alarm systems.
  - Electrically powered fire pumps.

403.5.1 Remoteness of exit stairway enclosures

- Where ≥ 2 required exit stairway enclosures exist, 2 of them must be separated as follows:
  - By the shorter distance of the following:
    - 30'
    - 1/4 the distance of the largest ovearall diagonal of the area served.
  - Separation distance is to be measured as follows:
    - On a straight line between nearest points of stairway enclosure walls.
- The following stairs count as a single exit stairway:
  - Interlocking stairs.
  - Scissor stairs.

403 High-Rise Buildings

403.5.2 Additional exit stairway

- This section governs buildings with both of the following characteristics:
  - In occupancies other than R-2.
  - > 420' in height.
  - Without occupant-controlled evacuation elevators.

Note: Section 3008, "Occupant Evacuation Elevators," is cited as governing the occupant-controlled evacuation elevators.

- This section requires the following for the buildings governed:
  - An exit stairway is required in excess of the number dictated by occupant load as follows:
    Scissor stairs cannot serve as the additional stairway required.
  - The sum of widths of all stairways less any single one must be ≥ that required by occupant load.

Note: The following are cited as governing the additional stairway required by this section:
Section 1009, "Stairways."
Section 1022, "Exit Enclosures."
1021.1, "Exits from stories," is cited as requiring the number of exits to which one additional exit stairway is added by this section.
1005.1, "Minimum required egress width," is cited as dictating the mininum total width of exit stairways based on occupant load.

403.5.3 Stairway door operation

- This section addresses stairway doors that are not exit discharge doors.
- Such doors may be locked so as to prevent opening from the stairway side in the following case:
  - Where the doors can be unlocked as follows:
    Simultaneously without being unlatched.
    By a signal from the fire command station.

403.5.3.1 Stairway communication system

- This section addresses stairways where doors are locked from the stairway side.
- A 2-way communication system is required in the stairway as follows:
  - To be connected to an approved and continuously attended station.
  - To be located at every 5th floor or at more frequent intervals.

403.5.4 Smokeproof exit enclosures

- The following required stairways must meet requirements for smokeproof enclosures:
  - Stairways serving floors > 75' above lowest access level for fire department vehicles.

Note: The following sections are cited as having applicable requirements:
909.20, "Smokeproof enclosures."
1022.9, "Smokeproof enclosures and pressurized stairways."

403 High-Rise Buildings

403.5.5 Luminous egress path markings

- Luminous egress path markings are required as follows:
  - In certain locations of occupancies A, B, E, I M, and R-1 such as the following:
    Buildings with an occupied floor > 75' above the following level:
      The lowest level with access for fire department vehicles.

Note: Section 1024, "Luminous Egress Path Markings," is cited as governing these markings and required locations. An introduction to the reference is provided above.

403.5.6 Emergency escape and rescue

- Such openings are not required in high-rise buildings.

Note: Section 1029, "Emergency Escape and Rescue," is cited as requiring the openings waived by this section.

403.6.1 Fire service access elevator

- ≥ 1 fire service access elevator is required in the following buildings:
  - Those with an occupied floor > 120' above the lowest level of access for fire department vehicles.

Note: Section 3007, "Fire Service Access Elevator," is cited as governing these elevators.

403.6.2 Occupant evacuation elevators

- This section applies to the following elevators:
  - Passenger elevators as follows:
    Used by the general public.
- Elevators governed by this section may be used for occupant self-evacuation in the following case:
  - Where they meet occupant evacuation requirements.

Note: Section 3008, "Occupant Evacuation Elevators," is cited as governing these elevators.

404 Atriums

404.1 General

- This section governs atrium-type vertical openings through floors as follows:
  - In other than Occupancy H.
  - Where the opening complies with requirements for atriums.

Note: 708.2, "Shaft enclosure required," Exception 5 is cited as governing floor openings to which this section applies. This exception requires that floor openings comply with Section 404, "Atriums."

404.1.1 Definition

- **Atrium**
  - The following are not considered to be stories for this definition:
    - Balconies in assembly occupancies.
    - Mezzanines.

Note: Section 505, "Mezzanines," is cited as the source of mezzanine requirements.

  - Openings through floors for the following are not included:
    - Air conditioning.
    - Electrical services.
    - Elevators.
    - Enclosed stairways.
    - Escalators.
    - Hoistways.
    - Plumbing.
    - Other equipment.
  - Openings through floors defined as a mall are not included.
  - Openings through floor(s) with both the following characteristics are included:
    - Openings connecting $\geq$ 2 floor levels.
    - Openings are closed at the top.

404.2 Use

- Atrium floors may serve any use where both of the following apply:
  - Use is approved.
  - Individual space is sprinklered.

Note: 903.3.1.1, "NFPA 13 sprinkler systems," is cited as governing the sprinklers.

- Otherwise, atrium floors may serve only the following:
  - Low-fire-hazard uses.
- The following components of an atrium must be approved:
  - Materials.
  - Decorations.

Note: The International Fire Code *is cited as governing materials and decorations.*

404 Atriums

404.3 Automatic sprinkler protection

- The following areas are not required to be sprinklered with the condition listed below:
  - Areas:
    Above an atrium.
    On levels adjacent to an atrium.
  - Condition:
    Where separated from the atrium by one or both of the following:
      Fire barriers with a fire-resistance rating ≥ 2 hrs.
      Horizontal assemblies with a fire-resistance rating ≥ 2 hrs.
- Sprinklers at the ceiling of an atrium are not required where the ceiling is > 55' above the floor.
- Otherwise, an approved sprinkler system is required as follows:
  - Required throughout a building containing an atrium.

Note: The following are cited as governing the separating assemblies:
Section 707, "Fire Barriers."
Section 712, "Horizontal Assemblies."

404.5 Smoke control

- Smoke control is not required for the following atriums:
  - Connecting ≤ 2 stories.
- Smoke control is required in other atriums.

Note: Section 909, "Smoke Control Systems," is cited as governing such systems.

404 Atriums

404.6 Enclosure of atriums

- Space adjacent to an atrium need not be separated from the atrium as follows:
  - Such space is limited to that on any 3 floors.
  - Such space must be accounted for in the smoke-control system design.
- In other cases, atriums must be separated from adjacent spaces by any of the following:
  - Fire-barrier with a fire-resistance rating of 1 hr.
  - A horizontal assembly with a fire-resistance rating of 1 hr.
  - Glass block wall with a 3/4-hr fire-protection rating.

Note: Section 707, "Fire Barriers," is cited as governing fire barriers.
Section 712, "Horizontal Assemblies," is cited as governing horizontal assemblies.
Section 2110, "Glass Unit Masonry," is cited as governing glass block.

  - A glass wall having all the following characteristics:
    Glass wall must function as a smoke partition.
    Sprinklers with all the following characteristics must be provided along the glass:
      Located in one of the following arrangements:
        On both the atrium side and the room side of the glass wall.
        On only the room side of the glass wall in the following case:
          Where there is no walkway on the atrium side.
      Spaced ≤ 6' apart.
      Located ≥ 4" and ≤ 12" from the glass wall.
      Upon activation sprinklers must wet the entire surface of the glass.
      Obstructions to sprinkler coverage are prohibited.
    Glass must be set in a gasketed frame which can deflect as follows:
      Prior to sprinkler activation.
      Without breaking the glass.

404.7 Standby power

- Smoke control equipment must be connected to standby power.

Note: 909.11, "Power systems," is cited as governing smoke control system power supply.

404.8 Interior finish

- ≥ Class B finishes are required for walls in atriums as follows:
  - Sprinklers do not warrant a reduction of this class.

404.9 Travel distance

- Means of egress travel through an atrium is governed as follows:
  - On levels other than the lowest level:
    Travel distance within the atrium is limited to ≤ 200'.

Note: Section 1016, "Exit Access Travel Distance," is cited as governing travel distance.

405 Underground Buildings

405.1 General

- Spaces governed as underground buildings have both the following characteristics:
  - Floors > 30' below the finished floor of the lowest exit discharge.
  - Floors occupied by humans.
- The following are not underground buildings:
  - 1- and 2-family dwellings as follows:
    Sprinklered.

Note: 903.3.1.3, "NFPA 13D sprinkler systems," is cited as governing the sprinkler system.

  - Parking garages as follows:
    With automatic fire-suppression systems.

Note: 405.3, "Automatic sprinkler system," is cited as governing the fire-suppression systems.

  - Transit systems as follows:
    With fixed guideways.
  - The following facilities:
    Arenas.
    Bleachers.
    Grandstands.
    Similar facilities.
    Stadiums.
  - A building story with all of the following characteristics:
    Lowest story of a building.
    The only story that otherwise qualifies as an underground building.
    An area ≤ 1,500 sf.
    < 10 occupants.
  - The following facilities where occupancy is intermittent involving service personnel:
    Mechanical spaces similar to pumping stations.
    Pumping stations.

405.2 Construction requirements

- The construction type required for the underground part of the building is as follows:
  - Type I.

405.3 Automatic sprinkler system

- Underground portions of the building must be sprinklered as follows:
  - At the highest level of exit discharge.
  - At all levels below the highest level of exit discharge.

Note: The following are cited as governing the sprinkler system:
903.3.1.1, "NFPA 13 sprinkler systems."
903.4, "Sprinkler system supervision and alarms," for water-flow switches and control valves.

405 Underground Buildings

405.4.1 Number of compartments

- A building with a floor > 60' below the finished floor of the lowest exit discharge is governed as follows:
  - It must be divided into ≥ 2 compartments as follows:
    - Compartments are to be about equal in size.
    - Compartments must extend as follows:
      - Through all underground levels.
      - To the highest level of exit discharge serving the underground levels.
  - The lowest story is not required to be compartmentalized as follows:
    - Where it has both the following characteristics:
      - ≤ 1,500 sf.
      - < 10 occupants.

405.4.2 Smoke barrier penetration

- Compartments must be separated by smoke barriers.

Note: Section 710, "Smoke Barriers," is cited as governing these barriers.

- Penetrations in smoke barriers are limited to the following:
  - Fire-stopped plumbing piping and electrical conduit.

Note: Section 713, "Penetrations," is cited as governing fire-stopped penetrations.

  - Doorways must be protected as follows:
    - By automatic-closing devices activated by smoke detection.

Note: The following are cited as governing door assembly protection:
715.4.8.3, "Smoke-activated doors."
NFPA 105, "Standard for the Installation of Smoke Door Assemblies."
715.4.3, "Door assemblies in corridors and smoke barriers."

- Air systems for compartments must be separate from each other.

405.4.3 Elevators

- Where elevators are provided to compartments the following applies:
  - Direct access to an elevator is required for each compartment.
  - The following is required where an elevator serves > 1 compartment:
    - An elevator lobby is required with all the following characteristics:
      - A smoke barrier must separate the lobby from each compartment.
      - Doors require gaskets and drop sills.
      - Doors must close automatically upon a signal from a smoke detection system.

Note: Section 710, "Smoke Barriers," is cited as governing these elements.
715.4.8.3, "Smoke-activated doors," is cited as governing the doors addressed above.

405.5.1 Control system

- A smoke control system is required for underground buildings as follows:
  - The system must contain smoke within the area of the fire origin.
  - The system must limit smoke in the means of egress to maintain its viability.

Note: Section 909, "Smoke Control Systems," is cited as governing smoke control systems.

405 Underground Buildings

405.5.2 Compartment smoke control system

- Compartments in underground buildings require smoke control systems as follows:
  - Each compartment must have its own system.
  - The system must be activated automatically.
  - Manual control of the system must be possible.

Note: The following are cited as governing the smoke control system:
907.2.18, "Underground buildings with smoke control systems."
907.2.19, "Deep underground buildings."

405.6 Fire alarm systems

- A fire alarm system is required as dictated by the references.

Note: The following are cited as governing the need for a fire alarm system:
907.2.18, "Underground buildings with smoke control systems."
907.2.19, "Deep underground buildings."

405.7.1 Number of exits

- Each floor requires ≥ 2 exits.
- Where compartments exist, each requires the following:
  - ≥ 1 exit.
  - An exit access doorway to the adjacent compartment.

Note: 405.4, "Compartmentation," is cited as the source of criteria for requiring compartments.

405.7.2 Smokeproof enclosure

- Stairways for floors > 30' below the finished floor of their exit discharge have the following requirement:
  - They must meet smokeproof enclosure requirements.

Note: 1022.9, "Smokeproof enclosures and pressurized stairways," is cited as the source of requirements for smokeproof enclosures.

405.8 Standby power

- Standby power is required for the following systems:
  - Smoke control.
  - Ventilation for smokeproof enclosures.
  - Fire detection for smokeproof enclosures.
  - Fire pumps.
  - Elevators.

Note: The following are cited as sources of requirements for standby power:
Chapter 27, "Electrical."
405.8.1, "Standby power loads," lists the systems needing standby power.

405 Underground Buildings

405.8.1 Standby power loads

- The following systems are standby power loads:
  - Smoke control.
  - Ventilation for smokeproof enclosures.
  - Fire detection for smokeproof enclosures.
  - Fire pumps.
  - Elevators.

Note: Section 3003, "Emergency Operations," is cited governing elevator standby power.

405.8.2 Pick-up time

- Standby power must supply power to connected loads as follows:
  - ≤ 60 seconds after power failure.

405.9 Emergency power

- Emergency power is required for the following systems:
  - Emergency voice communication.
  - Emergency alarms.
  - Fire alarms.
  - Fire detection.
  - Lighting for elevator cars.
  - Lighting for means of egress.
  - Illumination of exit signs.

Note: The following are cited as sources of requirements for emergency power:
Chapter 27, "Electrical."
405.9.1, "Emergency power loads," lists systems needing emergency power.

405.9.1 Emergency power loads

- The following are emergency power loads:
  - Emergency voice communication.
  - Emergency alarms.
  - Fire alarms.
  - Fire detection.
  - Lighting for elevator cars.
  - Lighting for means of egress.
  - Illumination of exit signs.

Note: Chapter 10, "Means of Egress," is cited as the source of requirements for illumination of means of egress and exit signs.

405.10 Standpipe system

- Standpipe systems are required in the following:
  - Underground buildings.

Note: Section 905, "Standpipe Systems," is cited as governing these systems.

406 Motor-Vehicle-Related Occupancies

406.1.1 Classification

- Occupancy U buildings or parts of buildings are limited as follows:
  - Area must be ≤ 1,000 sf where area increases are not permitted.
  - Height must be ≤ 1 story.

Note: 406.1.2, "Area increase," is cited as the source of increases permitted to the area limit.

406.1.2 Area increase

- An Occupancy U storage of the following vehicles may be ≤ 3,000 sf where the conditions below apply:
  - Vehicles:
    - Private motor vehicles.
    - Pleasure-type motor vehicles.
  - Conditions:
    - No repair work is done.
    - No fuel is dispensed.
    - For a mixed-occupancy building, the following is required:
      - The exterior wall for the Occupancy U area is governed as follows:
        - It must meet requirements for the major occupancy.
        - Openings must be protected as per requirements of the major occupancy.
      - The floor area permitted for the building is governed as follows:
        - The area is that permitted by the major occupancy.
    - For a building housing only Occupancy U, the following applies:
      - The exterior wall is governed as follows where the fire separation distance is ≥ 5':
        - It need not have a fire-resistance rating.
        - The area of openings is not limited.
- More than one Occupancy U area ≤ 3,000 sf is allowed in the same building as follows:
  - Where each 3,000 sf of Occupancy U is separated by fire walls.

Note: Section 706, "Fire Walls," is cited as governing the fire walls noted above.

406.1.3 Garages and carports

- Carports must meet the following requirements:
  - Open on ≥ 2 sides.
  - Floor surface must be one of the following:
    - Approved noncombustible material.
    - Asphalt on grade.
  - Parking surface must be sloped toward one of the following:
    - A drain.
    - The main vehicle entry.
- Carports open on < 2 sides must comply with garage requirements.

406 Motor-Vehicle-Related Occupancies

406.1.4 Separation

- A private garage must be separated from the following spaces by gypsum board as indicated:
  - Space to be separated from garage:
    Dwelling.
    Attic of dwelling.
  - Gypsum board to be 1/2" and applied to garage side of wall.
- A private garage must be separated from habitable space overhead as follows:
  - By 5/8" Type X gypsum board or equivalent.
- Doors between a private garage and a dwelling must be one of the following types:
  - Any of the following materials ≥ $1^{3}/_{8}$" thick:
    Solid wood.
    Solid-core steel.
    Honeycomb-core steel.
  - 20-minute fire-protection-rated door.

Note: 715.4.3, "Door assemblies in corridors and smoke barriers," is cited as governing the fire-protection-rated door option.

- Doors between a private garage and a dwelling must be both of the following:
  - Self-closing.
  - Self-latching.
- Openings are not permitted between a private garage and a sleeping room.
- A carport with both of the following characteristics is not required to be separated from Occupancy R-3:
  - Completely open on ≥ 2 sides.
  - No enclosed area above the carport.
- Ducts in the following locations must have the characteristics shown below:
  - Duct locations:
    In private garages.
    Penetrating a wall between a private garage and a dwelling unit.
    Penetrating a ceiling between a private garage and a dwelling unit.
  - Required characteristics:
    Sheet steel ≥ 0.019" thick.
    No openings into the garage.

406.1.5 Automatic garage door openers

- Automatic garage door openers where used must be UL listed.

Note: UL 325, "Door, Drapery, Gate, Louver, and Window Operators and Systems," is cited as governing garage door openers.

406 Motor-Vehicle-Related Occupancies

406.2.1 Classification

- Parking garages are classified as one of the following:
  - Open parking garage.
  - Enclosed parking garage.

Note: The following are cited as sources of requirements for parking garages:
406.3, "Open parking garages."
406.4, "Enclosed parking garages."
Section 509, "Special Provisions."

406.2.2 Clear height

- A clear height of ≥ 7' is required at each floor level of a parking garage in the following areas:
  - Vehicle traffic areas.
  - Pedestrian traffic areas.
- Areas serving required van-accessible parking must comply with accessibility requirements.

Note: The following are cited as sources of requirements for van-accessible parking:
1106.5, "Van spaces."
ICC/ANSI A117.1, "Accessible and Usable Buildings and Facilities."

406.2.3 Guards

- This section addresses guards.

Note: Section 1013, "Guards," is cited as governing guards.
406.2.4, "Vehicle barrier systems," is cited as governing guards acting as vehicle barriers.

406.2.4 Vehicle barrier systems

- This section does not apply to the following parking garage:
  - Where vehicle storage compartments are used and access is mechanical.
- The following vehicle barriers must be provided in the locations shown below:
  - Barriers ≥ 33" high.
  - Locations:
    - Where the floor level drops > 12" as follows:
      - At the ends of drive lanes.
      - At the ends of parking spaces.

Note: 1607.7.3, "Vehicle barrier systems," is cited as governing the structural design of vehicle barriers.

406.2.5 Ramps

- Vehicle ramps in parking garages cannot serve as a required exit in the following case:
  - Where pedestrian facilities are not provided.
- The following ramps in parking garages must have a slope ≤ 1 in 15:
  - Ramps serving as both vertical circulation and parking area.

406 Motor-Vehicle-Related Occupancies

Case study: Fig. 406.2.2. There are two levels of parking under the living units of this building. The upper parking level has headroom dictated by the significant floor-to-floor height of the first floor which houses retail shops. Here clear height in the parking garage is never less than 13'-1". The lower level of parking has a reduced floor to floor dimension with retail space spanning over the ramp. The lowest clear height on the ramp under the retail shop mezzanine is 7'-1 5/8". 7'-11" is provided under the lowest beam elsewhere. Headroom under the ramp drops to 7' at the end of 2 parking spaces beyond which vehicles may not pass due to the presence of wheel-stops. The parking garage complies with the code requirement to provide 7' clear height in vehicle and pedestrian traffic areas.

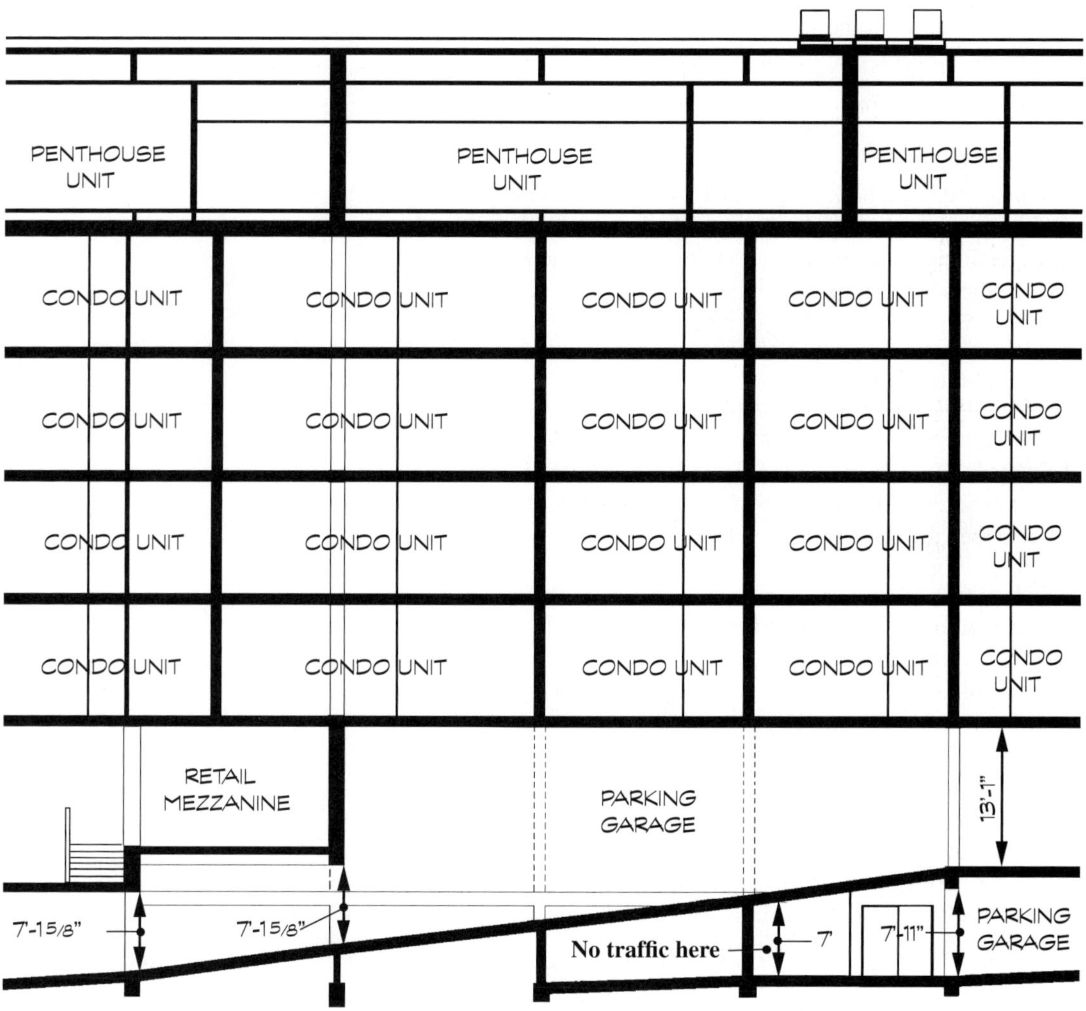

Fig. 406.2.2. Partial building section. McKenzie Lofts. Portland, Oregon. Ankrom Moisan Associated Architects. Portland, Oregon.

406 Motor-Vehicle-Related Occupancies

406.2.6 Floor surface

- The floor surface of a parking garage must be one of the following:
  - Concrete.
  - Material similar to concrete as follows:
    Noncombustible.
    Nonabsorbent.
  - Asphalt on grade.
- In other than S-2 garages, the parking surface must be sloped toward one of the following:
  - A drain.
  - The main vehicle entry.

406.2.7 Mixed occupancy separation

- Parking garages are to be separated from other occupancies according to the fire-resistance separation requirement for each individual occupancy.

Note: 508.1, "General," is cited as governing the separation of parking garages from other occupancies.

406.2.8 Special hazards

- The connection of a parking garage to a room containing a fuel-fired appliance is governed as follows:
  - Where the source of appliance ignition is $<$ 1'- 6" above the floor:
    A vestibule providing 2 doors between the spaces is required.
  - Where the source of appliance ignition is $\geq$ 1'- 6" above the floor:
    1 door is permitted in lieu of a vestibule.

406.2.9 Attached to rooms

- A parking garage may not open directly to a sleeping room.

406.3.1 Scope

- Parking garages must comply with other requirements of the code as follows:
  - Where they are not superseded by this section.

406.3.2 Definitions *(part 1 of 2)*

- **Mechanical-access open parking garages**
  - Machines similar to the following move vehicles to and from street level:
    Parking machines.
    Lifts.
    Elevators.
    Mechanical devices.
  - The public is not permitted above street level.
- **Open parking garage**
  - Used for parking or storage of private vehicles.
  - Has a specified area of openings for natural ventilation on two sides.

Note: 406.3.3.1, "Openings," is cited as governing the required openings of an open parking garage.
406.3.4, "Uses," is cited as governing the use of open parking garages.

406 Motor-Vehicle-Related Occupancies

406.3.2 Definitions *(part 2 of 2)*

- **Ramp-access open parking garages**
  - An open parking garage with one of the following configurations:
    Parking floors are sloped and serve as ramps between levels.
    Ramps provide access between floors for vehicles.
  - Vehicles travel between floors and the street by their own power.

406.3.3 Construction

- Open parking garages must be one of the following types of construction:
  - Type I, Type II, Type IV.

Note: The following are cited as having applicable requirements:
Chapter 16, "Structural Design."
406.2.4, "Vehicle barrier systems."

406.3.3.1 Openings

- Openings in exterior walls for natural ventilation at each level are required as follows:
  - To be on ≥ 2 sides of the building.
  - To be uniformly distributed.
  - Area of openings to be ≥ 20% of the total building perimeter wall area at each tier.
  - Total length of openings at each level to be ≥ 40% of the building perimeter as follows:
    Where they are not evenly distributed on opposite sides of the building.
  - Total length of openings at each level does not have a minimum in the following case:
    Where they are evenly distributed on opposite sides of the building.
- Interior walls are to be open as follows:
  - Open area to be > 20% of the wall area.
  - Open area to be evenly distributed.

406.3.4 Uses

- Limited mixed uses are permitted in an open parking garage building.

Note: The following sections are cited as governing the mixed uses permitted:
402.7.1, "Attached garage."
406.3.13, "Prohibitions."
508.1, "General."
509.3, "Group S-2 enclosed parking garage with Group S-2 open parking garage above."
509.4, "Parking beneath Group R."
509.7, "Open parking garage beneath Groups A, I, B, M and R."

406.3.5 Area and height

- The following is limited for open parking garages:
  - Height.
  - Area.

Note: Chapter 5, "General Building Heights and Areas," is cited as governing area and height in S-2.
508.1, "General," is cited as having applicable requirements.

406 Motor-Vehicle-Related Occupancies

406.3.5.1 Single use

- Only the following are permitted to be housed in garages governed by this section:
  - Private vehicles.
  - Any of the following functions restricted as indicated below:

    Functions:

    Office.

    Waiting room.

    Toilet rooms.

    Restrictions:

    At grade level only.

    Sum of all areas must be ≤ 1,000 sf.

    Separation from the garage is not required.
- In garages with a spiral floor or sloped floor, area is limited as follows:
  - The area of the projected plan at any horizontal section is limited to that for a parking tier.
- In garages with a continuous spiral type floor, a tier is defined as follows:
  - Each 9'-6" height or portion thereof constitutes a parking tier.
- Parking tier clear heights are governed as follows:
  - Height must be ≥ 7' where mechanical parking-access devices are not used.
  - Height < 7' is permitted where approved and where mechanical parking-access devices are used.
- Area in Type I construction is not limited.
- Area per tier in permitted construction types other than Type I is as follows:
  - Limited to ≤ 50,000 sf.
- Heights of garages are governed as follows:
  - Not limited in Type IA construction.
  - In other types of construction height is limited as indicated in the following table:

Table 406.3.5.1 Height Limits of Open Parking Garages

Type of construction	Height of garages with ramp access	Height of garages with mechanical access without sprinklers	Height of garages with mechanical access with sprinklers
IB	12 tiers	12 tiers	18 tiers
IIA	10 tiers	10 tiers	15 tiers
IIB	8 tiers	8 tiers	12 tiers
IV	4 tiers	4 tiers	4 tiers

Source: IBC Table 406.3.5.

Note: The following are cited as governing garage height and area:

IBC Table 406.3.5, "Open Parking Garages Area and Height."

406.3.6, "Area and height increases."

406 Motor-Vehicle-Related Occupancies

406.3.6 Area and height increases *(part 1 of 2)*

- Area is unlimited in garages with the following characteristics:
  - Type II construction.
  - Opening on each side must be ≥ 50% of the inside wall area uniformly distributed on each side.
  - Building height ≤ 75'.
  - Every point on the tier must be ≤ 200' horizontally from any of the following openings:
    - Openings as defined above.
    - Certain natural ventilation openings including the following:
      - Such openings may be in courts that are ≥ 20' wide for the full width of the opening.

Note: 406.3.3.1, "Openings," is cited as defining qualifying natural ventilation openings.

- For garages with the following openings, height and area limitations are indicated below:
  - Openings:
    - ≥ 3/4 of the building perimeter must have sides open.
    - A side that is open must have ≥ 50% of its area open as measured inside the wall.
    - An open side must have its openings uniformly distributed along its length.
  - Area per tier:
    - Area is unlimited for garages of Type I construction.
    - Area per tier for other permitted construction types is limited as follows:
      - Where 3/4 of the building perimeter is open, area must be ≤ 62,500 sf.
      - Where all the building perimeter is open, area must be ≤ 75,000 sf.
  - Height:
    - Height is unlimited for garages of Type IA construction.
    - Height limitations for other construction types are listed below:

Table 406.3.6a Increased Height Limits of Open Parking Garages

Type of construction	Height of garages with ramp access	Height of garages with mechanical access without sprinklers	Height of garages with mechanical access with sprinklers
IB	13 tiers	13 tiers	19 tiers
IIA	11 tiers	11 tiers	16 tiers
IIB	9 tiers	9 tiers	13 tiers
IV	5 tiers	5 tiers	5 tiers

- For the following conditions, tier area limits may be increased as indicated below:
  - Conditions:
    - Garage height must be less than the maximum otherwise permitted.
    - Openings must be located on ≥ 3 sides of the garage.
    - Openings must a clear height ≥ 2'-6".
    - Openings must extend for ≥ 80% of the length of each side where they are located.
    - Every point on the tier must be ≤ 200' measured horizontally from such an opening.
    - Openings must face one of the following:
      - Street ≥ 30' wide for the full opening length.
      - Yard ≥ 30' wide for the full opening length and have access to a street.
    - Standpipes are required on each such tier.

Note: IBC Table 406.3.5, "Open Parking Garages Area and Height," is cited as listing area and height.

406 Motor-Vehicle-Related Occupancies

406.3.6 Area and height increases *(part 2 of 2)*

- Tier areas are limited only by the total garage areas indicated in the following tables:

Table 406.3.6b Limitations for Total Garage Area with Ramp Access

Type of construction	Height of garage with ramp access	Total garage area
IIA	≤ 9 tiers	≤ 500,000 sf
IIB	≤ 7 tiers	≤ 400,000 sf
IV	≤ 3 tiers	≤ 200,000 sf

Table 406.3.6c Limitations for Total Garage Area with Mechanical Access

Type of construction	Height of garage with mechanical access	Total garage area
Without sprinklers:		
IIA	≤ 9 tiers	≤ 500,000 sf
IIB	≤ 7 tiers	≤ 400,000 sf
IV	≤ 3 tiers	≤ 200,000 sf
With sprinklers:		
IIA	≤ 14 tiers	≤ 750,000 sf
IIB	≤ 11 tiers	≤ 600,000 sf
IV	≤ 3 tiers	≤ 200,000 sf

406.3.7 Fire separation distance

- Exterior walls and openings must have fire resistance as required for other building types based on the following conditions:
  - Construction type.
  - Occupancy.
  - Fire separation distance.

Note: The following are cited as the source of requirements for the components above:
IBC Table 601, "Fire-Resistance Rating Requirements for Building Elements."
IBC Table 602, "Fire-Resistance Rating Requirements for Exterior Walls Based on Fire Separation Distance."

- The distance required between a garage and its property lines is similar to that for any building type based on the following:
  - The fire resistance of exterior walls and openings.
  - Construction type.
  - Occupancy.
  - Fire separation distance.

Note: The following are cited as sources for determining distance to property line:
IBC Table 602, "Fire-Resistance Rating Requirements for Exterior Walls Based on Fire Separation Distance."
Section 705, "Exterior Walls."

406 Motor-Vehicle-Related Occupancies

406.3.8 Means of egress

- Exit stairways are governed as follows:
  - Where only parking attendants have access:
    ≥ 2 exit stairways are required.
    Exit stairways must be ≥ 3' wide.
  - Where persons other than parking attendants have access:
    Means of egress requirements applicable to the occupancy apply.

Note: Chapter 10, "Means of Egress," is cited as governing open parking garages.

- Lifts are governed as follows:
  - May be provided for employee use only.
  - Must be enclosed by noncombustible materials.

406.3.9 Standpipes

- Standpipes are required in open parking garages as for any building type based on the following:
  - Building height.
  - Building area.
  - Occupant load.
  - Nature of building.

Note: Chapter 9, "Fire Protection Systems," is cited as governing standpipes.

406.3.10 Sprinkler systems

- Where sprinklers are required in open parking garages, the following applies:
  - They must comply with requirements similar to those for most other occupancies.

Note: Chapter 9, "Fire Protection Systems," is cited as governing sprinklers.

406.3.11 Enclosure of vertical openings

- This section addresses open parking garages.
- Where the public has access, the following applies:
  - Vertical openings must be enclosed as required for means of egress.
- Where only parking attendants have access:
  - ≥ 2 exit stairways are required.
  - Exit stairways must be ≥ 3' wide.
- Lifts must be enclosed by noncombustible materials.
- Other vertical openings are not required to be enclosed.

Note: 406.3.8, "Means of egress," is cited as governing vertical openings and is partially summarized above.

406.3.12 Ventilation

- In open parking garages more ventilation than that provided by required openings is not required.

Note: 406.3.3.1, "Openings," is cited as the source of requirements for openings.

406 Motor-Vehicle-Related Occupancies

406.3.13 Prohibitions

- The following uses are not permitted in open parking garages:
  - Vehicle repairs.
  - Parking of the following vehicles:
    - Buses.
    - Trucks.
    - Similar vehicles.
  - Partial or complete closing of required openings in exterior walls as follows:
    - By tarpaulins.
    - By any other means.
  - Dispensing fuel.

406.4.1 Heights and areas

- This section addresses enclosed parking garages.
- Garages and parts of garages that do not qualify as an open parking garage are governed as follows:
  - They must meet height and area limitations based on construction type.
- Parking is permitted on the roofs of enclosed parking garages.

Note: IBC Table 503, "Allowable Building Heights and Areas," is cited as governing the heights and areas of these garages as modified by Section 504, "Building Height," Section 506, "Building Area Modifications," and Section 507, "Unlimited Area Buildings," as applicable.

406.4.2 Ventilation

- Mechanical ventilation is required for enclosed parking garages.

Note: The International Mechanical Code *is cited as the source governing such ventilation.*

406.5.1 Construction

- Motor fuel-dispensing facilities are governed by the following:
  - This section series.
  - Other standards.

Note: The following are cited as governing motor fuel-dispensing facilities:

The International Fire Code.
406.5.1, "Construction."
406.5.2, "Vehicle fueling pad."
406.5.3, "Canopies."
406.5.3.1, "Canopies used to support gaseous hydrogen systems."

406.5.2 Vehicle fueling pad

- Vehicle fueling pads must be of one of the following materials:
  - Concrete without a coating.
  - Other approved paving with the following property:
    - With a resistivity ≤ 1 megohm as follows:
      - Resistivity to be measured by the specified method.

Note: EN 1081, "Resilient Floor Coverings — Determination of the Electrical Resistance," is cited as the specified method for determining resistivity.

406 Motor-Vehicle-Related Occupancies

406.5.3 Canopies

- This section addresses canopies over fuel dispensing pumps.
- Canopies and canopy supports must be constructed of one or more of the following:
  - Noncombustible materials.
  - Fire-retardant-treated wood.
  - Heavy timber as follows:
    - Sizes must comply with Type IV construction.
  - Construction having a 1-hr fire-resistance rating.

  Note: Chapter 23, "Wood," is cited as the source of requirements for fire-retardant-treated wood.

- Combustible materials at canopies must comply with one of the following:
  - They must be shielded from the pumps by one or more of the following:
    - Noncombustible materials.
    - Heavy timber as follows:
      - Sizes must comply with Type IV construction.
  - Plastics are governed as follows:
    - Must be covered by one of the following:
      - Aluminum ≥ 0.010" thick.
      - Corrosion-resistant steel with a base metal thickness ≥ 0.016".
    - Must be tested in the same form as installed as follows:
      - Must have a flame spread ≤ 25.
      - Must have a smoke-developed index ≤ 450.
      - Must have a self-ignition temperature ≥ 650° F.

  Note: The following are cited as the standards with which the plastics must comply:
  - *ASTM D 1929, "Standard Test Method for Determining Ignition Temperature of Plastics."*
  - *One of the following requires compliance:*
    - *UL 723, "Standard for Test for Surface Burning Characteristics of Building Materials."*
    - *ASTM E 84, "Standard Test Method for Surface Burning Characteristics of Building Materials."*

  - Light-transmitting plastics are governed as follows:
    - Panels must be located ≥ 10' from any building on the same property.
    - Panels must face yards or streets ≥ 40' wide on other sides.
    - Total area of all panels must be ≤ 1,000 sf.
    - Area of a single panel must be ≤ 100 sf.
- Canopy height is governed as follows:
  - Clear height in the drive area must be ≥ 13'-6".

407 Group I-2

407.2 Corridors

- The enclosure of Occupancy I-2 corridors is governed as follows:
  - Each corridor must be continuous to an exit.
  - Corridors may be open to the spaces indicated below where design and construction meet minimum requirements for fire safety:
    - Waiting areas.
    - Nurses' stations.
    - Mental health treatment areas.
    - Gift shops.
  - Otherwise, corridors must be separated from other spaces for purposes of smoke protection.

Note: The following are cited as sources of requirements for space opening to a corridor:
407.2.1, "Waiting and similar areas," which addresses waiting rooms.
407.2.2, "Nurses' stations."
407.2.3, "Mental health treatment areas."
407.2.4, "Gift shops."
407.3, "Corridor walls," for walls required to separate corridors from other spaces.

407.2.1 Waiting and similar areas

- Waiting and similar areas may be open to a corridor only where all of the following apply:
  - The areas may not be used for the following:
    - Patient sleeping rooms.
    - Treatment rooms.
    - Hazardous uses.
    - Incidental uses having their own separation requirements.

Note: 508.2, "Accessory occupancies," is cited as a source of requirements for incidental use areas and identifies areas that may not open to a corridor.

  - The area must be constructed as corridors are required to be constructed.
  - The area must be protected by a fire detection system.
  - One of the following is required for a smoke compartment open to a corridor:
    - The corridor must be protected by a fire detection system.
    - The entire smoke compartment must be protected by quick-response sprinklers.
  - Access to required exits may not be obstructed by the layout of the area.

Note: The following are cited as sources of requirements for this section:
Section 907, "Fire Alarm and Detection Systems," governs the fire detection systems.
903.3.2, "Quick-response and residential sprinklers," governs these sprinklers.

407 Group I-2

407.2.2 Nurses' stations

- The following areas may be open to corridors where they have all the characteristics listed below:
  - Areas:
    - Doctors' spaces.
    - Nurses' spaces.
  - Required characteristics:
    - Use of areas:
      - Charting.
      - Communications.
      - Related clerical work.
    - Construction of areas:
      - Areas must be constructed according to requirements for the corridor.

407.2.3 Mental health treatment areas

- Where any of the following spaces are open to the corridor, they must comply with all the requirements listed below:
  - Spaces:
    - Spaces housing patients who are not capable of self-preservation.
    - Group meeting spaces.
    - Multipurpose therapeutic spaces.

Note: 508.2.5, "Separation of incidental accessory occupancies," is cited as identifying spaces which are prohibited from being open to corridors and, thus, are not addressed by this section.

  - Requirements:
    - Each smoke compartment is limited to 1 such space open to the corridor.
    - Space must have continuous supervision by staff.
    - Each space must be ≤ 1,500 sf.
    - Space layout does not obstruct access to required exits.
    - Walls and ceilings are constructed as corridors are required to be constructed.
    - Space must have a fire detection system.

Note: Section 907.2, "Where required—new buildings and structures," is cited as the source of requirements for the fire detection system required by this section.

407.2.4 Gift shops

- Gift shops may be open to the corridor where all of the following apply:
  - The gift shop must be < 500 sf.
  - The gift shop must be sprinklered.
  - The gift shop storage area must be sprinklered.

Note: 508.2.5, "Separation of incidental accessory occupancies," is cited as having applicable requirements.

407.3 Corridor walls

- Corridor walls must be smoke partitions.

Note: Section 711, "Smoke Partitions," is cited as governing these partitions.

407 Group I-2

407.3.1 Corridor doors

- This section governs corridor doors in Occupancy I-2.
- Doors as follows must meet protection requirements consistent with the fire-resistance rating of the walls:
  - Doors in walls enclosing the following:
    - Incidental use spaces.
    - Vertical opening.
    - Exits.

Note: 508.2.5, "Separation of incidental accessory occupancies," is cited as the source listing incidental areas requiring protected doors.
715.4, "Fire door and shutter assemblies," is cited as the source of protection requirements for doors in fire-resistance-rated walls.

- Doors in other corridor walls are governed as follows:
  - A fire-protection rating is not required.
  - Self-closing devices are not required.
  - Automatic closing devices are not required.
  - Doors must limit the transfer of smoke.
  - Positive latching is required.
  - Roller latches are prohibited.

407.3.2 Locking devices

- The following locking devices in Occupancy I-2 are governed as indicated below:
  - Locking devices:
    - Those restricting access to the patient room from the corridor.
    - Those operable only by staff from the corridor side.
  - Requirements:
    - In mental health facilities:
      - Such locks may restrict the means of egress for patient rooms.
    - In other facilities:
      - Such locks may not restrict the means of egress for patient rooms.

407.4 Smoke barriers

- This section addresses the requirement for smoke barriers in Occupancy I-2.
- The following stories must be divided as indicated below:
  - Stories:
    - Where patients sleep.
    - Where patients receive treatment.
    - Other stories where the occupant load is ≥ 50.
  - Requirements:
    - Story must be divided by smoke barriers.
    - Story must have ≥ smoke compartments:
      - Area of compartment must be ≤ 22,500 sf.
      - Travel in the compartment is limited as follows:
        - From any point to a smoke-barrier door must be ≤ 200'.

Note: Section 710, "Smoke Barriers," is cited as the source of requirements for such components.

407 Group I-2

407.4.1 Refuge area

- This section addresses refuge area requirements in smoke compartments of Occupancy I-2.
- The area required for refuge in a smoke compartment is computed as follows:
  - For floors where patients are confined to a bed or litter:
    - Refuge area required per patient is ≥ 30 sf:
      - Patient count includes only those from the adjoining compartment.
  - For floors where patients are not confined to a bed or litter:
    - Refuge area required per occupant is ≥ 6 sf:
      - Occupant count includes only those from the adjoining compartment.
- The refuge area required in smoke compartments is to be distributed in one or more of the following locations:
  - Corridors.
  - Patient rooms.
  - Treatment rooms.
  - Lounges.
  - Dining areas.
  - Other low-hazard areas.

Note: The area required for refuge is in addition to the area required to meet means of egress minimums.

407.4.2 Independent egress

- This section addresses means of egress from smoke compartments in Occupancy I-2.
- Each smoke compartment requires a means of egress as follows:
  - Egress may not return to the compartment of origin.

407.4.3 Horizontal assemblies

- Smoke barriers that are required by this section are governed as follows:
  - Horizontal assemblies supporting the barriers must inhibit the passage of smoke.

Note: 712.9, "Smoke barrier," is cited as governing horizontal assemblies supporting smoke barriers.

407.5 Automatic sprinkler system

- Smoke compartments with patient sleeping units must be sprinklered as follows:
  - Using one of the following approved systems:
    - Quick-response sprinklers.
    - Residential sprinklers.

Note: 903.3.1.1, "NFPA 13 sprinkler systems," is cited as governing the sprinklers.
903.3.2, "Quick-response and residential sprinklers," is cited as governing this type sprinkler.

407 Group I-2

407.6 Fire alarm system

- The following locations require an automatic fire detection system:
  - Corridors in intermediate-care nursing homes.
  - Corridors in skilled nursing facilities.
  - Corridors in detoxification facilities.
  - Spaces open to corridors.

Note: 407.2, "Corridors," is cited as defining spaces permitted to be open to corridors.

- Corridor smoke detection is not required for either of the following conditions:
  - Condition 1:
    - Where patient sleeping units have smoke detectors as follows:
      - With an audible alarm at the nurses' station.
      - With a visual alarm at the nurses' station.
      - With a visual display on the corridor side of each sleeping unit.

Note: UL 268, "Smoke Detectors for Fire Protective Signaling Systems," is cited as the standard governing these smoke detectors.

  - Condition 2:
    - Where patient sleeping units have doors equipped with the following:
      - Automatic closing devices as follows:
        - With integral smoke detectors:
          - Located on the room side.
          - Installed according to their listing.
          - That perform the necessary alert.
- Certain hospital corridors and spaces open to corridors require smoke detection.

Note: 407.2, "Corridors," is cited as identifying the locations where smoke detection is required.

407.8 Secured yards

- Yard areas may be secured as follows where the conditions listed below exist:
  - Security:
    - Surrounded by fencing.
    - Gates provided with locks.
  - Required conditions:
    - The following safe dispersal area must be provided:
      - The area must be located as follows:
        - Between the building and the fencing.
        - ≥ 50' from the building.
      - The following net area must be provided for each patient:
        - 30 sf for the following:
          - Bed patients.
          - Litter patients.
        - 6 sf for the following:
          - Ambulatory patients.
          - Occupants who are not patients.

408 Group I-3

408.1 General

- Occupancy I-3 must comply with the following:
  - This section.
  - Other applicable sections of the code.

Note: 308.4, "Group I-3," is cited as a section applicable to this occupancy.

408.2 Other occupancies

- The following applies where an I-3 space occurs in an area with a different occupancy designation:
  - The larger area must meet requirements for its occupancy designation.
  - The following applies where means of egress must be locked for security purposes:
    The release of occupants must be possible at all times.
- Where means of egress from the following I-3 occupancies pass through other use designations, the requirements indicated below apply:
  - I-3 occupancies:
    Detention.
    Correctional.
  - Requirements:
    Egress through a horizontal exit into an occupancy that does not meet I-3 egress requirements is permitted as follows:
      The other occupancy must meet its own egress requirements.
      The other occupancy may not be an H occupancy.
    In all other cases the means of egress must comply with I-3 requirements.

408.3 Means of egress

- Means of egress for I-3 occupancies are governed by the following:
  - This section governs those aspects addressed.
  - Other code requirements govern aspects not addressed in this section.

Note: Chapter 10, "Means of Egress," is cited as governing egress issues not covered in this section.

408.3.1 Door width

- Resident sleeping-unit doors must have a clear width of ≥ 2'-4".

408.3.2 Sliding doors

- Horizontal sliding doors in an Occupancy I-3 means of egress are governed as follows:
  - Doors must fully open under the following conditions:
    With an opening force ≤ 50 lbs in either of the following circumstances:
      Simultaneously with a force ≤ 50 lbs applied ⊥ to the door.
      With no other force applied to the door.

408.3.3 Guard tower doors

- Guard towers may be accessed through the floor by the following:
  - An access panel with the following dimensions:
    Area ≥ 16 sf.
    All dimensions ≥ 2'.

408 Group I-3

408.3.4 Spiral stairways

- Spiral stairs may be used in Occupancy I-3 for staff operational purposes.

  Note: 1009.9, "Spiral stairways," is cited as governing such stairs.

408.3.5 Ship ladders

- Ship ladders may serve in a means of egress from the following:
  - Control rooms.
  - Elevated facility observation rooms.

  Note: 1009.11, "Ship ladders," is cited as governing these components.

408.3.6 Exit discharge

- Exits may discharge from an I-3 occupancy into the following area:
  - Courtyard as follows:
    - Fenced or walled.
    - Enclosed yard must accommodate all occupants in a zone as follows:
      - Zone to be located ≥ 50' from the building.
      - Size of zone must provide ≥ 15 sf per person.

408.3.7 Sallyports

- Sallyports may be located in an Occupancy I-3 means of egress as follows:
  - Where unobstructed travel through them is possible during emergencies.

408.3.8 Exit enclosures

- An exit enclosure of an I-3 occupancy may have glazing where all the following conditions apply:
  - Only one required exit enclosure per building may have glazing.
  - Glazing is permitted only in the following:
    - In doors at landings.
    - In interior walls at landings serving enclosure access.
  - Stairway is limited to serving ≤ 4 floor levels.
  - Doors must be fire doors with the larger fire-protection rating of the following:
    - ≥ that required for the fire-resistance rating of their wall.
    - ≥ 3/4 hour.
  - Total glazed area at each floor must be ≤ 5,000 sq in.
  - Individual panels of glazing must be ≤ 1,296 sq in.
  - Sprinklers must protect both sides of the glazing by wetting the entire surfaces.
  - Glazing must be in a gasketed frame as follows:
    - Frame must be able to deflect prior to sprinkler activation without breaking the glass.
  - Obstructions such as the following are not allowed between the sprinklers and the glazing:
    - Curtains and curtains rods.
    - Drapes and drapery traverse rods.
    - Similar obstructions.

Note: 715.4, "Fire door and shutter assemblies," is cited as the source of fire-protection ratings for fire doors as based on the fire-resistance rating of their walls.

408 Group I-3

408.4.1 Remote release

- In Occupancy I-3, remote control of locks is not required where all the following apply:
  - In restraint condition 4.
  - Locks that must be opened to move occupants to an area of refuge from a smoke compartment are governed as follows:
    - They must number ≤ 10.
    - The number of separate keys needed for all the locks is limited to ≤ 2.
    - Movement of all occupants to the area of refuge must be possible in ≤ 3 minutes.
- Otherwise, locks on required doors in a means of egress of Occupancy I-3 must have a remote release capability as follows:
  - Devices activating lock releases must be in a location remote from the resident living areas.
  - Locks preventing egress in the following conditions of restraint must meet the requirement indicated below:
    - Conditions of restraint:
      - Condition 3.
      - Condition 4.
    - Requirement:
      - Locks must be releasable in ≤ 2 minutes as follows:
      - By the minimum staff available.
      - At any time.

408.4.2 Power-operated doors and locks

- Emergency power for the following door mechanisms is not required in Occupancy I-3 for the conditions indicated below:
  - Mechanisms:
    - Power-operated sliding doors.
    - Power-operated locks for swinging doors.
  - Conditions:
    - In Occupancy I-3 restraint condition 4:
      - ≤ 10 locks must be opened to move occupants to an area of refuge.
      - ≤ 2 separate keys are needed for all the locks.
      - Movement of all occupants to the area of refuge must be possible in ≤ 3 minutes.

Note: The exception to 408.4.1, "Remote release," is cited as applicable and is summarized above.

- Otherwise, the following door mechanisms must comply with requirements indicated below:
  - Mechanisms:
    - Power-operated sliding doors.
    - Power-operated locks for swinging doors.
  - Requirements:
    - Manual release mechanism at the door must be provided.
    - One of the following must be provided:
      - Remote mechanical release for the door mechanisms.
      - Emergency power to door mechanisms.

408 Group I-3

408.4.3 Redundant operation

- In Occupancy I-3, the following mechanisms must have the redundant systems listed below:
  - Locks:
    - Remote release, mechanically operated sliding doors.
    - Remote release, mechanically operated locks.
  - Redundant systems:
    - Mechanically operated release mechanism at each door.
    - Redundant remote release mechanism.

408.4.4 Relock capability

- Doors in Occupancy I-3 that are unlocked remotely in an emergency are governed as follows:
  - Doors may not relock automatically upon closing without the following action:
    - Specific action is required at the control location to permit relocking.

408.5.1 Floor openings

- Openings in floors in housing units are governed as follows:
  - They need not have an enclosure where all the following conditions exist:
    - Staff can supervise the entire occupied area due to its openness.
    - The means of egress can accommodate all occupants from the following:
      - All interconnected cell tiers and areas.
    - The vertical distance between the following is ≤ 23':
      - Lowest and highest cell tier floor levels.
    - Egress between the following may not require travel on > 1 additional floor level in the housing unit:
      - From any part of a cell tier to an exit or exit access door.

408.5.2 Shaft openings in communicating floor levels

- This section addresses plumbing chases serving the following:
  - Cells within a residential housing area stacked vertically.
- Such chases do not require enclosure in the following case:
  - Where floor openings are permitted between floor levels open to each other.

Note: 408.5.1, "Floor openings," is cited as permitting certain floor openings.

408.6 Smoke barrier *(part 1 of 2)*

- This section addresses the division of Occupancy I-3 stories into smoke compartments.
- Spaces with the following characteristics are not required to be protected by smoke barriers:
  - Spaces must exit directly to one of the following locations:
    - A public way.
    - A building separated from resident housing by one of the following:
      - Construction with a fire-resistance rating ≥ 2 hours.
      - 50' of open space.
    - A secure yard or court with a holding space as follows:
      - Holding space provides ≥ 6 sf per occupant including the following:
        - Staff.
        - Residents.
        - Visitors.
      - Holding space is located ≥ 50' from resident housing.

408 Group I-3

408.6 Smoke barrier *(part 2 of 2)*

- The locking methods for doors in the exit system must meet the following:
  Restraint-condition requirements for the space.
- Otherwise, each of the following stories must be subdivided as indicated below:
  - Stories:
    Where residents sleep.
    With an occupant load ≥ 50 persons.
  - Division requirements:
    Each story must be divided into ≥ 2 smoke compartments by smoke barriers.

Note: The following are cited as governing these components.
408.8, "Subdivision of resident housing areas."
Section 710, "Smoke Barriers."

408.6.1 Smoke compartments

- Smoke compartments in Occupancy I-3 must have the following characteristics:
  - Number of residents in each compartment must be ≤ 200.
  - Travel distance between the following must be ≤ 150':
    Any room door required for exit access.
    The nearest door in a smoke barrier.
  - Travel distance between the following must be ≤ 200':
    Any point in the smoke compartment.
    The nearest door in a smoke barrier.

408.6.2 Refuge area

- This section addresses refuge areas in Occupancy I-3 smoke compartments.
- A refuge area is required in each smoke compartment as follows:
  - ≥ 6 sf is required for each occupant seeking refuge.
  - The occupant count for computing refuge area is based on the following:
    The occupant load of the adjacent smoke compartment.
  - The occupant count for computing refuge area does not include the following:
    Original occupants in the smoke compartment housing the refuge area.
  - The refuge area must be immediately available for occupation in a fire emergency.

Note: The required refuge area is in addition to the area required for the original occupants of the smoke compartment.

408.6.3 Independent egress

- This section addresses smoke compartments formed by smoke barriers in Occupancy I-3.
- A means of egress is required from each smoke compartment as follows:
  - Means of egress may not require occupants to return to the smoke compartment which they departed.

408 Group I-3

408.7 Security glazing

- This section addresses the following assemblies in Occupancy I-3 having a fire-resistance rating of 1 hr:
  - Fire barriers.
  - Fire partitions.
  - Smoke barriers.

Note: The following are cited as governing the assemblies addressed by this section.
Section 707, "Fire Barriers."
Section 709, "Fire Partitions."
Section 710, "Smoke Barriers."

- Security glazing is allowed in windows and doors of the assemblies where all of the following exist:
  - Each glazed area is ≤ 1,296 sq in.
  - Glazing is protected by sprinklers on both sides as follows:
    Upon activation sprinklers spray water on the whole surface subject to fire.
  - Glazing is mounted in a gasketed frame as follows:
    The system will deflect without breaking the glass before the sprinklers activate.
  - Obstructions, such as the following, are not located between the sprinklers and the glazing:
    Curtain rods.
    Drapery traverse rods.
    Curtains.
    Drapes.
    Similar elements.

408.8.1 Occupancy Conditions 3 and 4

- This section addresses the separation required in Occupancy I-3 restraint conditions 3 and 4.
- Where travel distance on the following routes is > 50', the requirement indicated applies:
  - Route:
    Beginning in a sleeping area.
    Passing through the common space.
    To a corridor.
  - Requirement:
    Common spaces and adjacent sleeping areas must be separated as follows:
    By a smoke-tight partition.

408.8.2 Occupancy Condition 5

- This section addresses separations required in Occupancy I-3 restraint condition 5.
- Each sleeping area must be separated from the following adjacent spaces by smoke-tight partitions:
  - Other sleeping areas.
  - Corridors.
  - Common spaces.
- A smoke-tight partition must separate the following areas from each other:
  - Corridor.
  - Common space.

408 Group I-3

408.8.3 Openings in room face

- This section addresses sleeping rooms in Occupancy I-3 restraint conditions 2, 3, 4, and 5.
- Openings in the solid face of sleeping rooms must meet the following requirements:
  - Area of openings includes the following:
    Grilles.
    Food passages.
    Door undercuts.
    All other openings.
  - Area of all openings combined is limited to ≤ 120 sq in.
  - Openings must be located ≤ 3' above the floor.
  - In restraint condition 5, openings must be closeable from the room side.

408.8.4 Smoke-tight doors

- Doors in smoke-tight partitions are governed as follows:
  - Doors must be substantial.
  - Doors must resist the passage of smoke.
  - The following are not required on cell doors:
    Latches.
    Door closers.

Note: Section 408.8, "Subdivision of resident housing areas," is cited as identifying partitions that must be smoke-tight.

408.9 Windowless buildings

- A windowless building or part of a building is defined as having one or more of the following characteristics:
  - The building has windows that do not open.
  - The building has windows that are not readily breakable.
  - The building does not have windows.
- Each windowless building requires an engineered smoke control system as follows:
  - Must be capable of removing smoke from the windowless smoke compartment where the fire occurred.

Note: Section 909, "Smoke Control Systems," is cited as the source of requirements for ventilating smoke control compartments.

410 Stages and Platforms

410.2 Definitions

- **Fly gallery**
  - A floor level above a stage:
    - For the movement of scenery.
    - For controlling other stage effects.
- **Gridiron**
  - Structural framing over a stage supporting equipment:
    - For hanging and flying scenery.
    - For supporting other stage effects.
- **Pinrail**
  - A rail on or above a stage:
    - For holding belaying pin to which lines are fastened.
- **Platform**
  - A raised area within a building.
  - Used for any of the following purposes:
    - Worship.
    - Music.
    - Plays.
    - Entertainment.
    - Head table for special guests.
    - Lecturers or speakers.
    - Boxing ring.
    - Wrestling ring.
    - Theater-in-the-round stage.
    - Similar activities.
  - None of the following devices are present:
    - Overhead hanging curtains.
    - Drops.
    - Scenery.
    - Stage effects other than lighting and sound.
  - A platform installed for ≤ 30 days is a temporary platform.
- **Proscenium wall**
  - A wall between the stage and either of the following:
    - Auditorium.
    - Assembly seating area.
- **Stage**
  - A space in a building used for either of the following:
    - Entertainment.
    - Presentations.
  - Includes the following elements:
    - Overhead hanging curtains.
    - Drops.
    - Scenery.
    - Stage effects besides the following:
      - Lighting.
      - Sound.

410 Stages and Platforms

410.3.1 Stage construction

- Stage floors may be constructed as follows where all the conditions indicated below apply:
  - Construction:
    - Wood deck.
    - Nominal thickness ≥ 2".
  - Conditions:
    - Stage is separated from other areas.
    - Stage floor construction is one of the following types:
      - Type IIB, Type IV.

Note: 410.3.4, "Proscenium wall," is cited as governing the separation of the stage from other areas in this case.

- A stage floor is not required to have a fire-resistance rating where all of the following conditions apply:
  - Building is one of the following construction types:
    - Type IIA, Type IIIA, Type VA.
  - The space below the stage is provided with the following:
    - An automatic fire-extinguishing system.

Note: The following are cited as governing the options for the fire-extinguishing system required above:
Section 903, "Automatic Sprinkler Systems."
Section 904, "Alternative Automatic Fire-Extinguishing Systems."

- Stage finished floors may be constructed out of one of the following materials where the condition below applies:
  - Materials:
    - Wood.
    - Approved noncombustible materials.
  - Condition:
    - Openings in stage floors must have trap doors with the following characteristics:
      - Tight fitting with approved safety locks.
      - Solid wood.
- In all other cases, stages must be constructed as follows:
  - Using materials required for the building construction type.

410.3.1.1 Stage height and area

- Stage area is measured including the following:
  - Performance area.
  - Spaces adjacent to performance area as follows:
    - Not separated by fire-resistance-rated construction.
    - Backstage.
    - Support areas.
- Stage height is measured from the lowest point on the stage floor to the following:
  - The highest point on the underside of one of the following:
    - Roof deck or floor deck above the stage.

410 Stages and Platforms

410.3.2 Galleries, gridirons, catwalks and pinrails

- Floors of the following may be constructed out of any approved material:
  - Fly galleries.
  - Catwalks.
- Materials for beams supporting only the following elements are governed as indicated below:
  - Elements:
    - Theater equipment:
      - Portable.
      - Fixed.
    - Gridirons.
    - Galleries.
    - Catwalks.
  - Beam materials:
    - Must be approved.
    - Must meet requirements of the construction type for the building.
    - Are not required to have a fire-resistance rating.
- For application of code requirements, the following elements do not constitute any of the components listed below:
  - Elements:
    - Fly galleries.
    - Gridirons.
    - Catwalks.
  - Components:
    - Floors
    - Mezzanines
    - Stories
    - Levels

410.3.3 Exterior stage doors

- Where protection is required for the following openings, the requirement below applies:
  - Openings:
    - Exterior exit doors from the stage.
  - Requirement:
    - Such doors must be fire doors.

Note: Section 715, "Opening Protectives," is cited as governing fire doors as required above.

- The following exterior openings from a stage are governed as indicated below:
  - Openings:
    - Exterior exit doors.
    - Exterior doors for loading and unloading.
  - Requirements:
    - Vestibules required as follows:
      - Where doors may be open while the theater is occupied.
      - To prevent air drafts into the auditorium.

410 Stages and Platforms

410.3.4 Proscenium wall

- The following applies where stage height is > 50':
  - All areas of the stage must be separated from the seating area as follows:
    - By a proscenium wall as follows:
      - Wall must have a fire-resistance rating ≥ 2 hr.
      - Wall must be continuous between the foundation and roof.

410.3.5 Proscenium curtain

- This section applies to assembly areas with a proscenium wall required to have a fire-resistance rating.
- One of the following is required:
  - An approved water curtain for the stage opening.
  - A fire curtain of approved material for the stage opening.
  - One of the following where audience seating design is not based on a smoke protection system:
    - A smoke control system.
    - A natural ventilation system designed to maintain smoke ≥ 6' above the means of egress floor.

Note: 903.3.1.1, "NFPA 13 sprinkler systems," is cited as governing water curtains.
NFPA 80, "Fire Doors and Other Opening Protectives," is cited as governing fire curtains.
1028.6.2, "Smoke-protected seating," is cited as describing smoke-protected seating design.
Section 909, "Smoke Control Systems," is cited as governing smoke control systems.

410.3.6 Scenery

- Materials for stage sets and scenery are governed as follows:
  - Combustible materials must be flame resistant.
  - Foam plastics are not addressed in this section.

Note: The following are cited as governing the above listed materials as indicated:
NFPA 701, "Standard Methods of Fire Tests for Flame-Propagation of Textiles and Films," for combustible materials.
Section 806, "Decorative Materials and Trim," for combustible materials.
Section 2603, "Foam Plastic Insulation," for foam plastics.
The International Fire Code *for combustible materials and foam plastics.*

410 Stages and Platforms

410.4 Platform construction

- Fire-retardant-treated wood may be used for platforms where all the following conditions are present:
  - The building is of one of the following types of construction:
    - Type I, Type II, Type IV, and the following apply:
      - The platform is ≤ 30" above the main floor.
      - The platform is ≤ 1/3 the room area.
      - The platform is ≤ 3,000 sf.
- In other cases, materials must meet requirements for the building construction type.
- Platform floors must have a fire-resistance rating ≥ 1 hr in the following case:
  - Where the space under the floor is used for any of the following:
    - Storage.
    - Purpose other than the following:
      - Equipment.
      - Wiring.
      - Plumbing.
  - The underside of the platform floor need not be protected in the following case:
    - Where the space under the floor is used only for the following:
      - Equipment.
      - Wiring.
      - Plumbing.

410.4.1 Temporary platforms

- This section addresses platforms installed for ≤ 30 days.
- Any material allowed by the code may be used for temporary platforms.
- The space under a temporary platform and above the building floor is not permitted to be used for any purposes except the following service to platform equipment:
  - Plumbing.
  - Electrical wiring.

410.5.1 Separation from stage

- Stages must be separated from the following spaces according to stage height as indicated below:
  - Spaces:
    - Dressing rooms.
    - Scene docks.
    - Property rooms and storerooms.
    - Workshops.
    - Compartments appurtenant to the stage.
    - Other parts of the building.
  - Stage height > 50' requires separation by the following:
    - Fire barriers and/or horizontal assemblies with a fire-resistance rating ≥ 2 hrs.
  - Stage height ≤ 50' requires separation by the following:
    - Fire barriers and/or horizontal assemblies with a fire-resistance rating ≥ 1 hr.

Note: The following are cited as governing the separating assemblies:
Section 707, "Fire Barriers."
Section 712, "Horizontal Assemblies."

410 Stages and Platforms

410.5.2 Separation from each other

- The following spaces must be separated from each other as indicated below:
  - Spaces:

Compartments appurtenant to the stage	Property rooms	Storerooms
Dressing rooms	Scene docks	Workshops

  - Separation:
    Fire barriers and/or horizontal assemblies with a fire-resistance rating ≥ 1 hr.

Note: The following are cited as governing the separating assemblies:
Section 707, "Fire Barriers."
Section 712, "Horizontal Assemblies."

410.5.3 Stage exits

- ≥ 1 approved means of egress is required from the following locations:
  - Each side of the stage.
  - Each side of the space under the stage.
- ≥ 1 means of escape is required from the following locations:
  - Each fly gallery.
  - The gridiron.
- The gridiron may have one of the following access devices to a scuttle in the stage roof:
  - Steel ladder.
  - Alternating tread stairway.
  - Spiral stairway.

410.6 Automatic sprinkler system

- Sprinklers are not required under stages areas with all of the following characteristics:
  - The space has a clear height < 4'.
  - The space is used only for storage of tables and chairs.
  - Concealed space is separated from other spaces by Type X gypsum board ≥ 5/8" thick.
- Sprinklers are not required for stages with all of the following characteristics:
  - Stages with ≤ 1,000 sf.
  - Stages ≤ 50' in height as follows:
    Where the following are not retractable vertically:
    Curtains and scenery.
    Other combustible hangings.
  - Combustible hangings are limited to the following:
    One main curtain.
    Borders and legs.
    One backdrop.
- Sprinklers are not required for stages for portable orchestra enclosures.
- Otherwise, stages require an automatic fire-extinguishing system in the following locations:
  - Under the roof and gridiron.
  - Under all catwalks and galleries over the stage.
  - In spaces accessory to the stage as follows:
    Dressing rooms and performer lounges.
    Shops and storerooms.

NOTES

5

General Building Heights and Areas

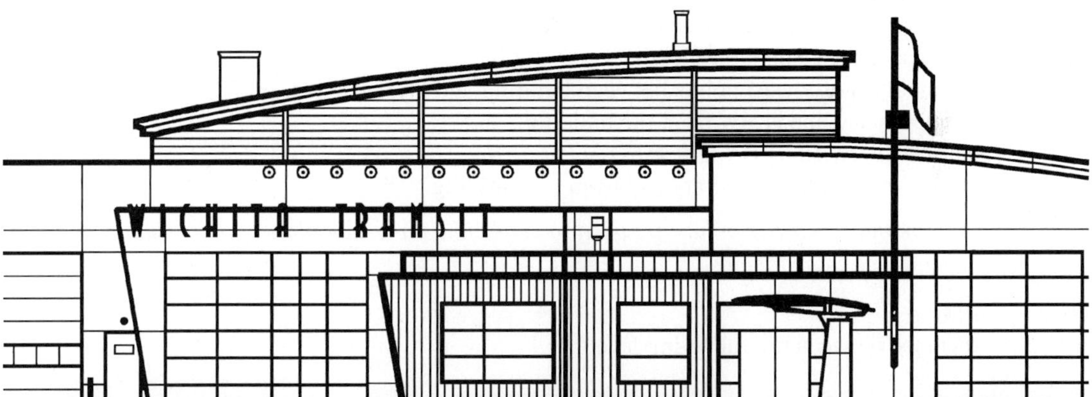

Wichita Transit Storage, Administration, and Maintenance Facility.
Wichita, Kansas. *(partial elevation)*
Wilson Darnell Mann, P.A., Architects. Wichita, Kansas.

502 Definitions

502.1 Definitions

- **Area, building**
  - Area within the inside surfaces of the following perimeters as applicable:
    Exterior walls.
    Exterior walls and fire walls.
  - Areas of vent shafts and courts are not included.
  - Usable area outside the exterior walls is included where it is under the following:
    Roof or floor above.
- **Equipment platform**
  - An unoccupied platform and associated components as follows:
    Elevated walkways.
    Stairs.
    Alternating tread devices.
    Ladders.
  - Used only for either of the following:
    Mechanical systems.
    Industrial process equipment.

Note: 505.5, "Equipment platforms," is cited as governing equipment platforms.

- **Grade plane**
  - Average level of finished grade at the building.
  - Points of elevation are taken at the building in the following cases:
    Where grade is level to a building façade.
    Where grade slopes down to a building façade.
  - Where grade slopes away from a building façade, the following apply:
    Where the property line is ≤ 6' from the building:
      Grade plane is based on the following:
        Average level of lowest points between building and property line.
    Where the property line is > 6' from the building:
      Grade plane is based on the following:
        Average level of lowest points ≤ 6' from the building.

Note: When more than 4 corner points are used to determine the grade plane, points should be evenly distributed along any façade.

- **Height, building**
  - Vertical distance between the following levels:
    Grade plane.
    Roof as follows:
      The top of a flat roof.
      A level halfway between the highest and lowest points of a sloped roof.
- **Mezzanine**
  - A level or levels between the following:
    Floor and ceiling of a story.

Note: Section 505, "Mezzanines," is cited as governing these levels.

503 General Building Height and Area Limitations

503.1 General *(part 1 of 4)*

- Building height and area are limited by this section according to construction type and occupancy.
- Each part of a building enclosed is considered to be a separate building as follows:
  - Within the exterior walls.
  - Within the exterior walls and fire walls where applicable.
- Height and area limitations are subject to modification in subsequent code sections.

Note: *The following are sections potentially modifying the requirements herein:*

406.1, "Private garages and carports," for Occupancy U facilities of this type.
406.3, "Open parking garages," for Occupancy S-2 facilities of this type.
415.5, "Special provisions for Group H-2 and H-3 occupancies."
504.2, "Automatic sprinkler system increase."
506.2, "Frontage increase."
506.3, "Automatic sprinkler system increase."
Section 507, "Unlimited Area Buildings."

IBC Table 503, "Allowable Building Heights and Areas," is cited as listing building height and area limits, which are shown in the tables on the following pages.
Section 302, "Classification," is cited as identifying occupancies.
Section 602, "Construction Classification," is cited as defining construction type.
Section 706, "Fire Walls," is cited as governing fire walls.

Case study: Fig. 503.1A. The Occupancy B building is Type IIB construction and is not sprinklered. A maximum height of 55' and 4 stories is permitted for this category of building. The building is 2 stories and 38'-4" high, measured to the average height of the highest roof; thus, it is in compliance with the code regarding height.

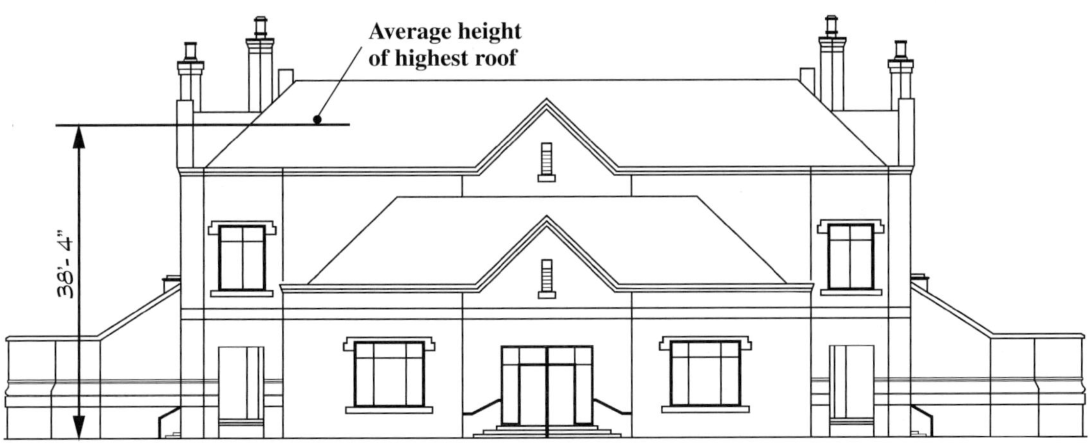

Fig. 503.1A. East elevation. Country Club Park Building One. Wichita, Kansas. Gossen Livingston Associates, Inc., Architecture. Wichita, Kansas.

503 General Building Height and Area Limitations

503.1 General *(part 2 of 4)*

Table 503.1 **Maximum Building Height and Area per Story**

		Type of construction								
Occ.	Height	IA	IB	IIA	IIB	IIIA	IIIB	IV	VA	VB
Occupancy A: maximum height in stories and feet										
A	Feet	UL	160	65	55	65	55	65	50	40
A-1	Stories	UL	5	3	2	3	2	3	2	1
A-2	"	UL	11	3	2	3	2	3	2	1
A-3	"	UL	11	3	2	3	2	3	2	1
A-4	"	UL	11	3	2	3	2	3	2	1
A-5	"	UL	UL	UL	UL	UL	UL	UL	UL	UL
Occupancy A: maximum area per story in square feet										
A-1	SF	UL	UL	15,500	8,500	14,000	8,500	15,000	11,500	5,500
A-2	"	UL	UL	15,500	9,500	14,000	9,500	15,000	11,500	6,000
A-3	"	UL	UL	15,500	9,500	14,000	9,500	15,000	11,500	6,000
A-4	"	UL	UL	15,500	9,500	14,000	9,500	15,000	11,500	6,000
A-5	"	UL	UL	UL	UL	UL	UL	UL	UL	UL
Occupancy B: maximum height in stories and feet										
B	Feet	UL	160	65	55	65	55	65	50	40
B	Stories	UL	11	5	3	5	3	5	3	2
Occupancy B: maximum area per story in square feet										
B	SF	UL	UL	37,500	23,000	28,500	19,000	36,000	18,000	9,000
Occupancy E: maximum height in stories and feet										
E	Feet	UL	160	65	55	65	55	65	50	40
E	Stories	UL	5	3	2	3	2	3	1	1
Occupancy E: maximum area per story in square feet										
E	SF	UL	UL	26,500	14,500	23,500	14,500	25,500	18,500	9,500

(Continued on next page)

Source: IBC Table 503.

UL = unlimited, Occ. = occupancy, SF = square feet.

503 General Building Height and Area Limitations

503.1 General *(part 3 of 4)*

Table 503.1—*Continued* **Maximum Building Height and Area per Story**

Occ.	Height	IA	IB	IIA	IIB	IIIA	IIIB	IV	VA	VB
		Type of construction								
Occupancy F: maximum height in stories and feet										
F	Feet	UL	160	65	55	65	55	65	50	40
F-1	Stories	UL	11	4	2	3	2	4	2	1
F-2	Stories	UL	11	5	3	4	3	5	3	2
Occupancy F: maximum area per story in square feet										
F-1	SF	UL	UL	25,000	15,500	19,000	12,000	35,500	14,000	8,500
F-2	SF	UL	UL	37,500	23,000	28,500	18,000	50,500	21,500	13,000
Occupancy H: maximum height in stories and feet										
H	Feet	UL	160	65	55	65	55	65	50	40
H-1	Stories	1	1	1	1	1	1	1	1	NP
H-2	"	UL	3	2	1	2	1	2	1	1
H-3	"	UL	6	4	2	4	2	4	2	1
H-4	"	UL	7	5	3	5	3	5	3	2
H-5	"	4	4	3	3	3	3	3	3	2
Occupancy H: maximum area per story in square feet										
H-1	SF	21,000	16,500	11,000	7,000	9,500	7,000	10,500	7,500	NP
H-2	"	21,000	16,500	11,000	7,000	9,500	7,000	10,500	7,500	3,000
H-3	"	UL	60,000	26,500	14,000	17,500	13,000	25,500	10,000	5,000
H-4	"	UL	UL	37,500	17,500	28,500	17,500	36,000	18,000	6,500
H-5	"	UL	UL	37,500	23,000	28,500	19,000	36,000	18,000	9,000
Occupancy I: maximum height in stories and feet										
I	Feet	UL	160	65	55	65	55	65	50	40
I-1	Stories	UL	9	4	3	4	3	4	3	2
I-2	"	UL	4	2	1	1	NP	1	1	NP
I-3	"	UL	4	2	1	2	1	2	2	1
I-4	"	UL	5	3	2	3	2	3	1	1
Occupancy I: maximum area per story in square feet										
I-1	SF	UL	55,000	19,000	10,000	16,500	10,000	18,000	10,500	4,500
I-2	"	UL	UL	15,500	11,000	12,000	NP	12,000	9,500	NP
I-3	"	UL	UL	15,000	10,000	10,500	7,500	12,000	7,500	5,000
I-4	"	UL	60,500	26,500	13,000	23,500	13,000	25,500	18,500	9,000

(Continued)

Source: IBC Table 503.

UL = unlimited, Occ. = occupancy, SF = square feet, NP = not permitted.

503 General Building Height and Area Limitations

Case study: Fig. 503.1B. The clothing factory is designated as Occupancy F-1. It is construction Type IIB and is not sprinklered. Consequently, IBC Table 503 limits the building to 2 stories and 55' in height. The building is 1 story and 19' to the average height of the roof surface; thus, it is in compliance with code requirements.

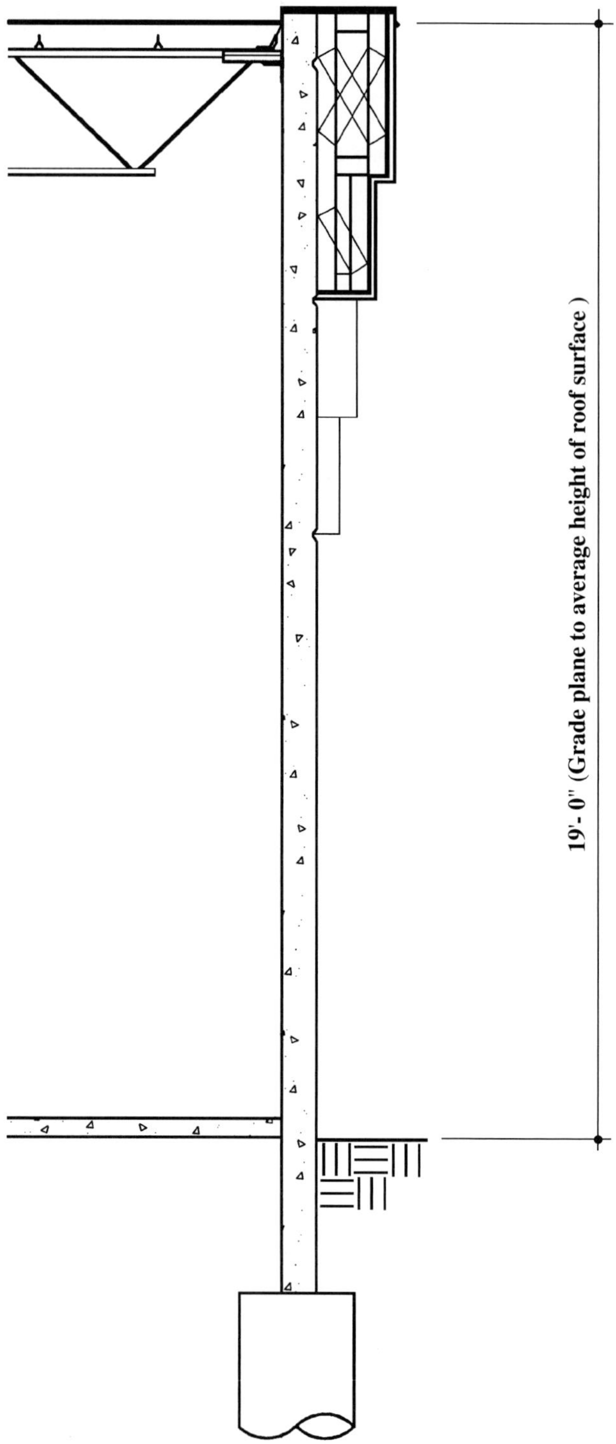

Fig. 503.1B. Wall section. Garments to Go. Bastrop, Texas. Spencer Godfrey Architects. Round Rock, Texas.

503 General Building Height and Area Limitations

503.1 General *(part 4 of 4)*

Table 503.1—*Continued* **Maximum Building Height and Area per Story**

Occ.	Height	Type of construction IA	IB	IIA	IIB	IIIA	IIIB	IV	VA	VB
Occupancy M: maximum height in stories and feet										
M	Feet	UL	160	65	55	65	55	65	50	40
M	Stories	UL	11	4	2	4	2	4	3	1
Occupancy M: maximum area per story in square feet										
M	SF	UL	UL	21,500	12,500	18,500	12,500	20,500	14,000	9,000
Occupancy R: maximum height in stories and feet										
R	Feet	UL	160	65	55	65	55	65	50	40
R-1	Stories	UL	11	4	4	4	4	4	3	2
R-2	"	UL	11	4	4	4	4	4	3	2
R-3	"	UL	11	4	4	4	4	4	3	3
R-4	"	UL	11	4	4	4	4	4	3	2
Occupancy R: maximum area per story in square feet										
R-1	SF	UL	UL	24,000	16,000	24,000	16,000	20,500	12,000	7,000
R-2	"	UL	UL	24,000	16,000	24,000	16,000	20,500	12,000	7,000
R-3	"	UL	UL	UL	UL	UL	UL	UL	UL	UL
R-4	"	UL	UL	24,000	16,000	24,000	16,000	20,500	12,000	7,000
Occupancy S: maximum height in stories and feet										
S	Feet	UL	160	65	55	65	55	65	50	40
S-1	Stories	UL	11	4	2	3	2	4	3	1
S-2	Stories	UL	11	5	3	4	3	5	4	2
Occupancy S: maximum area per story in square feet										
S-1	SF	UL	48,000	26,000	17,500	26,000	17,500	25,500	14,000	9,000
S-2	SF	UL	79,000	39,000	26,000	39,000	26,000	38,500	21,000	13,500
Occupancy U: maximum height in stories and feet										
U	Feet	UL	160	65	55	65	55	65	50	40
U	Stories	UL	5	4	2	3	2	4	2	1
Occupancy U: maximum area per story in square feet										
U	SF	UL	35,500	19,000	8,500	14,000	8,500	18,000	9,000	5,500

Source: IBC Table 503.
UL = unlimited, Occ. = occupancy, SF = square feet.

503 General Building Height and Area Limitations

503.1.1 Special industrial occupancies

- The following buildings and structures are not required to meet height and area limitations based on occupancy and construction type:
  - Those containing special industrial processes as follows:
    - Processes requiring large areas and heights to accommodate the following:
      - Craneways.
      - Special machinery and equipment.
      - Rolling mills.
      - Structural metal fabrication shops and foundries.
      - Production and distribution of power as follows:
        - Electric.
        - Gas.
        - Steam.

Note: IBC Table 503, "Allowable Building Heights and Areas," is cited as listing the limitations from which these buildings are exempt.

503.1.2 Buildings on same lot

- Multiple buildings on the same lot may be considered to be either of the following:
  - Separate buildings.
  - A single building where all of the following apply:
    - Height of each building meets the following:
      - Height limits based on occupancy and construction type as modified.
    - Sum of areas of the buildings meets the following:
      - Area limits based on occupancy and construction type as modified.
    - The individual structures meet applicable code requirements.
    - The group of structures meet applicable code requirements as a single building.

Note: The following are cited as providing limits for the buildings described above:
IBC Table 503, "Allowable Building Heights and Areas," provides the base limits.
Section 504, "Building Height," alters the limits based on conditions.
Section 506, "Building Area Modifications," alters the limits based on conditions.

503.1.3 Type I construction

- Buildings of Type I construction, which are allowed unlimited height or area based on occupancy and construction type, are governed as follows:
  - These buildings are not subject to the conditions that are required for unlimited or increased height or area in other cases or construction types.

Note: 504.3, "Roof structures," is cited as having limitations for unlimited-height buildings which do not apply here.
503.1.1, "Special industrial occupancies," is cited as having limitations for unlimited-height buildings which do not apply here.
Section 507, "Unlimited Area Buildings," is cited as a source of conditions for unlimited-area buildings which do not apply here.

504 Building Height

504.1 General

- Height of the following buildings is unlimited where all of the conditions indicated apply:
  - Buildings:
    Aircraft hangars.
    Aircraft paint hangars.
    Buildings in which aircraft are manufactured.
  - Conditions:
    Automatic fire-extinguishing system is required.
    Building is surrounded by public ways or yards as follows:

    Width of yard is ≥ 1 1/2 × building height.

  Note: Chapter 9, "Fire Protection Systems," is cited as governing the fire-extinguishing system above.

- Height limits of other buildings are increased according to this section.

  Note: IBC Table 503, "Allowable Building Heights and Areas," is cited as defining the height limits subject to increase by this section.

504.2 Automatic sprinkler system increase *(part 1 of 4)*

- This section does not apply to the following locations:
  - Occupancy I-2 buildings or parts of buildings in the following construction types:
    Type IIB, Type III, Type IV, Type V.
  - Occupancy H-1, H-2, H-3, or H-5 buildings or parts of buildings.
  - Where sprinklers are used to achieve a 1-hr fire-resistance rating reduction.

  Note: Table 601, "Fire-resistance Rating Requirements for Building Elements," footnote "e" is cited as defining cases not eligible for height increases specified by this section due to the use of sprinklers to achieve a 1-hr fire-resistance rating reduction.

- In other locations, an increase in height limit is permitted as follows:
  - Height limits of buildings spinklered with approved systems may be increased 20' and 1 story.

  Note: IBC Table 503, "Allowable Building Heights and Areas," is cited as the source of height limits. 903.3.1.1, "NFPA 13 sprinkler systems," is cited as governing the sprinklers indicated above.

  - These height increases are allowed in addition to building area increases afforded by sprinkler systems.

  Note: The following are cited as providing building area increases afforded by sprinkler systems:
  506.2, "Frontage increase."
  506.3, "Automatic sprinkler system increase."

- In Occupancy R buildings, an increase in height limit is permitted as follows:
  - For buildings spinklered with approved systems:
    Height limits ≤ 40' may be increased 20'.
    Height limits ≤ 3 stories may be increased 1 story.

  Note: IBC Table 503, "Allowable Building Heights and Areas," is cited as the source of height limits. 903.3.1.2, "NFPA 13R sprinkler systems," is cited as governing the sprinklers for Occupancy R.

- Increased height limits due to sprinklers are shown in the tables on the following pages.
- The tables on the following pages show original height limits where increases are not permitted.

504 Building Height

504.2 Automatic sprinkler system increase *(part 2 of 4)*

- This table lists maximum heights of buildings sprinklered as per NFPA 13.

Table 504.2a Maximum Height of Sprinklered Buildings, NFPA 13

		Type of construction								
Occ.	Height	IA	IB	IIA	IIB	IIIA	IIIB	IV	VA	VB
Occupancy A:										
A	Feet	UL	180	85	75	85	75	85	70	60
A-1	Stories	UL	6	4	3	4	3	4	3	2
A-2	"	UL	12	4	3	4	3	4	3	2
A-3	"	UL	12	4	3	4	3	4	3	2
A-4	"	UL	12	4	3	4	3	4	3	2
A-5	"	UL	UL	UL	UL	UL	UL	UL	UL	UL
Occupancy B:										
B	Feet	UL	180	85	75	85	75	85	70	60
B	Stories	UL	11	6	4	6	4	6	4	3
Occupancy E:										
E	Feet	UL	180	85	75	85	75	85	70	60
E	Stories	UL	6	4	3	4	3	4	2	2
Occupancy F:										
F	Feet	UL	180	85	75	85	75	85	70	60
F-1	Stories	UL	12	5	3	4	3	5	3	2
F-2	Stories	UL	12	6	4	5	4	6	4	3
Occupancy H:										
H	Feet	UL	160	65	55	65	55	65	50	40
H-1	Stories	1	1	1	1	1	1	1	1	NP
H-2	"	UL	3	2	1	2	1	2	1	1
H-3	"	UL	6	4	2	4	2	4	2	1
H-5	"	4	4	3	3	3	3	3	3	2
Occupancy H-4:										
H-4	Feet	UL	180	85	75	85	75	85	70	60
H-4	Stories	UL	8	6	4	6	4	6	4	3

(Continued)

Based on IBC Table 503.
UL = unlimited, Occ. = occupancy, NP = not permitted.

504 Building Height

Case study: Fig. 504.2A. The Occupancy B building is sprinklered and is Type VB construction. This qualifies the building for an increase in height from limits of 40' and 2 stories for buildings that are unsprinklered to limits of 60' and 3 stories. The building is 2 stories and 30'-8" high, measured to the average height of the roof behind the parapet (just below the top of the spandrel glass). Thus, the building complies with the code requirements for height.

Fig. 504.2A. Partial elevation. AmberGlen Business Center. Hillsboro, Oregon. Ankrom Moisan Associated Architects. Portland, Oregon.

504 Building Height

504.2 Automatic sprinkler system increase *(part 3 of 4)*

- This table lists maximum heights of buildings sprinklered as per NFPA 13.

Table 504.2a—*Continued* **Maximum Height of Sprinklered Buildings, NFPA 13**

		Type of construction								
Occ.	Height	IA	IB	IIA	IIB	IIIA	IIIB	IV	VA	VB
Occupancy I:										
I	Feet	UL	180	85	75	85	75	85	70	60
I-1	Stories	UL	10	5	4	5	4	5	4	3
I-3	"	UL	5	3	2	3	2	3	3	2
I-4	"	UL	6	4	3	4	3	4	2	2
Occupancy I-2:										
I-2	Feet	UL	180	85	55	65	55	65	50	40
I-2	Stories	UL	5	3	1	1	NP	1	1	NP
Occupancy M:										
M	Feet	UL	180	85	75	85	75	85	70	60
M	Stories	UL	12	5	3	5	3	5	4	2
Occupancy R:										
R	Feet	UL	180	85	75	85	75	85	70	60
R-1	Stories	UL	12	5	5	5	5	5	4	3
R-2	"	UL	12	5	5	5	5	5	4	3
R-3	"	UL	12	5	5	5	5	5	4	4
R-4	"	UL	12	5	5	5	5	5	4	3
Occupancy S:										
S	Feet	UL	180	85	75	85	75	85	70	60
S-1	Stories	UL	12	5	3	4	3	5	4	2
S-2	Stories	UL	12	6	4	5	4	6	5	3
Occupancy U:										
U	Feet	UL	180	85	75	85	75	85	70	60
U	Stories	UL	6	5	3	4	3	5	3	2

Based on IBC Table 503.

UL = unlimited, NP = not permitted, Occ. = occupancy.

504 Building Height

504.2 Automatic sprinkler system increase *(part 4 of 4)*

- This table lists maximum heights of buildings sprinklered as per NFPA 13R.

Table 504.2b Maximum Height of Sprinklered Buildings, NFPA 13R

		Type of construction								
Occ.	Height	IA	IB	IIA	IIB	IIIA	IIIB	IV	VA	VB
Occupancy R:										
R	Feet	UL	160	65	65	65	60	65	60	60
R-1	Stories	UL	11	4	4	4	4	4	4	3
R-2	"	UL	11	4	4	4	4	4	4	3
R-3	"	UL	11	4	4	4	4	4	4	4
R-4	"	UL	11	4	4	4	4	4	4	3

Based on IBC Table 503.
UL = unlimited, Occ. = occupancy.

504 Building Height

Case study: Fig. 504.2B. The Occupancy B building is sprinklered and is Type VB construction. This qualifies the building for an increase in height from limits of 40' and 2 stories for buildings that are unsprinklered to limits of 60' and 3 stories. The building is 2 stories and 30'-8" high, measured to the average height of the roof behind the parapet (just below the top of the spandrel glass). Thus, the building complies with the code requirements for height.

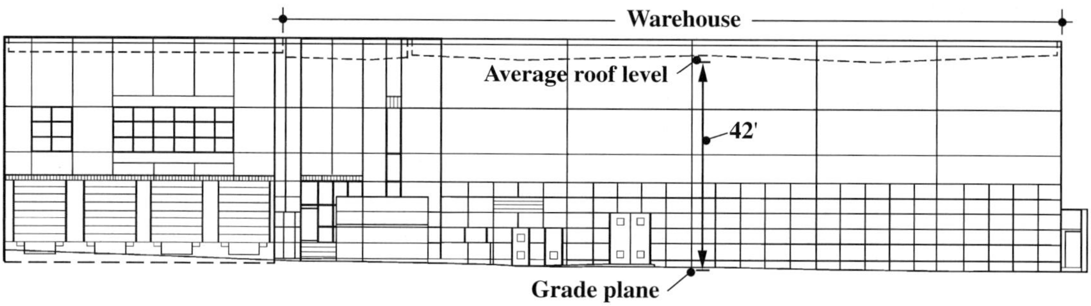

Fig. 504.2B. Elevation. New Warehouse Addition. Los Angeles, California. Stephen Wen + Associates, Architects, Inc. Pasadena, California.

504.3 Roof structures

- The following elements are governed as indicated below:
  - Elements:
    - Towers.
    - Spires.
    - Steeples.
    - Other roof structures.
  - Requirements:
    - Elements must be constructed with materials according to construction type.
    - Elements may not be used for the following:
      - Habitation.
      - Storage.
    - Elements are unlimited in height as follows:
      - Where constructed of noncombustible materials.
    - Elements must be ≤ 20' above allowable building height as follows:
      - Where constructed of combustible materials.

Note: 1509.2.4, "Type of construction," is cited as providing alternatives to materials requirements. Chapter 15, "Roof Assemblies and Rooftop Structures," is cited as the source for additional requirements.

505 Mezzanines

505.1 General

- Mezzanines are considered to be a part of the story within which it is located.
- The areas of mezzanines are included in the computation of fire areas.
- The clear height above a mezzanine floor must be ≥ 7'.
- The clear height below a mezzanine floor construction must be ≥ 7'.
- The areas of mezzanines are not included in building area.
- Mezzanines are not considered to be a story when computing the number of stories in a building.

Note: The following are cited as sources of requirements applicable as indicated:
503.1, "General," which addresses area and number of stories.
Section 505, "Mezzanines."
Section 902, "Definitions," defines fire area as referred to above.

505.2 Area limitation

- Mezzanine areas in special industrial occupancies are governed as follows:
  - In Type I or Type II construction:
    The total area of all mezzanines in a space is limited as follows:
    Area must be ≤ 2/3 the floor area of the space.

Note: 503.1.1, "Special industrial occupancies," is cited as defining these types of spaces.

- Mezzanine areas in the following locations have the area limitation listed below:
  - Locations where all the following apply:
    In Type I or Type II construction.
    Sprinklered throughout.
    Equipped with an approved emergency voice/alarm system.

Note: The following are cited as governing components listed above:
903.3.1.1, "NFPA 13 sprinkler systems."
907.5.2.2, "Emergency voice/alarm communication systems."

  - Area limitation:
    The total area of all mezzanines in a space must be ≤ 1/2 the floor area of the space.
- Mezzanine areas in other locations are governed as follows:
  - Total area of all mezzanines in a space must be ≤ 1/3 the floor area of that space.
- The following are not included in the floor area of a space for purposes of determining the maximum mezzanine area in the space:
  - Enclosed areas within the space.
  - The mezzanine area.

505 Mezzanines

505.3 Egress

- This section does not address certain locations based on the following:
  - Occupancy type.
  - Common path travel distance.
  - Occupant load.

Note: 1015.1, "Exits or exit access doorways from spaces," is cited as governing locations not addressed in this section.

- In other locations ≥ 2 means of egress are required from a mezzanine in either of the following cases:
  - Where its occupant load exceeds that listed below.
  - Where its common path of travel would otherwise exceed that listed below:

Table 505.3 Length Limits for Common Path of Egress Travel

		Common path	
Occupancy	Occupant load	with sprinklers	without sprinklers
B, F	> 49	100'	75'
A without seating, E, M, U	> 49	75'	75'
E day care	> 10	75'	75'
H-1, H-2, H-3	> 3	25'	not permitted
H-4, H-5	> 10	75'	not permitted
I-3	> 10	100'	100'
I-1, I-4, R-1, R-3, R-4	> 10	75'	75'
I-2	any	75'	75'
R-2	> 10	125'	75'
S	> 29	100'	75'
Tenant space in B, S, or U	< 31	100'	100'
Tenant space in B, S, or U	> 30	75'	75'

Note: 1014.3, "Common path of egress travel," is cited as the source of limits for egress travel, a partial summary of which is provided in the table above.

1015.1, "Exits or exit access doorways from spaces," is cited as identifying spaces requiring only 1 means of egress. The occupant load thresholds from that section are included in the table above.

- Length of egress travel on a mezzanine stairway is included in travel distance as follows:
  - As measured on a line connecting the tread nosings.

505 Mezzanines

505.4 Openness

- Mezzanines are not required to be open to the space in which they are located in any of the following cases:
  - Where the total occupant load of all the enclosed mezzanine areas totals ≤ 10.
  - Where both of the following apply:
    - The mezzanine has ≥ 2 means of egress.
    - ≥ 1 means of egress has direct access to an exit from the mezzanine.
  - Where the total of enclosed portions are ≤ 10% of the total of the mezzanine areas.
- Mezzanines with both of the following conditions may be glazed on all sides:
  - Located in an industrial facility.
  - Used for control of equipment.
- Mezzanines need not be open to the surrounding space where all of the following conditions apply:
  - In any of the following occupancies:
    - A, B, E, F, M, R, S, U.
  - ≤ 2 stories above the grade plane.
  - Sprinklered throughout the building.
  - With ≥ 2 means of egress.

Note: 903.3.1.1, "NFPA 13 sprinkler systems," is cited as governing the sprinklers above.

- All other mezzanines must be open to the surrounding space as follows:
  - The following are permitted:
    - Walls ≤ 42" above the mezzanine floor.
    - Columns.
    - Posts.

505.5 Equipment platforms

- Platforms are not considered to be a part of the floor below.
- The areas of platforms are not included in the area limit per floor.
- Platforms are not considered to be a story for purposes of building height limits.

Note: 503.1, "General," which includes IBC Table 503, "Allowable Building Heights and Areas," is cited as governing building heights and area limits as indicated above.

- The areas of platforms are not included in the computation of fire areas.

Note: Section 903, "Automatic Sprinkler Systems," is cited with regard to fire areas. The square footage in fire areas act as thresholds for certain sprinkler requirements in this section.

- Platforms may not be a part of a mezzanine.
- Platforms and the following components may not serve as a means of egress from a building:
  - Components providing access to platforms:
    - Walkways.
    - Stairs.
    - Alternating tread devices.
    - Ladders.

505 Mezzanines

505.5.1 Area limitations

- The combined areas of equipment platforms in a space is limited as follows:
  - Area must be ≤ 2/3 the area of the space.
- The combined areas of platforms and mezzanines in the same space is limited as follows:
  - Area must be ≤ 2/3 the area of the space.

Note: 505.2, "Area limitation," is cited as governing mezzanine area, which also applies when mezzanines are in the same space as an equipment platform.

505.5.2 Fire suppression

- Sprinklers are required above and below equipment platforms as follows:
  - Where the building is required to be sprinklered.
  - Where required by referenced standards.

Note: 903.3, "Installation requirements," is cited as referencing standards that govern the need for sprinkler protection.

505.5.3 Guards

- Guards are required along the open sides of equipment platforms as follows:
  - Where an open side is > 30" above the floor below.

Note: 1013.1, "Where required," is cited as governing guards. One requirement among others is listed above.

506 Building Area Modifications

506.1 General

- Area limits per story may be increased as follows:

Limit of area per story
\+ **Limit of area per story × frontage factor** *(see 506.2 this handbook)*
\+ **Additional area allowed due to sprinkler protection** *(see 506.3 this handbook)*

Increased limit of area per story

Note: The following are cited as sources of applicable requirements:
IBC Table 503, "Allowable Building Heights and Areas," defines limits of area per floor prior to any increases.
506.2, "Frontage increase," defines area increases permitted based on frontage. (Tables of frontage factors used to calculate these increases are provided on the next 3 pages.)
506.3, "Automatic sprinkler system increase," defines area increases based on sprinklers. (Tables of permitted increases are provided on subsequent pages of this handbook.)

506.2 Frontage increase *(part 1 of 4)*

- This section addresses increases to area limits per floor as permitted by frontage.
- The area limit may be increased where the following frontage conditions exist:
  - Where > 25% of the building perimeter is "open" as in either of the following cases:
    Abuts a public way ≥ 20' wide.
    Abuts an open space ≥ 20' wide.
- Area increase due to frontage is determined by the following equation:

Area increase = limit of area per story × frontage factor

- Frontage factors to determine the permitted increase in area are listed in the following tables:
  - These tables are based on the following equation:

Frontage factor = (% of open perimeter – 25%)/100 × (width of open area ÷ 30)

Open perimeter = perimeter abutting a public way or open space.
"Width of open area" ÷ 30" is limited to ≤ 2 where the following applies:
The building would have unlimited area except it has < the 60' open space required.
"Width of open area" ÷ 30" is limited to ≤ 1 for other buildings.

Note: 506.2.1, "Width limits," is cited as governing the width of the frontage open area.

506 Building Area Modifications

506.2 Frontage increase *(part 2 of 4)*

- The table below lists frontage factors as follows:
  - Based on % of open perimeter from 26% to 49%.
  - For frontage open space widths of 20' to 30'.

Table 506.2 Frontage Factors for Increases in Limits of Area per Story

% open	Width (in feet) of open space at frontage										
	20	21	22	23	24	25	26	27	28	29	30
26	0.0067	0.0070	0.0073	0.0077	0.0080	0.0083	0.0087	0.0090	0.0093	0.0097	0.0100
27	0.0133	0.0140	0.0147	0.0153	0.0160	0.0167	0.0173	0.0180	0.0187	0.0193	0.0200
28	0.0200	0.0210	0.0220	0.0230	0.0240	0.0250	0.0260	0.0270	0.0280	0.0290	0.0300
29	0.0267	0.0280	0.0293	0.0307	0.0320	0.0333	0.0347	0.0360	0.0373	0.0387	0.0400
30	0.0333	0.0350	0.0367	0.0383	0.0400	0.0417	0.0433	0.0450	0.0467	0.0483	0.0500
31	0.0400	0.0420	0.0440	0.0460	0.0480	0.0500	0.0520	0.0540	0.0560	0.0580	0.0600
32	0.0467	0.0490	0.0513	0.0537	0.0560	0.0583	0.0607	0.0630	0.0653	0.0677	0.0700
33	0.0533	0.0560	0.0587	0.0613	0.0640	0.0667	0.0693	0.0720	0.0747	0.0773	0.0800
34	0.0600	0.0630	0.0660	0.0690	0.0720	0.0750	0.0780	0.0810	0.0840	0.0870	0.0900
35	0.0667	0.0700	0.0733	0.0767	0.0800	0.0833	0.0867	0.0900	0.0933	0.0967	0.1000
36	0.0733	0.0770	0.0807	0.0843	0.0880	0.0917	0.0953	0.0990	0.1027	0.1063	0.1100
37	0.0800	0.0840	0.0880	0.0920	0.0960	0.1000	0.1040	0.1080	0.1120	0.1160	0.1200
38	0.0867	0.0910	0.0953	0.0997	0.1040	0.1083	0.1127	0.1170	0.1213	0.1257	0.1300
39	0.0933	0.0980	0.1027	0.1073	0.1120	0.1167	0.1213	0.1260	0.1307	0.1353	0.1400
40	0.1000	0.1050	0.1100	0.1150	0.1200	0.1250	0.1300	0.1350	0.1400	0.1450	0.1500
41	0.1067	0.1120	0.1173	0.1227	0.1280	0.1333	0.1387	0.1440	0.1493	0.1547	0.1600
42	0.1133	0.1190	0.1247	0.1303	0.1360	0.1417	0.1473	0.1530	0.1587	0.1643	0.1700
43	0.1200	0.1260	0.1320	0.1380	0.1440	0.1500	0.1560	0.1620	0.1680	0.1740	0.1800
44	0.1267	0.1330	0.1393	0.1457	0.1520	0.1583	0.1647	0.1710	0.1773	0.1837	0.1900
45	0.1333	0.1400	0.1467	0.1533	0.1600	0.1667	0.1733	0.1800	0.1867	0.1933	0.2000
46	0.1400	0.1470	0.1540	0.1610	0.1680	0.1750	0.1820	0.1890	0.1960	0.2030	0.2100
47	0.1467	0.1540	0.1613	0.1687	0.1760	0.1833	0.1907	0.1980	0.2053	0.2127	0.2200
48	0.1533	0.1610	0.1687	0.1763	0.1840	0.1917	0.1993	0.2070	0.2147	0.2223	0.2300
49	0.1600	0.1680	0.1760	0.1840	0.1920	0.2000	0.2080	0.2160	0.2240	0.2320	0.2400

(Continued on next page)

Data is rounded down to the 4th decimal place.

506 Building Area Modifications

506.2 Frontage increase *(part 3 of 4)*

- The table below lists frontage factors as follows:
  - Based on % of open perimeter from 50% to 73%.
  - For frontage open space widths of 20' to 30'.

Table 506.2—*Continued* **Frontage Factors for Increases in Limits of Area per Story**

% open	Width (in feet) of open space at frontage										
	20	21	22	23	24	25	26	27	28	29	30
50	0.1667	0.1750	0.1833	0.1917	0.2000	0.2083	0.2167	0.2250	0.2333	0.2417	0.2500
51	0.1733	0.1820	0.1907	0.1993	0.2080	0.2167	0.2253	0.2340	0.2427	0.2513	0.2600
52	0.1800	0.1890	0.1980	0.2070	0.2160	0.2250	0.2340	0.2430	0.2520	0.2610	0.2700
53	0.1867	0.1960	0.2053	0.2147	0.2240	0.2333	0.2427	0.2520	0.2613	0.2707	0.2800
54	0.1933	0.2030	0.2127	0.2223	0.2320	0.2417	0.2513	0.2610	0.2707	0.2803	0.2900
55	0.2000	0.2100	0.2200	0.2300	0.2400	0.2500	0.2600	0.2700	0.2800	0.2900	0.3000
56	0.2067	0.2170	0.2273	0.2377	0.2480	0.2583	0.2687	0.2790	0.2893	0.2997	0.3100
57	0.2133	0.2240	0.2347	0.2453	0.2560	0.2667	0.2773	0.2880	0.2987	0.3093	0.3200
58	0.2200	0.2310	0.2420	0.2530	0.2640	0.2750	0.2860	0.2970	0.3080	0.3190	0.3300
59	0.2267	0.2380	0.2493	0.2607	0.2720	0.2833	0.2947	0.3060	0.3173	0.3287	0.3400
60	0.2333	0.2450	0.2567	0.2683	0.2800	0.2917	0.3033	0.3150	0.3267	0.3383	0.3500
61	0.2400	0.2520	0.2640	0.2760	0.2880	0.3000	0.3120	0.3240	0.3360	0.3480	0.3600
62	0.2467	0.2590	0.2713	0.2837	0.2960	0.3083	0.3207	0.3330	0.3453	0.3577	0.3700
63	0.2533	0.2660	0.2787	0.2913	0.3040	0.3167	0.3293	0.3420	0.3547	0.3673	0.3800
64	0.2600	0.2730	0.2860	0.2990	0.3120	0.3250	0.3380	0.3510	0.3640	0.3770	0.3900
65	0.2667	0.2800	0.2933	0.3067	0.3200	0.3333	0.3467	0.3600	0.3733	0.3867	0.4000
66	0.2733	0.2870	0.3007	0.3143	0.3280	0.3417	0.3553	0.3690	0.3827	0.3963	0.4100
67	0.2800	0.2940	0.3080	0.3220	0.3360	0.3500	0.3640	0.3780	0.3920	0.4060	0.4200
68	0.2867	0.3010	0.3153	0.3297	0.3440	0.3583	0.3727	0.3870	0.4013	0.4157	0.4300
69	0.2933	0.3080	0.3227	0.3373	0.3520	0.3667	0.3813	0.3960	0.4107	0.4253	0.4400
70	0.3000	0.3150	0.3300	0.3450	0.3600	0.3750	0.3900	0.4050	0.4200	0.4350	0.4500
71	0.3067	0.3220	0.3373	0.3527	0.3680	0.3833	0.3987	0.4140	0.4293	0.4447	0.4600
72	0.3133	0.3290	0.3447	0.3603	0.3760	0.3917	0.4073	0.4230	0.4387	0.4543	0.4700
73	0.3200	0.3360	0.3520	0.3680	0.3840	0.4000	0.4160	0.4320	0.4480	0.4640	0.4800

(Continued on next page)

Data is rounded down to the 4th decimal place.

506 Building Area Modifications

506.2 Frontage increase *(part 4 of 4)*

- The table below lists frontage factors as follows:
  - Based on % of open perimeter from 74% to 100%.
  - For frontage open space widths of 20' to 30'.

Table 506.2—*Continued* **Frontage Factors for Increases in Limits of Area per Story**

% open	Width (in feet) of open space at frontage										
	20	21	22	23	24	25	26	27	28	29	30
74	0.3267	0.3430	0.3593	0.3757	0.3920	0.4083	0.4247	0.4410	0.4573	0.4737	0.4900
75	0.3333	0.3500	0.3667	0.3833	0.4000	0.4167	0.4333	0.4500	0.4667	0.4833	0.5000
76	0.3400	0.3570	0.3740	0.3910	0.4080	0.4250	0.4420	0.4590	0.4760	0.4930	0.5100
77	0.3467	0.3640	0.3813	0.3987	0.4160	0.4333	0.4507	0.4680	0.4853	0.5027	0.5200
78	0.3533	0.3710	0.3887	0.4063	0.4240	0.4417	0.4593	0.4770	0.4947	0.5123	0.5300
79	0.3600	0.3780	0.3960	0.4140	0.4320	0.4500	0.4680	0.4860	0.5040	0.5220	0.5400
80	0.3667	0.3850	0.4033	0.4217	0.4400	0.4583	0.4767	0.4950	0.5133	0.5317	0.5500
81	0.3733	0.3920	0.4107	0.4293	0.4480	0.4667	0.4853	0.5040	0.5227	0.5413	0.5600
82	0.3800	0.3990	0.4180	0.4370	0.4560	0.4750	0.4940	0.5130	0.5320	0.5510	0.5700
83	0.3867	0.4060	0.4253	0.4447	0.4640	0.4833	0.5027	0.5220	0.5413	0.5607	0.5800
84	0.3933	0.4130	0.4327	0.4523	0.4720	0.4917	0.5113	0.5310	0.5507	0.5703	0.5900
85	0.4000	0.4200	0.4400	0.4600	0.4800	0.5000	0.5200	0.5400	0.5600	0.5800	0.6000
86	0.4067	0.4270	0.4473	0.4677	0.4880	0.5083	0.5287	0.5490	0.5693	0.5897	0.6100
87	0.4133	0.4340	0.4547	0.4753	0.4960	0.5167	0.5373	0.5580	0.5787	0.5993	0.6200
88	0.4200	0.4410	0.4620	0.4830	0.5040	0.5250	0.5460	0.5670	0.5880	0.6090	0.6300
89	0.4267	0.4480	0.4693	0.4907	0.5120	0.5333	0.5547	0.5760	0.5973	0.6187	0.6400
90	0.4333	0.4550	0.4767	0.4983	0.5200	0.5417	0.5633	0.5850	0.6067	0.6283	0.6500
91	0.4400	0.4620	0.4840	0.5060	0.5280	0.5500	0.5720	0.5940	0.6160	0.6380	0.6600
92	0.4467	0.4690	0.4913	0.5137	0.5360	0.5583	0.5807	0.6030	0.6253	0.6477	0.6700
93	0.4533	0.4760	0.4987	0.5213	0.5440	0.5667	0.5893	0.6120	0.6347	0.6573	0.6800
94	0.4600	0.4830	0.5060	0.5290	0.5520	0.5750	0.5980	0.6210	0.6440	0.6670	0.6900
95	0.4667	0.4900	0.5133	0.5367	0.5600	0.5833	0.6067	0.6300	0.6533	0.6767	0.7000
96	0.4733	0.4970	0.5207	0.5443	0.5680	0.5917	0.6153	0.6390	0.6627	0.6863	0.7100
97	0.4800	0.5040	0.5280	0.5520	0.5760	0.6000	0.6240	0.6480	0.6720	0.6960	0.7200
98	0.4867	0.5110	0.5353	0.5597	0.5840	0.6083	0.6327	0.6570	0.6813	0.7057	0.7300
99	0.4933	0.5180	0.5427	0.5673	0.5920	0.6167	0.6413	0.6660	0.6907	0.7153	0.7400
100	0.5000	0.5250	0.5500	0.5750	0.6000	0.6250	0.6500	0.6750	0.7000	0.7250	0.7500

Data is rounded down to the 4th decimal place.

506 Building Area Modifications

Case study: Fig. 506.2. The sports and fine arts center is divided into three buildings, A, B, and C by fire walls. Building A contains A-1, A-3, and B occupancies. Building B is Occupancy A-3. Building C contains A-3 and B occupancies. The occupancies within buildings A and C are not separated by 2-hr walls as per IBC Table 508.4 so the buildings have "nonseparated uses" governed by 508.3.2. In this case, the most restrictive area limitation of the occupancies in each building dictates the area permitted.

Building A is construction Type IIB and is sprinklered. IBC Table 503 permits 8,500 sf per floor for Occupancy A-1 (the most restrictive case in this building). 506.1 allows this limit to be increased due to open area around the building and to the presence of sprinklers. Building A has an open frontage ≥ 30' deep at 90% of its perimeter. 506.2 permits an area increase of 65% or 5,525 sf based on the frontage. 506.3 permits an area increase equal to twice the original limit, or 17,000 sf. The new limit of area per floor is, therefore, *8,500 sf + 5,525 sf + 17,000 sf = 31,025 sf*. Building A has 31,010 sf on the 1st floor and less on the 2nd floor; thus, it complies with the limit of 31,025 sf per floor.

Building B is construction Type IIB and is sprinklered. IBC Table 503 permits 9500 sf for Occupancy A-3. 506.1 allows this limit to be increased due to open area around the building and to the presence of sprinklers. Building B has an open frontage ≥ 30' deep at 58% of its perimeter. 506.2 permits an area increase of 33% or 3,135 sf based on the frontage. 506.3 permits an area increase equal to twice the original limit, or 19,000 sf. The new limit of area per floor is, therefore, *9,500 sf + 3,135 sf + 19,000 sf = 31,635 sf*. Building B has 7,176 sf on the 1st floor and less on the 2nd floor; thus, it complies with the limit of 31,635 sf per floor.

Building C is construction Type IIA and is not sprinklered. IBC Table 503 permits 15,500 sf per floor for Occupancy A-3 (the more restrictive case in this building). Building C has an open frontage ≥ 30' deep at 85% of its perimeter. 506.2 permits an area increase of 60% or 9,300 sf based on the frontage. The new limit of area per floor is, therefore, *15,500 sf + 9,300 sf = 24,800 sf*. Building C has 23,365 sf on the 1st floor and less on the 2nd floor; thus, it complies with the limit of 24,800 sf per floor.

506 Building Area Modifications

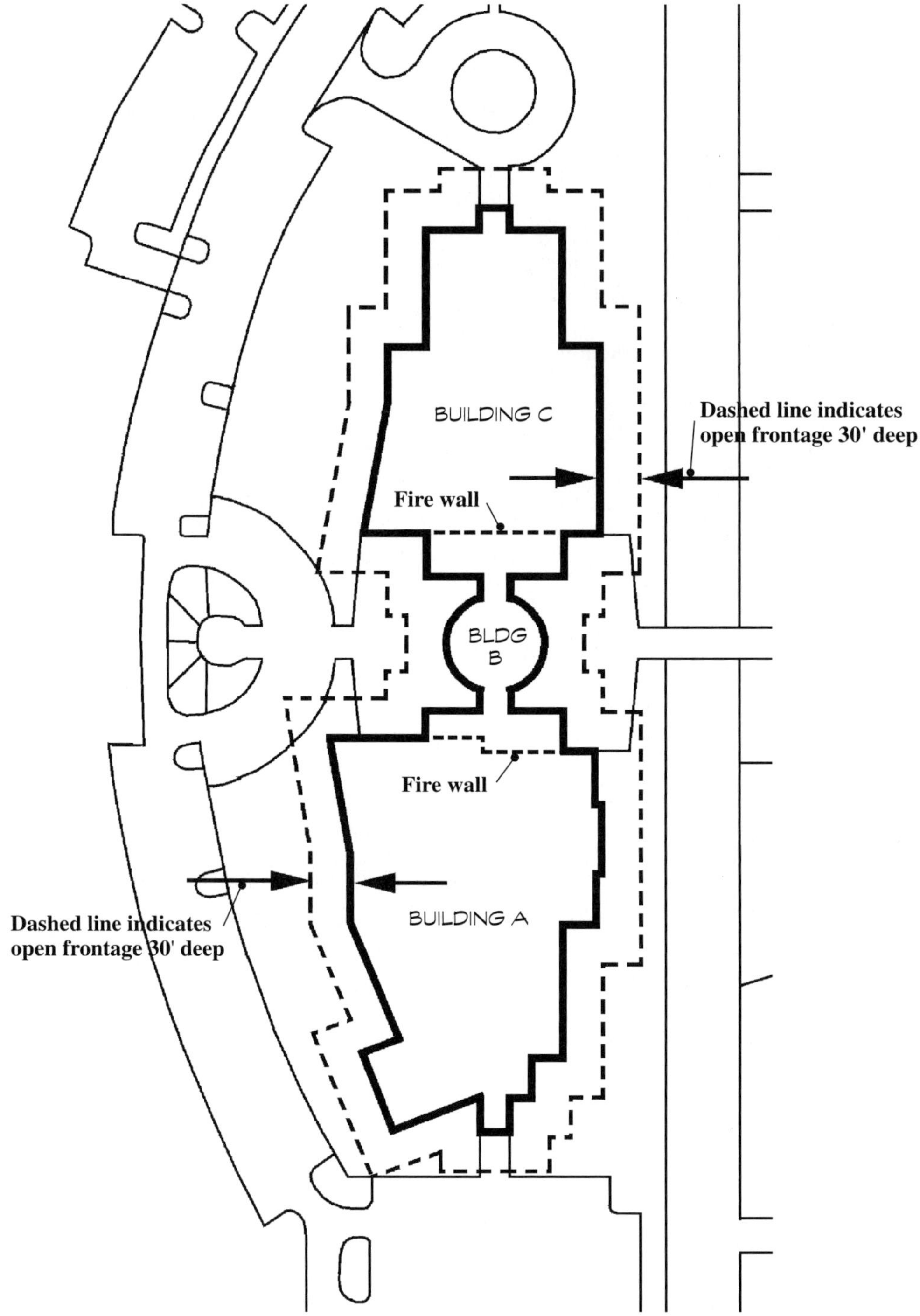

Fig. 506.2. Site plan. Newman University Sports and Fine Arts Center. Wichita, Kansas. Gossen Livingston Associates, Inc., Architecture. Wichita, Kansas.

506 Building Area Modifications

506.2.1 Width limits

- *W* as used in the equation for increased area limits based on frontage is as follows:
  - *W* is the width of public way or open area at the perimeter of a building.
  - *W* must be ≥ 20'.
  - *W* ÷ 30 is limited to a value ≤ 2.0 where both of the following apply:
    The building has < the 60' of the open area required to qualify for unlimited area.
    The building qualifies for unlimited area in all other respects.
  - Where the width of open space varies along the building perimeter, the following applies:
    W is the weighted average of widths > 20'.
    30' is used in place of any widths > 30' to determine the weighted average for *W*.
  - W is measured between the exterior faces of 2 buildings on the same lot.

Note: Section 507, "Unlimited Area Buildings," is cited as the source of requirements for buildings to qualify for unlimited area.

506.2.2 Open space limits

- This section addresses the frontage used as a basis for increasing the limits of area per floor.
- Open space qualifying as frontage is governed as follows:
  - It must be one of the following:
    Located on the same lot.
    Dedicated for public use.
  - It must be accessed by one of the following:
    From a street.
    By an approved fire lane.

506.3 Automatic sprinkler system increase *(part 1 of 3)*

- This section does not apply to the following cases:
  - Occupancy H-1.
  - H-2 and H-3 building areas.
  - Where sprinklers are used to achieve a 1-hr fire-resistance rating reduction.

Note: Table 601, "Fire-Resistance Rating Requirements for Building Elements," footnote "e" is cited as defining cases not eligible for height increases specified by this section due to the use of sprinklers to achieve a 1-hr fire-resistance rating reduction.

- Buildings with H-2 or H-3 areas are governed as follows:
  - Where H-2 or H-3 areas are separated from the rest of the building, the following applies:
    The rest of the building is eligible for area increases due to sprinklers.

Note: 508.4.2, "Allowable building area," is cited as governing mixed use area calculations.

- In buildings of other occupancies, area per floor limits are increased as follows:
  - Where the building is sprinklered.
  - Area increases are listed in the tables on the following pages.

Note: IBC Table 503, "Allowable Building Heights and Areas," is cited as the source of area limits that may be increased by the presence of sprinklers.
903.3.1.1, "NFPA 13 sprinkler systems," is cited as governing the sprinklers.

506 Building Area Modifications

506.3 Automatic sprinkler system increase *(part 2 of 3)*

- Area increases permitted in this section are in addition to the following:
  - Height and story count increases permitted due to the presence of sprinklers.

Note: 504.2, "Automatic sprinkler system increase," is cited as permitting height and story count increases due to the presence of sprinklers.

- This table lists area increases for multistory buildings as follows:
  - Sprinklered buildings.
  - Increase to area is calculated by multiplying the original permitted area per floor × 2.

Table 506.3a Added Area per Story in SF for Sprinklered Multistory Buildings

	Type of construction								
Occ.	IA	IB	IIA	IIB	IIIA	IIIB	IV	VA	VB
A-1	UL	UL	31,000	17,000	28,000	17,000	30,000	23,000	11,000
A-2	UL	UL	31,000	19,000	28,000	19,000	30,000	23,000	12,000
A-3	UL	UL	31,000	19,000	28,000	19,000	30,000	23,000	12,000
A-4	UL	UL	31,000	19,000	18,000	19,000	30,000	23,000	12,000
A-5	UL	UL	UL	UL	UL	UL	30,000	UL	UL
B	UL	UL	75,000	46,000	57,000	38,000	72,000	36,000	18,000
E	UL	UL	53,000	29,000	47,000	29,000	51,000	37,000	19,000
F-1	UL	UL	50,000	31,000	38,000	24,000	71,000	28,000	17,000
F-2	UL	UL	75,000	46,000	57,000	36,000	101,000	43,000	26,000
H-4	UL	UL	75,000	35,000	57,000	35,000	72,000	36,000	13,000
H-5	UL	UL	75,000	46,000	57,000	38,000	72,000	36,000	18,000
I-1	UL	110,000	38,000	20,000	33,000	20,000	36,000	21,000	9,000
I-2	UL	UL	31,000	22,000	24,000	NP	24,000	19,000	NP
I-3	UL	UL	30,000	20,000	21,000	15,000	24,000	15,000	10,000
I-4	UL	121,000	53,000	26,000	47,000	26,000	51,000	37,000	18,000
M	UL	UL	43,000	25,000	37,000	25,000	41,000	28,000	18,000
R-1	UL	UL	48,000	32,000	48,000	32,000	41,000	24,000	14,000
R-2	UL	UL	48,000	32,000	48,000	32,000	41,000	24,000	14,000
R-3	UL	UL	UL	UL	UL	UL	UL	UL	UL
R-4	UL	UL	48,000	32,000	48,000	32,000	41,000	24,000	14,000
S-1	UL	96,000	52,000	35,000	52,000	35,000	51,000	28,000	18,000
S-2	UL	158,000	78,000	52,000	78,000	52,000	77,000	42,000	27,000
U	UL	71,000	38,000	17,000	28,000	17,000	36,000	18,000	11,000

Based on IBC Table 503.
SF = square feet, NP = not permitted, UL = unlimited, Occ. = occupancy.

506 Building Area Modifications

506.3 Automatic sprinkler system increase *(part 3 of 3)*

- This table lists area increases for 1-story buildings as follows:
  - Sprinklered buildings.
  - Increase to area is calculated by multiplying the original permitted area per floor × 3.

Table 506.3b Added Area per Story in SF for Sprinklered 1-Story Buildings

	Type of construction								
Occ.	IA	IB	IIA	IIB	IIIA	IIIB	IV	VA	VB
A-1	UL	UL	46,500	25,500	42,000	25,500	45,000	34,500	16,500
A-2	UL	UL	46,500	28,500	42,000	28,500	45,000	34,500	18,000
A-3	UL	UL	46,500	28,500	42,000	28,500	45,000	34,500	18,000
A-4	UL	UL	46,500	28,500	42,000	28,500	45,000	34,500	18,000
A-5	UL	UL	UL	UL	UL	UL	45,000	UL	UL
B	UL	UL	112,500	69,000	85,500	57,000	108,000	54,000	27,000
E	UL	UL	79,500	43,500	70,500	43,500	76,500	55,500	28,500
F-1	UL	UL	75,000	46,500	57,000	36,000	106,500	42,000	25,500
F-2	UL	UL	112,500	69,000	57,000	54,000	151,500	64,500	39,000
H-4	UL	UL	112,500	52,500	85,500	52,500	108,000	54,000	19,500
H-5	UL	UL	112,500	69,000	85,500	57,000	108,000	54,000	27,000
I-1	UL	165,000	57,000	30,000	49,500	30,000	54,000	31,500	13,500
I-2	UL	UL	46,500	33,000	36,000	NP	36,000	28,500	NP
I-3	UL	UL	45,000	30,000	31,500	22,500	36,000	22,500	15,000
I-4	UL	181,500	79,500	39,000	70,500	39,000	76,500	55,500	27,000
M	UL	UL	64,500	37,500	55,500	37,500	61,500	42,000	27,000
R-1	UL	UL	72,000	48,000	72,000	48,000	61,500	36,000	21,000
R-2	UL	UL	72,000	48,000	72,000	48,000	61,500	36,000	21,000
R-3	UL	UL	UL	UL	UL	UL	UL	UL	UL
R-4	UL	UL	72,000	48,000	72,000	48,000	61,500	36,000	21,000
S-1	UL	144,000	78,000	52,500	78,000	52,500	76,500	42,000	27,000
S-2	UL	237,000	117,000	78,000	117,000	78,000	115,500	63,000	49,500
U	UL	106,500	57,000	25,500	42,000	25,500	54,000	27,000	16,500

Based on IBC Table 503.
SF = square feet, NP = not permitted, UL = unlimited, Occ. = occupancy.

506 Building Area Modifications

506.4 Single occupancy buildings with more than one story

- This section addresses single occupancy buildings with > 1 story above grade plane.
- The sum of actual floor areas in a building is governed as follows:
  - The sum is limited to the total allowable building area.
  - A single basement does not contribute to the sum in the following case:
    Where the basement area is not > that allowed for a building with ≤ 1 story above grade plane.

506.4.1 Area determination

Note: IBC Table 503, "Allowable Building Heights and Areas," limits the area per story based on occupancy and construction type.
506.1, "General," is cited as governing the method used to determine the adjusted area per story.

- This section does not apply to buildings of unlimited area.

Note: Section 507, "Unlimited Area Buildings," is cited as the source of requirements for the unlimited area buildings excluded from this section.

- Buildings sprinklered as per NFPA 13R are governed as follows:
  - The total area allowed for the building is calculated by the following equation:

Total building area = Allowable area/story × number of stories above grade plane

Note: 903.3.1.2, "NFPA 13R sprinkler systems," is cited as governing the sprinklers.

- The total area allowed for other single occupancy buildings > 1 story above grade plane is calculated as follows:

Area per story as limited by occupancy and construction type
+ Additional area permitted due to the presence of frontage
+ Additional area permitted due to the presence of sprinklers

---

Adjusted area permitted per story

Total area permitted for buildings 2-stories above grade plane =
Adjusted area permitted per floor × 2

Total area permitted for buildings ≥ 3-stories above grade plane =
Adjusted area permitted per floor × 3

- No story may have a greater area than the adjusted area permitted per story.

506.5 Mixed occupancy area determination

- This section governs the total building area permitted for mixed occupancies considering the following:
  - A single basement is not included in total allowable building area in the following case:
    Where the basement area is not > that allowed for a building with ≤ 1 story above grade plane.

506 Building Area Modifications

506.5.1 No more than one story above grade plane

- The following buildings are not governed by this section:
  - Buildings with ≤ 1 story above grade plane as follows:
    With mixed occupancies.

Note: 508.1, "General," is cited as governing total building area for buildings with ≤ 1 story above grade plane.

506.5.2 More than one story above grade plane

- The following are governed by other sections of this code addressing mixed occupancies:
  - Buildings with mixed occupancies with > 1 story above grade plane.
  - Buildings with > 3 stories above grade plane as follows:
    Sum of ratios of actual area of each story ÷ allowable area must be ≤ 3.

Note: 508.1, "General," is cited as governing individual stories of buildings with mixed occupancies with > 1 story and allowable area of each floor in buildings with > 3 stories as listed above.

507 Unlimited Area Buildings

507.1 General

- The areas of buildings as defined in this section are not limited.

507.2 Nonsprinklered, one story

- F-2 and S-2 buildings are not limited in area where they meet both of the following requirements:
  - The building must be ≤ 1 story.
  - One or more of the following open areas must surround the entire perimeter of the building:
    Public way ≥ 60' wide.
    Yard ≥ 60' wide.

507.3 Sprinklered, one story

- Rack storage facilities of any height meeting all the following criteria are not limited in area:
  - Building must be sprinklered.
  - Building must be Type I or Type II construction.
  - Public does not have access to the building.
  - The building must conform to requirements of this section.

  Note: The following are cited as governing the facilities described above:
  507.2, "Nonsprinklered, one story."
  903.3.1.1, "NFPA 13 sprinkler systems."
  Chapter 23, "High-Piled Combustible Storage," of the International Fire Code.

- Occupancy A-4 areas housing participant sports such as follows have the requirements listed below:
  - Sports:
    Tennis.
    Skating.
    Swimming.
    Equestrian activities.
  - Requirements:
    Sprinklers are not required in sports areas.
    Sports areas must have exit doors opening directly to the outdoors.
    The building must have manual fire alarm activation switches.

  Note: Section 907, "Fire Alarm and Detection Systems," is cited as applicable.

- In other cases, buildings may have unlimited area if they meet all of the following conditions:
  - Buildings must be among the following occupancies:
    B, F, M, S.
    A-4 with any of the following construction types:
      Type I, Type II, Type III, Type IV.
  - Building must be 1 story.
  - One or more of the following open areas must surround the entire perimeter of the building:
    Public way ≥ 60' wide.
    Yard ≥ 60' wide.
  - Building must be sprinklered.

  Note: 903.3.1.1, "NFPA 13 sprinkler systems," is cited as governing the sprinklers.

507 Unlimited Area Buildings

507.3.1 Mixed occupancy buildings with Group A-1 and A-2

- Occupancies A-1 and A-2 are permitted in unlimited area buildings where all conditions apply as follows:
  - Where construction is not Type V.
  - Where these occupancies are separated from other spaces as follows:
    Without reduction due to sprinklers to the required separation fire-resistance rating.

Note: 508.4.4 "Separation," is cited as governing the separation.

  - Where these occupancies have areas within code limits for the occupancies.

Note: 503.1, "General," is cited as governing area limits and does so by referencing Table 503, "Allowable Building Heights and Areas,"

  - Where all required exits discharge directly outside.

507.4 Two story

- 2-story buildings are not limited in areas if they meet all of the following conditions:
  - Buildings must be among the following occupancies:
    B, F, M, S.
  - One or more of the following open areas must surround the entire perimeter of the building:
    Public way ≥ 60' wide.
    Yard ≥ 60' wide.
  - Building must be sprinklered.

Note: 903.3.1.1, "NFPA 13 sprinkler systems," is cited as governing the sprinklers.

507.5 Reduced open space

- This section addresses the reduction of width for part of the open area required at the perimeter of unlimited area buildings.
- Such open area may be reduced from ≥ 60' to ≥ 40' if all of the following conditions are met:
  - Reduced width is limited to ≤ 75% of the building perimeter.
  - A fire-resistance rating of ≥ 3 hrs is provided as follows:
    For exterior walls facing the reduced width.
  - A fire-protection rating of ≥ 3 hrs is provided as follows:
    For protectives at openings in exterior walls facing the reduced width.

Note: The following are cited as requiring ≥ 60' of open area at the perimeter of unlimited area buildings:
507.2, "Nonsprinklered, one story."
507.3, "Sprinklered, one story."
507.4, "Two story."
507.6, "Group A-3 buildings of Type II construction."
507.11, "Motion picture theaters."

507 Unlimited Area Buildings

Case study: Fig. 507.4. The 2-story office building is not limited in area based on the facts that it is Occupancy B, it is sprinklered, and it has the necessary open space around it. The dashed line shown is 60' from the structure, thus, indicating that the yards and public ways surrounding the building are all larger than the 60' minimum required.

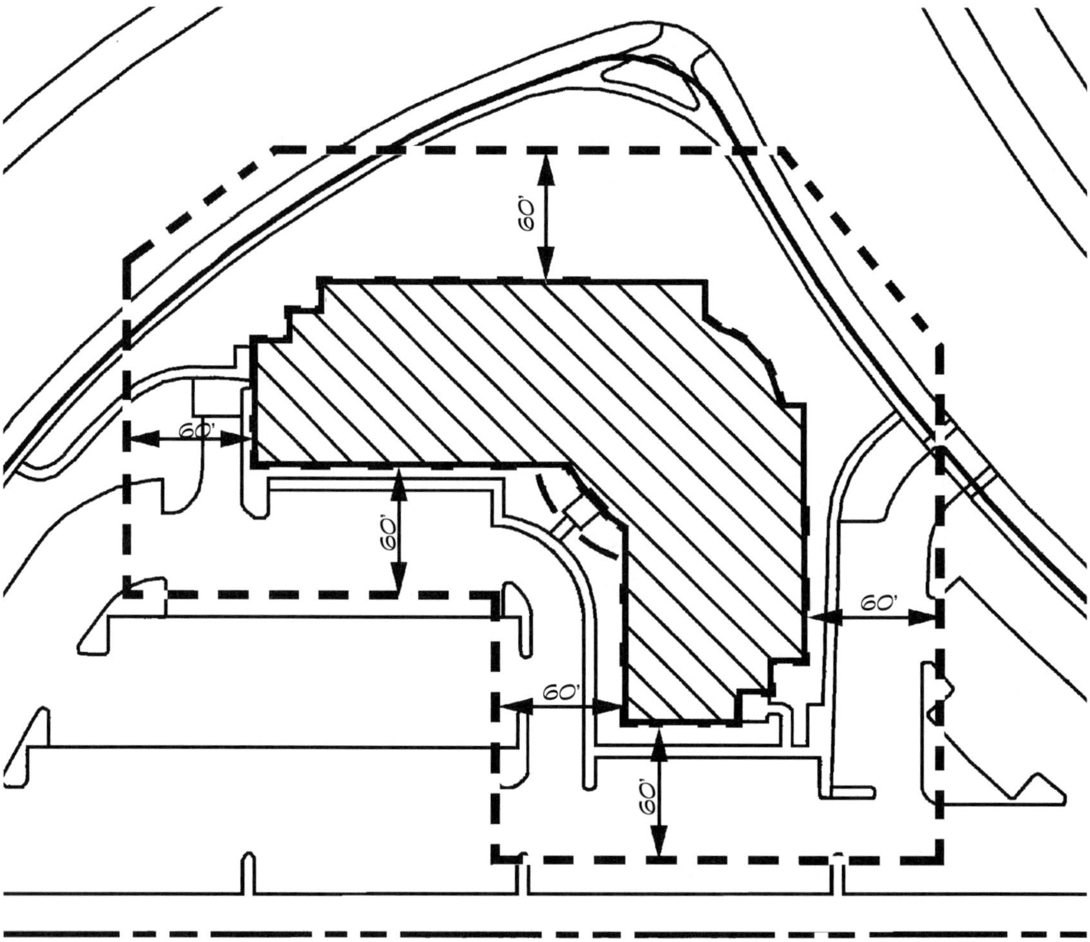

Fig. 507.4. Site plan. AmberGlen Business Center. Hillsboro, Oregon. Ankrom Moisan Associated Architects. Portland, Oregon.

507 Unlimited Area Buildings

507.6 Group A-3 buildings of Type II construction

- A building is not limited in area if it meets all of the following conditions:
  - Building must be one of the following:
    - Place of religious worship.
    - Community hall.
    - Dance hall.
    - Exhibition hall.
    - Gymnasium.
    - Lecture hall.
    - Indoor swimming pool.
    - Indoor tennis court.
  - Building must meet the following requirements:
    - Must be ≤ 1 story above grade plane.
    - Must be in Occupancy A-3.
    - Must have Type II construction.
    - May have a platform but not a stage.
    - One or more of the following open areas must surround the entire perimeter of the building:
      - Public way or yard ≥ 60' wide.
    - Building must be sprinklered.

Note: 903.3.1.1, "NFPA 13 sprinkler systems," is cited as governing the sprinklers.

507.7 Group A-3 buildings of Type III and IV construction

- A building is not limited in area if it meets all of the following conditions:
  - Building must be one of the following:
    - Place of religious worship.
    - Community hall.
    - Dance hall.
    - Exhibition hall.
    - Gymnasium.
    - Lecture hall.
    - Indoor swimming pool.
    - Indoor tennis court.
  - Building must meet the following requirements:
    - Must be ≤ 1 story above grade plane.
    - Must be in Occupancy A-3.
    - Must have Type III or IV construction.
    - May have a platform but not a stage.
    - Assembly floor must be ≤ 21" above street or grade level.
    - One or more of the following open areas must surround the entire perimeter of the building:
      - Public way or yard ≥ 60' wide.
    - Building must be sprinklered.
    - All exits must have ramps to street or grade level.

Note: 903.3.1.1, "NFPA 13 sprinkler systems," is cited as governing the sprinklers.
1010.1, "Scope," is cited as the source of requirements for the specified ramps.

507 Unlimited Area Buildings

507.8 Group H occupancies

- This section addresses unlimited area buildings containing F or S occupancies.

Note: The following are cited as having requirements for unlimited area Occupancy F or S buildings.
507.3, "Sprinklered, one story."
507.4, "Two story."

- H-2, H-3, and H-4 occupancies are permitted in the cases specified by this section.
- Where H-2, H-3 or H-4 occupancies are at the building perimeter, the following applies:
  - The sum of the H-occupancy areas is limited to both of the following:
    ≤ 10% of the building area.
    ≤ Code limits for H-occupancy areas as follows:
    As modified where a frontage increase is permitted.

Note: IBC Table 503, "Allowable Building Heights and Areas," is cited as limiting H-occupancy areas.
506.2, "Frontage increase," is cited as governing frontage increases to the area limits of IBC Table 503. Any increase is based upon the amount of H-occupancy area meeting frontage requirements.

- Where H-2, H-3 or H-4 occupancies are not at the building perimeter, the following applies:
  - The sum of the H-occupancy areas is limited to the following:
    ≤ 25% of the code limits for H-occupancy areas.

Note: IBC Table 503, "Allowable Building Heights and Areas," is cited as limiting H-occupancy areas.

- H-occupancy fire areas must be separated as follows:
  - From the rest of the building.
  - From each other.

Note: IBC Table 508.4, "Required Separation of Occupancies," is cited as governing the fire-resistance ratings of occupancy separations.

- 2-story unlimited area buildings are governed as follows:
  - H occupancies can be located above the 1st story above grade plane only in the following case:
    Where permitted by the construction type of an unlimited area building.

Note: IBC Table 503, "Allowable Building Heights and Areas," is cited as limiting H-occupancy heights based on construction type.

507.9 Aircraft paint hangar

- Aircraft hangers are not limited in area where all of the following requirements are met:
  - Building must be 1 story.
  - Building must be Occupancy H-2.
  - Building must have one of the following on each side:
    Public way ≥ 1.5 × building height.
    Yard ≥ 1.5 × building height.
  - Building must comply with other code requirements for aircraft paint hangars.

Note: 412.6, "Aircraft paint hangars," is cited as a source of requirements.

507 Unlimited Area Buildings

507.10 Group E buildings

- Occupancy E buildings may have unlimited area where all the following apply:
  - Building is 1 story.
  - Building is one of the following types of construction:
    Type II, Type IIIA, Type IV.
  - Each classroom has ≥ 2 means of egress.
  - Each classroom has ≥ 1 means of egress with a direct exit to the exterior.

Note: Section 1020, "Exits," is cited as governing exits.

  - One or more of the following open areas must surround the entire perimeter of the building:
    Public way ≥ 60' wide.
    Yard ≥ 60' wide.
  - Building must be sprinklered.

Note: 903.3.1.1, "NFPA 13 sprinkler systems," is cited as governing the sprinklers.

507.11 Motion picture theaters

- Motion picture theaters meeting all the following requirements are not limited in area:
  - Building must have Type II construction.
  - Theater must be on the 1st story above the grade plane.
  - One or more of the following open areas must surround the entire perimeter of the building:
    Public way ≥ 60' wide.
    Yard ≥ 60' wide.
  - Building must be sprinklered.

Note: 903.3.1.1, "NFPA 13 sprinkler systems," is cited as governing sprinklers.

507.12 Covered mall buildings and anchor stores

- The area is not limited for the following buildings:
  - Certain covered mall buildings ≤ 3 stories.
  - Certain anchor stores ≤ 3 stories.

Note: 402.6, "Types of construction," is cited as qualifying the buildings listed above for unlimited area. Type V construction is excluded and ≥ 60' of open area is required around the buildings among other limitations.

508 Mixed Use and Occupancy

508.1 General

- Each part of a building must be classified as to occupancy category.

  Note: *302.1, "General," is cited as governing the classification of occupancy.*

- This section governs certain buildings and parts of buildings containing ≥ 2 occupancies.

  Note: *The following are cited as governing buildings* containing ≥ 2 occupancies covered by this section:
  508.2, "Accessory occupancies."
  508.3, "Nonseparated occupancies."
  508.4, "Separated occupancies."

- This section does not govern certain occupancy combinations.

  Note: *Section 509, "Special Provisions," is cited as governing certain occupancy combinations not governed by this section.*

- H-1, H-2, and H-3 are not governed by this section where they house large amounts of hazardous materials:

  Note: *IBC Table 415.3.2, "Detached Building Required," is cited as listing quantity thresholds which require H-1, H-2, and H-3 to be housed in buildings detached from other occupancies.*

- Live/Work units are not governed by this section.

  Note: *Section 419, "Live/Work Units," is cited as governing this type of occupancy which is considered to be a single occupancy.*

508.2 Accessory occupancies

- Accessory occupancies are spaces supporting or serving a main occupancy.

  Note: *The following are cited as governing accessory occupancies:*
  508.2.1, "Area limitations" through 508.2.5.3, "Protection."

508.2.1 Area limitations

- The sum of accessory space areas in any story is governed as follows:
  - Limited to ≤ 10% of the area of the story they occupy.
  - Must be ≤ area limitations based on occupancy and construction type as follows:
    Area limits are without benefit of an increase due to frontage or sprinklers.

  Note: *IBC Table 503, "Allowable Building Heights and Areas," is cited as listing area limits with which accessory spaces must comply.*
  Section 506, "Building Area Modifications," is cited as defining area increases based on frontage and/or sprinklers that may not be used for increasing area limits for accessory spaces.

508.2.2 Occupancy classification

- The occupancy of each accessory space must be classified.

  Note: *302.1, "General," is cited as governing the classification of spaces.*

- Each space must meet code requirements based on its occupancy classification.

508 Mixed Use and Occupancy

508.2.3 Allowable building area and height

- The height and area limits of a building are based on its main occupancy and construction type.

Note: 503.1, "General," is cited as governing height and area. The section does so by reference to IBC Table 503, "Allowable Building Heights and Areas."

- The height of any accessory occupancy is limited based on occupancy and construction type as follows:
  - Height limits for accessory occupancies are without benefit of increases afforded by sprinklers.

Note: IBC Table 503, "Allowable Building Heights and Areas," is cited as limiting height. The limits are based on occupancy and construction type.
Section 504, "Building Height," is cited as the section permitting height increases which cannot be used.

- The sum of accessory space areas in any story is governed as follows:
  - Limited to ≤ 10% of the area of the story they occupy.
  - Must be ≤ area limitations based on occupancy and construction type as follows:
    Area limits are without benefit of an increase due to frontage or sprinklers.

Note: 508.2.1, "Area limitations," is cited as governing the area of accessory occupancies, part of which is included above.

508.2.4 Separation of occupancies

- The following occupancies must be separated from all other occupancies:
  - H-2, H-3, H-4, H-5.

Note: 508.4, "Separated occupancies," is cited as governing the separation of H-2, H-3, H-4, and H-5.

- Incidental accessory occupancies must comply with one or both of the following as specified:
  - They must be separated from other occupancies
  - They must be protected with a fire-suppression system.

Note: 508.2.5, "Separation of incidental accessory occupancies," is cited as specifying the separation or fire suppression required for these spaces.

- The following occupancies must be separated from other dwelling units, sleeping units, and occupancies:
  - I-1, R-1, R-2, R-3 dwelling units and sleeping units.

Note: Section 420, "Groups I-1, R-1, R-2, R-3," is cited as governing the separation of the occupancies above.

- In other cases, accessory occupancies are not required to be separated from the main occupancy.

508.2.5 Separation of incidental accessory occupancies *(part 1 of 3)*

- This section does not govern incidental accessory spaces in and serving a dwelling unit.
- This section requires one or both of the following for the incidental accessory occupancies listed:
  - A fire-rated separation from the rest of the building.
  - An automatic fire-extinguishing system.
- Refrigerant machinery rooms require one of the following:
  - 1-hr separation.
  - Sprinklers in the space.

508 Mixed Use and Occupancy

508.2.5 Separation of incidental accessory occupancies *(part 2 of 3)*

- Incinerator rooms require both of the following:
  - 2-hr separation.
  - Sprinklers in the space.
- Paint shops meeting both of the following conditions have the requirements listed below:
  - Conditions:
    - Where not classified as Occupancy H.
    - Where located in an occupancy that is not F.
  - Requirements:
    - One of the following is required:
      - 2-hr separation.
      - 1-hr separation with an automatic fire-extinguishing system in the space.
- The following spaces require a 1-hr separation or an automatic fire-extinguishing system:
  - Furnace rooms with any equipment > 400,000 Btu/h input.
  - Rooms with boilers having equipment with both of the following characteristics:
    - > 15 psi snf > 10 horsepower.
  - The following spaces in Occupancy E or I-2:
    - Laboratories and vocational shops not classified as H.
  - Laundry rooms and linen collection rooms > 100 sf.
  - Waste collection rooms > 100 sf.
- The following areas require a 1-hr separation:
  - Hydrogen cut-off rooms meeting all of the following conditions:
    - Not classified as Occupancy H.
    - Located in occupancies B, F, M, S, U.
  - Padded cells in Occupancy I-3.
  - Waste collection rooms and linen collection rooms in Occupancy I-2.
- Battery systems meeting all of the following conditions have the requirements listed below:
  - Conditions:
    - Serving as stationary power storage with one of the following capacities:
      - Liquid electrolyte capacity > 50 gallons.
      - Lithium capacity = 1,000 lbs.
    - Used for any of the following:
      - Facility standby power or emergency power.
      - Uninterrupted power supply.
  - Requirements:
    - 1-hr separation in occupancies B, F, M, S, U.
    - 2-hr separation in occupancies A, E, I, R.
- The following areas require a 2-hr separation:
  - Hydrogen cut-off rooms meeting all of the following conditions:
    - Not classified as Occupancy H.
    - Occupancies A, E, I, and R.
- Rooms with fire pumps require one of the following where located in buildings other than high-rise:
  - 2-hr separation.
  - 1-hr separation plus a fire-extinguishing system in the entire building.

508 Mixed Use and Occupancy

508.2.5 Separation of incidental accessory occupancies *(part 3 of 3)*

- Rooms with fire pumps require the following where located in high-rise buildings:
  - 2-hr separation.

Note: IBC Table 508.2.5, "Incidental Accessory Occupancies," is cited as governing the separation and/or fire protection of these areas. The table is summarized above.

508.2.5.1 Fire-resistance-rated separation

- This section does not require the following to be fire-resistance rated:
  - Construction supporting 1-hr fire barriers or horizontal assemblies used for the following:
    Incidental accessory occupancy separations in the following construction types:
    IIB, IIIB, VB.
- Separation of incidental accessory occupancies, where required, must be one or both of the following:
  - A fire barrier.
  - A fire-rated horizontal assembly.

Note: IBC Table 508.2.5, "Incidental Accessory Occupancies," is cited as specifying where the occupancies must be separated from the rest of the building.
Section 707, "Fire Barriers," is cited as governing these assemblies.
Section 712, "Horizontal Assemblies," is cited as governing these assemblies.

508.2.5.2 Nonfire-resistance-rated separation and protection *(part 1 of 2)*

- This section addresses incidental accessory occupancies where both of the following apply:
  - A fire-extinguishing system is required.
  - A fire barrier is not required to separate the occupancy from the rest of the building.

Note: IBC Table 508.2.5, "Incidental Accessory Occupancies," is cited as specifying where the occupancies must have fire protection or separation.

- The following applies to incidental accessory occupancies governed by this section:
  - The occupancy must be separated from the rest of the building as follows:
    With construction that prevents the passage of smoke as follows:
    The bottom of walls must extend to one of the following as applicable:
    The top of the foundation.
    The floor/ceiling assembly below.
    The tops of walls must extend to the underside of one of the following as applicable:
    The fire-resistance rated floor/ceiling asssembly above.
    The fire-resistance rated roof/ceiling assembly above.
    The floor or roof sheathing above.
    The deck or slab above.
    Doors must close automatically when smoke is detected.

Note: 715.4.8.3, "Smoke-activated doors," is cited as governing the doors addressed in this section.

Doors may not have openings for the purpose of transferring air.
Doors may not be undercut more than indicated in the Table 508.2.5.2:

508 Mixed Use and Occupancy

508.2.5.2 Nonfire-resistance rated separation and protection *(part 2 of 2)*

Table 508.2.5.2 Maximum Undercut of Doors

Door type	Material below door	Clearance below door
Swinging, builders hardware	Rigid floor tile	5/8"
All types	Raised noncombustible sill	3/8"
All types	Floor with no sill	3/4"
All types	Floor covering	1/2"

Source: NFPA 80, "Fire Doors and Other Opening Protectives," Table 1-11.4.

Note: NFPA 80, "Fire Doors and Other Opening Protectives," is cited as governing undercuts for doors.

Walls may not have openings for the purpose of transferring air without a smoke damper.

Note: 711.7, "Ducts and air transfer openings," is cited as governing smoke dampers as noted above.

508.2.5.3 Protection

- The following applies to fire-suppression systems required in incidental accessory occupancies:
  - The fire-suppression systems need be located only in the incidental accessory occupancy space.

Note: IBC Table 508.2.5, "Incidental Accessory Occupancies," is cited as specifying where fire-suppression systems are required.

508.3 Nonseparated occupancies

- The following are categorized as nonseparated occupancies:
  - Buildings meeting the requirements of this section.
  - Parts of buildings meeting the requirements of this section.

508.3.1 Occupancy classification

- Each nonseparated occupancy is assigned an occupancy classification based on occupancy criteria.

Note: 302.1, "General," is cited as providing criteria for occupancy designations.

- Each nonseparated occupancy is subject to code requirements based on its occupancy and as follows:
  - High-rise building requirements supersede less restrictive requirements where applicable.
  - Fire-protection requirements supersede less restrictive requirements where applicable.

Note: Section 403, "High-Rise Buildings," is cited as governing these buildings.
Chapter 9, "Fire Protection Systems," is cited as governing these systems.

508.3.2 Allowable building area and height

- The height and area limits of a building or parts of a building are based on the following:
  - The most restrictive limits among the occupancies being considered.

Note: 503.1, "General," is cited as governing building height and area limits.

508 Mixed Use and Occupancy

508.3.3 Separation

- The following occupancies must be separated from all other occupancies:
  - H-2, H-3, H-4, H-5.

Note: 508.4, "Separated occupancies," is cited as governing the separation of the H-occupancies.

- The following occupancies must be separated from other dwelling units, sleeping units, and occupancies:
  - I-1, R-1, R-2, R-3 dwelling units and sleeping units.

Note: Section 420, "Groups I-1, R-1, R-2, R-3," is cited as governing the separation of the occupancies above.

- Other occupancies designated as nonseparated need not be separated from the main occupancy.

508.4 Separated occupancies

- The following are categorized as separated occupancies:
  - Buildings meeting the requirements of this section.
  - Parts of buildings meeting the requirements of this section.

508.4.1 Occupancy classification

- Each separated occupancy is assigned an occupancy classification based on occupancy criteria.
- Each separated occupancy must comply with code requirements based on its occupancy classification.

Note: 302.1, "General," is cited as providing criteria for occupancy designations.

508.4.2 Allowable building area

- The sum of the following calculations must be ≤ 1 for each story:
  - Actual area of each separated occupancy ÷ the allowable area for each separated occupancy.

508.4.3 Allowable height

- Certain parking garages and certain Occupancy R buildings are not governed by this section.
- Each separated occupancy must conform to height limitations based on building construction type.

Note: Section 509, "Special Provisions," is cited as governing buildings not governed by this section. 503.1, "General," is cited as governing the height of buildings based on construction type. It does so by reference to IBC Table 503, "Allowable Building Heights and Areas."

508 Mixed Use and Occupancy

508.4.4 Separation *(part 1 of 11)*

- Separated occupancies must be separated by fire-resistance-rated assemblies as specified in this section.

  Note: *IBC Table 508.4, "Required Separation of Occupancies," is cited as governing fire-resistance ratings for separations between occupancies and is summarized on the following pages.*

- The tables on the following pages list required fire-resistance ratings for separations as follows:
  - Fire-resistance ratings are listed for the following sprinkler categories:
    Sprinklers provided throughout the building.
    Sprinklers not provided throughout the building.
    Sprinklers for Occupancy H-5 have requirements in addition to the tabular listing.

  Note: *903.3.1.1, "NFPA 13 sprinkler systems," is cited as governing the sprinklers.*
  903.2.5.2, "Group H-5," is cited as governing sprinkler design for Occupancy H-5.

  - Commercial kitchens are governed as follows:
    They are not required to be separated from the seating area served.
  - Separations for S-2 parking are governed as follows:
    Separation requirements ≥ 2 hr may be 1 hr less than the tabular listing as follows:
    Where vehicles are private or for pleasure.
  - Separations between parking and dwellings have requirements in addition to the tabular listing.

  Note: *406.1.4, "Separation," is cited as listing additional requirements for parking associated with dwellings.*

508 Mixed Use and Occupancy

508.4.4 Separation *(part 2 of 11)*

- This part governs separations for A and E occupancies.
- Fire-resistance ratings as required for separation from other occupancies are shown below:

Table 508.4.4a Occupancy A or E: Fire-Resistance Ratings for Occupancy Separations

With sprinklers throughout building:

Occupancy	Ratings (hrs)
A	0
B	1
E	0
F-1	1
F-2	0
H-1	NP
H-2	3
H-3	2
H-4	2
H-5	2
I-1, I-3, I-4	1
I-2	2
M	1
R	1
S-1	1
S-2	0
U	0

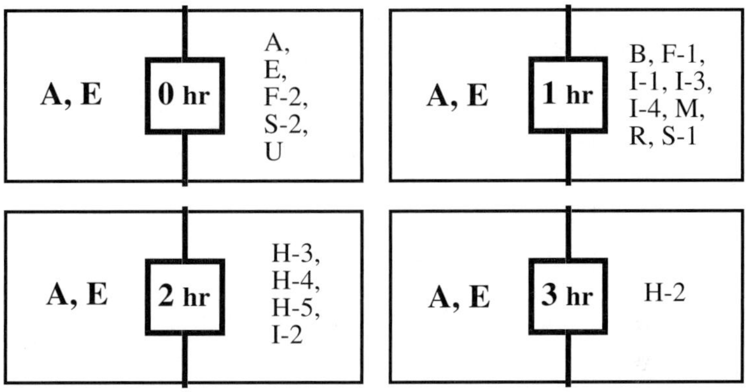

Without sprinklers throughout building:

Occupancy	Ratings (hrs)
A	0
B	2
E	0
F-1	2
F-2	1
H-1	NP
H-2	4
H-3	3
H-4	3
H-5	3
I-1, I-3, I-4	2
I-2	NP
M	2
R	2
S-1	2
S-2	1
U	1

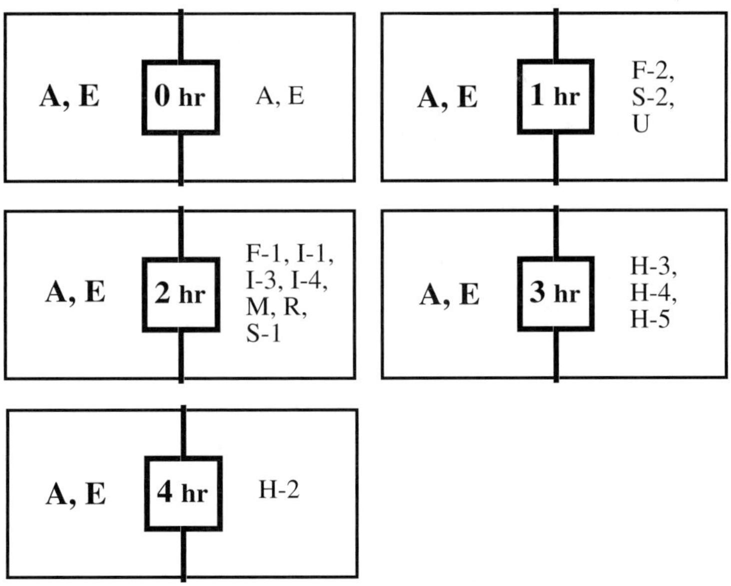

Source: IBC Table 508.4.

NP = not permitted.

508 Mixed Use and Occupancy

508.4.4 Separation *(part 3 of 11)*

- This part governs separations for I-1, I-3, and I-4 occupancies.
- Fire-resistance ratings as required for separation from other occupancies are shown below:

Table 508.4.4b Occupancy I-1, I-3, or I-4: Fire-Resistance Ratings for Occupancy Separations

With sprinklers throughout building:

Occupancy	Ratings (hrs)
A	1
B	1
E	1
F-1	1
F-2	1
H-1	NP
H-2	3
H-3	2
H-4	2
H-5	2
I-1, I-3, I-4	0
I-2	2
M	1
R	1
S-1	1
S-2	1
U	1

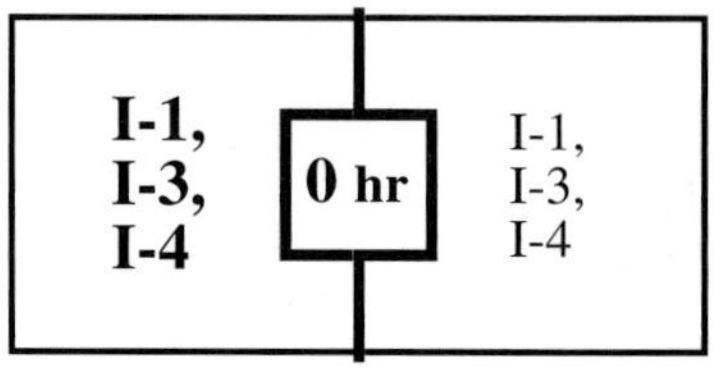

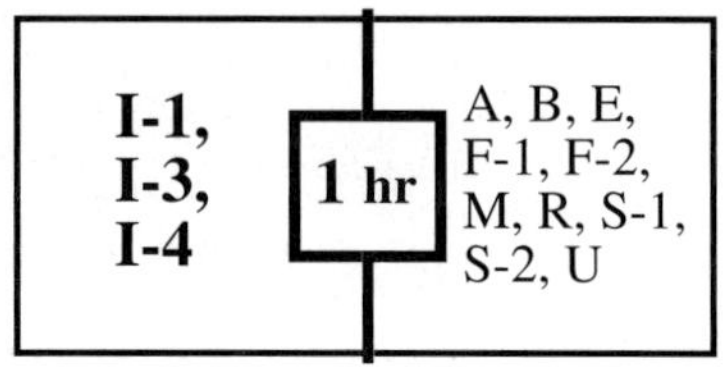

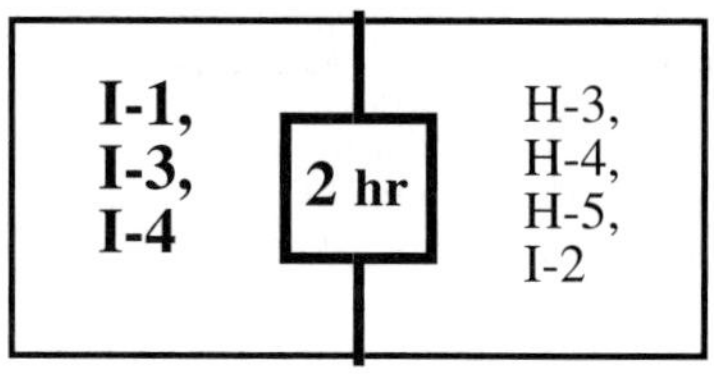

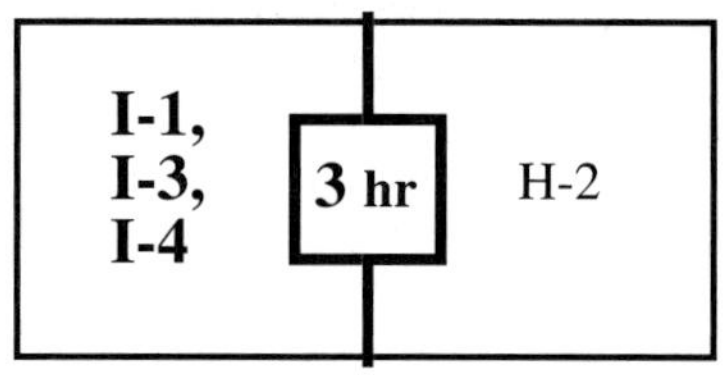

Without sprinklers throughout building:

Occupancy	Ratings (hrs)
A	2
B	2
E	2
F-1	2
F-2	2
H	NP
I-1, I-3, I-4	0
I-2	NP
M	2
R	NP
S-1	2
S-2	2
U	2

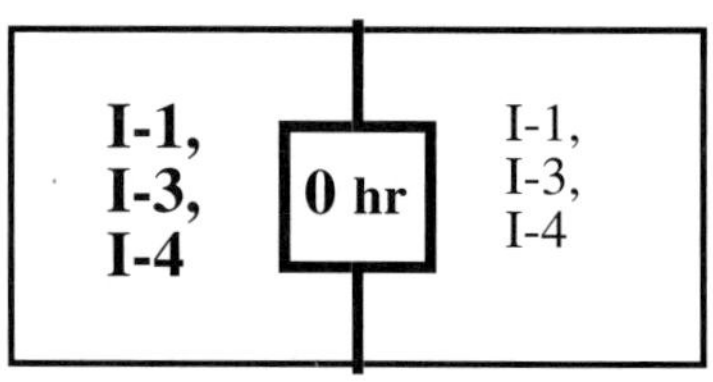

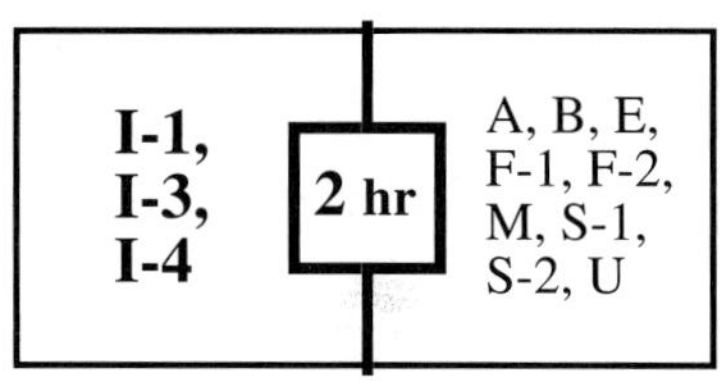

Source: IBC Table 508..4.

NP = not permitted.

508 Mixed Use and Occupancy

508.4.4 Separation *(part 4 of 11)*

- This part governs separations for the I-2 occupancy.
- Fire-resistance ratings as required for separation from other occupancies are shown below:

Table 508.4.4c Occupancy I-2: Fire-Resistance Ratings for Occupancy Separations

With sprinklers throughout building:

Occupancy	Ratings (hrs)
A	2
B	2
E	2
F-1	2
F-2	2
H-1	NP
H-2	3
H-3	2
H-4	2
H-5	2
I-1, I-3, I-4	2
I-2	0
M	2
R	2
S-1	2
S-2	2
U	2

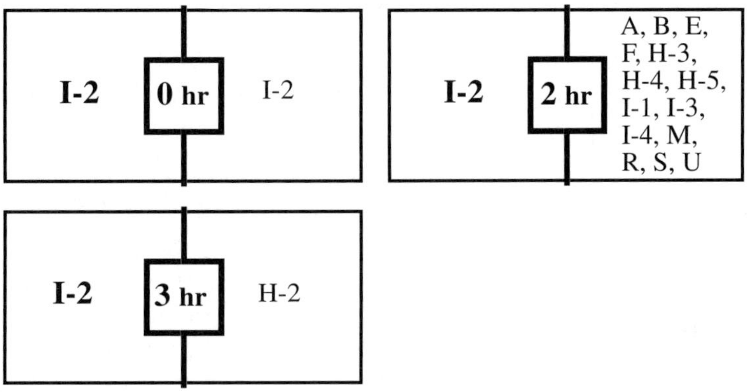

Without sprinklers throughout building:

Occupancy	Ratings (hrs)
A	NP
B	NP
E	NP
F	NP
H	NP
I-1, I-3, I-4	NP
I-2	0
M	NP
R	NP
S	NP
U	NP

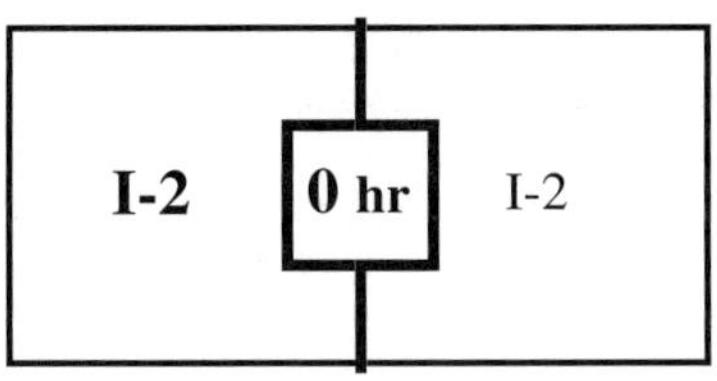

Source: IBC Table 508.4.

NP = not permitted.

508 Mixed Use and Occupancy

508.4.4 Separation *(part 5 of 11)*

- This part governs separations for R occupancies.
- Fire-resistance ratings as required for separation from other occupancies are shown below:

Table 508.4.4d Occupancy R: Fire-Resistance Ratings for Occupancy Separations

With sprinklers throughout building:

Occupancy	Ratings (hrs)
A	1
B	1
E	1
F-1	1
F-2	1
H-1	NP
H-2	3
H-3	2
H-4	2
H-5	2
I-1, I-3, I-4	1
I-2	2
M	1
R	0
S-1	1
S-2	1
U	1

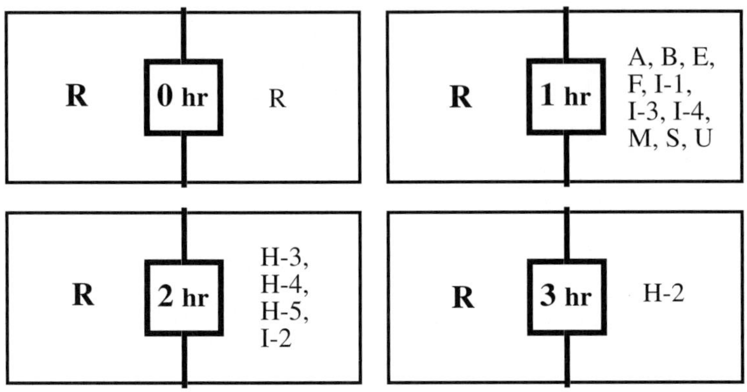

Without sprinklers throughout building:

Occupancy	Ratings (hrs)
A	2
B	2
E	2
F-1	2
F-2	2
H	NP
I	NP
M	2
R	0
S-1	2
S-2	2
U	2

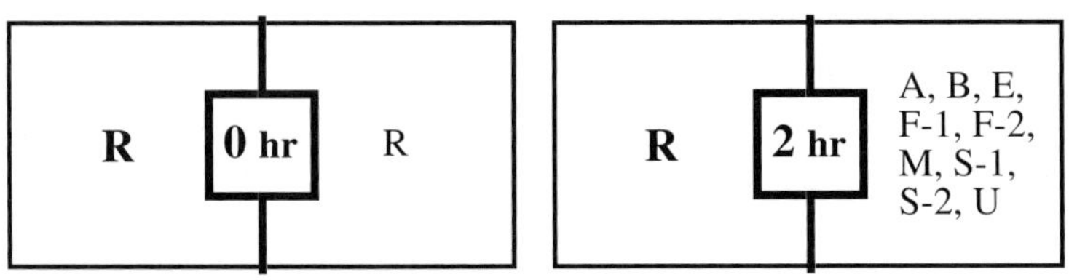

Source: IBC Table 508.4.
NP = not permitted.

508 Mixed Use and Occupancy

508.4.4 Separation *(part 6 of 11)*

- This part governs separations for F-2, S-2, and U occupancies.
- Fire-resistance ratings as required for separation from other occupancies are shown below:

Table 508.4.4e Occupancy F-2, S-2, or U: Fire-Resistance Ratings for Occupancy Separations

With sprinklers throughout building:

Occupancy	Ratings (hrs)
A	0
B	1
E	0
F-1	1
F-2	0
H-1	NP
H-2	3
H-3	2
H-4	2
H-5	2
I-1, I-3, I-4	1
I-2	2
M	1
R	1
S-1	1
S-2	0
U	0

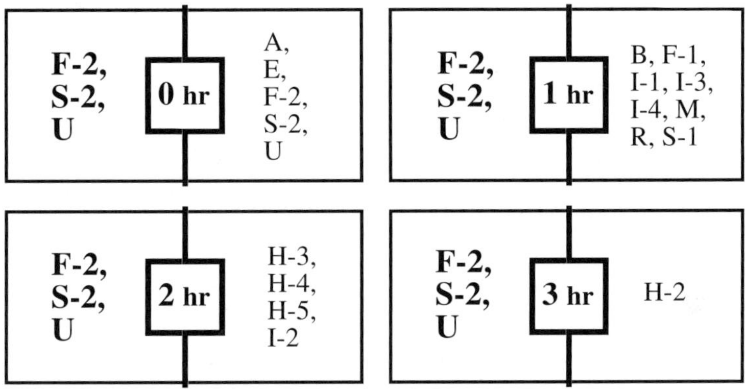

Without sprinklers throughout building:

Occupancy	Ratings (hrs)
A	1
B	2
E	1
F-1	2
F-2	0
H-1	NP
H-2	4
H-3	3
H-4	3
H-5	3
I-1, I-3, I-4	2
I-2	NP
M	2
R	2
S-1	2
S-2	0
U	0

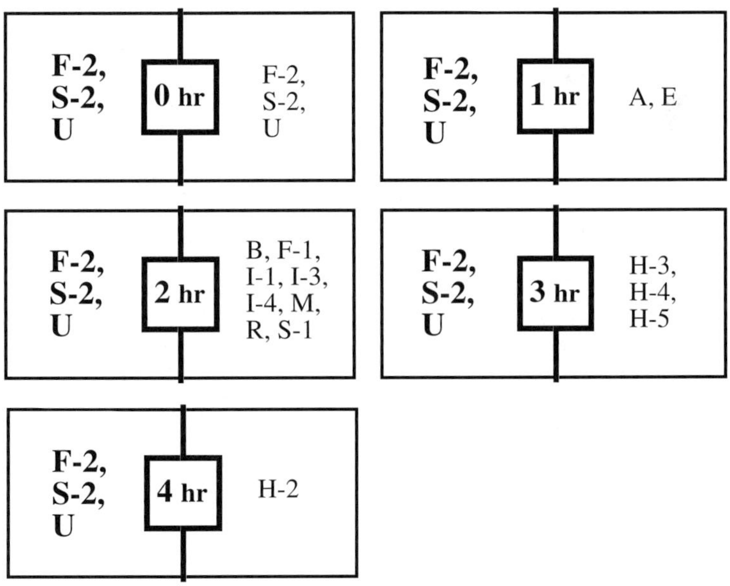

Source: IBC Table 508.4.

NP = not permitted.

508 Mixed Use and Occupancy

508.4.4 Separation *(part 7 of 11)*

- This part governs separations for B, F-1, M, and S-1 occupancies.
- Fire-resistance ratings as required for separation from other occupancies are shown below:

Table 508.4.4f **Occupancy B, F-1, M, or S-1: Fire-Resistance Ratings for Occupancy Separations**

With sprinklers throughout building:

Occupancy	Ratings (hrs)
A	1
B	0
E	1
F-1	0
F-2	1
H-1	NP
H-2	2
H-3	1
H-4	1
H-5	1
I-1, I-3, I-4	1
I-2	2
M	0
R	1
S-1	0
S-2	1
U	1

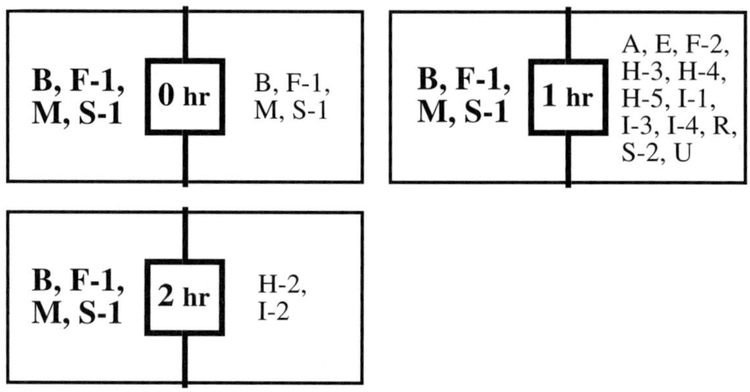

Without sprinklers throughout building:

Occupancy	Ratings (hrs)
A	2
B	0
E	2
F-1	0
F-2	2
H-1	NP
H-2	3
H-3	2
H-4	2
H-5	2
I-1, I-3, I-4	2
I-2	NP
M	0
R	2
S-1	0
S-2	2
U	2

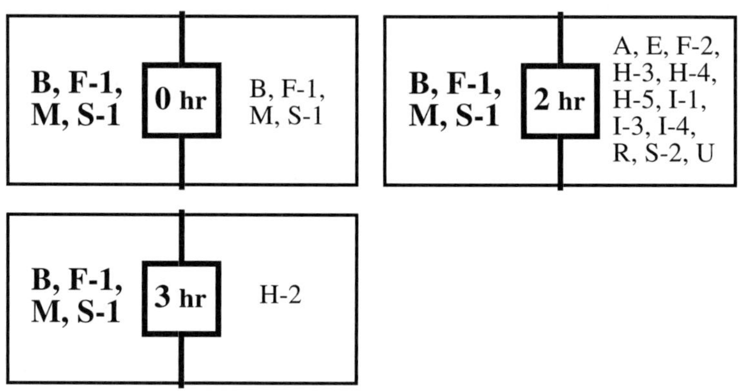

Source: IBC Table 508.4.

NP = not permitted.

508 Mixed Use and Occupancy

508.4.4 Separation *(part 8 of 11)*

- This part governs separations for H-2 occupancies.
- Fire-resistance ratings as required for separation from other occupancies are shown below:

Table 508.4.4g Occupancy H-2: Fire-Resistance Ratings for Occupancy Separations

With sprinklers throughout building:

Occupancy	Ratings (hrs)
A	3
B	2
E	3
F-1	2
F-2	3
H-1	NP
H-2	0
H-3	1
H-4	1
H-5	1
I	3
M	2
R	3
S-1	2
S-2	3
U	3

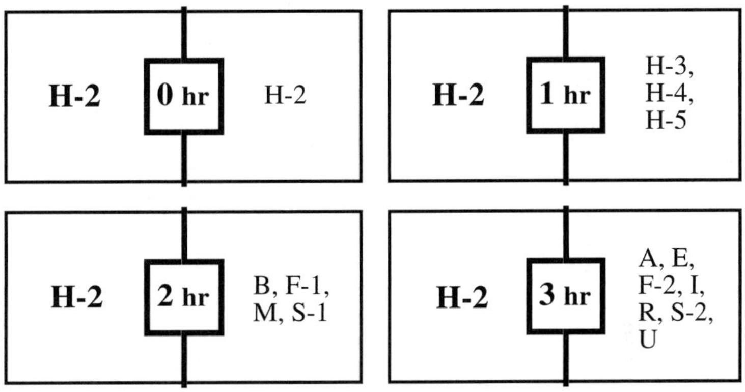

Without sprinklers throughout building:

Occupancy	Ratings (hrs)
A	4
B	3
E	4
F-1	3
F-2	4
H	NP
I	NP
M	3
R	NP
S-1	3
S-2	4
U	4

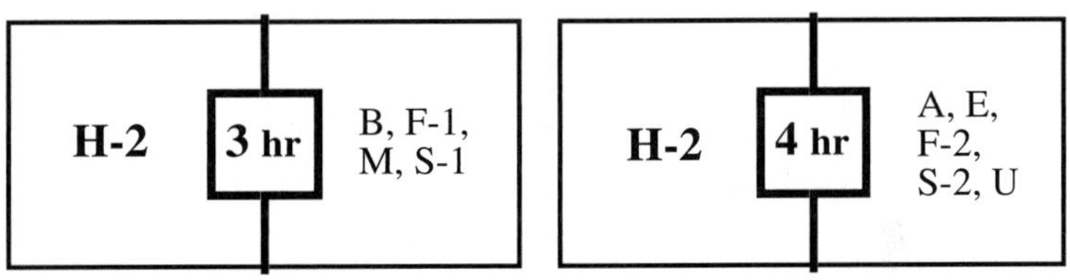

Source: IBC Table 508.4.
NP = not permitted.

508 Mixed Use and Occupancy

508.4.4 Separation *(part 9 of 11)*

- This part governs separations for H-3 occupancies.
- Fire-resistance ratings as required for separation from other occupancies are shown below:

Table 508.4.4h Occupancy H-3: Fire-Resistance Ratings for Occupancy Separations

With sprinklers throughout building:

Occupancy	Ratings (hrs)
A	2
B	1
E	2
F-1	1
F-2	2
H-1	NP
H-2	1
H-3	0
H-4	1
H-5	1
I	2
M	1
R	2
S-1	1
S-2	2
U	2

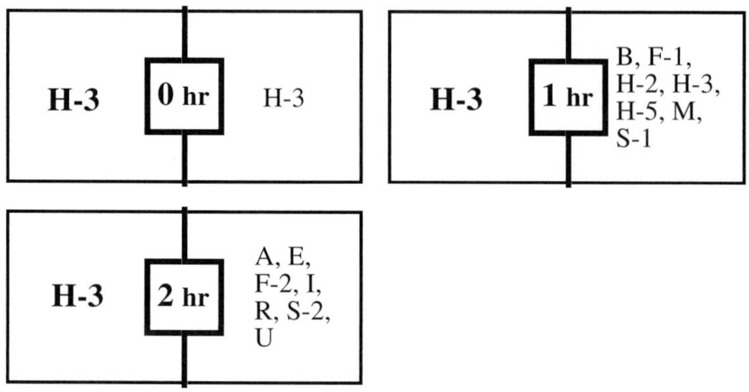

Without sprinklers throughout building:

Occupancy	Ratings (hrs)
A	3
B	2
E	3
F-1	2
F-2	3
H	NP
I	NP
M	2
R	NP
S-1	2
S-2	3
U	3

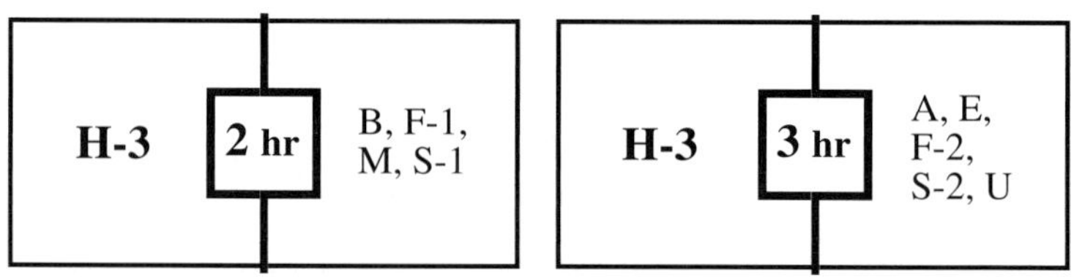

Source: IBC Table 508.4.
NP = not permitted.

508 Mixed Use and Occupancy

508.4.4 Separation *(part 10 of 11)*

- This part governs separations for H-4 occupancies.
- Fire-resistance ratings as required for separation from other occupancies are shown below:

Table 508.4.4i Occupancy H-4: Fire-Resistance Ratings for Occupancy Separations

With sprinklers throughout building:

Occupancy	Ratings (hrs)
A	2
B	1
E	2
F-1	1
F-2	2
H-1	NP
H-2	1
H-3	1
H-4	0
H-5	1
I	2
M	1
R	2
S-1	1
S-2	2
U	2

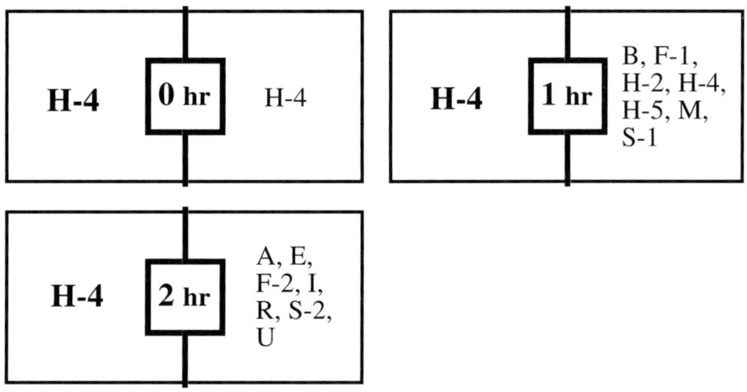

Without sprinklers throughout building:

Occupancy	Ratings (hrs)
A	3
B	2
E	3
F-1	2
F-2	3
H	NP
I	NP
M	2
R	NP
S-1	2
S-2	3
U	3

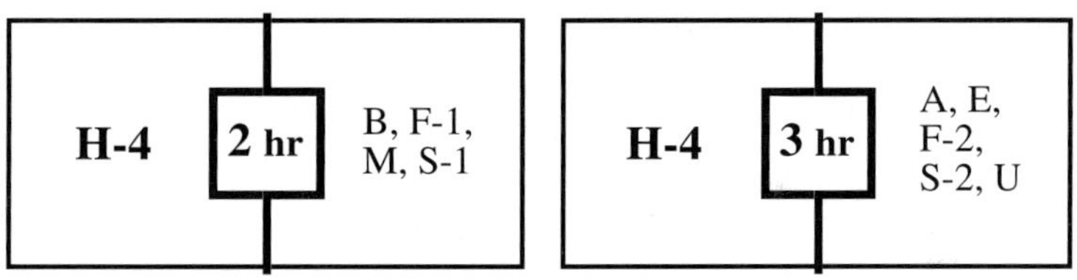

Source: IBC Table 508.4.
NP = not permitted.

508 Mixed Use and Occupancy

508.4.4 Separation *(part 11 of 11)*

- This part governs separations for H-5 occupancies.
- Fire-resistance ratings as required for separation from other occupancies are shown below:

Note: 415.8.2.2, "Separation," is cited as governing additional separation requirements for H-5.

Table 508.4.4j Occupancy H-5: Fire-Resistance Ratings for Occupancy Separations

With sprinklers throughout building:

Occupancy	Ratings (hrs)
A	2
B	1
E	2
F-1	1
F-2	2
H-1	NP
H-2	1
H-3	1
H-4	1
H-5	0
I	2
M	1
R	2
S-1	1
S-2	2
U	2

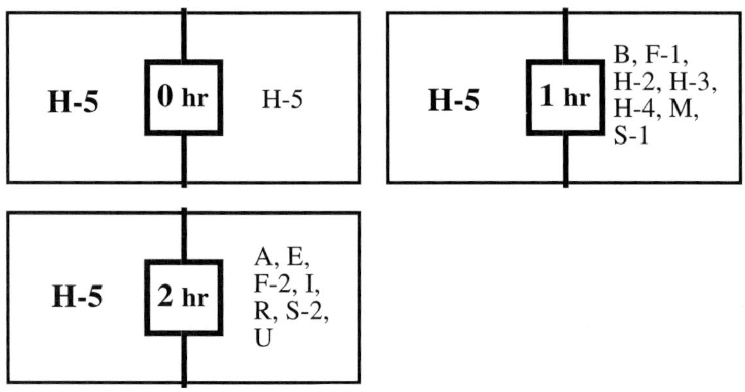

Without sprinklers throughout building:

Occupancy	Ratings (hrs)
A	3
B	2
E	3
F-1	2
F-2	3
H	NP
I	NP
M	2
R	NP
S-1	2
S-2	3
U	3

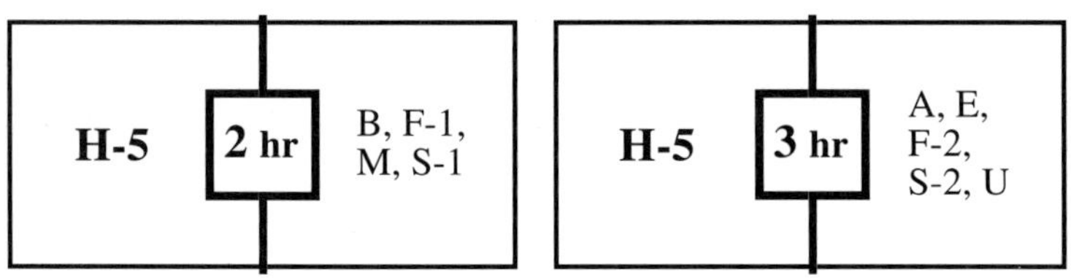

Source: IBC Table 508.4.
NP = not permitted.

508.4.4.1 Construction

- Required separations between occupancies must be one or both of the following:
  - Fire barriers.
  - Horizontal assemblies.

Note: Section 707, "Fire Barriers," is cited as governing these assemblies.
Section 712, "Horizontal Assemblies," is cited as governing these assemblies.

6

Types of Construction

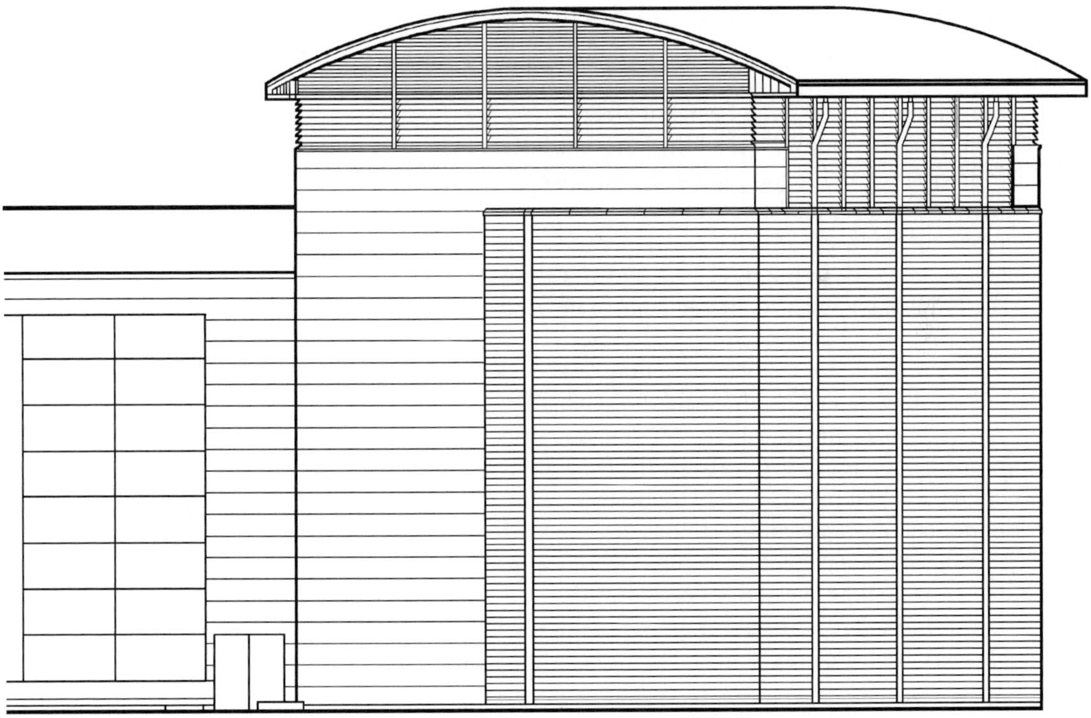

Montachusett Regional Vocational-Technical High School.
Fitchburg, Massachusetts. *(partial elevation)*
HKT Architects, Inc. Somerville, Massachusetts.

602 Construction Classification

602.1 General *(part 1 of 11)*

- The following buildings and structures are governed by this section:
  - New construction and alterations.
  - Additions to height and area.
- Buildings must be classified in 1 of the 5 construction types described in this section.

Note: The following are cited as defining the 5 construction types of this section:
602.2, "Types I and II."
602.3, "Type III."
602.4, "Type IV."
602.5, "Type V."

- Buildings components must have the fire-resistive ratings listed in this section as follows:
  - The primary structural frame includes the following:
    Columns.
    Members connecting directly to columns as follows:
    Beams Spandrels Other members
    Girders Trusses
    Bracing members necessary for vertical stability when primary frame is subjected to gravity loads.
  - The structural frame does not include the following secondary members:
    Floor and roof framing not connected directly to columns.

Note: Section 202, "Definitions," is cited as defining primary and secondary structural members.

  - The following applies to Type I construction:
    The fire-resistive rating for the following is 1 hr less than where other than a roof is supported:
    Primary structural frame supporting only a roof.
    Interior bearing walls supporting only a roof.
  - The following applies to Types I, II, III, and VA construction:
    Occupancies A, B, E, F-2, I, R, S-2, and U are governed as follows:
    The following applies to all elements of the roof construction ≥ 20' above the floor below:
    The following do not require fire protection:
    Roof structural members, framing, and decking.
    The following may be fire-retardant-treated wood:
    Roof structural members, framing, and decking.
  - The following applies to Types IB, II, III, IV, and VA construction:
    Roof construction may be heavy timber where the required fire-resistance rating is ≤ 1 hr.
  - Interior nonbearing walls and partitions are governed as follows:
    Higher fire-resistance ratings required by other sections of the code supersede those of this section.
  - The following applies to exterior bearing walls:
    Where differing fire-resistance ratings are required by this section, the highest governs.
    Occupancy U has requirements in addition to those specified in this section.

Note: 406.1.2, "Area increase," is cited as a source of additional requirements for Occupancy U.

  - Party walls are not governed by this section.

Note: 706.1.1, "Party walls," is cited as governing these walls.

602 Construction Classification

602.1 General *(part 2 of 11)*

- The following applies to interior walls of Types IIA, IIIA, and VA construction:
  - Sprinklers can be used in lieu of 1-hr fire-resistance-rated construction as follows:
    - Where sprinklers are not used to justify the following:
      - An increase in height or area limits.

Note: 903.3.1.1, "NFPA 13 sprinkler systems," is cited as governing the sprinklers.
504.2, "Automatic sprinkler system increase," is cited as governing increases to height limits due to the presence of sprinklers.
506.3, "Automatic sprinkler system increase," is cited as governing increases to area limits due to the presence of sprinklers.

- Fire-resistance ratings are not required for certain open parking garages:

Note: Section 406, "Motor-Vehicle-Related Occupancies," is cited as governing open parking garages which qualify for no fire-resistance ratings.

- Fire-resistance ratings of exterior walls are to be the higher of the following:
  - That required by the fire separation of the exterior wall and the story in which it is located.
  - That required for the components in the wall based on building construction type.

Note: 704.10, "Exterior structural members," is cited as referencing fire-resistance rating requirements for exterior walls.

- Fire-resistance ratings of the primary structural frame members are to be the highest of the following:
  - That required based on building construction type.
  - That required for the exterior wall based on fire separation distance as applicable.
  - That required for the exterior wall based on type of construction as applicable.

Note: 704.10, "Exterior structural members," is cited as referencing fire-resistance rating requirements.

- Occupancy H has certain requirements not addressed in the following tables.

Note: 415.3, "Fire separation distance," is cited as having additional requirements for Occupancy H.

- Occupancy S aircraft hangers have certain requirements not addressed in the following tables.

Note: 412.4.1, "Exterior walls," is cited as having additional requirements for Occupancy S.

- Fire-resistance ratings required by this section are listed in the tables on the following pages:
  - Protection for the following is required only if required by other sections of the IBC:
    - Openings.
    - Ducts.
    - Air-transfer openings.
  - Required fire-resistance ratings must be determined by methods specified by the IBC.

Note: 703.2, " Fire-resistance ratings," is cited as specifying methods for determining such ratings.
The following are cited as governing fire-resistance ratings for building elements:
IBC Table 601, "Fire-Resistance Rating Requirements for Building Elements," which is summarized above and on the following pages.
IBC Table 602, "Fire-Resistance Rating Requirements for Exterior Walls Based on Fire Separation Distance," which is summarized above and on the following pages.

602 Construction Classification

602.1 General *(part 3 of 11)*

- Type IA buildings and structures must have the fire-resistance ratings listed below:

Table 602.1a Fire-Resistance Ratings for Type IA Buildings and Structures

Construction Type IA components	Fire-resistance rating
Primary structural frame supporting a floor:	
Columns	≥ 3 hr
All members connected to columns:	
Girders, trusses, beams, spandrels, other	≥ 3 hr
Bracing necessary for vertical stability when primary frame is subjected to gravity loads	≥ 3 hr
Primary structural frame supporting only a roof:	
Columns	≥ 2 hr
All members connected to columns:	
Girders, trusses, beams, spandrels, other	≥ 2 hr
Bracing necessary for vertical stability when primary frame is subjected to gravity loads	≥ 2 hr
Exterior load-bearing walls other than party walls	≥ 3 hr
Interior load-bearing walls:	
Supporting a floor	≥ 3 hr
Supporting only a roof	≥ 2 hr
Exterior nonload-bearing walls other than party walls:	
All occupancies:	
Fire separation ≥ 30'	≥ 0 hr
Occupancies A, B, E, F-2, I, R:	
Fire separation distance < 30'	≥ 1 hr
Occupancy U:	
Fire separation distance < 5'	≥ 1 hr
Fire separation distance ≥ 5'	≥ 0 hr
Occupancy S-2 open parking garage:	
Fire separation distance ≥ 10'	≥ 0 hr
Occupancy S-2 other than open parking garage:	
Fire separation distance < 30'	≥ 1 hr
Occupancies F-1, M, S-1:	
Fire separation distance < 10'	≥ 2 hr
Fire separation distance ≥ 10' < 30'	≥ 1 hr
Occupancy H:	
Fire separation distance < 10'	≥ 3 hr
Fire separation distance ≥ 10' < 30'	≥ 2 hr
Interior nonload-bearing walls and partitions	≥ 0 hr
Floor construction and secondary members	≥ 2 hr
Structural members not connected to columns	
Floor construction not connected to columns	
Bracing not part of the primary structural frame	
Roof construction and secondary members	≥ 1 1/2 hr
Structural members not connected to columns	
Roof construction not connected to columns	
Bracing not part of the primary structural frame	

Source: IBC Tables 601 and 602.

602 Construction Classification

602.1 General *(part 4 of 11)*

- Type IB buildings and structures must have the fire-resistance ratings listed below:

Table 602.1b Fire-Resistance Ratings for Type IB Buildings and Structures

Construction Type IB components	Fire-resistance rating
Primary structural frame supporting a floor:	
Columns	≥ 2 hr
All members connected to columns:	
Girders, trusses, beams, spandrels, other	≥ 2 hr
Bracing necessary for vertical stability when primary frame is subjected to gravity loads	≥ 2 hr
Primary structural frame supporting only a roof:	
Columns	≥ 1 hr
All members connected to columns:	
Girders, trusses, beams, spandrels, other	≥ 1 hr
Bracing necessary for vertical stability when primary frame is subjected to gravity loads	≥ 1 hr
Exterior load-bearing walls other than party walls:	
Occupancy H:	
Fire separation distance < 5'	≥ 3 hr
Fire separation distance ≥ 5'	≥ 2 hr
All other occupancies	≥ 2 hr
Interior load-bearing walls:	
Supporting a floor	≥ 2 hr
Supporting only a roof	≥ 1 hr
Exterior nonload-bearing walls other than party walls:	
All occupancies:	
Fire separation ≥ 30'	≥ 0 hr
Occupancies A, B, E, F-2, I, R:	
Fire separation distance < 30'	≥ 1 hr
Occupancy U:	
Fire separation distance < 5'	≥ 1 hr
Fire separation distance ≥ 5'	≥ 0 hr
Occupancy S-2 open parking garage:	
Fire separation distance ≥ 10'	≥ 0 hr
Occupancy S-2 other than open parking garage:	
Fire separation distance < 30'	≥ 1 hr
Occupancies F-1, M, S-1:	
Fire separation distance < 5'	≥ 2 hr
Fire separation distance ≥ 5' < 30'	≥ 1 hr
Occupancy H:	
Fire separation distance < 5'	≥ 3 hr
Fire separation distance ≥ 5' < 30'	≥ 2 hr
Interior nonload-bearing walls and partitions	≥ 0 hr
Floor construction and secondary members	≥ 2 hr
Structural members not connected to columns	
Floor construction not connected to columns	
Bracing not part of the primary structural frame	
Roof construction and secondary members	≥ 1 hr
Structural members not connected to columns	
Roof construction not connected to columns	
Bracing not part of the primary structural frame	

Source: IBC Tables 601 and 602.

602 Construction Classification

Case study: Fig. 602.1A. Fire-resistance ratings for various elements of the construction type IB building are indicated in the section. The fire separation distance is > 30'. The building meets code requirements for this type of construction.

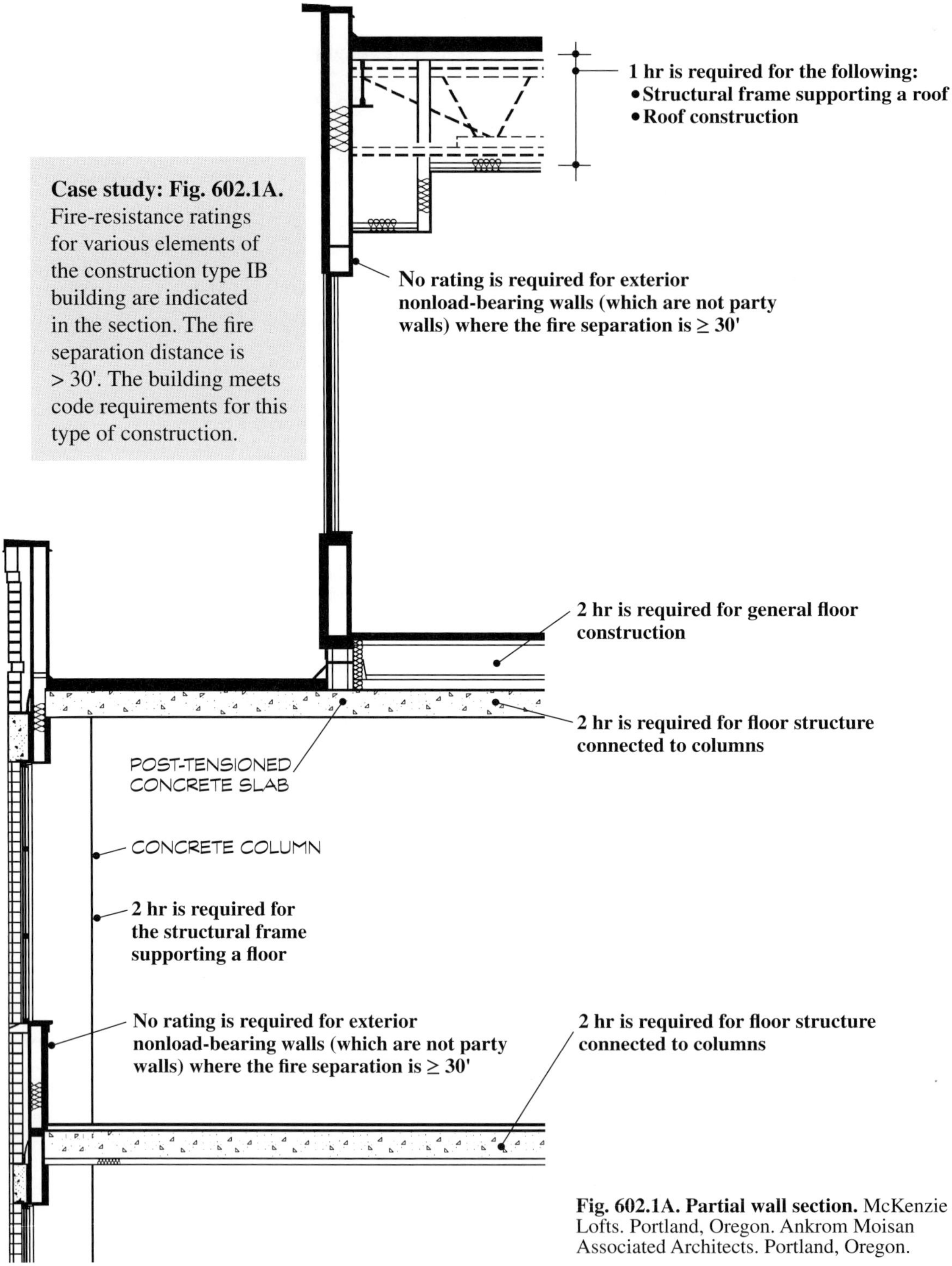

Fig. 602.1A. Partial wall section. McKenzie Lofts. Portland, Oregon. Ankrom Moisan Associated Architects. Portland, Oregon.

602 Construction Classification

602.1 General *(part 5 of 11)*

- Type IIA buildings and structures must have the fire-resistance ratings listed below:
  - Reduction based on sprinklers is for sprinklers as follows:
    Sprinklers not otherwise required or used to increase area or height.

Table 602.1c Fire-Resistance Ratings for Type IIA Buildings and Structures

Construction Type IIA components	With sprinklers	No sprinklers
Primary Structural frame supporting a floor or roof:		
Columns	≥ 0 hr	≥ 1 hr
All members connected to columns:		
Girders, trusses, beams, spandrels, other	≥ 0 hr	≥ 1 hr
Bracing necessary for vertical stability when primary frame is subjected to gravity loads	≥ 0 hr	≥ 1 hr
Exterior load-bearing walls other than party walls:		
Occupancies A, B, E, F-2, I, R, S-2, U	≥ 1 hr	≥ 1 hr
Occupancies F-1, M, S-1:		
Fire separation distance < 5'	≥ 2 hr	≥ 2 hr
Fire separation distance ≥ 5' < 30'	≥ 1 hr	≥ 1 hr
Occupancy H:		
Fire separation distance < 5'	≥ 3 hr	≥ 3 hr
Fire separation distance ≥ 5' < 10'	≥ 2 hr	≥ 2 hr
Fire separation distance ≥ 10' < 30'	≥ 1 hr	≥ 1 hr
Interior load-bearing walls	≥ 0 hr	≥ 1 hr
Exterior nonload-bearing walls other than party walls:		
All occupancies:		
Fire separation distance ≥ 30'	≥ 0 hr	≥ 0 hr
Occupancies A, B, E, F-2, I, R:		
Fire separation distance < 30'	≥ 1 hr	≥ 1 hr
Occupancy U:		
Fire separation distance < 5'	≥ 1 hr	≥ 1 hr
Fire separation distance ≥ 5'	≥ 0 hr	≥ 0 hr
Occupancy S-2 open parking garage:		
Fire separation distance ≥ 10'	≥ 0 hr	≥ 0 hr
Occupancy S-2 other than open parking garage:		
Fire separation distance < 30'	≥ 1 hr	≥ 1 hr
Occupancies F-1, M, S-1:		
Fire separation distance < 5'	≥ 2 hr	≥ 2 hr
Fire separation distance ≥ 5' < 30'	≥ 1 hr	≥ 1 hr
Occupancy H:		
Fire separation distance < 5'	≥ 3 hr	≥ 3 hr
Fire separation distance ≥ 5' < 10'	≥ 2 hr	≥ 2 hr
Fire separation distance ≥ 10' < 30'	≥ 1 hr	≥ 1 hr
Interior nonload-bearing walls and partitions	≥ 0 hr	≥ 0 hr
Floor construction and secondary members	≥ 0 hr	≥ 1 hr
Structural members not connected to columns		
Floor construction not connected to columns		
Bracing not part of the primary structural frame		
Roof construction and secondary members	≥ 0 hr	≥ 1 hr
Structural members not connected to columns		
Roof construction not connected to columns		
Bracing not part of the primary structural frame		

Source: IBC Tables 601 and 602.

602 Construction Classification

602.1 General *(part 6 of 11)*

- Type IIB buildings and structures must have the fire-resistance ratings listed below:

Table 602.1d Fire-Resistance Rating for Type IIB Buildings and Structures

Construction Type IIB components	Fire-resistance rating
Primary structural frame supporting a floor or roof:	
Columns	≥ 0 hr
All members connected to columns:	
Girders, trusses, beams, spandrels, other	≥ 0 hr
Bracing necessary for vertical stability when primary frame is subjected to gravity loads	≥ 0 hr
Exterior walls other than party walls:	
Load-bearing walls and nonload-bearing walls:	
Occupancies A, B, E, F-2, I, R, S-2:	
Fire separation distance < 10'	≥ 1 hr
Fire separation distance ≥ 10'	≥ 0 hr
Occupancy U:	
Fire separation distance < 5'	≥ 1 hr
Fire separation distance ≥ 5'	≥ 0 hr
Occupancies F-1, M, S-1:	
Fire separation distance < 5'	≥ 2 hr
Fire separation distance ≥ 5' <10'	≥ 1 hr
Fire separation distance ≥ 10'	≥ 0 hr
Occupancy H:	
Fire separation distance < 5'	≥ 3 hr
Fire separation distance ≥ 5' < 10'	≥ 2 hr
Fire separation distance ≥ 10' < 30'	≥ 1 hr
Fire separation distance ≥ 30'	≥ 0 hr
Interior load-bearing walls	≥ 0 hr
Interior nonload-bearing walls and partitions	≥ 0 hr
Floor construction and secondary members	≥ 0 hr
Structural members not connected to columns	
Floor construction not connected to columns	
Bracing not part of the primary structural frame	
Roof construction and secondary members	≥ 0 hr
Structural members not connected to columns	
Floor construction not connected to columns	
Bracing not part of the primary structural frame	

Source: IBC Tables 601 and 602.

602 Construction Classification

602.1 General *(part 7 of 11)*

- Type IIIA buildings and structures must have the fire-resistance ratings listed below:
  - Reduction based on sprinklers is for sprinklers as follows:
    Sprinklers not otherwise required or used to increase area or height.

Table 602.1e Fire-Resistance Ratings for Type IIIA Buildings and Structures

Construction Type IIIA components	With sprinklers	Other conditions
Primary structural frame supporting a floor or roof:		
Columns	≥ 0 hr	≥ 1 hr
All members connected to columns:		
Girders, trusses, beams, spandrels, other	≥ 0 hr	≥ 1 hr
Bracing necessary for vertical stability when primary frame is subjected to gravity loads	≥ 0 hr	≥ 1 hr
Exterior load-bearing walls other than party walls:		
Occupancy H:		
Fire separation distance < 5'	≥ 3 hr	≥ 3 hr
Fire separation distance ≥ 5'	≥ 2 hr	≥ 2 hr
All other occupancies	≥ 2 hr	≥ 2 hr
Interior load-bearing walls	≥ 0 hr	≥ 1 hr
Exterior nonload-bearing walls other than party walls:		
All occupancies:		
Fire separation distance ≥ 30'	≥ 0 hr	≥ 0 hr
Occupancies A, B, E, F-2, I, R:		
Fire separation distance < 30'	≥ 1 hr	≥ 1 hr
Occupancy U:		
Fire separation distance < 5'	≥ 1 hr	≥ 1 hr
Fire separation distance ≥ 5'	≥ 0 hr	≥ 0 hr
Occupancy S-2 open parking garage:		
Fire separation distance ≥ 10'	≥ 0 hr	≥ 0 hr
Occupancy S-2 other than open parking garage:		
Fire separation distance < 30'	≥ 1 hr	≥ 1 hr
Occupancies F-1, M, S-1:		
Fire separation distance < 5'	≥ 2 hr	≥ 2 hr
Fire separation distance ≥ 5' < 30'	≥ 1 hr	≥ 1 hr
Occupancy H:		
Fire separation distance < 5'	≥ 3 hr	≥ 3 hr
Fire separation distance ≥ 5' < 10'	≥ 2 hr	≥ 2 hr
Fire separation distance ≥ 10' < 30'	≥ 1 hr	≥ 1 hr
Interior nonload-bearing walls and partitions	≥ 0 hr	≥ 0 hr
Floor construction and secondary members	≥ 0 hr	≥ 1 hr
Structural members not connected to columns		
Floor construction not connected to columns		
Bracing not part of the primary structural frame		
Roof construction and secondary members	≥ 0 hr	≥ 1 hr
Structural members not connected to columns		
Roof construction not connected to columns		
Bracing not part of the primary structural frame		

Source: IBC Tables 601 and 602.

602 Construction Classification

Case study: Fig. 602.1B. The sprinklered construction Type IIIA Occupancy S-1 warehouse meets the fire-resistance rating requirements of IBC Table 601 as indicated in the illustration.

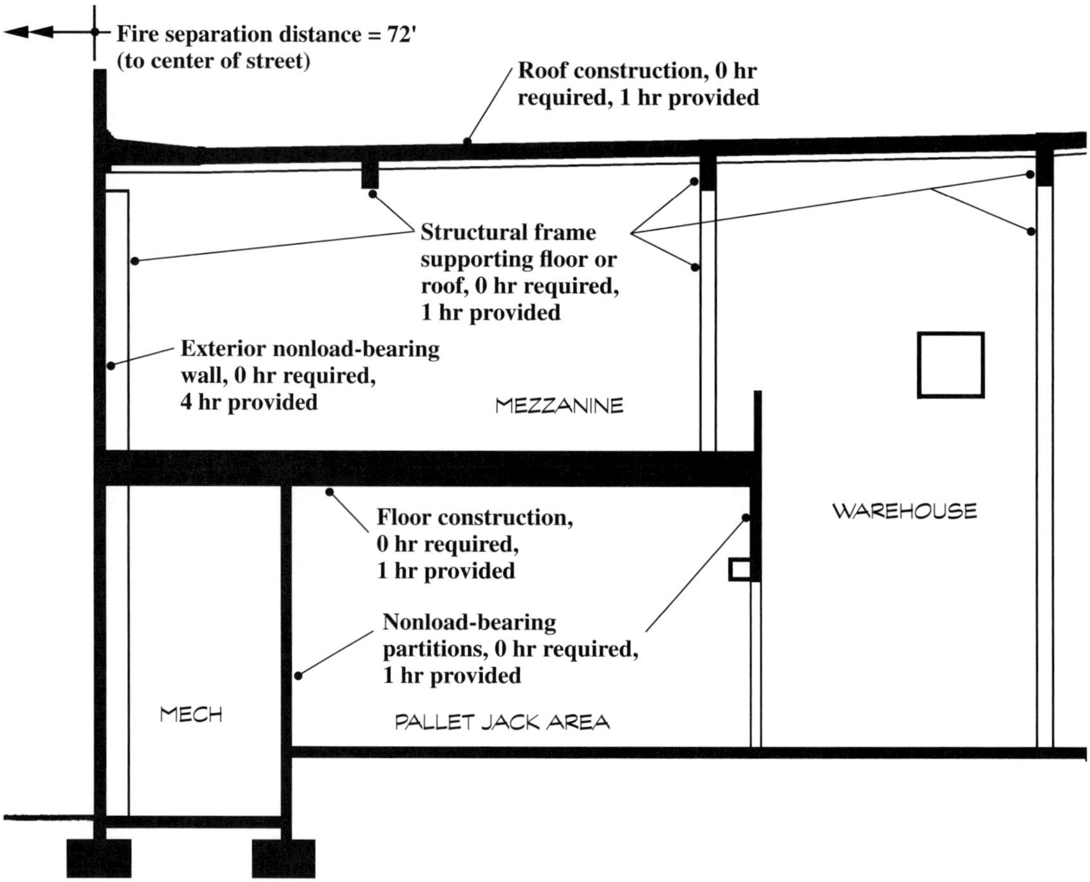

Fig. 602.1B. Partial section. New Warehouse Addition. Los Angeles, California. Stephen Wen + Associates, Architects, Inc. Pasadena, California.

602 Construction Classification

602.1 General *(part 8 of 11)*

- Type IIIB buildings and structures must have the fire-resistance ratings listed below:

Table 602.1f Fire-Resistance Rating for Type IIIB Buildings and Structures

Construction Type IIIB components	Fire-resistance rating
Primary structural frame supporting a floor or roof:	
Columns	≥ 0 hr
All members connected to columns:	
Girders, trusses, beams, spandrels, other	≥ 0 hr
Bracing necessary for vertical stability when primary frame is subjected to gravity loads	≥ 0 hr
Exterior load-bearing walls other than party walls:	
Occupancy H:	
Fire separation distance < 5'	≥ 3 hr
Fire separation distance ≥ 5'	≥ 2 hr
All other occupancies	≥ 2 hr
Interior load-bearing walls	≥ 0 hr
Exterior nonload-bearing walls other than party walls:	
All occupancies:	
Fire separation distance ≥ 30'	≥ 0 hr
Occupancies A, B, E, F-2, I, R:	
Fire separation distance < 30'	≥ 1 hr
Occupancy U:	
Fire separation distance < 5'	≥ 1 hr
Fire separation distance ≥ 5'	≥ 0 hr
Occupancy S-2 open parking garage:	
Fire separation distance ≥ 10'	≥ 0 hr
Occupancy S-2 other than open parking garage:	
Fire separation distance < 30'	≥ 1 hr
Occupancies F-1, M, S-1:	
Fire separation distance < 5'	≥ 2 hr
Fire separation distance ≥ 5' < 30'	≥ 1 hr
Occupancy H:	
Fire separation distance < 5'	≥ 3 hr
Fire separation distance ≥ 5' < 10'	≥ 2 hr
Fire separation distance ≥ 10' < 30'	≥ 1 hr
Interior nonload-bearing walls and partitions	≥ 0 hr
Floor construction and secondary members	≥ 0 hr
Structural members not connected to columns	
Floor construction not connected to columns	
Bracing not part of the primary structural frame	
Roof construction and primary members	≥ 0 hr
Structural members not connected to columns	
Roof construction not connected to columns	
Bracing not part of the primary structural frame	

Source: IBC Tables 601 and 602.

602 Construction Classification

602.1 General *(part 9 of 11)*

- Type IV buildings and structures must have the fire-resistance ratings or meet heavy timber detailing requirements as listed below:

Table 602.1g Fire-Resistance Rating or Heavy Timber Requirements for Type IV Buildings and Structures

Construction Type IV components	Fire-resistance rating or heavy timber requirements
Primary structural frame supporting a floor or roof:	
Columns	heavy timber
All members connected to columns:	
Girders, trusses, beams, spandrels, other	heavy timber
Bracing necessary for vertical stability when primary frame is subjected to gravity loads	heavy timber
Exterior load-bearing walls other than party walls:	
Occupancy H:	
Fire separation distance < 5'	≥ 3 hr
Fire separation distance ≥ 5'	≥ 2 hr
All other occupancies	≥ 2 hr
Interior load-bearing walls	≥ 1 hr or heavy timber
Exterior nonload-bearing walls other than party walls:	
All occupancies:	
Fire separation distance ≥ 30'	≥ 0 hr
Occupancies A, B, E, F-2, I, R:	
Fire separation distance < 30'	≥ 1 hr
Occupancy U:	
Fire separation distance < 5'	≥ 1 hr
Fire separation distance ≥ 5'	≥ 0 hr
Occupancy S-2 open parking garage:	
Fire separation distance ≥ 10'	≥ 0 hr
Occupancy S-2 other than open parking garage:	
Fire separation distance < 30'	≥ 1 hr
Occupancies F-1, M, S-1:	
Fire separation distance < 5'	≥ 2 hr
Fire separation distance ≥ 5' < 30'	≥ 1 hr
Occupancy H:	
Fire separation distance < 5'	≥ 3 hr
Fire separation distance ≥ 5' < 10'	≥ 2 hr
Fire separation distance ≥ 10' < 30'	≥ 1 hr
Interior nonload-bearing walls and partitions	≥ 1 hr or code alternatives
Floor construction and secondary members	heavy timber
Structural members not connected to columns	
Floor construction not connected to columns	
Bracing not part of the primary structural frame	
Roof construction and secondary members	heavy timber
Structural members not connected to columns	
Roof construction not connected to columns	
Bracing not part of the primary structural frame	

Source: IBC Tables 601 and 602.

602 Construction Classification

602.1 General *(part 10 of 11)*

- Type VA buildings and structures must have the fire-resistance ratings listed below:
  - Reduction based on sprinklers is for sprinklers as follows:
    Sprinklers not otherwise required or used to increase area or height.

Table 602.1h Fire-Resistance Ratings for Type VA Buildings and Structures

Construction Type VA components	With sprinklers	Other conditions
Primary structural frame supporting a floor or roof:		
Columns	≥ 0 hr	≥ 1 hr
All members connected to columns:		
Girders, trusses, beams, spandrels, other	≥ 0 hr	≥ 1 hr
Bracing necessary for vertical stability when primary frame is subjected to gravity loads	≥ 0 hr	≥ 1 hr
Exterior load-bearing walls other than party walls:		
Occupancies A, B, E, F-2, I, R, S-2, U	≥ 1 hr	≥ 1 hr
Occupancies F-1, M, S-1:		
Fire separation distance < 5'	≥ 2 hr	≥ 2 hr
Fire separation distance ≥ 5'	≥ 1 hr	≥ 1 hr
Occupancy H:		
Fire separation distance < 5'	≥ 3 hr	≥ 3 hr
Fire separation distance ≥ 5' < 10'	≥ 2 hr	≥ 2 hr
Fire separation distance ≥ 10'	≥ 1 hr	≥ 1 hr
Interior load-bearing walls	≥ 0 hr	≥ 1 hr
Exterior nonload-bearing walls other than party walls:		
All occupancies:		
Fire separation distance ≥ 30'	≥ 0 hr	≥ 0 hr
Occupancies A, B, E, F-2, I, R:		
Fire separation distance < 30'	≥ 1 hr	≥ 1 hr
Occupancy U:		
Fire separation distance < 5'	≥ 1 hr	≥ 1 hr
Fire separation distance ≥ 5'	≥ 0 hr	≥ 0 hr
Occupancy S-2 open parking garage:		
Fire separation distance ≥ 10'	≥ 0 hr	≥ 0 hr
Occupancy S-2 other than open parking garage:		
Fire separation distance < 30'	≥ 1 hr	≥ 1 hr
Occupancies F-1, M, S-1:		
Fire separation distance < 5'	≥ 2 hr	≥ 2 hr
Fire separation distance ≥ 5' < 30'	≥ 1 hr	≥ 1 hr
Occupancy H:		
Fire separation distance < 5'	≥ 3 hr	≥ 3 hr
Fire separation distance ≥ 5' < 10'	≥ 2 hr	≥ 2 hr
Fire separation distance ≥ 10' < 30'	≥ 1 hr	≥ 1 hr
Interior nonload-bearing walls and partitions	≥ 0 hr	≥ 0 hr
Floor construction and secondary members	≥ 0 hr	≥ 1 hr
Structural members not connected to columns		
Floor construction not connected to columns		
Bracing not part of the primary structural frame		
Roof construction not connected to columns	≥ 0 hr	≥ 1 hr
Structural members not connected to columns		
Roof construction not connected to columns		
Bracing not part of the primary structural frame		

Source: IBC Tables 601 and 602.

602 Construction Classification

602.1 General *(part 11 of 11)*

- Type VB buildings and structures must have the fire-resistance ratings listed below:

Table 602.1i Fire-Resistance Rating for Type VB Buildings and Structures

Construction Type VB components	Fire-resistance rating
Primary structural frame supporting a floor or roof:	
Columns	≥ 0 hr
All members connected to columns:	
Girders, trusses, beams, spandrels, other	≥ 0 hr
Bracing necessary for vertical stability when primary frame is subjected to gravity loads	≥ 0 hr
Exterior walls other than party walls:	
Load-bearing walls and nonload-bearing walls:	
Occupancies A, B, E, F-2, I, R, S-2:	
Fire separation distance < 10'	≥ 1 hr
Fire separation distance ≥ 10'	≥ 0 hr
Occupancy U:	
Fire separation distance < 5'	≥ 1 hr
Fire separation distance ≥ 5'	≥ 0 hr
Occupancies F-1, M, S-1:	
Fire separation distance < 5'	≥ 2 hr
Fire separation distance ≥ 5' < 10'	≥ 1 hr
Fire separation distance ≥ 10'	≥ 0 hr
Occupancy H:	
Fire separation distance < 5'	≥ 3 hr
Fire separation distance ≥ 5' < 10'	≥ 2 hr
Fire separation distance ≥ 10' < 30'	≥ 1 hr
Fire separation distance ≥ 30'	≥ 0 hr
Interior load-bearing walls	≥ 0 hr
Interior nonload-bearing walls and partitions	≥ 0 hr
Floor construction and secondary members	≥ 0 hr
Structural members not connected to columns	
Floor construction not connected to columns	
Bracing not part of the primary structural frame	
Roof construction and secondary members	≥ 0 hr
Structural members not connected to columns	
Roof construction not connected to columns	
Bracing not part of the primary structural frame	

Source: IBC Tables 601 and 602.

602 Construction Classification

602.1.1 Minimum requirements

- The following applies to detailing that complies with a construction type higher than required:
  - Other components of the occupancy need not comply with the higher construction type.

602.2 Types I and II

- In special cases combustible materials are permitted in construction Types I and II.

Note: Section 603, "Combustible Material in Type I and II Construction," and other parts of the code are cited as permitting certain combustible materials in construction Type I and II.

- In other than special cases, construction Types I and II require noncombustible materials for the following:
  - Structural frame:
    - Columns.
    - Members connected to columns:
      - Girders.
      - Trusses.
      - Spandrels.
    - Bracing members for gravity loads.
  - Load-bearing walls:
    - Exterior.
    - Interior.
  - Nonload-bearing walls and partitions:
    - Exterior.
    - Interior.
  - Floor construction including the following:
    - Beams.
    - Joists.
  - Roof construction including the following:
    - Beams.
    - Joists.

Note: Table 601, "Fire-Resistance Rating Requirements for Building Elements," is cited as listing the building elements that are required to be noncombustible in Types I and II construction and is partially summarized above.

602.3 Type III

- In construction Type III, building elements are of the following materials:
  - Noncombustible materials are required for exterior walls as follows:
    - Where exterior wall assemblies have fire-resistance rated at ≤ 2 hrs:
      - Fire-retardant-treated wood is allowed therein.
  - The following materials are allowed for interior building elements:
    - Any material permitted by the code.

Note: 2303.2, "Fire-retardant-treated wood," is cited as governing this material.

602 Construction Classification

602.4 Type IV

- Construction Type IV (Heavy Timber, HT) construction consists of the following:
  - Noncombustible materials are required for exterior walls as follows:
    Fire-retardant-treated wood is allowed in exterior walls with a ≥ 2-hr fire-resistance rating.
  - Interior building elements are required as follows:
    Solid or laminated wood.
    Containing no concealed spaces.
  - Details are governed by this section.

Note: 2303.2, "Fire-retardant-treated wood," is cited as the source of requirements for this material.

- The pairs of wood sizes listed below are considered to be equivalent as follows:
  - For purposes of meeting minimum size requirements for Type IV construction.

Wood type	Wood sizes (in)				
Sawn timber (nominal)	4×6	6×6	6×8	6×10	8×8
Glued laminated timber (in.)	3×6 7/8	5×6	5×8 1/4	5×10 1/2	6 3/4×8 1/4

Source: IBC Table 602.4.

602.4.1 Columns

- Construction Type IV wood columns are governed as follows:
  - Columns must be continuous or stacked with approved connections.
  - Columns must sawn as a single piece or be glue-laminated.
  - Column dimensions are required as follows:

Loads supported	Width	Depth
Floor loads	≥ 8"	≥ 8"
Roof and ceiling loads only	≥ 6"	≥ 8"

602.4.2 Floor framing

- Construction Type IV wood beams and girders are governed as follows:
  - Members must be sawn as a single piece or be glue-laminated.
  - Minimum nominal member dimensions are as follows:

Component	Width	Depth
Beams and girders	≥ 6"	≥ 10"
Arches springing from floor line and supporting floor loads	≥ 8"	≥ 8"
Members of trusses supporting floor loads	≥ 8"	≥ 8"

602 Construction Classification

602.4.3 Roof framing

- Construction Type IV roof framing members are governed as follows:
  - || members spaced on either side of another member must have all of the following characteristics:
    - Assembled of ≥ 2 members.
    - Each member must have a nominal thickness ≥ 3".
    - Open space between spaced members requires the following:
      - Space must be closed by one of the following means:
        - Continuous blocking between spaced members with a nominal thickness ≥ 2".
        - Continuous wood cover plate with a nominal thickness ≥ 2":
          - Applied to underside of members.
  - Splice plates must have a nominal thickness ≥ 3".
  - Other roof framing members are governed as follows:
    - Required widths may be reduced with sprinklers as follows:
      - Sprinklers must be located under the roof deck.
    - Individual framing members must have the nominal dimensions listed below:

Table 602.4.3 Nominal Dimensions of Individual Roof Framing Members

Components supporting no floor loads	Width with sprinklers	Width with no sprinklers	Depth
Framed or glue-laminated arches:			
Where they spring from floor line or grade:			
In the upper half of height	≥ 6"	≥ 6"	≥ 6"
In the lower half of height	≥ 6"	≥ 6"	≥ 8"
Where they spring from the following:			
Top of walls or wall abutments	≥ 3"	≥ 4"	≥ 6"
Framed timber trusses	≥ 3"	≥ 4"	≥ 6"
Other roof framing	≥ 3"	≥ 4"	≥ 6"

602.4.4 Floors *(part 1 of 2)*

- No concealed spaces in floors are permitted in Type IV construction.
- The wood decking system must be one of the following:
  - Structural decking laid flat as follows:
    - Nominal thickness in the vertical dimension must be ≥ 3".
    - Either sawn or glue-laminated is required.
    - One of the following edge details for decking members is required:
      - Splined.
      - Tongue-and-groove.
    - One of the following types of subflooring is required:
      - Nominal 1" tongue-and-groove flooring laid ⊥ to or diagonally across decking.
      - 1/2" particle board.

602 Construction Classification

602.4.4 Floors *(part 2 of 2)*

- Structural decking laid on edge as follows:
  - Nominal thickness in the vertical dimension must be ≥ 4".
  - Horizontal dimension must be < the vertical dimension.
  - Adjacent members must have continuous contact.
  - Members must be securely spiked together.
  - One of the following types of sub-flooring is required:
    - Nominal 1" wood flooring.
    - 15/32" wood structural panel.
    - 1/2" particle board.
- Butt-joints of lumber must be staggered as follows:
  - So joints are not aligned at locations other than on supports.
- A gap ≥ 1/2" is required between floors and walls as follows:
  - Gaps must be closed by one of the following means:
    - Moulding attached to the wall as follows:
      - Moulding must permit movement in the floor due to expansion and contraction.
    - Masonry wall corbeling below the floor edge.

602.4.5 Roofs

- This section addresses roofs in Type IV construction.
- Concealed spaces are not permitted in roof systems.
- Wood roof decks must be one of the following types:
  - Decking laid flat as follows:
    - Either sawn or glue-laminated is required.
    - One of the following edge details for plank decking is required:
      - Splined.
      - Tongue-and-groove.
    - Nominal thickness in the vertical dimension must be ≥ 2".
  - Decking laid on edge as follows:
    - Nominal thickness in the vertical dimension must be ≥ 3".
    - Horizontal dimension must be < the vertical dimension.
    - Adjacent members must have continuous contact.
    - Members must be securely spiked together.
  - Wood structural panels with both of the following characteristics:
    - Thickness ≥ 1 1/8".
    - Exterior glue.
  - Other types of decking with both of the following characteristics:
    - ≥ the fire resistance as the other options above.
    - ≥ structural properties as the other options above.

602.4.6 Partitions

- Partitions in Type IV construction must be one of the following:
  - Solid wood with ≥ 2 layers of one of the following:
    - 1" thick matched boards.
    - 4" thick laminated construction.
  - Construction with a fire-resistance rating of 1 hr.

602 Construction Classification

602.4.7 Exterior structural members

- Construction Type IV requires exterior wood columns and arches to have both of the following:
  - A horizontal fire-separation distance ≥ 20'.
  - Sizes complying with heavy timber size requirements.

602.5 Type V

- Construction Type V permits any material otherwise allowed by the code for the following:
  - Structural elements.
  - Exterior and interior walls.

602 Construction Classification

Case study: Fig. 602.5. The building is an example of Type V construction. In this category, materials are neither restricted nor specified so long as they are permitted by the code for building construction. This applies to both exterior and interior walls as well as structural components. Type V construction is not limited to the materials in this example. This represents only one particular combination.

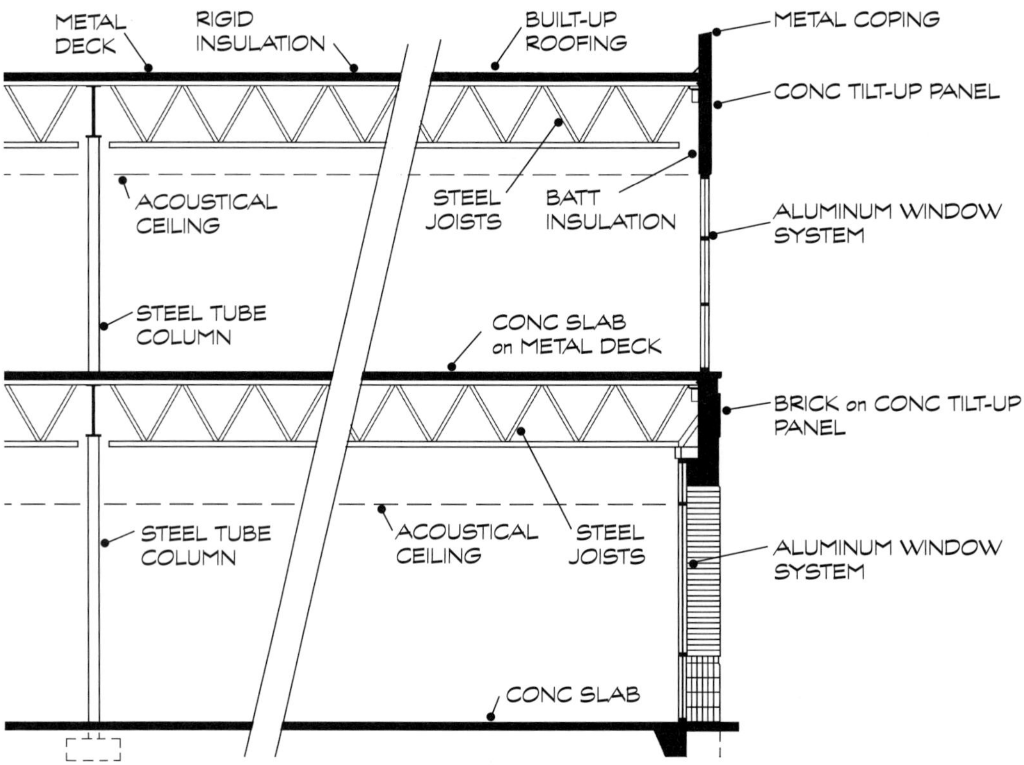

Fig. 602.5. Partial building section. AmberGlen Business Center. Hillsboro, Oregon. Ankrom Moisan Associated Architects. Portland, Oregon.

603 Combustible Material in Type I and II Construction

603.1 Allowable materials *(part 1 of 4)*

- This section lists combustible materials permitted in Type I and Type II construction.

  Note: The following are cited as requiring compliance:
  603.1.1, "Ducts."
  603.1.2, "Piping."
  603.1.3, "Electrical."

- Fire-retardant-treated wood is permitted in the following applications:
  - Nonload-bearing construction as follows:
    - Partitions with a required fire-resistance rating ≤ 2 hrs.
    - Exterior walls with no fire-resistance rating required.
  - In the following roof construction where all of the conditions indicated below apply:
    - Roof construction:
      - Decking
      - Framing
      - Girders
      - Trusses
    - Conditions:
      - The building must be one of the following:
        - Type IA construction ≤ 2 stories.
        - Type IA construction > 2 stories with the following condition:
          - Roof must be ≥ 20' above the top floor.
        - Type IB construction.
        - Type II construction.
- Insulation is permitted with the following conditions:
  - Layered between noncombustible materials as follows:
    - No airspace.
    - Flame spread index is ≤ 100.
  - Layered between the following components with the conditions indicated below:
    - Components:
      - Finished floor.
      - Solid decking.
    - Conditions:
      - No air space.
      - Flame spread index is ≤ 200.
  - Other insulation as follows:
    - Other than foam.
    - With a flame spread index ≤ 25.
    - Thermal or acoustical insulation.
- Roof coverings are permitted in one of the following classifications:
  - Class A.
  - Class B.
  - Class C.
- Plastics are permitted as regulated by the code as follows:
  - Foam plastics.
  - Light-transmitting plastics.

  Note: Chapter 26, "Plastic," is cited as the source of requirements for the plastics.

603 Combustible Material in Type I and II Construction

603.1 Allowable materials *(part 2 of 4)*

- The following interior materials are permitted:
  - Floor finishes and coverings.

  Note: Section 804, "Interior Floor Finish," is cited as governing floor finishes and coverings.

  - Millwork such as the following:
    - Doors and door frames.
    - Window sashes and window frames.
  - Wall and ceiling finishes.

  Note: The following are cited as governing wall and ceiling finishes:
  Section 801, "General."
  Section 803, "Wall and Ceiling Finishes."

  - Trim.

  Note: Section 806, "Decorative Materials and Trim," is cited as governing trim.

- Combustible materials are permitted in the following applications located ≥ 15' above grade:
  - Show windows and related elements as follows:
    - Nailing or furring strips.
    - Wood bulkheads below show windows.
    - Frames.
    - Aprons.
    - Show cases.
- Finish flooring.

  Note: Section 805, "Combustible Materials in Types I and II Construction," is cited as governing finish flooring.

- Partitions with the following characteristics may be constructed of the materials listed below:
  - Characteristics:
    - Used to subdivide the following of a single tenant:
      - Store.
      - Offices.
      - Similar spaces.
    - Partitions may not create a corridor serving ≥ 30 occupants.
  - Materials:
    - Any of the following are permitted for partitions ≤ 6' high:
      - Fire-retardant-treated wood.
      - 1-hr fire-resistant-rated construction.
      - Wood panels.
      - Similar light construction.

- Certain stages and platforms are permitted to have combustible construction.

  Note: The following are cited as governing stages and platforms:
  410.3, "Stages."
  410.4, "Platform construction."

603 Combustible Materials in Type I and II Construction

603.1 Allowable materials *(part 3 of 4)*

- The following combustible materials are permitted at exterior walls:
  - Exterior wall coverings.
  - Projections such as follows:
    - Balconies.
    - Bay windows.
    - Oriel windows.
    - Similar projections.

Note: Chapter 14, "Exterior Walls," is cited as governing combustible materials at exterior walls.

- Blocking such as for the following is permitted:
  - Handrails.
  - Millwork.
  - Cabinets.
  - Window frames.
  - Door frames.
- Certain light-transmitting plastics are permitted.

Note: Chapter 26, "Plastic," is cited as governing these plastics.

- Sealing materials between components of exterior walls is permitted as follows:
  - Mastics.
  - Caulking.
- Exterior plastic veneer.

Note: 2605.2, "Exterior use," is cited as governing these plastics.

- Certain nailing or furring strips are permitted.

Note: 803.11, "Application of interior finish materials to fire-resistance-rated structural elements," is cited as a source of requirements for furring.

- Heavy timber is permitted in certain applications.

Note: The following are cited as defining permitted applications of heavy timber:
IBC Table 601, "Fire-Resistance Rating Requirements for Building Elements," footnote "d".
602.4.7, "Exterior structural members."
1406.3, "Balconies and similar projections."

- Combustible ingredients for cement are permitted as follows:
  - Aggregates in gypsum concrete mixtures.
  - Aggregates in portland cement concrete mixtures.
  - Approved materials in assemblies meeting required fire-resistance ratings as follows:
    - Admixtures.
    - Component materials.

Note: 703.2.2, "Combustible components," is cited as the source establishing the acceptability of combustible ingredients, a summary of which is provided above.

603 Combustible Materials in Type I and II Construction

603.1 Allowable materials *(part 4 of 4)*

- The following materials are permitted where they are determined to be fire-resistant by tests:
  - Sprayed materials.
  - Intumescent and mastic coatings.

Note: The following are cited as governing sprayed, intumescent, or mastic materials:
703.2, "Fire-resistance ratings."
1704.12, "Sprayed fire-resistant materials."
1704.13, "Mastic and intumescent fire-resistant coatings."

- Certain materials protecting penetrations in fire-resistance rated assemblies are permitted.

Note: Section 713, "Penetrations," is cited as governing materials sealing penetrations.

- Materials in joints of assemblies with fire-resistance ratings.

Note: Section 714, "Fire-Resistant Joint Systems," is cited as governing these joint materials.

- Certain combustible materials are permitted as follows:
  - Class A interior finish materials.
  - Piping.
  - Piping insulation.

Note: 717.5, "Combustible materials in concealed spaces in Type I or II construction," is cited as listing materials permitted in concealed spaces, a partial summary of which is provided above.

- Certain materials exposed in plenums are permitted.

Note: The International Mechanical Code *Section 602, "Plenums," is cited as the source of combustible materials permitted in plenums.*

603.1.1 Ducts

- Certain nonmetallic ducts are permitted.

Note: The International Mechanical Code *is cited as defining conditions permitting nonmetallic ducts.*

603.1.2 Piping

- Certain combustible piping is permitted.

Note: The following are cited as governing combustible piping:
International Mechanical Code.
International Plumbing Code.

603.1.3 Electrical

- The following combustible electrical components are permitted:
  - Wiring insulation.
  - Tubing.
  - Raceways.
  - Related components.

7

Fire and Smoke Protection Features

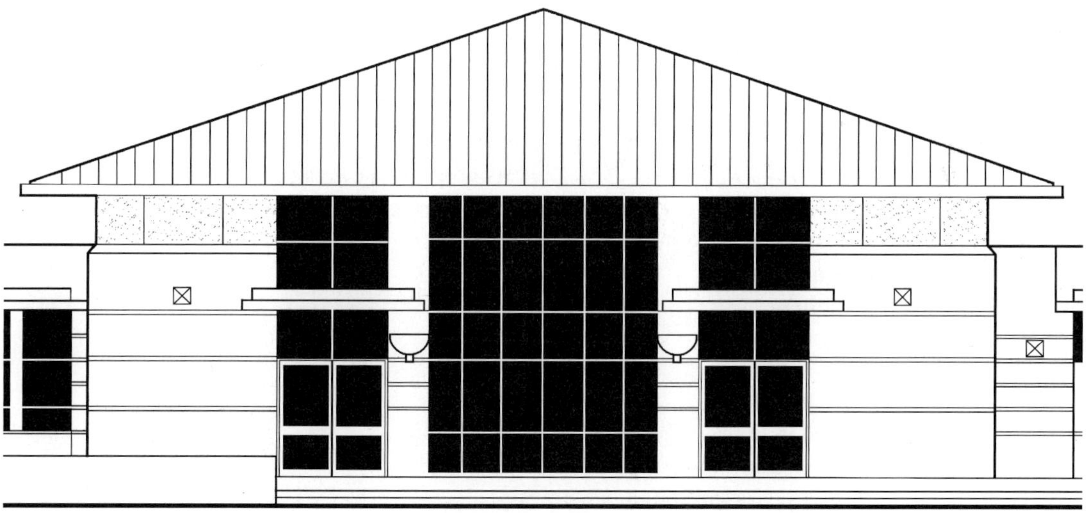

Lee's Summit Police and Court Facility. Lee's Summit, Missouri. *(partial elevation)*
The Hollis and Miller Group, Inc. Lee's Summit, Missouri.

702 Definitions

702.1 Definitions *(part 1 of 7)*

- **Annular space**
  - The gap around a component that is penetrating an assembly.
- **Building element**
  - A building construction component as follows:
    - Can be fire-resistance rated.
    - Not required to be fire-resistance rated.
    - Consists of materials related to a construction type category.

Note: IBC Table 601, "Fire-Resistance Rating Requirements for Building Elements," is cited as the table where building elements are listed.

- **Combination fire/smoke damper**
  - A listed device.
  - Installed in any of the following locations:
    - Ducts.
    - Air transfer openings.
  - Controlled by a smoke detection system.
  - Closes when the following is detected:
    - Heat.
    - Smoke.
  - Inhibits the flow of the following:
    - Flame.
    - Smoke.
  - Adjustable where required as follows:
    - From a fire command center.
- **Draftstop**
  - One of the following:
    - A material.
    - A device.
    - A construction.
  - Installed to limit the movement of air within the following types of concealed spaces:
    - Crawl spaces.
    - Floor-ceiling assemblies.
    - Roof-ceiling assemblies.
    - Attics.
    - Similar spaces.
- **F Rating**
  - A time period expressed in hours.
  - Applies to a through-penetration firestop system as follows:
    - The time it takes fire to pass through a penetration.

Note: The following are cited as alternatives specifying procedures for determing an F Rating:
ASTM E 814, "Test Method for Fire Tests of Through-Penetration Fire Stops."
UL 1479, "Fire Tests of Through-Penetration Firestops."

702 Definitions

702.1 Definitions *(part 2 of 7)*

Case study: Fig. 702.1A. TJI joists rest on 2"× 4" bearing plates which sit on a concrete slab. An air space results between the slab and each joist through which air can flow in the concealed space. Draft stop materials placed under periodic joists isolate small areas of continuous air space between which no air can move.

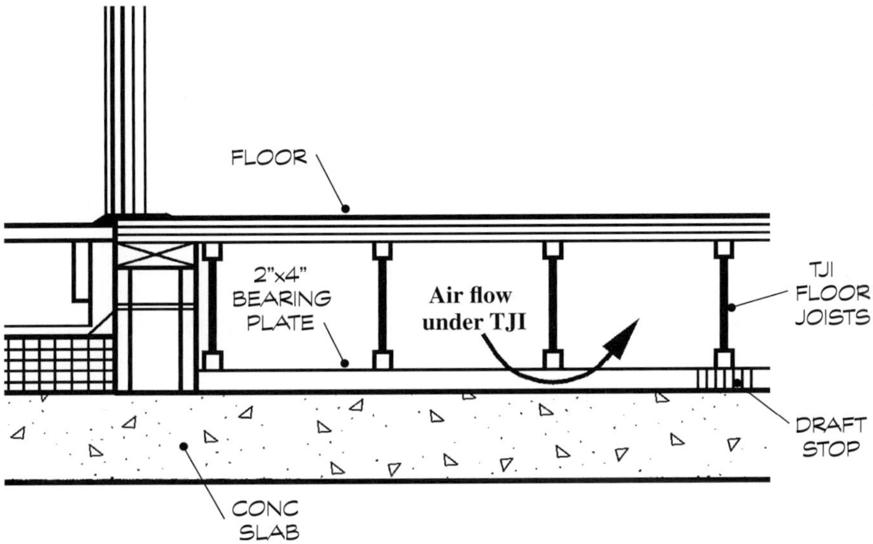

Fig. 702.1A. Detail at raised floor. McKenzie Lofts. Portland, Oregon. Ankrom Moisan Associated Architects. Portland, Oregon.

- **Fire barrier**
  - A wall as follows:
    - With a fire-resistance rating.
    - Designed to limit the spread of fire.
    - Continuity is required.
- **Fire damper**
  - A listed device.
  - Installed in any of the following locations:
    - Ducts.
    - Air transfer openings.
  - Closes when heat is detected.
  - Inhibits the passage of flame.
  - Categorized as one of two types:
    - A static system closes in case of fire.

702 Definitions

702.1 Definitions *(part 3 of 7)*

- **Fire door assembly**
  - An assembly of the following:
    - Fire door.
    - Door frame.
    - Hardware.
    - Accessories.
  - Provides fire protection to an opening at a defined level.
- **Fire partition**
  - A vertical assembly.
  - Limits the spread of fire.
  - Any openings in the partition are protected.
- **Fire-protection rating**
  - Pertains to a protective for an opening.
  - Indicates the length of time a protective can contain a fire as follows:
    - Measured in one of the following units:
      - Hours.
      - Minutes.

Note: Section 715, "Opening Protectives," is cited as describing tests to determine fire-protection rating.

- **Fire resistance**
  - Ability to resist the transmission of the following:
    - Excessive heat.
    - Hot gases.
    - Flames.
- **Fire-resistance rating**
  - The length of time an assembly or component can function in a fire as follows:
    - Confine a fire.
    - Perform assigned structural task.

Note: Section 703, "Fire-Resistance Ratings and Fire Tests," is cited as the source of methods for determining fire-resistance ratings.

- **Fire-resistant joint system**
  - Assembly of elements is as follows:
    - Retards the passage of fire through joints as follows:
      - In fire-resistance-rated assemblies.
      - Between fire-resistance-rated assemblies.
      - For a specified length of time.
    - System is fire-resistance-rated.

Note: The following are cited as alternative standards governing these systems:
ASTM E 1966, "Standard Test Method for Fire-Resistant Joint Systems."
UL 2079, "Tests for Fire Resistance of Building Joint Systems."

702 Definitions

Case study: Fig. 702.1B. The selection of walls are among several types used at the hospital. The fire-resistance-rated walls shown are similar to those tested by Underwriters Laboratories, Inc.® and described in their publication, *Fire Resistance Directory,* or those listed in the Gypsum Association's *Fire Resistance Design Manual,* which are tested by several agencies. The numbers under the fire-resistance ratings shown in the wall sections indicate the index number under which descriptions of the walls are provided in the reference publications. Wall assemblies and horizontal assemblies are not considered to have a fire-resistance rating unless they have been tested by a recognized agency. Two walls without fire-resistance ratings are also shown.

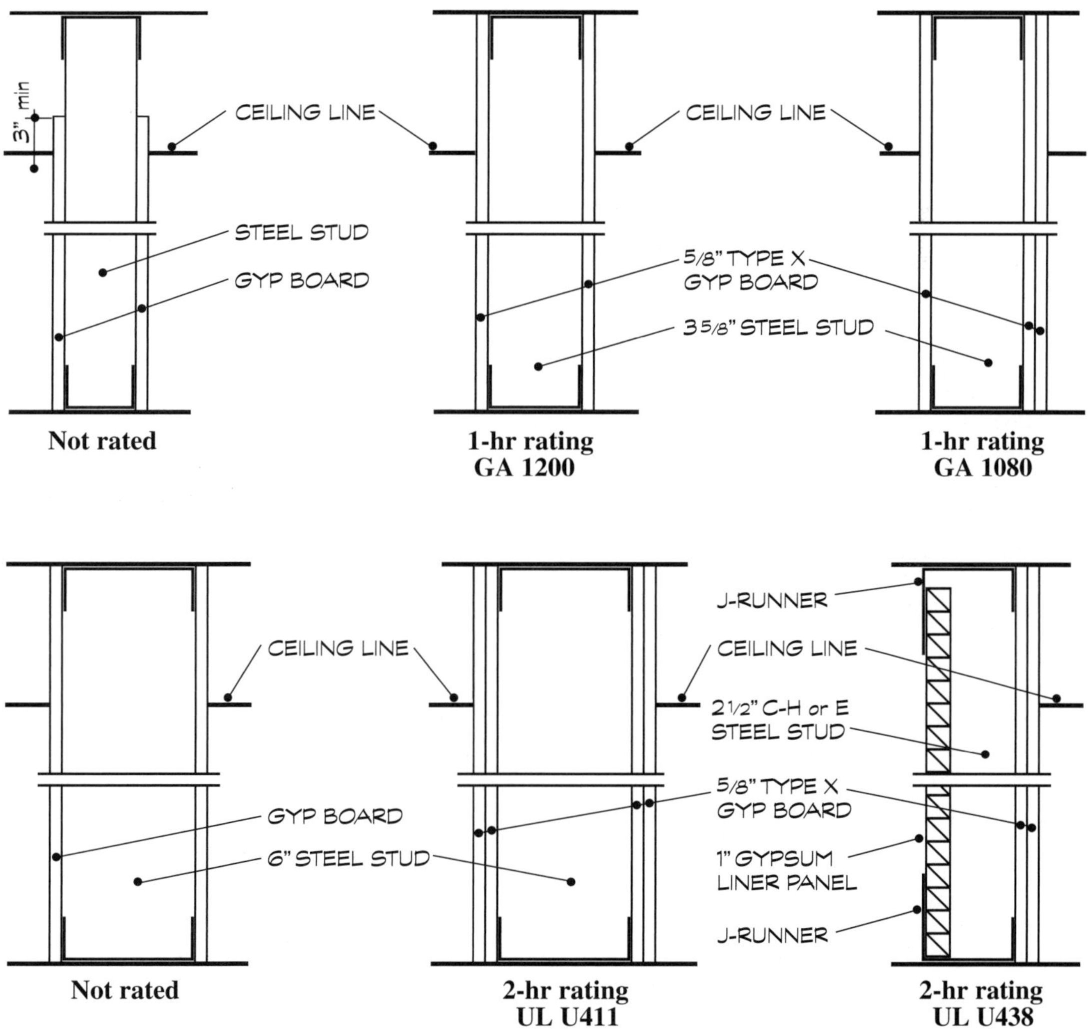

Fig. 702.1B. Selected wall sections. Methodist Community Health Center. Sugar Land, Texas. HKS, Inc., Architects, Engineers, Planners. Dallas, Texas.

702 Definitions

702.1 Definitions *(part 4 of 7)*

- **Fire separation distance**
  - The distance measured from the building face to any of the following:
    Closest interior lot line.
    Centerline of a street.
    Centerline of an alley.
    Centerline of a public way.
    An imaginary line between two buildings in the following case:
      Buildings on the same lot.
  - The distance is measured ⊥ to the building wall.

Case study: Fig. 702.1C. Measurement of fire separation distances in the example are ⊥ to the face of each exterior wall of the building extending to interior lot lines or to the center of the street.

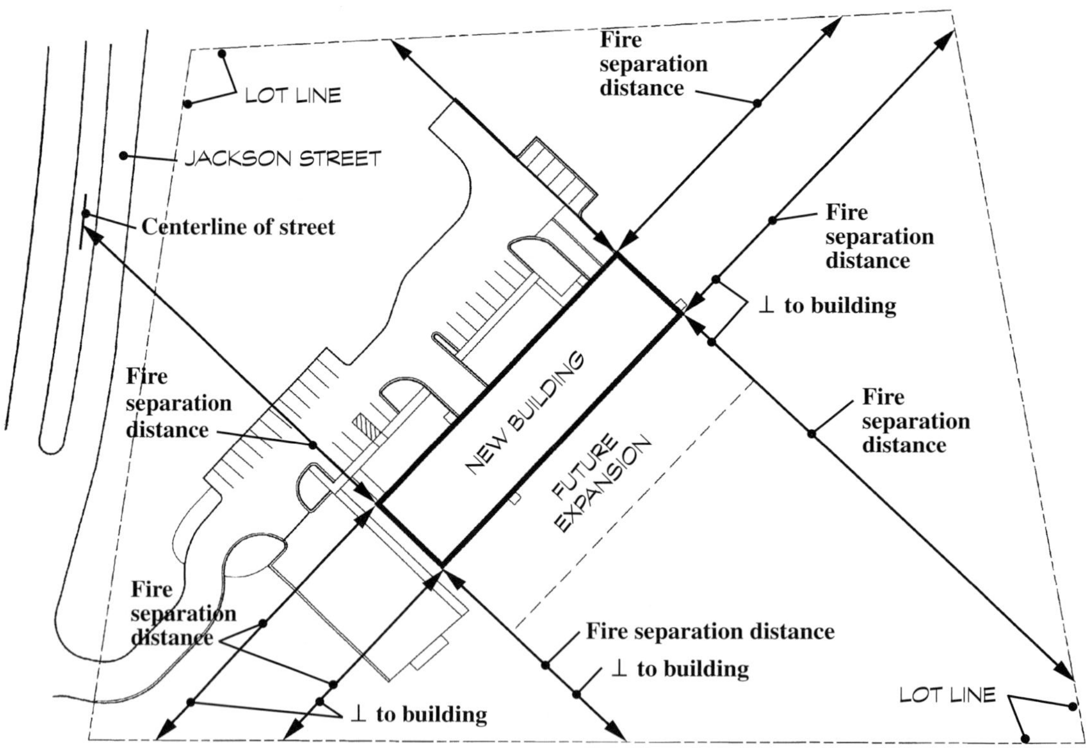

Fig. 702.1C. Site plan. Garments to Go. Bastrop, Texas. Spencer Godfrey Architects. Round Rock, Texas.

702 Definitions

702.1 Definitions *(part 5 of 7)*

- **Fire wall**
  - Has a fire-resistance rating.
  - Openings in wall are protected.
  - Wall retards the spread of fire.
  - Wall extends from foundation to or through roof.
  - Wall is detailed so as to remain standing as follows:
    In case of construction collapse on either side.
- **Fire window assembly**
  - A window that resists the passage of fire due to the following:
    Its construction.
    Its glazing.
- **Fireblocking**
  - One of the following materials installed in concealed spaces to prevent the spread of fire:
    Building materials.
    Fireblocking materials.

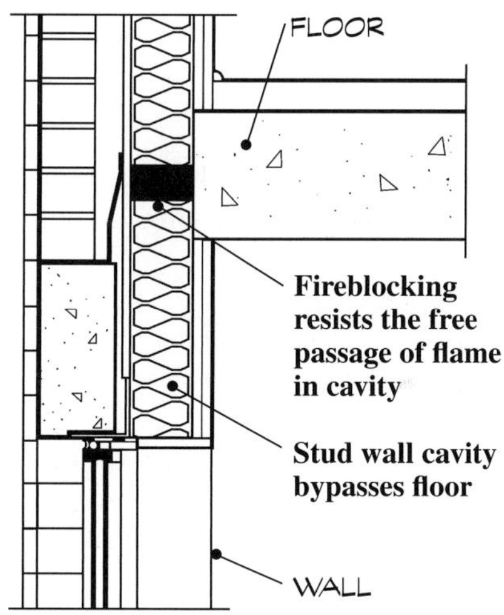

Fig. 702.1D. Detail at floor-wall intersection. McKenzie Lofts. Portland, Oregon. Ankrom Moisan Associated Architects. Portland, Oregon.

- **Floor fire door assembly**
  - An assembly including the following:
    Fire door.
    Frame.
    Hardware.
    Accessories.
  - Installed horizontally.
  - Provides fire protection at a defined level as follows:
    To an opening through a floor with a fire-resistance rating.

Note: 712.8, "Floor fire door assemblies," is cited as governing these doors. Tests and other requirements are included, some of which are summarized above.

702 Definitions

702.1 Definitions *(part 6 of 7)*

- **Horizontal assembly**
  - A floor or roof as follows:
    - With a fire-resistance rating.
    - Designed to limit the spread of fire.
    - Continuity is required.
- **Joint**
  - A linear gap in fire-resistance-rated construction.
  - Allows independent movement in any plane resulting from any of the following:
    - Thermal expansion and contraction.
    - Seismic activity.
    - Wind or other loading.
- **Membrane penetration**
  - An opening through any of the following surface membranes:
    - A wall.
    - A floor.
    - A ceiling.
- **Membrane-penetration firestop**
  - Any of the following:
    - A material.
    - A device.
    - A construction.
  - Prohibits the passage of flame and heat as follows:
    - Through membrane openings serving the following:
      - Cables or cable trays.
      - Conduit, tubing, or pipes.
      - Similar items.
  - Is effective for a specified length of time.
- **Penetration firestop**
  - A material or assembly protecting either of the following openings:
    - An opening passing through an entire assembly.
    - An opening through a membrane on one side of an assembly.
- **Self-closing**
  - A door equipped with a device as follows:
    - Device closes the door after it is opened.
    - Device must be approved.
- **Shaft**
  - An enclosed space.
  - Extends through ≥ 1 stories.
  - Connects vertical openings in any of the following:
    - Floors.
    - Floor and roof.
- **Shaft enclosure**
  - Any of the following elements surrounding a shaft:
    - Walls.
    - Other construction.

702 Definitions

702.1 Definitions *(part 7 of 7)*

- **Smoke barrier**
  - A continuous membrane.
  - Oriented vertically or horizontally.
  - Examples include walls, floors, and ceilings.
  - Limits the movement of smoke.
- **Smoke compartment**
  - A space surrounded by smoke barriers as follows:
    - All sides.
    - Above.
    - Below.
- **Smoke damper**
  - A listed device installed in any of the following locations:
    - Air ducts.
    - Openings for the transfer of air.
  - Inhibits the flow of smoke.
  - Closes when smoke is detected.
  - Adjustable where required as follows:
    - From a fire command center.
- **Splice**
  - Connection of fire-resistant joint systems as follows:
    - To form a continuous system by either of the following methods:
      - Factory process.
      - Field process.
- **T rating**
  - Length of time in hours that a penetration firestop system is able to limit temperature rise as follows:
    - The rise above initial temperature on the nonfire side of a penetration is limited to the following:
      - 325° F.

Note: The following are cited as alternatives specifying procedures for determining a T Rating:
ASTM E 814, "Test Method for Fire Tests of Through-Penetration Firestop Systems."
UL 1479, "Fire Tests of Through-Penetration Firestops."

- **Through penetration**
  - An opening completely through an assembly.
- **Through-penetration firestop system**
  - Either of the following that prevents the spread of fire through penetrations:
    - Materials.
    - Products.
  - The system is fire-resistance-rated.
  - The system is effective for a specified length of time indicated by the following ratings:
    - F rating.
    - T rating.

Note: The following are cited as alternatives specifying procedures for determining F and T Rating:
ASTM E 814, "Test Method for Fire Tests of Through-Penetration Firestop Systems."
UL 1479, "Fire Tests of Through-Penetration Firestops."

702 Definitions

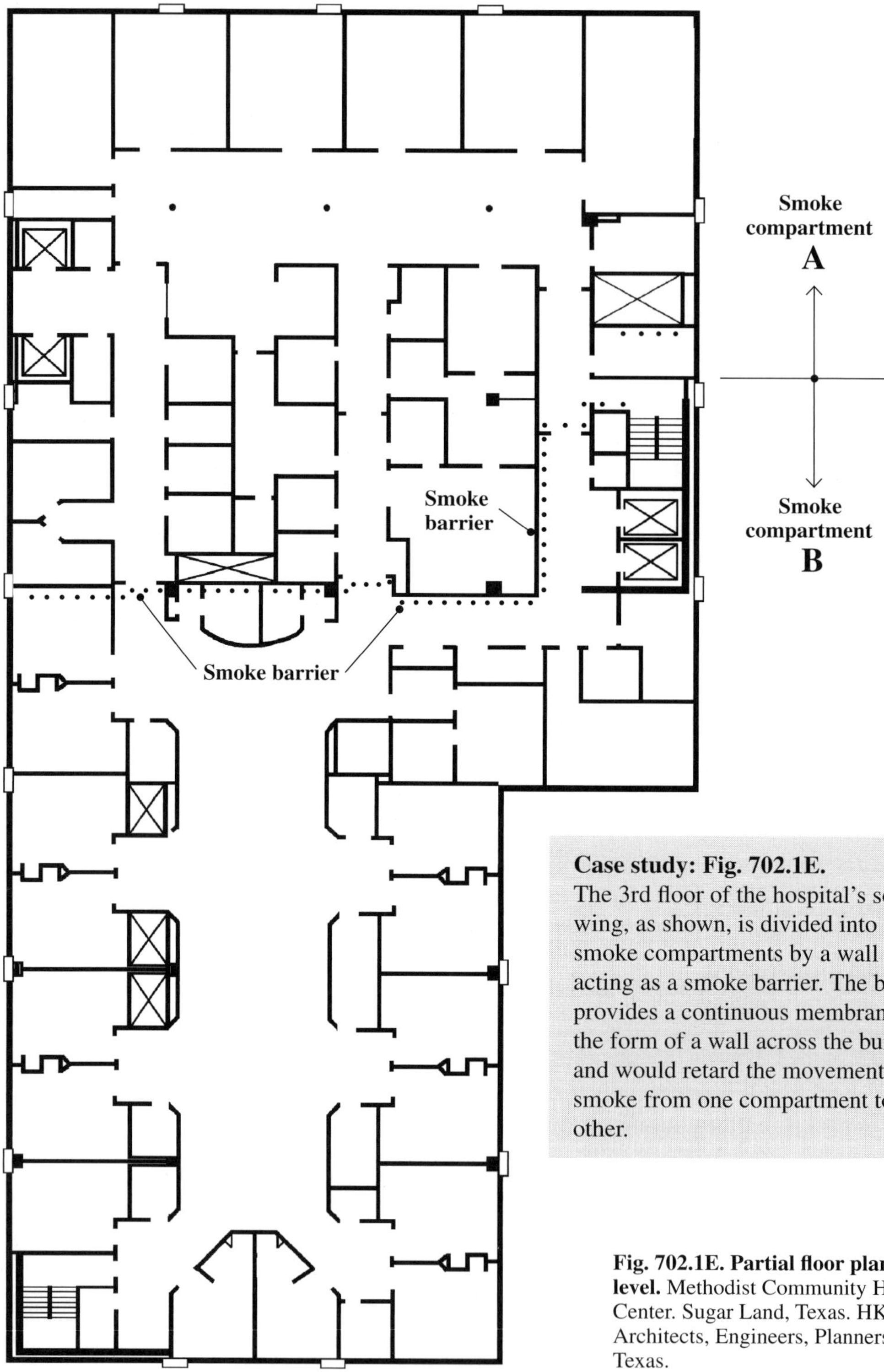

Case study: Fig. 702.1E.
The 3rd floor of the hospital's south wing, as shown, is divided into 2 smoke compartments by a wall acting as a smoke barrier. The barrier provides a continuous membrane in the form of a wall across the building and would retard the movement of any smoke from one compartment to the other.

Fig. 702.1E. Partial floor plan at 3rd level. Methodist Community Health Center. Sugar Land, Texas. HKS, Inc., Architects, Engineers, Planners. Dallas, Texas.

705 Exterior Walls

705.2 Projections

- This section does not govern buildings on the same lot as follows:
  - Where considered to be parts of a single building.

Note: 705.3, "Buildings on the same lot," is cited as governing these buildings.

- This section governs building projections extending beyond the exterior wall as follows:
  - Cornices and eave overhangs.
  - Exterior balconies and stairways.
  - Similar projections.
- Combustible projections must comply with requirements for combustible materials.

Note: Section 1406, "Combustible Materials on the Exterior Side of Exterior Walls," is cited as governing the projections listed above.
Section 1019, "Egress Balconies," is cited as providing additional requirements.
Section 1026, "Exterior Exit Ramps and Stairways," is cited as providing additional requirements.

- Projections may not extend closer to the lot line than any of the following points:
  - A point within 1/3 the distance from the face of the exterior wall to the property line in either of the following cases:
    Where protected openings are required.
    Where protected openings are required and unprotected openings are permitted.
  - A point within 1/2 the distance from the face of the exterior wall to the property line in either of the following cases:
    Where all openings can be unprotected.
    Where the building is sprinklered.
  - A point > 12" into the zone in which openings are not permitted.

Note: 705.8.2, "Protected openings," is cited as addressing sprinkler requirements for openings.

705.2.1 Type I and II construction

- Projections beyond the face of building may include the following materials:
  - Noncombustible materials.
  - Certain combustible materials.

Note: 1406.3, "Balconies and similar projections," is cited as governing projections where certain combustible materials are permitted.
1406.4, "Bay windows and oriel windows," is cited governing projections where certain combustible materials are permitted.

705.2.2 Type III, IV or V construction

- Any approved material is permitted for projections as follows:
  - From walls of Types III, IV, and V construction.

705 Exterior Walls

Case study: Fig. 705.2. A projection from the warehouse faces a street. The fire separation distance is 67'- 6". IBC Table 705.8 indicates that for distances > 30', openings are not regulated by the table; thus, the extent of this projection on the second floor is not limited. Because the building is Type III construction, the projection may be constructed of any approved material. It is constructed of noncombustible materials.

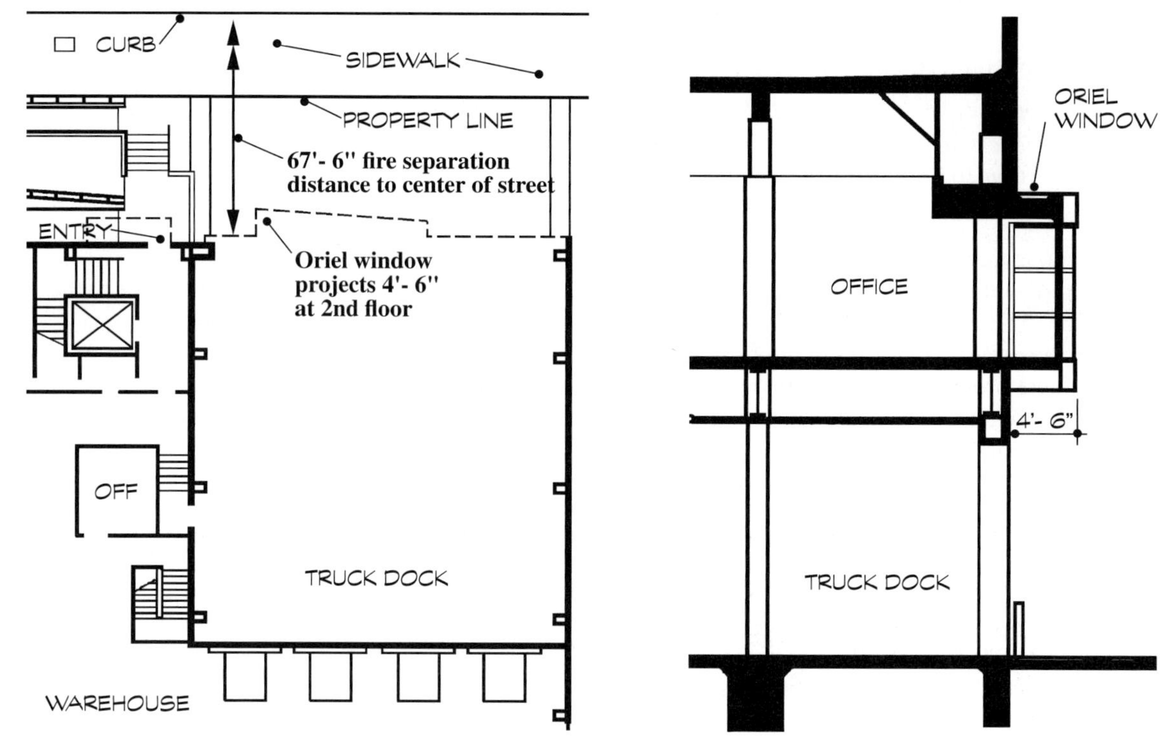

Fig. 705.2. Partial plan and section at truck dock. New Warehouse Addition. Los Angeles, California. Stephen Wen + Associates, Architects, Inc. Pasadena, California.

705.2.3 Combustible projections

- This section applies to projections of combustible construction in the following locations:
  - Where openings are not allowed.
  - Where openings are required to be protected.
- Such projections must comply with one of the following:
  - Have ≥ 1-hour fire-resistance rating.
  - Be heavy timber construction.
  - Be fire-retardant-treated wood.
  - Meet requirements for balconies and similar projections.

Note: 1406.3, "Balconies and similar projections," is cited governing combustible construction in these elements.

705 Exterior Walls

705.3 Buildings on the same lot

- ≥ 2 buildings may be regulated as 1 building where they meet all of the following conditions:
  - They are located on the same lot.
  - The sum of their areas is ≤ the area limit for 1 building as follows:
    - The smallest area limit governs where the following differ among buildings:
      - Occupancy classification.
      - Construction type.

Note: Chapter 5, "General Building Heights and Areas," is cited as governing area limits.

- In other cases where buildings are on the same lot, the following applies:
  - A line (a surrogate lot line) is assumed to be located between buildings as follows:
    - So that requirements for the following can be determined:
      - Wall protection.
      - Opening protection.
      - Roof covering.
    - Where a new building is adjacent to an existing building the following applies:
      - The assumed line must be located so the existing building complies as follows:
        - Adjacent wall and openings of the existing building must meet the following requirements:
          - Fire resistance based on construction type.
          - Fire resistance based on fire separation distance.
          - Opening size limitations based on the following:
            - Protection.
            - Fire separation distance.

Note: The following are cited as governing exterior walls and openings such as those of the existing building indicated above:
705.5, "Fire-resistance ratings."
705.8, "Openings."

705.4 Materials

- Exterior walls are required to be of materials as follows:
  - Materials within the designated construction type.

705.5 Fire-resistance ratings

- Exterior walls are rated with regard to fire-separation distance as follows:
  - Where fire-separation distance > 10' the following applies:
    - Walls are rated for fire exposure to the inside.
  - Where fire-separation distance ≤ 10' the following applies:
    - Walls are rated for fire exposure to both sides.

Note: The following tables are cited as governing fire-resistance ratings of exterior walls:
IBC Table 601, "Fire-Resistance Rating Requirements for Building Elements."
IBC Table 602, "Fire-Resistance Rating Requirements for Exterior Walls Based on Fire Separation Distance."

705 Exterior Walls

705.6 Structural stability

- Exterior walls must be detailed to remain standing during a fire as follows:
  - For a length of time equal to its fire-resistance rating.
- Exterior walls must extend above the roof or to a lower height as per the fire hazard.

Note: 705.11, "Parapets," is cited as governing exterior wall height.

705.7 Unexposed surface temperature *(part 1 of 4)*

- This section addresses a surface of an exterior wall as follows:
  - Surface is subject to a rise in surface temperature due to fire on the other side of the wall.
  - Surface is not directly exposed to fire.
- In the following cases the rise of temperature on the unexposed surface is not limited to 250° F as otherwise required:
  - Where the fire separation distance is > 20'.
  - Where the fire separation distance is ≤ 20' and the following reduction is applied:

    The allowable area of protected openings is reduced by subtracting the following amount:

    Amount subtracted = Wall area not including openings × Equivalent Opening Factor

    Equivalent Opening Factors are based on the following equation:

$$\textbf{Factor} = \frac{(\textbf{Average °F of surface not exposed to fire} + 460° \textbf{ F})^4}{(\textbf{Fire-resistance temperature coefficient} + 460° \textbf{ F})^4}$$

Fire-resistance temperature coefficients used in the equation above are as follows:

Wall fire-resistance rating	Fire-resistance temperature coefficient
1 hr	1,700° F
2 hr	1,850° F
3 hr	1,925° F
4 hr	2,000° F

Equivalent Opening Factors as derived from the equation above are provided as follows:

In parts 2, 3, and 4 of this section.

For every 10° F of unexposed surface temperature.

From 410° F to 2,000° F.

Note: The following are cited as standards which require a 250° F limit of temperature rise on an unexposed surface where either condition outlined by this section does not apply:

ASTM E 119, "Standard Test Methods for Fire Tests of Building Construction and Materials."

UL 263, "Fire Tests of Building Construction and Materials."

705.8, "Openings," is cited as a factor in identifying protected windows that are subject to the temperature rise limit of ASTM E 119 or UL 263. Key requirements therein are integrated into this section.

705 Exterior Walls

705.7 Unexposed surface temperature *(part 2 of 4)*

- Reduced allowable area of protected openings is calculated as follows:
  Allowable area – (Wall area not including openings × Equivalent Opening Factor)
- Equivalent Opening Factors are provided in the table below as follows:
  - From 410° F to 940° F.
  - Based on the following:
    Average temperature in °F of the unexposed wall surface.
    Fire-resistance rating of the wall.

Table 705.7a Equivalent Opening Factors for Exterior Walls (410° F–940° F)

Surface Temp.,°F	Wall fire-resistance rating 1 hr	2 hr	3 hr	4 hr	Surface Temp.,°F	Wall fire-resistance rating 1 hr	2 hr	3 hr	4 hr
410	0.026	0.020	0.018	0.016	680	0.078	0.059	0.052	0.046
420	0.028	0.021	0.019	0.016	690	0.080	0.061	0.054	0.048
430	0.029	0.022	0.019	0.017	700	0.083	0.064	0.056	0.049
440	0.030	0.023	0.020	0.018	710	0.086	0.066	0.058	0.051
450	0.032	0.024	0.021	0.019	720	0.089	0.068	0.060	0.053
460	0.033	0.025	0.022	0.020	730	0.092	0.070	0.062	0.055
470	0.034	0.026	0.023	0.020	740	0.095	0.073	0.064	0.057
480	0.036	0.027	0.024	0.021	750	0.098	0.075	0.066	0.059
490	0.037	0.029	0.025	0.022	760	0.102	0.078	0.068	0.060
500	0.039	0.030	0.026	0.023	770	0.105	0.080	0.071	0.063
510	0.041	0.031	0.027	0.024	780	0.109	0.083	0.073	0.065
520	0.042	0.032	0.029	0.025	790	0.112	0.086	0.075	0.067
530	0.044	0.034	0.030	0.026	800	0.116	0.089	0.078	0.069
540	0.046	0.035	0.031	0.027	810	0.120	0.091	0.080	0.071
550	0.048	0.037	0.032	0.028	820	0.123	0.094	0.083	0.073
560	0.050	0.038	0.033	0.030	830	0.127	0.097	0.086	0.076
570	0.052	0.040	0.035	0.031	840	0.131	0.100	0.088	0.078
580	0.054	0.041	0.036	0.032	850	0.135	0.103	0.091	0.080
590	0.056	0.043	0.038	0.033	860	0.139	0.107	0.094	0.083
600	0.058	0.044	0.039	0.034	870	0.144	0.110	0.097	0.085
610	0.060	0.046	0.041	0.036	880	0.148	0.113	0.100	0.088
620	0.063	0.048	0.042	0.037	890	0.153	0.117	0.103	0.091
630	0.065	0.050	0.044	0.039	900	0.157	0.120	0.106	0.093
640	0.068	0.051	0.045	0.040	910	0.162	0.124	0.109	0.096
650	0.070	0.053	0.047	0.041	920	0.167	0.127	0.112	0.099
660	0.072	0.055	0.049	0.043	930	0.171	0.131	0.115	0.102
670	0.075	0.057	0.050	0.045	940	0.176	0.135	0.119	0.105

705 Exterior Walls

705.7 Unexposed surface temperature *(part 3 of 4)*

- Reduced allowable area of protected openings is calculated as follows:

  Allowable area – (Wall area not including openings × Equivalent Opening Factor)
- Equivalent Opening Factors are provided in the table below as follows:
  - From 950° F to 1,480° F.
  - Based on the following:

    Average temperature in °F of the unexposed wall surface.

    Fire-resistance rating of the wall.

Table 705.7b Equivalent Opening Factors for Exterior Walls (950° F–1,480° F)

Surface Temp.,°F	Wall fire-resistance rating 1 hr	2 hr	3 hr	4 hr	Surface Temp.,°F	Wall fire-resistance rating 1 hr	2 hr	3 hr	4 hr
950	0.182	0.139	0.122	0.108	1,220	0.366	0.280	0.246	0.218
960	0.187	0.143	0.126	0.111	1,230	0.375	0.286	0.252	0.223
970	0.192	0.147	0.129	0.114	1,240	0.384	0.293	0.258	0.228
980	0.198	0.151	0.133	0.117	1,250	0.393	0.300	0.264	0.233
990	0.203	0.155	0.137	0.121	1,260	0.402	0.307	0.270	0.239
1,000	0.209	0.160	0.140	0.124	1,270	0.411	0.315	0.277	0.245
1,010	0.215	0.164	0.144	0.128	1,280	0.421	0.322	0.283	0.250
1,020	0.220	0.168	0.148	0.131	1,290	0.431	0.329	0.290	0.256
1,030	0.226	0.173	0.152	0.135	1,300	0.441	0.337	0.297	0.262
1,040	0.233	0.178	0.156	0.138	1,310	0.451	0.345	0.303	0.268
1,050	0.239	0.183	0.161	0.142	1,320	0.461	0.352	0.310	0.274
1,060	0.245	0.187	0.165	0.146	1,330	0.472	0.361	0.317	0.280
1,070	0.252	0.192	0.169	0.150	1,340	0.482	0.369	0.324	0.287
1,080	0.258	0.198	0.174	0.154	1,350	0.493	0.377	0.332	0.293
1,090	0.265	0.203	0.178	0.158	1,360	0.504	0.385	0.339	0.300
1,100	0.272	0.208	0.183	0.162	1,370	0.515	0.394	0.347	0.306
1,110	0.279	0.213	0.189	0.166	1,380	0.527	0.403	0.354	0.313
1,120	0.286	0.219	0.193	0.170	1,390	0.538	0.411	0.362	0.320
1,130	0.294	0.224	0.198	0.175	1,400	0.550	0.420	0.370	0.327
1,140	0.301	0.230	0.203	0.179	1,410	0.562	0.429	0.378	0.334
1,150	0.309	0.236	0.208	0.183	1,420	0.574	0.439	0.386	0.341
1,160	0.316	0.242	0.213	0.188	1,430	0.586	0.448	0.394	0.348
1,170	0.324	0.248	0.218	0.193	1,440	0.599	0.458	0.403	0.356
1,180	0.332	0.254	0.224	0.198	1,450	0.611	0.467	0.411	0.363
1,190	0.341	0.260	0.229	0.202	1,460	0.624	0.477	0.420	0.371
1,200	0.349	0.267	0.235	0.207	1,470	0.637	0.487	0.429	0.379
1,210	0.357	0.273	0.240	0.212	1,480	0.651	0.497	0.438	0.387

705 Exterior Walls

705.7 Unexposed surface temperature *(part 4 of 4)*

- Reduced allowable area of protected openings is calculated as follows:

  Allowable area – (Wall area not including openings × Equivalent Opening Factor)
- Equivalent Opening Factors are provided in the table below as follows:
  - From 1,490° F to 2,000° F.
  - Based on the following:

    Average temperature in °F of the unexposed wall surface.

    Fire-resistance rating of the wall.

Table 705.7c Equivalent Opening Factors for Exterior Walls (1,490° F–2,000° F)

Surface Temp.,°F	Wall fire-resistance rating 1 hr	2 hr	3 hr	4 hr	Surface Temp.,°F	Wall fire-resistance rating 1 hr	2 hr	3 hr	4 hr
1,490	0.664	0.508	0.447	0.395	1,750	NA	0.838	0.737	0.651
1,500	0.678	0.518	0.456	0.403	1,760	NA	0.853	0.751	0.663
1,510	0.692	0.529	0.465	0.411	1,770	NA	0.869	0.764	0.675
1,520	0.706	0.540	0.475	0.420	1,780	NA	0.884	0.778	0.687
1,530	0.720	0.551	0.485	0.428	1,790	NA	0.900	0.792	0.700
1,540	0.735	0.562	0.495	0.437	1,800	NA	0.916	0.806	0.712
1,550	0.750	0.573	0.504	0.446	1,810	NA	0.933	0.821	0.725
1,560	0.765	0.585	0.515	0.455	1,820	NA	0.949	0.835	0.738
1,570	0.780	0.596	0.525	0.464	1,830	NA	0.966	0.850	0.751
1,580	0.796	0.608	0.535	0.473	1,840	NA	0.983	0.865	0.764
1,590	0.811	0.620	0.546	0.482	1,850	NA	1.000	0.880	0.778
1,600	0.827	0.632	0.557	0.492	1,860	NA	NA	0.895	0.791
1,610	0.843	0.645	0.567	0.501	1,870	NA	NA	0.911	0.805
1,620	0.860	0.657	0.578	0.511	1,880	NA	NA	0.927	0.819
1,630	0.877	0.670	0.590	0.521	1,890	NA	NA	0.943	0.833
1,640	0.893	0.683	0.601	0.531	1,900	NA	NA	0.959	0.847
1,650	0.911	0.696	0.613	0.541	1,910	NA	NA	0.975	0.861
1,660	0.928	0.709	0.624	0.552	1,920	NA	NA	0.992	0.876
1,670	0.946	0.723	0.636	0.562	1,930	NA	NA	NA	0.891
1,680	0.963	0.737	0.648	0.573	1,940	NA	NA	NA	0.906
1,690	0.982	0.750	0.660	0.583	1,950	NA	NA	NA	0.921
1,700	1.000	0.764	0.673	0.594	1,960	NA	NA	NA	0.937
1,710	NA	0.779	0.685	0.605	1,970	NA	NA	NA	0.952
1,720	NA	0.793	0.698	0.617	1,980	NA	NA	NA	0.968
1,730	NA	0.808	0.711	0.628	1,990	NA	NA	NA	0.984
1,740	NA	0.823	0.724	0.640	2,000	NA	NA	NA	1.000

705 Exterior Walls

705.8.1 Allowable area of openings *(part 1 of 3)*

- The following openings qualify for unlimited area:
  - Openings in exterior walls that meet the following conditions:
    - Located in any of the following occupancies:
      - A, B, E, F, I, M, R, S, U.
    - Located in the 1st story above grade.
    - The wall faces either of the following open spaces:
      - A street with a fire separation distance >15'.
      - An unoccupied open space with the following characteristics:
        - Space is ≥ 30' wide.
        - Space is one of the following:
          - Located on the same lot as the wall.
          - Dedicated to public use.
        - A posted fire lane provides access to the space from a street.

Note: The International Fire Code *is cited as governing fire lanes.*

  - Openings in exterior walls of buildings as follows:
    - Where the following are not required to have a fire-resistance rating:
      - Exterior bearing walls.
      - Exterior nonbearing walls.
      - Exterior primary structural frame.
- Other area limits for openings in exterior walls are listed in the tables of this section:
  - The following types of openings are included:
    - Protected openings.
    - Unprotected openings.

Note: IBC Table 705.8, "Maximum Area of Exterior Wall Openings Based on Fire Separation Distance and Degree of Opening Protection," is cited as listing the maximum areas for windows.

  - Openings in firewalls of buildings having different heights are not included.

Note: 706.6.1, "Stepped buildings," is cited as governing openings in such firewalls.

  - Openings in firewalls of buildings on the same lot are not included.

Note: 706.8, "Openings," is cited as governing openings in such firewalls.

705 Exterior Walls

705.8.1 Allowable area of openings *(part 2 of 3)*

Table 705.8.1a — % of an Exterior Wall in Buildings Not Sprinklered That May Be Occupied by Openings Where All Openings Are Unprotected

Occupancy	Fire separation distance	Unprotected openings limit
R-3	< 3'	Not permitted
R-3	≥ 3' < 5'	≤ 25%
R-3	≥ 5'	Not limited
H-2, H-3	< 15'	Not permitted
"	≥ 15' < 20'	25%
"	≥ 20' < 25'	45%
"	≥ 25' < 30'	70%
"	≥ 30'	Not limited
Open parking garages	< 5'	Not permitted
"	≥ 5' < 10'	10%
"	≥ 10'	Not limited
Other occupancies	< 5'	Not permitted
Other occupancies (including accessory to R-3)	≥ 5' < 10'	10%
Other occupancies (including accessory to R-3)	≥ 10' < 15'	15%
Other occupancies	≥ 15' < 20'	25%
"	≥ 20' < 25'	45%
"	≥ 25' < 30'	70%
"	≥ 30'	Not limited

Source: IBC Table 705.8.

- The following Table 705.8.1b does not apply to H-1, H-2, or H-3 occupancies.

Table 705.8.1b — % of an Exterior Wall in Sprinklered Buildings That May Be Occupied by Openings Where All Openings Are Unprotected

Occupancy	Fire separation distance	Unprotected openings limit
R-3	< 3'	Not permitted
R-3	≥ 3' < 5'	25%
R-3	≥ 5'	Not limited
Open parking garages	< 3'	Not permitted
"	≥ 3' < 5'	15%
"	≥ 5' < 10'	25%
"	≥ 10'	Not limited
Other occupancies	< 3'	Not permitted
"	≥ 3' < 5'	15%
"	≥ 5' < 10'	25%
"	≥ 10'	45%
"	≥ 15' < 20'	75%
"	≥ 20'	Not limited

Source: IBC Table 705.8.

705 Exterior Walls

705.8.1 Allowable area of openings *(part 3 of 3)*

Table 705.8.1c % of an Exterior Building Wall That May Be Occupied by Openings Where All Openings Are Protected

Occupancy	Fire separation distance	Protected openings limit
R-3	< 3'	Not permitted
R-3	≥ 3' < 5'	25%
R-3	≥ 5'	Not limited
Open parking garages	< 3'	Not permitted
"	≥ 3' < 5'	15%
"	≥ 5' < 10'	25%
"	≥ 10'	Not limited
Other occupancies	< 3'	Not permitted
"	≥ 3' < 5'	15%
"	≥ 5' < 10'	25%
"	≥ 10'	45%
"	≥ 15' < 20'	75%
"	≥ 20'	Not limited

Source: IBC Table 705.8.

Note: 705.8.2, "Protected openings," is cited as governing protected openings.

705.8.2 Protected openings

- Where openings are required to be protected, any of the following protectives may be used:
  - Fire doors.
  - Fire shutters.
  - Fire window assemblies.
  - Water curtains as follows:
    Where the building is sprinklered throughout.
    With water curtain sprinklers protecting the outside of the glass.

Note: The following are cited as governing the protective assemblies:
715.4, "Fire door and shutter assemblies."
715.5, "Fire-protection-rated glazing," for fire window assemblies.
903.3.1.1, "NFPA 13 sprinkler systems."

705.8.3 Unprotected openings

- Unprotected windows and doors are to be constructed as follows:
  - With any approved material.
  - With any glass or plastic as applicable meeting code requirements.

Note: The following are cited as governing glazing in unprotected openings:
Chapter 24, "Glass and Glazing."
Chapter 26, "Plastic."

705 Exterior Walls

705.8.4 Mixed openings

- The following applies where protected and unprotected openings are used in an exteior wall of any story:
  - Maximum area permitted for the combination is determined by the following equation:

$$\frac{\textbf{Actual area (or equivalent) of protected openings}}{\textbf{Allowable area of protected openings}} + \frac{\textbf{Actual area of unprotected openings}}{\textbf{Allowable area of unprotected openings}} \leq 1$$

Note: *705.7, "Unexposed surface temperature," is cited as the source of the method for determining equivalent area of protected openings.*

705.8.5 Vertical separation of openings

- This section does not apply to the following:
  - Buildings ≤ 3 stories.
  - Sprinklered buildings as per NFPA 13 or NFPA 13R.
  - Open parking garages.

Note: *The following are cited as governing the specified sprinkler systems:*
903.3.1.1, "NFPA 13 sprinkler systems."
903.3.1.2, "NFPA 13R sprinkler systems."

- Other buildings with both of the following conditions have requirements as indicated below:
  - Conditions:
    The horizontal distance between an upper and lower opening is ≤ 5'.
    The opening in the lower story is as follows:
    Without protection with a fire-protection rating ≥ 3/4 hr.
  - Requirements:
    Vertical separation of upper and lower windows is required by one of the following:
    A vertical assembly as follows:
    Provides ≥ 3' between openings.
    Has a fire-resistance rating ≥ 1 hr.
    Is one of the following components:
    Spandrel girder.
    Exterior wall.
    Other similar assembly.
    Flame barrier as follows:
    Extends ≥ 30" horizontally beyond the exterior wall plane.
    Has a fire-resistance rating ≥ 1 hr.
    The rise of temperature on the barrier surface not exposed to a fire is not limited to 250° F by this section.

Note: *The following are cited as alternatives limiting temperature rise on unexposed surfaces:*
ASTM E 119, "Test Methods for Fire Tests of Building Construction and Materials."
UL 263, "Fire Tests of Building Construction and Materials."

705 Exterior Walls

705.8.6 Vertical exposure

- This section does not apply to buildings on the same lot as follows:
  - Where they are considered to be parts of the same building.

Note: 705.3, "Buildings on the same lot," is cited as governing conditions permitting buildings on the same lot to be considered as one building.

- This section addresses buildings on the same lot as follows:
  - An imaginary reference line of the following type is to be located between the two buildings:
    - The imaginary line is similar to those used to determine fire separation distance.
- The following applies to exterior wall openings higher than the roof of an adjacent building:
  - Openings do not require protectives in any of the following cases:
    - Where bottom edges of openings are ≥ 15' above the lower roof.
    - Where the lower roof has both of the following characteristics:
      - The roof assembly has a fire-resistance rating ≥ 1 hr in the following area:
        - An area ≥ 10' wide bordering the following exterior wall of the lower building:
          - The exterior wall facing the imaginary reference line.
      - Supporting structure for the rated assembly complies with the following:
        - It has a fire-resistance rating ≥ 1 hr for its full length.
  - Openings require a fire protection rating ≥ 3/4 hr where both of the following apply:
    - Where bottom edges of openings are < 15' above the lower roof.
    - Where the exterior wall of the lower building is < 15' from the imaginary reference line as follows:
      - This exterior wall is the wall facing the imaginary reference line.

706 Fire Walls

706.5 Horizontal continuity

- Fire walls must be continuous between exterior walls.
- Fire walls must terminate at exterior walls in one of the following configurations:
  - Fire walls may extend past the exterior surface of the exterior as follows:
    For a distance ≥ 18".
  - Fire walls may terminate at the interior surface of the following where all of the conditions listed apply to the exterior wall:
    Surfaces:
      Combustible exterior sheathing or siding.
    Conditions:
      Wall has a fire-resistance rating ≥ 1 hr as follows:
        Extending ≥ 4' on both sides of the fire wall.
        Openings in this 4' length must have fire assemblies as follows:
          With a fire-protection rating ≥ 3/4 hr.
  - Fire walls may terminate at the interior surface of the following where the condition listed exists:
    Surface:
      Noncombustible sheathing or siding.
      Noncombustible exterior material.
    Condition:
      The noncombustible extends ≥ 4' on both sides of the fire wall.
  - Fire walls may terminate at the following interior surface where the condition listed exists:
    Surface:
      Noncombustible exterior sheathing.
    Condition:
      Where the building is sprinklered on each side of the fire wall.

Note: The following are cited as alternatives to govern the sprinklers.
903.3.1.1, "NFPA 13 sprinkler systems."
903.1.2, "NFPA 13R sprinkler systems."

706.5.1 Exterior walls *(part 1 of 2)*

- This section addresses 2 exterior walls that come together at the end of a fire wall as follows:
  - Only exterior walls forming an exterior angle < 180° with each other are governed by this section.
- Walls governed by this section must meet one of the following 2 sets of conditions:
  - Condition 1: exterior wall construction within 4' of the fire wall in both directions has the following requirements:
    It must have a fire-resistance rating ≥ 1 hr.
    Openings must have a fire-protection rating ≥ 3/4 hr where required by fire-separation distance.

Note: *705.8, "Openings," is cited as the source of requirements for opening protection based on fire-separation distance.*

706 Fire Walls

706.5.1 Exterior walls *(part 2 of 2)*

- Condition 2: exterior walls on both sides of the fire wall must have the following:
  - An imaginary reference line between them to determine fire-separation distance as follows:
    - The line extends from the fire wall to a position between the exterior walls on either side.
    - The line is similar to an imaginary lot line between two buildings on the same lot.
  - A fire-resistance rating required by fire-separation distance, occupancy, and construction type.
  - Opening protection as required by fire-separation distance.

Note: *705.5, "Fire-resistance ratings," is cited as providing fire-separation distance guidelines and listing sources governing fire-resistance ratings based on occupancy and construction type.*
705.8, "Openings," is cited as the source of requirements for opening protection based on fire-separation distance.

706.5.2 Horizontal projecting elements *(part 1 of 2)*

- This section addresses the relationship between fire walls and horizontal projecting elements in the location as follows:
  - Projections:
    - Balconies.
    - Roof overhangs.
    - Canopies.
    - Marquees.
    - Similar projections.
  - Location:
    - Where they occur ≤ 4' from a fire wall.
- Fire walls need not extend to the outer edge of projecting elements in the following cases:
  - Where projections without concealed spaces comply with the following:
    - Exterior wall behind and below the projection must be as follows:
      - The wall must have a fire-resistance rating ≥ 1 hr:
        - Rating must extend along both sides of the fire wall as follows:
          - For a distance ≥ the depth of the projection.
        - Openings in this rated zone must have the following:
          - A fire-protection rating ≥ 3/4 hr.
  - Where noncombustible projections with concealed spaces comply with the following:
    - A wall with a fire resistance rating ≥ 1 hr must extend through the concealed space.
    - The projecting element must be separated from the building as follows:
      - Separating construction must have a fire-resistance rating ≥ 1 hr:
        - Rating must extend along both sides of the fire wall as follows:
          - For a distance ≥ the depth of the projection.
    - The 1-hr-rated wall need not extend under the projected element in the following case:
      - Where the exterior wall has a fire resistance rating ≥ 1 hr:
        - Rating must extend along both sides of the fire wall as follows:
          - For a distance ≥ the depth of the projection.
        - Openings in this rated zone must have the following:
          - A fire-protection rating ≥ 3/4 hr.

706 Fire Walls

706.5.2 Horizontal projecting elements *(part 2 of 2)*

- ◦ Where combustible projections with concealed spaces comply with the following:
  - Fire wall must extend through the concealed space to outer edge of projection.
  - Exterior wall behind and below the projecting element must comply with the following:
    - The wall must have a fire-resistance rating ≥ 1 hr:
      - Rating must extend along both sides of the fire wall as follows:
        - For a distance ≥ the depth of the projection.
      - Openings in this rated zone must have the following:
        - A fire-protection rating ≥ 3/4 hr.
- • In other cases fire walls must extend to the outer edge of projecting elements.

706.6 Vertical continuity *(part 1 of 2)*

- • This section does not address the upper termination fire walls in the following case:
  - ◦ Where fire walls separate certain buildings with different roof levels.

Note: 706.6.1, "Stepped buildings," is cited as the source governing the tops of fire walls separating such buildings.

- • The top of 2-hr fire-resistance-rated fire walls may terminate as follows, if the listed conditions are met:
  - ◦ Terminations:
    - At the underside of any of the following:
      - Roof sheathing.
      - Roof deck.
      - Roof slab.
  - ◦ Conditions:
    - Elements of the lower roof assembly must have a fire-resistance rating ≥ 1 hr as follows:
      - The roof assembly for a distance ≥ 4' from the wall.
      - The entire length of components supporting the rated roof assembly.
    - No openings are permitted in the roof as follows:
      - ≤ 4' from the fire wall.
    - The buildings on both sides of the fire wall require roofing as follows:
      - ≥ Class B.
- • Fire walls may terminate as follows, if all the listed conditions are met:
  - ◦ Terminations:
    - At the underside of any of the following:
      - Noncombustible roof sheathing.
      - Noncombustible roof deck.
      - Roof slab.
  - ◦ Conditions:
    - No openings are permitted in the roof ≤ 4' from the fire wall.
    - The buildings on both sides of the fire wall require ≥ Class B roofing.
- • Fire walls may terminate as follows, if all of the listed conditions are met:
  - ◦ Terminations:
    - At the underside of combustible roof sheathing.
    - At the underside of a combustible roof deck.

706 Fire Walls

706.6 Vertical continuity *(part 2 of 2)*

- Conditions:
  - Buildings must be one of the following construction types:
    - Type III, IV, V.
  - No openings are permitted in the roof as follows:
    - ≤ 4' from the fire wall.
  - The buildings on both sides of the fire wall require roofing as follows:
    - ≥ Class B.
  - One of the following applies to an area extending to ≥ 4' from both sides of the fire wall:
    - The roof sheathing or deck is constructed of the following:
      - Fire-retardant-treated wood.
    - 5/8" Type X gypsum board is applied as follows:
      - Underneath the roof sheathing or deck.
      - Supported on ≥ 2" ledgers attached to roof framing.

- The following applies where a 3-hr horizontal assembly separates certain buildings:
  - The bottom of a fire wall may terminate at the horizontal assembly between the buildings.

Note: 509.2, "Horizontal building separation allowance," is cited as the source of requirements qualifying buildings for this type of fire wall termination.

- In cases other than those addressed above in this section, the following applies:
  - The bottoms of fire walls must terminate at the foundations.
  - Fire walls must extend to a level ≥ 30" above the roofs on each side of the fire wall.

706.6.1 Stepped buildings

- This section addresses fire walls with both of the following characteristics:
  - The fire wall serves as an exterior wall.
  - The fire wall separates buildings with roofs at different levels.
- Such fire wall must terminate at its top in one of the following ways:
  - The fire wall must terminate at a level ≥ 30" above the lower roof as follows:
    - The exterior wall above the lower roof is has the following requirements:
      - It must have a fire-resistance rating ≥ 1 hr from both sides as follows:
        - This rating must extend for a height ≥ 15' above the lower roof.
        - Openings in this rated zone must have a fire-protection rating ≥ 3/4 hr.
- Such a fire wall must terminate at the underside of one of the following elements of the lower roof with other conditions meeting all of the requirements listed below:
  - Elements:
    - Roof sheathing.
    - Roof deck.
    - Roof slab.
  - Conditions:
    - The lower roof assembly must have a fire-resistance rating ≥ 1 hr as follows:
      - The rating must extend for a distance ≥ 10' from the fire wall.
    - The support system for the rated roof assembly is governed as follows:
      - It must have a fire-resistance rating ≥ 1 hr for its full length.
    - No openings are permitted in the lower roof ≤ 10' from the fire wall.

706 Fire Walls

706.7 Combustible framing in fire walls

- The following applies where combustible members penetrate a fire wall from opposite sides:
  - Fire walls of concrete or masonry are governed as follows:
    - A wall thickness ≥ 4" between embedded ends of members is required.
  - Fire walls that are hollow or have hollow units must have hollow spaces filled solid as follows:
    - For the full thickness of the wall.
    - For a distance ≥ 4" from the embedded ends of members in the following locations:
      - Above members.
      - Below members.
      - Between members.
    - Filler materials are required to be as follows:
      - Noncombustible.
      - Approved for fireblocking.

706.8 Openings

- Openings are not allowed in fire walls serving as party walls.

Note: 706.1.1, "Party walls," is cited as governing fire walls acting as party walls.

- Openings through other fire walls are governed as follows:
  - The sum of opening widths at a floor level must be ≤ 25% of the wall length.
  - Area of each opening must be ≤ 156 sf where either building is without sprinklers.
  - Area of each opening is not limited to 156 sf where the building is sprinklered.
  - Openings must be protected.

Note: 715.4, "Fire door and shutter assemblies," is cited as governing opening protection in fire walls.
903.3.1.1, "NFPA 13 sprinkler systems," is cited as governing the sprinklers.

707 Fire Barriers

707.2 Materials

- Fire barriers may be constructed of the following materials:
  - Conforming to the construction type.

707.3.1 Shaft enclosures

- Fire barriers isolating shafts are governed as follows:
  - Basements are included in the number of stories connected by a shaft.
  - Mezzanines are not included in the number of stories connected by a shaft.
  - Where shafts connect < 4 stories:

    The greater of the following fire-resistance is required up to a maximum of 2 hrs:

    ≥ 1 hr.

    ≥ the fire-resistance rating of the floor penetrated.
  - Where shafts connect ≥ 4 stories:

    A fire-resistance rating ≥ 2 hrs is required.

Note: 708.4, "Fire-resistance rating," is cited as governing shaft enclosures and is summarized above.

707.3.2 Exit enclosures

- Fire barriers isolating exits are governed as follows for most cases:
  - Basements are included in the number of stories connected by a shaft.
  - Mezzanines are not included in the number of stories connected by a shaft.
  - Where exits connect < 4 stories:

    A fire-resistance rating of ≥ 1 hr is required.
  - Where exits connect ≥ 4 stories:

    A fire-resistance rating ≥ 2 hrs is required.

Note: 1022.1, "Enclosures required," is cited as the source of requirements for exit enclosures. A partial summary is provided above which excludes the 9 exceptions listed in the section.

707.3.3 Exit passageway

- Exit passageways must be separated from other building areas with fire barriers as follows:
  - Enclosure must have a fire-resistance rating ≥ the larger of the following:

    1 hr.

    The rating required for any connecting exit enclosure.

Note: 1023.3, "Construction" is cited as the source of enclosure requirements for exit passageways.

707.3.4 Horizontal exit

- The separation between areas provided by a horizontal exit must have a fire-resistance rating ≥ 2 hr.

Note: 1025.1, "Horizontal exits," is cited as governing. 1025.2, "Separation," requires the 2-hr rating indicated above. An exception for the fire-resistance rating is also provided.

707.3.6 Incidental accessory occupancies

- Fire barriers isolating incidental use areas are governed as follows:
  - They must have fire-resistance ratings based on the function of the incidental space.

Note: IBC Table 508.2.5, "Incidental Accessory Occupancies," is cited as the source of fire-resistance ratings for the areas.

707 Fire Barriers

707.3.7 Control areas

- Fire barriers separating control areas must meet minimum fire-resistance rating requirements based on various conditions.

Note: 414.2.4, "Fire-resistance-rating requirements," provides minimum fire-resistance ratings for control areas via IBC Table 414.2.2, "Design and Number of Control areas." Various conditions are specified for rating requirements of 1 and 2 hr.

707.3.8 Separated occupancies

- Where fire barriers are required for separating mixed occupancies, the following applies:
  - They must have a fire-resistance rating based on the classification of the separated occupancies.

Note: 508.4, "Separated occupancies," is cited as the source of requirements for fire barriers separating mixed occupancies.
IBC Table 508.4, "Required Separation of Occupancies," is cited listing fire-resistance ratings required for the fire barriers.

707.3.9 Fire areas*

- One or both of the following constructions are required to divide occupancies addressed in this section into separate fire areas:
  - Fire barriers.
  - Horizontal assemblies.
- A fire-resistance rating $\geq$ 4 hr is required for the separation of the following occupancies into separate fire areas:
  - H-1, H-2.
- A fire-resistance rating $\geq$ 3 hr is required for fire barriers dividing the following occupancies into separate fire areas:
  - F-1, H-3, S-1.
- A fire-resistance rating $\geq$ 2 hr is required for fire barriers dividing the following occupancies into separate fire areas:
  - A, B, E, F-2, H-4, H-5, I, M, R, S-2.
- A fire-resistance rating $\geq$ 1 hr is required for fire barriers dividing the following occupancy into separate fire areas:
  - U.
- Where fire areas of mixed occupancies are required to be separated, the following applies:
  - The fire-resistance rating of the separation assembly must be as follows:
    The higher rating of that required for either occupancy separated.

Note: IBC Table 707.3.9, "Fire-Resistance Rating Requirements for Fire Barrier Assemblies or Horizontal Assemblies between Fire Areas," is cited as governing fire-resistance ratings as required above.

* *Source*: IBC Table 707.3.9.

707 Fire Barriers

707.4 Exterior walls

- The following exterior walls where required to be fire-resistance-rated are not governed by this section:
  - At exterior egress balconies.

  Note: Section 1019, "Egress Balconies," is cited as governing adjacent exterior walls.

  - At exit ramps.

  Note: 1026.6, "Exterior ramps and stairway protection," is cited as governing adjacent exterior walls at exit ramps.

  - At exit stairways.

  Note: 1026.6, "Exterior ramps and stairway protection," is cited as governing adjacent exterior walls at exit stairways.

  - At exit enclosures.

  Note: 1022.6, "Exit enclosure exterior walls," is cited as governing these walls.

- Exterior walls at the following locations have the requirements listed below:
  - Locations:
    - Fire-resistance-rated shaft enclosures.
    - Fire-resistance-rated exit enclosures.
  - Requirements:
    - Fire-resistance ratings required for the enclosures or separations do not apply as follows:
      - The walls must comply with fire-resistance rating requirements for exterior walls.

  Note: Section 705, "Exterior Walls," is cited as governing exterior walls.

707.5 Continuity

- Fire barriers must extend continuously between the following levels:
  - From the top of the floor/ceiling assembly located below the fire barrier wall.
  - To the underside of one of the following elements as the upper terminal as applicable:
    - Floor sheathing, slab or deck above the fire barrier wall.
    - Roof sheathing, slab or deck above the fire barrier wall.
- Fire barriers must be securely attached at upper and lower terminations.
- Fire barriers must be continuous through concealed spaces such as follows:
  - Space above a suspended ceiling.

707 Fire Barriers

707.5.1 Supporting construction

- Shaft enclosures are not required to extend to the underside of a roof deck in the following case:
  - Where the fire-resistance rating of the shaft's top closure is ≥ the larger of the following:
    The fire-resistance rating of the highest floor penetrated by the shaft.
    The fire-resistance rating of the shaft walls.

Note: 708.12, "Enclosure at the top," is cited as governing the top of a shaft, a partial summary of which is provided above.

- This section does not require that the following construction have the fire-resistance rating noted below:
  - Construction:
    Where supporting fire barriers with all of the following characteristics:
      Fire barriers that are required to separate incidental use areas from other building areas.
      Fire barriers have a fire-resistance rating = 1 hr.
      Fire barriers are located in any of the following types of construction:
        Type IIB, Type IIIB, Type VB.
  - Fire-resistance rating:
    ≥ the fire barrier supported.

Note: IBC Table 508.2.5, "Incidental Accessory Occupancies," is cited as the source listing incidental use areas requiring 1-hr rated walls.

- Applications that separate tank storage areas of H-2 occupancies from other uses must have a fire-resistance rating ≥ the larger of the following:
  - ≥ 2 hr.
  - ≥ fire-resistance rating required for building elements based on construction type.

Note: 415.6.2.1, "Mixed occupancies," is cited as governing tank storage separation.
IBC Table 601, "Fire-Resistance Rating Requirements for Building Elements," is cited as the source for ratings based on construction type.

- All other applications must have the following fire-resistance rating:
  - ≥ the fire barrier supported.
- Fire blocking is required at each floor as follows:
  - In any hollow vertical spaces within a fire barrier.

Note: 717.2, "Fireblocking," is cited as governing the fireblocking in the fire barrier walls.

707 Fire Barriers

707.6 Openings

- This section governs openings in a fire barrier.
- Openings in any of the following cases are not limited to 156 sf each:
  - Where floor areas on both sides of the fire barrier are sprinklered.
  - Where openings to exit enclosures are protected fire doors.
  - Where the openings have both of the following:
    An opening protective tested by applicable standards.
    A fire-resistance rating ≥ the wall.
  - Where the opening protective is a fire door assembly in a fire barrier separating the following:
    An exit enclosure and an exit passageway.

Note: 903.3.1.1, "NFPA 13 sprinkler systems," is cited as governing the sprinklers.
1022.2.1, "Extension," is cited as governing separations of exit enclosures and passageways.
The following are cited as alternative applicable standards for testing opening protectives:
ASTM E 119, "Test Methods for Fire Tests of Building Construction and Materials."
UL 263, "Fire Tests of Building Construction and Materials."

- The sum of the opening widths in any of the following cases is not limited to 25% of the wall length:
  - Where openings to exit enclosures are protected by fire doors.
  - Where the openings in atrium separation walls are protected by fire window assemblies.
  - Where the openings have both of the following:
    A fire-resistance rating ≥ the wall.
    An opening protective tested by applicable standards.
  - Where the opening protective is a fire door assembly in a fire barrier separating the following:
    An exit enclosure and an exit passageway.

Note: The following are cited as alternative applicable standards for testing opening protectives:
ASTM E 119, "Test Methods for Fire Tests of Building Construction and Materials."
UL 263, "Fire Tests of Building Construction and Materials."

- In other cases openings are limited as follows:
  - Each opening limited to 156 sf.
  - The sum of all widths of openings are limited to 25% of wall length.
  - All openings require protectives.
  - Openings in exit enclosures must meet additional requirements specific to their location.
  - Openings in exit passageways must meet additional requirements specific to their location.

Note: Section 715, "Opening Protectives," is cited as governing opening protectives.
The following are cited as providing additional requirements for specific locations:
1022.3, "Openings and penetrations," for openings in exit enclosures.
1023.5, "Openings and penetrations," for openings in exit passageways.

707.7.1 Prohibited penetrations

- The following are limited in size, function, and detailing as well as in other ways:
  - Penetrations into an exit enclosure or exit passageway.

Note: 1022.4, "Penetrations," is cited as governing penetrations into exit enclosures.
1023.6, "Penetrations," is cited as governing penetrations into exit passageways.

708 Shaft Enclosures

708.1 General

- This section governs shafts as follows:
  - Where required to protect openings and penetrations through the following:
    - Floor/ceiling assemblies.
    - Roof/ceiling assemblies.
  - Shaft enclosures must be fire barriers and/or horizontal assemblies.

Note: The following are cited as governing these assemblies as applicable:
Section 707, "Fire Barriers."
Section 712, "Horizontal Assemblies."

708.2 Shaft enclosure required *(part 1 of 3)*

- Grease ducts are not governed by this section.

Note: The International Mechanical Code *is cited as governing grease ducts.*

- A shaft enclosure is not required for the following cases:
  - For an opening through a floor/ceiling assembly with both of the following characteristics:
    - The opening is contained entirely within an individual dwelling unit.
    - The opening connects ≤ 4 stories.
  - For an opening through a floor/ceiling assembly meeting all of the following conditions:
    - The building is sprinklered.
    - The opening serves one of the following:
      - An escalator.
      - A stairway not required for means of egress.
    - Power-operated automatic shutters protect the opening at every floor as follows:
      - Shutters are approved.
      - Shutters are of noncombustible materials.
      - Shutters have a fire-resistance rating ≥ 1 1/2 hr.
      - Shutters close immediately upon detection of smoke.
      - Shutters completely surround and seal the opening well.
      - Shutters move ≤ 30 fpm.
      - Shutters have a leading edge that controls movement as follows:
        - Shutters stop upon meeting an obstacle.
        - Shutters resume movement when the obstacle is removed.
      - Escalator stops when shutters begin to close.

Note: 907.3, "Fire safety functions," is cited as governing shutter function upon the detection of smoke.
903.3.1.1, "NFPA 13 sprinkler systems," is cited as governing the sprinklers.

  - For penetrations as follows through a floor/ceiling assembly meeting the requirement indicated below:
    - Penetrations:
      - Pipe, tube, conduit.
      - Wire, cable.
      - Vents.
    - Requirement:
      - Penetrations must meet requirements for protection against the passage of fire.

Note: 713.4, "Horizontal assemblies," is cited as governing the penetrations listed above.

708 Shaft Enclosures

708.2 Shaft enclosure required *(part 2 of 3)*

- For penetrations by certain ducts.

Note: 716.6, "Horizontal assemblies," is cited as governing ducts that do not require a shaft enclosure.

- For an opening through a floor/ceiling assembly meeting all of the following conditions:
  The building is sprinklered.
  The opening serves one of the following:
    An escalator.
    A stairway not required for means of egress.
  The area of the opening is limited to one of the following:
    ≤ 2 × the projected area of the escalator.
    ≤ 2 × the projected area of the stairway.
  The opening is protected by a draft curtain.
  The opening is protected by closely spaced sprinklers.
  The opening ≤ 4 stories in occupancies A, E, F, H, I, R, S, U.

Note: 903.3.1.1, "NFPA 13 sprinkler systems," is cited as governing the sprinklers.

- For a floor opening where all of the following apply:
  Where the opening complies with requirements for atriums.
  Where not located in Occupancy H.

Note: Section 404, "Atriums," is cited as governing atriums.

- For masonry chimneys as follows:
  Where approved.
  Where the annular space is protected at each floor.

Note: 717.2.5, "Ceiling and floor openings," is cited as the source of requirements for fireblocking at openings around chimneys.

- For a floor opening connecting a floor and a mezzanine above.
- For joints with a fire-resistant joint protection system.

Note: Section 714, "Fire-Resistant Joint Systems," is cited as governing such joints.

- For automobile ramps as follows:
  In open parking garages.
  In enclosed closed parking garages.

Note: The following are cited as governing the construction of the garages listed above:
406.3, "Open parking garages."
406.4, "Enclosed parking garages."

- Where other sections of the code permit its omission.
- For floor openings serving the following:
  Unenclosed stairs.
  Unenclosed ramps.

Note: 1016.1, "Travel distance limitations," exceptions 3 and 4 are cited as governing the unenclosed stairs or ramps noted above.

708 Shaft Enclosures

708.2 Shaft enclosure required *(part 3 of 3)*

- For the following openings where all of the conditions listed below are met:
  - Openings:
    - Floor opening.
    - Air transfer opening.
  - Conditions:
    - Opening is located in an occupancy other than I-2, 1-3.
    - Opening connects ≤ 2 stories.
    - Opening does not serve a required means of egress.
    - Opening is not concealed in a wall or floor/ceiling assembly.
    - Opening is not open to a corridor as follows:
      - In occupancies I, R.
      - On a nonsprinklered floor in any occupancy.
    - Opening is separated from the following by construction meeting shaft enclosure requirements:
      - Floor and air transfer openings serving other floors.
    - Opening does not connect different smoke compartments.
- For floor openings as follows:
  - Where protected by floor fire door assemblies.

Note: 712.8, "Floor fire door assemblies," is cited as governing these doors.

- A shaft enclosure is not required for certain floor openings in Occupancy I-3.

Note: 408.5, "Protection of vertical openings," is cited as the floor openings noted above.

- A shaft enclosure is not required for mechanical supply or exhaust ducts meeting both of the following:
  - Located in an open or enclosed parking garage.
  - Ducts serve only the parking garage and are located within it.
- A shaft enclosure is not required for elevators in the following case:
  - Where located in an open or enclosed parking garage and serving only the garage.
- In all other cases, the following applies:
  - Openings through floor/ceiling assemblies require the following:
    - A shaft enclosure meeting the requirements of this section.

708.4 Fire-resistance rating

- Shaft enclosures are governed as follows:
  - Basements are included in the number of stories connected by a shaft.
  - Mezzanines are not included in the number of stories connected by a shaft.
  - Where they connect < 4 stories, the following applies:
    - The greater of the following fire-resistance ratings is required up to a maximum of 2 hrs:
      - ≥ 1 hr.
      - ≥ the fire-resistance rating of the floor penetrated.
  - Where they connect ≥ 4 stories the following applies:
    - A fire-resistance rating ≥ 2 hrs is required.

Note: 703.2.1, "Nonsymmetrical wall construction," is cited as governing the shaft enclosures addressed in this section.

708 Shaft Enclosures

708.5 Continuity

- Shaft enclosures must be one or both of the following:
  - Fire barrier.
  - Horizontal assembly.

Note: The following are cited as governing the shaft enclosure construction:
Section 707, "Fire Barriers."
Section 712, "Horizontal Assemblies."

- Shaft enclosures must have continuity.

Note: 707.5, "Continuity," is cited as governing continuity for fire barriers.
712.4, "Continuity," is cited as governing continuity for horizontal assemblies.

708.6 Exterior walls

- The following exterior walls are not governed by this section:
  - Where required to be fire-resistance rated at the following locations:
    Exterior egress balconies.
    Exit enclosures.
    Exit ramps.
    Exit stairways.

Note: The following are cited as governing fire-resistance ratings for exterior walls as noted above:
1019.2, "Wall separation," for egress balconies.
1022.6, "Exit enclosure exterior walls."
1026.6, "Exterior ramps and stairway protection."

- Exterior walls on a required shaft enclosure are governed as follows:
  - Fire-resistance-rated enclosure requirements do not apply to the walls.
  - The walls must comply with fire-resistance rating requirements for exterior walls.

Note: Section 705, "Exterior Walls," is cited as governing these walls.

708.7 Openings

- Doors in shaft enclosures must be one of the following:
  - Self-closing.
  - Automatic closing by smoke detection.

Note: 715.4.8.3, "Smoke-activated doors," governs these doors.

- Openings in shaft enclosures must be protected as for fire barriers.

Note: Section 715, "Opening Protectives," is cited as the source of requirements for opening protectives.

708.7.1 Prohibited openings

- The only openings permitted in a shaft enclosure are those required for the function of the shaft.

708.8 Penetrations

- Penetrations in shaft enclosures must be protected as required for fire barriers.

Note: Section 713, "Penetrations," is cited as governing these penetrations.

708 Shaft Enclosures

708.8.1 Prohibited penetrations

- The only penetrations permitted in a shaft enclosure are as follows:
  - Those required for the function of the shaft.

708.11 Enclosure at the bottom *(part 1 of 2)*

- This section addresses shafts that terminate at a point above the bottom of the building.
- The room in which a shaft terminates is not required to have the following characteristics where the shaft meets all the requirements indicated below:
  - Characteristics:
    - The room need not be separated from the building as follows:
      - By fire-resistance-rated construction.
    - The bottom of the shaft need not have opening protectives.
  - Requirements:
    - The shaft must contain no combustibles.
    - There may be no openings in the shaft enclosure as follows:
      - To the interior of the building.
    - There may be no penetrations through the shaft enclosure as follows:
      - To the interior of the building.
- The room in which a shaft terminates is not required to have the following characteristics where the shaft meets all the requirements indicated below:
  - Characteristics:
    - The room need not be separated from the building as follows:
      - By fire-resistance-rated construction.
  - Requirements:
    - There may be no openings into the shaft enclosure as follows:
      - To the interior of the building other than at the bottom.
    - There may be no penetrations of the shaft enclosure as follows:
      - To the interior of the building other than at the bottom.
    - One of the following conditions must be provided:
      - The room must have an approved automatic fire-suppression system.
      - Draftstopping materials as follows must be provided in the location indicated below:
        - Materials:
          - 1/2" gypsum board.
          - 3/8" wood structural panel.
          - 3/8" particleboard.
          - Other approved materials.
        - Location:
          - Materials must seal around penetrating items at the bottom of the shaft.

Note: 717.3.1, "Draftstopping materials," is cited as the source of materials permitted for this function, a summary of which is provided above.

- A shaft enclosure containing either of the following functions must comply with the requirements indicated below:
  - Functions:
    - Laundry chute.
    - Refuse chute.

708 Shaft Enclosures

708.11 Enclosure at the bottom *(part 2 of 2)*

- Requirements:
  - The shaft may not be used for any other purpose.
  - The shaft must terminate at the bottom in a room as follows:
    - Room must be separated from the rest of the building as follows:
      - By construction with a fire-resistance rating ≥ 1 hr.
      - Openings into the room must be protected as follows:
        - By protectives with a fire-protection rating ≥ 3/4 hr.
      - Openings into the room must be one of the following:
        - Self-closing.
        - Automatic-closing activated by smoke detection.
  - Refuse chutes may not terminate in the following room:
    - An incinerator room.

Note: 708.13.4, "Termination room," is cited as the source of requirements for the termination room serving laundry and refuse chutes.

- Other shafts must comply with one of the following requirements:
  - The shaft must be enclosed at the lowest level as follows:
    - By construction with the greater of the following fire-resistance ratings:
      - That of the lowest floor through which the shaft penetrates.
      - That required for the shaft enclosure.
  - The shaft must terminate in a room as follows:
    - The room function must relate to the function of the shaft.
    - The room must be separated from the rest of the building as follows:
      - By a fire barrier and/or horizontal assembly with the following:
        - A fire-resistance rating ≥ that required for the shaft enclosure.
        - Opening protectives providing the following fire protection:
          - ≥ than the fire protection required for the shaft enclosure.
  - The shaft must be protected with an approved fire damper as follows:
    - Dampers must be installed as per their listing.
    - Dampers must be installed in the shaft enclosure at the lowest floor level penetrated.

Note: Section 707, "Fire Barriers," is cited as governing fire barriers.
Section 712, "Horizontal Assemblies," is cited as governing horizontal assemblies.

708.12 Enclosure at the top

- Shaft enclosures that do not reach any of the following elements have the requirements listed below:
  - Elements:
    - Roof sheathing.
    - Roof deck.
    - Roof slab.
  - Requirements:
    - They must be enclosed at the top with construction as follows:
      - Construction having the greater of the following:
        - Fire-resistance rating ≥ that of the highest floor penetrated.
        - Fire-resistance rating ≥ that required for the shaft enclosure.

708 Shaft Enclosures

708.13.1 Refuse and laundry chute enclosures

- This section addresses shaft enclosures for the following:
  - Refuse chute.
  - Laundry chute.
- Such shaft enclosure may not be used for other purposes.
- Fire-resistance ratings required for such shaft enclosures are as follows:
  - Where they connect < 4 stories:
    - The greater of the following fire-resistance ratings is required up to a maximum of 2 hrs:
      - ≥ 1 hr.
      - ≥ the fire-resistance rating of the floor penetrated.
  - Where they connect ≥ 4 stories:
    - A fire-resistance rating ≥ 2 hrs is required.

Note: 708.4, "Fire-resistance rating," is as cited governing. A partial summary is provided above.

- Openings into such shafts require fire protection as follows:
  - Openings from access rooms.
  - Openings from termination rooms.
  - All other openings.

Note: Section 715, "Opening Protectives," and this section are cited as governing opening protection.

- Corridors may not have openings into chutes.
- Doors for shaft openings must be one of the following:
  - Self-closing.
  - Automatic-closing by one of the following means:
    - Activated by heat or smoke as follows:
      - Between the shaft and termination room.

Note: 715.4.8.3, "Smoke-activated doors," is cited as governing smoke activation.

708.13.2 Materials

- Shaft enclosures for laundry or refuse chutes require the following:
  - Materials must be consistent with the building construction type.

708.13.3 Refuse and laundry chute access rooms

- Chute access openings must be in rooms enclosed as follows:
  - By one or both of the following with a fire-resistance rating ≥ 1 hr:
    - Fire barriers.
    - Horizontal assemblies.
  - Openings to access rooms must have protectives as follows:
    - With a fire-protection rating ≥ 3/4 hr.
    - Which close by one of the following means:
      - Self-closing.
      - Automatic-closing when activated by smoke.

Note: Section 707, "Fire Barriers," is cited as governing fire barriers.
Section 712, "Horizontal Assemblies," is cited as governing horizontal assemblies.
715.4.8.3, "Smoke-activated doors," is cited as governing smoke activation.

708 Shaft Enclosures

708.13.4 Termination room

- This section governs the following rooms:
  - Laundry rooms.
  - Refuse rooms.
- Rooms receiving chute discharge have the following requirements:
  - Room must be separated from the rest of the building as follows:
    - By fire barriers and/or horizontal assemblies having a fire-resistance rating ≥ 1 hr.
    - Openings to the room must have protectives with the following rating:
      - A fire-protection rating ≥ 3/4 hr.
    - Doors must close by one of the following means:
      - Self-closing.
      - Automatic-closing activated by smoke.

Note: Section 707, "Fire Barriers," is cited as governing fire barriers.
Section 712, "Horizontal Assemblies," is cited as governing horizontal assemblies.
715.4.8.3, "Smoke-activated doors," is cited as governing these doors.

- Incinerator rooms may not receive chute discharge.
- Rooms that do not receive chute discharge are governed as follows:
  - They must be separated from the rest of the building as incidental use areas.

Note: IBC Table 508.2.5, "Incidental Accessory Occupancies," is cited as governing separation of these rooms.

708.13.6 Automatic sprinkler system

- The following chutes must be sprinklered:
  - Rubbish chutes.
  - Linen chutes.
- Sprinklers for chutes must comply with the following:
  - Characteristics:
    - Must be approved.
    - Must be automatic.
    - Must have access for servicing.
  - Locations:
    - Required at the top of chutes.
    - Required at alternate floors in the following case:
      - Where chutes pass through ≥ 3 floors.

Note: 903.2.11.2, "Rubbish and linen chutes," is cited as governing and is summarized above.

708.14 Elevator, dumbwaiter and other hoistways

- The following are governed by shaft requirements as noted:
  - Elevator hoistways.
  - Dumbwaiter hoistways.
  - Other hoistways.

Note: The following are cited as governing the above listed hoistways:
Section 708, "Shaft Enclosures."
Chapter 30, "Elevators and Conveying Systems."

708 Shaft Enclosures

708.14.1 Elevator lobby *(part 1 of 2)*

- This section addresses enclosed lobbies as separations of elevator shaft doors from each floor.
- An enclosed elevator lobby is not required in the following cases:
  - On ground floors of office buildings where the entire ground floor is sprinklered.

  Note: 903.3.1.1, "NFPA 13 sprinkler systems," is cited as governing the required sprinklers.

  - Where an elevator shaft is not required.

  Note: 708.2, "Shaft enclosure required," is cited as identifying these locations.

  - Where doors are provided at the hoistway opening in addition to the elevator door system.

  Note: 3002.6, "Prohibited doors," is cited as governing additional doors.
  UL 1784, "Air Leakage Tests of Door Assemblies," is cited as describing tests required for the additional doors. Doors are to be tested without an artificial bottom seal.

  - In sprinklered buildings where any of the following conditions apply:
    The building occupancy is not I-2.
    The building occupancy is not I-3.
    The building has no occupied floors ≥ 75' above the lowest fire department vehicle access level.

  Note: The following are cited as governing the sprinklers as applicable:
  903.3.1.1, "NFPA 13 sprinkler systems."
  903.3.1.2, "NFPA 13R sprinkler systems."

  - Where the elevator shaft is pressurized.

  Note: 708.14.2, "Enclosed elevator lobby," is cited as governing.

  - Where the elevator serves only an open parking garage.

  Note: 406.3, "Open parking garages," is cited as governing these garages.

- Smoke partitions may substitute for fire partitions where the building is sprinklered.

  Note: The following are cited as governing the sprinklers as applicable:
  903.3.1.1, "NFPA 13 sprinkler systems."
  903.3.1.2, "NFPA 13R sprinkler systems."
  The following are cited as governing smoke partitions, their doors, and duct penetrations:
  Section 711, "Smoke Partitions."
  711.5.2, "Smoke and draft control doors."
  711.5.3, "Self- or automatic-closing doors."
  715.4.8, "Door closing."
  716.5.4.1, "Corridors."
- In other buildings, an enclosed elevator lobby is required as follows:
  - To separate shaft doors from the rest of the building.
  - Where the elevator shafts connects ≥ 3 stories.
  - Must be located at each floor.
  - Must have the necessary opening protection.
  - Must have ≥ 1 means of egress complying with the IBC.

708 Shaft Enclosures

708.14.1 Elevator lobby *(part 2 of 2)*

- Must provide a separation from the connecting corridors as follows:
  - Where the corridor has a fire-resistance rating, the following applies:
    - Separation must be fire partitions with a fire-resistance rating = that of the corridor.
  - Where the corridor is not fire-resistance rated, the separation must be smoke partitions.

Note: Section 709, "Fire Partitions," is cited as governing fire partitions.
715.4.3, "Door assemblies in corridors and smoke barriers," is cited as governing lobby doors.
716.5.4.1, "Corridors," is cited as governing ducts and air transfer openings in lobby walls.
Chapter 10, "Means of Egress," is cited, in addition to other sections of the code, as governing means of egress.

708.14.2.6 Activation of pressurization system

- The pressurization system for an elevator shaft must activate in response to either of the following:
  - Sounding of the building alarm system.
  - Activation of the smoke detectors of the elevator lobby.
- Where they coexist, the following must be able to independently activate elevator pressurization:
  - Building alarm system.
  - Smoke detectors of the elevator lobby.

709 Fire Partitions

709.1 General

- This section series governs the following walls:
  - Walls between dwelling units in the same building.
  - Walls separating sleeping units in the same building.

  Note: 420.2, "Separation walls," is cited as governing walls between dwelling units or sleeping units in the same building.

  - Walls in covered malls between tenant spaces.

  Note: 402.7.2, "Tenant separations," is cited as governing these walls.

  - Corridor walls.

  Note: 1018.1, "Construction," is cited as governing corridor walls. The section excludes certain corridor walls in occupancies E and R, in open parking garages, in a limited range of B occupancies, and in many occupancies with a sprinkler system.

  - Elevator lobby separation walls.

  Note: 708.14.1, "Elevator lobby," is cited as governing the separation walls.

709.2 Materials

- Fire partitions must be constructed of materials as follows:
  - Consistent with building construction type.

709.3 Fire-resistance rating

- ≥ 1/2 hr fire-resistance rating is required for walls where all of the following apply:
  - The wall is a separation for any of the following:
    Dwelling unit.
    Sleeping unit.
  - The construction type of the building is any of the following:
    Type IIB.
    Type IIIB.
    Type VB.
  - The building is sprinklered.

  Note: 903.3.1.1, "NFPA 13 sprinkler systems," is cited as governing the required sprinkler system.

- ≥ 1/2 hr fire-resistance rating is required for corridor walls with all the following characteristics:
  - Located in Occupancy R.
  - Protected by sprinklers.
  - Corridor serves an occupant load > 10.

  Note: Table 1018.1, "Corridor Fire-Resistance Rating," is cited as listing exceptions to the 1-hr requirement for corridor walls.

- ≥ 1 hr fire-resistance rating is required for all other fire partitions.

709 Fire Partitions

709.4 Continuity *(part 1 of 2)*

- This section addresses the upper and lower points of termination for a fire partition.
- A fire partition is not required to extend into a crawl space as follows:
  - Where the floor above the crawl space has the following:
    - A fire-resistance rating ≥ 1 hr.
- Requirements listed below apply to the corridor ceiling adjacent to the following membrane:
  - Membrane:
    - Membrane is located on the room side of the corridor wall.
    - Membrane is fire-resistance-rated.
    - Membrane extends to the underside of one of the following as applicable:
      - Fire-resistance-rated floor above.
      - Fire-resistance-rated roof above.
  - Requirements:
    - Corridor ceiling may be protected by ceiling materials as follows:
      - As required for a fire-resistance rating ≥ 1 hr for one of the following:
        - A roof system.
        - A floor system.
- Corridor walls may terminate as follows:
  - At the upper membrane of the corridor ceiling in the following case:
    - Where the corridor ceiling complies with the same requirements as for the walls.
- Fire partitions between tenant spaces in a covered mall building are governed as follows:
  - They are not required to extend above the following height:
    - The underside of a ceiling in the following case:
      - Where the ceiling is not part of a fire-resistance-rated assembly.
  - A wall is not required in the following spaces:
    - In the attic above tenant separation walls.
    - In the ceiling spaces above tenant separation walls.

Note: 402.7.2, "Tenant separations," is cited as the source requiring tenant spaces to be separated by fire partitions.

- Fireblocking or draftstopping is not required in the following locations:
  - At the line of a fire partition in either of the following cases:
    - In Occupancy R-2 buildings where all of the following conditions apply:
      - In buildings ≤ 4 stories.
      - Where the attic has draftstops isolating areas ≤ the smaller of the following:
        - Area above every two dwelling units.
        - 3,000 sf.
    - In sprinklered buildings as follows:
      - Sprinklers are also installed in the following spaces:
        - Combustible floor/ceiling spaces.
        - Combustible roof/ceiling spaces.

Note: The following are cited as governing sprinkler systems that meet the sprinkler requirement:
903.3.1.1, "NFPA 13 sprinkler systems."
903.3.1.2, "NFPA 13R sprinkler systems."

709 Fire Partitions

709.4 Continuity *(part 2 of 2)*

- Construction supporting a fire partition is governed as follows:
  - Supporting construction need not have a fire-resistance rating ≥ that of the partition supported in the following cases:
    - In types IIB, IIIB, and VB construction as follows:
      - Tenant separation walls in covered mall buildings.
      - Sleeping unit separation walls.
      - Dwelling unit separation walls.
      - Corridor walls.
  - Supporting construction must have a fire-resistance rating ≥ that of the partition supported in the following types of construction:
    - Types I, IIA, IIIA, IV, VA.
- In all other cases, fire-partition continuity is required as follows:
  - Partitions must extend to the following termination levels:
    - Either of the following:
      - Top of foundation.
      - Top of floor assembly below.
    - Underside of one of the following:
      - Floor deck, slab or sheathing (subsequently referred to as deck).
      - Roof deck, slab or sheathing (subsequently referred to as deck).
      - Fire-resistance-rated floor/ceiling assembly.
      - Fire-resistance-rated roof/ceiling assembly.
  - Partitions must be securely attached to the termination points.
  - One of the following is required for combustible partitions that do not reach the deck above:
    - Fireblocking as follows:
      - In the plane of the partition.
      - Between the ceiling and deck above.
    - Draftstopping as follows:
      - In the plane of the partition.
      - Between the ceiling and deck above.

Note: The following are cited as governing the required fireblocking and draftstopping:
717.2, "Fireblocking."
717.3, "Draftstopping in floors."

709.5 Exterior walls

- This section does not apply to certain exterior walls.

Note: The following are cited as governing fire-resistance ratings for walls not subject to this section:
1019.2, "Wall separation," for egress balconies.
1022.6, "Exit enclosure exterior walls."
1026.6, "Exterior ramps and stairway protection."

- Other exterior walls of a required fire-resistance-rated separation are governed as follows:
  - Such walls must comply with fire-resistance requirements for exterior walls.
  - Fire-resistance-rated separation requirements do not apply.

Note: Section 705, "Exterior Walls," is cited as governing fire-resistance requirements.

710 Smoke Barriers

710.2 Materials

- Materials for smoke barriers must conform to the construction type for the building.

710.3 Fire-resistance rating

- Smoke barriers with both the following characteristics need not have a fire-resistance rating ≥ 1 hr:
  - Located in Occupancy I-3.
  - Constructed of ≥ 0.10" thick steel.
- Other smoke barriers must have a fire-resistance rating ≥ 1 hr.

710.4 Continuity

- Smoke barriers must form a membrane continuous between the following points:
  - From exterior wall to exterior wall.
  - Between the following points:
    - One of the following at the bottom of the barrier as applicable:
      - Top of foundation.
      - Top of floor/ceiling assembly.
    - One of the following at the top of the barrier as applicable:
      - Underside of floor sheathing, deck, or slab.
      - Underside of roof sheathing, deck, or slab.
- Smoke barriers must be continuous through the following spaces where applicable:
  - Concealed spaces such as above suspended ceilings.
  - Interstitial structural or mechanical space in the following case:
    - Where ceilings do not resist the passage of fire and smoke to the following degree:
      - To a degree equal that of the smoke barrier walls.
- Construction supporting smoke barriers must have a fire-resistance rating ≥ that of the wall or floor supported in the following construction types:
  - Type I, Type IIA, Type IIIA, Type IV, and Type VA.

710 Smoke Barriers

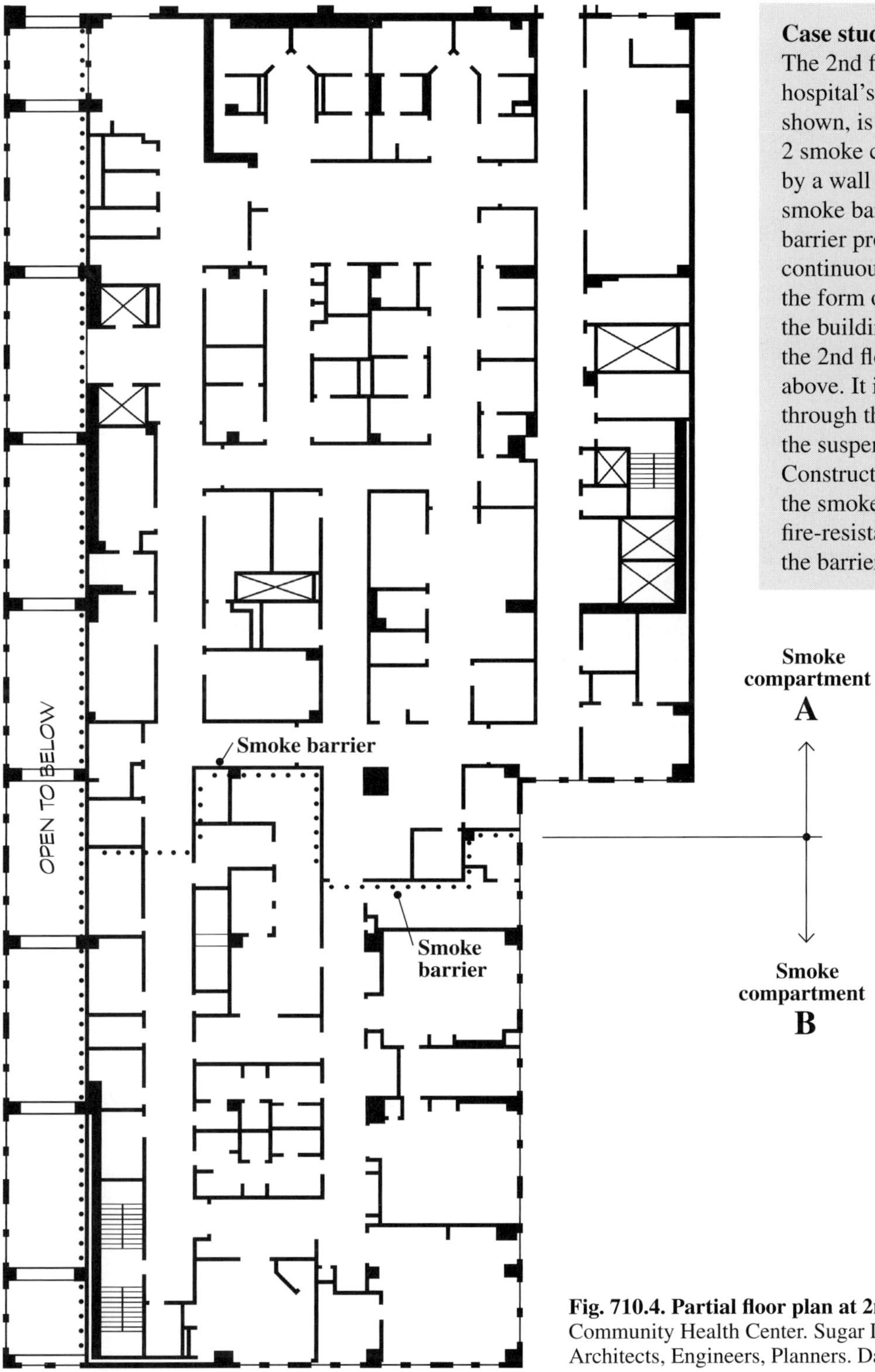

Case study: Fig. 710.4. The 2nd floor of the hospital's south wing, as shown, is separated into 2 smoke compartments by a wall acting as a smoke barrier. The barrier provides a continuous membrane in the form of a wall across the building and from the 2nd floor to the floor above. It is continuous through the space above the suspended ceiling. Construction supporting the smoke barrier has a fire-resistance rating ≥ the barrier as required.

Fig. 710.4. Partial floor plan at 2nd level. Methodist Community Health Center. Sugar Land, Texas. HKS, Inc., Architects, Engineers, Planners. Dallas, Texas.

710 Smoke Barriers

710.5 Openings

- Doors with all of the following characteristics need not meet the requirements indicated below:
  - Characteristics:
    - Occupancy I-2.
    - Located across a corridor.
    - In a smoke barrier.
    - Opposite swinging with no center mullion.
    - With vision panels as follows:
      - Fire-protection-rated glazing and frames.
      - With an area ≤ that tested.
    - Close fitting within operational tolerances.
    - With none of the following:
      - Undercuts > 3/4".
      - Louvers or grilles.
    - With all of the following:
      - Head stops.
      - Jamb stops.
      - Astragals at meeting edges.
      - Rabbets at meeting edges.
      - Automatic-closing devices activated by smoke detection.

Note: 715.4.8.3, "Smoke-activated doors," is cited as governing smoke-activation.

  - Requirements:
    - Doors need not have meet other opening protection requirements.
    - Doors need not have a fire-protection rating of 20 min.
    - Doors need not have positive-latching devices as follows:
      - Where the door manufacturer's listing allows their omission.
- Horizontal sliding doors are permitted in smoke barriers as follows:
  - Where located in Occupancy I-2.

Note: 1008.1.4.3, "Horizontal sliding doors," is cited as governing these doors.
Section 715, "Opening Protectives," is cited as governing the opening protective requirements for the horizontal sliding doors when installed in smoke barriers.

- All other openings in smoke barriers must meet both the following requirements:
  - Openings must comply with opening protective requirements.
  - Openings must have a fire-protection rating ≥ 20 min.

Note: Section 715, "Opening Protectives," is cited as governing the opening protective requirements for openings in smoke barriers.

711 Smoke Partitions

711.1 General

- This series of sections governs smoke partitions required elsewhere in the code.

711.2 Materials

- Smoke partition materials are governed by the construction type of the building.

711.3 Fire-resistance rating

- The following smoke partitions do not require a fire-resistance rating:
  - Where not required by other sections of the code.

711.4 Continuity

- Smoke partitions must extend to the following termination points:
  - The top of either of the following for the bottom of the smoke partition:
    The foundation.
    The floor.
  - The underside of one of the following overhead for the top of the smoke partition:
    Roof deck, sheathing, or slab.
    Floor deck, sheathing, or slab.
    Ceiling where built to limit the passage of smoke.

711.5 Openings

- Windows in smoke partitions must comply with one of the following:
  - Sealed to prevent the passage of smoke.
  - Close automatically when smoke is detected.
- Doors are governed in subsequent entries of this section series.

711.5.1 Louvers

- Louvers are prohibited smoke-partition doors.

711.5.2 Smoke and draft control doors

- Where other sections of the code require doors in smoke partitions to control smoke and drafts, all the following apply:
  - They must meet smoke and draft control door assembly requirements.
  - They must be tested for air leakage as per all the following requirements:
    Air leakage may not be > 3 cfm/sf of door opening as follows:
      At 0.10" of water for both the following:
        Ambient temperature test.
        Elevated-temperature exposure test.

Note: UL 1784, "Air Leakage Tests of Door Assemblies," is cited as the required test.
NFPA 105, "Standard for the Installation of Smoke Door Assemblies and Other Opening Protectives," is cited as governing the installation of smoke doors.

711 Smoke Partitions

711.5.3 Self- or automatic-closing doors

- This section addresses self-closing and automatic-closing doors.
- Where either of these door types are required in smoke partitions, the following applies:
  - Doors must be one of the following:
    - Self-closing.
    - Automatic-closing.
  - Door closing must be activated by smoke detection.

Note: 715.4.8.3, "Smoke-activated doors," is cited as governing these doors.

711.6 Penetrations and joints

- Gaps at the following locations must be filled with a material as indicated below:
  - Locations:
    - Around elements penetrating a smoke partition.
    - At joints in smoke partitions.
  - Fill material must comply with both of the following:
    - Must be approved.
    - Must retard the transfer of smoke.

711.7 Ducts and air transfer openings

- Openings for the passage of air through smoke partitions are governed as follows:
  - Where a smoke damper will not interfere with a required smoke control system:
    - A smoke damper is required.

Note: 716.3.2.2, "Smoke damper ratings," is cited as governing the smoke dampers and addresses class and temperature ratings, as well as actuation methods.

  - Where a smoke damper would interfere with a required smoke control system the following applies:
    - An approved protection system as an alternative to a smoke damper is required.

Note: Section 909, "Smoke Control Systems," is cited as governing the system which must be compatible with air passage openings in smoke partitions.

  - The space around a penetrating duct is governed as follows:
    - It must be filled with an approved material with the following purpose:
      - To limit the passage of smoke.

712 Horizontal Assemblies

712.2 Materials

- Construction for the following must be consistent with the building construction type:
  - Roof and floor assemblies.

712.3 Fire-resistance rating

- Floor and roof assemblies must have fire-resistance ratings as follows:
  - ≥ that dictated by the building construction type.
- Floor assemblies separating occupancies are governed as follows:
  - They must have the fire-resistance rating required for mixed occupancy separations.

Note: 508.4, "Separated occupancies," governs fire-resistance ratings required for separating occupancies.

- Floor assemblies dividing an occupancy into fire areas are governed as follows:
  - They must have the following fire-resistance ratings:
    - ≥ 1 hr in occupancy U.
    - ≥ 3 hr in occupancies F-1, H-3, S-1.
    - ≥ 4 hr in occupancies H-1, H-2.
    - ≥ 2 hr for all other occupancies.

Note: 707.3.9, "Fire areas," is cited as the source of requirements for dividing an occupancy into fire areas. The section is summarized above with data from IBC Table 707.3.9.

- Horizontal assemblies separating the following units require fire-resistance ratings as indicated below:
  - Units:
    - Dwelling units in the same building.
    - Sleeping units in the same building.
  - Fire-resistance ratings required:
    - ≥ 1/2 hr in any of the following construction types where sprinklers are provided:
      - Type IIB.
      - Type IIIB.
      - Type VB.
    - ≥ 1 hr in other locations.

Note: 903.3.1.1, "NFPA 13 sprinkler systems," is cited as governing the sprinkler system.

712.3.1 Ceiling panels

- Lay-in ceiling panels must resist upward displacement as follows:
  - From a force ≤ 1 psf by one of the following methods:
    - Self-weight.
    - One of the following restraining systems above the panel:
      - Wire system.
      - Other approved system.

712 Horizontal Assemblies

712.3.2 Access doors

- Access doors in a fire-resistance-rated ceiling must have the following characteristics:
  - They must be tested as follows:
    - By an approved agency.
    - As a horizontal assembly.
  - They must be labeled for their purpose.

Note: The following are cited as alternative applicable standards for testing the access doors:
ASTM E 119, "Test Methods for Fire Tests of Building Construction and Materials."
UL 263, "Fire Tests of Building Construction and Materials."

712.3.3 Unusable space

- The following membranes may be omitted from 1-hr fire-resistance-rated horizontal assemblies:
  - The ceiling membrane as follows:
    - At the floor assembly directly over an unusable crawl space.
  - The floor membrane as follows:
    - At the ceiling assembly directly below unusable attic space.

712.4 Continuity *(part 1 of 2)*

- Horizontal assemblies must be continuous as follows:
  - Discontinuity is permitted only where indicated by the code as follows:
    - Openings.
    - Penetrations.
    - Joints.
  - Discontinuity is governed as follows:
    - Shaft enclosures are required for most openings through a horizontal assembly.
    - Penetrations must be detailed to maintain the integrity of the horizontal assembly.
    - Joints must be detailed to maintain the integrity of the horizontal assembly.

Note: The following are cited as sources of applicable requirements:
708.2, "Shaft enclosure required," governs openings in horizontal assemblies.
713.4, "Horizontal assemblies," governs penetrations.
Section 714, "Fire-Resistant Joint Systems," governs joints.
1022.1, "Enclosures required," governs required enclosures for stairs and ramps.

  - Skylights and other penetrations in fire-resistance-rated roof assemblies are governed as follows:
    - Unprotected skylights and other penetrations are permitted in the following case:
      - Where the structural integrity of the roof assembly is maintained as follows:
        - Upon collapse of the unprotected penetrating assemblies due to fire.
    - Unprotected skylights are not permitted where they are a fire threat to an adjacent higher building.

Note: 704.10, "Exterior structural members," is cited as defining conditions prohibiting unprotected skylights.

712 Horizontal Assemblies

712.4 Continuity *(part 2 of 2)*

- Supporting construction is governed as follows:
  - Supporting construction is not required to be fire-resistance rated in the following construction types:
    - Type IIB, IIIB, or VB in the following cases:
      - Where separating incidental uses are as follows:
        - Where otherwise the required fire-resistance rating is ≤ 1hr.
      - Where separating sleeping units.
      - At smoke barriers.
  - In other cases, supporting construction must have a fire-resistance rating as follows:
    - ≥ the horizontal assembly supported.

Note: IBC Table 508.2.5, "Incidental Accessory Occupancies," is cited as requiring separations for these areas.
420.3, "Horizontal separation," is cited as governing these assemblies separating sleeping units.
Section 710, "Smoke Barriers," is cited as governing smoke barriers.

712.8 Floor fire door assemblies

- Fire door assemblies in floors with a fire-resistance rating must comply with the following:
  - Doors must have a fire-resistance rating ≥ the floor penetrated.
  - Doors must be tested.
  - Doors must be labeled by an approved agency as follows:
    - Label must be permanent and include the following:
      - Manufacturer.
      - Test standard.
      - Fire-resistance rating.

Note: NFPA 288, "Standard Methods of Fire Tests of Floor Fire Door Assemblies in Fire-resistance-rated Floor Systems," is cited as the required test.

712.9 Smoke barrier

- This section addresses horizontal assemblies where required by the code as follows:
  - To resist the flow of smoke as follows:
    - Per the definition of a smoke barrier.
- The following must be protected as required for smoke barriers in horizontal assemblies:
  - Penetrations and joints.

Note: 713.5, "Penetrations in smoke barriers," is cited as governing penetrations.
714.6, "Fire-resistant joint systems in smoke barriers," is cited as governing joints.

- Doors at elevator shafts that penetrate the horizontal assembly are governed as follows:
  - They must be proteced by enclosed elevator lobbies as follows:
    - For any number of stories connected by the elevator shaft.

Note: 708.14.1, "Elevator lobby," is cited as governing enclosed elevator lobbies.

- Shaft enclosures are required for openings through horizontal assemblies.

Note: Section 708, "Shaft Enclosures," is cited as governing shaft enclosures.

- Vertical openings in horizontal assemblies must be protected.

713 Penetrations

713.1 Scope

- This section governs the protection of the following penetrations in the assemblies listed:
  - Penetrations:
    - Passing entirely through assemblies.
    - Passing through membranes of assemblies.
  - Assemblies:
    - Horizontal assemblies.
    - Wall assemblies with fire-resistance ratings.

713.1.1 Ducts and air transfer openings

- This paragraph identifies requirements for penetrations of fire-resistance-rated assemblies as follows:
  - The following penetrations through walls are addressed:
    - Ducts without dampers.

Note: *The following sections are cited as governing ducts without dampers:*
713.2, "Installation details."
713.3, "Fire-resistance-rated walls."
713.3.1, "Through penetrations."
713.3.1.1, "Fire-resistance-rated assemblies."
713.3.1.2, "Through-penetration firestop system."
713.3.2, "Membrane penetrations."
713.3.3, "Dissimilar materials."

- Ducts with dampers.
- Air transfer openings.

Note: *Section 716, "Ducts and Air Transfer Openings." is cited as governing air transfer openings and ducts with dampers.*

- The following penetrations through horizontal assemblies are addressed:
  - Certain applications where there is no shaft as follows:
    - Where no damper is required elsewhere in the code.

Note: *708.2, "Shaft enclosure required," Exception 4, is cited as defining these penetrations.*
The following sections are cited as governing these penetrations:
713.4, "Horizontal assemblies."
713.4.1, "Fire-resistance-rated assemblies."
713.4.1.1 "Through penetrations."
713.4.1.1.1, "Installation."
713.4.1.1.2, "Through-penetration firestop system."
713.4.1.2, "Membrane penetrations."
713.4.1.3, "Ducts and air transfer openings."
713.4.1.4, "Dissimilar materials."
713.4.2, "Nonfire-resistance-rated assemblies."
713.4.2.1, "Noncombustible penetrating items."
713.4.2.2, "Penetrating items."

713 Penetrations

713.2 Installation details

- Sleeves in penetrations are governed as follows:
  - Sleeves must be secured to the assembly penetrated.
  - The following spaces must be protected as per requirements of this section:
    - Between a sleeve and the component inside the sleeve.
    - Between a sleeve and the opening through which the sleeve passes.
- Insulation or covering as part of a penetrating element is governed as follows:
  - The materials may not penetrate an assembly except for the following case:
    - Where they have been tested with the assembly as per requirements of this section.

713.3.1 Through penetrations

- This section addresses protection of penetrations through fire-resistive-rated walls.
- The following penetrating items must comply with one of the protection methods indicated:
  - Penetrating items:
    - Steel pipes, tubes, or conduits.
    - Ferrous pipes, tubes, or conduits.
    - Copper pipes, tubes, or conduits.
  - Protection methods:
    - Where concrete or masonry walls are penetrated the following applies:
      - Annular space may be filled with concrete, mortar, or grout in the following conditions:
        - The penetrating item must be ≤ 6" nominal diameter.
        - The opening through the wall must be ≤ 144 sq in.
        - Annular space must be filled to one of the following extents:
          - Filled to the full thickness of the wall.
          - Filled to a thickness adequate to maintain the fire-resistance rating of the wall.
    - Where other walls are penetrated, the following applies:
      - The annular space may be filled with a substance able to perform as follows:
        - Prevents passage of the following fire hazards for the time period indicated below:
          - Fire hazards:
            - Flame.
            - Gases as follows:
              - Hot enough to ignite cotton waste in the following conditions:
                - With a positive pressure differential at penetration ≥ 0.01" of water.
          - Time period:
            - For a length of time = the fire-resistance rating of the wall penetrated.

Note: The following are cited as alternative required standards for testing the substance filling the annular space as described above:
ASTM E 119, "Test Methods for Fire Tests of Building Construction and Materials."
UL 263, "Fire Tests of Building Construction and Materials."

- Other through penetrations must meet certain test standards.

Note: The following citations identify tests required for the penetrations:
713.3.1.1, "Fire-resistance-rated assemblies."
713.3.1.2, "Through-penetration firestop system."

713 Penetrations

713.3.1.1 Fire-resistance-rated assemblies

- Penetrations must be installed as follows:
  - In an approved fire-resistance-rated assembly.
  - In the same detailing as tested.

713.3.1.2 Through-penetration firestop system

- Through penetrations must be protected as follows:
  - By a penetration fire-stop system with the following conditions:
    - Must be approved.
    - Must be installed as tested.
    - Must be tested with a positive pressure differential ≥ 0.01" of water.
    - Must have an F rating ≥ the required fire-resistance rating of the wall penetrated.

Note: The required test must conform to one of the following cited standards:
ASTM E 814, "Standard Test Method for Fire Tests of Through-Penetration Fire Stops."
UL 1479, "Standard for Fire Tests of Through-Penetration Firestops."

713.3.2 Membrane penetrations *(part 1 of 2)*

- Steel electrical boxes may penetrate a membrane where all of the following apply:
  - In walls or partitions with a fire-resistance rating ≤ 2 hr.
  - The area of the box must be ≤ 16 sq in.
  - The sum of opening areas in any 100 sf of wall must be ≤ 100 sq in.
  - The annular space between membrane and box is limited to ≤ 1/8".
  - Such boxes on opposite faces of a wall or partition must be separated by one of the following:
    - By one of the following horizontal distances:
      - ≥ the depth of the wall cavity where filled with any of the following insulation:
        - Cellulose loose-fill.
        - Rockwool or slag mineral wool.
      - ≥ 2' in walls or partitions having individual spaces (not interconnected) between studs.
    - By solid fire-blocking.

Note: 717.2.1, "Fireblocking materials," is cited as governing the fireblocking.

    - By using putty pads to protect both boxes as follows:
      - Putty pads must be listed.
    - By other listed materials and methods.
- Electrical boxes of any material may penetrate a membrane where all of the following conditions are met:
  - Boxes must be listed.
  - Boxes must be tested for use in a fire-resistance-rated wall.
  - Boxes must be installed according to their listing.
  - The annular space between membrane and box is limited to ≤ 1/8".
  - Such boxes on opposite faces of a wall or partition must be separated by one of the following:
    - By the distance required by the electrical box listing, measured horizontally.
    - By solid fire-blocking.

Note: 717.2.1, "Fireblocking materials," is cited as governing the fireblocking.

    - By using listed putty pads to protect both boxes.
    - By other materials and methods that are listed.

713 Penetrations

713.3.2 Membrane penetrations *(part 2 of 2)*

- Electrical boxes as follows may penetrate a membrane where all of the conditions listed below are met:
  - Electrical boxes:
    - Any size and type.
  - Conditions:
    - Boxes must be listed as follows:
      - As part of a wall opening protective material system.
      - For use in fire-resistance-rated assemblies.
    - Boxes must be installed per listing instructions.
- Boxes that are not electrical boxes may penetrate a membrane if all of the following conditions are met:
  - The following elements must have protection as listed below:
    - Elements:
      - The box.
      - The annular ring between the box and the wall membrane.
    - Protection required:
      - A membrane penetration firestop system as follows:
        - Must be approved.
        - Must be tested as follows:
          - With a minimum positive pressure differential of 0.01" water.
        - Must be installed as tested.
        - Must have F and T ratings as follows:
          - ≥ the fire-resistance rating of the wall penetrated.
        - Must be installed according to the listing.

Note: The following are cited as required alternative tests for the membrane firestop system:
ASTM E 814, "Test Method for Fire Tests of Penetration Firestop Systems."
UL 1479, "Fire Tests of Through-Penetration Firestops."

- The gap around a fire-sprinkler pipe penetration must be covered by a metal collar.
- Recessed fixtures may not reduce the fire-resistance rating of walls and partitions.
- Other membrane penetrations must be protected as are through penetrations.

Note: 713.3.1, "Through penetrations," is cited as governing membrane penetrations.

713.3.3 Dissimilar materials

- The following connection is permitted only with the condition indicated below:
  - Connection:
    - Between the following components at a point beyond the fire-stopping:
      - Noncombustible penetrating component.
      - Combustible component.
  - Condition:
    - It must be shown that the required fire resistance of the wall is not diminished.

713 Penetrations

713.4 Horizontal assemblies

- This section addresses penetrations permitted to occur without a shaft as follows:
  - Through floors.
  - Through floor/ceiling assemblies.
  - Through ceiling membrane of a roof/ceiling assembly.

Note: 708.2, "Shaft enclosure required," is cited as permitting penetrations to occur without a shaft. The following sections are cited as governing these penetrations:
713.4.1, "Fire-resistance-rated assemblies."
713.4.1.1, "Through penetrations."
713.4.1.1.1, "Installation."
713.4.1.1.2, "Through-penetration firestop system."
713.4.1.2, "Membrane penetrations."
713.4.1.3, "Ducts and air transfer openings."
713.4.1.4, "Dissimilar materials."
713.4.2, "Nonfire-resistance-rated assemblies."
713.4.2.1, "Noncombustible penetrating items."
713.4.2.2, "Penetrating items."

713.4.1.1 Through penetrations *(part 1 of 2)*

- This section addresses penetrations passing entirely through fire-resistance-rated horizontal assemblies.
- The following penetrations of single fire-resistance-rated floors have the requirements listed:
  - Penetrating items:
    The following components in the materials indicated below:
      Components:
        Conduits, tubes, and pipes.
        Vents.
      Materials:
        Steel, ferrous metals, copper.
    Concrete and masonry elements.
  - Requirements:
    The annular space must be filled with a substance able to perform as follows:
      Prevents passage of the following fire hazards for the time period listed:
        Fire hazards:
          Flame.
          Gases as follows:
            Hot enough to ignite cotton waste in the following condition:
              With a positive pressure differential at penetration ≥ 0.01" of water.
        Time period:
          For a length of time = the fire-resistance rating of the wall penetrated.

Note: The following are cited as alternative required standards for governing the substance filling the annular space:
ASTM E 119, "Test Methods for Fire Tests of Building Construction and Materials."
UL 263, "Fire Tests of Building Construction and Materials."

713 Penetrations

713.4.1.1 Through penetrations *(part 2 of 2)*

- Penetrations are not limited to a single floor where both of the following conditions are met:
  - The penetrating component is ≤ 6" nominal diameter.
  - The sum of the areas of openings through the assembly ≤ 144 sq in as follows:
    - Within any 100 sf of floor.
- Requirements are indicated below for penetrations of the following floors by the items listed:
  - Floors:
    - A single concrete floor.
    - Multiple concrete floors as follows:
      - Where the sum of areas of openings ≤ 144 sq in on any floor.
  - Penetrating items:
    - The following components in the materials indicated below:
      - Components:
        - Conduits, tubes, and pipes.
        - Vents.
      - Materials:
        - Steel, ferrous metals, copper.
- Electrical boxes of any material may be used as follows:
  - They must be tested for use in such assemblies.
  - They must be installed as per listing instructions.
  - Requirements:
    - Diameter of penetrating item must be ≤ 6".
    - The annular space must be filled with one of the following:
      - Concrete.
      - Grout.
      - Mortar.
    - The extent of the fill in the annular space must be one of the following:
      - For the full thickness of the floor.
      - ≥ a thickness which does not diminish the required fire-resistance rating.
- Other penetrations must comply with one of the following:
  - They must be installed as they were tested in the approved fire-resistance-rated assembly.
  - They must be protected by an approved through-penetration fire-stop system.

Note: The following are cited governing other penetrations, partial summaries of which are provided above:

713.4.1.1.1, "Installation."

713.4.1.1.2, "Through-penetration firestop system."

713.4.1.1.1 Installation

- Through penetrations must be installed as follows:
  - In the manner tested in the approved fire-resistance-rated assembly.

713 Penetrations

713.4.1.1.2 Through-penetration firestop system

- Through penetrations of fire-resistance-rated horizontal assemblies must be protected by an approved through-penetration fire-stop system tested and installed as follows:
  - System must be tested as follows:
    - With a positive pressure differential of ≥ 0.01" of water.
  - System must be installed as tested.
- Through penetrations of fire-resistance-rated horizontal assemblies must be protected by a fire-stop system with the following F and T ratings:
  - System must have an F rating ≥ the larger of the following:
    - 1 hr.
    - The rating of the floor penetrated.
  - System T rating is governed as follows:
    - No T rating is required for a floor penetration in a wall cavity in either of the following locations:
      - Above the floor.
      - Below the floor.
    - Otherwise, T rating must be ≥ the larger of the following:
      - 1 hr.
      - The rating of the floor penetrated.

Note: The following as cited as alternative standards for the fire-stop system.
ASTM E 814, "Test Method for Fire Tests of Penetration Firestop Systems."
UL 1479, "Fire Tests of Through-Penetration Firestops."

713.4.1.2 Membrane penetrations *(part 1 of 2)*

- This section addresses the following:
  - Membrane penetrations in horizontal assemblies.
- Requirements for penetrations by the following elements are listed below:
  - Penetrating elements:
    - Concrete or masonry components.
    - The following components made of the materials listed:
      - Components:
        - Conduits, tubes, and pipes.
        - Vents.
      - Materials:
        - Steel and ferrous metals.
        - Copper.
  - Requirements:
    - The gap around the penetrating item must comply with one of the following:
      - Must be protected.
      - Must be protected to prevent the passage of the following:
        - Flame.
        - Products of combustion.
    - Where assemblies were tested without penetrations the following applies:
      - The sum of openings through the membrane is limited as follows:
        - ≤ 100 sq in within any 100 sf of ceiling area.

Note: 713.4.1.1, "Through penetrations," is cited as governing the protection gaps around penetrations.

713 Penetrations

713.4.1.2 Membrane penetrations *(part 2 of 2)*

- Steel electrical boxes may penetrate a ceiling membrane where all of the following apply:
  - In horizontal assemblies with a fire-resistance rating ≤ 2 hr.
  - The area of the box must be ≤ 16 sq in.
  - The sum of opening areas in any 100 sf of ceiling must be ≤ 100 sq in.
  - The annular space between membrane and box is limited to ≤ 1/8".
- Penetrations by electrical boxes with all of the following characteristics are permitted:
  - May be any size or type.
  - Must be listed as part of a wall opening protective material system.
  - Must be for use in a horizontal assembly.
  - Must be installed according to listing instructions.
- Penetrations by electrical boxes of any material are allowed as follows:
  - Boxes must be listed.
  - Boxes must be tested for use in assemblies with fire-resistance ratings.
  - Boxes must be installed as per instructions dictated by their listing.
  - The annular space between membrane and box is limited to ≤ 1/8".
- A fire-sprinkler pipe penetration is governed as follows:
  - A metal collar must cover the gap around the pipe at the penetration.
- Other penetrations must comply with one of the following:
  - Must be as tested in the approved fire-resistance-rated assembly.
  - Must be protected by an approved through-penetration fire-stop system.

Note: The following are cited as governing the requirements for other penetrations, a partial summary of which is provided above:
712.4.1.1.1, "Installation."
712.4.1.1.2, "Through-penetration firestop system."

- Recessed fixtures may not reduce the fire-resistance rating of the following assembly:
  - Floor/ceiling assemblies with a required fire-resistance rating.

713.4.1.4 Dissimilar materials

- The following connection is permitted only with the condition listed:
  - Connection:
    Between the following components at a point beyond the fire-stopping:
    Noncombustible penetrating component.
    Combustible component.
  - Condition:
    It must be shown that the required fire-resistance of the assembly is not diminished.

713 Penetrations

713.4.2.1 Noncombustible penetrating items

- Penetrations as follows must meet the requirements listed:
  - Penetrations:
    - Noncombustible items.
    - Connecting ≤ 3 stories.
  - Requirement:
    - Annular space must be filled with one of the following materials:
      - An approved noncombustible material as follows:
        - Must inhibit the passage of flame and combustion products.
      - A fill, void, or cavity substance as follows:
        - Must be tested and classified for use in through-penetration firestop systems.
        - Must inhibit the passage of flame and combustion products.

713.4.2.2 Penetrating items

- Penetrations connecting ≤ 2 stories are governed as follows:
  - Annular space must be filled with an approved material as follows:
    - Must restrict the passage of flame and combustion products.

715 Opening Protectives

715.2 Fire-resistance-rated glazing

- This paragraph addresses fire-resistance-rated glazing with both of the following characteristics:
  - It must be labeled.
  - It must be tested as part of a wall with a fire-resistance rating.

Note: The following are cited as alternative required standards for testing fire-resistance-rated glazing:
ASTM E 119, "Test Methods for Fire Tests of Building Construction and Materials."
UL 263, "Fire Tests of Building Construction and Materials."

- Such glazing is governed only as follows without additional requirements by this section series:
  - It is permitted in fire doors according to its listing.
  - It is permitted in fire window assemblies according to its listing.

715.4 Fire door and shutter assemblies *(part 1 of 2)*

- The following are not governed by this section:
  - Labeled tin-clad fire doors.

Note: The following are cited as governing tin-clad fire doors:
UL 10A, "Standard for Tin Clad Fire Doors."
UL 14B, "Standard for Sliding Hardware for Standard Horizontally Mounted Tin-Clad Fire Doors."
UL 14C, "Standard for Swinging Hardware for Standard Tin-Clad Fire Doors Mounted Singly and in Pairs."

  - Floor fire door assemblies.

Note: 712.8, "Floor fire door assemblies," is cited as governing these doors.

- Other approved fire door and shutter assemblies are governed as follows:
  - They may be of any material as follows:
    Materials must comply with specified fire tests.
  - They must be installed as per the standard cited for fire doors and windows.

Note: NFPA 80, "Fire Doors and Fire Windows," is cited as governing fire door installation.
The following are cited as listing required tests for doors governed by this section:
715.4.1, "Side-hinged or pivoted swinging doors."
715.4.2, "Other types of assemblies."
715.4.3, "Door assemblies in corridors and smoke barriers."
IBC Table 715.4, "Fire Door and Fire Shutter Fire Protection Ratings," is cited as governing the fire protection ratings for fire doors and fire shutters.

715 Opening Protectives

715.4 Fire door and shutter assemblies *(part 2 of 2)**

- They must have the following fire-protection ratings:
  - ≥ 20 minutes where located in any of the following where tested as specified:
    - 1/2-hr fire-resistance-rated fire partition as follows:
      - A corridor wall.
      - Other fire partition.
    - 1-hr fire-resistance-rated fire partition corridor wall.
  - ≥ 3/4 hr where located in the following:
    - 1-hr fire-resistance-rated walls in any of the following applications:
      - Exterior walls.
      - Fire partitions other than corridor walls.
      - Fire barriers other than the following:
        - Shaft exit enclosure walls.
        - Exit passageway walls.
  - ≥ 1 hr where located in the following:
    - 1-hr fire-resistance-rated fire barriers in either of the following applications:
      - Shaft exit enclosure walls.
      - Exit passageway walls.
  - ≥ 1 1/2 hr where located in one of the following:
    - 1 1/2-hr fire-resistance-rated walls of either of the following types:
      - Fire walls or fire-barriers.
    - 2-hr fire-resistance-rated walls of any of the following types:
      - Fire walls.
      - Fire barriers.
      - Exterior walls.
  - ≥ 1 1/2 hr where located in the following:
    - 3-hr fire-resistance-rated exterior wall.
  - ≥ 1 1/2 hr each where 2 doors or shutters meet all of the following conditions:
    - Located at the same opening.
    - One door is on each side of a wall with all of the following characteristics:
      - 3-hr fire-resistance-rated.
      - Wall is one of the following types:
        - Fire wall or fire barrier.
  - ≥ 3 hr where located in a fire wall or fire barrier with one of the following ratings:
    - 3-hr fire-resistance rating.
    - 4-hr fire-resistance rating.
  - ≥ 1/3 hr where located in a smoke barrier with the following rating:
    - 1-hr fire-resistance rating.
- Fire door frames with transom lights and/or sidelights must meet code protection requirements.

Note: 715.4.5, "Fire door frames with transom lights and sidelights," is cited as governing required protection for these assemblies.

**Source*: IBC Table 715.4.

715 Opening Protectives

715.4.4.1 Glazing in doors

- Glazing > 100 sq in is allowed in fire door assemblies as follows:
  - Glazing must have a fire-protection rating.
  - Glazing must be tested as follows:
    - As a component of the door assembly, not as a glass light.
    - The transmitted temperature rise is governed as follows:
      - Limited to ≤ 450° F above ambient after 1/2 hr of testing.
      - Not required as follows:
        - Where the building is sprinklered.

Note: 715.4.4, "Doors in exit enclosures and exit passageways," is cited as governing fire testing for doors with glazing.
The following are cited as governing the referenced sprinklers:
903.3.1.1, "NFPA 13 sprinkler systems."
903.3.1.2, "NFPA 13R sprinkler systems."

715.4.5 Fire door frames with transom lights and sidelights

- This section addresses transom lights or sidelights in fire door frames.
- Such lights are permitted as follows:
  - Where a fire protection rating of ≤ 3/4 hr is permitted.
  - Where a fire protection rating of > 3/4 hr is required, fire-resistance-rated glazing is required as follows:
    - Glazing must meet testing requirements as an assembly.

Note: IBC Table 715.4, "Fire Door and Fire Shutter Fire Protection Ratings," is cited as governing required protection ratings.
The following are cited as alternative tests for the glazing:
ASTM E119, "Test Methods for Fire Tests of Building Construction and Materials."
UL 263, "Fire Tests of Building Construction and Materials."

715.4.7 Glazing material

- Fire door assemblies may have glazing that conforms to the following:
  - Glazing must have a fire-protection rating.
  - Glazing must conform to opening protection requirements.

Note: 715.4, "Fire door and shutter assemblies," is cited as governing opening protection.

715.4.7.1 Size limitations *(part 1 of 2)*

- Glazing in fire doors in fire walls is governed as follows:
  - Where fire doors are serving as a horizontal exit, the following applies:
    - A glazed vision panel is permitted as follows:
      - Door must be swinging type.
      - Door must be self-closing.
      - Glazing must have a fire-protection rating.
      - Glazing area is limited to ≤ 100 sq in.
      - Glazing dimensions are limited to ≤ 10 in.
  - In other cases, glazing is not permitted.

715 Opening Protectives

715.4.7.1 Size limitations *(part 2 of 2)*

- Glazing in fire doors in fire barriers is governed as follows:
  - Glazing must have a fire-protection rating.
  - Glazing area is limited to ≤ 100 sq in. where installed in the following doors:
    Doors with a 1 1/2-hr rating.
- In other cases, glazing in fire doors must comply with size requirements of the applicable standard.

Note: NFPA 80, "Fire Doors and Fire Windows," is cited as the applicable standard.

Case study: Fig. 715.4.7.1. Wired glass is provided in the 1-hr fire door at the exit enclosure. IBC Table 715.5.4 limits the area of such glass to 100 sq in where the fire-protection rating of the opening is 1 hr. The height of the glass is limited to 33", and the maximum width permitted by the table is 10". The 10" × 10" glass meets these requirements.

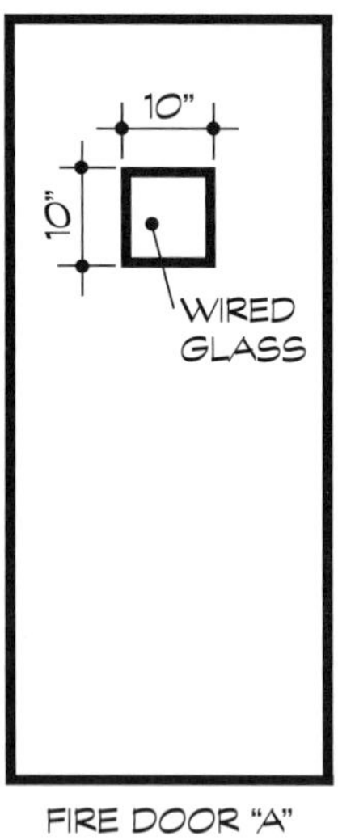

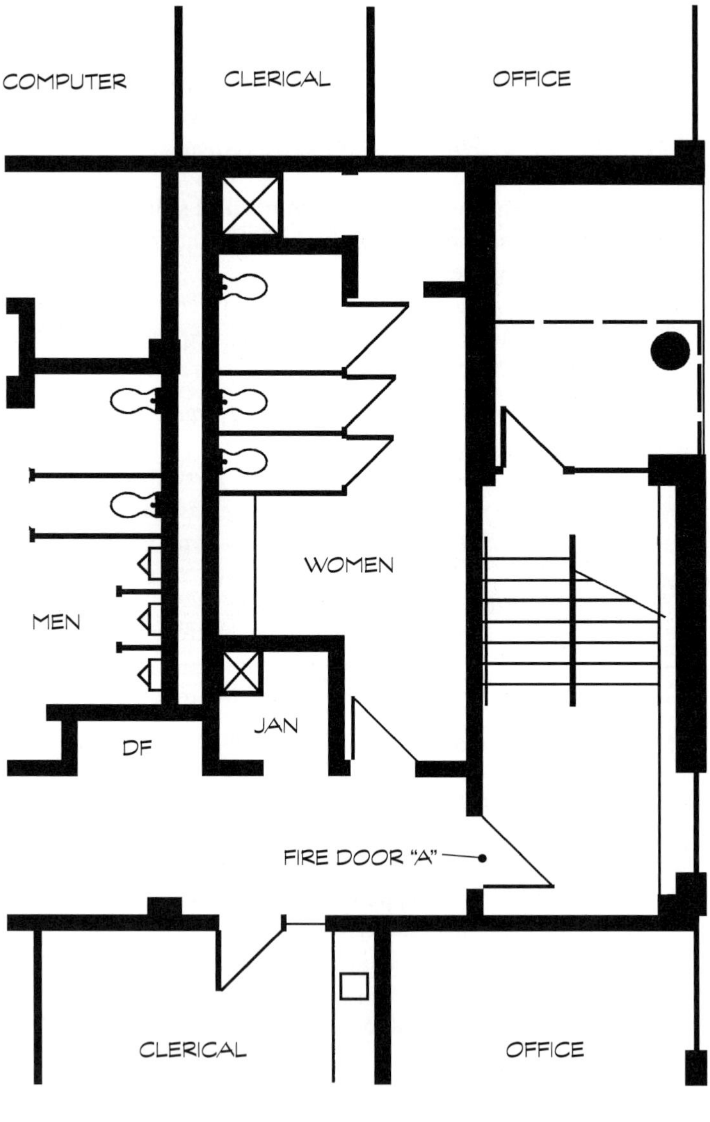

Fig. 715.4.7.1. Fire door elevation and plan. Lubrication Engineers, Inc. Wichita, Kansas. Gossen Livingston Associates, Inc., Architecture. Wichita, Kansas.

715 Opening Protectives

715.5.4 Wired glass

- A 3/4-hr fire window assembly has the following characteristics:
  - The window frame is one of the following materials:
    - ≥ 0.125" thick solid steel.
    - ≥ 0.048" thick formed sheet steel.
  - The window frame is fabricated by any of the following methods:
    - Pressing.
    - Mitering.
    - Riveting.
    - Interlocking.
    - Welding.
  - The window frame is able to receive 1/4" thick wired glass.
  - The window frame is secured into the building construction.
  - The glazing is 1/4" thick wired glass.
  - The glazing is labeled.
- Wired glass size limitations are as follows:*
  - Where located in openings requiring the following:
    - 3-hr fire-protection rating:
      - Not permitted.
  - Where located in doors of exterior walls requiring the following:
    - 1 1/2-hr fire-protection rating:
      - Not permitted.
  - Where located in openings requiring the following:
    - 1-hr fire-protection rating:
      - Area must be ≤ 100 sq in.
      - Height must be ≤ 33".
      - Width must be ≤ 10".
  - Where located in openings requiring the following:
    - 1 1/2-hr fire-protection rating:
      - Area must be ≤ 100 sq in.
      - Height must be ≤ 33".
      - Width must be ≤ 10".
  - Where located in openings requiring the following:
    - 3/4-hr fire-protection rating:
      - Area must be ≤ 1296 sq in.
      - Each dimension must be ≤ 54".
  - Where located in openings requiring the following:
    - 20-minute fire-protection rating:
      - Area not limited.
      - Dimensions not limited.
  - Fire window assemblies:
    - Area must be ≤ 1296 sq in.
    - Each dimension must be ≤ 54".

**Source*: IBC Table 715.5.4.

715 Opening Protectives

715.5.5 Nonwired glass

- Glazing other than wired glass where used in fire window assemblies is governed as follows:
  - It must be fire-protection-rated.
  - It must comply with standards other than this code for installation and size.

Note: NFPA 80, "Fire Doors and Fire Windows," is cited as governing glazing other than wired glass used in fire window assemblies.

715.5.6 Installation

- Glazing with a fire-protection rating must comply with the following:
  - Frames must be approved.
  - Glazing must be fixed or automatic-closing.

715.5.7 Window mullions

- The following metal mullions must comply with the requirement indicated below:
  - Mullions:
    Serving glazing with a fire-protection rating.
    > 12' tall.
  - Requirement:
    Mullion fire-resistance rating must be ≥ that of the wall.

715.5.8 Interior fire window assemblies

- This section governs glazing with a fire-protection rating as used in fire window assemblies.
- Such glazing is limited to use of the following components:
  - Fire partitions with a fire-resistance rating ≤ 1 hr.
  - Fire barriers with a fire-resistance rating ≤ 1 hr.

715.5.8.1 Where 3/4-hour fire protection window assemblies permitted

- This section governs glazing with a required fire-protection rating of 3/4 hr.

Note: IBC Table 715.5, "Fire Window Assembly Fire Protection Ratings," is cited as governing protection ratings for glazing.

- Such glazing may be used only in the following applications:
  - Fire partitions with a fire-resistance rating of 1 hr.
  - Fire barriers as follows:
    With a fire-resistance rating of 1 hr.
    Separating the following:
      Incidental use areas from the rest of the building.
      Mixed occupancies from each other.
      A single occupancy into different fire areas.

Note: The following are cited as governing the partitions and barriers indicated above:
Section 709, "Fire Partitions."
707.3.6, "Incidental accessory occupancies," addresses fire barriers.
707.3.8, "Separated occupancies," addresses fire barriers.

717 Concealed Spaces

717.1 General

- The following are required in combustible concealed spaces:
  - Fire-blocking at the following locations:
    - In walls at floor and ceilings.
    - In stair construction.
    - At penetrations.
    - In exterior wall finish systems.
    - Between sleepers.
  - Draft-stopping at the following locations:
    - Floor/ceiling spaces.
    - Attic spaces.
  - In Type I or II construction the following applies:
    - The use of combustible materials in concealed spaces is limited as follows:
      - To locations with conditions specified by the code.

Note: The following are cited as sources of applicable requirements:
717.2, "Fireblocking."
717.3, "Draftstopping in floors."
717.4, "Draftstopping in attics."
717.5, "Combustible materials in concealed spaces in Type I or II construction."

717.2 Fireblocking

- Fire-blocking is required to seal openings to restrict drafts as follows:
  - In combustible construction in concealed spaces as follows:
    - Vertical openings.
    - Horizontal openings.
- Fire-blocking is required to form a barrier between the following elements:
  - Between floors.
  - Between the top story and a roof.
  - Between the top story and an attic.

Note: The following are cited as the source for specific locations requiring fireblocking:
717.2.2, "Concealed wall spaces."
717.2.3, "Connections between horizontal and vertical spaces."
717.2.4, "Stairways."
717.2.5, "Ceiling and floor openings."
717.2.6, "Architectural trim."
717.2.7, "Concealed sleeper spaces."

717 Concealed Spaces

717.2.1 Fireblocking materials

- Fireblocking must be one of the following materials:
  - 2" nominal lumber.
  - 2 layers of 1" nominal lumber with staggered joints.
  - 0.719" wood structural panels as follows:
    - Joints to be backed by the same material.
  - 0.75" particle board as follows:
    - Joints to be backed by the same material.
  - 1/2" gypsum board.
  - 1/4" cement-based millboard.
  - Any of the following batts or blankets where securely fixed in place:
    - Mineral wool.
    - Glass fiber.
    - Other approved material.

717.2.1.1 Batts or blankets of mineral wool or mineral fiber

- This section applies where fireblocking is required as follows:
  - In concealed spaces of the following:
    - Stud walls.
    - Partitions with studs.
    - Furring.
  - Spaced ≤ 10' measured horizontally.
- The following materials are permitted to serve as fireblocking:
  - Batts or blankets of the following:
    - Mineral wool.
    - Mineral fiber.
    - Other approved nonrigid material.

717.2.1.2 Unfaced fiberglass

- Unfaced fiberglass batt insulation fireblocking must be detailed as follows:
  - It must fill the entire cross section of the wall cavity.
  - It must have a vertical dimension ≥ 16".
  - It must be packed tightly around any obstructions such as the following:
    - Pipes.
    - Conduits.
    - Similar obstructions.

717.2.1.3 Loose-fill insulation material

- This paragraph applies to the following materials where used as fireblocking:
  - Loose-fill insulation.
  - Insulating foam sealants.
  - Caulking materials.
- Such materials must be tested for use as fireblocking as follows:
  - Materials must be tested in the detailing to be used as fireblocking.
  - Materials must demonstrate the abililty to stay in place.
  - Materials must demonstrate the ability to obstruct the flow of fire and hot gasses.

717 Concealed Spaces

717.2.1.4 Fireblocking integrity

- Fireblocking integrity must be sustained.

717.2.1.5 Double stud walls

- The following nonrigid materials may serve as fireblocking in the walls indicated below:
  - Materials:
    - Batts or blankets of mineral fiber.
    - Batts or blankets of glass fiber.
    - Other approved nonrigid materials.
  - Walls:
    - Double walls with parallel rows of studs.
    - Walls with cavity space wider than stud width as follows:
      - Alternate studs are flush with opposite sides of the wall.

717.2.2 Concealed wall spaces

- Fireblocking is required as shown below in the following construction:
  - Construction:
    - Stud walls and partitions.
    - Associated furred spaces.
    - Parallel rows of studs.
    - Staggered studs.
  - Fireblocking:
    - Required in concealed spaces of the construction as follows:
      - At ceiling level.
      - At floor level.
      - At ≤ 10' intervals measured horizontally.

717.2.3 Connections between horizontal and vertical spaces

- Fireblocking is required to seal through connections as follows:
  - Between the concealed spaces in stud walls or partitions and the following:
    - Concealed spaces created by floor joists.
    - Concealed spaces created by trusses.
  - Between vertical and horizontal spaces occurring at the following and similar locations:
    - Soffits.
    - Drop ceilings.
    - Cove ceilings.

717.2.4 Stairways

- Fireblocking is required in concealed spaces as follows:
  - Between stair stringers at the top of a stair run.
  - Between stair stringers at the bottom of a stair run.
  - Enclosed space under a stair requires the greater of the following:
    - A fire-resistance rating ≤ 1 hr.
    - A fire-resistance rating ≤ the fire-resistance rating of the stairway enclosure.

Note: 1009.6.3, "Enclosures under stairways," is cited as governing enclosed space under a stair and is partially summarized above. The section includes other requirements and exceptions.

717 Concealed Spaces

717.2.5 Ceiling and floor openings

- Fireblocking is required at annular space of the following penetrations at locations indicated:
  - Penetrations:
    - Vents.
    - Pipes.
    - Chimneys.
    - Ducts.
    - Conduits.
    - Fireplaces.
  - Locations:
    - Ceilings.
    - Floors.

Note: The following are cited as sources listing penetrations requiring fireblocking, a partial summary of which is provided above:
708.2, "Shaft enclosure required," Exception 6.
713.4.1.2, "Membrane penetrations," Exception 1.
713.4.2, "Nonfire-resistance-rated assemblies."

- Fireblocking must be material tested in the format of the actual installation.
- Fireblocking must resist the passage of the following:
  - Flame and combustion products.

717.2.5.1 Factory-built chimneys and fireplaces

- Prefabricated fireplaces and chimneys must have fireblocking.

Note: The following are cited as governing prefabricated fireplaces and chimneys:
UL 103, "Standard for Factory-Built Chimneys for Residential Type and Building Heating Appliances."
UL 127, "Standard for Factory-Built Fireplaces."

717.2.6 Architectural trim *(part 1 of 2)*

- Fireblocking is required in cornices at the party wall of duplexes.
- Fireblocking is not required in the following cases:
  - At locations in cornices other than at the party wall of duplexes.
  - In cornices of single-family dwellings.
  - Where both of the following conditions are present:
    - The architectural trim is installed on noncombustible framing.
    - The exterior wall finish exposed to the concealed space is one of the following:
      - Aluminum ≥ 0.019" thick.
      - Corrosion-resistant steel as follows:
        - Base metal thickness is ≥ 0.016" at thinnest point.
      - Other approved noncombustible materials.
- Fireblocking is required in concealed spaces of the following components as indicated:
  - Components:
    - With combustible construction or framing as follows:
      - Exterior wall finish.
      - Exterior architectural elements.

717 Concealed Spaces

717.2.6 Architectural trim *(part 2 of 2)*

- Requirements:
  - Continuous concealed spaces must be fireblocked as follows:
    - At a spacing ≤ 20'.
    - So as to limit open space to ≤ 100 sf.
  - Noncontinuous concealed spaces must have both of the following:
    - Closed ends.
    - Separation between sections ≥ 4".

Note: Section 1406, "Combustible Materials on the Exterior Side of Exterior Walls," is cited as governing the combustible materials at exterior walls indicated above.

- Wood furring where used is governed as follows:
  - Must be of approved wood.
  - Must be one of the following:
    - Wood with a natural resistance to decay.
    - Preservative-treated wood.

717.2.7 Concealed sleeper spaces

- Slab-on-grade gymnasium floors do not require fireblocking:
- Bowling alley lanes require blocking only at the following locations:
  - At the juncture of every other lane.
  - At the ends of each lane.
- In other cases, space between wood sleepers on floors as follows are governed as indicated:
  - Floors:
    - Masonry or concrete floors with a fire-resistance rating.
  - Sleeper space:
    - Space between flooring and floor slab must be detailed in one of the following ways:
      - Fire-blocked to limit the airspace to ≤ 100 sf.
      - Filled with an approved material as follows:
        - That will prevent passage of the following:
          - Flame and combustion products.
    - Space between flooring and slab must be completely filled in the following location:
      - Under permanent partitions between rooms.

717.3 Draftstopping in floors

- This section addresses draftstopping in the following locations:
  - In combustible construction.
  - In floor/ceiling assemblies.

Note: The following are cited as sources of specific locations where draftstopping is required:
717.3.2, "Groups R-1, R-2, R-3 and R-4."
717.3.3, "Other groups."

717 Concealed Spaces

717.3.1 Draftstopping materials

- The following qualify as draftstopping materials:
  - Gypsum board ≥ 1/2" thick.
  - Wood structural panel ≥ 3/8" thick.
  - Particleboard ≥ 3/8" thick.
  - Nominal 1" thickness of lumber.
  - Cement fiberboard.
  - Batts or blankets of mineral wool.
  - Batts or blankets of glass fiber.
  - Other approved materials.
- Draftstopping materials must be adequately supported.
- Integrity of draftstops must be maintained.

717.3.2 Groups R-1, R-2, R-3 and R-4

- This section applies to draft-stopping in concealed spaces as follows:
  - Combustible floor/ceiling assemblies in the following occupancies:
    R-1.
    R-2 with ≥ 3 dwelling units.
    R-3 with 2 dwelling units.
    R-4.
- Draftstopping is not required as follows:
  - In buildings sprinklered as per NFPA 13.
  - In buildings sprinklered as per NFPA 13R as follows:
    Sprinklers provided in combustible concealed spaces.

Note: The following are cited as governing the referenced sprinklers:
903.3.1.1, "NFPA 13 sprinkler systems."
903.3.1.2, "NFPA 13R sprinkler systems."

- In other cases, draftstopping is required in the following locations:
  - On the line of separation between dwelling units.
  - On the line of separation between sleeping units.

717.3.3 Other groups

- This section applies to draftstopping in concealed spaces as follows:
  - Combustible floor/ceiling assemblies in the following:
    Occupancies other than R.
- Draftstopping is not required in sprinklered buildings.

Note: 903.3.1.1, "NFPA 13 sprinkler systems," is cited as governing the sprinklers.

- In other buildings, draftstopping is required as follows:
  - Horizontal airspace must be limited to ≤ 1,000 sf of floor area.

717 Concealed Spaces

717.4 Draftstopping in attics

- This section series addresses draftstopping in the following locations of combustible construction:
  - In attics and concealed roof spaces.

Note: The following sections are cited as sources of specific locations required for draftstopping:
717.4.2, "Groups R-1 and R-2."
717.4.3, "Other groups."

- Ventilation of concealed spaces under the roof must be provided.

Note: 1203.2, "Attic spaces," is cited as the source for ventilation requirements.

717.4.1 Draftstopping materials

- The following qualify as draftstopping materials in attic spaces:
  - Gypsum board ≥ 1/2" thick.
  - Wood structural panel ≥ 3/8" thick.
  - Particleboard ≥ 3/8" thick.
  - Nominal 1" thickness of lumber.
  - Cement fiberboard.
  - Batts or blankets of mineral wool.
  - Batts or blankets of glass fiber.
  - Other approved materials.
- Draftstopping materials must be adequately supported.
- Integrity of draftstops must be maintained.

Note: 717.3.1, "Draftstopping materials," is cited as governing these materials in attics, the content of which is reproduced above.

717.4.1.1 Openings

- Openings in draftstopping attic partitions must be protected as follows:
  - With self-closing doors.
  - Doors must have automatic latches.

717.4.2 Groups R-1 and R-2 *(part 1 of 2)*

- This section applies to draftstopping where required as follows:
  - In combustible concealed spaces of the following occupancies in locations listed below:
    - Occupancies:
      - R-1.
      - R-2 with ≥ 3 dwellings.
    - Locations:
      - General locations:
        - Attics.
        - Mansards.
        - Overhangs.
        - Other concealed roof spaces.
      - Specific locations:
        - On line with separation walls which do not reach the roof sheathing as follows:
          - Between dwelling units.
          - Between sleeping units.

717 Concealed Spaces

717.4.2 Groups R-1 and R-2 *(part 2 of 2)*

- Draftstopping is required above only one of the two corridor walls in the following case:
  - Where corridor walls provide the following separations:
    - Between dwelling units.
    - Between sleeping units.
- Draftstopping is not required as follows:
  - In buildings sprinklered as per NFPA 13.
  - In buildings sprinklered as per NFPA 13R as follows:
    - Sprinklers provided in combustible concealed spaces.

  Note: The following are cited as governing the referenced sprinklers:
  903.3.1.1, "NFPA 13 sprinkler systems."
  903.3.1.2, "NFPA 13R sprinkler systems."

- Draftstopping is required as indicated below in the following occupancy:
  - Occupancy:
    - R-2 ≤ 4 stories above grade plane.
  - Requirement:
    - Draft-stops must divide attic space into the smaller of the following areas:
      - ≤ 3,000 sf.
      - ≤ area above every two dwellings.
- Draftstopping is otherwise required in the general and specific locations specified in this section.

717.4.3 Other groups

- Draftstopping is not required in sprinklered buildings.

  Note: 903.3.1.1, "NFPA 13 sprinkler systems,"is cited as governing the referenced sprinklers.

- Draftstopping is otherwise required in the following locations as indicated below:
  - Locations:
    - Attics.
    - Concealed roof spaces.
  - Requirement:
    - Horizontal areas must be limited to ≤ 3,000 sf.

717 Concealed Spaces

Case study: Fig. 717.4.3. The attic of the Occupancy E building is divided by walls providing draftstops. One draftstopping wall separates the new attic (not sprinklered) from the existing attic. A second draftstopping wall separates the new attic into areas of 2,780 sf and 1,670 sf, both within the 3,000 sf limit.

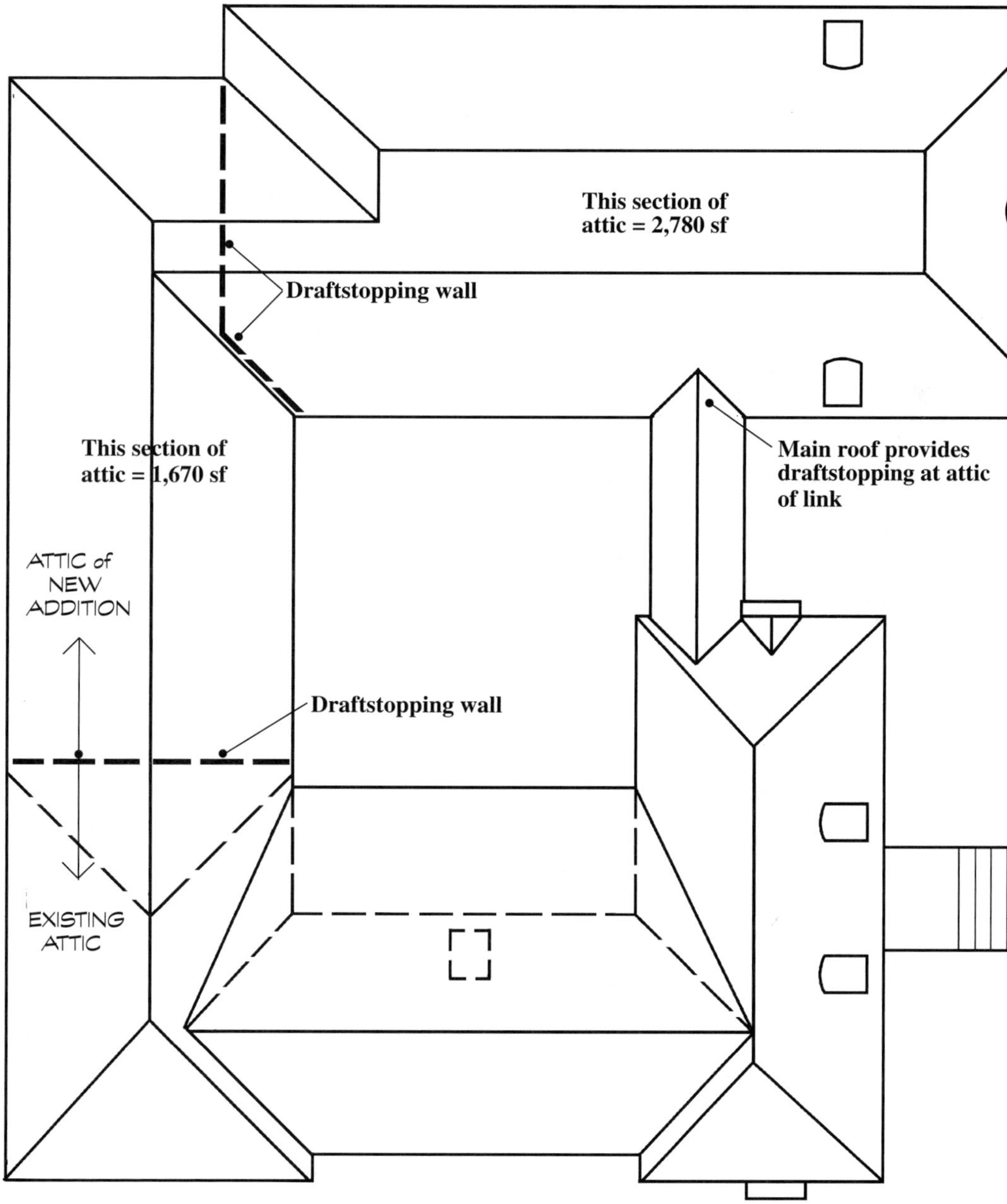

Fig. 717.4.3. Roof plan. Multipurpose Building Addition to Children's Home. Wilkes-Barre, Pennsylvania. C. Allen Mullins, Architect. Bear Creek, Pennsylvania.

717 Concealed Spaces

717.5 Combustible materials in concealed spaces in Type I or II construction

- The following combustible materials may be used in concealed spaces of Type I and II construction:
  - Certain materials identified elsewhere.

  Note: The following sources are cited as identifying combustible materials permitted in concealed spaces in Type I or II construction:
  Section 603, "Combustible Material in Type I and II Construction."
  International Mechanical Code, *Section 602, "Plenums."*

  - Class A finish materials.

  Note: Section 803, "Wall and Ceiling Finishes," is cited as defining the Class A materials.

  - Combustible piping as follows:
    In the following locations as otherwise governed by this code:
    Partitions.
    Shaft enclosures.
    In concealed spaces where installed as per applicable codes.

  Note: The following are cited as governing combustible piping in concealed spaces:
  International Mechanical Code.
  International Plumbing Code.

  - In other than plenums the following are permitted:
    Combustible insulation.
    Combustible pipe and tube covering.

  Note: 719.7, "Insulation and covering on pipe and tubing," is cited as governing these types of insulation and coverings.

- Otherwise, combustible materials are not allowed in the following locations:
  - In concealed spaces of Type I and II construction.

718 Fire-Resistance Requirements for Plaster

718.1 Thickness of plaster

- The thickness required for the following types of plaster is determined by specified fire tests:
  - Gypsum plaster and portland cement plaster.
- Where applied to the following, plaster thickness is measured to the face of the lath:
  - Gypsum lath and metal lath.

718.2 Plaster equivalents

- The following are considered to have equal fire resistance:
  - 0.5" unsanded gypsum plaster.
  - 0.75" 1:3 gypsum sand plaster.
  - 1" of portland cement sand plaster.

718.3 Noncombustible furring

- In Type I and II construction, plaster must be applied directly to one of the following:
  - Concrete or masonry.
  - Approved noncombustible plastering base and furring.

718.4 Double reinforcement

- The following plaster systems do not require supplementary reinforcement:
  - Solid plaster partitions.
  - Assemblies where it is deemed by fire test to be unnecessary.
- Other plaster as follows requires supplementary reinforcement as indicated below:
  - Plaster:
    - With both the following characteristics:
      - Plaster used as fire protection.
      - Plaster > 1" thick.
  - Requirement:
    - An additional layer of approved lath is required as follows:
      - Lath must be embedded ≥ 0.75" from outer surface of plaster.
      - Lath must be fastened securely in place.

718.5 Plaster alternatives for concrete

- This section addresses concrete cover protection in reinforced concrete construction.
- The following plaster types may substitute for concrete cover as indicated below:
  - Plaster types:
    - Gypsum plaster.
    - Portland cement plaster.
  - Substitution:
    - Up to 1/2" of concrete cover may be replaced with plaster as follows:
      - Concrete cover may not be reduced below the following:
        - 3/8" for poured reinforced concrete floors in addition to the plaster.
        - 1" for reinforced concrete columns in addition to the plaster.

Note: 2510.7, "Preparation of masonry and concrete," is cited as the source of requirements for the concrete base.

719 Thermal- and Sound-Insulating Materials

719.1 General

- This section does not apply to the following materials:
  - Fiberboard insulation.
  - Foam plastic insulation.
  - Duct insulation and coverings.
  - Pipe insulation and coverings.
  - Linings in plenums.

Note: Chapter 23, "Wood," is cited as governing fiberboard insulation.
Chapter 26, "Plastic," is cited as governing foam plastic insulation.
The International Mechanical Code *is cited as governing ducts, pipes, and plenums.*
Section 2613, "Reflective Plastic Core Insulation," is cited as governing all layers of single and multilayer reflective plastic core insulation.

- This section governs the following materials:
  - Other insulating materials.
  - Facings such as follows:
    - Vapor retarders.
    - Vapor-permeable membranes.
    - Similar coverings.
  - All layers of the following:
    - Single-layer reflective foil insulation.
    - Multilayer reflective foil insulation.
- Materials are not allowed as follows:
  - Where the following factors increase indexes as indicated below:
    - Factors:
      - Age.
      - Moisture.
      - Other atmospheric conditions.
    - Indexes:
      - Where either of the following increase to surpass prescribed limits:
        - Flame spread index.
        - Smoke-developed index.

Note: The following are cited as alternative required tests for flame spread index and smoke-developed:
ASTM E 84, "Test Method for Surface Burning Characteristics of Building Materials."
UL 723, "Test for Surface Burning Characteristics of Building Materials."

719 Thermal- and Sound-Insulating Materials

719.2 Concealed installation

- Reflective plastic core insulation is not governed by this section.

  Note: Section 2613, "Reflective Plastic Core Insulation," is cited as governing all layers of single and multilayer reflective plastic core insulation.

- Concealed cellulose insulation as follows must have a smoke-developed index ≤ 450:
  - Loose-fill insulation.
  - Insulation that is not spray-applied.
  - Complies with applicable third-party standards.

  Note: 719.6, "Cellulose loose-fill insulation," is cited as the source of required standards.

- Other concealed insulating materials must comply with the following indexes:
  - Flame spread index ≤ 25.
  - Smoke-developed index ≤ 450.

719.2.1 Facings

- Reflective plastic core insulation is not governed by this section.

  Note: Section 2613, "Reflective Plastic Core Insulation," is cited as governing all layers of single and multilayer reflective plastic core insulation.

- This section applies to the following facings installed as indicated below:
  - Facings:
    - As follows on insulation installed in concealed spaces governed by this section:
      - Facings and coverings.
      - Layers of reflective foil insulation.
  - Installation:
    - Where installed in the following construction types:
      - Type III, Type IV, and Type V.
    - Where concealed behind and in substantial contact with the following elements:
      - Ceiling finish.
      - Wall finish.
      - Floor finish.
- The following indexes are not limited for facings addressed in this section:
  - Flame spread index.
  - Smoke-developed index.

719.3 Exposed installation

- Exposed cellulose insulation as follows must have a smoke-developed index ≤ 450:
  - Loose-fill insulation.
  - Insulation that is not spray-applied.
  - Complies with applicable third-party standards.

  Note: 719.6, "Cellulose loose-fill insulation," is cited as the source of required standards.

- Other exposed insulating materials must comply with the following indexes:
  - Flame spread index ≤ 25.
  - Smoke-developed index ≤ 450.

719 Thermal- and Sound-Insulating Materials

719.3.1 Attic floors

- Insulation materials exposed on attic floors must have the following property:
  - A critical radiant flux ≥ 0.12 watt/sq cm.

Note: ASTM E 970, "Test Method for Critical Radiant Flux of Exposed Attic Floor Insulation Using a Radiant Heat Energy Source," is cited as the required test method.

719.4 Loose-fill insulation

- This section addresses loose-fill insulation that cannot be mounted in a test apparatus as follows:
  - The apparatus of either specified surface burning test without a special device such as the following:
    A screen.
    Artificial support.

Note: The following are the surface burning tests referenced above:
ASTM E 84, "Test Method for Surface Burning Characteristics of Building Materials."
UL 723, "Test for Surface Burning Characteristics of Building Materials."

- Two categories of loose-fill insulation are governed by this section as follows:
  - Loose-fill insulation other than cellulose must be tested by an alternate surface burning test as follows:
    Flame spread limits set by the IBC must be met.
    Smoke-developed limits as set by the IBC must be met.

Note: CAN/ULC S102.2, "Test for Surface Burning Characteristics of Flooring, Floor Covering and Miscellaneous Materials and Assemblies," is cited as the alternate surface burning test.
The following are cited as listing IBC limits required for compliance in the test identified above:
719.2, "Concealed installation," for concealed applications.
719.3, "Exposed installation," for exposed applications.

  - Cellulose loose-fill insulation need not comply with surface burning tests in this section as follows:
    Where it complies with applicable sections of the IBC.

Note: The following IBC sections are cited as listing requirements that may be met in lieu of being tested by any of the surface burning tests listed in this section:
719.2, "Concealed installation," for concealed applications.
719.3, "Exposed installation," for exposed applications.
719.6, "Cellulose loose-fill insulation," for all applications.

719.5 Roof insulation

- Combustible roof insulation need not meet the following limits where conditions listed below apply:
  - Limits:
    Flame spread index limit ≤ 25.
    Smoke-developed index limit ≤ 450.
  - Conditions:
    Where insulation is covered with an approved roof covering.
    Where the roof covering is applied directly to the insulation.

Note: The following are cited as the sources of requirements which are waived for the insulation addressed in this section, a partial summary of which is provided above:
719.2, "Concealed insulation," which limits flame spread and smoke-developed index.
719.3, "Exposed insulation," which limits flame spread and smoke-developed index.

719 Thermal- and Sound-Insulating Materials

719.6 Cellulose loose-fill insulation

- Cellulose loose-fill insulation must comply with the following:
  - Meet standards of the Consumer Product Safety Commission.
  - Have packaging labeled as per the standards.

Note: The following Consumer Product Safety Commission standards are cited as governing cellulose loose-fill insulation:
CPSC 16 CFR, 1209, "Interim Safety Standard for Cellulose Insulation."
CPSC 16 CFR, 1404, "Cellulose Insulation."

719.7 Insulation and covering on pipe and tubing

- The following insulation and coverings on pipe and tubing are not governed by this section:
  - That installed in plenums.

Note: The International Mechanical Code *is cited as governing this installation.*

- Other insulation and coverings on pipe and tubing must comply with the following standards:
  - Flame spread index must be ≤ 25.
  - Smoke-developed index must be ≤ 450.

720 Prescriptive Fire Resistance

720.1 General

- This section provides details with assigned fire-resistance ratings as shown on the following pages.

Note: The following are cited as sources of details with fire-resistance ratings:
IBC Table 720.1(1), "Minimum Protection of Structural Parts Based on Time Periods for Various Noncombustible Insulating Materials."
IBC Table 720.1(2), "Rated Fire-Resistance Periods for Various Walls and Partitions."
IBC Table 720.1(3), "Minimum Protection for Floor and Roof Systems."

- Where changes are made to the details provided herein, the following applies:
  - Where changes affect the heat dissipation potential of the detail:
    Documentation must be made available as follows:
      Type:
        Fire tests.
        Other data.
      Content:
        Verifying that the fire-resistance period of the detail is not reduced.
      Availability:
        Must be made available to the building official.

720.1.1 Thickness of protective coverings

- Thickness of fire-resistant materials protecting structure must be one of the following:
  - As indicated in details provided on subsequent pages as follows:
    Detail drawings are in the format of typical construction drawing details.
    Detail drawings do not include typical specification information such as the following:
      Test requirements as listed in the IBC descriptions or footnotes.
      Fastening requirements as listed in the IBC descriptions.
  - As otherwise indicated in this section.

Note: IBC Table 720.1(1), "Minimum Protection of Structural Parts Based on Time Periods for Various Noncombustible Insulating Materials," is cited as the source of details.

- Protective covering thickness indicated in this section is defined as follows:
  - Net thickness of protecting materials.
  - Thickness does not include air space behind the protecting material.

720.1.2 Unit masonry protection

- Where required, metal ties must be installed as follows:
  - In bed joints of masonry protecting steel columns.
  - Ties must be one of the following:
    As shown in details provided in this section.
    Equivalent to that shown in details of this section.

Note: IBC Table 720.1(1), "Minimum Protection of Structural Parts Based on Time Periods for Various Noncombustible Insulating Materials," is cited as the source of details.

720 Prescriptive Fire Resistance

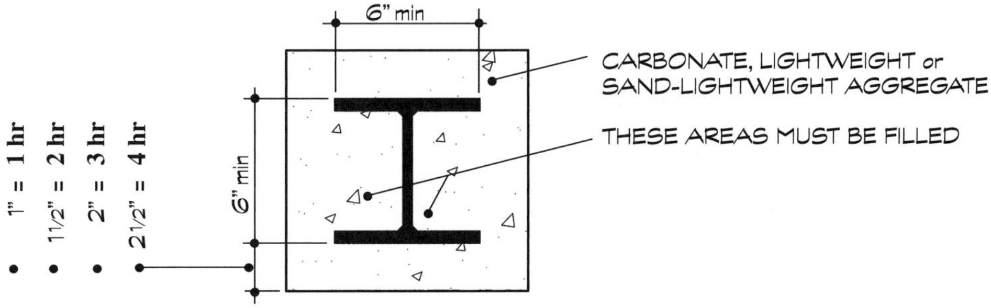

1-1.1

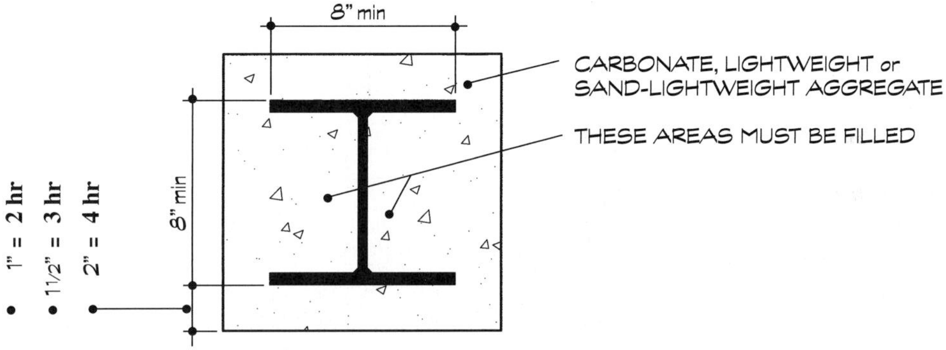

1-1.2

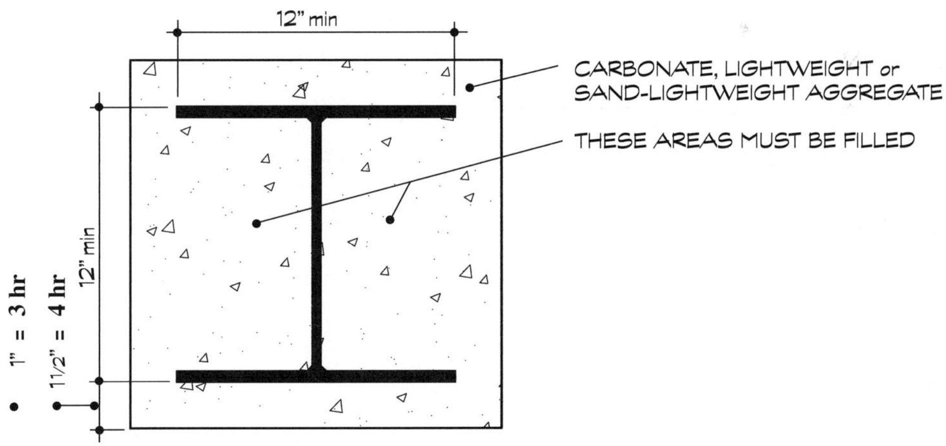

1-1.3

Fig. 720.1(1) 1A. Minimum protection of steel columns and all members of primary trusses. Minimum thicknesses of noncombustible insulating materials are indicated as required for the fire-resistance ratings shown. Such thicknesses are the same on all sides where insulating materials occur. *[Source: IBC Table 720.1(1).]*

720 Prescriptive Fire Resistance

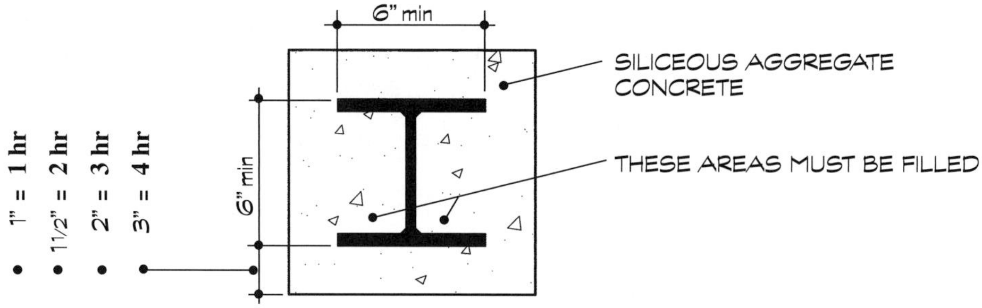

1-1.4

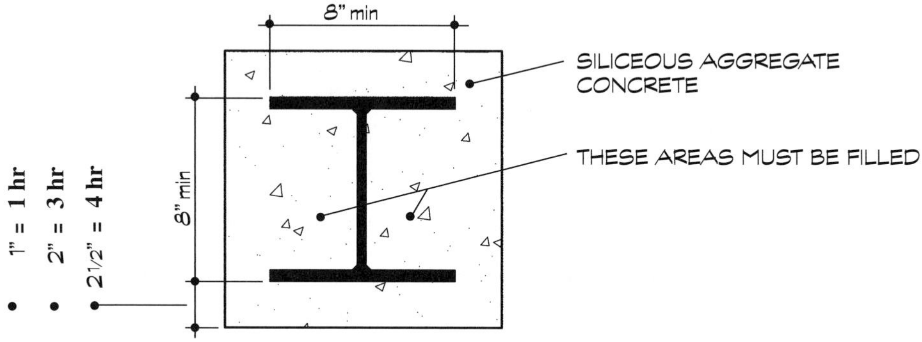

1-1.5

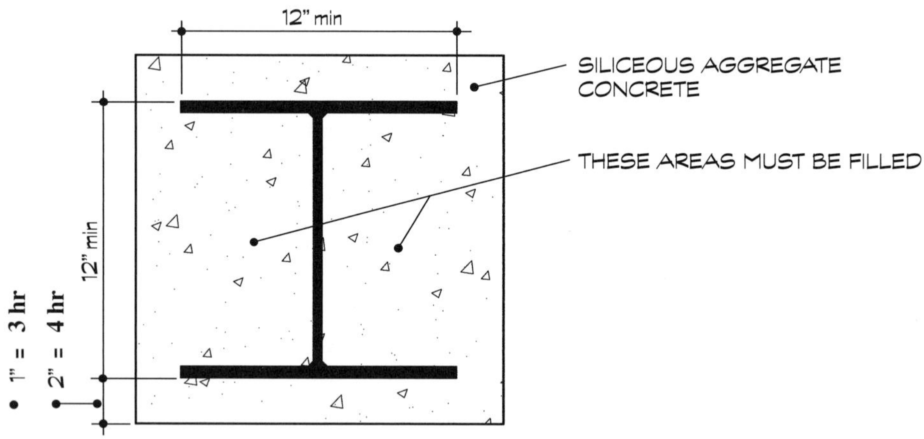

1-1.6

Fig. 720.1(1) 1B. Minimum protection of steel columns and all members of primary trusses. Minimum thicknesses of noncombustible insulating materials are indicated as required for the fire-resistance ratings shown. Such thicknesses are the same on all sides where insulating materials occur. *[Source: IBC Table 720.1(1).]*

720 Prescriptive Fire Resistance

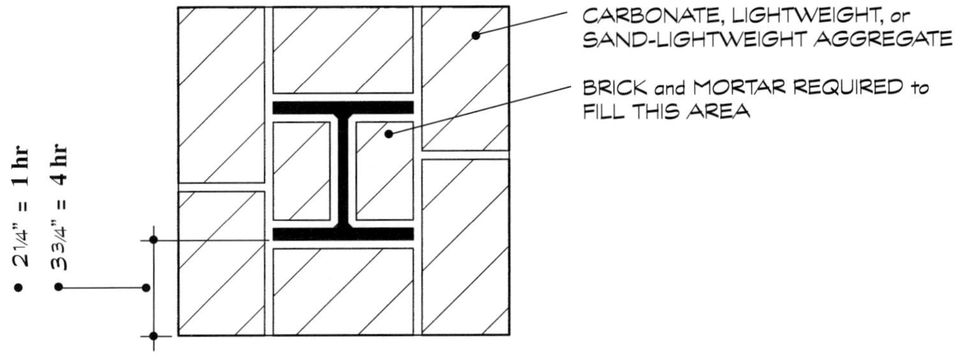

1-2.1

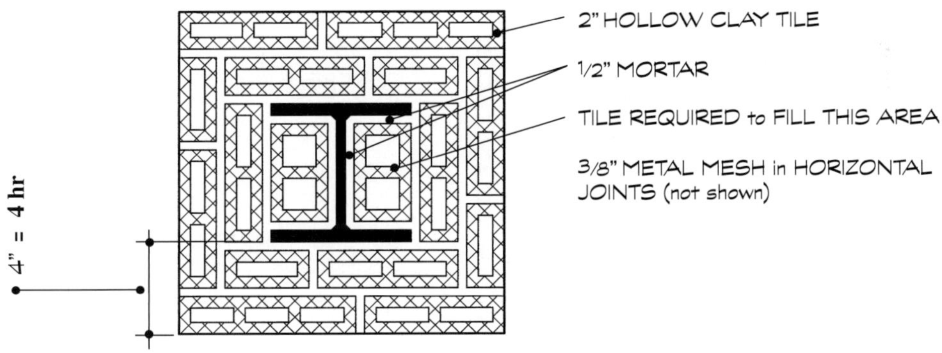

1-3.1

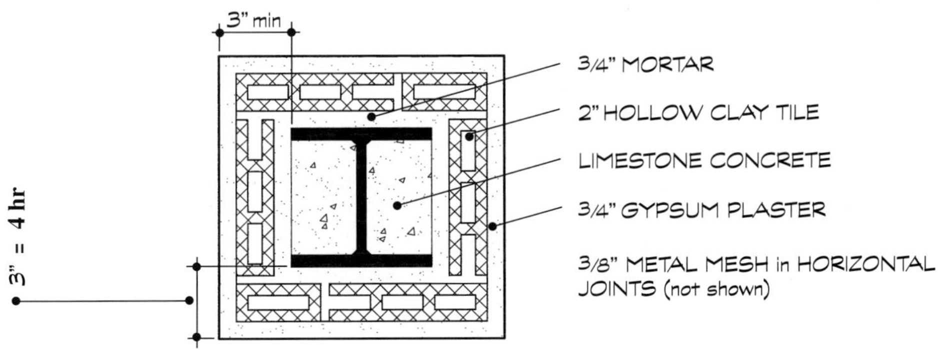

1-3.2

Fig. 720.1(1) 1C. Minimum protection of steel columns and all members of primary trusses. Minimum thicknesses of noncombustible insulating materials are indicated as required for the fire-resistance ratings shown. Such thicknesses are the same on all sides where insulating materials occur. *[Source: IBC Table 720.1(1).]*

720 Prescriptive Fire Resistance

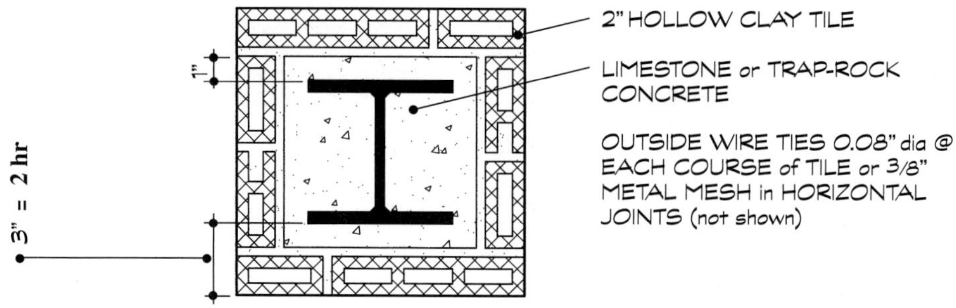

1-3.3

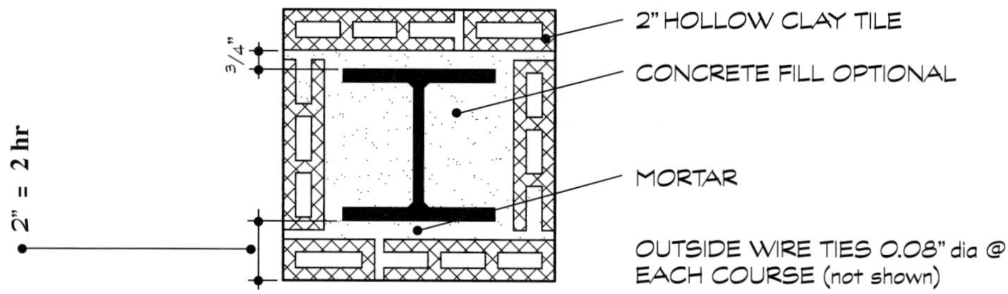

1-3.4

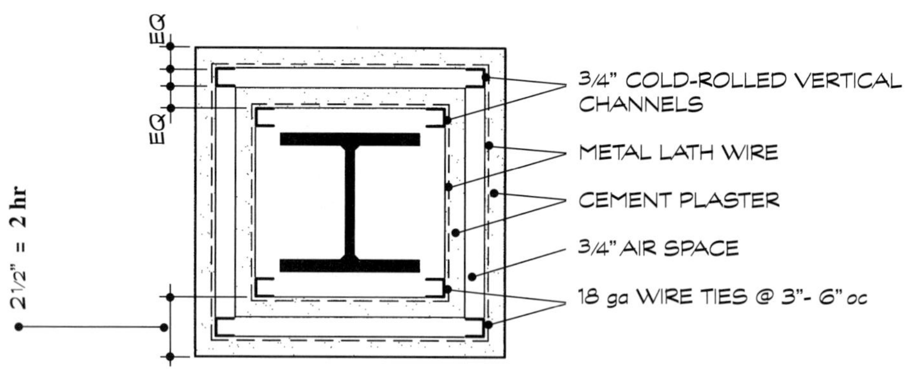

1-4.1a

Fig. 720.1(1) 1D. Minimum protection of steel columns and all members of primary trusses. Minimum thicknesses of noncombustible insulating materials are indicated as required for the fire-resistance ratings shown. Such thicknesses are the same on all sides where insulating materials occur. *[Source: IBC Table 720.1(1).]*

720 Prescriptive Fire Resistance

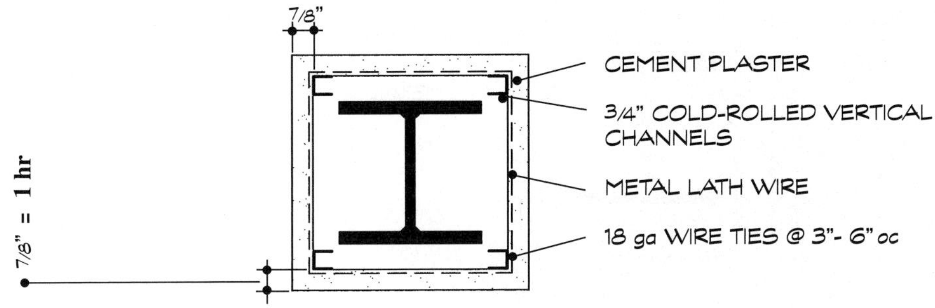

1-4.1b

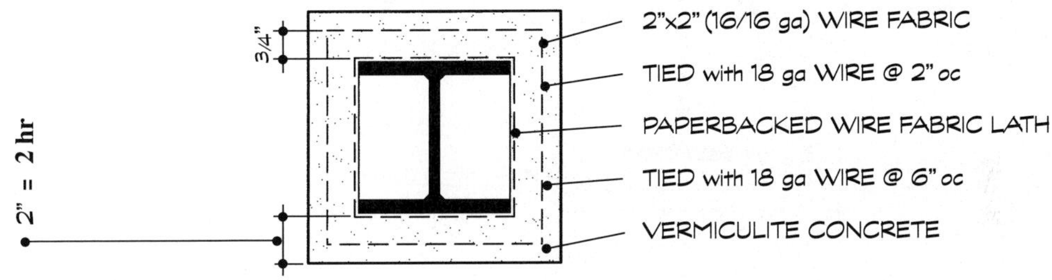

1-5.1

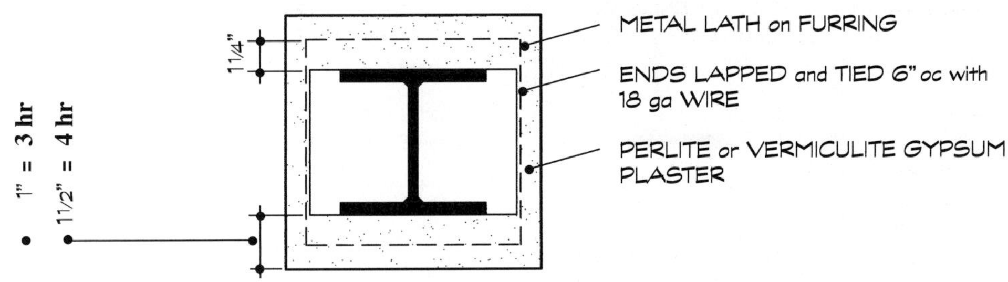

1-6.1

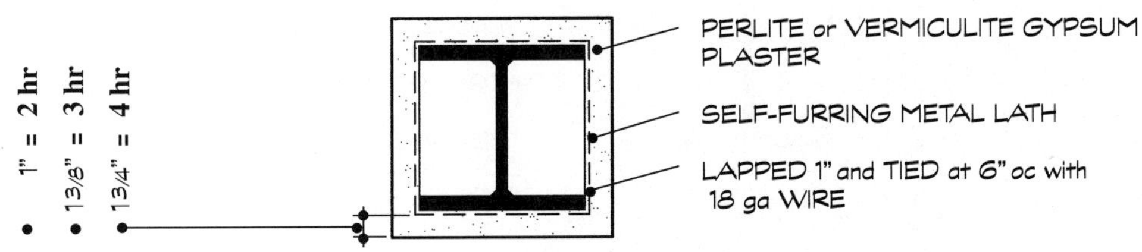

1-6.2

Fig. 720.1(1) 1E. Minimum protection of steel columns and all members of primary trusses. Minimum thicknesses of noncombustible insulating materials are indicated as required for the fire-resistance ratings shown. Such thicknesses are the same on all sides where insulating materials occur. *[Source: IBC Table 720.1(1).]*

720 Prescriptive Fire Resistance

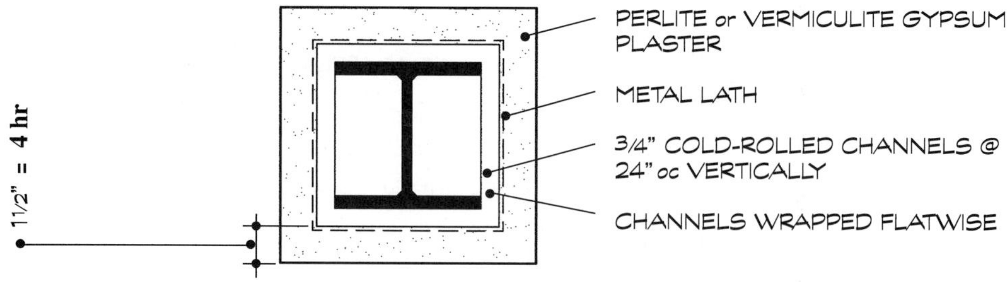

1-6.3

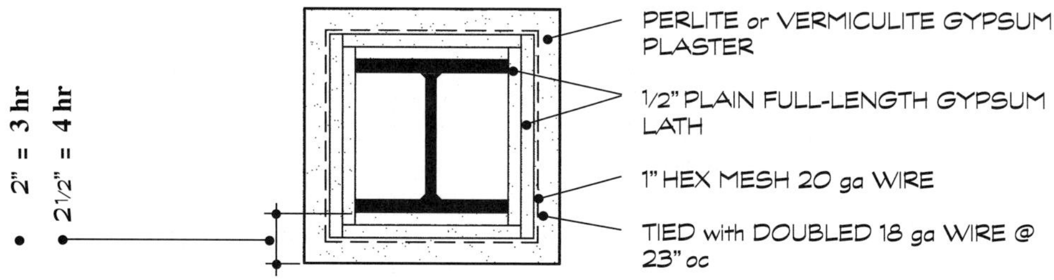

1-6.4

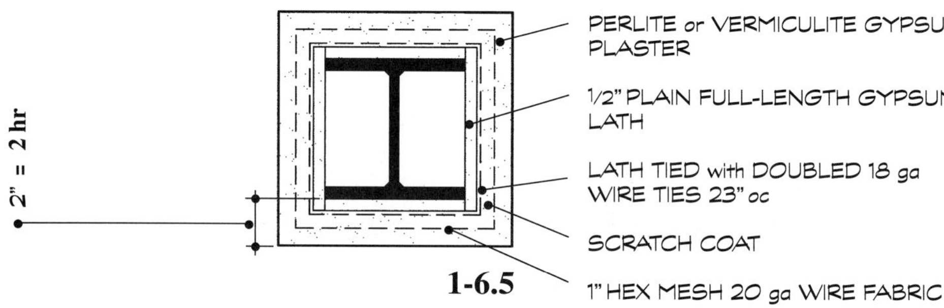

1-6.5

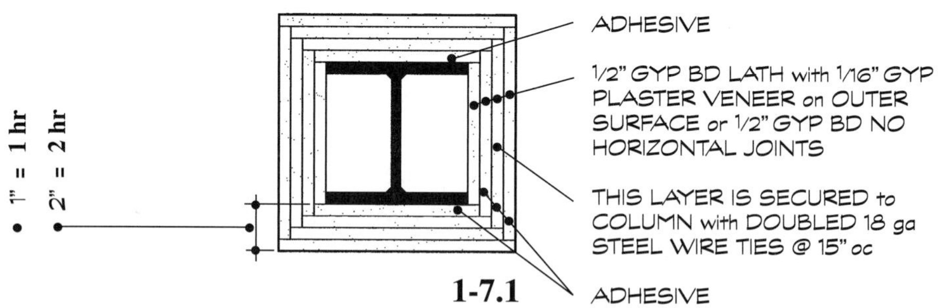

1-7.1

Fig. 720.1(1) 1F. Minimum protection of steel columns and all members of primary trusses. Minimum thicknesses of noncombustible insulating materials are indicated as required for the fire-resistance ratings shown. Such thicknesses are the same on all sides where insulating materials occur. *[Source: IBC Table 720.1(1).]*

720 Prescriptive Fire Resistance

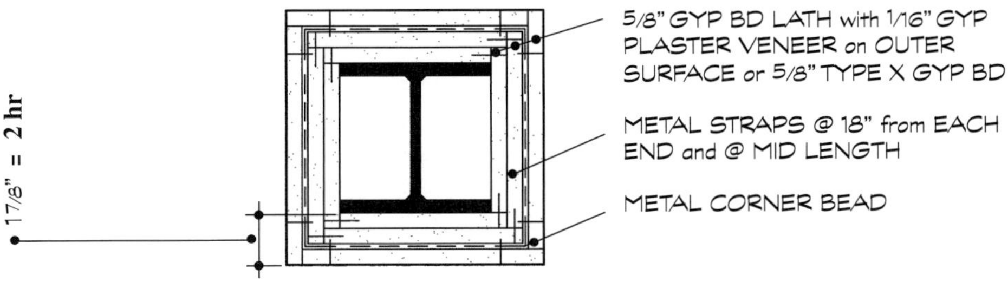

1-7.2

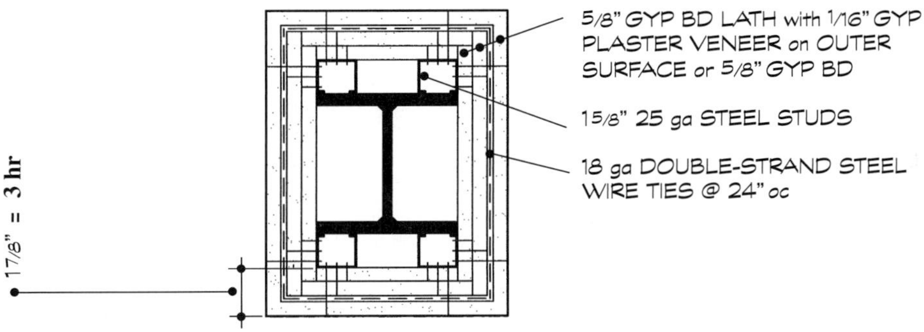

1-7.3

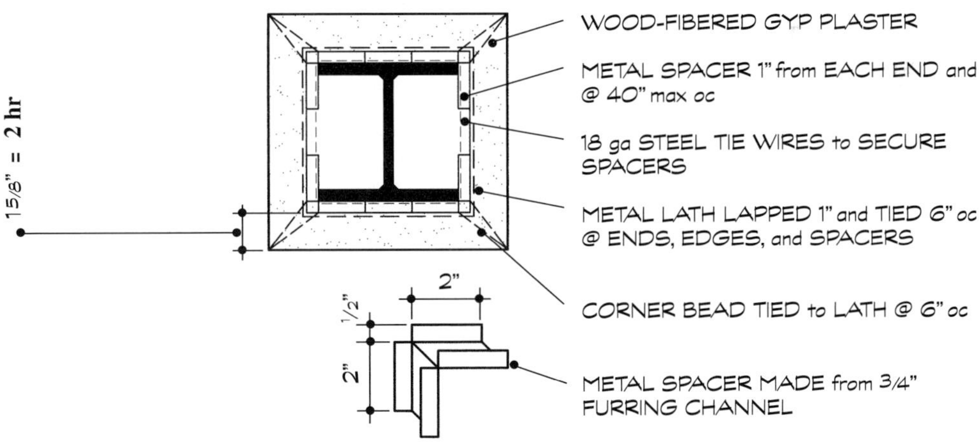

1-8.1

Fig. 720.1(1) 1G. Minimum protection of steel columns and all members of primary trusses. Minimum thicknesses of noncombustible insulating materials are indicated as required for the fire-resistance ratings shown. Such thicknesses are the same on all sides where insulating materials occur. *[Source: IBC Table 720.1(1).]*

720 Prescriptive Fire Resistance

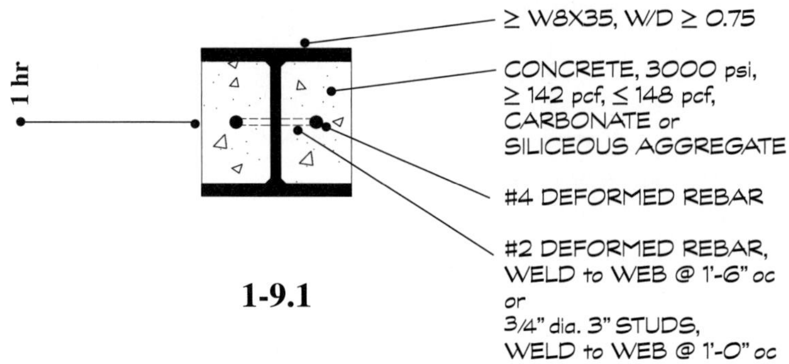

1-9.1

Fig. 720.1(1) 1H. Minimum protection of steel columns and all members of primary trusses. Minimum thicknesses of noncombustible insulating materials are indicated as required for the fire-resistance ratings shown. Such thicknesses are the same on all sides where insulating materials occur. *[Source: IBC Table 720.1(1).]*

720 Prescriptive Fire Resistance

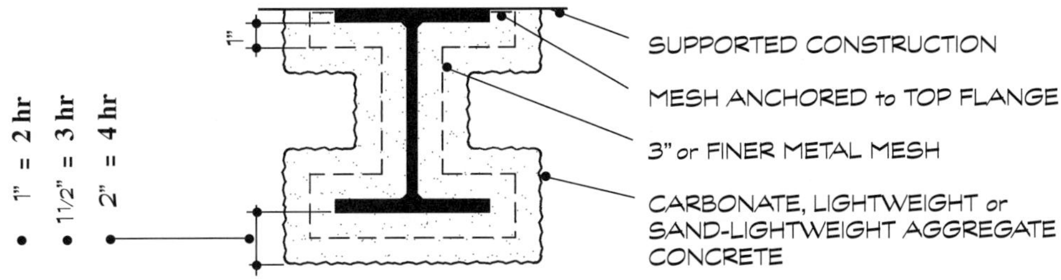

2-1.1

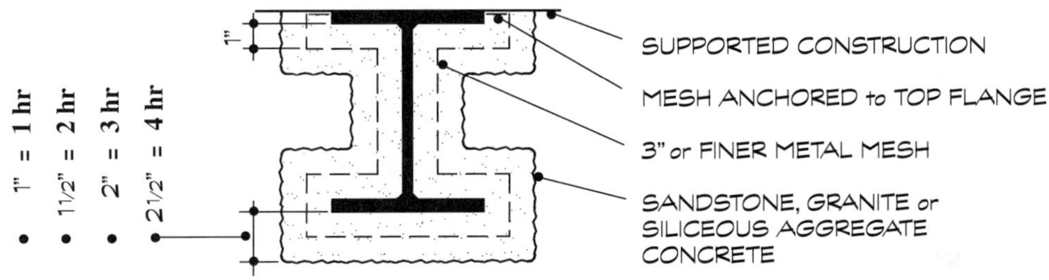

2-1.2

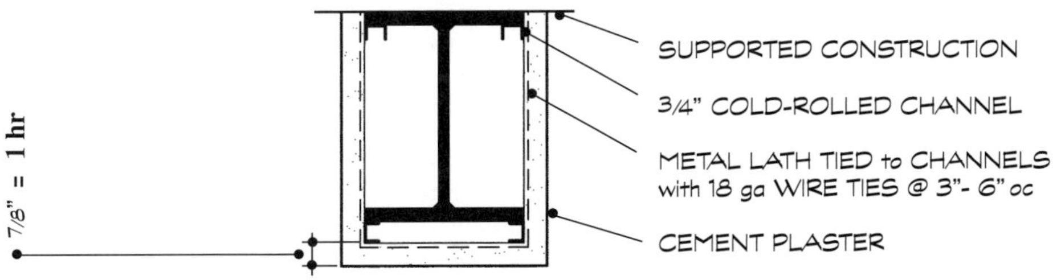

2-2.1a

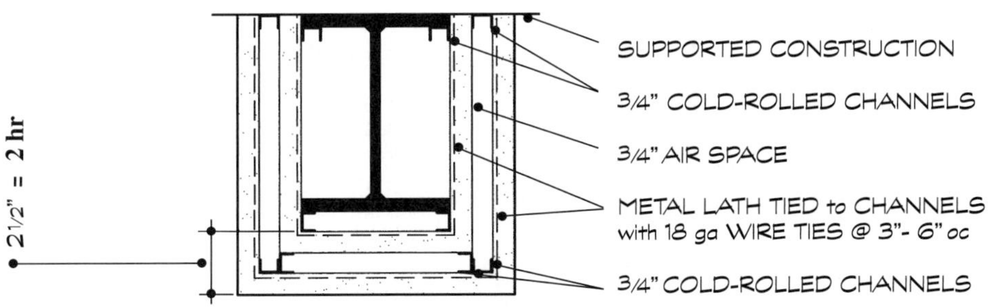

2-2.1b

Fig. 720.1(1) 2A. Minimum protection of webs and flanges of steel beams and girders. Minimum thicknesses of noncombustible insulating materials are indicated as required for the fire-resistance ratings shown. Such thicknesses are the same on all sides where insulating materials occur. *[Source: IBC Table 720.1(1).]*

720 Prescriptive Fire Resistance

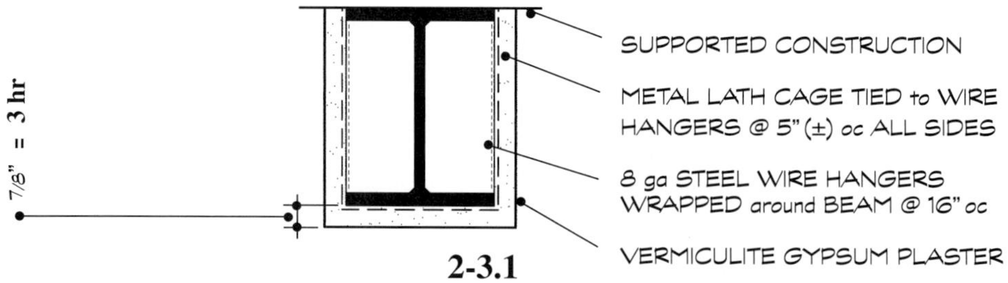

2-3.1

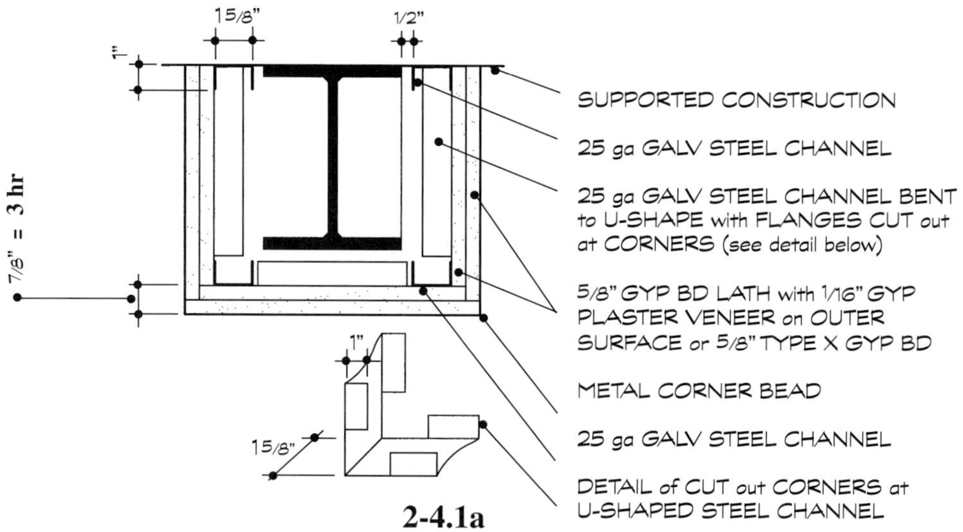

2-4.1a

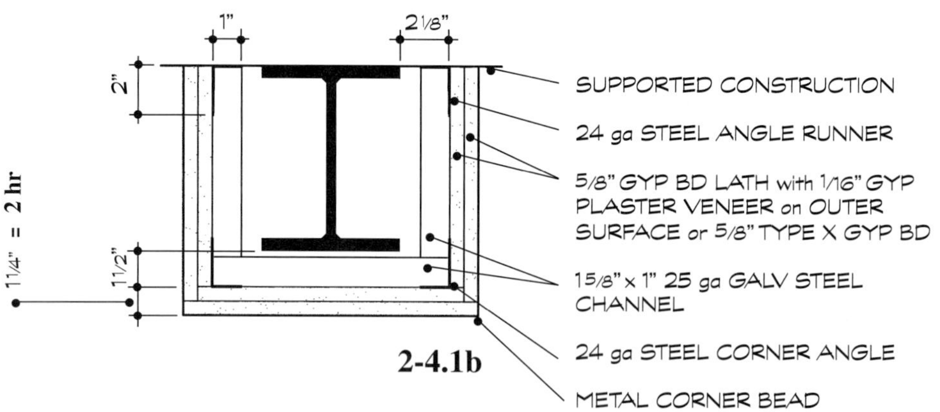

2-4.1b

Fig. 720.1(1) 2B. Minimum protection of webs and flanges of steel beams and girders. Minimum thicknesses of noncombustible insulating materials are indicated as required for the fire-resistance ratings shown. Such thicknesses are the same on all sides where insulating materials occur. *[Source: IBC Table 720.1(1).]*

720 Prescriptive Fire Resistance

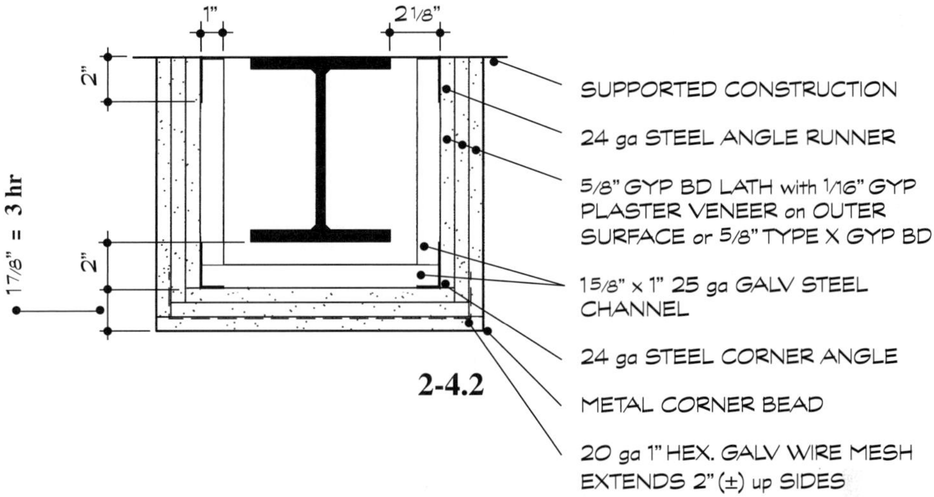

Fig. 720.1(1) 2C. Minimum protection of webs and flanges of steel beams and girders. Minimum thicknesses of noncombustible insulating materials are indicated as required for the fire-resistance ratings shown. Such thicknesses are the same on all sides where insulating materials occur. *[Source: IBC Table 720.1(1).]*

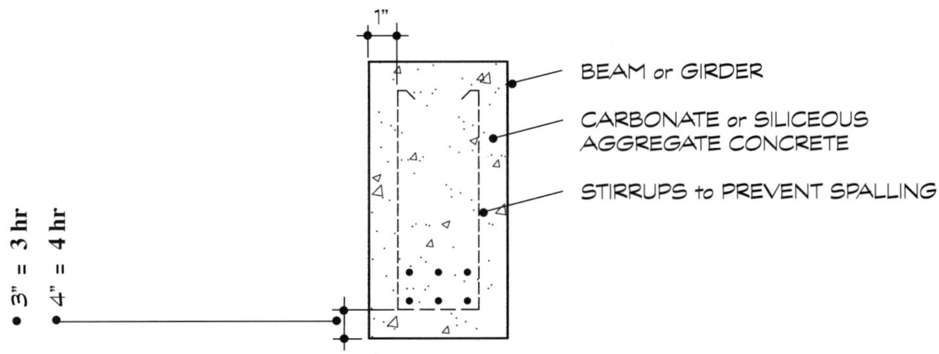

3-1.1a

Fig. 720.1(1) 3A. Minimum protection of bonded pretensioned reinforcement in prestressed concrete. Minimum thicknesses of concrete cover are indicated as required for the fire-resistance ratings shown. Such minimum thicknesses are the same on all sides. *[Source: IBC Table 720.1(1).]*

720 Prescriptive Fire Resistance

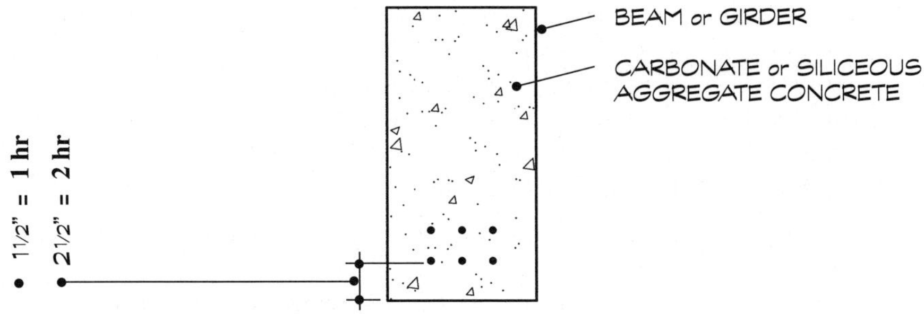

3-1.1b

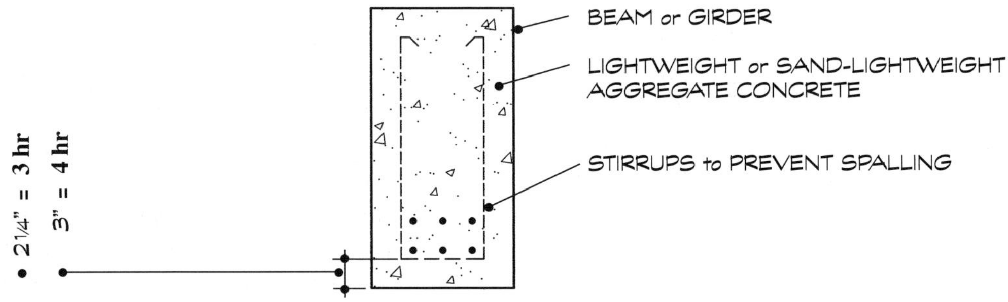

3-1.1c

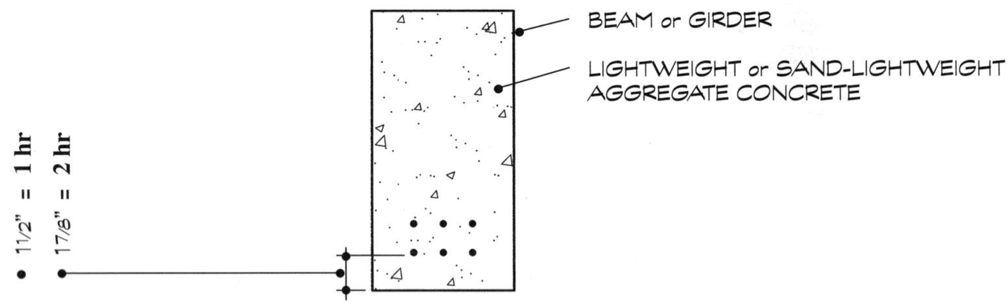

3-1.1d

Fig. 720.1(1) 3B. Minimum protection of bonded pretensioned reinforcement in prestressed concrete. Minimum thicknesses of concrete cover are indicated as required for the fire-resistance ratings shown. Such minimum thicknesses are the same on all sides. *[Source: IBC Table 720.1(1).]*

720 Prescriptive Fire Resistance

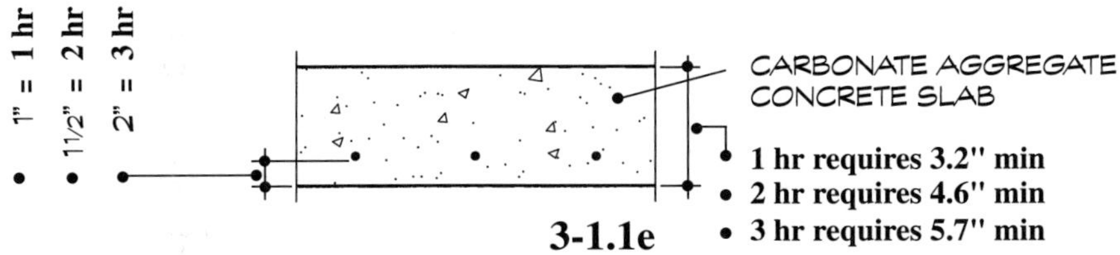

3-1.1e

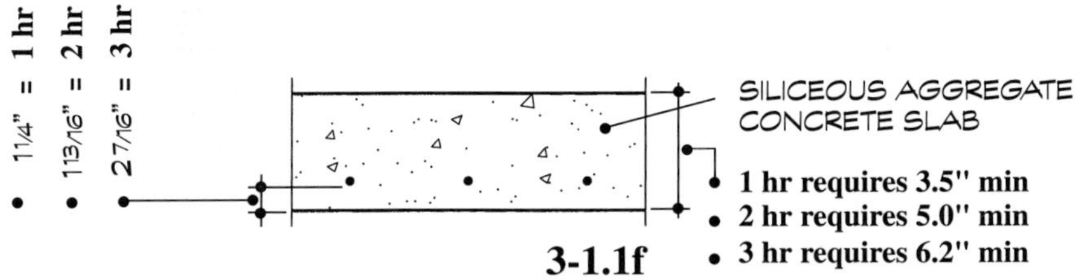

3-1.1f

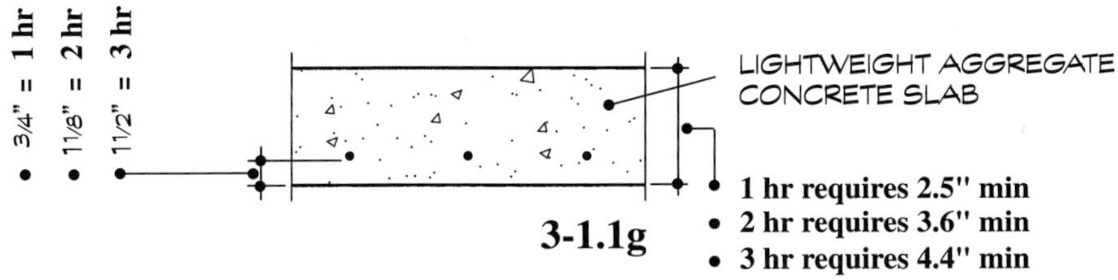

3-1.1g

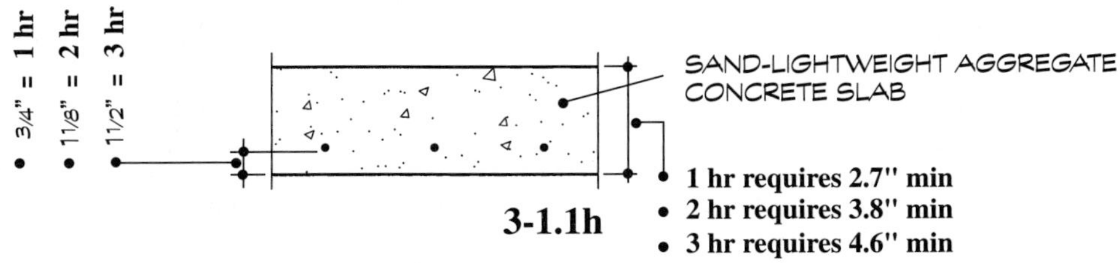

3-1.1h

Fig. 720.1(1) 3C. Minimum protection of bonded pretensioned reinforcement in prestressed concrete. Minimum thicknesses of concrete cover are indicated as required for the fire-resistance ratings shown. Such minimum thicknesses are the same on all sides. *[Source: IBC Table 720.1(1).]*

720 Prescriptive Fire Resistance

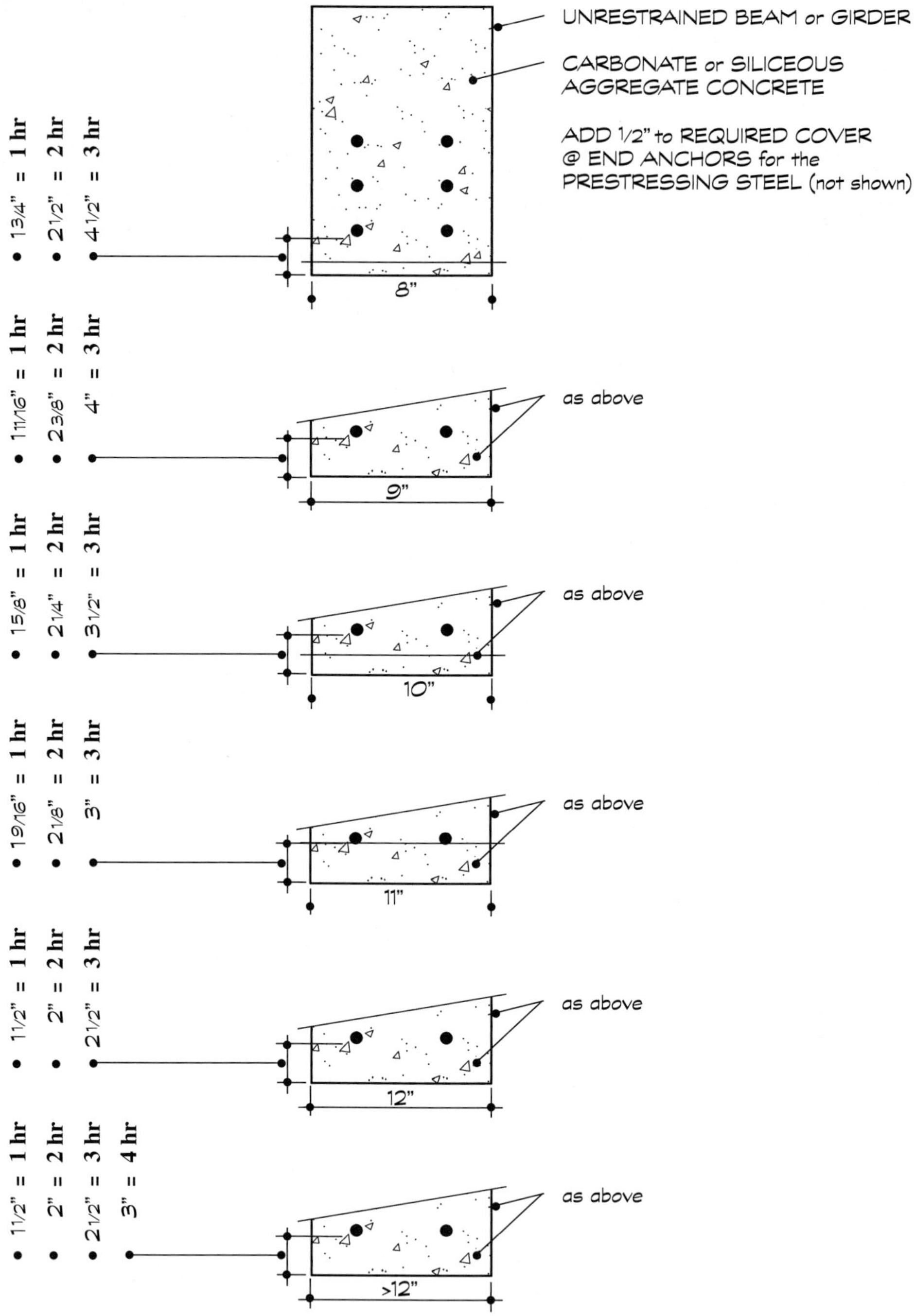

4-1.1a

Fig. 720.1(1) 4A. Minimum protection of bonded or unbonded post-tensioned tendons in prestressed concrete. Minimum thicknesses of concrete cover are indicated as required for the fire-resistance ratings shown. Such minimum thicknesses are the same on all sides. *[Source: IBC Table 720.1(1).]*

720 Prescriptive Fire Resistance

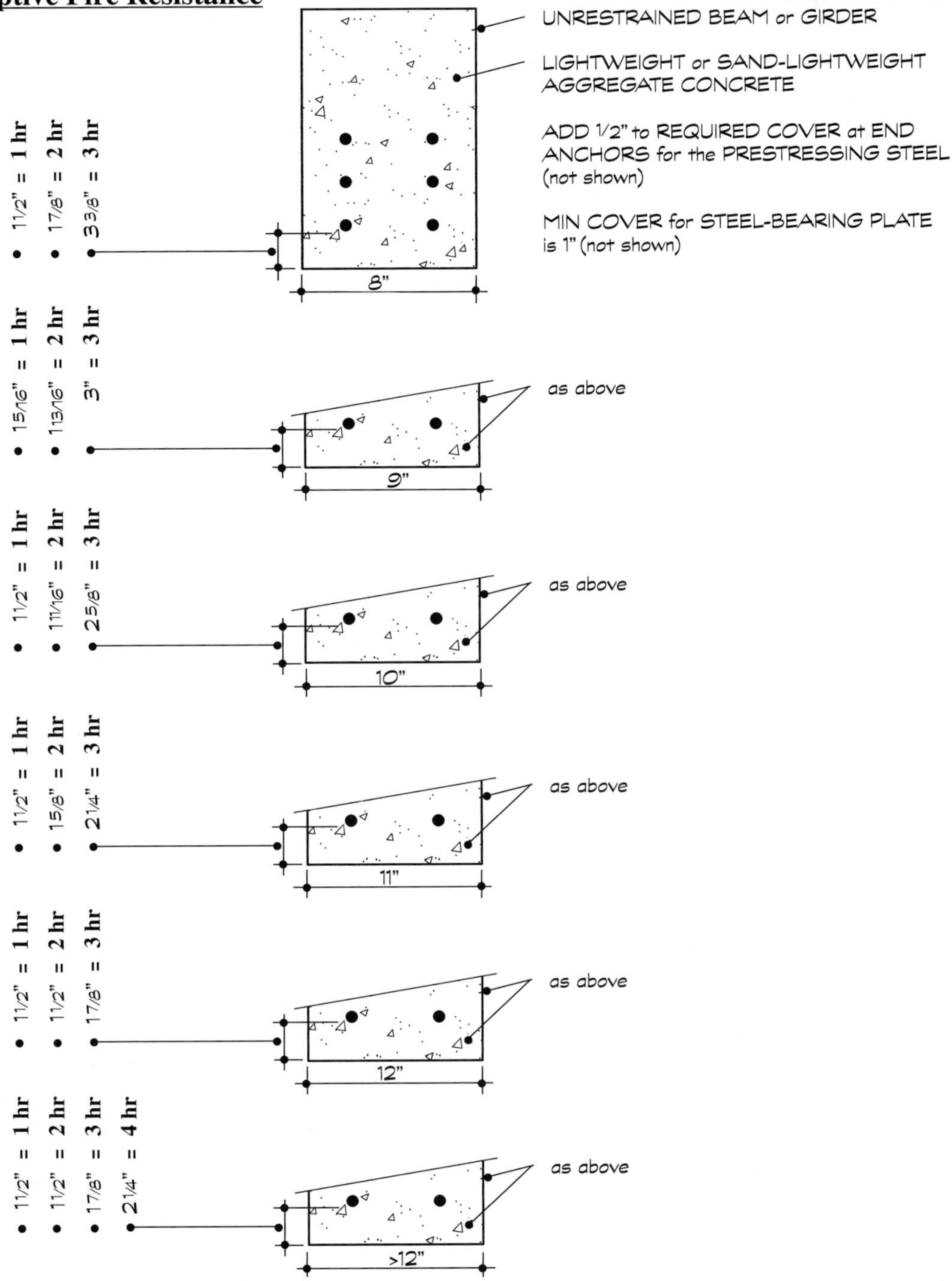

4-1.1b

Fig. 720.1(1) 4B. Minimum protection of bonded or unbonded post-tensioned tendons in prestressed concrete. Minimum thicknesses of concrete cover are indicated as required for the fire-resistance ratings shown. Such minimum thicknesses are the same on all sides. *[Source: IBC Table 720.1(1).]*

720 Prescriptive Fire Resistance

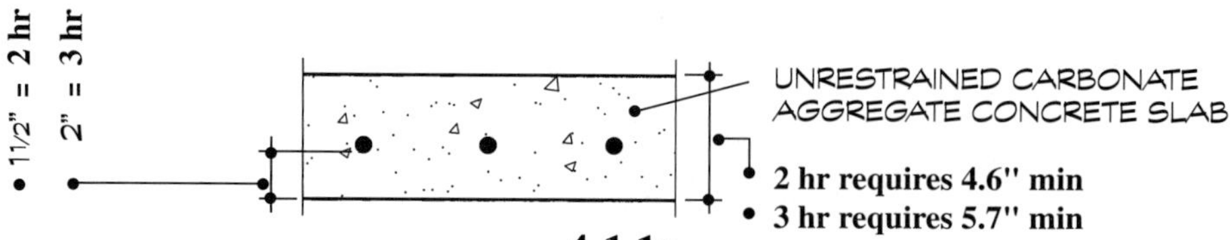

4-1.1c

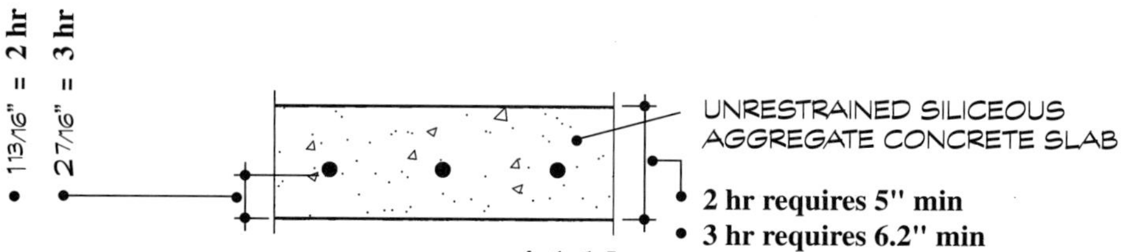

4-1.1d

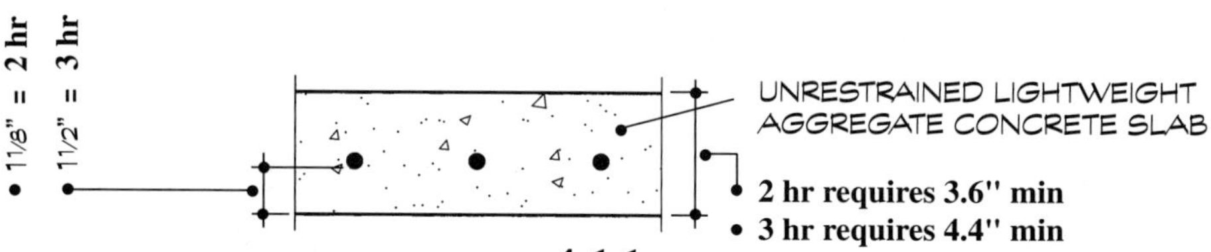

4-1.1e

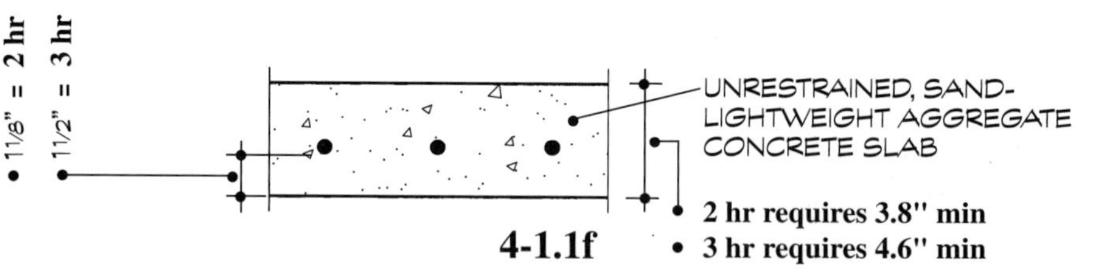

4-1.1f

Fig. 720.1(1) 4C. Minimum protection of bonded or unbonded post-tensioned tendons in prestressed concrete. Minimum thicknesses of concrete cover are indicated as required for the fire-resistance ratings shown. Such minimum thicknesses are the same on all sides. *[Source: IBC Table 720.1(1).]*

720 Prescriptive Fire Resistance

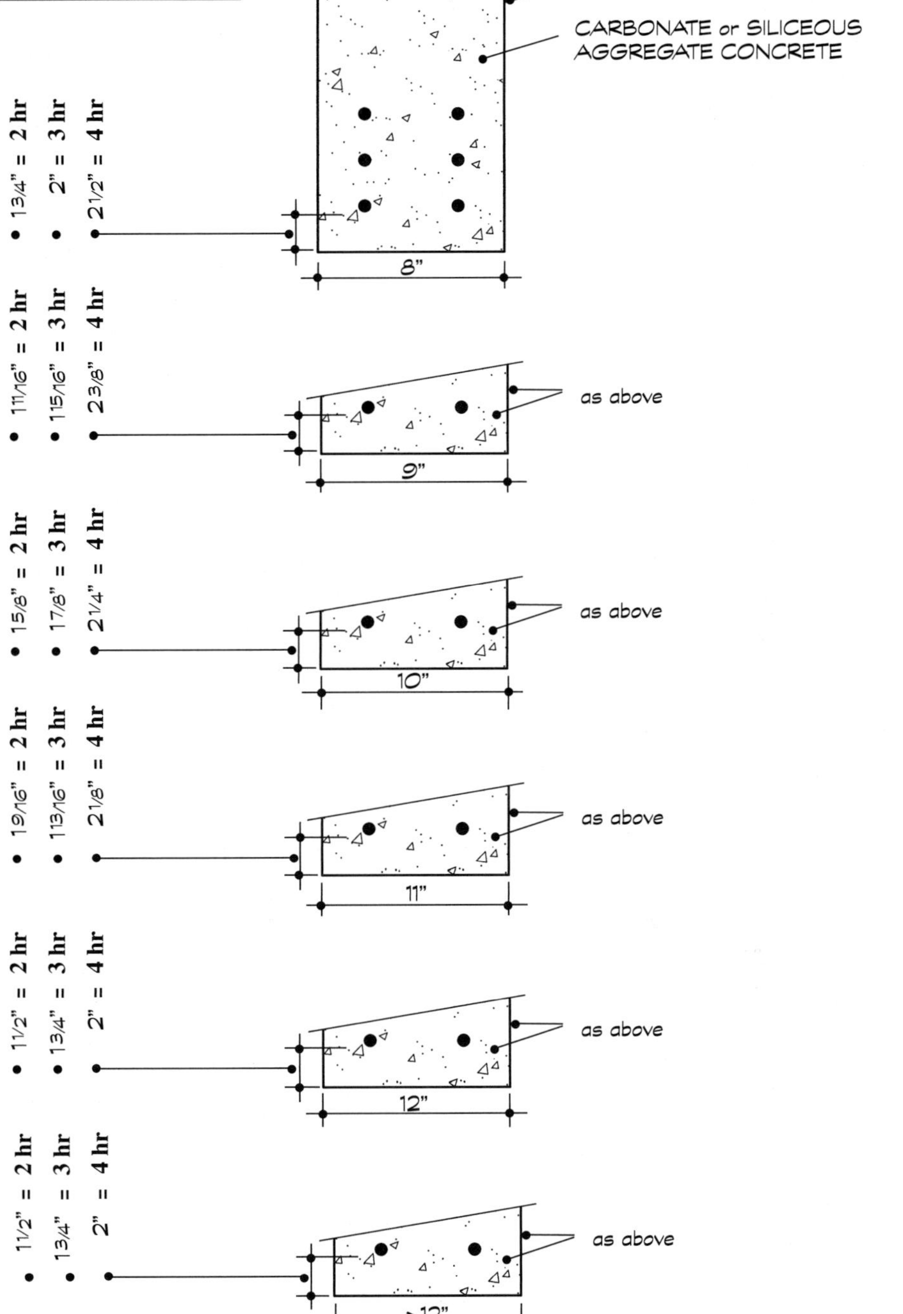

4-1.2a

Fig. 720.1(1) 4D. Minimum protection of bonded or unbonded post-tensioned tendons in prestressed concrete. Minimum thicknesses of concrete cover are indicated as required for the fire-resistance ratings shown. Such minimum thicknesses are the same on all sides. *[Source: IBC Table 720.1(1).]*

720 Prescriptive Fire Resistance

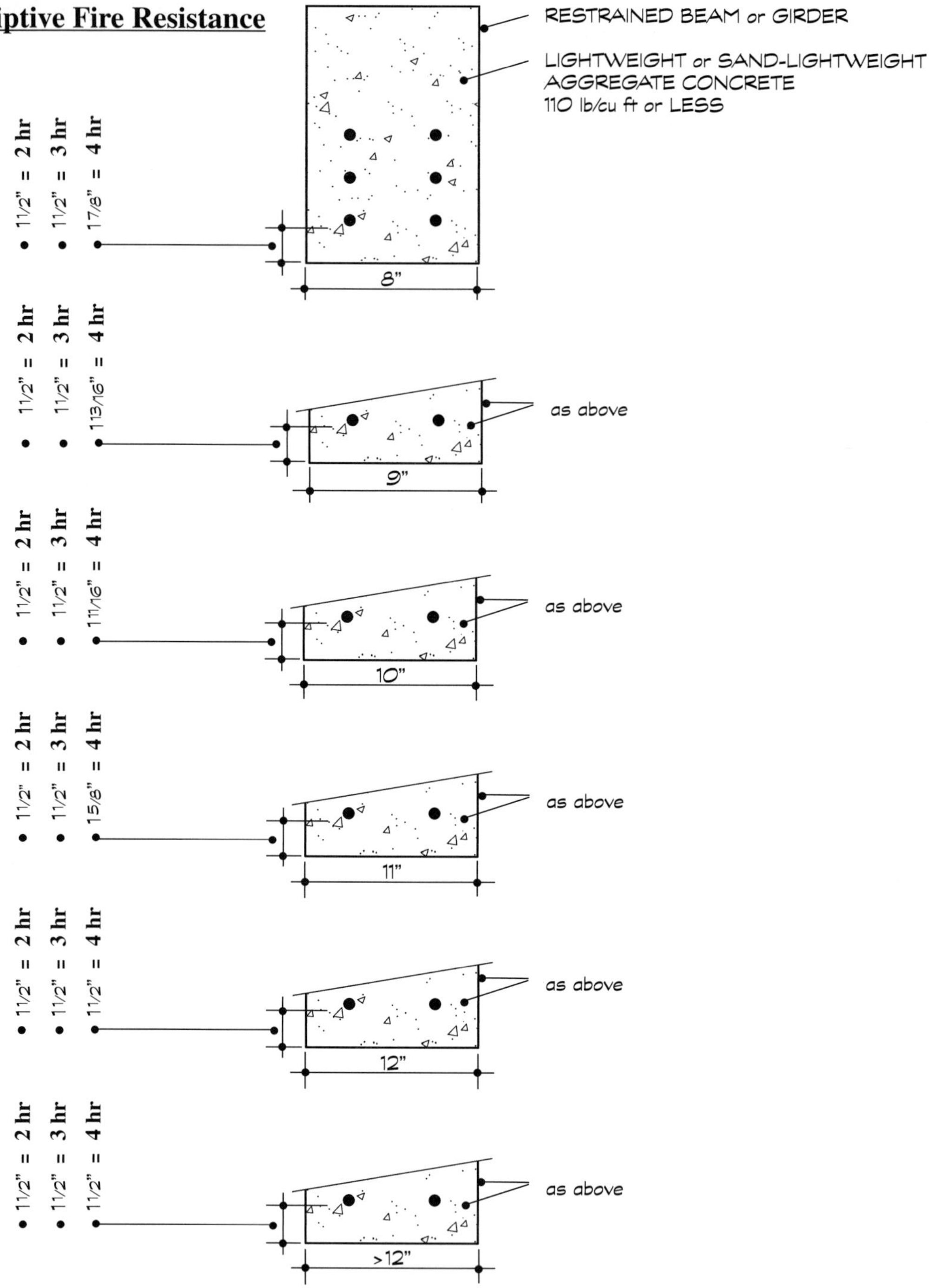

4-1.2b

Fig. 720.1(1) 4E. Minimum protection of bonded or unbonded post-tensioned tendons in prestressed concrete. Minimum thicknesses of concrete cover are indicated as required for the fire-resistance ratings shown. Such minimum thicknesses are the same on all sides. *[Source: IBC Table 720.1(1).]*

720 Prescriptive Fire Resistance

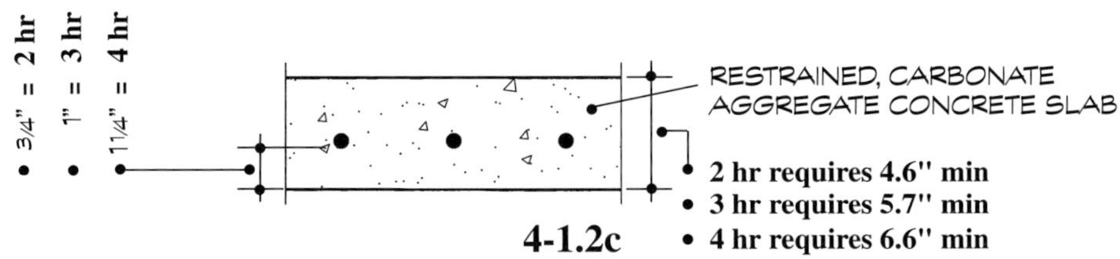

4-1.2c

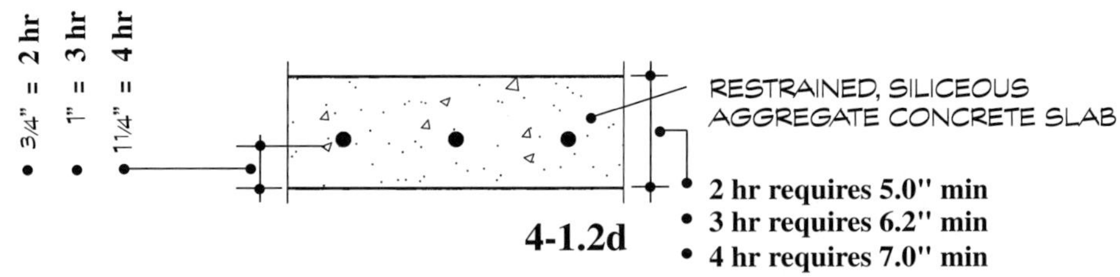

4-1.2d

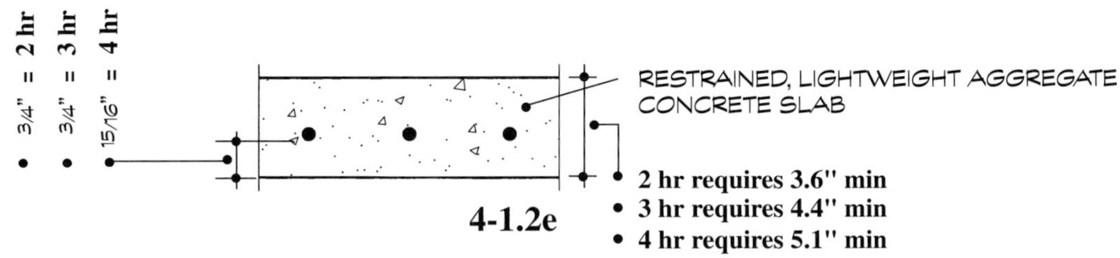

4-1.2e

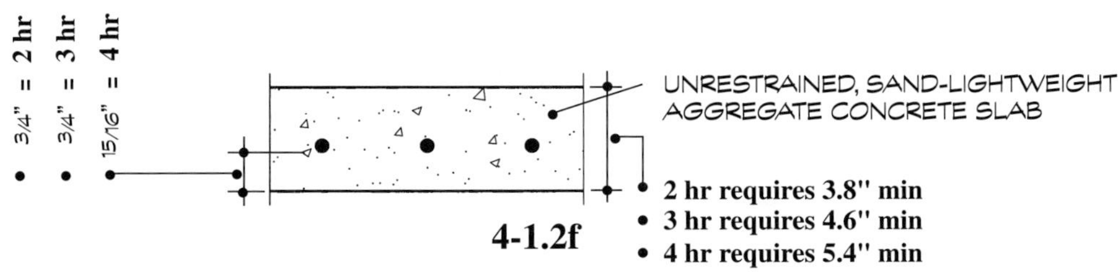

4-1.2f

Fig. 720.1(1) 4F. Minimum protection of bonded or unbonded post-tensioned tendons in prestressed concrete. Minimum thicknesses of concrete cover are indicated as required for the fire-resistance ratings shown. Such minimum thicknesses are the same on all sides. *[Source: IBC Table 720.1(1).]*

720 Prescriptive Fire Resistance

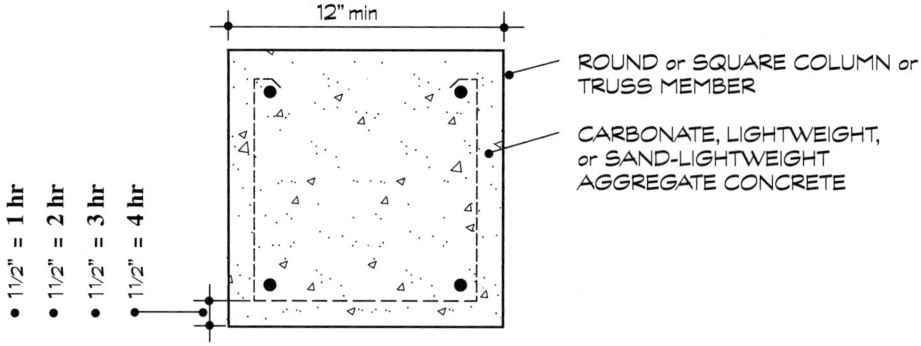

5-1.1a

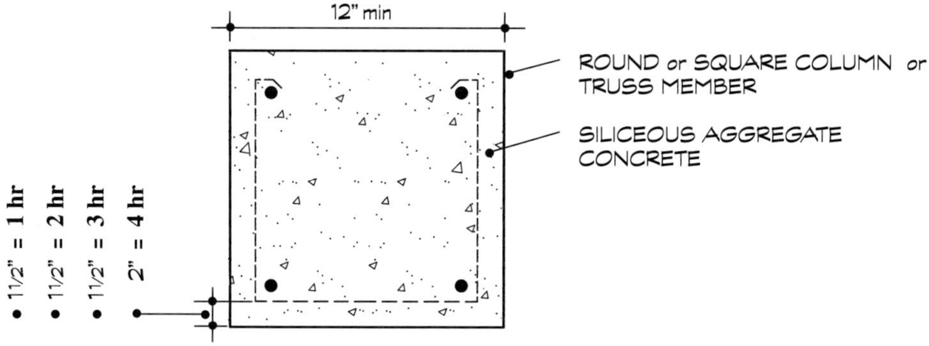

5-1.1b

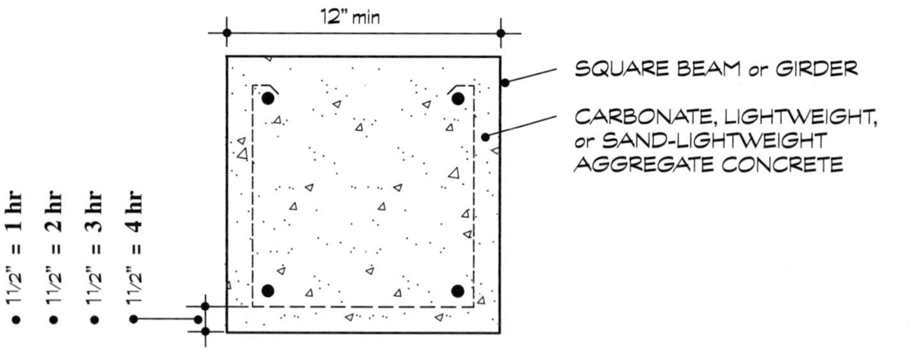

5-1.1c

Fig. 720.1(1) 5A. Minimum protection of reinforcing steel in reinforced concrete columns, beams, girders, and trusses. Minimum thicknesses of concrete cover are indicated as required for the fire-resistance ratings shown. Such minimum thicknesses are the same on all sides. *[Source: IBC Table 720.1(1).]*

720 Prescriptive Fire Resistance

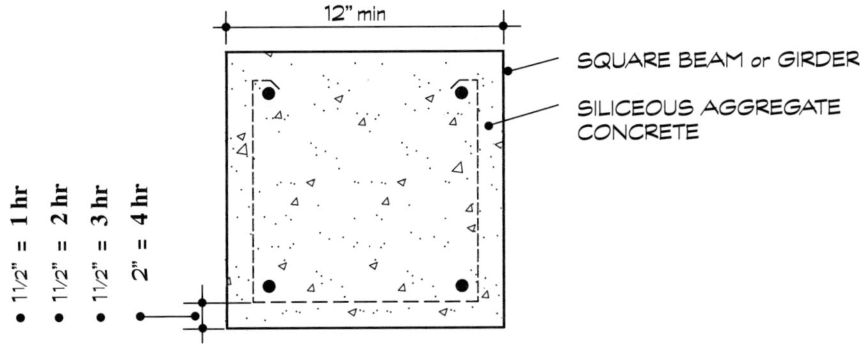

5-1.1d

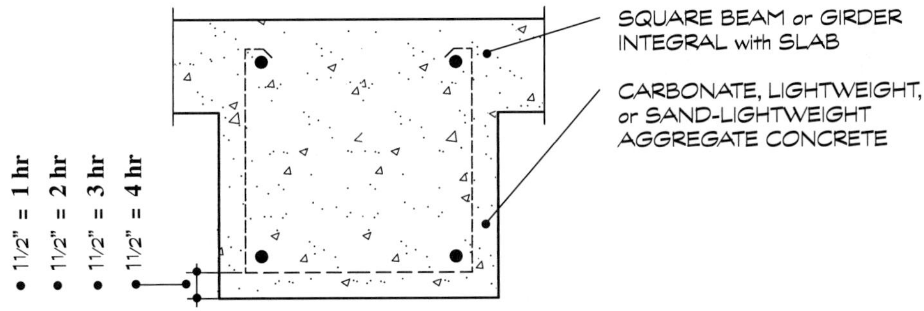

5-1.1e

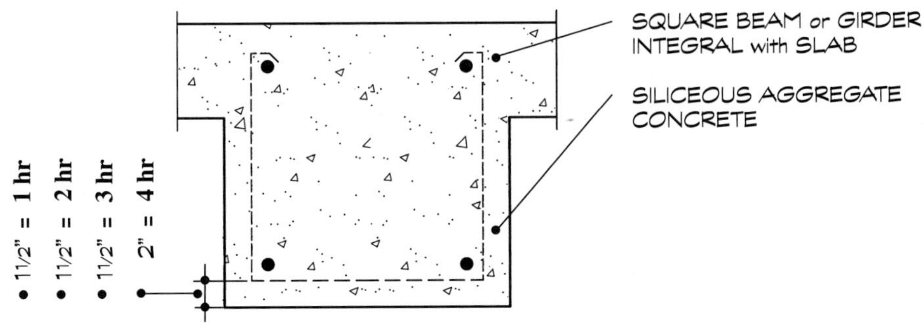

5-1.1f

Fig. 720.1(1) 5B. Minimum protection of reinforcing steel in reinforced concrete columns, beams, girders, and trusses. Minimum thicknesses of concrete cover are indicated as required for the fire-resistance ratings shown. Such minimum thicknesses are the same on all sides. *[Source: IBC Table 720.1(1).]*

720 Prescriptive Fire Resistance

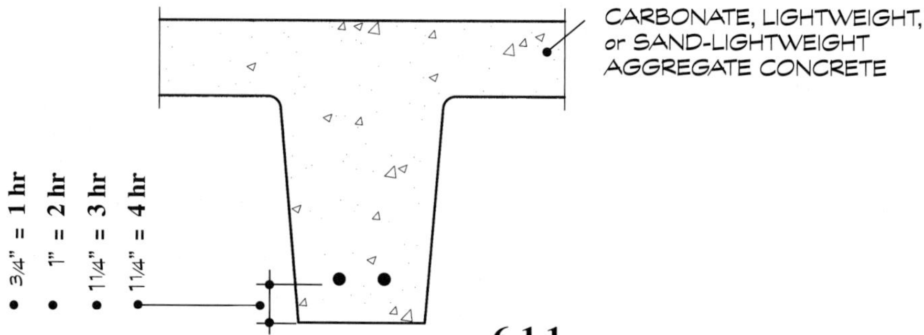

6-1.1

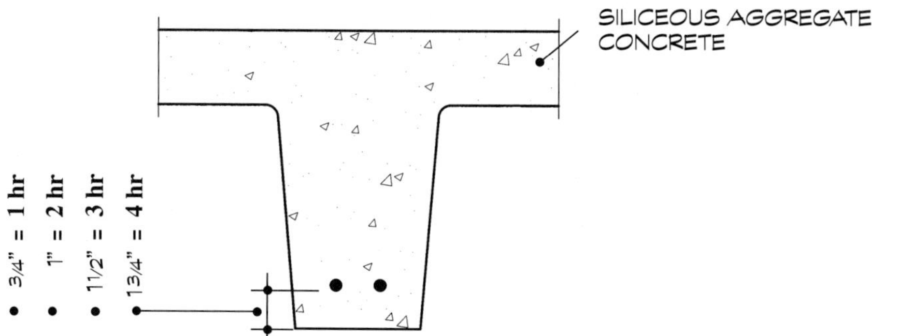

6-1.2

Fig. 720.1(1) 6A. Minimum protection of reinforcing steel in reinforced concrete joists. Minimum thicknesses of concrete cover are indicated as required for the fire-resistance ratings shown. Such minimum thicknesses are the same on all sides. *[Source: IBC Table 720.1(1).]*

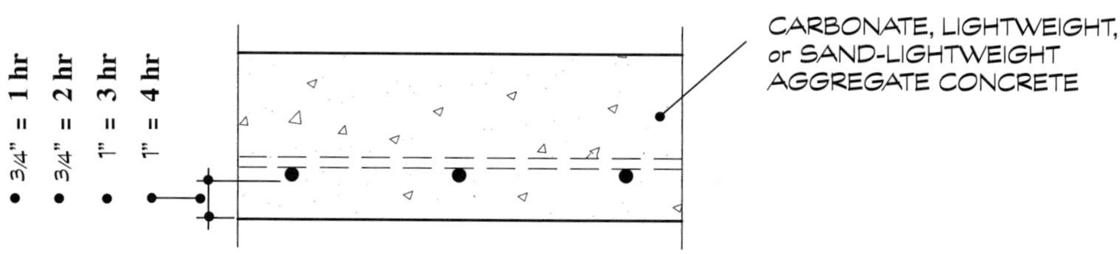

7-1.1

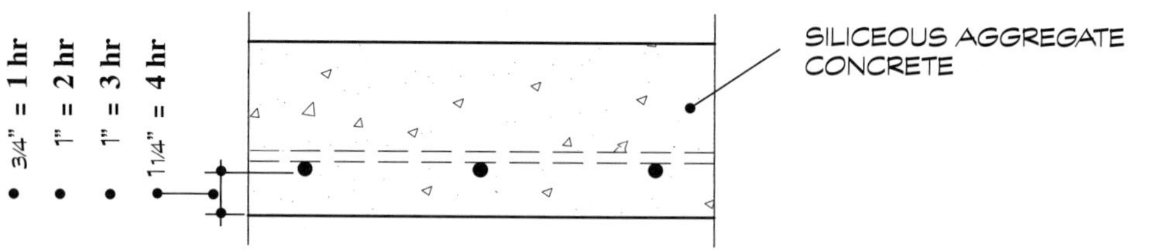

7-1.2

Fig. 720.1(1) 7A. Minimum protection of reinforcing and tie rods in floor and roof slabs. Minimum thicknesses of concrete cover are indicated as required for the fire-resistance ratings shown. Such minimum thicknesses are the same on all sides. *[Source: IBC Table 720.1(1).]*

720 Prescriptive Fire Resistance

Note: Acceptable fill materials for voids in details 1-1.1 through 1-1.3 are as follows:

Silicone-treated loose-fill insulation	*Expanded shale lightweight aggregate*
Vermiculite loose-fill insulation	*Expanded lightweight aggregate*
Expanded clay lightweight aggregate	*Grout*

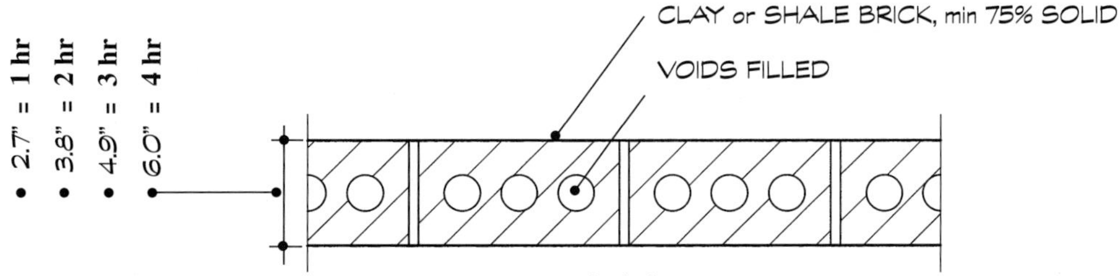

1-1.1

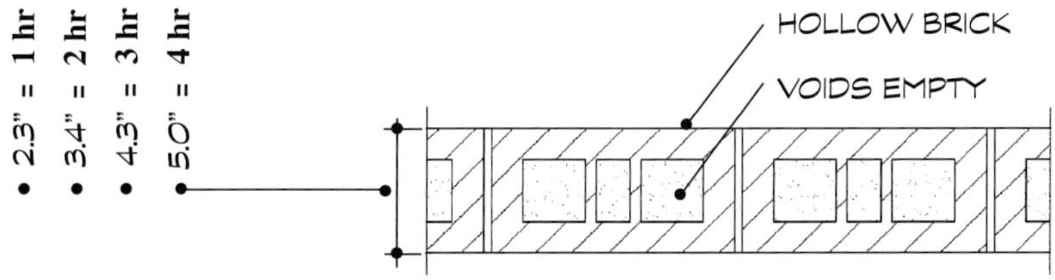

1-1.2

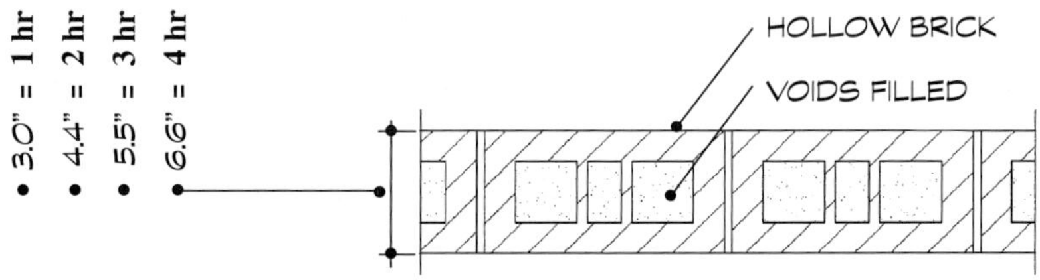

1-1.3

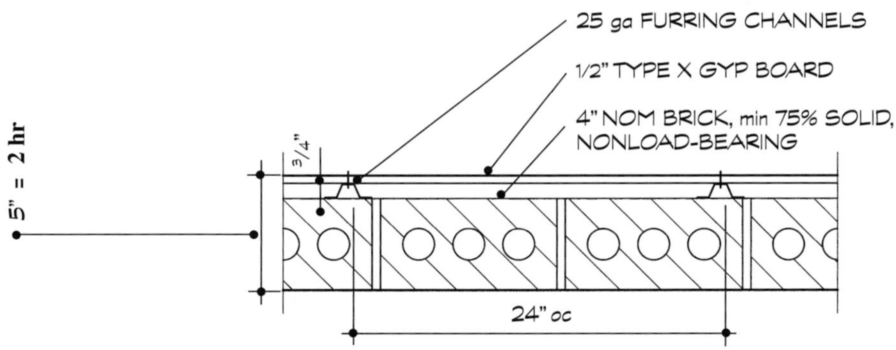

1-2.1

Fig. 720.1(2) 1A. Fire-resistance ratings for clay or shale brick walls and partitions. Minimum thickness of assembly is indicated as required for the fire-resistance rating shown. *[Source: IBC Table 720.1(2).]*

720 Prescriptive Fire Resistance

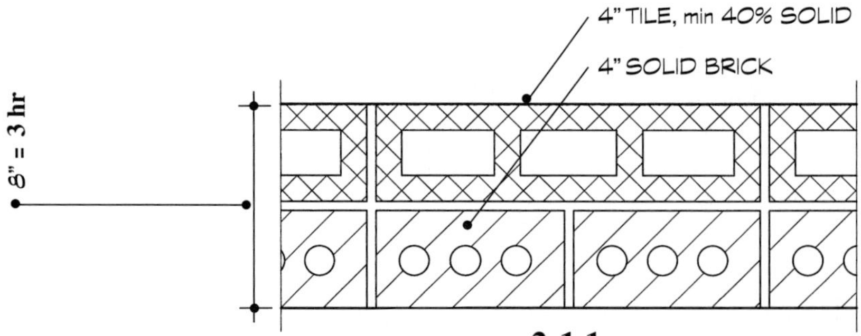

2-1.1

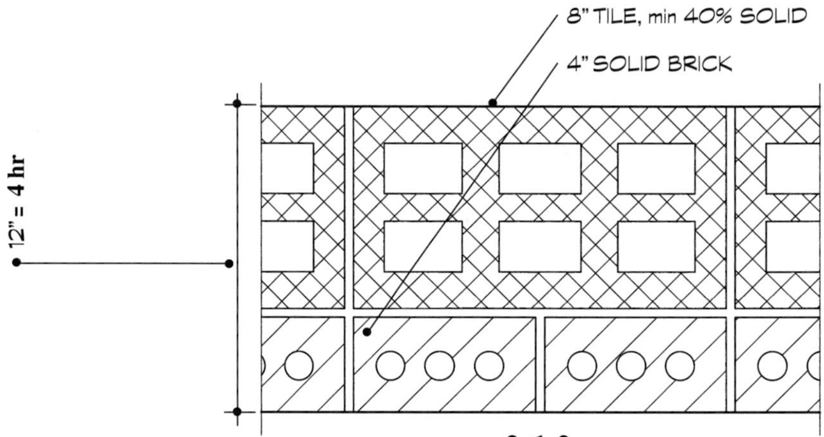

2-1.2

Note: Acceptable fill materials for voids in details 3-1.1 are as follows:

Silicone-treated loose-fill insulation	*Expanded shale lightweight aggregate*
Vermiculite loose-fill insulation	*Expanded lightweight aggregate*
Expanded clay lightweight aggregate	*Grout*

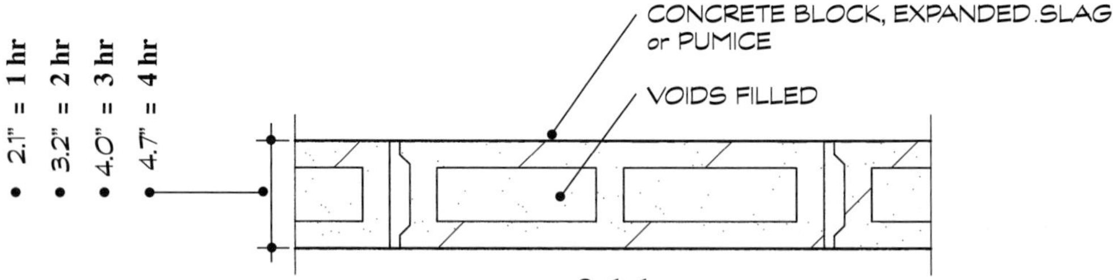

3-1.1

Fig. 720.1(2) 2A. Fire-resistance ratings for clay brick and load-bearing hollow clay tile walls and partitions. Minimum thickness of assembly is indicated as required for the fire-resistance rating shown. *[Source: IBC Table 720.1(2).]*

720 Prescriptive Fire Resistance

Note: Acceptable fill materials for voids in details 3-1.2 through 3-1.4 are as follows:

Silicone-treated loose-fill insulation	*Expanded shale lightweight aggregate*
Vermiculite loose-fill insulation	*Expanded lightweight aggregate*
Expanded clay lightweight aggregate	*Grout*

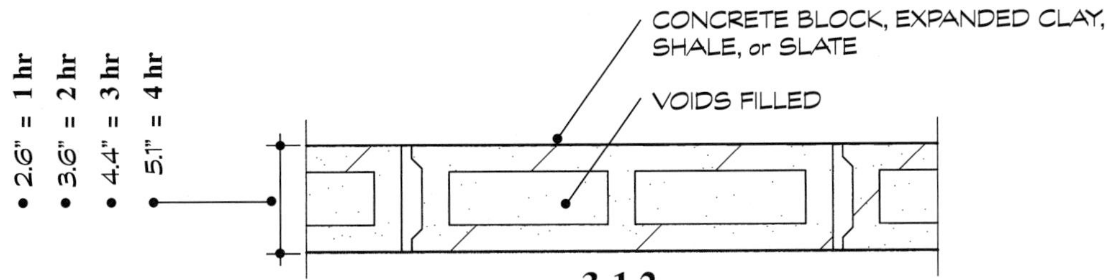

3-1.2

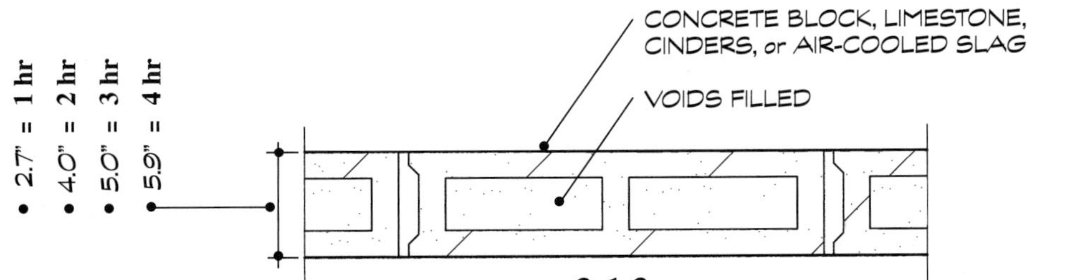

3-1.3

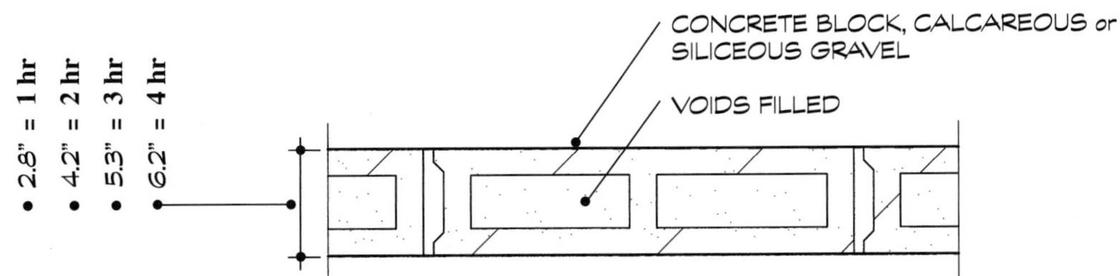

3-1.4

Fig. 720.1(2) 3A. Fire-resistance ratings for concrete masonry walls and partitions. Minimum thickness of assembly is indicated as required for the fire-resistance rating shown. *[Source: IBC Table 720.1(2).]*

720 Prescriptive Fire Resistance

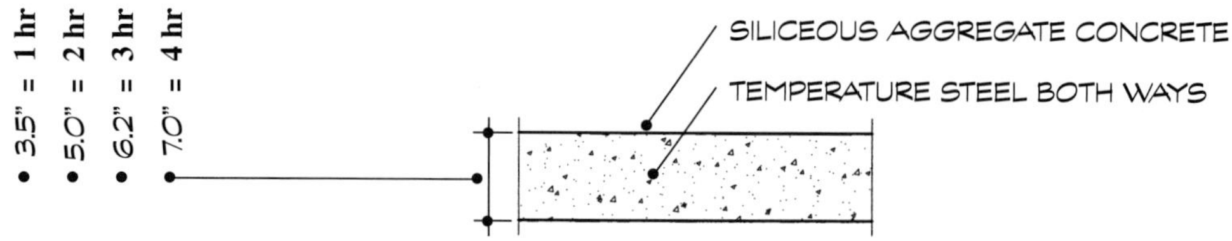

4-1.1a

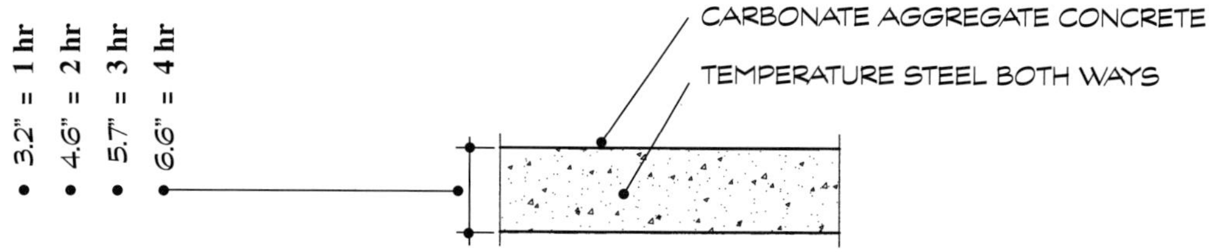

4-1.1b

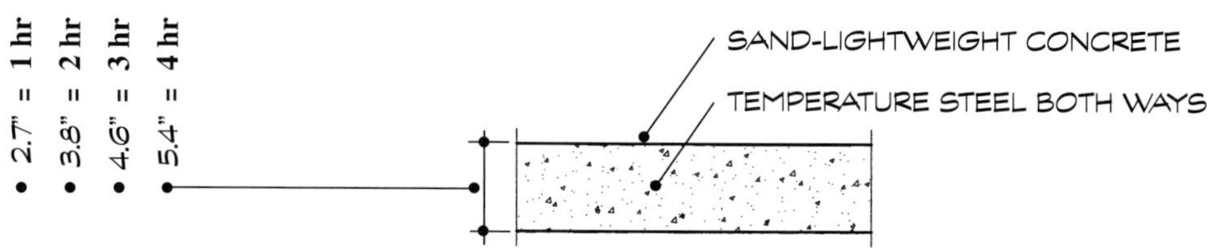

4-1.1c

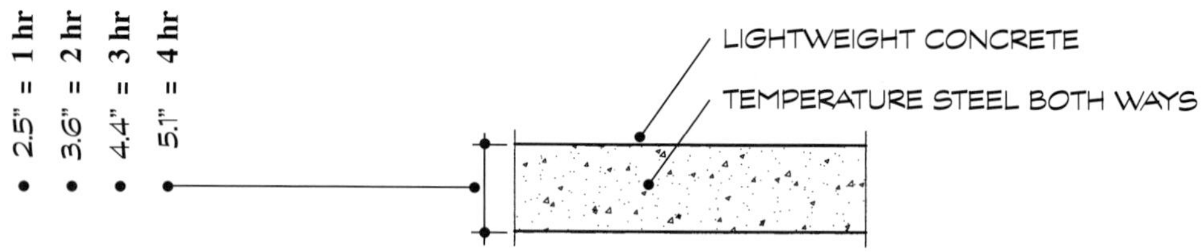

4-1.1d

Fig. 720.1(2) 4A. Fire-resistance ratings for solid concrete walls and partitions. Minimum thickness of assembly is indicated as required for the fire-resistance rating shown. *[Source: IBC Table 720.1(2).]*

720 Prescriptive Fire Resistance

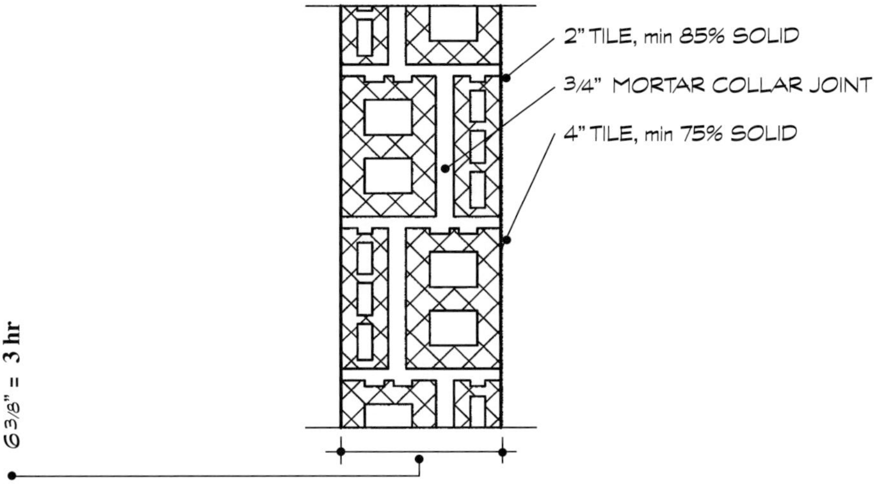

5-1.1

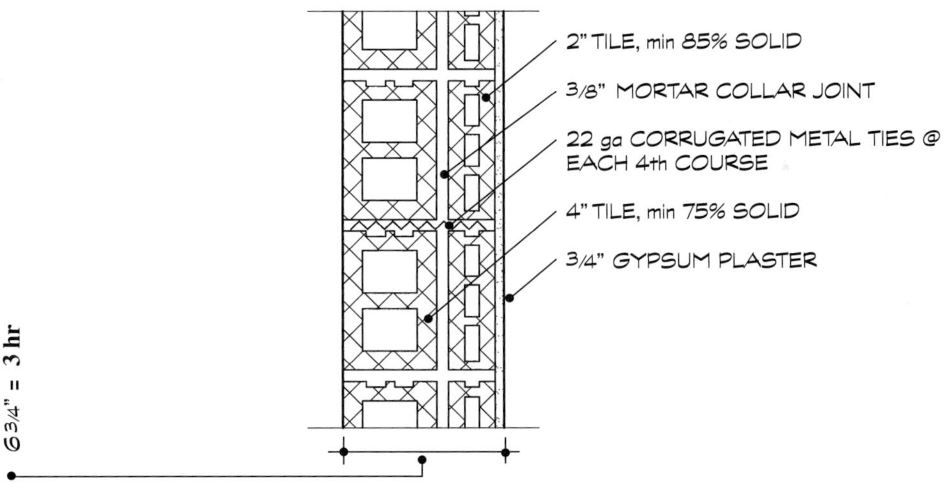

5-1.2

Fig. 720.1(2) 5A. Fire-resistance ratings for glazed or unglazed nonload-bearing facing tile walls and partitions. Minimum thickness of assembly is indicated as required for the fire-resistance rating shown. *[Source: IBC Table 720.1(2).]*

720 Prescriptive Fire Resistance

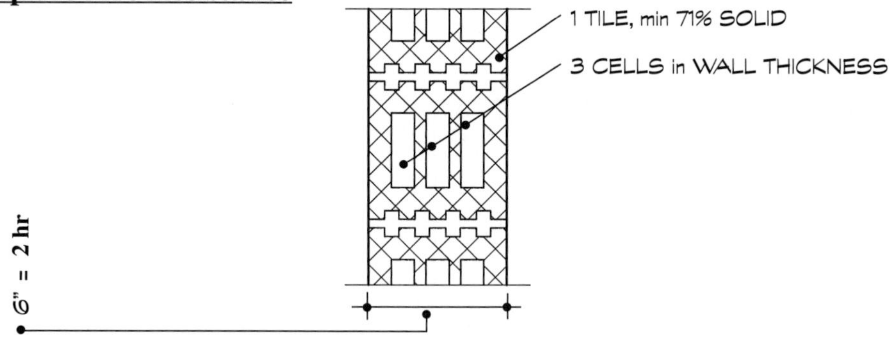

5-1.3

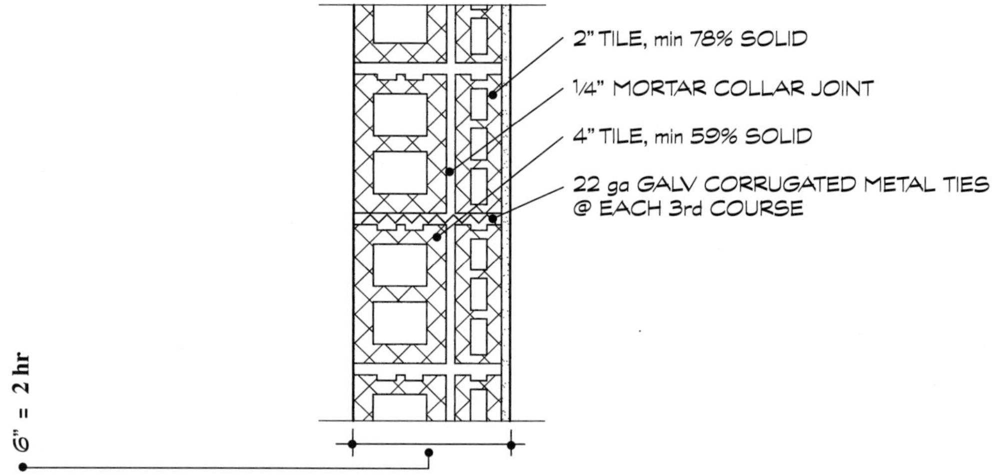

5-1.4

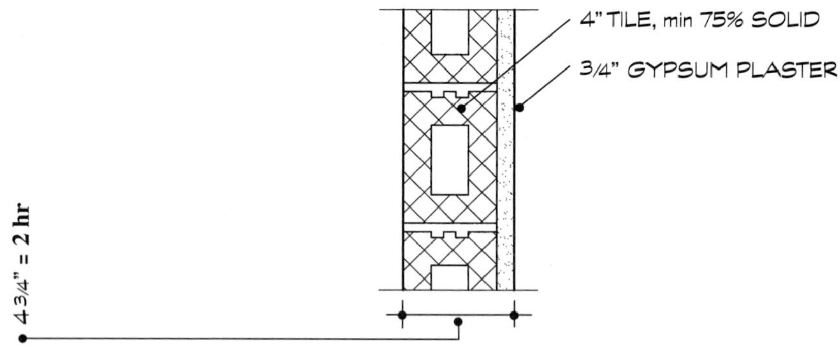

5-1.5

Fig. 720.1(2) 5B. Fire-resistance ratings for glazed or unglazed nonload-bearing facing tile walls and partitions. Minimum thickness of assembly is indicated as required for the fire-resistance rating shown. *[Source: IBC Table 720.1(2).]*

720 Prescriptive Fire Resistance

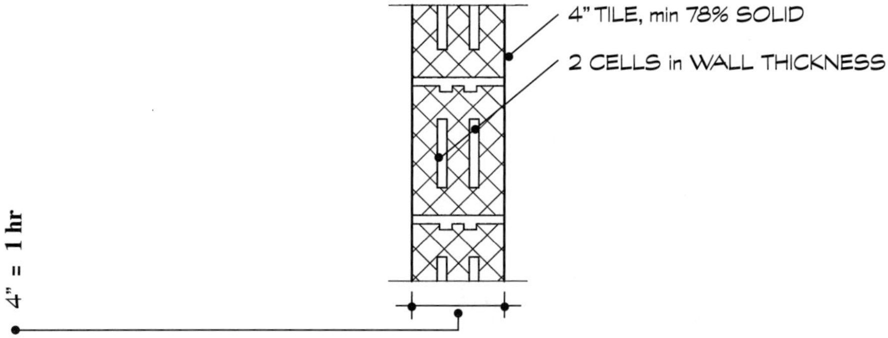

5-1.6

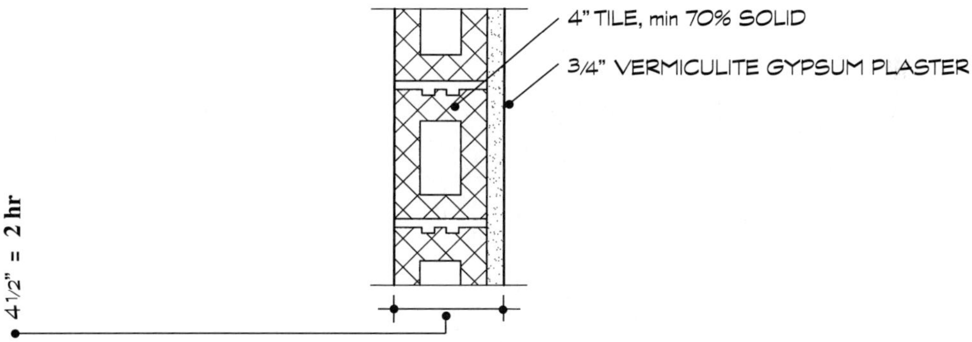

5-1.7

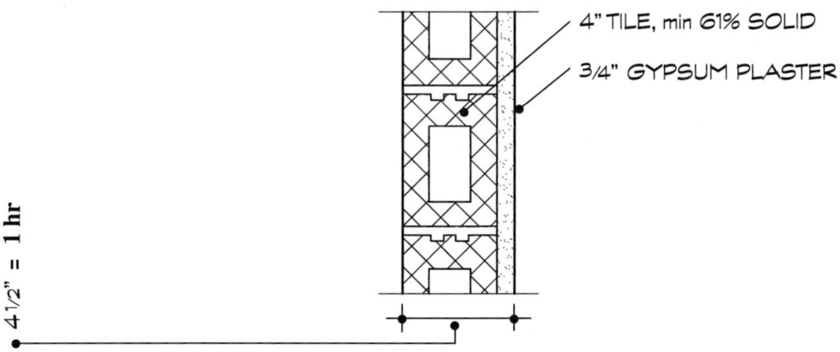

5-1.8

Fig. 720.1(2) 5C. Fire-resistance ratings for glazed or unglazed nonload-bearing facing tile walls and partitions. Minimum thickness of assembly is indicated as required for the fire-resistance rating shown. *[Source: IBC Table 720.1(2).]*

720 Prescriptive Fire Resistance

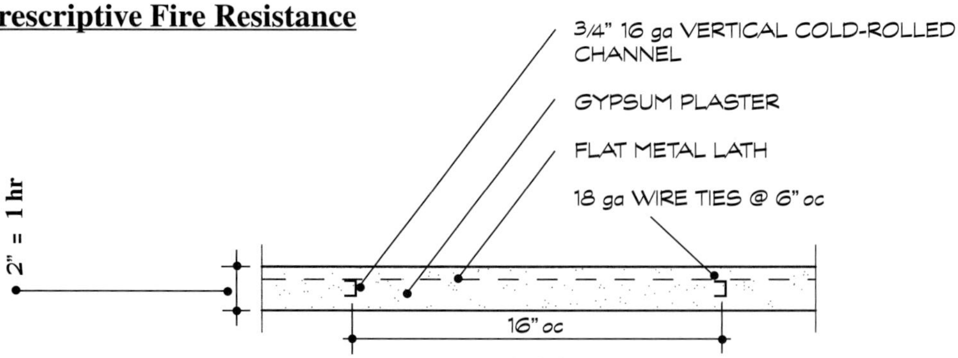

6-1.1

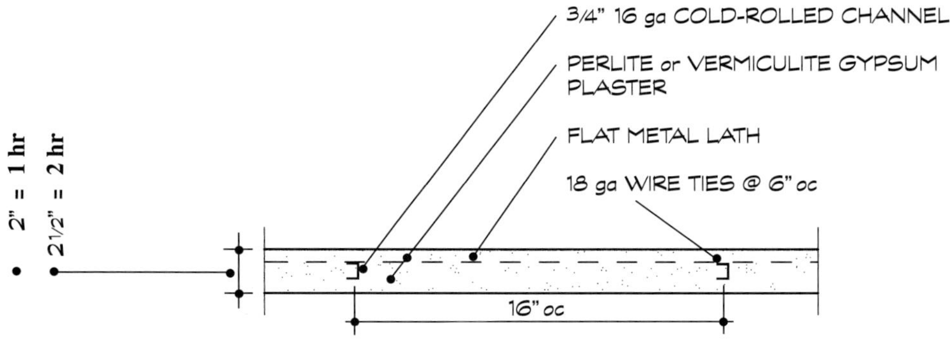

6-1.2

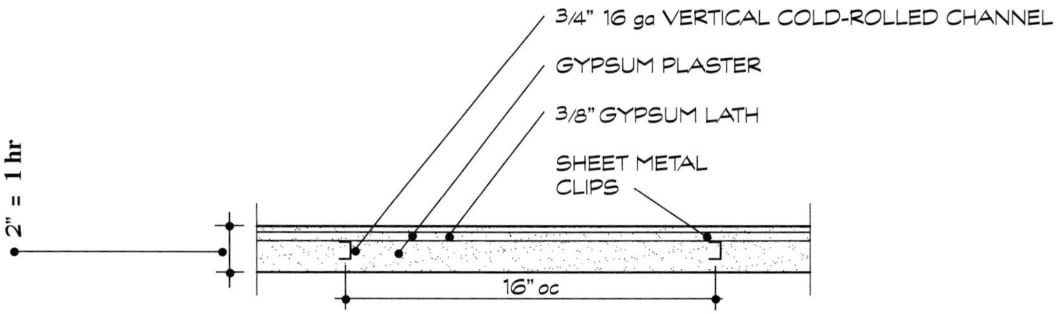

6-1.3

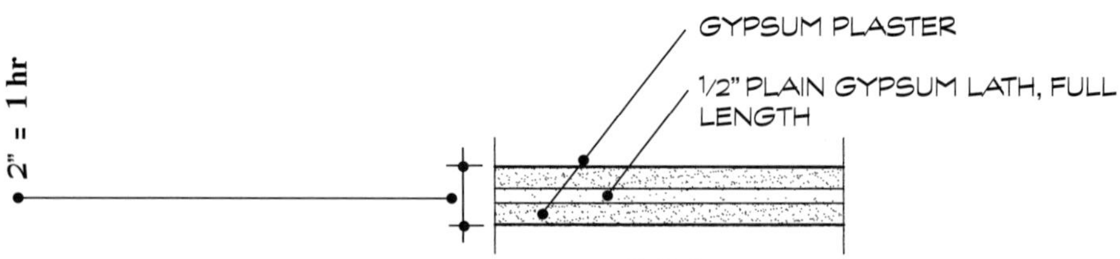

6-2.1

Fig. 720.1(2) 6A. Fire-resistance ratings for solid gypsum plaster nonload-bearing walls and partitions. Minimum thickness of assembly is indicated as required for the fire-resistance rating shown. *[Source: IBC Table 720.1(2).]*

720 Prescriptive Fire Resistance

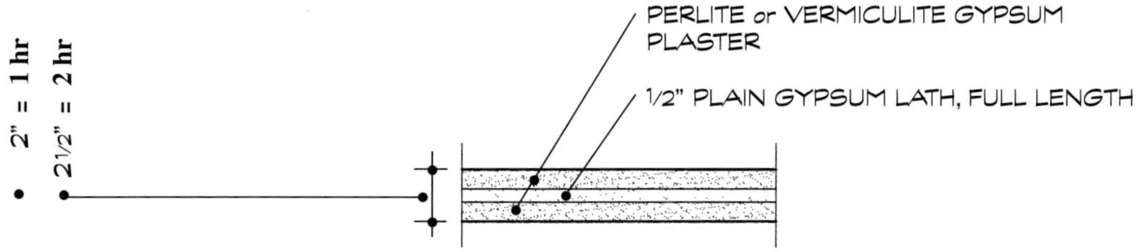

6-2.2

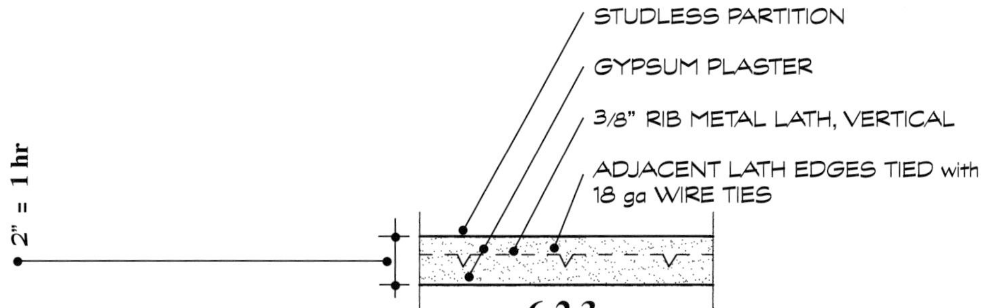

6-2.3

Fig. 720.1(2) 6B. Fire-resistance ratings for solid gypsum plaster nonload-bearing walls and partitions. Minimum thickness of assembly is indicated as required for the fire-resistance rating shown. *[Source: IBC Table 720.1(2).]*

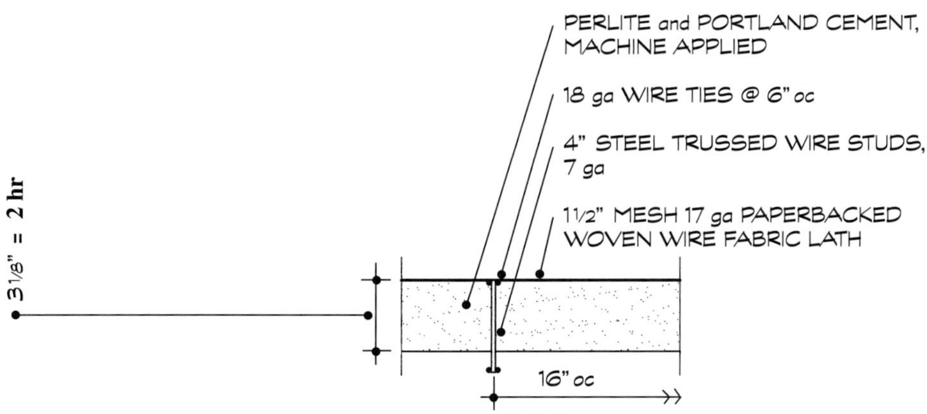

7-1.1

Fig. 720.1(2) 7A. Fire-resistance ratings for solid perlite and portland cement nonload-bearing walls and partitions. Minimum thickness of assembly is indicated as required for the fire-resistance rating shown. *[Source: IBC Table 720.1(2).]*

720 Prescriptive Fire Resistance

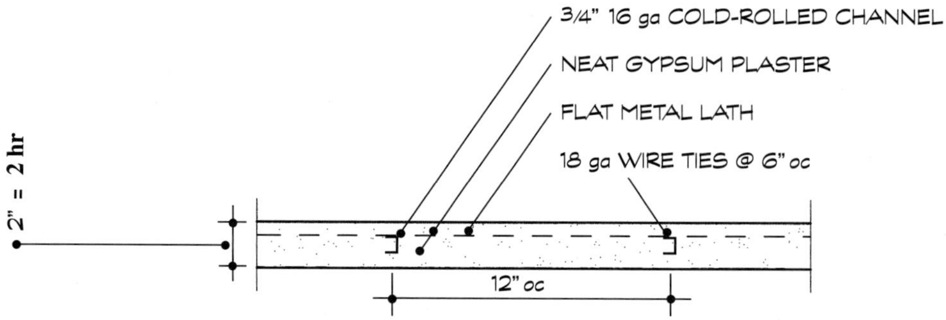

8-1.1

Fig. 720.1(2) 8A. Fire-resistance ratings for solid neat wood fibered gypsum plaster nonload-bearing walls and partitions. Minimum thickness of assembly is indicated as required for the fire-resistance rating shown. *[Source: IBC Table 720.1(2).]*

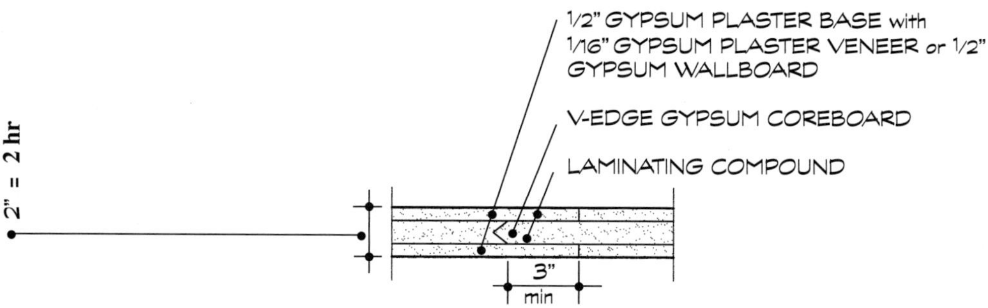

9-1.1

Fig. 720.1(2) 9A. Fire-resistance ratings for solid gypsum wallboard nonload-bearing walls and partitions. Minimum thickness of assembly is indicated as required for the fire-resistance rating shown. *[Source: IBC Table 720.1(2).]*

720 Prescriptive Fire Resistance

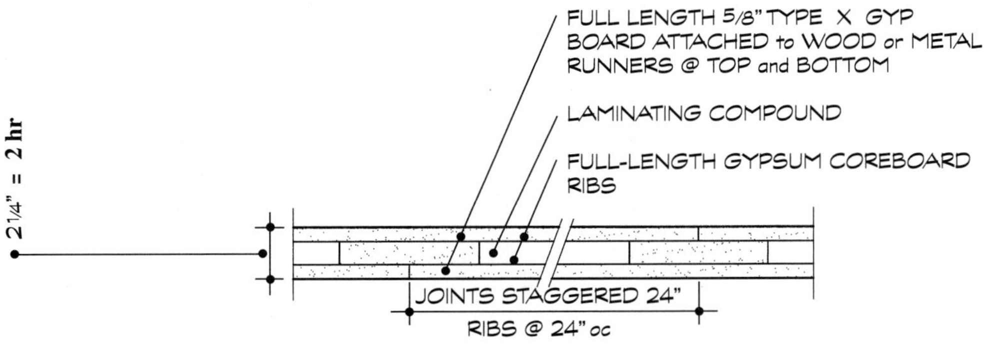

10-1.1

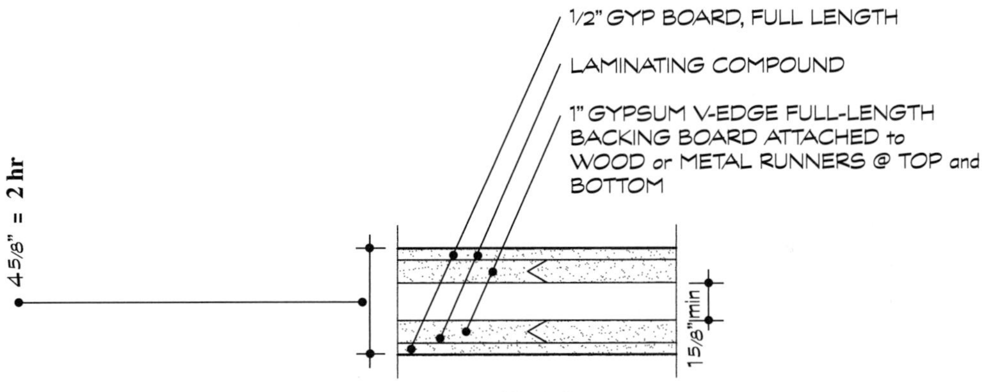

10-1.2

Fig. 720.1(2) 10A. Fire-resistance ratings for hollow (studless) gypsum wallboard nonload-bearing walls and partitions. Minimum thickness of assembly is indicated as required for the fire-resistance rating shown. *[Source: IBC Table 720.1(2).]*

720 Prescriptive Fire Resistance

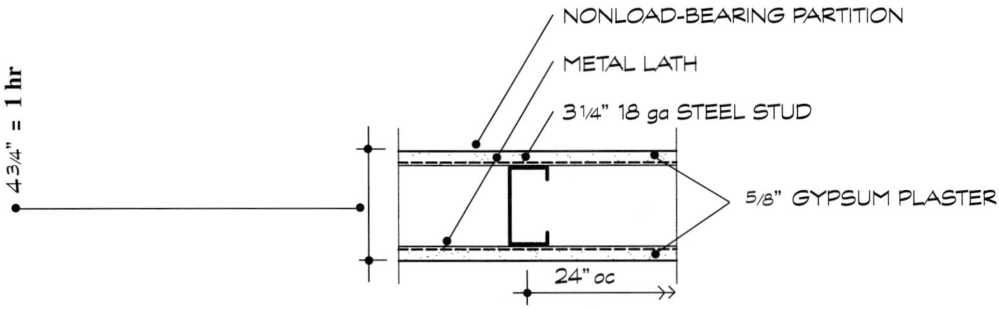

11-1.1

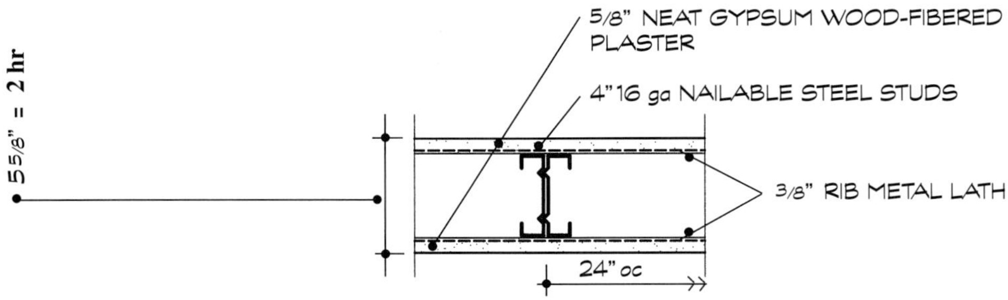

11-1.2

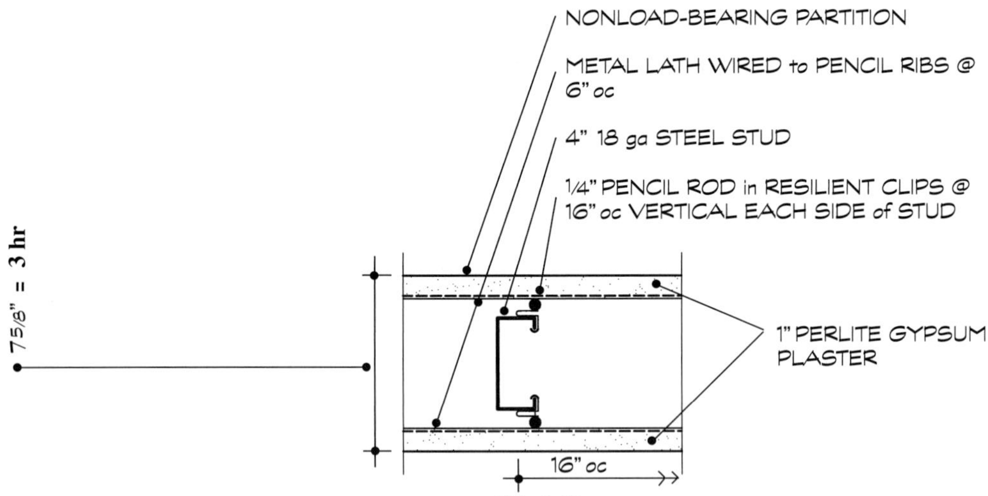

11-1.3

Fig. 720.1(2) 11A. Fire-resistance ratings for interior partitions with noncombustible studs and plaster. Minimum thickness of assembly is indicated as required for the fire-resistance rating shown. *[Source: IBC Table 720.1(2).]*

720 Prescriptive Fire Resistance

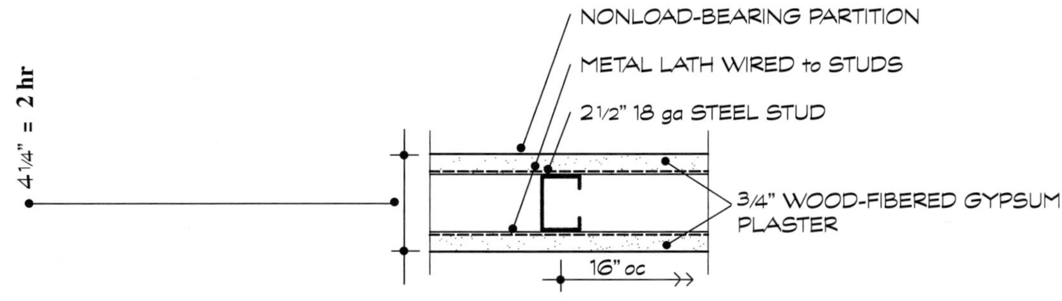

11-1.4

Fig. 720.1(2) 11B. Fire-resistance ratings for interior partitions with noncombustible studs and plaster. Minimum thickness of assembly is indicated as required for the fire-resistance rating shown. *[Source: IBC Table 720.1(2).]*

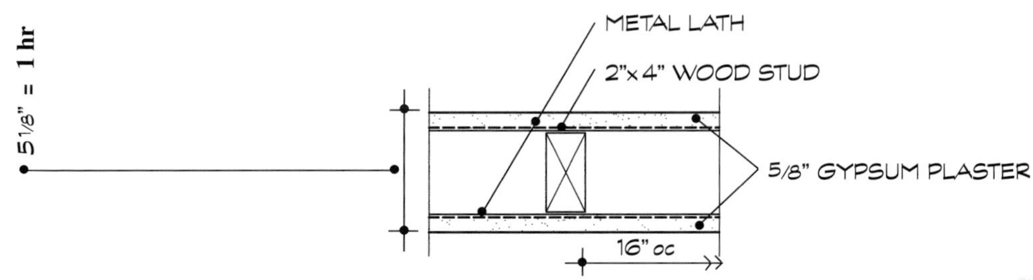

12-1.1

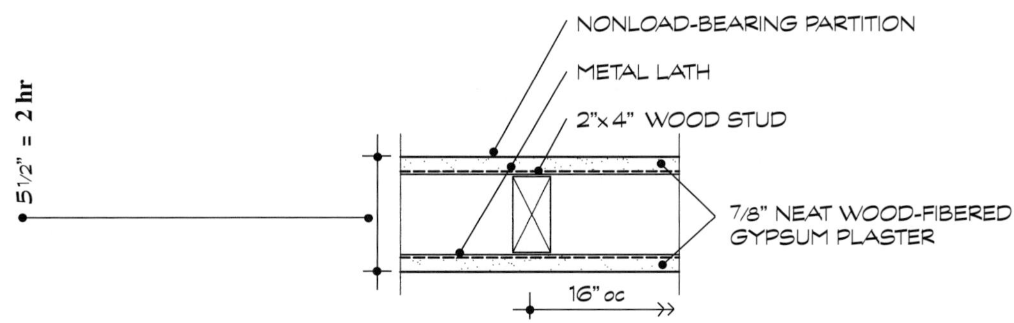

12-1.2

Fig. 720.1(2) 12A. Fire-resistance ratings for interior partitions with wood studs and plaster. Minimum thickness of assembly is indicated as required for the fire-resistance rating shown. *[Source: IBC Table 720.1(2).]*

720 Prescriptive Fire Resistance

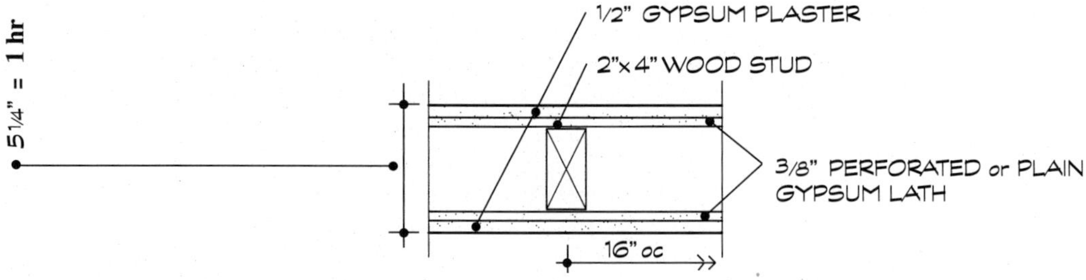

12-1.3

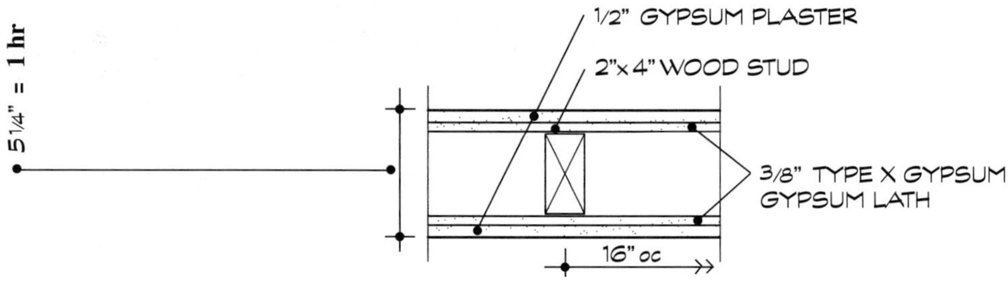

12-1.4

Fig. 720.1(2) 12B. Fire-resistance ratings for interior partitions with wood studs and plaster. Minimum thickness of assembly is indicated as required for the fire-resistance rating shown. *[Source: IBC Table 720.1(2).]*

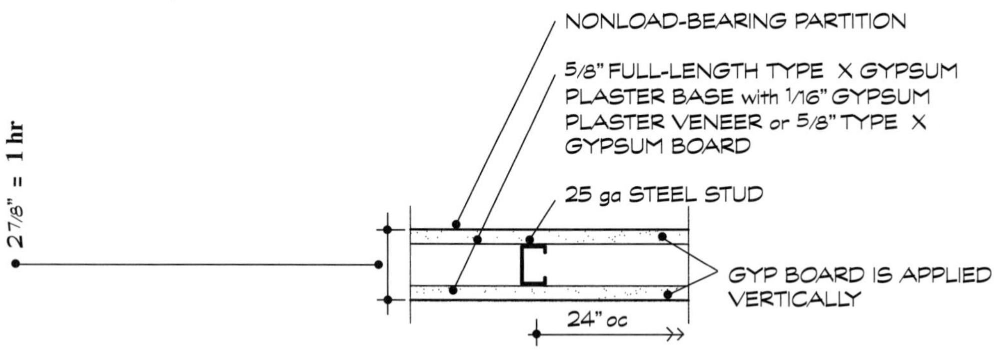

13-1.1a

Fig. 720.1(2) 13A. Fire-resistance ratings for interior partitions with noncombustible studs and gypsum board. Minimum thickness of assembly is indicated as required for the fire-resistance rating shown. *[Source: IBC Table 720.1(2).]*

720 Prescriptive Fire Resistance

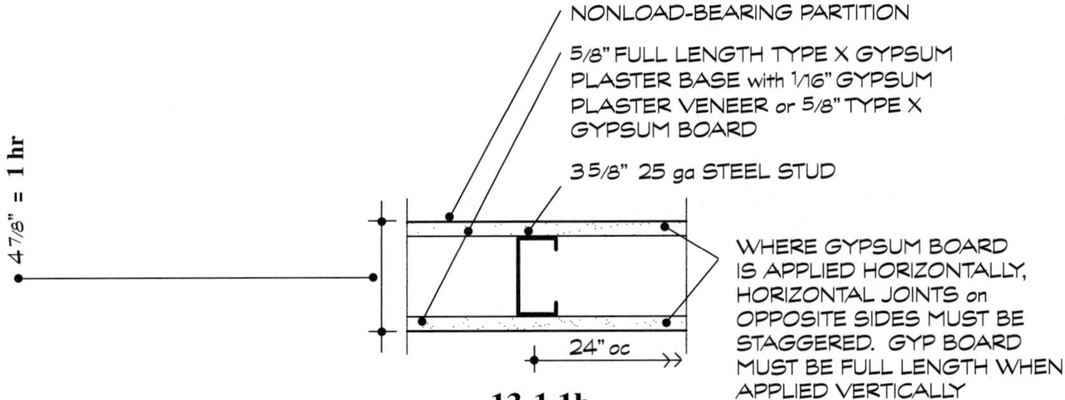

13-1.1b

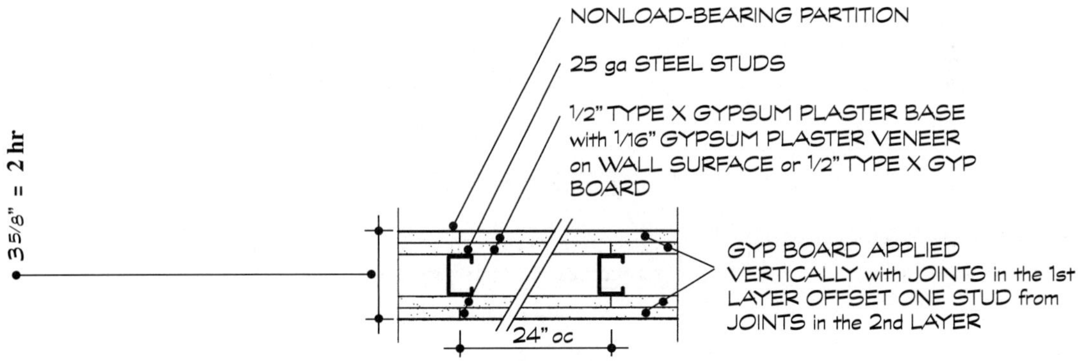

13-1.2

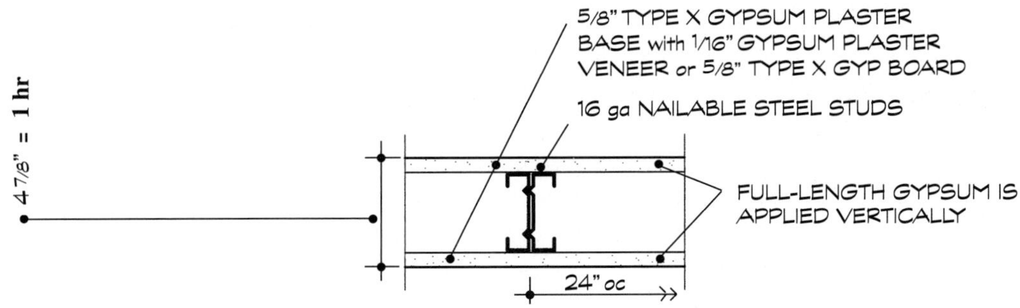

13-1.3

Fig. 720.1(2) 13B. Fire-resistance ratings for interior partitions with noncombustible studs and gypsum board. Minimum thickness of assembly is indicated as required for the fire-resistance rating shown. *[Source: IBC Table 720.1(2).]*

720 Prescriptive Fire Resistance

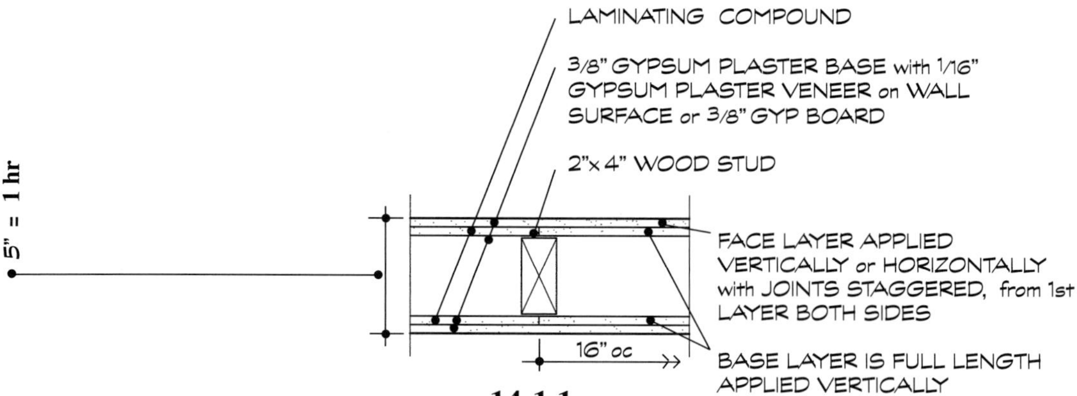

14-1.1

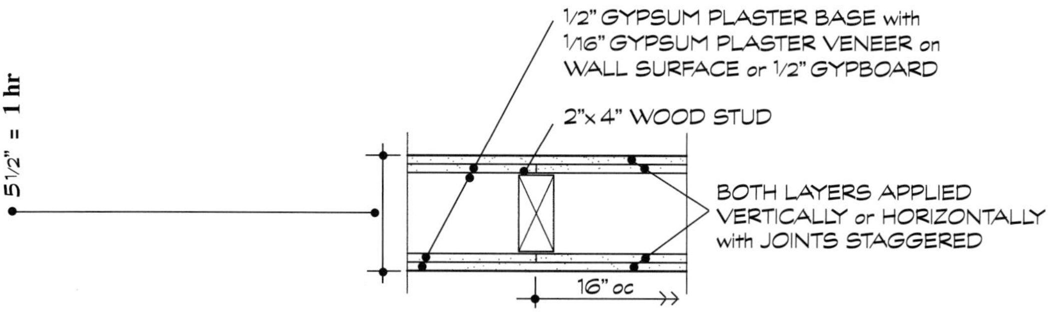

14-1.2

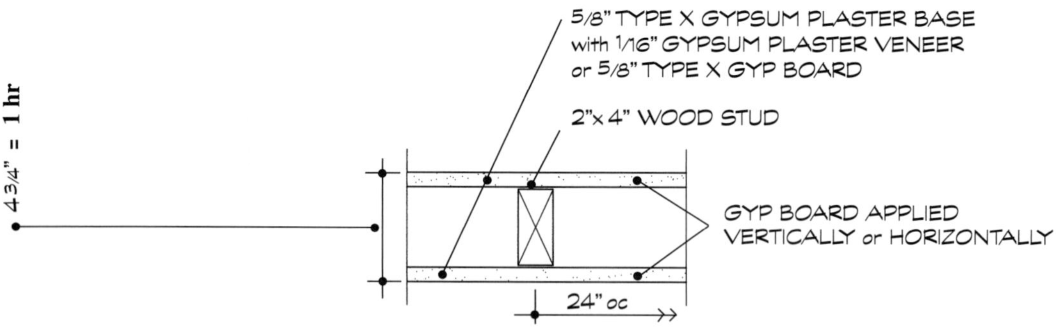

14-1.3

Fig. 720.1(2) 14A. Fire-resistance ratings for interior partitions with wood studs and gypsum board. Minimum thickness of assembly is indicated as required for the fire-resistance rating shown. *[Source: IBC Table 720.1(2).]*

720 Prescriptive Fire Resistance

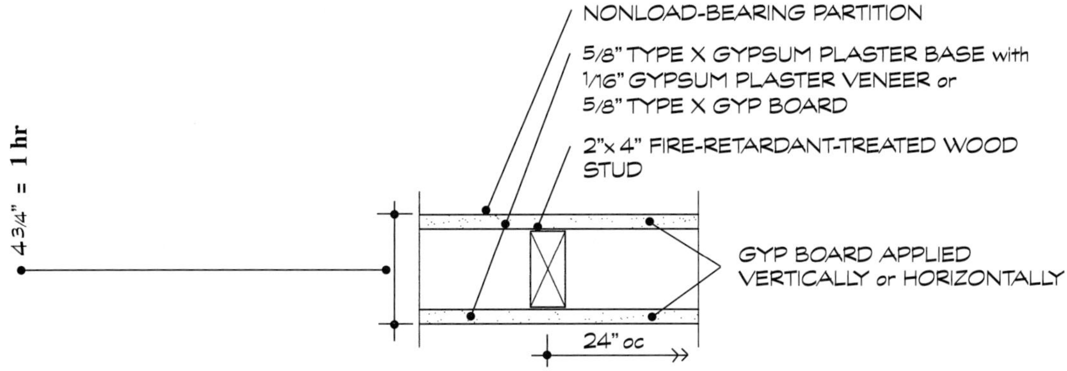

14-1.4

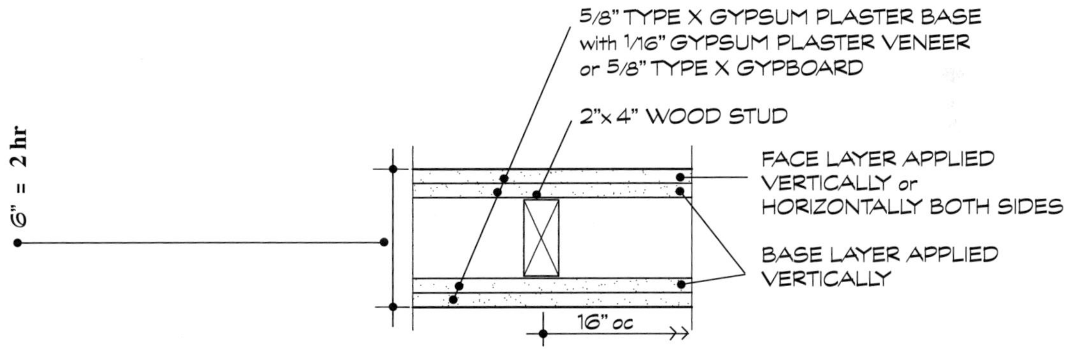

14-1.5

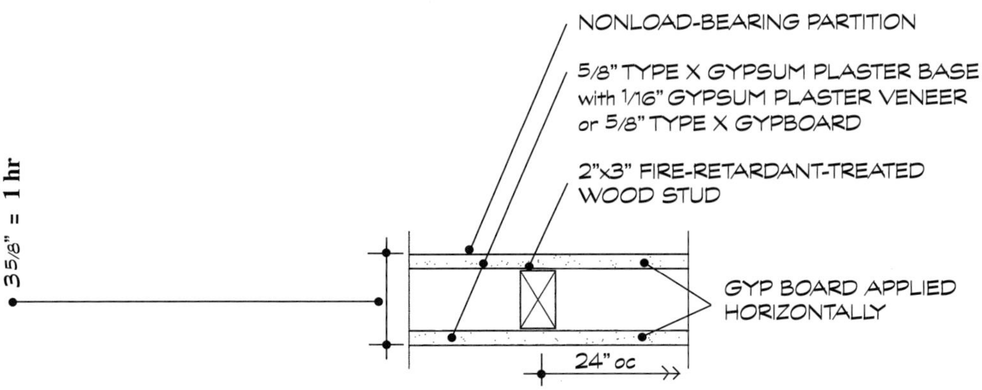

14-1.6

Fig. 720.1(2) 14B. Fire-resistance ratings for interior partitions with wood studs and gypsum board. Minimum thickness of assembly is indicated as required for the fire-resistance rating shown. *[Source: IBC Table 720.1(2).]*

720 Prescriptive Fire Resistance

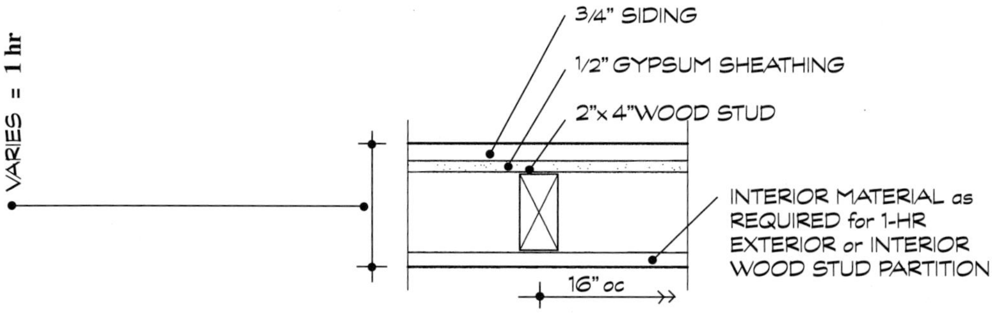

15-1.1

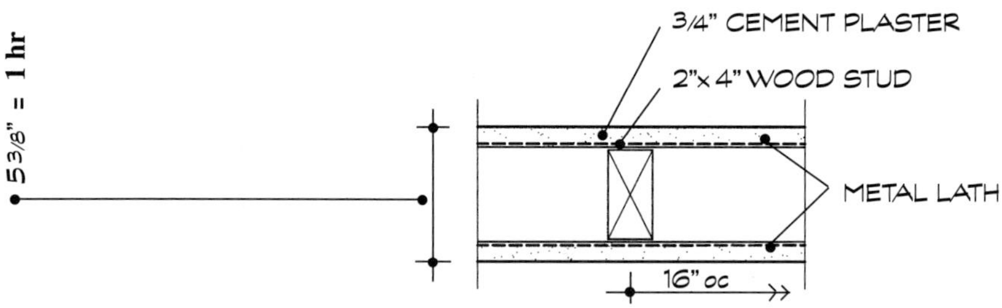

15-1.2

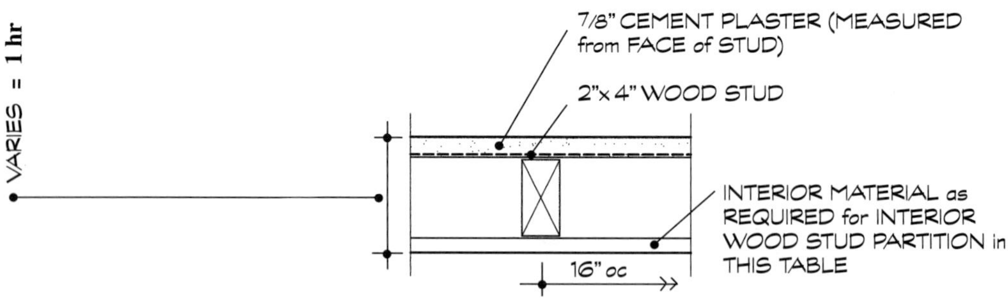

15-1.3

Fig. 720.1(2) 15A. Fire-resistance ratings for exterior or interior walls. Minimum thickness of assembly is indicated as required for the fire-resistance rating shown. *[Source: IBC Table 720.1(2).]*

720 Prescriptive Fire Resistance

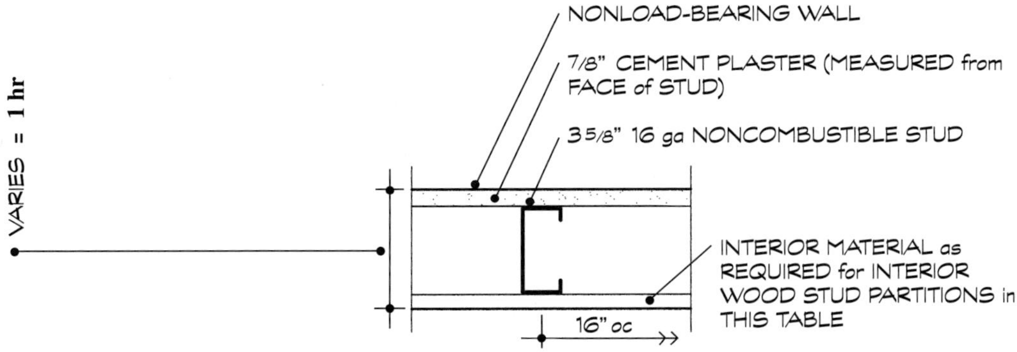

15-1.4

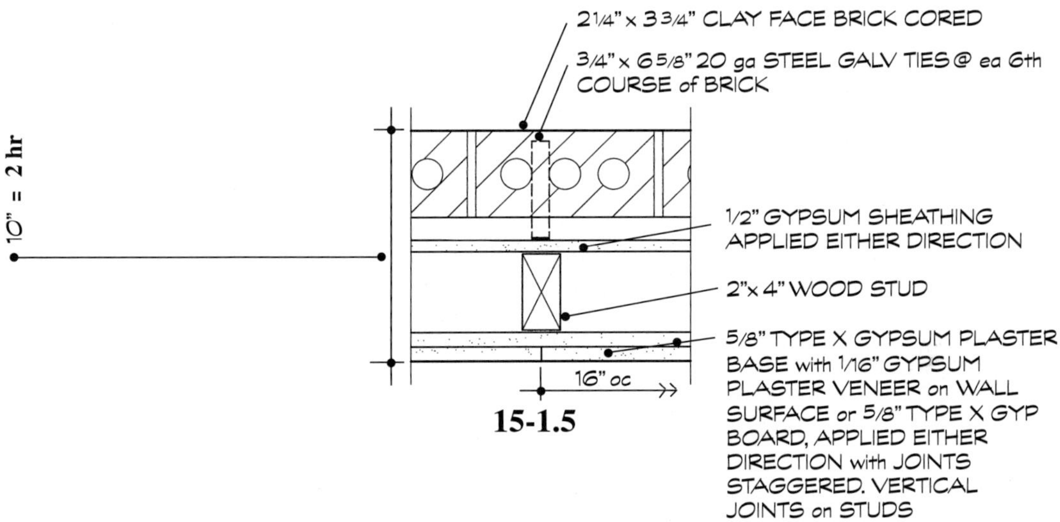

15-1.5

Fig. 720.1(2) 15B. Fire-resistance ratings for exterior or interior walls. Minimum thickness of assembly is indicated as required for the fire-resistance rating shown. *[Source: IBC Table 720.1(2).]*

720 Prescriptive Fire Resistance

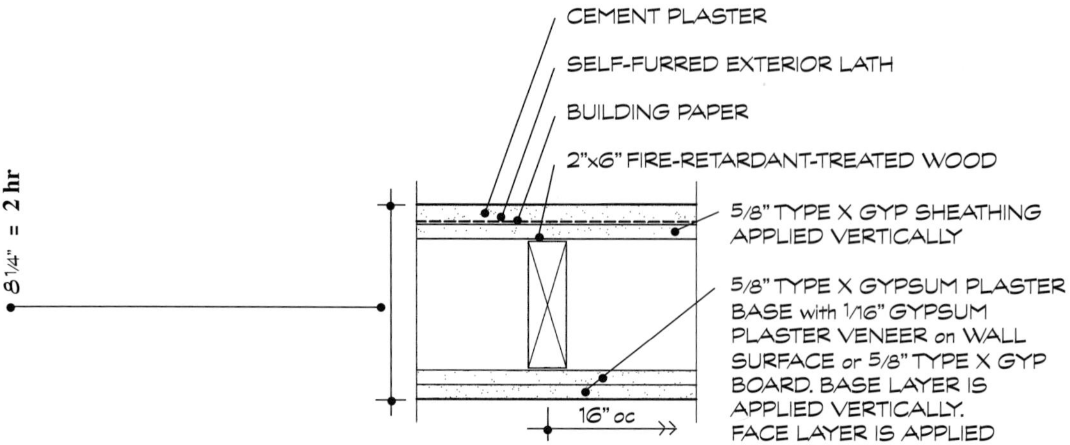

15-1.6

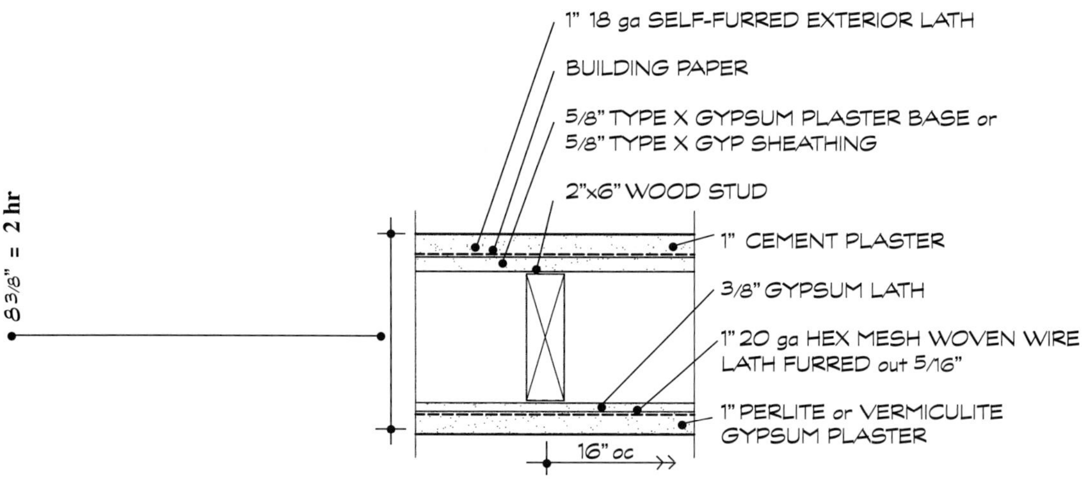

15-1.7

Fig. 720.1(2) 15C. Fire-resistance ratings for exterior or interior walls. Minimum thickness of assembly is indicated as required for the fire-resistance rating shown. *[Source: IBC Table 720.1(2).]*

720 Prescriptive Fire Resistance

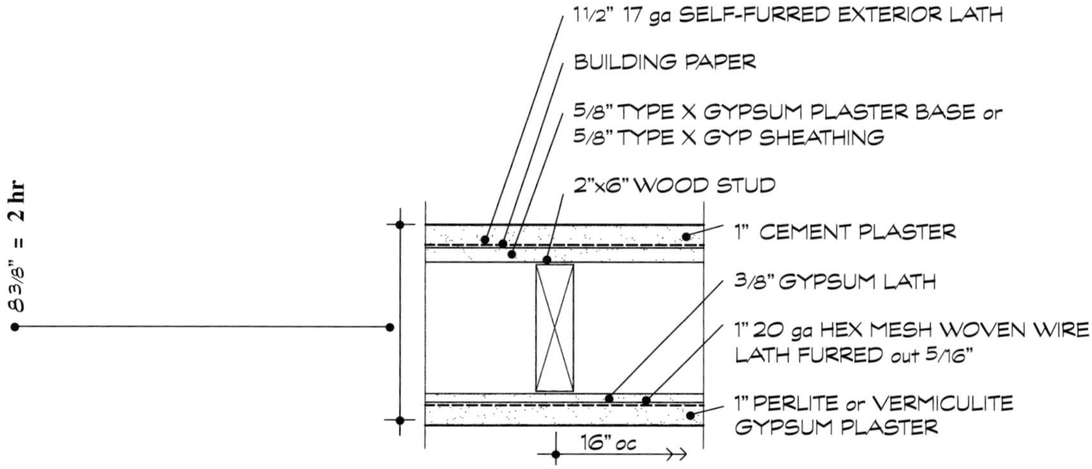

15-1.8

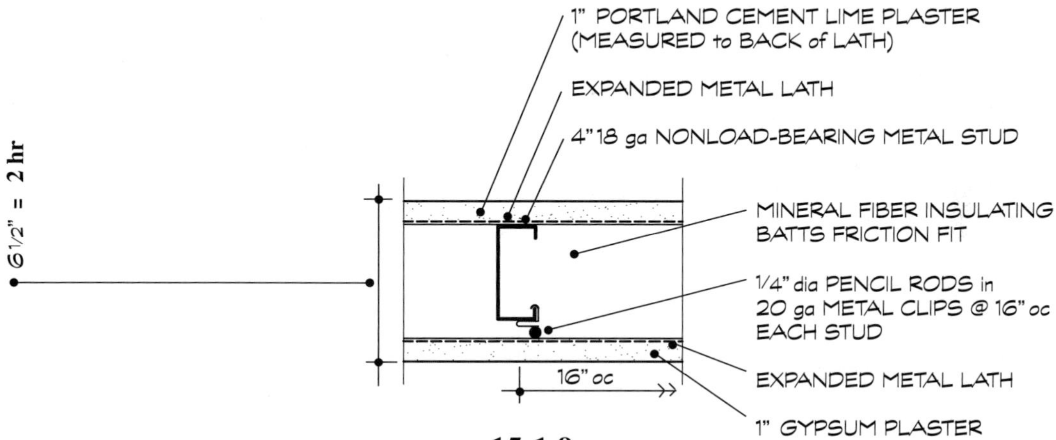

15-1.9

Fig. 720.1(2) 15D. Fire-resistance ratings for exterior or interior walls. Minimum thickness of assembly is indicated as required for the fire-resistance rating shown. *[Source: IBC Table 720.1(2).]*

720 Prescriptive Fire Resistance

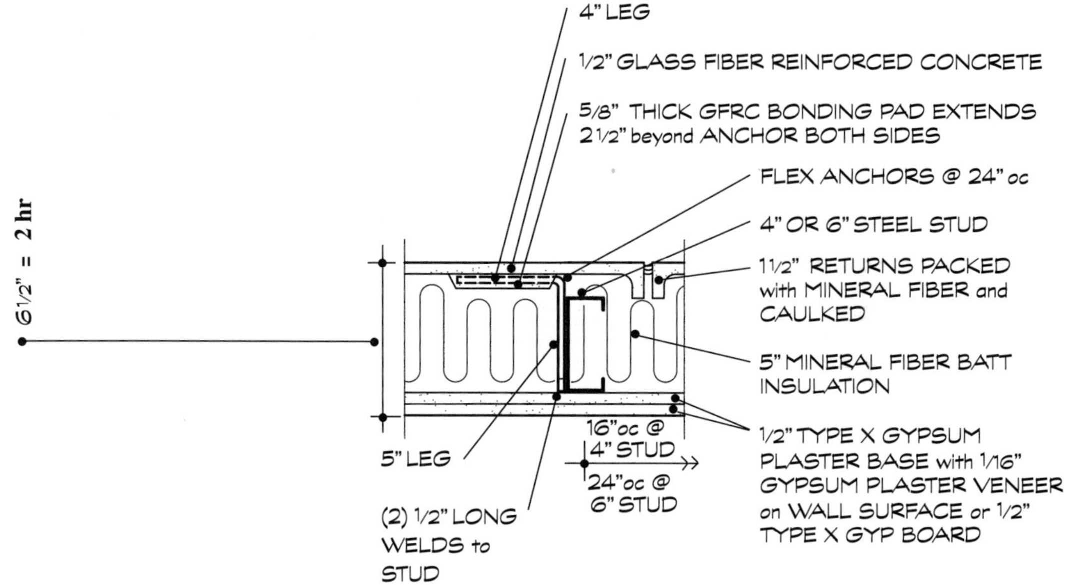

15-1.10

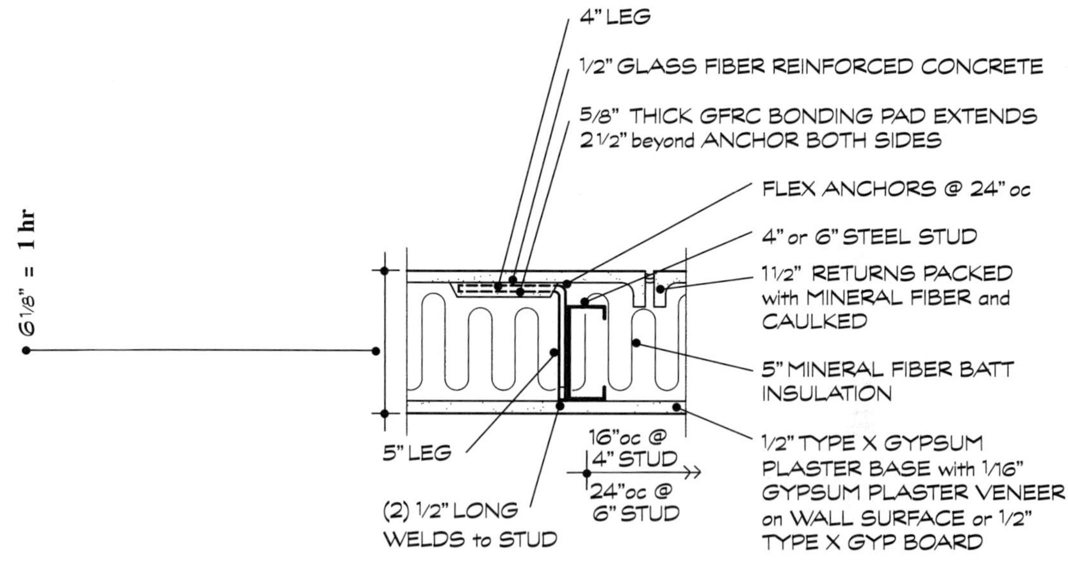

15-1.11

Fig. 720.1(2) 15E. Fire-resistance ratings for exterior or interior walls. Minimum thickness of assembly is indicated as required for the fire-resistance rating shown. *[Source: IBC Table 720.1(2).]*

720 Prescriptive Fire Resistance

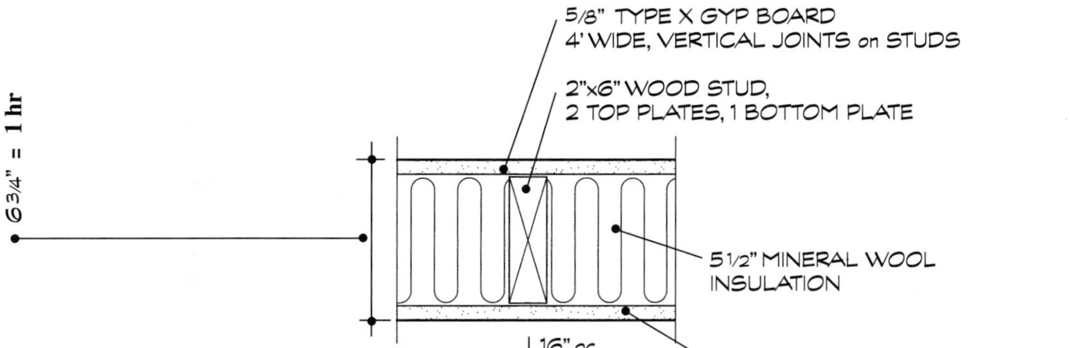

Note: Section 2306, "Allowable Stress Design," is cited as governing calculation of allowable stress.

15-1.12

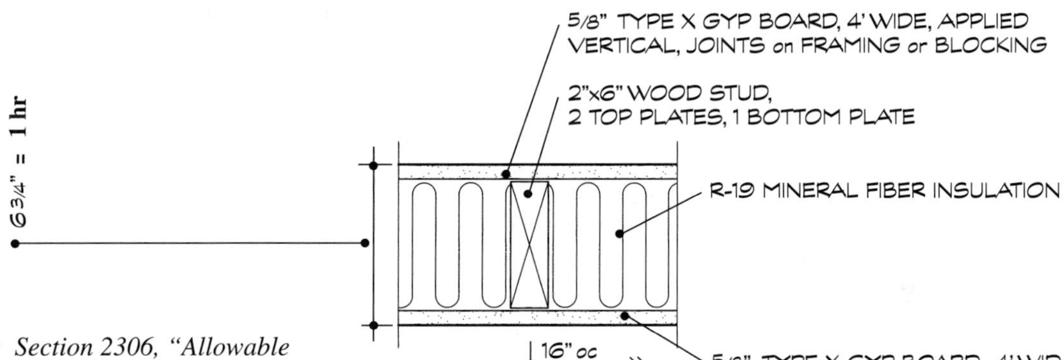

Note: Section 2306, "Allowable Stress Design," is cited as governing calculation of allowable stress.

15-1.13

Fig. 720.1(2) 15F. Fire-resistance ratings for exterior or interior walls. Minimum thickness of assembly is indicated as required for the fire-resistance rating shown. These fire-resistance ratings are established for walls loaded to 100% of their design load. *[Source: IBC Table 720.1(2).]*

720 Prescriptive Fire Resistance

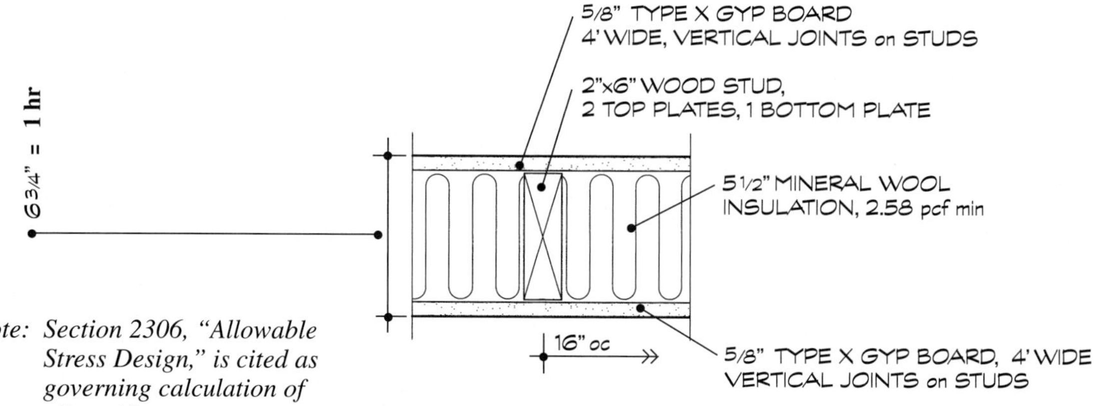

Note: Section 2306, "Allowable Stress Design," is cited as governing calculation of allowable stress.

15-1.14

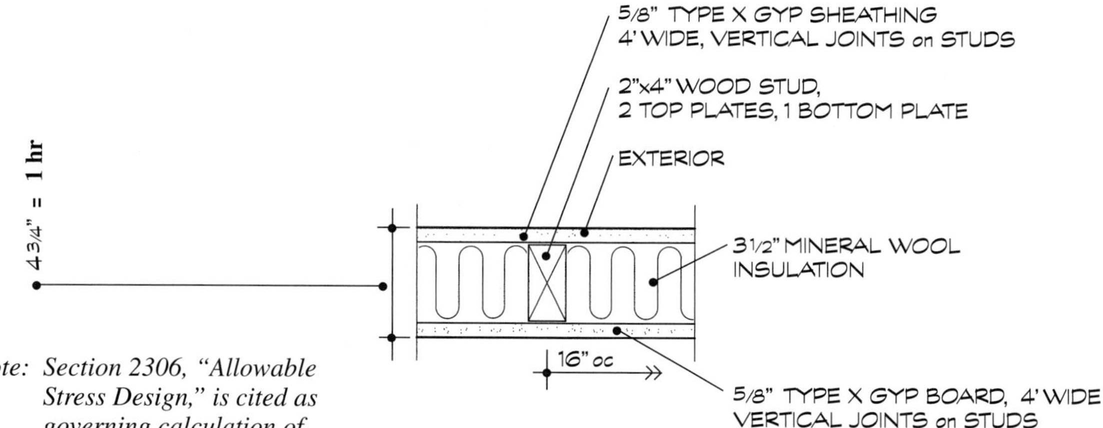

Note: Section 2306, "Allowable Stress Design," is cited as governing calculation of allowable stress.

15-1.15

Fig. 720.1(2) 15G. Fire-resistance ratings for exterior or interior walls. Minimum thickness of assembly is indicated as required for the fire-resistance rating shown. These fire-resistance ratings are established for walls loaded to 100% of their design load. *[Source: IBC Table 720.1(2).]*

720 Prescriptive Fire Resistance

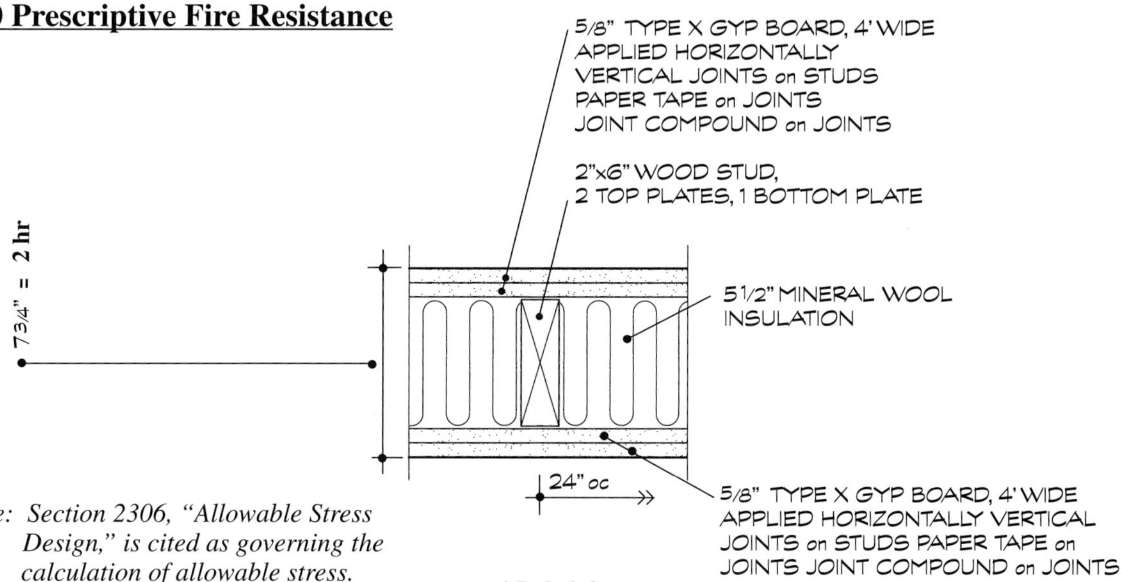

Note: Section 2306, "Allowable Stress Design," is cited as governing the calculation of allowable stress.

15-1.16

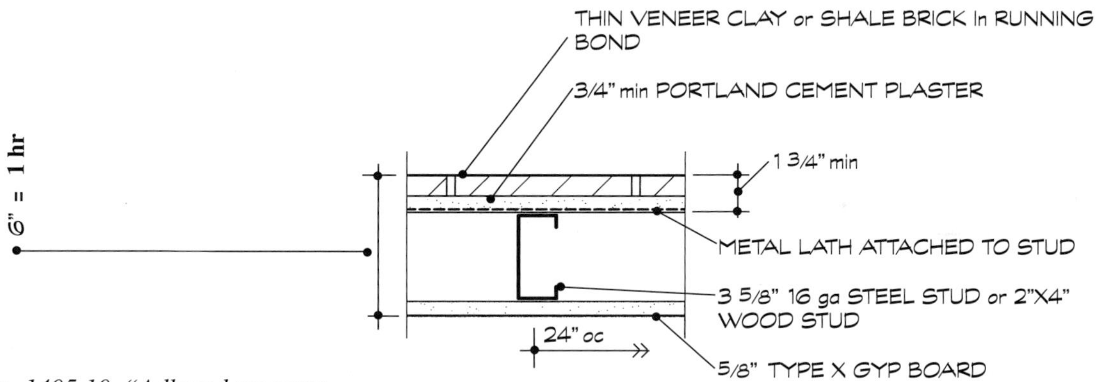

Note: 1405.10, "Adhered masonry veneer," is cited as governing installation of the brick veneer.

15-2.1

Fig. 720.1(2) 15H. Fire-resistance ratings for exterior or interior walls. Minimum thickness of assembly is indicated as required for the fire-resistance rating shown. These fire-resistance ratings are established for walls loaded to 100% of their design load. *[Source: IBC Table 720.1(2).]*

720 Prescriptive Fire Resistance

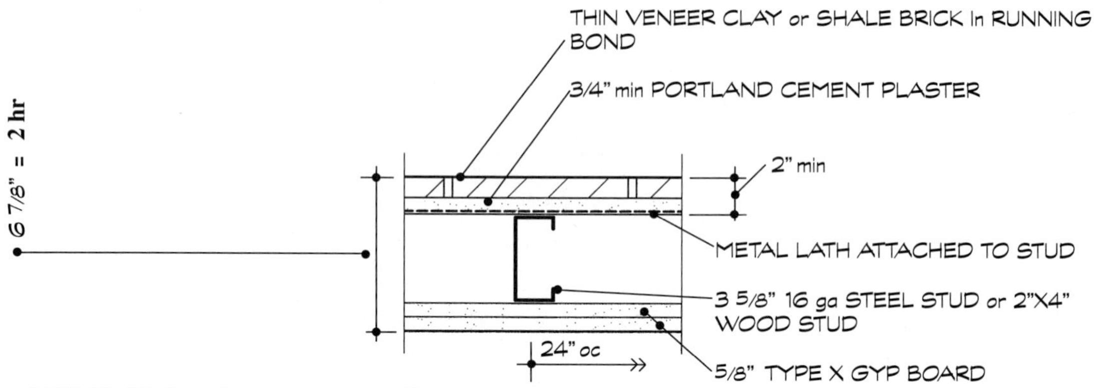

Note: 1405.10, "Adhered masonry veneer," is cited as governing installation of the brick veneer.

15-2.2

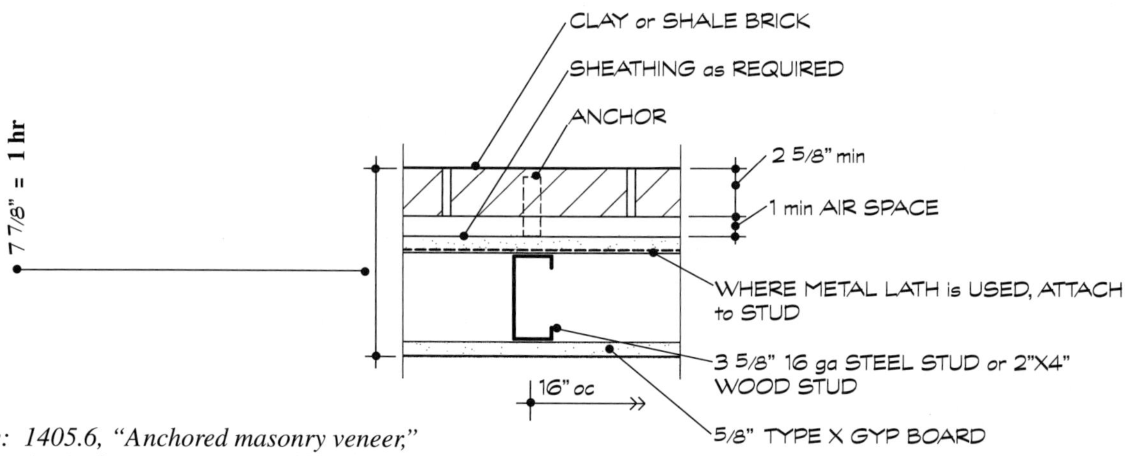

Note: 1405.6, "Anchored masonry veneer," is cited as governing installation of the brick.

15-2.3

Fig. 720.1(2) 15I. Fire-resistance ratings for exterior or interior walls. Minimum thickness of assembly is indicated as required for the fire-resistance rating shown. These fire-resistance ratings are established for walls loaded to 100% of their design load. *[Source: IBC Table 720.1(2).]*

720 Prescriptive Fire Resistance

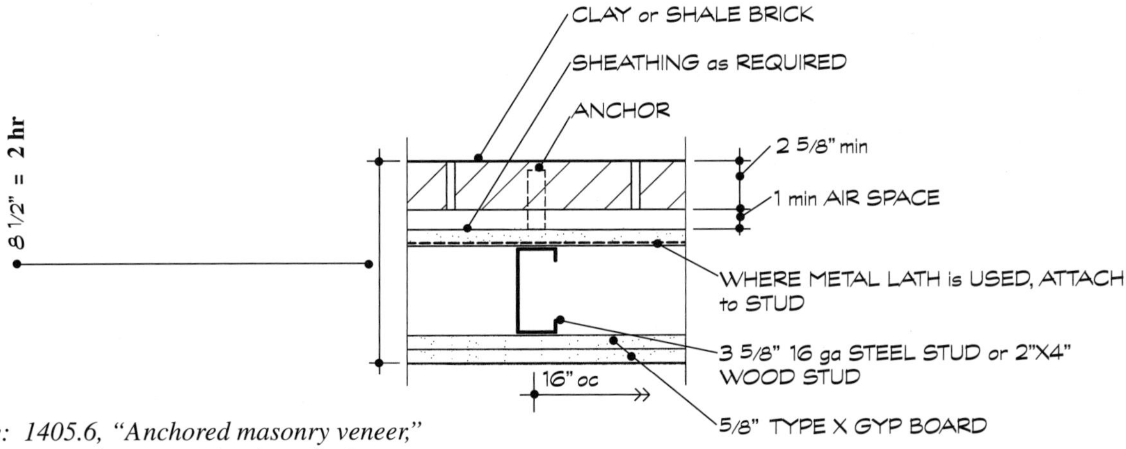

Note: 1405.6, "Anchored masonry veneer," is cited as governing installation of the brick.

15-2.4

Fig. 720.1(2) 15J. Fire-resistance ratings for exterior or interior walls. Minimum thickness of assembly is indicated as required for the fire-resistance rating shown. These fire-resistance ratings are established for walls loaded to 100% of their design load. *[Source: IBC Table 720.1(2).]*

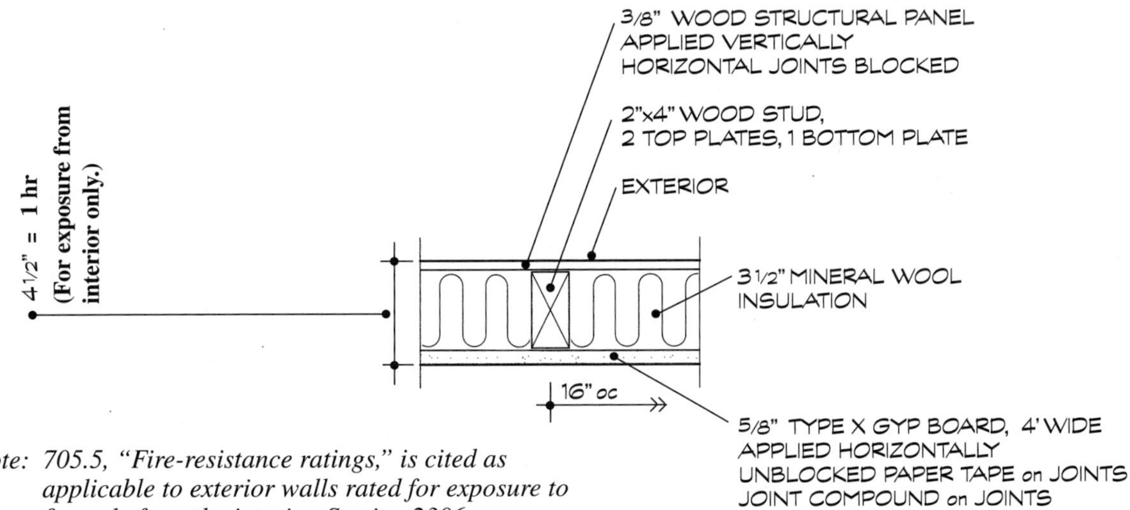

Note: 705.5, "Fire-resistance ratings," is cited as applicable to exterior walls rated for exposure to fire only from the interior. Section 2306, Allowable Stress Design," is cited as governing the calculation of allowable stress.

16-1.1

Fig. 720.1(2) 16A. Fire-resistance ratings for exterior walls. Minimum thickness of assembly is indicated as required for the fire-resistance rating shown. These fire-resistance ratings are established for walls loaded to 100% of their design load. *[Source: IBC Table 720.1(2).]*

720 Prescriptive Fire Resistance

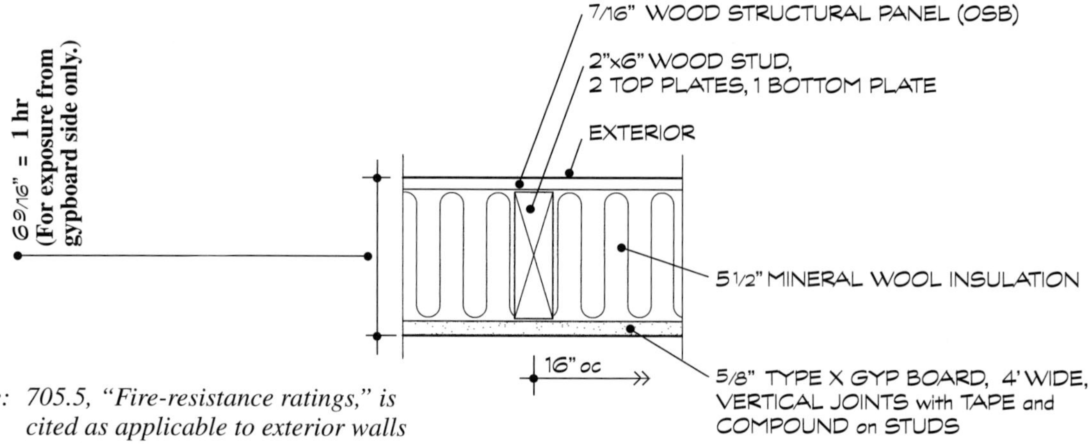

Note: 705.5, "Fire-resistance ratings," is cited as applicable to exterior walls rated for exposure to fire only from the interior. Section 2306, "Allowable Stress Design," is cited as governing the calculation of allowable stress.

16-1.2

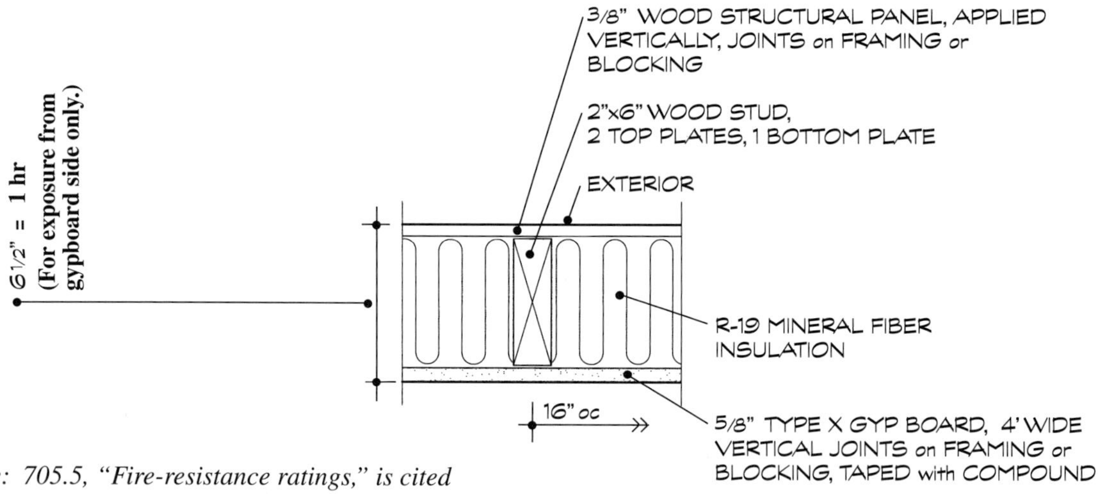

Note: 705.5, "Fire-resistance ratings," is cited as applicable to exterior walls rated for exposure to fire only from the interior. Section 2306, "Allowable Stress Design," is cited as governing the calculation of allowable stress.

16-1.3

Fig. 720.1(2) 16B. Fire-resistance ratings for exterior walls. Minimum thickness of assembly is indicated as required for the fire-resistance rating shown. These fire-resistance ratings are established for walls loaded to 100% of their design load. *[Source: IBC Table 720.1(2).]*

720 Prescriptive Fire Resistance

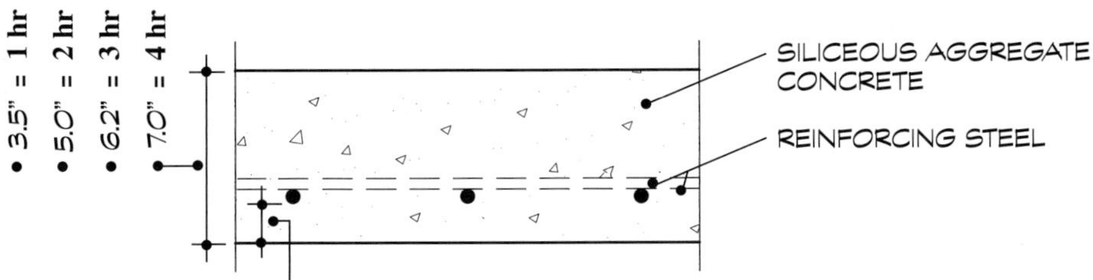

Min for nonprestressed steel

Slab unrestrained

- 3/4" = **1 hr**
- 1" = **2 hr**
- 1 1/4" = **3 hr**
- 1 5/8" = **4 hr**

Slab restrained

- 3/4" = **1-4 hr**

1-1.1

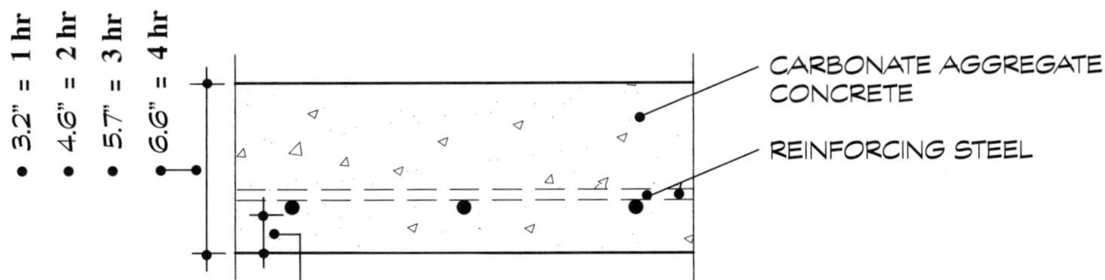

Min for nonprestressed steel

Slab unrestrained

- 3/4" = **1 hr**
- 3/4" = **2 hr**
- 1 1/4" = **3 hr**
- 1 1/4" = **4 hr**

Slab restrained

- 3/4" = **1-4 hr**

2-1.1

Fig. 720.1(3) 1–5A. Minimum protection for concrete floor and roof systems. Minimum thickness of assembly is indicated as required for the fire-resistance rating shown. *[Source: IBC Table 720.1(3).]*

720 Prescriptive Fire Resistance

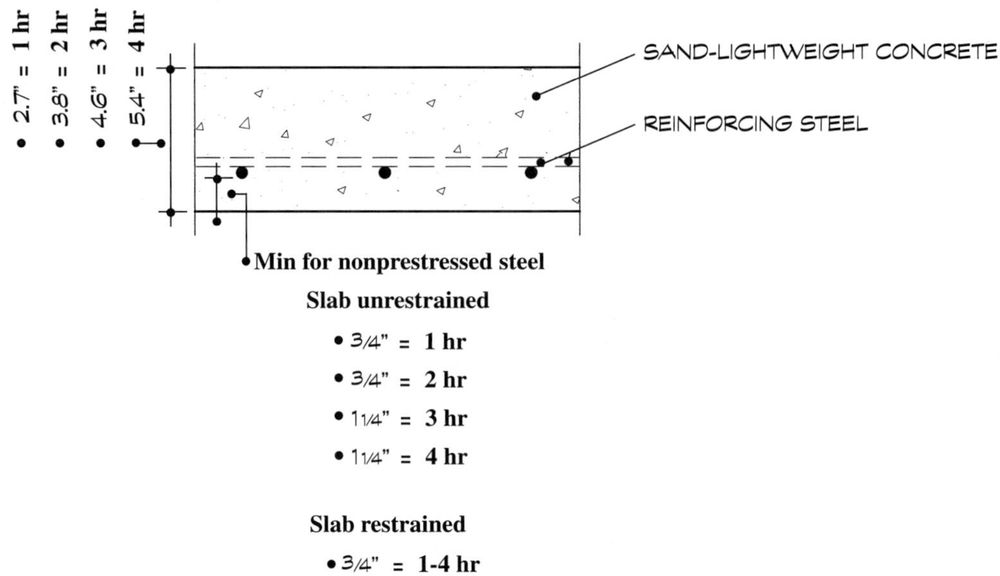

3-1.1

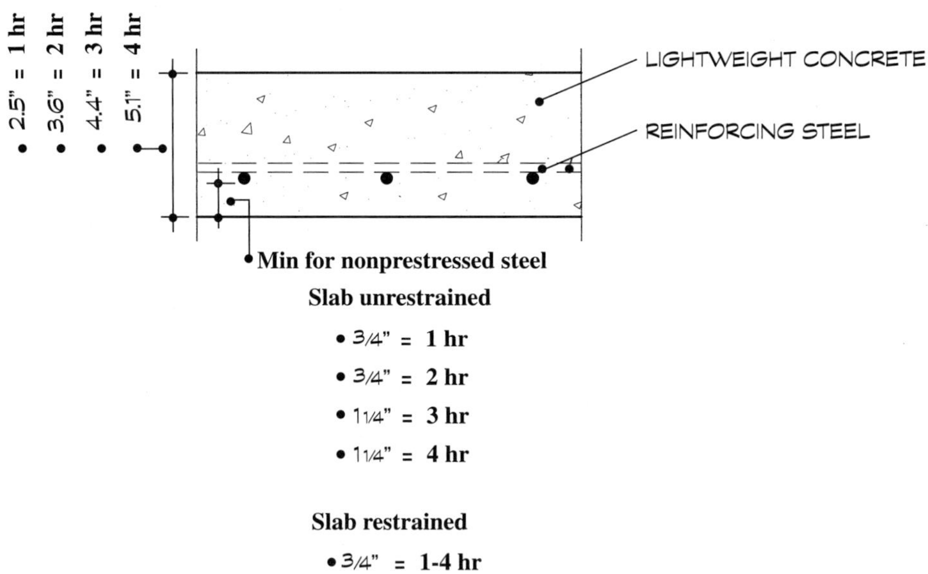

4-1.1

Fig. 720.1(3) 1–5B. Minimum protection for concrete floor and roof systems. Minimum thickness of assembly is indicated as required for the fire-resistance rating shown. *[Source: IBC Table 720.1(3).]*

720 Prescriptive Fire Resistance

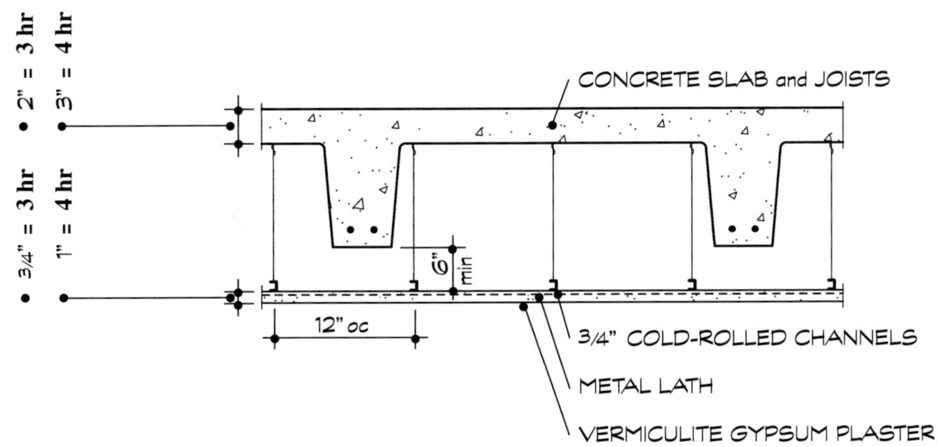

5-1.1

Fig. 720.1(3) 1–5C. Minimum protection for concrete floor and roof systems. Minimum thickness of assembly is indicated as required for the fire-resistance rating shown. *[Source: IBC Table 720.1(3).]*

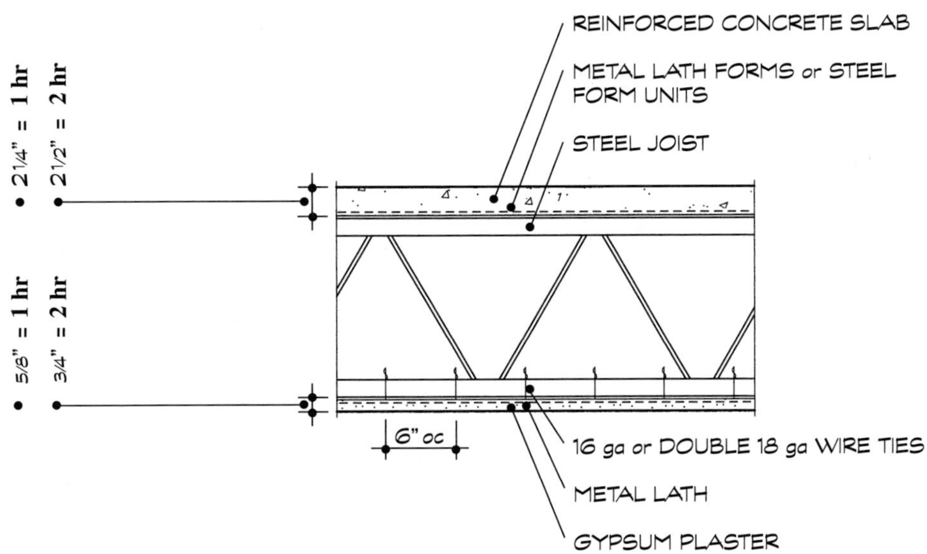

6-1.1

Fig. 720.1(3) 6–12A. Minimum protection for concrete and steel floor and roof systems. Minimum thickness of assembly is indicated as required for the fire-resistance rating shown. *[Source: IBC Table 720.1(3).]*

720 Prescriptive Fire Resistance

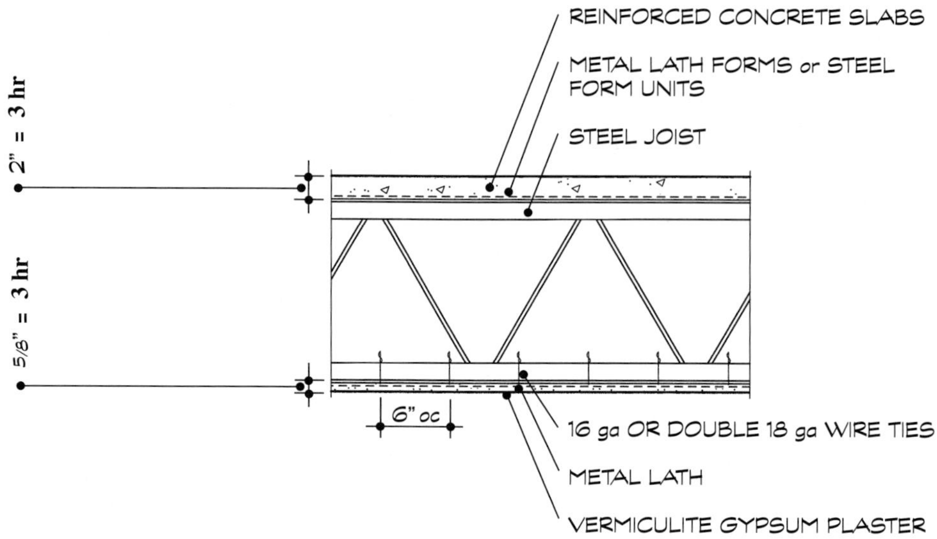

6-2.1

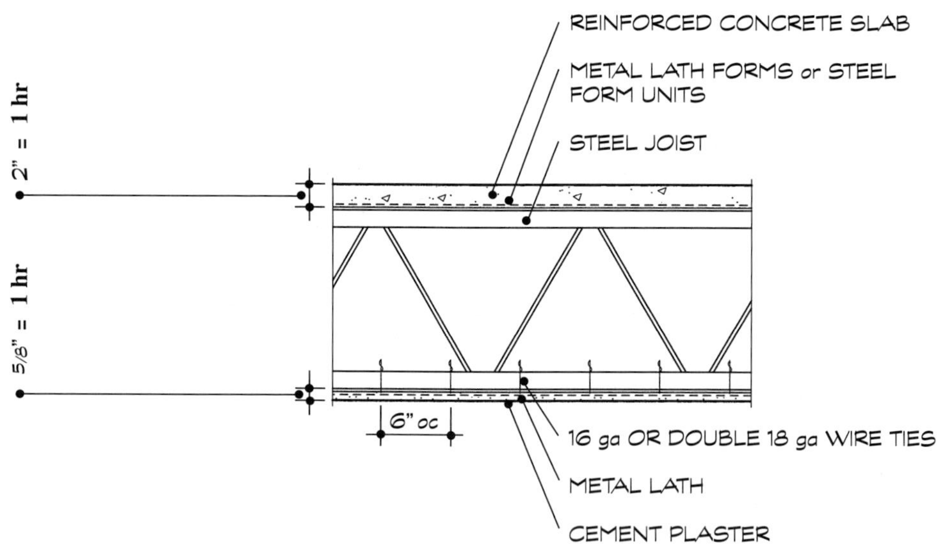

6-3.1

Fig. 720.1(3) 6–12B. Minimum protection for concrete and steel floor and roof systems. Minimum thickness of assembly is indicated as required for the fire-resistance rating shown. *[Source: IBC Table 720.1(3).]*

720 Prescriptive Fire Resistance

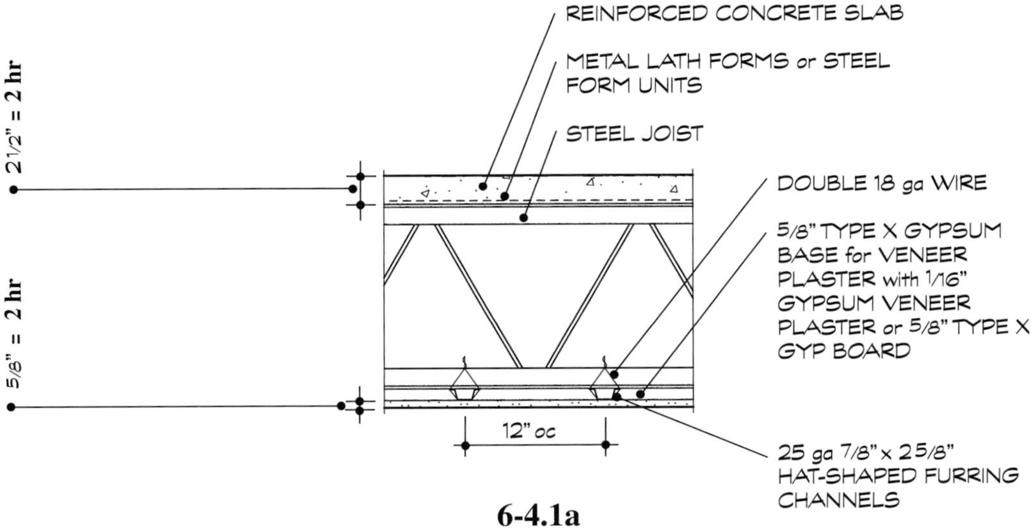

6-4.1a

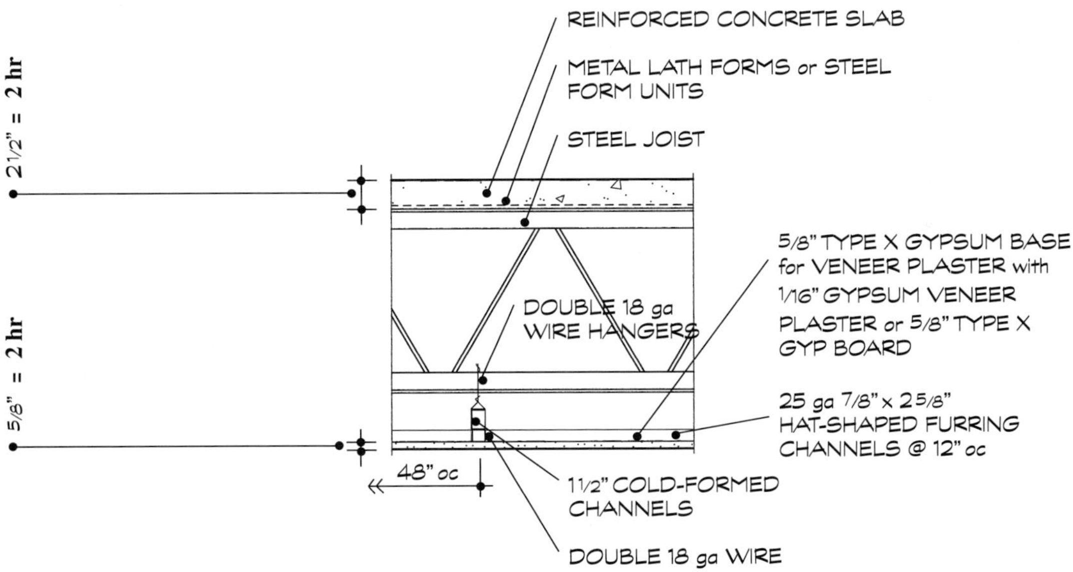

6-4.1b

Fig. 720.1(3) 6–12C. Minimum protection for concrete and steel floor and roof systems. Minimum thickness of assembly is indicated as required for the fire-resistance rating shown. *[Source: IBC Table 720.1(3).]*

720 Prescriptive Fire Resistance

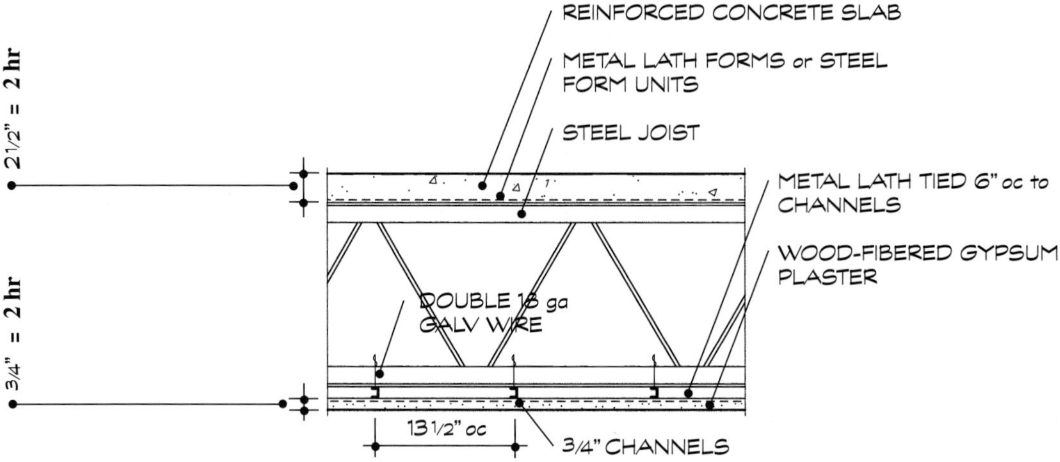

6-5.1

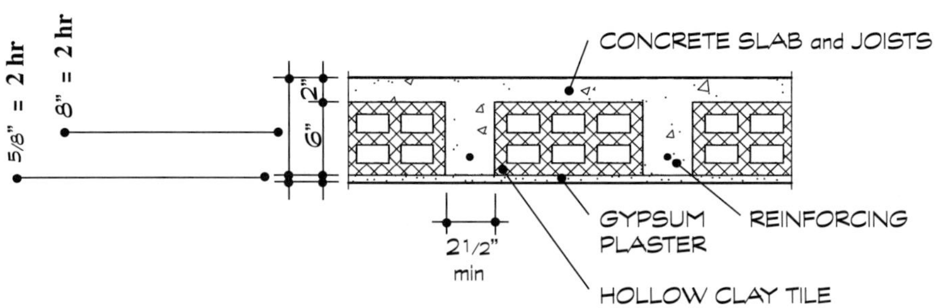

7-1.1

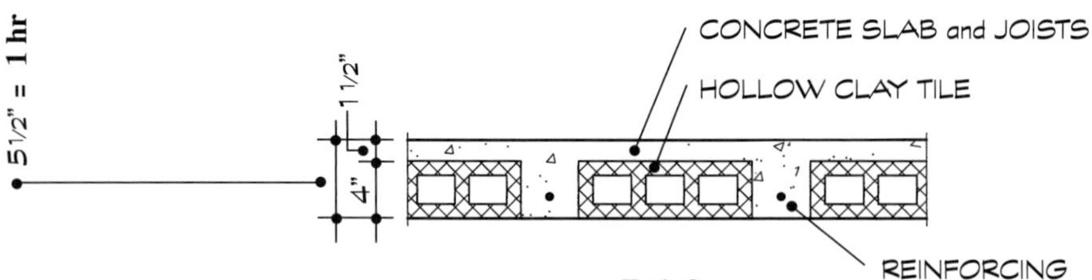

7-1.2

Fig. 720.1(3) 6–12D. Minimum protection for concrete and steel floor and roof systems. Minimum thickness of assembly is indicated as required for the fire-resistance rating shown. *[Source: IBC Table 720.1(3).]*

720 Prescriptive Fire Resistance

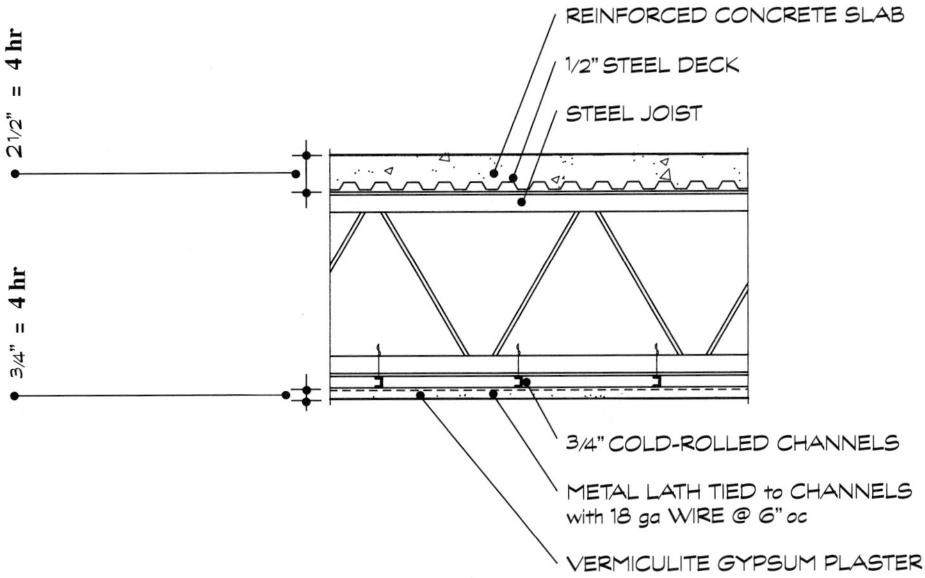

8-1.1

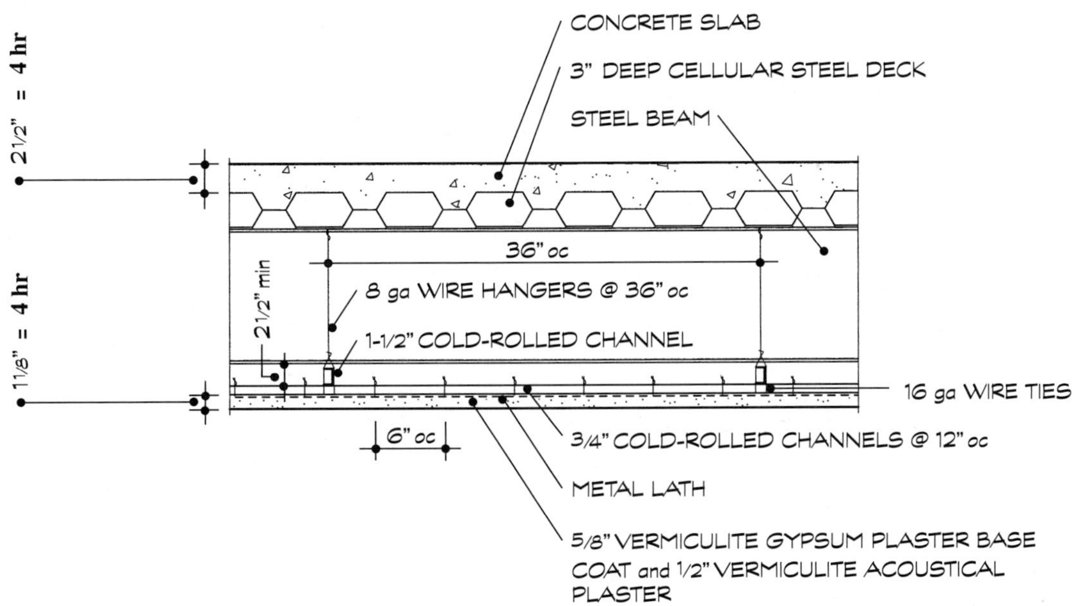

9-1.1

Fig. 720.1(3) 6–12E. Minimum protection for concrete and steel floor and roof systems. Minimum thickness of assembly is indicated as required for the fire-resistance rating shown. *[Source: IBC Table 720.1(3).]*

720 Prescriptive Fire Resistance

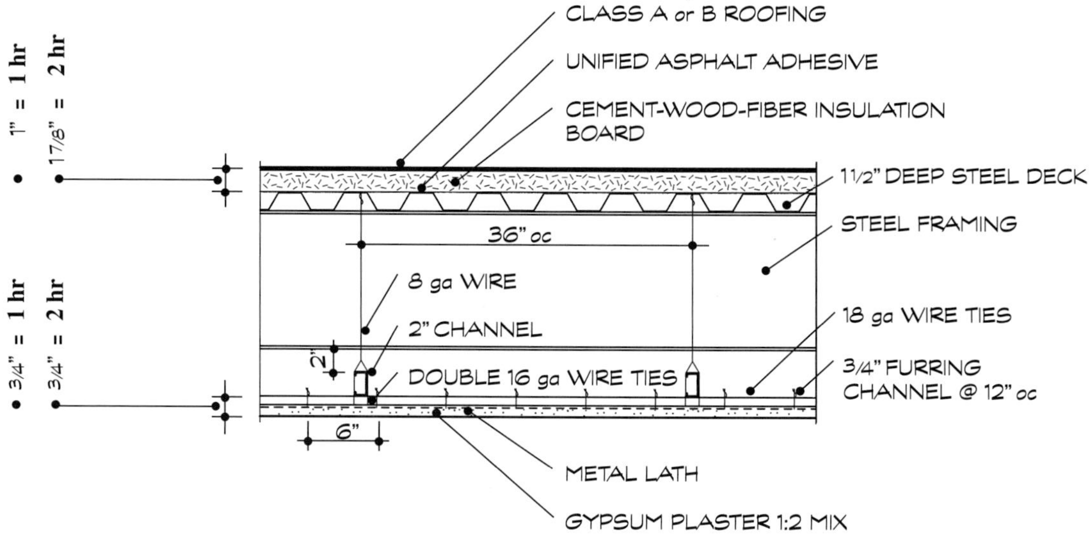

10-1.1

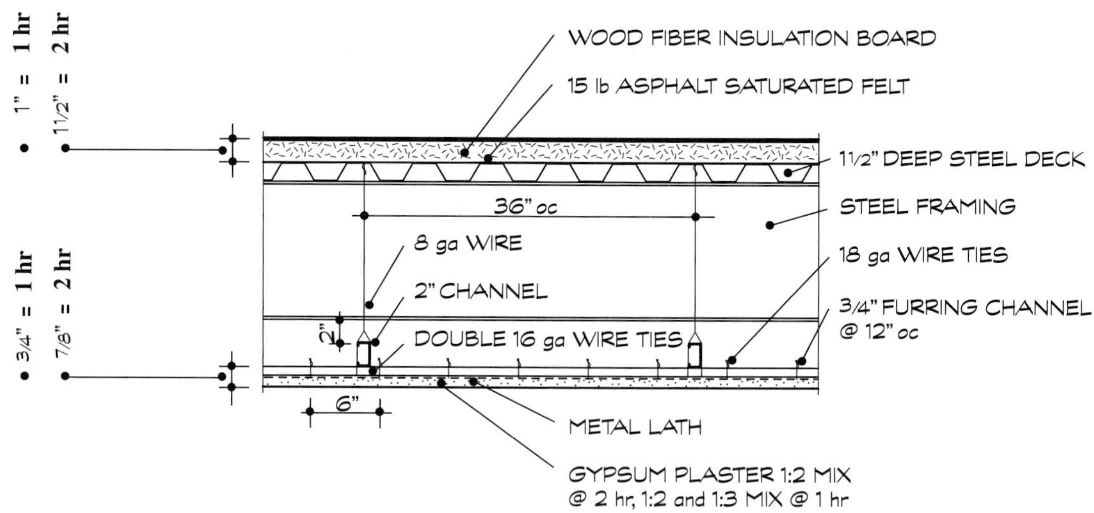

11-1.1

Fig. 720.1(3) 6–12F. Minimum protection for concrete and steel floor and roof systems. Minimum thickness of assembly is indicated as required for the fire-resistance rating shown. *[Source: IBC Table 720.1(3).]*

720 Prescriptive Fire Resistance

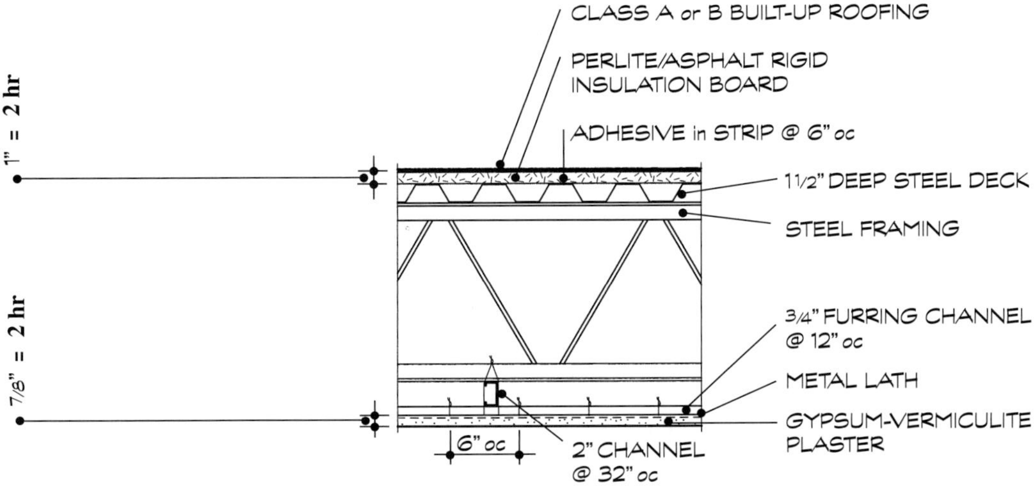

12-1.1

Fig. 720.1(3) 6–12G. Minimum protection for concrete and steel floor and roof systems. Minimum thickness of assembly is indicated as required for the fire-resistance rating shown. *[Source: IBC Table 720.1(3).]*

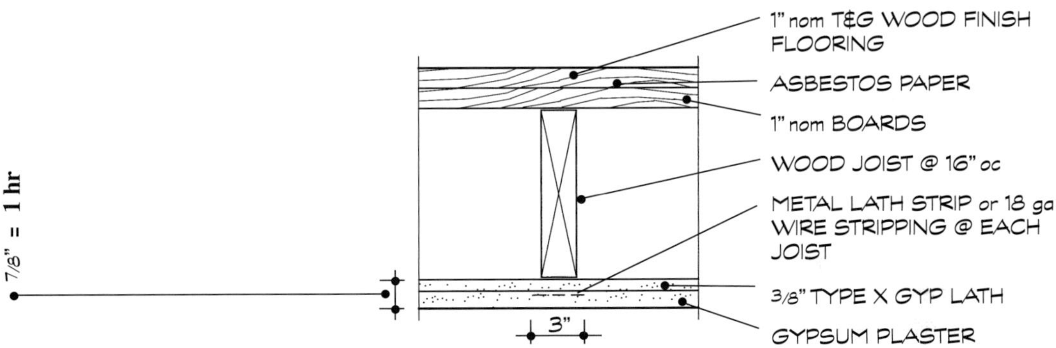

13-1.1a

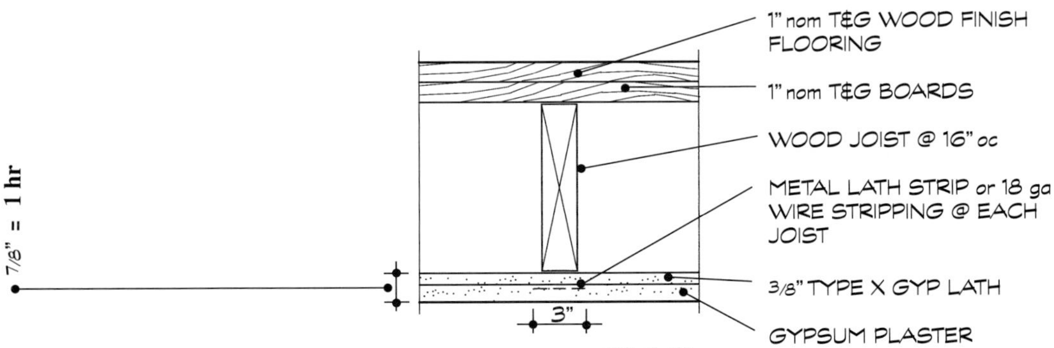

13-1.1b

Fig. 720.1(3) 13–14A. Minimum protection for wood floor and roof systems. Minimum thickness of assembly is indicated as required for the fire-resistance rating shown. *[Source: IBC Table 720.1(3).]*

720 Prescriptive Fire Resistance

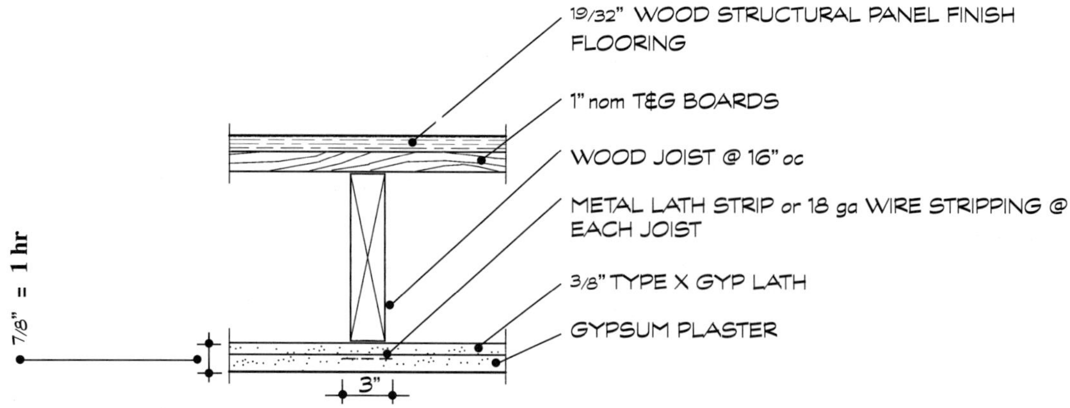

13-1.1c

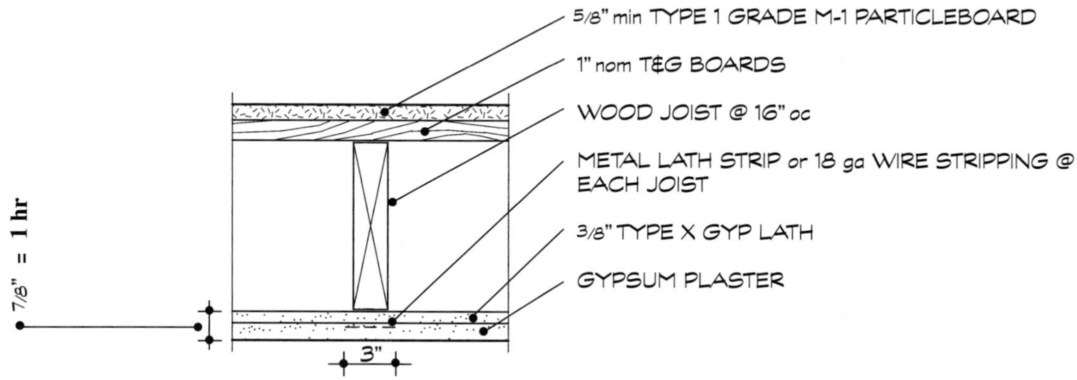

13-1.1d

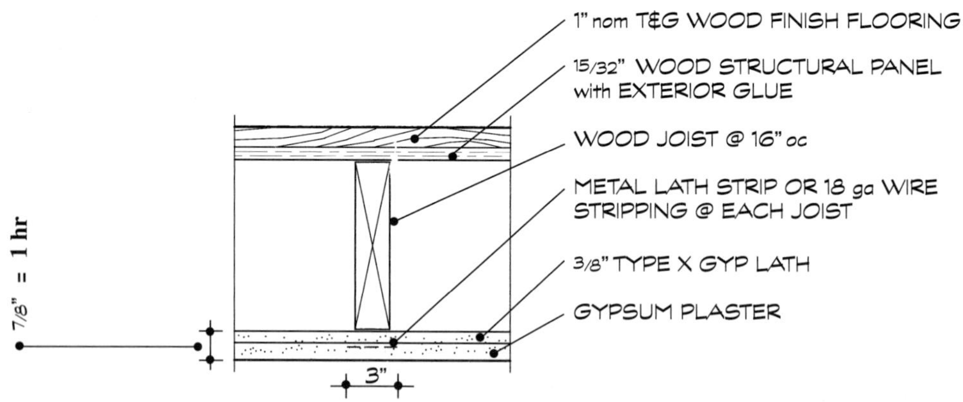

13-1.1e

Fig. 720.1(3) 13–14B. Minimum protection for wood floor and roof systems. Minimum thickness of assembly is indicated as required for the fire-resistance rating shown. *[Source: IBC Table 720.1(3).]*

720 Prescriptive Fire Resistance

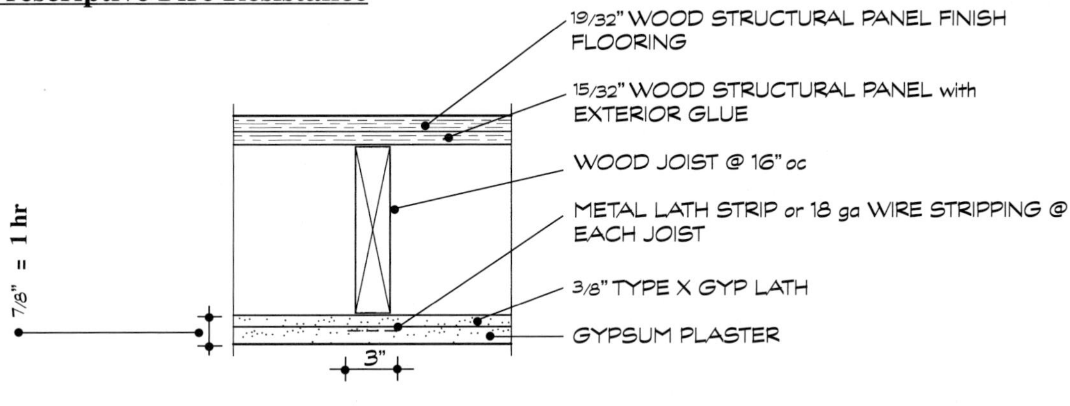

13-1.1f

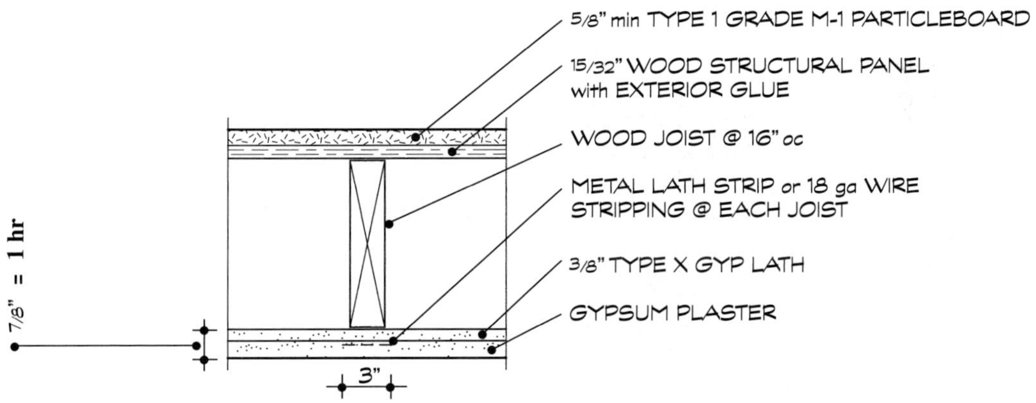

13-1.1g

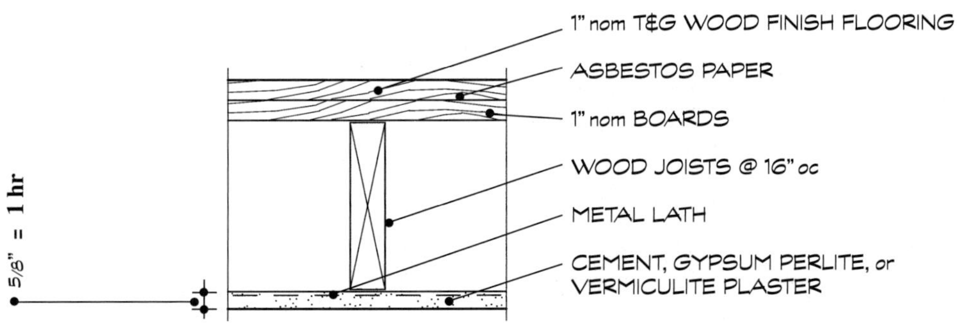

13-1.2+1.3a

Fig. 720.1(3) 13–14C. Minimum protection for wood floor and roof systems. Minimum thickness of assembly is indicated as required for the fire-resistance rating shown. *[Source: IBC Table 720.1(3).]*

720 Prescriptive Fire Resistance

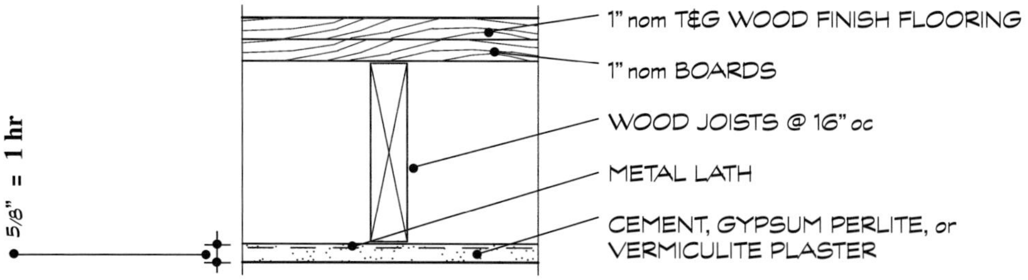

13-1.2+1.3b

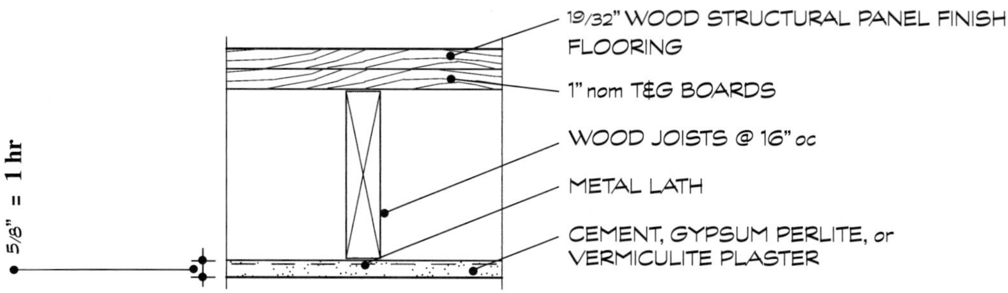

13-1.2+1.3c

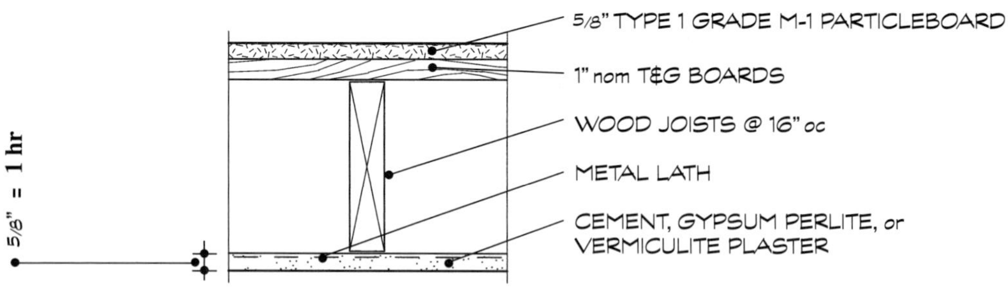

13-1.2+1.3d

Fig. 720.1(3) 13–14D. Minimum protection for wood floor and roof systems. Minimum thickness of assembly is indicated as required for the fire-resistance rating shown. *[Source: IBC Table 720.1(3).]*

720 Prescriptive Fire Resistance

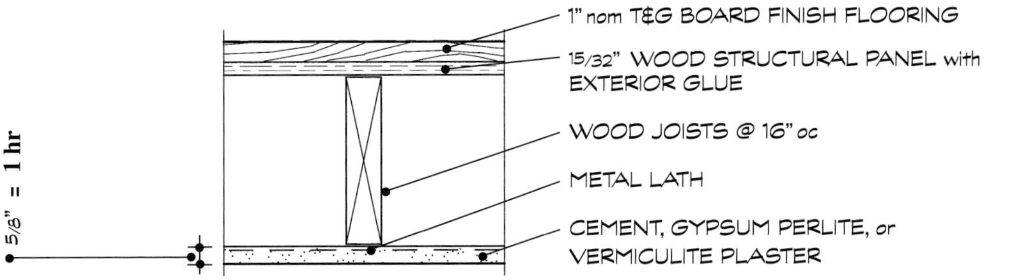

13-1.2+1.3e

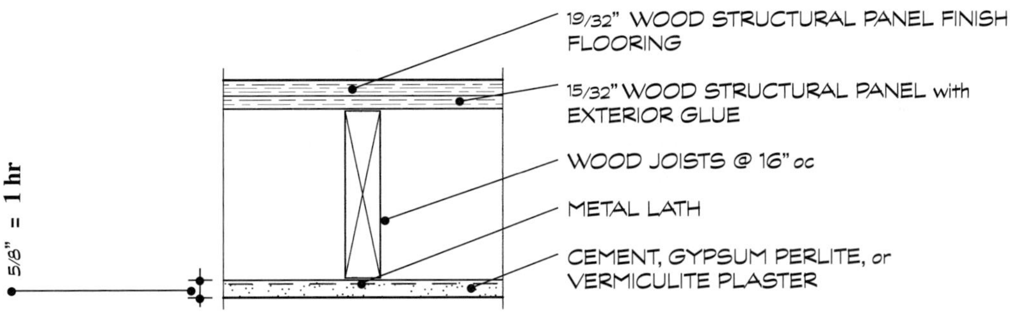

13-1.2+1.3f

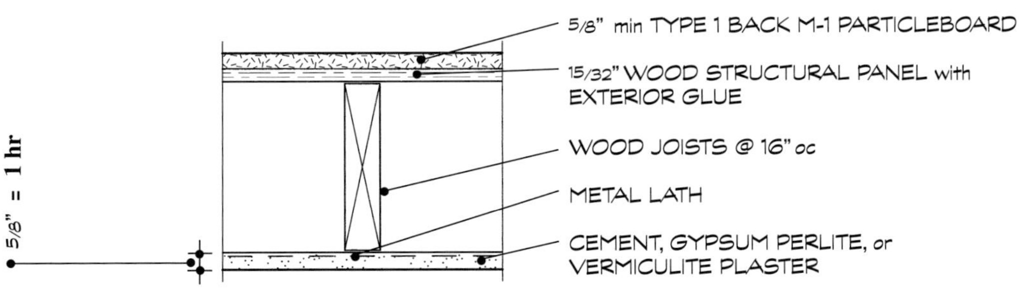

13-1.2+1.3g

Fig. 720.1(3) 13–14E. Minimum protection for wood floor and roof systems. Minimum thickness of assembly is indicated as required for the fire-resistance rating shown. *[Source: IBC Table 720.1(3).]*

720 Prescriptive Fire Resistance

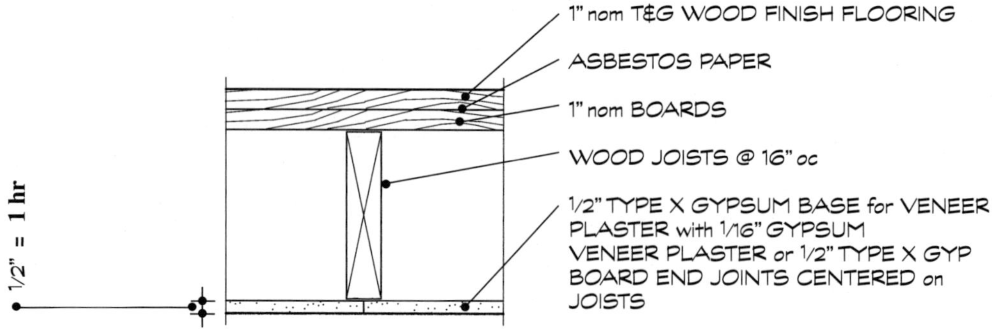

13-1.4a

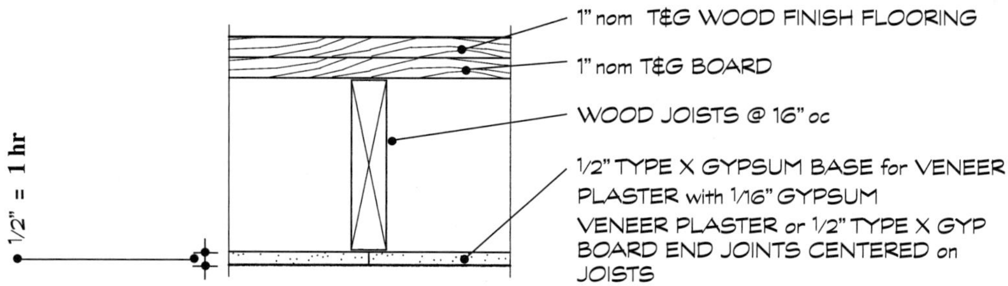

13-1.4b

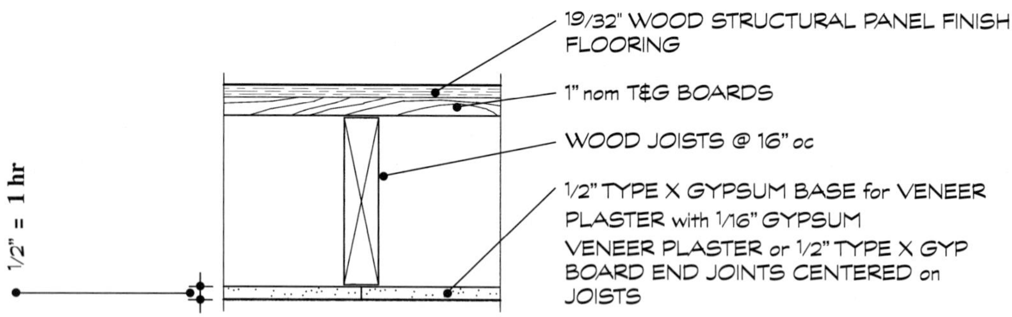

13-1.4c

Fig. 720.1(3) 13–14F. Minimum protection for wood floor and roof systems. Minimum thickness of assembly is indicated as required for the fire-resistance rating shown. *[Source: IBC Table 720.1(3).]*

720 Prescriptive Fire Resistance

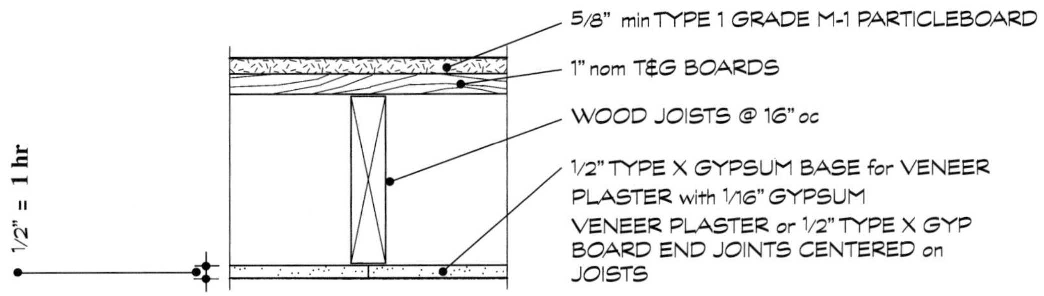

13-1.4d

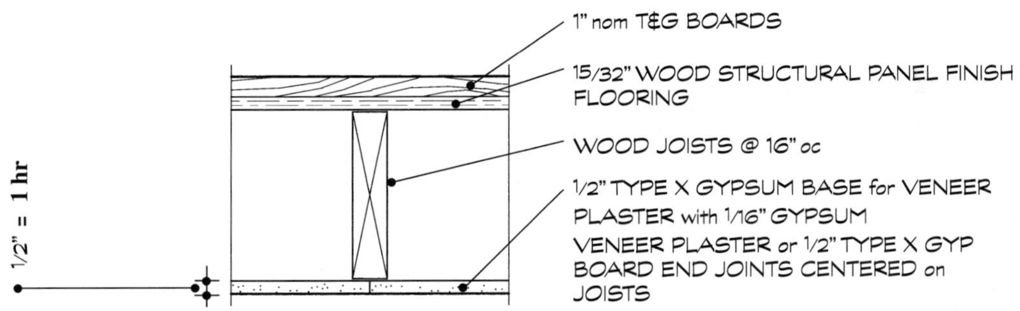

13-1.4e

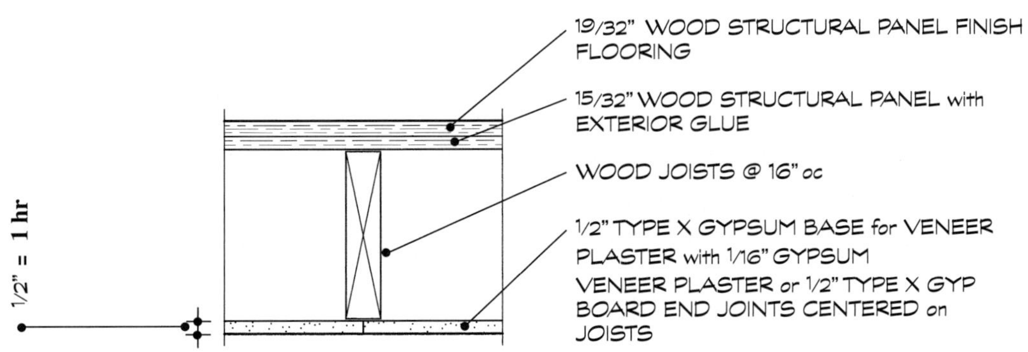

13-1.4f

Fig. 720.1(3) 13–14G. Minimum protection for wood floor and roof systems. Minimum thickness of assembly is indicated as required for the fire-resistance rating shown. *[Source: IBC Table 720.1(3).]*

720 Prescriptive Fire Resistance

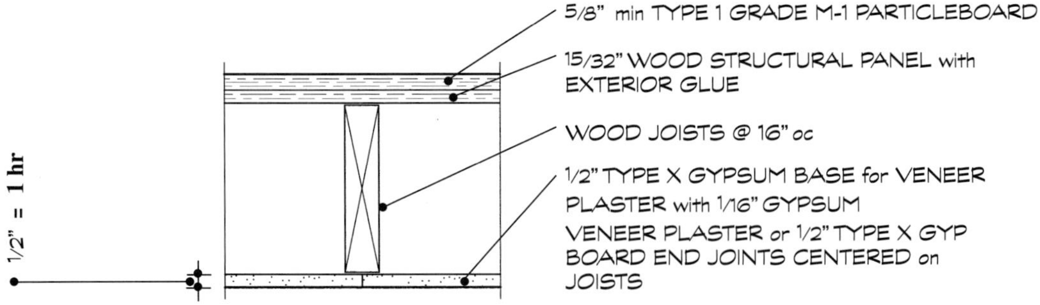

13-1.4g

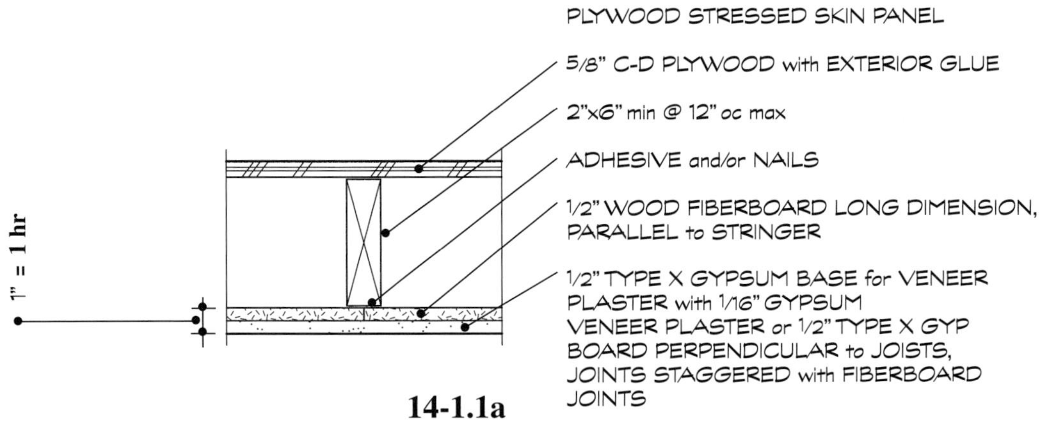

14-1.1a

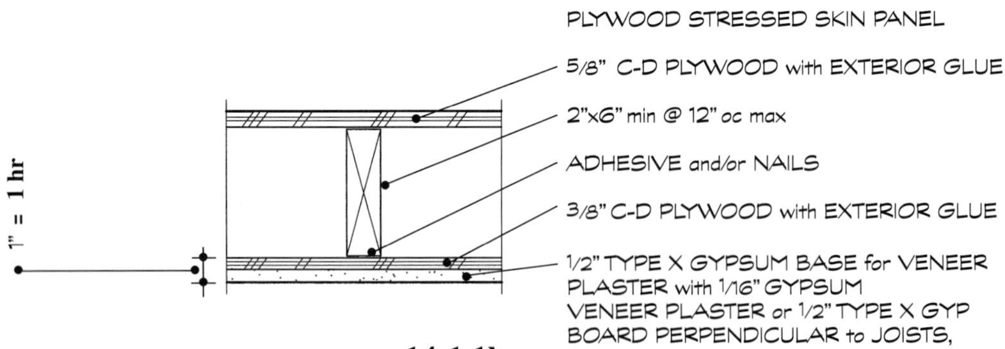

14-1.1b

Fig. 720.1(3) 13–14H. Minimum protection for wood floor and roof systems. Minimum thickness of assembly is indicated as required for the fire-resistance rating shown. *[Source: IBC Table 720.1(3).]*

720 Prescriptive Fire Resistance

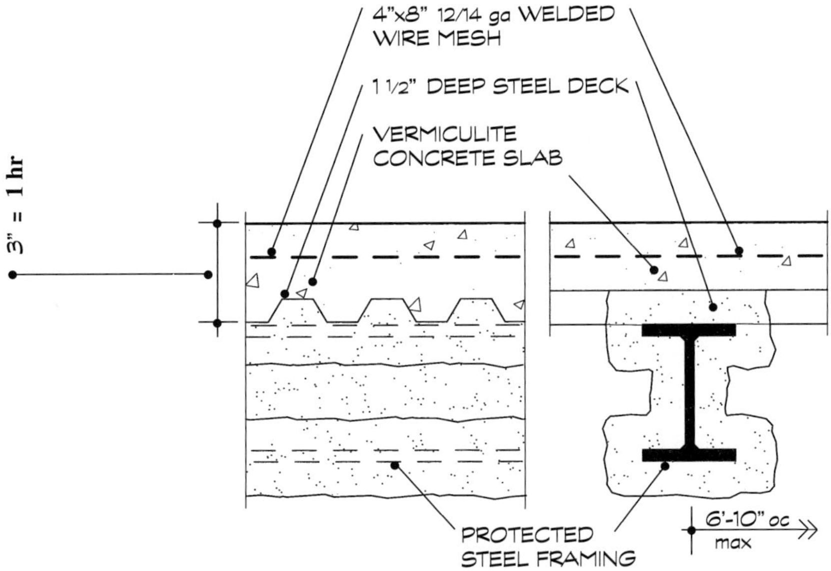

15-1.1

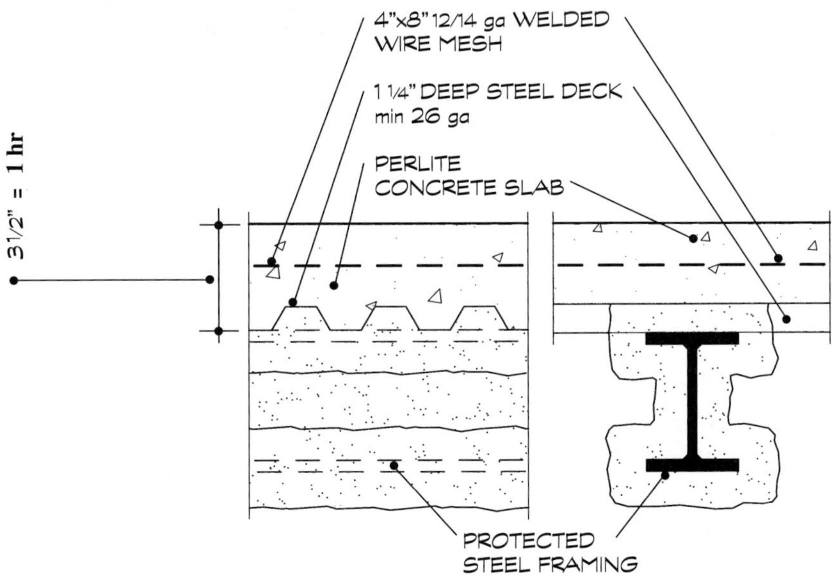

16-1.1

Fig. 720.1(3) 15–20A. Minimum protection for concrete and steel floor and roof systems. Minimum thickness of assembly is indicated as required for the fire-resistance rating shown. *[Source: IBC Table 720.1(3).]*

720 Prescriptive Fire Resistance

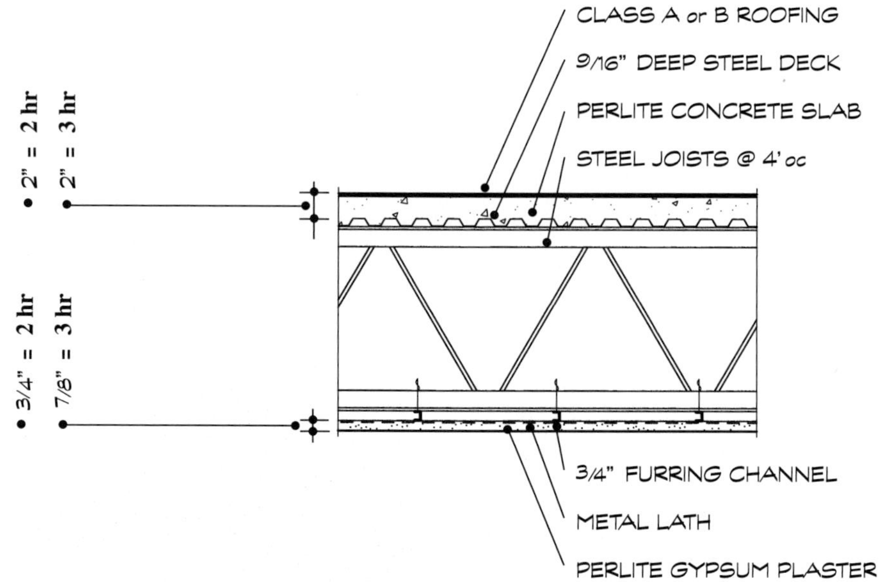

17-1.1

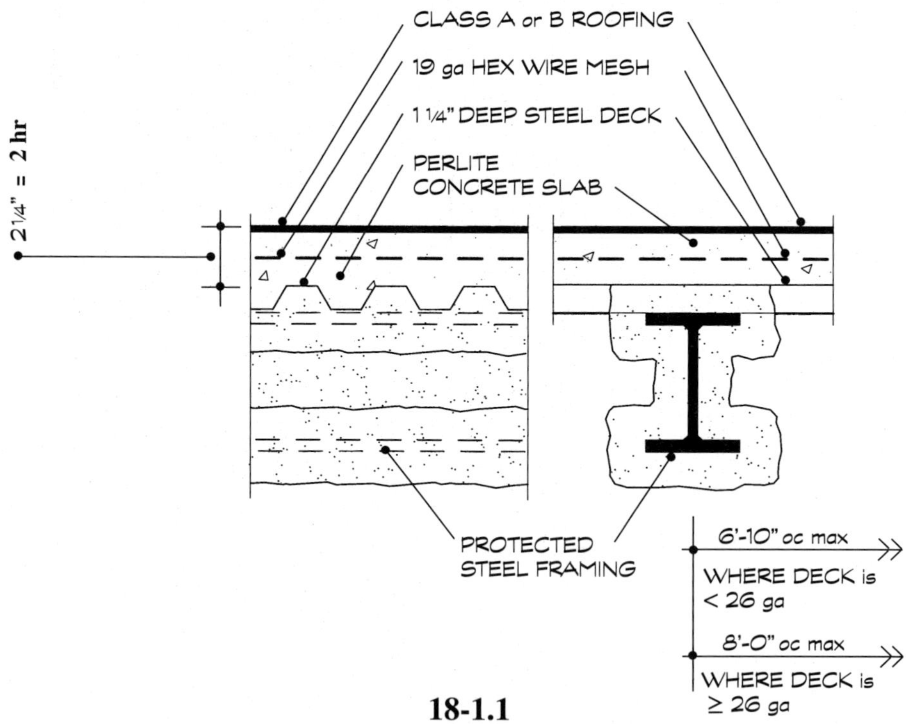

18-1.1

Fig. 720.1(3) 15–20B. Minimum protection for concrete and steel floor and roof systems. Minimum thickness of assembly is indicated as required for the fire-resistance rating shown. *[Source: IBC Table 720.1(3).]*

720 Prescriptive Fire Resistance

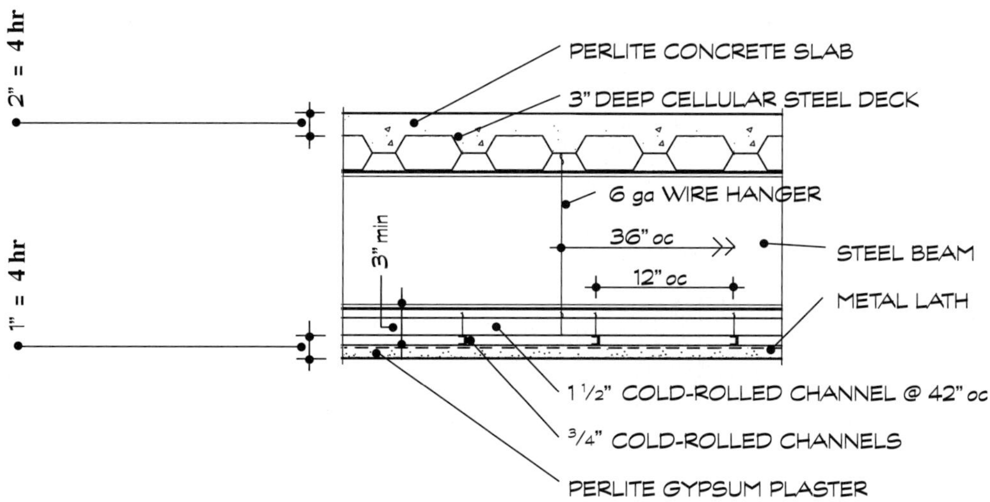

19-1.1

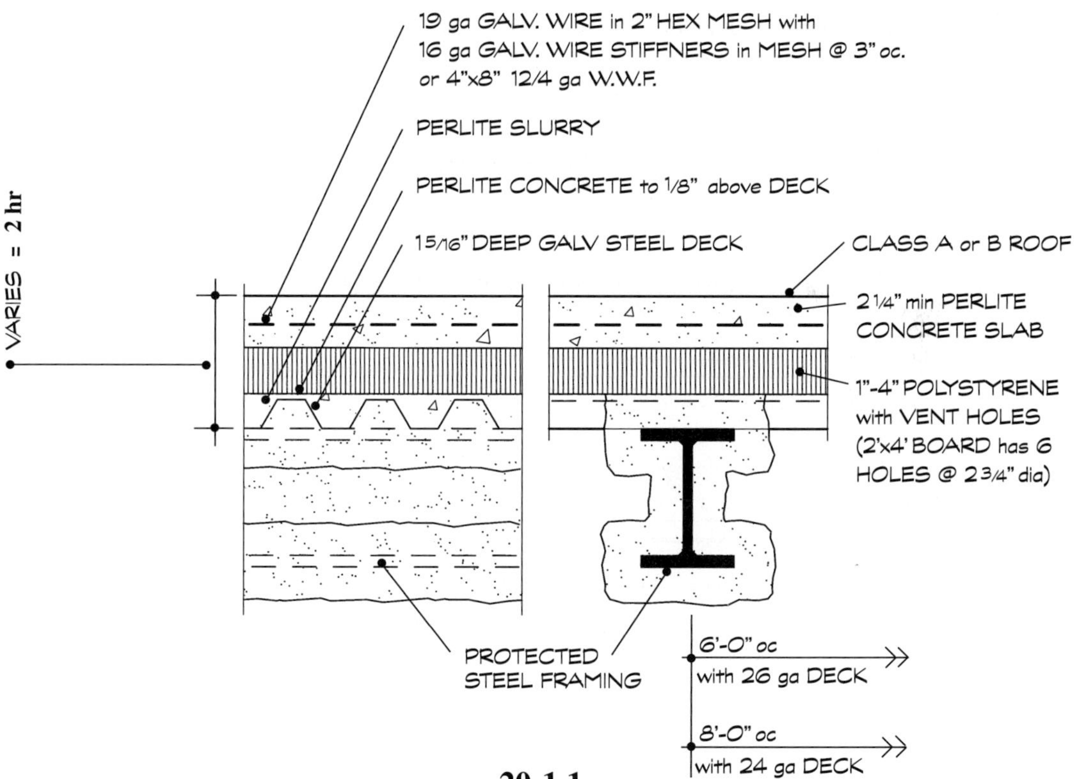

20-1.1

Fig. 720.1(3) 15–20C. Minimum protection for concrete and steel floor and roof systems. Minimum thickness of assembly is indicated as required for the fire-resistance rating shown. *[Source: IBC Table 720.1(3).]*

720 Prescriptive Fire Resistance

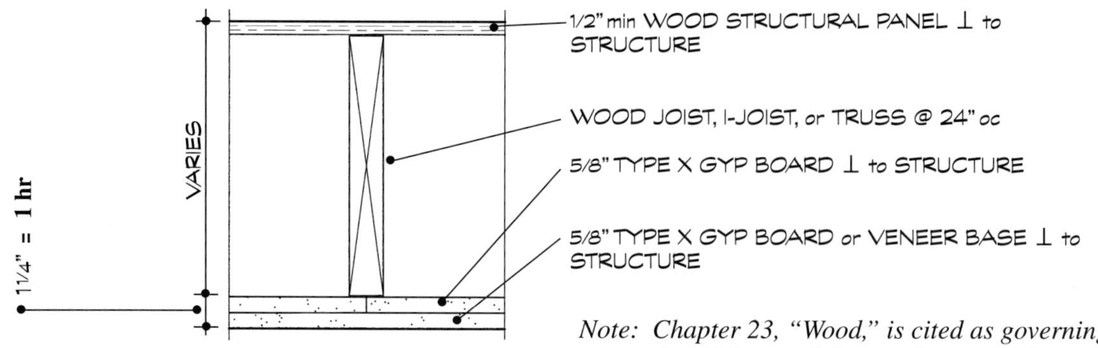

Note: Chapter 23, "Wood," is cited as governing the thickness of the wood structural panel, which must not be < 1/2" of the thickness specified in the chapter.

21-1.1

Fig. 720.1(3) 21A. Minimum protection for wood floor and roof systems. Minimum thickness of assembly is indicated as required for the fire-resistance rating shown. *[Source: IBC Table 720.1(3).]*

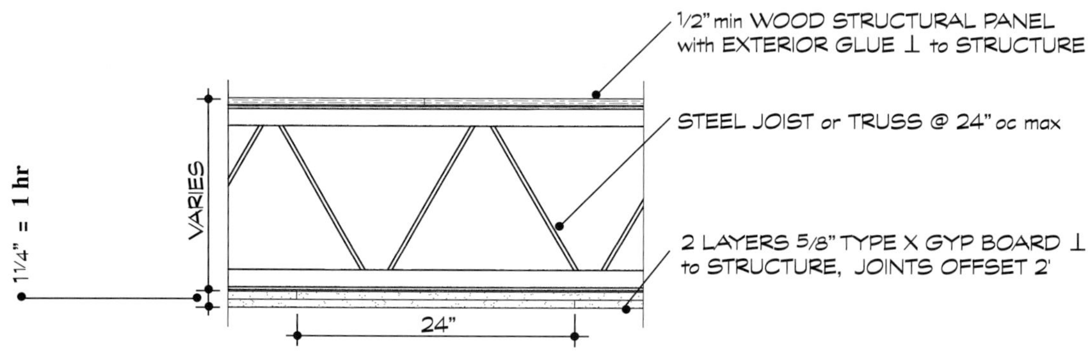

Note: Chapter 23, "Wood," is cited as governing the thickness of the wood structural panel, which must not be < 1/2" of the thickness specified in the chapter.

22-1.1

Fig. 720.1(3) 22A. Minimum protection for steel and wood floor and roof systems. Minimum thickness of assembly is indicated as required for the fire-resistance rating shown. *[Source: IBC Table 720.1(3).]*

720 Prescriptive Fire Resistance

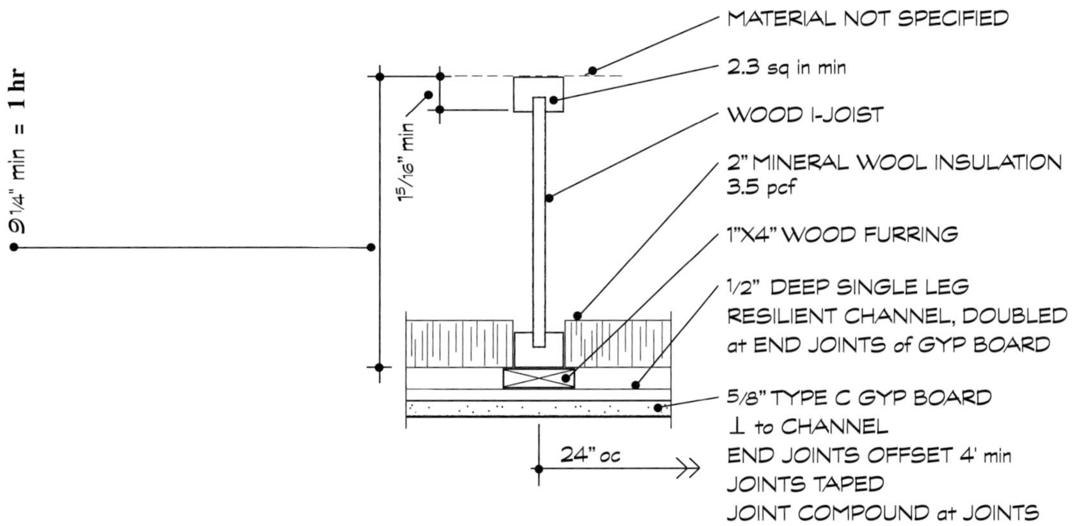

23-1.1

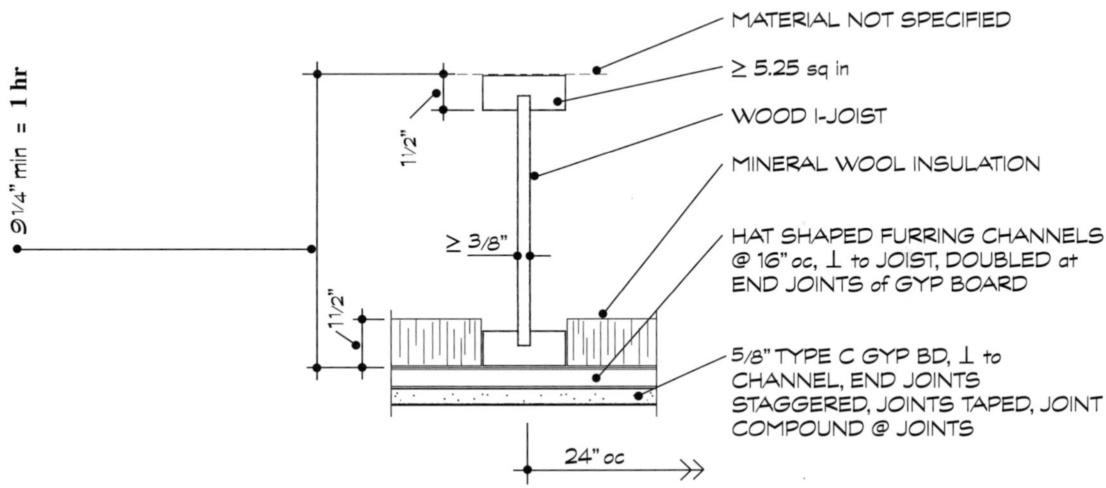

24-1.1

Fig. 720.1(3) 23-28A. Minimum protection for wood floor and roof systems. Dimensions and components required for the fire-resistance rating shown. *[Source: IBC Table 720.1(3).]*

720 Prescriptive Fire Resistance

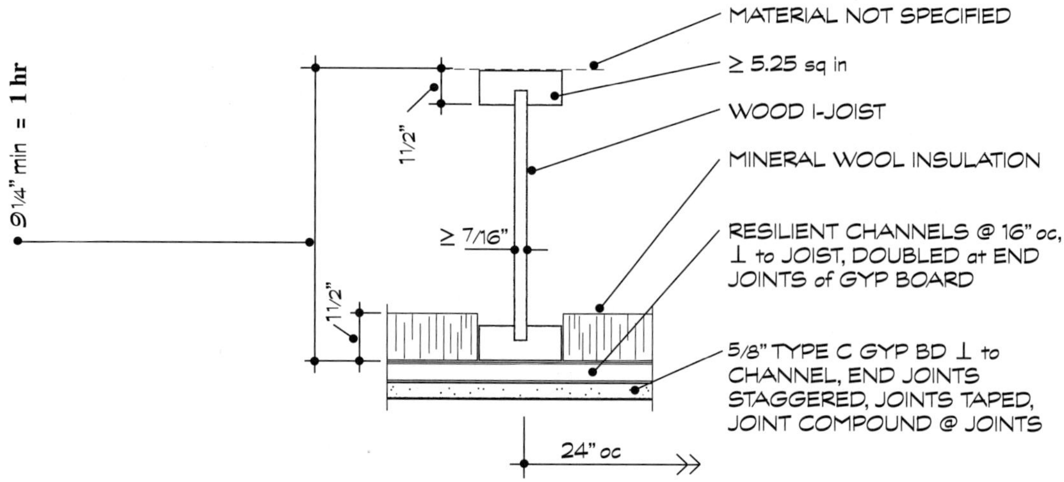

25-1.1

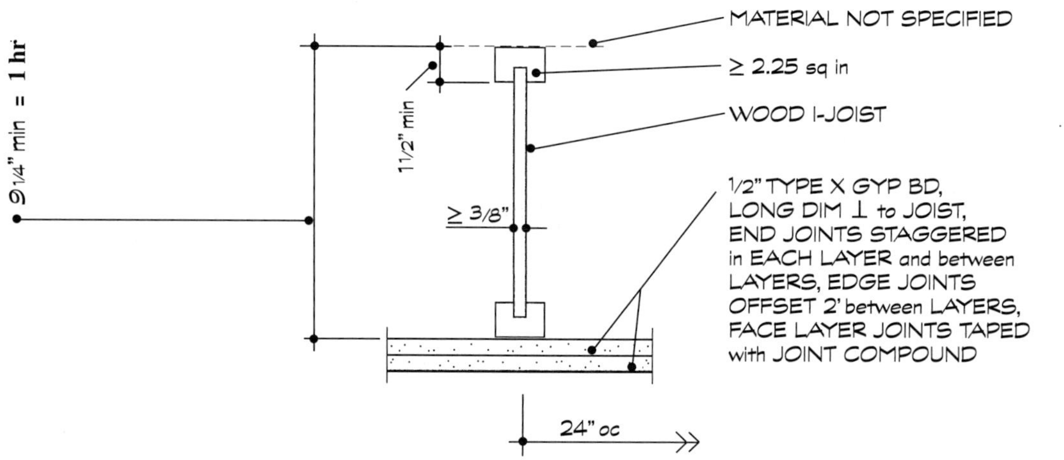

26-1.1

Fig. 720.1(3) 23-28B. Minimum protection for wood floor and roof systems. Dimensions and components required for the fire-resistance rating shown. *[Source: IBC Table 720.1(3).]*

720 Prescriptive Fire Resistance

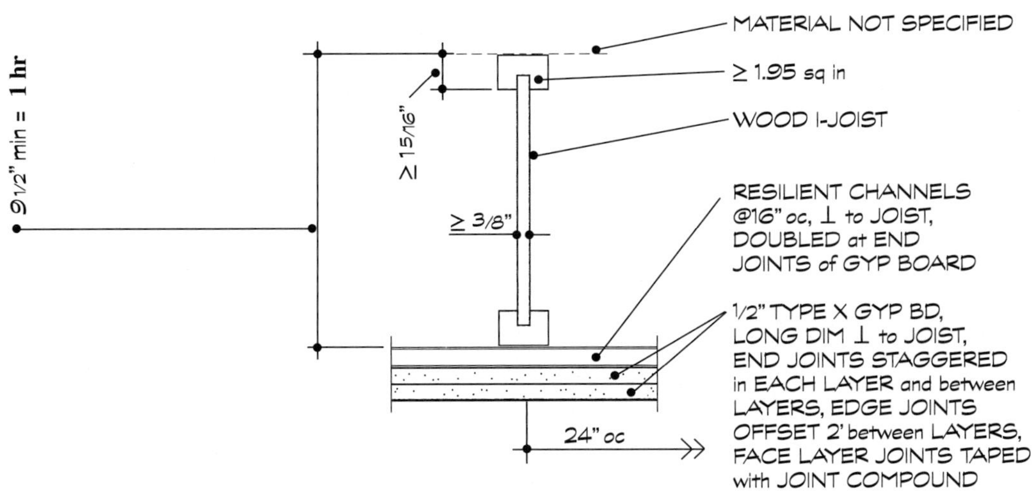

27-1.1

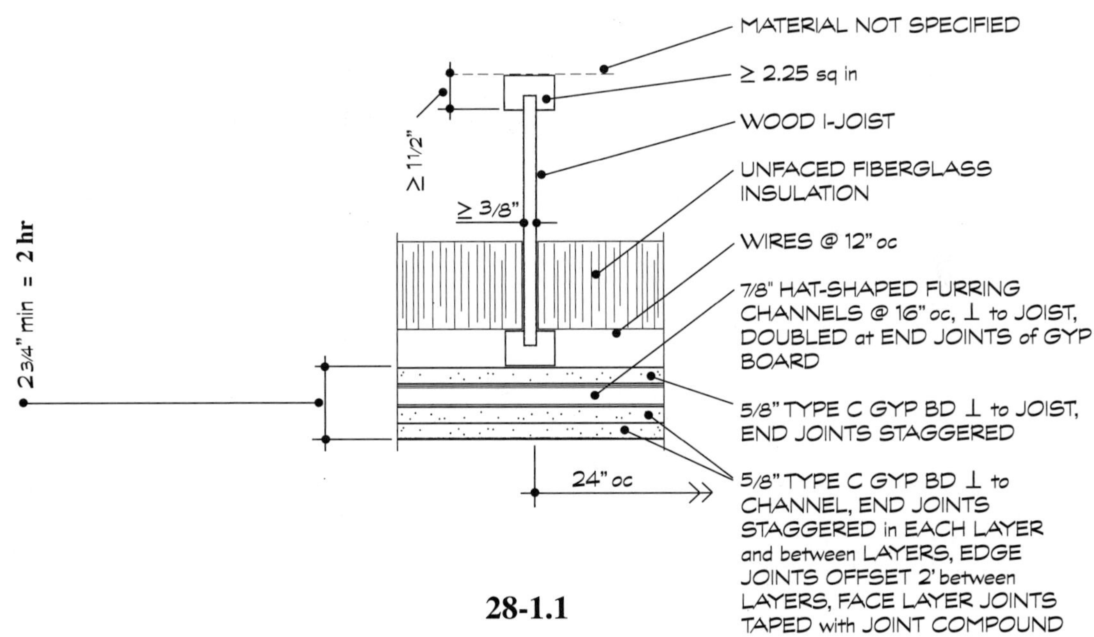

28-1.1

Fig. 720.1(3) 23-28C. Minimum protection for wood floor and roof systems. Dimensions and components required for the fire-resistance rating shown. *[Source: IBC Table 720.1(3).]* *(Note: Test data submitted to the ICC with the approved proposal for this entry indicates that the insulation is to be ≥ 3 1/2" thick.)*

720 Prescriptive Fire Resistance

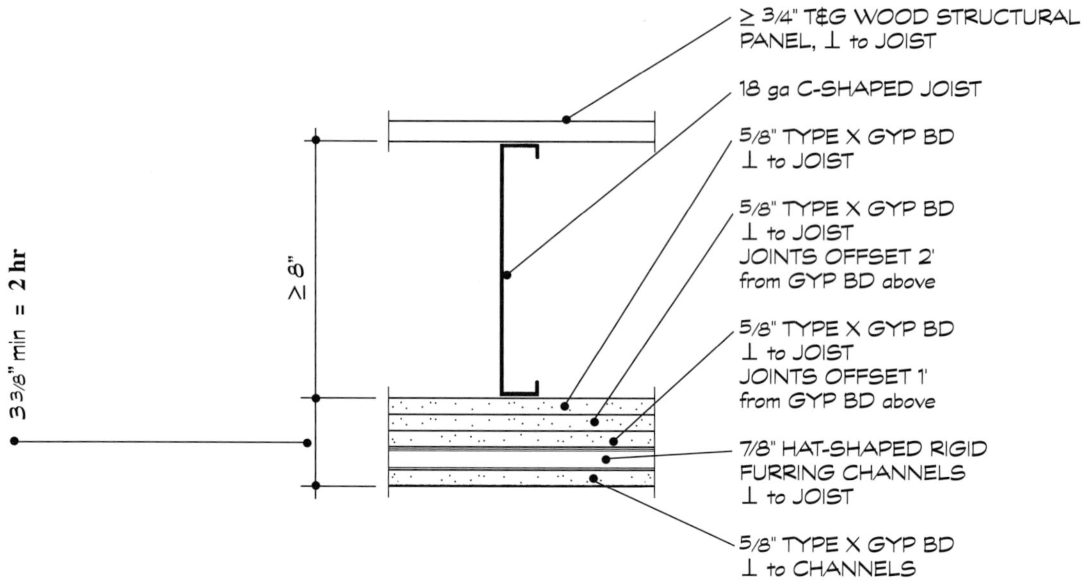

29-1.1

Fig. 720.1(3) 29A. Minimum protection for steel and wood floor and roof systems. Dimensions and components required for the fire-resistance rating shown. *[Source: IBC Table 720.1(3).]*

720 Prescriptive Fire Resistance

720.1.3 Reinforcement for cast-in-place concrete column protection

- Cast-in-place concrete protection for steel columns requires the following or equivalent:
  - Wire ties as follows:
    To have a diameter ≥ 0.18".
    To be located a the edges of the column.
    To be wound around the column in a spiral path with a pitch ≤ 8".

720.1.4 Plaster application

- A finish coat of a protective plaster cover is not required in the following case:
  - Where the following meet the requirements of details provided in this section:
    Plaster design mix.
    Total plaster thickness indicated.

Note: *The following tables are cited as listing minimum requirements:*
IBC Table 720.1(1), "Minimum Protection of Structural Parts Based on Time Periods for Various Noncombustible Insulating Materials."
IBC Table 720.1(2), "Rated Fire-Resistance Periods for Various Walls and Partitions."
IBC Table 720.1(3), "Minimum Protection for Floor and Roof Systems."

720.1.5 Bonded prestressed concrete tendons

- The concrete cover for prestressed tendons must be as required by details provided in this section as follows:
  - Where there are single tendons:
    Cover is measured from the nearest surface.
- Multiple tendons with different concrete covers are governed for fire protection as follows:
  - The average cover must be ≥ that required by details provided in this section as follows:
    The average cover is based the following:
      The clear distance from each tendon to the nearest surface.
  - The clear cover required for any tendon is as follows:
    ≥ 1/2 that specified by details provided in this section.
    Required cover for slabs with any type aggregate is ≥ 3/4".
    Required cover for beams with any type aggregate is ≥ 1".
    Tendons with cover less than required are governed as follows:
      They must provide ≤ 50% of the ultimate moment capacity in the following case:
        Where members have a cross-sectional area < 350 sq in.
      They must provide ≤ 65% of the ultimate moment capacity in the following case:
        Where members have a cross-sectional area ≥ 350 sq in.
    - The following assumption is made regarding reduced cover permitted for fire protection:
      Structural integrity is not affected.

Note: *IBC Table 720.1(1), "Minimum Protection of Structural Parts Based on Time Periods for Various Noncombustible Insulating Materials," is cited as governing the above requirements.*

721 Calculated Fire Resistance

721.1.1 Definitions *(part 1 of 2)*

- **Ceramic fiber blanket**
  - Mineral wool insulation as follows:
    - Alumina-silica fibers.
    - ≥ 4 lb/cu ft and ≤ 10 lb/cu ft.
- **Concrete, carbonate aggregate**
  - Aggregates are mainly one or both of the following substances:
    - Calcium carbonate.
    - Magnesium carbonate.
  - Examples of aggregates include the following:
    - Limestone.
    - Dolomite.
  - Aggregates consist of ≤ 40% of the following substances:
    - Quartz.
    - Chert.
    - Flint.
- **Concrete, cellular**
  - Insulating concrete as follows:
    - Preformed foam and portland cement slurry mixture.
    - Dry weight ≃ 30 pcf.
- **Concrete, lightweight aggregate**
  - Aggregates are one or more of the following:
    - Expanded clay.
    - Expanded shale.
    - Expanded slag.
    - Expanded slate.
    - Sintered fly ash.
    - Natural lightweight aggregates as follows:
      - With the same fire-resistive properties as those listed above.
      - ≥ 85 and ≤ 115 pcf.

Note: ASTM C 330, "Standard Specification for Lightweight Aggregates for Structural Concrete," is cited as governing aggregate properties listed above.

- **Concrete, perlite**
  - Insulating concrete as follows:
    - Dry weight ≃ 30 pcf.
    - With perlite aggregate as follows:
      - From volcanic rock.
      - Expanded with heat.
      - A glasslike substance.
      - Cellular in nature.

721 Calculated Fire Resistance

721.1.1 Definitions *(part 2 of 2)*

- **Concrete, sand-lightweight**
  - Aggregates are one or more of the following mixed with natural sand:
    - Expanded clay.
    - Expanded shale.
    - Expanded slag.
    - Expanded slate.
    - Sintered fly ash.
    - Natural lightweight aggregates as follows:
      - With the same fire-resistive properties as those listed above.
      - ≥ 105 and ≤ 120 pcf.

  Note: ASTM C 330, "Standard Specification for Lightweight Aggregates for Structural Concrete," is cited as governing aggregate properties.

- **Concrete, siliceous aggregate**
  - Aggregates are normal weight.
  - Aggregates are mainly in one of the following substances:
    - Silica.
    - Compounds other than the following:
      - Calcium carbonate.
      - Magnesium carbonate.
  - Aggregates contain > 40% of the following substances:
    - Quartz.
    - Chert.
    - Flint.
- **Concrete, vermiculite**
  - Insulating concrete as follows:
    - Dry weight ≃ 30 pcf.
    - With vermiculite aggregate as follows:
      - Laminated micaceous substance.
      - Ore is expanded with heat.
- **Glass fiberboard**
  - Roof insulation as follows:
    - Inorganic glass fibers with binder.
    - Formed into rigid boards.
  - Top surface is faced as follows:
    - With asphalt and glass fiber reinforced kraft paper.
- **Mineral board**
  - Rigid insulation in flat rectangular boards.
  - One of the following substances:
    - Felted mineral fiber.
    - Cellular beads of expanded aggregate.

NOTES

8

Interior Finishes

McKenzie Lofts. Portland, Oregon.
Ankrom Moisan Associated Architects. Portland, Oregon.

801 General

801.1 Scope

- This chapter governs materials used for the following:
  - Interior finishes.
  - Interior trim.
  - Interior decorative materials.

801.4 Decorative materials and trim

- The following properties of decorative materials and trim are governed as noted:
  - Combustibility.
  - Flame propagation,

Note: The following are cited as governing decorative materials and trim:
NFPA 701, "Standard Methods of Fire Tests for Flame-propagation of Textiles and Films."
Section 806, "Decorative Materials and Trim."

801.5 Applicability

- This section addresses the following materials in a flood hazard area:
  - Interior finishes.
  - Trim.
  - Decorative materials.

Note: 1612.3, "Establishment of flood hazard areas," is cited as the source of guidelines for establishing such an area.

- Materials governed by this section must be resistant to flood damage as follows:
  - Where located below the design flood elevation.

801.6 Application

- Combustible materials may be used as a finish for the following interior surfaces:
  - Walls.
  - Ceilings.
  - Floors.
  - Other interior surfaces.

801.7 Windows

- 1st story above grade show windows may be of the following materials:
  - Wood.
  - Unprotected metal framing.

801.8 Foam plastics

- This section governs the following:
  - Exposed foam plastics.
  - Textile or vinyl covered foam plastics.
- Foam plastics used as interior finish or trim must meet requirements listed elsewhere in the code.

Note: 803.4, "Foam plastics," is cited as governing foam plastics used as interior finishes.
The following are cited as governing foam plastics used as interior trim:
806.3, "Foam plastic."
2604.2, "Interior trim."

802 Definitions

802.1 General *(part 1 of 2)*

- **Expanded vinyl wall covering**
  - Woven textile backing.
  - Expanded vinyl base coat layer:
    - A homogeneous vinyl layer.
    - Contains a blowing agent:
      - Agent decomposes during processing, causing the layer to expand:
        - Closed cells are formed in this process.
  - Nonexpanded vinyl skin coat.
  - Total thickness is ≥ 0.055"(±), ≤ 0.070"(±).
- **Flame spread**
  - The expansion of flame over a surface.
- **Flame spread index**
  - A comparative measure as follows:
    - A dimensionless number assigned to a material.
    - Based on visual measurements of the following:
      - Spread of flame vs. time.
    - Based on a test.

Note: The following are cited as alternative standards governing the test:
ASTM E 84, "Standard Test Method for Surface Burning Characteristics of Building Materials."
UL 723, "Standard for Test for Surface Burning Characteristics of Building Materials."

- **Interior finish**
  - Interior wall finish.
  - Interior ceiling finish.
  - Interior floor finish.
- **Interior floor finish**
  - Exposed floor surfaces.
  - Coverings applied over the following:
    - A finished floor.
    - Stair treads.
    - Stair risers.
    - Stair landings.
- **Interior floor-wall base**
  - Interior finish trim.
  - Located where the wall meets the floor.
  - Serves either or both of the following purposes:
    - Functional border.
    - Decorative border.

802 Definitions

802.1 General *(part 2 of 2)*

- **Interior wall and ceiling finish**
  - Interior exposed surfaces of a building including but not limited to the following:
    Walls:
      Fixed.
      Mobile.
    Partitions:
      Fixed.
      Mobile.
    Toilet room privacy partitions.
    Columns.
    Ceilings.
    Wainscoting.
    Paneling.
    Other finish applied structurally.
    Other finish applied decoratively.
    Acoustical materials.
    Surface insulation.
    Structural fire resistance or similar function.
  - Trim is not included.
- **Smoke-developed index**
  - A comparative measure as follows:
    A dimensionless number assigned to a material.
    Based on measurements of the following:
      Smoke obscuration vs. time.
    Based on a test.

  Note: ASTM E 84, "Test Method for Surface Burning Characteristics of Building Materials," is cited as the governing test.

- **Trim**
  - The following components used in fixed applications:
    Picture molding.
    Chair rails.
    Baseboards.
    Handrails.
    Door frames.
    Window frames.
    Similar decorative materials.
    Similar protective materials.

803 Wall and Ceiling Finishes

803.1.1 Interior wall and ceiling finish materials

- Interior wall and ceiling finishes are classified as shown below where tested as cited in this section:

Class	Flame spread index	Smoke-developed index
A	0 – 25	0 – 450
B	26 – 75	0 – 450
C	76 – 200	0 – 450

Note: One of the following tests must be used to determine classification criteria:
ASTM E 84, "Test Method for Surface Burning Characteristics of Building Materials."
UL 723, "Standard for Test for Surface Burning Characteristics of Building Materials."

803.1.2 Room corner test for interior wall or ceiling finish materials

- This section addresses the following:
  - Interior wall and ceiling materials tested with the alternative method cited in this section.
- Materials governed by this section must comply test criteria listed in the next section.

Note: NFPA 286, "Standard Methods of Fire Tests for Evaluating Contribution of Wall and Ceiling Interior Finish to Room Fire Growth," is cited as the alternative test method.
803.1.2.1, "Acceptance criteria for NFPA 286," is cited as listing required criteria for materials tested by the NFPA 286 alternative test.

803.1.2.1 Acceptance criteria for NFPA 286

- The interior finish must perform as follows where the alternative fire test is used:
  - Flames may not spread to the ceiling during the 40 kW exposure.
  - Flames must not spread to the edges of the sample during the 160 kW exposure as follows:
    Where mounted on the wall or ceiling.
  - Flashover is not permitted during the 160 kW exposure.
  - Peak rate of heat release must be ≤ 800 kW during the entire test.
  - Total smoke released during the fire test is limited to ≤ 1000 m^2.

Note: NFPA 286, "Standard Methods of Fire Test for Evaluating Contribution of Wall and Ceiling Interior Finish to Room Fire Growth," is cited as the alternative fire test used and as the standard that defines "flashover."

803.1.3 Room corner test for textile wall coverings and expanded vinyl wall coverings

- The following materials are addressed by this section:
  - Textile and expanded vinyl wall coverings tested in the format to be used in their building application.
- Where materials addressed in this section are tested by the cited method, the following applies:
  - The materials must meet the test criteria listed in the next section.
  - Testing must use the following methods as specified in the testing instructions:
    The product-mounting system specified.
    The adhesive specified.

Note: 803.1.3.1, "Acceptance criteria for NFPA 265," is cited as listing criteria required for materials tested as specified in this section.
Method B of NFPA 265, "Standard Methods of Fire Tests for Evaluating Room Fire Growth Contribution of Textile Coverings on Full Height Panels and Walls," is cited as the test requiring compliance with 803.1.3.1.

803 Wall and Ceiling Finishes

803.1.3.1 Acceptance criteria for NFPA 265

- During the 40 kW exposure part of the test, the following applies:
  - Flames must not spread to the ceiling.
- During the 150 kW exposure part of the test, the following applies:
  - Flames must not spread to the outer edges of the materials on the 8' x 12' walls.
  - Flashover must not occur.
- Total smoke released during the test is limited to ≤ 1000 m^2.

Note: NFPA 265, "Standard Methods of Fire Tests for Evaluating Room Fire Growth Contribution of Textile Coverings on Full Height Panels and Walls," is cited as defining "flashover."

803.1.4 Acceptance criteria for textile and expanded vinyl wall or ceiling coverings tested to ASTM E 84 or UL 723

- This section addresses the following materials:
  - Textile wall coverings.
  - Textile ceiling coverings.
  - Expanded wall coverings.
  - Expanded ceiling coverings.
- Materials governed by this section must meet the following requirements:
  - Materials must have a Class A flame spread index.
  - Materials must be protected by sprinklers.

Note: The following are cited as alternative tests for establishing the flame spread index:
ASTM E 84, "Test Method for Surface Burning Characteristics of Building Materials."
UL 723, "Standard for Test for Surface Burning Characteristics of Building Materials."
ASTM E 2404, "Practice for Specimen Preparation and Mounting of Textile, Paper or Vinyl Wall or Ceiling Coverings to Assess Surface Burning Characteristics," is cited as governing preparation and mounting of specimans for the burning test.
The following are cited as alternatives for governing the required sprinklers:
903.3.1.1, "NFPA 13 sprinkler systems."
903.3.1.2, "NFPA 13R sprinkler systems."

803.2 Thickness exemption

- Materials with all of the following characteristics need not be tested:
  - < 0.036" thick.
  - Applied directly to either of the following surfaces:
    - Wall.
    - Ceiling.

803.3 Heavy timber exemption

- The following are not required to meet the requirements for interior finish materials:
  - Exposed parts of Type IV construction structural members.

Note: 602.4, "Type IV," is cited as governing Type IV construction.

803 Wall and Ceiling Finishes

803.4 Foam plastics

- This section addresses the following materials:
  - Exposed foam plastics.
  - Foam plastics used with textiles.
  - Foam plastics used with vinyl facings.
  - Foam plastics used with vinyl coverings.
- Foam plastics addressed in this section are governed as follows:
  - They must comply with the applicable section of this code when used as an interior finish.

Note: 2603.9, "Special approval," is cited as governing foam plastics used as an interior finish.

803.5 Textile wall coverings

- The following materials where used as an interior finish for walls are addressed in this section:
  - Textile wall coverings including the following types:
    - Those having any of the following types of surface:

      Looped, Nonwoven, Woven, Napped, Tufted
    - Carpet.
    - Textile materials similar to those listed above.
- Materials governed by this section must be tested as follows:
  - In the format to be used in the final building application as follows:
    - Incorporating the same material mounting system.
    - Incorporating the same adhesive system as applicable.
  - In compliance with applicable code testing requirements.

Note: The following are cited as governing testing of materials addressed by this section as applicable:
803.1.2, "Room corner test for interior wall or ceiling finish materials."
803.1.3, "Room corner test for textile wall coverings and expanded vinyl wall coverings."
803.1.4, "Acceptance criteria for textile and expanded vinyl wall or ceiling coverings tested to ASTM E 84 or UL 723."

803.6 Textile ceiling coverings *(part 1 of 2)*

- The following materials where used as an interior finish for ceilings are addressed in this section:
  - Textile ceiling coverings including the following types:
    - Those having any of the following types of surface:

      Looped, Nonwoven, Woven, Napped, Tufted
    - Carpet.
    - Textile materials similar to those listed above.

803 Wall and Ceiling Finishes

803.6 Textile ceiling coverings *(part 2 of 2)*

- Materials governed by this section must be tested as follows:
  - In the format to be used in the final building application as follows:
    Incorporating the same material mounting system.
    Incorporating the same adhesive system as applicable.
  - In compliance with applicable code testing requirements.

Note: The following are cited as governing testing of materials addressed by this section as applicable:
803.1.2, "Room corner test for interior wall or ceiling finish materials."
803.1.4, "Acceptance criteria for textile and expanded vinyl wall or ceiling coverings tested to ASTM E 84 or UL 723."

803.7 Expanded vinyl wall coverings

- Expanded vinyl wall coverings are addressed by this section as follows:
  - Where used as an interior wall finish.
- Expanded vinyl wall coverings must be tested as follows:
  - In the format to be used in the final building application as follows:
    Incorporating the same material mounting system.
    Incorporating the same adhesive system as applicable.
  - In compliance with applicable code testing requirements.

Note: The following are cited as governing testing of materials addressed by this section as applicable:
803.1.2, "Room corner test for interior wall or ceiling finish materials."
803.1.3, "Room corner test for textile wall coverings and expanded vinyl wall coverings."
803.1.4, "Acceptance criteria for textile and expanded vinyl wall or ceiling coverings tested to ASTM E 84 or UL 723."

803.8 Expanded vinyl ceiling coverings

- Expanded vinyl ceiling coverings are addressed by this section as follows:
  - Where used as an interior ceiling finish.
- Expanded vinyl wall coverings must be tested as follows:
  - In the format to be used in the final building application as follows:
    Incorporating the same material mounting system.
    Incorporating the same adhesive system as applicable.
  - In compliance with applicable code testing requirements.

Note: The following are cited as governing testing of materials addressed by this section as applicable:
803.1.2, "Room corner test for interior wall or ceiling finish materials."
803.1.4, "Acceptance criteria for textile and expanded vinyl wall or ceiling coverings tested to ASTM E 84 or UL 723."

803 Wall and Ceiling Finishes

803.9 Interior finish requirements based on group *(part 1 of 5)**

- This section dictates the minimum flame-spread class of finish materials required for interior walls and ceilings.

Note: IBC Table 803.9, "Interior Wall and Ceiling Finish Requirements by Occupancy," is cited as the source of flame-spread limitations. The table is summarized here and on the following pages.

- Where Class A materials are required, the following materials are acceptable:
  - Those tested by the alternative fire test with associated acceptance criteria.

Note: NFPA 286, "Standard Methods of Fire Test for Evaluating Contribution of Wall and Ceiling Interior Finish to Room Fire Growth," is cited as the alternative fire test that has associated acceptance criteria. This test is an alternative to ASTM E 84 or UL 723 which defines the flame spread index of Class A materials as well as others.
803.1.2.1, "Acceptance criteria for NFPA 286," is cited as the source of test criteria required for the NFPA 286 test.
The following are cited as the tests for establishing the flame-spread index:
ASTM E 84, "Test Method for Surface Burning Characteristics of Building Materials."
UL 723, "Standard for Test for Surface Burning Characteristics of Building Materials."
903.3.1.1, "NFPA 13 sprinkler systems," governs sprinklers where applicable.
903.3.1.2, "NFPA 13R sprinkler systems," governs sprinklersr where applicable.

- This section does not restrict the flame-spread class for materials in Occupancy U.
- Exit enclosures and exit passageways require the following flame-spread class:
  - ≥ Class C finish materials are required for the following conditions:
    - Materials are ≤ 1,000 sf of surface area of wainscot or paneling.
    - Materials are used in the grade-level lobby.
    - Materials are applied to one of the following:
      - Directly to a noncombustible base.
      - To furring on a noncombustible base and fireblocked by one of the following methods:
        - Airspaces filled with inorganic or noncombustible materials.
        - Airspaces filled with Class A materials.
        - Airspaces fireblocked so as to isolate airspaces ≤ 8' in length in any direction.

Note: 803.11.1, " Direct attachment and furred construction," is cited as governing the application of materials noted above.

  - Flame-spread class for other conditions are listed with occupancy designations in this section.
- Exit enclosures in other than Occupancy I-3 require materials with the following flame spreads:
  - ≥ Class C in buildings ≤ 3 stories in sprinklered buildings.
  - ≥ Class B in buildings ≤ 3 stories in buildings not sprinklered.
  - Flame-spread class for other buildings are reported with occupancy designations listed in this section.
- Rooms and enclosed spaces are defined by partitions that run from the floor to the ceiling as follows:
  - Where the structure requires a fire-resistance rating.

Note: A room or enclosed space does not terminate at any partition that does not reach the ceiling, but continues into the adjacent area to a point where a partition reaches the ceiling. Where more than one occupancy occupies such spaces, the most restrictive governs flame-spread class.

**Source*: IBC Table 803.9.

803 Wall and Ceiling Finishes

803.9 Interior finish requirements based on group *(part 2 of 5)**

Occupancies A-1 and A-2

- Exit enclosures and exit passageways:
  - ≥ Class B finish materials are required where spaces are sprinklered.
  - ≥ Class A finish materials are required where spaces are not sprinklered.
- Corridors:
  - ≥ Class B finish materials are required where spaces are sprinklered.
  - Where spaces are not sprinklered, the following applies:
    - ≥ Class B finish materials are required in lobbies.
    - ≥ Class A finish materials are required in other spaces.
- Rooms and enclosed spaces:
  - ≥ Class C finish materials are required where spaces are sprinklered.
  - Where spaces are not sprinklered the following applies:
    - ≥ Class C finish materials are required for occupancy loads ≤ 300.
    - ≥ Class B finish materials are required for occupancy loads > 300.

Occupancies A-3, A-4, and A-5

- The following applies to Occupancy A-3:
  - In places of worship, wood may be used for the following:
    - Ornamental purposes.
    - Trusses.
    - Paneling.
    - Chancel furnishing.
  - ≥ Class B finish materials are required in nonsprinklered lobbies.
  - Other spaces and buildings in A-3 are governed by other requirements of this section.
- Exit enclosures and exit passageways:
  - ≥ Class B finish materials are required where spaces are sprinklered.
  - ≥ Class A finish materials are required where spaces are not sprinklered.
- Corridors:
  - ≥ Class B finish materials are required where spaces are sprinklered.
  - Where spaces are not sprinklered:
    - ≥ Class B finish materials are required in lobbies.
    - ≥ Class A finish materials are required in other spaces.
- Rooms and enclosed spaces:
  - ≥ Class C finish materials are required.

Occupancies B, E, M, R-1

- Exit enclosures and exit passageways:
  - ≥ Class B finish materials are required where spaces are sprinklered.
  - ≥ Class A finish materials are required where spaces are not sprinklered.
- Corridors:
  - ≥ Class C finish materials are required where spaces are sprinklered.
  - ≥ Class B finish materials are required where spaces are not sprinklered.
- Rooms and enclosed spaces:
  - ≥ Class C finish materials are required.

**Source*: IBC Table 803.9.

803 Wall and Ceiling Finishes

803.9 Interior finish requirements based on group *(part 3 of 5)**

Occupancy R-4

- Exit enclosures and exit passageways:
  - ≥ Class B finish materials are required where spaces are sprinklered.
  - ≥ Class A finish materials are required where spaces are not sprinklered.
- Corridors:
  - ≥ Class C finish materials are required where spaces are sprinklered.
  - ≥ Class B finish materials are required where spaces are not sprinklered.
- Rooms and enclosed spaces:
  - ≥ Class B finish materials are required.

Occupancy F

- Exit enclosures and exit passageways:
  - ≥ Class C finish materials are required where spaces are sprinklered.
  - ≥ Class B finish materials are required where spaces are not sprinklered.
- Corridors:
  - ≥ Class C finish materials are required.
- Rooms and enclosed spaces:
  - ≥ Class C finish materials are required.

Occupancy H

- Exit enclosures and exit passageways:
  - ≥ Class B finish materials are required where spaces are sprinklered.
  - ≥ Class A finish materials are required where spaces are not sprinklered.
- Corridors:
  - ≥ Class B finish materials are required where spaces are sprinklered.
  - ≥ Class A finish materials are required where spaces are not sprinklered.
- Rooms and enclosed spaces:
  - ≥ Class C finish materials are required where both of the following conditions are present:
    Where the spaces are sprinklered.
    Where the building is ≤ 2 stories.
  - ≥ Class B finish materials are required in either of the following cases:
    Where the building is not sprinklered.
    Where the building is > 2 stories.

Occupancy I-1

- Exit enclosures and exit passageways:
  - ≥ Class B finish materials are required where spaces are sprinklered.
  - ≥ Class A finish materials are required where spaces are not sprinklered.
- Corridors:
  - ≥ Class C finish materials are required where spaces are sprinklered.
  - ≥ Class B finish materials are required where spaces are not sprinklered.
- Rooms and enclosed spaces:
  - ≥ Class C finish materials are required where spaces are sprinklered.
  - ≥ Class B finish materials are required where spaces are not sprinklered.

**Source*: IBC Table 803.9.

803 Wall and Ceiling Finishes

803.9 Interior finish requirements based on group *(part 4 of 5)**

Occupancy I-2

- Exit enclosures and exit passageways:
  - ≥ Class B finish materials are required where spaces are sprinklered.
  - ≥ Class A finish materials are required where spaces are not sprinklered.
- Corridors:
  - ≥ Class B finish materials are required where spaces are sprinklered.
  - ≥ Class A finish materials are required where spaces are not sprinklered.
- Rooms and enclosed spaces:
  - Where spaces are sprinklered:
    - ≥ Class C finish materials are required in administrative spaces.
    - ≥ Class C finish materials are required in rooms having a capacity ≤ 4 persons.
    - ≥ Class B finish materials are required for other conditions.
  - ≥ Class B finish materials are required where spaces are not sprinklered.

Occupancy I-3

- Exit enclosures and exit passageways:
  - ≥ Class A finish materials are required.
- Corridors:
  - ≥ Class B finish materials are required in sprinklered spaces as follows:
    Where used as a wainscot ≤ 48" above the finished floor in exit access corridors.
  - ≥ Class A finish materials are required for all other conditions.
- Rooms and enclosed spaces:
  - ≥ Class C finish materials are required where spaces are sprinklered.
  - ≥ Class B finish materials are required where spaces are not sprinklered.

Occupancy I-4

- Exit enclosures and exit passageways:
  - ≥ Class B finish materials are required where spaces are sprinklered.
  - ≥ Class A finish materials are required where spaces are not sprinklered.
- Corridors:
  - ≥ Class B finish materials are required where spaces are sprinklered.
  - ≥ Class A finish materials are required where spaces are not sprinklered.
- Rooms and enclosed spaces:
  - Where spaces are sprinklered:
    - ≥ Class C finish materials are required in administrative spaces.
    - ≥ Class C finish materials are required in rooms having a capacity ≤ 4 persons.
    - ≥ Class B finish materials are required for other conditions.
  - ≥ Class B finish materials are required where spaces are not sprinklered.

**Source*: IBC Table 803.9.

803 Wall and Ceiling Finishes

803.9 Interior finish requirements based on group *(part 5 of 5)**

Occupancies R-2 and S

- Exit enclosures and exit passageways:
  - ≥ Class C finish materials are required where spaces are sprinklered.
  - ≥ Class B finish materials are required where spaces are not sprinklered.
- Corridors:
  - ≥ Class C finish materials are required where spaces are sprinklered.
  - ≥ Class B finish materials are required where spaces are not sprinklered.
- Rooms and enclosed spaces:
  - ≥ Class C finish materials are required.

Occupancy R-3

- Exit enclosures and exit passageways:
  - ≥ Class C finish materials are required.
- Corridors:
  - ≥ Class C finish materials are required.
- Rooms and enclosed spaces:
  - ≥ Class C finish materials are required.

803.10 Stability

- Interior finish materials governed by this chapter must comply with the following:
  - They must remain securely attached under the following conditions:
    In room temperatures ≤ 200° F as follows:
      For a time ≥ 30 min.

803.11 Application of interior finish materials to fire-resistance-rated structural elements

- Interior finish materials must comply with this section series as follows:
  - Where applied to walls, ceilings, or structural elements required to have any of the following:
    A fire-resistance rating.
    Noncombustible materials.

803.11.1 Direct attachment and furred construction *(part 1 of 2)*

- This section governs walls and ceilings that are required to have either of the following:
  - A fire-resistance rating.
  - To be noncombustible.
- Interior finish materials applied to walls and ceilings governed by this section must comply with one of the following:
  - It must be applied directly against the element.
  - It must be applied to furring strips as follows:
    The furring strips must be ≤ 1 3/4" thick.
    The furring strips must be applied directly to the element.

**Source*: IBC Table 803.9.

803 Wall and Ceiling Finishes

803.11.1 Direct attachment and furred construction *(part 2 of 2)*

- Spaces between furring strips must be detailed in one of the following ways:
  - Spaces must be filled with one of the following materials:
    - Inorganic.
    - Noncombustible.
    - Class A or equivalent.

Note: Class A or equivalent materials must comply with the following as applicable:
803.1.1, "Interior wall and ceiling finish materials."
803.1.2, "Room corner test for interior wall or ceiling finish materials."

  - Spaces must be fireblocked so as to isolate air space ≤ 8' in length in any direction.

Note: Section 717, "Concealed Spaces," is cited as governing fireblocking.

803.11.2 Set-out construction

- This section governs the following interior building elements where they are required to have any of the properties listed below:
  - Building elements:
    - Walls where the finished surface is set out from the wall structure > 1 3/4".
    - Ceilings that are dropped > 1 3/4" from the ceiling structure.
  - Properties:
    - A fire-resistance rating.
    - To be noncombustible.

Note: 803.11.1, "Direct attachment and furred construction," is cited as specifying the distance for set-out construction to which this section applies. This dimension is given above.

- Walls and ceilings governed by this section require one of the following:
  - Class A or equivalent finish materials.
  - Finish materials protected on both sides by sprinklers.
  - Finish materials attached to noncombustible backing or furring strips.

Note: Class A or equivalent materials must comply with the following as applicable:
803.1.1, "Interior wall and ceiling finish materials."
803.1.2, "Room corner test for interior wall or ceiling finish materials,"
803.11.1, "Direct attachment and furred construction," is cited as governing the backing or furring strips noted above.

- In dropped ceilings the following components are governed as indicated below:
  - Components below the main ceiling line:
    - Hangers.
    - Assembly members.
  - The components below the main ceiling line must be one of the following:
    - Noncombustible materials.
    - Fire-retardant-treated wood in the following types of construction only:
      - Type III and Type V.
- Set-out wall construction must be fire-resistance-rated as applicable per other code requirements.

803 Wall and Ceiling Finishes

803.11.3 Heavy timber construction

- This section addresses wall and ceiling finish materials as follows:
    - Finish materials permitted by this chapter.
    - Located in Type IV construction.
    - Where applied directly to any of the following:
        Wood decking.
        Wood planking.
    - Where attached to wood furring strips that are applied directly to any of the following:
        Wood decking.
        Wood planking.
- Finish materials addressed by this section must be fireblocked.

Note: 803.11.1, "Direct attachment and furred construction," is cited as governing the fireblocking required by this section.

803.11.4 Materials

- This section applies to materials ≤ 1/4" thick used as follows:
    - Used as interior wall finish.
    - Used as interior ceiling finish.
- The following materials are not required to be applied directly to a noncombustible backing:
    - Noncombustible materials.
    - Materials successfully tested while set out from the noncombustible backing.
- All other materials must be applied directly to a noncombustible backing.

804 Interior Floor Finish

804.1 General

- The following traditional floor finishes and coverings are not governed by this section:
  - Wood.
  - Vinyl.
  - Linoleum.
  - Terrazzo.
  - Resilient coverings not composed of fibers.
- Other interior floor finishes and coverings are governed by this section.

Note: The following are cited as applicable to the other floor finishes and coverings above.
804.2, "Classification."
804.3, "Testing and identification."
804.4, "Interior floor finish requirements."
804.4.1, "Minimum critical radiant flux."

804.2 Classification

- Materials required by this section to be Class I or Class II must have a heat threshold preventing the advent of flame spread as follows:
  - Class I materials must have a critical radiant flux ≥ 0.45 watts/sq cm.
  - Class II materials must have a critical radiant flux ≥ 0.22 watts/sq cm.

Note: NFPA 253, "Test for Critical Radiant Flux of Floor Covering Systems Using a Radiant Heat Energy Source," is cited as the applicable standard.
804.4.1, "Minimum critical radiant flux," is cited as requiring minimum heat thresholds for interior floor finishes.

804.3 Testing and identification

- Reports of testing must be provided to the building official upon request.
- Interior floor finish and covering materials must be tested as follows:
  - By an approved agency.
  - To determine classification according to critical radiant flux.
  - Carpet-type coverings must be tested in the manner installed, including underlayment.

Note: NFPA 253, "Test for Critical Radiant Flux of Floor Covering Systems Using a Radiant Heat Energy Source," is cited as the applicable standard.

- Identification of floor covering materials is governed as follows:
  - Method of identification required:
    Hang tag or other suitable method.
  - Identification information required:
    Manufacturer or supplier.
    Style.
    Classification.

Note: 804.2, "Classification," is cited as governing classification.

804 Interior Floor Finish

804.4 Interior floor finish requirements

- This section applies to all occupancies.
- Interior floor finishes and coverings in the following locations have the requirement listed below:
  - Locations:
    - Exit enclosures.
    - Exit passageways.
    - Corridors.
    - Spaces as follows:
      - That are not separated from corridors by the following:
        - Partitions spanning from the floor to the underside of the ceiling.
  - Requirement:
    - Such finishes must withstand a minimum critical radiant flux.

Note: 804.4.1, "Minimum critical radiant flux," is cited as specifying the necessary critical radiant flux.

804.4.1 Minimum critical radiant flux

- The minimum critical radiant flux for interior floor finishes in the following locations is required for specific occupancies as shown below:
  - Floor locations:
    - Exit enclosures.
    - Exit passageways.
    - Corridors.
    - Rooms or spaces not separated from exit access corridors by the following:
      - Partitions running from the floor to the underside of the ceiling.
  - Buildings not sprinklered:
    - ≥ Class I finish materials are required for the following occupancies:
      - I-1, I-2, I-3.
    - ≥ Class II finish materials are required for the following occupancies:
      - A, B, E, H, I-4, M, R-1, R-2, S.
  - Sprinklered buildings:
    - ≥ Class II finish materials are permitted for the following occupancies:
      - I-1, I-2, I-3.
    - ≥ Class II materials and those meeting the "pill test" are permitted in the following:
      - Occupancies A, B, E, H, I-4, M, R-1, R-2, S.

Note: DOC FF-1, "Pill test" (CPSC 16 CFR 1630), "Standard for the Surface Flammability of Carpets and Rugs," is cited as the required test for coverings.

The following are cited as governing sprinklered buildings as noted above as applicable:

903.3.1.1, "NFPA 13 sprinkler systems."

903.3.1.2, "NFPA 13R sprinkler systems."

- In all other areas, the interior floor finish must meet the requirements of the "pill test" as noted above.

805 Combustible Materials in Type I and II Construction

805.1 Application

- This section does not apply to stages or platforms.

Note: 410.3, "Stages," is cited as governing stages.
410.4, "Platform construction," is cited as governing platforms.

- This section governs other combustible floor materials Type I and Type II construction.

Note: The following are cited as governing the floor materials indicated above:
805.1.1, "Subfloor construction."
805.1.2, "Wood finish flooring."
805.1.3, "Insulating boards."

805.1.1 Subfloor construction

- This section applies to the following floor components in Type I and Type II construction:
  - Sleepers.
  - Bucks.
  - Nailing blocks.
- Such floor components must be noncombustible unless the following is provided:
  - The space between the flooring and the floor assembly with a fire-resistance rating must be detailed by one of the following methods:
    Space to be filled solid with approved noncombustible materials.
    Space must be fireblocked.

Note: Section 717, "Concealed Spaces," is cited as the source for fireblocking requirements.

  - Open space between flooring and floor assembly may not pass by the following:
    Permanent walls or partitions.

805.1.2 Wood finish flooring

- Wood finish flooring in Type I and Type II construction may be attached as follows:
  - Directly to wood sleepers of the following type:
    Embedded wood sleepers.
    Fire-blocked wood sleepers.
  - Cemented directly to the top surface of the floor assembly of the following type:
    Approved.
    With a fire-resistance rating.
  - Directly to a wood subfloor that is attached to sleepers.

Note: 805.1.1, "Subfloor construction," is cited as the source of requirements for sleepers.

805.1.3 Insulating boards

- Combustible insulating boards may be used in flooring of Type I and Type II construction as follows:
  - Boards must be ≤ 0.5" thick.
  - Boards must be covered with an approved finish flooring.
  - Boards must be attached by one of the following details:
    Directly to a noncombustible floor assembly.
    To wood subflooring attached to sleepers.

Note: 805.1.1, "Subfloor construction," is cited as the source for sleeper requirements.

806 Decorative Materials and Trim

806.1 General requirements

- This section addresses the following materials hung from walls or ceilings:
  - Curtains and draperies.
  - Hangings.
  - Other decorative materials.
- Decorative materials must have certain flame propagation properties or be noncombustible in the following locations:
  - Occupancies A, E, I, R-1.
  - Dormitories in Occupancy R-2.
  - Occupancies B and M as follows:
    - Includes fabric partitions with both of the following characteristics:
      - Suspended from the ceiling.
      - Not supported by the floor.

Note: The following are cited defining flame propagation properties:
NFPA 701, "Standard Methods of Fire Tests for Flame-propagation of Textiles and Films."
806.2, "Acceptance criteria and reports."

- Combustible decorations are governed in occupancies I-1 and I-2 as follows:
  - The following decorations are not governed if their quantity is too small to be a hazard:
    - Photographs.
    - Paintings.
    - Similar decorations.
  - Other combustible decorations must be flame retardant.
- Combustible decorations in Occupancy I-3 are not permitted.
- The following elements with the characteristics listed below are designated as indicated:
  - Elements:
    - Walls and partitions.
    - Panelling.
    - Wall pads.
    - Crash pads.
  - Characteristics:
    - Fixed or movable.
    - Serving any of the following purposes:
      - Structure.
      - Decoration.
      - Acoustical.
      - Surface insulation.
      - Other.
    - Covering ≥ 10% of either of the following:
      - Walls.
      - Ceiling.
  - Designation:
    - Considered to be interior finish.
    - Not considered to be either of the following:
      - Decorations.
      - Furnishings.

806 Decorative Materials and Trim

806.1.1 Noncombustible materials

- Noncombustible decorative materials are not limited in quantity.

806.1.2 Combustible decorative materials

- In Occupancy A auditoriums of sprinklered buildings the following material is limited as listed below:
  - Material:
    Decorative.
    Meeting required flame propagation properties.

  Note: NFPA 701, "Standard Methods of Fire Tests for Flame-Propagation of Textiles and Films," is cited as defining the required flame propagation properties.

  - Limits:
    May cover ≤ half the sum of wall and ceiling areas.
    Must meet wall and ceiling application requirements.

  Note: 803.11, "Application of interior finish materials to fire-resistance-rated structural elements," is cited as the source for application requirements.
  903.3.1.1, "NFPA 13 sprinkler systems," is cited as governing the sprinklers.

- The following materials are not limited in occupancies B and M:
  - Fabric partitions with both of the following characteristics:
    Suspended from the ceiling.
    Not supported by the floor.
- In other locations, the following material is limited as listed below:
  - Material:
    Decorative.
    Meeting flame propagation properties as noted.

  Note: NFPA 701, "Standard Methods of Fire Tests for Flame-Propagation of Textiles and Films," is cited as defining the required flame propagation properties.

  - Limits:
    May cover ≤ 10% of the sum of wall and ceiling areas.

806.2 Acceptance criteria and reports

- Where required, decorative materials must comply with one of the following:

  Note: 806.1, "General requirements," is cited as listing requirements for decorative materials.

  - They must be noncombustible.
  - They must be tested by an approved agency as follows:
    They must meet flame propagation requirements.

  Note: NFPA 701, "Standard Methods of Fire Tests for Flame-Propagation of Textiles and Films," is cited as determining flame propagation properties. Test results are to be available to the fire code official.

806 Decorative Materials and Trim

806.3 Foam plastic

- Foam plastic trim must meet the following requirements for interior trim:
  - Minimum density.
  - Maximum thickness.
  - Maximum area.
  - Maximum flame spread.

Note: 2604.2, "Interior trim," is cited as governing foam plastic and lists requirements addressing the properties listed above.

806.4 Pyroxylin plastic

- The following is prohibited in Occupancy A:
  - Materials involving pyroxylin plastic such as imitation leather and other materials.
  - Materials involving a substance equally as hazardous as pyroxylin plastic.

806.5 Interior trim

- This section does not require that foam plastic used as interior trim equal the following standard:
  - Class C flame spread.
  - Class C smoke-developed index.
- Other interior trim must have a flame spread and smoke-developed index ≥ Class C.

Note: The following are cited as alternative standards establishing the classification of materials:
ASTM E 84, "Standard Test Method for Surface Burning Characteristics of Building Materials."
UL 723, "Standard for Test for Surface Burning Characteristics of Building Materials."
803.1.1, "Interior wall and ceiling finish materials," is cited as further governing interior trim.

- Combustible trim is governed as follows:
  - Handrails or guardrails are not limited by this section.
  - Other combustible trim is limited to the following:
    - ≤ 10% of the sum of surface areas as follows:
      - Ceiling surface in the space where located as applicable.
      - Wall surface in the space where located as applicable.

806.6 Interior floor-wall base

- This section does not govern materials meeting the requirements of the previous section.

Note: 806.5, "Interior trim," is cited as listing requirements that, if met, remove a material from the jurisdiction of this section.

- This section governs other interior floor-wall base materials as follows:
  - That are ≤ 6" high.
- Floor-wall base materials governed by this section have the following requirements:
  - They must be tested.
  - They must be ≥ Class II.
  - They must be Class I in the following case:
    - Where a Class I floor finish is required.

Note: 804.2, "Classification," is cited as addressing required testing for materials governed by this section.

808 Acoustical Ceiling Systems

808.1 Acoustical ceiling systems

- The following aspects of the ceiling systems listed below must comply with the requirements indicated:
  - Aspects:
    Quality.
    Design.
    Fabrication.
  - Systems:
    Metal suspension systems for the following:
      Acoustical tile ceiling systems.
      Acoustical lay-in panel ceiling systems.
  - Requirements:
    Generally accepted engineering practice.
    Requirements of this chapter.
    Other applicable requirements of the code.

808.1.1 Materials and installation

- Acoustical materials for ceiling systems must comply with the following:
  - Manufacturer's instructions.
  - Provisions of this section series regarding the application of interior finish.

808.1.1.1 Suspended acoustical ceilings

- Suspended acoustical ceilings must be installed according to the following standards:
  - Standards of the American Society for Testing and Materials.

Note: The following standards are cited as governing suspended acoustical ceilings:
ASTM C 635, "Specification for the Manufacture, Performance, and Testing of Metal Suspension Systems for Acoustical Tile and Lay-in Panel Ceilings."
ASTM C 636, "Standard Practice for Installation of Metal Ceiling Suspension Systems for Acoustical Tile and Lay-in Panels."

808.1.1.2 Fire-resistance-rated construction

- The following acoustical ceiling systems must comply with the requirements listed below:
  - Ceiling systems:
    Those that are part of construction which is fire-rated.
  - Requirements:
    Such systems must be installed in the same format as tested.
    Such systems must comply with fire-resistance-rated requirements.

Note: Chapter 7, "Fire and Smoke Protection Features," is cited as the source for requirements for acoustical ceiling systems in fire-rated construction.

9

Fire Protection Systems

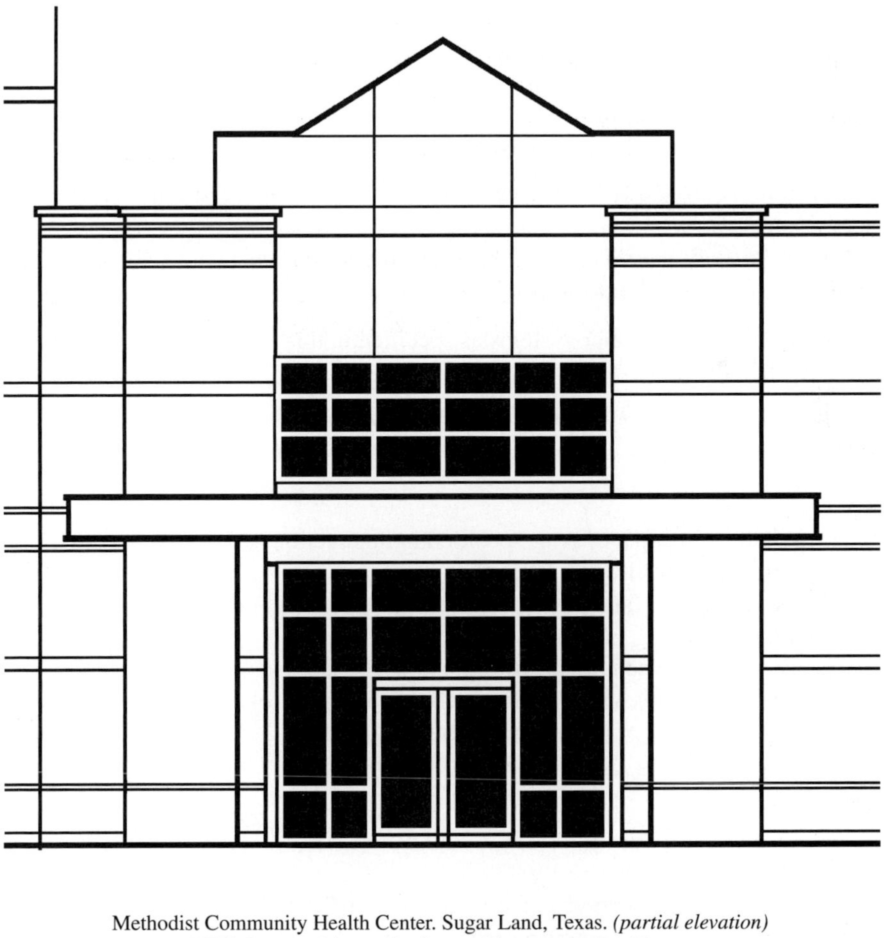

Methodist Community Health Center. Sugar Land, Texas. *(partial elevation)*
HKS, Inc., Architects, Engineers, Planners. Dallas, Texas.

903 Automatic Sprinkler Systems

903.1.1 Alternative protection

- Other automatic fire-protection systems may be substituted for automatic sprinklers where the following apply:
  - Must be approved by the fire code official.
  - Must meet applicable standards.

Note: Section 904, "Alternative Automatic Fire-Extinguishing Systems," is cited as the source of requirements for such systems.

903.2 Where required

- The following spaces with the qualifications listed below do not require sprinklers:
  - Locations:
    - In telecommunications buildings used only for the following:
      - Telecommunications equipment.
      - Associated electrical power distribution equipment.
      - Batteries.
      - Standby engines.
  - Qualifications:
    - Spaces must have an automatic smoke detection system.
    - Spaces must be isolated from other areas with one or both of the following as applicable:
      - Fire barriers with a fire-resistance rating ≤ 1 hr.
      - Horizontal assemblies with a fire-resistance rating ≤ 2 hr.

Note: 907.2, "Where required—new buildings and structures," is cited as governing the smoke detection system noted above.
Section 707, "Fire Barriers," is cited as governing these elements.
Section 712, "Horizontal Assemblies," is cited as governing these elements.

- Other new buildings require sprinkler systems as specified in this section and as follows:
  - To be approved and automatic.

Note: Sections 903.2.1, "Group A," through 903.2.12, "During construction," are cited as governing the required sprinkler systems.

903.2.1 Group A

- Sprinklers are required in Occupancy A as specified in this section series.
- Sprinklers are required in the following locations:
  - In occupancies A-1, A-2, A-3, and A-4 as follows:
    - Throughout the floor area.
    - Throughout all floors between the following:
      - The Group A occupancies and the nearest exit discharge level serving Occupancy A areas.
  - In Occupancy A-5 where any of the following are > 1,000 sf:
    - Concession stands.
    - Retail areas.
    - Press boxes.
    - Other accessory use areas.

Note: 903.2.1.5, "Group A-5," is cited as listing areas in A-5 requiring sprinklers and is partially summarized above. The section also includes an exception.

903 Automatic Sprinkler Systems

903.2.1.1 Group A-1

- Sprinklers are required in an A-1 occupancy fire area with any of the following characteristics:
  - Area is > 12,000 sf.
  - Area occupant load is ≥ 300.
  - Area is not on the exit discharge level.
  - Area contains multiple motion picture theaters.

903.2.1.2 Group A-2

- Sprinklers are required in an A-2 occupancy fire area with any of the following characteristics:
  - Area is > 5,000 sf.
  - Area occupant load is ≥ 100.
  - Area is not on the exit discharge level.

903.2.1.3 Group A-3

- Sprinklers are required in an A-3 occupancy fire area with any of the following characteristics:
  - Area is > 12,000 sf.
  - Area occupant load is ≥ 300.
  - Area is not on the exit discharge level.

903.2.1.4 Group A-4

- Sprinklers are required in an A-4 occupancy fire area with any of the following characteristics:
  - Area is > 12,000 sf.
  - Area occupant load is ≥ 300.
  - Area is not on the exit discharge level.

903.2.1.5 Group A-5

- Sprinklers are required in the following functions where areas are > 1,000 sf:
  - Concession stands.
  - Retail areas.
  - Press boxes.
  - Other accessory use areas.

903.2.2 Group B ambulatory health care facilities

- Ambulatory health care facilities are to be sprinklered where either of the following cases exist:
  - Where ≥ 4 patients are not capable of self-preservation at any time.
  - Where ≥ 1 patients are not capable of self-preservation as follows:
    Are not located on the level of exit discharge.

903.2.3 Group E

- Sprinklers are not required in areas below the level of exit discharge as follows:
  - Where every classroom throughout the building has ≥ 1 exterior exit door at ground level.
- Otherwise, sprinklers are required as follows:
  - Throughout all Occupancy E fire areas > 12,000 sf.
  - Throughout all parts of Occupancy E buildings as follows:
    Below the lowest exit discharge level serving the E occupancy areas.

903 Automatic Sprinkler Systems

903.2.4 Group F-1

- Sprinklers are required throughout buildings in all of the following cases:
  - Where a fire area has an F-1 occupancy with any of the following characteristics:
    - Fire area is > 12,000 sf.
    - Fire area is > 3 stories above grade.
  - Where the sum of all F-1 fire areas in a building including mezzanines is as follows:
    - Total area is > 24,000 sf.

903.2.4.1 Woodworking operations

- Sprinklers are required in all F-1 fire areas that have all of the following characteristics:
  - Woodworking operations are present.
  - Woodworking operations are > 2,500 sf.
  - Fine combustible particles are present in either of the following formats:
    - As waste.
    - As materials used in the process.

903.2.5.1 General

- An automatic sprinkler system is required in Occupancy H facilities.

903.2.5.2 Group H-5

- Where an H-5 occupancy is present, automatic sprinklers are required throughout the building as follows:
  - System design must comply with both of the following:
    - Code requirements according to the following hazard classifications:
      - Ordinary Hazard Group 2 applies to the following:
        - Fabrication areas.
        - Corridors and service corridors.
        - Storage rooms without dispensing functions.
      - Extra Hazard Group 2 applies to the following:
        - Storage rooms with dispensing functions.

Note: IBC Table 903.2.5.2, "Group H-5 Sprinkler Design Criteria," is cited as the source of requirements based on hazard level, the content of which is summarized above.

  - Corridors with one row of sprinklers are governed as follows:
    - ≤ 13 sprinklers are required based on calculations.

903.2.5.3 Pyroxylin plastics

- Sprinklers are required where the following substances are present in amounts > 100 lbs in any of the activities indicated below:
  - Substances:
    - Cellulose nitrate film.
    - Pyroxylin plastics.
  - Activities:
    - The plastics are manufactured.
    - The plastics are stored or handled.

903 Automatic Sprinkler Systems

903.2.11.1.1 Opening dimensions and access

- This section addresses stories and basements in occupancies other than R-3 and U.
- Openings provided in lieu of sprinklers must have all of the following characteristics:
  - Dimensions must be ≥ 2'-6".
  - Must provide access from the outside for the fire department.
  - Must provide the possibility of the following from the outside:
    - Fire fighting.
    - Rescue.

903.2.11.1.2 Openings on one side only

- This section addresses stories in occupancies other than R-3 and U.
- Sprinklers are required where openings are not provided as follows:
  - Openings, as required in lieu of sprinklers, must meet one of the following conditions:
    - Openings are on ≥ 2 sides of the story.
    - Openings are on 1 side of the story as follows:
      - The wall opposite the side with openings is ≤ 75' away.

903.2.11.1.3 Basements

- This section addresses occupancies other than R-3 and U.
- An approved sprinkler system is required in basements as follows:
  - Where any point is > 75' from the following opening:
    - An opening qualifying to be provided in lieu of sprinklers.

Note: 903.2.11.1, "Stories without openings," is cited as the source of requirements for openings qualifying to substitute for sprinklers.

903.2.11.2 Rubbish and linen chutes

- This section addresses rubbish and linen chutes in occupancies other than R-3 and U.
- Sprinklers are required as follows:
  - At the top of chutes.
  - In the terminal rooms of chutes.
  - In chutes through ≥ 3 stories as follows:
    - At alternate floors.
    - At the top of chutes.
    - In the terminal rooms of chutes.
  - Sprinklers must be accessible for servicing.

903.2.11.3 Buildings 55 feet or more in height

- This section addresses occupancies other than R-3 and U.
- The following high-rise occupancies are not governed by this section:
  - Airport control towers.
  - Open parking garages.
  - Occupancy F-2.
- Other buildings as follows require sprinklers throughout:
  - Buildings with a floor level as follows:
    - Occupant load ≥ 30.
    - Located ≥ 55' above lowest level of fire department vehicle access.

903 Automatic Sprinkler Systems

903.2.11.4 Ducts conveying hazardous exhausts

- The following ducts are not governed by this section:
  - Ducts with a largest diameter of < 10".
- In certain cases, sprinklers may be required in ducts carrying the following:
  - Hazardous exhaust.
  - Flammable materials.
  - Combustible materials.

Note: The International Mechanical Code *is cited as the standard governing the need for sprinklers in the ducts noted above.*

903.2.11.5 Commercial cooking operations

- Sprinklers protecting commerical cooking systems must be as follows:
  - Automatic.
  - Installed in the kitchen exhaust hood and duct system.

Note: The sprinklers are those specified in order to comply with:
Section 904, "Alternative Automatic Fire-Extinguishing Systems."

903.2.11.6 Other required suppression systems *(part 1 of 2)**

- The following require fire suppression systems in certain cases:

Note: 903.2, "Where required," is cited as specifying locations requiring sprinklers.

  - Covered malls.

Note: 402.9, "Automatic sprinkler system," is cited as the source of requirements.

  - High-rise buildings.

Note: 403.2, "Construction," is cited as a source of requirements.
403.3, "Automatic sprinkler system," is cited as a source of requirements.

  - Atriums.

Note: 404.3, "Automatic sprinkler protection," is cited as the source of requirements.

  - Underground structures.

Note: 405.3, "Automatic sprinkler system," is cited as the source of requirements.

  - Group I-2.

Note: 407.5, "Automatic sprinkler system," is cited as the source of requirements.

  - Stages.

Note: 410.6, "Automatic sprinkler system," is cited as the source of requirements.

  - Special amusement buildings.

Note: 411.4, "Automatic sprinkler system," is cited as the source of requirements.

**Source*: IBC Table 903.2.11.6

903 Automatic Sprinkler Systems

903.2.11.6 Other required suppression systems *(part 2 of 2)**

- Aircraft hangars.

Note: The following are cited as sources of requirements:
412.4.6, "Fire suppression."
412.4.6.1, "Hazardous operations."
412.6.5, "Fire suppression."

- Group H-2.

Note: 415.6.2.4, "Suppression," is cited as the source of requirements.

- Flammable finishes.

Note: 416.4, "Spray booths," is cited as the source of requirements.

- Drying rooms.

Note: 417.4, "Fire protection," is cited as the source of requirements.

- Unlimited area buildings.

Note: Section 507, "Unlimited Area Buildings," is cited as the source of requirements.

- Incidental accessory occupancies.

Note: 508.2.5, "Separation of incidental accessory occupancies," is cited as the source of requirements.

- Smoke-protected assembly seating.

Note: 1028.6.2.3, "Automatic sprinklers," is cited as the source of requirements.

- Other sprinkler requirements.

Note: International Fire Code, *903.2.11.6, "Other required suppression systems," is cited as the source of additional requirements.*

903.2.12 During construction

- In certain cases sprinklers may be required during the following:
  - Construction.
  - Alteration.
  - Demolition.

Note: The International Fire Code, *Chapter 14, "Fire Safety During Construction and Demolition," is cited as governing sprinklers in the above cases.*

**Source*: IBC Table 903.2.11.6.

903 Automatic Sprinkler Systems

903.3.1.1 NFPA 13 sprinkler systems

- Sprinklers governed by this section must comply with sprinkler standards.

  Note: NFPA 13, "Installation of Sprinkler Systems," is cited as the applicable standard.

- Requirements for buildings to have sprinklers do not include the following spaces:
  - Certain spaces with one of the following fire detection systems:
    - Detects invisible particles of combustion.
    - Detects visible particles of combustion.

  Note: 903.3.1.1.1, "Exempt locations," is cited as listing spaces not required to have a sprinkler system governed by this section where the space is located in a building which does have such a requirement.

903.3.1.1.1 Exempt locations *(part 1 of 2)*

- Sprinklers are not required in any of the following spaces:
  - Where protected by a fire detection system with all of the following characteristics:
    - Approved.
    - Automatic.
    - Detects one of the following products of combustion:
      - Visible.
      - Invisible.

  Note: 907.2, "Where required—new buildings and structures," is cited as governing the fire detection systems.
  - Where the application of the following substances results in the hazards indicated below:
    - Substances:
      - Water.
      - Flame and water.
    - Hazards:
      - Hazard to life.
      - Fire hazard.
  - Where both of the following conditions apply:
    - Nature of contents render sprinklers undesirable.
    - Omission of sprinklers is approved by the fire code official.
  - In both of the following rooms where the condition indicated below applies:
    - Rooms:
      - Generator rooms.
      - Transformer rooms.
    - Condition:
      - Rooms are separated from rest of building as follows:
        - By the following assemblies as applicable with a fire-resistance rating ≥ 2 hrs:
          - Walls.
          - Floor/ceiling assemblies.
          - Roof/ceiling assemblies.

903 Automatic Sprinkler Systems

903.3.1.1.1 Exempt locations *(part 2 of 2)*

- In spaces of telecommunications buildings as follows with the conditions indicated below:
  - Spaces used only for the following:
    - Telecommunications equipment.
    - Related electrical power distribution equipment.
    - Batteries.
    - Standby engines.
  - Conditions:
    - Spaces have an automatic fire alarm system.
    - Spaces are separated from rest of building as follows:
      - Wall has a fire-resistance rating ≥ 1 hr.
      - Floor/ceiling assembly has a fire-resistance rating ≥ 2 hr.
- Where spaces are as follows:
  - Constructed of noncombustible construction.
  - Has noncombustible contents.
- In the following areas serving fire access service elevators:
  - Machine rooms.
  - Machinery spaces.

- The following conditions do not justify the omission of sprinklers from a space:
  - Space is damp.
  - Space is of construction with a fire-resistance rating.
  - Space contains electrical equipment.

903.3.1.2 NFPA 13R sprinkler systems

- Sprinklers governed by this section must meet sprinkler standards as follows:
  - In Occupancy R buildings ≤ 4 stories.

Note: NFPA 13R, "Installation of Sprinkler Systems in Residential Occupancies up to and Including Four Stories in Height," is cited as the standard for the sprinklers.

903.3.1.2.1 Balconies and decks

- This section governs the following construction with the requirements listed:
  - Construction:
    - Type V construction as follows:
      - Occupancy R.
  - Requirements:
    - The following elements require sprinkler protection:
      - Ground-floor patios with a roof or deck above.
      - Exterior balconies and decks with a roof or deck above.
    - Sidewall sprinklers where protecting open wood joist construction may be located as follows:
      - Deflectors may be ≥ 1" and ≤ 6" below structural members.
      - Deflectors may be ≤ 14" below the deck.

903 Automatic Sprinkler Systems

903.3.1.3 NFPA 13D sprinkler systems

- Sprinklers installed in the following locations must meet the referenced standard:
  - In 1- and 2-family dwellings.
  - In townhouses.

Note: NFPA 13D, "Installation of Sprinkler Systems in One- and Two-family Dwellings and Manufactured Homes," is cited as the standard for the sprinklers.

903.3.2 Quick-response and residential sprinklers

- Where sprinklers are required, one of the following types must be located as indicated:
  - Types:
    - Quick-response.
    - Residential.
  - Locations required:
    - In Occupancy I-2 as follows:
      - Throughout all spaces within a smoke compartment with sleeping units.
    - In the following rooms in the occupancies R and I-1:
      - Dwelling units.
      - Sleeping units.
    - In light-hazard occupancies as follows:
      - Buildings with the following characteristics:
        - Quantity of combustible material is low.
        - Combustibility of contents is low.
        - Low rates of heat are anticipated from any fire.
      - Typical building types:
        - Churches.
        - Libraries excluding large stack rooms.
        - Clubs.
        - Educational buildings.
        - Offices.
        - Nursing homes, convalescent homes, and hospitals.
        - Institutional buildings.
        - Restaurant seating.
        - Auditoriums and theaters excluding stages and prosceniums.
        - Museums.
        - Unused attics.
        - Residential buildings.

Note: NFPA 13, "Installation of Sprinkler Systems," is cited as the source of light-hazard occupancies, a partial summary of which is provided above.

903.3.1, "Standards," is cited as governing the sprinklers addressed in this section.

903 Automatic Sprinkler Systems

903.3.3 Obstructed locations

- Kitchen equipment located under exhaust hoods that are protected by a fire-extinguishing system are not governed by this section.

Note: *Section 904, "Alternative Automatic Fire-Extinguishing Systems," is cited as the source of requirements for the facilities indicated above.*

- Sprinklers must be installed so that obstructions will not hinder their function as follows:
  - Water distribution may not be obstructed.
  - Activation may not be delayed.
- Sprinklers must be installed in or under the following covered locations:
  - Kiosks and concession stands.
  - Displays and booths.
  - Equipment > 4' wide.
- Clearance of ≥ 3' is required between sprinklers and the top of stacks of combustible fibers.

903.3.4 Actuation

- Sprinkler systems must be activated automatically unless otherwise specifically permitted.

903.3.5 Water supplies

- Water supplies for sprinklers must comply with the following:
  - This section series.
  - Referenced standards.

Note: *903.3.1, "Standards," is cited as the source of applicable standards.*

  - The potable water supply must be protected as follows:
    Against backflow as per this section.

Note: *The* International Plumbing Code *is cited as governing backflow protection.*

NOTES

10

Means of Egress

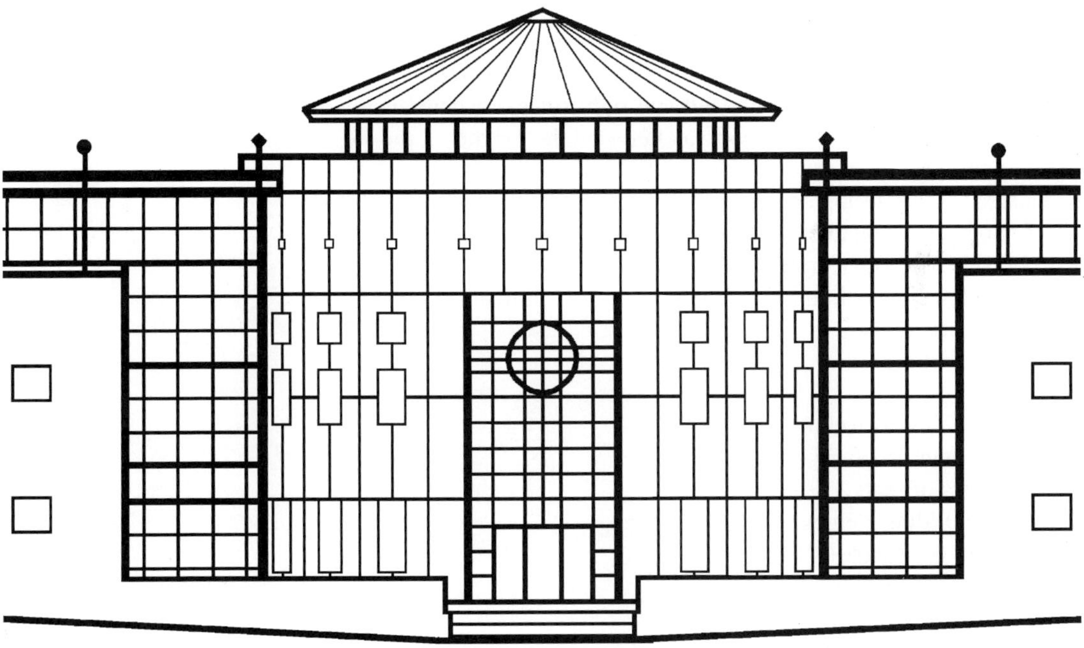

Newman University Sports and Fine Arts Center. Wichita, Kansas. *(partial elevation)*
Gossen Livingston Associates, Inc., Architecture. Wichita, Kansas

1002 Definitions

1002.1 Definitions *(part 1 of 7)*

- **Accessible means of egress**
  - A route of travel as follows:
    - In either of the following:
      - A building.
      - A facility.
    - Route is continuous.
    - Route is unobstructed.
    - Route is from any accessible location to the following:
      - A public way.
- **Aisle**
  - An unenclosed exit access component as follows:
    - Defines a path of egress travel.
- **Aisle accessway**
  - A segment of an exit access path as follows:
    - The path leads to an aisle.
- **Alternating tread device**
  - A component with a series of steps as follows:
    - Angle of travel is as follows:
      - $\geq 50°$ and $\leq 70°$.
      - Angle is measured from the horizontal.
    - Steps are typically designed as follows:
      - Step width is half the width of the component.
      - Individual steps are provided for each foot of the user.
      - Steps are positioned on the left and right sides of a central support as follows:
        - Steps on one side are halfway between the heights of the steps on the other side.
- **Area of refuge**
  - An area for use during emergency evacuation as follows:
    - For occupants unable to use stairways.
    - Affords temporary protection:
      - Used while waiting for one or both of the following:
        - Instructions.
        - Assistance.
- **Bleachers**
  - Seating in tiers as follows:
    - On its own structural system.
    - ≥ 2 rows of seats
    - Not a building element
- **Common path of egress travel**
  - A route of travel as follows:
    - Toward an exit.
    - Provides the only option for a route of travel.
    - Terminates at the following location:
      - At the point where more than one option of travel toward an exit is provided.
  - Routes that are initially separate, but which merge at some point, are common paths.

1002 Definitions

1002.1 Definitions *(part 2 of 7)*

- **Corridor**
  - An exit access component as follows:
    - Enclosed.
    - Provides a route to an exit.
- **Egress court**
  - One of the following elements providing access to a public way:
    - Court.
    - Yard.
- **Emergency escape and rescue opening**
  - The following openings providing a means of escape or rescue:
    - Operable window.
    - Door.
    - Other similar element.

Case study: Fig. 1002.1A. A common path of egress travel is measured on the most direct route available to occupants. The rectangular paths shown in the figure approximate the distance around objects such as furniture and fixtures, the locations of which may vary over time. Some jurisdictions don't require egress paths to follow the rectangular geometry.

Retail 107: Path AD is a common path, there being only one choice of travel. Some jurisdictions may not include toilet rooms in common path measurements. Path BDE and BC are not common paths but "exit access travel distance." This is due to the fact that from point B, there are two choices of travel.

Retail 106A and 106B: All paths of egress travel in these rooms are common paths since all routes merge at point B and since there is a single choice of path from B to the exit door.

Storage 108: All paths in the room are common paths since they must merge at some point prior to leaving the room. The common path extends to point C, which is the first opportunity for more than one choice of travel to an exit. The one diagonal segment is shown since this area is for circulation only, permitting no furnishings to be located there.

1002 Definitions

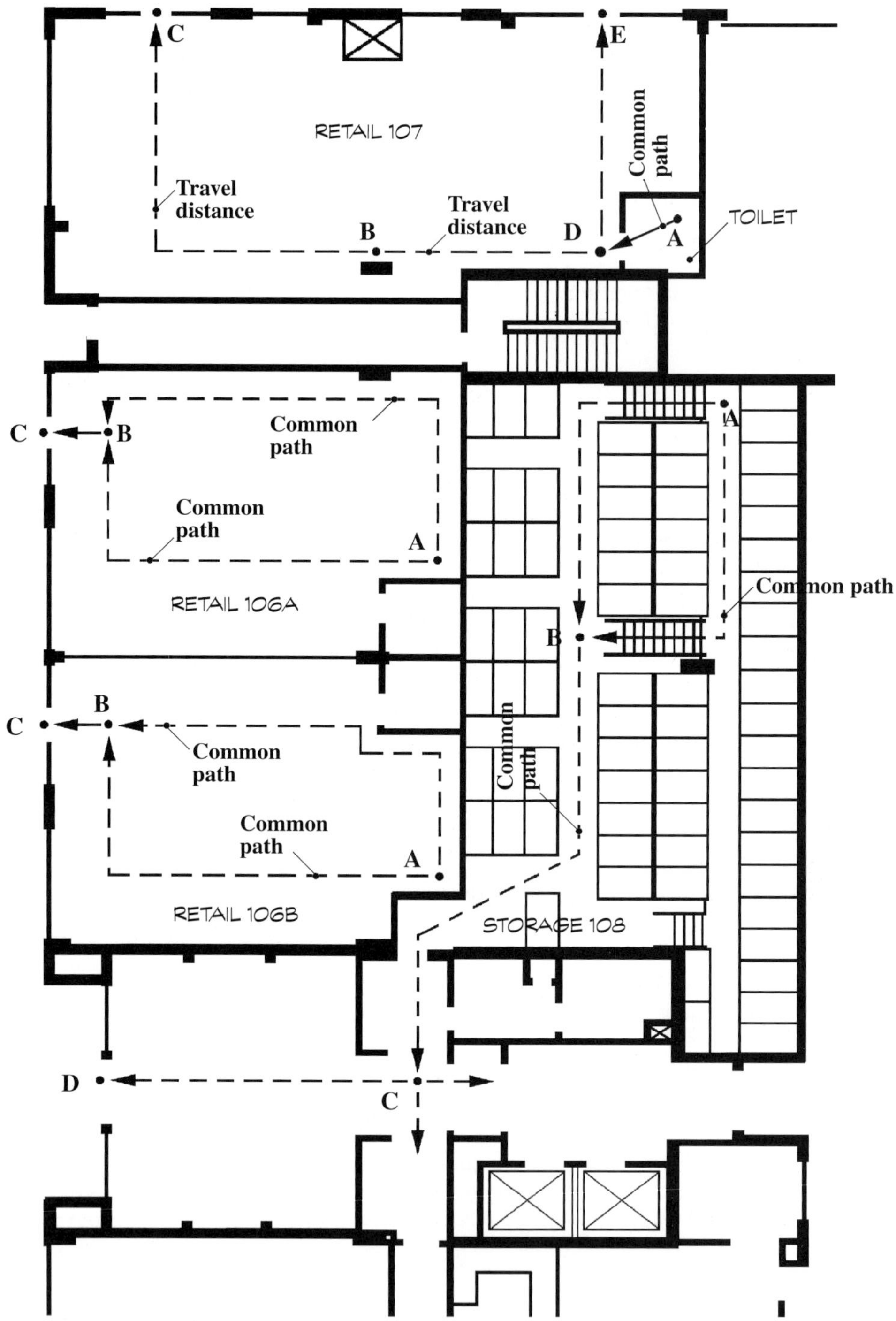

Fig. 1002.1A. Partial ground floor plan, east building. Hoyt Street Properties, Portland, Oregon. Ankrom Moisan Associated Architects. Portland, Oregon.

1002 Definitions

1002.1 Definitions *(part 3 of 7)*

- **Exit**
  - A route of travel from an exit access to an exit discharge.
  - Separated from interior building spaces:
    - By fire-resistance-rated construction.
    - By opening protectives.
  - May include the following elements:
    - Exterior exit doors at level of exit discharge.
    - Vertical exit enclosures.
    - Exit passageways.
    - Exterior exit stairways.
    - Exterior exit ramps.
    - Horizontal exits.
  - A part of a means of egress.
- **Exit access**
  - A route of travel from any occupiable place in a building to an exit.
  - A part of a means of egress.
- **Exit access doorway**
  - A doorway in the path of egress travel from any of the following:
    - An occupiable room, area, or space as follows:
      - Where the egress path enters any of the following:
        - An intervening room.
        - A corridor.
        - An open exit access stairway.
        - An open exit access ramp.
- **Exit discharge**
  - A route between the following:
    - The termination of an exit.
    - A public way.
  - A part of a means of egress.
- **Exit discharge, level of**
  - A horizontal plane.
  - Located at the termination of an exit.
  - Located at the beginning of an exit discharge.
- **Exit enclosure**
  - A component of an exit.
  - Separated from interior building spaces:
    - By fire-resistance-rated construction.
    - By opening protectives.
  - A protected route of travel to one of the following:
    - Exit discharge.
    - Public way.
  - May be in either of the following directions:
    - Horizontal.
    - Vertical.

1002 Definitions

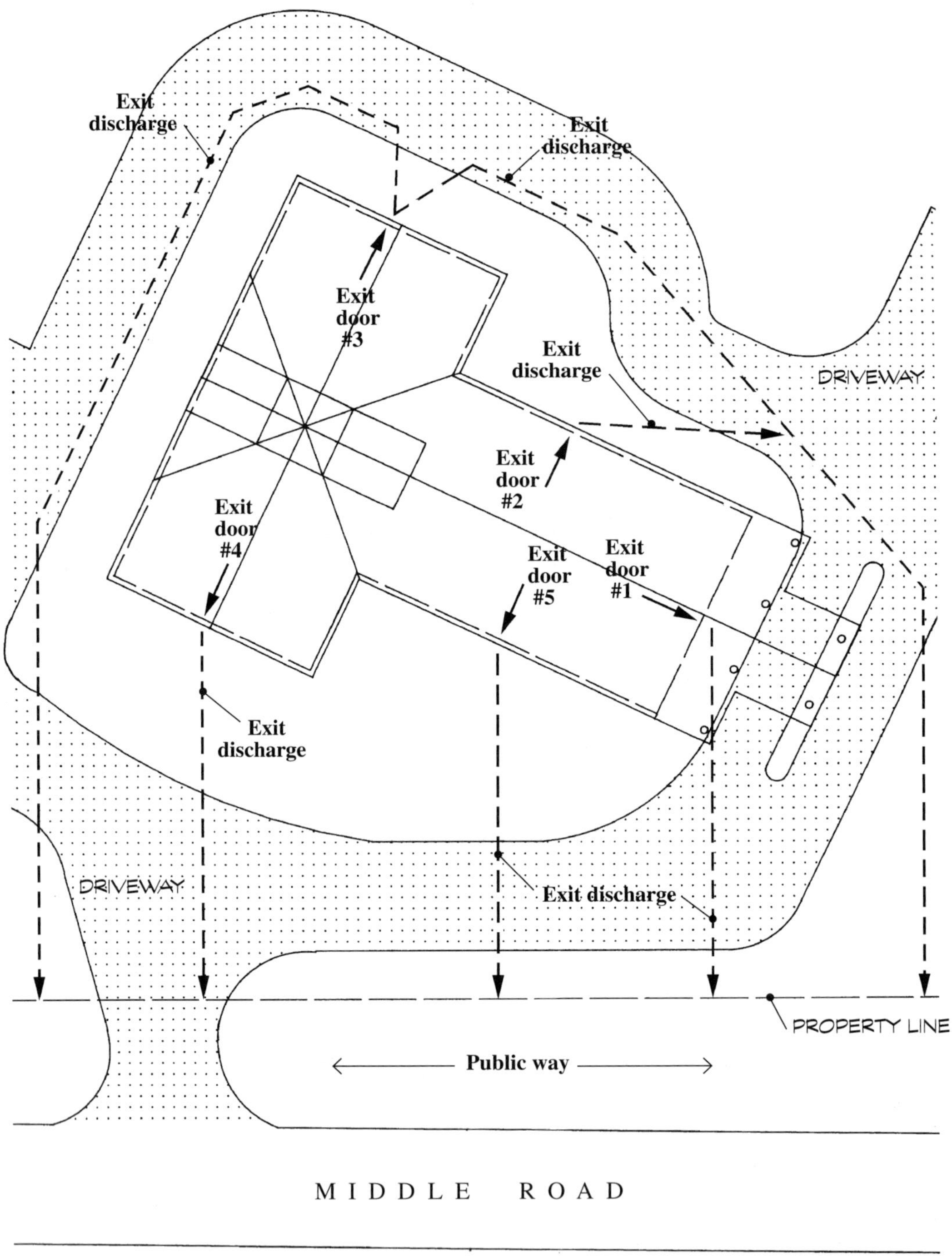

Fig. 1002.1B. Site plan. Glad Tidings Assembly of God Church. Naticoke, Pennsylvania. Mullins and Weida, Architect and Associate. Bear Creek, Pennsylvania.

1002 Definitions

1002.1 Definitions *(part 4 of 7)*

- **Exit, horizontal**
  - A route of egress travel.
  - Between either of the following:
    Two buildings on approximately the same level.
    Two areas in the same building approximately on the same level.
  - Through an assembly providing protection from fire and smoke.

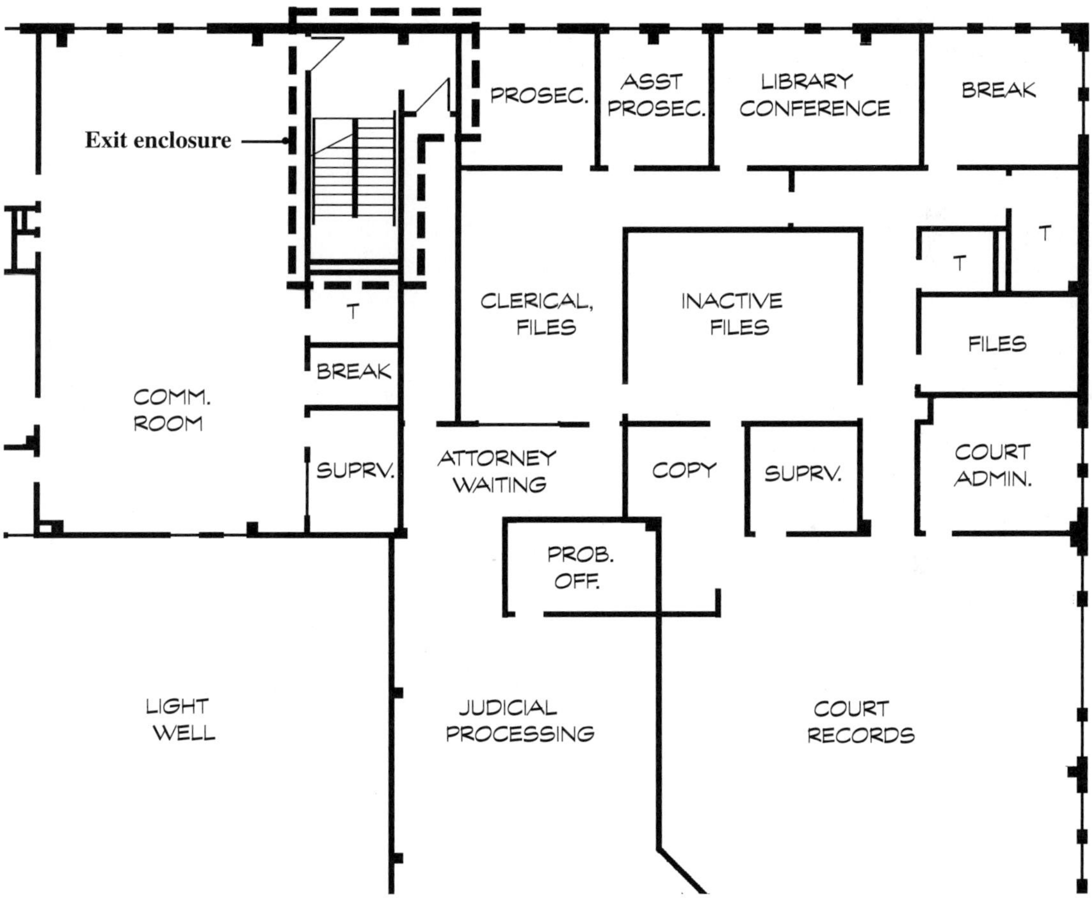

Fig. 1002.1C. Partial floor plan. Lee's Summit Police and Court Facility. Lee's Summit, Missouri. The Hollis and Miller Group, Inc. Lee's Summit, Missouri.

1002 Definitions

1002.1 Definitions *(part 5 of 7)*

- **Exit passageway**
  - A component of an exit.
  - Separated from interior building spaces:
    - By fire-resistance-rated construction.
    - By opening protectives.
  - A protected route of travel to one of the following:
    - Exit discharge.
    - Public way.
  - A horizontal route of travel.
- **Fire exit hardware**
  - Panic hardware as follows:
    - Listed for fire door assemblies.
- **Floor area, gross**
  - Area within the inside perimeter of the exterior walls.
  - Does not include the following:
    - Vent shafts.
    - Shafts with no openings.
    - Interior courts.
  - Includes space occupied by the following:
    - Corridors.
    - Stairways.
    - Closets.
    - Interior walls.
    - Columns.
    - Other interior elements.
  - Space not enclosed with exterior walls is included as follows:
    - Usable area under the following horizontal projections:
      - Roof above.
      - Floor above.
- **Floor area, net**
  - Occupied area.
  - Unoccupied accessory areas, such as the following, are not included:
    - Corridors.
    - Stairways.
    - Toilet rooms.
    - Mechanical rooms.
    - Closets.
- **Folding and telescopic seating**
  - A structure for seating as follows:
    - Seating is tiered.
    - Size can be reduced for the following purposes:
      - For moving.
      - For storing.
    - Not a building element.

1002 Definitions

1002.1 Definitions *(part 6 of 7)*

- **Grandstand**
  - Seating in tiers as follows:
    - On its own structural system.
    - ≥ 2 rows of seats.
    - Not a building element.
- **Guard**
  - A barrier as follows:
    - Located at the open edge of walking surfaces where there is an elevation change.
    - Minimizes chances for falling from the walking surface to the lower level.
- **Handrail**
  - A rail for grasp by the hand as follows:
    - Rail can be in either of the following positions:
      - Horizontal.
      - Sloping.
    - For either of the following purposes:
      - Guidance.
      - Support.
- **Means of egress**
  - A route of travel from any occupied place in a building to a public way:
    - Route is continuous.
    - Route is unobstructed.
  - Consists of 3 separate segments:
    - Exit access.
    - Exit.
    - Exit discharge.
- **Nosing**
  - The leading edge of the following stair components:
    - Treads.
    - Landings at the top of stair runs.
- **Occupant load**
  - The number of people for which a means of egress is designed.
- **Panic hardware**
  - A door-latching assembly as follows:
    - Latch releases upon application of a force in the direction of egress.
- **Public way**
  - Any of the following that leads to a street dedicated to public use:
    - Street.
    - Alley.
    - Parcel of land.
  - Has clear dimensions as follows:
    - ≥ 10' in height.
    - ≥ 10' in width.
- **Ramp**
  - A walking surface sloping > 1:20 in the direction of travel.

1002 Definitions

1002.1 Definitions *(part 7 of 7)*

- **Scissor stair**
  - Two stairways with all of the following characteristics:
    - Interlocking.
    - Two separate egress routes.
    - Located in the same stairway enclosure.
- **Smoke-protected assembly seating**
  - Seating with all of the following characteristics:
    - Served by a means of egress.
    - Where smoke cannot accumulate.
- **Stair**
  - A change in elevation as follows:
    - Has $\geq$ 1 riser.
- **Stairway, exterior**
  - Open to the exterior on at least one side:
    - Where required, the following are permitted at open sides:
      - Structural columns and beams.
      - Guards and handrails.
  - Open to one of the following:
    - Yard.
    - Court.
    - Public way.
- **Suite**
  - A group of patient rooms in Occupancy I-2 as follows:
    - Rooms are for treament or sleeping.
    - Staff are within the rooms supervising patients.
    - The rooms meet certain code requirements.

  Note: The following are cited as governing suites:
  - *1014.2.2, "Group I-2."*
  - *1014.2.3, "Suites in patient sleeping areas."*
  - *1014.2.3.1, "Area."*
  - *1014.2.3.2, "Exit access."*
  - *1014.2.3.3, "Travel distance."*
  - *1014.2.4, "Suites in areas other than patient sleeping areas."*
  - *1014.2.4.1, "Area."*
  - *1014.2.4.2, "Exit access."*
  - *1014.2.4.3, "One intervening room."*
  - *1014.2.4.4, "Two intervening rooms."*
  - *1014.2.5, "Exit access through suites."*
  - *1014.2.6, "Travel distance."*
  - *1014.2.7, "Separation."*

- **Winder**
  - A stair tread as follows:
    - Long edges are not parallel.

1002 Definitions

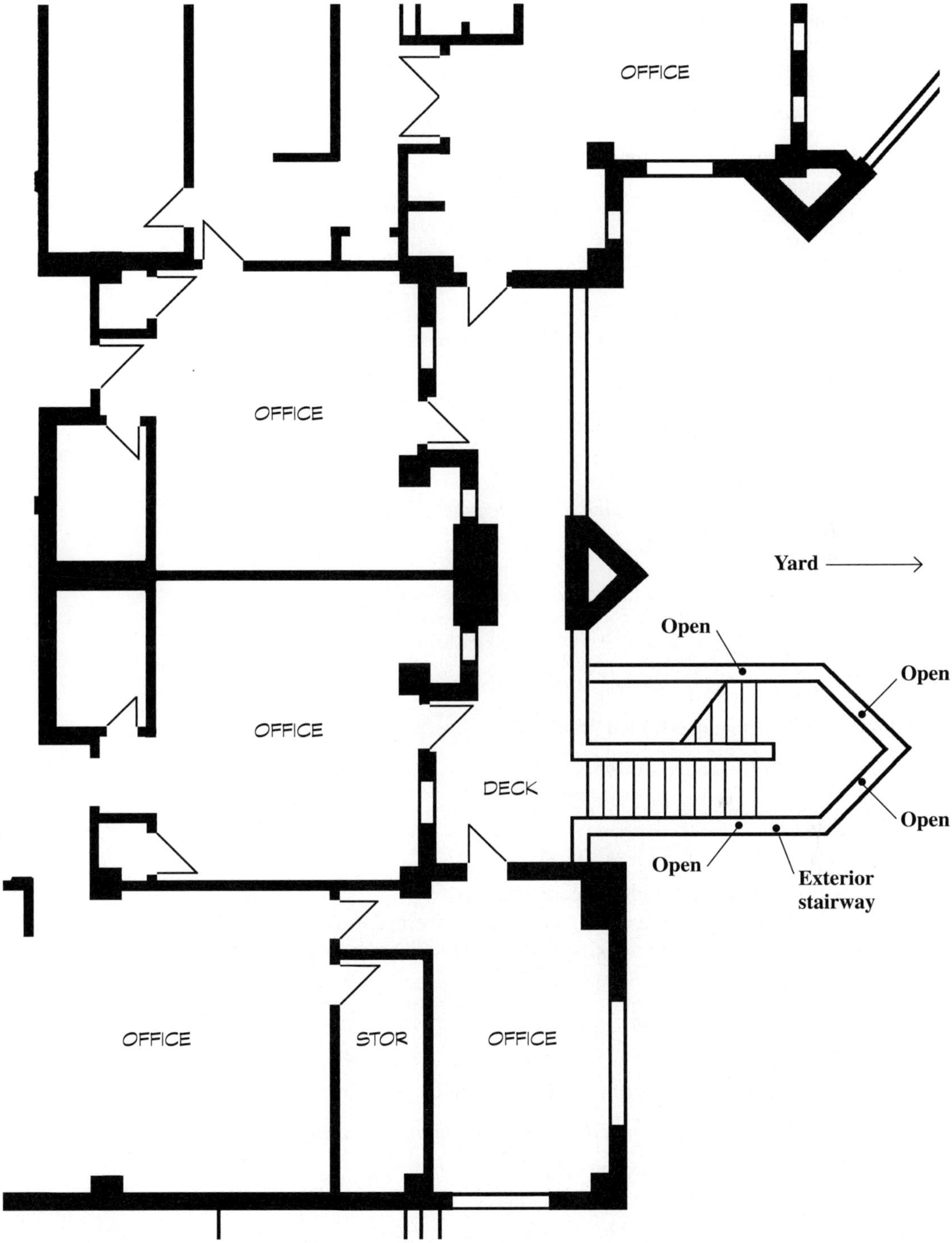

Fig. 1002.1D. Partial second-floor plan showing an exterior stairway. Country Club Park Building One. Wichita, Kansas. Gossen Livingston Associates, Inc., Architecture. Wichita, Kansas.

1003 General Means of Egress

1003.2 Ceiling height

- The height minimum of this section does not apply to the following:
  - Sloped ceilings.

  Note: 1208.2, "Minimum ceiling heights," is cited as applicable.

  - Ceilings in Occupancy R as follows:
    Dwelling units.
    Sleeping units.

  Note: 1208.2, "Minimum ceiling heights," is cited as applicable.

  - Projections where permitted.

  Note: 1003.3, "Protruding objects," is cited as applicable.

  - Headroom at stairs.

  Note: 1009.2, "Headroom," is cited as applicable.

  - Height at doors.

  Note: 1008.1.1, "Size of doors," is cited as applicable.

  - Headroom at ramps.

  Note: 1010.5.2, "Headroom," is cited as applicable.

  - Clear height in parking garages for vehicles and pedestrians.

  Note: 406.2.2, *"Clear height," is cited as applicable.*

  - Height above and below mezzanine floors.

  Note: 505.1, *"General," is cited as applicable.*

- Otherwise, ceiling height must be ≥ 7'-6" in a means of egress.

1003.3.1 Headroom

- Door stops and closers may not reduce headroom to < 6'-6".
- Other protrusions below minimum ceiling height are governed as follows:
  - ≥ 6'-8" headroom is required at walking surfaces including the following:
    Walks.
    Corridors.
    Aisles.
    Passageways.
  - ≤ 1/2 the ceiling area in a means of egress may be reduced below required height.

  Note: 1003.2, "Ceiling height," is cited as governing minimum ceiling height.

- A barrier is required where headroom is reduced to < 6'-8" as follows:
  - The leading edge must be ≤ 2'-3" above the floor.
  - At locations other than doorways.

1003 General Means of Egress

Case study: Fig. 1003.3.1A. The soffit of the stair landing at 7'-4" is below the 7'-6" minimum ceiling height in a means of egress as now required. The 7'-4" headroom meets the 7' minimum that was in effect at the time the building was constructed. A wall flush with the stringer of the lower flight provides the barrier required where protrusions below the 7' minimum ceiling height would otherwise reduce headroom below 6'-8". Doors B5 and B6 at 7'-2" meet the 6'-8" minimum headroom. Their 5/8" stops are above the 6'-6" minimum in doorways.

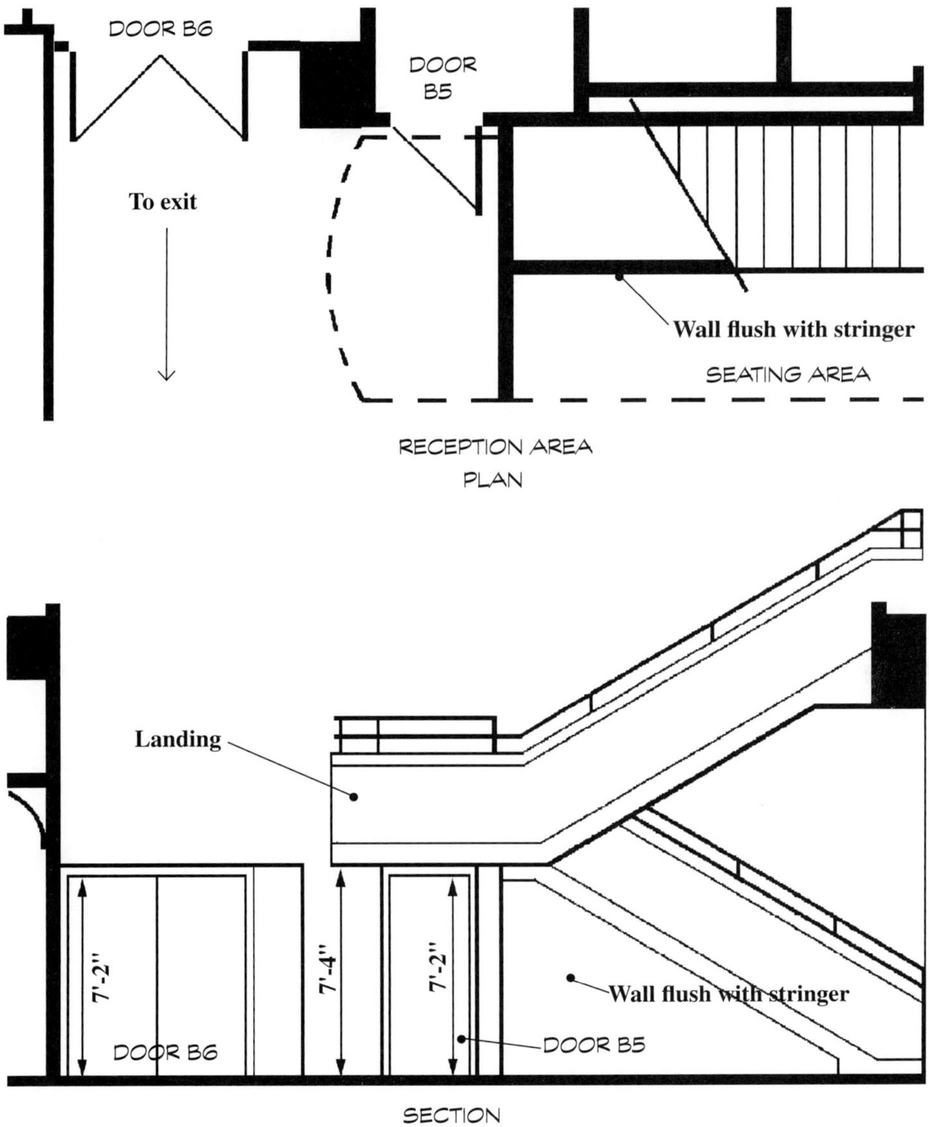

Fig. 1003.3.1A. Partial floor plan and section at reception area. Lubrication Engineers, Inc. Wichita, Kansas. Gossen Livingston Associates, Inc., Architecture. Wichita, Kansas.

1003 General Means of Egress

Case study: Fig. 1003.3.1B. The detailing complies with the requirement to provide a barrier ≤ 2'-3" high to protect a protrusion which reduces headroom to < 6'-8". In this case, the protrusion is the lower flight of the stairway, and the barrier is a rail.

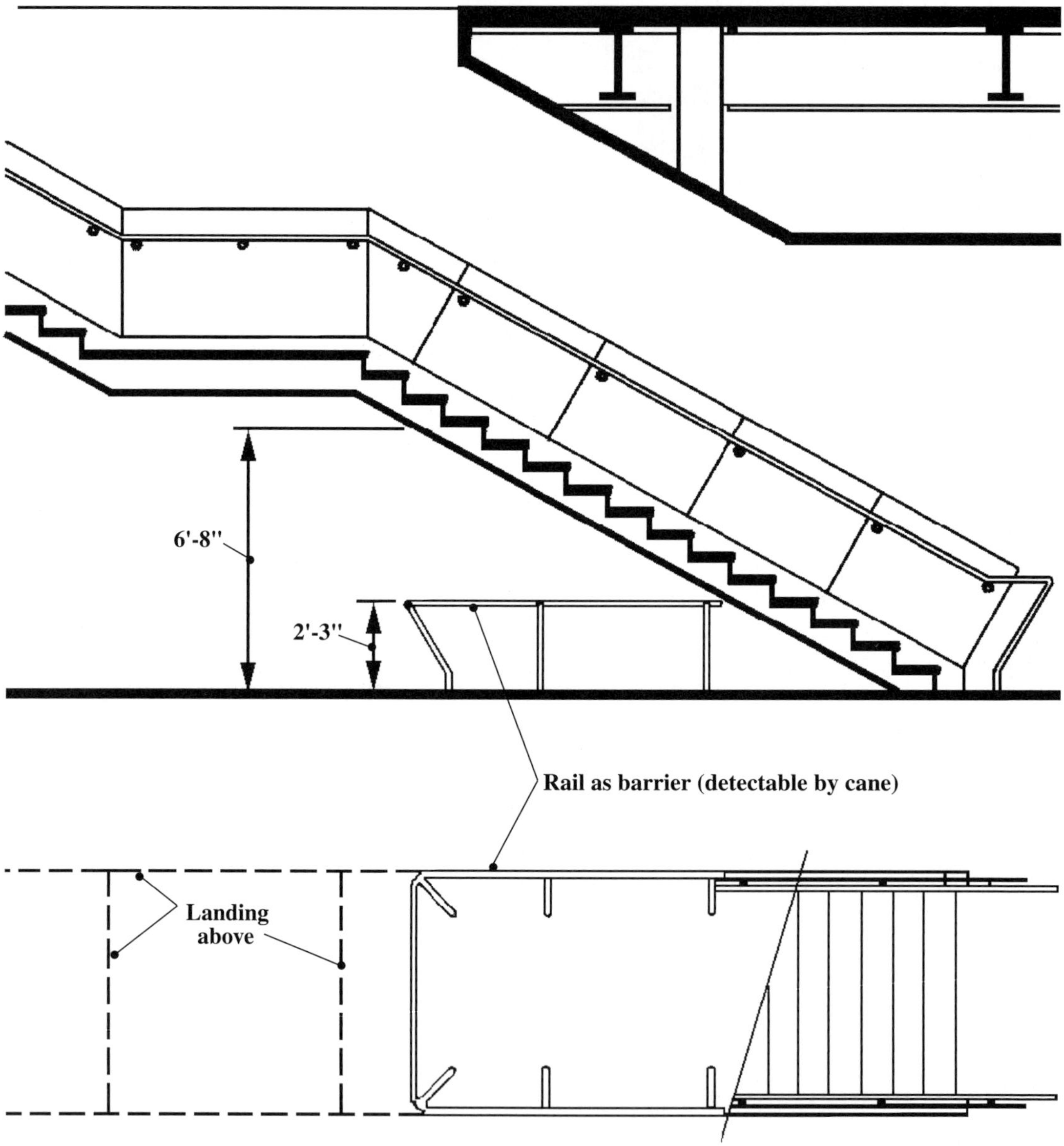

Fig. 1003.3.1B. Partial plan and section of stair in the first-floor concourse. University of Connecticut New Downtown Campus at Stamford, Connecticut. Perkins Eastman Architects, P.C. New York, New York.

1003 General Means of Egress

Case study: Fig. 1003.3.1C. The ceiling height of the corridor is 11'-4", well over the minimum. The light alcoves at the walls and those spanning across the corridor at the columns have soffits at 7'-4". Thus, the required 6'-8" minimum is maintained for headroom under objects extending below the ceiling.

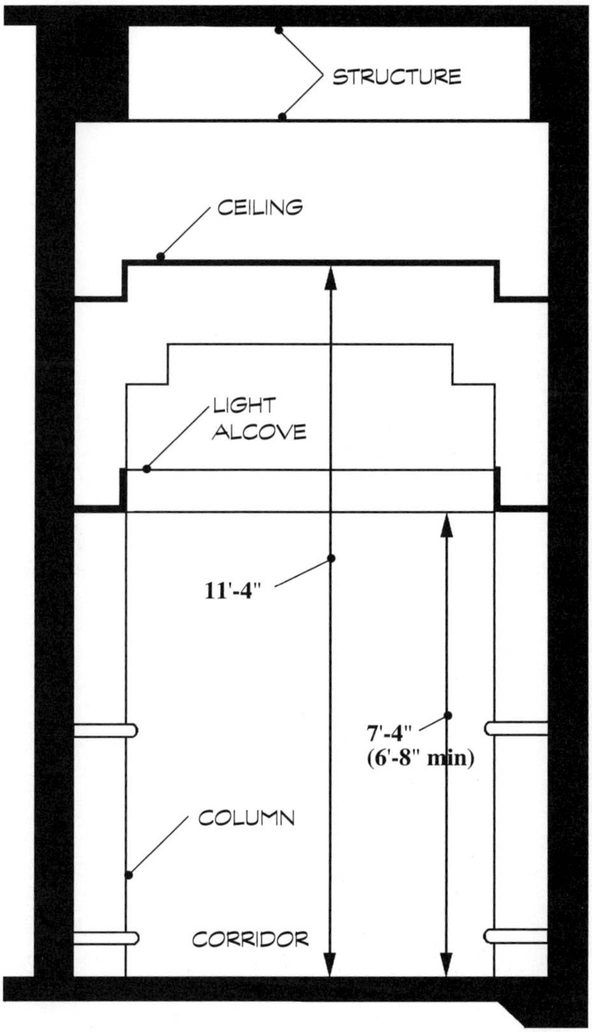

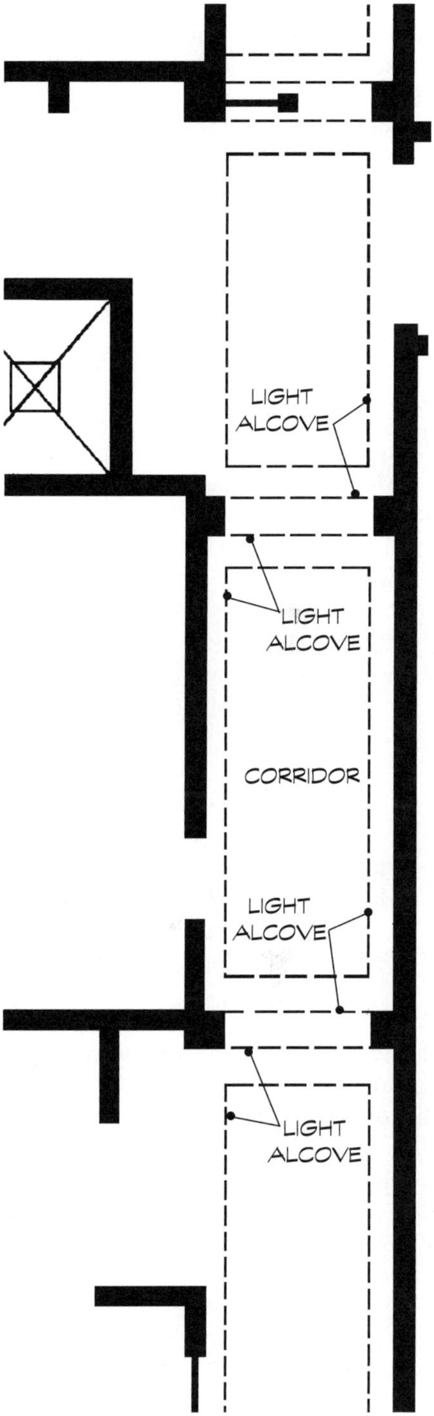

Fig. 1003.3.1C. Partial floor plan and section at corridor. Lubrication Engineers, Inc. Wichita, Kansas. Gossen Livingston, Associates, Inc., Architecture. Wichita, Kansas.

1003 General Means of Egress

1003.3.2 Post-mounted objects

- This section does not apply to the following:
  - Sloping parts of handrails located as follows:
    - Between top and bottom risers of stairs.
    - Above ramp runs.
- The following freestanding object has the requirement indicated below:
  - Object:
    - Mounted on one of the following:
      - Post.
      - Pylon.
    - Mounted so that its lowest leading edge is between the following heights:
      - > 27" above floor.
      - < 80" above floor.
  - Requirement:
    - May not overhang > 4".
- The following object has the requirements indicated below:
  - Object:
    - Any of the following:
      - Sign.
      - Other obstruction.
    - Mounted as follows:
      - Between any of the following supports:
        - Posts.
        - Pylons.
      - Where supports are spaced as follows:
        - > 12" clear space apart.
  - Requirements:
    - Height of lowest edge cannot be within the following range:
      - > 27" above floor.
      - < 80" above floor.

1003.3.3 Horizontal projections

- This section limits the projection of elements over a walking surface.
- Handrails are limited to ≤ 4 1/2".
- The following elements are limited as shown below:
  - Elements:
    - Structural elements.
    - Fixtures.
    - Furnishings.
  - Limitation:
    - ≤ 4" from either side in the following zone:
      - From 2'-3" to 6'-8" above the walking surface:

1003 General Means of Egress

Case study: Fig. 1003.3.3A. The fire hose cabinet and fire extinguisher cabinet comply with the limit of protrusions to 4" over a walking surface, at a height > 2'-3" and < 6'-8".

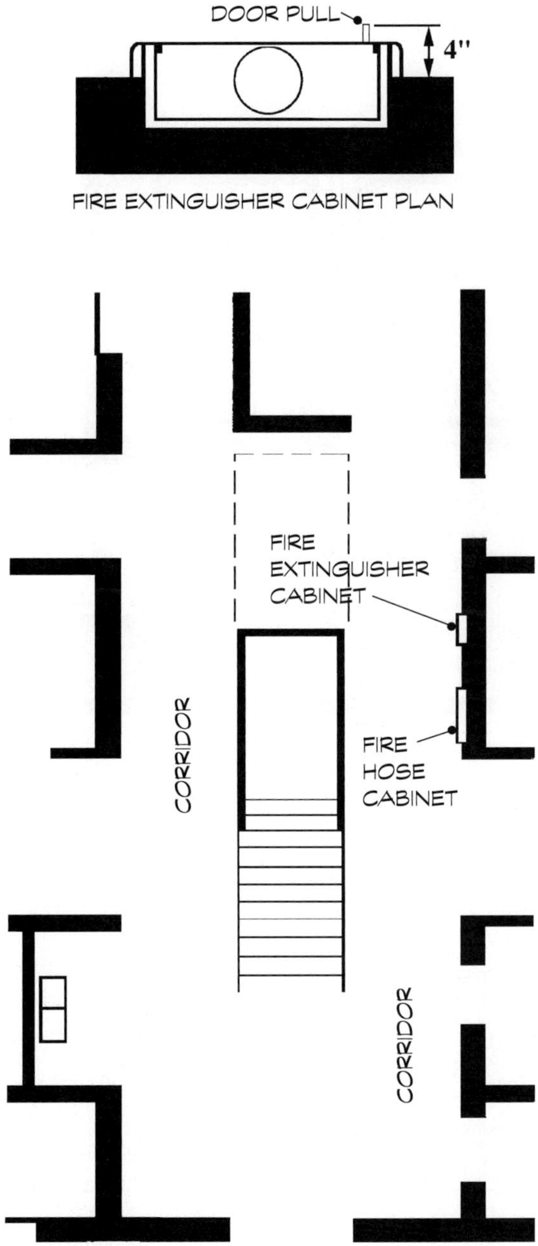

Fig. 1003.3.3A. Partial plan and detail. Newman University Sports and Fine Arts Center. Wichita, Kansas. Gossen Livingston Associates, Inc., Architecture. Wichita, Kansas.

1003 General Means of Egress

Case study: Fig. 1003.3.3B. The wall-mounted light fixtures protrude only 4" into the corridor, thus, complying with the 4" limitation for horizontal projections at heights > 2'-3" and < 6'-8". The wall mass projects 6" into the corridor at point A, but since it extends below the 2'-3" height, it is detectable by a cane and complies with the code.

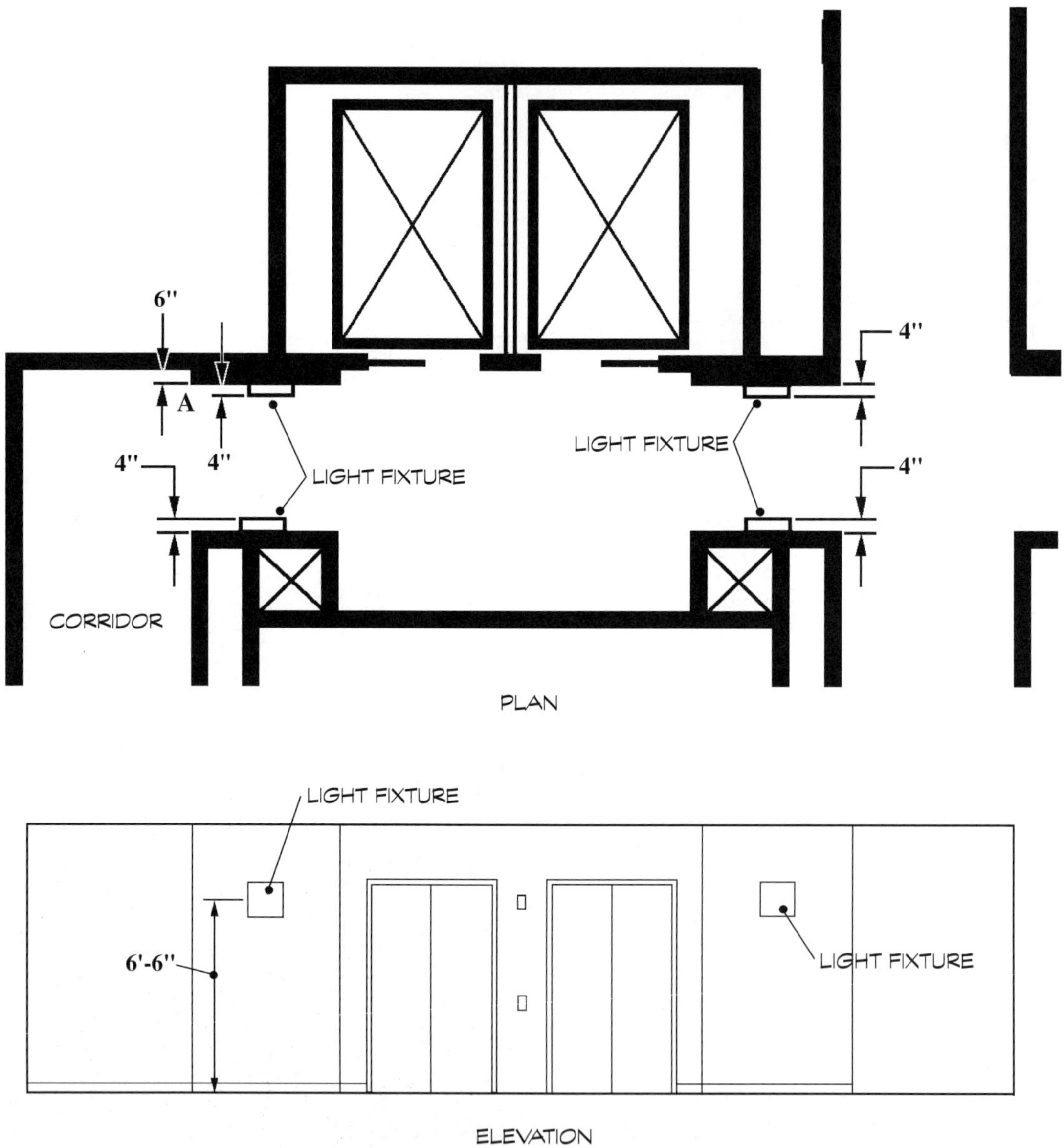

Fig. 1003.3.3B. Partial floor plan and interior elevation at elevators. Methodist Community Health Center. Sugar Land, Texas. HKS, Inc., Architects, Engineers, Planners. Dallas, Texas.

1003 General Means of Egress

1003.3.4 Clear width

- Protruding objects in accessible routes are governed as follows:
  - They may not reduce the minimum required clear width.

1003.4 Floor surface

- Means of egress walking surface must have the following characteristics:
  - Surface must be slip resistant.
  - Surface must be securely attached.

1003.5 Elevation change *(part 1 of 2)*

- A single riser ≤ 7" may be used in a means of egress as follows:
  - At exterior doors not required to be accessible in the following locations:
    In occupancies F, H, R-2, R-3, S, U.

Note: Chapter 11, "Accessibility," is cited as the source for accessibility requirements throughout this section.

- The following stair has the requirements indicated below:
  - Stair:
    Located in routes not required to be accessible.
    ≤ 2 risers.
  - Requirements:
    Risers must be ≥ 4" and ≤ 7".
    Treads must be ≥ 13".
    ≥ 1 handrail must be provided as follows:
      Located ≤ 2'-6" of the center of anticipated egress travel on the stair.

Note: The following are cited as sources for other requirements as noted:
1009.4, "Stair treads and risers," for step requirements, some data from which is listed above.
Section 1012, "Handrails," for handrail requirements.

- Aisles with all of the following characteristics may have a step:

Note: 1028.11, "Assembly aisle walking surfaces," is cited as governing the steps.

  - For seating that is not required to be accessible.
  - Where seating has a difference in elevation < 12".
  - Where an aisle handrail is provided.

Note: 1028.13, "Handrails," is cited as governing the handrails.

1003 General Means of Egress

1003.5 Elevation change *(part 2 of 2)*

- In other locations, a sloped floor with the requirements listed below is required for the following conditions:
  - Conditions:
    - Changes in elevation with both of the following characteristics:
      - < 12".
      - Located in a means of egress.
  - Requirements:
    - Where the slope is > 1:20 the following applies:
      - A ramp is required as follows:
        - For elevation changes ≤ 6" one of the following is required for the ramp:
          - Handrails.
          - A surface that contrasts with adjacent floor surfaces.

Note: Section 1010, "Ramps," is cited as governing required ramps.

- A sloped floor is required in the following cases:
  - For changes in elevation where both of the following apply:
    - In Occupancy I-2.
    - Paths of travel serving nonambulatory occupants.

1003.6 Means of egress continuity

- Continuity of the travel route in a means of egress is required as follows:
  - Only means of egress building components are permitted to interrupt the travel route.
  - The only obstructions permitted in the route are protrusions allowed by this chapter.
- The capacity required for a means of egress may not diminish in the direction of egress.

1003.7 Elevators, escalators, and moving walks

- Elevators may serve as an accessible means of egress.

Note: 1007.4, "Elevators," is cited as the source of requirements.

- The following may not serve as a required means of egress for persons not already using them at the advent of an emergency:
  - Elevators and escalators.
  - Moving walks.

1003 General Means of Egress

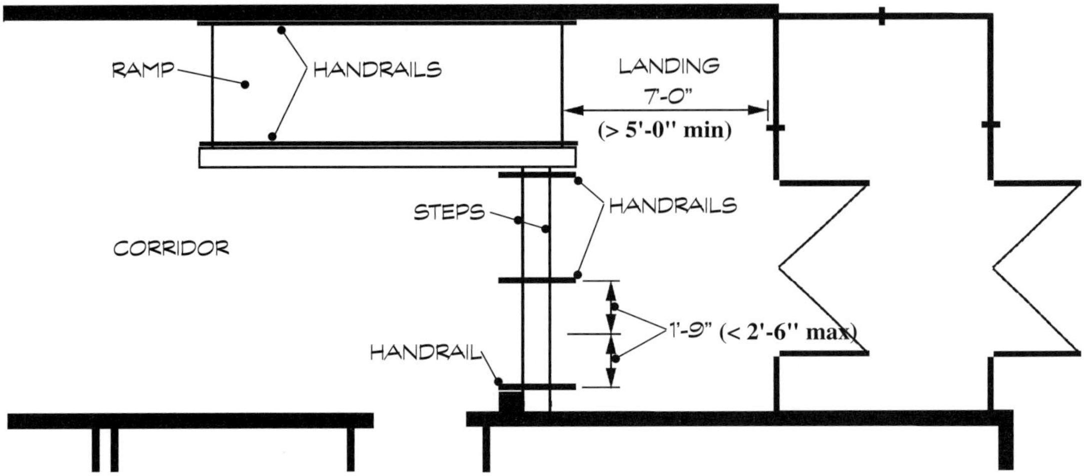

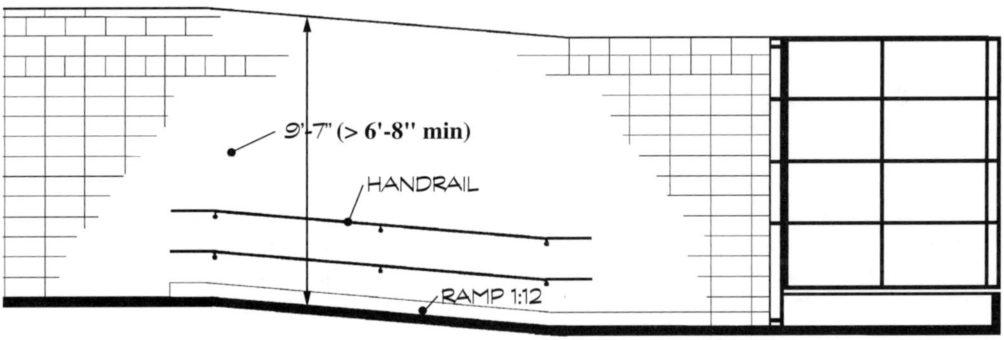

Case study: Fig. 1003.5. The ramp meets requirements for a sloped surface at a change in level < 12". Since its slope is > 1:20, it also must comply with requirements for all ramps. These include among others, having a slope ≤ 1:12, a cross slope ≤ 1:48, a rise ≤ 2'-6", headroom ≥ 6'-8", handrails on both sides, and landings ≥ 5' in length. Requirements for stairs at changes of elevation < 12" are shown in the stair section. The intermediate handrail complies with the need for a railing within 2'-6" of the center line of travel.

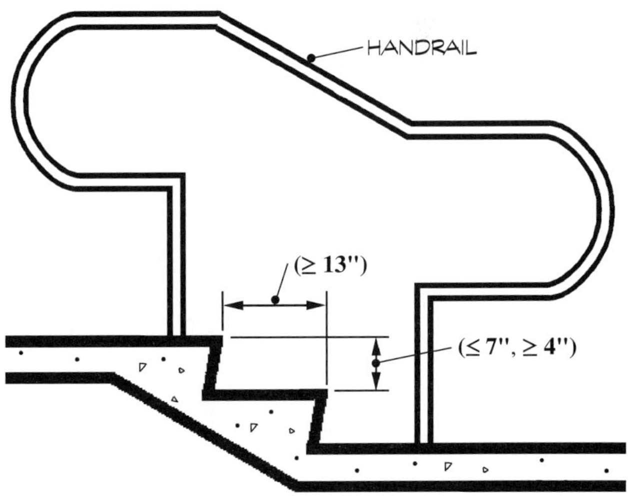

Fig. 1003.5. Partial plan and sections at corridor. Montachusett Regional Vocational-Technical High School. Fitchburg, Massachusetts. HKT Architects, Inc. Somerville, Massachusetts.

1003 General Means of Egress

Case study: Fig. 1003.7. The escalators shown on this hospital floor do not serve as a means of egress as per code limitations. Exits are provided with exit access travel distances in compliance with code requirements. The elevators qualify as an accessible means of egress by virtue of the building being sprinklered and by their size, which meets required standards.

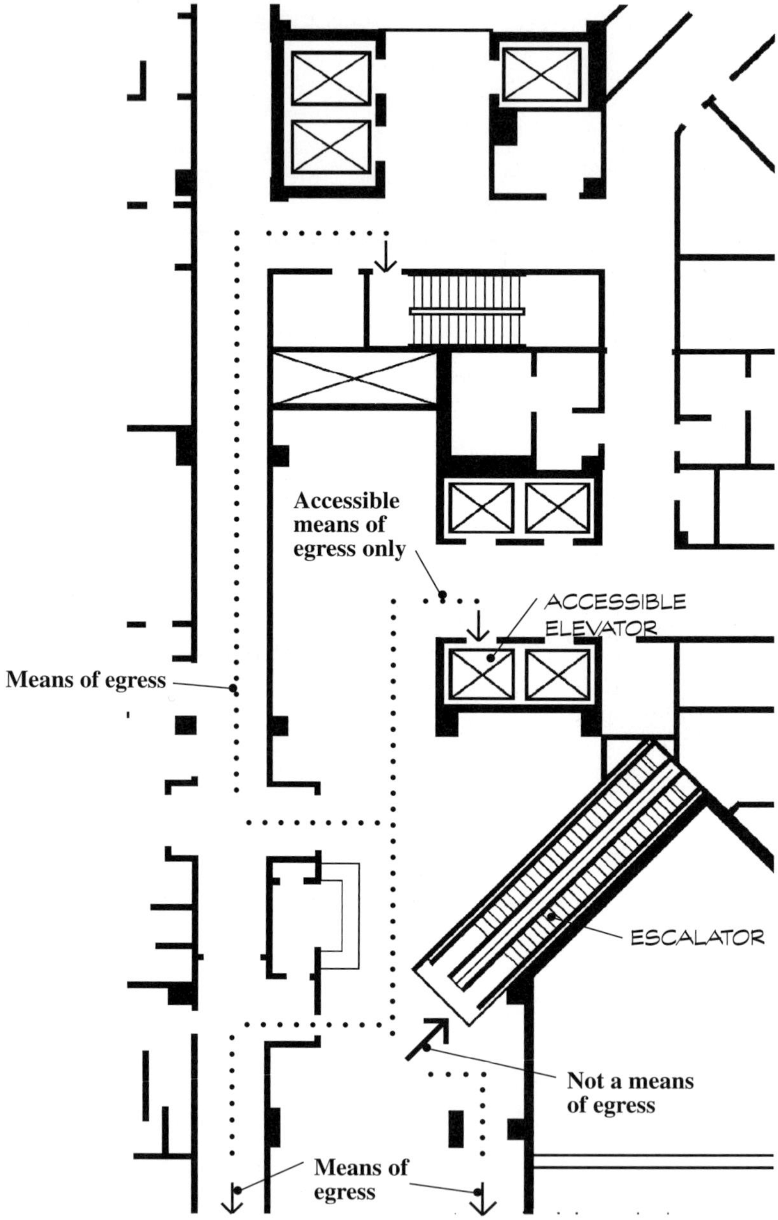

Fig. 1003.7. Partial floor plan. Christus St. Michael Health Care Center. Texarkana, Texas. Watkins Hamilton Ross Architects, Inc. Houston, Texas.

1004 Occupant Load

1004.1 Design occupant load

- This section governs the number of occupants assigned to a means of egress.
- The occupant load of a space is the sum of the following:
  - Occupant load assigned to the space.
  - Occupant load assigned to any accessory area egressing through the space.

1004.1.1 Areas without fixed seating

Note: 1004.7, "Fixed seating," is listed as governing the occupant load in assembly fixed seating.

- Where the actual number of occupants for which an area is designed is < that calculated by this section, the following applies:
  - The actual number of occupants can be used to determine egress requirements in the following case: Where approved.
- The calculation of occupant load used to determine egress requirements is as follows:
  - Occupant load ≥ floor area served ÷ the square feet per occupant listed in the table below:
    Where a use is not listed, the following applies:
    The building official will identify a similar use in the table:

Table 1004.1.1 Maximum Floor Area per Occupant

Function	SF/ occupant	Function	SF/occupant
Agricultural building	300 gross	**H-5 Fabrication and**	
Aircraft hangars	500 gross	**manufacturing areas**	200 gross
Airport terminal		**Industrial areas**	100 gross
Concourse	100 gross	**Institutional areas**	
Waiting area	15 gross	In patient treatment	240 gross
Baggage claim	20 gross	Outpatient areas	100 gross
Baggage handling	300 gross	Sleeping areas	120 gross
Assembly		**Kitchens, commercial**	200 gross
Gambling areas	11 gross	**Library**	
No fixed seats		Reading rooms	50 net
Concentrated		Stack area	100 gross
(chairs only)	7 net	**Locker rooms**	50 gross
Standing space	5 net	**Mercantile**	
Not concentrated		Basement	30 gross
(tables, chairs)	15 net	Grade floor	30 gross
Bowling centers		Other floors	60 gross
5 occupants per		Storage, shipping	300 gross
lane including 15'		**Mechanical room**	300 gross
of lane-approach	7 net	**Parking garages**	200 gross
Business areas	100 gross	**Residential**	200 gross
Courtrooms		**Skating rink**	
Areas with no fixed seats	40 net	Rink	50 gross
Day care	35 net	Deck	15 gross
Dormitories	50 gross	**Swimming pool**	
Educational		Pool	50 gross
Classroom	20 net	Deck	15 gross
Shops	50 net	**Stages and platforms**	15 net
Vocational areas	50 net	**Storage, accessory**	300 gross
Exercise rooms	50 gross	**Warehouses**	500 gross

Source: IBC Table 1004.1.1.

1004 Occupant Load

Case study: Fig. 1004.1.1A. The office area is designated as Occupancy B for purposes of determining occupant load. IBC Table 1004.1.1 indicates that the area allowance for occupants is 100 sf (gross) per person. Since the allowance is in gross sf, interior wall thicknesses, toilets, closets, and hall are included in the area computation. The area allowance for the 2 storage rooms is 300 sf (gross) per person. In this example, computing occupants for the storage rooms separately from the business area would result in 1 less occupant for the complex. Since the means of egress serving this occupant load is larger than required, treating the storage rooms separately is not justified. The occupant count for the business area, including storage, is, therefore, determined by the conservative following calculation:
3,210 gross sf ÷ 100 sf/occupant = 33 occupants.

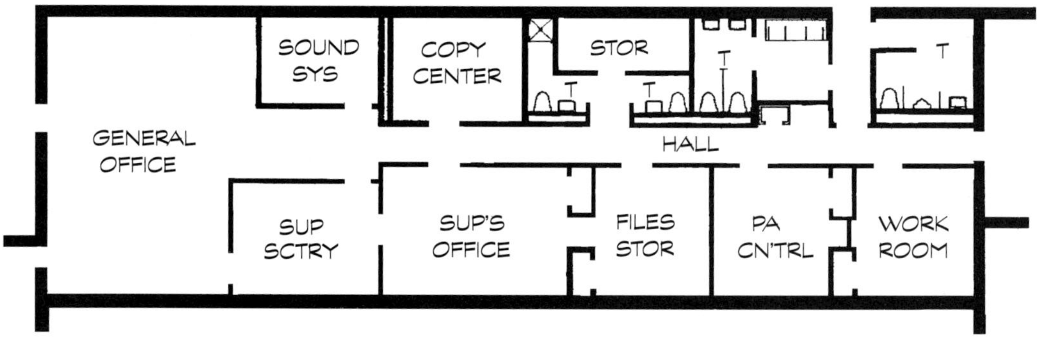

Fig. 1004.1.1A. Partial floor plan. Montachusett Regional Vocational-Technical High School. Fitchburg, Massachusetts. HKT Architects, Inc. Somerville, Massachusetts.

Case study: Fig. 1004.1.1B. Gross (gr) and net square foot (sf) allowances from IBC Table 1004.1.1 are applied to the case study example as follows:

415 Electric room	Mechanical:	985 gr sf	÷ 300 gr sf/occ	=	4 occupants
416 Storage	Storage:	675 gr sf	÷ 300 gr sf/occ	=	3 occupants
417 Utility and storage	Storage:	390 gr sf	÷ 300 gr sf/occ	=	2 occupants
418 Emergency generator	Mechanical:	300 gr sf	÷ 300 gr sf/occ	=	1 occupant
419 Maintenance office	Office:	480 gr sf	÷ 100 gr sf/occ	=	5 occupants
423 Storage	Storage:	1642 gr sf	÷ 300 gr sf/occ	=	6 occupants
424 Incinerator	Mechanical:	350 gr sf	÷ 300 gr sf/occ	=	2 occupants
426 Office	Office:	470 gr sf	÷ 100 gr sf/occ	=	5 occupants
431 Restaurant	Assembly:	2440 net sf	÷ 15 net sf/occ	=	163 occupants
433 Culinary arts	Vocational:	3870 net sf	÷ 50 net sf/occ	=	78 occupants
440 Solarium	Assembly:	184 net sf	÷ 15 net sf/occ	=	13 occupants
448 Nonfood storage	Storage:	144 gr sf	÷ 300 gr sf/occ	=	1 occupant
449 Food storage	Storage:	172 gr sf	÷ 300 gr sf/occ	=	1 occupant
450 Classroom	Classroom:	750 net sf	÷ 20 net sf/occ	=	38 occupants
451 Receiving	Storage:	920 gr sf	÷ 300 gr sf/occ	=	4 occupants

1004 Occupant Load

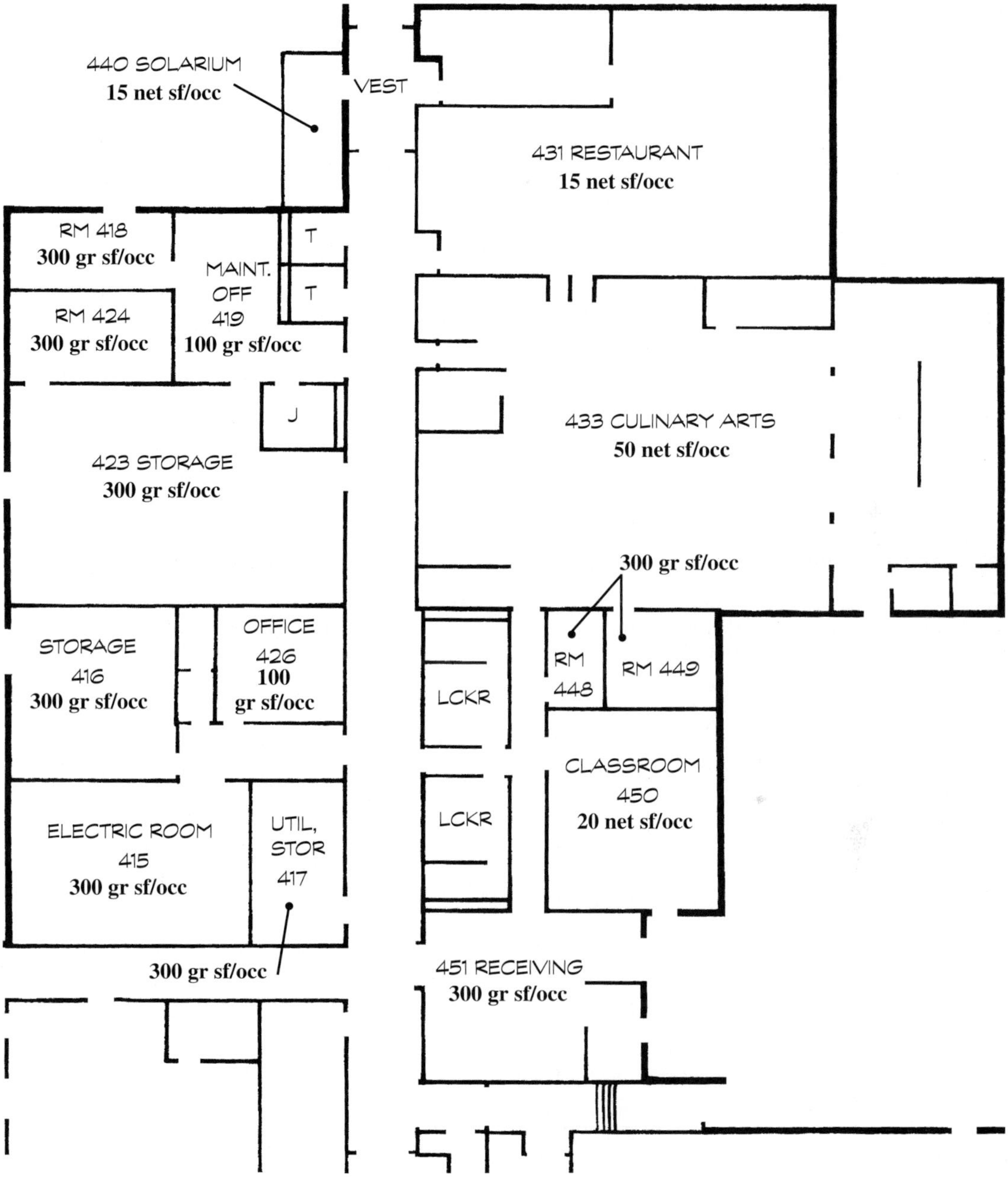

Fig. 1004.1.1B. Partial floor plan. Montachusett Regional Vocational-Technical High School. Fitchburg, Massachusetts. HKT Architects, Inc. Somerville, Massachusetts.

1004 Occupant Load

1004.4 Exiting from multiple levels

- This section addresses exits serving ≥ 1 floor.
- The capacity of the exit at each floor is serves occupants of each floor only.
- Occupant loads are not summed from each floor in the direction of egress.
- The capacity of the exit may not decrease along the route of egress.

Case study: Fig. 1004.4. The occupant load entering the exit stairway at each floor above grade level is 38. The occupant load used to determine the minimum requirements for the stairway down to the ground floor is, therefore, 38. It is assumed that the occupants from each floor enter the stairway at the same time and vacate each segment of the stairway as the following group enters it behind them. Thus, no segment of the stair has more than 38 occupants in it at any time. Should a larger number have been assigned to any floor, all stairway segments below that floor would have to serve the larger number.

The occupant load entering the stairway from the basement parking area is 61. These occupants merge with the 38 from the 2nd floor, placing both groups in the means of egress at the 1st floor (the intermediate floor) at the same time. The means of egress at the point of convergence must, therefore, serve *38 + 61 = 99 occupants*.

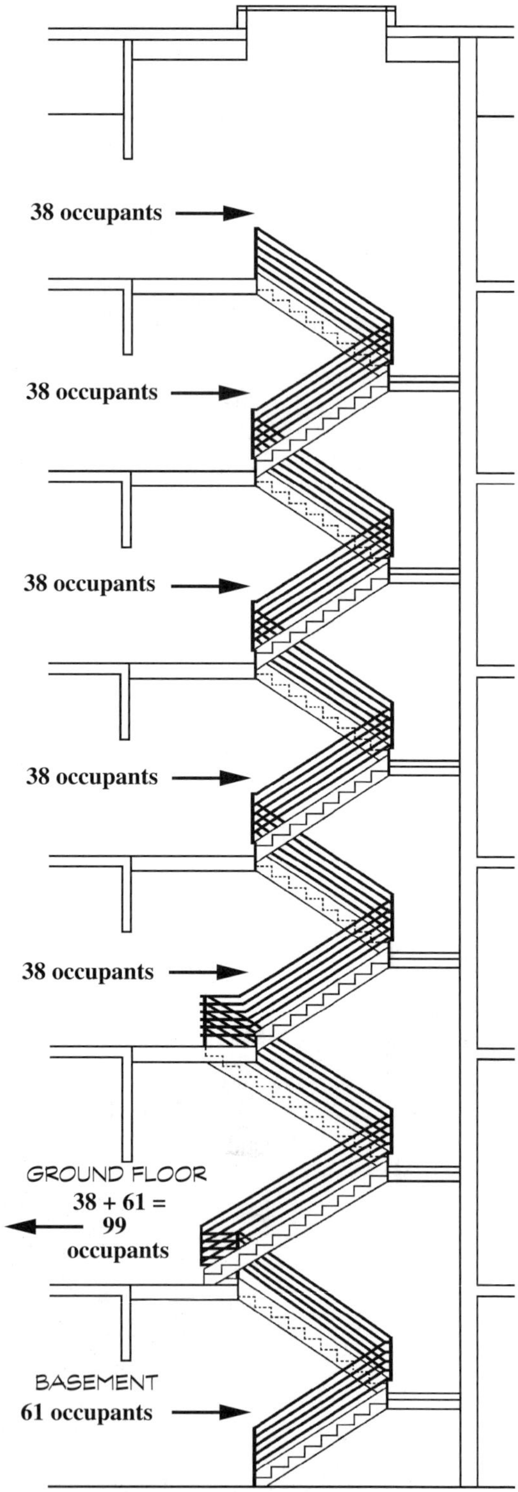

Fig. 1004.4. Stairway section. Hoyt Street Properties, Portland, Oregon. Ankrom Moisan Associated Architects, Portland, Oregon.

1004 Occupant Load

1004.5 Egress convergence

- Where the means of egress from a floor above, converges with that of the floor below, at a level in between, the following applies:
  - Egress capacity in the direction of travel from the point of convergence is as follows:
    Capacity must be ≥ the sum of the capacities of the converging means of egress.

1004.6 Mezzanine levels

- This section governs the exit capacity of a space with mezzanine egress through it.
- The exit capacity of such space is based on the sum of occupant loads of the following:
  - The mezzanine.
  - The space.

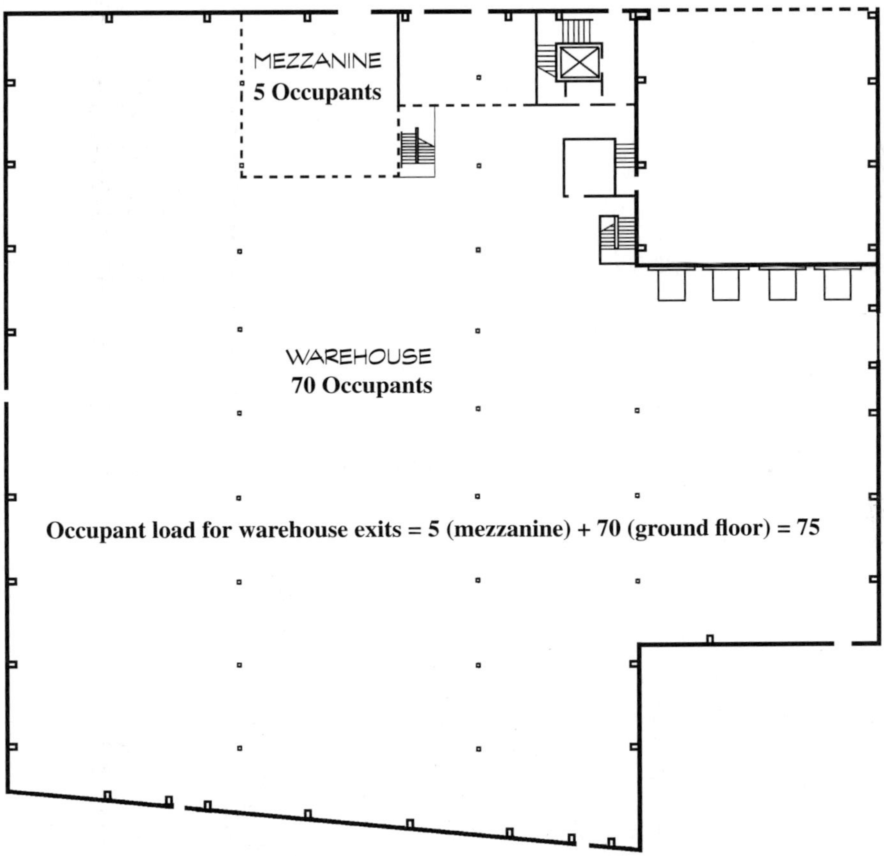

Fig. 1004.6. Floor plan. New Warehouse Addition. Los Angeles, California. Stephen Wen + Associates, Architects, Inc. Pasadena, California.

1004 Occupant Load

1004.7 Fixed seating *(part 1 of 2)*

- Occupant load in fixed seating areas is determined in one of the following ways:
  - For individual seats, *occupant load = number of seats.*
  - For continuous seating without arms, *occupant load = length of seating ÷ 18".*
  - For booth seating, *occupant load = length of seating ÷ 24",* as follows:
    Length is measured at the backrest.

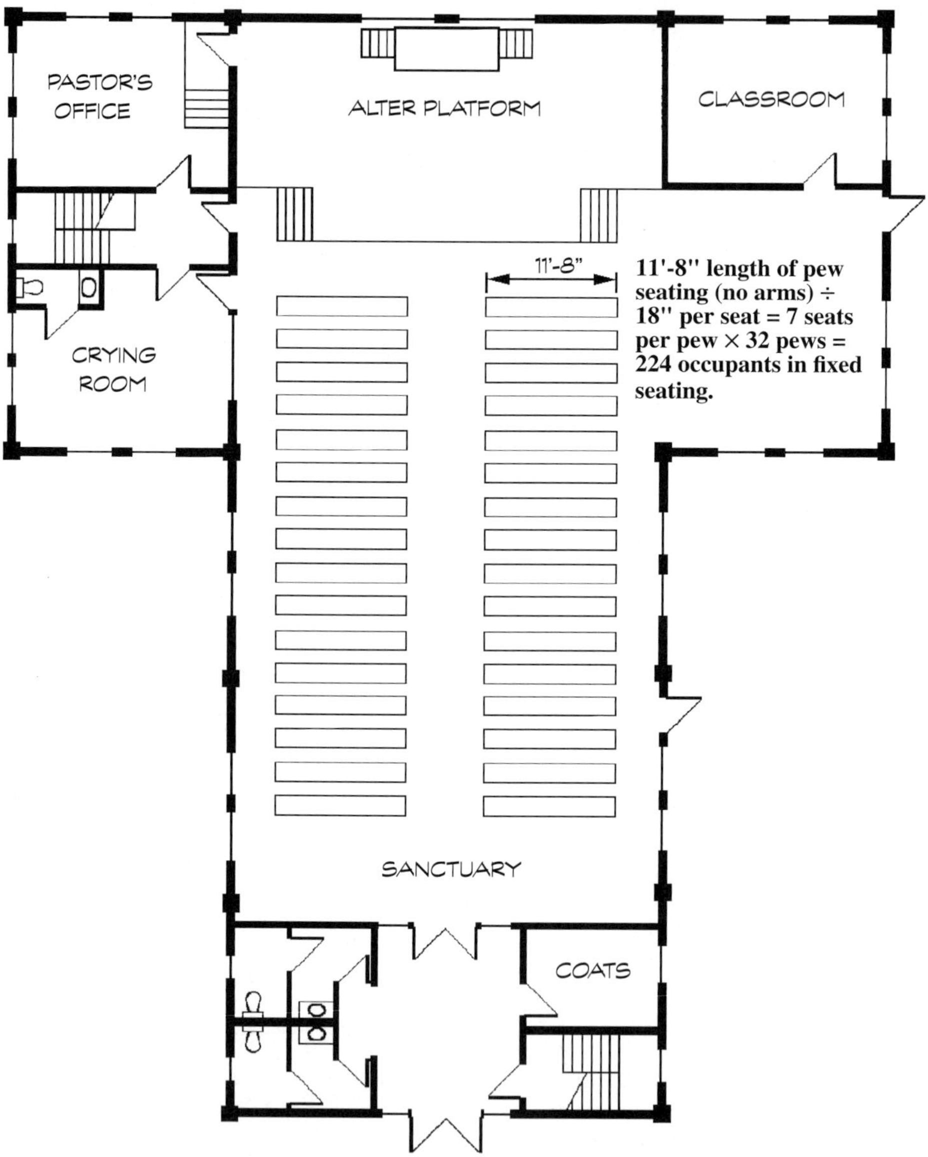

Fig. 1004.7. First-floor plan. Glad Tidings Assembly of God Church. Nanticoke, Pennsylvania. Mullins and Weida, Architect and Associate. Bear Creek, Pennsylvania.

1004 Occupant Load

1004.7 Fixed seating *(part 2 of 2)*

- Areas such as the following, without fixed seating, have the requirements indicated below:
  - Areas integral with the fixed seating function:
    - Waiting spaces.
    - Wheelchair spaces.
    - Other similar spaces.
  - Requirements:
    - The occupant load of such an area is added to the fixed seating occupant load.

Note: 1004.1.1, "Areas without fixed seating," is cited as defining the method for calculating the occupant load of these spaces.

1004.8 Outdoor areas

- This section does not apply to the following outdoor areas:
  - Outdoor areas of Occupancy R-3.
  - Outdoor areas of individual dwelling units of Occupancy R-2.
- Outdoor areas used only for building service are required to have only one means of egress.
- Other outdoors areas usable by building occupants, such as the following, must have a means of egress complying with this chapter:
  - Yards.
  - Patios.
  - Courts.
  - Similar outdoor areas.
- The occupant load of outdoor areas is assigned by the building official as per expected use.
- Means of egress of outdoor areas passing through a building are governed as follows:
  - Where an outdoor area serves persons other than building occupants, the following applies:
    - Requirements for the means of egress based on the sum of the following occupants:
      - Occupant load from the building.
      - Occupant load from outside the building.

Case study: Fig. 1004.8. If 102 occupants from adjoining rooms egress through those rooms, 1,125 of the courtyard occupant load of 1,227 will pass through the auditorium lobby. Total capacity of doors to the lobby is 1,440. 566 of the auditorium occupants also egress through the lobby, thus, its means of egress serves 1,691. Total capacity of exit doors serving the lobby is 2,400. To omit the courtyard load from the lobby means of egress, it would have to be shown that the auditorium and courtyard would not be occupied simultaneously. This scenario reflects one set of assumptions about use of the courtyard.

1004 Occupant Load

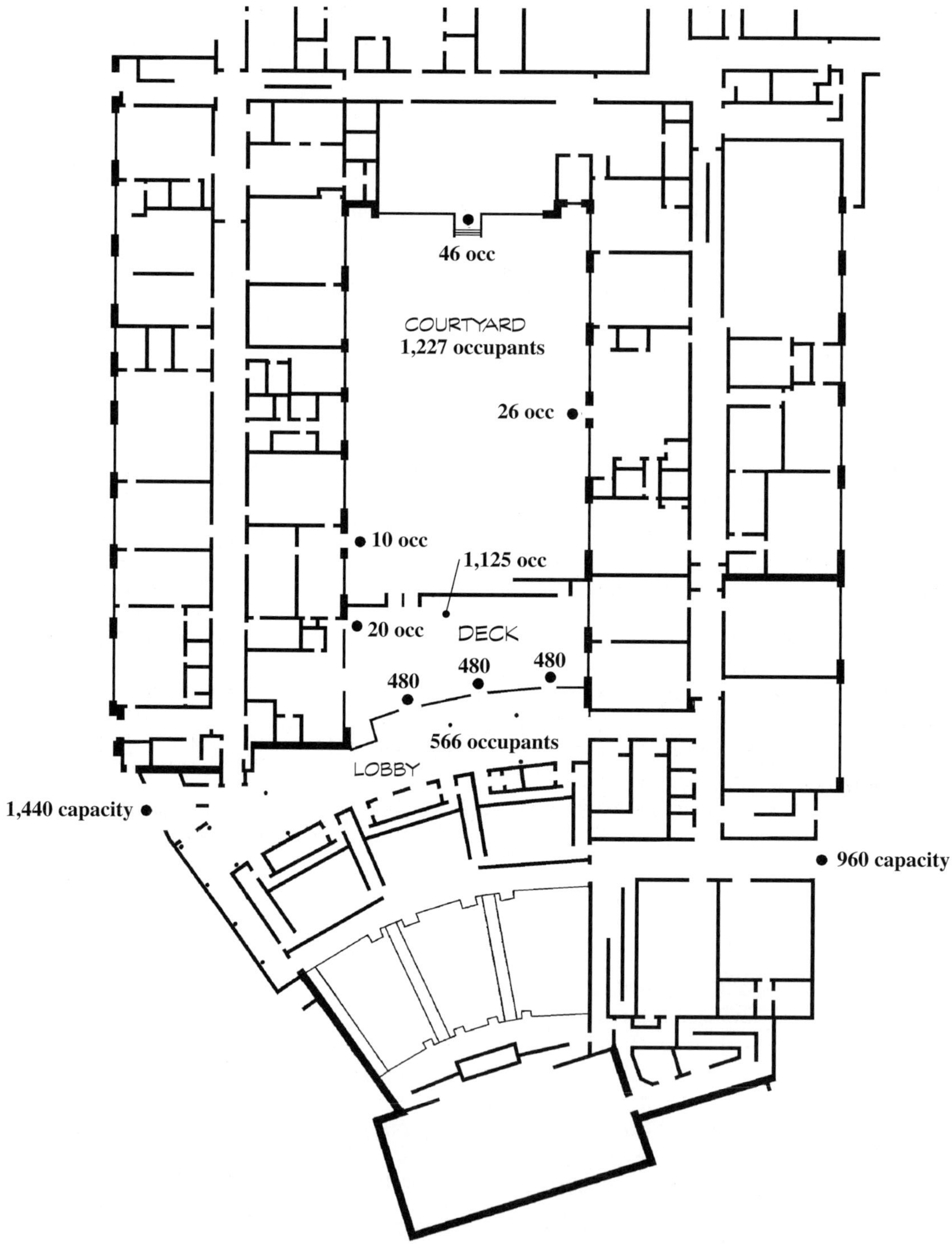

Fig. 1004.8. Partial floor plan at courtyard. Montachusett Regional Vocational-Technical High School. Fitchburg, Massachusetts. HKT Architects, Inc. Somerville, Massachusetts.

1004 Occupant Load

1004.9 Multiple occupancies

- This section applies to buildings containing more than one occupancy.
- Means of egress within an occupancy are governed by the occupancy requirements.
- Where more than one occupancy shares parts of the same means of egress, shared portions are governed by the more restrictive requirements of the occupancies served.

Case study: Fig. 1004.9A. IBC Table 1018.1 requires that corridors serving I-3 occupancies with sprinklers have walls with a fire-resistance rating of 1 hr, although the walls of corridors serving A and B occupancies with sprinklers need have no fire-resistance rating. The corridor between the I-3 and A or B occupancies in the case study serves all these occupancies and, therefore, must have a minimum of 1-hr-rated walls in compliance with the more restrictive I-3 requirement. (In this case, the stair walls and the I-3 corridor wall are rated at 2 hrs, although the other corridor walls are 1 hr.)

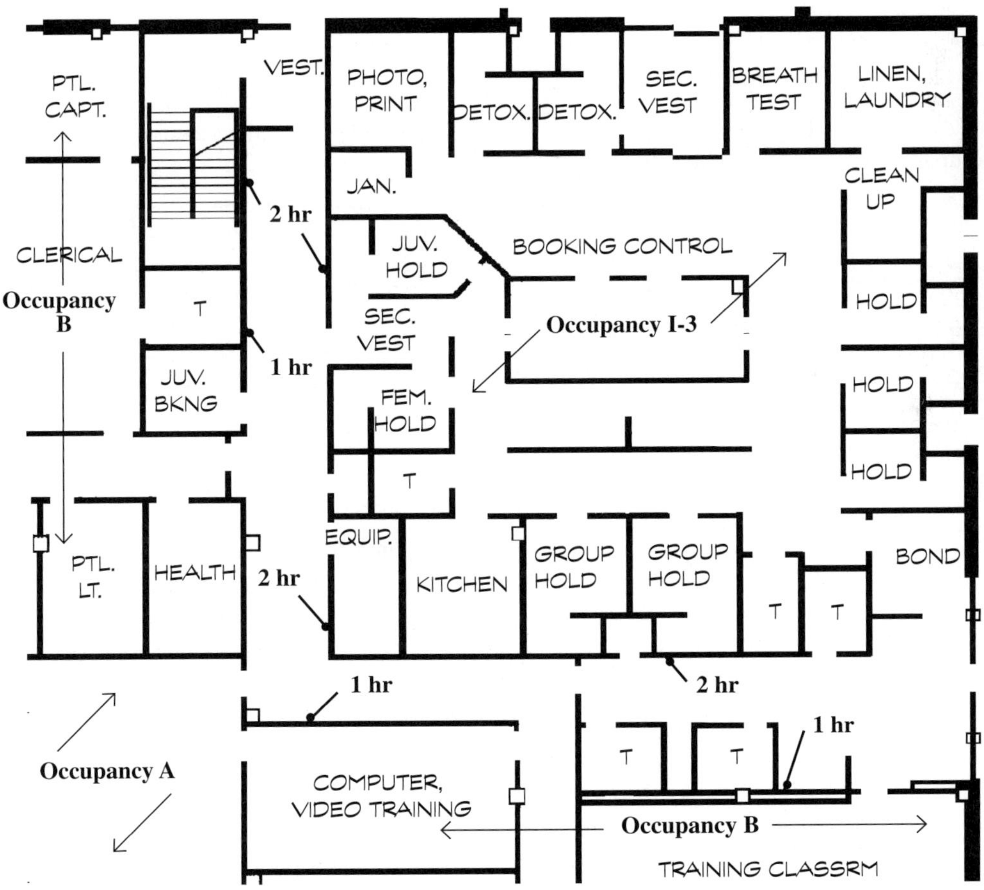

Fig. 1004.9A. Partial floor plan. Lee's Summit Police and Court Facility. Lee's Summit, Missouri. The Hollis and Miller Group, Inc. Lee's Summit, Missouri.

1004 Occupant Load

Case study: Fig. 1004.9B. Each exit door in the plan is indicated by a dot with its maximum egress capacity listed in number of occupants. The doors in occupancies B and S have the same capacity as doors in the H-2 occupancy for the same width. 1005.1 indicates that each occupant requires 0.2" of egress door width (without regard to occupancy). Consequently, the 36" exterior doors have capacities of 180 occupants in the B and S occupancies as well as in the H occupancy. The egress width per occupant is the same regardless of whether the building is sprinklered or not sprinklered.

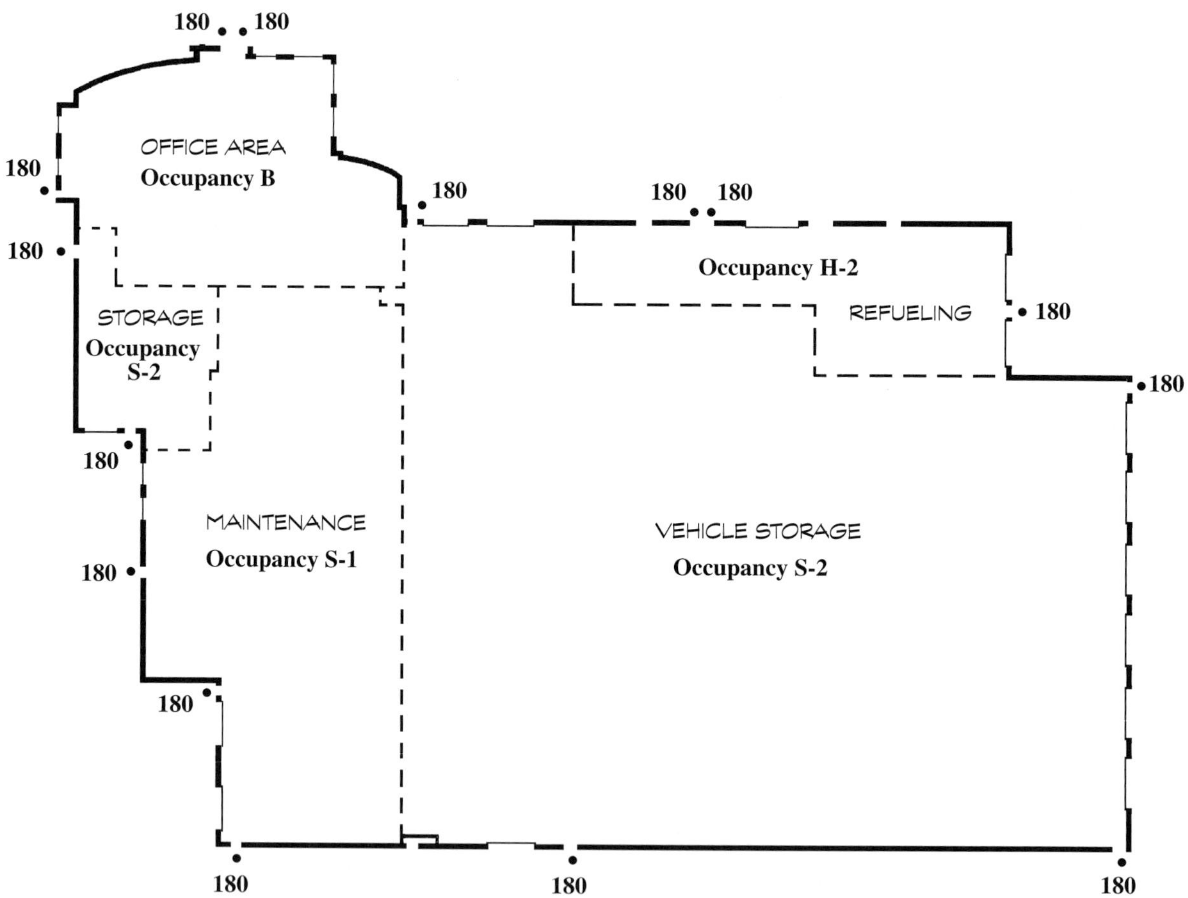

Fig. 1004.9B. Floor plan. Wichita Transit Storage, Administration, and Maintenance Facility. Wichita, Kansas. Wilson Darnell Mann, P.A., Architects. Wichita, Kansas.

1005 Egress Width

1005.1 Minimum required egress width

- Assembly occupancies complying with assembly egress requirements are not governed by this section.

  Note: Section 1028, "Assembly," is cited as governing means of egress in assembly occupancies.

- In other cases, means of egress widths must be ≥ the greater of the following:
  - Minimums required elsewhere in the code.
  - Number of occupants served × 0.3 for egress stairways in any occupancy.
  - Number of occupants served × 0.2 for egress components other than stairways in any occupancy.
- A loss of a means of egress may not reduce the total egress capacity by > 1/2 that required.
- Means of egress capacity must be ≥ the largest required from any point to its termination.

1005.2 Door encroachment

- Encroachment restrictions of this section do not apply to the following doors:
  - In dwelling units of the following occupancies:
    R-2 and R-3.
  - In sleeping units of the following occupancy:
    R-2.
- Otherwise, door encroachment into a required means of egress width is limited to the following:
  - 1/2 the required egress width when the door is in any position.
  - 7" when the door is fully open as follows:
    Includes certain door hardware.
- Handrails may not encroach into a required means of egress width > 7" total.
- Encroachment of nonstructural elements such as the following are limited to 1 1/2" on each side:
  - Trim.
  - Similar decorative elements.

1005.3 Door hardware encroachment

- This section addresses the following door hardware:
  - Surface-mounted latch release type as follows:
    Located on the door side facing the corridor when the door is open.
    Mounted at the following height:
      Between 34" and 48" above the finished floor.
- Door hardware addressed by this section is governed as follows:
  - Its encroachment into a required means of egress is not limited to 7".

  Note: The previous section, 1005.2, "Door encroachment," is cited as listing the 7" encroachment rule to which the hardware of this section is not limited.

1005 Egress Width

Case study: Fig. 1005.1A. When the building was built containing this stairway, it met the required egress width as follows: minimum width per occupant for a stairway in Occupancy E with sprinklers = 0.2" from 1005.1. Stairway occupant load = 435. Actual width of stairway is 7'-3 1/2". *Minimum width of stairway = 0.2" × 435 = 7'-3".* 1005.1 no longer distinguishes between sprinklered buildings and buildings without sprinklers thus, the minimum width per occupant is now 0.3". If this stairway were built today, the required width would be 0.3" x 435 = 10'-10 1/2". The stairway continues to comply with the code under which it was built.

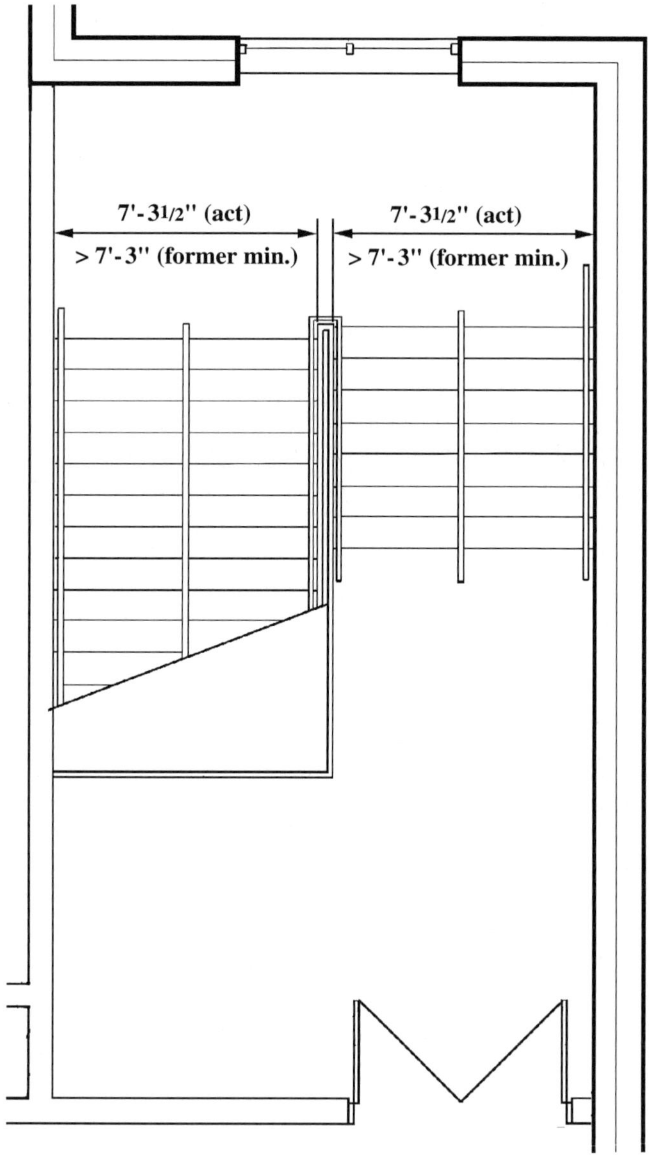

Fig. 1005.1A. Stairway plan. Montachusett Regional Vocational-Technical High School. Fitchburg, Massachusetts. HKT Architects, Inc. Somerville, Massachusetts.

1005 Egress Width

Case study: Fig. 1005.1B. The auditorium requires a total means of egress width ≥ 102", based on 1005.1. While other requirements are also applicable, this section prohibits a layout wherein the loss of one means of egress would reduce the total width to less than 51". The actual means of egress width available (as limited by exit access doors) is 276". The auditorium meets this requirement because the loss of the largest doorway (72") would leave 204" available. 204" > 51", the minimum permitted to remain upon the loss of a doorway.

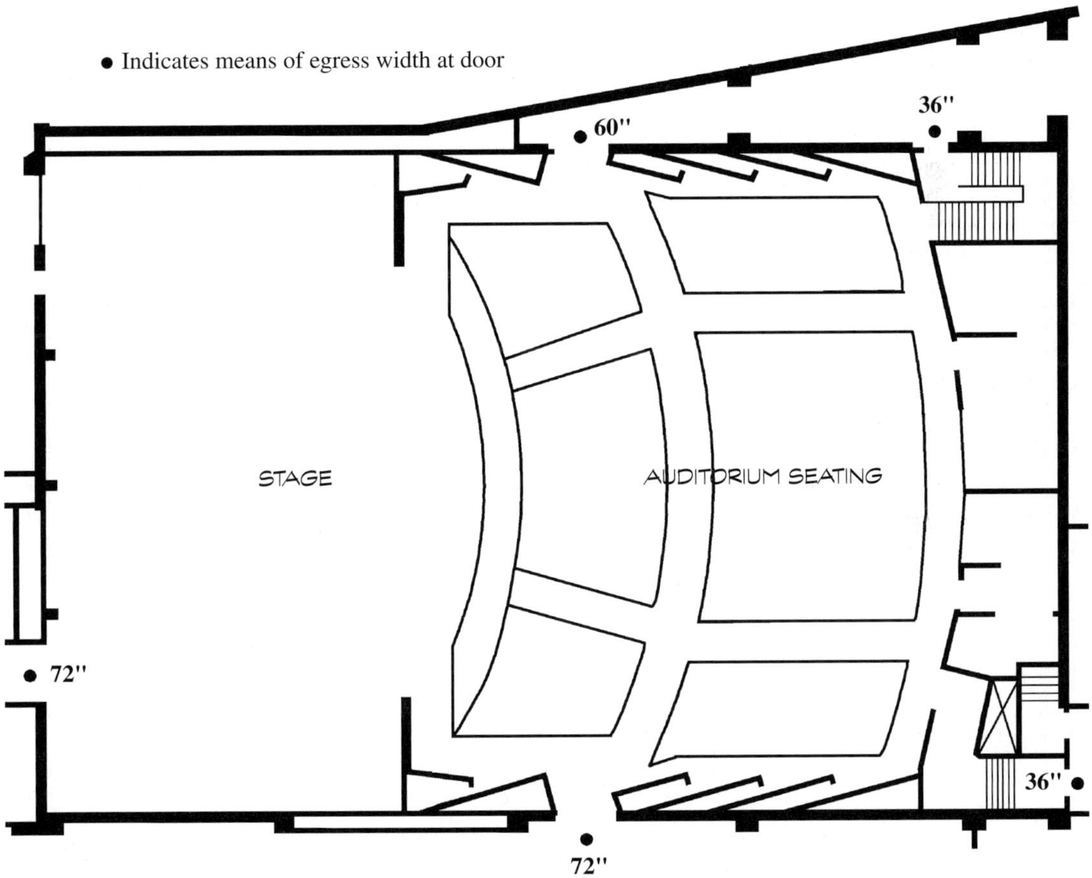

Fig. 1005.1B. Partial floor plan. Newman University Sports and Fine Arts Center. Wichita, Kansas. Gossen Livingston Associates, Inc., Architecture. Wichita, Kansas.

1005 Egress Width

Case study: Figure 1005.2. The required egress width of corridor 368 is 6' based on occupant load. The actual width is 8'. It is assumed that the required width is centered in the corridor, although other locations are feasible. When door E05.1 swings 90°, it can open no further. Therefore, the two limitations of its projection into required egress width are applied to the door fully open. When fully open it projects 6" into the required egress width, which is < the 7" maximum and < 1/2 the required width. The greatest projection of double doors 368.2 into required egress width during their swing is 6", which is < 1/2 the required width. Although, due to the column, one door is not flush to the wall when fully open, neither door encroaches on the required egress width in the open position.

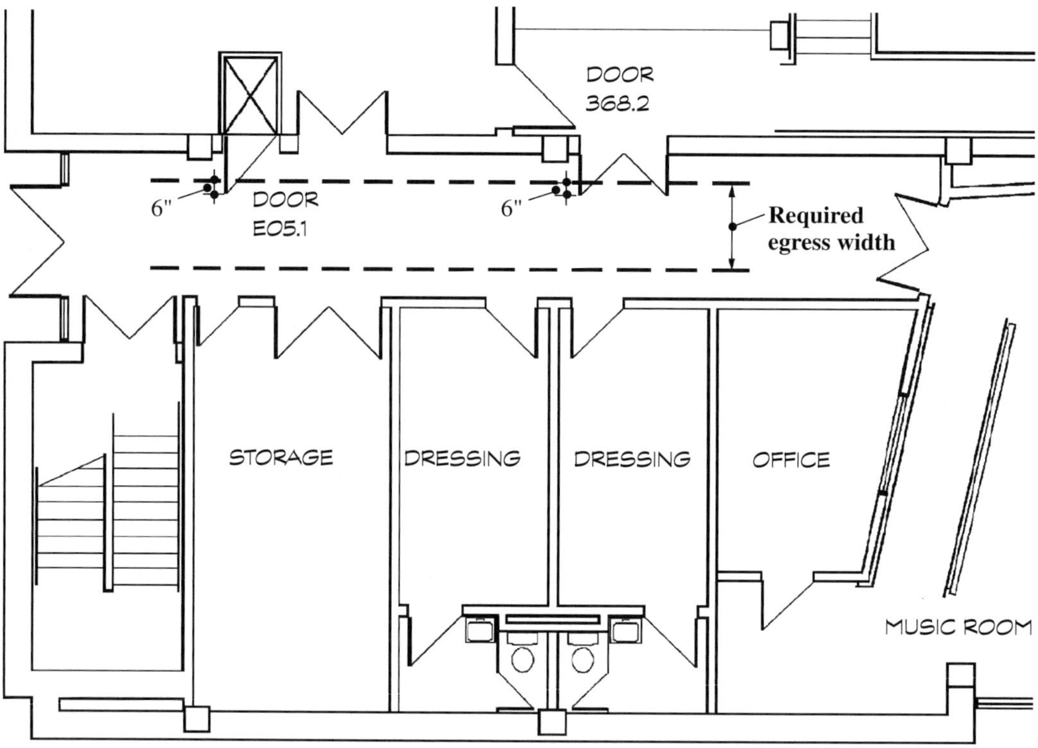

Fig. 1005.2. Partial floor plan. Newman Elementary School Renovations. Needham, Massachusetts. HKT Architects, Inc. Somerville, Massachusetts.

1006 Means of Egress Illumination

1006.1 Illumination required

- Egress illumination requirements during the following period are shown in the table below:
  - Egress elements included are as follows:
    Means of egress.
    Exit discharge.
  - Period requiring illumination is as follows:
    When the part of the building using the means of egress is occupied:

Table 1006.1 Egress Illumination

Occupancy	Illumination required
U	no
Aisle accessways in A	no
R-1, R-2, R-3 dwelling units	no
R-1, R-2, R-3 sleeping units	no
Sleeping units in I	no
All other	yes

1006.2 Illumination level

- Requirements for illumination at floor level in a means of egress are as follows:

Table 1006.2 Egress Illumination at Walking Surface Level

Location and time	Illumination required if no fire alarm is activated	Illumination required if a fire alarm is activated
During performances in auditoriums, theaters, opera halls, and similar assembly occupancies	≥ 0.2 footcandle	≥ 1.0 footcandle
All other	≥ 1.0 footcandle	≥ 1.0 footcandle

1006 Means of Egress Illumination

1006.3 Illumination emergency power

- Normal building power is required to illuminate the means of egress.
- An emergency power system must illuminate the following areas in case of power failure:
  - In areas requiring ≥ 2 means of egress as follows:
    Unenclosed egress stairways.
    Aisles.
  - In buildings requiring ≥ 2 means of egress as follows:
    Corridors.
    Exit enclosures.
    Exit passageways.
    Exterior egress elements at the following levels:
      At levels other than their level of exit discharge as follows:
        Continuous to the end of exit discharge.
    Exterior landings.
    Exit discharge elements that are within the building interior.

Note: 1008.1.6, "Landings at doors," is cited as governing these elements.
1027.1, "General," is cited as governing interior exit discharge elements.

- The emergency power system is governed as follows:
  - It must provide power for ≥ 90 minutes.
  - It must consist of one of the following:
    Storage batteries.
    Unit equipment.
    On-site generator.

Note: Chapter 27, "Electrical," is cited as governing the installation of emergency power systems.

Case study: Fig. 1006.3. The example shows a typical emergency lighting fixture and the symbol used on floor plans to indicate location. The many fixtures available for this purpose are similar in their surface mounting and two adjustable lamps. The battery-powered units meet code performance requirements.

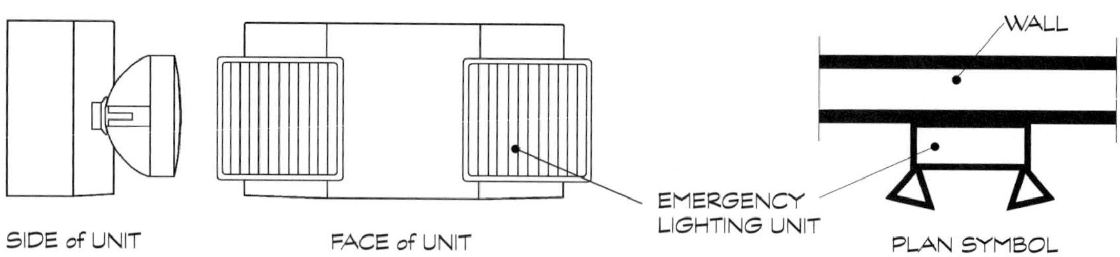

Fig. 1006.3. Elevations and symbol. Emergency lighting unit. Emergi-Lite. St. Matthews, South Carolina.

1007 Accessible Means of Egress

1007.1 Accessible means of egress required

- Accessible means of egress are not required as follows:
  - In alterations to existing buildings.
- One accessible means of egress is required in the following case:
  - From an accessible mezzanine.

Note: One of the following governs this egress as applicable:
1007.3, "Stairways."
1007.4, "Elevators."
1007.5, "Platform lifts."

- One accessible means of egress is permitted in the following assembly areas:
  - A space with a sloped or stepped aisle as follows:
    The common path of travel meets both of the following conditions:
      The path is accessible.
      The path meets aisle width and other requirements.

Note: 1028.8, "Common path of egress travel," is cited as governing the aisles.

- In other cases accessible means of egress are governed as follows:
  - They must meet the requirements of this section.
  - ≥ 1 accessible means of egress is required as follows:
    From each accessible space.
  - ≥ 2 accessible means of egress are required as follows:
    From each accessible part of an accessible space where the following is required:
      Where > 1 means of egress is required from the accessible space.

Note: Either of the following is cited as defining where >1 means of egress is required:
1015.1, "Exits or exit access doorways from spaces."
1012.1, "Where required."

1007.2 Continuity and components *(part 1 of 2)*

- An exterior area for assisted rescue is required in the following case:
  - Where exit discharge is not accessible.

Note: 1007.7, "Exterior area for assisted rescue," is cited as the source of requirements for this feature.

- Where the exit stairway is open to the exterior, the accessible means of egress must include one of the following:
  - An area of refuge.

Note: 1007.6, "Areas of refuge," is cited as governing the area.

  - An exterior area for assisted rescue.

Note: 1007.7, "Exterior area for assisted rescue," is cited as governing the area.

1007 Accessible Means of Egress

Case study: Fig. 1007.1 The 1,242 occupants of the gymnasium require 4 exits. Therefore, ≥ 2 accessible means of egress are required. The gymnasium is in compliance, as 4 accessible exits are provided.

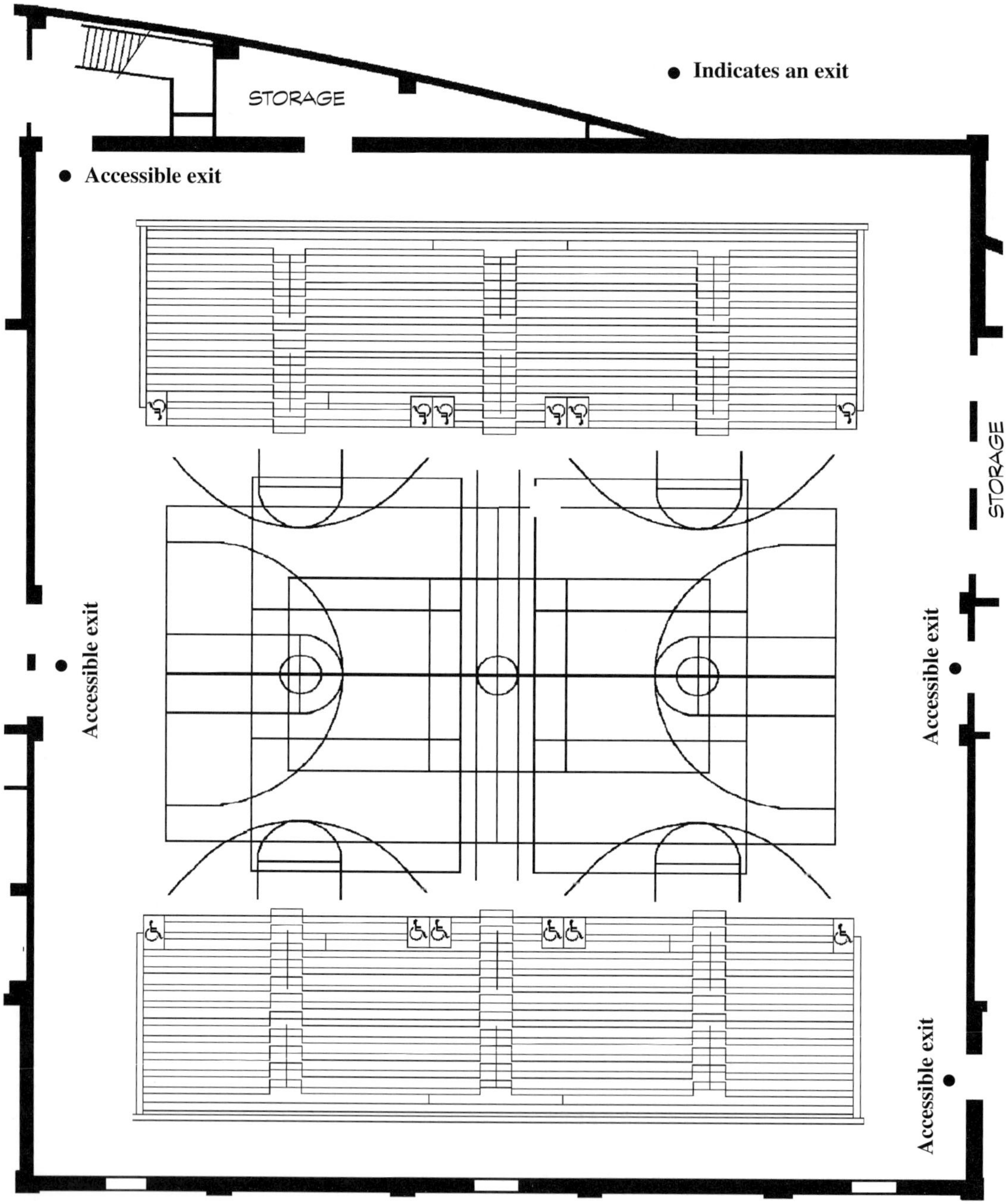

Fig. 1007.1. Gymnasium floor plan. Newman University Sports and Fine Arts Center. Wichita, Kansas. Gossen Livingston, Associates, Inc., Architecture. Wichita, Kansas.

1007 Accessible Means of Egress

1007.2 Continuity and components *(part 2 of 2)*

- Otherwise, required accessible means of egress must comply with the following:
  - They must be continuous to a public way.
  - They must consist of ≥ 1 of the following:

    Accessible routes.
    Areas of refuge.
    Elevators.
    Exterior exit stairways.
    Horizontal exits.
    Interior exit stairways.
    Platform lifts.
    Ramps.

Note: The following are cited as governing the above-listed elements as applicable:
Section *1010, "Ramps."*
Section 1022, "Exit Enclosures."
Section 1025, "Horizontal Exits."
Section 1026, "Exterior Exit Ramps and Stairways."
Section 1104, "Accessible Route."
1007.3, "Stairways."
1007.4, "Elevators."
1007.5, "Platform lifts."
1007.6, "Areas of Refuge."

1007.2.1 Elevators required

- This section applies to buildings as follows:
  - In which a required accessible floor has either of the following locations:

    ≥ 4 stories above the level of exit discharge.
    ≥ 4 stories below the level of exit discharge.
- An elevator serving as an accessible means of egress is not required where all of the following apply:
  - The building is sprinklered.
  - The floor has a horizontal exit.
  - The floor is at or above the levels of exit discharge.
- An elevator serving as an accessible means of egress is not required where all of the following apply:
  - The building is sprinklered.
  - The floor has a ramp serving as an accessible means of egress.

Note: Section 1010, "Ramps," is cited as governing ramps as means of egress.
The following are cited as governing all sprinklers identified in this section where applicable:
903.3.1.1, "NFPA 13 sprinkler systems."
903.3.1.2, "NFPA 13R sprinkler systems."

- In other cases ≥ 1 accessible means of egress must be an elevator.

Note: 1007.4, "Elevators," is cited as governing these elevators.

1007 Accessible Means of Egress

1007.3 Stairways

- An area of refuge is not required in the following locations:
  - In sprinklered buildings as follows:
    At open exit or exit access stairways.

  Note: *The following are cited as governing locations where an open exit access or exit stairway is permitted:*
  1016.1, "Travel distance limitations."
  1022.1, "Enclosures required."
  The following are cited as alternative requirements for sprinklers as applicable:
  903.3.1.1, "NFPA 13 sprinkler systems."
  903.3.1.2, "NFPA 13R sprinkler systems."

  - At exit stairways in opening parking garages.
  - At smoke-protected seating.

  Note: *1028.6.2, "Smoke-protected seating," is cited as governing the smoke-protected seating where an area of refuge is not required.*

  - In R-2 occupancies.
- A clear width between handrails of ≥ 4' is not required in either of the following cases:
  - Where the exit stairway is accessed from a horizontal exit.
  - Where the building is sprinklered.

  Note: *1016.1, "Travel distance limitations," is cited as governing exit access stairways.*
  The following are cited as governing the sprinklers where applicable:
  903.3.1.1, "NFPA 13 sprinkler systems."
  903.3.1.2, "NFPA 13R sprinkler systems."

- In other cases, stairways serving an accessible means of egress must have the following:
  - A clear width ≥ 4' between handrails.
  - One of the following is required:
    An area of refuge within a floor-level landing.
    To be accessed from a horizontal exit.
    To be accessed from an area of refuge.

  Note: *1007.6, "Areas of refuge," is cited as governing areas of refuge from which the stairway is to be accessed where this option is employed.*

1007.4 Elevators *(part 1 of 2)*

- The following elevators are not required to be accessed from an area of refuge or horizontal exit:
  - In open parking garages.
  - In sprinklered buildings.

  Note: *The following are cited as governing the sprinklers where applicable:*
  903.3.1.1, "NFPA 13 sprinkler systems."
  903.3.1.2, "NFPA 13R sprinkler systems."

  - Where the elevator is not required to be in a shaft.

  Note: *708.2, "Shaft enclosure required," is cited as specifying where shaft enclosures are required.*

1007 Accessible Means of Egress

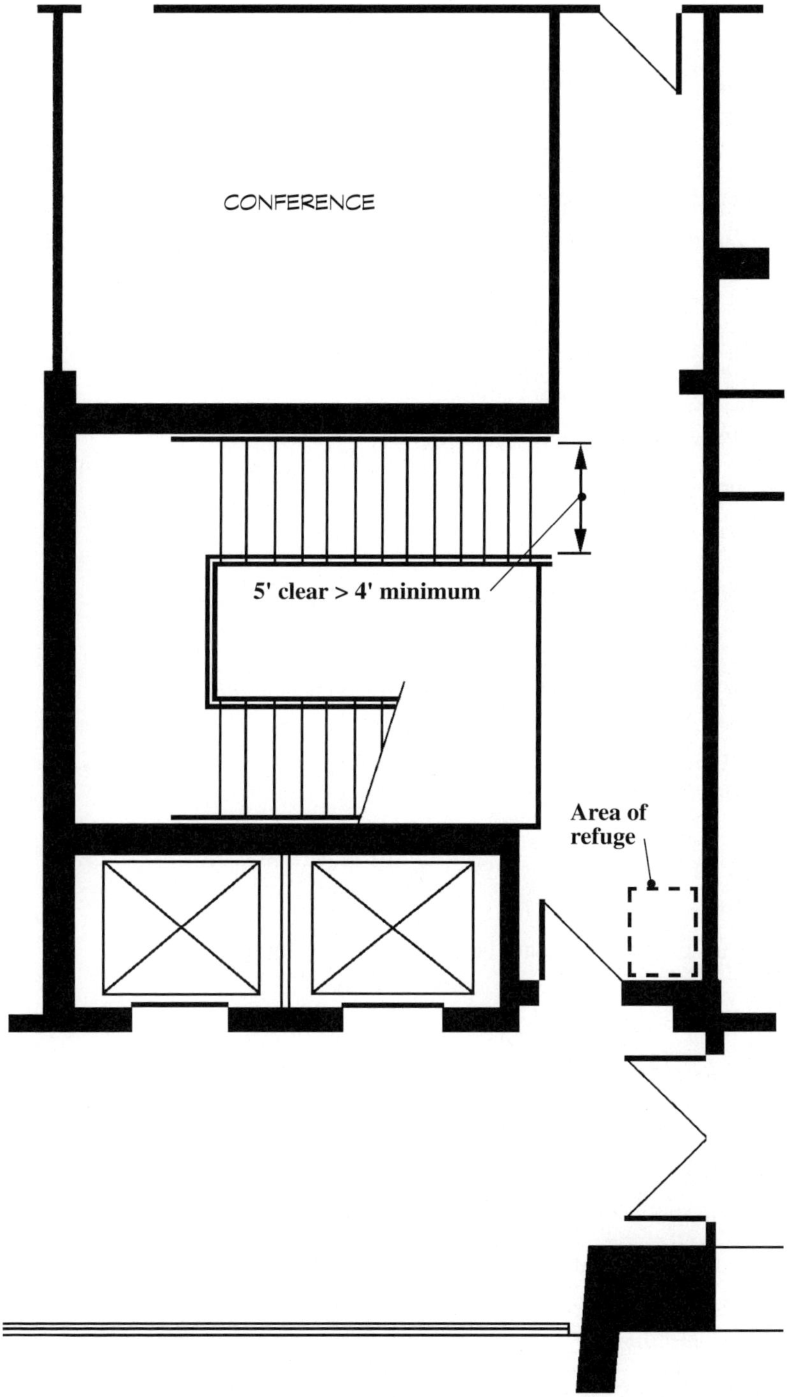

Case study:
Fig. 1007.3. The enclosed stairway qualifies as an accessible means of egress by two criteria in 1007.3. First, a means of refuge is provided and the clear width of the stair between the handrails is 5', > the 4' minimum. Without these aspects, the stairway could serve as an accessible means of egress since the building is sprinklered.

The elevators qualify as an accessible means of egress according to 1007.4, due to their adequate size, to the fact that the building is sprinklered, and to the provision of the necessary emergency operation and signaling systems.

Fig. 1007.3. Partial plan at stairway and elevators. University of Connecticut New Downtown Campus at Stamford, Connecticut. Perkins Eastman Architects, P.C. New York, New York.

1007 Accessible Means of Egress

1007.4 Elevators *(part 2 of 2)*

- In smoke-protected seating.

Note: 1028.6.2, "Smoke-protected seating," is cited as governing the smoke-protected seating where an area of refuge is not required.

- In other cases, elevators must comply with the following to serve as an accessible means of egress:
  - They must meet emergency operation and signaling device requirements.

Note: Section 2.27 of ASME A17.1, "Safety Code for Elevators and Escalators," is cited as governing the emergency systems.

  - They must have standby power.

Note: The following are cited as governing standby power:
Chapter 27, "Electrical."
Section 3003, "Emergency Operations."

  - They must be accessed from one of the following:
    Area of refuge.
    Horizontal exit.

Note: 1007.6, "Areas of refuge," is cited as governing these components.

1007.5 Platform lifts

- Platform lifts for wheelchairs may serve as an accessible means of egress as follows:
  - Where standby power is supplied.

Note: Chapter 27, "Electrical," is cited as governing standby power.

  - Where permitted in a required accessible route.

Note: 1109.7, "Lifts," items 1 through 9, is cited as permitting lifts in certain accessible routes.

1007.5.1 Openess

- Platform lifts in an accessible means of egress may not be installed in the following:
  - A fully enclosed shaft.

1007.6 Areas of refuge *(part 1 of 2)*

- A required area of refuge must be accessible from the space it serves as follows:
  - Access must be provided by an accessible means of egress.
  - The travel distance between an accessible space and the nearest area of refuge is limited as follows:
    Route must be ≤ that limited by the occupancy.

Note: 1016.1, "Travel distance limitations," is cited as governing travel distance.

- Certain stairways serving an area of refuge need not be enclosed.

Note: The following are cited as permitting certain stairways to be open:
1016.1, "Travel distance limitations."
1022.1, "Enclosures required."

1007 Accessible Means of Egress

1007.6 Areas of refuge *(part 2 of 2)*

- The following elevator lobbies are not required to have a smokeproof enclosure:
  - Where they serve as an area of refuge that is not required to be enclosed.
  - Where they are located in an area of refuge formed by one of the following:
    - A horizontal exit.
    - A smoke barrier.
- In other cases, the following applies:
  - Elevator shaft and lobby must meet smokeproof enclosure requirements in the following case:
    - Where the elevator lobby serves as an area of refuge.

  Note: 1022.9, "Smokeproof enclosures and pressurized stairways," is cited as governing the shaft and lobby.

  - A required area of refuge must have direct access to one of the following:
    - A stairway within an exit enclosure.

  Note: The following are cited as governing the enclosed stairway:
  1007.3, "Stairways."
  Section 1022, "Exit Enclosures."

    - An elevator.

  Note: 1007.4, "Elevators," is cited as governing elevators in this case.

Case study: Fig. 1007.6. Access to the area of refuge is provided by an accessible means of egress. The area of refuge has direct access to the elevator. The elevator lobby, including the telephone area, which serves as the area of refuge, meets smokeproof enclosure requirements. The area of refuge is, therefore, in compliance with the code.

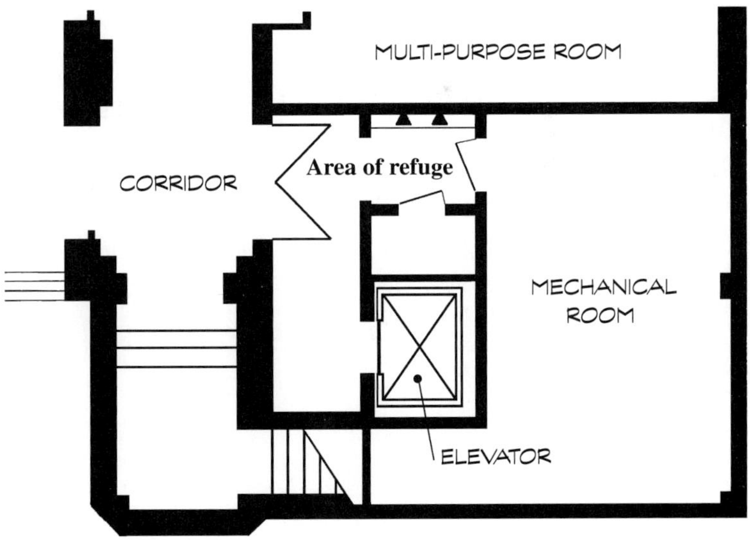

Fig. 1007.6. Plan of elevator with refuge area. Lady Bird Johnson Wildflower Center. Austin, Texas. Overland Partners, Inc. San Antonio, Texas.

1007 Accessible Means of Egress

1007.6.1 Size

- This section addresses wheelchair spaces in an area of refuge.
- Wheelchair spaces may not protrude into the minimum required width of the means of egress.
- Wheelchairs may be positioned no more than two deep from the path of travel.
- Occupant load for determining the number of wheelchairs required is as follows:
  - The sum of the occupant load of the area of refuge plus that of the area served.
- The minimum size of a wheelchair space is 30" × 48".
- Wheelchair spaces required, as listed below, are based on the following equation:

Occupant load ÷ 200 = number of wheelchair spaces required

Table 1007.6.1 Wheelchair Spaces Required

Occupant load	Spaces required	Occupant load	Spaces required
1 – 200	1	1,001 – 1,200	6
201 – 400	2	1,201 – 1,400	7
401 – 600	3	1,401 – 1,600	8
601 – 800	4	1,601 – 1,800	9
801 – 1,000	5	etc.	etc.

1007.6.2 Separation

- This section does not apply to the following:
  - Areas of refuge in an exit enclosure.
- Other areas of refuge must be separated from the rest of the story by one of the following:
  - A smoke barrier.
  - A horizontal exit.

Note: Section 710, "Smoke Barriers," is cited as governing these elements.
Section 1025, "Horizontal Exits," is cited as governing these elements.

- Areas of refuge must be designed to minimize smoke penetration.

1007.6.3 Two-way communication

- Areas of refuge must have a two-way communication system.

Note: The following are cited as governing the communication system required above:
1007.8.1, "System requirements."
1007.8.2, "Directions."

1007 Accessible Means of Egress

Case study: Fig. 1007.6.1. The areas of refuge in the enclosed stairway do not encroach on the minimum width required for egress. They are no more than 2 deep from the path of travel. They are 30" × 48", thus, meeting size requirements. The areas of refuge are in compliance.

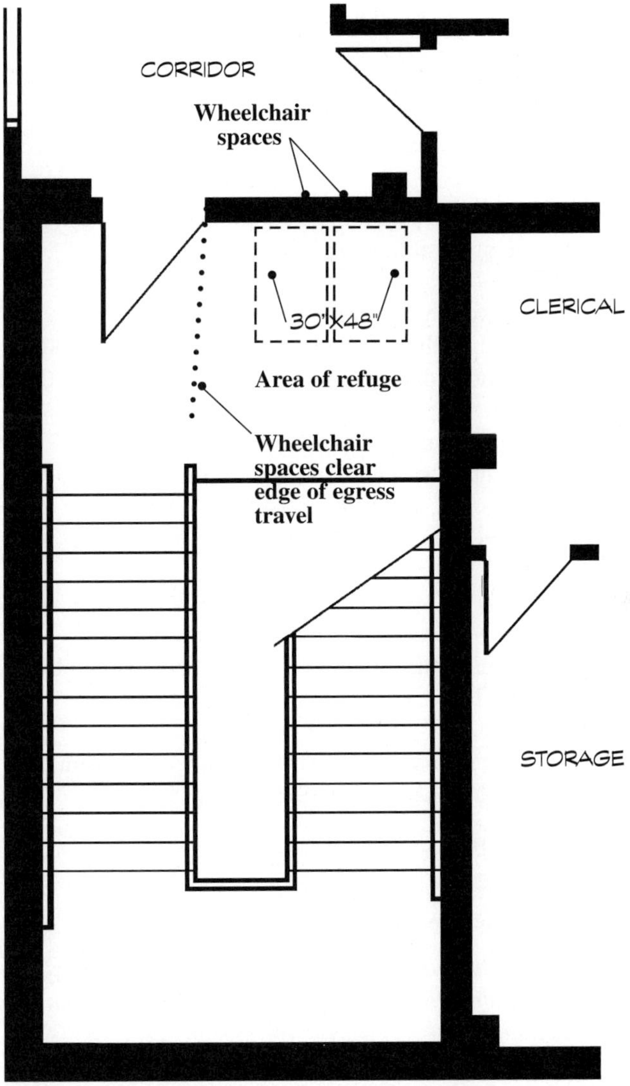

Fig. 1007.6.1. Plan of stair with area of refuge. University of Connecticut New Downtown Campus at Stamford, Connecticut. Perkins Eastman Architects, P.C. New York, New York.

1007 Accessible Means of Egress

1007.8 Two-way communication

- This section governs two-way communication systems at elevator landings as follows:
  - Such systems are not required in the following locations:
    On elevator landings where such a system is provided in areas of refuge.
    On floors with exit ramps.

Note: 1007.6.3, "Two-way communication," is cited as governing these systems.
Section 1010, "Ramps," is cited as governing these ramps.

  - Such are required in the following locations:
    At accessible elevator landings as follows:
    On each accessible floor ≥ 1 story above or below exit discharge.

Note: The following are cited as governing the required two-way communication systems:
1007.8.1, "System requirements."
1008.8.2, "Directions."

1007.8.1 System requirements

- The required two-way communication system must communicate between the following locations:
  - The required location and one of the following:
    The fire command center.
    A central control point approved by the fire department as follows:
    Where the point is not continually attended, the following is required:
    A timed automatic telephone dial-out system capability to one of the following:
    A monitored location.
    The 911 emergency telephone number.
- The required two-way communication system must include the following:
  - Audible signals.
  - Visual signals.

1007.9 Signage

- Signage showing accessibility provisions is required as follows:
  - Doors to areas of refuge from the adjacent floor area must be identified by a sign as follows:
    Sign must state "AREA OF REFUGE."
    Tactile signage is required.
  - Doors to an exterior area for assisted rescue must be identified by a sign as follows:
    Sign must state "EXTERIOR AREA FOR ASSISTED RESCUE."
    Tactile signage is required.
  - Signage must have the International Symbol of Accessibility.
  - Signage must comply with the reference standard for visual characters.
  - Exit signs must be illuminated where required.

Note: 1011.2, "Illumination," is cited as identifying locations requiring exit sign illumination.
ICC A117.1, "Accessible and Usable Buildings and Facilities," is cited as governing the signage required by this section.

1008 Doors, Gates and Turnstiles

1008.1 Doors

- Means of egress doors must be apparent as follows:
  - They must be easily distinguished from adjacent surfaces.
  - They must be readily recognizable as doors.
  - They may not be covered with reflective materials such as mirrors.
  - They may not be concealed by the following:
    Curtains.
    Drapes.
    Decorations.
    Similar materials.

  Note: 1020.2, "Exterior exit doors," is cited as providing additional requirements.

- Doors provided that are in excess of the minimum required must meet the same requirements as those of required doors.

1008.1.1 Size of doors *(part 1 of 2)*

- Clear width for swinging doors is measured as follows:
  - From the stop to the face of the door when it is open at 90°.
- The following doorway must have a width ≥ 32" when only one leaf is open:
  - A door with two leafs.
  - A door required to have a width ≥ 32".
- Minimum door width is the greater of the following:
  - Minimum width established by occupant load.
  - Minimum width listed below:

Table 1008.1.1a Minimum Width of Doors

Doors	Minimum clear width	Maximum width of swinging door
Serving other than means of egress in R-2 and R-3	NA	NA
To storage closets < 10 sf other than R-2 and R-3	NA	48"
To resident sleeping units in I-3	28"	48"
For moving beds in means of egress in I-2	41 1/2"	48"
Revolving doors	NA	NA
Interior egress doors in dwelling units and sleeping units other than R-1 not required to be an accessible unit, Type A unit, or Type B unit	NA	48"
Doors required to be accessible in Type B units	31 3/4"	48"
Other means of egress doors	32"	48"

NA = not applicable, not governed by this section.

Note: 1008.1.4.1, "Revolving doors," is cited as governing revolving doors.

1008 Doors, Gates and Turnstiles

1008.1.1 Size of doors *(part 2 of 2)*

- Minimum required door heights are as listed below:

Table 1008.1.1b Minimum Door Height

Doors	Minimum height
Doors within dwelling units or sleeping units	6'-6"
Exterior doors in dwelling units and sleeping units other than the required exit	6'-4"
Other doors	6'-8"

1008.1.1.1 Projections into clear width

- Projections permitted into the clear width of a door opening are as shown below:

Height in door opening	Projection allowed	Object
≥ 2'-10" and ≤ 6'-6"	≤ 4"	Door closer, door stop
≥ 2'-10" and ≤ 6'-8"	≤ 4"	Other objects
< 2'-10"	none	Any object

1008.1.2 Door swing

- Egress doors are not required to be side-hinged swinging or pivoted as follows:
  - In the following locations with an occupant load ≤ 10:
    Private garages.
    Office areas.
    Factory areas.
    Storage areas.
  - In an Occupancy I-3 detention function.
  - In health care facility suites as follows:
    Critical care patient rooms.
    Intensive care patient rooms.
  - Within or serving a single dwelling unit in occupancies R-2 and R-3.
  - Revolving doors in occupancies other than H.

  Note: 1008.1.4.1, "Revolving doors," is cited as governing these doors.

  - Horizontal sliding doors in occupancies other than H.

  Note: 1008.1.4.3, "Horizontal sliding doors," is cited as governing these doors.

  - Power-operated doors.

  Note: 1008.1.4.2, "Power-operated doors," is cited as governing these doors.

  - Bathroom doors in an R-1 sleeping unit.
  - Manual horizontal sliding doors serving an occupant load < 10 as follows:
    In other than Occupancy H.
- Other egress doors are required to be side-hinged swinging or pivoted as follows:
  - Doors must swing in the direction of travel in either of the following cases:
    Where the occupant load ≥ 50.
    In Occupancy H.

1008 Doors, Gates and Turnstiles

1008.1.3 Door opening force

- The force necessary to open means of egress doors is limited as follows:

Table 1008.1.3 Force Required at Door

Door type	Door action	Force necessary for door action
Swinging, interior (not fire doors)	To open	≤ 5 lbs
Other swinging, sliding, folding	Release of latch	≤ 15 lbs
"	Start door in motion	≤ 30 lbs
"	Swing to full-open position	≤ 15 lbs

1008.1.3.1 Location of applied forces

- The forces limited by code to activate means of egress doors are applied to the latch side.

1008.1.4.1 Revolving doors

- Revolving doors may not be part of an accessible route.

Note: The following are cited as governing accessible routes:
Section 1007, "Accessible Means of Egress."
Chapter 11, *"Accessibility."*

- Revolving doors must have all of the following characteristics:
  - Be collapsible into a bookfold position as follows:
    Provide parallel paths of travel.
    Travel paths to total 36" in width.
  - Be ≤ 10' from stairs and escalators.
  - Be separated from stairs and escalators by a dispersal area.
  - Be within 10' of a side-swinging door in the same wall.

Note: 1008.1, "Doors," is cited as governing side-swinging doors.

  - Have a maximum speed in revolutions per minute as follows:

Table 1008.1.4.1 Maximum Speed for Revolving Doors

	Inside diameter of door							
Speed control type	6.5'	7.0'	7.5'	8.0'	8.5'	9.0'	9.5'	10.0'
RPM for power-driven speed control	11	10	9	9	8	8	7	7
RPM for manual speed control	12	11	11	10	9	9	8	8

Source: IBC Table 1008.1.4.1.

1008.1.4.1.1 Egress component

- This section addresses revolving doors serving in a means of egress.

Note: 1008.1.4.1, "Revolving doors," is cited as requiring compliance.

- The doors may not be assigned more than half the required egress capacity.
- The number of occupants assigned to a revolving door must be ≤ 50.
- The doors must be collapsible into a bookfold position as follows:
  - With a force of ≤ 130 lbs.
  - With the force applied ≤ 3" from the outer edge of a door leaf.

1008 Doors, Gates and Turnstiles

1008.1.4.1.2 Other than egress component

- This section addresses revolving doors not serving in a means of egress.
- Such revolving doors must comply with the following:
  - Have collapsible doors of adequate size.
  - Have the necessary relationship to the following:
    - Stairs and escalators.
    - A swinging door.
  - Have RPM limits as appropriate.

Note: 1008.1.4.1, "Revolving doors," is cited as requiring compliance. A partial summary of its requirements is shown above.

- The force necessary to collapse the doors may be >180 lbs as follows:
  - Where it is reduced to ≤ 130 lbs in at least one of the following circumstances:
    - Loss of power that holds doors in position.
    - Activation of sprinklers.
    - Activation of smoke detection system as follows:
      - System covers the area inside the building within 75' of the revolving doors.
    - Activation of manual control switch reducing the necessary force to < 130 lbs:
      - The switch must be in an approved location.
      - The switch must be clearly defined.

Note: Section 907, "Fire Alarm and Detection Systems," is cited as governing the smoke detection system indicated above.

- In other cases, the force necessary to collapse a revolving door must be ≤ 180 lbs.

1008.1.4.2 Power-operated doors *(part 1 of 2)*

- Doors in I-3 are not subject to this section.
- Certain horizontal sliding doors are not subject to this section.

Note: 1008.1.4.3, "Horizontal sliding doors," is cited as defining sliding doors that are not governed by this subsection.

- This section addresses other doors serving in a means of egress as follows:
  - Operated by power.
  - Doors with a photoelectric-actuated mechanism as follows:
    - Opens door upon approach of a person.
  - Manual doors with power-assistance.

Note: The following are cited as governing the types of doors indicated:
BHMA A156.10, "American National Standard for Power Operated Pedestrian Doors."
BHMA A156.19, "Standard for Power Assist and Low Energy Operated Doors."

- Such doors must have the following capability upon loss of power:
  - Able to be opened or closed manually.
- The force required to start moving the door manually is limited to the following:
  - ≤ 50 lbs.

Note: 1008.1.3, "Door opening force," is cited as governing other forces required to open these doors manually.

1008 Doors, Gates and Turnstiles

1008.1.4.2 Power-operated doors (*part 2 of 2)*

- Upon application of force in the direction of egress, the door must perform as follows:
  - Swing to full opening width from any position.
- The following applies to biparting doors in the emergency break-out position:
  - The requirement for a 32" width applies to the opening with both doors open.

Note: 1008.1.1, "Size of doors," is cited as requiring the 32" width when one door of a double door is open in other cases.

1008.1.4.3 Horizontal sliding doors

- This section addresses horizontal sliding doors serving in a means of egress.
- Such doors may serve in a means of egress in occupancies other than H.

Note: 1008.1.2, "Door swing," exception 6, is cited as permitting this use of sliding doors. The doors are required to meet other code requirements for sliding doors.

- Power requirements are as follows:
  - Doors must be power operated.
  - Doors must have an integrated standby power supply.
  - Door power must be supervised electronically.
- Manual requirements are as follows:
  - Doors must operate manually if power is lost as follows:
    By a simple method from both sides.
    Without special knowledge or effort.
  - The force necessary to operate the doors manually is limited as follows:
    ≤ 30 lbs to set door in motion.
    ≤ 15 lbs to open door to minimum width required for opening.
    ≤ 15 lbs to close door.
    ≤ 15 lbs to open door in the following case:
    When 250-lb force is applied ⊥ to door at a point adjacent to operating device.
- Where fire-protection rating is required, the door is governed as follows:
  - Must have applicable fire-protection rating.
  - Must be automatically closed by smoke detection system.

Note: NFPA 80, "Fire Doors and Fire Windows," is cited as governing the door assembly.
Section 715, "Opening Protectives," is cited as governing the door assembly.
715.4.8.3, "Smoke-activated doors," is cited as governing smoke activation.

- Doors must open to required width as follows:
  - Within 10 seconds of activating operating device.

1008 Doors, Gates and Turnstiles

Case study: Fig. 1008.1.4.3. Power-operated sliding doors become manually operated doors upon a loss of power to the doors. As manually operated doors, they are required in many locations by 1008.1.2, "Door swing," to be side-hinged doors that swing in the direction of egress travel. The horizontal sliding doors in the hospital of the example below are such a case. They comply with the power and force requirements, as well as having the capability to swing from the side in an emergency. They are in compliance with all code requirements.

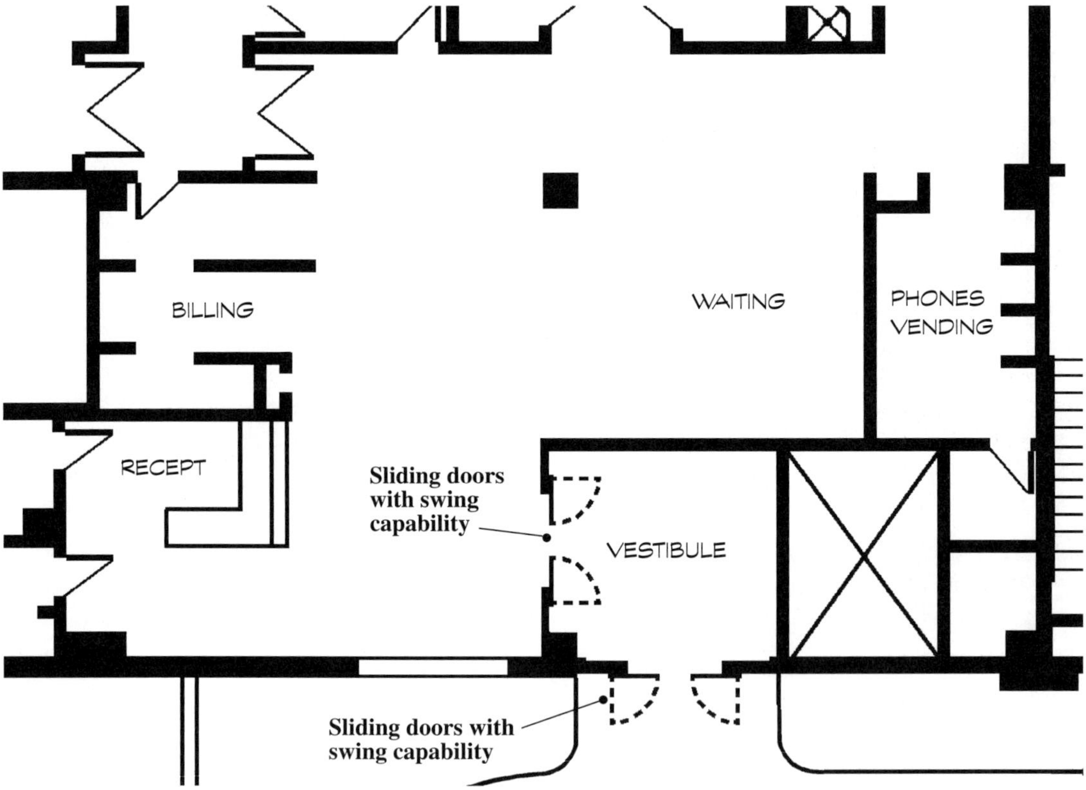

Fig. 1008.1.3.3. Partial floor plan. Christus St. Michael Health Care Center. Texarkana, Texas. Watkins Hamilton Ross Architects, Inc. Houston, Texas.

1008 Doors, Gates and Turnstiles

1008.1.4.4 Access-controlled egress doors

- The section addresses entrance doors in a means of egress.
- The following doors may have an entrance and egress access control system:
  - Doors in occupancies A, B, E, I-2. M, R-1, R-2 as follows:
    - Entrance doors.
    - Entrance doors to tenant spaces.
- Access control systems are governed as follows:
  - They must be approved.
  - A sensor is required on the egress side as follows:
    - Sensor detects an approaching person.
    - Doors unlock upon a signal from the sensor.
    - Doors unlock upon loss of power to the sensor.
  - Doors must unlock upon loss of power to door locks.
  - A manual unlocking device must be provided as follows:
    - Located ≥ 3'-4" and ≤ 4' above floor.
    - Located ≤ 5' from locked doors.
    - Must be readily accessible.
    - Must be identified by a sign that reads as follows:
      - PUSH TO EXIT.
    - Device activation must interrupt power to lock.
    - Device must operate independently of access control electronics.
    - Device activation must leave doors unlocked for ≥ 30 seconds.
  - Doors must unlock upon activation of a fire-protection system as follows:
    - Building fire alarm system.
    - Building sprinkler system.
    - Building fire detection system.
    - Doors to remain unlocked until the fire alarm system is reset.
- The following applies to occupancies A, B, E, and M:
  - Entrance doors may not be locked as follows:
    - From the egress side.
    - During times the building is open to the public.

1008.1.4.5 Security grilles

- Security grilles are permitted at the main exit of the following occupancies:
  - B, F, M, S.
- The following types of security grilles are permitted:
  - Horizontal sliding.
  - Vertical.
- Grilles must be operable, as follows, during the time the space is occupied:
  - From inside without the use of a key.
  - Without special knowledge or effort.
- Grilles must remain in full-open position as follows:
  - During times the space is occupied by the general public.

1008 Doors, Gates and Turnstiles

1008.1.5 Floor elevation

- This section addresses floor elevations on each side of a door.
- The following locations have requirements listed below:
  - Locations:
    - Individual units of Occupancy R-3 > 3 stories.
    - Individual units of Occupancy R-2 > 3 stories.
  - Requirements:
    - Door at interior stairs:
      - Door may open at the top step as follows:
        - Door may not swing over the top step.
    - The following doors may swing over stairs and landings:
      - Screen doors.
      - Storm doors.
- A single step is permitted at exterior doors where all of the following apply:
  - In occupancies F, H, R-2, R-3, S, U.
  - Where the door is not required to be accessible.
  - Where the step is ≤ 7" high.

Note: 1003.5, "Elevation change," exception 1, is cited as governing this circumstance, a partial summary of which is provided above.
1020.2, "Exterior exit doors," is cited as defining the door to which the above requirements are applicable.

- The requirement listed below governs the following locations:
  - Locations where all of the following apply:
    - Occupancy R-3.
    - Where units are not required to be accessible units, Type A units, or Type B units.
    - Where the door other than screen or storm doors does not swing over the landing.
  - Requirement:
    - Exterior landing must be ≤ 7 3/4" below the threshold top.
- Variation in elevation on each side of a door is permitted where both of the following apply:
  - Where caused by differences in finish materials.
  - Where the difference is ≤ 1/2".
- Variation in elevation on each side of certain doors have the limitations listed below for the following locations:
  - Locations:
    - In Type B dwelling units as follows:
      - At exterior decks and patios.
      - At exterior balconies.
  - Limitations for exterior surface:
    - Must be impervious.
    - Must be ≤ 4" below adjacent interior finished floor of the unit.
- The following applies to all other cases:
  - One of the following is required on each side of a door at the same elevation:
    - A floor or landing as follows:
      - Interior surface must be level.
      - Exterior surface may slope ≤ 1:48.

1008 Doors, Gates and Turnstiles

1008.1.6 Landings at doors

- Landing length in the direction of travel is governed as follows:
  - In the following locations such length is not required to be > 36":
    - Occupancy R-3.
    - Occupancy U.
    - Within individual units of R-2.
  - Landings in other locations must have a length ≥ 44" in the direction of travel.
- Landings must have a width equal to the larger of the following:
  - Equal to the width of the stairway.
  - Equal to the width of the door.
- A door may protrude ≤ 7" into the required width of a landing as follows:
  - Where the door is fully open.
  - Door hardware is included.
- A door may protrude ≤ 1/2 the required width of a landing where all of the following apply:
  - The occupant load is ≥ 50.
  - With the door in any position.
  - Door hardware is included.

Case study: Fig. 1008.1.6. The landing width at the turn is measured as a radius which must be ≥ the 3'-11" width of the stair. In this case, the landing at the door is 4'-0" wide. The length of the landing in the direction of travel is > than the 44" minimum. The width of the landing is greater than the 3' width of the door. The door (including hardware) protrudes 7" into the landing radius when fully open, thus, complying with the maximum permitted. The door does not protrude into the required landing width more than half the required width, as indicated by the arc in the illustration. The landings comply with the code.

1008 Doors, Gates and Turnstiles

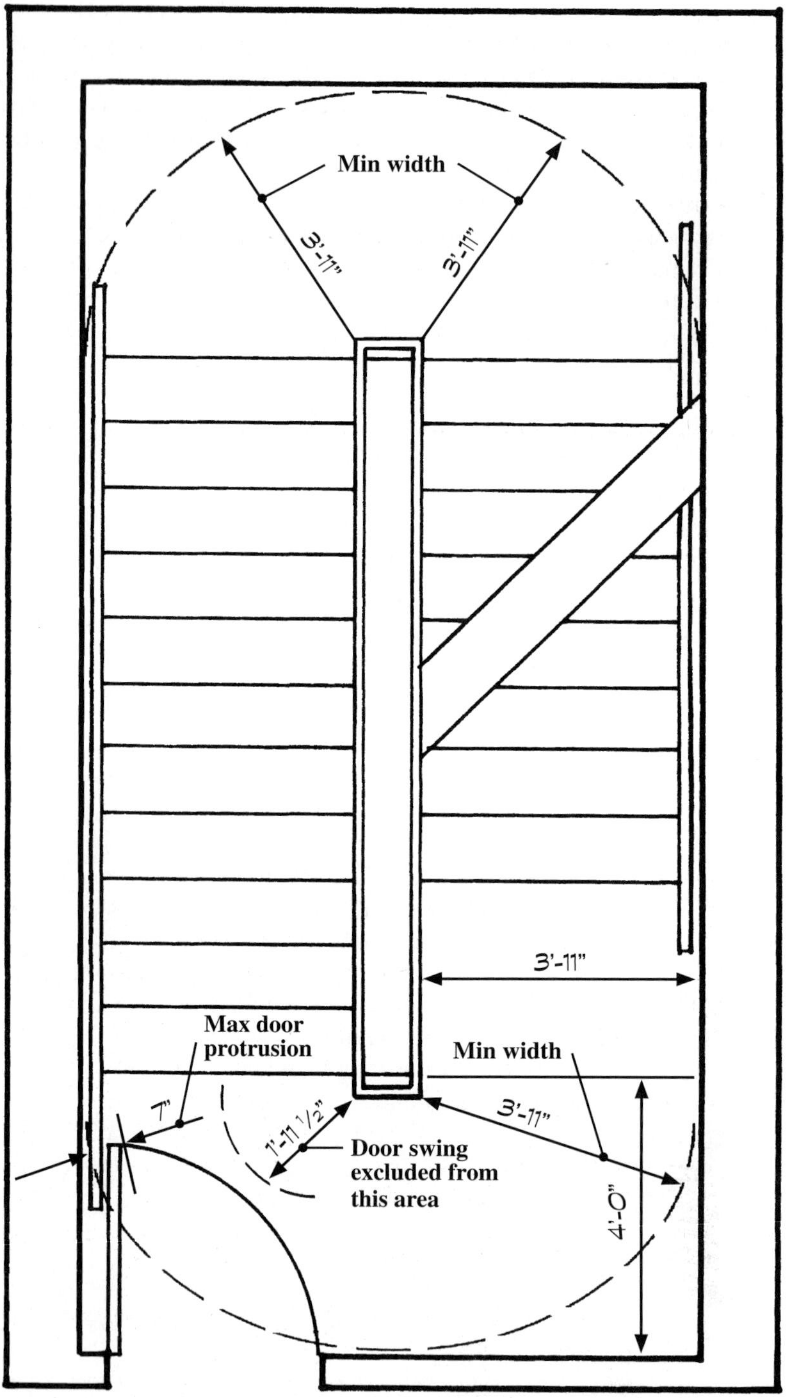

Fig. 1008.1.6. Plan at stairway. Hoyt Street Properties. Portland, Oregon. Ankrom Mosian Associated Architects. Portland, Oregon.

1008 Doors, Gates and Turnstiles

1008.1.7 Thresholds

- Doorway thresholds are limited in height as follows:
  - ≤ 7 3/4" where all of the following apply:
    - Located in Occupancy R-2 or R-3.
    - Exterior door as follows:
      - Where not in a required means of egress.
      - Where not on a required accessible route.
      - Where not in any of the following:
        - An accessible unit.
        - A Type A unit.
        - A Type B unit.
      - Where it does not swing over the landing or step as follows:
        - Where the door is not a storm door.
        - Where the door is not screen door.

Note: Chapter 11, "Accessibility," is cited as the source requiring accessible routes.

  - ≤ 3/4" for dwelling unit sliding doors.
  - ≤ 1/2" for other doors.
- Changes in floor level > 1/4" have the following requirement:
  - Beveled edge(s) at a slope ≤ 1:2 for the following:
    - Thresholds.
    - Floor elevation variations.

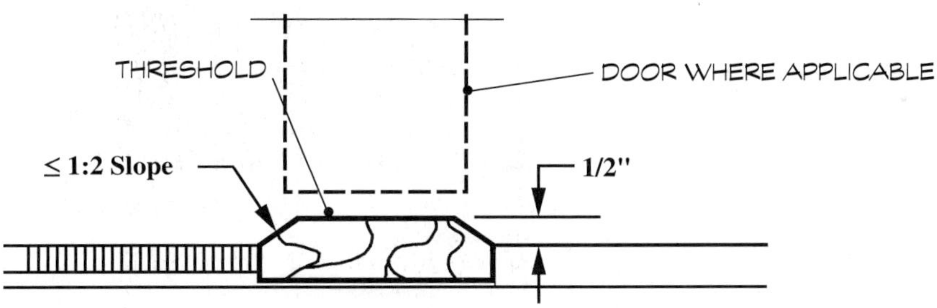

Fig. 1008.1.7. Threshold detail. Hot Springs Police Department New Headquarters. Hot Springs National Park, Arkansas. Cromwell Architects Engineers. Little Rock, Arkansas.

1008 Doors, Gates and Turnstiles

1008.1.8 Door arrangement

- This section addresses the distance between multiple doors in sequence in the direction of travel.
- Space required between horizontal sliding power-operated doors in a series is ≥ 4'.
- Spacing of multiple doors in the direction of travel is governed as follows:
  - The following doors are not required to be spaced ≥ 4' apart:
    - Exterior doors with storm doors or screen doors for individual dwelling units as follows:
      - In Occupancy R-2.
      - In Occupancy R-3.
    - Doors within dwelling units that are not Type A as follows:
      - Occupancy R-2.
      - Occupancy R-3.
- Other swinging doors in a series must be spaced the following distance apart:
  - 4' + the width of the door swinging into the space.
- Doors in a series must swing in one of the following patterns:
  - In the direction of travel.
  - Away from the space between the doors.

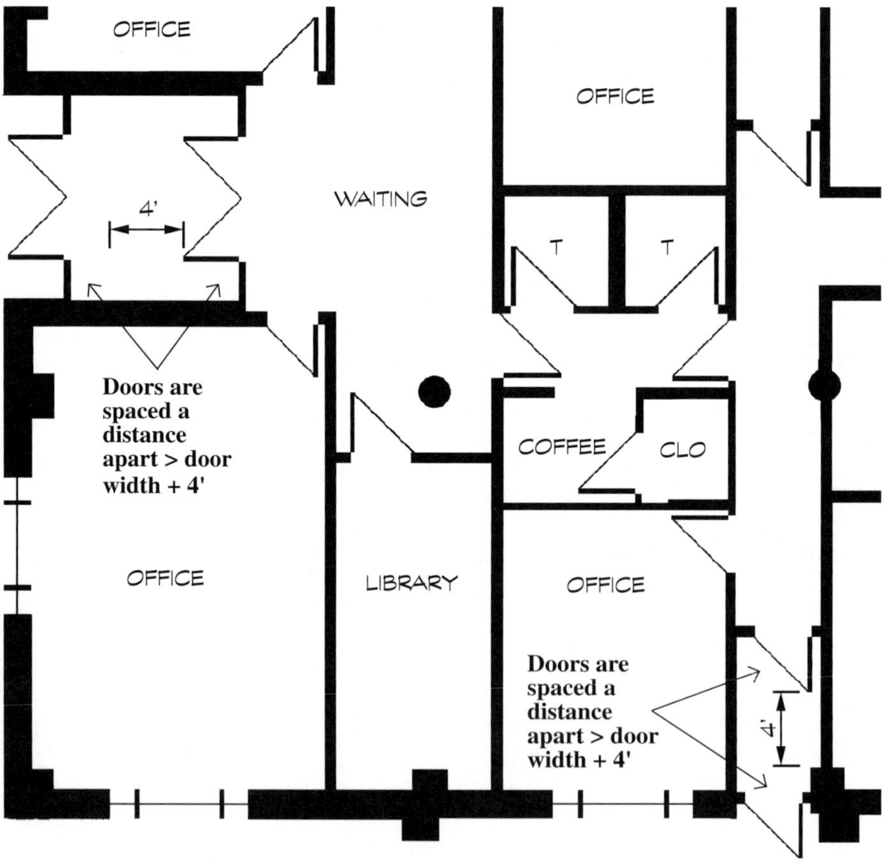

Fig. 1008.1.8. Partial first-floor plan. Country Club Park Building One. Wichita, Kansas. Gossen Livingston Associates, Inc., Architecture. Wichita, Kansas.

1008 Doors, Gates and Turnstiles

1008.1.9 Door operations

- Unless otherwise permitted by this section the following requirement applies to egress doors:
  - Egress doors must be operable as follows:
    - From the egress side.
    - Without the use of a key.
    - Without the requirement of special knowledge.
    - Without special effort.

1008.1.9.1 Hardware

- The following hardware has the requirements listed below:
  - Hardware:
    - For doors where required to be accessible as follows:
      - Handles.
      - Pulls.
      - Latches.
      - Locks.
      - Other operating devices.

Note: Chapter 11, "Accessibility," is cited as defining doors required to be accessible.

  - Requirements:
    - Hardware must not require the following to operate:
      - Tight grasping.
      - Tight pinching.
      - Twisting of the wrist.

1008.1.9.2 Hardware height *(part 1 of 2)*

- The following locks may be mounted at any height:
  - Locks not used for routine operation but only for security purposes.
- The requirements listed below govern doors or gates in the following locations:
  - Locations:
    - Barrier walls or fences protecting the following:
      - Pools.
      - Spas.
      - Hot tubs.
  - Requirements:
    - The following operable mechanisms may be ≤ 4'-6" above the walking surface:
      - Self-latching devices other than the following type:
        - Self-locking operated as follows:
          - By a key.
          - By an electronic opener.
          - By an integral combination lock.

1008 Doors, Gates and Turnstiles

1008.1.9.2 Hardware height *(part 2 of 2)*

- Other door hardware, as follows, must be positioned at the height indicated below:
  - Door hardware:
    Handles.
    Pulls.
    Latches.
    Locks.
    Other operating devices.
  - Required height:
    ≥ 2'-10" and ≤ 4'.

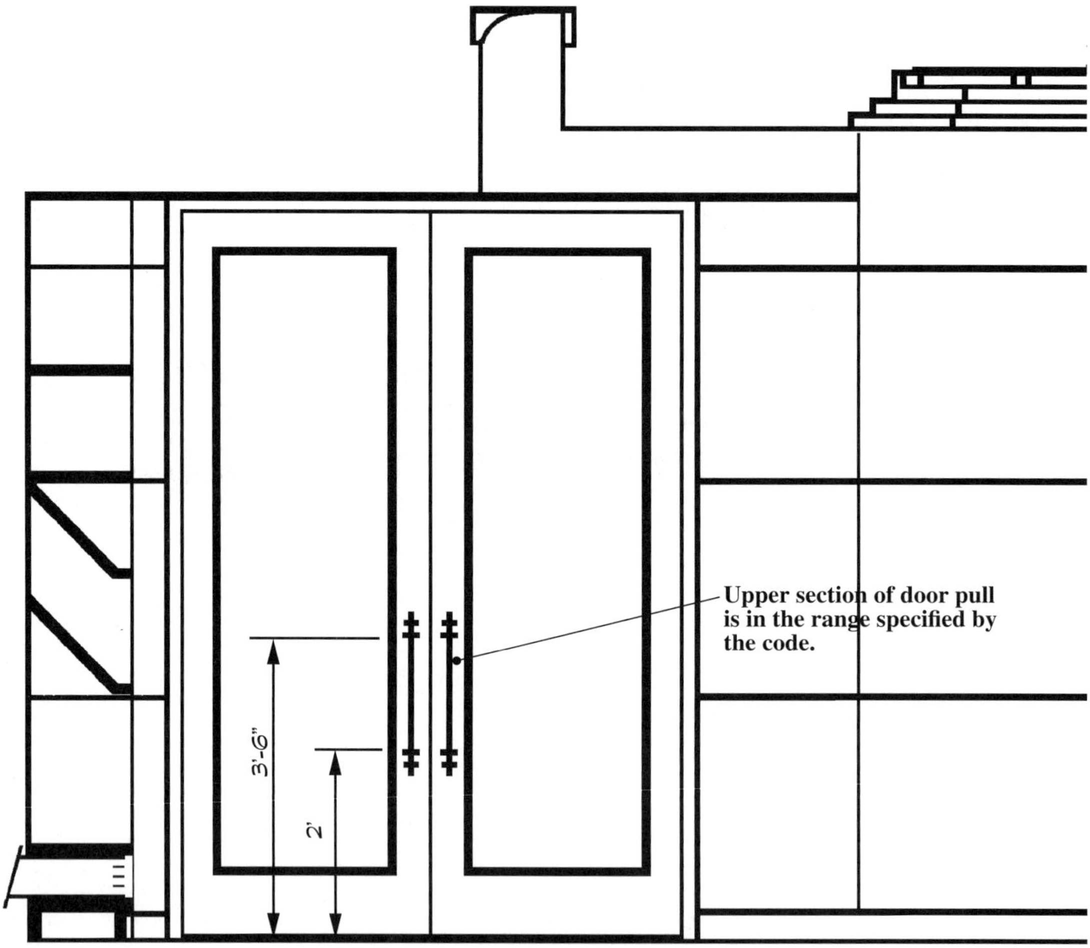

Fig. 1008.1.9.2. Door elevation. Christus St. Michael Health Care Center. Texarkana, Texas. Watkins Hamilton Ross Architects, Inc. Houston, Texas.

1008 Doors, Gates and Turnstiles

1008.1.9.3 Locks and latches *(part 1 of 2)*

- Doors may be secured by locks and latches in any of the following cases:
  - For detention purposes.
  - For restraint purposes.
  - Main exterior door(s) may have the locks specified at the following locations if all the requirements listed below are met:
    - Locks:
      - Key-operated locks.
    - Locations:
      - In Occupancy A as follows:
        - Occupant load is ≤ 300.
      - In the following occupancies:
        - B, F, M, S.
      - At places of religious worship.
    - Requirements:
      - Keys operate from the egress side.
      - It is obvious that the device is locked.
      - A sign is posted with all of the following characteristics:
        - Readily visible.
        - Durable.
        - Posted on egress side of door.
        - Posted in one of the following locations:
          - On the door.
          - Adjacent to the door.
        - Has the following message:
          - THIS DOOR TO REMAIN UNLOCKED WHEN BUILDING IS OCCUPIED.
          - Message is to have all of the following characteristics:
            - 1" high letters.
            - Letters to be on a contrasting background.
        - Building official has not disapproved the use of the lock as follows:
          - Based on due cause.
- Fire doors where all of the following conditions exist:
  - After the unlatching device has been disabled as follows:
    - By elevated temperature.
    - According to listed fire door test procedures.
- Pairs of egress doors may have the following devices if all of the requirements below are met:
  - Devices:
    - Flush bolts.
  - Requirements:
    - Must be approved.
    - Must be automatic.
    - Door has no doorknob or hardware mounted on its surface.
- Any of the following devices are permitted on the specified doors where requirements below are met:
  - Devices:
    - Night latch or dead bolt.
    - Security chain.

1008 Doors, Gates and Turnstiles

1008.1.9.3 Locks and latches *(part 2 of 2)*

- Doors:
  - In Occupancy R from any of the following:
    - Individual dwelling units and sleeping units.
- Requirements:
  - Occupant load must be ≤ 10.
  - Devices must be operable as follows:
    - From the inside.
    - Without a key.
    - Without a tool.

1008.1.9.4 Bolt locks

- Manual flush bolts or surface bolts are permitted at the following locations:
  - On doors not required for egress at either of the following:
    - Individual dwelling units.
    - Individual sleeping units.
- Manual edge- or surface-mounted bolts are permitted on the following doors:
  - On the inactive door of double doors to the following types of rooms:
    - Storage room.
    - Equipment room.
  - On the inactive door of double doors meeting all of the following conditions:
    - In B, F, S occupancies.
    - Doors serves < 50 occupants.
    - Where the inactive door has none of the following:
      - Door knobs or panic hardware.
      - Similar operating hardware.
  - On the inactive door of double doors meeting all of the following conditions:
    - In B, F, S occupancies.
    - Where the inactive door is not needed to meet egress width requirements.
    - The building is sprinklered.
    - Where the inactive door has none of the following:
      - Door knobs or panic hardware.
      - Similar operating hardware.

  Note: 903.3.1.1, "NFPA 13 sprinkler systems," is cited as governing the sprinkler system.

  - On the inactive door of double doors meeting all of the following conditions:
    - In Occupancy I-2.
    - Where the inactive door is not needed to meet egress width requirements.
    - Where the inactive door has none of the following:
      - Door knobs or panic hardware.
      - Similar operating hardware.
- In other cases, the following devices are not permitted:
  - Manual flush bolts.
  - Manual surface bolts.

1008 Doors, Gates and Turnstiles

1008.1.9.5 Unlatching

- Unlatching a door may require more > 1 action in the following cases:
  - At doors for detention or restraint.
  - At doors where a manual bolt lock is allowed.

Note: 1008.1.9.4, "Bolt locks," is cited as defining locations where these devices are permitted.

  - At doors where automatic flush bolts are allowed.

Note: 1008.1.9.3, "Locks and latches," exception 3, is cited as defining permitted locations.

  - At doors in any of the following Occupancy R doors:
    From individual dwelling units.
    From individual guest rooms.

Note: 1008.1.9.3, "Locks and latches," exception 4, is cited as defining permitted locations.

- In other locations, unlatching the following must not require > 1 action:
  - A door or leaf of a door.

1008.1.9.5.1 Closet and bathroom doors in Group R-4 occupancies

- Closet doors must be openable from inside the closet in R-4 occupancies.
- Lockable bathroom doors must be unlockable from outside the bathroom in R-4 occupancies.

1008.1.9.6 Special locking arrangements in Group I-2

- Delayed egress locks are permitted in I-2 occupancies where all of the following conditions apply:
  - Locks are approved.
  - Clinical needs of patients require such locks.
  - One of the following systems is provided:
    Building is sprinklered.
    Building has an approved smoke or heat detection system.

Note: 903.3.1.1, "NFPA 13 sprinkler systems," is cited as governing the sprinklers.
Section 907, *"Fire Alarm and Detection Systems," is cited as governing these systems.*

  - Occupants can enter an exit by passing through ≤ 1 delayed egress locked door.
  - In other than mental hospitals where patients require restraint, the locks unlock as follows:
    By actuation of a sprinkler or fire detection system.
    By any of the following:
      A signal from the fire command center.
      A signal from a nursing station.
      A signal from an approved location.
  - Approved procedures for unlocking the locks are described as follows:
      In the emergency planning and preparedness plan.

Note: Chapter 4, "Emergency Planning and Preparedness," in the International Fire Code *is cited as governing emergency planning and preparedness.*

  - Clinical staff has the means such as the following to open the locks:
    Keys and/or codes.
  - Emergency lighting is provided at the door with the delayed lock.

1008 Doors, Gates and Turnstiles

1008.1.9.7 Delayed egress locks

- Delayed locks may be used on egress doors where the following requirements are met:
  - Locks must be approved.
  - Locks must be listed.
  - Doors are limited to the following occupancies:
    B, F, I, M, R, S, U as follows:
    Where the building has one of the following:
    Sprinklers.
    Smoke detection system.
    Heat detection system.

Note: 903.3.1.1, "NFPA 13 sprinkler systems," is cited as governing the sprinklers noted above. Section 907, "Fire Alarm and Detection Systems," is cited as governing smoke and heat detection systems.

  - Doors must unlock as follows:
    Within one of the following periods:
    15 seconds.
    30 seconds where approved.
    Due to a force on the release mechanism as follows:
    ≤ 15 lb.
    Applied for 1 second.
    Lock release must activate a signal audible near the door.
    Lock must require manual relocking.
    The unlocking process cannot be reversible once initiated.
  - Doors with delayed locks may not be located as follows:
    So that travel into an exit requires passing through > 1 door with a delayed lock.
  - Doors must unlock as follows:
    When sprinklers activate.
    When a fire detection system activates.
    By a signal from the fire command center.
  - The following sign must posted:
    Located as follows:
    Above the release device.
    Within 12" of the release device.
    Sign must have one of the following messages as applicable:
    PUSH UNTIL ALARM SOUNDS. DOOR CAN BE OPENED IN 15 SECONDS.
    PUSH UNTIL ALARM SOUNDS. DOOR CAN BE OPENED IN 30 SECONDS.
  - Doors require emergency lighting.

1008.1.9.8 Electromagnetically locked egress doors *(part 1 of 2)*

- This section addresses means of egress doors in the following categories:
  - Doors not required to have panic hardware in the following locations:
    Occupancy A, B, E M. R-1, R-2.
  - Doors to tenant spaces in the following locations:
    Occupancy A, B, E M. R-1, R-2.

1008 Doors, Gates and Turnstiles

1008.1.9.8 Electromagnetically locked egress doors *(part 2 of 2)*

- Doors governed by this section may have electromagnetic locks if all of the following conditions are met:
  - The hardware is listed.
  - The hardware has a built-in switch that unlocks the door as follows:
    - Mounted on the door leaf.
    - Has an obvious method of operation.
    - Readily operated in all lighting conditions.
    - Can be operated with one hand.
    - Operation of the hardware unlocks the door immediately.
    - The door unlocks upon loss of power to the hardware.

1008.1.9.9 Locking arrangements in correctional facilities

- This section addresses means of egress doors in the following locations:
  - Correctional and detention facilities also containing any of the following occupancies:
    - A-2, A-3, A-4, B, E, F, I-2, I-3, M, S.
- Doors may be locked to areas containing persons whose movements are restricted as follows:
  - Locks must be unlockable by both of the following means:
    - Manually.
    - Automatically by ≥ 1 of the following methods:
      - Upon activation of sprinklers.
      - Upon activation of an approved alarm box.
      - Upon a signal from a continuously attended location.

Note: 903.3.1.1, "NFPA 13 sprinkler systems," is cited as governing the sprinklers noted above.

1008.1.9.10 Stairway doors

- This section governs the following means of egress doors:
  - At interior stairways.
- This section does not govern the following doors:
  - Certain stairway doors permitted to be locked from the stairway side.

Note: 403.5.3, "Stairway door operation," is cited as governing this type of door.

- Stairway discharge doors are governed as follows:
  - They must be openable from the stairway side.
  - They may be locked only from outside the stairway.
- Doors in stairways serving ≤ 4 stories are governed as follows:
  - They may be locked from the stairway side where both of the following apply:
    - Where they are openable from the side outside the stairway.
    - Where they all can be unlocked at the same time without unlatching by one of the following:
      - From the fire command station.
      - From a location inside the building main entrance.
- In other cases, the stairway doors are governed as follows:
  - They must be operable from both sides without the following:
    - Use of a key.
    - Special knowledge.
    - Special effort.

1008 Doors, Gates and Turnstiles

1008.1.10 Panic and fire exit hardware

- Certain main exits of Occupancy A may have locks and latches.

  Note: 1008.1.9.3, Item 2, "Locks and latches," is cited as defining such locations in Occupancy A.

- In other cases panic hardware is the only latch or lock type permitted on the following doors:
  - Doors in a means of egress in the following occupancies:
    - Occupancies A and E with an occupant count ≥ 50.
    - Occupancy H.
- Exit access doors in the following electrical rooms have the requirements listed below:
  - Electrical rooms:
    - With both of the following:
      - Equipment rated ≥ 1,200 amps.
      - Width > 6' *(measured along the egress path).*
    - With any of the following:
      - Overcurrent devices.
      - Switching devices.
      - Control devices.
  - Requirements:
    - Must have panic hardware.
    - Must swing in the direction of egress.
- Panic hardware in balanced doors is governed as follows:
  - It must be the push-pad type as follows:
    - The pad must be on the latch side of the door.
    - The pad may not extend past the center of the door.

1008.1.10.1 Installation

- Panic and fire exit hardware must comply with the following requirements:
  - The component that releases the door latch must extend ≥ half the door width.
  - The force required to release the door latch is limited to ≤ 15 lb.

  Note: UL 305, "Panic Hardware," is cited as governing panic hardware.
  The following are cited as alternatives governing fire exit hardware:
  UL 10C, "Positive Pressure Fire Tests of Door Assemblies."
  UL 305, "Panic Hardware."

1008.2 Gates

- Gates > 4' wide are allowed in fences and walls around a stadium as follows:
  - Horizontal sliding gates.
  - Swinging gates.
- Other gates in a means of egress must comply with applicable requirements for doors.

1008 Doors, Gates and Turnstiles

1008.2.1 Stadiums

- Panic hardware is not required on gates where the following conditions exist:
  - Gates are around stadiums.
  - Gates are continually supervised when the public is present.
  - A dispersal area is provided as follows:
    - Area is located between the fence and enclosed space.
    - Area is ≥ 50' from enclosed space.
    - Area must be ≥ the following size:

Occupant load × 3 sf (per occupant)

Note: 1027.6, "Access to a public way," is cited as governing means of egress from a safe dispersal area.

1008.3 Turnstiles

- A turnstile may serve in a means of egress as follows:
  - Where assigned ≤ 50 occupants for egress.
  - Turnstile must rotate freely as follows:
    - In the direction of egress.
    - When released manually by staff.
  - Where assigned ≤ half of the required capacity for egress.
  - Turnstile dimensions must be as follows:
    - Height ≤ 39".
    - Clear width ≥ 16 1/2" @ ≤ 39" height.
    - Clear width ≥ 22" @ > 39" height.
- A turnstile may serve in an accessible route where it meets all of the following conditions:
  - Where it does not have a revolving mechanism.
  - Where it has the following dimensions:
    - Clear width ≥ 36" @ ≤ 34" height.
    - Clear width ≥ 32" @ ≥ 34" and ≤ 80" height.
- In other cases the following may not obstruct a means of egress:
  - Turnstiles.
  - Similar devices as follows:
    - That permit travel only in one direction.

1008.3.1 High turnstile

- Turnstiles > 39" in height are governed as follows:
  - They must meet revolving door requirements.

1008.3.2 Additional door

- The following turnstiles have the requirements listed below:
  - Turnstiles:
    - Serving an occupant load > 300.
    - Other than portable.
  - Requirements:
    - Must have a side-hinged swinging door as follows:
      - Located ≤ 50' from the turnstile.

Note: 1008.1, "Doors," is cited as governing these swinging doors.

1009 Stairways

1009.1 Stairway width

- Spiral stairways are not governed by this section.

  Note: 1009.9, "Spiral stairways," is cited as the source of requirements for this type of stairway.

- Assembly aisle stairs are not governed by this section.

  Note: Section 1028, "Assembly," is cited as governing aisle stairs.

- A clear passage ≥ 1'-8" is required on stairways where both of the following apply:
  - Where either of the following is installed:
    - An incline platform lift.
    - A stairway chairlift as follows:
      - Where the lift folds, the clearance is measured to the following point:
        - To the lift in the folded-up position.
  - In R-3 occupancies.
  - Within dwelling units of R-2 occupancies.
- In other cases, stairway width must be the larger of 0.3" × occupants served or one of the following:
  - ≥ 4' clear width between handrails in certain cases where required to be accessible.
  - For stairs not required to be accessible, the following widths apply:
    - ≥ 3' for occupant loads < 50.
    - ≥ 3'-8" for occupant loads ≥ 50.

Note: 1005.1, "Minimum required egress width," is cited as governing stairway width and is the source of the 0.3" required per occupant as shown above.
1007.3, "Stairways," is cited as governing stairways in an accessible means of egress.

1009.2 Headroom

- Spiral stairways may have a headroom clearance of 6'-6".

  Note: 1009.9, "Spiral stairways," is cited as governing these stairs.

- The following stairways have an encroachment into stairway headroom as noted below:
  - Stairways:
    - Located in Occupancy R-3.
    - Located in dwelling units in Occupancy R-2.
    - Located in Occupancy U that are accessory to R-3.
    - Located in Occupancy U that are accessory to individual dwelling units in R-2.
  - Encroachment:
    - The side of the floor opening for a stairway may encroach into the stairway headroom as follows:
      - ≤ 4 3/4" in the following case:
        - Where the treads extend beyond the edge of the opening in the lower part of the stairway.
- Other stairways require a headroom clearance ≥ 6'-8" as follows:
  - Measured vertically from a line tangent to tread nosings.
  - Required headroom must be continuous to a point directly above the following:
    - The point where the line tangent to the nosings intersects the following:
      - The landing below the stair run as follows:
        - The point is located a distance away from the bottom riser equal to 1 tread depth.
  - Minimum headroom clearance is required for the full width of the landing and stairway.

1009 Stairways

1009.3 Walkline

- The walkline on winder stairs is positioned as follows:
  - Concentric to the curved line of travel.
  - 12" from the narrow side measured as follows:
    From the edge of the walking surface at the side of the stairs toward the center of the stairway width.

1009.4.1 Dimension reference surfaces

- In this section the following are not included in dimensions:
  - Carpets.
  - Rugs.
  - Runners.

1009.4.2 Riser height and tread depth *(part 1 of 2)*

- The following stairways are not governed by this section:
  - Alternating tread devices.

Note: 1009.10, "Alternating tread devices," is cited as governing these devices.

  - Ship ladders.

Note: 1009.11, "Ship ladders" is cited as governing these components.

  - Spiral stairways.

Note: 1009.9, "Spiral stairways," is cited as governing these stairs.

  - Stairways replacing existing stairways.

Note: 3404.1, "General," is cited as governing stairway replacement. Relief from new stairway requirements is provided where existing construction prohibits the reduction of pitch or slope.

- In assembly seating areas, the following applies:
  - Where aisle gradient is dictated by sightlines, stair dimensions are required as follows:
    Risers ≥ 4" and ≤ 8".

Note: 1028.11.2, "Risers," is cited as governing these stairs and is partially summarized above.

- Occupancy I-3 stairways to the following facilities ≤ 250 sf have requirements as indicated below:
  - Facilities:
    Guard towers.
    Observation stations.
    Control rooms.
  - Requirements:
    Risers to be ≤ 8".
    Tread depth to be ≥ 9".
- For the following locations and conditions, the requirements indicated below apply:
  - Locations:
    R-3 and within dwelling units of R-2.
    U where serving R-3 or dwelling units within R-2.

1009 Stairways

1009.4.2 Riser height and tread depth *(part 2 of 2)*

- Conditions:
  - Stairs with solid risers.
  - Treads < 11" in depth.
- Requirements:
  - A nosing ≥ 3/4" and ≤ 1 1/4".

- Winder tread depth is measured as follows:
  - ⊥ to leading edge.
  - Horizontally between vertical planes as follows:
    - At the leading edges of adjacent treads.
- The requirements indicated below apply to winders in the following locations:
  - Locations:
    - R-3.
    - Within dwelling units of R-2.
    - U where serving R-3.
  - Requirements:
    - Tread depth must be ≥ 6" at all points.
    - Tread depth must be ≥ 10" as follows:
      - Measured at 12" from the narrow end.
- Other winder treads have the following requirements:
  - Tread depth must be ≥ 10" within the clear width of the stair.
  - Tread depth must be ≥ 11" measured as follows:
    - At at the walk line.
- Other tread and riser sizes requirements are as follows:

Table 1009.4.2 Tread and Riser Heights

Occupancy	Riser	Tread
R-3	≥ 4" and ≤ 7 3/4"	≥ 10"
Within dwelling units of R-2	≥ 4" and ≤ 7 3/4"	≥ 10"
U where serving R-3	≥ 4" and ≤ 7 3/4"	≥ 10"
All other stairs with rectangular treads	≥ 4" and < 7"	≥ 11"

- Riser height is measured vertically between leading edges of adjacent treads.
- Tread depth is measured horizontally as follows:
  - Horizontally between vertical planes as follows:
    - At the leading edges of adjacent treads.
  - ⊥ to the leading edge of the tread.

1009.4.3 Winder treads

- The following are not governed by this section:
  - Curved stairways.
  - Spiral stairways.

Note: 1009.8, "Curved stairways," is cited as governing this type of stairway.
1009.9, "Spiral stairways," is cited as governing this type of stairway.

- Winder treads may be used in a means of egress stairway in a dwelling unit.
- Otherwise, winder treads may not be used in a means of egress stairway.

1009 Stairways

1009.4.4 Dimensional uniformity

- Aisle risers serving assembly seating where the gradient is dictated by sightlines are not required to be uniform within certain limits.

  Note: 1028.11.2, "Risers," is cited as governing aisle risers.

- Winders may differ from rectangular treads in the same stairway flight as follows:
  - Winders must be consistent in shape with each other.

  Note: 1009.4.2, "Riser height and tread depth," is cited as governing these stairs.

- Top or bottom stairway risers abutting a sloped surface are governed as follows:
  - Sloped surfaces as follows permitted to serve as landings must have a fixed gradient:
    - Public way.
    - Walkway.
    - Driveway.
  - The abutting riser may follow the slope of the landing to a height < 4" as follows:
    - Slope must be ≤ 1:12 across the stairway width.
  - The leading edge of the abutting riser must be identified with a marking stripe as follows:
    - Marking stripe to be distinctive.
    - Marking stripe to differ from any other nosing marking in the flight.
    - Marking stripe to be visible when traveling down the stairs.
    - Marking stripe to have a slip-resistant surface.
    - Marking stripe to have a width ≥ 1" and ≤ 2".
- Other stair treads and risers are governed as follows:
  - They must be uniform in size and shape.
  - The difference in depth between the largest and smallest tread in a flight must be ≤ 3/8".
    - Tread depth at winders is measured as follows:
      - At the walk line.
  - The difference in height between the largest and smallest riser in a flight must be ≤ 3/8".

1009.4.5 Profile *(part 1 of 2)*

- Solid risers are not required for the following stairways:
  - Located in Occupancy I-3 as follows:
    - The size of the open area in the riser is not limited.
  - Located in F, H, and S occupancies where not available to the public as follows:
    - The size of the open area in the riser is not limited.
  - Stairways not required to be an enclosed accessible means of egress as follows:
    - Where the opening between treads will not pass a 4" sphere.

  Note: 1007.3, "Stairways," is cited as the source for these and other requirements governing accessible stairways.

  - Spiral stairways.

  Note: 1009.9, "Spiral stairways," is cited as governing these stairs.

  - Alternating tread devices.

  Note: 1009.10, "Alternating tread devices," is cited as governing these devices.

1009 Stairways

1009.4.5 Profile *(part 2 of 2)*

- Other stairways must have solid risers in one of the following positions:
  - Vertical.
  - Sloped at ≤ 30° from the vertical to a point under the tread above as follows:
    - Adjoins the underside of the nosing above.
- Other aspects of stairways are governed as follows:
  - The leading edge of the treads may not have a radius > 9/16".
  - Nosings may not be beveled > 1/2".
  - Nosings may not project > 1 1/4" over the tread below.
  - Nosings must be uniform in size, including the following:
    - The nosing of the floor at the top of a stair run.

1009.5 Stairway landings

- Aisle stairs are not governed by this section where they comply with requirements for aisles in assembly occupancies.

  Note: Section 1028, "Assembly," is cited as the source of requirements for aisle stairs.

- Other stairway landings are governed as follows:
  - One of the following is required at the top and bottom of a stairway:
    - Floor.
    - Landing.
  - The landing width must be ≥ the stairway width.
  - Landings located in other than straight runs must have the following dimension in the direction of travel:
    - A dimension ≥ the stair width.
  - Landings located in straight runs must have the smaller of the following dimensions in the direction of travel:
    - A dimension ≥ the stair width.
    - ≥ 48".
  - Doors opening onto landings may protrude into the required width as follows:
    - A distance ≤ 1/2 of the required width while opening.
    - A distance ≤ 7" when the door is fully open as follows:
      - The protrusion of a door includes its hardware.
  - Required wheelchair spaces may not be in the required width of the landing.
  - Doors may not swing through required wheelchair spaces.

  Note: 1007.6.1, "Size," is cited as governing wheelchair spaces.

1009.6 Stairway construction

- Wood handrails are allowed in stairways in all types of building construction.
- Otherwise, stairway materials must conform to the construction-type requirements for the building.

1009 Stairways

1009.6.1 Stairway walking surface

- The walking surface of the following stairway components must not slope > 1:48 in any direction:
  - Landings.
  - Treads.
- The surfaces of treads and landings are governed as follows:

Table 1009.6.1 Surfaces of Treads and Landings

Locations	Tread and landing surfaces
Occupancies A, B, E, I, M, R, U, and parking structure public areas.	Openings allowed which cannot pass a 1/2" sphere. Long dimension of elongated openings to be ⊥ to the dominant direction of travel.
Occupancies F, H.	Openings allowed which cannot pass a 1 1/8" sphere.
Occupancy S other than public areas of parking structures.	Openings allowed which cannot pass a 1 1/8" sphere.

1009.6.2 Outdoor conditions

- The design of following walking surfaces has the requirement indicated below:
  - Surfaces:
    Outdoor stairways.
    Outdoor approaches to stairways.
  - Requirement:
    Design must not permit water to accumulate.

Case study: Fig. 1009.6.2. The surfaces of the stairway slope 2% to shed water. Snow melting mats are embedded in the landing and the steps, to prevent the accumulation of snow or ice. The stairway meets requirements for exterior conditions as required for a B occupancy.

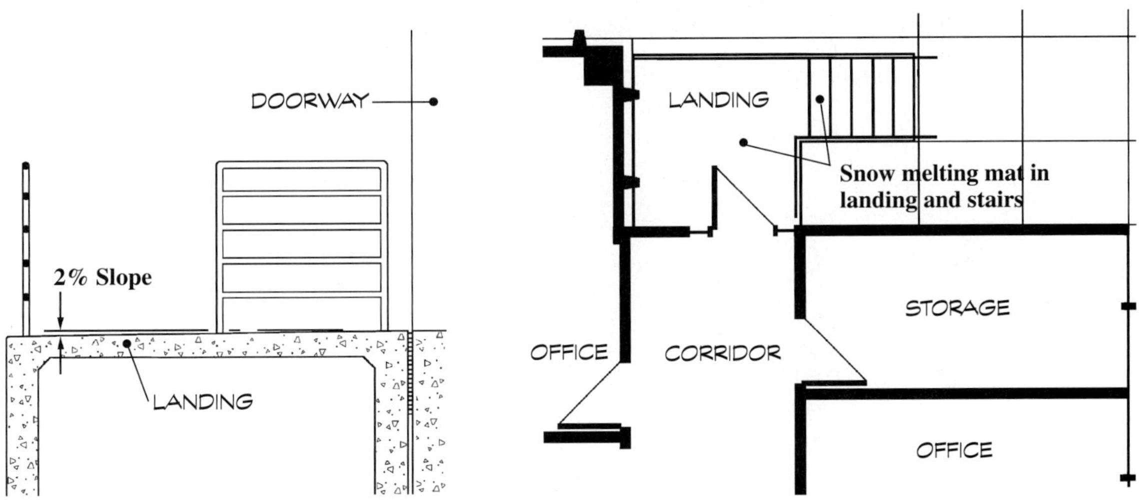

Fig. 1009.6.2. Plan and section at exterior stair. Lubrication Engineers, Inc. Wichita, Kansas. Gossen Livingston, Associates, Inc., Architecture. Wichita, Kansas.

1009 Stairways

1009.6.3 Enclosures under stairways

- This section addresses space under stairways as follows:
  - Stairways are one of the following:
    - Enclosed.
    - Open.
  - Space under the stairway is as follows:
    - Enclosed.
    - Usable.
- The space under a stairway in the following locations has the requirement indicated:
  - Locations:
    - In dwelling units of occupancies R-2 and R-3.
  - Requirement:
    - Walls and soffit must be covered as follows:
      - With 1/2" gypsum board.
      - On the enclosed side.
- In other locations, space under stairways is governed as follows:
  - Walls and soffits must have the larger of the following:
    - ≥ 1-hr fire-resistance rating.
    - Fire-resistance rating ≥ that of the stairway enclosure.
  - Access to the space must be from outside the stair enclosure.
- Space under exterior exit stairways is governed as follows:
  - Where enclosed the following applies:
    - Enclosure must have a fire-resistance rating ≥ 1 hr.
  - Where open the following applies:
    - The space may not be used for any purpose.

1009.7 Vertical rise

- Aisle stairs are not governed by this section.

  Note: Section 1028, "Assembly," is cited as the source of requirements for aisle stairs.

- Alternating tread devices have the following rise limitation:
  - 20' as follows:
    - Where located in a means of egress.
    - Measured between the following:
      - Floor levels.
      - Landings.
- Other stairs are limited in their vertical rise to ≤ 12'.

1009 Stairways

Case study: Fig. 1009.6.3. A meter room is located under an interior stairway as shown in the illustration. The walls of the space and the soffit of the landing above it have a 1-hr fire-resistance rating. The room is accessed from outside the stairway enclosure. The meter room is in compliance with code requirements.

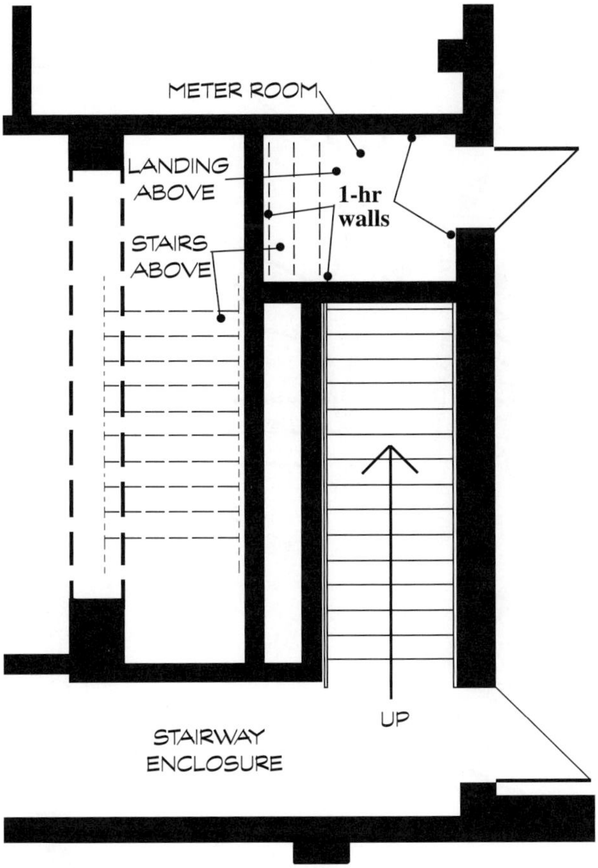

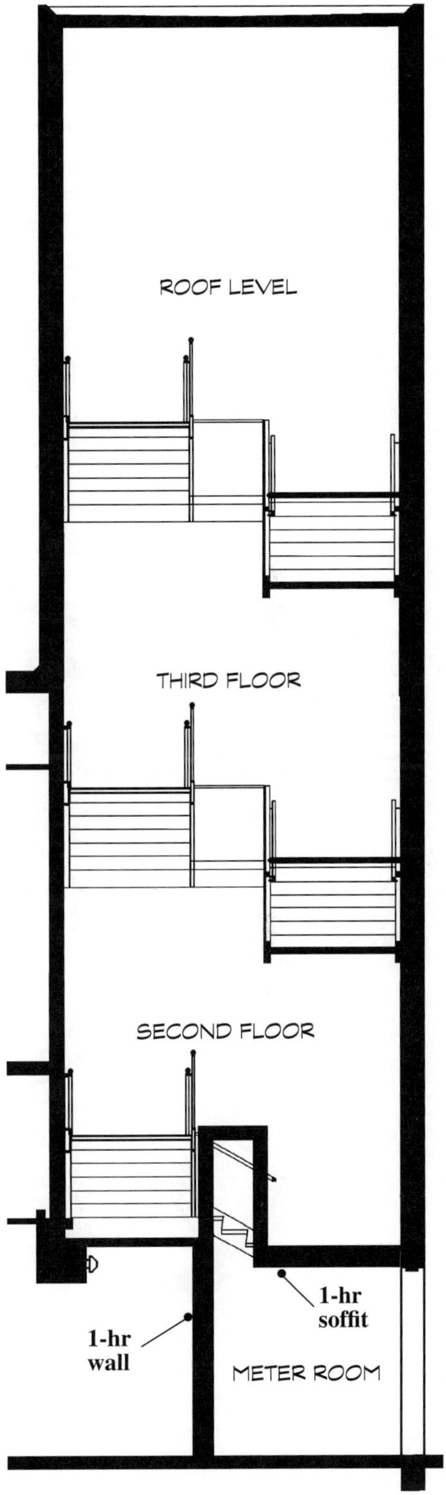

Fig. 1009.6.3. Plan and section of meter room.
University of Connecticut New Downtown Campus at Stamford, Connecticut. Perkins Eastman Architects, P.C. New York, New York.

1009 Stairways

1009.8 Curved stairways

- The radii of the following circular stairs are not governed by this section:
  - In Occupancy R-3.
  - In individual dwelling units of R-2.
- Other curved stairs with winder treads are governed as follows:
  - The smallest stairway radius has the following size requirements:
    It must be ≥ 2 × the stairway required width.

Note: 1009.4, "Stair treads and risers," is cited as governing curved stair tread width and riser height.

Case study: Fig. 1009.8. The tower stairway meets requirements for radius as well as tread depth at the narrow end and at a point 12" from the narrow end as indicated in the illustration. The adjacent steps are not circular in the strictest sense, but meet the same requirements.

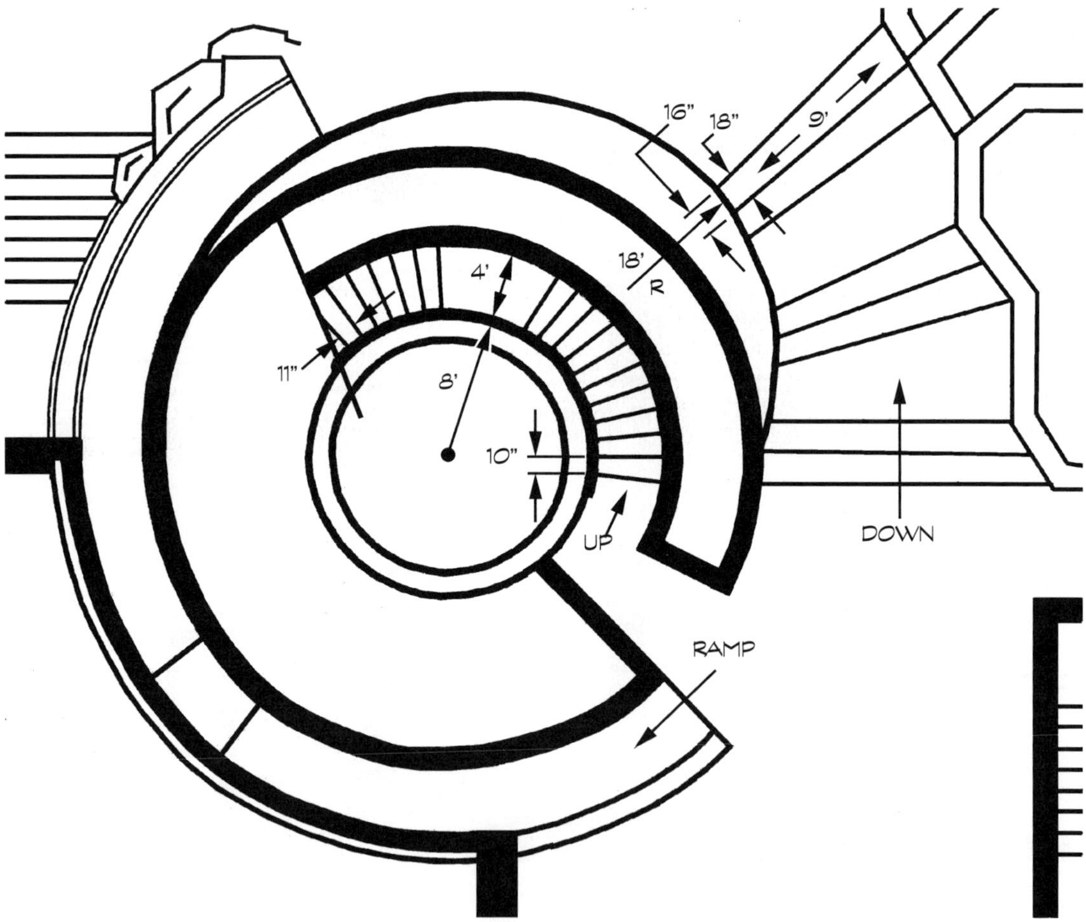

Fig. 1009.8. Plan at tower. Lady Bird Johnson Wildflower Center. Austin, Texas. Overland Partners, Inc. San Antonio, Texas.

1009 Stairways

1009.9 Spiral stairways

- Use of spiral stairways in a means of egress is governed as follows:
  - Permitted in the following locations:
    - Within a dwelling unit:
      - From a space with both of the following characteristics:
        - Area ≤ 250 sf.
        - Occupant load ≤ 5 persons.
    - For stage-related areas as follows:
      - Galleries.
      - Catwalks.
      - Gridirons.

Note: 1015.6, "Stage means of egress," is cited as governing galleries, catwalks, and gridirons.

  - Not permitted in other locations.
- Dimensions required of a spiral stairway are as follows:
  - Tread width must be ≥ 7 1/2" measured in the following location:
    - At 12" from narrow end.
  - Headroom must be ≥ 6'-6".
  - Riser height must be ≤ 9 1/2".
  - Clear width of stairway must be ≥ 2'-2" as follows:
    - Measured at the handrail.
    - Measured below the handrail.

1009.10 Alternating tread devices

- Use of alternating tread devices in a means of egress is governed as follows:
  - Permitted in the following locations:
    - From mezzanines in occupancies F, H, and S as follows:
      - Mezzanine area ≤ 250 sf.
      - Mezzanine occupants ≤ 5.
    - In I-3 areas ≤ 250 sf as follows:
      - Guard towers.
      - Observation stations.
      - Control rooms.
      - For access to unoccupied roofs.
  - Not permitted in other locations.

1009.10.1 Handrails of alternating tread devices

- Handrails for alternating tread devices must meet the following requirements:
  - Must be located on both sides of the stair system.
  - Must meet the same requirements as for stairway handrails.

Note: Section 1012, "Handrails," is cited as governing the handrails.

1009 Stairways

1009.10.2 Treads of alternating tread devices

- The initial tread of an alternating tread device must be at the same elevation as the following:
  - Landing or floor surface.
  - Platform.
- Alternating tread device dimensions are governed as follows and in the table below:
  - Projected tread depth is measured horizontally as follows:
    Between vertical planes at nosings of adjacent treads on opposite sides of the center line.
  - Riser height is measured vertically as follows:
    Between horizontal planes at the top surfaces of adjacent treads on opposite sides of the center line.

Table 1009.10.2 Alternating Tread Devices

Device location	Projected tread	Tread depth	Tread width	Rise to next tread
Means of egress from a mezzanine of ≤ 250 sf and ≤ 5 occupants	≥ 8 1/2"	≥ 10 1/2"	≤ 7"	≤ 8"
Other locations	≥ 5"	≥ 8 1/2"	≥ 7"	≤ 9 1/2"

1009.11 Ship ladders

- Ship ladders may serve in a means of egress for the following:
  - For the following spaces ≤ 250 sf with ≤ 3 occupants:
    Control rooms.
    Elevated observation stations.
  - For unoccupied roofs.
- The following is required for ship ladders:
  - Tread depth must be ≥ 5".
  - Tread depth + tread projected nosing must be ≥ 8.5" as follows:
    The projected nosing extends in front of the plane of the nosing edge of the next tread above.
  - Riser height must be ≥ 9.5".
  - Hand rails are required on both sides as follows:
    Clear space between handrails must be ≥ 20".

1009.12 Handrails *(part 1 of 2)*

- Handrails are not required for aisle stairs with certain conditions.

Note: 1028.13, "Handrails" is cited as governing aisles noted above.

- The following may have a handrail on one side only:
  - Stairways within dwelling units.
  - Spiral stairways.
  - Aisle stairs with seating only on one side.
- Handrails are not required for the following components where conditions indicated apply:
  - Components:
    Decks and patios.
    Walkways.
  - Conditions:
    At a single change in elevation.
    Landing depth on each side of elevation change is greater than required.

1009 Stairways

1009.12 Handrails *(part 2 of 2)*

- Handrails are not required at the following:
  - At a single riser at the following Occupancy R-3 locations:
    - Entrance and egress doors.
  - At a change of room elevation with ≤ 3 risers as follows:
    - In a dwelling unit of the following occupancies:
      - R-2, R-3.
    - In a sleeping unit of the following occupancies:
      - R-2, R-3.
- Other stairways require a handrail on each side as follows:
  - Handrails must have adequate strength and attachment.

Note: Section 1012, "Handrails," is cited as governing handrails.
Section 2407, "Glass in Handrails and Guards," is cited as governing glass used for handrails.

1009.13 Stairway to roof

- A stairway is required to extend to the roof where both of the following apply:
  - In buildings ≥ 4 stories in height.
  - Where the roof slopes ≤ 4:12.
- Access to unoccupied roof from the top floor may be by the following:
  - An alternating tread device.

1009.13.1 Roof access

- Access to unoccupied roofs may be through a roof hatch as follows:
  - Area of hatch must be ≥ 16 sf.
  - Dimensions of hatch must be ≥ 2'.
- Access provided to other roofs must be by way of a penthouse.

Note: 1509.2, "Penthouses," is cited as the source of applicable requirements.

1009.13.2 Protection at roof hatch openings

- Guards are required for the following roof hatch:
  - Where located ≤ 10' from the edge of the roof.
- The guards must protect the following as applicable:
  - The roof access.
  - The roof edge.

Note: Section 1013, "Guards," is cited as governing the installation of guards.

1009.14 Stairway to elevator equipment

- A stairway is required to access the following where elevator equipment requires maintenance:
  - Roofs.
  - Penthouses.

1010 Ramps

1010.1 Scope

- This section series does not govern the following ramps:
  - Ramped aisles with both of the following characteristics:
    In assembly spaces.
    That do not provide access to wheelchair spaces.

Note: 1028.11, "Assembly aisle walking surfaces," is cited as governing ramped aisles.
The following are cited as governing access for wheelchairs:
1108.2, "Assembly area seating," through 1108.2.4, "Dispersion of wheelchair spaces in multilevel assembly seating areas."
1108.2.6, "Lawn seating."

  - Curb ramps.

Note: ICC A117.1, "Accessible and Usable Buildings and Facilities," is cited as governing curb ramps.

- This section series governs only the following aspects for the ramps listed below:
  - Aspects:
    Running slope.
    Guards.
  - Ramps:
    Vehicle ramps with all of the following characteristics:
    In parking garages.
    For pedestrian exit access.
    Not providing an accessible route.
    Not serving accessible parking spaces.
    Not serving other accessible elements.
    Not part of an accessible means of egress.

Note: 1010.3, "Cross slope," through 1010.9, "Edge protection," are cited as not applicable to the ramps described above.

- Other ramps as follows are governed by this section series:
  - Ramps used in a means of egress.

1010.2 Slope

- Aisle ramp slope in the following locations are not governed by this section:
  - Occupancy A.
  - Assembly areas accessory to Occupancy E.

Note: 1028.11, "Assembly aisle walking surfaces," is cited as governing aisle ramp slope.

- Other ramps are governed as follows:
  - Ramps in a means of egress must have a slope $\leq$ 1:12.
  - Other pedestrian ramps must have a slope $\leq$ 1:8.

1010.3 Cross slope

- The slope of a ramp $\perp$ to direction of travel must be $\leq$ 1:48.

1010 Ramps

1010.4 Vertical rise

- Rise of any ramp is limited to ≤ 2'-6".

1010.5.1 Width

- The width required for a means of egress ramp is the same as that required for corridors.

  Note: 1018.2, "Corridor width," is cited as the source governing applicable widths.

- ≥ 3' clear width is required for ramps in a means of egress as follows:
  - Between handrails.
  - Between permissible projections into the ramp width.

1010.5.2 Headroom

- Headroom required for all parts of a ramp in a means of egress is ≥ 6'-8".

1010.5.3 Restrictions

- Means of egress ramps are governed as follows:
  - Width may not reduce in direction of egress travel.
  - Projections are not permitted into required width of ramps and landings.
  - Door swing onto a landing must leave ≥ 3'-6" clear width unobstructed.

1010.6 Landings

- Landings are required for ramps at the following locations:
  - Bottom of ramp.
  - Top of ramp.
  - At changes of direction.
  - At doors.
  - At entrances.
  - At exits.

  Note: 1010.6.1, "Slope," through 1010.6.5, "Doorways," are cited as governing landings.

1010.6.1 Slope

- Changes of level on a ramp landing are not permitted.
- Ramp landings may not slope ≥ 1:48 in any direction.

1010.6.2 Width

- The width of a landing must be ≥ the width of adjoining ramps.

1010.6.3 Length

- Length in the direction of travel required for ramp landings are as follows:
  - Landing length must be ≥ 3' in the following locations:
    - In dwelling and sleeping units as follows:
      - In R-2 and R-3 where not required to be Accessible, Type A or Type B units.

  Note: Section 1107, "Dwelling Units and Sleeping Units," is cited as governing these facilities.

  - Where not on an accessible route, the following applies:
    - Landing length is not required to be > 4'.
  - Landing length must be ≥ 5' in other locations.

1010 Ramps

1010.6.4 Change in direction

- Dimensions of ramp landings at changes in direction are required as follows:
  - ≥ 3' in all dimensions at the following locations:
    In dwelling and sleeping units as follows:
      In R-2 and R-3 where not required to be Accessible, Type A or Type B units.

  Note: Section 1107, "Dwelling Units and Sleeping Units," is cited as governing these facilities.

  - ≥ 5' in all dimensions at other locations.

1010.6.5 Doorways

- Clearances required for accessibility at doorways may overlap required landing dimensions.

  Note: ICC A117.1, "Accessible and Usable Buildings and Facilities," is cited as the source of maneuvering requirements at doorways.

1010.7 Ramp construction

- Ramp materials are governed as follows:
  - Wood handrails are allowed at ramps in all building construction types.
  - Otherwise, all ramp construction must conform to that required for the building construction type.
- Ramps used as exits must meet the same requirements as do exit enclosures.

  Note: 1022.1, "Enclosures required," through 1022.6, "Exit enclosure exterior walls," are cited as governing requirements for exit enclosures.

1010.7.1 Ramp surface

- Materials on surfaces must comply with the following:
  - Must be slip-resistant.
  - Must be securely attached.

1010.7.2 Outdoor conditions

- The design of following walking surfaces has the requirement listed:
  - Surfaces:
    Outdoor ramps.
    Outdoor approaches to ramps.
  - Requirement:
    Design must not permit water to accumulate.

1010.8 Handrails

- Handrails are not required for certain ramped aisles.

  Note: 1028.13, "Handrails," is cited as governing ramped aisles that do not require handrails.

- Other ramps with a rise > 6" must meet the following requirements:
  - Handrails are required on both sides as follows:
    Handrails must meet the same requirements as do stairways.

  Note: Section 1012, "Handrails," is cited as the source of stairway handrail requirements.

1010 Ramps

1010.9 Edge protection

- Edge protection is not required at the following ramps:
  - At ramps where both of the following conditions apply:
    - Handrails are not required.
    - Ramps have flared sides meeting curb ramp requirements.

Note: ICC A117.1, "Accessible and Usable Buildings and Facilities," is cited as governing the pertinent curb ramps.

  - At the sides of ramp landings that meet either of the following:
    - A ramp.
    - A stairway.
  - At the sides of ramp landings with the following condition:
    - Vertical drop at the ramp edge is ≤1/2" over the following distance:
      - A 10" distance measured horizontally from the edge of the required landing area.
  - At the sides of ramps in assembly areas with fixed seating as follows:
    - Where serving adjacent seating.
    - Where serving aisle accessways.
- Other ramps and landings must have one of the following edge protections:
  - Curb or barrier.
  - Extended floor or ground surface.
  - Railing.

Note: The following are cited as governing edge protection listed above as applicable:
1010.9.1, "Curb, rail, wall, or barrier."
1010.9.2, "Extended floor or ground surface."

1010.9.1 Curb, rail, wall or barrier

- One of the following types of edge protection is required with the characteristics noted below:
  - Edge protection:
    - Barrier.
    - Curb.
  - Characteristics:
    - Curb must be ≥ 4" high.
    - Barrier must prevent the passage of a 4" sphere as follows:
      - So that no part of the sphere can pass ≤ 4" from the floor or ground surface.

1010.9.2 Extended floor or ground surface

- The floor or ground surface of the following has the requirement indicated:
  - Surface:
    - Ramp.
    - Ramp landing.
  - Requirement:
    - Surface must extend past the handrail as follows:
      - ≥ 1' from the inside face of the handrail.

Note: Section 1012, "Handrails," is cited as governing the handrail.

1010 Ramps

1010.10 Guards

- Guards are required along ramps and landings more than 2'-6" above floor or grade.
- Guards for ramps and landings must comply with the same construction requirements as for stairways.

Note: Section 1013, "Guards," is cited as the source of requirements for the location and construction of guards for ramps, stairways, and other locations.

1011 Exit Signs

1011.1 Where required

- Exit signs are not required in the following locations:
  - In spaces requiring only one exit or exit access.
  - At main exterior exit doors or gates as follows:
    - Where their identity as exits is obvious.
    - Where approved by the building official.
  - In individual sleeping or dwelling units of the following occupancies:
    - R-1, R-2, R-3.
  - In Occupancy U.
  - In Occupancy I-3 as follows:
    - Dayrooms.
    - Sleeping rooms.
    - Dormitories.
  - On seating side of vomitories or openings to seating areas, where all of the following apply:
    - In occupancies A-4 and A-5.
    - Where there is grandstand seating.
    - Where exit signs are provided as follows:
      - In the concourse.
      - Exit signs are readily apparent from vomitories.
    - Where vomitories or openings are identified by emergency egress lighting.
- In other locations, approved exit signs are required as follows:
  - At exits and exit access doors as follows:
    - Signs must be readily visible from any direction of egress travel.
  - Where exit access is not readily apparent to occupants as follows:
    - Exit signs must indicate the path of egress travel as follows:
      - To exits.
      - Within exits.
    - Exit signs must indicate the direction of egress travel.
  - Exit signs must be located on intervening egress doors within exits.
  - Exit signs must be located as follows in an exit access corridor or exit passageway:
    - So that the distance between any occupant position and the nearest sign within sight is as follows:
      - The distance is to be the shorter of the following:
        - ≤ 100'.
        - ≤ the viewing distance listed for the sign.

1011.2 Illumination

- Lighting for tactile signs is not required.

  Note: 1011.3, "Tactile exit signs," is cited as the source requiring these signs.

- Other exit signs require lighting by one of the following methods:
  - Internal.
  - External.

1011 Exit Signs

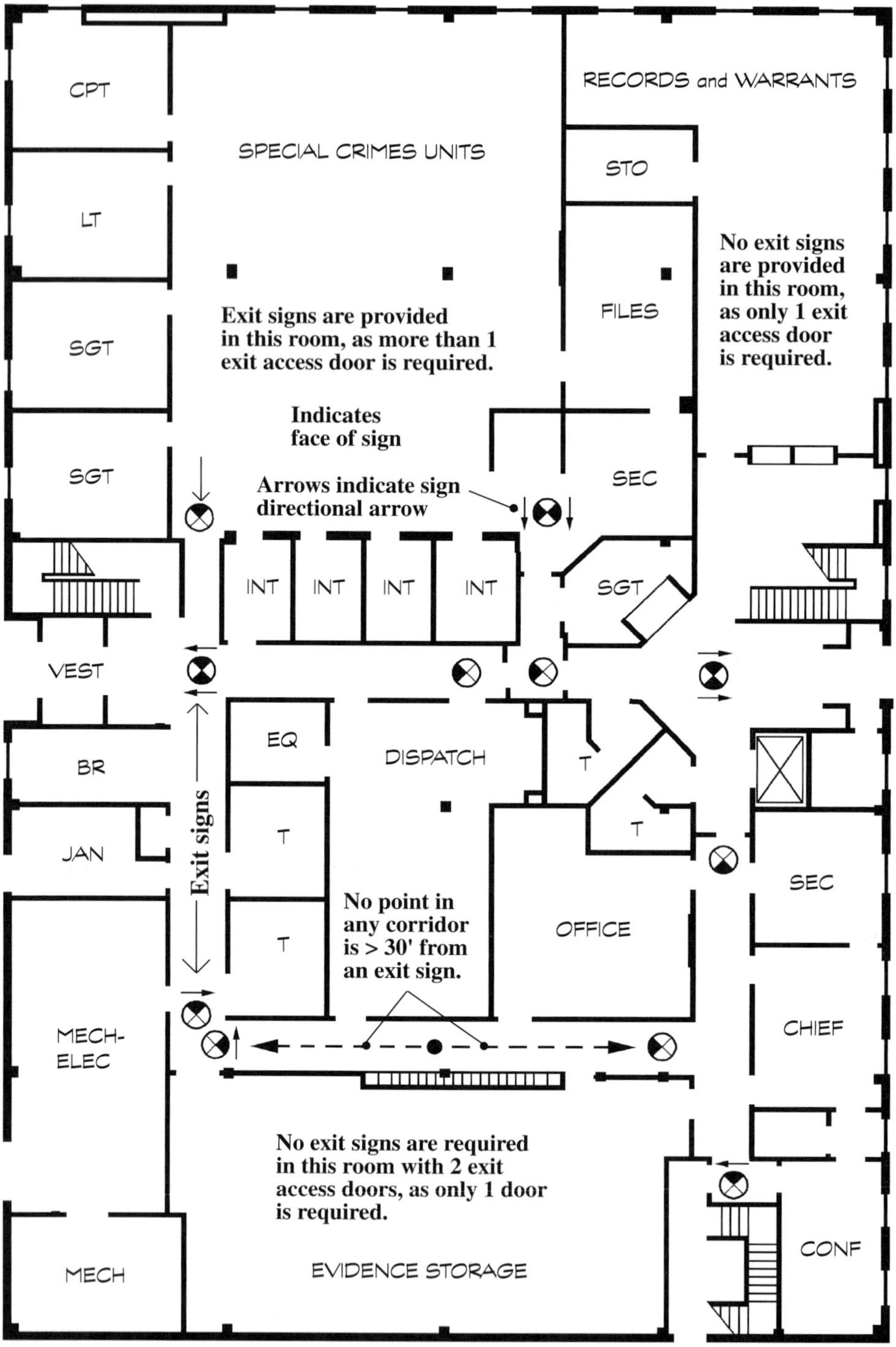

Fig. 1011.1. First-floor plan. Hot Springs Police Department New Headquarters. Hot Springs National Park, Arkansas. Cromwell Architects Engineers. Little Rock, Arkansas.

1011 Exit Signs

1011.3 Tactile exit signs

- Tactile exit signs are required as follows:
  - Sign must say EXIT.
  - Must be adjacent to each door to the following:
    Exit discharge.
    Exit passageway.
    Exit ramp.
    Exit stairs.

Note: ICC A117.1, "Accessible and Usable Buildings and Facilities," is cited as governing these signs.

1011.4 Internally illuminated exit signs

- This section governs the following exit signs:
  - Electrically powered.
  - Self-luminous.
  - Photoluminescent.
- Exit signs governed by this section must be listed and labeled.

Note: UL 924, "Emergency Lighting and Power Equipment," is cited as governing exit signs.

  - Must be installed as per manufacturer's instructions.
  - Must be lit at all times.

Note: Chapter 27, "Electrical," is cited as another source of requirements for these signs.

1011.5.1 Graphics

- Graphics for exit signs is governed as follows:
  - Direction of any chevron direction indicator must not be readily changeable.
  - Letters in the word "EXIT" must have high contrast with background.
  - Must be clearly discernible when illuminated or not.
  - Proportions of letters larger than minimum must be the same as required for minimum size.
- Required sizes of letters in the word "EXIT" are as follows:

Table 1011.5.1 Exit Sign Graphics

Graphics	Dimension
Height	≥ 6"
Width of letter strokes	≥ 3/4"
Width of letters E, X, and T	≥ 2"
Width of letter I	≥ 3/4"
Spacing between letters	≥ 3/8"

1011.5.2 Exit sign illumination

- Exit signs that are illuminated from the outside have the following requirements:
  - The face must have the following amount of light:
    ≥ 5 footcandles.

1011 Exit Signs

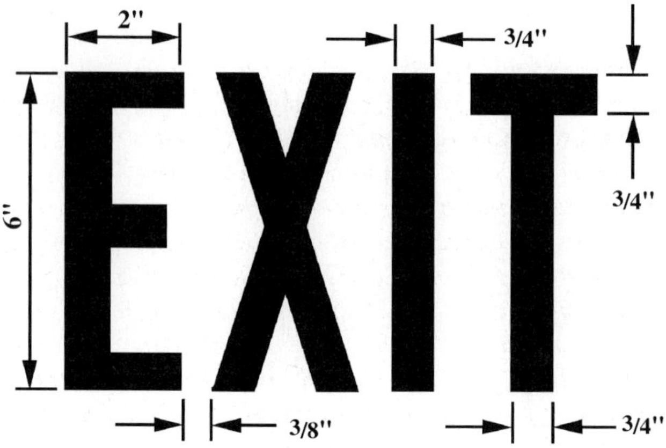

Fig. 1011.5.1A. Exit sign letter dimensions. Required minimums for letter sizes and spacing are shown for exit signs.

Case study: Fig. 1011.5.1B. The signs meet letter size and spacing minimums and other requirements. The arrow direction is not readily changeable, as it is available as a "snap-out" section in the sign face. As required, sign letters (in red) contrast with the sign face (aluminum) when the LED lights are on or off.

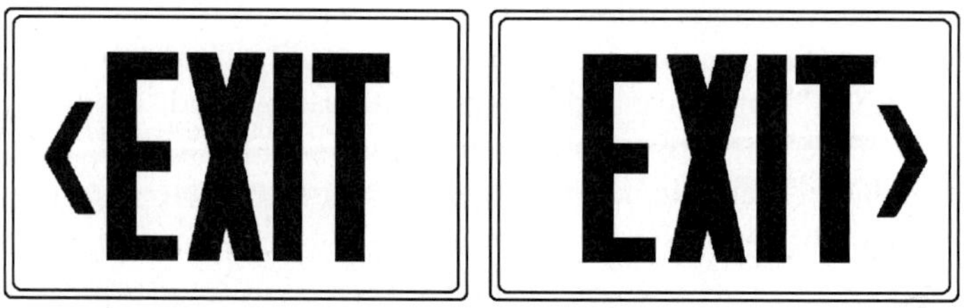

Fig. 1011.5.1B. Typical exit signs. Emergi-Lite. St. Matthews, South Carolina.

1011.5.3 Power source

- Exit signs must be lit at all times.
- Upon power failure signs must remain illuminated ≥ 90 minutes.
- Sign lighting that is not dependent on the primary power source need not be connected to an emergency power supply.
- Sign lighting that is dependent on the primary power source must also be connected to one of the following:
  - Storage battery.
  - Unit equipment.
  - On site generator.

Note: Chapter 27, "Electrical," is cited as governing these signs.

1012 Handrails

1012.2 Height

- Handrail height must be uniform.
- The height of a handrail for stairs and ramps must be in the following range:
  - ≥ 2'-10" and ≤ 3'-2" measured as follows:
    - Above stair tread nosings.
    - Above ramp finished surface.
- The height of a handrail for alternating tread devices and ship ladders must be in the following range:
  - ≥ 2'-6" and ≤ 2'-10" measured as follows:
    - Above stair tread nosings.

1012.3 Handrail graspability

- Required handrails for the following locations are to be Type I or Type II or provide equal graspability:
  - Occupancy R-3.
  - Within R-2 dwelling units.
  - In Occupancy U where accessory to R-3.
  - In Occupancy U where accessory to R-2 individual dwelling units.

  Note: 1012.3.1, "Type I," is cited as governing this type of handrail.
  1012.3.2, "Type II," is cited as governing this type of handrail.

- Other required handrails must be Type I or provide equal graspability.

  Note: 1012.3.1, "Type I," is cited as governing this type of handrail.

1012.3.1 Type I

- Type I handrails must have the following characteristics:
  - Circular handrails must have an outside diameter ≥ 1 1/4" and ≤ 2".
  - Other handrail shapes must comply with the following:
    - Have a perimeter ≥ 4" and ≤ 6 1/4".
    - Have all cross-section dimensions ≤ 2 1/4".
    - Have edges with a radius ≥ 0.01".

1012.3.2 Type II

- Type II handrails must have the following characteristics:
  - Where the perimeter is > 6 1/4" the following applies:
    - A graspable finger recess is required on both sides as follows:
      - Top of recess must begin ≤ 3/4" from the tallest part of the profile (measured vertically).
      - Recess must reach a depth ≤ 5/16" as follows:
        - Measured from the widest part of the profile.
        - At a location ≤ 7/8" below the widest part of the profile.
        - Required depth to continue vertically ≤ 3/8" as follows:
          - To a level ≥ 1 3/4" below the tallest part of the profile.
      - Handrail width above the recess must be ≥ 1 1/4" and ≤ 2 3/4".
    - Edges must have a radius ≥ 0.01".

1012 Handrails

1012.4 Continuity

- Handrail-gripping surface continuity is governed as follows:
  - Within a dwelling unit, the following applies:
    - A newel post may interrupt a handrail at a turn or stair landing.
    - The following details are permitted over the lowest tread:
      - Volute or turnout.
      - Starting easing.
      - Starting newel.
  - The following applies to handrails by walkways that slope ≤ 1:20:
    - The entire length of the bottom of the handrail gripping surface may be obstructed as follows:
      - Where the handrail is integral to either of the following:
        - Bumper guards.
        - Crash rails.
  - The following handrail attachments are allowed where they comply with the restrictions listed below:
    - Attachments:
      - Balusters.
      - Brackets.
    - Restrictions:
      - They must be attached to the handrail's bottom surface.
      - Clearance below the handrail is required as listed in the following table:

Table 1012.4 Clearance Required at Handrail

Circular handrail diameter	Noncircular handrail diameter	Circumference	Clearance below handrail required
1.25"	NA	3.93"	1.500"
1.26" – 1.28"	NA	3.94" – 4.00"	1.500"
1.29" – 1.44"	NA	4.10" – 4.50"	1.375"
1.45" – 1.60"	NA	4.60" – 5.00"	1.250"
1.61" – 1.76"	NA	5.10" – 5.50"	1.125"
1.77" – 1.91"	NA	5.60" – 6.00"	1.000"
1.92" – 2.00"	NA	6.10" – 6.28"	1.000"

  - Otherwise, handrails must be continuous as follows:
    - Without interruptions by the following:
      - Newel posts.
      - Other obstructions.

1012.5 Fittings

- Handrails may not rotate.

1012 Handrails

1012.6 Handrail extensions

- Handrails serving aisles in occupancies A and E are not governed by this section.

  Note: 1028.13, "Handrails," is cited as defining the pertinent handrails.

- Handrails in dwelling units not required to be accessible must extend between the following locations:
  - Between the top and bottom risers.
- The following applies to alternating tread devices and ship ladders:
  - Handrails are permitted to terminate above the top and bottom risers.
  - Handrails are not required to be continuous between flights.
  - Handrails are not required to extend beyond the top or bottom risers.
- Other handrails are governed as follows:
  - They must return to one of the following points:
    - A wall.
    - A guard.
    - The walking surface.
    - Be continuous to the handrail of the adjacent stair or ramp run.
  - Handrails that are not continuous between stair runs must terminate as follows:
    - They must extend horizontally ≥12" beyond the top riser.
    - They must continue to slope beyond the bottom riser equal to a distance of 1 tread.
  - Handrails that are not continuous between ramp runs must terminate as follows:
    - They must extend horizontally above the landing ≥12" as follows:
      - Beyond the top and bottom of the sloped run.
  - Handrail extensions must be in the same direction as the following components as applicable:
    - Stair flights.
    - Ramp runs.

1012.7 Clearance

- Clear space ≥ 1 1/2" is required between a handrail and the following:
  - Wall.
  - Other surface.
- The following surfaces must be without sharp or abrasive elements:
  - Handrail surfaces.
  - Surfaces adjacent to handrail.

1012.8 Projections

- Clear width required between ramp handrails is ≥ 3'.
- Projections into the required width of a stairway or ramp are limited as follows:

Height	Projection permitted
At handrail height	≤ 4 1/2"
Below handrail height	≤ 4 1/2"
Above 6'-8"	Not limited

Note: 1009.2, "Headroom," is cited as specifying minimum headroom which is listed in the above table.

1012 Handrails

1012.9 Intermediate handrails

- Intermediate handrails are required at stairways where necessary as follows:
  - So that all points in the required width are ≤ 2'-6" away from a handrail.
  - Intermediate handrails necessary for selected required widths are indicated below:

Table 1012.9 Intermediate Handrails Required

Required egress width	Intermediate handrails required	Required egress width	Intermediate handrails required
≤ 5'	0	> 15' and ≤ 20'	3
> 5' and ≤ 10'	1	> 20' and ≤ 25'	4
> 10' and ≤ 15'	2	> 25' and ≤ 30'	5

- Handrails must be located on monumental stairs as follows:
  - Along the most direct path of egress travel.

1013 Guards

1013.1 Where required

- This section does not govern certain guards in assembly seating as follows:
  - Where guards are providing meeting requirements specific to the occupancy.

Note: 1028.14, "Assembly guards," is cited as the source of requirements for guards in assembly seating that are not governed by this section.

- Guards are not required in the following locations:
  - On the loading side of loading docks or piers.
  - At stages or raised platforms as follows:
    - On the audience side.
    - At steps.
    - At openings in the floors of performance areas.
    - At elevated walkways providing access to special lighting or equipment.
  - On the following where used for entertainment or presentations:
    - Runways.
    - Ramps.
    - Side stages.
  - Along vehicle service pits where not accessed by the public.
- Guards are required in other locations as follows:
  - Along the following which are > 2'-6" above an adjacent level within 3' of the open side:
    - Open-sided walking surfaces such as follows:
      - Mezzanines.
      - Equipment platforms.
      - Stairs.
      - Ramps.
      - Landings.
  - With adequate strength.

Note: 1607.7, "Loads on handrails, guards, grab bars, seats and vehicle barrier systems," is cited as governing the strength of guards.

1013.1.1 Glazing

- Glazing on the open side of walkways where guards are required is governed as follows:
  - Where the glazing does not meet strength requirements, guards are required along the glazing.

Note: Section 2407, "Glass in Handrails and Guards," is cited as governing glass used for guards.
1607.7, "Loads on handrails, guards, grab bars, seats and vehicle barrier systems," is cited as governing structural requirements for guards.

1013 Guards

1013.2 Height

- This section does not govern the heights of certain guards in assembly seating as follows:

  Note: 1028.14, "Assembly guards," is cited as governing guards in assembly seating.

- Guards in Occupancy R-3 and withing dwelling units of Occupancy R-2 are governed as follows:
  - The following heights are required:
    - ≥ 34" where adjacent to an open side of stairs:
      - Measured vertically from a line connecting the leading edges of the treads.
    - ≥ 34" and ≤ 38" where the top of the guard serves as a handrail as follows:
      - Measured vertically from a line connecting the leading edges of the treads.
- The following applies to alternating tread devices and ship ladders:
  - Guards with a top rail serving as a handrail are governed as follows:
    - The following height is required for the handrail:
      - ≥ 30" and ≤ 34" as follows:
        - Measured vertically from the leading edge of the tread nosing.
- Other guards must have a height ≥ 42" as follows:
  - Measured vertically from the following lower points as applicable:
    - Adjacent walking surface.
    - Adjacent fixed seating.
    - A line connecting the leading edges of the treads.

1013.3 Opening limitations

- Opening sizes in guards must restrict the passage of spheres of the sizes as indicated below.

  Note: This method of measure accounts for space between elements that might not be in a vertical plane.

Table 1013.3 Sizes of Spheres that May Not Pass through a Guard

Location	Sphere size	Height within guard
Individual dwelling units in occupancies R-2 and R-3	4 3/8"	Required height to walking surface
Individual sleeping units in occupancies R-2 and R-3	4 3/8"	Required height to walking surface
Assembly seating at ends of aisles terminating at fascias of boxes, balconies, galleries	4"	< 26" above walking surface
"	8"	≥ 26", ≤ 42" above walking surface
Walking surfaces for access to mechanical electrical, and plumbing systems or equipment	21"	Required height to walking surface
In nonpublic areas of occupancies I-3, F, H, S and all alternating tread devices	21"	Required height to walking surface
Stairs	6"	In triangular space between bottom rail and treads and risers
Other locations	4"	< 36" above the walking surface
"	4 3/8"	≥ 36", ≤ 42" above the walking surface

1013 Guards

Case study:
Fig. 1013.3. 4", 6", and 8" spheres cannot pass through the guard at heights specified by 1013.3. Circles superimposed on the illustration are to scale and illustrate their relationship to openings. The centerline of the top rail is at 42" above tread nosings, thus meeting the minimum height requirement of 42" to the top of the rail, as specified in 1013.2. The guard is in full compliance with the code.

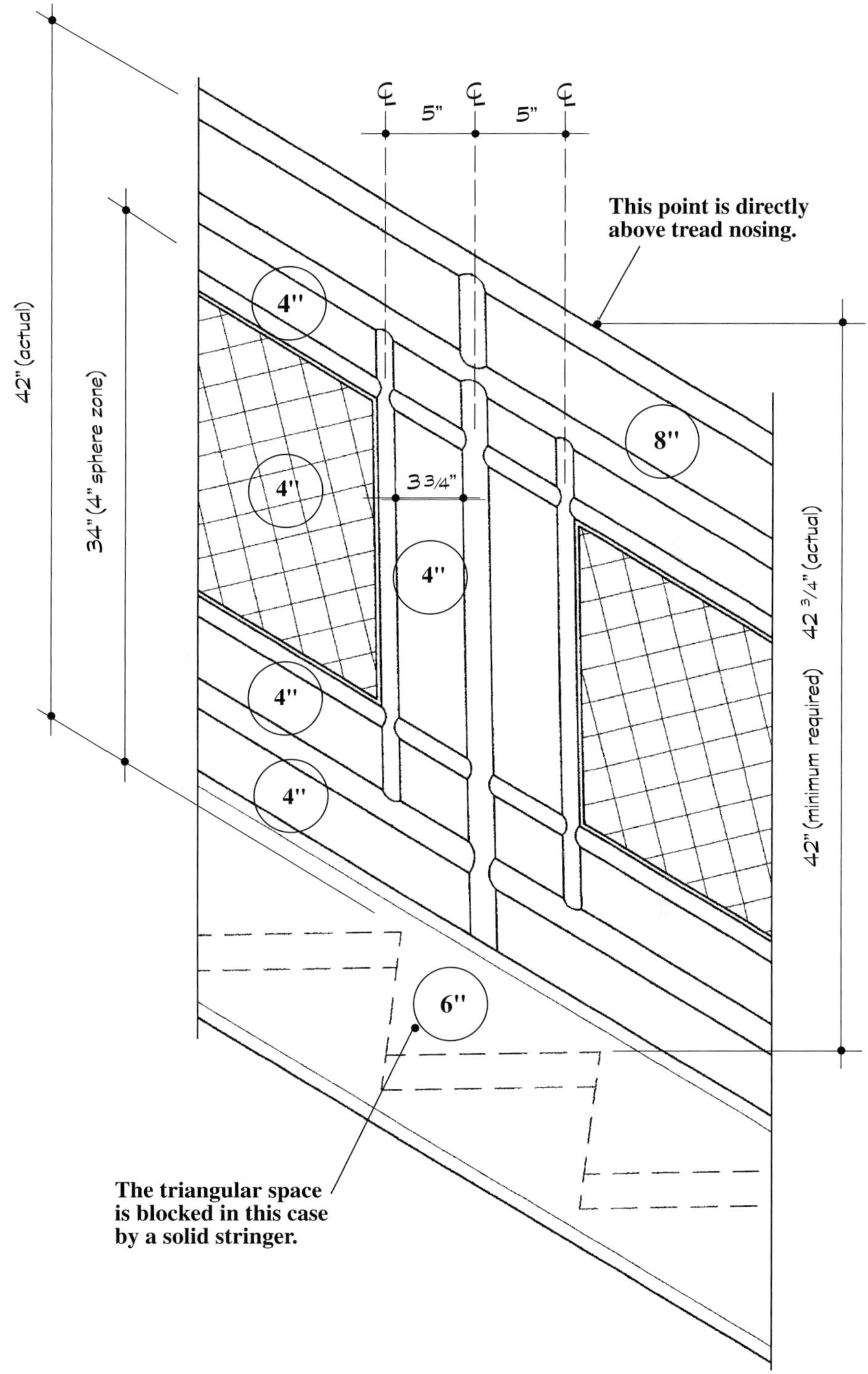

Fig. 1013.3. Elevation of guard at stairs. Montachusett Regional Vocational-Technical High School. Fitchburg, Massachusetts. HKT Architects, Inc. Somerville, Massachusetts.

1013 Guards

1013.4 Screen porches

- Screened-in porches or decks are governed as follows:
  - Guards are required in the following case:
    - Where the floor is > 2'-6" above the adjacent floor or grade.

1013.5 Mechanical equipment

- The following elements have the requirements listed below:
  - Elements:
    - Appliances.
    - Equipment.
    - Fans.
    - Other devices needing service.
    - Roof hatch.
  - Requirements:
    - A guard is required as follows:
      - Where the elements are ≤ 10' from an edge that is > 2'-6" above the surface below.
      - Guard must not pass a 21" sphere.

1013.6 Roof access

- Guards are required for a roof hatch as follows:
  - Where the hatch is ≤ 10' from an edge that is > 2'-6" above the surface below.
  - Guard must not pass a 21" sphere.

1014 Exit Access

1014.2 Egress through intervening spaces

- Egress through an intervening space in the following occupancies has the requirements listed below:
  - Occupancies:
    - H, S, F.
  - Requirements:
    - Such egress is permitted where the intervening space is one of the following:
      - In an occupancy of equal hazard to the space where egress is begun.
      - In an occupancy of lower hazard to the space where egress is begun.
- Egress through storage rooms is governed as follows:
  - Egress through Occupancy M stockrooms is permitted if all of the following conditions exist:
    - The hazard of the stock in storage is the same as that of the stock in the retail area.
    - This route constitutes ≤ half of the exit access.
    - The stockroom cannot be locked from the egress side.
    - A path must exist through the stockroom as follows:
      - The path must be distinct and obstruction free.
      - The path must be ≥ 3'-8" wide.
  - In other locations, egress may not pass through a storage room.
- Egress from an adjoining space may not pass through the following spaces:
  - Closets.
  - Similar spaces.
- Egress though a kitchen from a adjoining space is governed as follows:
  - Egress is permitted in the following locations:
    - Within a dwelling unit.
    - Within a guest room.
  - In other cases, egress through a kitchen is not permitted.
- Egress may not pass through a room that can be locked so as to obstruct egress.
- Egress from dwelling units or sleeping areas may not pass through the following:
  - Other sleeping areas.
  - Toilet rooms.
  - Bathrooms.
- In other cases, the following applies:
  - Egress from a space may pass through an intervening space where all of the following conditions exist:
    - Where either the space or the adjoining space is accessory to the other.
    - Where the adjoining space is not an H occupancy.
    - Where adjoining space provides a readily apparent route to an exit.
  - Otherwise, egress through an intervening space is not permitted.

1014 Exit Access

Case study: Fig. 1014.2A–F. Examples on the following pages indicate a variety of intervening rooms in several occupancies though which egress travel may pass or may not pass. Those rooms through which egress travel is not permitted cannot be counted as a required means of egress in order to meet the number required, the total width required, in measuring minimum travel distances, or for other purposes. In all cases illustrated, means of egress requirements are met by the plans with appropriate routes some of which are marked and some of which are not indicated. Arbitrary starting points are marked with a dot, and each means of egress analysis applies only to the room so marked.

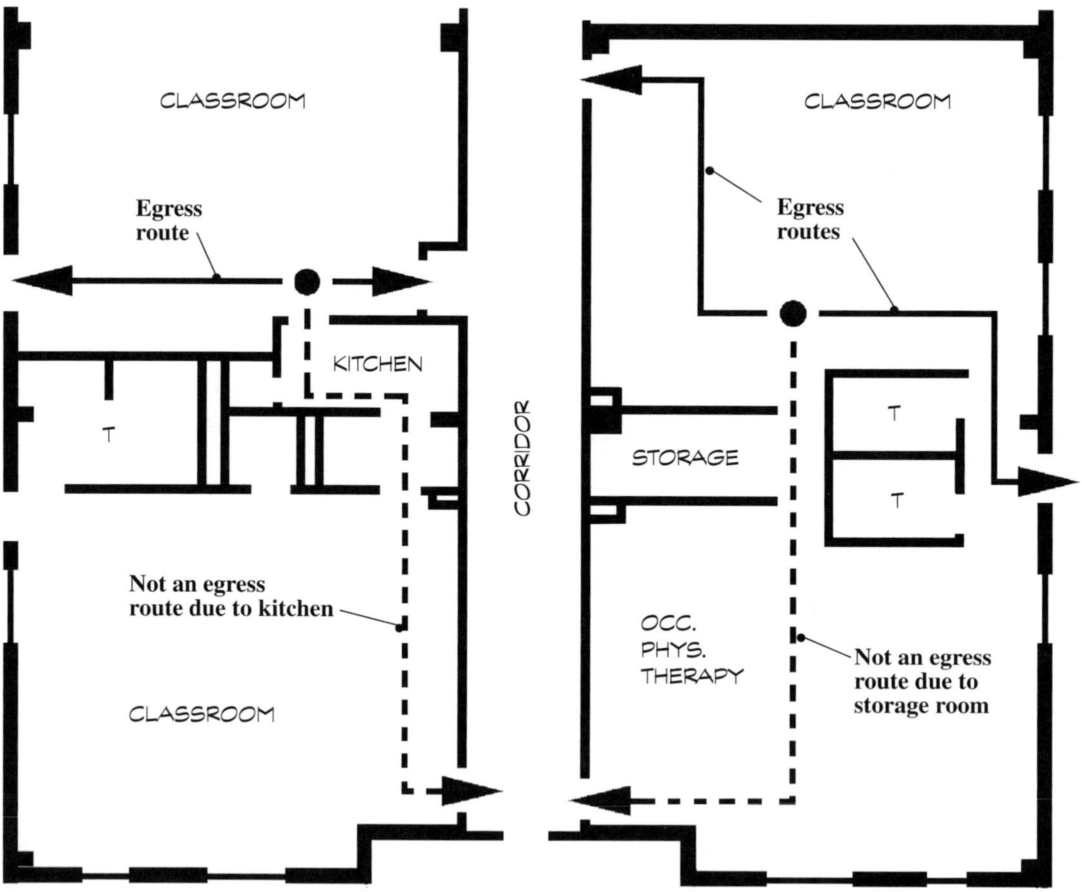

Fig. 1014.2A. Partial floor plan. New Jasper Pre-K – 2nd Grade School. Jasper, Texas. PBK Architects, Inc. Houston, Texas.

1014 Exit Access

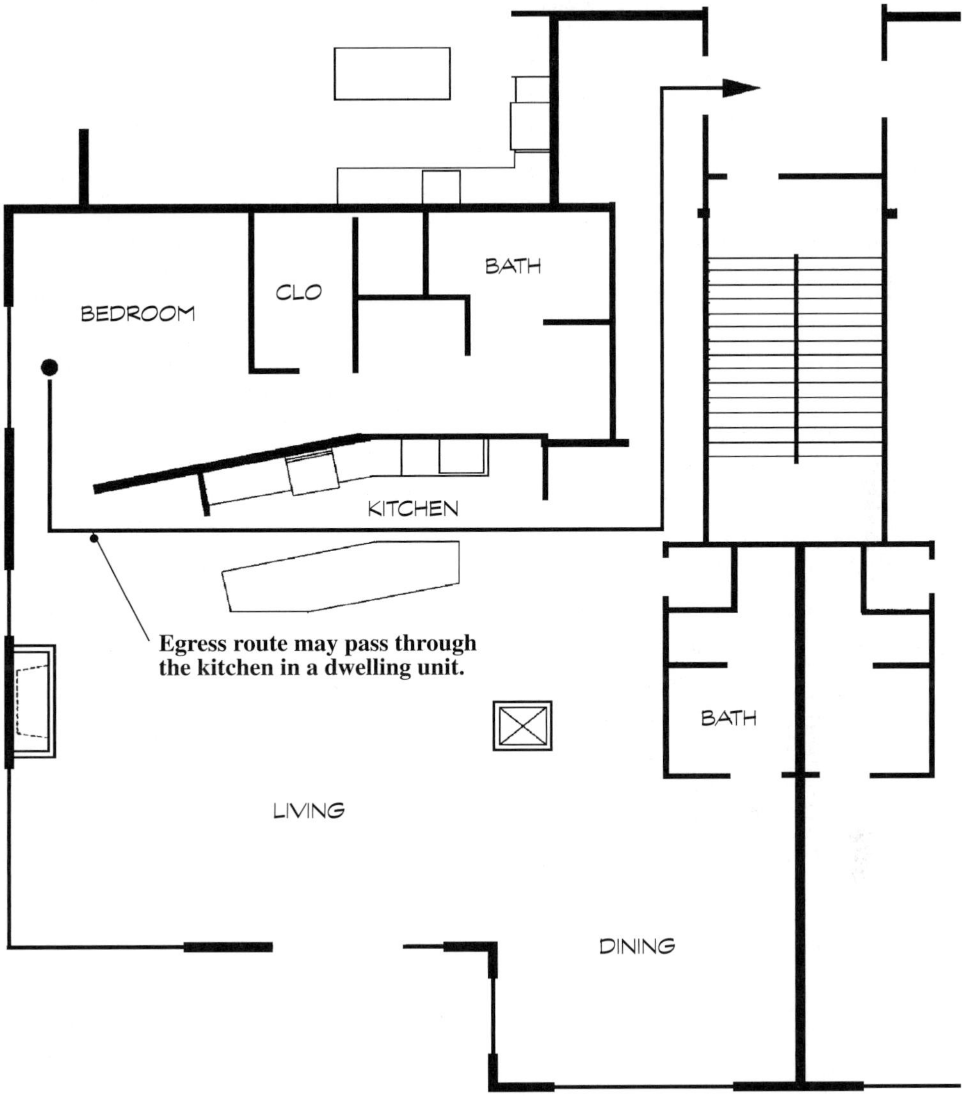

Fig. 1014.2B. Apartment floor plan. McKenzie Lofts**.** Portland, Oregon. Ankrom Moisan Associated Architects. Portland, Oregon.

1014 Exit Access

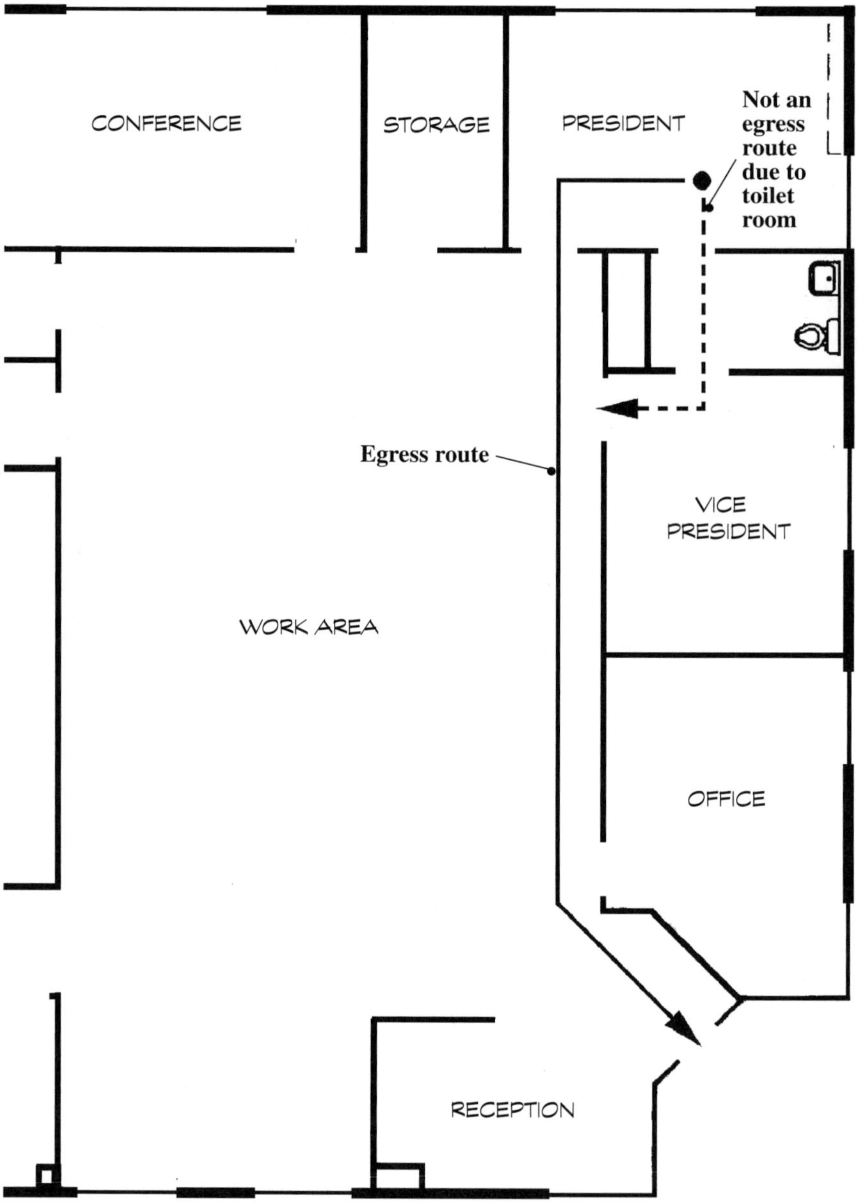

Fig. 1014.2C. Partial floor plan. Garments to Go. Bastrop, Texas. Spencer Godfrey Architects, Round Rock, Texas.

1014 Exit Access

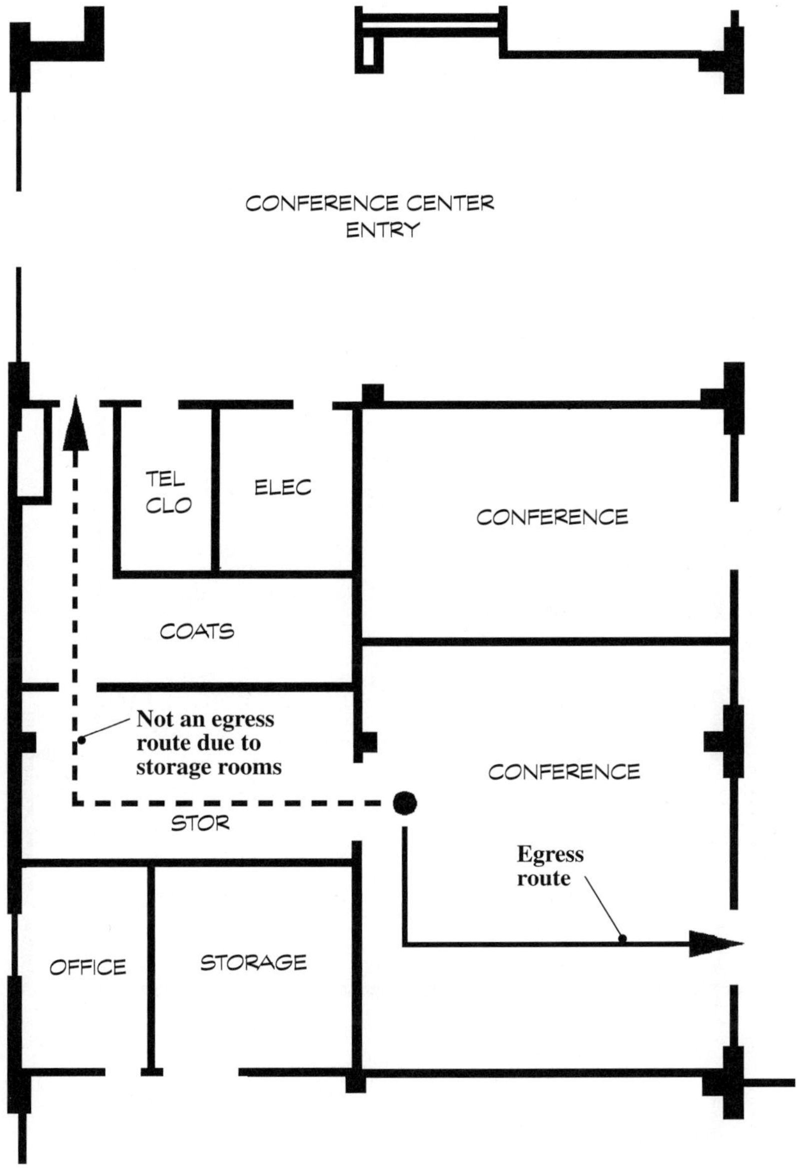

Fig. 10142D. Partial floor plan at conference center. University of Connecticut New Downtown Campus at Stamford, Connecticut. Perkins Eastman Architects, New York, New York.

1014 Exit Access

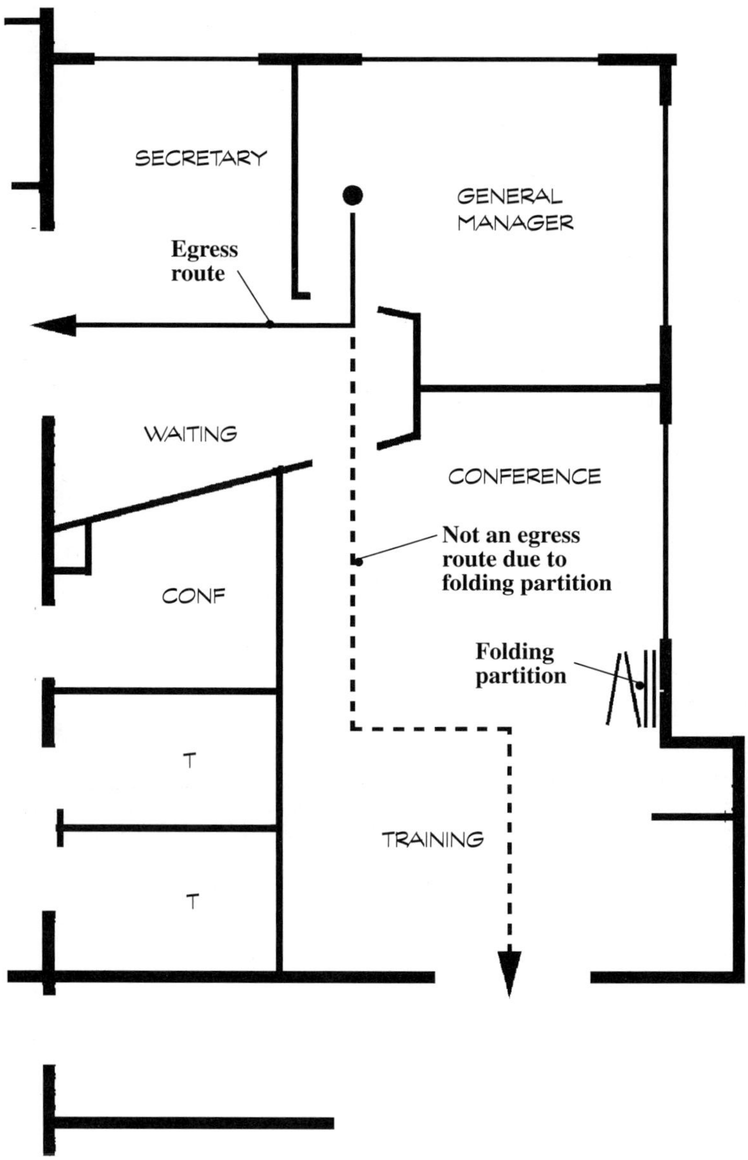

Fig. 1014.2E. Partial floor plan. Wichita Transit Storage Administration, and Maintenance Facility. Wichita, Kansas. Wilson Darnell Mann, P.A., Architects. Wichita, Kansas.

1014 Exit Access

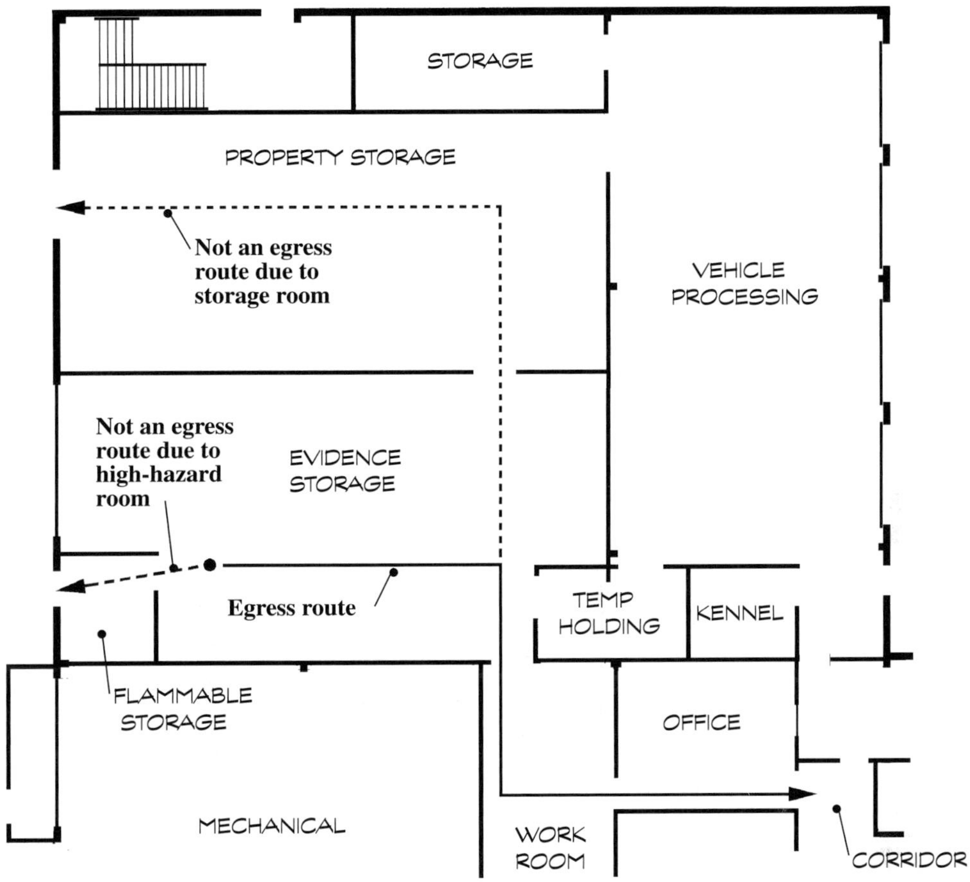

Fig. 1014.2F. Partial floor plan. Lee's Summit Police and Court Facility. Lee's Summit, Missouri. The Hollis and Miller Group, Inc. Lee's Summit, Missouri.

1014 Exit Access

1014.2.1 Multiple tenants

- This section addresses locations where more than one tenant occupies a floor.
- Means of egress from a smaller tenant space may pass through an adjoining larger tenant space where all of the following conditions exist:
  - The smaller space is ≤ 10% of the larger adjoining space.
  - The two tenant spaces are the same or similar occupancy.
  - An identifiable egress path exists through the larger adjoining space.
  - A barrier cannot be locked so as to prevent egress from the smaller space.
- A required means of egress from a larger tenant space is governed as follows:
  - It may not pass through a smaller adjoining tenant space.
- Otherwise the following spaces have the requirements listed below:
  - Spaces:
    - Tenant space.
    - Dwelling units.
    - Sleeping units.
  - Requirements:
    - Access to required exits may not pass through the following adjacent spaces:
      - Tenant space.
      - Dwelling units.
      - Sleeping units.

1014.2.2 Group I-2

- Exit access doors leading directly to a corridor are not required as follows:
  - For rooms with exit doors as follows:
    - Opening directly to the exterior.
    - Opening at ground level.
- In other habitable rooms or suites in Occupancy I-2, the following is required:
  - An exit access door leading directly to a corridor.

1014.2.3 Suites in patient sleeping areas

- I-2 patient sleeping areas may be divided into suites as follows:
  - They may have 1 intervening room where the following conditions exist:
    - Intervening room does not serve as exit access for > 8 patient beds.
    - Direct continual visual supervision by nursing staff is possible.

1014.2.3.1 Area

- Sleeping room suites are limited to ≥ 5,000 sf ea.

1014.2.3.2 Exit access

- The following are required to have ≥ 2 exit access doors:
  - Patient sleeping rooms > 1,000 sf.
  - Patient sleeping room suites > 1,000 sf.
- Exit access doors required by this section must be located remote from each other.

1014.2.3.3 Travel distance

- Travel distance in a sleeping room suite must be ≤ 100 ft as follows:
  - From any point to an exit access door.

1014 Exit Access

1014.2.4 Suites in areas other than patient sleeping areas

- This section governs areas in I-2 other than patient sleeping areas.
- Area governed by this section may be divided into suites.

1014.2.4.1 Area

- I-2 suites that are not for patient sleeping are limited to the following:
  - 10,000 sf.

1014.2.4.2 Exit access

- This section governs I-2 areas that are not for patient sleeping.
- Areas governed by this section are required to have ≥ 2 exit access doors as follows:
  - Rooms > 2,500 sf.
  - Suites of rooms > 2,500 sf.
- Exit access doors required by this section must be located remote from each other.

1014.2.4.3 One intervening room

- This section governs I-2 suites that are not for patient sleeping.
- Suites governed by this section may have 1 intervening room as follows:
  - Where travel distance is ≤ 100 ft as follows:
    From any point in the suite to an exit access door.

1014.2.4.4 Two intervening rooms

- This section governs I-2 suites that are not for patient sleeping.
- Rooms in suites governed by this section may have 2 intervening rooms as follows:
  - Where travel distance is ≤ 50 ft as follows:
    From any point in the room to an exit access door.

1014.2.6 Travel distance

- This section governs I-2 patient sleeping rooms.
- Travel distance in a sleeping room must be ≤ 50 ft as follows:
  - From any point in the room to an exit access door in the room.

1014.2.7 Separation

- Suites in I-2 must be isolated by smoke partitions.

Note: Section 711, "Smoke Partitions," is cited as governing these partitions.

1014 Exit Access

1014.3 Common path of egress travel

- The length of common path egress travel is limited as shown below:

Table 1014.3 Common Path Distance Limits

Tenant spaces:

		Common path distance	
Occupancy	Occupant load	Buildings sprinklered	Buildings not sprinklered
B, S, U	≤ 30	≤ 100'	≤ 100'
B, S	> 30	≤ 100'	≤ 75'
U	> 30	≤ 75'	≤ 75'

Locations other than tenant spaces:

	Common path distance	
Occupancy	Buildings sprinklered	Buildings not sprinklered
A without fixed seating	≤ 75'	≤ 75'
E, I-1, I-2, I-4, M, R-1, R-3, R-4, U	≤ 75'	≤ 75'
R-2	≤ 125'	≤ 75'
B, F, S	≤ 100'	≤ 75'
H-1, H-2, H-3	≤ 25'	—
H-4, H-5	≤ 75'	—
I-3	≤ 100'	≤ 100'

Note: 1028.8, "Common path of egress travel," is cited as governing Occupancy A and assembly accessory to Occupancy E with fixed seating.
903.3.1.1, "NFPA 13 sprinkler systems," is cited as governing sprinkler systems as applicable.
903.3.1.2, "NFPA 13R sprinkler systems," is cited as governing sprinkler systems as applicable.

1014 Exit Access

Case study: Fig. 1014.3A–B. The shortest common paths of egress travel are shown from remote points in each area. Where a room or area has a single exit door or a single exit access door, all routes in the space are common paths. The termination of a common path arrow indicates the point at which the common path ends. Where this occurs prior to reaching an exit, it indicates the first point encountered in the path at which the occupant has a choice of two routes to separate exits. Diagonal paths are through spaces where no furnishings will block such a route. Otherwise, travel is measured on a rectangular pattern.

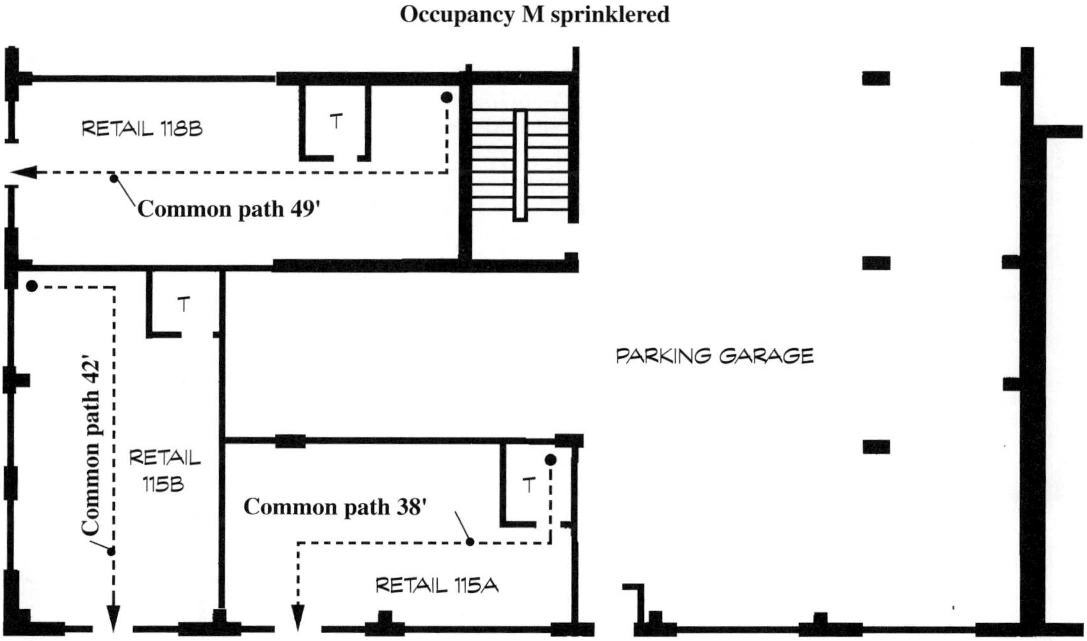

Fig.1014.3A. Partial floor plan at shops. McKenzie Lofts. Portland, Oregon. Ankrom Moisan Associated Architects. Portland, Oregon.

1014 Exit Access

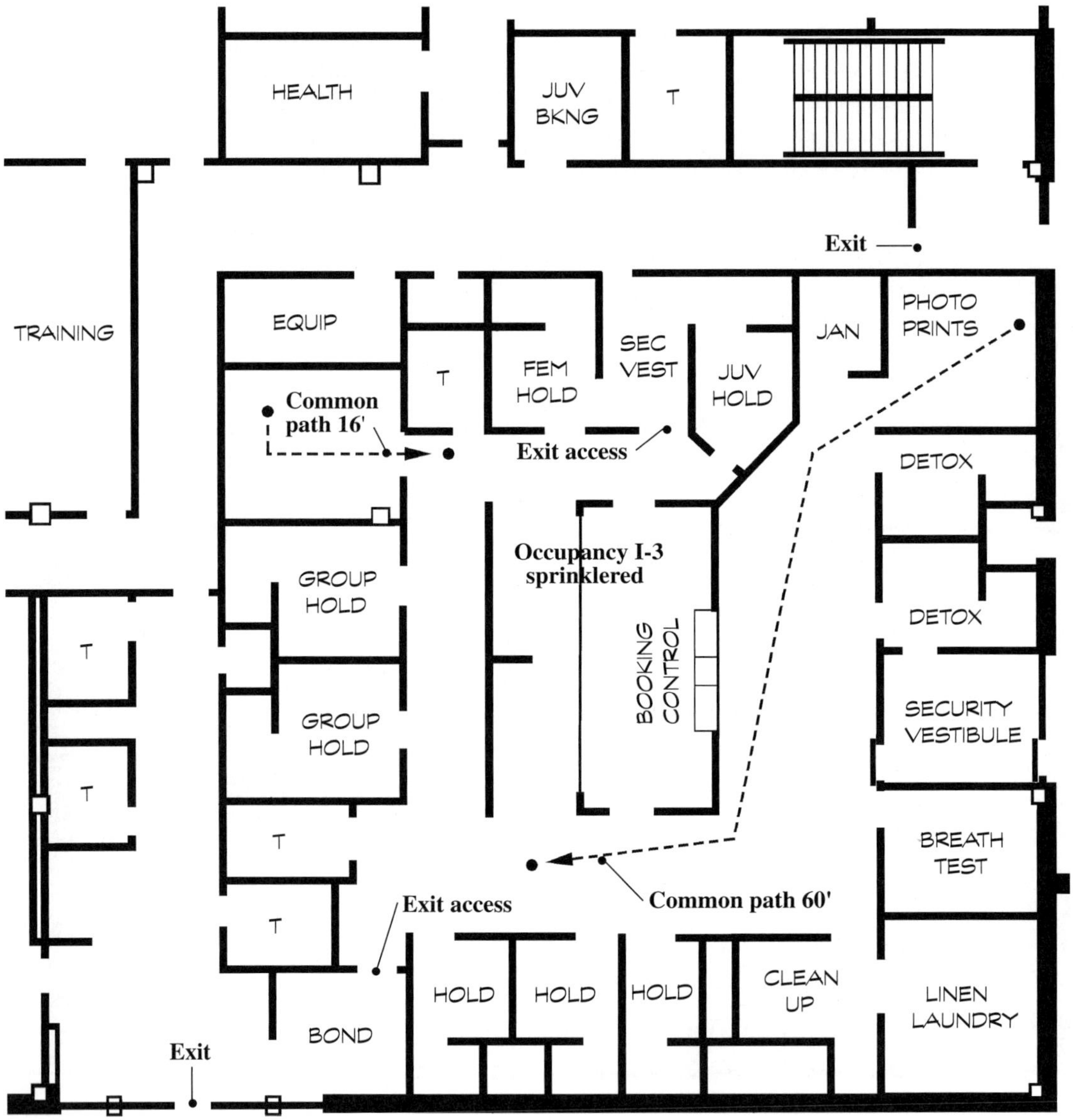

Fig.1014.3B. Partial floor plan. Lee's Summit Police and Court Facility. Lee's Summit, Missouri. The Hollis and Miller Group, Inc. Lee's Summit, Missouri.

1015 Exit and Exit Access Doorways

1015.1 Exit or exit access doorways from spaces

- Occupancy I-2 is not governed by this section.

  Note: 1014.2.2, "Group I-2," through 1014.2.7, "Separation," are cited as governing I-2.

- The following locations permit 1 means of egress:
  - Within and from R-2 and R-3 dwelling units where all of the following conditions are met:
    Occupant load ≤ 20.
    Dwelling is sprinklered.

  Note: 903.3.1.1, "NFPA 13 sprinkler systems," is cited as governing sprinkler systems as applicable.
  903.3.1.2, "NFPA 13R sprinkler systems," is cited as governing sprinkler systems as applicable.

- Other exit or exit access doorways are governed as follows:
  - ≥ 2 exit or exit access doorways are required in the following cases:
    For the following spaces given certain conditions:

Boiler rooms	Refrigerated rooms or spaces	Stage galleries
Furnace rooms	Refrigeration machinery rooms	Stage grid irons
Incinerator rooms	Stage catwalks	Stages

  Note: The following are cited as requiring 2 exits or exit access doorways for certain conditions:
  1015.3, "Boiler, incinerator and furnace rooms."
  1015.4, "Refrigeration machinery rooms."
  1015.5, "Refrigerated rooms or spaces."
  1015.6, "Stage means of egress."
  1015.6.1, "Gallery, gridiron and catwalk means of egress."

    Where the common path of egress travel would otherwise exceed that permitted.

  Note: 1014.3, "Common path of egress travel," is cited as defining limits of common path travel.

    In other occupancies where the occupant load is as follows:

Table 1015.1 Spaces Requiring ≥ 2 Exits or Exit Access Doors

Occupancy	Occupants
A, B, E (no day care), F, M, U	> 49
E (day care)	> 10
H-1, H-2, H-3	> 3
H-4, H-5, I-1, I-3, I-4, R	> 10
S	> 29

Source: IBC Table 1015.1

- Each individual occupancy in mixed occupancy buildings must comply with the following:
  - Requirements for each individidual occupancy.
- Occupant load for determining number of exits includes the following:
  - The load of occupancies exiting through a subsequent occupancy in the path of egress.

  Note: 1004.1, "Design occupant load," is cited as governing occupant count.

1015 Exit and Exit Access Doorways

Case study: Fig. 1015.1. The church organ loft has an occupant load of 5 based on gross area. Since this is < 50 and the common path distance is < the 75' maximum for the Occupancy B space, only 1 exit or exit door is required.

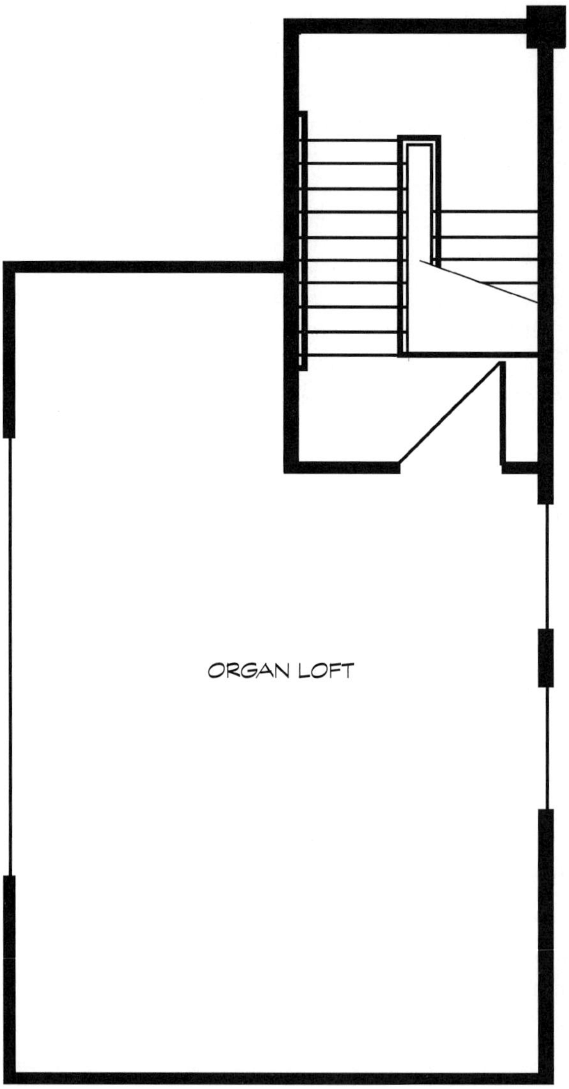

Fig. 1015.1. Floor plan of organ loft. Glad Tidings Assembly of God Church. Naticoke, Pennsylvania. Mullins and Weida, Architect and Associate. Bear Creek, Pennsylvania.

1015 Exit and Exit Access Doorways

1015.1.1 Three or more exits or exit access doorways

- ≥ 3 exit or exit access doorways are required from spaces as follows:
  - Where the occupant load is ≥ 501 and ≤ 1,000.
- ≥ 4 exit or exit access doorways are required from spaces as follows:
  - Where the occupant load is > 1,000.

1015.2 Exit or exit access doorway arrangement

- Required exits must be clearly available for their purpose.
- Exits may not be obstructed at any time.
- Exits and exit access doorways must be separated by the following distance:
  - A distance large enough to prevent the loss of more than one in an emergency.

Note: The following are cited as sources of requirements for separation distances:
1015.2.1, "Two exits or exit access doorways."
1015.2.2, "Three or more exits or exit access doorways."

1015.2.1 Two exits or exit access doorways

- Where there are 2 exit enclosures meeting the following conditions, separation is required as listed below:
  - Exit enclosures connected to each other by the following:
    A corridor with a fire-resistance rating of 1 hr.

Note: Section 1018, "Corridors," is cited as governing these corridors.

  - Requirements:
    The required distance separating the exits is measured as follows:
      On the shortest line of travel in the corridor.
- In sprinklered buildings the following applies:
  - The distance between 2 exits or exit access doors is required to be as follows:
    ≥ 1/3 the greatest overall diagonal dimension of the area served.

Note: 903.3.1.1, "NFPA 13 sprinkler systems," is cited as governing sprinkler systems as applicable.
903.3.1.2, "NFPA 13R sprinkler systems," is cited as governing sprinkler systems as applicable.

- In other cases, where 2 exits or exit access doorways are required, they must be separated as follows:
  - By a distance ≥ 1/2 the greatest overall diagonal dimension of the building or area served:
    The distance is measured on a straight line between the exits or exit access doorways.
- The following stairways are counted as one exit:
  - Interlocking stairs.
  - Scissor stairs.

1015.2.2 Three or more exits or exit access doorways

- The following apply where ≥ 3 exits are required:
  - The distance required between at least 2 of the exits or exit access doorways is as follows:
    The same distance that is required between exit or exit access doorways where 2 are required.

Note: 1015.2.1, "Two exits or exit access doorways," is cited as governing the separation.

1015 Exit and Exit Access Doorways

Case study: Fig. 1015.2.1. Since the space on each side of the folding partition in the multipurpose room requires an exit access door, the whole space with the open partition will have 2 doors regardless of its occupant load. To determine whether or not the distance between the 2 doors must meet the code minimum, the whole space is analyzed as a single room with the partition open. As an assembly space with chairs and tables, the occupant load would be 98. The storage room adds 1 additional occupant. Consequently, 2 exit access doors are required and must be spaced a minimum distance apart. In this sprinklered building, the minimum distance is 1/3 the diagonal of the area or 28'-4". Since double doors are provided and, in this example only a single door is required at each location, an argument could be made for measuring between the most remote doors. The closest doors comply with the code, in any case, as they are 38' apart.

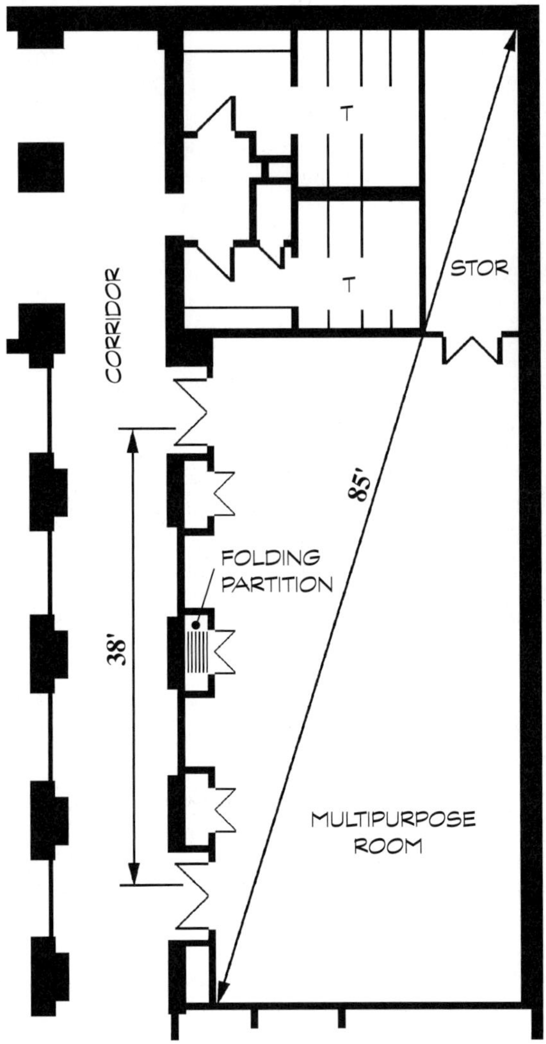

Fig. 1015.2.1. Partial floor plan. Lady Bird Johnson Wildflower Center. Austin, Texas. Overland Partners, Inc. San Antonio, Texas.

1015 Exit and Exit Access Doorways

1015.3 Boiler, incinerator and furnace rooms

- The following rooms require ≥ 2 exit access doorways where both of the conditions listed below apply:
  - Rooms:
    Boiler.
    Incinerator.
    Furnace.
  - Conditions:
    Area > 500 sf.
    Fuel-fire equipment > 400,000 Btu input capacity.
- Where ≥ 2 exit access doorways are required, the following applies:
  - 1 exit access doorway may be accessed by the following:
    By a fixed ladder.
    By an alternating tread stair.
  - Exit access doorways must be separated by a horizontal distance as follows:
    By a distance ≥ 1/2 the greatest diagonal dimension of the room.

1015.4 Refrigeration machinery rooms

- ≥ 2 exits or exit access doors are required in machinery rooms > 1,000 sf.
- Where ≥ 2 exit access doorways are required, the following apply:
  - 1 may be accessed by the following:
    By a fixed ladder.
    By an alternating tread stair.
  - Doorways must be separated by a horizontal distance as follows:
    By a distance ≥ 1/2 the greatest horizontal dimension of the room.
- Travel distance in the room to an exit or exit access doorway is limited as follows:
  - ≤ 150'.

Note: 1016.1, "Travel distance limitations," is cited as the source of potential increase to the travel distance limit.

- Doors must swing in the direction of egress travel for any occupant load.
- Doors must fit tightly and be self-closing.

1015.5 Refrigerated rooms or spaces

- Room with limited amounts of refrigerants are not governed by this section.

Note: The International Mechanical Code *is cited governing these rooms.*

- Otherwise, ≥ 2 exits or exit access doors are required in rooms with all of the following characteristics:
  - > 1,000 sf.
  - Contains a refrigerant evaporator.
  - Has a room temperature < 68° F.
- Egress travel may pass through adjoining refrigerated spaces.
- Where not sprinklered, travel distance in the room to an exit or exit access doorway is ≤ the smaller of the following:
  - 150'.
  - As required for building occupancy.

Note: 1016.1, "Travel distance limitations," is cited as governing travel distance based on occupancy.

1015 Exit and Exit Access Doorways

1015.6 Stage means of egress

- A means of egress must be located on each side of a stage as follows:
  - Where ≥ 2 are required based on either of the following:
    - Stage size.
    - Occupant load.

1015.6.1 Gallery, gridiron, and catwalk means of egress

- This section addresses the means of egress from the following:
  - Lighting catwalks.
  - Access catwalks.
  - Galleries.
  - Gridirons.
- Lighting and access catwalks must have a width ≥ 22".
- The following are permitted in the means of egress:
  - Spiral stairs.
  - Open stairs.
  - Stairs with width ≥ 22".
  - Ladders.
- > 1 means of egress are not required as follows:
  - Where a means of escape is provided to the following locations by the devices indicated below:
    - Locations:
      - To a roof.
      - To a floor.
    - Devices:
      - Ladders.
      - Alternating tread stairs.
      - Spiral stairs.
- Otherwise, means of egress requirements are the same as those for Occupancy F-2.

1016 Exit Access Travel Distance

1016.1 Travel distance limitations *(part 1 of 3)*

- All exit access travel distance addressed in this section is governed as follows:
  - It is measured along a normal unobstructed route of circulation.
  - Where applicable and where an open stair or ramp is in the exit access path, the following applies:
    - Travel distance includes the length along any open stair or ramp in the exit access path as follows:
      - Measured on the centerline of a stair.
      - Measured on a line tangent to the tread nosings of a stair.
    - Travel distance on connecting floors is included.
  - Travel distance is affected by the following as summarized in this section:
    - Occupancy.
    - Sprinkler availability.

Note: IBC Table 1016.1, "Exit Access Travel Distance," is cited as limiting travel distance.

- Travel distance is measured as follows in the cases cited:
  - To the nearest riser of an open exit stairway at the following:
    - An open parking garage.
    - An outdoor facility with the following components:
      - Open exit access components.
      - Exterior exit stair.
  - To the nearest slope of an exterior exit ramp where located at the following:
    - An outdoor facility with open exit access components.
- The travel distance below applies to ≤ 1/2 the exits in the following facilities:
  - Facilities:
    - Occupancies A, B, E, F, M, R, S, U.
    - With open exit stairways or ramps in the exit access path as follows:
      - Connecting two stories as follows:
        - That have ≥ 2 means of egress.
        - Not open to other stories.
  - Travel distance:
    - Measured between the following two points:
      - The most remote point on the upper floor.
      - An exit on the lower floor.
- The travel distance below applies to the exits in the following facilities:
  - Facilities:
    - Occupancies A, B, E, F, M, R, S, U.
    - Where the entire building is sprinklered.

Note: 903.3.1.1, "NFPA 13 sprinkler systems," is cited as governing the sprinklers.

    - With unenclosed stairways or ramps in the exit access path as follows:
      - Connecting the 1st and 2nd stories above grade plane as follows:
        - That have ≥ 2 means of egress.
        - Not open to other stories.
  - Travel distance:
    - Measured between the following two points:
      - The most remote point on the upper story.
      - An exit on the lower story.

1016 Exit Access Travel Distance

1016.1 Travel distance limitations *(part 2 of 3)*

- In other cases exit access travel distance is measured between the following two points:
  - The most remote location in a story and an exit.
- Travel distances are limited by other sections of the code for certain locations.

Note: The following are cited in footnotes to IBC Table 1016.1, "Exit Access Travel Distance," as governing travel distance in certain locations:

402.4, " Means of egress," limits travel distance by way of the following sections:

402.4.2, "Number of means of egress," limits travel distance to 75' where there is only 1 means of egress for other than employees from any point in a tenant space to the mall.

402.4.4, "Distance to exits," limits travel distance to 200' from any point in a mall tenant space to a mall exit or entrance. It also limits travel distance to 200' from any point in a mall to an exit.

404.9, "Travel distance," limits travel distance to 200' in a means of egress through an atrium above the lowest level is 200'.

407.4, "Smoke barriers," limits travel distance to 200' from any point in a smoke barrier protected area (for patient sleeping or patient treatment or floors with 50 or more occupants) in Occupancy I-2 to a smoke barrier door.

408.6.1, "Smoke compartments," limits travel distance to 150' from any room exit access door in a smoke compartment required by this section to a door to the smoke compartment in Occupancy I-3. Maximum distance from any point in a room to a door to the smoke compartment is 200'.

408.8.1, "Occupancy conditions 3 and 4," 50' travel distance from a sleeping area through a common space to a corridor in I-3 is the threshold for a requirement for smoke-tight partitions between the sleeping area and adjacent common spaces.

411.4, "Automatic sprinkler system." A travel distance > 50' from any point to an exit in temporary special amusement buildings < 1,000 sf is the threshold for requiring sprinklers.

1014.2.2, "Group I-2," requires that habitable rooms or suites have doors directly to corridors and/or to the exterior. Note also that 1014.2.4.3, "One intervening room," says that in rooms other than patient sleeping rooms in I-2 suites, the path of travel is permitted through one intervening room where travel distance to an exit access door is ≤ 100' and 1014.2.4.4, "Two intervening rooms," adds that egress travel through two intervening rooms is permitted where the travel distance is ≤ 50'.

1015.4, "Refrigeration machinery room," limits travel distance to 150' from any point in a refrigeration machinery room > 1,000 sf to an exit or exit access door is 150'. The travel distance can be increased to that specified for sprinklered buildings in IBC Table 1016.1 according to the occupancy.

1015.5, "Refrigerated rooms or spaces," limits travel distance from any point in a refrigeration area of 1,000 sf or more (and other specifications) to an exit or exit access door is 150' for unsprinklered rooms. The travel distance can be increased to that specified for sprinklered buildings in IBC Table 1016.1 according to the occupancy.

1016 Exit Access Travel Distance

1016.1 Travel distance limitations *(part 3 of 3)*

1021.2, "Single exits." In the first story or basement of buildings qualifying for one exit: In A, B, E, F, M, U, H-4, H-5, I, R travel distance is ≤ 75' where not sprinklered. In B, F travel distance is ≤ 100' where sprinklered. In S travel distance is ≤ 75' or ≤ 100' depending on number of occupancts and whether it is sprinklered or not. In H-2, H-3 travel distance is 25'. In 2nd stories qualifying for one exit: in B, F, M, S travel distance is ≤ 75'. In R-2 travel distance is 50' where there are ≤ 4 dwelling units. In 3rd stories qualifying for 1 exit: in R-2 travel distance is 50' where there are ≤ 4 dwelling units. All travel distances are based on various limits to the number of occupants or number of dwelling units per IBC Table 1021.2.

1028.7, "Travel distance." In smoke-protected seating travel distance from a seat to the nearest vomitory entrance is 200' and from there to a stair, ramp, or walk outside of the building is 200'. The travel distance from a seat in an open air structure to the exterior of the building is 400'. In type I and II construction of open air seating, there is no limit to travel distance.

3103.4, "Means of egress," limits travel distance to 100' in temporary structures.

3104.9, "Exit access travel." Unsprinklered pedestrian walkways have a 200' travel distance. Sprinklered pedestrian walkways have a 250' travel distance. Unsprinklered pedestrian walkways, both sides at least 50% open have a 300' travel distance. Sprinklered pedestrian walkways with both sides at least 50% open have a 400' travel distance.

- In other locations, exit access travel distance is limited as follows:

Table 1016.1 Exit Access Travel Distance Limits

Buildings sprinklered per NFPA 13 and where permitted NFPA 13R:

Occupancy	Travel distance	Occupancy	Travel distance
A, E, F-1, I-1, M, R, S-1	≤ 250'	H-2	≤ 100'
B	≤ 300'	H-3	≤ 150'
F-2, S-2, U	≤ 400'	H-4	≤ 175'
H-1	≤ 75'	H-5, I-2, I-3, I-4	≤ 200'

Buildings not sprinklered:

Occupancy	Travel distance	Occupancy	Travel distance
A, B, E, F-1, M, R, S-1	≤ 200'	F-2, S-2, U	≤ 300'

Source: IBC Table 1016.1.

- I and H occupancies are not permitted to be without sprinkers.

1016.2 Exterior egress balcony increase

- Certain travel distance limitations are increased ≤ 100' in the following case:
  - Where the last segment of travel to an exit occurs as follows:

    On an exterior egress balcony.

    The increase may not be longer than the egress balcony.

Note: 1016.1, "Travel distance limitations," is cited as listing the travel distances that may be increased. Section 1019, "Egress Balconies," is cited as governing these components.

1016 Exit Access Travel Distance

Case study: Fig. 1016.1A–D. Exit access travel distances are shown in the following examples for various occupancies. Rectangular paths of travel are followed as they simulate actual routes available around possible furnishing or other obstacles. Diagonal paths are followed through areas where no furnishings will occur. The most remote beginning point for travel is indicated in each case by a dot. All examples are in compliance with the code.

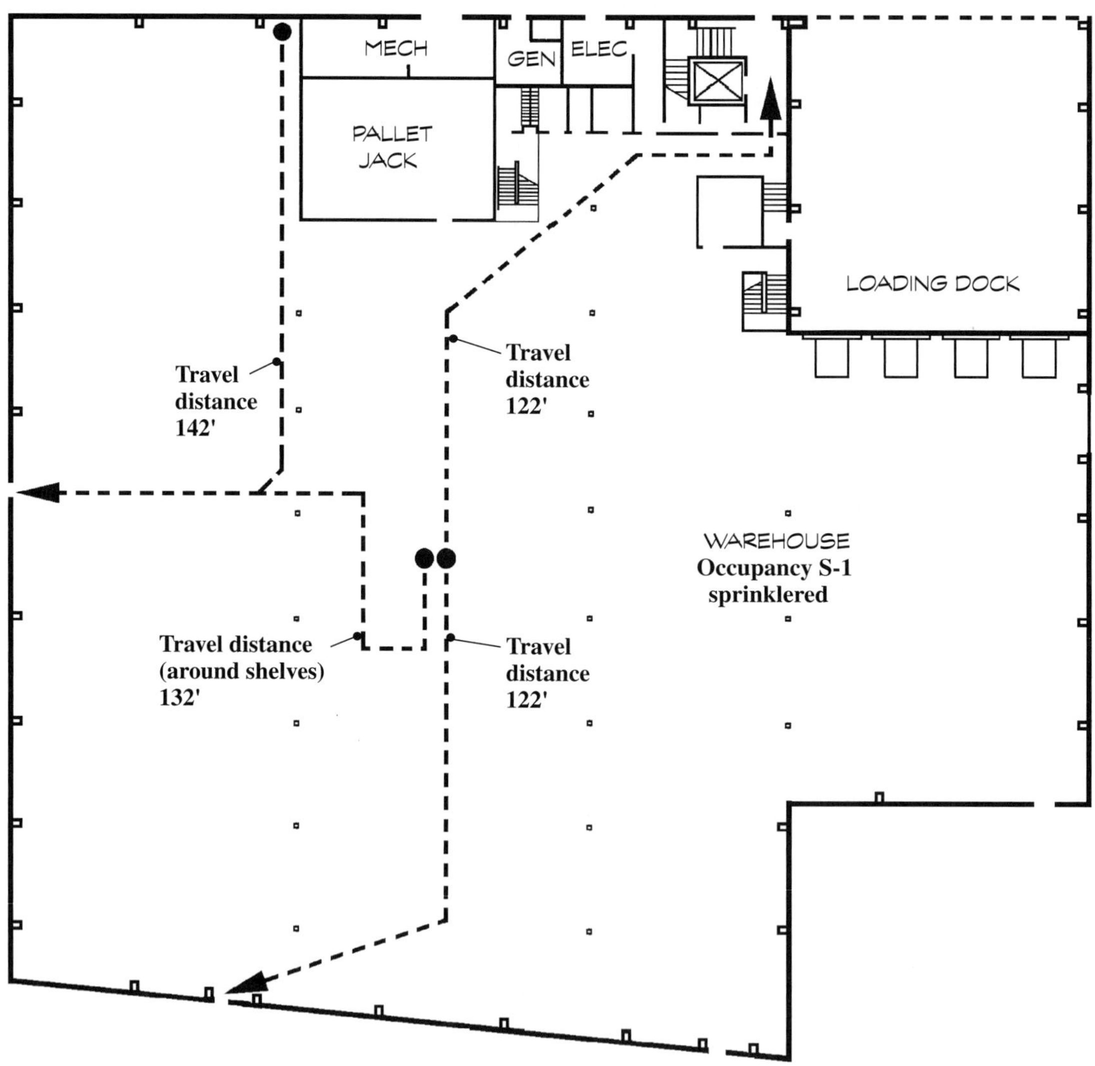

Fig.1016.1A. Floor plan. New Warehouse Addition. Los Angeles, California. Stephen Wen + Associates, Architects, Inc. Pasadena, California.

1016 Exit Access Travel Distance

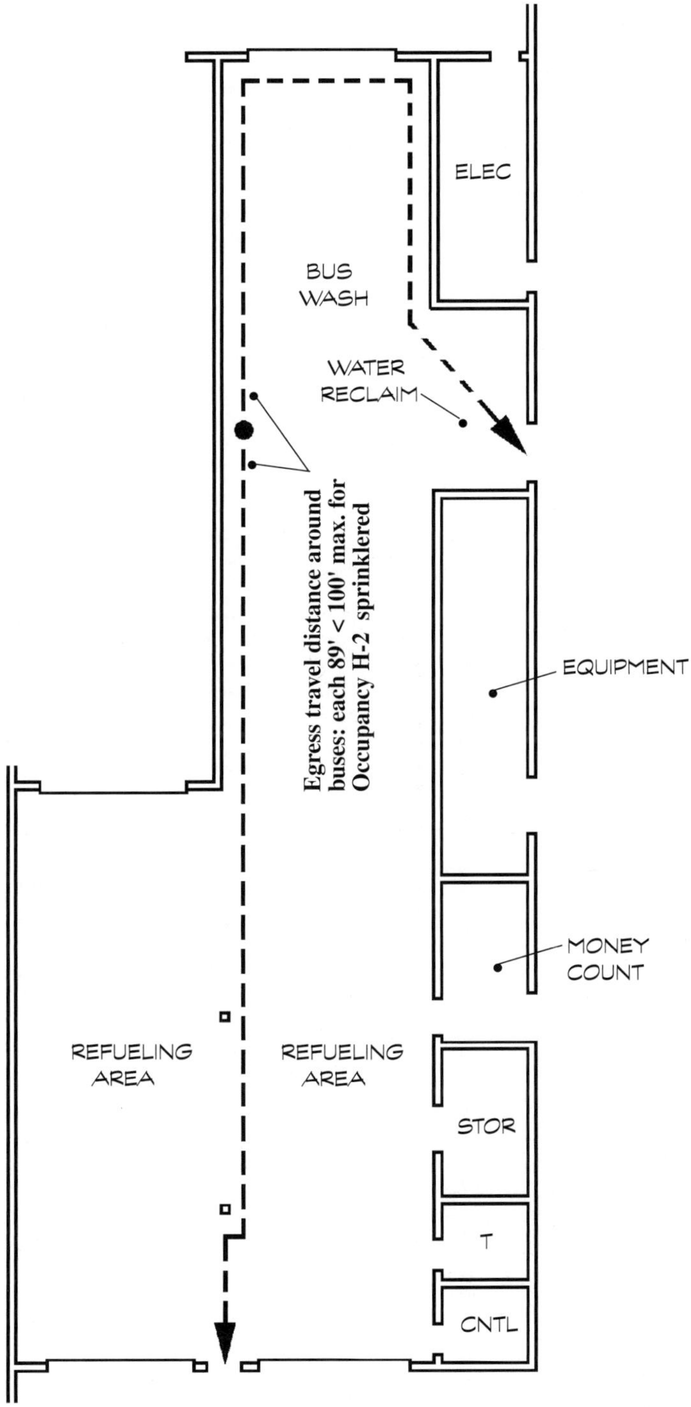

Fig. 1016.1B. Partial floor plan. Wichita Transit Storage, Administration, and Maintenance Facility. Wichita, Kansas. Wilson Darnell Mann, P.A. Wichita, Kansas.

1016 Exit Access Travel Distance

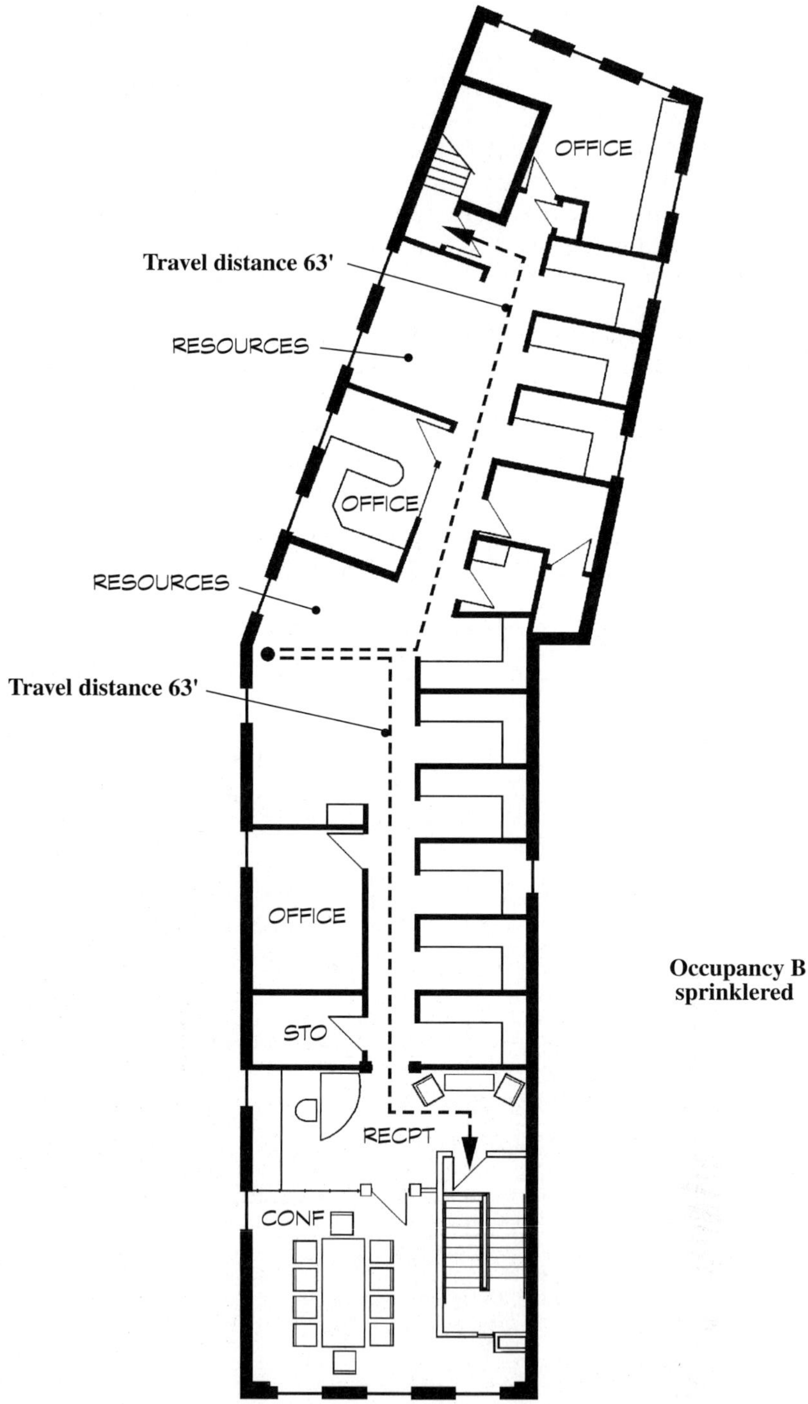

Fig. 1016.1C. 2nd-floor plan. Alterations to 209 Main Street. Annapolis, Maryland. Alt Breeding Schwarz Architects, LLC. Annapolis, Maryland.

1016 Exit Access Travel Distance

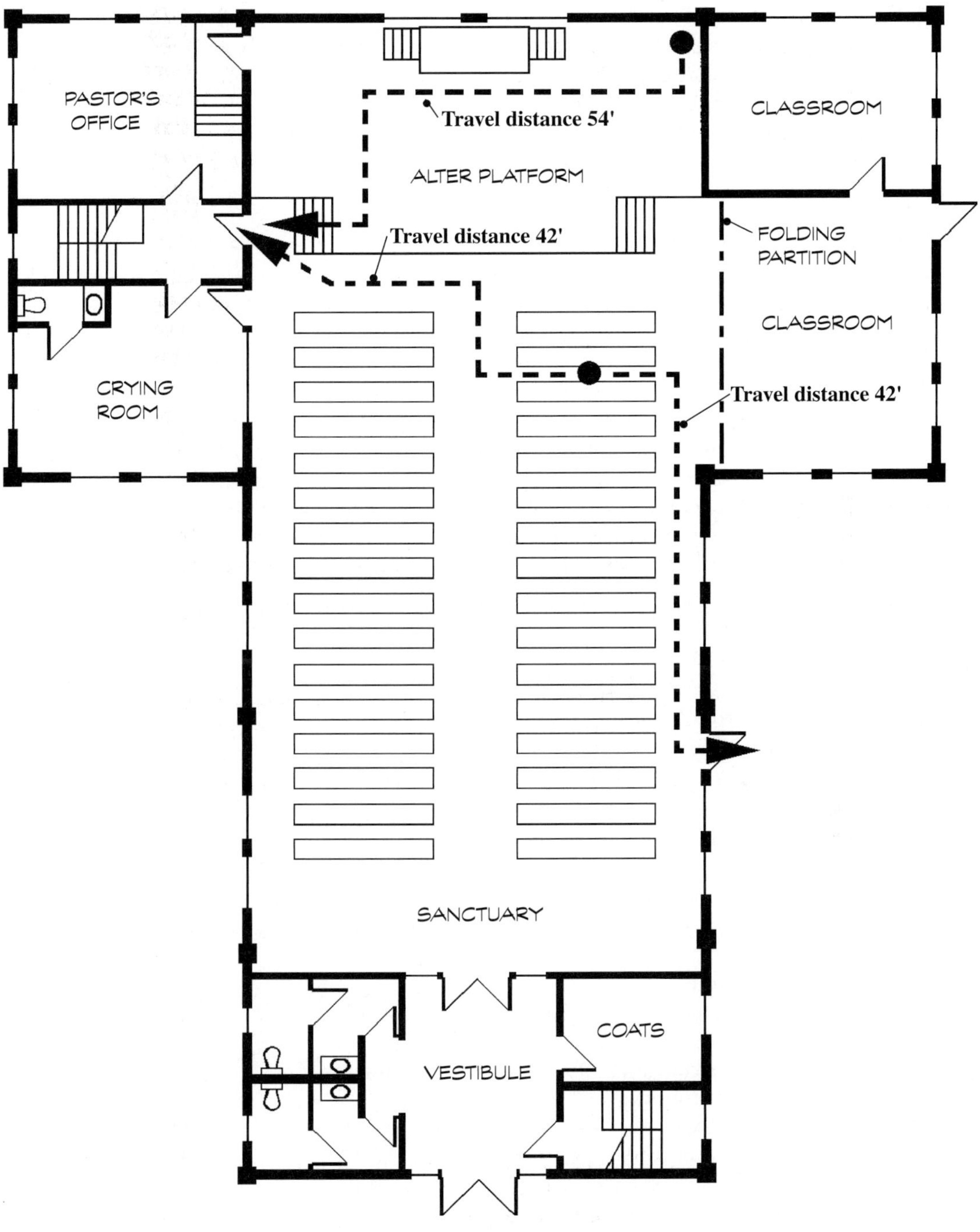

Occupancy A-3 not sprinklered

Fig. 1016.1D. Floor plan. Glad Tidings Assembly of God Church. Naticoke, Pennsylvania. Mullins and Weida, Architect and Associate. Bear Creek, Pennsylvania.

1017 Aisles

1017.1 General

- This section addresses aisles serving in a means of egress.
- The following aisles are not addressed by this section:
  - In assembly.
  - In reviewing stands.
  - In grandstands.
  - In bleachers.

Note: Section 1028, "Assembly," is cited as governing these aisles.

- Aisles are required for occupied areas containing the following:
  - Seats.
  - Tables.
  - Furnishing.
  - Displays.
  - Similar fixtures or equipment.
- Obstructions in required aisle width are limited to the following:
  - Fully opened doors may protrude a total ≤ 7" into the required width.
  - Doors in any position may not reduce the required aisle width to < 1/2.
  - The following nonstructural items may protrude into required width ≤ 1 1/2" from each side:
    - Trim and similar decorations.

Note: 1005.2, "Door encroachment," is cited as governing door obstructions and is summarized above.

1017.2 Aisles in Groups B and M

- In occupancies B and M, aisle widths are limited as follows:
  - Minimum width varies with occupant load but cannot be less than the following:
    - ≥ 28" where all of the following conditions exist:
      - Aisles are not public.
      - Occupant load ≤ 50.
      - Aisle is not required to be accessible.
    - ≥ 36" for other aisles.

Note: 1005.1, "Minimum required egress width," is cited as governing aisle width.
Chapter 11, "Accessibility," is cited as defining aisles required to be accessible.

1017.3 Aisle accessways in Group M

- Aisle accessways in sales display fixture areas are governed as follows:
  - An aisle accessway is required on ≥ 1 side of the following:
    - Nonfixed sales display fixtures ≤ 5'-9" high such as follows:
      - Cases
      - Partitions
      - Shelving
      - Counters
      - Racks
  - Where not required to be accessible, clear width must be ≥ 2'-6".
  - Clear width must extend to an adjacent aisle or aisle accessway.
- Common path of travel from any point in the sales display fixture area is limited as follows:
  - ≤ 75' for occupant loads ≤ 50.
  - ≤ 30' in other cases.

1017 Aisles

1017.4 Seating at tables

- The clear width of an aisle or aisle accessway along the following seating is measured as indicated below:
  - Movable seating:
    - At tables.
    - At counters.
  - Width measurement:
    - Aisle width is measured to a line 19" away from and || to the edge of the following:
      - Table.
      - Counter.
    - The 19" distance is measured ⊥ to the edge of the table or counter.
- The clear width of an aisle or aisle accessway along the following seating is measured as indicated below:
  - Fixed seating:
    - At tables.
    - At counters.
  - Width measurement is made to the back of the fixed seating.
- Other aisle or aisle accessway width measurements are to bypass any handrails and continue to the bordering element such as follows:
  - Walls.
  - Edges of seating.
  - Tread edges.

1017.4.1 Aisle accessway for tables and seating

- An aisle accessway, as follows, has the requirement listed below:
  - Aisle accessway:
    - For seating at tables or counters.
  - Requirement:
    - Must have a width ≥ the larger of the following:
      - Egress width required for means of egress.
      - Accessway width required for seating at tables and counters.

Note: 1005.1, "Minimum required egress width," is cited as the source of width requirements for means of egress.
1017.4.2, "Table and seating accessway width," is cited as a source of width requirements for accessways for seating at tables and counters.

1017.4.2 Table and seating accessway width *(part 1 of 2)*

- Aisle accessway length is measured to the center of the seat most remote from the aisle.
- Aisle accessway segments with both of the following characteristics are not governed by this section:
  - Length ≤ 6'.
  - Occupant load ≤ 4.
- Other aisle accessways must have a width as follows:
  - ≥ 12" where accessways are ≤ 12' in length.
  - The following table lists widths for accessways > 12' and ≤ 30' in length.

1017 Aisles

1017.4.2 Table and seating accessway width *(part 2 of 2)*

Table 1017.4.2 Aisle Accessway Widths

Length	Width	Length	Width	Length	Width
> 12' ≤ 13'	≥ 12.5"	> 18' ≤ 19'	≥ 15.5"	> 24' ≤ 25'	≥ 18.5"
> 13' ≤ 14'	≥ 13.0"	> 19' ≤ 20'	≥ 16.0"	> 25' ≤ 26'	≥ 19.0"
> 14' ≤ 15'	≥ 13.5"	> 20' ≤ 21'	≥ 16.5"	> 26' ≤ 27'	≥ 19.5"
> 15' ≤ 16'	≥ 14.0"	> 21' ≤ 22'	≥ 17.0"	> 27' ≤ 28'	≥ 20.0"
> 16' ≤ 17'	≥ 14.5"	> 22' ≤ 23'	≥ 17.5"	> 28' ≤ 29'	≥ 20.5"
> 17' ≤ 18'	≥ 15.0"	> 23' ≤ 24'	≥ 18.0"	> 29' ≤ 30'	≥ 21.0"

The table above is based on the following equation:

Minimum clear width = 12" + 0.5" × [(length in ft – 12') rounded up to next foot]

1017.4.3 Table and seating aisle accessway length

- Travel distance in an aisle accessway is limited to ≤ 30' as follows:
  - The distance is measured between the following points:
    From any seat.
    To the point providing a choice of ≥ 2 routes to separate exits.

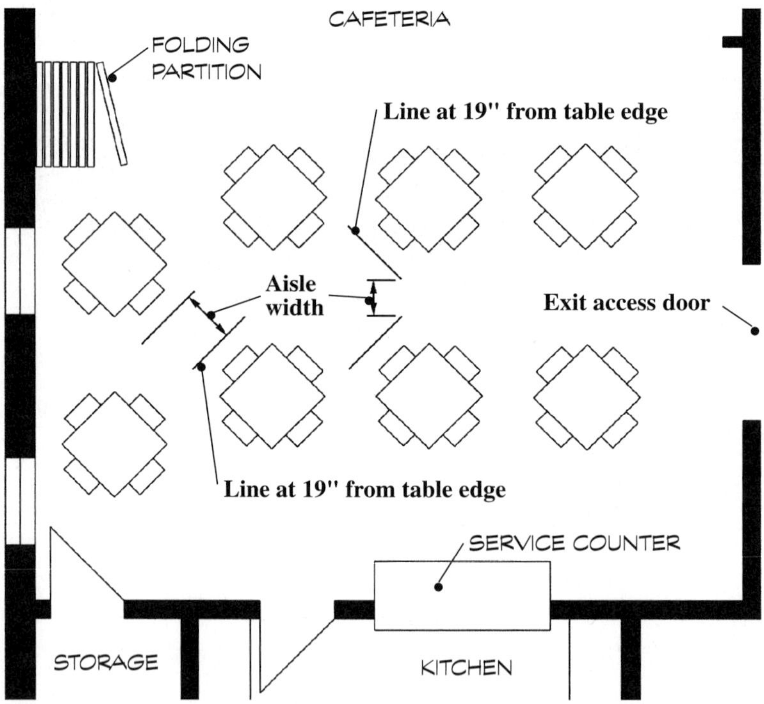

Fig. 1017.4.1. Partial floor plan of cafeteria. Multipurpose Building Addition to Children's Home. Wilkes-Barre, Pennsylvania. C. Allen Mullins, Architect. Bear Creek, Pennsylvania.

1017 Aisles

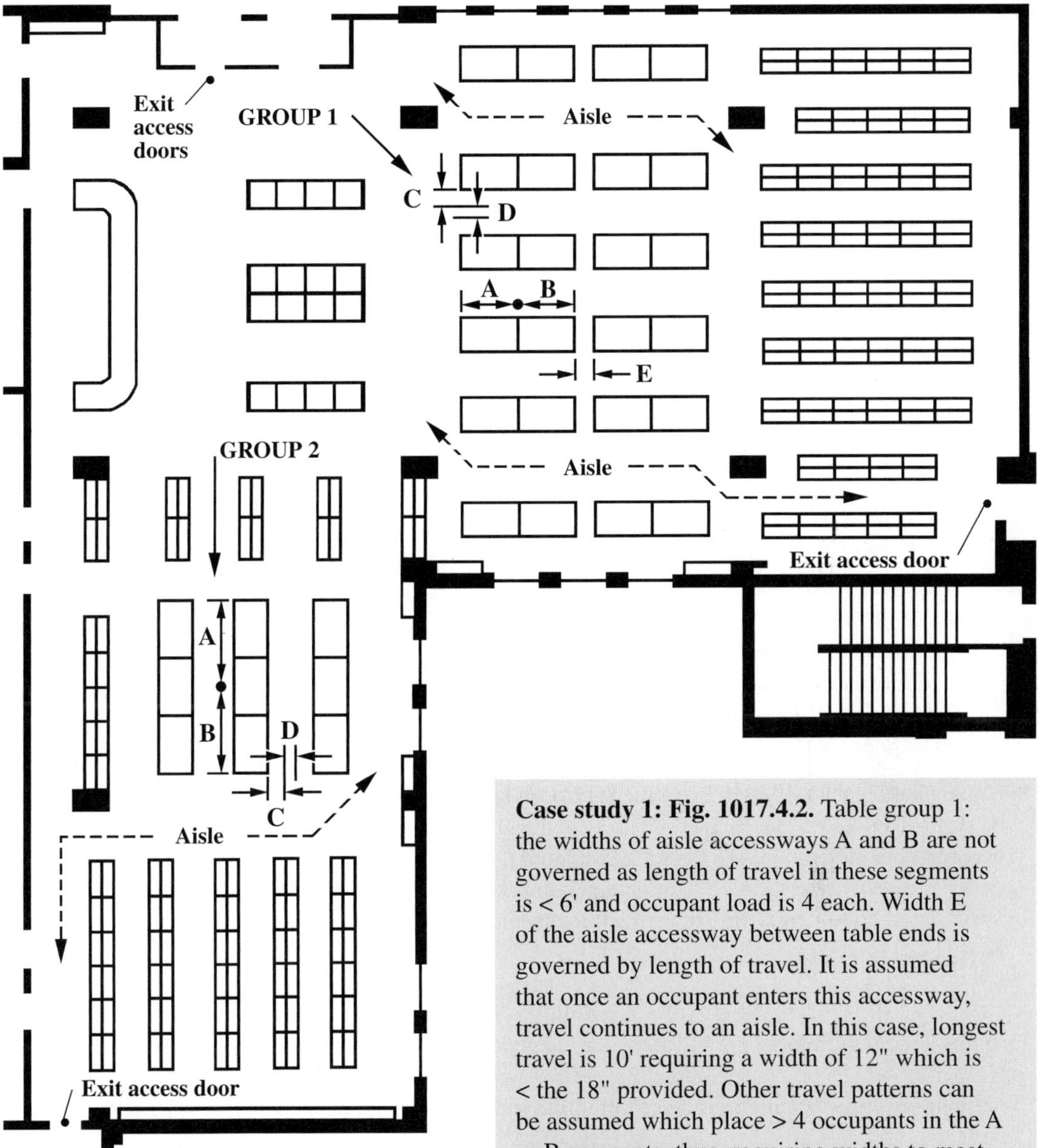

Case study 1: Fig. 1017.4.2. Table group 1: the widths of aisle accessways A and B are not governed as length of travel in these segments is < 6' and occupant load is 4 each. Width E of the aisle accessway between table ends is governed by length of travel. It is assumed that once an occupant enters this accessway, travel continues to an aisle. In this case, longest travel is 10' requiring a width of 12" which is < the 18" provided. Other travel patterns can be assumed which place > 4 occupants in the A or B segments, thus, requiring widths to meet minimums based on length. Such width D is measured between lines that are 19" (dimension C) from the table edges. Width D is 12", which is adequate for travel up to 12'. Travel in this scenario is 10'.

Case study 2: Fig. 1017.4.2. Table group 2: aisle accessways A and B are 7'-6" in length, thus, requiring a minimum width D of 12" which is provided. Width D is measured between lines at 19" (dimension C) from the table edges.

Fig. 1017.4.2. Partial floor plan of library. High School 6, Cypress-Fairbanks Independent School District. Harris County, Texas. PBK Architects, Inc. Houston, Texas.

1018 Corridors

1018.1 Construction

- A fire-resistance rating is not required for the following corridors:
  - In Occupancy E as follows:
    - Each instruction room must have ≥ 1 door meeting the following condition:
      - It must discharge directly to the outside at grade level.
    - Each assembly room must have the following:
      - ≥ half the required means of egress doors must meet the following condition:
        - They must discharge directly to the outside at grade level.
  - In Occupancy R as follows:
    - In a dwelling unit or sleeping unit.
  - In open parking garages.
  - In Occupancy B as follows:
    - Where only 1 means of egress is required.

Note: 1015.1, "Exits or exit access doorways from spaces," is cited as governing locations requiring only one means of egress.

  - In the following occupancies where sprinklered:*
    - A, B, E, F, I-2, I-4, M, S, U.

Note: The following are cited as governing corridors in occupancy I-2:
407.2, "Corridors."
407.3, "Corridor walls."
The following are cited as governing sprinklers as applicable in this section:
903.3.1.1, "NFPA 13 sprinkler systems."
903.3.1.2, "NFPA 13R sprinkler systems."

  - Where the occupant load is ≤ 30 in the following sprinklered occupancies:*
    - A, B, E, F, H-4, H-5, M, S, U.
  - Where the occupant load is ≤ 10 in a sprinklered Occupancy R.*
- A corridor fire-resistance rating ≥ 1 hr is required as follows:*
  - In sprinklered buildings of the following occupancies:
    - H-1, H-2, H-3, I-1, I-3.

Note: 408.8, "Subdivision of resident housing areas," is cited as governing I-3.

  - In sprinklered buildings of the following occupancies where the corridor load is > 30:
    - H-4, H-5.
  - Where the corridor occupant load is > 30 in the following occupancies:
    - A, B, E, F, M, S, U.
  - Where the corridor occupant load is > 10 in Occupancy R.
- A corridor fire-resistance rating ≥1/2 hr is required in the following case:*
  - In sprinklered buildings of Occupancy R as follows:
    - Where the corridor load is > 10.
- Corridor walls required to have a fire-resistance rating must comply with the following:
  - Requirements for fire partitions.

Note: Section 709, "Fire Partitions," is cited as governing these walls.

**Source*: IBC Table 1018.1.

1018 Corridors

1018.2 Corridor width

- The following widths are required for corridors serving the locations noted:

Table 1018.2 Minimum Corridor Widths

Width	Occupants	Location
≥ 24"	Not limited	Access to building service equipment
≥ 36"	< 50	All locations
≥ 36"	Not limited	Within a dwelling unit
≥ 72"	≥ 100	Occupancy E
≥ 72"	Not limited	Corridors and areas with gurney traffic serving outpatients incapable of self-preservation
≥ 96"	Not limited	Occupancy I-2 where bed movement is required

- In other locations, corridors must be the larger of the following widths:
  - ≥ 3'-8".
  - ≥ occupant load × 0.2" for each occupant served.

Note: 1005.1, "Minimum required egress width," is cited as governing corridor width based on occupant load.

1018.3 Corridor obstruction

- The required width of a corridor must not be obstructed.

1018.4 Dead ends

- Where > 1 exit or exit access doorway is required, the following applies:
  - Dead-end corridor length is limited as follows:

Table 1018.4 Dead-End Length Limits

Occupancy	Conditions	Length
B, E, F, I-1, M, R-1, R-2, R-4, S, U	Sprinklered	≤ 50'
I-3	Condition 2, 3, or 4	≤ 50'
All occupancies	Width measured at narrowest point	< 2.5 × width of dead-end corridor

Note: 308.4, "Group I-3," is cited as the source of characteristics for I-3 conditions. 903.3.1.1, "NFPA 13 sprinkler systems," is cited as governing the sprinklers.

  - In all other locations, dead-end corridors are limited to ≤ 20'.

1018 Corridors

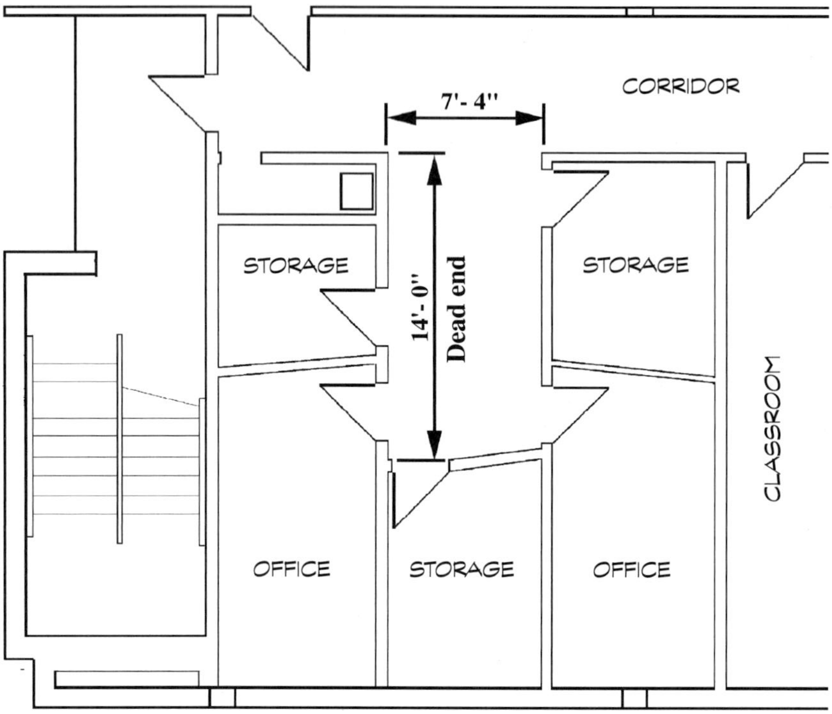

Case study: Fig. 1018.4A. The dead-end corridor at the storage room in the Occupancy E building is 14'-0" long and 7'-4" wide. The length of a dead end must be < 2.5 × its width or ≤ 20', whichever is greater. Since *2.5 × 7'- 4" = 18'- 4"* which is < 20', maximum length is 20'. The 14'-0" length is < 20'; thus, the dead end is in compliance with the code.

Fig. 1018.4A. Partial floor plan. Newman Elementary School Renovations. Needham, Massachusetts. HKT Architects, Inc. Somerville, Massachusetts.

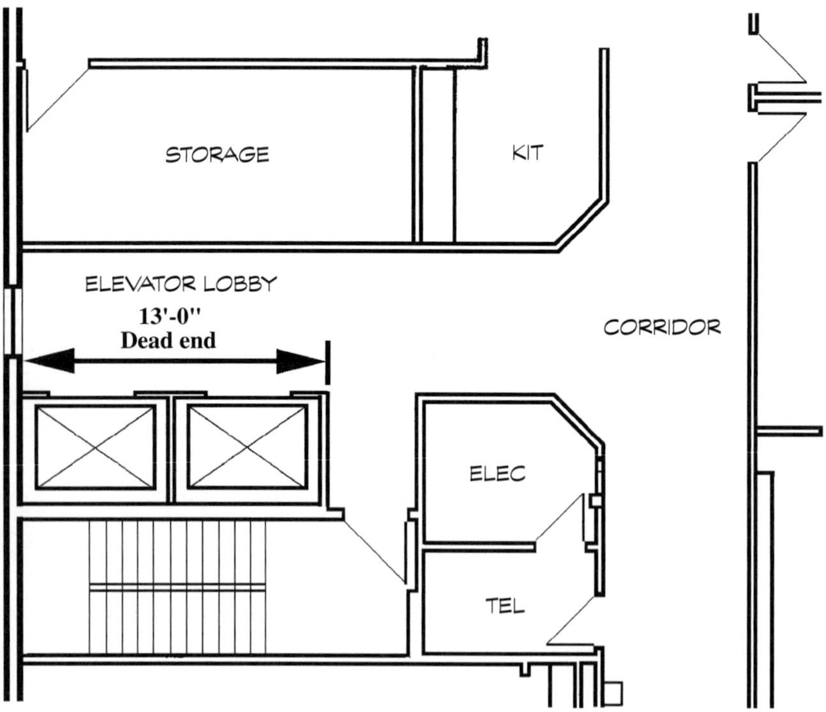

Case study: Fig. 1018.4B. The elevator lobby of the Occupancy B sprinklered building is a dead end 13' in length. This is in compliance with the 50' maximum for sprinklered buildings of this occupancy.

Fig. 1018.4B. Partial floor plan at elevator lobby. South Texas Blood and Tissue Center. San Antonio, Texas. Overland Partners, Inc. San Antonio, Texas.

1018 Corridors

1018.5 Air movement in corridors

- A corridor may be used for exhaust system makeup air for the following rooms where the requirements listed below are met:
  - Rooms:
    - Toilet rooms.
    - Bathrooms.
    - Dressing rooms.
    - Smoking lounges.
    - Janitor closets.
  - Requirements:
    - The rooms must open directly onto the corridor.
    - The corridor must receive outside air at the following rate:
      - > the rate at which makeup air is taken out.
- Corridors may be used for HVAC return air as follows:
  - In dwelling units.
  - In tenant spaces as follows:
    - Of an area ≤ 1,000 sf.
- Corridors may be used for incidental air movement as follows:
  - From pressurized rooms in healthcare facilities as follows:
    - Corridor may not be the primary vehicle for the following:
      - Supply air.
      - Return air.
- In other cases, corridors may not convey the following:
  - Supply air.
  - Return air.
  - Exhaust air.
  - Relief air.
  - Ventilation air.

1018.5.1 Corridor ceiling

- The space above a corridor ceiling may be used for HVAC return air in any of the following cases:
  - Where the corridor construction does not have a fire-resistance rating.
  - Where the return air zone is separated from the corridor as follows:
    - By fire-resistance-rated construction.
  - Where the fans serving the corridor are shut off in one of the following ways:
    - By a sprinkler flow switch where the building is sprinklered.
    - By smoke detectors at the fan unit.

  Note: The International Mechanical Code *is cited as governing this aspect.*

  - Where the space above the corridor ceiling is used as follows:
    - As part of a smoke control system with both the following characteristics:
      - Approved.
      - Engineered.

1018 Corridors

1018.6 Corridor continuity

- This section addresses corridors with a fire-resistant rating.
- Corridors may pass through the following rooms with the condition indicated below:
  - Rooms:
    Foyers
    Lobbies.
    Reception rooms.
  - Condition:
    Rooms to be constructed with fire resistance ≥ required for the corridor.
- Otherwise, corridors must comply with both of the following:
  - They must be continuous between the following points:
    Their beginning.
    An exit.
  - They may not pass through the following:
    Intervening rooms.

1019 Egress Balconies

1019.1 General

- Balconies in a means of egress must comply with corridor requirements as follows:
  - Width.
  - Headroom.
  - Dead ends.
  - Projections.

1019.2 Wall separation

- This section addresses the separation of a means of egress balcony from the building interior.
- Separation is not required where both of the following conditions apply:
  - The egress balcony is served by ≥ 2 stairs.
  - Travel from a dead end does not pass an unprotected opening en route to a stair.
- Otherwise, such a balcony must be separated from the interior of the building as follows:
  - By the walls as required for corridors.
  - By opening protectives as required for corridors.

1019.3 Openness

- The area on the long side of the egress balcony must be open to exterior as follows:
  - ≥ 50% openness is required.
- Open area above guards must be distributed to minimize the collection of smoke and gases.

1020 Exits

1020.1 General

- Any use of an exit as follows is prohibited:
  - In any way that interferes with its function.
- The level of protection at any point in an exit may not diminish as follows:
  - Until the exit discharge.

Note: Section 1003, "General Means of Egress," through Section 1013, "Guards," are cited as governing exits.
Section 1020, "Exits," through Section 1026, "Exterior Exit Ramps and Stairways," (applicable parts) are cited as governing exits.

1020.2 Exterior exit doors

- ≥ 1 exterior exit door is required as follows:
  - In any building for human occupancy.
  - Must meet size requirements for means of egress doors.

Note: 1008.1.1, "Size of doors," is cited as the source of size requirements for means of egress doors.

1020.2.1 Detailed requirements

- Exterior exit doors must meet the applicable requirements for means of egress doors.

Note: 1008.1, "Doors," is cited as the source of requirements for means of egress doors.

1020.2.2 Arrangement

- Exterior exit doors must open directly to one of the following:
  - An exit discharge.
  - A public way.

1021 Number of Exits and Continuity

1021.1 Exits from stories

- Occupied roofs have the same requirements for exits as do stories.
- The number of exits required for certain tall buildings is determined as follows:
  - Occupant load determines the base number.
  - Certain occupancies and height add one exit to the base number required.

  Note: 403.5.2, "Additional exit stairway," is cited as requiring an additional stairway for certain tall buildings.

- Certain small buildings require one exit based on the following:
  - Occupancy.
  - Number of occupants.
  - Number of stories.
  - Travel distance.

  Note: 1021.2, "Single exits," is cited as governing the number of exits required for certain (small) buildings.

- Open exit access stairways and ramps can meet minimum exit requirements in certain locations.

  Note: 1016.1, "Travel distance limitations," Exceptions 3 and 4 are cited as governing open exit access stairways and ramps that meet exit access requirements in certain locations.

- Buildings with all of the following characteristics require only one means of egress:
  - In occupancies R-2, R-3.
  - Individual dwelling unit.
  - Occupant load ≤ 20.
  - Sprinklered.

  Note: 903.3.1.1, "NFPA 13 sprinkler systems," is cited as governing sprinkler systems as applicable. 903.3.1.2, "NFPA 13R sprinkler systems," is cited as governing sprinkler systems as applicable.

- One exit is permitted to serve rooms and spaces within a story as follows:
  - Exits must discharge directly to the exterior at the level of exit discharge.
  - Rooms and spaces must meet certain criteria regarding travel, occupancy, occupant load, and use.

  Note: 1015.1, "Exits or exit access doorways from spaces," is cited as governing criteria for one exit.

- In other buildings, all spaces in a story must have access to the required number of exits as follows:
  - Exits must be approved.
  - Exits must be independent.
  - Number of exits are based on occupant load as follows:*
    ≤ 500 occupants require ≥ 2 exits per story.
    ≥ 501 and ≤ 1,000 occupants require ≥ 3 exits per story.
    > 1,000 occupants require ≥ 4 exits per story.

  Note: IBC Table 1021.1, "Minimum Number of Exits for Occupant Load," is cited as governing the number of exits based on occupant load.

**Source*: IBC Table 1021.1.

1021 Number of Exits and Continuity

Case study:
Fig. 1021.1. The 646 occupants in this Occupancy E building require 3 exits. The building complies with the code by providing 8 exits. Exits are numbered in the illustration.

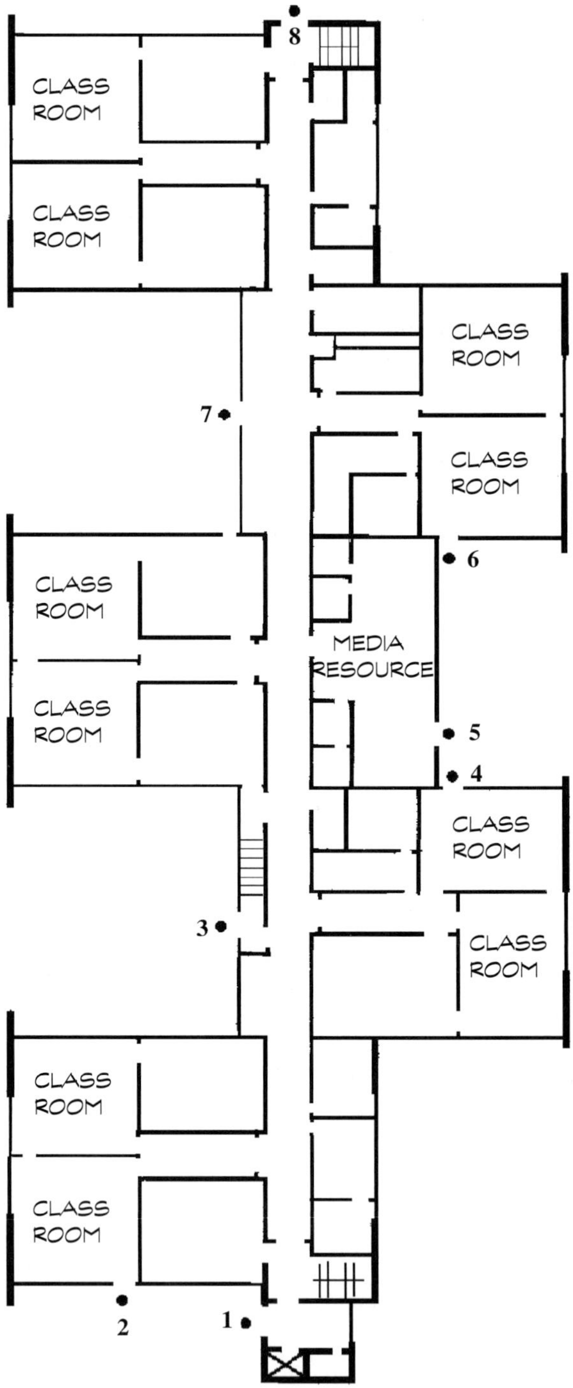

Fig. 1021.1. Floor plan of classroom wing. Newman Elementary School Renovations. Needham, Massachusetts. HKT Architects, Inc. Somerville, Massachusetts.

1021 Number of Exits and Continuity

1021.1.1 Exits maintained

- The required number of exits from any story must be maintained to one of the following points:
  - Grade.
  - Public way.

1021.1.2 Parking structures

- Where vehicles are mechanically parked, exits required are as follows:
  - Only 1 exit is required from each parking level.
- In other parking structures, exits are required as follows:
  - ≥ 2 exits are required from each parking level.
- Vehicle ramps without pedestrian facilities are not considered exits.
- Vehicle ramps with pedestrian facilities are considered exits.

1021.1.3 Helistops

- Helistops require 2 means of egress as follows:
  - Means of egress must comply with this chapter.
  - One means of egress may be either of the following components where the conditions below exist:

    Components:

      A fire escape to the floor below.

      A ladder to the floor below.

      Alternating tread device to the floor below.

    Conditions:

      Where the landing platform or roof area have either of the following dimensions:

        < 60' in length.

        < 2,000 sf.

Case study: Figure 1021.2A. The modular classroom building is Occupancy E, 1 story, and has no basement. Its 45 occupants and exit access travel distance ≤ 58' require only 1 exit.

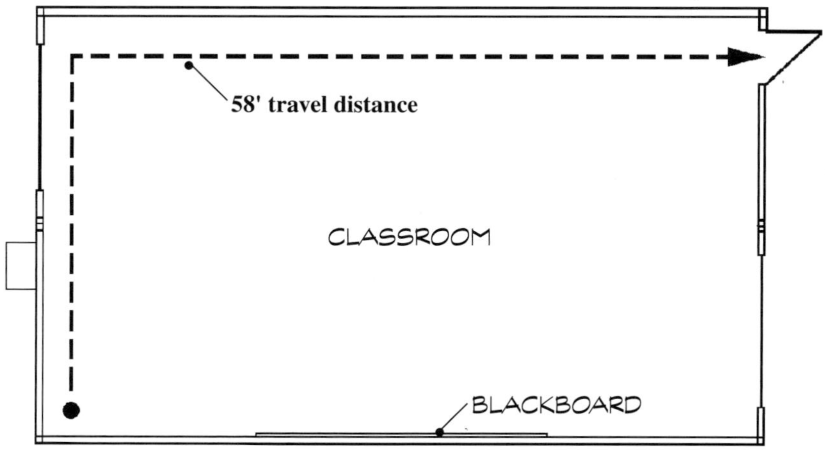

Fig. 1021.2A. Floor plan. Modular Classroom Building, Creston Elementary School. Creston, California. Phillips Metsch Sweeney Moore Architects. Santa Barbara, California.

1021 Number of Exits and Continuity

1021.2 Single exits

- Air traffic control towers are not governed by this section.

  Note: 412.3, "Airport traffic control towers," is cited as providing exit requirements for these structures.

- Parking garages are not governed by this section.

  Note: 1021.1.2, "Parking structures," is cited as providing requirements for exits from parking garages.

- Only 1 exit is required in Occupancy R-3.
- In other cases, only 1 exit is required where the criteria in the table below are met and as follows:
  - Individual occupancies that meet the criteria of the table below may have 1 exit as follows:
    Where they are part of mixed occupancies.
    Where they are located in a building that requires more than one exit.
  - Occupant load for determining number of exits includes the following:
    The load of occupancies exiting through a subsequent occupancy in the path of egress.

Note: 1004.1, "Design occupant load," is cited as governing the determination of occupant load.

Table 1021.2 Conditions Permitting 1 Exit from a Story

Occupancy	Occupants per floor	Travel distance
1st story or basement (≤ 1 story below grade plane):		
A, E (no day care), M, U	≤ 49	≤ 75'
E (day care)	≤ 10	≤ 75'
B, F, S sprinklered	≤ 49	≤ 100'
B, F, S not sprinklered	≤ 49	≤ 75'
H-2, H-3	≤ 3	≤ 25'
H-4, H-5, I, R	≤ 10	≤ 75'
S (no parking garages)	≤ 29	≤ 100'
2nd story:		
B, F, M, S	≤ 29	≤ 75'

2nd or 3rd story:

Occupancy	Sprinklered	Stories	Dwelling units	Travel distance
R-2	no	2nd	≤ 4	≤ 50'
R-2	yes	2nd	≤ 4	≤ 50'
R-2 with escape & rescue openings	yes	3rd	≤ 4	≤ 50'

Source: IBC Table 1021.2.

Note: Section 1029, "Emergency Escape and Rescue," is cited as governing escape and rescue openings.
903.3.1.1, "NFPA 13 sprinkler systems," is cited as governing sprinklers for the occupancies B, F, S, and R-2 as applicable.
903.3.1.2, "NFPA 13R sprinkler systems," is cited as governing sprinklers for Occupancy R-2 as applicable.

1021 Number of Exits and Continuity

Case study: Fig. 1021.2B. The Occupancy B pavilion is 1 story with no basement. The space is used by staff only for refreshment preparation. Customers are served at pass-through windows and do not enter the building. With 10 occupants and no sprinklers, the building qualifies for 1 exit since no exit access travel distance is > 54' from any point to either of the 2 exits provided. Since only 1 exit is required, the 2 provided do not have to be located a minimum distance apart. Travel distances shown are those applicable, if either door was omitted. With 2 exits, the distances are shorter.

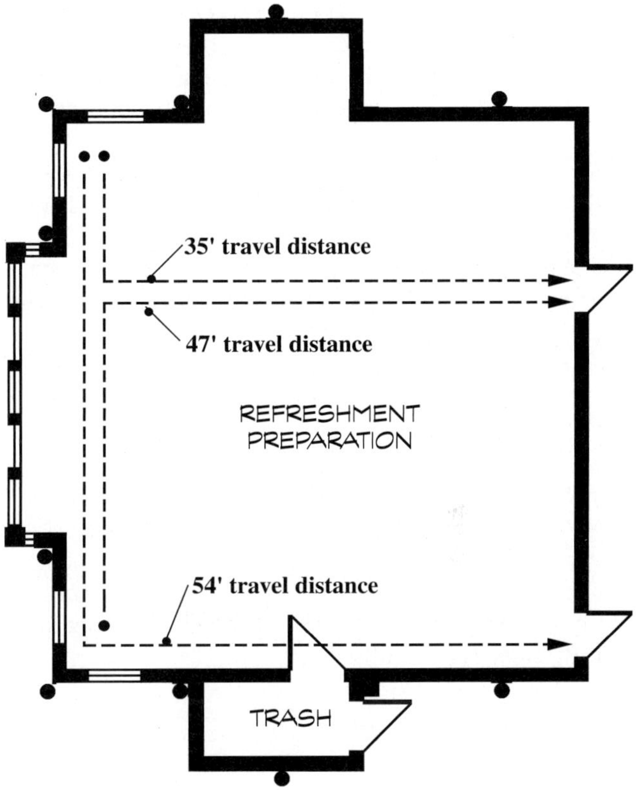

Fig. 1021.2B. Floor plan. Visitor Services Pavilions at Clarence Buckingham Memorial Fountain. Chicago Park District. Chicago, Illinois. David Woodhouse, Architects. Chicago, Illinois.

1021.3 Exit continuity

- Exits must be continuous between the following points:
  - The beginning of the exit.
  - The exit discharge.

1021.4 Exit door arrangement

- 2 exit doors must be separated by a distance based on the following:
  - In buildings not sprinklered, separation is required as follows:
    ≥1/2 the diagonal dimension of the space served.
  - In sprinklered buildings separation is required as follows:
    ≥1/3 the diagonal dimension of the space served.

Note: 1015.2, "Exit or exit access doorway arrangement," through 1015.2.2, "Three or more exits or exit access doorways," are cited as governing the separation of the exits, a partial summary of which is provided above.

1022 Exit Enclosures

1022.1 Enclosures required

- Exits do not require an enclosure where all of the following conditions are present:
  - In Occupancy A-5 where all parts of the means of egress are essentially open to the exterior.
- Stairways do not require an enclosure in occupancies A, B, E, F, M, R, S, U in the following case:
  - Where all of the following conditions are present:
    - Occupant load < 10.
    - Stairway connects only the story at the level of exit discharge and one of the following:
      - 1 story above or below its level of exit discharge.
  - In any dwelling or sleeping unit of R-1, R-2, R-3.
  - Where the stairway is in an open parking structure that serves only the parking structure.
  - Where the stairway is serving a stage area, gallery, gridiron, or catwalk.

Note: 410.5.3, "Stage exits," is cited governing these elements.
1015.6.1, "Gallery, gridiron and catwalk means of egress," is cited as governing these elements.

- 1 required stairway enclosure in a building need not meet all the enclosure requirements of this section where both of the following conditions apply:
  - Where located in Occupancy I-3.
  - Where complying with requirements permitting glazing in a stairway enclosure.

Note: 408.3.8, "Exit enclosures," is cited as governing modifications for egress stairways in I-3.

- Certain means of egress stairways at the following facilities need not be enclosed:
  - Balconies and galleries.
  - Press boxes.

Note: 1028.5.1, "Enclosure of openings," is cited as governing open stairways from the facilities noted.

- All other interior exit stairways and ramps must be enclosed as follows:
  - Enclosures must be one or both of the following:
    - Fire barriers.
    - Horizontal assemblies.

Note: Section 707, "Fire Barriers," is cited as governing these elements.
Section 712, "Horizontal Assemblies," is cited as governing these elements.

  - Enclosure fire-resistance ratings are required as follows:
    - ≥ 1 hr where < 4 stories counting basements but not mezzanines.
    - ≥ 2 hr where ≥ 4 stories counting basements but not mezzanines.
- An exit enclosure must have a fire resistance rating = the lesser of the following:
  - ≥ the floor assembly penetrated.
  - ≥ 2 hr.
- An exit enclosure may be used only as a means of egress.
- An exit enclosure must discharge to the exterior by one of the following means:
  - Directly.
  - By way of an exit passageway.
  - As specified elsewhere in the code.

Note: Section 1023, "Exit passageways," is cited as governing these elements.
1027.1, "General," is cited as listing exit enclosures permitted to discharge to the interior.

1022 Exit Enclosures

1022.2 Termination

- Exit enclosures may terminate at an exit passageway as follows:
  - Where the exit passageway terminates at one of the following:
    - Exit discharge.
    - Public way.

Note: Section 1023, "Exit Passageways," is cited as governing these elements.

- Other exit enclosures are required to terminate at one of the following:
  - Exit discharge.
  - Public way.

1022.2.1 Extension

- This section governs the following exit enclosures:
  - Where connected to an exit discharge or public way by the following:
    - An exit passageway.
- Exit enclosures connected to exit passageways must be separated by the following:
  - A fire barrier as follows:
    - Constructed per one or both of the following:
      - Code requirements for a fire barrier.
      - Code requirments for a horizontal assembly.
    - With a fire-resistance rating ≥ that of the exit enclosure.
    - With access to the exit passageway by a fire door assembly.
    - Without openings other than the fire door assembly.
    - Limited to certain penetrations.

Note: 1022.4, "Penetrations," is cited as governing the only penetrations permitted in the fire barrier.
Section 707, "Fire Barriers," is cited as governing these elements.
Section 712, "Horizontal Assemblies," is cited as governing these elements.
715.4, "Fire door and shutter assemblies," is cited as governing these elements.

1022.3 Openings and penetrations

- Unprotected exterior openings in exit enclosures are not governed by this section.
- Openings into an exit enclosure are limited to the following:
  - Openings required for exit access as follows:
    - From normally occupied spaces.
  - Openings required for egress out of the enclosure.

Note: Section 715, "Opening Protectives," is cited as governing openings to exit enclosures.

- Elevators are not permitted to open into an exit enclosure.

1022 Exit Enclosures

1022.4 Penetrations

- Only the following components are permitted to penetrate an exit enclosure:
  - Standpipes.
  - Exit doors.
  - Components serving the exit enclosure as follows:
    - The following components used to pressurize the enclosure:
      - Equipment.
      - Ductwork.
      - Sprinkler piping.
    - The following components terminating as noted below:
      - Components:
        - Electrical raceways as follows:
          - Including those for fire department communication.
      - Termination:
        - Raceways must terminate in a steel box as follows:
          - Side facing the enclosure must be ≤ 16 sq in.

Note: Section 713, "Penetrations," is cited as governing penetrations into an exit enclosure.

- No other penetrations or openings through an exit enclosure are permitted.
- None of the following are permitted between adjacent exit enclosures:
  - Penetrations as follows:
    - Protected or unprotected.
  - Openings as follows:
    - Protected or unprotected.

1022.5 Ventilation *(part 1 of 2)*

- Only the following openings into fire-resistance-rated construction are permitted:
  - Openings needed for the following:
    - Maintenance.
    - Operation.
  - Opening protectives must be provided.

Note: Section 715, "Opening Protectives," is cited as governing these devices.

- Ventilation systems for exit enclosures are restricted as follows:
  - They must be independent of other ventilating systems.
- The following components must comply with at least one of the requirements listed:
  - Components:
    - Equipment for exit enclosure ventilation.
    - Ductwork for exit enclosure ventilation.

Note: 1022.4, "Penetrations," is cited as defining the equipment and ductwork governed by this section.

1022 Exit Enclosures

1022.5 Ventilation *(part 2 of 2)*

- Requirements:
  - Where located outside the building, the following applies:
    - Must be connected directly to the exit enclosure as follows:
      - By construction meeting shaft requirements.
  - Where located inside the exit enclosure, the following applies:
    - Intake air must be taken directly from outside.
    - Exhaust air must be directly to the outside.
  - Where located inside the exit enclosure the following applies:
    - The following air must be transported through ducts as listed below:
      - Air:
        - Intake air.
        - Exhaust air.
      - Ducts:
        - Within construction meeting shaft requirements.
  - Where located inside the building the following applies:
    - Must be isolated with construction meeting shaft requirements as follows:
      - From the building.
      - From other equipment.

1022.6 Exit enclosure exterior walls

- Exit enclosure walls that are exterior walls must meet fire-resistance requirements for exterior walls.

  Note: Section 705, "Exterior Walls," is cited as the source of fire-resistance requirements.

- In the following circumstances, the conditions listed below are required:
  - Circumstances applying to the exterior wall of a stairway enclosure:
    - The wall has one or both the following characteristics:
      - It does not have a fire-resistance rating.
      - It has unprotected openings.
    - The wall is exposed to another exterior wall of the building at an angle < 180°.
  - Conditions required:
    - The building exterior wall so exposed must be constructed as follows:
      - It must have a fire-resistance rating ≥ 1 hr.
      - It must have opening protectives > 3/4 hr.
    - Required fire-resistant construction must cover an area as follows:
      - ≤ 10' horizontally from the stairway wall as defined above.
      - From grade to the lower of the two upper limits described as follows:
        - A level 10' above the highest landing of the stairway.
      - The roof line.

1022.7 Discharge identification

- Enclosed exit stairways continuing to below their level of exit discharge are governed as follows:
  - An approved barrier is required at the level of exit discharge as follows:
    - Barrier must prevent accidental egress travel to below the level of exit discharge.
  - Directional exit signs are required.

  Note: Section 1011, "Exit Signs," is cited as the source of exit sign requirements.

1022 Exit Enclosures

1022.8 Floor identification signs

- Signs are required in exit enclosures connecting > 3 stories with the following information:
  - Floor level.
  - Identifying the termination of the stairway enclosure at the top and bottom.
  - Stairway or ramp identification.
  - Story of the exit discharge and direction to the exit discharge.
  - Availability of roof access for the fire department.
- Required signs are to be positioned as follows:
  - 5' above the floor landing.
  - To be visible when the doors are open or closed.
- Signs with tactile characters are required as follows:
  - Located at each floor level landing as adjacent to the following door:
    The door between the enclosure and the corridor.
  - With information identifying each floor.

Note: ICC A117.1, "Accessible and Usable Buildings and Facilities," is cited as governing tactile-character signs.

1022.8.1 Signage requirements

- Requirements for stairway identification signs are as follows:
  - Size of sign must be ≥ 18" × ≥ 16".
  - Floor number designation must be positioned in the center of the sign.
  - Characters on the sign have the following requirements:
    They must have a nonglare finish.
    They must contrast with the background in one of the following ways:
      Dark characters on a light background.
      Light characters on a dark background.
  - Sign background must have a nonglare finish.
  - Size of sign characters are required as follows:

Sign character	Height of character
Letters identifying stairway	≥ 1 1/2"
Number(s) identifying floor level	≥ 5"
All other characters	≥ 1"

  - Signs in certain interior exit enclosures must be luminescent.

Note: Section 1024, "Luminous Egress Path Markings," is cited as governing signage in the interior exit enclosures addressed above.
1024.4, "Self-luminous and photoluminescent," is cited as governing the materials required for signs in interior exit enclosures addressed above.

1022 Exit Enclosures

1022.9 Smokeproof enclosures and pressurized stairways

- This section addresses exits serving the following floor levels:
  - Serving stories in high-rise buildings as follows:
    Floor surface > 75' above lowest level of fire department vehicle access.
  - Serving stories in underground buildings as follows:
    Floor surface > 30' below a level of exit discharge.

Note: Section 403, "High-Rise Buildings," defines these buildings governed by this section.
Section 405, "Underground Buildings," defines these buildings governed by this section.

- Exits serving such stories must be one of the following:
  - A smokeproof enclosure.
  - A pressurized stairway.

Note: 909.20, "Smokeproof enclosures," is cited as governing these elements.

1022.9.1 Termination and extension *(part 1 of 2)*

- This section addresses smokeproof enclosures and pressurized stairways.
- Such enclosures are permitted to egress through the following in certain cases:
  - Areas on the level of exit discharge.
  - Vestibules.

Note: Section 1027, "Exit Discharge," is cited as governing discharge through areas or vestibules.

- Otherwise, such enclosures must discharge to one of the following:
  - A public way.
  - An exit discharge.
  - An exit passageway leading to one of the following:
    A public way.
    An exit discharge.

Note: 1022.2, "Termination," is cited as governing the discharge of such enclosures into an exit passageway.

- An exit passageway serving such an enclosure is governed as follows:
  - Passageway must be separated from the rest of the building as follows:
    By one or both of the following with a 2-hr fire-resistant rating:
    Fire barrier.
    Horizontal assembly.

Note: Section 707, "Fire Barriers," is cited as governing these elements.
Section 712, "Horizontal Assemblies," is cited as governing these elements.

  - Where passageways are not protected and pressurized equaling the enclosure, the following applies:
    A fire barrier must separate the enclosure from the passageway.
    Passage between enclosure and passageway must be by way of a fire door assembly.

Note: 1022.2, "Termination," is cited as requiring a fire door assembly as the opening in the fire barrier.

1022 Exit Enclosures

1022.9.1 Termination and extension *(part 2 of 2)*

- ○ Openings in the passageway are permitted with the following conditions:
  - Where a passageway serves a smokeproof enclosure the following applies:
    - Passageway must have the same protection and pressurization as the smokeproof enclosure.
    - Openings are protected as required for access from other floors.
  - Where a passageway serves a pressurized stairway the following applies:
    - Passageway must have the same protection and pressurization as the smokeproof enclosure.

1022.9.2 Enclosure access

- This section governs access to a stairway in a smokeproof enclosure via the following elements:
  - ○ Vestibule.
  - ○ Exterior balcony.
- For such access the following applies:
  - ○ Where the stairway is pressurized access by a vestuble or exterior balcony is not required.
  - ○ Where the stairway is not pressurized access via the vestibule or balcony is required.

Note: 909.20.5, "Stair pressurization alternative," is cited as governing stairway pressurization.

1023 Exit Passageways

1023.1 Exit passageway

- An exit passageway may be used only for a means of egress.

1023.2 Width

- The width of an exit passageway must be the larger of the following:
  - One of the following dimensions as applicable:
    ≥ 44" for occupant load ≥ 50.
    ≥ 36" for occupant load < 50.
  - As calculated by occupant load.

Note: 1005.1, "Minimum required egress width," is cited as governing width per occupant load.

- Protrusions into the required width of an exit passageway are limited as follows:

Table 1023.2 Protrusions into Exit Passageways

Protruding elements	Protrusion permitted
Handrails	≤ 7"
Doors at full open position	≤ 7"
Doors in any position	≤ 1/2 required width
Nonstructural trim and similar decorative features	≤ 1 1/2" from each side

Note: 1005.2, "Door encroachment," is cited as permitting certain protrusions into the required width which is summarized in the table above.

1023.3 Construction

- The fire-resistance rating for the following exit passageway enclosure components must be as indicated below:
  - Components:
    Walls.
    Floors.
    Ceilings.
  - Fire-resistance rating must be the larger of the following:
    1 hr.
    The rating required for any connecting exit enclosure.
- An exit passageway must be constructed as one or both of the following:
  - A fire barrier.
  - A horizontal assembly.

Note: The following are cited as governing the exit passageway construction:
Section 707, "Fire Barriers."
Section 712, "Horizontal Assemblies."

1023.4 Termination

- Exit passage ways must discharge into one of the following:
  - An exit discharge.
  - A public way.

1023 Exit Passageways

1023.5 Openings and penetrations

- Openings permitted in exit passageways include the following:
  - Unprotected exterior openings.
  - Openings to certain building service areas.

Note: 402.4.6, "Service areas fronting on exit passageways," is cited as permitting these areas to open to an exit passageway with certain fire-resistance-rated construction and opening protectives. Section 715, "Opening Protectives," is cited as governing openings in exit enclosures.

- Other openings into an exit enclosure are limited to the following:
  - Openings required for exit access as follows:
    - From spaces which usually are occupied.
  - Openings required for egress out of the enclosure.
- Exit passageways that extend exit enclosures to the following have additional code requirements:
  - A public way.
  - An exit discharge.

Note: 1022.2.1, "Extension," is cited as governing exit passageways where serving exit enclosures.

- Elevators are not permitted to open into an exit passageway.

1023.6 Penetrations

- Only the following components are permitted to penetrate an exit passageway:
  - Standpipes.
  - Exit doors.
  - Components serving the exit passageway as follows:
    - The following components used to pressurize the passageway:
      - Equipment and ductwork.
    - Sprinkler piping.
    - The following components terminating as noted below:
      - Components:
        - Electrical raceways as follows:
          - Including those for fire department communication.
      - Termination:
        - Raceways must terminate in a steel box as follows:
          - Side facing the enclosure must be ≤ 16 sq in.

Note: Section 713, "Penetrations," is cited as governing these penetrations.

- No other penetrations or openings through an exit passageway are permitted.
- None of the following are permitted between adjacent exit passageways:
  - Protected penetrations.
  - Unprotected penetrations.
  - Protected openings.
  - Unprotected openings.

1024 Luminous Egress Path Markings

1024.1 General

- The following locations are not governed by this section:
  - Certain lobbies that are in the means of egress.

  Note: 1027.1, "General," Exception 1, is cited as governing lobbies in a means of egress where path markings are not required.

  - Certain open parking garages where both of the following apply:
    On the level of exit discharge.
    With exit stairs or ramps essentially open.

    Note: 1027.1, "General," Exception 3, is cited as governing parking garages where path markings are not required, a partial summary of which is provided above.

- Otherwise, this section addresses exit paths in the following occupancies:
  - A, B, E, I, M, R-1 as follows:
    Where occupied floors are > 75' above the following level:
    The lowest level of fire department vehicle access.
- Exit paths governed by this section require the following:
  - Approved luminous markings defining the paths.

  Note: The following sections are cited as governing the luminous exit path markings required above: 1024.1, "General," through 1024.5, "Illumination."

1024.2 Markings within exit enclosures

- Egress path markings are required in the following:
  - Exit enclosures.
  - Vertical exit enclosures.
  - Exit passageways.

  Note: The following are cited as governing egress path markings: 1024.2.1, "Steps," through 1024.2.6, "Doors from exit enclosures."

1024.2.1 Steps

- A solid and continuous stripe is required on each step in the following location:
  - On the horizontal leading edge as follows:
    For the full length of the step.
    With the leading edge of the stripe ≤ 1/2" from the edge.
    With any overlap on the vertical face of the leading edge ≤ 1/2" down the face.
- Required stripes that do not comply with the cited standard must have the following width:
  - ≥ 1" and ≤ 2".

  Note: UL 1994, "Low Level Path Marking and Lighting Systems," is cited as the standard with which stripes that comply are not required to have a minimum width of 1".

1024.2.2 Landings

- The leading edge of landings require the same stripe as required for steps.

1024 Luminous Egress Path Markings

1024.2.3 Handrails

- Handrails and extensions require a solid and continuous stripe as follows:
  - Stripe width to be ≥ 1" where not in compliance with the cited standard.

  Note: UL 1994, "Low Level Path Marking and Lighting Systems," is cited as the standard with which stripes that comply are not required to have a minimum width of 1".

  - Stripe to be located on top the handrail as follows:
    - On the full length of the handrail.
    - On extensions.
    - On newel post caps.
    - ≤ 4" gaps in the stripe are permitted only as follows:
      - Where handrails or extensions bend.
      - Where handrails or extensions turn corners.

1024.2.4 Perimeter demarcation lines

- The following surfaces in exit enclosures are governed by this section:
  - Landings.
  - Floor areas.
- The following are not governed by this section:
  - The sides of steps in exit enclosures.
- Solid and continuous demarcation lines are required as follows:
  - On one or both of the following surfaces:
    - Floors.
    - Walls.
  - With the following dimensions:
    - ≥ 1"wide and ≤ 2" wide.
    - With no gaps > 4".

  Note: UL 1994, "Low Level Path Marking and Lighting Systems," is cited as the standard with which stripes that comply are not required to have a minimum width of 1".

1024.2.4.1 Floor-mounted demarcation lines

- The following surfaces in exit enclosures are governed by this section:
  - Landings.
  - Floor areas.
- Perimeter demarcation lines must be located as follows:
  - Lines are to be ≥ 4" from the wall.
  - Lines are to extend to a point ≤ 2" from the line at the landing leading edge.
  - Lines passing by doorways are governed as follows:
    - They are not required at the following exit doors:
      - Those which lead out of an exit enclosure and require use to exit the building.
    - They are required at all other doors of exit enclosures.

1024 Luminous Egress Path Markings

1024.2.4.2 Wall-mounted demarcation lines

- This section governs demarcation lines placed on walls in exit enclosures.
- Demarcation lines must be placed as follows:
  - The bottom edge of the line must be ≤ 4" from the finished floor.
- Demarcation lines must turn down to meet the floor as follows:
  - At the tops or bottoms of stairs.
  - At a point ≤ 2" from the edge of the step or landing.
  - Once at the floor, the line must run across the floor in the following case:
    Where this is the only practical way of defining a path.
- Demarcation lines interrupted by doorways are governed as follows:
  - They are not to be extended on or past the following exit doors:
    Those which lead out of an exit enclosure and require use to exit the building.
  - At other doors, a demarcation line must take one of the following paths:
    The line must run across the door surface.
    The line must turn down to run across the doorway on the floor in front of the door.

1024.2.4.3 Transition

- This section governs the connection of wall demarcation lines to floor demarcation lines.
- Where the two lines connect to form a continuous line, the following applies:
  - The wall line must turn down to the floor.
  - The floor line must extend to meet the wall line where it is turned down.

1024.2.5 Obstacles

- This section addresses obstacles in a means of egress as follows:
  - Includes obstacles such as the following:

Hose cabinets	Standpipes
Obstacles restricting height	Wall projections

  - Only includes obstacles with either of the following dimensional characteristics:
    ≤ 6'-6" in height.
    Projection > 4" into the egress path.
- Such obstacles must be outlined with a pattern of markings as follows:
  - Alternating equal bands of luminescent luminous material and black.
  - In a pattern ≥ 1" wide with ≤ 2" thick bands at 45°.
  - Markings may not obscure information regarding the obstacles such as the following:
    Required information.
    Required indicators.
    Instruction to occupants for use of standpipes.

1024.2.6 Doors from exit enclosures

- This section governs doors within an exit enclosure through which occupants must pass to egress.
- Such doors require special markings.

Note: The following sections are cited as governing the markings of doors governed by this section:
1024.2.6.1, "Emergency exit symbol."
1024.2.6.2, "Door hardware markings."
1024.2.6.3, "Door frame markings."

1024 Luminous Egress Path Markings

1024.2.6.1 Emergency exit symbol

- Doors governed by this section must have a luminous emergency exit symbol as follows:
  - The symbol must be ≥ 4" in height.
  - The symbol must be mounted on the door as follows:
    - Centered horizontally.
    - The top of the symbol must be ≥ 18" above the floor.

Note: NFPA 170, "Standard for Fire Safety and Emergency Symbols," is cited as governing the exit symbol.

1024.2.6.2 Door hardware markings

- Door hardware governed by this section must have special markings as follows:
  - Hardware must have ≥ 16 sq in of luminous material marking it.
  - Hardware marking must be located in ≥ 1 of the following places:
    - Immediately adjacent to the hardware.
    - On the door handle or escutcheon plate.
    - On panic hardware where installed as follows:
      - Marking to be ≥ 1" wide for the full length of bar or touchpad.

1024.2.6.3 Door frame markings

- Doors governed by this section must have the following markings:
  - Solid and continuous stripe ≥ 1" and ≤ 2" wide located as follows:
    - On the top and side mouldings of the door where possible.
    - On the wall above and on each side of the door where not possible on the moulding.

1024.3 Uniformity

- The format of demarcation lines is to be uniform as follows:
  - Within an exit enclosure.

1024.4 Self-luminous and photoluminescent

- This section governs luminous egress path perimeter lines as follows:
  - The following materials are permitted to create the lines:
    - Paint.
    - Self-luminous materials.
    - Photoluminescent materials.
    - Other materials.
  - An electrical charge is not permitted to provide luminescence of the lines.
  - Materials for the lines must be applicable standards.

Note: The following standards are cited as alternatives governing luminescent path lines:
- *UL 1994, "Low Level Path Marking and Lighting Systems."*
- *ASTM E 2072, "Specification for Photoluminescent (Phosphorescent) Safety Markings," with the following variations:*
  - *The charging source is to be as follows:*
    - *1 foot candle of fluorescent illumination for 1 hr.*
  - *Luminance is to be as follows:*
    - *≥ 30 millicandelas/sq meter after 10 min.*
    - *≥ 5 millicandelas/sq meter after 1 1/2 hr.*

1024 Luminous Egress Path Markings

1024.5 Illumination

- This section governs exit enclosures as follows:
  - Where exit path perimeter photoluminescent lines are used.
- ≥ 1 foot candle of illumination is required as follows:
  - ≥ 1 hr before the building is occupied.

Note: Section 1006, "Means of Egress Illumination," is cited as governing illumination in exit enclosures where exit path perimeter photoluminescent lines are used and is partially summarized above.

1025 Horizontal Exits

1025.1 Horizontal exits

- All required exits may be horizontal exits in Occupancy I-3 as follows:
  - The space on each side of a horizontal exit must have the following occupant capacity:
    Area ≥ 6 sf × (sum of occupants from both sides of the horizontal exit).
  - A fire compartment defined by a horizontal exit is not required to have the following egress components where the conditions indicated below are present:
    Egress components:
      Stairway leading directly to the exterior.
      Door leading directly to the exterior.
    Conditions:
      An adjoining fire compartment must have one of the egress components listed above.
      Egress does not return to the compartment of origin.
- ≤ 2/3 of required exits in Occupancy I-2 may be horizontal exits as follows:
  - From a building or a floor.
- In other occupancies, horizontal exits are limited as follows:
  - They may not provide the only means of exit from any part of a building.
  - They may provide ≤ 1/2 the required exits.
  - They may provide ≤ 1/2 the required exit width.
- The compartment served by a horizontal exit need not have a stairway or door leading outside as follows:
  - Where the area of refuge accessed by the horizontal exit has the following:
    A stairway or door leading directly to the exterior.
    Means of egress that do not require occupants to return to the compartment of origin.

1025.2 Separation *(part 1 of 2)*

- This section governs separation walls between the following:
  - Buildings connected by a horizontal exit.
  - Areas of refuge connected by a horizontal exit.
- The separation wall must be one of the following types:
  - A fire wall.
  - One or both of the following:
    A fire barrier as follows:
      Walls must completely divide the floor served by the horizontal exit.
      Walls must be continuous between exterior walls.
    A horizontal assembly.

Note: Section 706, "Fire Walls," is cited as governing this type of wall.
Section 707, "Fire Barriers," is cited as governing this type of wall.
Section 712, "Horizontal Assemblies," is cited as governing this type of construction.

- The separation wall must have a fire-resistance rating ≥ 2 hr.
- Separation wall must extend vertically through the entire building in either of the following:
  - Where fire-resistance rating of floor assemblies is < 2 hr.
  - Where floor assemblies have unprotected openings.
- Openings in the separation wall must be protected.

Note: Section 715, "Opening Protectives," is cited as governing these openings.

1025 Horizontal Exits

1025.2 Separation *(part 2 of 2)*

- Any of the following in the required separations must be protected:
  - Duct openings.
  - Air transfer openings.

Note: Section 716, "Ducts and Air Transfer Openings," is cited as governing these openings.

- A horizontal exit does not require a fire-resistance rating where all of the following conditions are met:
  - Located between the following:
    - A building area.
    - A pedestrian walkway above grade.
  - The distance between the connected buildings must be > 20'.

Note: Section 3104, "Pedestrian Walkways and Tunnels," is cited as governing the walkway.

- Horizontal exit walls built as fire barriers have the following requirement:
  - They must run from exterior wall to exterior wall as follows:
    - So as to divide the floor completely.

1025.3 Opening protectives

- Fire doors in horizontal exits must close when activated by a smoke detector.
- Fire doors in a cross-corridor configuration must close when activated by a smoke detector.

Note: 715.4.8.3, "Smoke-activated doors," is cited as governing smoke activation.

1025.4 Capacity of refuge area

- Refuge areas of a horizontal exit must be one of the following:
  - Area occupied by the same tenant.
  - Public areas.
- A refuge area must be able to hold the sum of the following:
  - Its original occupants.
  - Occupants expected from the area connected by the horizontal exit.
- The number of occupants expected to travel to an area of refuge through a horizontal exit is based on the following:
  - By the capacity of the horizontal exit doors through which they must pass.
- The space required to house people in a refuge area is governed as follows:
  - Required area ≥ floor area per person × total number of occupants to be accommodated.
  - Floor areas required per person are based on occupancy as follows:

Table 1025.4 Refuge Area Capacity

Occupancy	Net floor area per person
I-2 housing nonambulatory occupants	30 sf
I-2 housing ambulatory occupants	15 sf
I-3	6 sf
All other occupancies	3 sf

- The refuge area accessed by a horizontal exit must have the following:
  - Exits complying with the code as follows:
    - Occupant load does not include those entering the area of refuge through a horizontal exit.
    - ≥ 1 exit leading directly to the exterior or an exit enclosure.

1026 Exterior Exit Ramps and Stairways

1026.1 Exterior exit ramps and stairways

- Exterior exit ramps and stairways at outdoor stadiums as follows are not governed by this section series:
  - Where all portions of the means of egress are essentially open to the exterior.

  Note: 1022.1, "Enclosures required," Exception 2 is cited as governing these elements.

- Other ramps and stairways are governed by this section series.

1026.2 Use in a means of egress

- The use of exterior exit stairways in a required means of egress is governed as follows:
  - Not permitted in Occupancy I-2.
- Exterior exit stairways and ramps are permitted in other occupancies where both of the following apply:
  - Building is ≤ 6 stories above grade plane.
  - Building has occupied floors ≤ 75' above the following:
    Lowest level of fire department vehicle access.

1026.3 Open side

- Exterior exit ramps and stairways serving in a means of egress must have ≥ 1 side open as follows in the locations listed below:
  - A total of ≥ 35 sf must be open.
  - Required open area must be located ≤ 42" above each of the following:
    Each adjacent floor level.
    Each adjacent intermediate landing.

1026.4 Side yards

- The required open side of an exterior exit ramp or stairway must adjoin one of the following:
  - Yard.
  - Court.
  - Public way.
- Closed sides of an exterior exit ramp or stairway may be exterior walls of the building.

1026.5 Location

- Exterior exit ramps and stairways must be positioned as follows:
  - ≥ 10' from lot lines.
  - ≥ 10' from other buildings on the same lot as follows:
    Where adjacent, the exterior building walls and openings are not protected as follows:
    As per exterior wall requirements vs. fire separation distance.

Note: 1027.3, "Exit discharge location," is cited as the source of requirements for adjacent walls within 10' of the stairway according to fire separation distance.

1026 Exterior Exit Ramps and Stairways

Case study: Fig. 1026.3. 100% of the exterior stairway is open above the 42" level on both sides of the intermediate landing and on all sides of the 2nd-floor landing. The east side of the stairway is 100% open at the 1st-floor landing, thus, providing 38 1/2 sf of opening above the 42" level. This is > than the 35 sf minimum. The open sides face yards and are > 10' from lot lines. The stairway serves the power plant and is not part of the I-2 occupancy building. The stairway is in compliance with the code.

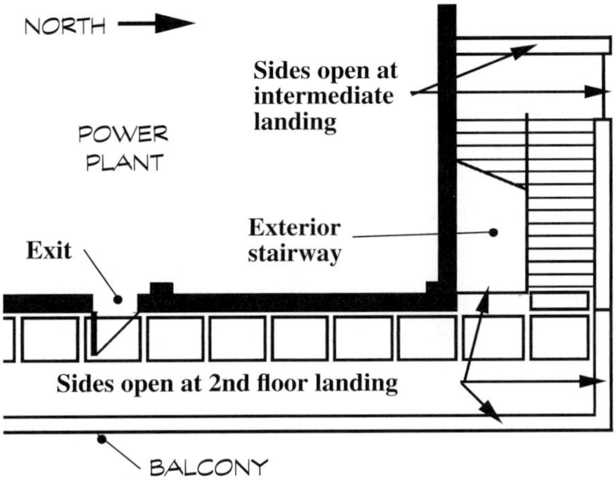

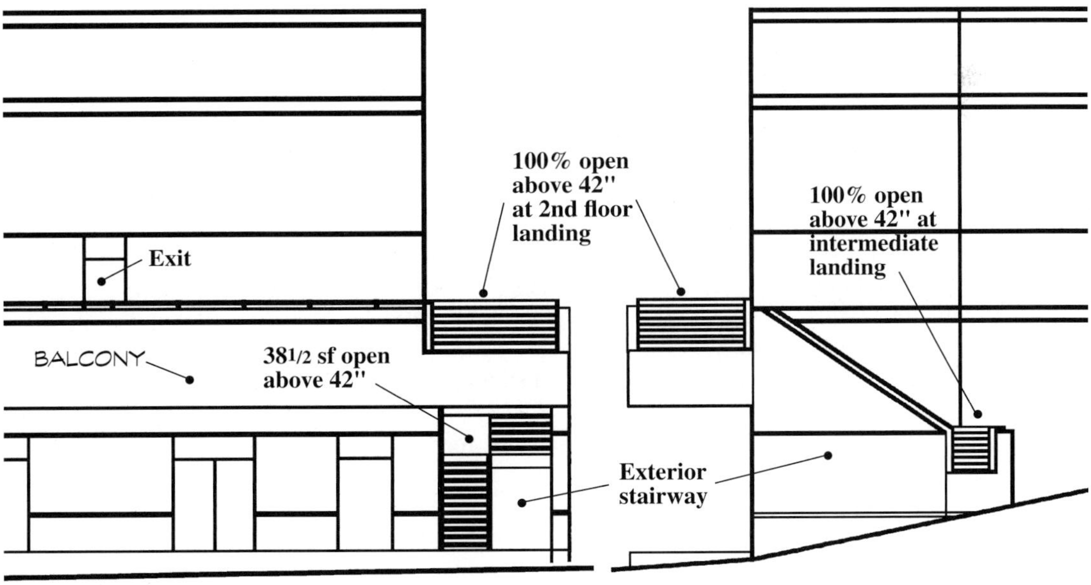

Fig. 1026.3. Plan and elevations of exterior stairway. Christus St. Michael Health Care Center. Texarkana, Texas. Watkins Hamilton Ross Architects, Inc. Houston, Texas.

1026 Exterior Exit Ramps and Stairways

1026.6 Exterior ramps and stairway protection

- This section addresses the separation of an exterior exit ramp or stairway from the building interior.
- Separation is not required where all of the following conditions apply:
  - In occupancies A, B, E, F, H, I, M, R-3, R-4, S, and U.
  - Buildings are ≤ 2 stories above the grade plane.
  - Level of exit discharge serving such occupancies is the 1st story above the grade plane.
- Separation is not required where the stairway is served by an exterior balcony meeting all of the following conditions:
  - Balcony connects 2 remote exits that are of the following types:
    - Exterior stairways.
    - Other approved exits.
  - Balcony has a perimeter that is ≥ 50% open as follows:
    - Open area is ≥ 1/2 the height of the enclosing wall.
    - Top of the open area ≥ 7' above the floor of the balcony.
- Separation is not required at buildings where enclosures are not required for interior stairways.

Note: 1022.1, "Enclosures required," is cited as the source of requirements permitting open interior stairs.

- Separation is not required where the exterior exit stairway is connected to an open-ended corridor meeting all of the following conditions:
  - Building is sprinklered including the following:
    - The corridors.
    - The stairs.

Note: The following are cited as alternatives to governing the sprinkler system as applicable:
903.3.1.1, "NFPA 13 sprinkler systems."
903.3.1.2, "NFPA 13R sprinkler systems."

  - Open-ended corridor complies with requirements for interior corridors.

Note: Section 1018, "Corridors," is cited as governing interior corridors.

  - Each end of the open-ended corridor connects to an exterior exit ramp or stairway.

Note: Section 1026, "Exterior Exit Ramps and Stairways," is cited as governing.

  - One of the following is provided at any change of direction > 45°:
    - A clear opening to the exterior ≥ 35 sf as follows:
      - Opening minimizes accumulation of smoke and toxic gases.
    - An exterior exit ramp or stairway.
- For all other cases, the walls separating exterior exit stairways from the interior of the building must comply with the following:
  - They must meet fire-resistance and other requirements for vertical exit enclosures.

Note: 1022.1, "Enclosures required," is cited as the source of applicable requirements.

  - Only openings necessary for egress are permitted in the separating walls as follows:
    - Egress from spaces which are normally occupied.

1027 Exit Discharge

1027.1 General

- The following two egress configurations may not be used simultaneously for > 50% of a building's exits:
  - Egress from an exit enclosure may pass through an interior space only where all of the following apply:
    - The space is at the level of exit discharge.
    - Egress from ≤ 50% of the capacity of enclosures passes through interior space.
    - Egress from ≤ 50% of the number enclosures passes through interior space.
    - The space provides a path to an exterior door as follows:
      - Path must be unobstructed.
      - Path must be readily apparent from the terminal of the exit enclosure.
    - Level of discharge must be separated from areas below as follows:
      - Separation construction has a fire-resistance rating = that of the exit enclosure.
    - The egress path from the exit enclosure is sprinklered.
    - Areas at discharge level with access to the egress path must have one of the following:
      - A sprinkler system as per NFPA 13 or 13R.
      - Separation from the rest of the building as per exit enclosure requirements.

  Note: The following are cited as alternatives to governing the sprinkler system as applicable:
  903.3.1.1, "NFPA 13 sprinkler systems."
  903.3.1.2, "NFPA 13R sprinkler systems."

  - Egress from an exit enclosure may pass through a vestibule only where all of the following apply:
    - Egress from ≤ 50% of the capacity of enclosures passes through interior space.
    - Egress from ≤ 50% of the number of enclosures passes through interior space.
    - The vestibule must have all of the following characteristics:
      - It is separated from areas below it as follows:
        - Separation construction has a fire-resistance rating = that of the exit enclosure.
      - Depth of vestibule measured from exterior of building is ≤ 10'.
      - It is separated from rest of exit discharge level as follows:
        - Separation construction is equivalent to the following:
          - To that provided by wire glass in steel frames:
            - Wire glass must be approved.
      - Vestibule is used as a means of egress only.
      - Vestibule discharges directly to the exterior.
- Stairways in open parking garages are governed as follows:
  - They may egress through the open parking garage as follows:
    - At their exit discharge level.

  Note: 1022.1, "Enclosures required," Exception 5, is cited as permitting this condition.

- Horizontal exits are not required to discharge directly to the outside.

  Note: Section 1025, " Horizontal Exits," is cited as governing these exits.

- All other exit discharges must comply with all of the following:
  - Discharge must be directly to the outside.
  - Discharge must occur in one of the following ways:
    - At grade.
    - At direct access to grade.
  - Discharge may not return to the building.

1027 Exit Discharge

1027.2 Exit discharge capacity

- Exit discharge capacity must be ≥ the required capacity of exits being discharged.

1027.3 Exit discharge location

- The following elements must be located as indicated below:
  - Elements:
    - Exterior balconies.
    - Exterior stairways.
    - Exterior ramps.
  - Location requirements:
    - ≥ 10' from lot lines.
    - ≥ 10' from other buildings on the same lot in the following circumstance:
      - Where adjacent exterior building walls and openings are not protected as follows:
        - As per exterior wall requirements based on fire separation distance.

Note: Section 704, "Fire-Resistance Rating of Structural Members," is cited as governing exterior walls and openings that are < 10' from lot lines and other buildings on the same lot.

1027.4 Exit discharge components

- Exit discharge components must be open to the exterior as follows:
  - To a degree that minimizes the accumulation of smoke and toxic gases.

1027.5 Egress courts

- An egress court as follows has the requirement listed below:
  - Egress court:
    - Acting as a component of the exit discharge as follows:
      - In a means of egress.
  - Requirement:
    - The egress court must comply with exit discharge requirements.

Note: Section 1027, "Exit Discharge," is cited as the source of applicable requirements.

1027.5.1 Width *(part 1 of 2)*

- The width of an egress court must be ≥ the larger of the following:
  - The width required for means of egress.

Note: 1005.1, "Minimum required egress width," is cited as governing width.

  - Widths as follows:

Occupancy	Width
R-3, U	≥ 3'
Other	≥ 3'-8"

- Unobstructed height of an egress court within required width must be ≥ 7'.

1027 Exit Discharge

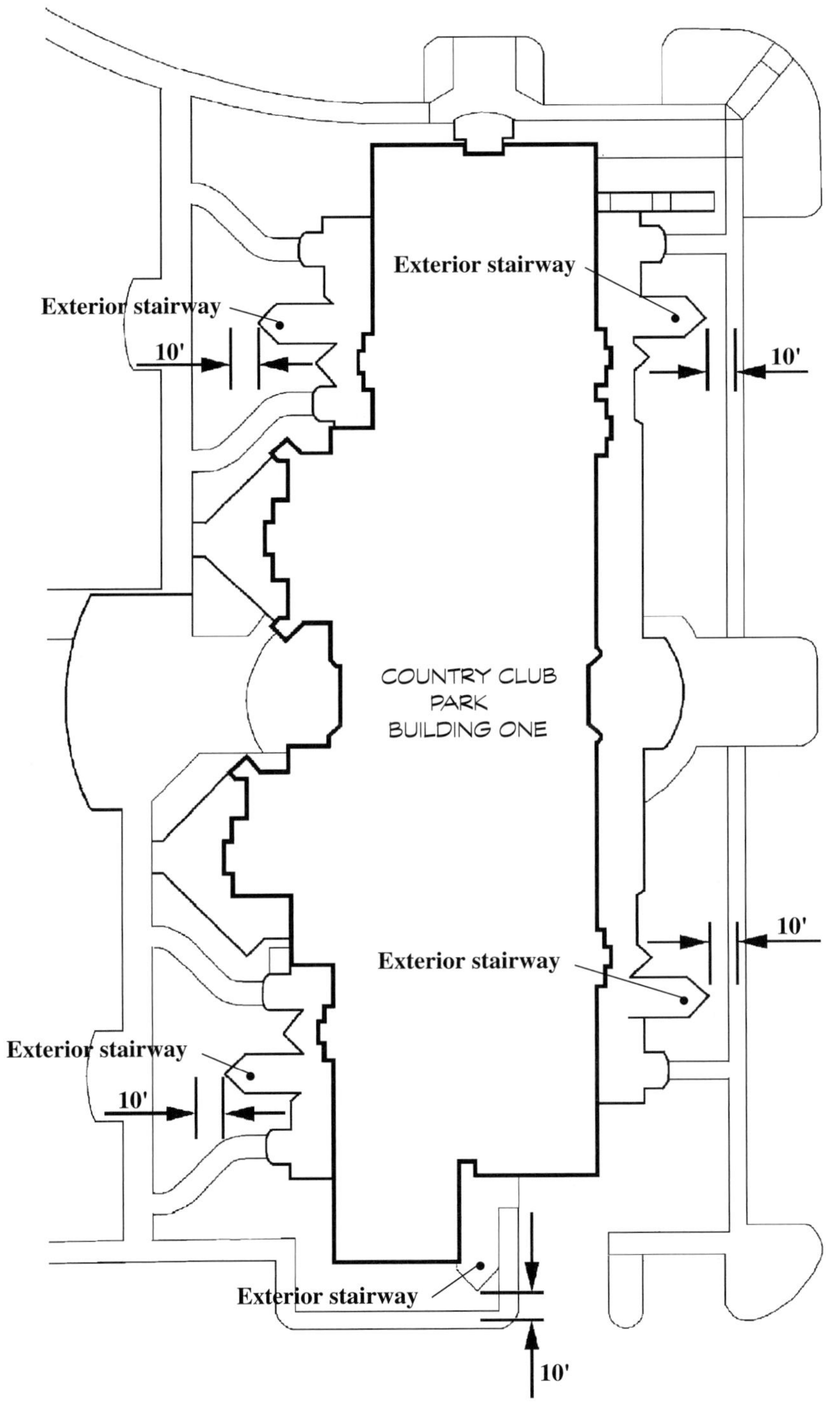

Case study:
Fig. 1027.3. The 5 exterior stairways are > 10' from lot lines ranging from 160' to 240' away. The 10' minimum distance is indicated at each stairway. The 6 exterior balconies are further from the lot lines. There are no other buildings on the site.

Fig. 1027.3. Partial site plan. Country Club Park Building One. Wichita, Kansas. Gossen Livingston Associates, Inc. Architecture. Wichita, Kansas.

1027 Exit Discharge

1027.5.1 Width *(part 2 of 2)*

- Only the following protrusions are permitted into the required width of an egress court:

Table 1027.5.1 Protrusions into Egress Courts

Protruding elements	Protrusion permitted
Handrails	≤ 7"
Doors at full open position	≤ 7"
Doors in any position	≤ 1/2 required width
Nonstructural trim and similar decorative features	≤ 1 1/2" from each side

Note: 1005.2, "Door encroachment," is cited as permitting certain protrusions into the required width which is summarized in the table above.

- An egress court may diminish in width in the direction of egress only as follows:
  - Width may not diminish to < the required width.
  - The transition must be gradual.
  - The transition may not form an angle > 30° with the line of travel.
  - The transition must be bordered by a guard ≥ 36" high.

1027.5.2 Construction and openings

- The following egress courts are not governed by this section:
  - Where the occupant load is < 10.
  - In Occupancy R-3.
- In other cases, where an egress court is < 10' wide, the following applies:
  - Court walls must have a fire-resistance rating ≥ 1 hr as follows:
    - Rated construction must extend ≥ 10' above the court floor.
    - Openings in the wall ≤ 10' high must have the following:
      - Opening protectives rated ≥ 3/4 hr.

1027.6 Access to a public way

- Where access to a public way is not possible, the following is permitted:
  - Provision of a safe dispersal area with all of the following characteristics:
    - Must provide ≥ 5 sf per person.
    - Must be located on the same lot as the building served.
    - Must be located ≥ 50' from the building served.
    - Must be permanently maintained as a safe dispersal area.
    - Must be identified as a safe dispersal area.
    - Must be accessed from the building served by the following:
      - A safe path as follows:
        - Unobstructed.
  - In all other cases, the exit discharge must provide the following:
    - Direct access to a public way as follows:
      - Unobstructed.

1028 Assembly

1028.1 General

- This section governs the following occupancies which have the furnishings listed below:
  - Occupancies:
    - Occupancy A, assembly.
    - Assembly that is accessory to Occupancy E.
  - Furnishings:
    - Displays.
    - Equipment.
    - Seats.
    - Tables.
    - Other material.

1028.1.1 Bleachers

- The following nonbuilding elements must comply with specified standards:
  - Bleachers.
  - Folding seating.
  - Grandstands.
  - Telescopic seating.

Note: *ICC 300, "ICC Standard on Bleachers, Folding and Telescopic Seating and Grandstands," is cited as governing this seating.*

1028.2 Assembly main exit

- This section governs buildings in the following occupancies:
  - Occupancy A, assembly.
  - Assembly that is accessory to Occupancy E.
- Exits may be distributed on the perimeter of buildings where all of the following conditions apply:
  - Exits have a total egress width ≥ the required width.
  - One of the following cases applies:
    - There are no well-defined main exits.
    - Multiple main exits are provided.

Note: *Stadiums and arenas are typical of this condition.*

- Where the following conditions apply, other buildings and spaces must have a main exit as indicated below:
  - Where the occupant load > 300.
  - Main exit requirement:
    - Width must accommodate ≥ 1/2 the total occupant load.
    - Width must accommodate all means of egress served by the exit.
- Where the building is designated as Occupancy A, the main exit must face one of the following:
  - A street.
  - An unoccupied space with both of the following characteristics:
    - ≥ 10' wide.
    - Adjoins a street or public way.

1028 Assembly

1028.3 Assembly other exits

- This section governs buildings in the following occupancies:
  - Occupancy A, assembly.
  - Assembly that is accessory to Occupancy E.
- Exits may be distributed on the perimeter of buildings where all of the following conditions apply:
  - Exits have a total egress width ≥ the required width.
  - One of the following cases applies:
    - There are no well-defined main exits.
    - Multiple main exits are provided.

  Note: Stadiums and arenas are typical of this condition.

- Other buildings and spaces with an occupant load > 300 are governed as follows:
  - Means of egress in addition to the main exit are required for each level of Occupancy A as follows:
    - Egress capacity must ≥ 1/2 the total occupant load of the level served.
    - Means of egress must comply with exit access location requirements.

  Note: 1015.2, "Exit or exit access doorway arrangement," is cited as the applicable requirements for exit access location.

1028.4 Foyers and lobbies

- This section addresses the following in Occupancy A-1:
  - Lobbies.
  - Similar spaces where people wait for seating.
- Such spaces must comply with the following:
  - Waiting spaces may not overlap into the means of egress required width.
  - Waiting spaces must have one of the following relationships to a public street:
    - Be connected directly by all the main entrances or exits.
    - Have a corridor or path of travel to every main entrance or exit as follows:
      - Travel to be straight.
      - Travel to be unobstructed.

1028.5 Interior balcony and gallery means of egress

- This section addresses the following assembly means of egress:
  - From balconies.
  - From galleries.
  - From press boxes.
- Where seating is provided for ≥ 50, the following applies:
  - ≥ 2 means of egress are required as follows:
    - ≥ 1 means of egress must be located at each side of the seating area.
    - ≥ 1 means of egress must lead directly to an exit.

1028 Assembly

Case study: Fig. 1028.5. The swimming pool balcony seats 86 occupants, thus, requiring ≥ 2 means of egress. The balcony meets the requirement with the provision of a stairway at each side of the seating as stipulated by the code. Both means of egress lead directly to an exit.

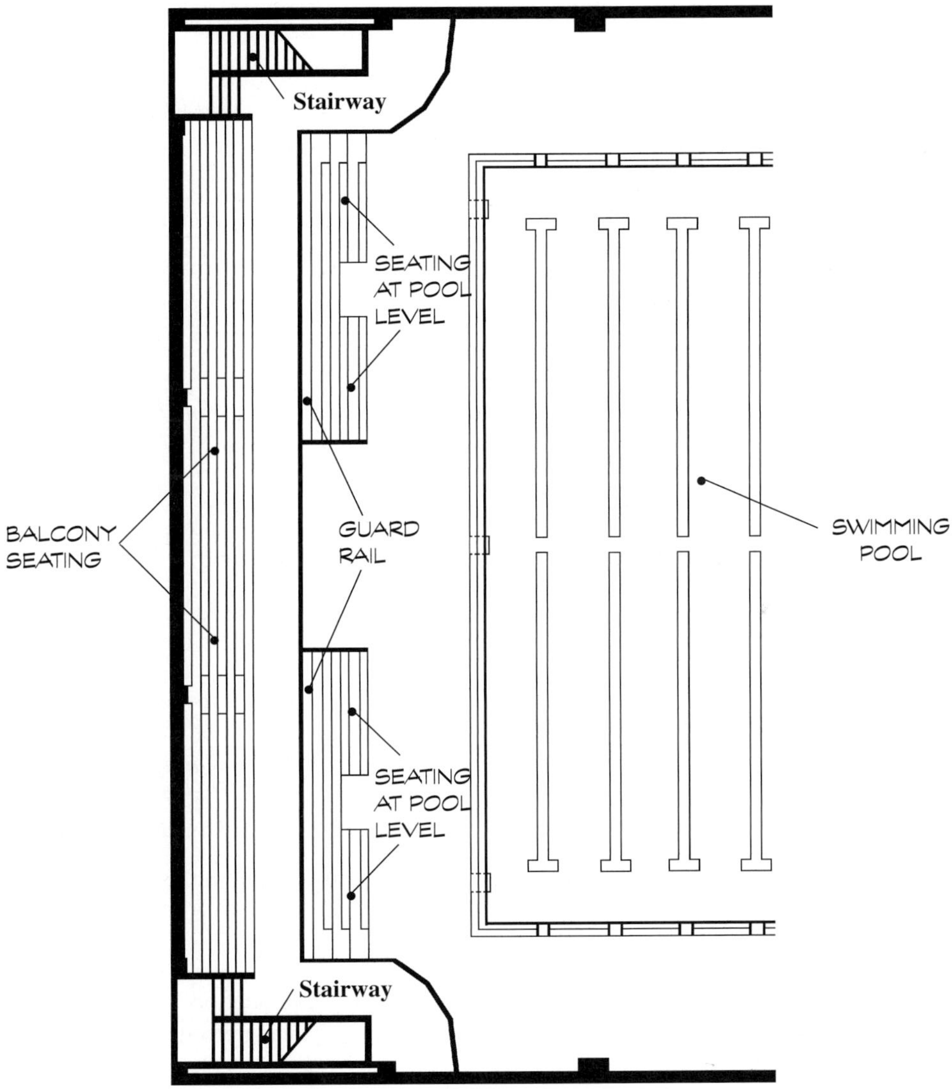

Fig. 1028.5. Partial floor plan at swimming pool. High School 6, Cypress-Fairbanks Independent School District. Harris County, Texas. PBK Architects, Inc. Houston, Texas.

1028 Assembly

1028.5.1 Enclosure of openings

- Stairways may be open between the following facilities and main assembly floors such as listed below:
  - Facilities:
    Balconies.
    Galleries.
    Press boxes.
  - Main assembly floors:
    Theaters.
    Places of religious worship.
    Auditoriums.
    Sports facilities.
- In other assembly spaces, the following openings at balconies must be enclosed in exit enclosures:
  - Interior stairways.
  - Other vertical openings.

Note: 1022.1, "Enclosures required," is cited as the source of applicable requirements.

- ≥ 1 accessible means of egress is required from the following levels:
  - A balcony with accessible seating.
  - A gallery with accessible seating.
  - A press box level with accessible seating.

Note: The following are cited as governing accessible means of egress as noted above:
1007.3, "Stairways."
1007.4, "Elevators."

1028.6 Width of means of egress for assembly

- Clear width of aisles and other means of egress is measured to the following:
  - Walls.
  - Edges of seating.
  - Edges of treads.

Note: 1028.6.1, "Without smoke protection," is cited as governing the widths noted above where seating is not smoke-protected.
The following are cited as governing widths as noted above where seating is smoke-protected:
1028.6.2, "Smoke-protected seating."
1028.6.3, "Width of means of egress for outdoor smoke-protected assembly."

- A requirement for a specified clear width does not preclude protrusions into the clear width in certain cases.

1028.6.1 Without smoke protection *(part 1 of 2)*

- This section addresses means of egress clear width with no smoke protection.
- A width per occupant ≥ 0.3" is required on stairways with the following profile:
  - Riser height ≤ 7".
  - Tread depth ≥ 11".

1028 Assembly

1028.6.1 Without smoke protection *(part 2 of 2)*

- Where the riser height is > 7", the required width per occupant is as follows:

  Width per occupant ≥ {[(riser height – 7") ÷ 0.1] × 0.005"} + 0.3"

  This equation yields the following required widths per occupant for riser heights up to 8".

Table 1028.6.1a Stairway Width Based on Riser Height

Riser height	Width per occupant	Riser height	Width per occupant	Riser height	Width per occupant	Riser height	Width per occupant
7.05"	≥ 0.3025"	7.30"	≥ 0.3150"	7.55"	≥ 0.3275"	7.80"	≥ 0.3400"
7.10"	≥ 0.3050"	7.35"	≥ 0.3175"	7.60"	≥ 0.3300"	7.85"	≥ 0.3425"
7.15"	≥ 0.3075"	7.40"	≥ 0.3200"	7.65"	≥ 0.3325"	7.90"	≥ 0.3450"
7.20"	≥ 0.3100"	7.45"	≥ 0.3225"	7.70"	≥ 0.3350"	7.95"	≥ 0.3475"
7.25"	≥ 0.3125"	7.50"	≥ 0.3250"	7.75"	≥ 0.3375"	8.00"	≥ 0.3500"

- A width per occupant ≥ 0.375" is required as follows:
  - Where egress requires going down stairs.
  - Where risers are ≤ 7".
  - Required for the segment of stairs > 30" from a handrail.
- The width indicated below is required where all of the following conditions apply:
  - Conditions:
    Where egress requires going down stairs.
    Where risers are > 7".
    Required for the segment of stairs > 30" from a handrail.
  - The width required per occupant is defined by the following equation:

  Width per occupant ≥ {[(riser height – 7") ÷ 0.1] × 0.005"} + 0.375" required

  This equation yields the following required widths per occupant for riser heights up to 8".

Table 1028.6.1b Descending Egress Stairway Width Based on Riser Height

Riser height	Width per occupant	Riser height	Width per occupant	Riser height	Width per occupant	Riser height	Width per occupant
7.05"	≥ 0.3775"	7.30"	≥ 0.3900"	7.55"	≥ 0.4025"	7.80"	≥ 0.4150"
7.10"	≥ 0.3800"	7.35"	≥ 0.3925"	7.60"	≥ 0.4050"	7.85"	≥ 0.4175"
7.15"	≥ 0.3825"	7.40"	≥ 0.3950"	7.65"	≥ 0.4075"	7.90"	≥ 0.4200"
7.20"	≥ 0.3850"	7.45"	≥ 0.3975"	7.70"	≥ 0.4100"	7.95"	≥ 0.4225"
7.25"	≥ 0.3875"	7.50"	≥ 0.4000"	7.75"	≥ 0.4125"	8.00"	≥ 0.4250"

- Ramped or level means of egress require clear widths per occupant as follows:

Slope	Width per occupant
> 1:12	≥ 0.22"
≤ 1:12	≥ 0.20"

1028 Assembly

1028.6.2 Smoke-protected seating *(part 1 of 3)*

- This section does not address the width of means of egress for the following:
  - Outdoor smoke-protected assembly as follows:
    With an occupant load ≤ 18,000.

Note: 1028.6.3, "Width of means of egress for outdoor smoke-protected assembly," is cited as governing this seating.

- This section addresses the clear width required for means of egress in smoke-protected assembly seating.
- Width required is calculated by the following equation:
  - *Total width = (width required per seat served) × (number of seats served)* as follows:
    The seats in a space where they are exposed to same smoke-protected environment.
  - Factors for inches of width required per seat served, as provided by the code, may be interpolated.
- A Life Safety Evaluation is required for seating utilizing egress widths stipulated by this section.

Note: NFPA 101, "Life Safety Code," is cited as the source of requirements for the life safety evaluation.

- Required widths for means of egress elements are provided in the tables below as follows:
  - Equations for interpolating factors for width per seat served as required by the code.
  - Tables of required total egress widths for selected quantities of seats.

Table 1028.6.2a Width of Stairs and Aisle Steps ≤ 30" from a Handrail

Method of calculating inches per seat served for the table of minimum widths below:

Seats	Calculation
1 – 5,000	0.200" – a fixed number
5,001 – 10,000	0.200" – [0.0000140” × (number of seats > 5,000)]
10,001 – 15,000	0.130" – [0.0000068" × (number of seats > 10,000)]
15,001 – 20,000	0.096" – [0.0000040" × (number of seats > 15,000)]
20,001 – 25,000	0.076" – [0.0000032" × (number of seats > 20,000)]
25,000 – up	0.060" – a fixed number

Minimum width per seat served:

Seats	Width	Seats	Width	Seats	Width	Seats	Width
500	≥ 0.200"	7,000	≥ 0.172"	14,000	≥ 0.103"	21,000	≥ 0.073"
1,000	≥ 0.200"	8,000	≥ 0.158"	15,000	≥ 0.096"	22,000	≥ 0.696"
2,000	≥ 0.200"	9,000	≥ 0.144"	16,000	≥ 0.092"	23,000	≥ 0.066"
3,000	≥ 0.200"	10,000	≥ 0.130"	17,000	≥ 0.088"	24,000	≥ 0.063"
4,000	≥ 0.200"	11,000	≥ 0.123"	18,000	≥ 0.084"	25,000	≥ 0.060"
5,000	≥ 0.200"	12,000	≥ 0.116"	19,000	≥ 0.080"	26,000	≥ 0.060"
6,000	≥ 0.186"	13,000	≥ 0.110"	20,000	≥ 0.076"	27,000	≥ 0.060"

Source: IBC Table 1028.6.2.

1028 Assembly

1028.6.2 Smoke-protected seating *(part 2 of 3)*

Table 1028.6.2b Width of Stairs and Aisle Steps > 30" from a Handrail

Method of calculating inches per seat served for the table of minimum widths below:

Seats	Calculation
1 – 5,000	0.250" – a fixed number
5,001 – 10,000	0.250" – [0.0000174" × (number of seats > 5,000)]
10,001 – 15,000	0.163" – [0.0000086" × (number of seats > 10,000)]
15,001 – 20,000	0.120" – [0.0000050" × (number of seats > 15,000)]
20,001 – 25,000	0.095" – [0.0000040" × (number of seats > 20,000)]
25,000 – up	0.075" – a fixed number

Minimum width per seat served:

Seats	Width	Seats	Width	Seats	Width	Seats	Width
500	≥ 0.250"	7,000	≥ 0.215"	14,000	≥ 0.129"	21,000	≥ 0.091"
1,000	≥ 0.250"	8,000	≥ 0.198"	15,000	≥ 0.120"	22,000	≥ 0.087"
2,000	≥ 0.250"	9,000	≥ 0.180"	16,000	≥ 0.115"	23,000	≥ 0.083"
3,000	≥ 0.250"	10,000	≥ 0.163"	17,000	≥ 0.110"	24,000	≥ 0.079"
4,000	≥ 0.250"	11,000	≥ 0.154"	18,000	≥ 0.105"	25,000	≥ 0.075"
5,000	≥ 0.250"	12,000	≥ 0.146"	19,000	≥ 0.100"	26,000	≥ 0.075"
6,000	≥ 0.233"	13,000	≥ 0.137"	20,000	≥ 0.095"	27,000	≥ 0.075"

Source: IBC Table 1028.6.2

Table 1028.6.2c Width of Passageways, Doorways, and Ramps ≤ 1:10 Slope

Method of calculating inches per seat served for the table of widths below:

Seats	Calculation
1 – 5,000	0.150" – a fixed number
5,001 – 10,000	0.150" – [0.0000100" × (number of seats > 5,000)]
10,001 – 15,000	0.100" – [0.0000060" × (number of seats > 10,000)]
15,001 – 20,000	0.070" – [0.0000028" × (number of seats > 15,000)]
20,001 – 25,000	0.056" – [0.0000024" × (number of seats > 20,000)]
25,000 – up	0.044" – a fixed number

Minimum width per seat served:

Seats	Width	Seats	Width	Seats	Width	Seats	Width
500	≥ 0.150"	7,000	≥ 0.130"	14,000	≥ 0.076"	21,000	≥ 0.054"
1,000	≥ 0.150"	8,000	≥ 0.120"	15,000	≥ 0.070"	22,000	≥ 0.051"
2,000	≥ 0.150"	9,000	≥ 0.110"	16,000	≥ 0.067"	23,000	≥ 0.049"
3,000	≥ 0.150"	10,000	≥ 0.100"	17,000	≥ 0.064"	24,000	≥ 0.046"
4,000	≥ 0.150"	11,000	≥ 0.094"	18,000	≥ 0.062"	25,000	≥ 0.044"
5,000	≥ 0.150"	12,000	≥ 0.088"	19,000	≥ 0.059"	26,000	≥ 0.044"
6,000	≥ 0.140"	13,000	≥ 0.082"	20,000	≥ 0.056"	27,000	≥ 0.044"

Source: IBC Table 1028.6.2.

1028 Assembly

1028.6.2 Smoke-protected seating *(part 3 of 3)*

Table 1028.6.2d Width of Ramps > 1:10 Slope

Method of calculating inches per seat served for the table of widths below:

Seats	Calculation
1 – 5,000	0.165" – a fixed number
5,001 – 10,000	0.165" – [0.0000110" × (number of seats > ,5,000)]
10,001 – 15,000	0.110" – [0.0000066" × (number of seats > 10,000)]
15,001 – 20,000	0.077" – [0.0000030" × (number of seats > 15,000)]
20,001 – 25,000	0.062" – [0.0000028" × (number of seats > 20,000)]
25,000 – up	0.048" – a fixed number

Minimum width per seat served:

Seats	Width	Seats	Width	Seats	Width	Seats	Width
500	≥ 0.165"	7,000	≥ 0.143"	14,000	≥ 0.084"	21,000	≥ 0.059"
1,000	≥ 0.165"	8,000	≥ 0.132"	15,000	≥ 0.077"	22,000	≥ 0.056"
2,000	≥ 0.165"	9,000	≥ 0.121"	16,000	≥ 0.074"	23,000	≥ 0.054"
3,000	≥ 0.165"	10,000	≥ 0.110"	17,000	≥ 0.071"	24,000	≥ 0.051"
4,000	≥ 0.165"	11,000	≥ 0.103"	18,000	≥ 0.068"	25,000	≥ 0.048"
5,000	≥ 0.165"	12,000	≥ 0.097"	19,000	≥ 0.065"	26,000	≥ 0.048"
6,000	≥ 0.154"	13,000	≥ 0.090"	20,000	≥ 0.062"	27,000	≥ 0.048"

Source: IBC Table 1028.6.2.

1028.6.2.1 Smoke control

- This section addresses means of egress from smoke-protected seating in an assembly area.
- The means of egress must be protected from smoke by one of the following methods:
  - A smoke control system.

Note: Section 909, "Smoke Control Systems," is cited as the source of requirements.

  - Natural ventilation as follows:
    Smoke must be held to ≥ 6' above the floor.

1028.6.2.2 Roof height

- This section addresses the roof height in a smoke-protected assembly area.
- In an outdoor stadium, the following applies:
  - A canopy < 15' above the highest aisle or aisle accessway is permitted as follows:
    Where a height < 6'-8" is clear of any object.
- In other locations, the lowest roof deck must be ≥ 15' above the following:
  - The highest aisle.
  - The highest aisle accessway.

1028 Assembly

1028.6.2.3 Automatic sprinklers

- This section addresses smoke-protected assembly seating enclosed with ceiling and walls.
- This section does not apply to outdoor seating facilities as follows:
  - Where seating is essentially open to the exterior.
  - Where means of egress in seating areas are essentially open to the exterior.
- Sprinklers are not required for the following floor areas where the conditions below apply:
  - Floor areas used for the following:
    - Competition.
    - Performance.
    - Entertainment.
  - Conditions:
    - Roof construction must be > 50' above the floor level.
    - Only low-hazard uses occur.
- Sprinklers are not required for the following where < 1,000 sf:
  - Press boxes.
  - Storage.
- In other cases the seating must be protected with sprinklers.

Note: 903.3.1.1, "NFPA 13 sprinkler systems," is cited as governing the sprinklers.

1028.6.3 Width of means of egress for outdoor smoke-protected assembly

- Means of egress in an outdoor-smoke-protected assembly area must comply with one of the following width requirements:
  - Width required for indoor smoke-protected assembly.

Note: 1028.6.2, "Smoke-protected seating," is cited as a source of width requirements.

  - Width = (number of occupants served) × (required width per occupant) as follows:

Table 1028.6.3 Egress Width in Outdoor Smoke-Protected Assembly

Means of egress	Width/occupant	Means of egress	Width/occupant
Aisles	≥ 0.08"	Stairs	≥ 0.08"
Ramps	≥ 0.06"	Corridors	≥ 0.06"
Tunnels	≥ 0.06"	Vomitories	≥ 0.06"

1028 Assembly

1028.7 Travel distance

- Travel distance is measured along aisles and aisle accessways as follows:
  - Without crossing over seats.
- Travel distance is limited in assembly occupancies as indicated below:

Table 1028.7 Travel Distance in Assembly Spaces

Location and description of egress route	Travel distance
From each seat to outside the building:	
In open-air seating on Type I or II construction	not limited
In open-air seating of Type III, IV, and V construction	≤ 400'
In smoke-protected seating:	
From each seat to nearest entrance to a vomitory	≤ 200'
From vomitory entrance to outside the building:	
To a stair, ramp, or walk	≤ 200'
From each seat to nearest entrance to the concourse	≤ 200'
From concourse entrance outside the building:	
To a stair, ramp, or walk	≤ 200'
In sprinklered buildings to exit door	≤ 250'
In unsprinklered buildings to an exit door	≤ 200'

1028.8 Common path of egress travel

- The following common path of egress travel is limited as shown below:
  - Path:
    Between the following locations:
      Any seat.
      A point where the occupant has a choice as follows:
        Between two routes to two exits.
  - Limits:
    The common path is limited as shown in the following table:

Table 1028.8 Common Path of Travel Distance in Assembly Spaces

Conditions	Common path distance
< 50 occupants	≤ 75'
Smoke-protected seating	≤ 50'
Other	≤ 30'

1028 Assembly

1028.8.1 Path through adjacent row

- For smoke-protected seating, the following applies:
  - Where 1 of the 2 paths of egress travel is located as follows, the requirements indicated below apply:

    Path:

    Passes across an aisle through a row to the aisle beyond.

    Requirements:

    Seats in the aisle must be ≤ 40.

    Clear width in row must be ≥ 12" + [0.3" × (total number of seats - 7)] as follows:

    The following table lists required clear width based on this requirement:

Table 1028.8.1a Required Width of Smoke-Protected Row Serving as Means of Egress

Seats in row	Clear width	Seats in row	Clear width	Seats in row	Clear width
1–7	≥ 12"	19	≥ 15.6"	30	≥ 18.9"
8	≥ 12.3"	20	≥ 15.9"	31	≥ 19.2"
9	≥ 12.6"	21	≥ 16.2"	32	≥ 19.5"
10	≥ 12.9"	22	≥ 16.5"	33	≥ 19.8"
11	≥ 13.2"	23	≥ 16.8"	34	≥ 20.1"
12	≥ 13.5"	24	≥ 17.1"	35	≥ 20.4"
13	≥ 13.8"	25	≥ 17.4"	36	≥ 20.7"
14	≥ 14.1"	26	≥ 17.7"	37	≥ 21"
15	≥ 14.4"	27	≥ 18"	38	≥ 21.3"
16	≥ 14.7"	28	≥ 18.3"	39	≥ 21.6"
17	≥ 15"	29	≥ 18.6"	40	≥ 21.9"
18	≥ 15.3"				

- For seating that is not smoke-protected the following applies:
  - Where 1 of the 2 paths of egress travel is located as follows, the requirements indicated below apply:

    Path:

    Passes across an aisle through a row to the aisle beyond.

    Requirements:

    Seats in the aisle must be ≤ 40.

    Clear width in row must be ≥ 12" + [0.6" × (total number of seats - 7)] as follows:

    The following table lists required clear width based on this requirement:

Table 1028.8.1b Required Width of Not Smoke-Protected Row Serving as Means of Egress

Seats in row	Clear width	Seats in row	Clear width	Seats in row	Clear width
1–7	≥ 12.0"	13	≥ 15.6"	19	≥ 19.2"
8	≥ 12.6"	14	≥ 16.2"	20	≥ 19.8"
9	≥ 13.2"	15	≥ 16.8"	21	≥ 20.4"
10	≥ 13.8"	16	≥ 17.4"	22	≥ 21.0"
11	≥ 14.4"	17	≥ 18.0"	23	≥ 21.6"
12	≥ 15.0"	18	≥ 18.6"	24	≥ 22.2"

1028 Assembly

1028.9 Assembly aisles are required

- This section addresses the following occupied areas:
  - Areas of Occupancy A.
  - Areas of assembly accessory to Occupancy E.
- Areas governed by this section with the following furnishings must comply with the requirements listed below:
  - Furnishings:

    Displays — Similar fixtures
    Seats — Tables
    Similar equipment

  - Requirements:

    Aisles must lead to one of the following:
      Exits.
      Exit access doorways.
    Must comply with this section series.
    Aisle accessways must comply with requirements listed elsewhere.

Note: 1017.4, "Seating at tables," is cited governing aisle accessways.

1028.9.1 Minimum aisle width

- Required aisle widths in assembly areas are required as follows:

Table 1028.9.1 Aisle Width in Assembly Spaces

Description	Width
Aisle stairs:	
With seats on both sides totaling ≥ 50	≥ 48"
With seats on both sides totaling < 50	≥ 36"
With seats on one side only	≥ 36"
With ≤ 5 rows of seats on one side	≥ 23" between handrail and seating
Aisle stairs divided by handrail:	≥ 23" between handrail or guard and seating
Level or ramped aisles:	
With seats on both sides totaling ≥ 50	≥ 42"
With seats on both sides totaling >14 and < 50	≥ 36"
With seats on both sides totaling ≤ 14	≥ 30"
With seats on one side only totaling > 14	≥ 36"
With seats on one side only totaling ≤ 14	≥ 30"

1028.9.2 Aisle width

- An aisle must have sufficient width as follows:
  - To serve as egress for its assigned portion of the space.
- Areas served by aisles must be sized as follows:
  - So that the capacity of aisles reflects the following:

    A balanced use of all means of egress.

1028.9.3 Converging aisles

- A means of egress receiving occupant loads of converging aisles must have the following capacity:
  - ≥ the sum of the required capacities of the converging aisles.

1028 Assembly

1028.9.4 Uniform width

- The following parts of aisles require a uniform width:
  - Where egress travel is in either direction.

1028.9.5 Assembly aisle termination

- A dead-end aisle may be > 20' only as follows:
  - Seats beyond a point 20' into a dead-end must comply with both of the following:
    - Seats must be ≤ 24 seats from another aisle counted along the row of seats.
    - Required clear width between rows is as follows:

$$\textbf{Width} \geq 12'' + [(\textbf{number of seats} - 7) \times 0.6'']$$

This equation yields the following required clear widths between rows:

Table 1028.9.5a Row Width at Long Dead-End Aisles

Seats	Width	Seats	Width	Seats	Width	Seats	Width	Seats	Width
1–7	≥ 12.0"	11	≥ 14.4"	15	≥ 16.8"	19	≥ 19.2"	23	≥ 21.6"
8	≥ 12.6"	12	≥ 15.0"	16	≥ 17.4"	20	≥ 19.8"	24	≥ 22.2"
9	≥ 13.2"	13	≥ 15.6"	17	≥ 18.0"	21	≥ 20.4"		
10	≥ 13.8"	14	≥ 16.2"	18	≥ 18.6"	22	≥ 21.0"		

- In smoke-protected seating, a dead-end vertical aisle (⊥ to rows) is governed as follows:
  - It may be > than 21 rows in length only:
    - Where seats beyond the 21 rows comply with both of the following:
      - Seats must be ≤ 40 seats from another aisle counted along the row of seats.
      - Required clear width between rows served by this part of the dead end is:

$$\textbf{Width} \geq 12'' + [(\textbf{number of seats} - 7) \times 0.3'']$$

This equation yields the following required clear widths between rows:

Table 1028.9.5b Clear Width between Rows of Assembly Seating

Seats	Width	Seats	Width	Seats	Width	Seats	Width	Seats	Width
1–7	≥ 12.0"	14	≥ 14.1"	21	≥ 16.2"	28	≥ 18.3"	35	≥ 20.4"
8	≥ 12.3"	15	≥ 14.4"	22	≥ 16.5"	29	≥ 18.6"	36	≥ 20.7"
9	≥ 12.6"	16	≥ 14.7"	23	≥ 16.8"	30	≥ 18.9"	37	≥ 21.0"
10	≥ 12.9"	17	≥ 15.0"	24	≥ 17.1"	31	≥ 19.2"	38	≥ 21.3"
11	≥ 13.2"	18	≥ 15.3"	25	≥ 17.4"	32	≥ 19.5"	39	≥ 21.6"
12	≥ 13.5"	19	≥ 15.6"	26	≥ 17.7"	33	≥ 19.8"	40	≥ 21.9"
13	≥ 13.8"	20	≥ 15.9"	27	≥ 18.0"	34	≥ 20.1"		

- All other dead-end vertical aisles in smoke-protected seating must be ≤ 21 rows in length.
- All other dead-end aisles must be ≤ 20' in length.
- All other aisles must terminate at both ends as follows:
  - Termination must be at one of the following having access to an exit:
    - Cross aisle.
    - Vomitory or concourse.
      - Doorway.
      - Foyer.

1028 Assembly

1028.9.6 Assembly aisle obstructions

- Handrails are the only obstructions permitted within the required aisle width.

  Note: 1028.13, "Handrails," is cited as the source of applicable requirements.

1028.10 Clear width of aisle accessways serving seating

- Clear aisle-accessway width is measured between the following points:
  - From the back of a row of seats to the following:
    The closest element of the row of seats behind it.
- Width is measured with the seats up as follows:
  - Where chairs have self-rising seats.
- Width is measured with the seat down as follows:
  - For any chair without a self-rising seat.
- Aisles are measured as follows for seats with folding tablet arms:
  - With the tablet arm in the stored position in the following case:
    Where gravity returns the tablet arm to the stored position as follows:
    When the arm is manually raised to a vertical position in one motion.
  - With the tablet arm in the usable position as follows:
    For seats with other types of tablet arms.
- For ≤ 14 seats in a row the following applies:
  - The clear aisle-accessway width required is ≥ 12".

Case study: Fig. 1028.10. The central area of the auditorium has 14 seats or fewer seats per row. Since they are self-rising the aisle accessway width is measured with the seat up at 1'-6¾". This meets the code minimum of 12" for this number of seats.

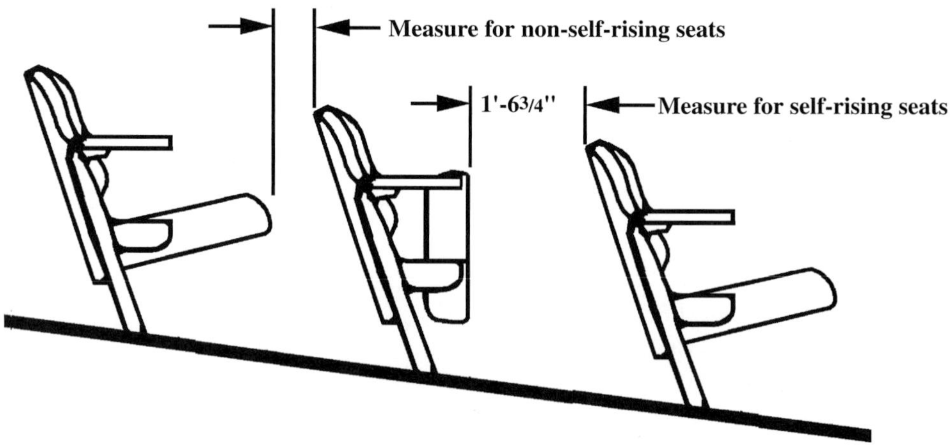

Fig. 1028.10. Elevation of auditorium seating. High School 6, Cypress-Fairbanks Independent School District. Harris County, Texas. PBK Architects, Inc. Houston, Texas.

1028 Assembly

1028.10.1 Dual access *(part 1 of 3)*

- Rows of seating served by aisles or doorways at each side are limited as follows:
    - Limited to ≤ 100 seats per row.
- A clear width of ≥ 12" is required for rows with 1–14 seats.
- Clear width between rows > 22" is not required for rows of any length.
- For rows with > 14 seats, required width is as follows:

Width ≥ 12" + [(number of seats – 14)] × 0.3"

This equation yields the following required clear widths up to 22":

Table 1028.10.1a Required Width between Rows for Seating Not Smoke-Protected, Access from 2 Sides

Seats	Width	Seats	Width	Seats	Width	Seats	Width	Seats	Width
1–14	≥ 12.0"	21	≥ 14.1"	28	≥ 16.2"	35	≥ 18.3"	42	≥ 20.4"
15	≥ 12.3"	22	≥ 14.4"	29	≥ 16.5"	36	≥ 18.6"	43	≥ 20.7"
16	≥ 12.6"	23	≥ 14.4"	30	≥ 16.8"	37	≥ 18.9"	44	≥ 21.0"
17	≥ 12.9"	24	≥ 15.0"	31	≥ 17.1"	38	≥ 19.2"	45	≥ 21.3"
18	≥ 13.2"	25	≥ 15.3"	32	≥ 17.4"	39	≥ 19.5"	46	≥ 21.6"
19	≥ 13.5"	26	≥ 15.6"	33	≥ 17.7"	40	≥ 19.8"	47	≥ 21.9"
20	≥ 13.8"	27	≥ 15.9"	34	≥ 18.0"	41	≥ 20.1"	48	≥ 22.0"

Source: IBC Table 1028.10.1.

- A clear width of ≥ 12" is required for rows with lengths varying from 14–17 seats and to which 0.3" per seat is added for longer rows as follows:

Table 1028.10.1b Required Width between Rows for Smoke-Protected Seats, Access from 2 Sides

Seats	Width	Seats	Width	Seats	Width	Seats	Width	Seats	Width
< 4,000 smoke-protected seats:									
1–14	≥ 12.0"	21	≥ 14.1"	28	≥ 16.2"	35	≥ 18.3"	42	≥ 20.4"
15	≥ 12.3"	22	≥ 14.4"	29	≥ 16.5"	36	≥ 18.6"	43	≥ 20.7"
16	≥ 12.6"	23	≥ 14.4"	30	≥ 16.8"	37	≥ 18.9"	44	≥ 21.0"
17	≥ 12.9"	24	≥ 15.0"	31	≥ 17.1"	38	≥ 19.2"	45	≥ 21.3"
18	≥ 13.2"	25	≥ 15.3"	32	≥ 17.4"	39	≥ 19.5"	46	≥ 21.6"
19	≥ 13.5"	26	≥ 15.6"	33	≥ 17.7"	40	≥ 19.8"	47	≥ 21.9"
20	≥ 13.8"	27	≥ 15.9"	34	≥ 18.0"	41	≥ 20.1"	48	≥ 22.0"
4,000–6,999 smoke-protected seats:									
1–15	≥ 12.0"	22	≥ 14.1"	29	≥ 16.2"	36	≥ 18.3"	43	≥ 20.4"
16	≥ 12.3"	23	≥ 14.4"	30	≥ 16.5"	37	≥ 18.6"	44	≥ 20.7"
17	≥ 12.6"	24	≥ 14.4"	31	≥ 16.8"	38	≥ 18.9"	45	≥ 21.0"
18	≥ 12.9"	25	≥ 15.0"	32	≥ 17.1"	39	≥ 19.2"	46	≥ 21.3"
19	≥ 13.2"	26	≥ 15.3"	33	≥ 17.4"	40	≥ 19.5"	47	≥ 21.6"
20	≥ 13.5"	27	≥ 15.6"	34	≥ 17.7"	41	≥ 19.8"	48	≥ 21.9"
21	≥ 13.8"	28	≥ 15.9"	35	≥ 18.0"	42	≥ 20.1"	49	≥ 22.0"

(Continued on next page)

Source: IBC Table 1028.10.1.

1028 Assembly

1028.10.1 Dual access *(part 2 of 3)*

Table 1028.10.1b—*Continued* **Required Width between Rows for Smoke-Protected Seats, Access from 2 Sides**

Seats	Width	Seats	Width	Seats	Width	Seats	Width	Seats	Width
7,000 –9,999 smoke-protected seats:									
1–16	≥ 12.0"	23	≥ 14.1"	30	≥ 16.2"	37	≥ 18.3"	44	≥ 20.4"
17	≥ 12.3"	24	≥ 14.4"	31	≥ 16.5"	38	≥ 18.6"	45	≥ 20.7"
18	≥ 12.6"	25	≥ 14.4"	32	≥ 16.8"	39	≥ 18.9"	46	≥ 21.0"
19	≥ 12.9"	26	≥ 15.0"	33	≥ 17.1"	40	≥ 19.2"	47	≥ 21.3"
20	≥ 13.2"	27	≥ 15.3"	34	≥ 17.4"	41	≥ 19.5"	48	≥ 21.6"
21	≥ 13.5"	28	≥ 15.6"	35	≥ 17.7"	42	≥ 19.8"	49	≥ 21.9"
22	≥ 13.8"	29	≥ 15.9"	36	≥ 18.0"	43	≥ 20.1"	50	≥ 22.0"
10,000 –12,999 smoke-protected seats:									
1–17	≥ 12.0"	24	≥ 14.1"	31	≥ 16.2"	38	≥ 18.3"	45	≥ 20.4"
18	≥ 12.3"	25	≥ 14.4"	32	≥ 16.5"	39	≥ 18.6"	46	≥ 20.7"
19	≥ 12.6"	26	≥ 14.4"	33	≥ 16.8"	40	≥ 18.9"	47	≥ 21.0"
20	≥ 12.9"	27	≥ 15.0"	34	≥ 17.1"	41	≥ 19.2"	48	≥ 21.3"
21	≥ 13.2"	28	≥ 15.3"	35	≥ 17.4"	42	≥ 19.5"	49	≥ 21.6"
22	≥ 13.5"	29	≥ 15.6"	36	≥ 17.7"	43	≥ 19.8"	50	≥ 21.9"
23	≥ 13.8"	30	≥ 15.9"	37	≥ 18.0"	44	≥ 20.1"	51	≥ 22.0"

Source: IBC Table 1028.10.1.

- A clear width of ≥ 12" is required for rows with lengths varying from 18–21 seats and to which 0.3" per seat is added for longer rows as follows:

Table 1028.10.1c **Required Width between Rows for Smoke-Protected Seats, Access from 2 Sides**

Seats	Width	Seats	Width	Seats	Width	Seats	Width	Seats	Width
13,000 –15,999 smoke-protected seats:									
1–18	≥ 12.0"	25	≥ 14.1"	32	≥ 16.2"	39	≥ 18.3"	46	≥ 20.4"
19	≥ 12.3"	26	≥ 14.4"	33	≥ 16.5"	40	≥ 18.6"	47	≥ 20.7"
20	≥ 12.6"	27	≥ 14.4"	34	≥ 16.8"	41	≥ 18.9"	48	≥ 21.0"
21	≥ 12.9"	28	≥ 15.0"	35	≥ 17.1"	42	≥ 19.2"	49	≥ 21.3"
22	≥ 13.2"	29	≥ 15.3"	36	≥ 17.4"	43	≥ 19.5"	50	≥ 21.6"
23	≥ 13.5"	30	≥ 15.6"	37	≥ 17.7"	44	≥ 19.8"	51	≥ 21.9"
24	≥ 13.8"	31	≥ 15.9"	38	≥ 18.0"	45	≥ 20.1"	52	≥ 22.0"
16,000 –18,999 smoke-protected seats:									
1–19	≥ 12.0"	26	≥ 14.1"	33	≥ 16.2"	40	≥ 18.3"	47	≥ 20.4"
20	≥ 12.3"	27	≥ 14.4"	34	≥ 16.5"	41	≥ 18.6"	48	≥ 20.7"
21	≥ 12.6"	28	≥ 14.4"	35	≥ 16.8"	42	≥ 18.9"	49	≥ 21.0"
22	≥ 12.9"	29	≥ 15.0"	36	≥ 17.1"	43	≥ 19.2"	50	≥ 21.3"
23	≥ 13.2"	30	≥ 15.3"	37	≥ 17.4"	44	≥ 19.5"	51	≥ 21.6"
24	≥ 13.5"	31	≥ 15.6"	38	≥ 17.7"	45	≥ 19.8"	52	≥ 21.9"
25	≥ 13.8"	32	≥ 15.9"	39	≥ 18.0"	46	≥ 20.1"	53	≥ 22.0"

(Continued on next page)

Source: IBC Table 1028.10.1.

1028 Assembly

1028.10.1 Dual access *(part 3 of 3)*

Table 1028.10.1c—*Continued* **Required Width between Rows for Smoke-Protected Seats Access from 2 Sides**

Seats	Width	Seats	Width	Seats	Width	Seats	Width	Seats	Width
19,000 –21,999 smoke-protected seats:									
1–20	≥ 12.0"	27	≥ 14.1"	34	≥ 16.2"	41	≥ 18.3"	48	≥ 20.4"
21	≥ 12.3"	28	≥ 14.4"	35	≥ 16.5"	42	≥ 18.6"	49	≥ 20.7"
22	≥ 12.6"	29	≥ 14.4"	36	≥ 16.8"	43	≥ 18.9"	50	≥ 21.0"
23	≥ 12.9"	30	≥ 15.0"	37	≥ 17.1"	44	≥ 19.2"	51	≥ 21.3"
24	≥ 13.2"	31	≥ 15.3"	38	≥ 17.4"	45	≥ 19.5"	52	≥ 21.6"
25	≥ 13.5"	32	≥ 15.6"	39	≥ 17.7"	46	≥ 19.8"	53	≥ 21.9"
26	≥ 13.8"	33	≥ 15.9"	40	≥ 18.0"	47	≥ 20.1"	54	≥ 22.0"
≥ 22,000 smoke-protected seats:									
1–21	≥ 12.0"	28	≥ 14.1"	35	≥ 16.2"	42	≥ 18.3"	49	≥ 20.4"
22	≥ 12.3"	29	≥ 14.4"	36	≥ 16.5"	43	≥ 18.6"	50	≥ 20.7"
23	≥ 12.6"	30	≥ 14.4"	37	≥ 16.8"	44	≥ 18.9"	51	≥ 21.0"
24	≥ 12.9"	31	≥ 15.0"	38	≥ 17.1"	45	≥ 19.2"	52	≥ 21.3"
25	≥ 13.2"	32	≥ 15.3"	39	≥ 17.4"	46	≥ 19.5"	53	≥ 21.6"
26	≥ 13.5"	33	≥ 15.6"	40	≥ 17.7"	47	≥ 19.8"	54	≥ 21.9"
27	≥ 13.8"	34	≥ 15.9"	41	≥ 18.0"	48	≥ 20.1"	55	≥ 22.0"

Source: IBC Table 1028.10.1.

1028.10.2 Single access *(part 1 of 2)*

- Clear width required between rows with an aisle or doorway at one side only, where not smoke protected, is as follows:
  - For rows ≤ 7 seats required width is ≥ 12".
  - Clear width between rows of any length is not required to be > 22".
  - For rows > 7 seats, required width is determined by the following equation:

    Width ≥ 12" + [(number of seats – 7) × 0.6"]

    This equation yields the clear widths in the table below:

Table 1028.10.2a **Required Clear Width between Rows for Seating Not Smoke-Protected, Access 1 Side**

Seats	Width	Seats	Width	Seats	Width	Seats	Width	Seats	Width
1–7	≥ 12.0"	11	≥ 14.4"	15	≥ 16.8"	19	≥ 19.2"	23	≥ 21.6"
8	≥ 12.6"	12	≥ 15.0"	16	≥ 17.4"	20	≥ 19.8"	24	≥ 22.0"
9	≥ 13.2"	13	≥ 15.6"	17	≥ 18.0"	21	≥ 20.4"	25	≥ 22.0"
10	≥ 13.8"	14	≥ 16.2"	18	≥ 18.6"	22	≥ 21.0"	26	≥ 22.0"

Source: IBC Table 1028.10.1.

1028 Assembly

1028.10.2 Single access *(part 2 of 2)*

- A clear width of ≥ 12" is required for rows with lengths varying from 7–11 smoke-protected seats and to which 0.6" per seat is added for longer rows as in the table below:

Table 1028.10.2b Required Clear Width between Rows for Smoke-Protected Seats, Access from 1 Side

Seats	Width	Seats	Width	Seats	Width	Seats	Width	Seats	Width
< 7,000 smoke-protected seats:									
1–7	≥ 12.0"	11	≥ 14.4"	15	≥ 16.8"	19	≥ 19.2"	23	≥ 21.6"
8	≥ 12.6"	12	≥ 15.0"	16	≥ 17.4"	20	≥ 19.8"	24	≥ 22.0"
9	≥ 13.2"	13	≥ 15.6"	17	≥ 18.0"	21	≥ 20.4"	25	≥ 22.0"
10	≥ 13.8"	14	≥ 16.2"	18	≥ 18.6"	22	≥ 21.0"	26	≥ 22.0"
7,000 – 12,999 smoke-protected seats:									
1–8	≥ 12.0"	12	≥ 14.4"	16	≥ 16.8"	20	≥ 19.2"	24	≥ 21.6"
9	≥ 12.6"	13	≥ 15.0"	17	≥ 17.4"	21	≥ 19.8"	25	≥ 22.0"
10	≥ 13.2"	14	≥ 15.6"	18	≥ 18.0"	22	≥ 20.4"	26	≥ 22.0"
11	≥ 13.8"	15	≥ 16.2"	19	≥ 18.6"	23	≥ 21.0"	27	≥ 22.0"
13,000 – 18,999 smoke-protected seats:									
1–9	≥ 12.0"	13	≥ 14.4"	17	≥ 16.8"	21	≥ 19.2"	25	≥ 21.6"
10	≥ 12.6"	14	≥ 15.0"	18	≥ 17.4"	22	≥ 19.8"	26	≥ 22.0"
11	≥ 13.2"	15	≥ 15.6"	19	≥ 18.0"	23	≥ 20.4"	27	≥ 22.0"
12	≥ 13.8"	16	≥ 16.2"	20	≥ 18.6"	24	≥ 21.0"	28	≥ 22.0"
19,000 – 21,999 smoke-protected seats:									
1–10	≥ 12.0"	14	≥ 14.4"	18	≥ 16.8"	22	≥ 19.2"	26	≥ 21.6"
11	≥ 12.6"	15	≥ 15.0"	19	≥ 17.4"	23	≥ 19.8"	27	≥ 22.0"
12	≥ 13.2"	16	≥ 15.6"	20	≥ 18.0"	24	≥ 20.4"	28	≥ 22.0"
13	≥ 13.8"	17	≥ 16.2"	21	≥ 18.6"	25	≥ 21.0"	29	≥ 22.0"
≥ 22,000 smoke-protected seats:									
1–11	≥ 12.0"	15	≥ 14.4"	19	≥ 16.8"	23	≥ 19.2"	27	≥ 21.6"
12	≥ 12.6"	16	≥ 15.0"	20	≥ 17.4"	24	≥ 19.8"	28	≥ 22.0"
13	≥ 13.2"	17	≥ 15.6"	21	≥ 18.0"	25	≥ 20.4"	29	≥ 22.0"
14	≥ 13.8"	18	≥ 16.2"	22	≥ 18.6"	26	≥ 21.0"	30	≥ 22.0"

Source: IBC Table 1028.10.1.

1028 Assembly

1028.11 Assembly aisle walking surfaces

- Aisles with a gradient must have the following configurations:

Aisle Slope	Aisle configuration required
$\leq$ 1:8	Ramp with slip-resistant surface
> 1:8	Treads and risers = to width of aisle

Note: The following are cited as governing the treads and risers cited above:
1028.11.1, "Treads."
1028.11.2, "Risers."
1028.11.3, "Tread contrasting marking stripe."

1028.11.1 Treads

- Treads in aisle steps must comply with the following:
  - Depth required is $\geq$ 11".
  - Variation in adjacent tread depth is limited to $\leq$ 3/16".

1028.11.2 Risers

- Risers of aisle stairs with the same gradient as adjacent seating are governed as follows:
  - Required riser height is $\geq$ 4".
  - Required riser height is $\leq$ 8" where sightlines permit.
  - Required riser height is $\leq$ 9" where required by sightlines.
- Variations in riser height are limited to those necessary for changes in the slope of seating.
- Riser height is to be uniform in each flight.
- Adjacent riser heights which vary more than 3/16" must be identified as follows:
  - A distinctive stripe must be on the nosing of a tread next to a nonuniform riser as follows:
    Width of stripe to be $\geq$ 1" and $\leq$ 2".
    Appearance of stripe to be distinctly different from strips marking other tread edges.

1028.11.3 Tread contrasting marking stripe

- Tread edges which are not readily apparent during descent must be marked as follows:
  - A stripe contrasting with the appearance of the tread is required on each nosing as follows:
    Width of marking stripe is to be $\geq$ 1" and $\leq$ 2".
    Marking must make treads readily apparent during descent.

1028 Assembly

1028.12 Seat stability

- Seats are not required to be fastened to the floor in the following applications:
  - ≤ 200 seats without ramped or tiered floors for the seating.
  - > 200 seats where both of the following apply:
    Where floors for seating are neither ramped nor tiered.
    Where seats are joined together in groups ≥ 3 seats.
  - Seating at tables without ramped or tiered floors for the seating.
  - ≤ 200 seats where all of the following apply:
    Where flexibility of seating layout is integral with the function of the space.
    Where seating is on tiered levels.
    Where plans for seating, tiers, and aisles are submitted for approval.
  - Groups ≤ 14 seats with both of the following conditions:
    Where the seats are separated from other seating by any of the following:
      Railings or guards.
      Low walls or similar barriers.
    Where floors are level.
  - Seats for musicians or other performers with the following conditions:
    Where the seats are separated from other seating by any of the following:
      Railings or guards.
      Low walls or similar barriers.
- In all other places of assembly, seats must be fastened to the floor.

1028.13 Handrails

- This section addresses the following in places of assembly:
  - Ramped aisles.
  - Stepped aisles.
- Handrails are not required as follows:
  - Where a guard meets handrail, the graspability requirement is as follows:
    The guard is located at the side of the aisle.
- Handrails are not required for ramped aisles where both of the following apply:
  - Where the slope is ≤ 1:8.
  - Where seating is on both sides.
- Handrail extensions are not required following locations:
  - At the tops and bottoms of aisle stairs.
  - At the tops and bottoms of aisle ramps.
- In other cases, handrails are required as follows:
  - For aisle stairs.
  - For ramped aisles sloping ≥ 1:15.

1028 Assembly

1028.13.1 Discontinuous handrails

- This section addresses aisles where handrails are required.
- The following applies where seating is on both sides of an aisle:
  - Handrails must be discontinuous as follows:
    - Gaps in handrails must be spaced as follows:
      - At intervals ≤ 5 rows of seating.
    - Clear width in handrail gaps is to be as follows:
      - ≥ 22" and ≤ 36".
        - Measured horizontally.
    - The following handrail details are to be rounded:
      - Terminations.
      - Bends.

1028.13.2 Intermediate handrails

- Where handrails occur in the center of aisle stairs, the following applies:
  - A second handrail is required at 12" (±) below the main handrail.

> **Case study: Fig. 1028.13.2.** The handrail complies with requirements of 1028.13, 1028.13.1, and 1028.13.2. It is provided in the center of stepped aisles where seats are on both sides; it has a 2nd handrail below the top; and it is discontinuous with rounded corners. The intermediate handrail is 12" lower than the main handrail as required.

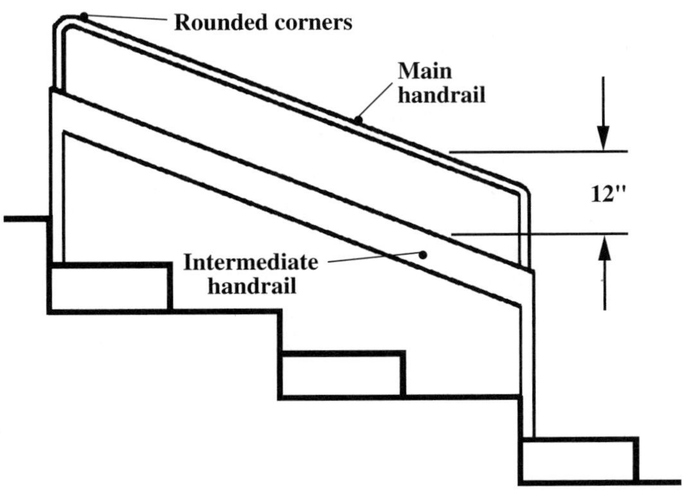

Fig. 1028.13.2. Elevation of handrail. Newman University Sports and Fine Arts Center. Wichita, Kansas. Gossen Livingston Associates, Inc., Architecture. Wichita, Kansas.

1028 Assembly

1028.14.1 Cross aisles

- Cross aisles > 30" above an adjacent level require guards as follows:
  - Guards must be ≥ 42" high.
  - Guards must meet opening and other requirements.
- Cross aisles ≤ 30" above an adjacent lower level are governed as follows:
  - Guards are not required in the following case:
    Where the backs of seats on the lower level extend ≥ 24" above the aisle.
  - In other cases, guards ≥ 26" high are required.

Note: Section 1013, "Guards," is cited as the source of requirements for guards required by this section.

1028.14.2 Sightline-constrained guard heights

- This section does not apply to the edge of a floor or footboard at the ends of aisles.

Note: 1028.14.3, "Guards at the end of aisles," is cited as governing the barriers at the end of aisles that are not addressed in this section.

- A guard is required at bleachers with certain conditions:

Note: ICC 300, "ICC Standard on Bleachers, Folding and Telescopic Seating and Grandstands," is cited as specifying bleacher conditions that required a guard.

- A fascia or guard is required in other locations as follows:
  - Where a floor or foot board is > 30" above the adjacent level.
  - Barriers must be ≥ 26" high in the following location:
    Where a higher barrier would obstruct sightlines of adjacent seating.

Note: Section 1013, "Guards," is cited as the source of requirements for guards required by this section.

1028.14.3 Guards at the end of aisles

- Where the end of an aisle is > 30" above the adjacent level, a barrier is required as follows:
  - Barrier must meet guard requirements.
  - Barrier must extend the full width of the aisle.
  - Barrier must be ≥ 36" high.
  - The diagonal distance between the following points must be ≥ 42":
    The top of the barrier and the nosing of the nearest tread.

Note: Section 1013, "Guards," is cited as governing these guards.

1028.15 Bench seating

- The capacity of bench seating is determined as follows:
  - Number of seats = bench length ÷ 18".

1028 Assembly

Case study: Fig. 1028.14.1. The floor of the balcony cross aisle is 8'-11" above the floor of the adjacent lower level seating, which is greater than the 30" maximum. The edge of the cross aisle, therefore, requires a guard ≥ 3'-6" high as provided.

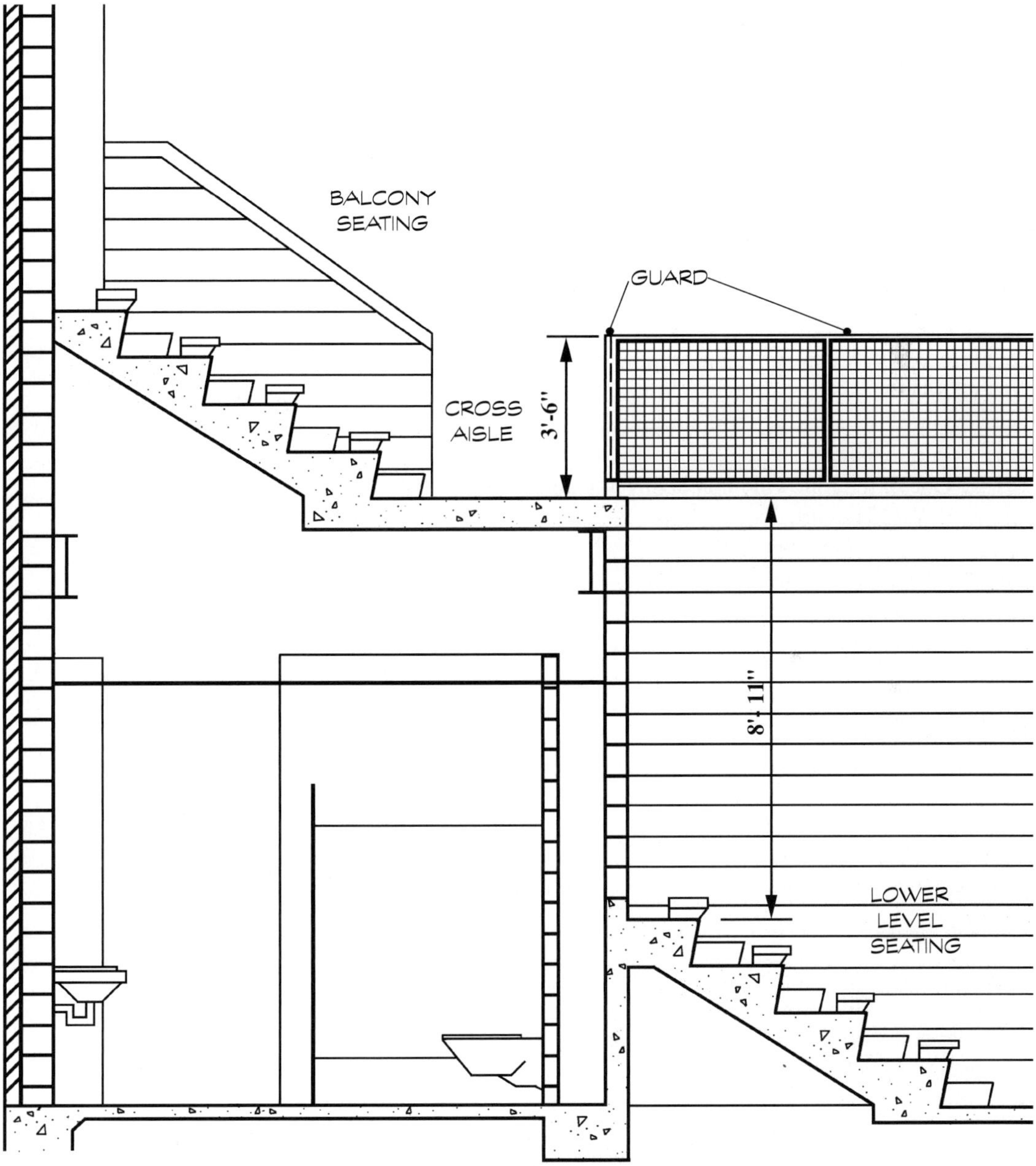

Fig. 1028.14.1. Section of seating at swimming pool. High School 6, Cypress-Fairbanks Independent School District. Harris County, Texas. PBK Architects, Inc. Houston, Texas.

1029 Emergency Escape and Rescue

1029.1 General

- This section series addresses emergency escape and rescue openings as follows:
  - In occupancies R and I-1.
- Such openings are not required in the following cases:
  - In buildings other than R-3, where either of the following conditions apply:
    - Where the building is sprinklered.
    - Where sleeping rooms have a door to a corridor as follows:
      - Corridor has a fire-resistance rating.
      - Corridor provides access in opposite directions to 2 remote exits.

    Note: The following are cited as governing the sprinklers as applicable:
    903.3.1.1, "NFPA 13 sprinkler systems."
    903.3.1.2, "NFPA 13R sprinkler systems."

  - In high-rise buildings.

    Note: Section 403, "High-Rise Buildings," is cited as describing buildings to which this section series does not apply.

  - In basements or sleeping rooms with the following doors:
    - Exit door or exit access door as follows:
      - Opens directly to one of the following:
        - Public way.
        - One of the following that opens to a public way:
          - Yard or court.
          - Exterior exit balcony.
  - In either of the following basements:
    - Where ceiling height is < 6'-8".
    - Where both the following apply:
      - There is no habitable space.
      - The area is ≤ 200 sf.
- Such openings may open to a balcony in an atrium where all of the following apply:
  - Where the balcony serves one of the following:
    - Dwelling unit.
    - Sleeping unit.
  - Where the balcony provides access to an exit.
  - Where a means of egress not open to the atrium is provided.

  Note: Section 404, "Atriums," is cited as governing atriums.

- In other cases ≥ 1 exterior escape and rescue opening is required as follows:
  - In sleeping rooms below the 4th story as follows:
    - Includes 4 stories above the grade plane.
  - In basements as follows:
    - In each sleeping room in a basement.
    - In basement space that does not adjoin a sleeping room.
  - Openings must open directly into one of the following:
    - Public way.
    - Yard or court that opens to a public way.

1029 Emergency Escape and Rescue

1029.2 Minimum size

- Sizes required for emergency escape and rescue openings are as follows:

Location	Clear area required
At grade floor	≥ 5.0 sf
Other locations	≥ 5.7 sf

1029.2.1 Minimum dimensions

- Dimensions required for emergency escape and rescue openings are as follows:

Clear height required	Clear width required
≥ 2'	≥ 1'-8"

- Normal operation of the opening must yield the required clear dimensions.

1029.3 Maximum height from floor

- Height of emergency escape and rescue openings is limited as follows:
  - The bottom of the clear opening must be ≤ 3'-8" from the floor.

1029.4 Operational constraints

- Emergency escape and rescue openings must be operable as follows:
  - From inside the room.
  - Without the use of keys or tools.
- The following devices may cover such openings where the conditions indicated apply:
  - Devices:
    Bars.
    Grilles.
    Grates.
    Similar devices.
  - Required conditions:
    The size requirements for the opening may not be reduced.
    The device must be openable from inside the room as follows:
      Without keys or tools.
      Without a force greater than that required for the opening itself.
    Smoke detectors must be provided in the following case:
      Where devices are installed in existing buildings for all alterations.

Note: The following are cited as the sources of additional requirements:
1029.2, "Minimum size," governs the size of the opening that must remain available after any covering device is installed.
907.2.11, "Single- and multiple-station smoke alarms," governs the smoke detectors required where covering devices are installed in existing buildings.

1029 Emergency Escape and Rescue

1029.5 Window wells

- This section addresses emergency escape and rescue openings below grade.
- Where the finished sill is below adjacent grade, a window well is required.

Note: 1029.5.1, "Minimum size," is cited as a source of requirements for window wells.
1029.5.2, "Ladders or steps," is cited as a source of requirements for window wells.

1029.5.1 Minimum size

- Window wells serving emergency escape and rescue openings must have the following size:
  - Window well must allow escape and rescue opening to fully open.
  - All horizontal dimensions of the window well to be ≥ 3'.
  - Window well must be ≥ 9 sf in plan.

1029.5.2 Ladders or steps

- This section addresses ladders or steps in window wells serving emergency escape and rescue openings.
- Ladders or steps required by this section are not governed by stair requirements.

Note: Section 1009, "Stairways," is cited as having the requirements that are waived for ladders and steps required by this section.

- Approved ladders or steps must be provided in window wells ≥ 3'-8" in depth as follows:
  - Must be permanently affixed.
  - Must have an inside width ≥ 12".
  - Must project ≥ 3" from the wall.
  - Rungs or steps must be spaced ≤ 1'-6" center to center vertically.
  - Must extend full height of well.
  - Must not protrude > 6" into required dimensions of well.
  - Must not be obstructed by the escape and rescue opening.

11

Accessibility

Hot Springs Police Department New Headquarters.
Hot Springs National Park, Arkansas. *(partial elevation)*
Cromwell Architects Engineers. Little Rock, Arkansas.

1102 Definitions

1102.1 Definitions *(part 1 of 3)*

- **Accessible**
  - A construction that meets the requirements of this chapter as follows:
    - Site.
    - Building.
    - Facility.
- **Accessible route**
  - A path that meets the requirements of this chapter as follows:
    - Continuous.
    - Unobstructed.
- **Accessible unit**
  - A unit as follows that meets the requirements of this chapter and other standards:
    - Dwelling unit.
    - Sleeping unit.

Note: ICC A117.1, "Accessible and Usable Buildings and Facilities," is cited as governing accessible units.

- **Circulation path**
  - A pedestrian route in one of the following locations:
    - Outside.
    - Inside.
- **Common use**
  - The following items with all of the characteristics listed:
    - Items:
      - Circulation paths.
      - Rooms.
      - Spaces.
      - Elements.
    - Characteristics:
      - Exterior or interior.
      - Not for the public.
      - Used by ≥ two persons.
- **Detectable warning**
  - A standardized feature warning of hazards:
    - On a walking surface.
    - On other elements.
  - Detectable by visually impaired persons.
- **Dwelling unit or sleeping unit, multistory**
  - Has habitable space on > 1 story.
- **Dwelling unit or sleeping unit, Type A**
  - Complies with fully with accessibility requirements of this code and other standards.

Note: ICC A117.1, "Accessible and Usable Buildings and Facilities," is cited as the source of accessibility requirements in addition to this code.

1102 Definitions

1102.1 Definitions *(part 2 of 3)*

- **Dwelling unit or sleeping unit, Type B**
  - Consistent with the Fair Housing Act.
  - Less accessible than a Type A dwelling unit.

  Note: ICC A117.1, "Accessible and Usable Buildings and Facilities," is cited as a source of accessibility requirements for Type B units.

- **Employee work area**
  - A space with all of the following characteristics:
    - Used by employees only.
    - Used for work only.
  - Does not include the following:
    - Corridors.
    - Toilet rooms.
    - Kitchenettes.
    - Break rooms.
- **Facility**
  - The following in whole or part:
    - Building.
    - Structure.
    - Site improvements.
    - Elements.
    - Pedestrian routes within a site.
    - Vehicular routes within a site.
- **Intended to be occupied as a residence**
  - A dwelling unit or sleeping unit as follows:
    - Usable as a place to live for either of the following time periods:
      - Full time.
      - Part time.
- **Multilevel assembly seating**
  - Either of the following seating arrangements:
    - Groups of multiple rows of seating as follows:
      - Where each group is at a level different from the other.
    - Box seats accessed as follows:
      - From a level different from the level on which the seats set.
- **Multistory unit**
  - Habitable space on > 1 story in the following types of units:
    - Dwelling unit.
    - Sleeping unit.
- **Public entrance**
  - An entrance other than either of the following:
    - Service entrance.
    - Restricted entrance.

1102 Definitions

1102.1 Definitions *(part 3 of 3)*

- **Public-use areas**
  - Space available to the public as follows:
    - Interior.
    - Exterior.
- **Restricted entrance**
  - An entrance with controlled common use.
  - Includes neither of the following:
    - Public entrance.
    - Service entrance.
- **Self-service storage facility**
  - A self-service construction for either of the following:
    - Rented for storage purposes.
    - Leased for storage purposes.
- **Service entrance**
  - An entrance mainly for deliveries.
- **Site**
  - Land defined by any of the following at its edges:
    - Property line.
    - Public right-of-way.
- **Type A unit**
  - A dwelling unit as follows:
    - Complies with fully with accessibility requirements of this code and other standards.

Note: ICC A117.1, "Accessible and Usable Buildings and Facilities," is cited as the source of accessibility requirements in addition to this code.

- **Type B unit**
  - A dwelling unit as follows:
    - Consistent with the Fair Housing Act.
    - Less accessible than a Type A dwelling unit.

Note: ICC A117.1, "Accessible and Usable Buildings and Facilities," is cited as a source of accessibility requirements for Type B units.

- **Wheelchair space**
  - A space designed to accommodate the following:
    - A wheelchair.
    - The occupant of a wheelchair.

1103 Scoping Requirements

1103.1 Where required

- Temporary and permanent constructions as follows are required to be accessible:
  - Buildings.
  - Elements.
  - Facilities.
  - Sites.
  - Spaces.
  - Structures.

1103.2 General exceptions

- This section identifies certain facilities in the following categories not required to be accessible:
  - Buildings.
  - Elements.
  - Facilities.
  - Sites.
  - Spaces.
  - Structures.

1103.2.2 Existing buildings

- Accessibility of existing buildings is governed elsewhere in the code.

Note: Section 3411, "Accessibility for Existing Buildings," is cited as governing.

1103.2.3 Employee work areas

- Work areas with all of the following characteristics have no accessibility requirements:
  - < 300 sf.
  - Raised above adjacent grade or floor ≥ 7" as follows:
    The height must be necessary for the function of the space.
  - Function of area is not that of a courtroom station.
- Other employee work areas must meet accessibility requirements only as follows:
  - For certain visible fire alarm requirements.

  Note: 907.5.2.3.2, "Employee work areas," is cited as governing the alarm requirements.

  - For certain accessible means of egress requirements.

  Note: 1007, "Accessible Means of Egress," is cited as governing egress requirements.

  - For certain circulation routes.

  Note: 1104.3.1, "Employee work areas," is cited as governing circulation routes.

  - Disabled persons must be able to do the following:
    Approach the work area.
    Enter the work area.
    Exit the work area.

1103 Scoping Requirements

1103.2.4 Detached dwellings

- The following dwellings are not required to be accessible:
  - Detached 1-family.
  - Detached 2-family.
  - Accessory structures.
  - Associated sites.
  - Associated facilities.

1103.2.5 Utility buildings

- Agricultural buildings require accessible access to the following areas:
  - Paved work areas.
  - Areas open to the public.
- Private garages are required to meet the requirements of this chapter as follows:
  - Where they are required to have accessible parking.
- Carports are required to meet the requirements of this chapter as follows:
  - Where they are required to have accessible parking.
- Other facilities in Occupancy U need not meet the requirements of this chapter.

1103.2.6 Construction sites

- The following elements of construction sites are not required to be accessible:
  - Structures.
  - Sites.
  - Equipment.
  - Materials storage.
  - Scaffolding.
  - Bridging.
  - Materials hoists.
  - Construction trailers.

1103.2.7 Raised areas

- Raised areas similar to the types listed below and used primarily for the following are not required to be accessible or on an accessible route:
  - Uses:
    - Security.
    - Life safety.
    - Fire safety.
  - Types:
    - Observation galleries.
    - Lifeguard stands.
    - Prison guard towers.
    - Fire towers.

1103 Scoping Requirements

1103.2.8 Limited access spaces

- Nonoccupiable spaces, with access only by the following, need not be accessible:
  - Catwalks.
  - Crawl spaces.
  - Freight elevators.
  - Ladders.
  - Very narrow passageways.

1103.2.9 Equipment spaces

- Spaces such as the types indicated below that are accessed only for the following functions are not required to be accessible:
  - Functions:
    - Maintenance.
    - Monitoring.
    - Repair.
  - Types:
    - Communication equipment rooms.
    - Electric substations.
    - Elevator penthouses.
    - Elevator pits.
    - Equipment catwalks.
    - Highway utility facilities.
    - Mechanical equipment rooms.
    - Piping catwalks.
    - Sewage treatment pump rooms.
    - Sewage treatment stations.
    - Transformer vaults.
    - Tunnel utility facilities.
    - Water treatment pump rooms.
    - Water treatment stations.

1103.2.10 Single-occupant structures

- Facilities occupied by a single person, such as the example indicated below and accessed only as follows, are not required to be accessible:
  - Access:
    - By routes elevated above grade.
    - By routes below grade.
  - Example:
    - Tollbooths.

1103.2.11 Residential Group R-1

- Buildings in Occupancy R-1 meeting both of the following conditions:
  - ≤ 5 sleeping units for rent.
  - Serve as the residence of the proprietor.

1103 Scoping Requirements

1103.2.12 Day care facilities

- This section applies to day care facilities of the following occupancies:
  - A-3, E, I-4, R-3.
- Where a dwelling unit has a day care component, the following applies:
  - Only the day care component must be accessible.

1103.2.13 Live/work units

- This section addresses live/work units.

Note: Section 419, "Live/Work Units," is cited as governing these types of units.

- The nonresidential part of the unit must be accessible.
- The residential part of the unit is evaluated for accessibility separately from the nonresidential part.

Note: The following are cited as the basis for evaluating the residential part of the live/work unit:
1107.6.2, "Group R-2."
1107.7, "General exceptions."

1103.2.14 Detention and correctional facilities

- Common use areas with all the following characteristics governed as indicated below:
  - Characteristics:
    Where utilized only by any of the following:
    Detainees.
    Inmates.
    Security personnel.
    Which are not holding cells.
    Which are not used for housing as follows:
    Where required to be accessible.
  - Requirements:
    The areas are not required to be accessible.
    The areas are not required to be on an accessible route.

1103.2.15 Walk-in coolers and freezers

- The following need not be accessible where meant for employee use only:
  - Walk-in freezers.
  - Walk-in coolers.

1104 Accessible Route

1104.1 Site arrival points

- No accessible route is required between the following locations in the circumstances noted:
  - Locations:
    Site arrival point and facility entrance.
  - Circumstances:
    Where no Type B units are present or served.
    Where only one access between the two locations is as follows:
    A vehicle route with no pedestrian accommodation.
- Otherwise, accessible routes to an accessible entrance are required from the following:
  - Accessible passenger loading areas.
  - Accessible parking.
  - Public streets.
  - Public transportation stops.
  - Public sidewalks.

Case study: Fig. 1104.1. An accessible route is provided as required from accessible parking to an accessible building entrance. Included is a ramp meeting slope requirements.

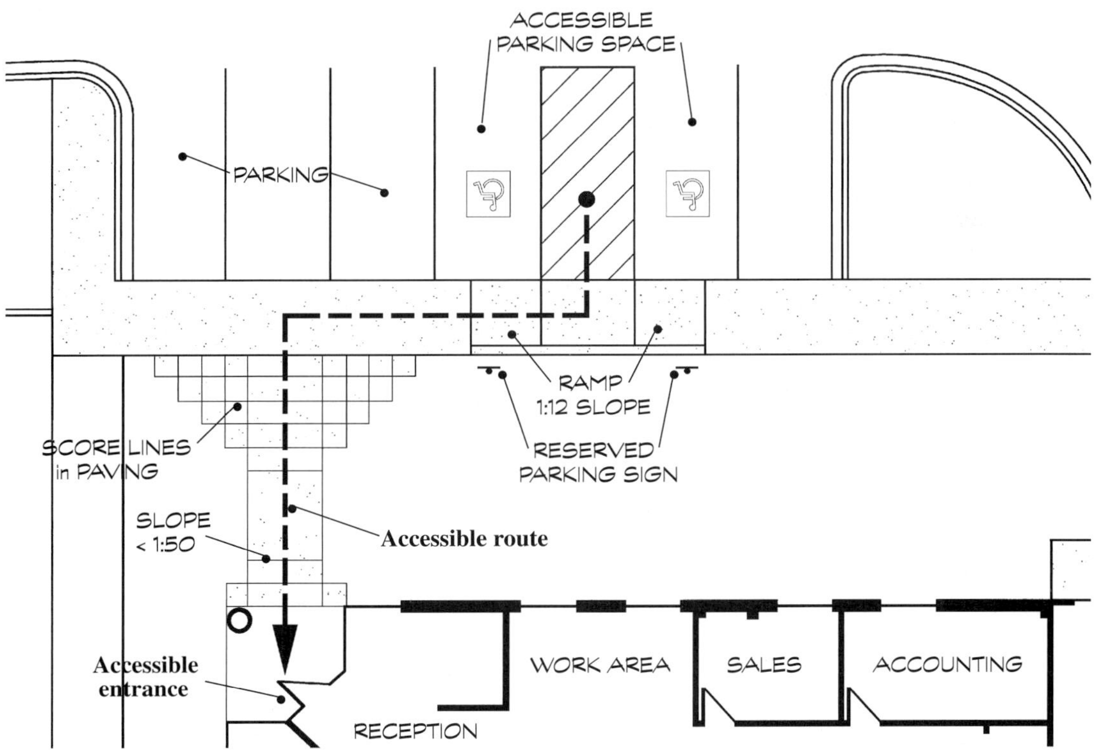

Fig. 1104.1. Partial site plan at entry. Garments to Go. Bastrop, Texas. Spencer Godfrey Architects. Round Rock, Texas.

1104 Accessible Route

1104.2 Within a site

- Accessible routes are not required between accessible facilities in the following case:
  - Where the only means of access between facilities is vehicular as follows:
    - No pedestrian access is provided.
- Otherwise, ≥ 1 accessible route is required connecting the following within a site:
  - Accessible buildings.
  - Accessible facilities.
  - Accessible elements.
  - Accessible spaces.

1104.3 Connected spaces

- Fixed seating in assembly areas is governed as follows:
  - Where wheelchair spaces are not provided, the following applies:
    - This seating is not required to be on an accessible route.
- The maneuvering clearance at the room-side of doors to sleeping units is not required in I-2 as follows:
  - Where the door is ≥ 44" wide.
- In other cases, an accessible route is required to connect the following:
  - Each part of a building required to be accessible.
  - Building entrances required to be accessible.
  - Pedestrian walkways required to be accessible.
  - The public way.

1104.3.1 Employee work areas

- The following common use routes need not be accessible routes:
  - Where the work area is < 300 sf as follows:
    - Perimeter is bounded by any of the following permanent fixtures:
      - Casework
      - Furnishings
      - Counters
      - Partitions
  - Where the work area is integral to equipment.
  - Where the work area is unprotected from the weather.
- In other cases, common use routes are governed as follows:
  - They must be accessible routes.

1104.3.2 Press boxes

- This section addresses press boxes in assembly areas.
- The following press boxes need not be on accessible routes:
  - Those in bleachers with both of the following characteristics:
    - With points of entry at 1 level only.
    - Total area of all press boxes must be ≤ 500 sf.
  - Where freestanding with both of the following characteristics:
    - Raised ≥ 12'.
    - Total area of all press boxes must be ≤ 500 sf.
- In other cases, press boxes are governed as follows:
  - They must be on accessible routes.

1104 Accessible Route

Case study: Fig. 1104.2. All buildings on the site are accessible and all are on accessible routes as indicated. The site plan complies with the code regarding accessibility.

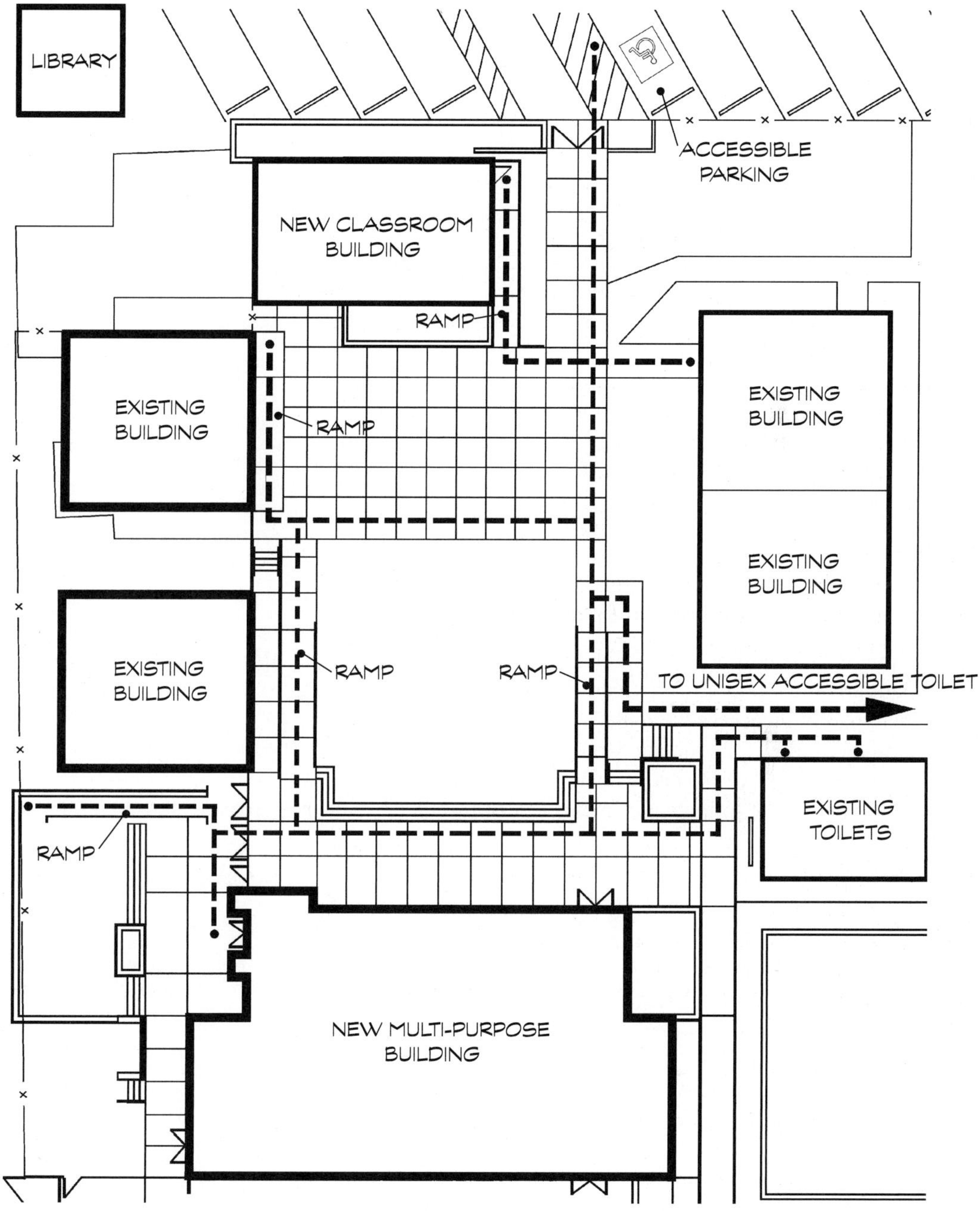

Fig. 1104.2. Site plan. Creston Elementary Multipurpose Building and New Classroom Building. Creston, California. Phillips Metsch Sweeney Moore Architects. Santa Barbara, California.

1104 Accessible Route

1104.4 Multilevel buildings and facilities

- Accessible routes are required for the following:
  - Occupancy M tenant spaces as follows:
    - ≥ 5 tenants.
  - Occupancy B or I as follows:
    - Health care provider offices.
  - Occupancy A-3 or B passenger facilities as follows:
    - Airports.
    - Other transportation.
- Otherwise, accessible routes are not required for the following:
  - For floors above and below accessible levels as follows:
    - Not required in the following case:
      - Total area of floors above and below an accessible level is ≤ 3,000 sf as follows:
        - Includes mezzanines.
  - In a 2-story building:
    - Not required between stories where 1 story has all of the following characteristics:
      - Occupant load is ≤ 5 people.
      - With no public space.
  - Levels without the following do not require an accesible route from an accessible level:
    - Accessible elements.
    - Accessible spaces required elsewhere in this chapter.

Note: The following are cited as requiring accessible spaces:
Section 1107, "Dwelling Units and Sleeping Units."
Section 1108, "Special Occupancies."

  - Not required for an airport control tower control room.
  - Not required on the floor directly below an airport control tower control room.
- The vertical part of an accessible route is not required as follows:
  - At the time of initial construction for the following:
    - Elevated employee work stations in a courtroom as follows:
      - Where one of the following elements can be installed per the requirement shown below:
        - Elements:
          - Ramp.
          - Lift.
          - Elevator.
        - Requirement:
          - Installation must be possible without reconfiguration or extension of either of the following:
            - The courtroom.
            - The electrical system.

Note: ICC A117.1, "Accessible and Usable Buildings and Facilities," is cited as governing the elements of vertical access to elevated work stations.

- In other cases the following applies to multiple levels including mezzanines:
  - ≥ 1 accessible route is required to connect each accessible level.

1104 Accessible Route

1104.5 Location

- Accessible routes must have one of the following relationships to general circulation:
  - Accessible routes must coincide with general circulation routes.
  - Accessible routes must be located in the same area as general circulation routes.
- The location of accessible routes is governed as follows:
  - Accessible routes are not required to be interior in the following case:
    From parking garages contained within and serving Type B units.
  - In other locations, the accessible route must be interior as follows:
    Where the general circulation path is interior.
- The following applies where only 1 accessible route is provided:
  - In an accessible unit, Type A unit, or Type B unit, the route may pass through the following:
    A kitchen.
    A storage room.
  - Otherwise, the route may not pass through the following or similar spaces:
    A kitchen.
    A storage room.
    A closet.
    A restroom.

1104.6 Security barriers

- The following security barriers must not obstruct the routes listed below:
  - Barriers:
    Bollards.
    Checkpoints.
    Others.
  - Routes:
    Accessible routes.
    Accessible means of egress.
- An accessible route must bypass the following obstructions as noted:
  - Obstructions:
    Security screening devices such as the following:
      Metal detectors.
      Fluoroscopes.
      Similar devices.
  - Requirements:
    Accessible route must provide users visual contact with personal items as follows:
      To the same degree as provided to persons passing through the screening devices.

1105 Accessible Entrances

1105.1 Public entrances

- Entrances are not required to be accessible in the following cases:
  - To spaces not required to be accessible.
  - Loading and service entrances that are not the sole entrance to the following:
    - A building.
    - A tenant space.
- In other locations, accessible entrances are required as follows:
  - As required by subsequent sections in this series.

  Note: The following are cited as requiring accessible entrances:
  1105.1.1, "Parking garage entrances."
  1105.1.2, "Entrances from tunnels or elevated walkways."
  1105.1.3, "Restricted entrances."
  1105.1.4, "Entrances for inmates or detainees."
  1105.1.5, "Service entrances."
  1105.1.6, "Tenant spaces, dwelling units and sleeping units."

  - ≥ 60% of the public entrances provided in addition to those noted above.

1105.1.1 Parking garage entrances

- The following building entrances are required to be accessible:
  - Any access provided directly from a parking structure.

1105.1.2 Entrances from tunnels or elevated walkways

- Where any of the following elements provide direct access to a building, the requirements indicated below apply:
  - Elements:
    - Pedestrian tunnels.
    - Elevated walkways.
  - Requirements:
    - ≥ 1 accessible entrance is required as follows:
      - From each of any of the elements that provide access.

1105.1.3 Restricted entrances

- ≥ 1 accessible restricted entrance to a building is required as follows:
  - Where ≥ 1 restricted entrance is provided.

1105 Accessible Entrances

1105.1.4 Entrances for inmates or detainees

- The following buildings used by any of the personnel listed have the requirement listed:
  - Buildings:
    - Judicial facilities.
    - Detention facilities.
    - Correctional facilities.
  - Personnel:
    - Inmates.
    - Detainees.
    - Security personnel.
  - Requirements:
    - ≥ 1 accessible entrance is required at each building as follows:
      - Among any entrances used only by the personnel listed above.

1105.1.5 Service entrances

- The service entrance must be accessible in the following case:
  - Where it is the only entrance to either of the following:
    - A building.
    - A tenant space.

1105.1.6 Tenant spaces, dwelling units and sleeping units

- Accessible entrances are not required for the following:
  - Tenant spaces that are not required to be accessible.
  - Dwelling units or sleeping units that are not required to be one of the following:
    - Accessible Type A unit.
    - Accessible Type B unit.
- In other cases:
  - ≥ 1 accessible entrance is required for the following:
    - Each tenant.
    - Each dwelling unit.
    - Each sleeping unit in a facility.

1106 Parking and Passenger Loading Facilities

1106.1 Required

- This section does not govern accessible parking for certain facilities where use by handicapped persons is greater than the norm for other buildings.

  Note: The following are cited as governing such facilities:
  1106.2, "Groups R-2 and R-3."
  1106.3, "Hospital outpatient facilities."
  1106.4, "Rehabilitation facilities and outpatient physical therapy facilities."

- This section does not govern accessible parking for the following:
  - Buses.
  - Trucks.
  - Other delivery vehicles.
  - Law enforcement vehicles.
  - Vehicular impound.
  - Motors pools as follows:
    Where parking accessed by the public has an accessible passenger loading zone.
- Where ≥ 1 parking lot exists on a site, the following applies:
  - The number of accessible parking spaces required are calculated separately for each lot.
- In other parking, the number required to be accessible is as follows:

Table 1106.1 Accessible Parking Spaces Required

Total spaces	Accessible required	Total spaces	Accessible required	Total spaces	Accessible required
1–25	≥ 1	651–700	≥ 14	1,501–1,600	≥ 26
26–50	≥ 2	701–750	≥ 15	1,601–1,700	≥ 27
51–75	≥ 3	751–800	≥ 16	1,701–1,800	≥ 28
76–100	≥ 4	801–850	≥ 17	1,801–1,900	≥ 29
101–150	≥ 5	851–900	≥ 18	1,901–2,000	≥ 30
151–200	≥ 6	901–950	≥ 19	2,001–2,100	≥ 31
201–300	≥ 7	951–1,000	≥ 20	2,101–2,200	≥ 32
301–400	≥ 8	1,001–1,100	≥ 21	2,201–2,300	≥ 33
401–500	≥ 9	1,101–1,200	≥ 22	2,301–2,400	≥ 34
501–550	≥ 11	1,201–1300	≥ 23	2,401–2,500	≥ 35
551–600	≥ 12	1,301–1400	≥ 24	2,501–2,600	≥ 36
601–650	≥ 13	1,401–1500	≥ 25	2,601–2,700	≥ 37

Source: IBC Table 1106.1.

- The number of accessible parking places required in parking > 2,700 total spaces is determined as indicated below:
  - Fractions of accessible spaces required are rounded up to the next higher whole number.

  Number of accessible parking spaces = [(Total spaces – 2,700) ÷ 100] + 37

1106 Parking and Passenger Loading Facilities

Case study: Fig. 1106.1. The parking lot in front of the Occupancy B building has 26 parking spaces. IBC Table 1106.1 requires 2 spaces to be accessible where 26 to 50 regular spaces are provided. 3 accessible spaces are provided, which meets the code requirement. Section 1106.5 requires that 1 of every 6 accessible spaces be van-accessible. 1 van-accessible space is provided in the front parking lot as required.

The parking lot in back of the building has 150 parking spaces. IBC Table 1106.1 requires that 5 spaces be accessible where 101 to 150 regular spaces are provided. 6 accessible parking spaces are provided, which supersede the minimum requirement. 1 van-accessible space is needed since 6 accessible spaces are provided. 2 van-accessible spaces are provided, which is > the minimum requirement.

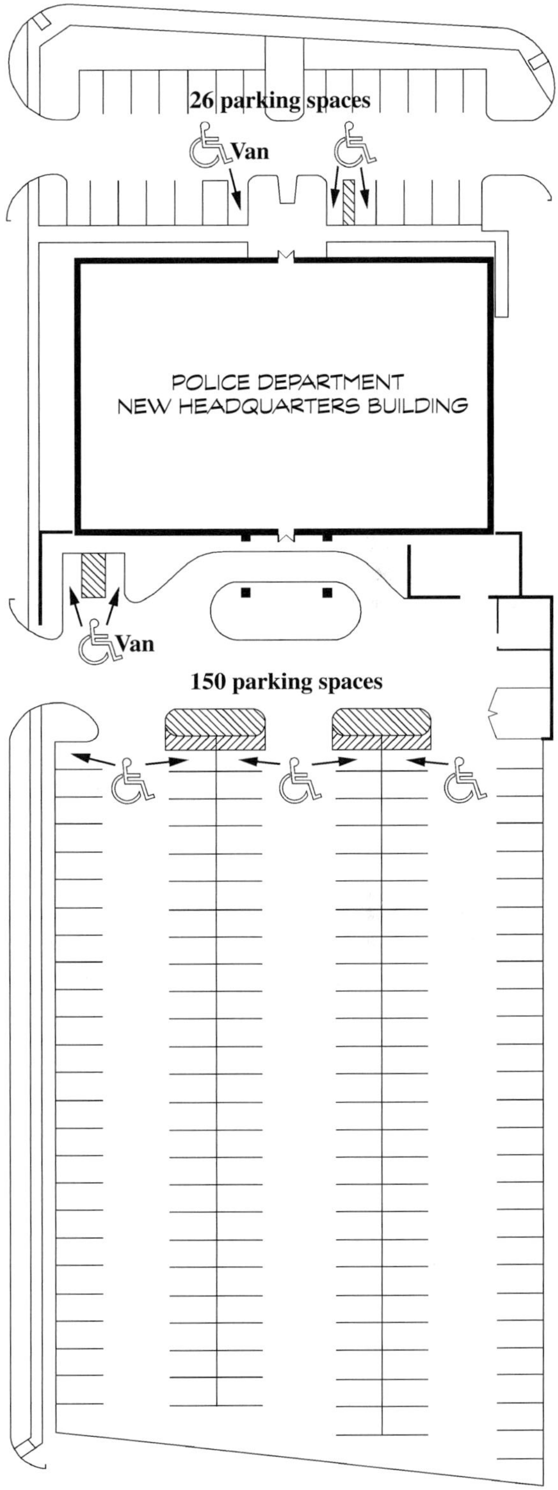

Fig. 1106.1. Site plan. Hot Springs Police Department New Headquarters. Hot Springs National Park, Arkansas. Cromwell Architects Engineers. Little Rock, Arkansas.

1106 Parking and Passenger Loading Facilities

1106.2 Groups R-2 and R-3

- Accessible parking as listed below is required in the following location:
  - Location:
    Where either of the following dwelling or sleeping units are required:
      Accessible Type A.
      Accessible Type B.
  - Requirements:
    ≥ 1 accessible parking place is required.
    ≥ 2% of the total parking spaces are required to be accessible as listed below:

Table 1106.2 Accessible Parking Required in R-2 and R-3

Total spaces	Accessible required	Total spaces	Accessible required	Total spaces	Accessible required
1–50	≥ 1	451–500	≥ 10	901–950	≥ 19
51–100	≥ 2	501–550	≥ 11	951–1,000	≥ 20
101–150	≥ 3	551–600	≥ 12	1,001–1,050	≥ 21
151–200	≥ 4	601–650	≥ 13	1,051–1,100	≥ 22
201–250	≥ 5	651–700	≥ 14	1,101–1,150	≥ 23
251–300	≥ 6	701–750	≥ 15	1,151–1,200	≥ 24
301–350	≥ 7	751–800	≥ 16	1,201–1,250	≥ 25
351–400	≥ 8	801–850	≥ 17	1,251–1,300	≥ 26
401–450	≥ 9	851–900	≥ 18	etc.	

- Accessible parking is required to be within or beneath a building in the following case:
  - Where general parking is provided within or beneath a building.

1106.3 Hospital outpatient facilities

- Patient and visitor parking must have the following accessible parking spaces provided:
  - ≥ 10% of the total number of spaces.
  - ≥ 1 parking space.
- A partial table of parking spaces required to be accessible is provided below:

Table 1106.3 Accessible Parking Required for Hospital Outpatient Facilities

Total spaces	Accessible required	Total spaces	Accessible required	Total spaces	Accessible required
1–10	≥ 1	81–90	≥ 9	161–170	≥ 17
11–20	≥ 2	91–100	≥ 10	171–180	≥ 18
21–30	≥ 3	101–110	≥ 11	181–190	≥ 19
31–40	≥ 4	111–120	≥ 12	191–200	≥ 20
41–50	≥ 5	121–130	≥ 13	201–210	≥ 21
51–60	≥ 6	131–140	≥ 14	211–220	≥ 22
61–70	≥ 7	141–150	≥ 15	221–230	≥ 23
71–80	≥ 8	151–160	≥ 16	etc.	

1106 Parking and Passenger Loading Facilities

Case study: Fig. 1106.2. This Occupancy R-2 building provides 136 parking spaces in the parking plan shown. At least 2% of these must be accessible. 4 accessible parking spaces are provided, thus, meeting the minimum requirement of 3. Section 1106.5 requires at least 1 van-accessible space for every 6 or fraction of 6 accessible spaces. 2 van-accessible spaces are provided, thus, meeting the minimum of 1 required. Since parking is provided beneath the building, accessible parking must be provided there as well. The parking garage also meets this requirement and is, therefore, in compliance with the code in all respects.

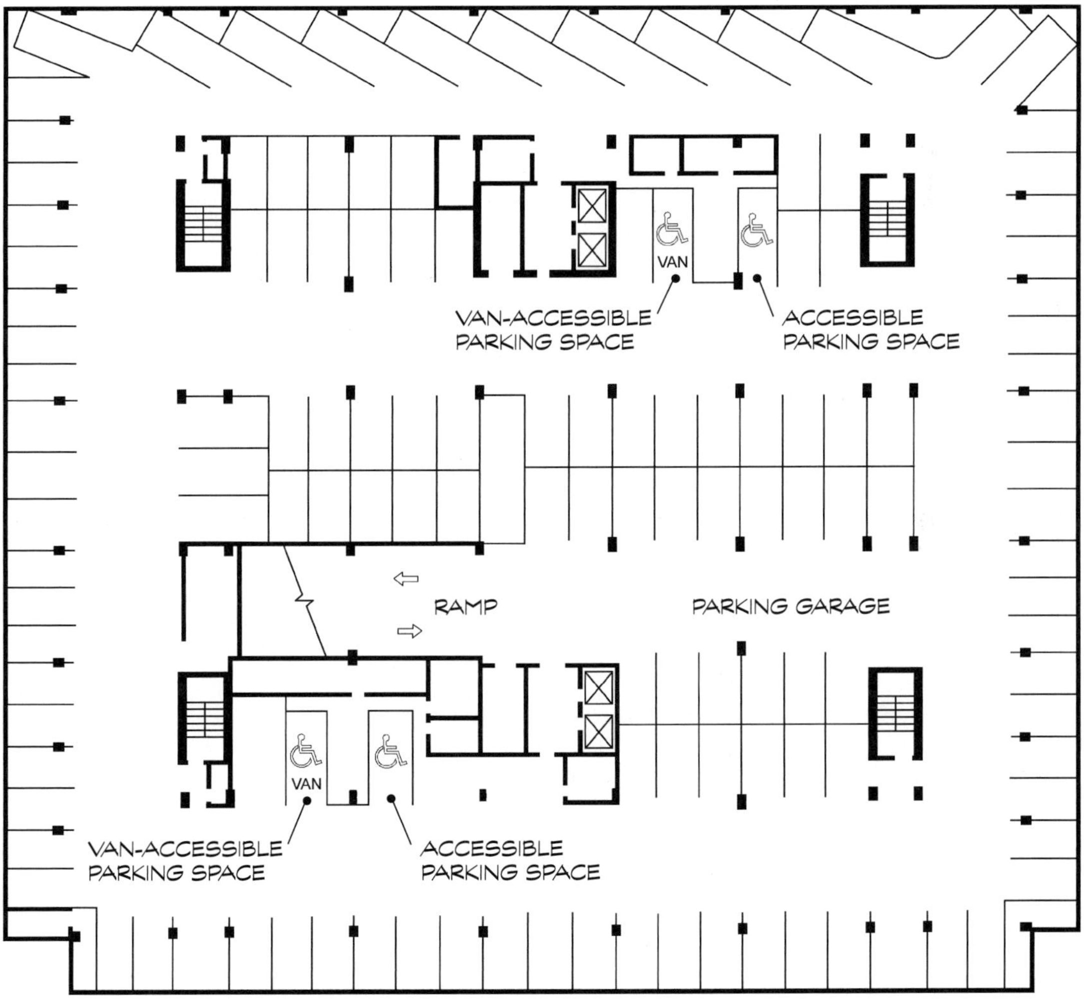

Fig. 1106.2. Parking garage floor plan. Hoyt Street Properties. Portland, Oregon. Ankrom Moisan Associated Architects. Portland, Oregon.

1106 Parking and Passenger Loading Facilities

1106.4 Rehabilitation facilities and outpatient physical therapy facilities

- The requirement listed below applies to the following types of parking:
  - Parking:
    - Patients of rehabilitation and physical therapy facilities specializing in patient mobility.
    - Visitors to rehabilitation and physical therapy facilities specializing in patient mobility.
  - Requirement:
    - The larger of the following numbers of parking places must be accessible:
      - One.
      - ≥ 20% of the parking as noted above.
- A partial table of parking spaces required to be accessible is provided below:

Table 1106.4 Accessible Parking at Rehabilitation and Outpatient Physical Therapy Facilities

Total spaces	Accessible required	Total spaces	Accessible required	Total spaces	Accessible required
1–5	≥ 1	31–35	≥ 7	61–65	≥ 13
6–10	≥ 2	36–40	≥ 8	66–70	≥ 14
11–15	≥ 3	41–45	≥ 9	71–75	≥ 15
16–20	≥ 4	46–50	≥ 10	76–80	≥ 16
21–25	≥ 5	51–55	≥ 11	81–85	≥ 17
26–30	≥ 6	56–60	≥ 12	etc.	

1106.5 Van spaces

- The following elements may have a vertical clearance ≥ 7' where the conditions below apply:
  - Elements:
    - Van vehicular routes.
    - Van entrances.
    - Van parking spaces.
    - Van access aisles.
  - Conditions:
    - In occupancies R-2 and R-3 as follows:
      - Where van-accessible spaces are in private garages.
- Otherwise, the larger of the following numbers of parking spaces must be van-accessible:
  - One.
  - ≥ 16 2/3% of accessible parking spaces.
- The required number of van-accessible spaces is also listed below:

Accessible spaces	Accessible van spaces required
1–6	≥ 1
7–12	≥ 2
13–18	≥ 3
19–24	≥ 4
etc.	

1106 Parking and Passenger Loading Facilities

1106.6 Location

- Where parking serves a particular building, the following applies:
  - Accessible parking must be located on the shortest accessible route to the following:
    - An accessible building entrance.
- Where parking does not serve a particular building, the following applies:
  - Accessible parking must be located on the shortest accessible route to the following:
    - An accessible pedestrian entrance to the parking facility.
- Distribution of accessible parking is governed as follows:
  - Accessible parking may be located in different facilities as follows:
    - Where there is equal or greater convenience for handicapped persons regarding the following:
      - Distance from an accessible entrance.
      - Parking fees.
      - General convenience.
  - Otherwise, the following applies:
    - All van-accessible spaces may be located on 1 level.
    - Other accessible spaces must be distributed near all accessible entrances.

1106.7 Passenger loading zones

- These facilities are governed by other standards.

Note: ICC A117.1, "Accessible and Usable Buildings and Facilities," is cited as governing passenger loading zones.

1106.7.1 Continuous loading zones

- Where passenger loading zones are provided, the following applies:
  - An accessible loading zone must be provided as follows:
    - In every continuous length of loading zone = 100'.

1106.7.2 Medical facilities

- An accessible passenger loading zone is required as follows:
  - At an accessible entrance to the following facilities with the functions listed:
    - Facilities:
      - Licensed medical facilities.
      - Licensed long-term care facilities.
    - Functions:
      - Where people stay > 24 hours per visit.
      - Where people receive the following care:
        - Physical.
        - Medical.

1106.7.3 Valet parking

- An accessible passenger loading zone is required where valet parking is provided.

1106.7.4 Mechanical access parking garages

- Mechanical access parking garages require the following:
  - ≥ 1 passenger loading area at the following locations:
    - Vehicle pick-up and drop-off.

1107 Dwelling Units and Sleeping Units

1107.3 Accessible spaces

- Recreational facilities are not governed by this section.

  Note: 1109.14, "Recreational and sports facilities," is cited as governing these areas.

- The maneuvering clearance at the room side of doors to sleeping units is not required as follows:
  - Where all of the following apply:
    - The occupancy is I-2.
    - The door is ≥ 44" wide.
- Other spaces with all of the following characteristics are required to be accessible:
  - Used by the following people:
    - Public.
    - Residents.
  - Serving the following units:
    - Accessible.
    - Type A units.
    - Type B units.
  - Including the following:
    - Toilet rooms.
    - Bathing rooms.
    - Kitchens.
    - Living areas.
    - Dining areas.
    - Patios or terraces.
    - Balconies.

1107.4 Accessible route *(part 1 of 2)*

- For any of the following conditions, an accessible route may be omitted in lieu of the substitution listed:
  - Conditions:
    - Where any of the following conditions are beyond the control of the owner:
      - Where the finished grade slopes > 1:12 between accessible facilities.
      - Where physical barriers prevent installation of an accessible route.
      - Where legal restrictions prevent installation of an accessible route.
  - Substitution:
    - A vehicular route with parking at each accessible facility or building.

  Note: Section 1106, "Parking and Passenger Loading Facilities," is cited as governing the parking as required above.

- The following facilities may have a floor surface ≤ 4" below the adjacent interior floor where the conditions below apply:
  - Facilities:
    - Exterior decks.
    - Patios.
    - Balconies.
  - Conditions:
    - At Type B dwelling units.
    - The facility must have an impervious surface.

1107 Dwelling Units and Sleeping Units

1107.4 Accessible route *(part 2 of 2)*

- In all other cases, ≥ 1 accessible route is required as follows:
  - Between accessible building entrances and the following elements of the units in the building as noted below:
    - Elements:
      - Primary entrances.
      - Exterior spaces serving the units.
      - Interior spaces serving the units.
    - Units:
      - Accessible dwelling units.
      - Type A units.
      - Type B units.

1107.5.1.1 Accessible units

- The following Occupancy I-1 facilities must be accessible in the quantity as listed:
  - Facilities:
    - Dwelling units and sleeping units.
  - Quantity:
    - The greater of the following quantities:
      - 1 unit.
      - ≥ 4% of the units as follows:
        - A partial list of units required to be accessible is listed in the table below:

Table 1107.5.1.1 Accessible Occupancy I-1 Dwelling Units and Sleeping Units

Total number of units	Units to be accessible	Total number of units	Units to be accessible
1–25	≥ 1	151–175	≥ 7
26–50	≥ 2	176–200	≥ 8
51–75	≥ 3	201–225	≥ 9
76–100	≥ 4	226–250	≥ 10
101–125	≥ 5	251–275	≥ 11
126–150	≥ 6	276–300	≥ 12

1107.5.1.2 Type B units

- The requirements listed below apply to Occupancy I-1 buildings as follows:
  - Buildings:
    - Where there are ≥ 4 units used as residences as follows:
      - Dwelling units and sleeping units.
  - Requirements:
    - Units must be Type B.

Note: 1107.7, "General exceptions," is cited as permitting the number of Type B units to be reduced based on the absence of elevators and other access problems.

1107 Dwelling Units and Sleeping Units

1107.5.2.1 Accessible units

- The larger number of the following dwelling units where provided must be accessible:
  - 1 of each type of unit.
  - ≥ half of the total number of each type of unit.
- The larger number of the following sleeping units where provided must be accessible:
  - 1 of each type of unit.
  - ≥ half of the total number of each type of unit.

1107.5.2.2 Type B units

- The following must be Type B units:
  - Where there are ≥ 4 of either or both of the following units used as residences in a building:
    Dwelling units.
    Sleeping units.

Note: 1107.7, "General exceptions," is cited as permitting the number of Type B units to be reduced based on the absence of elevators and other access problems.

1107.5.3.1 Accessible units

- The following units of the facilities in Occupancy I-2 listed below must be accessible in the quantity listed:
  - Units:
    Dwelling units.
    Sleeping units.
  - Facilities:
    General-purpose hospitals.
    Psychiatric facilities.
    Detoxification facilities.
    Residential care/assisted living facilities.
  - Quantity:
    The number of units that must be accessible is the larger of the following:
    1 unit.
    ≥ 10% of the total number of units.
- The following is a partial table showing the number of accessible units required by this section based on the total number of units.

Table 1107.5.3.1 Accessible Dwelling or Sleeping Units Required

Total units	Accessible required	Total units	Accessible required	Total units	Accessible required
1–10	≥ 1	81–90	≥ 9	161–170	≥ 17
11–20	≥ 2	91–100	≥ 10	171–180	≥ 18
21–30	≥ 3	101–110	≥ 11	181–190	≥ 19
31–40	≥ 4	111–120	≥ 12	191–200	≥ 20
41–50	≥ 5	121–130	≥ 13	201–210	≥ 21
51–60	≥ 6	131–140	≥ 14	211–220	≥ 22
61–70	≥ 7	141–150	≥ 15	221–230	≥ 23
71–80	≥ 8	151–160	≥ 16	etc.	

1107 Dwelling Units and Sleeping Units

1107.5.3.2 Type B units

- The following units of the facilities in Occupancy I-2 listed below have the noted requirement:
  - Units:
    - Where there are ≥ 4 of either or both of the following units used as residences in a building:
      - Dwelling units.
      - Sleeping units.
  - Facilities:
    - General-purpose hospitals.
    - Psychiatric facilities.
    - Detoxification facilities.
    - Residential care/assisted living facilities.
  - Requirement:
    - Units must be Type B.

Note: 1107.7, "General exceptions," is cited as permitting the number of Type B units to be reduced based on the absence of elevators and other access problems.

1107.5.4 Group I-2 rehabilitation facilities

- All of the following units must be accessible where they are in any of the Occupancy I-2 facilities listed below and having the function noted:
  - Units:
    - Dwelling units.
    - Sleeping units.
  - Facilities:
    - Hospitals.
    - Rehabilitation facilities.
  - Function:
    - Treatment of mobility problems.

1107.5.5.1 Group I-3 sleeping units *(part 1 of 2)*

- The following units must be accessible in the quantity indicated:
  - Units:
    - Dwelling units.
    - Sleeping units.
  - Quantity:
    - The number of units required to be accessible is the larger of the following:
      - 1 unit.
      - ≥ 2% of the total number of units as follows:
        - The partial table below lists the number of accessible units required.

1107 Dwelling Units and Sleeping Units

1107.5.5.1 Group I-3 sleeping units *(part 2 of 2)*

Table 1107.5.5.1 Accessible Dwelling Units and Sleeping Units in Occupancy I-3

Total units	Accessible required	Total units	Accessible required	Total units	Accessible required
1–50	≥ 1	451–500	≥ 10	901–950	≥ 19
51–100	≥ 2	501–550	≥ 11	951–1,000	≥ 20
101–150	≥ 3	551–600	≥ 12	1,001–1,050	≥ 21
151–200	≥ 4	601–650	≥ 13	1,051–1,100	≥ 22
201–250	≥ 5	651–700	≥ 14	1,101–1,150	≥ 23
251–300	≥ 6	701–750	≥ 15	1,151–1,200	≥ 24
301–350	≥ 7	751–800	≥ 16	1,201–1,250	≥ 25
351–400	≥ 8	801–850	≥ 17	1,251–1,300	≥ 26
401–450	≥ 9	851–900	≥ 18	etc.	

1107.5.5.2 Special holding cells and special housing cells or rooms

- The following Occupancy I-3 spaces are not required to have grab bars:
  - Spaces designed to prevent suicide as follows:
    Without protrusions.
- In other cases, one of each kind of the following I-3 spaces, where provided and used as listed below, must be accessible:
  - Spaces:
    Special holding cells.
    Special housing cells or rooms.
  - Uses:
    Orientation.
    Protective custody.
    Administrative detention.
    Disciplinary detention.
    Segregation.
    Detoxification.
    Medical isolation.

Note: 1107.5.5.1, "Group I-3 sleeping units," is cited as identifying required accessible spaces that are in addition to those listed in this section.

1107.5.5.3 Medical care facilities

- Where medical isolation cells are required, the following applies:
  - Any of the following that are required to be accessible must also be provided:
    Patient sleeping units.
    Patient sleeping cells.

Note: 1107.5.5.2, "Special holding cells and special housing cells or rooms," is cited as requiring medical isolation cells.

1107 Dwelling Units and Sleeping Units

1107.6.1.1 Accessible units *(part 1 of 2)*

- In Occupancy R-1, the following units must be provided as required below:
  - Units:
    - Accessible dwelling units.
    - Accessible sleeping units.
  - Requirements:
    - The number of accessible units required is based on the total number of R-1 units on the site.
    - Accessible units must be distributed among the various classes of units.
    - Roll-in showers in accessible units must have the following:
      - A permanent folding seat.
    - Where > 2500 units are on a site, the number of accessible units required is determined as follows and any fraction of a unit is rounded up:
      - Units to be accessible and requiring a roll-in shower:

        [(Total accommodations on site – 2500) ÷100] + 25

      - Units to be accessible and not requiring a roll-in shower:

        [(Total accommodations on site – 2500) ÷ 100] + 35

      - Total accessible units required:

        {[(Total accommodations on site – 2500) ÷ 100] × 2} + 60

    - Where ≤ 2500 units are on a site, the required accessible units are listed in the following tables:

Table 1107.6.1.1a Accessible Dwelling Units and Sleeping Units Required (1–23)

Total number of units on site	Required accessible units with roll-in shower	Required accessible units, no roll-in shower	Total accessible units required
1–25	≥ 0	≥ 1	≥ 1
26–50	≥ 0	≥ 2	≥ 2
51–75	≥ 1	≥ 3	≥ 4
76–100	≥ 1	≥ 4	≥ 5
101–150	≥ 2	≥ 5	≥ 7
151–200	≥ 2	≥ 6	≥ 8
201–300	≥ 3	≥ 7	≥ 10
301–400	≥ 4	≥ 8	≥ 12
401–500	≥ 4	≥ 9	≥ 13
501–533	≥ 6	≥ 11	≥ 16
534–550	≥ 6	≥ 11	≥ 17
551–566	≥ 6	≥ 12	≥ 17
567–600	≥ 6	≥ 12	≥ 18
601–633	≥ 7	≥ 13	≥ 19
634–650	≥ 7	≥ 13	≥ 20
651–666	≥ 7	≥ 14	≥ 20
667–700	≥ 7	≥ 14	≥ 21
701–733	≥ 8	≥ 15	≥ 22
734–750	≥ 8	≥ 15	≥ 23
751–766	≥ 8	≥ 16	≥ 23

1107 Dwelling Units and Sleeping Units

1107.6.1.1 Accessible units *(part 2 of 2)*

Table 1107.6.1.1b Accessible Dwelling Units and Sleeping Units Required (24–60)

Total number of units on site	Required accessible units with roll-in shower	Required accessible units, no roll-in shower	Total accessible units required
767–800	≥ 8	≥ 16	≥ 24
801–833	≥ 9	≥ 17	≥ 25
834–850	≥ 9	≥ 17	≥ 26
851–866	≥ 9	≥ 18	≥ 26
867–900	≥ 9	≥ 18	≥ 27
901–933	≥ 10	≥ 19	≥ 28
934–950	≥ 10	≥ 19	≥ 29
951–966	≥ 10	≥ 20	≥ 29
967–1,000	≥ 10	≥ 20	≥ 30
1,001–1,050	≥ 11	≥ 21	≥ 31
1,051–1,100	≥ 11	≥ 21	≥ 32
1,101–1,150	≥ 12	≥ 22	≥ 33
1,151–1,200	≥ 12	≥ 22	≥ 34
1,201–1,250	≥ 13	≥ 23	≥ 35
1,251–1,300	≥ 13	≥ 23	≥ 36
1,301–1,350	≥ 14	≥ 24	≥ 37
1,351–1,400	≥ 14	≥ 24	≥ 38
1,401–1,450	≥ 15	≥ 25	≥ 39
1,451–1,500	≥ 15	≥ 25	≥ 40
1,501–1,550	≥ 16	≥ 26	≥ 41
1,551–1,600	≥ 16	≥ 26	≥ 42
1,601–1,650	≥ 17	≥ 27	≥ 43
1,651–1,700	≥ 17	≥ 27	≥ 44
1,701–1,750	≥ 18	≥ 28	≥ 45
1,751–1,800	≥ 18	≥ 28	≥ 46
1,801–1,850	≥ 19	≥ 29	≥ 47
1,851–1,900	≥ 19	≥ 29	≥ 48
1,901–1,950	≥ 20	≥ 30	≥ 49
1,951–2,000	≥ 20	≥ 30	≥ 50
2,001–2,050	≥ 21	≥ 31	≥ 51
2,051–2,100	≥ 21	≥ 31	≥ 52
2,101–2,150	≥ 22	≥ 32	≥ 53
2,151–2,200	≥ 22	≥ 32	≥ 54
2,201–2,250	≥ 23	≥ 33	≥ 55
2,251–2,300	≥ 23	≥ 33	≥ 56
2,301–2,350	≥ 24	≥ 34	≥ 57
2,351–2,400	≥ 24	≥ 34	≥ 58
2,401–2,450	≥ 25	≥ 35	≥ 59
2,451–2,500	≥ 25	≥ 35	≥ 60

Source: IBC Table 1107.6.1.1.

1107 Dwelling Units and Sleeping Units

1107.6.1.2 Type B units

- The requirements listed below apply to Occupancy R-1 buildings as follows:
  - Buildings:
    - Where there are ≥ 4 of either or both of the following units used as residences:
      - Dwelling units.
      - Sleeping units.
  - Requirements:
    - Units must be Type B.

Note: 1107.7, "General exceptions," is cited as permitting the number of Type B units to be reduced based on the absence of elevators and other access problems.

1107.6.2.1.1 Type A units

- This section governs the following buildings in Occupancy R-2:
  - Apartment buildings.
  - Monasteries.
  - Convents.
- The requirements below apply where > 20 of any of the following units are provided:
  - Units:
    - Dwelling units.
    - Sleeping units.
  - Requirements:
    - Type A units must be distributed among the following:
      - All classes of units provided.
    - The number of Type A units required is the greater of the following:
      - 1 unit.
      - 2% of the total units on the site as follows:
        - Existing units on the site are not counted in the total units.
        - All other R-2 units on the site are counted in the total number of units.
    - The following table is a partial listing of the number of Type A units required:

Table 1107.6.2.1.1 Type A Units Required in Apt. Buildings, Monasteries, and Convents

Total units	Type A required	Total units	Type A required	Total units	Type A required
1–50	≥ 1	451–500	≥ 10	901–950	≥ 19
51–100	≥ 2	501–550	≥ 11	951–1,000	≥ 20
101–150	≥ 3	551–600	≥ 12	1,001–1,050	≥ 21
151–200	≥ 4	601–650	≥ 13	1,051–1,100	≥ 22
201–250	≥ 5	651–700	≥ 14	1,101–1,150	≥ 23
251–300	≥ 6	701–750	≥ 15	1,151–1,200	≥ 24
301–350	≥ 7	751–800	≥ 16	1,201–1,250	≥ 25
351–400	≥ 8	801–850	≥ 17	1,251–1,300	≥ 26
401–450	≥ 9	851–900	≥ 18	etc.	

Note: 1107.7, "General exceptions," is cited as permitting the number of Type A units to be reduced based on the absence of elevators and other access problems.

1107 Dwelling Units and Sleeping Units

1107.6.2.1.2 Type B units

- This section governs the following buildings in Occupancy R-2:
  - Apartment buildings.
  - Monasteries.
  - Convents.
- The requirements below apply where there are ≥ 4 of any of the following units used as residences:
  - Units:
    - Dwelling units.
    - Sleeping units.
  - Requirements:
    - Units must be Type B.

Note: 1107.7, "General exceptions," is cited as permitting the number of Type B units to be reduced based on the absence of elevators and other access problems.

1107.6.2.2.1 Accessible units

- The following buildings are governed by the requirements listed below:
  - Buildings:
    - Boarding houses.
    - Dormitories.
    - Fraternity houses.
    - Sorority houses.
  - Requirements:
    - These buildings require the following accessible units as noted below:
      - Units:
        - Dwelling units.
        - Sleeping units.
      - Requirements:
        - The number of accessible units required is as follows:
          - The same number as required for Occupancy R-1.

Note: IBC Table 1107.6.1.1, "Accessible Dwelling and Sleeping Units," is cited as listing the number of accessible units required.
The following tables of this handbook provide an expanded version of the IBC table:
Table 1107.6.1.1a, "Accessible Dwelling Units and Sleeping Units Required (1–23).
Table 1107.6.1.1b, "Accessible Dwelling Units and Sleeping Units Required (24–60).

1107 Dwelling Units and Sleeping Units

1107.6.2.2.2 Type B units

- The requirements listed below apply to Occupancy R-2 buildings as follows:
  - Buildings:
    Boarding houses.
    Dormitories.
    Fraternity houses.
    Sorority houses.
  - Requirements:
    Where there are ≥ 4 of either or both of the following units used as residences, the requirement indicated applies:
      Units:
        Dwelling units.
        Sleeping units.
      Requirement:
        Units must be Type B.

Note: 1107.7, "General exceptions," is cited as permitting the number of Type B units to be reduced based on the absence of elevators and other access problems.

1107.6.3 Group R-3

- This section applies to Occupancy R-3 buildings.
- Where there are ≥ 4 of either or both of the following units used as residences, the requirement listed below applies:
  - Units:
    Dwelling units.
    Sleeping units.
  - Requirement:
    Units must be Type B.

Note: 1107.7, "General exceptions," is cited as permitting the number of Type B units to be reduced based on the absence of elevators and other access problems.

1107.6.4.1 Accessible units

- The following is required in an Occupancy R-4 building:
  - ≥ 1 of either of the following units must be accessible:
    Dwelling unit.
    Sleeping unit.

1107 Dwelling Units and Sleeping Units

1107.6.4.2 Type B units

- This section applies to Occupancy R-4 buildings.
- Where there are ≥ 4 of either or both of the following units used as residences, the requirement listed below applies:
  - Units:
    Dwelling units.
    Sleeping units.
  - Requirement:
    Units must be Type B.

Note: 1107.7, "General exceptions," is cited as permitting the number of Type B units to be reduced based on the absence of elevators and other access problems.

1107.7 General exceptions

- This section provides for reductions in required numbers of the following units:
  - Type A units.
  - Type B units.

Note: The following sections are cited as permitting reductions of Type A units and Type B units:
1107.5, "Group I."
1107.6, "Group R."
The following are cited as modifying the required number of Type A units and Type B units where permitted:
1107.7.1, "Structures without elevator service."
1107.7.1.1, "One story with Type B units required."
1107.7.1.2, "Additional stories with Type B units."
1107.7.2, "Multistory units."
1107.7.3, "Elevator service to the lowest story with units."
1107.7.4, "Site impracticality."
1107.7.5, "Design flood elevation."

1107.7.1 Structures without elevator service

- This section applies to buildings without elevators and containing any of the following types of units:
  - Dwelling units.
  - Sleeping units
- Only certain stories must have the types of units indicated below:
  - Type A units.
  - Type B units.

Note: The following are cited as specifying which stories are required to have Type A and Type B units:
1107.7.1.1, "One story with Type B units required."
1107.7.1.2, "Additional stories with Type B units."
1107.6.2.1.1, "Type A units," is cited as governing the number of Type A units.

1107 Dwelling Units and Sleeping Units

1107.7.1.1 One story with Type B units required

- Where a building with any of the following units has no elevator, the requirement listed below applies:
  - Units:
    - Dwelling units to be used as residences.
    - Sleeping units to be used as residences.
  - Requirements:
    - ≥ 1 story with such units must have the following:
      - An accessible exterior entrance.
      - All Type B units as follows:
        - Where units are intended to be used as a residence.

1107.7.1.2 Additional stories with Type B units

- Building stories governed by this section are required to have all Type B units.
- This section governs each building story with the following characteristics:
  - The story does not have elevator service.
  - The story is not governed in the previous section of the code.
  - The story has any of the following units:
    - Dwelling units to be used as residences.
    - Sleeping units to be used as residences.
  - The story is served by a building entrance as follows:
    - Entrance is near the arrival point serving the story.
    - Entrance is served by the following grades in the locations listed below:
      - Grades:
        - Original site grade that slopes ≤ 1:10.
        - Design finished grade that slopes ≤ 1:10.
      - Locations:
        - Between the building entrance and the following points:
          - All vehicular arrival points ≤ 50' from the entrance.
          - All pedestrian arrival points ≤ 50' from the entrance.
          - Nearest arrival point to the entrance > 50' from it in the following case:
            - Where the nearest arrival point has both the following characteristics:
              - It is > 50' from the entrance.
              - It does not serve a story required to be all Type B units.

Note: 1107.7.1.1, "One story with Type B units required," is cited as the section requiring all units on one story to be Type B units.

1107 Dwelling Units and Sleeping Units

1107.7.2 Multistory units

- A multistory unit as follows has the requirements listed:
  - Unit:
    - Dwelling unit.
    - Sleeping unit.
  - Requirements:
    - The unit need not be a Type B unit in the following case:
      - Where the unit does not have its own internal elevator serving all levels.
    - Where an external elevator serves only 1 floor, the following applies:
      - The main entry floor is governed as follows:
        - It must be served by the elevator.
        - It must meet Type B unit requirements.
        - It must have a toilet.

1107.7.3 Elevator service to the lowest story with units

- The lowest story with the following units has the requirements listed:
  - Units:
    - Dwelling units to be used as residences.
    - Sleeping units to be used as residences.
  - Requirements:
    - Such units are required to be Type B units in the following case:
      - Where an elevator provides an accessible route to only that story.

1107.7.4 Site impracticality *(part 1 of 2)*

- This section applies to sites with > 1 building without an elevator.
- The following number of required Type B units may be reduced in certain cases:
  - The number required for buildings without elevators.

Note: 1107.7.1, "Structures without elevator service," is cited as the section requiring a minimum quantity of Type B units. 1107.7.1 provides the requirements for Type B units that are modified by the various entries of this section.

- The Type B units required may be reduced as follows where the conditions listed below are met:
  - Reduction:
    - The smaller number of units required is specified as a % of the following:
      - The number of units required without a reduction.
    - The % of units required is equal to the following:
      - The % of the site having original grade < 10% slope.

1107 Dwelling Units and Sleeping Units

1107.7.4 Site impracticality *(part 2 of 2)*

- Conditions:
  The number of Type B units on the site must total the following:
    ≥ 20% of the number required without a reduction.
  The following units with either circumstance listed below must be Type B units:
    Units:
      Dwelling units to be used as residences.
      Sleeping units to be used as residences.
    Circumstance 1:
      Grade serving the building entrance for the units noted above slopes as follows:
        It slopes ≤ 1:12 between the entrance and either of the following locations:
          A pedestrian arrival point.
          A vehicular arrival point.
    Circumstance 2:
      Grade serving the building entrance for the units noted above slopes as follows:
        It slopes ≤ 1:10 between the entrance and either of the following locations:
          A pedestrian arrival point.
          A vehicular arrival point.
      An elevated walkway is provided between the entrance and either of the above arrival points.
  Units served by an elevator are Type B units.

Note: 1107.7.3, "Elevator service to the lowest story with units," is cited as describing conditions wherein units served by elevators are Type B units.

1107.7.5 Design flood elevation

- This section applies to buildings without elevators on sites described herein.
- Sites addressed by this section do not require the following:
  - The number of Type A units and Type B units otherwise required.
- This section applies to sites where all of the following conditions apply:
  - The following must be at a level ≥ the required design flood elevation:
    Lowest floor.
    Lowest horizontal structural member.
  - The floor is at a height measured as follows with the grade conditions listed:
    Height measurement:
      Height is measured between the following two points:
        Minimum floor elevation required at primary entrances.
        One of the following elevations:
          Elevation of any of the following arrival points ≤ 50' away if any such points exist:
            Vehicular and/or pedestrian arrival points.
          Elevation at the closet arrival point in the following case:
            Where no arrival point is within 50' of a primary entrance.
    Grade conditions:
      Floor height is >30" above grade as specified above.
      Slope of grade is > 1:10 between the following two locations:
        Arrival point.
        Floor at primary entrance.

1108 Special Occupancies

1108.2.1 Services

- The following functions in assembly areas have the requirements listed:
  - Functions:
    Services in areas not required to be accessible.
    Facilities in areas not required to be accessible.
  - Requirements:
    Such functions must also be provided on an accessible level.
    Such functions must be accessible.

1108.2.2.1 General seating

- This section addresses wheelchair spaces in the following assembly areas with fixed seating:
  - Theaters.
  - Bleachers.
  - Grandstands.
  - Stadiums.
  - Arenas.
  - Other fixed seating.
- A partial listing of wheelchair spaces required is provided in the following table:

Table 1108.2.2.1 Wheelchair Spaces Required in Fixed Seating Assembly Areas

Total seats	Wheelchair spaces required	Total seats	Wheelchair spaces required	Total seats	Wheelchair spaces required
4–25	≥ 1	1,551–1,700	≥ 14	3,351–3,500	≥ 26
26–50	≥ 2	1,701–1,850	≥ 15	3,501–3,650	≥ 27
51–100	≥ 4	1,851–2,000	≥ 16	3,651–3,800	≥ 28
101–300	≥ 5	2,001–2,150	≥ 17	3,801–3,950	≥ 29
301–500	≥ 6	2,151–2,300	≥ 18	3,951–4,100	≥ 30
501–650	≥ 7	2,301–2,450	≥ 19	4,101–4,250	≥ 31
651–800	≥ 8	2,451–2,600	≥ 20	4,251–4,400	≥ 32
801–950	≥ 9	2,601–2,750	≥ 21	4,401–4,550	≥ 33
951–1,100	≥ 10	2,751–2,900	≥ 22	4,551–4,700	≥ 34
1,101–1,250	≥ 11	2,901–3,050	≥ 23	4,701–4,850	≥ 35
1,251–1,400	≥ 12	3,051–3,200	≥ 24	4,851–5,000	≥ 36
1,401–1,550	≥ 13	3,201–3,350	≥ 25	5,001–5,200	≥ 37

Source: IBC Table 1108.2.2.1.

- Number of wheelchair spaces required where total seats are > 5,000 is determined by the following equation:

  Wheelchair spaces = [(Total seats – 5,000) ÷ 200] + 36

  - Fractions of wheelchair spaces required are rounded up to the next higher whole number.

1108 Special Occupancies

Case study: Fig. 1108.2.2.1. The auditorium seats 120. IBC Table 1108.2.2.1 indicates that 5 wheelchair spaces are required where seating totals 101 through 300. The auditorium provides 8 wheelchair spaces, thus, meeting this requirement. The 4 wheelchair spaces in the front of the auditorium are provided by way of removable standard seats. As required, a companion seat is provided by every wheelchair space.

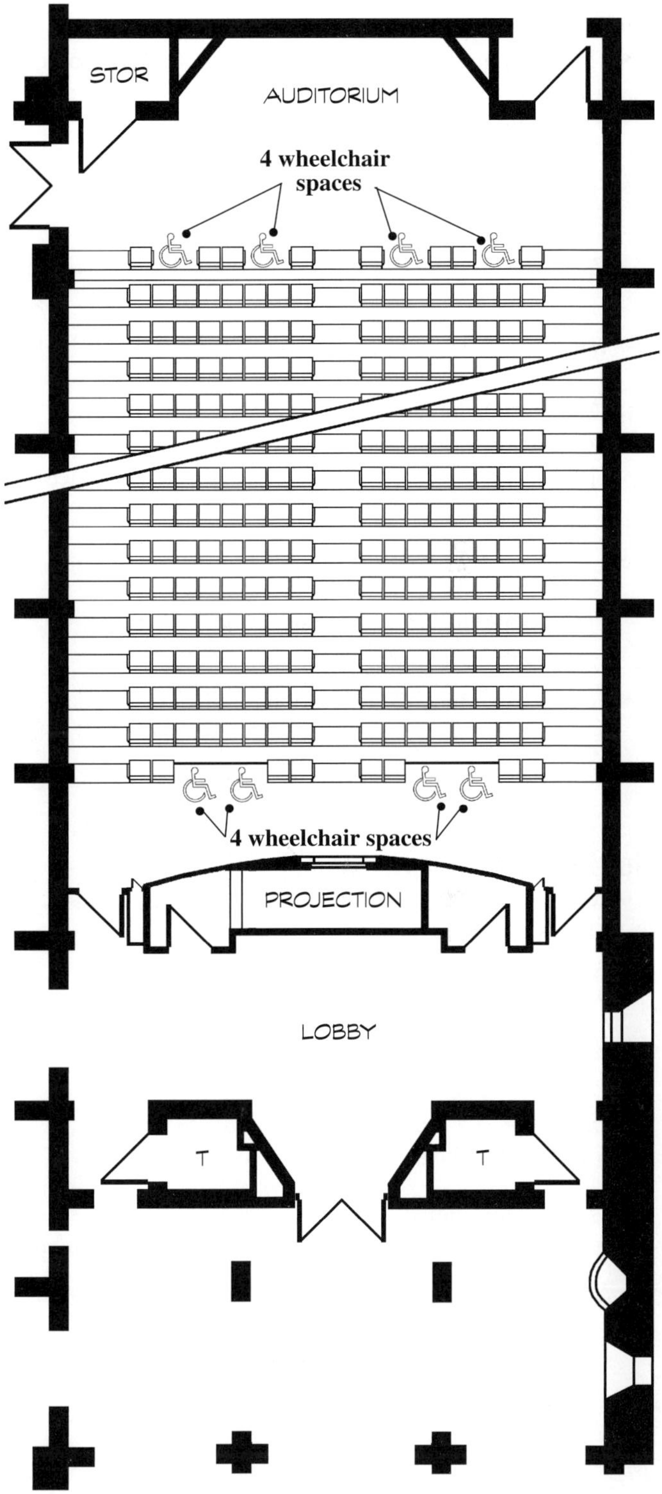

Fig. 1108.2.2.1. Partial floor plan at auditorium. Lady Bird Johnson Wildflower Center. Austin, Texas. Overland Partners, Inc. San Antonio, Texas.

1108 Special Occupancies

1108.2.2.2 Luxury boxes, club boxes and suites

- This section governs the following box seating in assembly areas having fixed seating:
  - Luxury boxes.
  - Club boxes.
  - Suites.
- Seating governed by this section requires wheelchair spaces as follows:
  - As per the table in the previous section.

Note: IBC Table 1108.2.2.1, "Accessible Wheelchair Spaces," is cited as governing the number of wheelchair spaces required by this section.

- Wheelchair spaces must be provided in ≥ 20% of all boxes.

1108.2.2.3 Other boxes

- Wheelchair spaces must be provided in ≥ 20% of all boxes.
- This section governs the number of wheelchair spaces in assembly box seating other than the following:
  - Luxury boxes.
  - Club boxes.
  - Suites.

Note: Section 1108.2.2.2, "Luxury boxes, club boxes and suites," is cited as governing wheelchair spaces in these spaces.

- Seating governed by this section requires wheelchair spaces as follows:
  - As per the previous table specifying required wheelchair spaces for fixed assembly seating.

Note: IBC Table 1108.2.2.1, "Accessible Wheelchair Spaces," is cited as governing the number of wheelchair spaces required by this section.

1108.2.2.4 Team or player seating

- Wheelchair spaces are not required where all of the following apply:
  - In team or player seating.
  - Seating area serving bowling lanes.
  - Not on an accessible route.

Note: 1109.14.4.1, "Bowling lanes," is cited as defining which bowling lanes must be on an accessible route.

- Otherwise, ≥ 1 wheelchair space must be provided as follows:
  - In team or player seating for a sport activity.

1108.2.3 Companion seats

- ≥ 1 companion seat is required for each wheelchair space in certain cases.

Note: The following are cited as requiring wheelchair spaces, each of which require a companion seat:
1108.2.2.1, "General seating."
1108.2.2.2, "Luxury boxes, club boxes and suites."
1108.2.2.3, "Other boxes."

1108 Special Occupancies

1108.2.4 Dispersion of wheelchair spaces in multilevel assembly seating areas

- Wheelchair spaces are required on the main floor level of all multilevel facilities.
- All wheelchair spaces may be on the main floor where both of the following apply:
  - The space is used for worship services.
  - The upper level contains ≤ 25% of the total seating capacity.
- All wheelchair spaces may be on the main floor where both of the following apply:
  - The upper level contains ≤ 25% of the total seating capacity.
  - The upper level contains ≤ 300 seats.
- Wheelchair spaces in the following areas need not be dispersed in the larger facility:
  - Team seating for a sport activity.
  - Player seating for a sport activity.
- For other conditions of multilevel assembly seating, the following applies:
  - Wheelchair spaces are required on upper levels as follows:
    Required on one level of every two levels.
- Wheelchair spaces are required in each of the following:
  - Luxury box.
  - Club box.
  - Suite.

1108.2.5 Designated aisle seats

- Designated aisle seats are not required in the following areas:
  - Team seating for a sport activity.
  - Player seating for a sport activity.
- Otherwise, the greater of the following in fixed seating assembly areas must be designated aisle seats:
  - 1 aisle seat.
  - ≥ 5% of the total number of aisle seats.
- Designated aisle seats are placed closest to an accessible route.

Note: Designated aisle seats have removable armrests, folding armrests, or no armrests on the aisle side.

1108.2.6 Lawn seating

- An accessible route shall connect to the following where fixed seats are not provided:
  - Lawn seating.
  - Overflow seating.

1108.2.7 Assistive listening systems

- An assistive listening system is required in both of the following types of courtrooms:
  - Where an audio amplification system is provided.
  - Where an audio amplification system is not provided.
- In other locations with no audio amplification system, the following applies:
  - An assistive listening system is not required.
- In areas with all of the following characteristics, an assistive listening system is required:
  - An assembly area with fixed seating.
  - An audio amplification system is provided.
  - Hearing is integral to the function of the space.

1108 Special Occupancies

1108.2.7.1 Receivers *(part 1 of 2)*

- The minimums required by this section do not apply in the following case:
  - In an assembly area where all seats are served by the following:
    An induction loop assistive listening system.
- Otherwise, receivers are required for assistive listening systems as indicated in the partial table below.
- Where there is more than one assembly area in a building requiring receivers, the following applies:
  - The total number of receivers required may be based on the following:
    The sum of the seats in all assembly areas where all of the following apply:
      All receivers must be usable in all assembly areas.
      All areas required to have receivers must be under the same management.

Table 1108.2.7.1 Number of Receivers Required for Assistive Listening

Total seats	Receivers required	Hearing-aid-compatible required
1–50	≥ 2	≥ 2
51–75	≥ 3	≥ 2
76–100	≥ 4	≥ 2
101–125	≥ 5	≥ 2
126–150	≥ 6	≥ 2
151–175	≥ 7	≥ 2
176–200	≥ 8	≥ 2
201–225	≥ 9	≥ 3
226–250	≥ 10	≥ 3
251–275	≥ 11	≥ 3
276–300	≥ 12	≥ 3
301–325	≥ 13	≥ 4
326–350	≥ 14	≥ 4
351–375	≥ 15	≥ 4
376–400	≥ 16	≥ 4
401–425	≥ 17	≥ 5
426–450	≥ 18	≥ 5
451–475	≥ 19	≥ 5
476–500	≥ 20	≥ 5
501–533	≥ 21	≥ 6
534–566	≥ 22	≥ 6
567–599	≥ 23	≥ 6
600–632	≥ 24	≥ 6
633–665	≥ 25	≥ 7
666–698	≥ 26	≥ 7
699–731	≥ 27	≥ 7
732–764	≥ 28	≥ 7
765–797	≥ 29	≥ 8
798–830	≥ 30	≥ 8
831–863	≥ 31	≥ 8
864–896	≥ 32	≥ 8
897–929	≥ 33	≥ 9
930–962	≥ 34	≥ 9
963–995	≥ 35	≥ 9
996–1,050	≥ 36	≥ 9
1,051–1,100	≥ 37	≥ 10
1,101–1,150	≥ 38	≥ 10
1,151–1,200	≥ 39	≥ 10
1,201–1,250	≥ 40	≥ 10
1,251–1,300	≥ 41	≥ 11
1,301–1,350	≥ 42	≥ 11
1,351–1,400	≥ 43	≥ 11
1,401–1,450	≥ 44	≥ 11
1,451–1,500	≥ 45	≥ 12
1,501–1,550	≥ 46	≥ 12
1,551–1,600	≥ 47	≥ 12
1,601–1,650	≥ 48	≥ 12
1,651–1,700	≥ 49	≥ 13
1,701–1,750	≥ 50	≥ 13
1,751–1,800	≥ 51	≥ 13
1,801–1,850	≥ 52	≥ 13
1,851–1,900	≥ 53	≥ 14
1,901–1,950	≥ 54	≥ 14
1,951–2,000	≥ 55	≥ 14
2,001–2,100	≥ 56	≥ 14
2,101–2,200	≥ 57	≥ 15
2,201–2,300	≥ 58	≥ 15
2,301–2,400	≥ 59	≥ 15
2,401–2,500	≥ 60	≥ 15
2,501–2,600	≥ 61	≥ 16

Source: IBC Table 1108.2.7.1.

1108 Special Occupancies

1108.2.7.1 Receivers *(part 2 of 2)*

- The number of receivers required for > 2,000 seats is determined as follows:

  Receivers required = [(Total seats – 2,000) ÷ 100] + 55

- The number of receivers that must be hearing-aid-compatible for > 200 seats is determined as follows:
  - ≥ 25% of the standard receivers required.
- Fractions of receivers required are rounded up to the next whole number.

1108.2.7.2 Public address systems

- The requirements below apply to the following facilities:
  - Facilities:
    - Stadiums.
    - Arenas.
    - Grandstands.
  - Requirements:
    - Where such facilities provide verbal public announcements, the following applies:
      - They must also provide equivalent messages in written form.

Note: The following are cited as providing specific requirements for required text messages:
1108.2.7.2.1, "Prerecorded text messages."
1108.2.7.2.2, "Real-time messages."

1108.2.7.2.1 Prerecorded text messages

- This section addresses announcements in assembly areas as follows:
  - Stadiums.
  - Arenas.
  - Grandstands.
- Verbal announcements that cannot be recorded prior to an event are governed as follows:
  - Such announcements are not required to be displayed in written form.
- Verbal announcements that are recorded prior to an event are governed as follows:
  - The equivalent of such announcements must be displayed as written messages in the following case:
    - Where electronic signs are available that can display such messages.

1108.2.7.2.2 Real-time messages

- Real-time verbal announcements at the following facilities are governed as listed:
  - Facilities:
    - Stadiums.
    - Arenas.
    - Grandstands.
  - Requirement:
    - The equivalent of such announcements must be displayed as written messages in the following case:
      - Where electronic signs are available that can display such messages.

1108 Special Occupancies

1108.2.8 Performance areas

- A direct accessible route must connect the following terminals in the circumstance indicated below:
  - Terminals:
    - Performance area.
    - Assembly seating.
  - Circumstance:
    - Where a circulation path connects the same terminals.
- An accessible route must connect the following:
  - Performance areas.
  - Related areas used by performers.

1108.2.9 Dining areas

- An accessible route (typically an elevator) to a mezzanine is not required where all of the following apply:
  - The mezzanine has < 1/4 of the total dining area.
  - The building has no accessible route (typically an elevator) between levels.
  - The same services provided on the mezzanine are provided in the accessible area.
- Tiered dining spaces in sports facilities are governed as follows:
  - Where the dining spaces are required to have accessible seating, the following applies:
    - Accessible routes are required to serve ≥ 1/4 of the dining area as follows:
      - Where both of the following conditions are present:
        - Accessible routes are provided for accessible seating.
        - All tiers have the same services.
- In other cases, the total dining area for tables and seating must be accessible.

1108.2.9.1 Dining surfaces

- This section addresses seating or standing spaces adjacent to dining surfaces as follows:
  - For eating.
  - For drinking.
- The greater number of the following adjacent spaces must comply with the requirements listed:
  - Spaces:
    - 1 standing and/or seating space as applicable.
    - ≥ 5% of all standing and/or seating spaces as applicable.
  - Requirements:
    - They must be accessible.
    - They must be dispersed throughout the dining area.
    - They must be accessed by an accessible route.

1108 Special Occupancies

1108.3 Self-service storage facilities

- The numbers of self-storage units required to be accessible are listed below:

Table 1108.3 Accessible Self-Storage Units

Total number of units provided	Number to be accessible	Total number of units provided	Number to be accessible	Total number of units provided	Number to be accessible
1–20	≥ 1	181–200	≥ 10	601–650	≥ 19
21–40	≥ 2	201–250	≥ 11	651–700	≥ 20
41–60	≥ 3	251–300	≥ 12	701–750	≥ 21
61–80	≥ 4	301–350	≥ 13	751–800	≥ 22
81–100	≥ 5	351–400	≥ 14	801–850	≥ 23
101–120	≥ 6	401–450	≥ 15	851–900	≥ 24
121–140	≥ 7	451–500	≥ 16		
141–160	≥ 8	501–550	≥ 17		
161–180	≥ 9	551–600	≥ 18		

Source: IBC Table 1108.3.

- The number of self-storage units required to be accessible in a facility with > 900 units is determined as indicated below:

  Required accessible units = [(Total number of units – 900) × 0.02] + 24

1108.3.1 Dispersion

- Where there are fewer accessible self-storage units required than there are classes of storage, the following applies:
  - The number of accessible units provided need not be > the number otherwise required for the following purpose:

    To distribute accessible units among all classes of storage.
- Otherwise, accessible self-storage units must be distributed among all classes of storage spaces provided.
- In a multibuilding storage facility, accessible self-storage units may be distributed in a single building.

1108.4.1 Courtrooms

- All courtrooms have the following requirements:
  - They must be accessible.
  - They must be on an accessible route.
  - They must comply with this section:

Note: The following are cited as governing courtrooms as part of this section:

1108.4.1.1, "Jury box."
1108.4.1.2, "Gallery seating."
1108.4.1.3, "Assistive listening systems."
1108.4.1.4, "Employee work stations."
1108.4.1.5, "Other work stations."

1108 Special Occupancies

1108.4.1.1 Jury box

- A wheelchair space is required in the jury box as follows:
  - A companion seat is not required.

Note: ICC A117.1, "Accessible and Usable Buildings and Facilities," is cited as governing wheelchair spaces.

1108.4.1.2 Gallery seating

- The following are required in gallery seating:
  - Wheelchair spaces.
  - Designated aisle seats.

Note: ICC A117.1, "Accessible and Usable Buildings and Facilities," is cited as governing wheelchair spaces.
IBC Table 1108.2.2.1, "Accessible Wheelchair Spaces," is cited as governing the number of wheelchair spaces required.
1108.2.5, "Designated aisle seats," is cited as governing this type of seat.

1108.4.1.3 Assistive listening systems

- An assistive listening system is required in each courtroom.

Note: 1108.2.7.1, "Receivers," is cited as governing the number of receivers required.

1108.4.1.4 Employee work stations

- The following are required to be on an accessible route:
  - Judge's bench.
  - Clerk's station.
  - Baliff's station.
  - Deputy clerk's station.
  - Court reporter's station.
- The vertical access to an elevated employee's work stations is not required as follows:
  - At the time of initial construction, given the following condition:
    Where one of the following elements can be installed per the requirement shown below:
    Elements:
    Ramp.
    Lift.
    Elevator.
    Requirement:
    Installation must be possible without reconfiguration or extension of either of the following:
    The courtroom.
    The electrical system.

Note: ICC A117.1, "Accessible and Usable Buildings and Facilities," is cited as governing the elements of vertical access to elevated work stations.

1108 Special Occupancies

1108.4.1.5 Other work stations

- The following work stations must be accessible:
  - Litigant's station including any lecturn.
  - Counsel's station including any lecturn.

Note: ICC A117.1, "Accessible and Usable Buildings and Facilities," is cited as governing accessibility.

1108.4.2.1 Central holding cells

- The following central holding cells must comply with the requirement listed:
  - Cells:
    Multipurpose cells not separated by the following categories:
      Age.
      Gender.
  - Requirement:
    ≥ 1 cell must be accessible.
- The following central holding cells must comply with the requirement listed:
  - Cells:
    Separate cells assigned to any of the following categories:
  - Requirement:
    ≥ 1 cell of each category must be accessible.

1108.4.2.2 Court-floor holding cells

- The following holding cells serving courtrooms must comply with the requirement listed:
  - Cells:
    Multipurpose cells at the court-floor level not separated by age or gender.
  - Requirement:
    ≥ 1 accessible cell must serve the courtrooms.
- The following holding cells serving courtrooms must comply with the requirements listed:
  - Cells:
    Separate cells at the court-floor level assigned to any of the following categories:
      Adult females.
      Adult males.
      Juvenile females.
      Juvenile males.
  - Requirement:
    ≥ 1 accessible cell of each category must serve each courtroom.
- More than one courtroom may be served by an accessible cell.

1108 Special Occupancies

1108.4.3.1 Cubicles and counters

- This section addresses the following elements in the visiting area of a judicial facility:
  - Detainee visiting cubicles.
  - Detainee visiting counters.
- This section does not govern the following elements in the visiting area of a judicial facility:
  - The detainee side of visiting cubicles where both of the following apply:
    - Where no contact between detainee and visitors is permitted.
    - Where the visiting area does not serve an accessible holding cell.
- In other cases, the following applies:
  - The greater number of the following visiting cubicles must meet the requirement listed:
    - Cubicles:
      - 1 visitor cubicle.
      - ≥ 5% of all visitor cubicles.
    - Requirement:
      - The cubicles must be accessible on both sides.
  - Any visiting counters provided are governed as follows:
    - ≥ 1 must be accessible on both sides.

1108.4.3.2 Partitions

- The following elements in the visiting area of a judicial facility must meet the requirements listed:
  - Elements:
    - Visiting cubicles with any of the following separating visitors from detainees:
      - Solid partitions.
      - Security glazing.
    - Visiting counters with any of the following separating visitors from detainees:
      - Solid partitions.
      - Security glazing.
  - Requirement:
    - ≥ 1 of each type of the following must be accessible:
      - Counter partition.
      - Cubicle.

1109 Other Features and Facilities

1109.2 Toilet and bathing facilities *(part 1 of 2)*

- Alternatives to accessibility requirements of this section for the following facilities are provided below:
  - Facilities:
    - Toilet rooms or bathing facilities where all of the following conditions apply:
      - Facilities accessed only through a private office.
      - Facilities not for public use.
      - Facilities intended for use by a single occupant.
  - Alternatives:
    - Doors may swing into the clear floor space in the following case:
      - Where it can be reversed to meet other accessibility standards.
    - Height requirements of other accessibility standards do not apply for the water closet.
    - Toilet room grab bars are not required in the following case:
      - Where backing is provided in walls suitable for their installation.
    - The following requirements at the lavatory do not apply:
      - Height minimum.
      - Knee clearance.
      - Toe clearance.

Note: ICC A117.1, "Accessible and Usable Buildings and Facilities," is cited as the accessibility standard that applies to the above alternatives or which is waived as indicated.

- This section does not apply to the following facilities:
  - Those serving the following units where not required to be accessible:
    - Dwelling units.
    - Sleeping units.

Note: Section 1107, "Dwelling Units and Sleeping Units," is cited as the source that does not require certain facilities to be accessible as listed above.

- Where there is a group of ≥ 2 toilet or bathing rooms as follows, the requirements listed below apply:
  - Rooms:
    - Each is for use by a single occupant.
  - Requirement:
    - The larger of the following number of rooms must be accessible:
      - For men's facilities:
        - 1 room.
        - ≥ 50% of the rooms.
      - For women's facilities:
        - 1 room.
        - ≥ 50% of the rooms.
      - For unisex facilities:
        - 1 room.
        - ≥ 50% of the rooms.

1109 Other Features and Facilities

1109.2 Toilet and bathing facilities *(part 2 of 2)*

- Other toilet rooms and bathing facilities are governed as follows:
  - Such rooms and facilities are required to be accessible.
  - ≥ 1 of the following in each accessible room or facility must be accessible:

    Control Element
    Dispenser Fixture
  - Where such rooms or facilities are the only ones in a building, the following applies:

    They may not be located on a floor without an accessible route.
- A urinal is not required to be accessible in the following cases:
  - Where it is the only one as follows:

    In a toilet room or bathing facility.
- The following toilet rooms are not required to be accessible:
  - Where in a critical-care patient sleeping room.
  - Where in an intensive-care patient sleeping room.

1109.2.1 Family or assisted-use toilet and bathing rooms

- A family or assisted-use bathing room is not required in the following case:
  - Where fixtures in each separate-sex bathing room are limited as follows:

    Where a bathtub is provided, there is only 1.
    Where a shower is provided, there is only 1.
- In other cases, the following requirements apply:
  - An accessible family or assisted-use toilet room is required where all of the following conditions apply:

    In occupancies A and M.
    Where the sum of required male and female water closets is ≥ 6.
  - Where A or M occupancies are mixed with others, the following applies:

    An accessible family or assisted-use toilet room is required for the following condition:
    Where the sum of required male and female water closets is ≥ 6 as follows:
    In the A or M occupancy only.
- An accessible family or assisted-use toilet room is required in recreational facilities as follows:
  - Where separate-sex bathing rooms are provided.
- Fixtures in family or assisted-use toilet and bathing rooms count toward the following minimum:
  - The total number of fixtures required.

1109.2.1.2 Family or assisted-use toilet rooms

- A family or assisted-use toilet room must have one of the following combinations of fixtures:
  - Combination 1:

    1 lavatory.
    1 water closet.
  - Combination 2:

    1 lavatory.
    1 water closet.
    1 urinal.

Note: 1109.2.1.3, "Family or assisted-use bathing rooms," is cited as listing bathing rooms that qualify as family or assisted-use toilets.

1109 Other Features and Facilities

1109.2.1.3 Family or assisted-use bathing rooms

- Family or assisted-use bathing rooms must include no more and no fewer of the following:
  - One bathtub or shower.
  - One water closet.
  - One lavatory.
- Accessible storage facilities are required for family or assisted-use bathing rooms in the following case:
  - Where storage is provided for separate-sex bathing rooms.

1109.2.1.4 Location

- Family or assisted-use toilet and bathing rooms must be on an accessible route.
- Family or assisted-use toilet rooms must be placed in one of the following locations:
  - ≤ 1 story above separate-sex toilet rooms.
  - ≤ 1 story below separate-sex toilet rooms.
- The accessible route between the following points is limited to ≤ 500':
  - Any separate-sex toilet room.
  - A family or assisted-use toilet room.

1109.2.1.5 Prohibited location

- This section addresses passenger transportation facilities and airports.
- The accessible route between the following facilities is governed as shown below:
  - Facilities:
    - Separate-sex toilet rooms.
    - A family or assisted-use toilet room.
  - Requirement:
    - The route may not pass through a security checkpoint.

1109.2.1.6 Clear floor space

- Where doors swing into the following rooms, the following clear space is required:
  - Rooms:
    - Family or assisted-use toilet room.
    - Family or assisted-use bathing room.
  - Requirement:
    - Clear floor space must be available in the room beyond the door swing as follows:
      - ≥ 30" × ≥ 48".

1109.2.1.7 Privacy

- Doors to the following rooms must be securable from inside the rooms:
  - Family or assisted-use rooms.
  - Family or assisted-use bathing rooms.

1109 Other Features and Facilities

1109.2.2 Water closet compartment

- This section addresses toilet rooms and bathing rooms.
- Where water closet compartments are provided, the following is required:
  - ≥ 1 wheelchair-accessible water closet compartment.
- Where the sum of the following facilities is ≥ 6, the requirement indicated below applies:
  - Facilities:
    Water closet compartments.
    Urinals.
  - Requirement:
    ≥ 1 wheelchair-accessible water closet compartment is required.
    ≥ 1 ambulatory-accessible water closet compartment is required.

Note: ICC A117.1, "Accessible and Usable Buildings and Facilities," is cited as governing wheelchair-accessible and ambulatory-accessible compartments.

1109.2.3 Lavatories

- Where lavatories are provided the larger of the following must be accessible:
  - 1.
  - ≥ 5%.
- Where there are ≥ 6 lavatories in a toilet room or bathing facility, the following applies:
  - ≥ 1 must be provided with enhanced-reach ranges.

Note: ICC A117.1, "Accessible and Usable Buildings and Facilities," is cited as governing enhanced-reach lavatories.

1109.3 Sinks

- The following sinks are not governed by this section:
  - Mop sinks.
  - Service sinks.
- In other cases where other sinks are provided, the number required to comply with accessibility requirements are listed below:

Total number of sinks provided	Sinks required to meet accessibility requirements
1–20	≥ 1
21–40	≥ 2
41–60	≥ 3

- Where > 60 sinks are provided, ≥ 5% must meet accessibility requirements.

Note: ICC A117.1, "Accessible and Usable Buildings and Facilities," is cited as governing the sinks listed above.

1109.4 Kitchens and kitchenettes

- Where provided in accessible spaces, the following must be accessible:
  - Kitchens and kitchenettes.

Note: ICC A117.1, "Accessible and Usable Buildings and Facilities," is cited as governing the facilities listed above.

1109 Other Features and Facilities

Case study: Fig. 1109.3. Where 1–20 sinks are provided, ≥ 1 must meet accessibility requirements. In this classroom laboratory, 4 sinks are provided. 1 meets accessibility requirements; thus, the sinks are in compliance with the code.

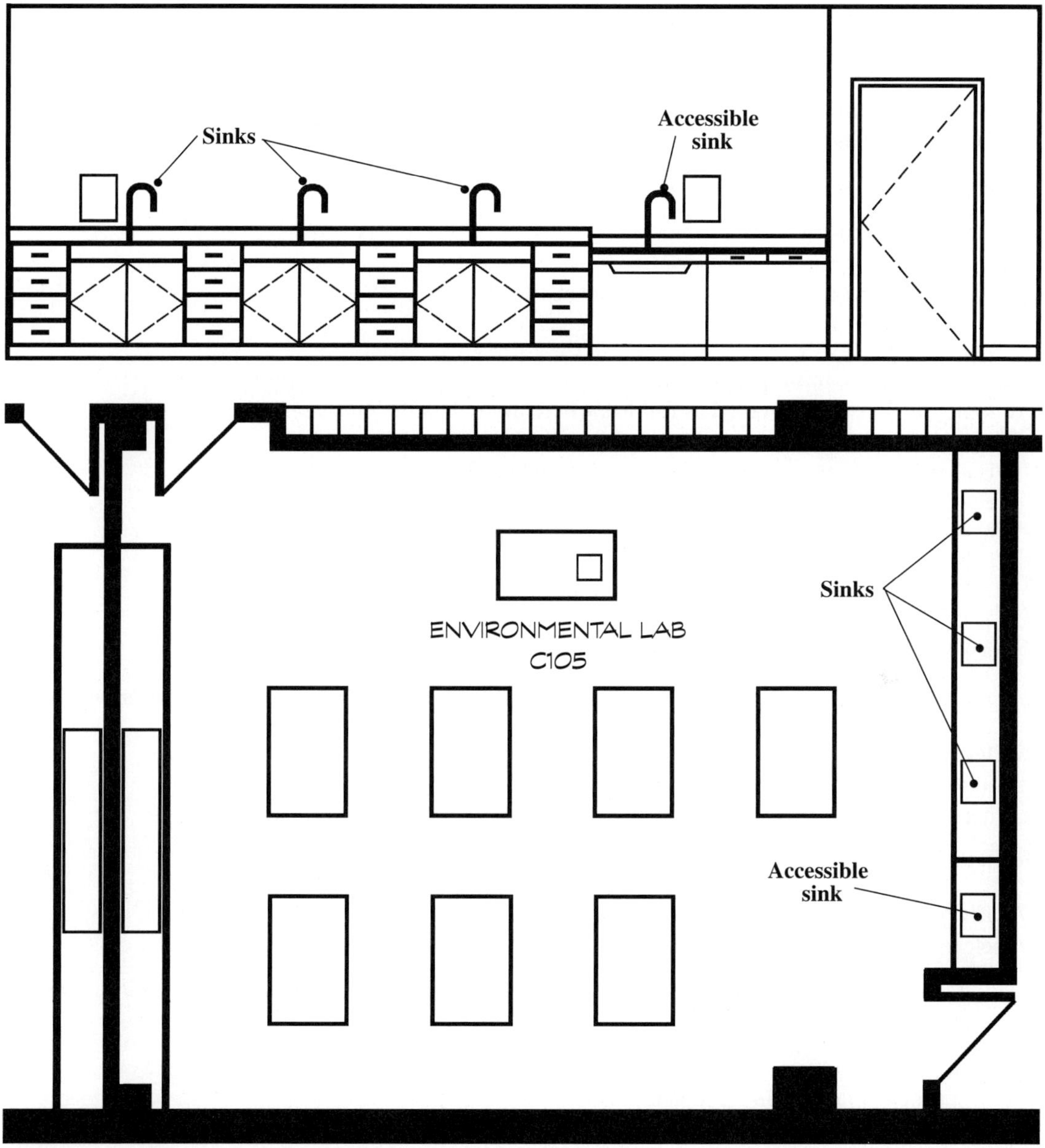

Fig. 1109.3. Plan and elevation at environmental lab C105. High School 6, Cypress-Fairbanks Independent School District. Harris County, Texas. PBK Architects, Inc. Houston, Texas.

1109 Other Features and Facilities

1109.5 Drinking fountains

- This section identifies requirements for drinking fountains in the following locations:
  - On an exterior site.
  - On a floor.
  - Within a secured area.

Note: The following sections are cited as governing these drinking fountains:
1109.5.1, "Minimum number."
1109.5.2, "More than the minimum number."

1109.5.1 Minimum number

- One of the following is required:
  - ≥ 1 drinking fountain meeting the following requirements:
    For persons in wheelchairs.
    For standing persons.
  - ≥ 2 drinking fountains as follows:
    One must meet requirements for persons in wheelchairs.
    One must meet requirements for standing persons.

1109.5.2 More than the minimum number

- Drinking fountains are governed as follows where more than the minimum are provided:
  - Where the total is an even number the following applies:
    Half must meet requirements for persons in wheelchairs.
    Half must meet requirements for standing persons.
  - Where the total is an odd number the following applies:
    One of the following distributions may be selected:
      There may be one more wheelchair fountain than standing fountain.
      There may be one more standing fountain than wheelchair fountain.
    All must meet the requirements of this section.

Note: 1109.5.1, "Minimum number," is cited as establishing the minimum number of drinking fountains.

1109.6 Elevators

- The following passenger elevators must be accessible:
  - Where on an accessible route.

Note: 3001.3, "Accessibility," is cited as governing these elevators.

1109.7 Lifts *(part 1 of 2)*

- Wheelchair lifts are permitted in the following accessible routes in new construction:

Note: ASME A18.1, "Safety Standard for Platform Lifts and Stairway Chairlifts," is cited as governing the installation of wheelchair lifts.

  - To a performing area in Occupancy A.
  - To a speaker's platform in Occupancy A.
  - To spaces meeting both of the following conditions:
    Not open to the general public.
    Having an occupant load ≤ 5.

1109 Other Features and Facilities

Case study: Fig. 1109.5.1 Two drinking fountains are provided on the 2nd floor of the building. Half of those provided or a minimum of 1 is required to be accessible. The drinking fountains meet this requirement.

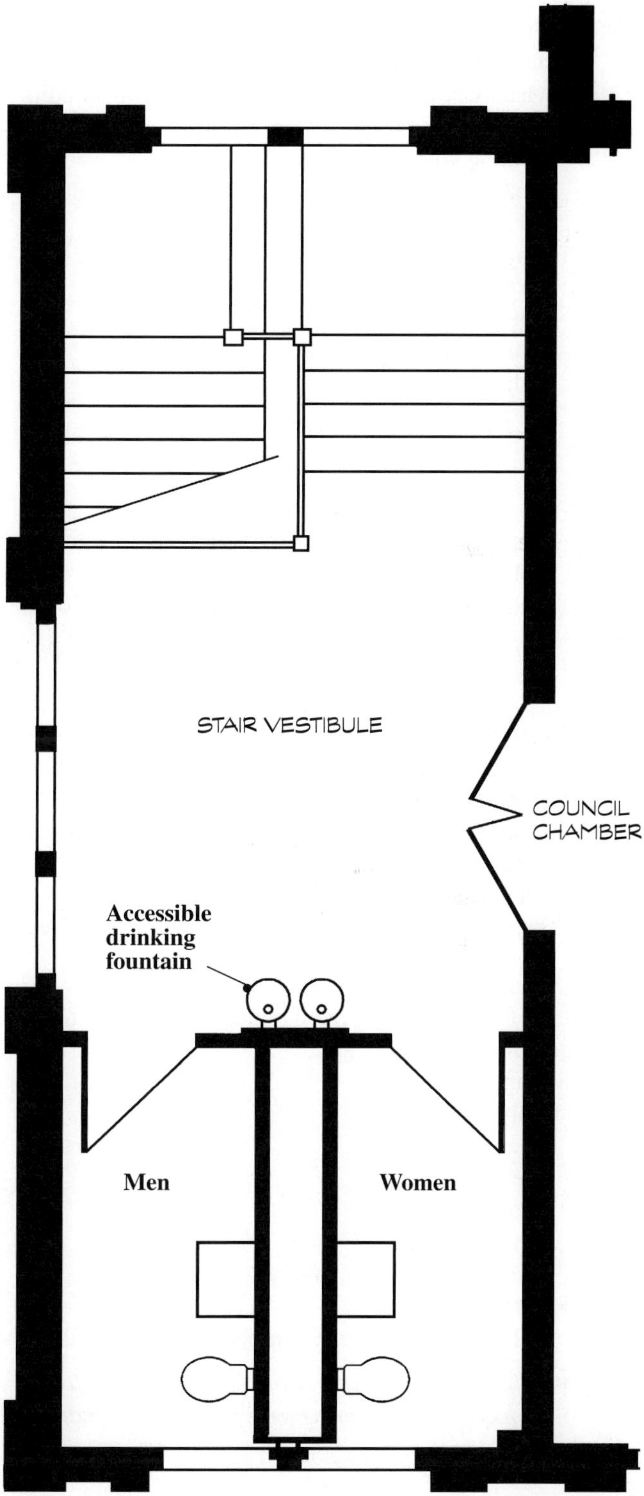

Fig. 1109.5.1. Plan at south stair. Lake Forest City Hall Renovation and Addition. Lake Forest, Illinois. David Woodhouse Architects. Chicago, Illinois.

1109 Other Features and Facilities

1109.7 Lifts *(part 2 of 2)*

- ◦ Within a dwelling unit or sleeping unit.
- ◦ In Occupancy A-5, with all of the following conditions:
  - Route is to wheelchair seating spaces.
  - Route is to outdoor dining.
  - Means of egress as follows is open to the outdoors:
    - From the outdoor dining to a public way.
- ◦ To any of the following areas in a courtroom:
  - Jury boxes.
  - Witness stands.
  - Raised courtroom stations.
  - Judges' benches
  - Clerks' stations.
  - Baliffs' stations.
  - Deputy clerks' stations.
  - Court reporters' stations.
  - Depressed areas such as the following:
    - The well of the court.
- ◦ To wheelchair spaces which are distributed according to code.

Note: 1108.2.2, "Wheelchair spaces," through 1008.2.6, "Lawn seating," are cited as governing distribution of wheelchair spaces.

- ◦ To loading and unloading areas of amusement rides.
- ◦ To play devices.
- ◦ To soft contained play structures.
- ◦ To either of the following where serving a sport activity:
  - Team member seating.
  - Player seating.
- ◦ Where a ramp or elevator are not feasible due to existing exterior site constraints:

1109.8 Storage

- Where fixed or built-in storage units such as the following are provided in a required accessible space, the requirement listed applies:
  - ◦ Storage units:
    - Cabinets.
    - Shelves.
    - Medicine cabinets.
    - Closets.
    - Drawers.
    - Similar storage components.
  - ◦ Requirement:
    - ≥ 1 of each type must comply with accessibility requirements.

Note: ICC A117.1, "Accessible and Usable Buildings and Facilities," is cited as the source of accessibility requirements for the units indicated above.

1109 Other Features and Facilities

1109.8.1 Lockers

- Where lockers are provided in accessible spaces, the following numbers must be accessible:

Total number of lockers provided	Lockers required to be accessible	Total number of lockers provided	Lockers required to be accessible
1–20	≥ 1	61–80	≥ 4
21–40	≥ 2	81–100	≥ 5
41–60	≥ 3	101–120	≥ 6

- Where > 120 lockers are provided, ≥ 5% are required to be accessible.

1109.8.2 Shelving and display units

- The following components are governed as indicated below:
  - Components:
    - Self-service shelves.
    - Self-service display units.
  - Requirements:
    - The components must be on an accessible route.
    - The components need not comply with reach-range requirements.

1109.8.3 Coat hooks and shelves

- Where the following devices are provided in the spaces listed below, the following requirements apply:
  - Devices:
    - Coat hooks.
    - Shelves.
  - Spaces:
    - Toilet rooms.
    - Toilet compartments.
    - Dressing rooms.
    - Fitting rooms.
    - Locker rooms.
  - Requirements:
    - ≥ 1 accessible version of each device is required as follows:
      - They must be provided for the following spaces:
        - Accessible toilet rooms without compartments.
        - Accessible toilet compartments.
        - Accessible dressing rooms.
        - Accessible fitting rooms.
        - Accessible locker rooms.

1109.9 Detectable warnings

- Detectable warnings are not required in the following locations:
  - At bus stops.
- Other passenger transit platforms, as follows, must have a detectable warning for the following hazard:
  - Any platform edge with a drop-off hazard as follows:
    - Where neither of the following is provided:
      - Protective screens.
      - Guards.

1109 Other Features and Facilities

1109.10 Seating at tables, counters and work surfaces

- This section does not apply to the following:
  - Where a checkout aisle is not accessible, the following applies:
    Check-writing surfaces are not required to be accessible.

  Note: 1109.11.2, "Check-out aisles," is cited as governing these aisles.

  - The following elements must meet the requirements listed below:
    Elements:
      Visiting cubicles in Occupancy I-3.
      Visiting counters in Occupancy I-3.
    Requirements:
      They are not required to be accessible on the detainee side as follows:
        In either of the following cases:
          Where the visiting area does not permit contact.
          Where the visiting area does not serve either of the following:
            Accessible holding cells.
            Accessible sleeping units.
- In other cases, the requirements listed below apply where there is seating and/or standing space at the following locations:
  - Locations:
    At fixed or built-in tables.
    At fixed or built-in counters.
    At fixed or built-in work surfaces.
  - Requirements:
    The larger of the following is required to be accessible:
      1 seat and standing space.
      ≥ 5% of the seats and standing spaces as indicated in the partial table below:

Total seats and spaces	Seats and spaces to be accessible	Total seats and spaces	Seats and spaces to be accessible
1–20	≥ 1	61–80	≥ 4
21–40	≥ 2	81–100	≥ 5
41–60	≥ 3	101–120	≥ 6

- Where the number of seats and standing spaces > 120, the following applies:
  - The number required to be accessible is ≥ 5% of the total provided.

1109.10.1 Dispersion

- Where accessible seating is provided at the following locations, it must be distributed throughout the seating area:
  - At fixed or built-in tables.
  - At fixed or built-in counters.
  - At fixed or built-in work surfaces.
- Such seating must be on an accessible route.

1109 Other Features and Facilities

1109.11.1 Dressing, fitting and locker rooms

- Where the following facilities are provided, the number required to be accessible in each group is as listed:
  - Dressing rooms.
  - Fitting rooms.
  - Locker rooms

Total of each facility type	Number of each to be accessible
1–20	≥ 1
21–40	≥ 2
41–60	≥ 3

- Where the number of such facilities is > 60, the following applies:
  - The number required to be accessible is ≥ 5% of the total number provided.

1109.11.2 Check-out aisles

- Accessible check-out aisles must be distributed throughout a facility in the following case:
  - Where inaccessible check-out aisles are distributed throughout the facility.
- The following devices located in accessible check-out aisles must be accessible:
  - Traffic control devices.
  - Security devices.
  - Turnstiles.
- Where check-out aisles serve different functions, the following applies:
  - ≥ 1 accessible check-out aisle must be provided for each function.
- Where check-out aisles are provided, the following number for each sales function must be accessible:

Table 1109.11.2 Accessible Check-out Aisles

Check-out aisles at each function	Check-out aisles at each function to be accessible	Check-out aisles at each function	Check-out aisles at each function to be accessible
1–4	≥ 1	16–20	≥ 4
5–8	≥ 2	21–25	≥ 5
9–15	≥ 3	26–30	≥ 6

Source: IBC Table 1109.11.2.

- Where check-out aisles are > 30, the number to be accessible is determined as follows:

Number to be accessible = [(Check-out aisles provided – 30) × 0.2] + 6

1109.11.3 Point of sale and service counters

- Counters for sales and distribution of goods and services are governed as follows:
  - Where such counters are provided, the following applies:
    ≥ 1 of each type must be accessible.
  - Where such counters are distributed throughout the facility, the following applies:
    The accessible counters must be distributed throughout the facility.

1109 Other Features and Facilities

1109.11.4 Food service lines

- Food service lines must be accessible.
- Where self-service shelves are provided, the following applies:
  - The greater of the following numbers of shelves must be accessible:
    - ≥ 1 of each type.
    - ≥ 50% of each type.

1109.11.5 Queue and waiting lines

- Queue and waiting lines for the following facilities must be accessible:
  - Accessible counters.
  - Accessible checkout aisles.

1109.12 Controls, operating mechanisms and hardware *(part 1 of 2)*

- The following are not required to be accessible:
  - Devices designed for use by service or maintenance personnel.
  - The following receptacles with a dedicated function:
    - Electrical.
    - Communication.
  - One of the following electrical outlets:
    - Where ≥ 2 outlets are over the following kitchen counter:
      - A counter without interruptions by a sink or appliance.
  - Electrical outlets located on the floor.
  - One of the following controls in each space:
    - Where redundant controls serve a single component as follows:
      - Where the controls are not light switches.
- The requirements listed below govern doors or gates in the following locations:
  - Locations:
    - Barrier walls or fences protecting the following:
      - Pools.
      - Spas.
      - Hot tubs.
  - Requirements:
    - The following operable mechanisms may be ≥ 4' and ≤ 4'-6" above the walking surface as follows:
      - On self-latching devices other than the following type:
        - Self-locking operated as follows:
          - By a key.
          - By an electronic opener.
          - By an integral combination lock.

1109 Other Features and Facilities

1109.12 Controls, operating mechanisms and hardware *(part 2 of 2)*

- In other cases, devices used by occupants, such as the following, must be accessible where located as listed:
  - Devices:
    - Controls.
    - Operating mechanisms.
    - Operating hardware.
    - Light switches.
    - Ventilation switches.
    - Switches for electrical convenience outlets.
  - Locations:
    - In accessible spaces.
    - Along accessible routes.
    - As parts of accessible elements.

1109.12.1 Operable window

- Accessible windows are not required in the following rooms:
  - Bathrooms.
  - Kitchens.
- In other rooms required to be accessible in the locations listed below, the following operable windows must be accessible:
  - Windows:
    - ≥ 1 operable window in each room where operable windows are provided.
    - Every required operable window.
  - Locations:
    - In accessible dwelling units and sleeping units as follows:
      - Occupancy I-3.
      - Occupancy I-2 nursing homes.
      - Occupancy I-2 hospitals.
      - Occupancy I-2 rehabilitation facilities.
      - Occupancy R-1.
      - Occupancy R-2.
      - Occupancy R-4.

Note: The following are cited as sources listing rooms that must be accessible, a partial summary of which is provided above:

1107.5.1.1, "Accessible units."
1107.5.2.1, "Accessible units."
1107.5.3.1, "Accessible units."
1107.5.4, "Group I-2 rehabilitation facilities."
1107.6.1.1, "Accessible units."
1107.6.2.1.1, "Type A units."
1107.6.2.2.1, "Accessible units."
1107.6.4.1, "Accessible units."

1109 Other Features and Facilities

1109.14.1 Facilities serving a single building

- Recreational facilities, where provided in the following locations, have requirements as listed:
  - Locations:
    - Occupancies R-2 and R-3 as follows:
      - Where recreational facilities serve 1 building with the following units:
        - Type A units and Type B units.
  - Requirements:
    - The larger of the following number must be accessible:
      - 1 facility of each type provided.
      - ≥ 25% of each type of facility provided.
  - The required accessible number of facilities is based on the following:
    - All facilities of every type on the site.

1109.14.2 Facilities serving multiple buildings

- Recreational facilities, where provided in the following locations, have requirements as listed:
  - Locations:
    - Occupancies R-2 and R-3 as follows:
      - Where recreational facilities serve > 1 building with the following units:
        - Type A units and Type B units.
  - Requirements:
    - The larger of the following numbers serving each building must be accessible:
      - 1 facility of each type provided.
      - ≥ 25% of each type of facility provided.
  - The required accessible number of facilities is based on the following:
    - All facilities of all types serving every building on the site.

1109.14.3 Other occupancies

- All of the following recreation and sports facilities must be accessible:
  - Those not governed by the previous two sections.

Note: The following are cited as governing the facilities not governed by this section:
1109.14.1, "Facilities serving a single building."
1109.14.2, "Facilities serving multiple buildings."

1109.14.4 Recreational and sports facilities exceptions

- This section specifies the following regarding recreational and sports facilities:
  - Certain facilities are not required to be accessible.
  - The extent to which certain other facilities are required to be accessible is specified.

1109.14.4.1 Bowling lanes

- The larger of the following numbers of bowling lanes must be on an accessible route:
  - 1 of each type of lane.
  - ≥ 5% of each type of lane.

1109.14.4.2 Court sports

- Both sides of a sports court must be directly connected by ≥ 1 accessible route.

1109 Other Features and Facilities

1109.14.4.3 Raised boxing or wrestling rings

- The following need not be accessible:
  - Raised boxing rings.
  - Raised wrestling rings.

1109.14.4.4 Raised refereeing, judging and scoring areas

- The following need not be accessible:
  - Raised structures used only for any of the following:
    Refereeing a sport.
    Judging a sport.
    Scoring a sport.

1109.14.4.5 Raised diving boards and diving platforms

- The following need not be accessible:
  - Raised diving boards.
  - Raised diving platforms.

1110 Signage

1110.1 Signs

- Required accessible elements require the International Symbol of Accessibility as follows:
  - For accessible parking spaces where ≥ 5 total parking spaces are provided.

Note: 1106.1, "Required," is cited as requiring accessible parking spaces as indicated above.

  - For accessible loading zones for passengers.
  - For accessible toilet and bathing rooms as follows:
    Single-user rooms grouped at a single location.
  - For accessible entrances in the following case:
    Where not all entrances are accessible.
  - For accessible check-out aisles where not all aisles are accessible as follows:
    Signage must be located above the aisle, as is other check-out aisle identification.
  - For unisex toilet rooms.
  - For unisex bathing rooms.
  - For the following facilities where not all similar facilities are accessible:
    Dressing rooms.
    Fitting rooms.
    Locker rooms.
  - For areas of refuge.
  - For exterior areas for assisted rescue.

Note: 1007.9, "Signage," is cited as governing signage in areas of refuge and in exterior areas for assisted rescue.

Case study: Fig. 1110.1. The International Symbol for Accessibility, as shown in the illustration, is painted on the pavement at each accessible parking space in this project and is applied to signs reserving the spaces. The symbol is also used on the sign beside the door of the unisex toilet room. In addition, the toilet room is identified as being unisex by the symbol on the door and by the wording on the accessibility sign, which includes Braille. The signs at the toilet room are mounted 5' above the floor. The signage complies with code requirements.

1110 Signage

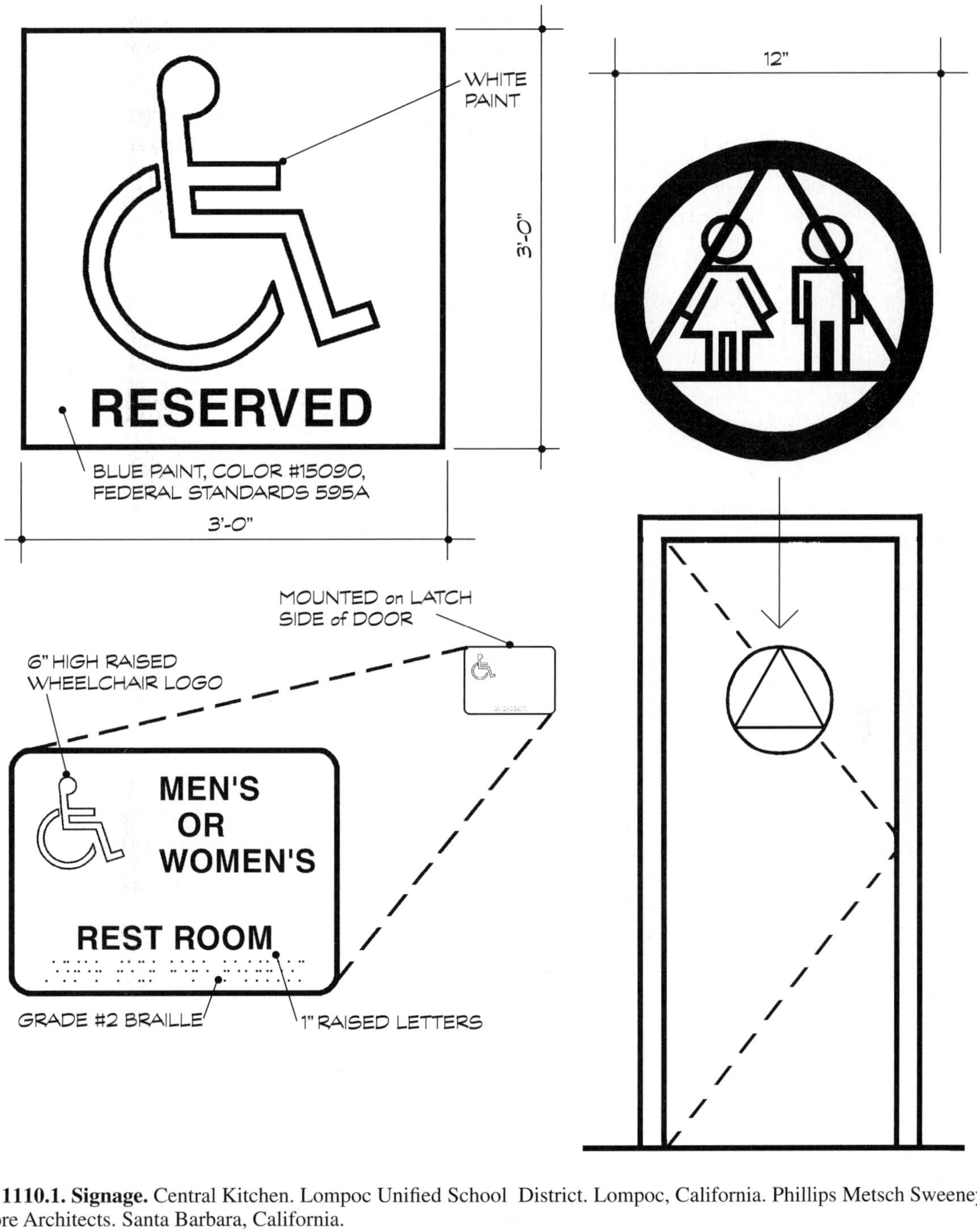

Fig. 1110.1. Signage. Central Kitchen. Lompoc Unified School District. Lompoc, California. Phillips Metsch Sweeney Moore Architects. Santa Barbara, California.

1110 Signage

1110.2 Directional signage

- Signage with the following information is required at the locations indicated below:
  - Information:
    - Indicates route to nearest similar facility that is accessible.
    - Exhibits the International Symbol of Accessibility.
  - Locations:
    - At inaccessible building entrances.
    - At inaccessible public toilets and bathing facilities.
    - At elevators not serving an accessible route.
    - At each separate-sex toilet and bathing room indicating the following:
      - The location of nearest unisex toilet or bathing room.

Note: 1109.2.1, "Family or assisted-use toilet and bathing rooms," is cited as governing such rooms.

At the following locations which serve a required accessible space:
- Exits which are not an accessible means of egress.
- Exit stairways which are not an accessible means of egress.

Note: 1007.10, "Directional signage," is cited as governing signs at exits and exit stairways that are not accessible.

1110.3 Other signs

- Signage reporting assistive listening system availability to patrons is required as follows:
  - Where assistive listening systems are required in assembly spaces.
  - Signage is to be placed in any of the following locations:
    - At ticket offices and ticket windows.
    - At the assembly area.

Note: 1108.2.7, "Assistive listening systems," is cited as the source requiring such systems.

- Tactile exit signage is required at each door to the following:
  - Area of refuge.
  - Exterior area for assisted rescue.
  - Egress stairways.
  - Exit passageways.
  - Exit discharges.

Note: 1011.3, "Tactile exit signs," is cited as governing these signs.

- Signage is required at areas of refuge and exterior areas for assisted rescue.

Note: 1007.11, "Instructions," is cited as governing signs at areas of refuge and exterior areas for assisted rescue.

- Signage is required at two-way communication systems.

Note: 1007.8.2, "Directions," is cited as governing the signage for two-way communication systems.

- Floor level identification signage is required at exit enclosures.

Note: 1022.8, "Floor identification signs," is cited as governing the signage at exit enclosures.

12

Interior Environment

Multipurpose Building Addition to Children's Home. Wilkes-Barre, Pennsylvania. *(partial elevation)*
C. Allen Mullins, Architects. Bear Creek, Pennsylvania.

1202 Definitions

1202.1 General

- **Sunroom**
  - A structure attached to a building as follows:
    - One story.
    - Glazing > 40% of the enclosure surface of the addition as follows:
      - Walls.
      - Roof.
- **Thermal isolation**
  - A separation between the following spaces by the elements listed:
    - Spaces:
      - Sunroom addition.
      - Dwelling unit.
    - Elements:
      - Existing or new elements among the following:
        - Walls.
        - Doors.
        - Windows.

1203 Ventilation

1203.1 General

- One of the following types of ventilation is required for buildings:
  - Natural ventilation as governed by this chapter.
  - Mechanical ventilation.

Note: The International Mechanical Code *is cited as governing mechanical ventilation.*

1203.2 Attic spaces

- Spaces between rafters enclosed as follows must be ventilated as indicated below:
  - Rafter spaces:
    - Enclosed by attaching a ceiling directly to the underside of rafters.
  - Requirements:
    - Each separate rafter space must be ventilated.
    - Air space required for ventilation is ≥ 1" as follows:
      - Between the insulation and roof sheathing.
    - Area of rafter space is measured on the plane of rafter slope.
- Enclosed attics and enclosed rafter spaces must be ventilated as follows:
  - Net clear open area for ventilation must be the following size:
    - ≥ 1/300 of the area ventilated.
  - 1/2 the required vents must be located as follows:
    - In the upper part of the ventilated space.
    - ≥ 3' above the eave or cornice vents.
  - 1/2 the required vents must be at the eave or cornice.
- Openings ventilation may not permit penetration by the following:
  - Rain.
  - Snow.
- The following may not interfere with ventilation airflow:
  - Blocking or bridging.

1203.2.1 Openings into attic

- This section addresses openings to the exterior from attics of buildings for human occupancy.
- Openings must prevent entry of the following and similar creatures:

Birds	Snakes
Rodents	Squirrels

- Openings for ventilation meet the following dimensional requirements:
  - The smallest opening size must be ≥ 1/16".
  - One of the following is required for the largest opening:
    - It must be ≤ 1/4".
    - Ventilation openings with a least dimension > 1/4" must be covered as follows:
      - With one of the following corrosion resistant materials having openings ≥ 1/16" and ≤ 1/4":

Hardware cloth	A similar material
Perforated vinyl	Wire cloth screening

- Combustion air taken from the attic is not governed by this section.

Note: The International Mechanical Code, *Chapter 7, is cited as governing combustion air taken from the attic.*

1203 Ventilation

Case study: Fig. 1203.2. Spaces, between rafters enclosed by the attachment of a ceiling to the bottom of the rafters, are ventilated. Metal soffit vents are provided under the eave and blocking is cut to permit the through-flow of air. A 1" airspace is provided between the top of the insulation and the bottom of the roof sheathing for airflow as required. Ventilation of the rafter spaces is in compliance with the code.

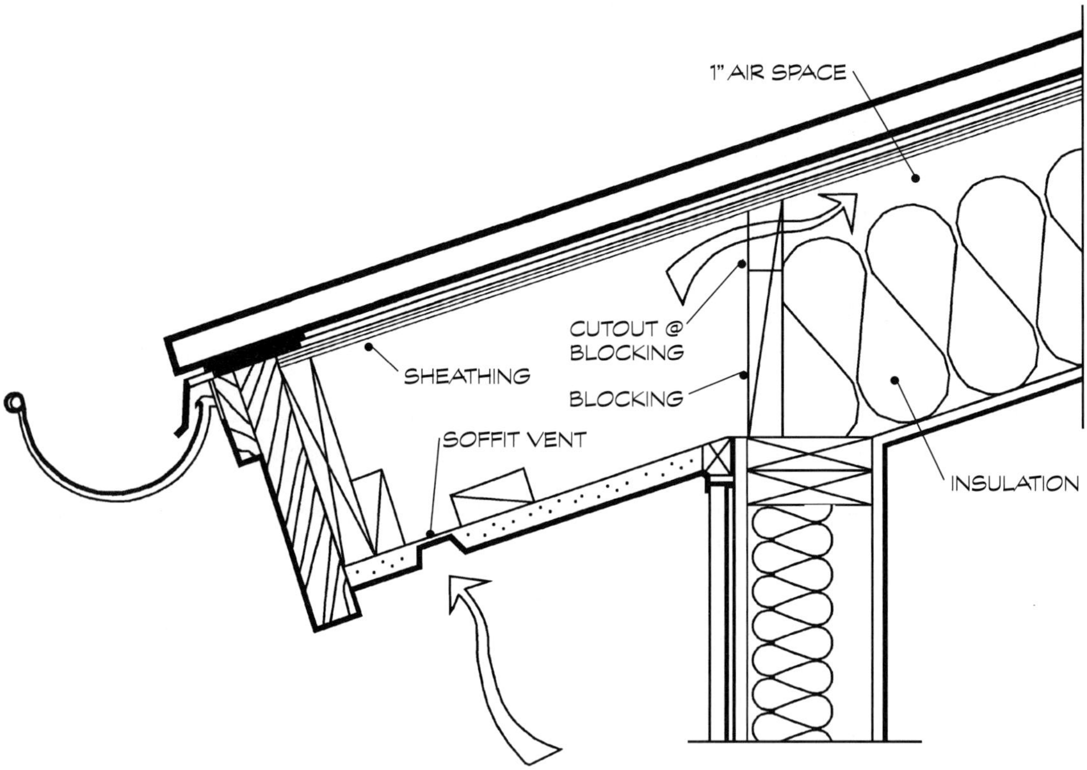

Fig. 1203.2. Detail at eave. Central Kitchen. Lompoc Unified School District. Lompoc, California. Phillips Metsch Sweeney Moore Architects. Santa Barbara, California.

1203 Ventilation

1203.3 Under-floor ventilation

- This section addresses crawl spaces for which conditions require natural ventilation.
- Space between the earth and floor joists, other than the following, must be ventilated as listed:
  - Spaces not governed by this section:
    - Basements.
    - Cellars.
  - Requirements:
    - Spaces must be ventilated by openings through the following as applicable:
      - Foundation walls.
      - Exterior walls.
    - Ventilation openings must be located to provide cross-ventilation of the space.

1203.3.1 Openings for under-floor ventilation

- This section addresses ventilation of crawl spaces where no vapor retarder is used.
- The net clear open area required for ventilating crawl spaces is as follows:
  - ≥ 1/150 of the crawl space area.
- Openings providing ventilation must be covered with one of the following:
  - Materials with openings ≤ 1/4" are as follows:
    - Perforated sheet metal ≥ 0.070" thick.
    - Expanded sheet metal ≥ 0.047" thick.
    - Cast iron grille or grating.
    - Extruded load-bearing vents.
    - Hardware cloth:
      - Wire diameter ≥ 0.035".
  - Corrosion-resistant wire mesh:
    - Openings ≤ 1/8".

Note: 1203.3.2, "Exceptions," lists conditions which warrant variations on these requirements.

1203.3.2 Exceptions *(part 1 of 2)*

- This section includes exceptions to the previous two sections, which addressed crawl-space ventilation.
- Openings providing ventilation are not required to the exterior in the following case:
  - Where the climate warrants vent openings to the interior of the building.
- The required net area of ventilation openings is as follows:
  - ≥ 1/1500 of the crawl-space area where the following are provided:
    - The ground is covered with a Class I vapor retarder.
      - Openings must be positioned to provide cross-ventilation of the space:
        - Openings with operable louvers are permitted.
- Openings providing ventilation are not required in the following case:
  - Where mechanical ventilation is provided as follows:
    - Mechanical ventilation must be continuous.
    - 1.0 cfm per 50 sf of crawl-space area must be provided.
    - The ground must be covered as follows:
      - With a Class I vapor retarder.

1203 Ventilation

Case study: Fig. 1203.2.1. A corrosion-resistant screen is provided at the attic opening to prevent the entry of birds and other creatures. Openings in the screen are between 1/8" and 1/4" as required. The sleeve surrounding the passage to the mechanical louvers is lined with sheet metal, and the bottom surface slopes to the exterior for drainage. Protection of the opening into the attic is in compliance with the code.

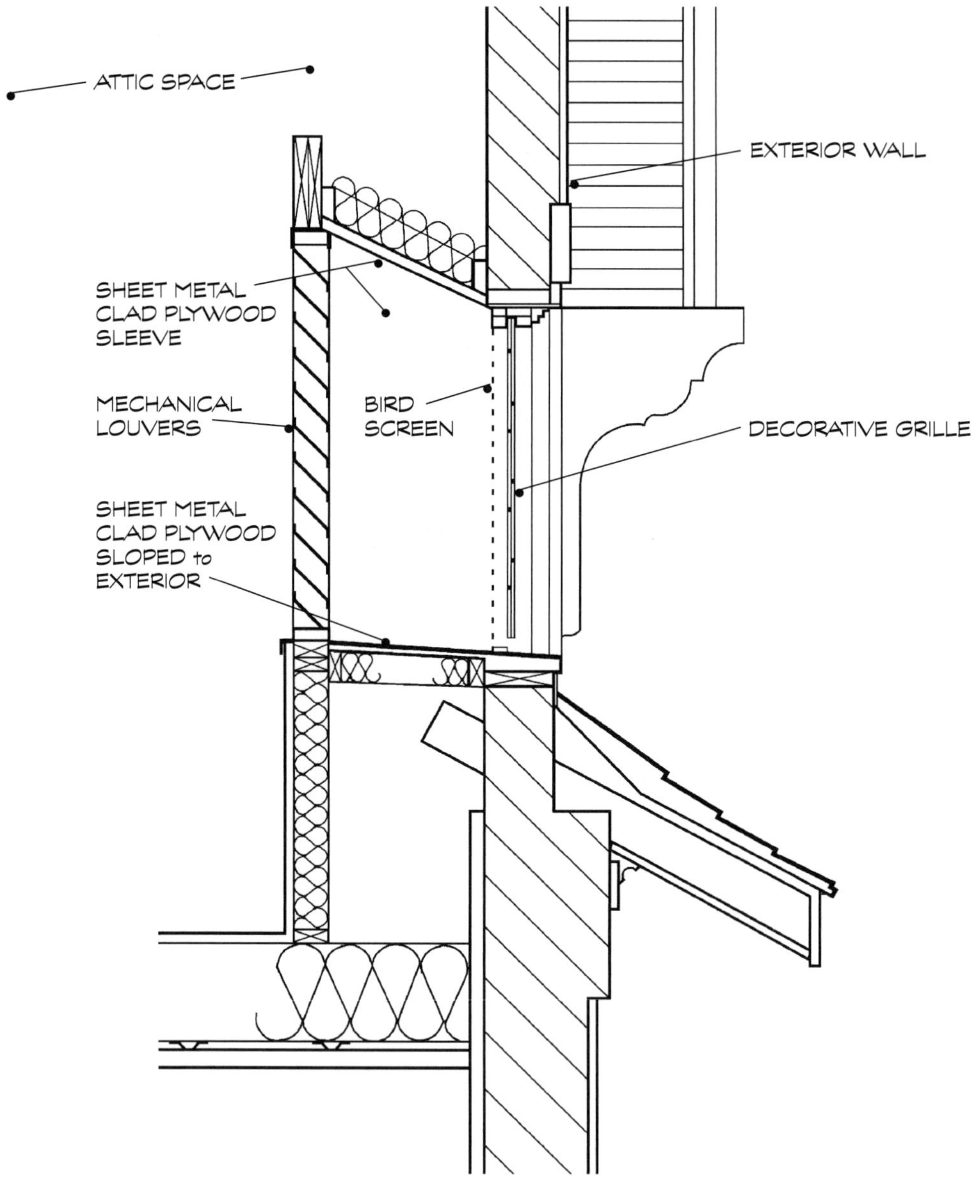

Fig. 1203.2.1. Detail at attic vent. Lake Forest City Hall Renovation and Addition. Lake Forest, Illinois. David Woodhouse Architects. Chicago, Illinois.

1203 Ventilation

1203.3.2 Exceptions *(part 2 of 2)*

- Openings providing ventilation are not required where both of the following apply:
  - Where the ground in the crawl space is covered as follows:
    With a Class I vapor retarder.
  - Where the surrounding walls are insulated.
  - Where the space is conditioned.

Note: The International Energy Conservation Code *is cited as governing conditioning of the insulated crawl space.*

- Openings providing ventilation in flood hazard areas are governed as follows:
  - They may be designed according to flood resistance principles.

Note: 1612.3, "Establishment of flood hazard areas," is cited as the source of requirements for establishing such areas.
ASCE 24, "Flood Resistant Design and Construction," is cited as governing the design of openings in flood hazard areas.

1203.4 Natural ventilation

- Natural ventilation provided to occupied space is governed as follows:
  - It must be through one or more of the following openings to the exterior:

Doors	Windows
Louvers	Other openings

1203.4.1 Ventilation area required

- Area of ventilating openings must be ≥ 4% of the area of the space ventilated.

1203.4.1.1 Adjoining spaces

- Required exterior ventilation openings may open to the following spaces where the requirements listed below are met:
  - Spaces:
    Thermally isolated sunroom addition.
    Thermally isolated patio cover.
  - Requirements:
    The opening between the inside space and the following spaces has the requirement listed:
      Spaces:
        Sunroom space.
        Patio cover.
      Requirement:
        The opening area must be equal to the greater of the following:
          ≥ 8% of the area of the ventilated interior space.
          20 sf.
- In other cases, space ventilated through an adjoining space is governed as follows:
  - The area of the opening between spaces must be the larger of the following:
    ≥ 8% of the area of the space being ventilated through an adjoining space.
    ≥ 25 sf.
  - The size of the ventilating openings to the exterior is based on the following:
    The total area of all spaces ventilated.

1203 Ventilation

1203.4.1.2 Openings below grade

- The following openings are governed as listed:
  - Openings:
    - Providing natural ventilation.
    - Located all or partly below grade.
  - Requirement:
    - Clear space outside the opening must have the following size:
      - Horizontal dimension ⊥ to the opening:
        - Must be ≥ 1.5 × opening depth:
          - Opening depth is measured as follows:
            - Vertically from average adjacent grade to opening bottom.

1203.4.2 Contaminants exhausted

- This section addresses spaces that are naturally ventilated.
- The removal of contaminants from such spaces is outside the scope of this code.

Note: The following are cited as governing the removal of such contaminants:
International Mechanical Code.
International Fire Code.

1203.4.2.1 Bathrooms

- Spaces with the following fixtures must be mechanically ventilated:
  - Bathtubs.
  - Showers.
  - Spas.
  - Similar bathing fixtures.

Note: The International Mechanical Code *is cited as governing such ventilation.*

1203.4.3 Openings on yards or courts

- The following may provide natural ventilation to an interior space where they meet size and other requirements:
  - Yards.
  - Courts.

Note: Section 1206, "Yards or Courts," is cited as providing requirements for these areas.

1203.5 Other ventilation and exhaust systems

- Ventilation for spaces with the following hazards is beyond the scope of this code:
  - Flammable hazards.
  - Combustible hazards.
  - Other contaminant sources.

Note: The following are cited as governing the removal of such hazards:
International Mechanical Code.
International Fire Code.

1204 Temperature Control

1204.1 Equipment and systems

- This section does not govern the following interior spaces:
  - Where human comfort is not required.
- In other spaces occupied by people the following applies:
  - Either of the following heating systems is required with the capability listed below:
    - Systems:
      - Passive systems.
      - Active systems.
    - Capability:
      - The system must be able to maintain a temperature in the space as follows:
        - ≥ 68° F as follows:
          - At 3' above floor level.
          - On the design heating day.

1205 Lighting

1205.1 General

- All spaces designed for people must have lighting by one of the following means:
  - Artificial.

Note: 1205.3, "Artificial light," is cited as governing this type of light.

  - Natural as follows:
    - Openings providing light must be as follows:
      - Exterior.
      - Glazed.
      - Open to one of the following:
        - Public way.
        - Yard.
        - Court.

Note: The following are cited as sources of applicable requirements:
1205.2, "Natural light."
Section 1206, "Yards or Courts."

1205.2 Natural light

- Net glazed area providing natural light must be the following size:
  - ≥ 8% of floor area served.

1205.2.1 Adjoining spaces

- Required exterior light openings may open to the following spaces where the requirements listed are met:
  - Spaces:
    - Thermally isolated sunroom addition.
    - Thermally isolated patio cover.
  - Requirements:
    - The opening in the common wall must have the characteristic listed:
      - Spaces:
        - Sunroom space.
        - Patio cover.
      - Characteristic:
        - The opening area must be equal to the greater of the following:
          - ≥ 10% of the area of the interior space.
          - 20 sf.
- In other cases, a space may receive required natural light through an adjoining space as follows:
  - ≥ 50% the common wall must be open.
  - The opening in the common wall must be the larger of the following:
    - ≥ 10% of the interior space floor area.
    - ≥ 25 sf.

1205 Lighting

1205.2.2 Exterior openings

- Exterior openings providing required natural light must comply with any of the following:
  - Openings must open directly to one of the following:
    - Public way.
    - Yard or court.
    - A roofed porch with all of the following characteristics:
      - Porch abuts one of the following:
        - Public way.
        - Yard or court.
      - Porch ceiling is ≥ 7' high.
      - Longer side of porch is ≥ 65% open.

  Note: The following are cited as sources of applicable requirements:
  1205.2, "Natural light."
  Section 1206, "Yards or Courts."

  - Openings must be skylights.

1205.3 Artificial light

- Artificial light must provide the following illumination:
  - An average intensity ≥ 10 footcandles is required as follows:
    - At a level 2'-6" high across the room.

1205.4 Stairway illumination

- Treads on the following stairways must be illuminated at a level ≥ 1 footcandle:
  - Within dwelling units.
  - Exterior dwelling unit stairways.
- Other stairways are not governed by this section.

Note: Chapter 10, "Means of Egress," is cited as governing other stairways.

1205.4.1 Controls

- The means for controlling stairway lighting is not governed by this code.

Note: NFPA 70, "National Electrical Code," is cited as governing these controls.

1205.5 Emergency egress lighting

- Means of egress illumination is not governed by this section.

Note: 1006.1, "Illumination required," is cited as governing means of egress lighting.

1206 Yards or Courts

1206.1 General

- This section addresses yards and courts as follows:
  - Next to openings for any of the following purposes:
    - Ventilation.
    - Natural light.
  - They must be located on the same property as the building served.

1206.2 Yards

- The width of a yard measured ⊥ to the building façade is required as listed:

Table 1206.2 Minimum Yard Width

Building height above grade plane	Yard width	Building height above grade plane	Yard width	Building height above grade plane	Yard width
1 story	≥ 3'	6 stories	≥ 7'	11 stories	≥ 12'
2 stories	≥ 3'	7 stories	≥ 8'	12 stories	≥ 13'
3 stories	≥ 4'	8 stories	≥ 9'	13 stories	≥ 14'
4 stories	≥ 5'	9 stories	≥ 10'	14 stories	≥ 15'
5 stories	≥ 6'	10 stories	≥ 11'	>14 stories	≥ 15'

1206.3 Courts *(part 1 of 2)*

- Courts with windows on any two opposite sides require widths as listed below:
  - Width is measured between walls with facing windows.
  - Where all walls have facing windows:
    - Either dimension is designated as width:
      - The other dimension is designated as length:
        - Length is governed by Table 1206.3c.

Table 1206.3a Courts with Windows on Opposite Sides

Building height above grade plane	Court width	Building height above grade plane	Court width	Building height above grade plane	Court width
1 story	≥ 6'	6 stories	≥ 10'	11 stories	≥ 15'
2 stories	≥ 6'	7 stories	≥ 11'	12 stories	≥ 16'
3 stories	≥ 7'	8 stories	≥ 12'	13 stories	≥ 17'
4 stories	≥ 8'	9 stories	≥ 13'	14 stories	≥ 18'
5 stories	≥ 9'	10 stories	≥ 14'	>14 stories	≥ 18'

- Courts without windows on opposite sides require widths as listed below:
  - Where courts are open on one end to a public way or yard:
    - Width is measured between facing walls.
  - Where courts are not open on one side to a public way or yard:
    - Either dimension is designated as width:
      - The other dimension is designated as length:
        - Length is governed by Table 1206.3c.

1206 Yards or Courts

1206.3 Courts *(part 2 of 2)*

Table 1206.3b Courts without Windows on Opposite Sides

Building height above grade plane	Court width	Building height above grade plane	Court width	Building height above grade plane	Court width
1 story	≥ 3'	6 stories	≥ 7'	11 stories	≥ 12'
2 stories	≥ 3'	7 stories	≥ 8'	12 stories	≥ 13'
3 stories	≥ 4'	8 stories	≥ 9'	13 stories	≥ 14'
4 stories	≥ 5'	9 stories	≥ 10'	14 stories	≥ 15'
5 stories	≥ 6'	10 stories	≥ 11'	>14 stories	≥ 15'

- Courts not open on one end to a public way or yard require lengths as listed below:
  - Length is ⊥ to the required widths governed by Table 1206.3a and 1206.3b.

Table 1206.3c Minimum Length of Courts

Building height above grade plane	Court length	Building height above grade plane	Court length	Building height above grade plane	Court length
1 story	≥ 10'	6 stories	≥ 18'	11 stories	≥ 28'
2 stories	≥ 10'	7 stories	≥ 20'	12 stories	≥ 30'
3 stories	≥ 12'	8 stories	≥ 22'	13 stories	≥ 32'
4 stories	≥ 14'	9 stories	≥ 24'	14 stories	≥ 34'
5 stories	≥ 16'	10 stories	≥ 26'	>14 stories	≥ 34'

1206.3.1 Court access

- Access is required to the bottom of a court for cleaning.

1206.3.2 Air intake

- Courts require a way to bring in air at the bottom as follows:
  - Air intake is required for the following courts:
    - Courts which do not abut a yard or public way.
    - Courts > 2 stories high.
  - Air intake requires the following characteristics:
    - Must be horizontal.
    - Must be located at the bottom of the court.
    - Must be ≥ 10 sf in area.
    - Must lead to the exterior.

1206.3.3 Court drainage

- The following is required for drainage of a court:
  - Court must be graded for adequate drainage.
  - Court must be drained to one of the following:
    - A public sewer.
    - Other approved disposal system.

Note: The International Plumbing Code *is cited as governing drainage disposal systems as required above.*

1206 Yards or Courts

Case study: Fig. 1206.3. All walls facing the courtyard of this 2-story building have windows. One corner of the courtyard is open to the public way on the ground level only. The minimum width of the courtyard is 6'. The minimum length of the courtyard between walls is 10'. The courtyard is 27'-9" wide and 33' long, thus, meeting size minimums. The courtyard is graded and drained as required by the code. The courtyard is in compliance with code requirements.

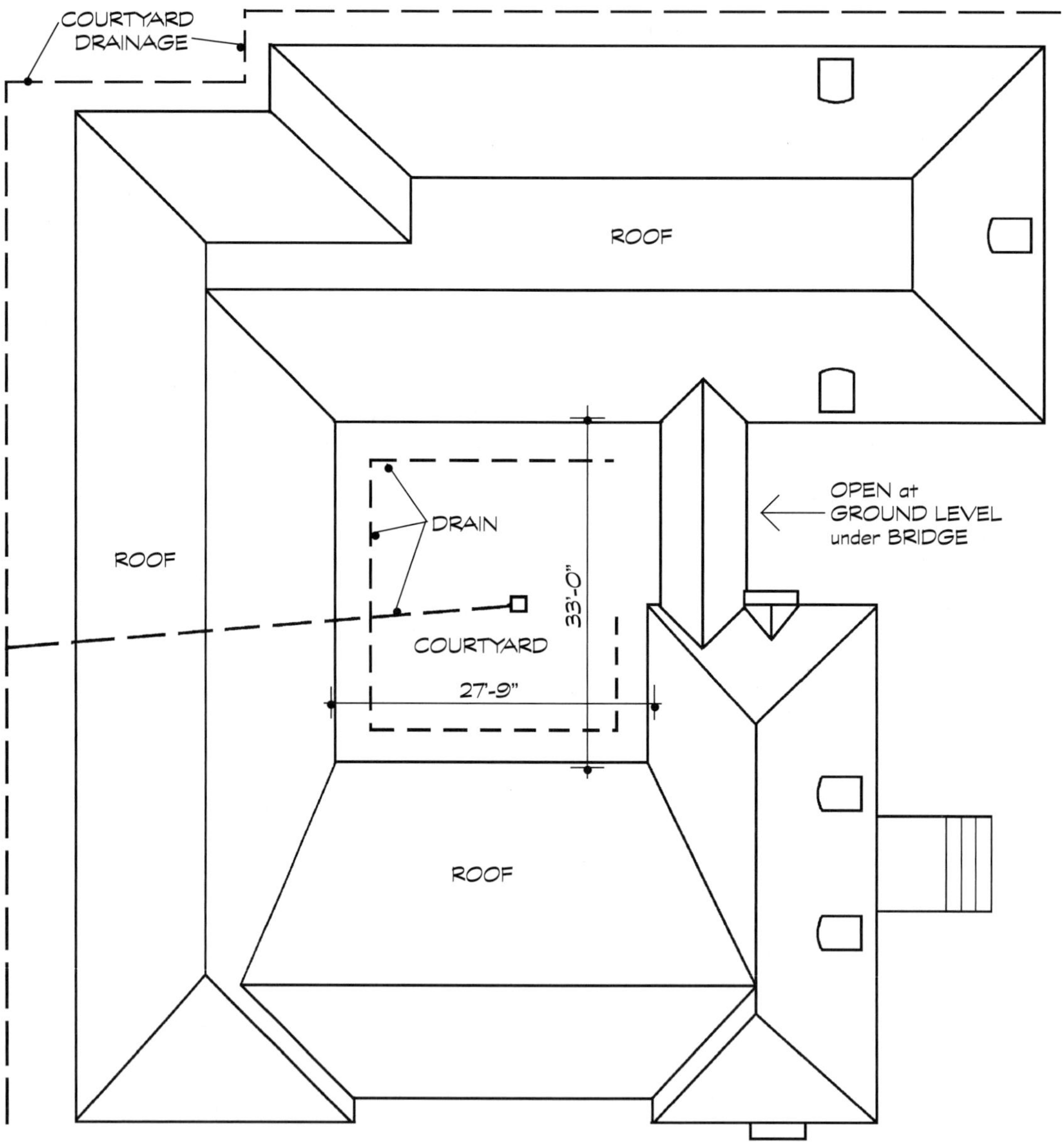

Fig. 1206.3. Courtyard plan. Multipurpose Building Addition to Children's Home. Wilkes-Barre, Pennsylvania. C. Allen Mullins, Architect. Bear Creek, Pennsylvania.

1207 Sound Transmission

1207.1 Scope

- This section provides sound transmission requirements indicated below as applicable to elements of dwelling units as follows:
  - Elements:
    - Common interior walls between adjacent units.
    - Interior walls between units and adjacent public areas:
      - Halls.
      - Corridors.
      - Stairs.
      - Service areas.
    - Floor/ceiling assemblies between adjacent units.
    - Floor/ceiling assemblies between units and adjacent public areas:
      - Halls.
      - Corridors.
      - Stairs.
      - Service areas.

1207.2 Air-borne sound

- Dwelling unit entrance doors are governed for air-borne sound as follows:
  - Tight fit to the frame and sill are required.
  - A required Sound Transmission Class (STC) rating is not specified.
- One of the following STC ratings is required for walls and floors governed by this section:
  - STC $\geq$ 50 is required where not verified by field test.
  - STC $\geq$ 45 is required where verified by field test.

Note: ASTM E 90, "Standard Test Method for Laboratory Measurement of Airborne Sound Transmission Loss of Building Partitions and Elements," is cited as governing the method for determining STC ratings.

1207.3 Structure-borne sound

- One of the following Impact Insulation Class (IIC) ratings is required for floor/ceiling assemblies governed by this section:
  - IIC $\geq$ 50 is required where not verified by field test.
  - IIC $\geq$ 45 is required where verified by field test.

Note: ASTM E 492, "Standard Test Method for Laboratory Measurement of Impact Sound Transmission Through Floor-Ceiling Assemblies Using the Tapping Machine," is cited as governing the method for determining IIC ratings.

1208 Interior Space Dimensions

1208.1 Minimum room widths

- Habitable spaces other than kitchens require the following size:
  - ≥ 7' for any room dimension in plan.
- Kitchens require ≥ 3' clear space as follows:
  - Between counter fronts and any of the following:
    - Appliances.
    - Other counter fronts.
    - Walls.

1208.2 Minimum ceiling heights

- Mezzanines are not governed by this section.

  Note: 505.1, "General," is cited as governing mezzanines.

- In 1- and 2-family dwellings, the following applies:
  - Beams and girders may project below the required the ceiling height as follows:
    - Where spaced ≥ 4' on center.
    - Where projecting ≤ 6" below required ceiling height.
- In all occupancies, the following applies:
  - ≥ 7'-0" is the required ceiling height in the following spaces:
    - Bathrooms.
    - Toilet rooms.
    - Kitchens.
    - Storage rooms.
    - Laundry rooms.
  - ≥ 7'-6" is the required ceiling height for other spaces as follows:
    - Habitable space.
    - Occupiable space.
    - Corridors.
- The following applies to sloped ceilings:
  - The required ceiling height must be provided for ≥ 1/2 the area.
  - Floor area counted in minimum area requirements must have the following height:
    - Ceiling height ≥ 5'.

1208.2.1 Furred ceiling

- Furred ceilings are governed as follows:
  - A furred ceiling must have the required ceiling height as follows:
    - In 2/3 the room area.
  - Otherwise, a furred ceiling may not have a height < 7'.

1208.3 Room area

- Rooms in dwelling units are governed as follows:
  - Kitchens in 1- and 2-family units must have the following size:
    - ≥ 50 sf of gross floor area.
  - A net floor area ≥ 120 sf is required in ≥ 1 room.
  - Other habitable rooms require a net floor area ≥ 70 sf.

1208 Interior Space Dimensions

Case study: Fig. 1208.1. The kitchen complies with the requirement that at least 3' of space be provided between counters and other objects and walls as indicated in the plan.

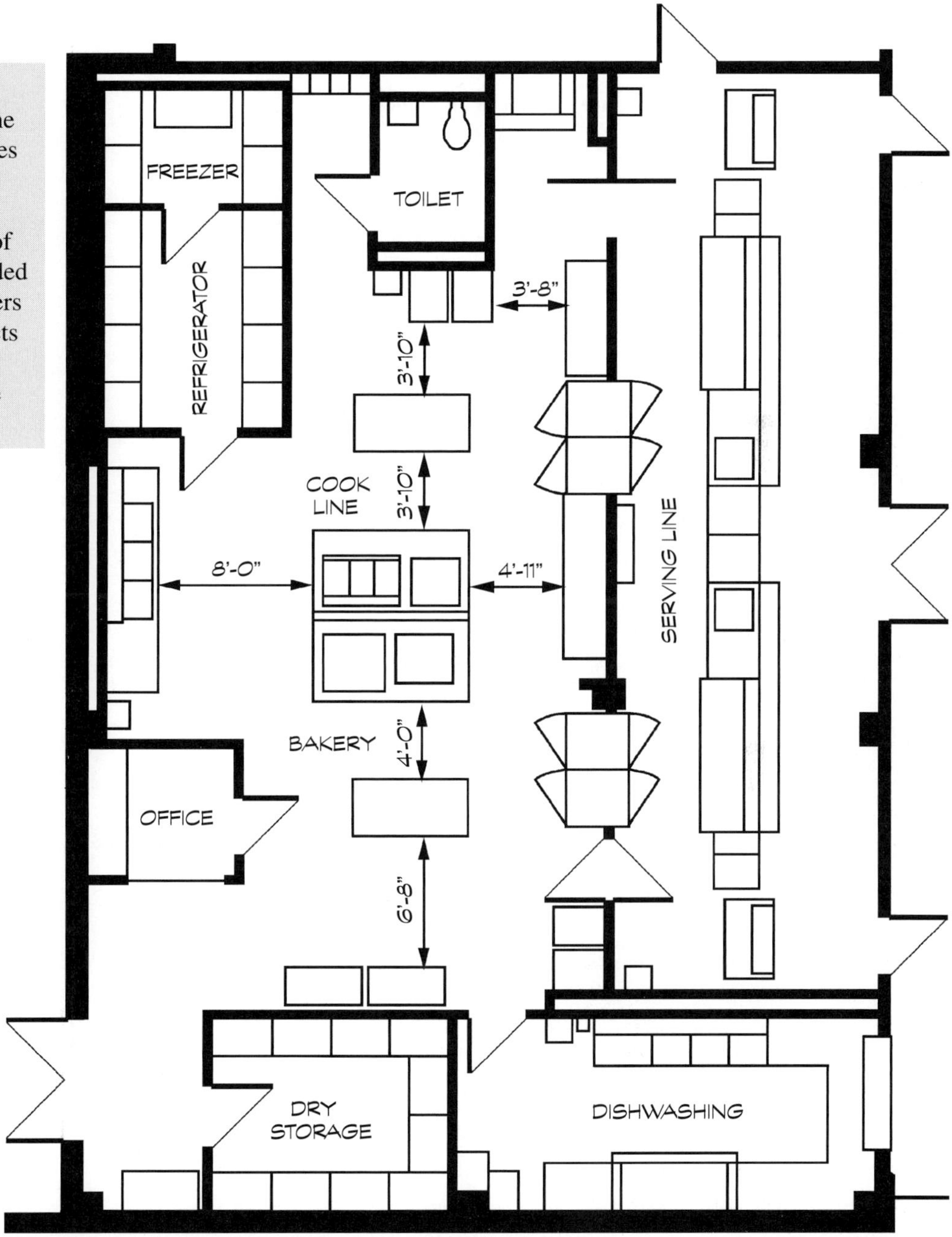

Fig. 1208.1. Partial floor plan at kitchen. New Jasper Pre-K–2nd Grade School. Jasper, Texas. PBK Architects, Inc. Houston, Texas.

1208 Interior Space Dimensions

1208.4 Efficiency dwelling units

- A living room ≥ 220 sf is required for units with ≤ 2 occupants:
  - 100 sf is added to this requirement for each additional occupant.
- A separate closet is required.
- The following appliances and clearances are required:

Appliance	Clearance in front of appliance
Cooking	≥ 2'-6"
Refrigerator	≥ 2'-6"

- A separate bathroom is required with the following:
  - Water closet.
  - Lavatory.
  - Bathtub or shower.
- Units must conform to the requirements of the code as follows:
  - For light.
  - For ventilation.
  - All other requirements as applicable.

1209 Access to Unoccupied Spaces

1209.1 Crawl spaces

- Crawl spaces required ≥ 1 access opening as follows:
  - Dimensions must be as follows:

    Width ≥ 1'-6".

    Length ≥ 2'.

1209.2 Attic spaces

- Attic space with a clear height ≥ 2'-6" has the following access requirements:
  - Dimensions of an access opening must be as follows:

    Width ≥ 1'-8".

    Length ≥ 2'-6".
  - Headroom ≥ 2'-6" is required at the access opening.

1209.3 Mechanical appliances

- Access to mechanical appliances in the following locations is not governed by this code:
  - In crawl spaces.
  - In attic spaces.
  - On roofs.
  - On elevated structures.

Note: The International Mechanical Code *is cited as governing such access.*

1209 Access to Unoccupied Spaces

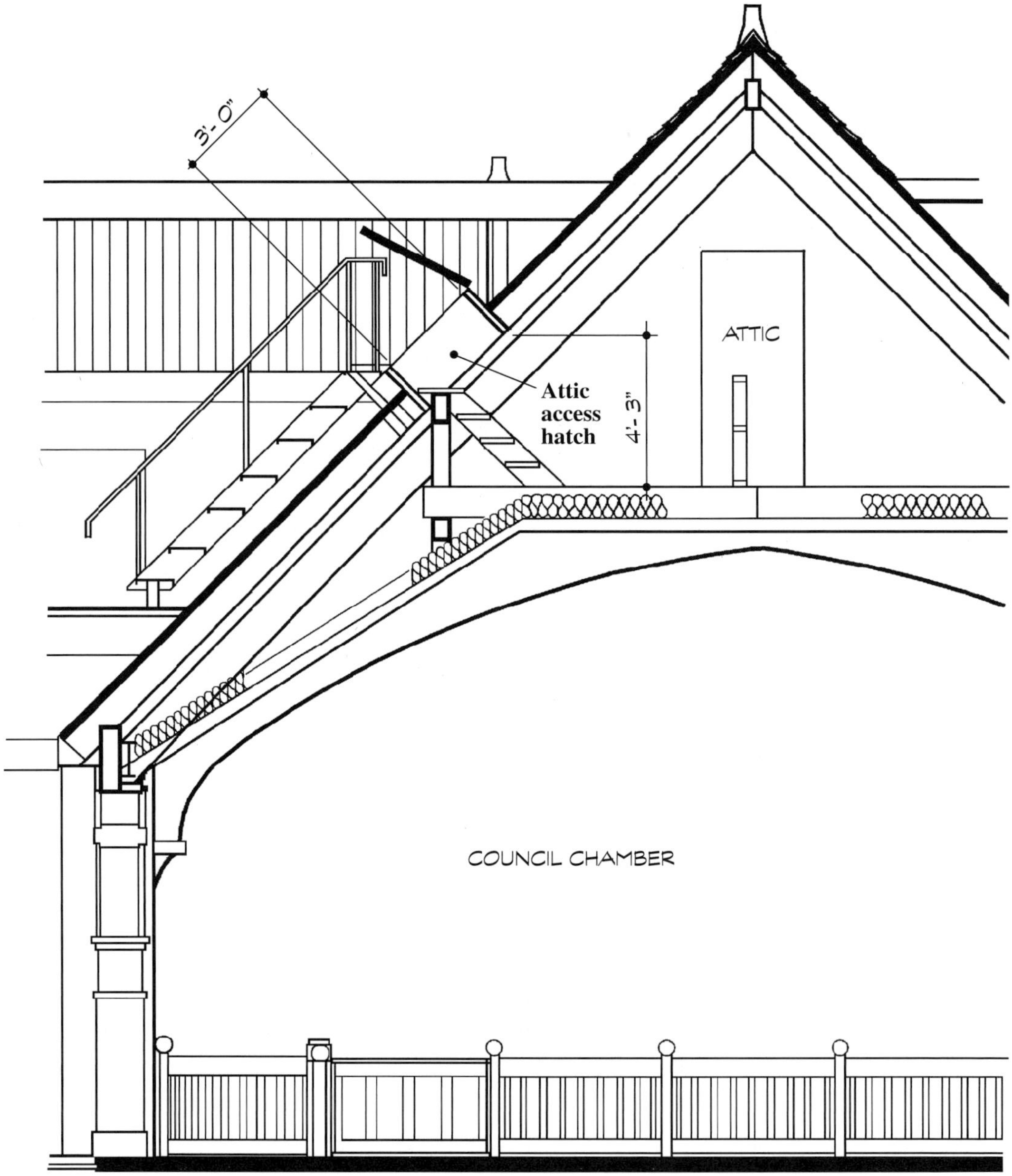

Case study: Fig. 1209.2. The clear height of the attic is > 2'-6"; thus, it must meet minimum access requirements. The headroom at the attic access is 4'-3", which is > than the 2'-6" minimum required. The access opening is 3' × 3', which complies with the minimum size requirements of 1'-8" × 2'-6". Access to the attic is in compliance with code requirements.

Fig. 1209.2. Partial section at attic. Lake Forest City Hall Renovation and Addition. Lake Forest, Illinois. David Woodhouse Architects. Chicago, Illinois.

1210 Surrounding Materials

1210.1 Floors and wall base finish materials

- This section governs the following rooms in other than dwelling units:
  - Toilets.
  - Bathing rooms.
  - Shower rooms.
- Floor surfaces of such rooms must be as follows:
  - Smooth.
  - Hard.
  - Nonabsorbent.
- A base material as follows must extend walls ≥ 4":
  - It must have the same surface characteristics as those of the floor.

1210.2 Walls and partitions

- Requirements indicated below do not apply to the following:
  - Toilet rooms in dwelling units and sleeping units.
  - Toilet rooms with both of the following characteristics:
    - Rooms not accessible to the public.
    - Rooms with < 2 water closets.
- Other toilet rooms are governed as follows:
  - Walls and partitions ≤ 2' of urinals and water closets have the following requirements:
    - Surface to a height ≥ 4' must have the following characteristics:
      - Smooth.
      - Hard.
      - Nonabsorbent.
    - Materials other than structure are governed as follows:
      - Must not be affected by moisture.
  - Accessories as follows must be installed as listed below:
    - Accessories:
      - Grab bars.
      - Towel bars.
      - Paper dispensers.
      - Soap dishes.
      - Similar accessories.
    - Requirement:
      - Installed and sealed to protect structure from moisture.

1210.3 Showers

- Finish materials in the following shower areas are governed as listed:
  - Shower areas:
    - Shower stalls.
    - Showers in bathtubs.
  - A surface as follows is required to a height ≥ 5'-10" above the drain inlet:
    - Smooth.
    - Hard.
    - Nonabsorbent.

1210 Surrounding Materials

1210.4 Waterproof joints

- The joint between the following components must be waterproof:
  - A built-in bathtub with a shower.
  - The adjacent wall.

Case study: Fig. 1210.2. The wall of the toilet room within 2' of the water closet is covered with tile to a height of 7'. The floor is tile, and tile is provided as a base at the wall. These surfaces meet the minimum requirement wherein a smooth, hard, and non-absorbent material must reach a height ≥ 4' on the wall, cover the floor, and reach ≥ 6" up the wall from the floor. As required, accessories such as grab bars and toilet paper holder are sealed where they connect to the wall to prohibit the penetration of moisture. This toilet room complies with all code requirements.

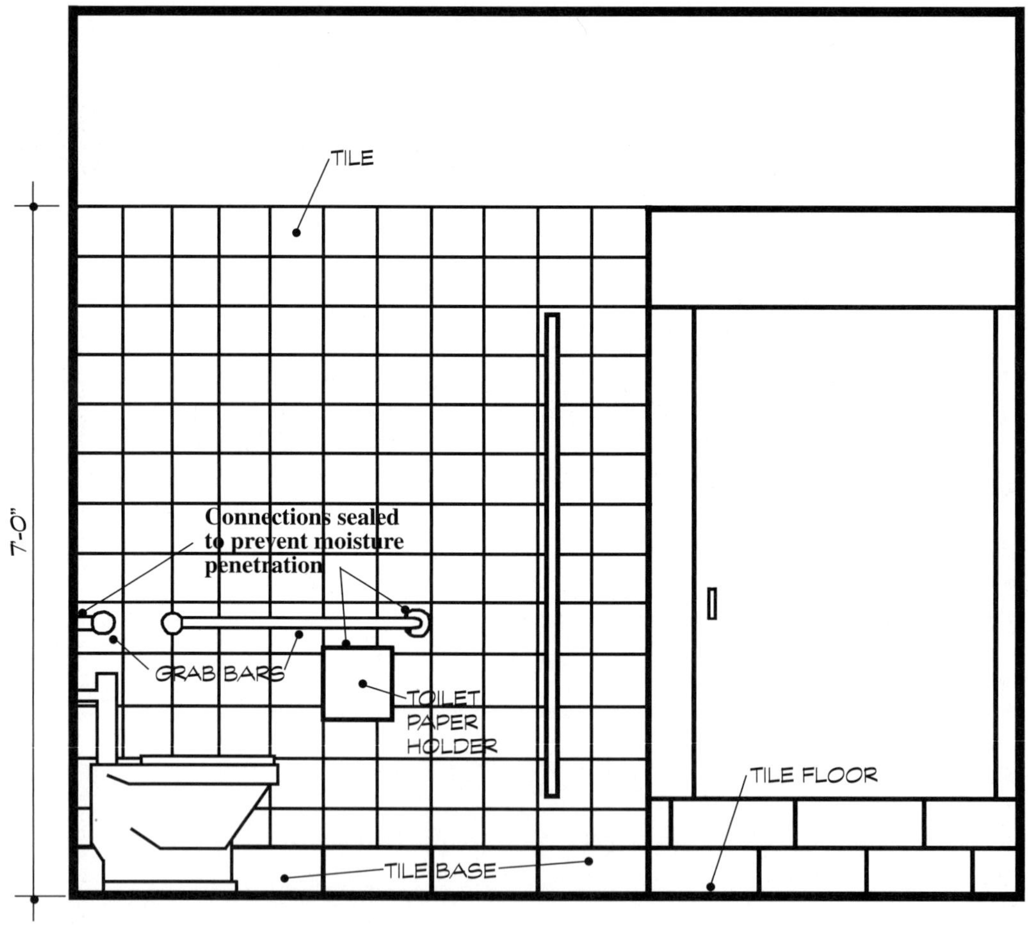

Fig. 1210.2. Toilet room elevation. AmberGlen Business Center. Hillsboro, Oregon. Ankrom Moisan Associated Architects. Portland, Oregon.

1210 Surrounding Materials

1210.5 Toilet rooms

- Toilet rooms may not open directly to the following space:
  - A space where food for service to the public is prepared.

Case study: Fig. 1210.5. Toilet rooms are prohibited from opening into a space where food is prepared for the public. The staff toilet room for this kitchen opens into an alcove, thus, complying with this code restriction.

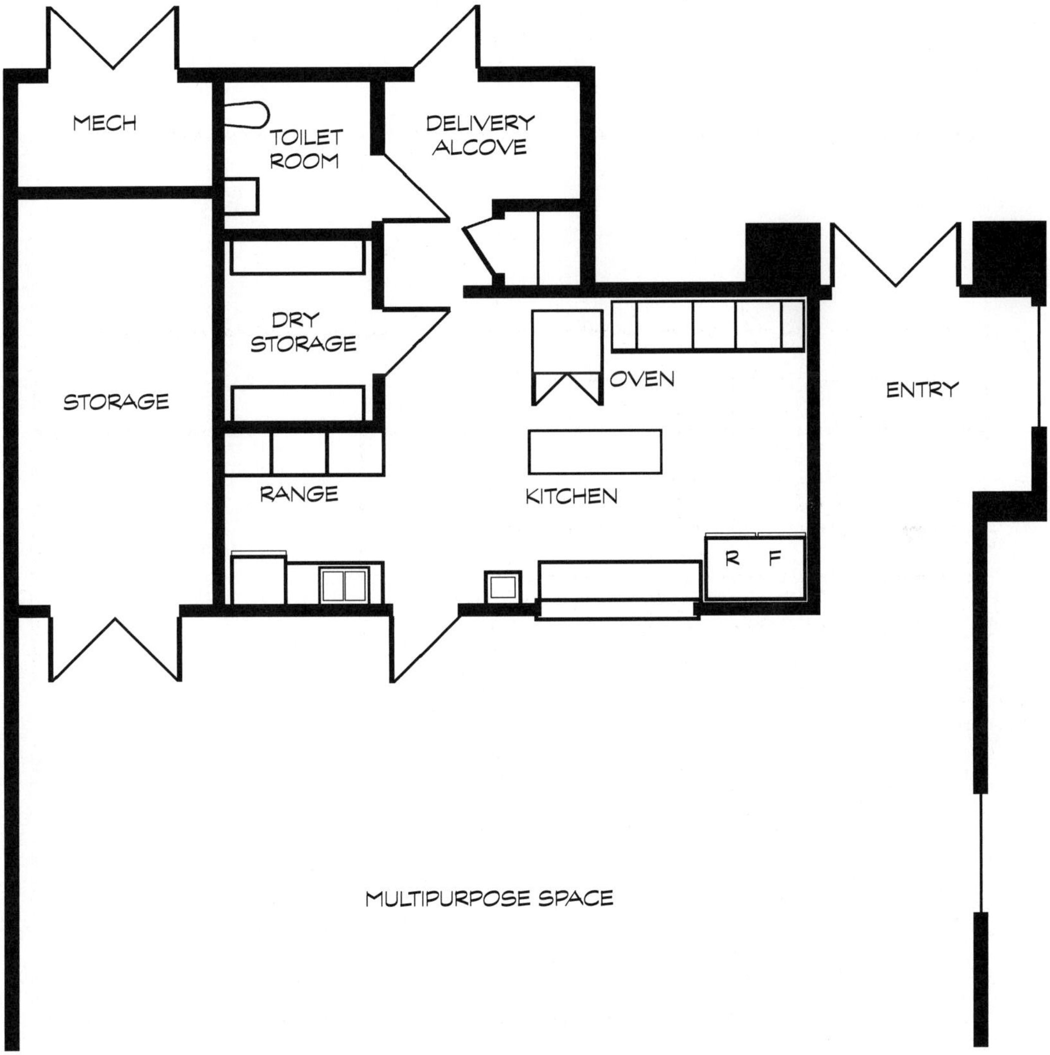

Fig. 1210.5. Partial floor plan. Creston Elementary Multipurpose Building. Creston, California. Phillips Metsch Sweeney Moore Architects. Santa Barbara, California.

NOTES

13

Energy Efficiency

Alterations to 209 Main Street. Annapolis, Maryland.
Alt Breeding Schwarz Architects, LLC. Annapolis, Maryland.

1301 General

1301.1.1 Criteria

- For purposes of energy efficiency, buildings must comply with the following code:
  - *International Energy Conservation Code.*

14

Exterior Walls

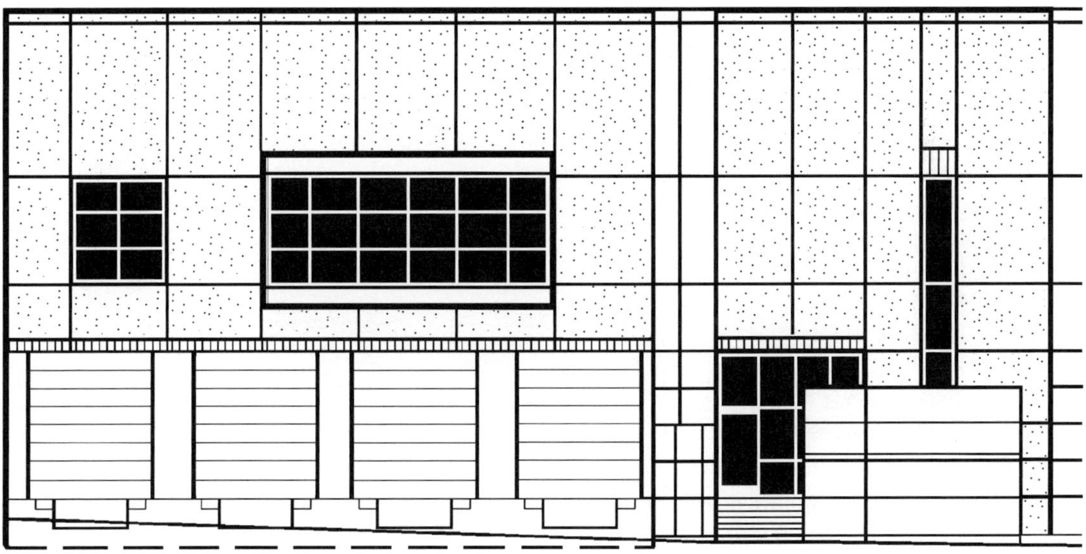

New Warehouse Addition. Los Angeles, California. *(partial elevation)*
Stephen Wen + Associates, Architects, Inc. Pasadena, California.

1403 Performance Requirements

1403.1 General

- This section applies to the following:
  - Exterior walls.
  - Exterior wall coverings.
  - Components of exterior walls.

1403.2 Weather protection *(part 1 of 2)*

- This section addresses detailing of the exterior wall.
- A weather-resistant exterior wall envelope is not required on the following:
  - Concrete walls.
  - Masonry walls.

Note: The following are cited as governing the walls indicated above:
Chapter 19, "Concrete."
Chapter 21, "Masonry."

- The following detailing may be omitted based on testing as indicated below:
  - Detailing as specified in this section for the wall envelope to address the following:
    Drainage.
    Water-resistive barrier.
    Flashing.

Note: The following are cited as governing the detailing specified in this section:
1404.2, "Water-resistive barrier."
1405.4, "Flashing."

  - Testing:
    Testing must demonstrate the following:
      Ability of alternative detailing to resist wind-driven rain including the following:
        At joints.
        At penetrations.
        At intersections with dissimilar materials.
    Testing must include the following:
      A test specimen representative of the actual wall envelope as follows:
        ≥ 1 opening.
        ≥ 1 control joint.
        ≥ 1 wall-eave interface.
        ≥ 1 wall sill.
      A test specimen ≥ 4' × ≥ 8'.
      A differential pressure on the test specimen ≥ 6.24 lb/sf.
      A duration ≥ 2 hr.

Note: ASTM E 331, "Standard Test Method for Water Penetration of Exterior Windows, Skylights, Doors, and Curtain Walls by Uniform Static Air Pressure Difference," is cited as governing the required testing.

- The wall must be protected against condensation.

Note: 1405.3, "Vapor retarders," is cited as providing requirements applicable to condensation.

1403 Performance Requirements

1403.2 Weather protection *(part 2 of 2)*

- Otherwise, the following is required to prevent the accumulation of water in exterior walls:
  - A weather-resistant exterior wall envelope as follows:
    Including a water-resistant barrier behind the exterior veneer.
  - Flashing.
  - A method for drainage of any water that enters the wall.

Note: 1404.2, "Water-resistive barrier," is cited as governing this component.
1405.4, "Flashing," is cited as governing this component.

1403.5 Flood resistance

- The following applies to buildings in flood hazard areas:
  - Exterior walls below design flood elevation are governed as follows:
    They must resist water damage.
    Wood must be one of the following:
      Pressure-treated with preservative.
      Decay-resistant heartwood of any of the following:
        Redwood.
        Black locust.
        Cedar.

Note: 1612.3, "Establishment of flood hazard areas," is cited as defining areas to which this section applies.
AWPA U1, "Use Category System: User Specification for Treated Wood Except Section 6, Commodity Specification H" is cited as governing pressure treatment regarding species, product, and end use. Preservatives permitted are listed in section 4.

1403.6 Flood resistance for high-velocity wave action areas

- Buildings in flood zones with high-velocity wave potential are governed as follows:
  - The following elements have the requirements listed below:
    Elements:
      Electrical, mechanical, and plumbing system elements.
    Requirements:
      Such elements may not be secured to or penetrate the following exterior walls:
        Those designed to collapse with wave impact.

Note: 1612.3, "Establishment of flood hazard areas," is cited as governing these zones.

1405 Installation of Wall Coverings

1405.2 Weather protection

- Exterior walls must provide the following for a building:
  - Protection from the weather.
- The following materials are acceptable as weather coverings on exterior walls:
  - Materials with the nominal thickness shown in details provided with this section as follows: Figs. 1405.2A through Fig. 1405.2G.

Note: IBC Table 1405.2, "Minimum Thickness of Weather Coverings," is cited as the source of acceptable materials as required above and is the source for the data with the details provided by this section.

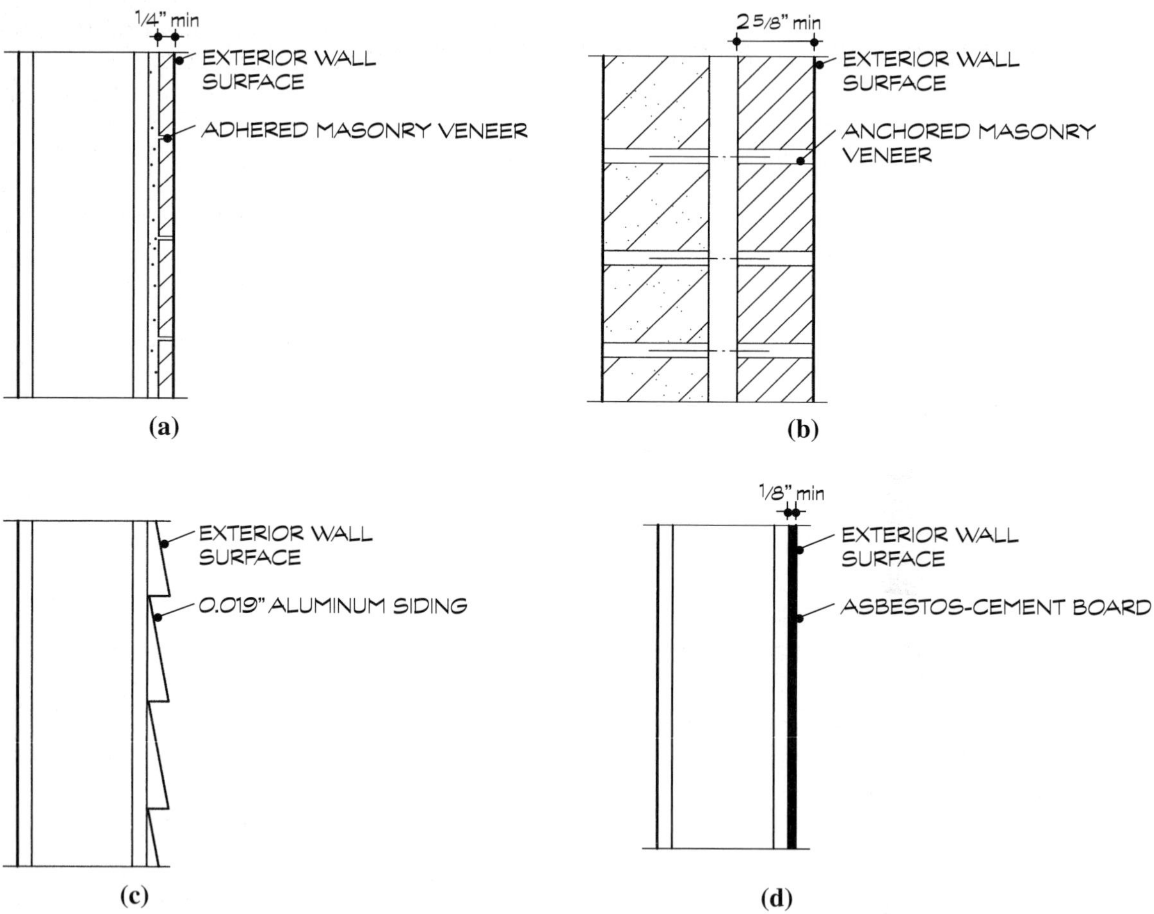

Fig. 1405.2A. Minimum thickness of weather coverings. Acceptable minimum nominal thickness of various types of cladding are shown in the wall sections.

1405 Installation of Wall Coverings

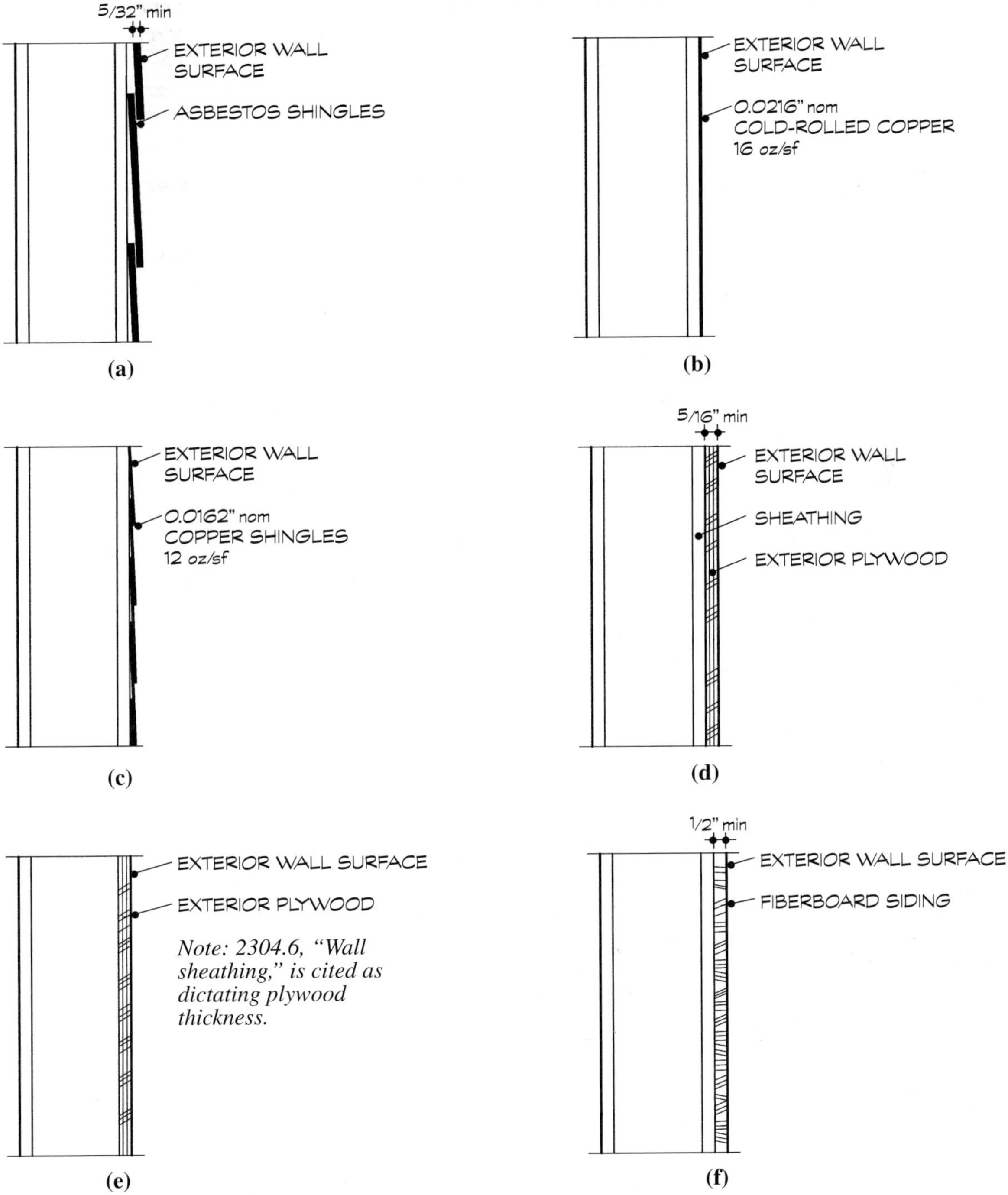

Fig. 1405.2B. Minimum thickness of weather coverings. Acceptable minimum nominal thickness of various types of cladding are shown in the wall sections.

1405 Installation of Wall Coverings

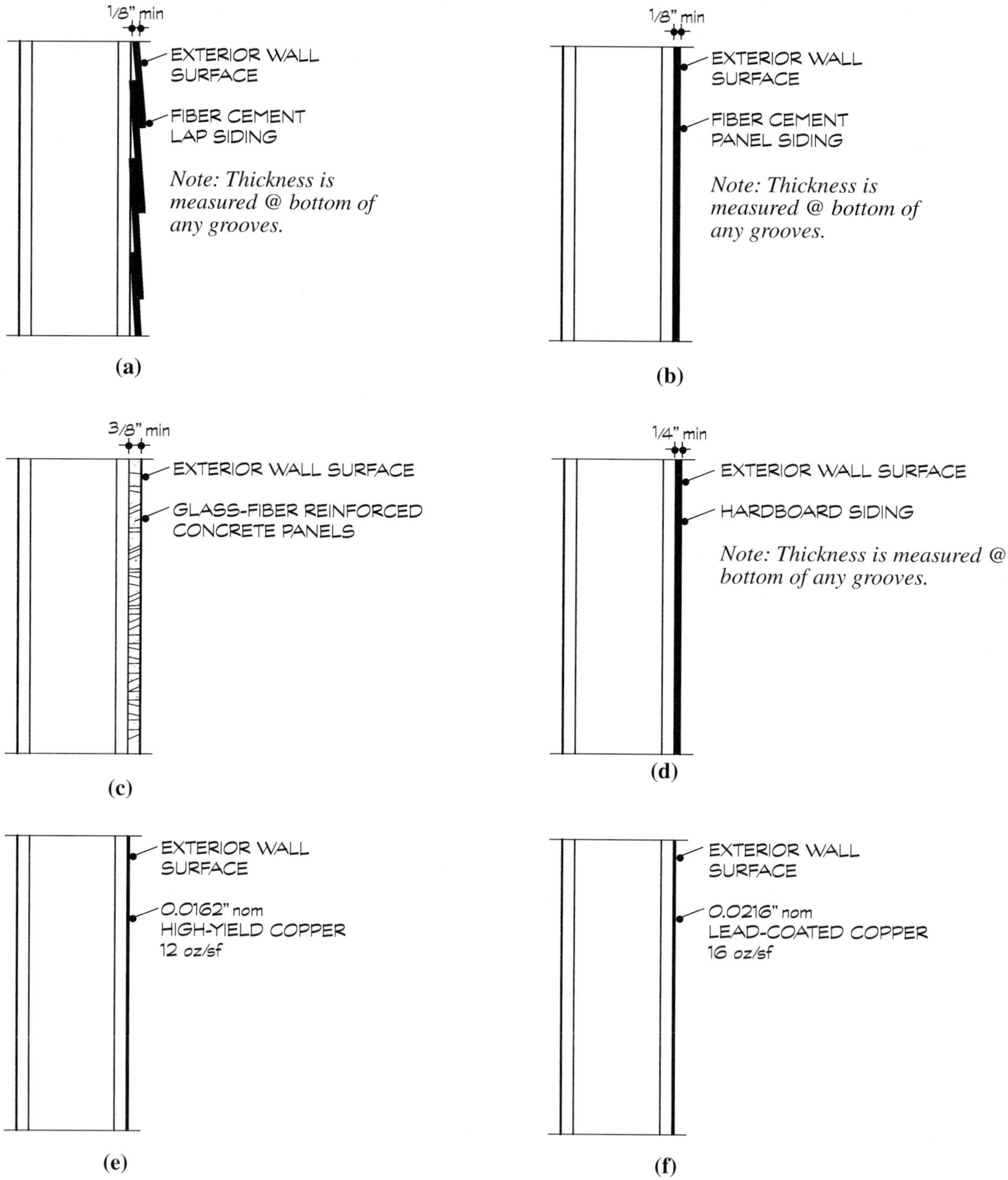

Fig. 1405.2C. Minimum thickness of weather coverings. Acceptable minimum nominal thickness of various types of cladding are shown in the wall sections.

1405 Installation of Wall Coverings

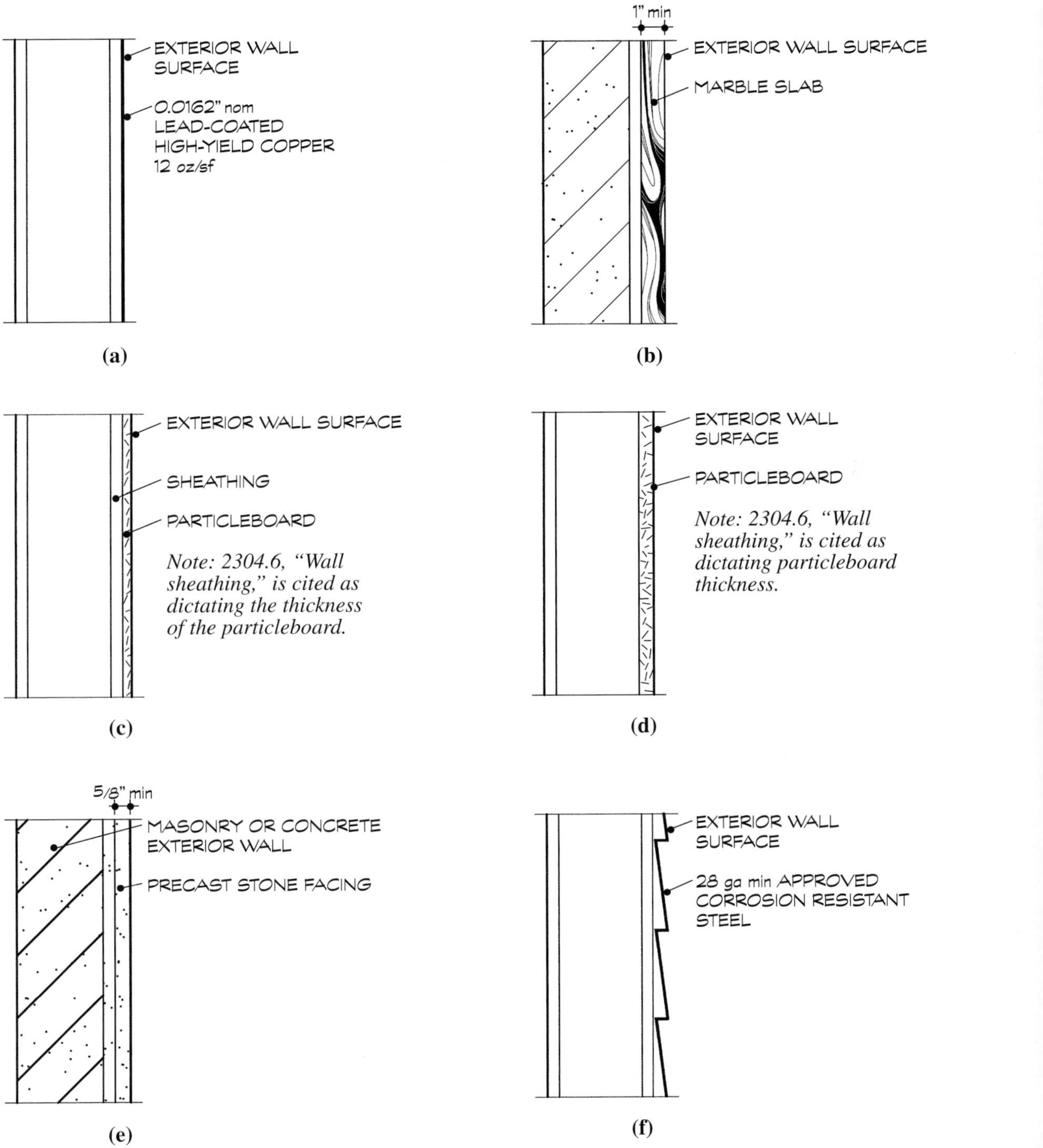

Fig. 1405.2D. Minimum thickness of weather coverings. Acceptable minimum nominal thickness of various types of cladding are shown in the wall sections.

1405 Installation of Wall Coverings

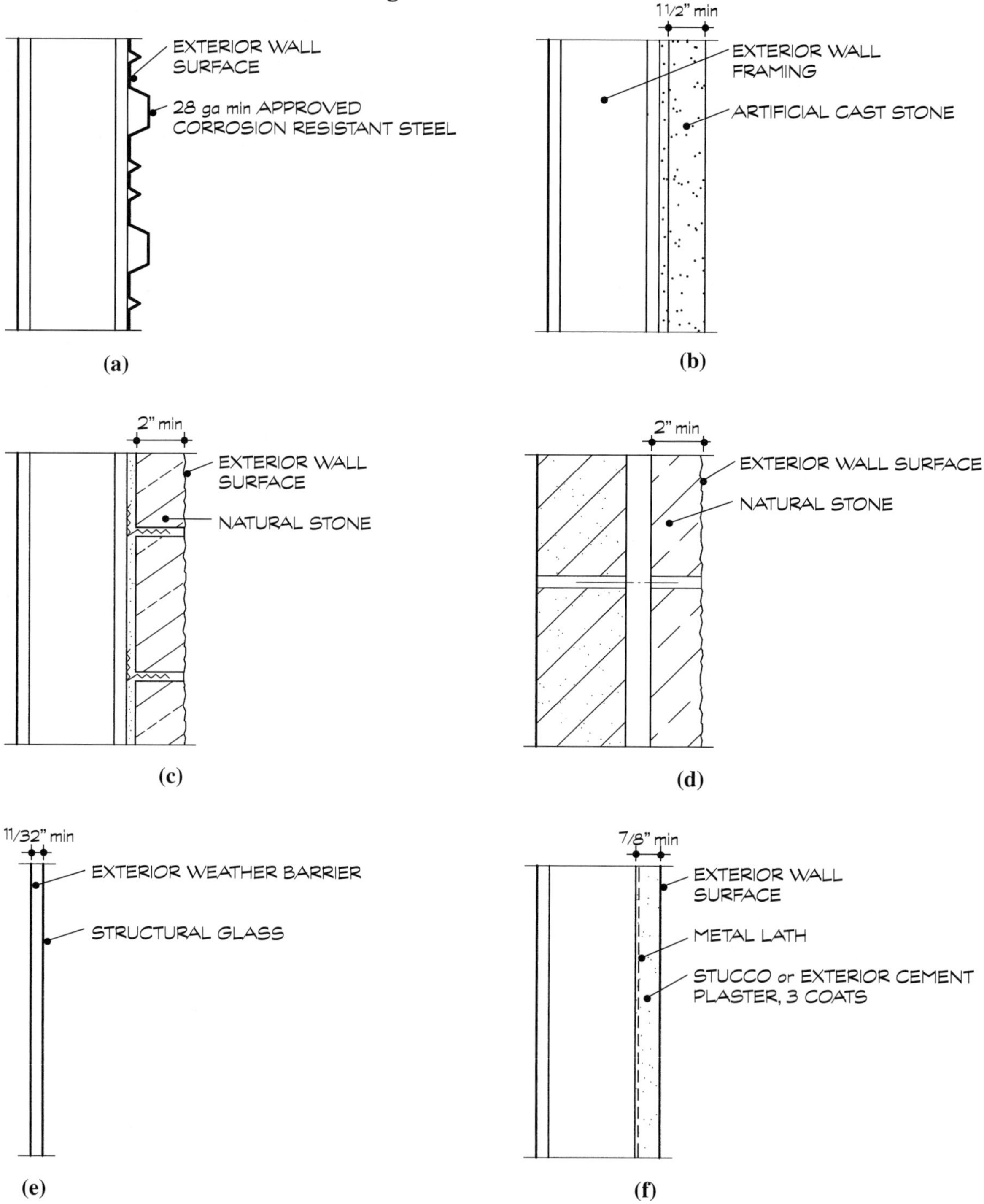

Fig. 1405.2E. Minimum thickness of weather coverings. Acceptable minimum nominal thickness of various types of cladding are shown in the wall sections.

1405 Installation of Wall Coverings

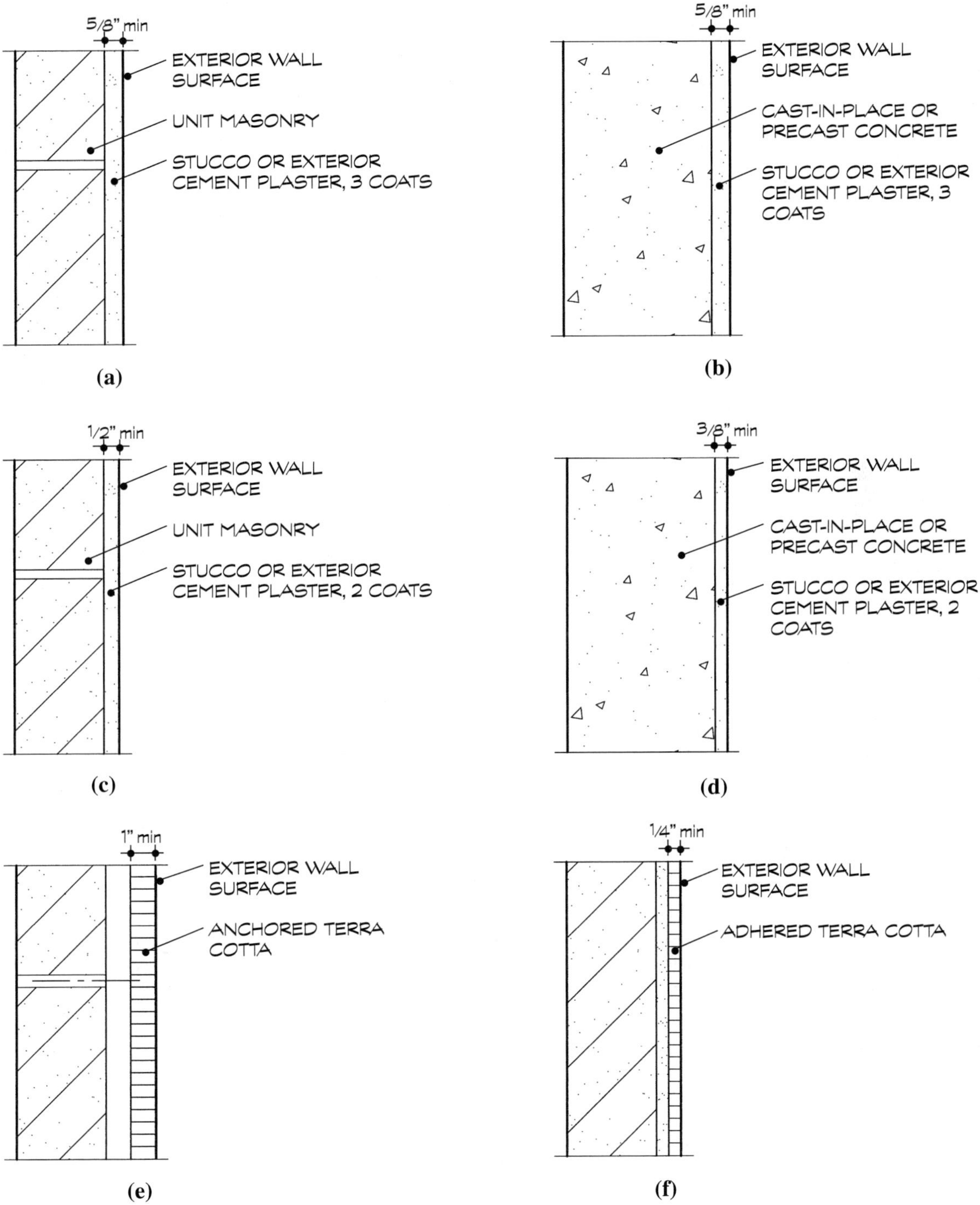

Fig. 1405.2F. Minimum thickness of weather coverings. Acceptable minimum nominal thickness of various types of cladding are shown in the wall sections.

1405 Installation of Wall Coverings

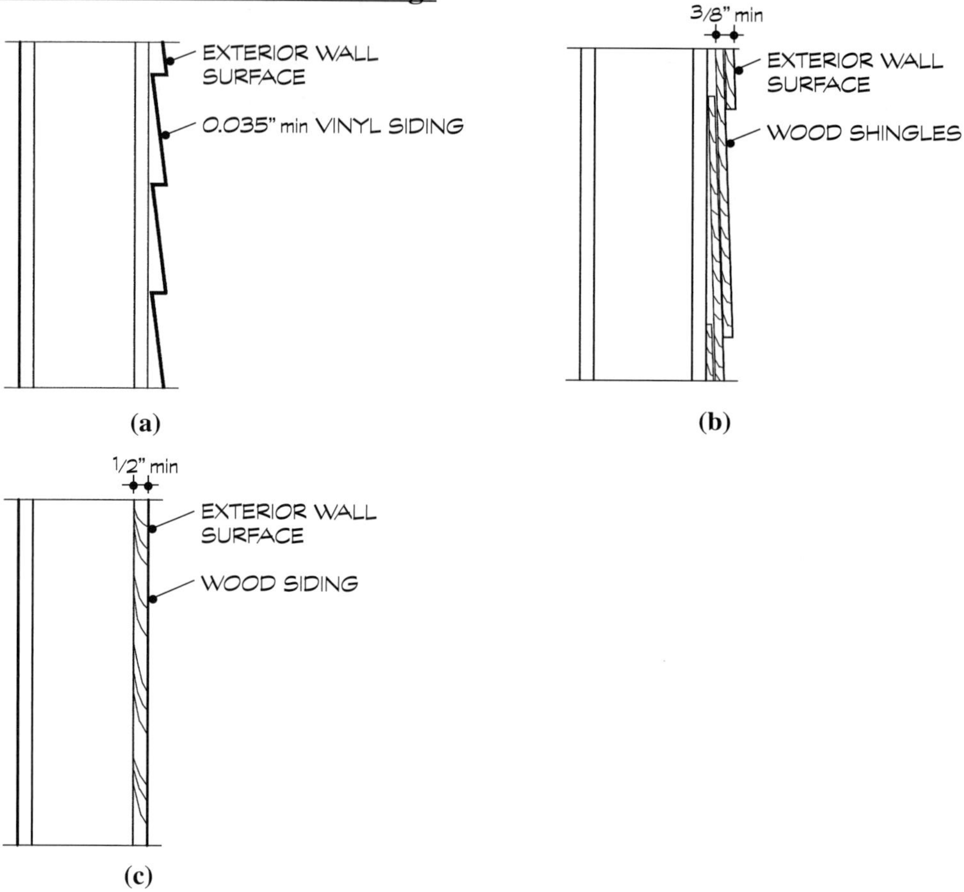

Fig. 1405.2G. Minimum thickness of weather coverings. Acceptable minimum nominal thickness of various types of cladding are shown in the wall sections.

1405.3 Vapor retarders

- This section addresses vapor retarders in the following zones:
  - 5, 6, 7, Marine 4.
- The following locations do not require Class I or II vapor retarders on the interior side of the frame:
  - Basement walls.
  - The part of any wall below grade.
  - Where the following will not damage the materials:
    Moisture.
    Freezing moisture.
- Other locations require Class I or II vapor retarders on the interior side of the frame.

1405 Installation of Wall Coverings

1405.3.1 Class III vapor retarders *(part 1 of 2)**

- Class III vapor retarders are permitted in the following zones:
  - In zone Marine 4, in one of the following applications:
    - Where there is vented cladding over any of the following:
      - Fiberboard OSB
      - Gypsum Plywood
    - On 2x4 walls in one of the following cases:
      - Where there is insulated sheathing with an R-value ≥ 2.5.
      - Where spray foam insulation is installed as follows:
        - Foam has a density ≥ 2 lbs/cu ft.
        - Foam is on the interior cavity side of one of the following:
          - Fiberboard Insulated sheathing Plywood
          - Gypsum OSB
        - Foam has an R-value ≥ 2.5.
    - On 2x6 walls in one of the following cases:
      - Where there is insulated sheathing with an R-value ≥ 3.75.
      - Where spray foam insulation is installed as follows:
        - Foam has a density ≥ 2 lbs/cu ft.
        - Foam is on the interior cavity side of one of the following:
          - Fiberboard Insulated sheathing Plywood
          - Gypsum OSB
        - Foam has an R-value ≥ 3.75.
  - In zone 5, in one of the following applications:
    - Where there is vented cladding over any of the following:
      - Fiberboard OSB
      - Gypsum Plywood
    - On 2x4 walls in one of the following cases:
      - Where there is insulated sheathing with an R-value ≥ 5.
      - Where spray foam insulation is installed as follows:
        - Foam has a density ≥ 2 lbs/cu ft.
        - Foam is on the interior cavity side of one of the following:
          - Fiberboard Insulated sheathing Plywood
          - Gypsum OSB
        - Foam has an R-value ≥ 5.
    - On 2x6 walls in one of the following cases:
      - Where there is insulated sheathing with an R-value ≥ 7.5.
      - Where spray foam insulation is installed as follows:
        - Foam has a density ≥ 2 lbs/cu ft.
        - Foam is on the interior cavity side of one of the following:
          - Fiberboard Insulated sheathing Plywood
          - Gypsum OSB
        - Foam has an R-value ≥ 7.5.

*Source: IBC Table 1405.3.1.

1405 Installation of Wall Coverings

1405.3.1 Class III vapor retarders *(part 2 of 2)**

- In zone 6, in one of the following applications:
  - Where there is vented cladding over any of the following:
    - Fiberboard.
    - Gypsum.
  - On 2x4 walls in one of the following cases:
    - Where there is insulated sheathing with an R-value ≥ 7.5.
    - Where spray foam insulation is installed as follows:
      - Foam has a density ≥ 2 lbs/cu ft.
      - Foam is on the interior cavity side of one of the following:

        Fiberboard, Insulated sheathing, Plywood, Gypsum, OSB

      - Foam has an R-value ≥ 7.5.
  - On 2x6 walls one of the following cases:
    - Where there is insulated sheathing with an R-value ≥ 11.25.
    - Where spray foam insulation is installed as follows:
      - Foam has a density ≥ 2 lbs/cu ft.
      - Foam is on the interior cavity side of one of the following:

        Fiberboard, Insulated sheathing, Plywood, Gypsum, OSB

      - Foam has an R-value ≥ 11.25.
- In zones 7 and 8, in one of the following applications:
  - On 2x4 walls in one of the following cases:
    - Where there is insulated sheathing with an R-value ≥ 10.
    - Where spray foam insulation is installed as follows:
      - Foam has a density ≥ 2 lbs/cu ft.
      - Foam is on the interior cavity side of one of the following:

        Fiberboard, Insulated sheathing, Plywood, Gypsum, OSB

      - Foam has an R-value ≥ 10.
  - On 2x6 walls in one of the following cases:
    - Where there is insulated sheathing with an R-value ≥ 15.
    - Where spray foam insulation is installed as follows:
      - Foam has a density ≥ 2 lbs/cu ft.
      - Foam is on the interior cavity side of one of the following:

        Fiberboard, Insulated sheathing, Plywood, Gypsum, OSB

      - Foam has an R-value ≥ 15.

*Source: IBC Table 1405.3.1.

1405 Installation of Wall Coverings

1405.3.2 Material vapor retarder class

- Vapor retarder class is to be based on the following:
  - Manufacturer's certified testing.
  - A tested assembly
- Vapor retarder class is assigned as follows:
  - Class I includes the following:
    - Sheet polyethyelene.
    - Nonperforated aluminum foil.
  - Class II includes the following:
    - Kraft-faced fiberglass batts.
    - Paint with a perm rating in the following range:
      - > 0.1 and ≤ 1.0.
  - Class III includes the following:
    - Latex paint.
    - Enamel paint.

1405.3.3 Minimum clear airspaces and vented openings for vented cladding

- Vented cladding, as cited in this section, requires the following clear air spaces:
  - Airspace as required elsewhere in this chapter for the following materials:
    - Siding of the following materials applied over a weather resistant barrier:
      - Vinyl.
      - Horizontal aluminum.
  - Airspace as required elsewhere in this code for the following material:
    - Brick.
  - Other vented claddings where approved.

1405.4 Flashing *(part 1 of 2)*

- Flashing must perform in one or both of the following ways:
  - Prevent water from penetrating a wall.
  - Deliver water from inside a wall to the exterior.
- Flashing is required at the following locations:
  - At the perimeters of door assemblies.
  - At the perimeters of window assemblies.
  - At penetrations of exterior wall assemblies.
  - At terminations of exterior wall assemblies.
- Flashing with flanges is required in the following locations:
  - At copings as follows:
    - On both sides.
    - On ends.
  - Under sills.
  - Above projecting trim as follows:
    - As continuous flashing.

1405 Installation of Wall Coverings

1405.4 Flashing *(part 2 of 2)*

- At exterior wall intersections with the following:
  - Roofs.
  - Chimneys.
  - Porches.
  - Decks.
  - Balconies.
  - Projections similar to balconies.
- At built-in gutters.
- At locations similar to built-in gutters as follows:
  - Where water might penetrate a wall.

Case study: Fig. 1405.4. The intersection of the entry canopy roof and the building wall is required to have flashing in the step-configuration. Such flashing is provided; thus, the detailing is in compliance with the code.

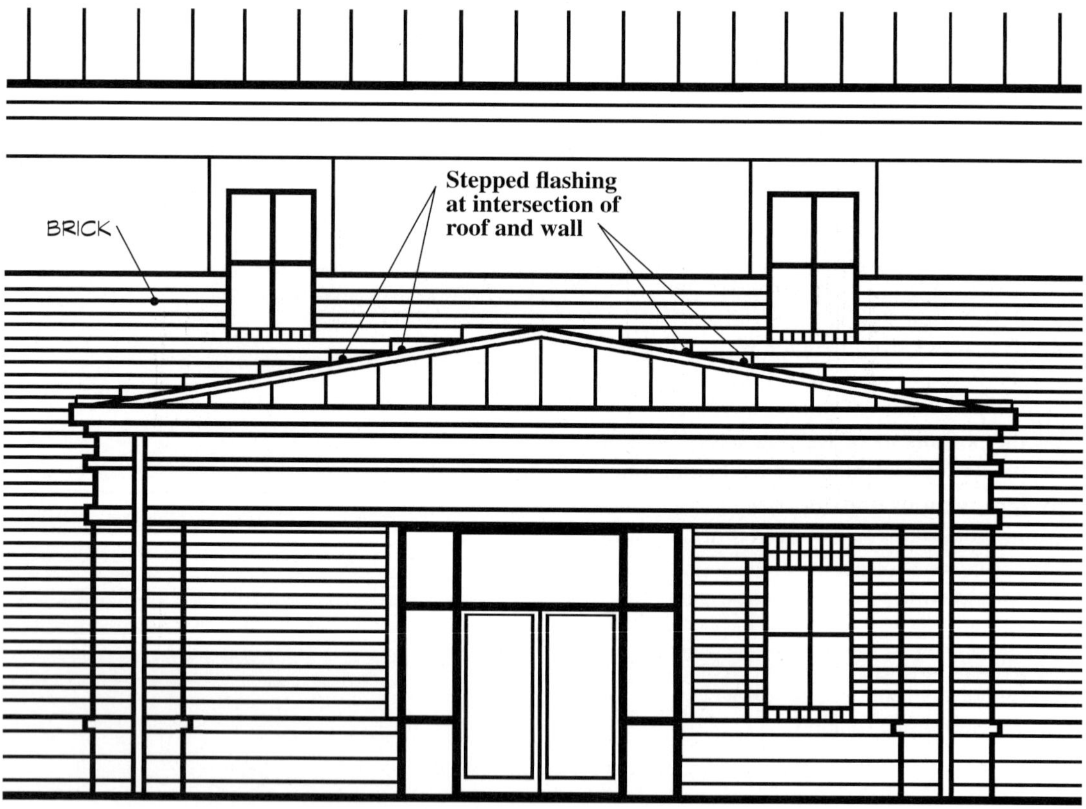

Fig. 1405.4. Partial elevation at east entry. Hot Springs Police Department New Headquarters. Hot Springs National Park, Arkansas. Cromwell Architects Engineers. Little Rock, Arkansas.

1405 Installation of Wall Coverings

1405.4.1 Exterior wall pockets

- Detailing at exterior walls that permits moisture to collect is governed as follows:
  - Such is not permitted without devices to prevent water damage such as follows:
    - Caps.
    - Drips.
    - Other approved protection.

1405.4.2 Masonry

- The following detailing is required in anchored masonry veneers at the locations listed:
  - Detailing:
    - Flashing and weep holes.
  - Locations:
    - Above grade and in the first course of masonry above the following:
      - Foundation wall.
      - Slab.
      - Points of masonry support such as the following:
        - Structural floors.
        - Shelf angles.
        - Lintels.

Note: 1405.6, "Anchored masonry veneer," is cited as a source of requirements.

1405.5 Wood veneers

- This section addresses exterior wood veneers in construction other than Type V.
- One of the following materials is required for such veneers:
  - ≥ 1" wood.
  - ≥ 7/16" exterior hardboard siding.
  - ≥ 3/8" exterior wood structural panels.
  - ≥ 3/8" exterior particleboard.
- Such wood veneers must comply with all of the following:
  - The height of the veneer measured from grade is limited as follows:
    - Fire-retardant-treated wood must be ≤ 60' above grade plane.
    - Other wood must be ≤ 40' above grade plane.
  - The veneer must be attached in one of the following ways to a backing as listed below:
    - Attachment:
      - To a noncombustible backing.
      - To furring on a noncombustible backing.
    - Backing:
      - Must have the fire-resistance rating as required elsewhere in the code.
  - Open or spaced veneers without concealed spaces are governed as follows:
    - Veneer not projecting > 2' from the building wall.

1405.7 Stone veneer *(part 1 of 2)*

- Stone veneer ≤ 10" thick must be anchored directly to one of the following types of construction:
  - Masonry.
  - Concrete.
  - Stud.

1405 Installation of Wall Coverings

1405.7 Stone veneer *(part 2 of 2)*

- Where backed by concrete or masonry the following applies:
  - One of the following anchor ties is required:
    - ≥ 0.1055" diameter (12 gage) corrosion-resistant wire.
    - Approved equal.
  - Anchor ties are to be formed in loops as follows:
    - Legs ≥ 6" are to be bent at right angles for embedding in the backing wall.
    - Loop must extend beyond the face of the backing wall.
    - Loops must be spaced ≤ 12" on center in both directions.
  - One of the following masonry ties is required as follows:
    - ≥ 0.1055" diameter (12 gage) corrosion-resistant wire.
    - Approved equal.
  - Masonry ties are to be formed into loops as follows:
    - They are to be threaded through anchor tie loops in the following manner:
      - ≥ 1 masonry tie is required for every ≤ 2 sf of stone.
      - Legs ≥ 15" bent are to lie in joint of stone veneer as follows:
        - The last 2" of each leg must be bent at a right angle.
  - ≥ 1" cement grout must be placed as follows:
    - Between stone and backing.
- Where backed by stud construction the following apply:
  - Wood studs to be spaced ≤ 16" on center.
  - 2 layers of waterproof paper backing are applied to the studs.

Note: 1404.2, "Water-resistive barrier," is cited as governing the waterproof paper.

  - Wire mesh is attached to the studs over the paper backing as follows:
    - Mesh to be corrosion-resistant wire ≥ 0.0625" diameter (16 gage).
    - Mesh to be 2" × 2".
    - Mesh is attached to studs as follows:
      - With 2" steel wire furring nails at 4" on center.
      - Nails to penetrate studs ≥ 1 1/8".
    - Mesh is attached to top and bottom plates as follows:
      - With the following 8d common nails or equivalent wire ties:
        - Nails at 8" on center.
  - One of the following masonry ties is required as follows:
    - Corrosion-resistant wire ≥ 0.1055" diameter (12 gage).
    - Approved equal.
  - Masonry ties to be formed into loops in the following manner:
    - Threaded through the wire mesh as follows:
      - ≥ 1 masonry tie is required for every ≤ 2 sf of stone.
      - Legs ≥ 15" bent to lie in joint of stone veneer as follows:
        - Last 2" of each leg must be bent at a right angle.
  - ≥ 1" cement grout must be placed as follows:
    - Between stone and backing.

1405 Installation of Wall Coverings

1405.8 Slab-type veneer

- Slab-shaped masonry veneer units must be anchored directly to the following construction:
  - Masonry.
  - Concrete.
  - Stud.
- Slab-shaped masonry veneer units must comply with the following size limits:
  - Thickness must be ≤ 2".
  - Face area must be ≤ 20 sf.
- The requirements indicated below apply to the following veneer units:
  - Units:
    - Marble.
    - Travertine.
    - Granite.
    - Other slab-shaped stone.
  - Requirements:
    - Corrosion-resistant dowels are required to secure ties to slabs.
    - ≥ 4 dowels and ties for each veneer unit are required.
    - Dowels are to be located as follows:
      - In holes drilled in the middle 1/3 of the edge thickness:
        - Holes to be spaced around the perimeter of the unit as follows:
          - At ≤ 2' on center.
        - Holes to be one of the following:
          - Tight-fitting around dowel.
          - With a diameter ≤ 1/16" > dowel diameter as follows:
            - Hole is countersunk as follows:
              - With a diameter 2 × that of the dowel.
              - To a depth 2 × that of the dowel diameter.
            - Hole is grouted with cement mortar around dowel.
    - Ties are to be as follows:
      - Corrosion-resistant metal.
      - Able to resist tensile or compressive forces as follows:
        - ≥ 2 × the weight of the veneer unit attached.
      - In one of the following shapes:
        - Sheet metal ≥ 0.0336" (22 gage) × 1" in cross section.
        - Wire ≥ 0.1483" diameter (9 gage).

1405.9 Terra cotta *(part 1 of 2)*

- This section addresses terra cotta or ceramic units anchored directly to backing walls.
- Veneer units specified in this section must be anchored directly to the following construction:
  - Masonry.
  - Concrete.
  - Stud.
- Veneer units must have the following physical properties:
  - Ribbed profile as follows:
    - Dovetail ribs at 8"(±) on center.
  - ≥ 15/8" thick including the ribs.

1405 Installation of Wall Coverings

1405.9 Terra cotta *(part 2 of 2)*

- Veneer units are tied to the backing as follows:
  - Ties are corrosion-resistant wire as follows:
    Diameter ≥ 0.162" (8 gage).
  - Ties are installed into the top of each unit:
    In horizontal bed joints as follows:
      ≥ 12" and ≤ 18" on center.
  - Ties are connected to 1/4" diameter steel rods as follows:
    Rods are vertical.
    Rods pass through loops of anchors secured in the backing wall.
  - Ties must have the following strength:
    To support the full weight of the veneer in tension.
- Veneer units must be installed as follows:
  - With ≥ 2" space behind the units as follows:
    Between the face of the ribs and the backing wall.
  - Space must be grouted solid with the following:
    Portland cement grout.
    Pea gravel.
  - Just prior to setting, the following must be thoroughly wetted with clean water:
    Veneer units.
    Backing wall.
  - The following must be visibly damp when the veneer units are set:
    Veneer units.
    Backing wall.

1405.10.1 Interior adhered masonry veneers

- Veneer weight is limited to ≤ 20 lbs/sf:
  - Where supported by wood construction, the following applies:
    Supporting members which have a span must be designed as follows:
      Deflection is limited to ≤ 1/600 of the span.

Note: 1405.10, "Adhered masonry veneer," is cited as the source of requirements for the installation of such components.

1405.11 Metal veneers

- Metal veneers must be one of the following:
  - Of approved corrosion-resistant materials.
  - Be protected with porcelain enamel as follows:
    On the front.
    On the back.
  - Be processed to provided corrosion resistance.
- Metal veneers must be as follows:
  - ≥ 0.0149" thick (28 gage) sheet steel.
  - Mounted on one of the following:
    Wood furring strips.
    Metal furring strips.
    Approved sheathing on wood construction.

1405 Installation of Wall Coverings

1405.11.1 Attachment

- Metal veneers must be attached to the following construction as indicated below:
  - Construction:
    - Masonry.
    - Framing.
  - Attachment:
    - One of the following attachment methods is required:
      - Corrosion-resistant fastenings.
      - Metal ties.
      - Other approved methods.
    - Spacing of attachments must be ≤ 2' as follows:
      - Vertically and horizontally.
    - Where units are > 4 sf:
      - ≥ 4 attachments per unit are required.
    - Attachment devices must have the following cross-sectional area:
      - ≥ that provided by W 1.7 wire (0.017 sq in).
    - Attachments and their supports must have strength as per the larger of the following:
      - They must be able to resist applicable wind loads.
      - They must be able to resist a horizontal load ≥ 20 lbs/sf.

Note: Section 1609, "Wind Loads," is cited as the source governing wind loading.

1405.11.2 Weather protection

- Metal supports for exterior metal veneer must be protected by one of the following:
  - Painting or galvanizing.
  - Other equivalent method.
- The following wood supports must be protected from moisture as indicated below:
  - Supports:
    - Studs and furring strips.
    - Other wood supports.
  - Moisture protection:
    - One of the following is required:
      - Supports must be approved pressure-treated wood.
      - Must be protected from moisture contact.
    - The following details must be protected as indicated below:
      - Details:
        - Joints and edges exposed to the weather.
      - Protection:
        - Must be caulked by one of the following to prevent moisture penetration:
          - With a durable waterproofing material.
          - Other approved material.

Note: 1403.2, "Weather protection," is cited as the source of methods to protect wood supports from moisture contact.

1405 Installation of Wall Coverings

1405.12 Glass veneer

- The area limitations of exterior glass veneer panels as follows is indicated below:
  - Structural glass veneer as follows:
    - Single sections of thin glass.

Height above grade	Glass area
≤ 15'	≤ 10 sf
> 15'	≤ 6 sf

1405.12.1 Length and height

- Thin exterior structural glass veneer dimensions are governed as follows:
  - All dimensions are limited to ≤ 4'.

1405.12.2 Thickness

- Thin exterior structural glass veneer thickness is governed as follows:
  - Thickness must be ≥ 0.344".

1405.12.3 Application

- Thin exterior structural glass veneer must be set in the following conditions:
  - Backing must be dry.
  - An approved bond coat applied to the backing as follows:
    - Bond coat must be evenly distributed.
    - Backing surface must be completely sealed.
  - Glass panels must be installed as follows:
    - With an approved mastic cement.
    - So that ≥ 50% of each glass panel is bonded to the backing as follows:
      - By mastic ≥ 1/4" and ≤ 5/8" thick.
      - Compatibility of mastic with bond coat must be verified.
      - Mastic must adhere securely to backing.

1405.12.4 Installation at sidewalk level

- Glass veneer at sidewalk level must be detailed as follows:
  - Glass panels must set in a metal molding as follows:
    - Must be approved.
    - Space between molding and sidewalk must be treated as follows:
      - Caulked for watertightness.
  - Glass panels must be set ≥ 1/4" above the following:
    - The highest point of the sidewalk.

1405.12.4.1 Installation above sidewalk level *(part 1 of 2)*

- This section applies to thin exterior structural glass veneer as follows:
  - Where installed ≥ higher than one of the following levels:
    - The level of the top of a bulkhead facing.
    - A level > 36" above the sidewalk.

1405 Installation of Wall Coverings

1405.12.4.1 Installation above sidewalk level *(part 2 of 2)*

- In such cases the following are required in addition to mastic behind the glass:
  - Shelf angles:
    - With all of the following characteristics:
      - Approved.
      - Nonferrous metal.
      - ≥ 0.0478" thick.
      - ≥ 2" long || to the glass.
    - Installed as follows:
      - Spaced at approved intervals.
      - ≥ 2 angles per glass panel.
      - Secured to the backing by one of the following:
        - Expansion or toggle bolts.
        - Other approved means.

1405.12.5 Joints

- The abutting edges of exterior structural glass veneer panels must be square as follows:
  - Miters are not permitted in the following case:
    - Where not specifically approved for obtuse-angled surfaces.
  - Other shapes are not permitted unless approved.
- An approved jointing compound must be applied uniformly to joints.
- Horizontal joints must be ≥ 1/16" as follows:
  - Spaced by approved nonrigid materials.
- Where glass panels abut rigid materials, the following is required:
  - Expansion joints ≥ 1/4" as follows:
    - At the sides and top of the glass.

1405.12.6 Mechanical fastenings

- Glass veneer panels installed at the following heights are governed as indicated below:
  - Heights:
    - Above the level of heads of show windows.
    - > 12' above the sidewalk.
  - Requirements:
    - Fastenings required include all of the following:
      - Mastic as required at lower levels.
      - Shelf angles as required at lower levels.
      - Mechanical fasteners are required as follows:
        - At one of the following locations of each glass panel:
          - At each horizontal or vertical edge.
          - At each corner.
        - Fasteners must be fixed to the backing by one of the following:
          - Expansion or toggle bolts.
          - Other methods.
        - Fasteners must support the glass as follows:
          - In a vertical plane without assistance from the mastic.
        - Shelf angles that also serve as mechanical fasteners are acceptable.

1405 Installation of Wall Coverings

1405.12.7 Flashing

- This section applies to thin exterior structural glass veneer.
- Exposed edges of the glass veneer panels must be detailed as follows:
  - Covered with overlapping flashing of the following material:
    - Corrosion-resistant metal.
  - Made moisture tight as follows:
    - By caulking with a waterproof substance.

1405.13 Exterior windows and doors

- Windows and doors in exterior walls must meet testing standards.

Note: 1715.5, "Exterior window and door assemblies" is cited as governing testing and performance requirements fot these elements.

1405.13.1 Installation

- The following components must meet the requirements listed below:
  - Components:
    - Windows.
    - Doors.
  - Requirements:
    - Components must be installed according to the manufacturer's instructions as follows:
      - Instructions must be approved.
      - Instructions must include fastener size and spacing as follows:
        - These must be calculated based on results where the following were tested:
          - Maximum loading.
          - Maximum fastener spacing.

1405.13.2 Window sills

- This section does not address windows with window guards meeting certain standards.

Note: The following alternatives are cited as governing the window guards mentioned above:
ASTM F 2006, "Safety Specification for Window Fall Prevention Devices for Non-Emergency Escape (Egress) and Rescue (Ingress) Windows."
ASTM F 2090, "Specification for Window Fall Prevention Devices with Emergency Escape (Egress) Release Mechanisms."

- This section addresses windows as follows:
  - In either of the following locations:
    - Occupancy R-2 multiple family dwellings.
    - Occupancy R-3 1- and 2- family dwellings.
  - Operable windows where the bottom of the clear opening is > 6' above either of the following:
    - Finished grade.
    - Other exterior horizontal surface below the window.
- The bottom of the clear opening must be ≥ 2' above the finished floor below the window.
- Any glazing within 2' above the finished floor must be one of the following:
  - Fixed glazing.
  - Glazing with any opening too small to pass a 4" sphere.

1405 Installation of Wall Coverings

1405.14 Vinyl siding

- Vinyl siding is permitted on exterior walls as follows:

Note: ASTM D 3679, "Specification for Rigid Poly (Vinyl Chloride) (PVC) Siding," is cited as the standard with which the vinyl siding must comply.

  - In Type V construction.
  - Where basic wind speed is ≤ 100 mph as follows:
    In Exposure C.

Note: Chapter 16, "Structural Design," is cited as providing the criteria for wind speed.

  - Where the building is < 40' tall.
  - Where test results or calculations are submitted as follows:
    Verifying compliance with structural requirements in either of the following cases:
      Where the building is > 40' tall.
      Where the wind speed is > 100 mph.
  - Vinyl siding must be installed so as to provide the following:
    Weather protection to exterior walls.

1405.14.1 Application

- Vinyl siding must be installed as follows:
  - Over sheathing or other materials.

Note: 2304.6, "Wall sheathing," is cited as the source of materials over which sheathing must be applied.

  - So as to meet weather-resistant barrier requirements.

Note: Section 1403, "Performance Requirements," is cited as governing.

  - In compliance with manufacturer's instructions as follows:
    Including accessories.
    Instructions must be approved.
  - Using one of the following types of nails:
    Nails specified in the manufacturer's instructions.
    Nails as follows where not specified in manufacturer's instructions:
      Head diameter = 0.313".
      Shank diameter = 0.125".
      Nails are to be corrosion-resistant.
      Nails must penetrate the nailing substrate ≥ 3/4".
      Nails are spaced as follows where siding is horizontal:
        ≤ 16" horizontally.
        ≤ 12" vertically.
      Nails are spaced as follows where siding is vertical:
        ≤ 12" horizontally.
        ≤ 12" vertically.

1405 Installation of Wall Coverings

1405.16 Fiber-cement siding

- Fiber-cement siding is permitted on exterior walls as follows:

  Note: 1404.10, "Fiber-cement siding," is cited as governing this material.

  - On all construction types.
  - For wind conditions cited by the following:
    - The manufacturer's listing and label.
    - Manufacturer's instructions as follows:
      - Where approved.
  - Where installed over sheathing or other materials.

    Note: 2304.6, "Wall sheathing," is cited as listing appropriate materials over which the siding may be installed.

  - Where meeting weather-resistant barrier requirements.

    Note: Section 1403, "Performance Requirements," is cited as governing these requirements.

  - Where installed according to manufacturer's instructions as follows:
    - Including accessories.
    - Instructions must be approved.
  - Using nails as follows where fastening to wood studs:
    - Nails specified in the manufacturer's instructions.
    - Nails as follows where not specified in manufacturer's instructions:
      - Nails are to be corrosion-resistant.
      - Nails are to have round heads.
      - Nails are to have smooth shanks.
      - Nails must penetrate the studs ≥ 1".
  - Using screws as follows where fastening to metal framing:
    - All-weather screws.
    - Screws must penetrate the metal framing ≥ 3 threads.

1405.16.1 Panel siding

- Fiber-cement panels must comply with applicable standards.

  Note: ASTM C 1186, "Standard Specification for Flat Fiber-Cement Sheets," Type A, ≥ Grade II, is cited as governing this product.

- Fiber-cement panels must be installed as follows:
  - Panels may be installed either vertically or horizontally.
  - All joints must fall on framing.
  - All joints must be sealed in one of the following ways:
    - Caulking.
    - Covered with battens.
    - Per the applicable standard.

  Note: 1403.2, "Weather protection," is cited as the standard applicable to joint treatment.

  Fasteners must comply with approved manufacturer's instructions.

1405 Installation of Wall Coverings

1405.16.2 Lap siding

- Fiber-cement lap siding is limited to 12" in width.

  Note: ASTM C 1186, "Standard Specification for Flat Fiber-Cement Sheets," Type A, ≥ Grade II, is cited as governing this product.

- Fiber-cement lap siding must be installed as follows:
  - Laps must be ≥ 11/4".
  - Lab siding without tongue-and-groove end joints must be installed by one of the following methods:
    - Method #1 as follows:
      - Caulk end joints.
      - Cover end joints with H-section joint covers over flashing.
    - Method #2 is per the applicable standard:

  Note: 1403.2, "Weather protection," is cited as the standard applicable to installation.

  - Fastener heads are to be exposed or covered per approved manufacturer's instructions.

1405.17 Fastening

- The following exterior wall materials must be fastened as indicated below:
  - Materials:
    - Weather boarding.
    - Wall coverings.
  - Fastening:
    - Any of the following fasteners must be used:
      - Aluminum.
      - Copper.
      - Zinc or zinc-coated.
      - Other corrosion-resistant fasteners where approved.
    - As per one of the following:
      - The manufacturer's installation instructions.
      - The nailing schedule provided by the code.

  Note: IBC Table 2304.9.1, "Fastening Schedule," is cited as governing the fasteners.

- The following exterior wall materials must be fastened as indicated below:
  - Materials:
    - Shingles.
    - Similar weather coverings.
  - Fastening:
    - One of the following fasteners is required:
      - Standard-shingle nails.
      - Mechanically bonded nails where approved.
    - Weather coverings are to be attached to one of the following:
      - Furring strips nailed to studs.
      - Wood sheathing ≥ 1" nominal thickness.
      - Wood structural panels as listed elsewhere in the code.

  Note: IBC Table 2308.9.3(3), "Wood Structural Panel Wall Sheathing," is cited as the source of appropriate wood structural panels.Wismod dolore dolorem doloreet, velis accummodo doloreet

1406 Combustible Materials on the Exterior Side of Exterior Walls

1406.1 General

- This section governs the following where constructed of combustible materials:
  - Exterior wall coverings.
  - Balconies.
  - Projections similar to balconies.
  - Bay windows.
  - Oriel windows.

1406.2 Combustible exterior wall coverings

- This section does not govern plastics.

  Note: Chapter 26, "Plastic," is cited as governing plastics.

- This section governs other exterior wall coverings as follows:
  - Where they are combustible.

1406.2.1 Ignition resistance

- The following exterior wall coverings need not be tested for ignition by radiant heat:
  - Wood.
  - Wood products.
  - Products that are covered with certain exterior materials other than vinyl sidings.

    Note: IBC Table 1405.2, "Minimum Thickness of Weather Coverings," is cited as listing coverings that preclude the need to be tested for ignition by radiant heat. Although vinyl sidings are listed in the table, they are not exempt from testing.

  - ≥ 0.019" thick aluminum.
  - Materials on Type V construction.
- Other exterior wall coverings must be tested for ignition by radiant heat.

  Note: NFPA 268, "Standard Test Method for Determining Ignitability of Exterior Wall Assemblies Using a Radiant Heat Energy Source," is cited as the required test.

1406.2.1.1 Fire separation of 5 feet or less

- Exterior wall materials installed where the fire separation distance is ≤ 5' must perform as follows:
  - They must not support sustained flaming.

  Note: NFPA 268, "Standard Test Method for Determining Ignitability of Exterior Wall Assemblies Using a Radiant Heat Energy Source," is cited as defining the sustained flaming noted above.

1406.2.1.2 Fire separation greater than 5 feet

- Exterior wall materials must resist ignition from radiant heat as follows:
  - Required resistance to radiant heat decreases with increased fire separation distance.

  Note: NFPA 268, "Standard Test Method for Determining Ignitability of Exterior Wall Assemblies Using a Radiant Heat Energy Source," is cited as governing.
  IBC Table 1406.2.1.2, "Minimum Fire Separation for Combustible Veneers," is cited as listing radiant heat tolerance required vs. fire separation distance.

1406 Combustible Materials on the Exterior Side of Exterior Walls

1406.2.2 Type I, II, III and IV construction

- This section applies to the following construction types:
  - Types I, II, III, IV.
- Exterior wall coverings may be of the following materials where the conditions indicated below apply:
  - Materials:
    - Wood.
    - Materials of equivalent combustibility.
  - Conditions:
    - Materials must meet code requirements for wood veneers.

Note: 1405.5, "Wood veneers," is cited as governing wood and materials of equivalent combustibility.

Other than fire-retardant treated wood the materials are governed as follows:
  Limited to ≤ 10% of the exterior wall surface in the following case:
    Where the fire separation distance is ≤ 5'.

- Combustible architectural trim is limited to 40' above grade plane.
- Fire-retardant treated wood in exterior applications is governed as follows:
  - It is limited to 60' above grade plane at any fire separation distance.
  - It is not limited in area at any fire separation distance.

Note: 2303.2, "Fire-retardant-treated wood," is cited as governing this material for exterior use.

1406.2.3 Location

- Combustible wall covering may not extend above the exterior wall.
- Combustible wall covering may not extend over the top of the exterior wall.
- Combustible wall covering at the top of an exterior wall is governed as follows:
  - The wall must be behind the combustible covering at every point.

1406.2.4 Fireblocking

- Furred space formed by combustible coverings on exterior walls is governed as follows:
  - The dimension between the back of the wall covering and the wall must be ≤ 15/8".
  - The furred space must be fireblocked in certain cases.

Note: Section 717, "Concealed Spaces," is cited specifying where the furred space must be fireblocked

1406.3 Balconies and similar projections *(part 1 of 2)*

- This section addresses the following elements where they are combustible construction:
  - Balconies and similar projections from buildings.
- Fire-retardant-treated wood may be used as follows:
  - For the following elements in construction Type I and Type II, where the conditions below apply:
    - Elements:
      - Balconies.
      - Porches.
      - Decks.
      - Exterior stairways.
    - Conditions:
      - Building must be ≤ 3 stories above grade plane.
      - Elements may not be required exits.

1406 Combustible Materials on the Exterior Side of Exterior Walls

1406.3 Balconies and similar projections *(part 2 of 2)*

- Wood that is not fire-retardant-treated may be used in guardrail systems that are ≤ 42" in height as follows:
  - As pickets.
  - As rails.
  - For similar components.
- Balconies and similar projections on buildings of the following construction types have requirements as indicated below:
  - Construction types:
    - III, IV, and V.
  - Requirements:
    - They may be Type V construction.
    - They are not required to have a fire-resistance rating as follows:
      - Where the element is protected by sprinklers.
- In other cases, balconies and similar projections are governed as follows:
  - They must comply with one of the following:
    - Must have the fire-resistance rating required floors in the building construction type.
    - Must be Type IV construction.
  - The total length of these on a building is limited as follows:
    - Must be ≤ 1/2 the building perimeter at each floor.

Note: IBC Table 601, "Fire-Resistance Rating Requirements for Building Elements," is cited as the source of such requirements for the floor construction noted above.
602.4, "Type IV," is cited as the source of requirements for Type IV construction.

1406.4 Bay windows and oriel windows

- The following windows are governed as indicated below:
  - Windows:
    - Bay windows.
    - Oriel windows.
  - Requirements:
    - Window assemblies may be fire-retardent-treated wood as follows:
      - In buildings ≤ 3 stories in the following types of construction:
        - Type I, Type II, Type III, and Type IV.
    - In other cases the window assemblies must conform to the following:
      - The type of construction in which they are installed.

Case study: Fig. 1406.3. Type V construction is permitted for the balcony of the Type IIIB building. The balcony must meet the fire-resistance requirements for floor construction as specified in IBC Table 601 for Type IIIB construction for which there is no fire-resistance-rating required. Wood that is not fire-retardant treated is permitted for the guardrail system. The balcony length is 8% of the building perimeter, which is less than the 50% maximum. The balcony is in compliance with the code.

1406 Combustible Materials on the Exterior Side of Exterior Walls

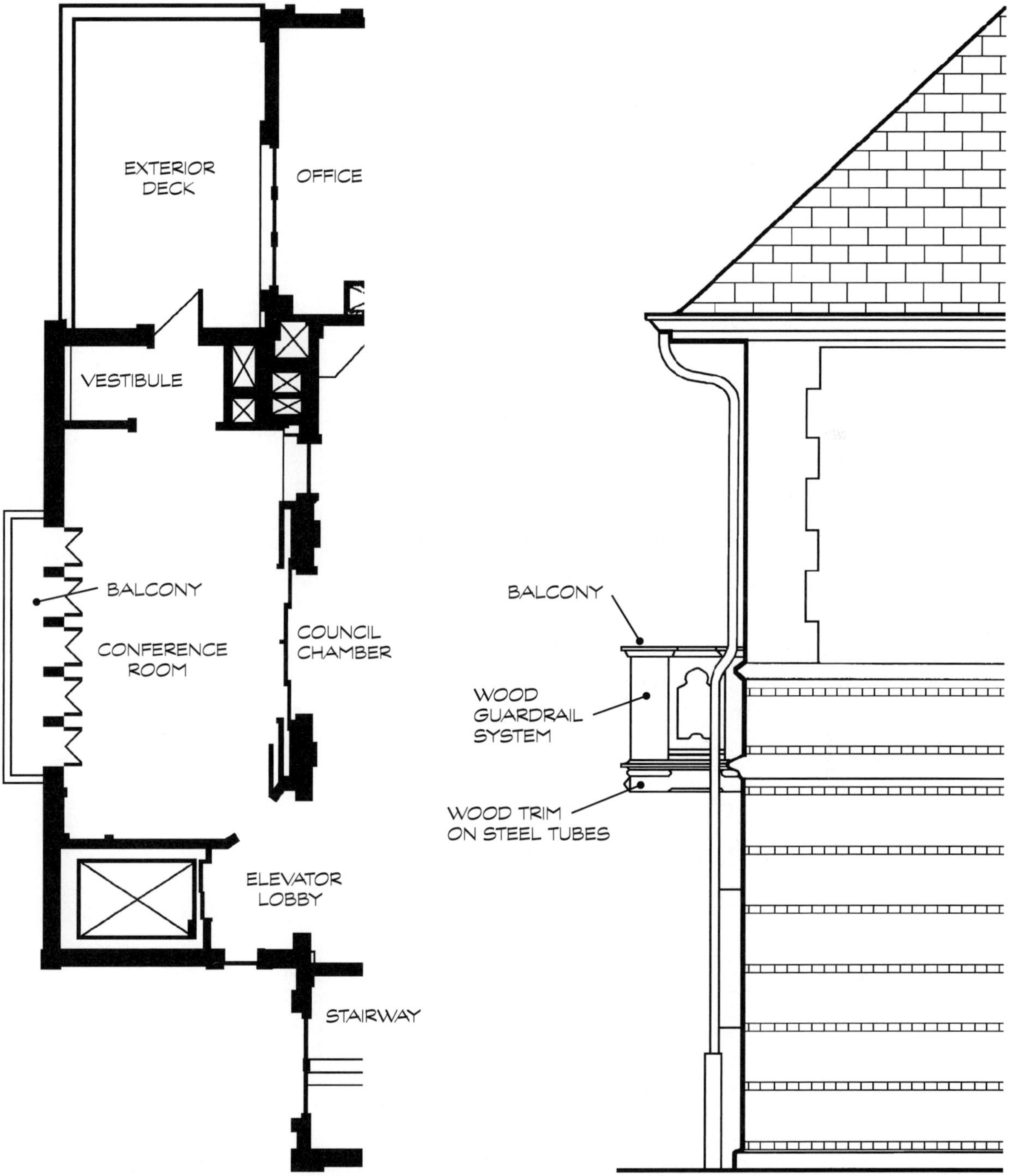

Fig. 1406.3. Partial elevation and plan at balcony. Lake Forest City Hall Renovation and Addition. Lake Forest, Illinois. David Woodhouse Architects. Chicago, Illinois.

NOTES

15

Roof Assemblies and Rooftop Structures

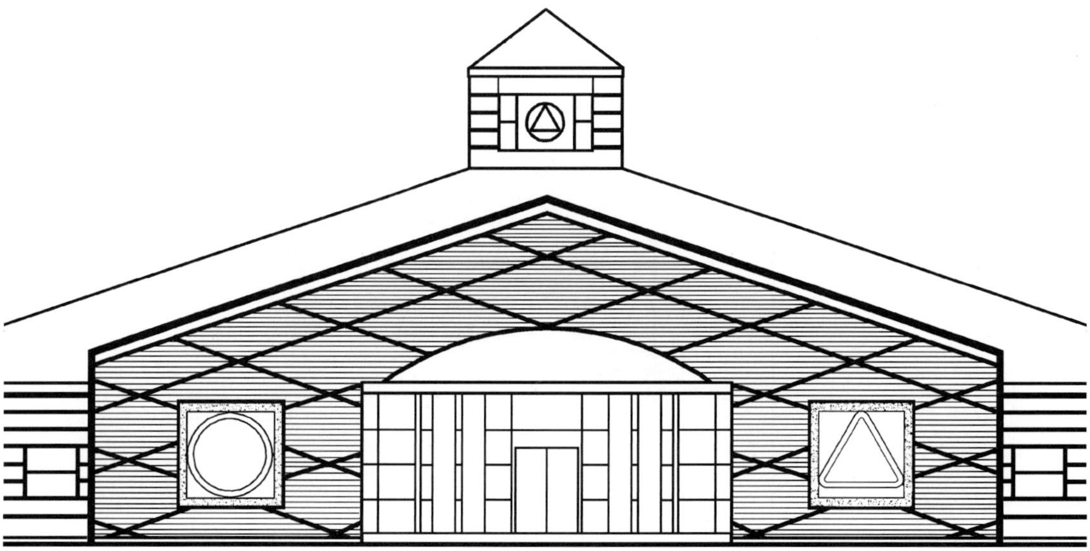

New Jasper Pre-K–2nd Grade School. Jasper, Texas. *(partial elevation)*
PBK Architects, Inc. Houston, Texas.

1503 Weather Protection

1503.2 Flashing

- Flashing is required to stop water penetration as follows:
  - Through the wall and roof at the following locations:
    Joints in copings.
    Through materials permeable to moisture.
    At parapet intersections with the roof.
    At other penetrations through the roof.

1503.2.1 Locations

- Flashing must be installed at the following locations:
  - At intersections of walls and roofs.
  - At gutters.
  - At changes in roof slope.
  - At changes in roof direction.
  - Around openings in the roof.
- Where flashing is metal, it must be as follows:
  - Corrosion resistant.
  - ≥ 0.019" thick.

Case study: Fig. 1503.2.1. Flashing is installed at the intersection of the wall and the roof as required by the code. It meets corrosion resistance and thickness requirements.

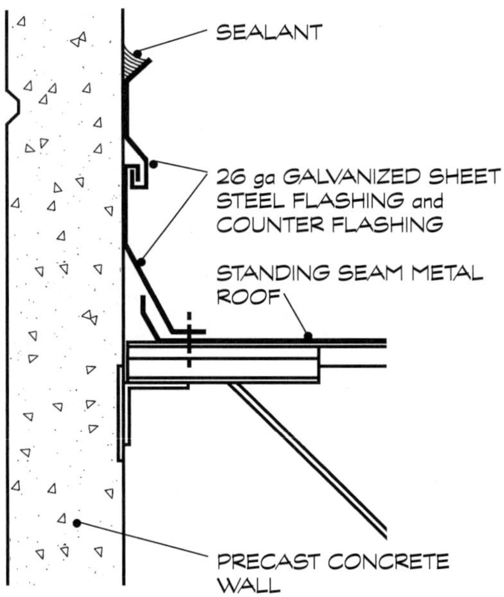

Fig. 1503.2.1. Wall detail. Garments to Go. Bastrop, Texas. Spencer Godfrey Architects. Round Rock, Texas.

1503 Weather Protection

1503.3 Coping

- Parapet walls must be coped with materials as follows:
  - Noncombustible.
  - Weatherproof.
  - With a width ≥ the thickness of the parapet.

1503.4.1 Secondary drainage required

- This section governs drainage for the following roofs:
  - Where roof perimeter construction extends above the roof as follows:
    - So as to trap water if the primary roof drains fail to drain properly.
- Such roofs require one of the following:
  - A secondary roof drain system.
  - Scuppers.

1503.4.2 Scuppers

- This section governs scuppers used as an emergency roof drain system.
- Scuppers must be designed as follows:
  - To prevent a water-ponding depth beyond the capability of the roof design as follows:
    - The following details must be designed to prevent excessive ponding during emergency use:
      - Quantity.
      - Size.
      - Location.
      - Inlet elevation.

Note: 1503.4.1, "Secondary drainage required," governs the design of roof drainage.

  - Scupper opening must be ≥ 4".
  - Scupper design must be without consideration of the drainage capability of the primary system.

1503.4.3 Gutters

- This section does not apply to the following:
  - Occupancy R-3.
  - Private garages.
  - Buildings of Type IV construction.
- Gutters and leaders must be one of the following materials:
  - Noncombustible.
  - Schedule 40 plastic pipe.

1503.6 Crickets and saddles

- A chimney or roof penetration requires a cricket in the following location:
  - On the side of the chimney or penetration where the roof is higher as follows:
    - Where the chimney or roof penetration is ≥ 2'-6" wide measured ⊥ to the slope.
  - Coverings must be one of the following:
    - Sheet metal.
    - Same material as roofing.

1503 Weather Protection

Case study: Fig. 1503.4.3. The gutters and leaders on this building are metal and, thus, meet the code requirements for these components by being noncombustible.

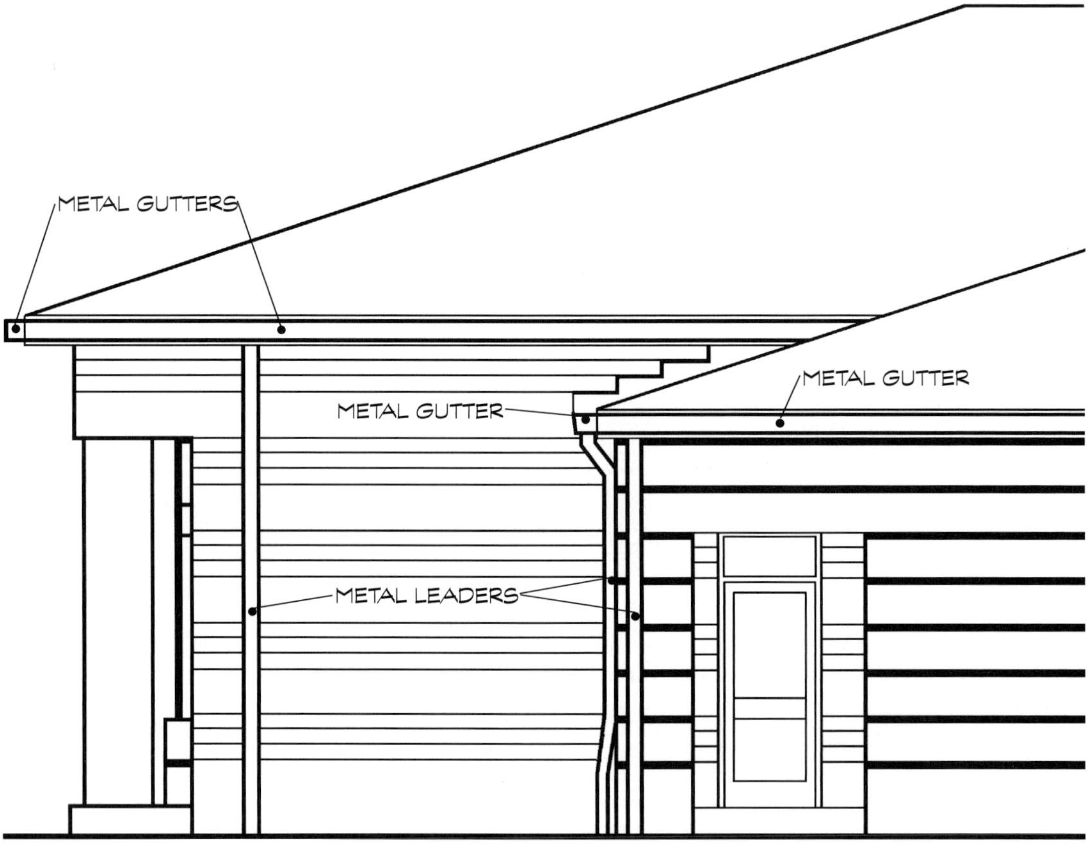

Fig. 1503.4.3. Partial elevation. New Jasper Pre-K–2nd Grade School. Jasper, Texas. PBK Architects, Inc. Houston, Texas.

1505 Fire Classification

1505.2 Class A roof assemblies

- Class A roofs may be used in all types of construction.
- Class A roofs include the following materials:
  - Brick.
  - Masonry.
  - Slate installed on noncombustible decks.
  - Clay roof tile installed on noncombustible decks.
  - Concrete roof tile installed on noncombustible decks.
  - Exposed concrete roof deck.
  - Ferrous shingles installed on noncombustible decks.
  - Copper shingles installed on noncombustible decks.
  - Metal shingles installed on noncombustible decks.
  - Ferrous sheet installed on noncombustible decks.
  - Copper sheet installed on noncombustible decks.
  - Metal sheet installed on noncombustible decks.
  - Ferrous sheet installed on noncombustible framing with no deck.
  - Copper sheet installed on noncombustible framing with no deck.
  - Metal sheet installed on noncombustible framing with no deck.
  - Other materials which perform as follows:
    They must be listed by an approved testing agency.
    They must be identified as Class A by an approved testing agency.
    They must be effective against severe fire testing.

1505.3 Class B roof assemblies

- Class B roof assemblies perform as follows:
  - Effective against moderate fire testing.
- Class B assemblies and roof coverings must have the following characteristics:
  - They must be listed by an approved testing agency.
  - They must be identified as Class B by an approved testing agency.

1505.4 Class C roof assemblies

- Class C roof assemblies perform as follows:
  - Effective against light fire testing.
- Such assemblies and roof coverings must have the following characteristics:
  - They must be listed by an approved testing agency.
  - They must be identified as Class C by an approved testing agency.

1505.5 Nonclassified roofing

- Materials with all of the following characteristics are designated as nonclassified roofing:
  - Roofing materials that are approved.
  - Roofing materials that are not classified as one of the following:
    Class A.
    Class B.
    Class C.

1505 Fire Classification

1505.6 Fire-retardant-treated wood shingles and shakes

- The following is required for fire-retardant-treated wood shingles and shakes:
  - They must be treated as follows:
    - By impregnation with chemicals using the following method:
      - Full-cell vacuum-pressure process.
      - As specified in the applicable standard.

  Note: AWPA C1, "All Timber Products—Preservative Treatment by Pressure Processes," is cited as the standard governing the process.

  - Bundles of the products must be labeled with the following information:
    - The product identification.
    - The manufacturer.
    - The company that treated the product.
    - The quality control agency.
    - The classification of the product as follows:
      - As established by fire tests.

  Note: 1505.1, "General," is cited as listing the necessary testing.

1505.7 Special purpose roofs

- The following special-purpose roofing is governed as listed below:
  - Roofing:
    - Wood shingles.
    - Wood shakes.
  - Requirements:
    - One of the following sheathings is required under the roofing:
      - ≥ 1/2" wood structural panel solid sheathing.
      - ≥ Nominal 1" boards spaced sheathing.
    - Gypsum backing board or sheathing is required under the wood sheathing as follows:
      - ≥ 5/8" Type X.
      - Water resistant.

  Note: The following are cited as sources of requirements for special purpose roofing:
  - *1507.8, "Wood shingles."*
  - *1507.9, "Wood shakes."*

1507 Requirements for Roof Coverings

1507.2.1 Deck requirements

- Asphalt shingles must be secured to the following:
  - Solidly sheathed deck.

Case study: Fig. 1507.2.1. A solid deck is required for the asphalt shingles on the roof of the example below. Such a deck is provided. The minimum slope of the shingles is 2:12. This roof slopes 3:12 and, thus, meets this code requirement. Since the slope is < 4:12, double underlayment is provided as required.

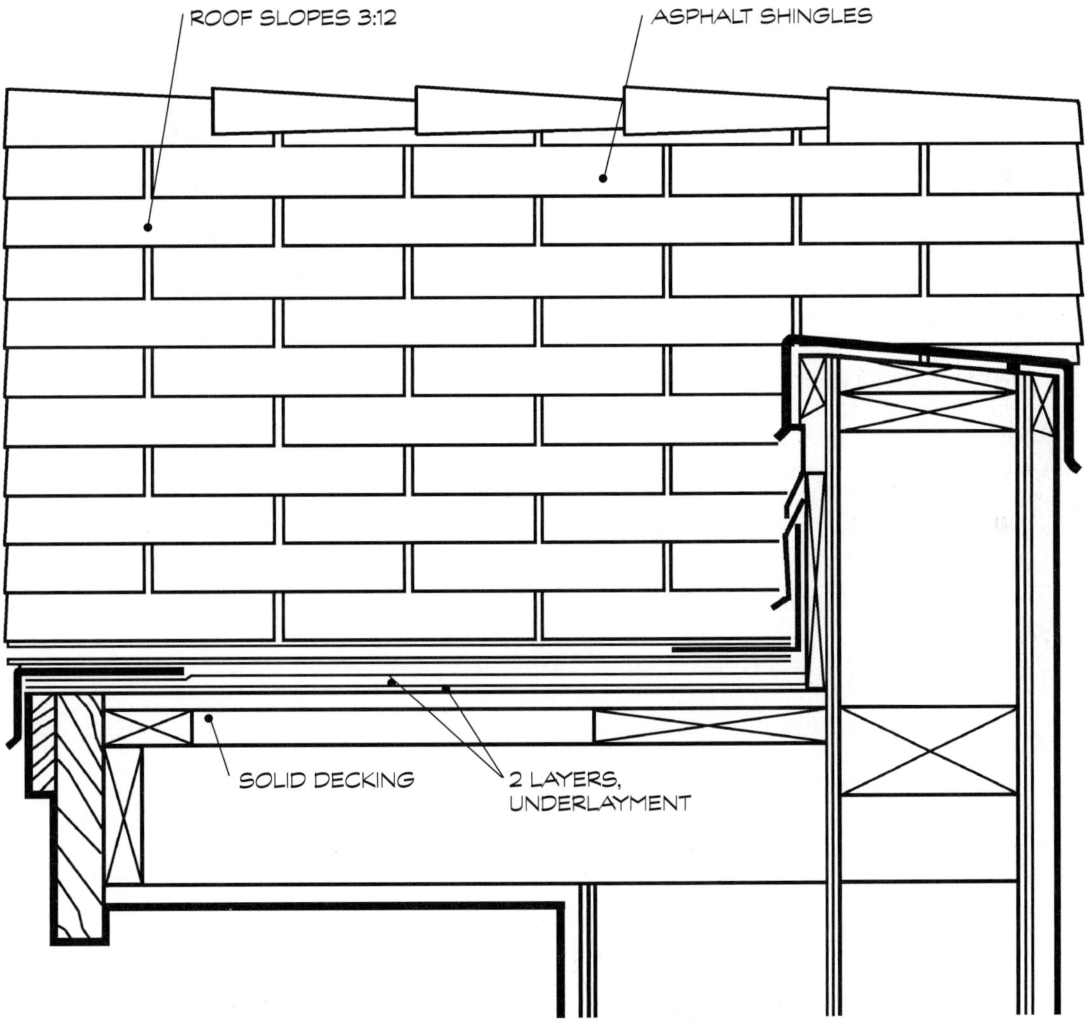

Fig. 1507.2.1. Detail at roof. Creston Elementary Multipurpose Building. Creston, California. Phillips Metsch Sweeney Moore Architects. Santa Barbara, California.

1507 Requirements for Roof Coverings

1507.2.2 Slope

- The slope required for asphalt shingles is as follows:
  - ≥ 2:12 slope.
- Double underlayment is required for asphalt shingles at the following range of slopes:
  - Slopes ≥ 2:12 and < 4:12.

Note: 1507.2.8, "Underlayment application," is cited as governing the application of two layers of underlayment.

1507.2.8 Underlayment application

- This section applies to asphalt shingles.
- For roof slopes ≥ 2:12 and < 4:12 underlayment is required as listed:
  - 2 layers must be applied as follows:
    - Apply ≥ 19" wide underlayment felt strip in the following way:
      - At the eaves.
      - || to the eaves.
    - Apply sheets of underlayment as follows:
      - 36" wide sheets in the following manner:
        - Starting at the eaves.
        - With 19" overlaps.
- For roof slopes ≥ 4:12 underlayment is required as indicated below:
  - Apply one layer in shingle format as follows:
    - Starting at the eaves.
    - With 2" overlaps.
- Any distortions in the underlayment must not prevent the shingles from sealing.

1507.2.8.2 Ice barrier

- This section applies to asphalt shingles.
- This section does not apply to accessory buildings without conditioned space.
- For the following condition, the underlayment indicated below is required:
  - Condition:
    - Where there is a history of ice forming along the eaves and accumulating water.
  - Underlayment:
    - One of the following ice barriers is required in lieu of standard underlayment:
      - ≥ 2 layers of underlayment cemented together.
      - Self-adhering polymer modified bitumen sheet.
    - Ice barriers must be applied as follows:
      - Barrier extends to the lowest edges of all roof surfaces.
      - Barrier extends over the interior of the building as follows:
        - ≥ 2' inside the exterior wall.

1507 Requirements for Roof Coverings

1507.2.9.1 Base and cap flashing

- This section applies to asphalt shingles.
- Base and cap flashing are required as follows:
  - Installed as per manufacturer's instructions.
  - Base flashing must be one of the following:
    - Corrosion-resistant metal ≥ nominal 0.019" thick.
    - Mineral-surfaced roll roofing ≥ 77 lbs/100 sf.
  - Cap flashing must be the following:
    - Corrosion-resistant metal ≥ nominal 0.019" thick.

1507.2.9.2 Valleys

- Valley linings for asphalt shingles are required as follows:
  - Installed as per manufacturer's instructions.
  - Installed prior to shingles.
  - The materials listed below are acceptable for the following valley types:
    - Valleys:
      - Where valley lining is exposed.
      - Where valley lining is covered with shingles.
    - Materials:
      - Metal as follows:
        - ≥ 2' wide.
        - Any of the following are permitted:
          - Copper.
          - Aluminum.
          - Stainless steel.
          - Painted terne.
          - Lead.
          - Zinc alloy.
          - Galvanized steel.
      - Mineral surface roll roofing as follows:
        - 2 plies required.
        - Bottom layer to be ≥ 1'-6" wide.
        - Top layer to be ≥ 3' wide.
- The following material may be used as a valley lining where the valley lining is covered with shingles:
  - 1-ply smooth roll roofing ≥ 3' wide.
  - Self-adhering polymer modified bituminous sheet materials.

Note: The following are cited as sources of applicable requirements:

IBC Table 1507.2.9.2, "Valley Lining Material," lists acceptable valley lining metals and metal thickness.

ASTM D 6380, "Specification for Asphalt Roll Roofing (Organic Felt)," as one alternative for smooth roll roofing.

ASTM D 3909, "Specification for Asphalt Roll Roofing (Glass Felt) Surfaced with Mineral Granules," as an alternative for smooth roll roofing.

ASTM D 1970, "Standard Specification for Self-Adhering Polymer Modified Bituminous Sheet Materials Used as Steep Roof Underlayment for Ice Dam Protection."

1507 Requirements for Roof Coverings

1507.2.9.3 Drip edge

- A drip edge for asphalt shingles is required on shingled roofs at the following locations:
  - Eaves.
  - Gables.
- A drip edge must be detailed as follows:
  - Extend ≥ 2" onto the roof under the shingles.
  - Extend downward past the bottom edge of sheathing ≥ 1/4".
  - Attached with mechanical fasteners ≤ 12" on center.

1507.3.1 Deck requirements

- The following roofing must be installed on one of the sheathing materials listed:
  - Roofing:
    - Concrete tile.
    - Clay tile.
  - Sheathing:
    - Solid sheathing.
    - Spaced structural sheathing boards.

1507.3.2 Deck slope

- Clay and concrete roof tile must be installed on slopes as follows:
  - ≥ 21/2:12 slope.
- Slopes < 4:12 require the following:
  - Double underlayment.

Note: 1507.3.3, "Underlayment," is cited as the source providing requirements applicable to double underlayment.

1507.3.3.1 Low-slope roofs

- This section addresses clay and concrete tile roofing.
- ≥ 2 layers of underlayment are required for roofs as follows:
  - Where the slope is 21/2:12 ≤ slope < 4:12, the following is required:
    - First strip of underlayment:
      - ≥ 1'-7" wide.
      - Applied || to and meeting the eave.
    - Second strip of underlayment:
      - ≥ 3' wide.
      - Applied over the 1st strip meeting the eave.
    - Subsequent strips of underlayment:
      - Applied overlapping previous strips as follows:
        - ≥ 1'-7" overlap.

1507.3.3.2 High-slope roofs

- This section addresses clay and concrete tile roofing.
- Roof slopes ≥ 4:12 require the following underlayment:
  - 1 layer of underlayment felt as follows:
    - First strip applied || to and meeting the eaves.
    - Subsequent strips lap previous strips ≥ 2".

1507 Requirements for Roof Coverings

1507.3.9 Flashing

- This section addresses flashing for clay and concrete tile roofing.
- Flashing and counterflashing is required as follows:
  - At the intersection of the roof with vertical surfaces.
- Where flashing is metal, the following is required:
  - Corrosion-resistant metal ≥ 0.019" thick.
- Flashing must be installed according to manufacturer's instructions.
- Valley flashing is governed as follows:
  - It must extend ≥ 11" into each side of the valley centerline.
  - It must have a splash diverter rib ≥ 1" high as follows:
    - At the centerline of the valley.
    - Integrally formed with the flashing.
- Flashing sections must overlap each other ≥ 4" where ends meet.
- For roof slopes ≥ 3:12, valley flashing is governed as follows:
  - The following valley underlayment is required:
    - ≥ 3' wide.
    - One of the following:
      - Type I underlayment.
      - Self-adhering polymer modified bitumen sheet.
    - Extending the full length of the valley.
    - Provided in addition to other underlayment required for the roof.

Note: ASTM D 1970, "Standard Specification for Self-Adhering Polymer Modified Bituminous Sheet Materials Used as Steep Roof Underlayment for Ice Dam Protection," is cited as governing polymer modified sheet products.

- Where all of the following conditions exist, the requirements listed below apply:
  - Conditions:
    - Roof slope is < 7:12.
    - Where one of the following applies:
      - Average daily January temperature is ≤ 25° F.
      - Where ice can form along the eaves to accumulate water.
  - Requirement:
    - The valley flashing underlayment must comply with one of the following:
      - It is cemented across its entire surface to the roofing underlayment.
      - Self-adhering polymer modified bitumen sheet is used for the underlayment.

1507.4.1 Deck requirements

- Metal roof panels must be applied to one of the following types of decks:
  - Solid deck.
  - Closely fitted deck.
  - Spaced decking as follows:
    - Where the metal roofing is specifically designed for such application.

1507 Requirements for Roof Coverings

1507.4.2 Deck slope

- The following metal roofing requires a slope ≥ 3:12:
  - Nonsoldered seams.
  - Lapped sections.
  - No lap sealant.
- The following metal roofing requires a slope ≥ 1/2:12:
  - Nonsoldered seams.
  - Lapped sections.
  - With lap sealant as follows:
    Applied according to approved manufacturer's instructions.
- The following metal roofing requires a slope ≥ 1/4:12:
  - Standing seam.

1507.5.1 Deck requirements

- Metal roof shingles must be applied to one of the following types of decks:
  - Solid deck.
  - Closely fitted deck.
  - Spaced decking as follows:
    Where the metal shingles are specifically designed for such application.

1507.5.2 Deck slope

- Metal roof shingles must slope ≥ 3:12.

1507.5.4 Ice barrier

- This section governs roof eaves where ice tends to accumulate causing the following:
  - Water backup.
- This section does not govern accessory structures with both of the following characteristics:
  - Detached.
  - Containing no conditioned space.
- At such eaves, an ice barrier is required as follows:
  - One of the following must be substituted for normal underlayment:
    ≥ 2 layers of underlayment cemented together.
    Self-adhering polymer-modified bituminous sheet.
  - Barrier must extend between the following:
    Lowest edges of all roof surfaces.
    A line ≥ 2' inside the exterior wall.

1507.5.7 Flashing *(part 1 of 2)*

- This section addresses metal roof shingles.
- Valley flashing must be one of the following materials:
  - Corrosion-resistant metal, same as the covering.
  - Other metal.

Note: IBC Table 1507.4.3(1), "Metal Roof Coverings," is cited as listing other acceptable metals for the valley flashing.

1507 Requirements for Roof Coverings

Case study: Fig. 1507.4.2A. The batten seam metal roofing requires a minimum slope of 3:12 according to manufacturer's recommendations and code requirements. The actual slope of the roofs is 4:12; thus, the roofing is in compliance with the code.

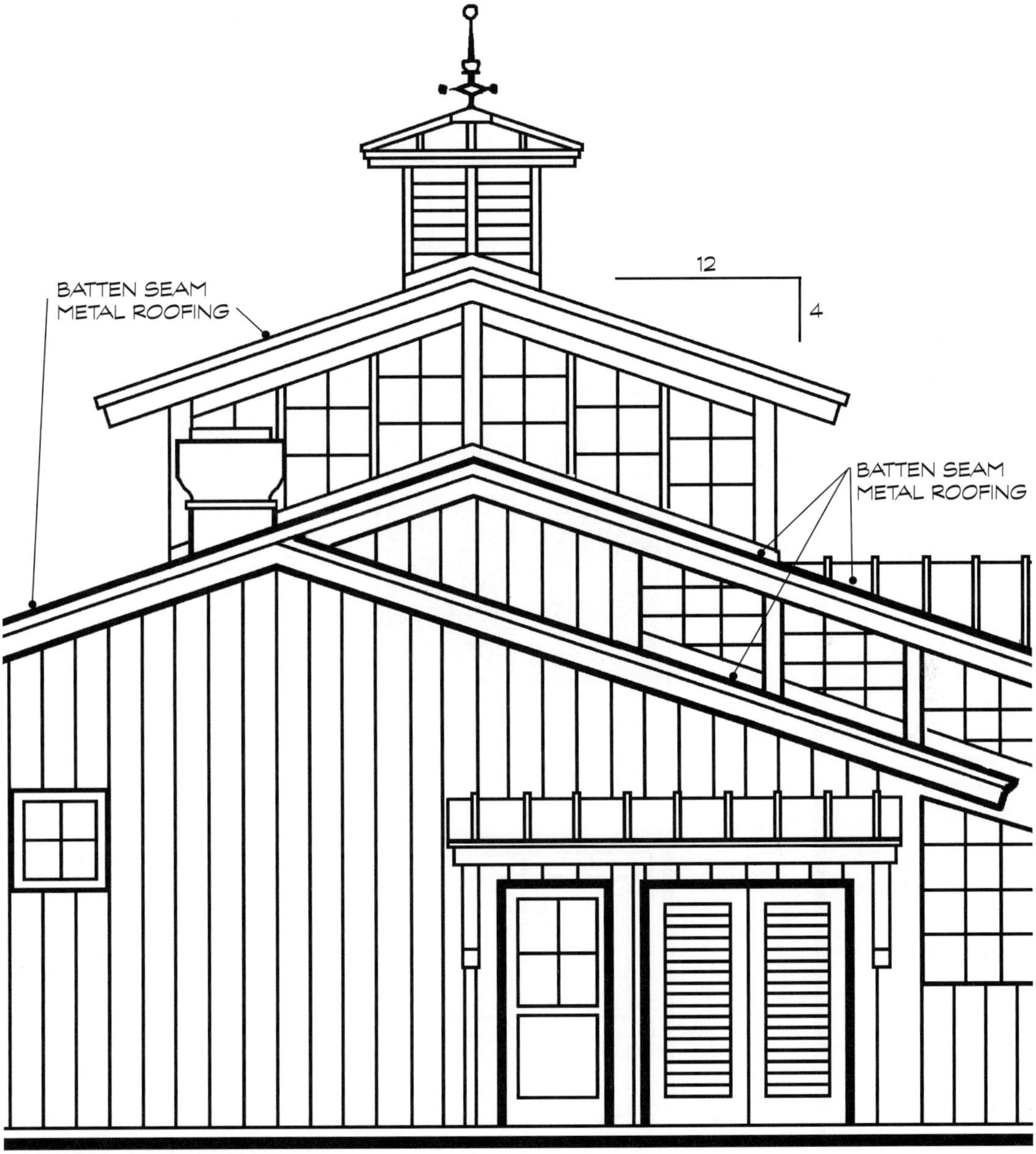

Fig. 1507.4.2A. Partial elevation. Central Kitchen. Lompoc Unified School District. Lompoc, California. Phillips Metsch Sweeney Moore Architects. Santa Barbara, California.

1507 Requirements for Roof Coverings

Case study: Fig. 1507.4.2B. The standing seam roof on the cupola of the tower slopes 2:12, which is > 1/4:12, the minimum required slope. The standing seam roof is, therefore, in compliance with code requirements for slope.

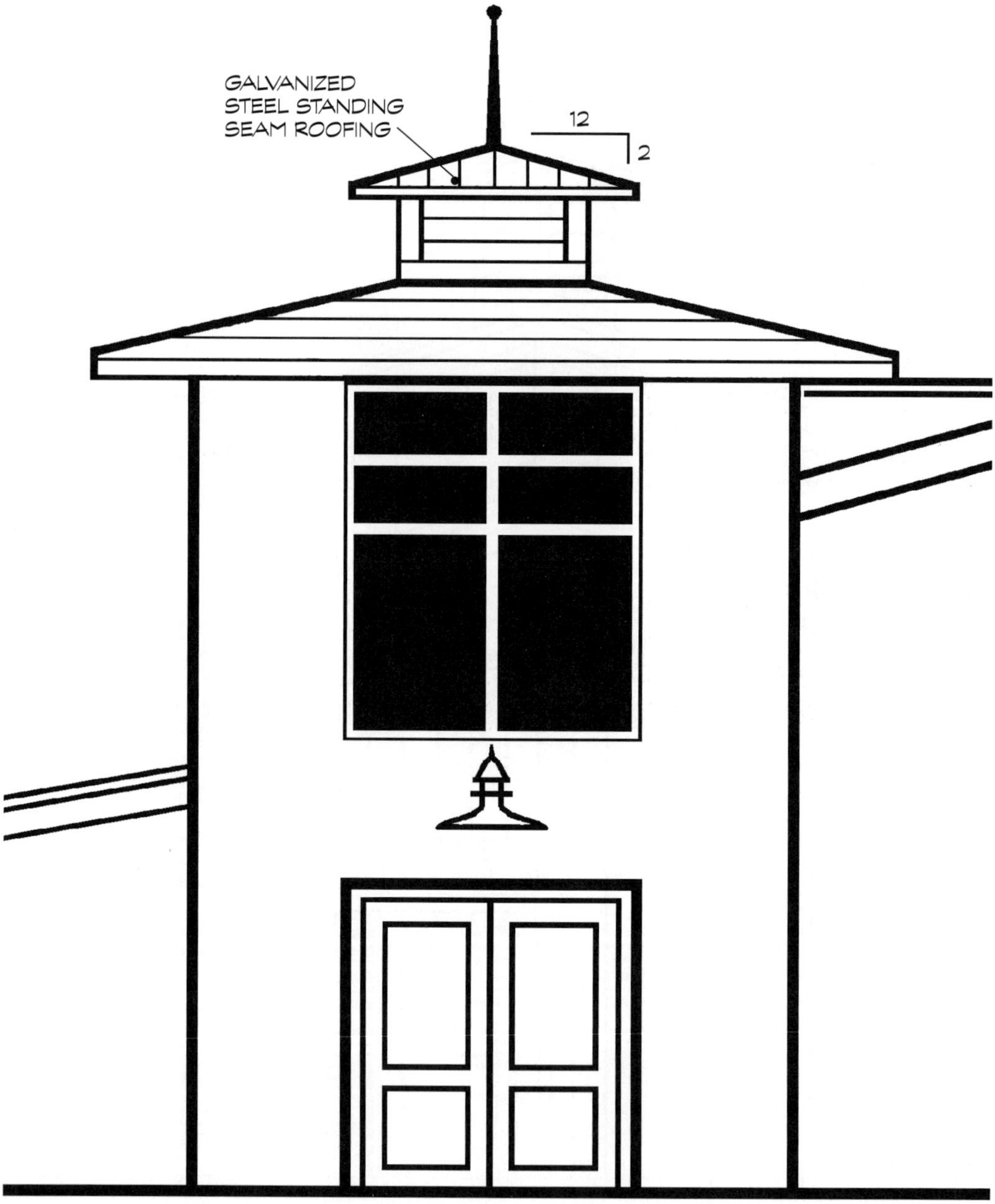

Fig. 1507.4.2B. Partial elevation at tower. Creston Elementary Multipurpose Building. Creston, California. Phillips Metsch Sweeney Moore Architects. Santa Barbara, California.

1507 Requirements for Roof Coverings

1507.5.7 Flashing *(part 2 of 2)*

- Valley flashing must be detailed as follows:
  - Extend ≥ 8" on both sides of the valley center line.
  - Have a splash diverter rib ≥ 3/4" high as follows:
    - At the center line of the valley.
    - Formed as an integral part of the flashing.
  - Flashing sections must lap ≥ 4" where ends meet.
- Where either of the following conditions exist, the requirements listed below apply:
  - Conditions:
    - Average daily January temperature is ≤ 25° F.
    - Where ice can form along the eaves to accumulate water.
  - Requirement:
    - The following valley lining is required under the flashing:
      - ≥ 3' wide extending the full length of the valley.
      - One of the following materials is required:
        - Underlayment.
        - Self-adhering polymer modified bitumen sheet.
      - Provided in addition to other underlayment required for the metal roof shingles.
    - Where the roof slope is < 7:12, the following is required:
      - The valley lining for the metal valley flashing must be one of the following:
        - Underlayment cemented across its entire surface to the roofing underlayment.
        - Self-adhering polymer modified bitumen sheet.

Note: ASTM D 1970, "Specification for Self-Adhering Polymer Modified Bituminous Sheet Materials Used as Steep Roof Underlayment for Ice Dam Protection," is cited as governing polymer modified sheet products.

1507.6.1 Deck requirements

- Mineral-surfaced roll roofing must be applied to a solidly sheathed surface.

1507.6.2 Deck slope

- Mineral-surface roll roofing must be applied to a slope ≥ 1:12.

1507.6.4 Ice barrier

- This section governs roof eaves where ice tends to accumulate causing the following:
  - Water backup.
- This section does not govern accessory structures with both of the following characteristics:
  - Detached.
  - Containing no conditioned space.
- At such eaves an ice barrier is required as follows:
  - One of the following must be substituted for normal underlayment:
    - ≥ 2 layers of underlayment cemented together.
    - Self-adhering polymer-modified bituminous sheet.
  - Barrier must extend between the following:
    - Lowest edges of all roof surfaces.
    - A line ≥ 2' inside the exterior wall.

1507 Requirements for Roof Coverings

1507.7.1 Deck requirements

- Slate shingles must be applied to a solidly sheathed surface.

1507.7.2 Deck slope

- Slate shingles must be applied to a slope ≥ 4:12.

1507.7.4 Ice barrier

- This section governs roof eaves where either of the following conditions exist:
  - Where the daily temperature in January averages ≤ 25 °F.
  - Where this is a possibility of ice accumulation causing water backup.
- This section does not govern accessory structures with both of the following characteristics:
  - Detached.
  - Containing no conditioned space.
- At such eaves an ice barrier is required as follows:
  - One of the following must be substituted for normal underlayment:
    - ≥ 2 layers of underlayment cemented together.
    - Self-adhering polymer-modified bituminous sheet.
  - Barrier must extend between the following:
    - Lowest edges of all roof surfaces.
    - A line ≥ 2' inside the exterior wall.

1507.7.6 Application

- Slate shingles must be applied to the roof as follows:
  - With 2 fasteners for each piece.
- Slate shingles must have the headlap shown in the details provided for this section.

Note: IBC Table 1507.7.6, "Slate Shingle Headlap," is cited as listing headlap requirements and is the source of data provided in Fig. 1507.7.6.

1507.7.7 Flashing

- Flashing and counterflashing for slate shingles must be of the following material:
  - Sheet metal as follows:
    - ≥ 0.0179" uncoated thickness zinc coated G90.
- Valley flashing must be sheet metal as follows:
  - ≥ 0.0179" uncoated thickness zinc coated G90.
  - ≥ 1'-3" wide.
- Intersections between the roof and the following vertical surfaces require the cap flashing as noted:
  - Vertical surfaces:
    - Chimneys.
    - Stucco walls.
    - Brick walls.
  - Cap flashing:
    - Two plies of felt are required as follows:
      - 4" wide strips of felt set in plastic cement.
      - Top felt extends 1" above the bottom felt.
      - Top coating of plastic cement.
      - Felts overlaps the base flashing 2".

1507 Requirements for Roof Coverings

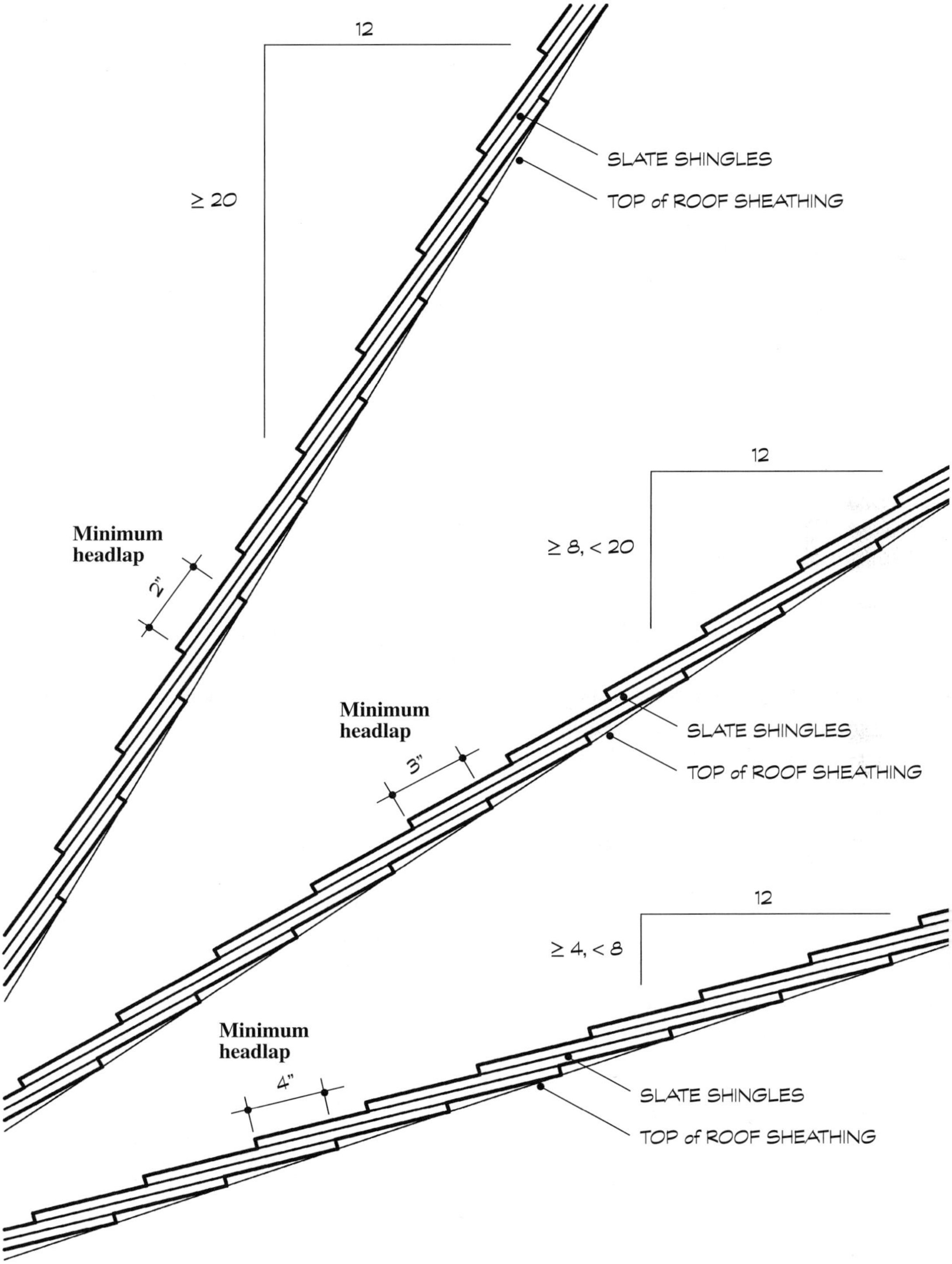

Fig. 1507.7.6. Minimum headlap for slate shingles. *[Source: IBC Table 1507.7.6.]*

1507 Requirements for Roof Coverings

1507.8.1 Deck requirements

- Wood shingles must be applied to one of the following:
  - Solidly sheathed surface.
  - Spaced sheathing as follows:
    - ≥ nominal 1"× 4" boards are required.
    - Spacing must equal the following:
      - Length of shingle exposed to the weather.

1507.8.1.1 Solid sheathing required

- Wood shingles require solid sheathing in either of the following cases:
  - Where average daily temperature in January is ≤ 25° F.
  - Where ice can form to accumulate water along the eaves.

1507.8.2 Deck slope

- Wood shingles must be applied to the following slope:
  - ≥ 3:12.

1507.8.4 Ice barrier

- This section governs roof eaves where ice tends to accumulate causing the following:
  - Water backup.
- This section does not govern accessory structures with both of the following characteristics:
  - Detached.
  - Containing no conditioned space.
- At such eaves an ice barrier is required as follows:
  - One of the following must be substituted for normal underlayment:
    - ≥ 2 layers of underlayment cemented together.
    - Self-adhering polymer-modified bituminous sheet.
  - Barrier must extend between the following:
    - Lowest edges of all roof surfaces.
    - A line ≥ 2' inside the exterior wall.

1507.8.7 Application

- Naturally durable wood shingles must be installed as follows:
  - With a lap between side joints in adjacent courses as follows:
    - ≥ 1 1/2" lap is required.
  - Shingles in alternate courses may not align.
  - Spacing between adjacent shingles in the same course must be as follows:
    - ≥ 1/2" and ≤ 3/8".
  - The length of shingles that may be exposed to the weather is as follows:
    - As indicated in the details provided for this section.

Note: IBC Table 1507.8.7, "Wood Shingle Weather Exposure and Roof Slope," is cited as governing weather exposure for shingles.

1507 Requirements for Roof Coverings

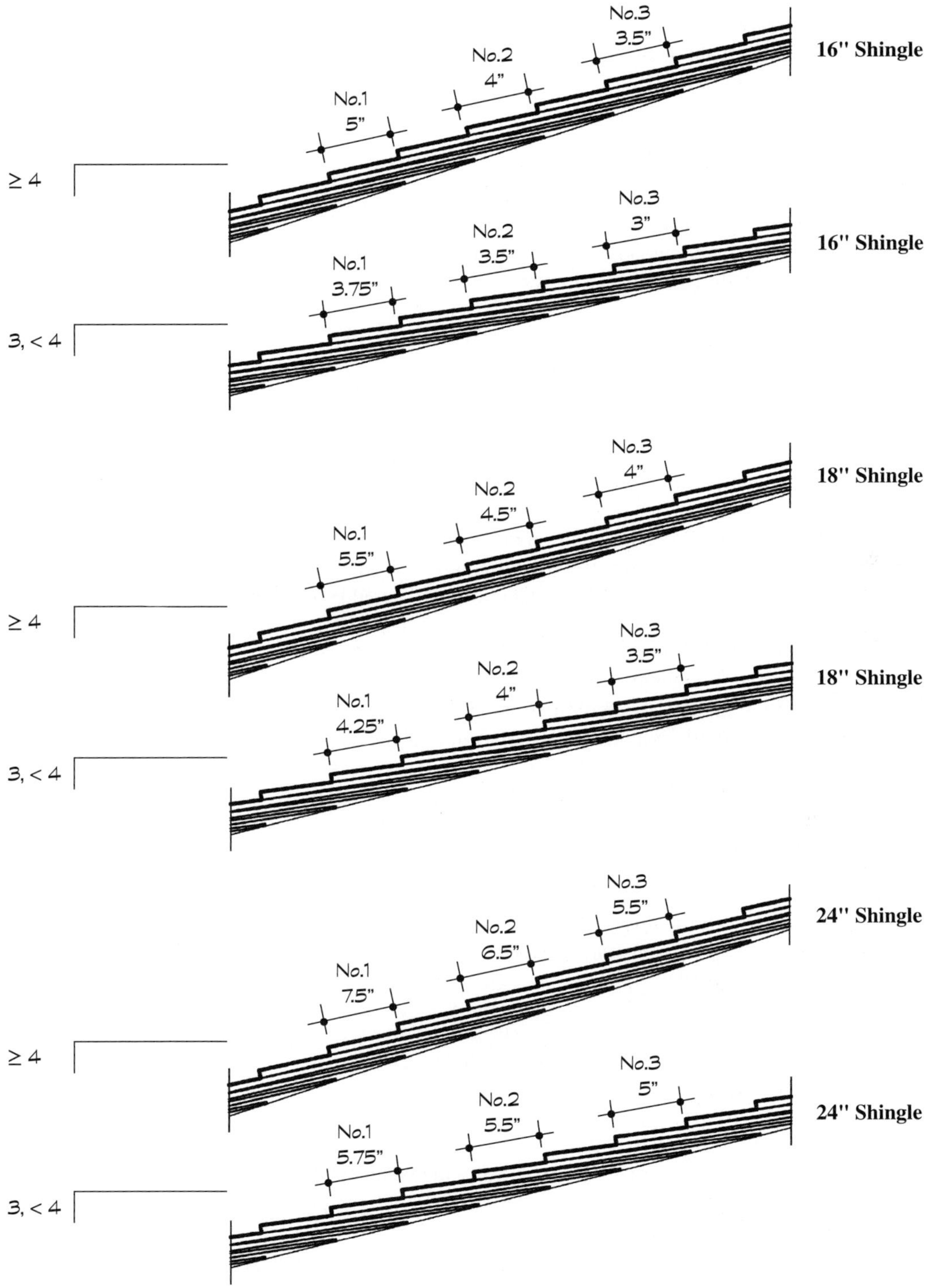

Fig. 1507.8.7. Maximum permitted exposure of wood shingles. *[Source: IBC Table 1507.8.7.]*

1507 Requirements for Roof Coverings

1507.8.8 Flashing

- Wood shingles require flashing and counterflashing as follows:
  - At the intersection of the roof and vertical surfaces.
  - Installed as per manufacturer's instructions.
  - Where flashing is metal, the following is required:
    - Must be corrosion-resistant.
    - Must be ≥ 0.019" thick (26 galvanized sheet gage).
- Valley flashing is governed as follows:
  - It must extend ≥ 11" on either side of the centerline of the valley.
  - It must have a splash diverter rib as follows:
    - ≥ 1" high.
    - Located at the centerline of the valley.
    - Formed as part of the valley flashing.
  - Flashing sections must lap ≥ 4" where ends meet.
  - Valley flashing requires an underlayment as follows:
    - ≥ 3' wide extending the full length of the valley.
    - One of the following materials:
      - Type I underlayment.
      - Self-adhering polymer modified bitumen sheet.
    - Provided in addition to other underlayment required for the roof shingles.
  - Where the following conditions apply, the detailing indicated below is required:
    - Conditions:
      - Roof slope is < 7:12.
      - Either one of the following:
        - Average daily January temperature is ≤ 25° F.
        - Where ice can form along the eaves to accumulate water.
    - Detailing:
      - One of the following is required under the metal valley flashing:
        - Self-adhering polymer bitumen sheet.
        - Valley flashing underlayment as follows:
          - Cemented across its entire surface to the roofing underlayment.

Note: ASTM D 1970, "Specification for Self-Adhering Polymer Modified Bituminous Sheet Materials Used as Steep Roof Underlayment for Ice Dam Protection," is cited as governing polymer modified sheet products.

1507.9.1 Deck requirements

- Wood shakes must be applied to one of the following:
  - Solidly sheathed surface.
  - Spaced sheathing as follows:
    - ≥ nominal 1" × 4" boards are required.
    - Spacing must = the length of shingle exposed to the weather.
- Where 1" × 4" sheathing is spaced 10" on center, the following applies:
  - 1" × 4" boards must be added between them.

1507 Requirements for Roof Coverings

1507.9.1.1 Solid sheathing required

- Wood shakes require solid sheathing in either of the following cases:
  - Where average daily temperature in January is ≤ 25° F.
  - Where ice can form to accumulate water along the eaves.

1507.9.2 Deck slope

- Wood shakes must be applied to the following slopes:
  - ≥ 4:12.

1507.9.4 Ice barrier

- This section governs roof eaves where ice tends to accumulate causing the following:
  - Water backup.
- This section does not govern accessory structures with both the following characteristics:
  - Detached.
  - Containing no conditioned space.
- At such eaves an ice barrier is required as follows:
  - One of the following must be substituted for normal underlayment:
    - ≥ 2 layers of underlayment cemented together.
    - Self-adhering polymer-modified bituminous sheet.
  - Barrier must extend between the following:
    - Lowest edges of all roof surfaces.
    - A line ≥ 2' inside the exterior wall.

1507.9.8 Application

- Wood shakes must be installed as follows:
  - With a lap ≥ 1 1/2" between side joints in adjacent courses.
  - Shakes in alternate courses may not align.
  - Spacing between adjacent shakes in the same course must be as follows:
    - ≥ 3/8" and ≤ 5/8" for the following:
      - Shakes.
      - Taper-sawn shakes of naturally durable wood.
    - ≥ 1/4" and ≤ 3/8" for the following:
      - Taper-sawn shakes that are preservative treated.
  - The length of shakes that may be exposed to the weather is as shown in the details provided for this section.

Note: IBC Table 1507.9.8, "Wood Shake Weather Exposure and Roof Slope," is cited as governing weather exposure and is the source of data shown in Fig. 1507.9.8.

1507 Requirements for Roof Coverings

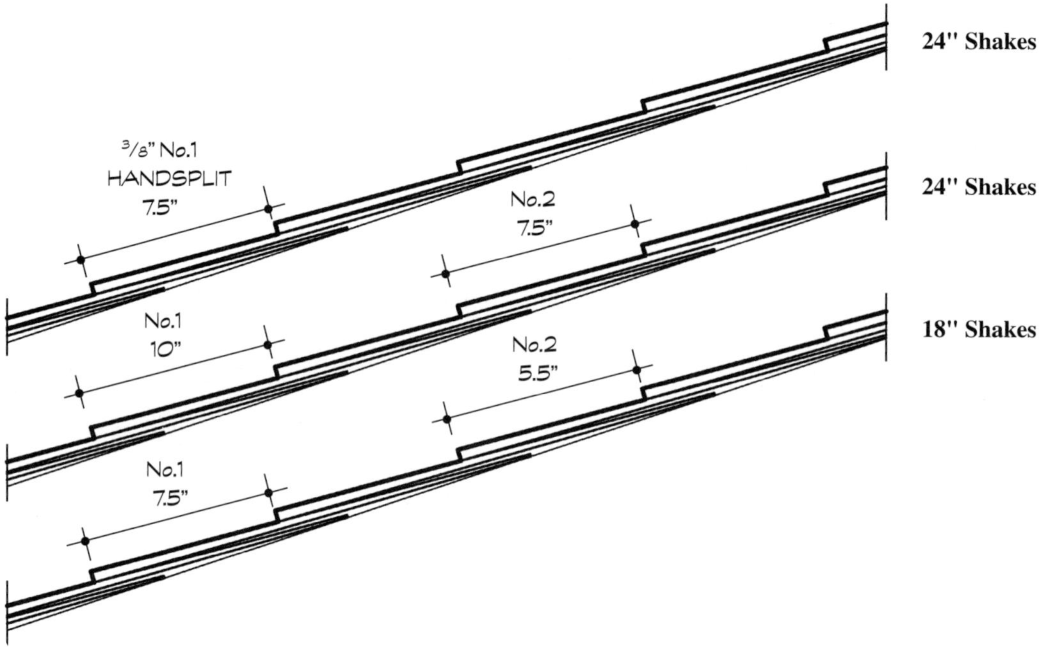

Fig. 1507.9.8. Maximum permitted exposure of wood shakes. *[Source: IBC Table 1507.9.8.]*

1507.9.9 Flashing *(part 1 of 2)*

- Wood shakes require flashing and counterflashing as follows:
  - At the intersection of the roof and vertical surfaces.
  - Installed as per manufacturer's instructions.
  - Where flashing is metal, the following is required:
    - Must be corrosion-resistant.
    - Must be ≥ 0.019" thick (26 galvanized sheet gage).
- Valley flashing for wood shakes is governed as follows:
  - It must extend ≥ 11" on either side of the centerline of the valley.
  - It must have a splash diverter rib as follows:
    - ≥ 1" high.
    - Located at the centerline of the valley.
    - Formed as part of the valley flashing.
  - Flashing sections must lap ≥ 4" where ends meet.
  - Valley flashing requires underlayment as follows:
    - ≥ 3' wide extending the full length of the valley.
    - One of the following materials:
      - Type I underlayment.
      - Self-adhering polymer modified bitumen sheet.
    - Provided in addition to other underlayment required for the roof shingles.

1507 Requirements for Roof Coverings

1507.9.9 Flashing *(part 2 of 2)*

- Where the following conditions apply, the detailing listed below is required:
  - Conditions:
    - Roof slope is < 7:12.
    - Either one of the following:
      - Average daily January temperature is ≤ 25° F.
      - Where ice can form at eaves to accumulate water.
  - Detailing:
    - One of the following is required under the metal valley flashing:
      - Self-adhering polymer bitumen sheet.
      - Valley flashing underlayment as follows:
        - Cemented across its entire surface to the roofing underlayment.

Note: ASTM D 1970, "Specification for Self-Adhering Polymer Modified Bituminous Sheet Materials Used as Steep Roof Underlayment for Ice Dam Protection," is cited as governing polymer modified sheet products.

1507.10.1 Slope

- Coal-tar built-up roofs must have a slope as follows:
  - ≥ 1/8:12.
- Other built-up roofs must have a slope as follows:
  - ≥ 1/4:12.

1507.12.1 Slope

- Thermoset single-ply membrane roofs must have a slope as follows:
  - ≥ 1/4:12.

1507.13.1 Slope

- Thermoplastic single-ply membrane roofs must have a slope as follows:
  - ≥ 1/4:12.

1507.14.1 Slope

- Sprayed polyurethane foam roofs must have the following slope:
  - ≥ 1/4:12.

1507.15.1 Slope

- Liquid-applied roofs must have the following slope:
  - ≥ 1/4:12.

1509 Rooftop Structures

1509.2.1 Height above roof

- This section addresses the following roof structures on buildings of Type II, III, IV or V construction:
  - Penthouses.
  - Other projections above the roof.
- Construction housing tanks and/or elevators stopping at roof level are governed as follows:
  - They may project above the roof ≤ 28'.
- Other constructions may project above the roof ≤ 18'.

1509.2.2 Area limitation

- This section addresses roof structures as follows:
  - Penthouses.
  - Other projections above the roof.
- Such roof structures are governed as follows:
  - They must not be < 1/3 the area of the supporting roof.
  - They do not count in the building area.
  - They do not count in the number of building stories.

  Note: 503.1, "General," is cited as governing building height and area.

  - They do not count in the fire area.

  Note: Section 902, "Definitions," is cited as defining fire area.

1509.2.3 Use limitations

- The following constructions above the roof are limited to the uses noted and the detailing listed:
  - Constructions:
    Penthouses.
    Bulkheads.
    Similar constructions.
  - Uses:
    Shelter of mechanical equipment.
    Shelter of vertical shaft openings in the roof.
  - Detailing:
    Such constructions are permitted for the following for protection of equipment and building interior:
      Louvers.
      Louver blades.
      Flashing.
      Similar provisions.
- The following constructions above the roof used for other purposes have the requirement listed:
  - Constructions:
    Penthouses.
    Bulkheads.
  - Requirement:
    Such constructions must meet code requirements for an additional story.
- The following devices are permitted above the roof:
  - Wood flagpoles.
  - Similar structures.

1509 Rooftop Structures

1509.2.4 Type of construction *(part 1 of 2)*

- The following applies to penthouses on buildings of Type I construction:
  - Where the fire separation distance is > 5' and < 20', the following applies:
    - The walls and roof require the following:
      - A fire-resistance rating of ≥ 1 hr.
      - Noncombustible construction.
  - Where the fire separation distance is ≥ 20', the following applies:
    - Noncombustible construction is required for walls and roof.
  - The following must be noncombustible construction:
    - Interior framing and walls.
- The requirements listed below apply to the following penthouses:
  - Penthouses:
    - On buildings of Type I construction and ≤ 2 stories above grade plane.
    - On buildings of Type II construction.
  - Requirements where the fire separation distance is > 5' and < 20':
    - Exterior walls and roofs must be one of the following:
      - Noncombustible construction with a fire-resistance rating of ≥ 1 hr.
      - Fire-retardant-treated wood construction with a fire-resistance rating of ≥ 1 hr.
  - Requirements where the fire separation distance is ≥ 20':
    - Exterior walls and roofs must be one of the following:
      - Noncombustible construction.
      - Fire-retardant-treated wood construction.
  - Requirements for interior framing and walls:
    - Walls and framing must be one of the following:
      - Noncombustible construction.
      - Fire-retardant-treated wood construction.
- The following applies to penthouses on buildings of Type III, IV, and V construction:
  - Where the fire separation distance is > 5' and < 20', the following applies:
    - Exterior walls must have a fire-resistance rating of ≥ 1 hr.
  - Where the fire separation distance is ≥ 20' from a common property line, the following applies:
    - Exterior walls must be ≥ one of the following:
      - Type IV construction.
      - Noncombustible construction.
      - Fire-retardant-treated wood construction.
  - Interior walls are required to be one of the following:
    - Type IV construction.
    - Noncombustible construction.
    - Fire-retardant-treated wood construction.
  - Roofs must have the following:
    - Materials as required for the construction type.
    - Fire-resistance ratings as required for the construction type.

Note: IBC Table 601, "Fire-Resistance Rating Requirements for Building Elements," is cited as the source of requirements for roofs as indicated above.

Section 603, "Combustible Material in Type I and II Construction," Item 25.3 (a numbered item in 603.1, "Allowable materials,") cited as governing roof materials.

1509 Rooftop Structures

1509.2.4 Type of construction *(part 2 of 2)*

- The following applies to buildings of Type I construction:
  - Unprotected noncombustible enclosures are permitted where both of the following conditions exist:
    - Where containing only mechanical equipment.
    - With a fire separation distance ≥ 20'.
- The requirements listed below apply to buildings of the following construction types:
  - Construction types:
    - Type I ≤ 2 stories above grade plane.
    - Type II.
    - Type III.
    - Type IV.
    - Type V.
  - Requirements:
    - Enclosures housing only mechanical equipment are governed as follows:
      - Where the fire separation distance is ≥ 20', the following are permitted:
        - Unprotected noncombustible construction.
        - Fire-retardant-treated wood construction.
- Mechanical equipment screens on the roof may be combustible where all of the following apply:
  - Building must be 1 story.
  - No roof may be present on the screens.
  - Screens must have a fire separation distance ≥ 20'.
  - Screens must be ≤ 4' in height above the roof surface.
- Dormers must be the same type construction as one of the following:
  - The roof of which they are a part.
  - The exterior walls of the building.
- Other penthouses must have the following constructed as required for the building:
  - Walls.
  - Floors.
  - Roof.

1509.3 Tanks

- Tanks > 500 gal in any of the following locations have the requirements listed:
  - Located in or on a building.
  - Requirements:
    - Such tanks must be supported by one of the following:
      - Masonry.
      - Reinforced concrete.
      - Steel.
      - Type IV construction.
    - Where tank supports in a building are above the lowest story, the following applies:
      - Support construction must meet fire-resistance ratings for Type IA construction.

1509 Rooftop Structures

1509.4 Cooling towers

- The following components of a cooling tower are permitted to be wood where all of the conditions listed below apply:
  - Components:
    - Drip boards.
    - Enclosing construction.
  - Requirements:
    - The wood must have a nominal thickness ≥ 1".
    - Wood must be covered on the outside of the tower with the following:
      - Noncombustible material.
- Otherwise, cooling towers must be of noncombustible construction in the follow cases:
  - Where either of the following applies to a building roof > 50' high:
    - Where a tower is > 250 sf at its base.
    - Where a tower is > 15' high.

1509.5 Towers, spires, domes and cupolas

- The following is required where the roof elements of this section are ≤ 85' above grade:
  - They must have a fire-resistance rating ≥ that of the building.
- Where roof elements addressed in this section have the following characteristics, the requirements listed below apply:
  - Characteristics:
    - Height is > 85' above the grade plane and either of the following also applies:
      - Area at any horizontal section is > 200 sf.
      - Where the elements are used for anything other than the following:
        - A belfry.
        - An architectural embellishment.
  - Requirements:
    - The following must be Type I or II construction:
      - Roof elements addressed in this section.
      - The construction on which such elements are supported.

1509 Rooftop Structures

1509.5.1 Noncombustible construction required

- This section addresses the following elements:
  - Towers.
  - Spires.
  - Domes.
  - Cupolas.
- Where such elements have any of the following characteristics, the requirements below apply:
  - Characteristics:
    - > 60' high above the following point:
      - Highest point of contact with the roof.
    - > 200 sf at any horizontal section.
    - To be used for any purpose other than the following:
      - A belfry.
      - An architectural embellishment.
  - Requirements:
    - Constructed entirely of noncombustible materials.
    - Supported by noncombustible materials.
    - Separated from the building below by construction as follows:
      - With a fire-resistance rating ≥ 1 1/2 hrs.
      - With openings having a fire-protection rating ≥ 1 1/2 hrs.
- The requirements listed below apply to elements on the roofs of buildings that are > 50' in building height, other than the following:
  - Elements:
    - The following elements are not included in the requirements below:
      - Aerial supports ≤ 12' high.
      - Flagpoles.
      - Water tanks.
      - Cooling towers.
  - Requirements:
    - Other types of elements must comply with the following:
      - Elements must be noncombustible.
      - Construction supporting the elements must be noncombustible.

1509.5.2 Towers and spires

- Where such structures are enclosed, the following applies:
  - The exterior walls must be constructed as required for the building.
- The roof covering of spires is governed as follows:
  - It must be the class of roofing required for the main roof of the building.

16

Structural Design

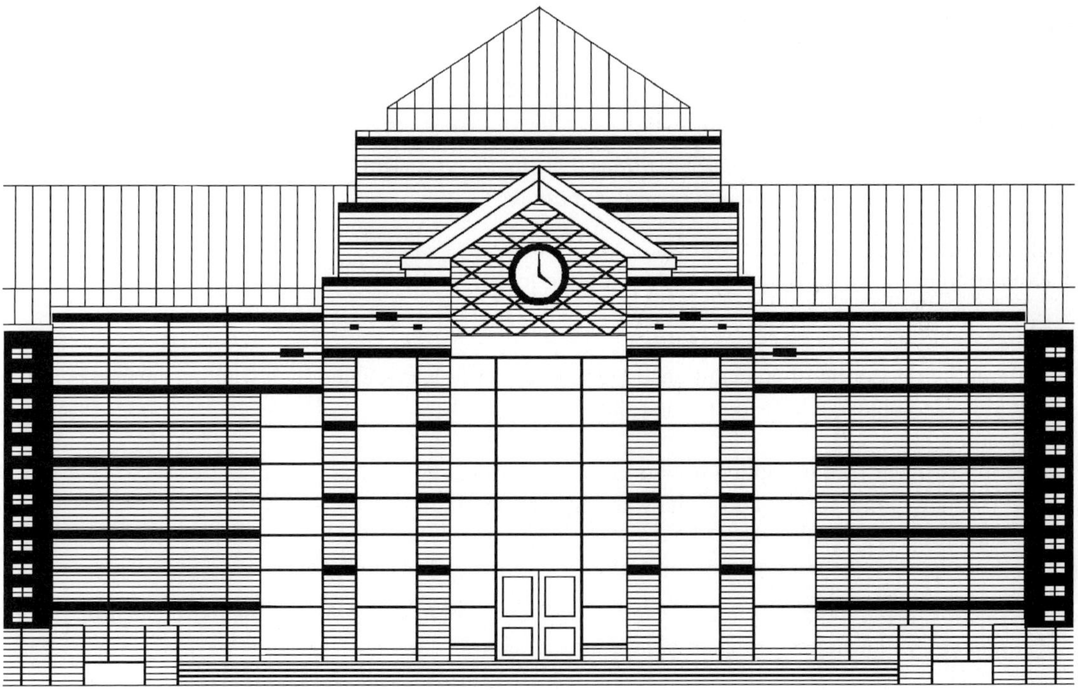

High School 6, Cypress-Fairbanks Independent School District.
Harris County, Texas. *(partial elevation)*
PBK Architects, Inc. Houston, Texas.

1604 General Design Requirements

1604.3.6 Limits *(part 1 of 3)*

- Deflection in metal building components is governed as follows:
  - Total load deflection of the following cladding is limited as listed below:
    - Structural roofing of formed metal sheets:
      - Deflection must be ≤ span ÷ 60.
    - Structural siding of formed metal sheets:
      - Deflection must be ≤ span ÷ 60.
  - Live load deflection of the following roof structure is limited as listed below:
    - Secondary roof structure supporting the following:
      - Formed metal roofing with no roof covering:
        - Deflection must be ≤ span ÷ 150.
  - Wind load deflection of the following wall structure is limited as listed below:
    - Secondary roof structure supporting formed metal siding:
      - Deflection must be ≤ span ÷ 90.
- Interior partitions are governed as follows:
  - The following interior partitions are not regulated by this section:
    - Partitions ≤ 6' high.
    - Flexible partitions.
    - Folding partitions.
    - Portable partitions.
  - Interior partitions > 6' high are governed as follows:
    - Actual deflection is based on the greater of the following lateral loads:
      - Actual loads to which the partition is subjected.
      - 5 psf.

Note: 1607.13, "Interior walls and partitions," is cited as the source of requirements for partition loading, a summary of which is provided above.

- Deflection of glass is limited as follows:
  - Deflection of the edge of a glass sheet is limited to the smaller of the following:
    - Length of glass edge ÷ 175.
    - 3/4".
  - Differential deflection of adjacent unsupported edges of glass located by a walking surface is limited to that indicated below, where loaded as follows:
    - Loading:
      - 50 lbs/linear foot is applied horizontally to one sheet as follows:
        - At any point ≤ 3'-6" above the walking surface.
    - Deflection:
      - Limited to ≤ the thickness of the glass.

Note: Section 2403, "General Requirements for Glass," is cited as governing glass deflection, a partial summary of which is provided above.

1604 General Design Requirements

1604.3.6 Limits *(part 2 of 3)*

- Wood structural members with all of the following characteristics are governed as listed below:
  - Characteristics:
    - Moisture content < 16% at time of installation.
    - Used in dry conditions.
  - Deflection:
    - Actual total deflection is permitted to be based on the following:
      - Live load + half the dead load.
- Where roof drainage is not assured, the possibility of ponding must be investigated for the following reason:
  - The deflection limits of this section do not necessarily prevent ponding.

Note: The following sections are cited as sources of applicable requirements:
Section 1611, "Rain Loads."
1503.4, "Roof drainage."

- In computing actual deflection, wind load may be taken as the following:
  - 0.7 × "component and cladding" loads.
- In computing actual total load deflection, dead load of steel structural members is taken as zero.
- The following aluminum members where used in the applications listed below have the requirements indicated below:
  - Members:
    - Structural members.
    - Panels.
  - Applications:
    - Roofs or walls of the following:
      - Sunroom additions.
      - Patio covers.
    - Skylights.
    - Framing for sloped glazing.
  - Requirements:
    - Where not supporting either of the following members, the deflection is limited as listed:
      - Members:
        - Edge of glass.
        - Aluminum sandwich panel.
      - Deflection:
        - Total load deflection must be ≤ span ÷ 60.
- The deflection of aluminum sandwich panels used in the following locations and applications is limited as listed below:
  - Locations:
    - Walls.
    - Roofs.
  - Applications:
    - Sunroom additions.
    - Patio covers.
  - Deflection:
    - Total load deflection must be ≤ span ÷ 120.

1604 General Design Requirements

1604.3.6 Limits *(part 3 of 3)*

- Continuous aluminum structural members supporting the edge of glass must meet both of the following:
  - Total load deflection must be ≤ L ÷ 175 for each glass pane.
  - Total load deflection must be ≤ L ÷ 60 for the whole member length.
- The following applies where determining deflection limits for cantilevers:
  - Span is taken as 2 × the length of the cantilever as follows:
    Where dividing span by the given denominator.

- Otherwise, deflection of structural members is limited to a fraction of span as indicated below:

Table 1604.3.6 Deflection Limit for Structural Members

Deflection of structural members	Fraction of span
Deflection based on live load:	
Roof members:	
Supporting a plaster ceiling	Span ÷ 360
Supporting a ceiling not plaster	Span ÷ 240
Not supporting a ceiling	Span ÷ 180
Floor members	Span ÷ 360
Deflection based on snow or wind load:	
Roof members:	
Supporting a plaster ceiling	Span ÷ 360
Supporting a ceiling not plaster	Span ÷ 240
Not supporting a ceiling	Span ÷ 180
Exterior walls and interior partitions:	
With brittle finishes	Span ÷ 240
With flexible finishes	Span ÷ 120
Deflection based on total load (dead load + live load):	
Roof members:	
Supporting a plaster ceiling	Span ÷ 240
Supporting a ceiling not plaster	Span ÷ 180
Not supporting a ceiling	Span ÷ 120
Floor members	Span ÷ 240
Farm buildings	Span ÷ 180
Greenhouses	Span ÷ 120

Source: IBC Table 1604.3.

1607 Live Loads

1607.3 Uniform live loads *(part 1 of 4)*

- Lives loads for structural design are to be the larger of the following:
  - The maximum loads anticipated.
  - The minimum uniformly distributed loads listed in this section.
- Passenger vehicle garage floors are to be designed for the greater of the following loads:
  - The minimum uniformly distributed load listed in this section.
  - The concentrated load described in the next section.

Note: IBC Table 1607.1, "Minimum Uniformly Distributed Live Loads, L_o, and Minimum Concentrated Live Loads," is cited as listing the applicable uniform load. The uniform loads of the table are summarized in this section.

- The library book stack loading minimum provided in this section is for the following:
  - A floor that supports book stacks as follows:
    - Nonmobile stacks.
    - Double-sided stacks.
    - Book-stack units with a nominal height ≤ 90".
    - Shelf depth on each side must have a nominal depth ≤ 12".
    - || rows of double-sided stacks must be separated by aisles with a width ≥ 36".
- Loading for the following seating is not governed by this section:
  - Reviewing stands.
  - Grandstands.
  - Bleachers.

Note: ICC 300, "ICC Standard on Bleachers, Folding and Telescopic Seating, and Grandstands," is cited as governing these types of seating.

- Loading for the following seating is governed as indicated below:
  - Seating:
    - Arena bleachers.
    - Stadium bleachers.
    - Arena fixed seats.
    - Stadium fixed seats.
  - Loading requirements provided by this section must be verified by other standards.

Note: ICC 300, "ICC Standard on Bleachers, Folding and Telescopic Seating, and Grandstands," is cited as governing these types of seating.

- Loading for the following elements is determined by one of the methods listed:
  - Elements:
    - The following where subject to truck loading:
      - Sidewalks.
      - Driveways.
      - Vehicular yards.
  - Methods:
    - Using loads specified in this section.
    - Using loads derived by other approved methods where appropriate.

1607 Live Loads

1607.3 Uniform live loads *(part 2 of 4)*

- Snow loading greater than that listed in this section must be accommodated as follows:
  - Where loads are elevated due to drifting.
  - Where required by the building official.

  Note: Section 1608, "Snow Loads," is cited as governing these loads.

- Roofs used for the following special purposes must be designed for the loads indicated:
  - For promenades, ≥ 60 psf.
  - For gardens, ≥ 100 psf.
  - For assembly, ≥ 100 psf.
  - Other special purposes, as directed by the building official.

  Note: 1607.11.2.2, "Special-purpose roofs," is cited as governing these roofs, a summary of which is provided above.

- Decks:
  - Live load to be the same as the occupancy served.

  Note: 1604.8.3, "Decks," is cited as a source of related requirements where decks are attached to exterior walls.

- Loading for the following is not provided in this section:
  - Trucks and buses.

  Note: 1607.6, "Truck and bus garages," is cited as governing this loading.

  - All of the following:
    - Handrails.
    - Guards.
    - Grab bars.

  Note: 1607.7, "Loads on handrails, guards, grab bars, seats and vehicle barrier systems," is cited as governing this loading.

  - Vehicle barrier systems.

  Note: 1607.7.3, "Vehicle barrier systms," is cited as governing this loading.

- Residential attics without storage are governed as follows:
  - They are defined as follows:
    - Rafter systems with the following characteristics:
      - Clear height between the following two levels is < 3'-6":
        - Top of ceiling joist.
        - Bottom of rafter.
    - Truss systems with the following characteristics:
      - Where the space within the webs of adjacent trusses could not contain a rectangular volume of the following cross section:
        - 3'-6" high.
        - 24" wide.
  - The required live load is not assumed to act as follows:
    - Simultaneously with any other live loads.

1607 Live Loads

1607.3 Uniform live loads *(part 3 of 4)*

- Residential attics with truss systems and with limited storage are governed as follows:
  - They are defined by the following characteristics:
    - The truss system has bottom chords at a slope ≤ 2:12.
    - The bottom chords are designed for the greater of the following dead loads:
      - Actual dead load.
      - 10 psf uniformly distributed across whole span.
    - Attic is accessible by one of the following:
      - Pull-down stairs.
      - A framed opening.

  Note: 1209.2, "Attic spaces," is cited as governing the framed opening listed above.

  - The required live load is applied only to the lengths of the bottom cord where the following applies:
    - Where the space within the webs of adjacent trusses has the following characteristics:
      - The space is large enough to contain a rectangular volume as follows:
        - With a cross section of the following size:
          - 3'-6" high and 24" wide.
        - Which will fit between the following members:
          - The top of the bottom chords and other truss members.
- The following roof system is not permitted to have its live load reduced:
  - Fabric supported as follows:
    - By a lightweight rigid frame structure.
- For other than special cases, minimum uniformly distributed live loads are to be as follows:

Table 1607.3 Minimum Uniformly Distributed Live Loads

Use	psf	Use	psf
Access floors:		**Balconies, exterior:**	
Computer	100	Other than 1- & 2-family dwelling	100
Office	50	**Ballrooms**	100
Apartments:		**Bowling alleys**	75
Corridors serving public rooms	100	**Canopies**	75
Private rooms	40	**Catwalks**	40
Public rooms	100	**Corridors:**	
Arenas:		Hospitals:	
Bleachers	100	1st floor	100
Fixed seats	60	Upper floors	80
Armories and drill rooms	150	Hotels, serving public rooms	100
Assembly:		Hotels, serving private rooms	40
Control rooms	50	Libraries:	
Fixed seat area	60	1st floor	100
Follow-spot floors	50	Upper floors	80
Lobbies	100	Office buildings:	
Movable seat area	100	1st floor	100
Platforms	125	Upper floors	80
Projection rooms	50	Penal institutions	100
Stages	125	*(continued on next page)*	
Other assembly areas	100		

Source: IBC Table 1607.1.

1607 Live Loads

1607.3 Uniform live loads *(part 4 of 4)*

Table 1607.3—*Continued* **Minimum Uniformly Distributed Live Loads**

Use	psf
Corridors *(continued)*:	
Schools:	
1st floor	100
Upper floors	80
Other locations	100
Dance halls	100
Dining rooms	100
Driveways subject to trucks	250
Fire escapes:	
1-family dwellings	40
Other locations	100
Garages, passenger cars	40
Grandstands	100
Gymnasiums:	
Main floor	100
Balconies	100
Hospitals:	
Corridors:	
1st floor	100
Upper floors	80
Laboratories	60
Operating rooms	60
Patient rooms	40
Hotels:	
Corridors serving private rooms	40
Corridors serving public rooms	100
Private rooms	40
Public rooms	100
Libraries:	
Corridors:	
Upper floors	80
Reading rooms	60
Stacks	150
Manufacturing:	
Light	125
Heavy	250
Marquees	75
Office buildings:	
Corridors:	
1st floor	100
Upper floors	80
Lobbies	100
Offices	50
Penal institutions:	
Cell blocks	40
Corridors	100
Platforms, elevated (not an exitway, not similar to stage)	60
Residential:	
1-family dwellings:	
Fire escapes	40
1- & 2-family dwellings:	
Attics served by fixed stairs	30
Balconies, exterior ≤ 100 sf	60
Balconies, exterior > 100 sf	100
Habitable attics	30
Sleeping areas	30
Stairs and exits	40
Uninhabitable attics:	0
No storage	10
With limited storage	20
Other areas not including decks	40
Multifamily dwellings:	
Corridors serving public rooms	100
Private rooms	40
Stairs and exits	100
Corridors serving private rooms	40
Restaurants	100
Schools:	
Classrooms	40
Corridors:	
1st floor	100
Upper floors	80
Sidewalks subject to trucks	250
Skating rinks	100
Stadium bleachers	100
Stadium fixed seats	60
Stairs and exits:	
Other than 1- & 2-family dwellings	40
Storage warehouses:	
Light	125
Heavy	250
Stores:	
Retail:	
1st floor	100
Upper floors	75
Wholesale, all floors	125
Terraces, pedestrian	100
Walkways (not exitways)	60
Yards:	
Pedestrian	100
Subject to trucks	250

Source: IBC Table 1607.1.

1607 Live Loads

1607.4 Concentrated loads *(part 1 of 2)*

- Floors and similar surfaces must be designed to support the following live loads:
  - The following load that produces greater stresses in the structure is required:
    - Uniformly distributed loads as specified in the previous section.
    - The following concentrated loads:
      - To be evenly distributed over an area 21/2' × 21/2'.
      - To be located to yield the greatest stresses in the structure.

Note: 1607.3, "Uniform Live Loads," is cited as the source of uniformly distributed live loads. IBC Table 1607.1, "Minimum Uniformly Distributed Live Loads, L_o, and Minimum Concentrated Live Loads," is cited as listing the applicable loads. The concentrated loads of the table are summarized in this section.

- Passenger vehicle garages are to be designed for the greater of the following loads:
  - The minimum uniformly distributed floor load listed in the previous section.
  - A concentrated floor load as follows:
    - Where a garage is limited to passenger vehicles carrying ≤ 9 people, the following applies:
      - Concentrated design live load must be ≥ 3,000 lb as follows:
        - Applied to an area 41/2" × 41/2".
    - For the following garages, the load listed below is required:
      - Garages:
        - Using only mechanical parking devices.
        - Without slab.
        - Without a deck.
      - Load:
        - The design load is to be ≥ 2,250 lb/wheel.
- The live load at each wheel for the following surfaces is to be distributed over an area 41/2" × 41/2":
  - Sidewalks subject to truck traffic.
  - Driveways subject to truck traffic.
  - Yards subject to truck traffic.
- The live load on a stair tread is to be ≥ 300 lbs as follows:
  - Distributed over 4 sq in.
- Loading for the following is not provided in this section:
  - Trucks and buses.

Note: 1607.6, "Truck and bus garages," is cited as governing this loading.

  - All of the following:
    - Handrails and guards.
    - Grab bars.
    - Vehicle barrier systems.

Note: 1607.7, "Loads on handrails, guards, grab bars, seats and vehicle barrier systems," is cited as governing this loading.

1607 Live Loads

1607.4 Concentrated loads *(part 2 of 2)*

- For other than special cases, minimum concentrated live loads are to be as follows:

Table 1607.4 Minimum Concentrated Live Loads

Concentrated use	Load in lbs	Concentrated use	Load in lbs
Access floors:		**Roofs:**	
Computer	2,000	Accessed by	
Office	2,000	maintenance people	300
Catwalks	300	Manufacturing	2,000
Ceilings:		Storage warehouses	2,000
Able to be accessed	200	Repair garages	2,000
Driveways:		Other occupancies	300
Subject to trucks	8,000	**Schools:**	
Exits:		Classrooms	1,000
On 4 sq in of tread	300	Corridors	1,000
Hospitals:		**Scuttles**	200
Corridors, upper floors	1,000	**Sidewalks:**	
Laboratories	1,000	Subject to trucks	8,000
Operating rooms	1,000	**Skylight ribs**	200
Patient rooms	1,000	**Stairs:**	
Libraries:		On 4 sq in of tread	300
Corridors, upper floors	1,000	**Stores:**	
Reading rooms	1,000	Retail	1,000
Stacks	1,000	Wholesale	1,000
Manufacturing:		**Yards:**	
Light	2,000	Subject to trucks	8,000
Heavy	2,000		
Office buildings:			
Corridors	2,000		
Lobbies	2,000		
Offices	2,000		

Source: IBC Table 1607.1.

1607.5 Partition loads

- In the following buildings, the additional live load must be added as listed below:
  - Buildings:
    - Office buildings.
    - Buildings where partition locations change.
  - Load:
    - Where the required live load is ≤ 80 psf:
      - 15 psf must be added to the required live load for partitions.

1607.7.1 Handrails and guards *(part 1 of 2)*

- This section does not apply to 1- and 2-family dwellings.

Note: 1607.7.1.1, "Concentrated load," is cited as governing handrails and guards in 1- and 2-family dwellings.

1607 Live Loads

1607.7.1 Handrails and guards *(part 2 of 2)*

- Handrails and guards in the following occupancies and locations have the requirements listed below:
  - Occupancies:
    I-3, F, H, S.
  - Locations:
    Locations not available to the general public.
    Where occupant load is < 50.
  - Requirements:
    Handrails and guards must resist a force of ≥ 20 plf as follows:
      Applied in any direction at the top.
      Load must be transferred to the structure by the supports.
- In other locations, handrails and guards have the following requirements:
  - Handrails and guards must resist a force of ≥ 50 plf as follows:
    Applied in any direction at the top.
    Load must be transferred to the structure by the supports.

Note: *Section 2407, "Glass in Handrails and Guards," is cited as governing glass in these elements.*

1607.7.1.1 Concentrated load

- Loading of handrail assemblies and guards is governed as follows:
  - They must be able to resist a single concentrated load ≥ 200 lb as follows:
    Applied in any direction at any point at the top.
    Load is not applied simultaneously with other loads.
  - They must be able to transfer the loading to the structure.

Note: *1607.7.1, "Handrails and guards," is cited as identifying the other loads not applied simultaneously.*

1607.7.1.2 Components

- Components of a handrail assembly or guard as follows must resist the load listed:
  - Components:
    All rails other than the top handrail.
    Balusters.
    Panel fillers.
  - ≥ 50 lb as follows:
    Applied horizontally ⊥ to the component.
    Applied to an area of 1 sf including the following:
      Openings and space between rails.
    Load is not applied simultaneously with loads specified in the previous 2 sections.

Note: *The following sections are cited as specifying the loads which are not to be applied simultaneously with the load specified in this section:*
1607.7.1, "Handrails and guards."
1607.7.1.1, "Concentrated load."

1607.7.2 Grab bars, shower seats and dressing room bench seats

- Such components must resist 250 lbs as follows:
  - Applied at any point in any direction.

NOTES

17

Structural Tests and Special Inspections

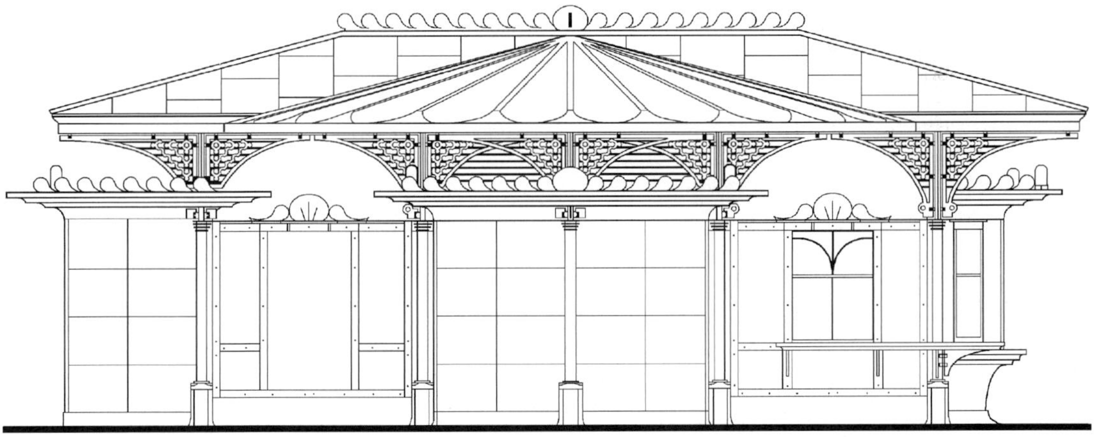

Visitor Services Pavilions at Clarence Buckingham Memorial Fountain.
Chicago Park District. Chicago, Illinois
David Woodhouse Architects. Chicago, Illinois.

1703 Approvals

1703.5 Labeling

- Where the following components are to be labeled, the requirements listed below apply:
  - Components:
    - Materials.
    - Assemblies.
  - Requirements:
    - Such components must be labeled by an approved agency.
    - Such components must be labeled according to the requirements of this section.

Note: Section 1703, "Approvals," is cited as governing the approved agency indicated.
The following are cited as governing the labeling of components indicated above:
1703.5.1, "Testing."
1703.5.2, "Inspection and identification."
1703.5.3, "Label information."

1703.5.1 Testing

- This section addresses a part of the labeling process.
- An approved agency must test an example of the component to be labeled as follows:
  - Tests must be based on relevant standards.
  - The agency must maintain a record of tests performed as follows:
    - Adequate information must be recorded to verify compliance with the relevant standards.

1703.5.2 Inspection and identification

- This section addresses a part of the labeling process.
- The approved agency must inspect the components subsequently labeled as follows:
  - Inspections must be periodic.
  - Inspections of the component fabrication must be done where necessary.
  - Inspections must verify that the labeled component is representative of that tested.

1703.5.3 Label information

- Labels must include the following information:
  - Identification of one of the following:
    - Manufacturer.
    - Distributor.
  - One of the following types of information:
    - Description of the components performance properties.
    - Both identification numbers as follows:
      - Model number.
      - Serial number.
  - Identification of the approved agency issuing the label.

18

Soils and Foundations

Lake Forest City Hall Renovation and Addition. Lake Forest, Illinois.
David Woodhouse Architects. Chicago, Illinois.

1804 Excavation, Grading and Fill

1804.1 Excavation near foundations

- Where excavation removes lateral support from a foundation, the following applies:
  - The following must be prevented by underpinning or other means:
    - Settlement of the foundation.
    - Lateral translation of the foundation.

1804.2 Placement of backfill

- This section addresses excavation outside the foundation.
- Excavation must be backfilled with one of the following:
  - Soil free of any of the following substances:
    - Organic material.
    - Construction debris.
    - Cobbles.
    - Boulders.
  - Controlled low-strength material (CLSM).
- Backfill is governed as follows:
  - CLSM is not required to be compacted.
  - Other backfill must be placed in lifts with the following process:
    - Compacted without damage to the following:
      - Foundation.
      - Waterproofing.
      - Dampproofing.

1804.3 Site grading

- Grade must slope away from the foundation in one of the following conditions:
  - At a slope ≥ 1:20 as follows:
    - For a distance ≥ 10' from the foundation measured as follows:
      - ⊥ perpendicular to the foundation.
  - At a slope ≥ 1:20 in the following case:
    - Using an approved alternate method directing water away from the foundation as follows:
      - Where a slope ≥ 1:20 is prevented from extending ≥ 10' by one of the following:
        - Physical obstructions.
        - Lot line.
      - A swale ≤ 10' from the foundation must slope ≥ 1:48.
- Impervious surfaces ≤ 10' from the foundation must slope ≥ 1:48.
- Settlement of backfill must be considered in determining final grade.

1805 Dampproofing and Waterproofing

1805.1 General

- This section addresses the following building elements :
  - Walls or parts of walls as follows:
    - That abut the earth and enclose space.
  - Floors below grade.
- Such elements need not be waterproofed or dampproofed where all of the following exist:
  - Where not among the following occupancies:
    - Institutional.
    - Residential.
  - Where the absence of waterproofing or damproofing does not adversely affect the building.
- In all other cases, the elements require waterproofing and dampproofing as required by this section.

Note: 1203.4, "Natural ventilation," is cited as governing ventilation for crawl spaces.

1805.1.1 Story above grade plane

- This section applies to a basement that qualifies as a story above the grade plane.
- Where finished grade is below the basement floor for ≥ 1/4 of the perimeter, the following applies:
  - The walls below grade must be dampproofed.
  - The floor must be dampproofed.
  - A foundation drain is required around the walls below grade.

Note: 1805.2, "Dampproofing," is cited as governing the damproofing for this condition above.
1805.4.2, "Foundation drain," is cited as governing the foundation drain.

- A subsurface soil investigation is not required as follows:
  - To determine the height of the water table.
- Such a basement need not be waterproofed.
- A base course is not required under such a basement floor.

Note: The following are cited as governing the soil investigation, the waterproofing, and the base course that are not required for basements governed by this section:
1803.5.4, "Ground-water table."
1805.3, "Waterproofing."
1805.4.1, "Floor base course."

1805.1.2 Under-floor space *(part 1 of 2)*

- Ground surface in a crawl space is governed as follows:
  - It may not be below footing-bearing level.
- One of the following details must be provided where either of the indicated conditions applies:
  - Details:
    - Ground surface in a crawl space must be at the following height:
      - ≥ the height of exterior finished grade.
    - An approved drainage system must be provided.
  - Conditions:
    - Ground-water table rises to ≤ 6" of finished grade at building perimeter.
    - Surface ground-water does not readily drain from the site.

1805 Dampproofing and Waterproofing

1805.1.2 Under-floor space *(part 2 of 2)*

- The following is not required for crawl spaces meeting the requirements of this section:
  - A subsurface soil investigation to determine the height of the water table.
  - Dampproofing.
  - Waterproofing.
  - A subsoil drainage system.

Note: The following are cited as governing the remedies noted above that are not required for crawl spaces meeting the requirements of this section:
1803.5.4, "Ground-water table."
1805.2, "Dampproofing."
1805.3, "Waterproofing."
1805.4, "Subsoil drainage system."

1805.1.2.1 Flood hazard areas

- Crawl spaces in Occupancy R-3 are not governed by this section:

Note: FEMA FIA-TB-11, "Crawlspace Construction for Buildings Located in Special Flood Hazard Areas," is cited as governing R-3 crawl spaces in flood hazard areas.

- Where located in a flood hazard area, the following applies:
  - Ground surface in a crawl space must be as follows:
    ≥ the height of exterior finished grade on ≥ 1 side.

Note: 1612.3, "Establishment of flood hazard areas," is cited as governing the establishment of flood hazard areas.

1805.1.3 Ground-water control

- The lowest floor and walls must be dampproofed in the following case:
  - Where the ground-water table is controlled as follows:
    Lowered and maintained at ≥ 6" below the bottom of lowest floor.

Note: 1805.2, "Dampproofing," is cited as the source of requirements for dampproofing.

- Systems utilized to lower the ground-water table must be in consideration of the following:
  - Accepted principles of engineering.
  - Permeability of the soil.
  - Rate of water flow into the drainage system.
  - Capacity of pumps.
  - Pressure against which pumps must perform.
  - Capacity of disposal area for the system.
  - Other applicable conditions.

1805 Dampproofing and Waterproofing

1805.2 Dampproofing

- This section does not apply to wood foundations.
- This section governs dampproofing for walls and floors as follows:
  - Where there is no hydrostatic pressure.

Note: The following are cited as sources of applicable requirements:
1803.5.4, "Ground-water table," describes conditions where no hydrostatic pressure will occur.
AF&PA PWF, "Permanent Wood Foundation Design Specification," provides requirements by which such foundations are designed.

1805.2.1 Floors

- This section applies to floors requiring dampproofing.
- Where there is no separate floor above a slab, dampproofing is installed as follows:
  - Dampproofing must be installed between the floor and base course.

Note: 1805.4.1, "Floor base course," is cited as governing the base course indicated above.

- Where located below a slab, dampproofing is governed as follows:
  - One of the following systems is required:
    - ≥ 6-mil polyethylene as follows:
      - With joints lapped ≥ 6".
    - Other approved materials or methods.
- Where located on top of a slab, dampproofing is governed as follows:
  - One of the following systems is required:
    - Mopped-on-bitumen.
    - ≥ 4-mil polyethylene.
    - Other approved materials or methods.
  - Joints in the dampproofing membrane are governed as follows:
    - To be lapped and sealed as per manufacturer's instructions.

1805.2.2 Walls *(part 1 of 2)*

- This section applies to walls requiring dampproofing.
- Dampproofing must be installed as follows:
  - On the exterior surface of the wall.
  - From the top of footing to above grade.
- Dampproofing must be one of the following materials:
  - Bituminous material.
  - Acrylic modified cement at 3 lbs/sq yd.

1805 Dampproofing and Waterproofing

1805.2.2 Walls *(part 2 of 2)*

- 1/8" coat of surface-bonding mortar.
- Any waterproofing material as follows:
  - Two-ply hot mopped felts.
  - ≥ 6-mil polyethylene.
  - ≥ 6-mil polyvinyl chloride.
  - ≥ 40-mil polymer-modified asphalt.
  - Other approved materials and methods.

Note: The following are cited as sources of applicable requirements:
ASTM C 887, "Standard Specification for Packaged, Dry, Combined Materials for Surface Bonding Mortar," governs surface bonding mortar as indicated above.
1805.3.2, "Walls," lists waterproofing materials as summarized above.

1805.2.2.1 Surface preparation of walls

- Walls to receive dampproofing must be prepared as follows:
  - Concrete walls:
    - Form tie-holes and recesses must be sealed with one of the following:
      - Bituminous material.
      - Other approved material or method.
  - Unit masonry:
    - Where dampproofing is not approved for direct application:
      - Walls must be parged as follows:
        - On the exterior surface below grade.
        - With ≥ 3/8" portland cement mortar.
        - Parging is to be coved at the footing.
    - Where dampproofing is approved for direct application to masonry, parging is not required.

1805.3 Waterproofing

- Walls and floors must be waterproofed where both of the following conditions apply:
  - Where there is hydrostatic pressure.
  - Where a ground-water control system is not provided.

Note: The following are cited as sources of applicable requirements:
1803.5.4, "Ground-water table," describes the conditions where hydrostatic pressure will occur.
1805.1.3, "Ground-water control," governs the control system indicated above.

1805.3.1 Floors

- Concrete floors are required to be waterproofed as follows:
  - They must be designed to resist the hydrostatic pressure against it.
- One of the following, or other approved waterproofing material, must be placed under the slab:
  - Rubberized asphalt membrane.
  - Butyl rubber membrane.
  - Fully adhered and fully bonded HDPE (high density polyethylene) membrane.
  - Fully adhered and fully bonded polyolefin composite membrane.
  - ≥ 60-mil polyvinyl chloride with joints lapped ≥ 6".
  - Other approved material.
  - Membrane joints must be lapped and sealed as per manufacturer's instructions.

1805 Dampproofing and Waterproofing

1805.3.2 Walls

- Walls and parts of walls required to be waterproofed are governed as follows:
  - They must be one of the following:
    - Concrete.
    - Masonry.
  - They must be designed to resist the following:
    - Hydrostatic pressure applied.
    - Lateral loads applied.
  - They must have waterproofing between the following points:
    - The bottom of the wall.
    - A point ≥ 12" above the highest ground-water table level.
  - They must have dampproofing as follows:
    - On the remainder of the wall to a point above grade.
  - Waterproofing must consist of one of the following:
    - Two-ply hot mopped felts.
    - ≥ 6-mil polyvinyl chloride.
    - ≥ 40-mil polymer-modified asphalt.
    - ≥ 6-mil polyethylene.
    - Other approved materials or methods.
  - Joints in the waterproofing membrane must be detailed as follows:
    - Lapped and sealed as per manufacturer's instructions.

Note: 1805.2.2, "Walls," is cited as governing dampproofing as required above.

1805.3.2.1 Surface preparation of walls

- Walls to receive dampproofing must be prepared as follows:
  - Concrete walls:
    - Form tie-holes and recesses must be sealed with one of the following:
      - Bituminous material.
      - Other approved material or method.
  - Unit masonry:
    - Where dampproofing is not approved for direct application:
      - Walls must be parged as follows:
        - On the exterior surface below grade.
        - With ≥ 3/8" portland cement mortar.
        - Parging is to be coved at the footing.
    - Where dampproofing is approved for direct application to masonry, the following applies:
      - Parging is not required.

Note: 1805.2.2.1, "Surface preparation of walls," is cited as governing wall preparation, a summary of which is provided above.

1805 Dampproofing and Waterproofing

1805.4 Subsoil drainage system

- Where hydrostatic pressure is not present, the following applies:
  - Dampproofing is required.
  - A base is required under the floor.
  - A drain is required at the perimeter of the foundation.
  - The water table may be lowered by the subsoil drainage system described in the referenced section.

Note: 1805.1.3, "Ground-water control," is cited as the source of requirements for the subsoil drainage system indicated above.

1805.4.1 Floor base course

- The following basement floors are not governed by this section:
  - Basements with both of the following characteristics:
    The basement qualifies as a story above grade plane.
    The finished grade is below ≥ 1/4 of the basement wall perimeter.
  - Where the site has either of the following conditions, a floor base is not required:
    Site soil is well-drained gravel mixture.
    Site soil is well-drained sand/gravel mixture.

Note: 1805.1.1, "Story above grade plane," is cited as governing floors not covered by this section.

- Other basement floors must be on a base course as follows:
  - ≥ 4" of gravel or crushed stone.
  - Base must have ≤ 10% material passing a #4 sieve.

1805.4.2 Foundation drain

- This section addresses foundation drains where required.
- One of the following foundation drains is required at the perimeter of the foundation:
  - Gravel or crushed stone as follows:
    With ≤ 10% material passing a #4 sieve.
    Drain must extend horizontally ≥ 12" beyond the outside edge of the footing.
    Bottom of drain must be ≤ the height of the bottom of the base under floor.
    Top of drain must be ≥ 6" above top of footing.
    Top of drain must be covered with an approved membrane filter.
  - Drain tile or perforated pipe as follows:
    Drain invert elevation must be ≤ that of the floor.
    Top of joints or perforations must be covered with an approved membrane filter.
    Drain must be placed on ≥ 2" of gravel or crushed stone.
    Drain must be covered with ≥ 6" of gravel or crushed stone.

Note: 1805.4.1, "Floor base course," is cited as governing the crushed gravel or stone listed above.

1807 Foundation Walls, Retaining Walls and Embedded Posts and Poles

1807.1.2 Unbalanced backfill height

- The height of unbalanced backfill is measured between the following two levels:
  - The finished exterior grade.
  - One of the following interior levels:
    - Top of concrete slab where provided.
    - The lower of the following two levels where there is no concrete slab:
      - Top of footing.
      - Interior grade.

1807.1.3 Rubble stone foundation walls

- Foundation walls of random rubble stone are governed as follows:
  - Thickness must be ≥ 1'-4".
  - Such walls may not be used in locations having the following Seismic Design Categories:
    - C, D, E, F.

1807.1.6 Prescriptive design of concrete and masonry foundation walls

- Concrete and masonry foundation walls may be designed according to this section as follows:
  - Walls must be laterally supported at top and bottom.

1807.1.6.1 Foundation wall thickness

- This section addresses concrete and masonry foundation walls.
- Foundation walls with nominal thickness ≥ 8" may support the following walls where meeting referenced code sections:
  - Walls:
    - Frame walls with brick veneers.
    - 10"-thick cavity walls.

Note: *The following are cited as governing foundation walls with nominal thickness ≥ 8" that support thicker walls as noted above:*
1807.1.6.2, "Concrete foundation walls."
1807.1.6.3, "Masonry foundation walls."

1807.1.6.2 Concrete foundation walls *(part 1 of 2)*

- The design of concrete foundation walls is governed by the table provided in this section as follows:
  - Required thickness is shown.
  - Spacing and size of vertical reinforcing bars is modified as follows:
    - For yield strength ≥ 60 ksi, spacing is as per tabular requirements.
    - For yield strength ≥ 50 ksi, spacing is 0.83 × tabular requirements.
    - For yield strength ≥ 40 ksi, spacing is 0.67 × tabular requirements.

Note: *Spacing for the three strengths of reinforcing listed above is shown in Figures 1807.1.6.2-1 through 1807.1.6.2-45b on the following pages, and is derived from IBC Table 1807.1.6.2, "Concrete Foundation Walls." Spacing for 50 ksi and 40 ksi reinforcing was calculated as being 5/6 and 2/3 respectively of that required for 60 ksi reinforcing. Resulting decimals were rounded down to the next 1/16". The distance* d *is shown in Figures 1807.1.6.2-1 through 1807.1.6.2-45b on the following pages and is derived from the IBC table listed in this note.*

1807 Foundation Walls, Retaining Walls and Embedded Posts and Poles

1807.1.6.2 Concrete foundation walls *(part 2 of 2)*

- Where vertical reinforcement is required, it must be positioned as follows:
  - Nearest the face of the wall opposite of grade as follows:
    - A distance *d* from the exterior-grade face of the wall as follows:
      - *d* = wall thickness minus 1/2 the reinforcing bar diameter and minus 1 1/4".
    - The tolerance for positioning reinforcing bars is as follows:
      - Where *d* is ≤ 8", tolerance is ± 3/8".
      - Where *d* is > 8", tolerance is ± 1/2".
- Reinforcing is permitted to be smaller than that specified in the tabular data as follows:
  - Where spacing is reduced so as to yield the following:
    - An average cross-sectional area of steel per foot of wall equal to that of the required steel.
- Concrete cover for reinforcing is required as follows:
  - ≥ 3/4" on the side of the wall opposite grade.
  - ≥ 1 1/2" on the grade-side face of the wall for bars ≤ #5.
  - ≥ 2" on the grade-side face of the wall for bars > #5.
- Compressive strength required for the concrete is ≥ 2,500 psi.
- Unfactored axial load/ft on the wall must be the following:
  - ≤ 1.2 × required wall thickness × required compressive strength.
- Where unbalanced backfill is ≥ 8', the following applies:
  - Requirements for 30 and 45 psf/ft of depth design lateral soil load do not apply in the following case:
    - Where the lateral soil load was determined by use of the referenced code table.

Note: IBC Table 1610.1, "Lateral Soil Load," is cited as the table that may not be used to identify lateral soil loads of 30 psf and 45 psf for walls with unbalanced backfill heights ≥ 8'.
Section 1610, "Soil Lateral Loads," is cited as applicable.

- The height of unbalanced backfill is measured between the following two levels:
  - The finished exterior grade.
  - One of the following interior levels:
    - Top of concrete slab where provided.
    - The lower of the following two levels where there is no concrete slab:
      - Top of footing.
      - Interior grade.

Note: 1807.1.2, "Unbalanced backfill height," is cited as defining this height and is summarized above.

1807.1.6.2.1 Seismic requirements

- The foundation walls illlustrated on the following pages have the following limitations:

Note: IBC Table 1807.1.6.2, "Concrete Foundation Walls," (as illustrated on the following pages) is cited as having the limitations dictated by this section.
Section 1613, "Earthquake Loads," is cited as governing seismic design categories.

- The addition of reinforcing around openings is required for seismic design categories A and B.
- Illustrated requirements are applicable in seismic design categories C, D, E, and F only as follows:
  - According to the guidelines of the code addressing these seismic design categories.

Note: 1909.6.3, "Openings in walls," is cited as governing reinforcing around openings in walls.
1908.1.8, "ACI 318, Section 22.10," is cited as providing guidelines for use of details provided in this section for plain concrete in sesimic design categories C, D, E, and F.

1807 Foundation Walls, Retaining Walls and Embedded Posts and Poles

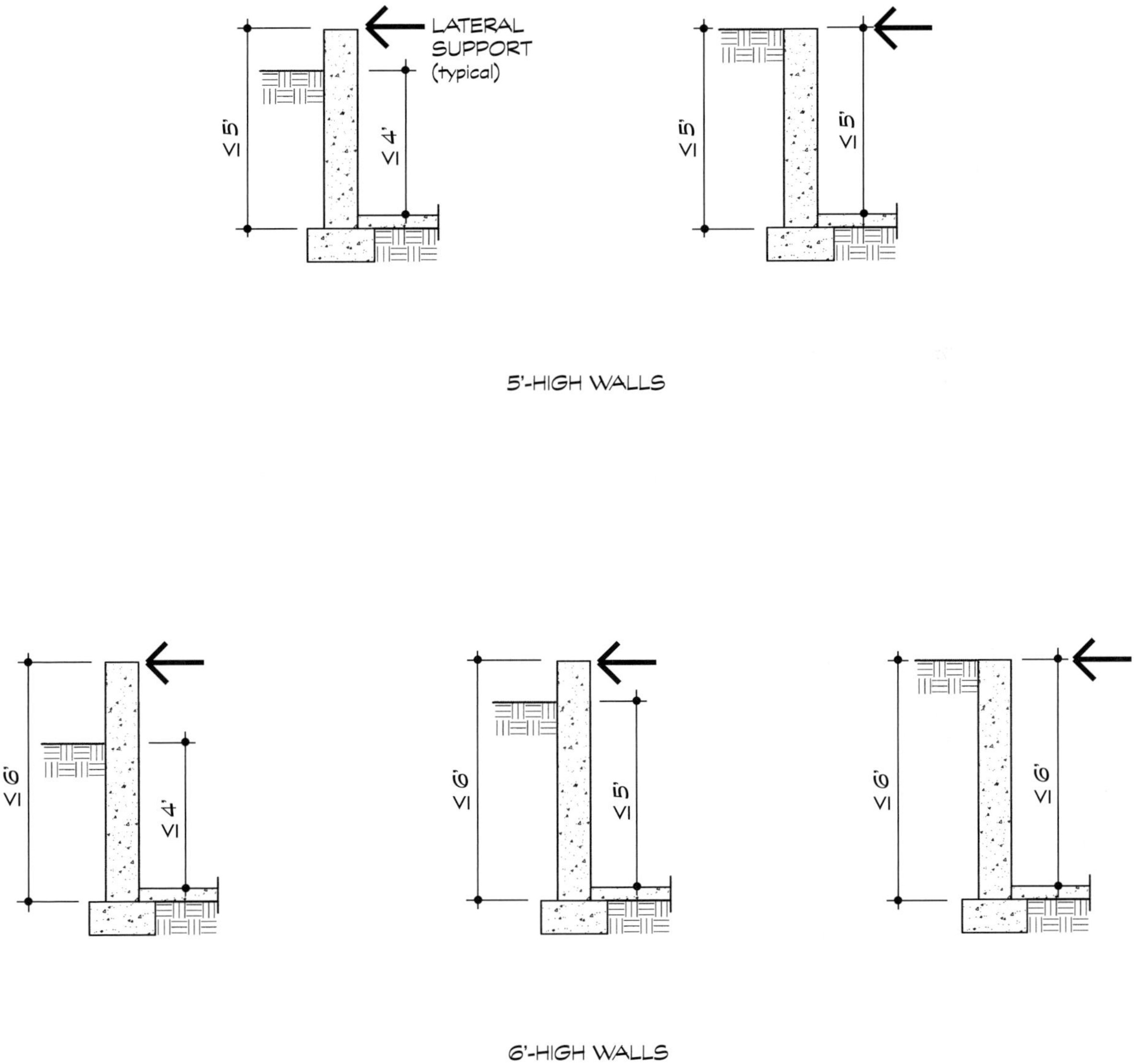

Fig. 1807.1.6.2-1. Plain concrete foundation wall, thickness ≥ 7 1/2" for a design lateral soil load of 30 psf/ft of depth. *[IBC Table 1807.1.6.2]*

1807 Foundation Walls, Retaining Walls and Embedded Posts and Poles

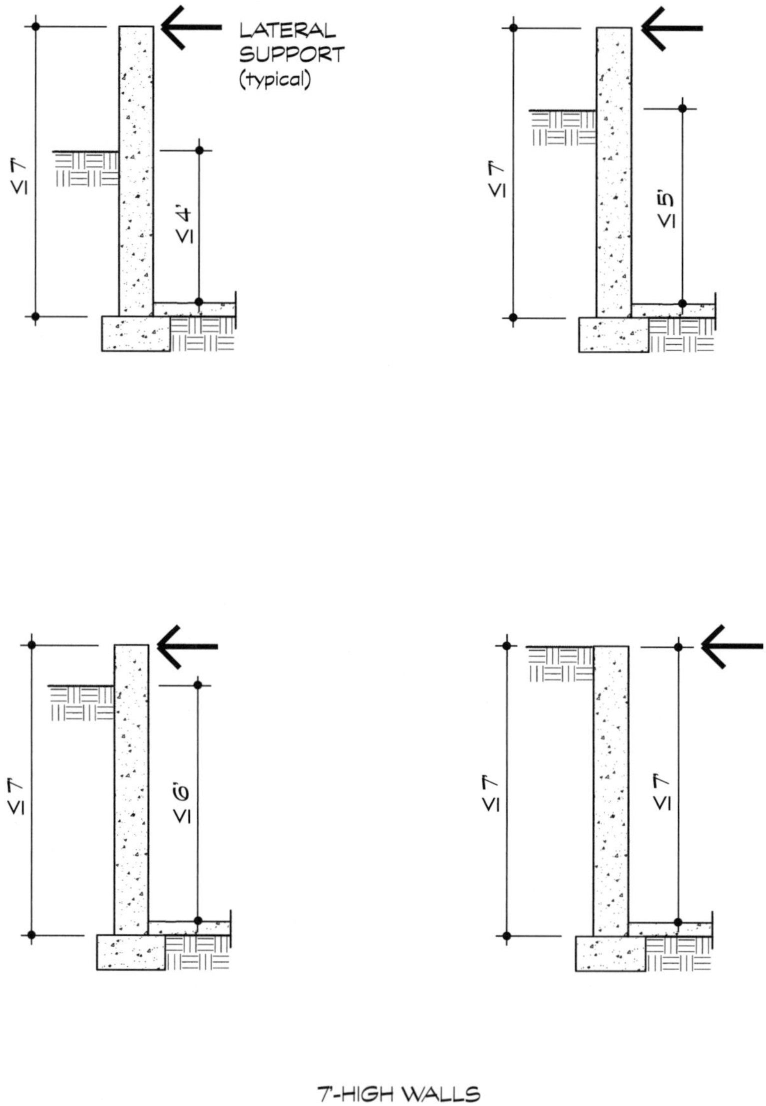

Fig. 1807.1.6.2-2. Plain concrete foundation wall, thickness ≥ 7 1/2" for a design lateral soil load of 30 psf/ft of depth. *[IBC Table 1807.1.6.2]*

1807 Foundation Walls, Retaining Walls and Embedded Posts and Poles

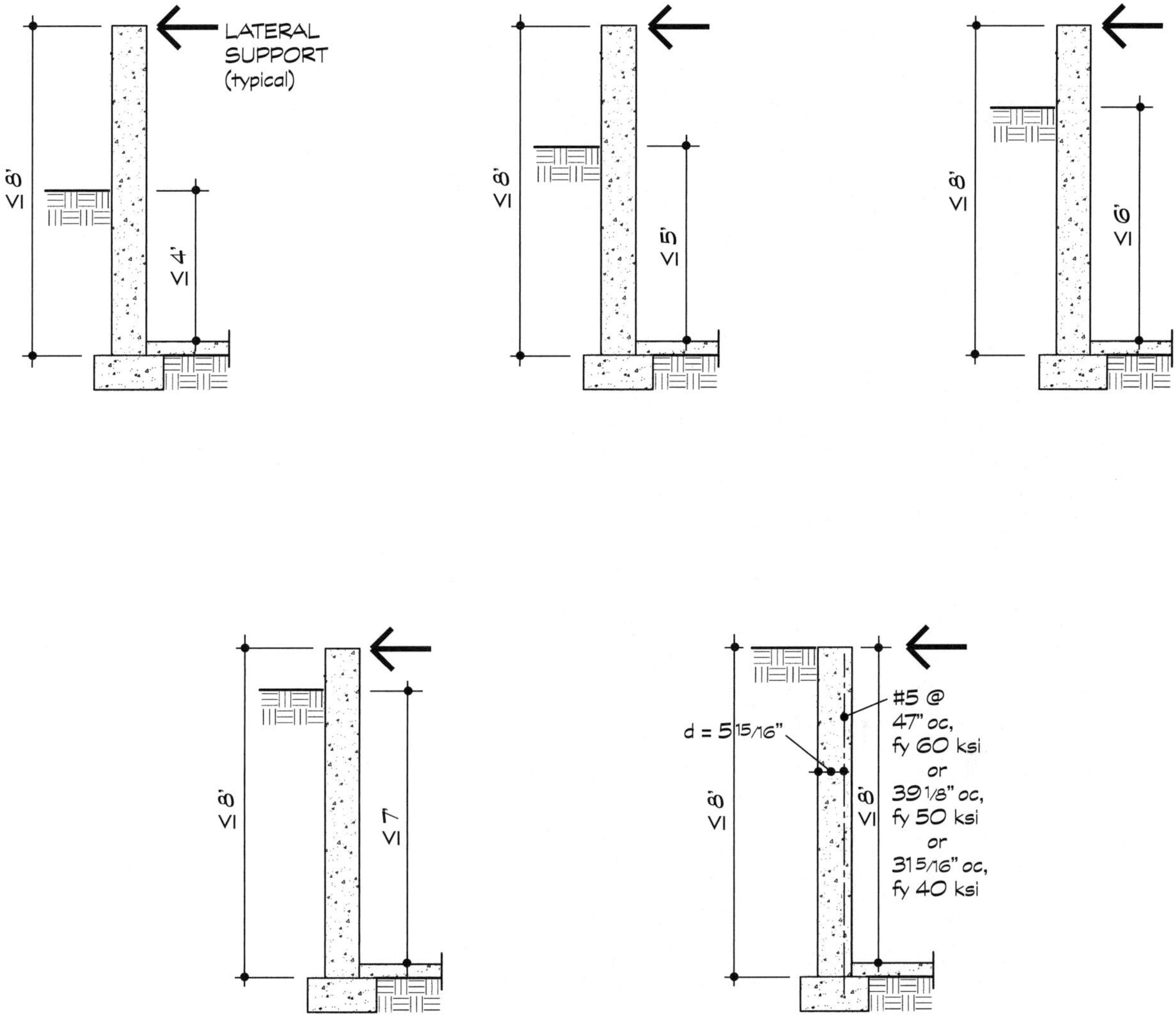

8'-HIGH WALLS

Fig. 1807.1.6.2-3. Plain concrete foundation wall, thickness ≥ 7 1/2" for a design lateral soil load of 30 psf/ft of depth. *[IBC Table 1807.1.6.2]*

1807 Foundation Walls, Retaining Walls and Embedded Posts and Poles

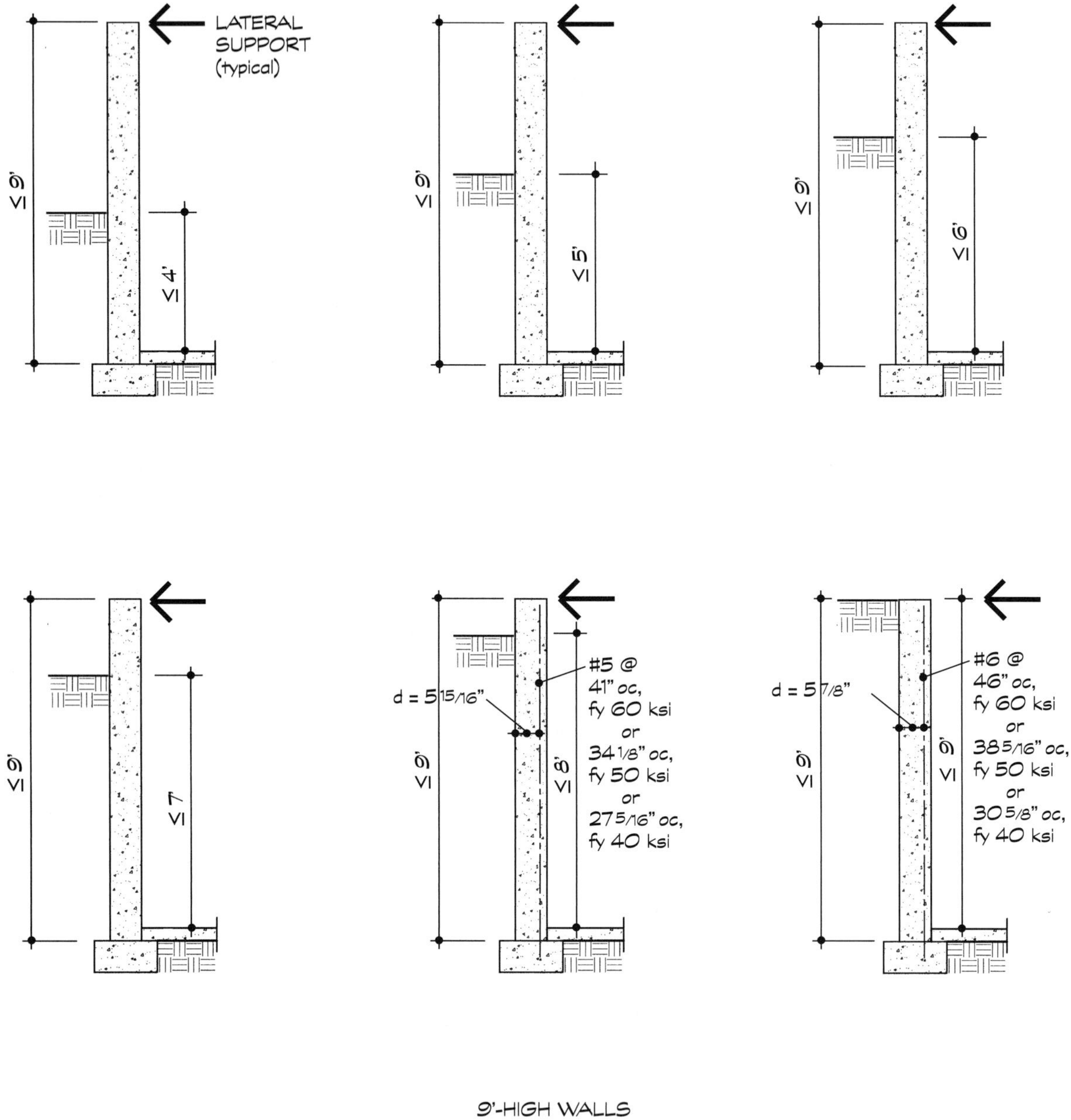

Fig. 1807.1.6.2-4. Plain concrete foundation wall, thickness ≥ 7 1/2" for a design lateral soil load of 30 psf/ft of depth. *[IBC Table 1807.1.6.2]*

1807 Foundation Walls, Retaining Walls and Embedded Posts and Poles

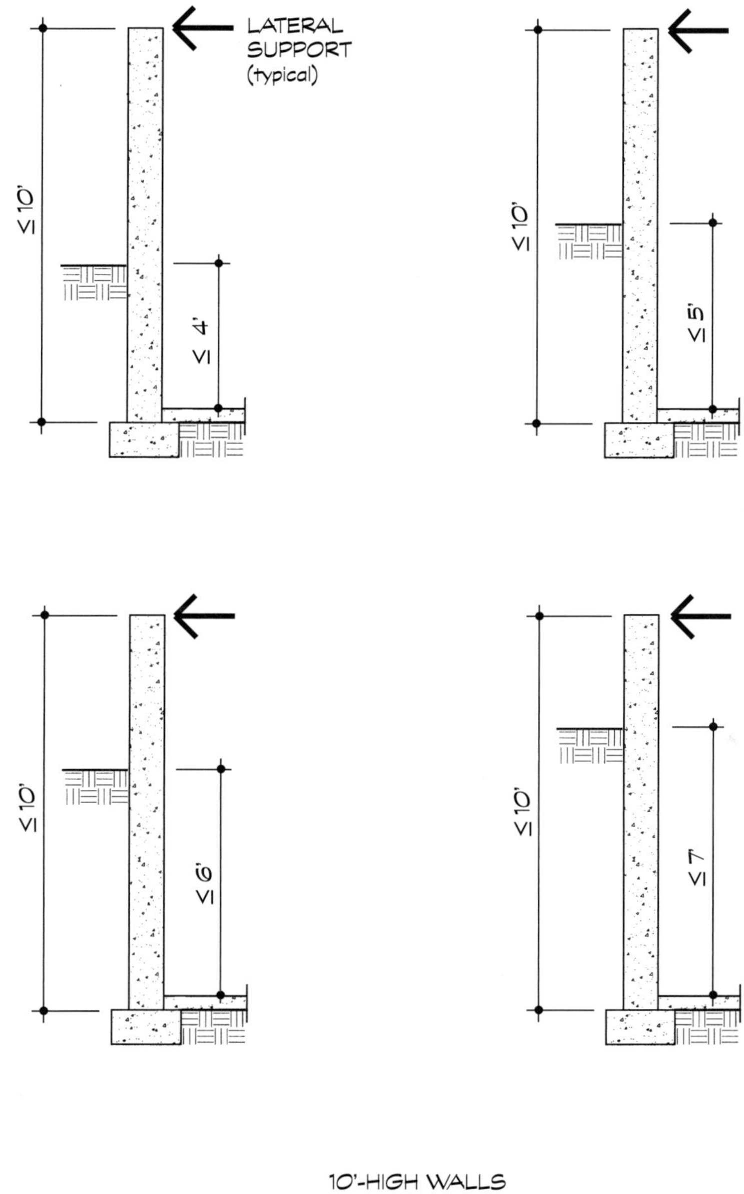

Fig. 1807.1.6.2-5a. Concrete foundation wall (plain concrete where reinforcing is not shown), thickness ≥ 7 1/2" for a design lateral soil load of 30 psf/ft of depth. *[IBC Table 1807.1.6.2]*

1807 Foundation Walls, Retaining Walls and Embedded Posts and Poles

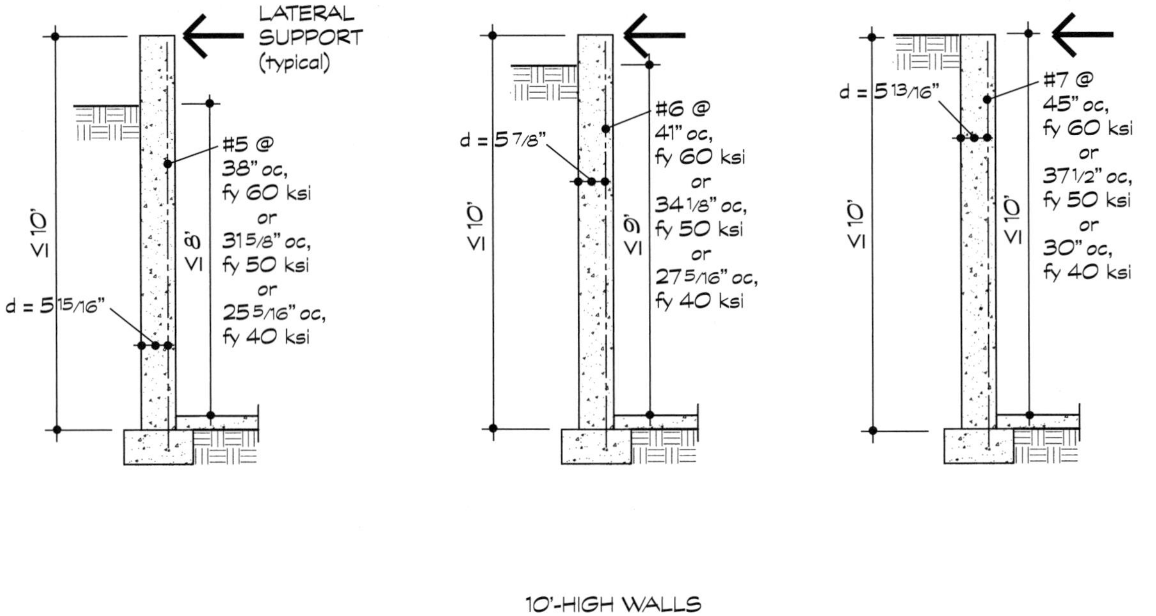

Fig. 1807.1.6.2-5b. Concrete foundation wall (plain concrete where reinforcing is not shown), thickness ≥ 7 1/2" for a design lateral soil load of 30 psf/ft of depth. *[IBC Table 1807.1.6.2]*

1807 Foundation Walls, Retaining Walls and Embedded Posts and Poles

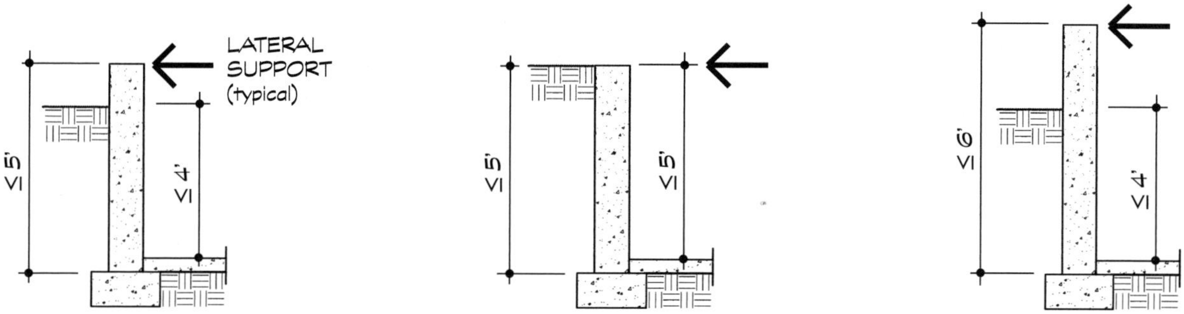

5'-HIGH WALLS

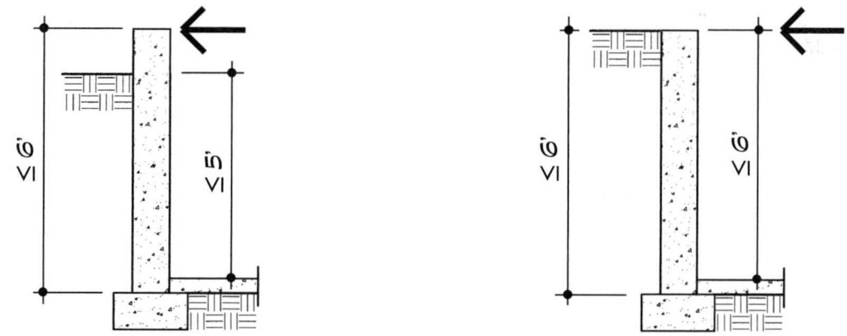

6'-HIGH WALLS

Fig. 1807.1.6.2-6. Plain concrete foundation wall, thickness ≥ 91/2" for a design lateral soil load of 30 psf/ft of depth. *[IBC Table 1807.1.6.2]*

1807 Foundation Walls, Retaining Walls and Embedded Posts and Poles

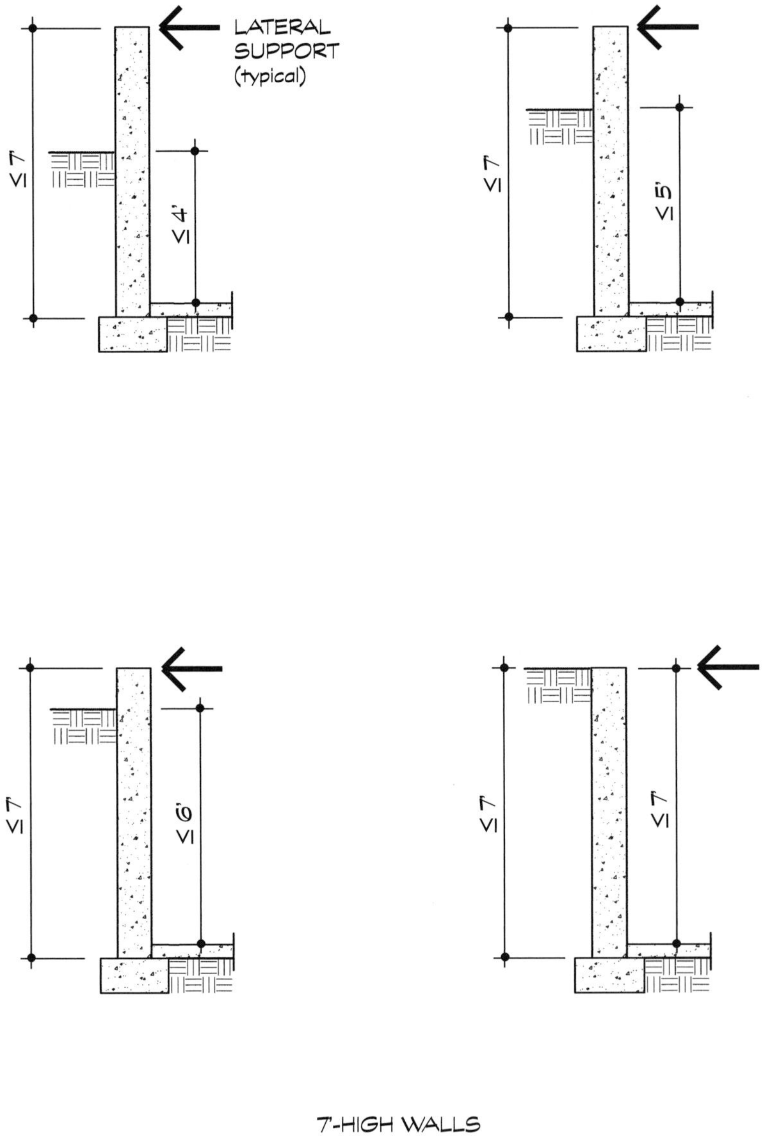

Fig. 1807.1.6.2-7. Plain concrete foundation wall, thickness ≥ 9 1/2" for a design lateral soil load of 30 psf/ft of depth. *[IBC Table 1807.1.6.2]*

1807 Foundation Walls, Retaining Walls and Embedded Posts and Poles

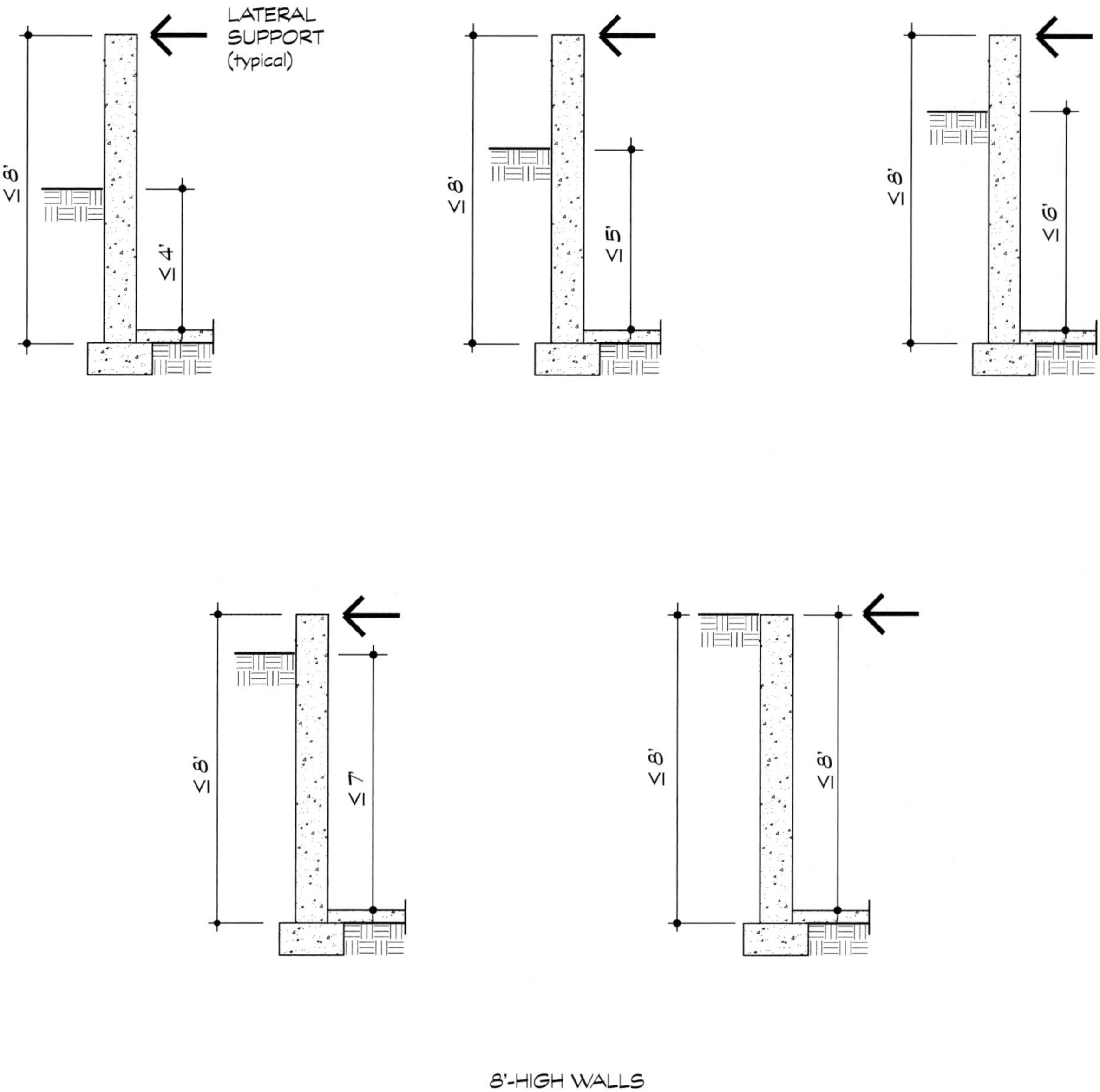

Fig. 1807.1.6.2-8. Plain concrete foundation wall, thickness ≥ 91/2" for a design lateral soil load of 30 psf/ft of depth. *[IBC Table 1807.1.6.2]*

1807 Foundation Walls, Retaining Walls and Embedded Posts and Poles

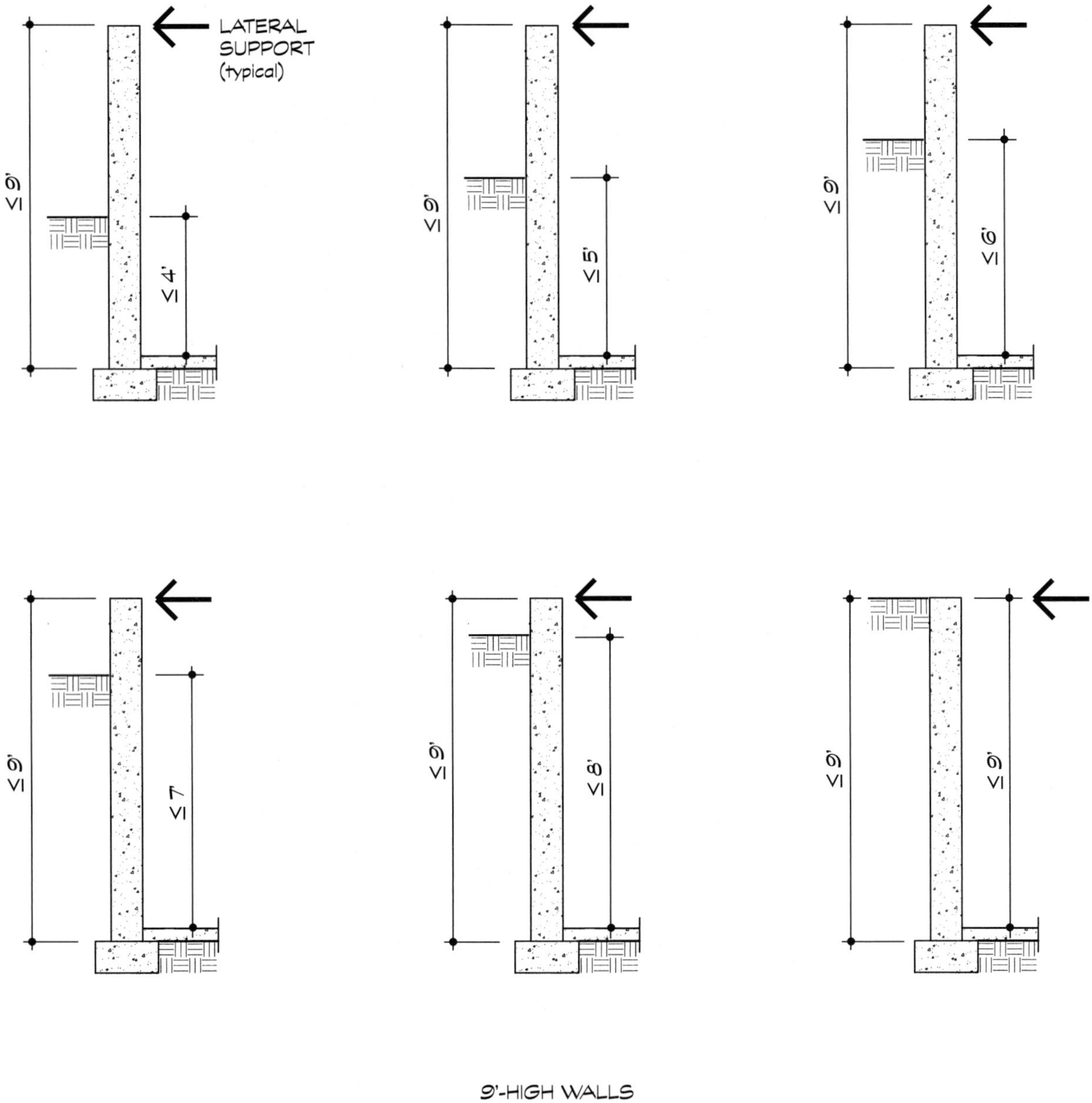

Fig. 1807.1.6.2-9. Plain concrete foundation wall, thickness ≥ 9 1/2" for a design lateral soil load of 30 psf/ft of depth. *[IBC Table 1807.1.6.2]*

1807 Foundation Walls, Retaining Walls and Embedded Posts and Poles

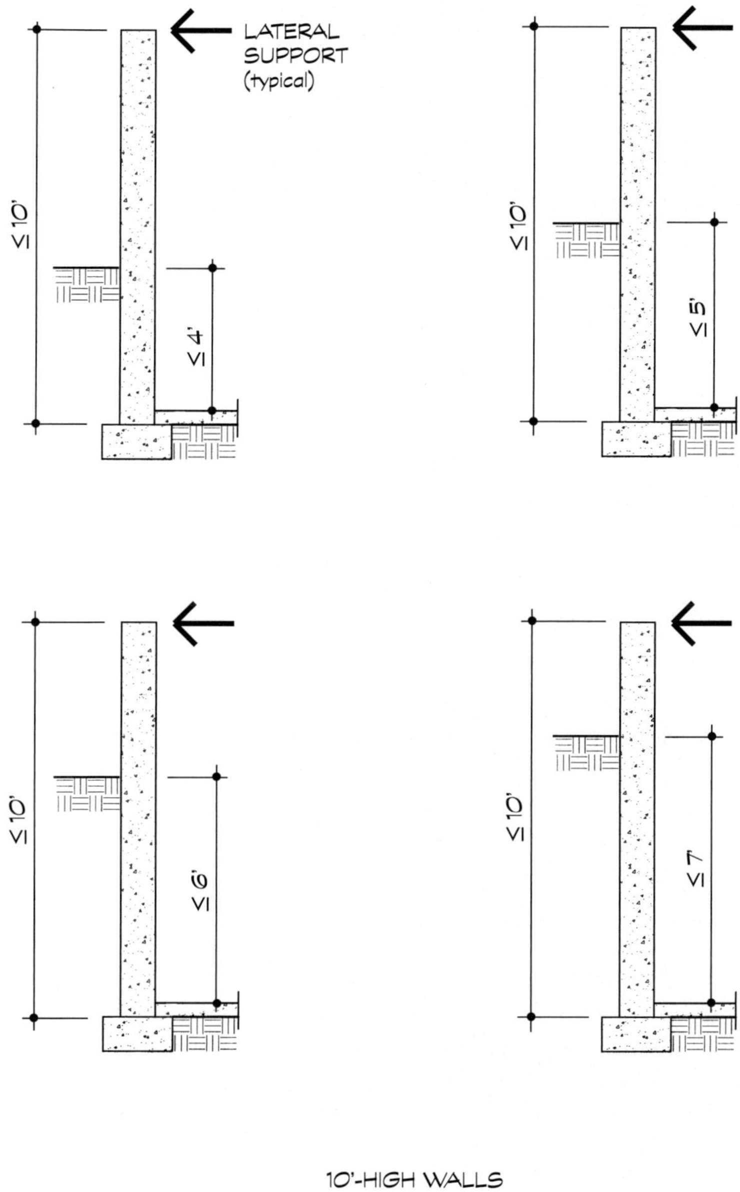

Fig. 1807.1.6.2-10a. Concrete foundation wall (plain concrete where reinforcing is not shown), thickness ≥ 9½" for a design lateral soil load of 30 psf/ft of depth. *[IBC Table 1807.1.6.2]*

1807 Foundation Walls, Retaining Walls and Embedded Posts and Poles

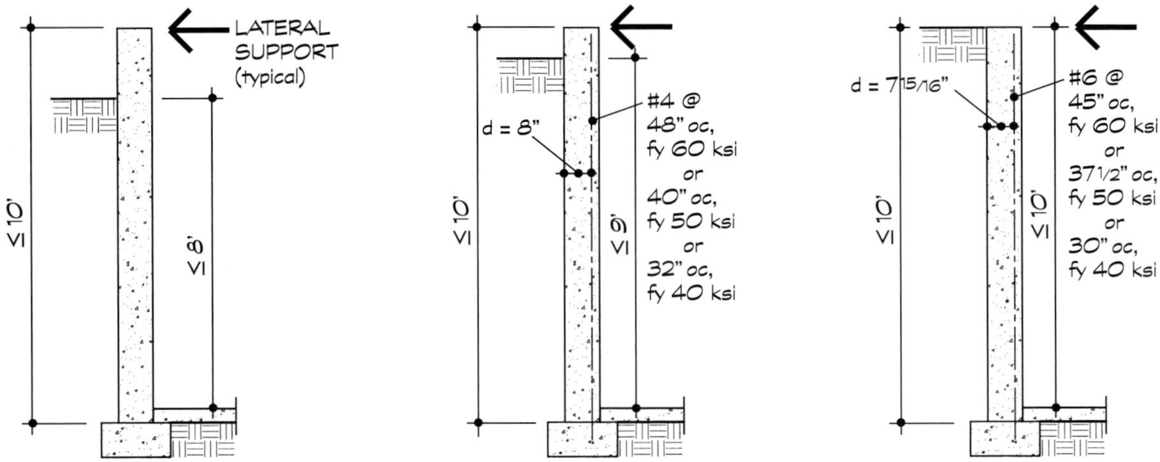

10'-HIGH WALLS

Fig. 1807.1.6.2-10b. Concrete foundation wall (plain concrete where reinforcing is not shown), thickness ≥ 9 1/2" for a design lateral soil load of 30 psf/ft of depth. *[IBC Table 1807.1.6.2]*

1807 Foundation Walls, Retaining Walls and Embedded Posts and Poles

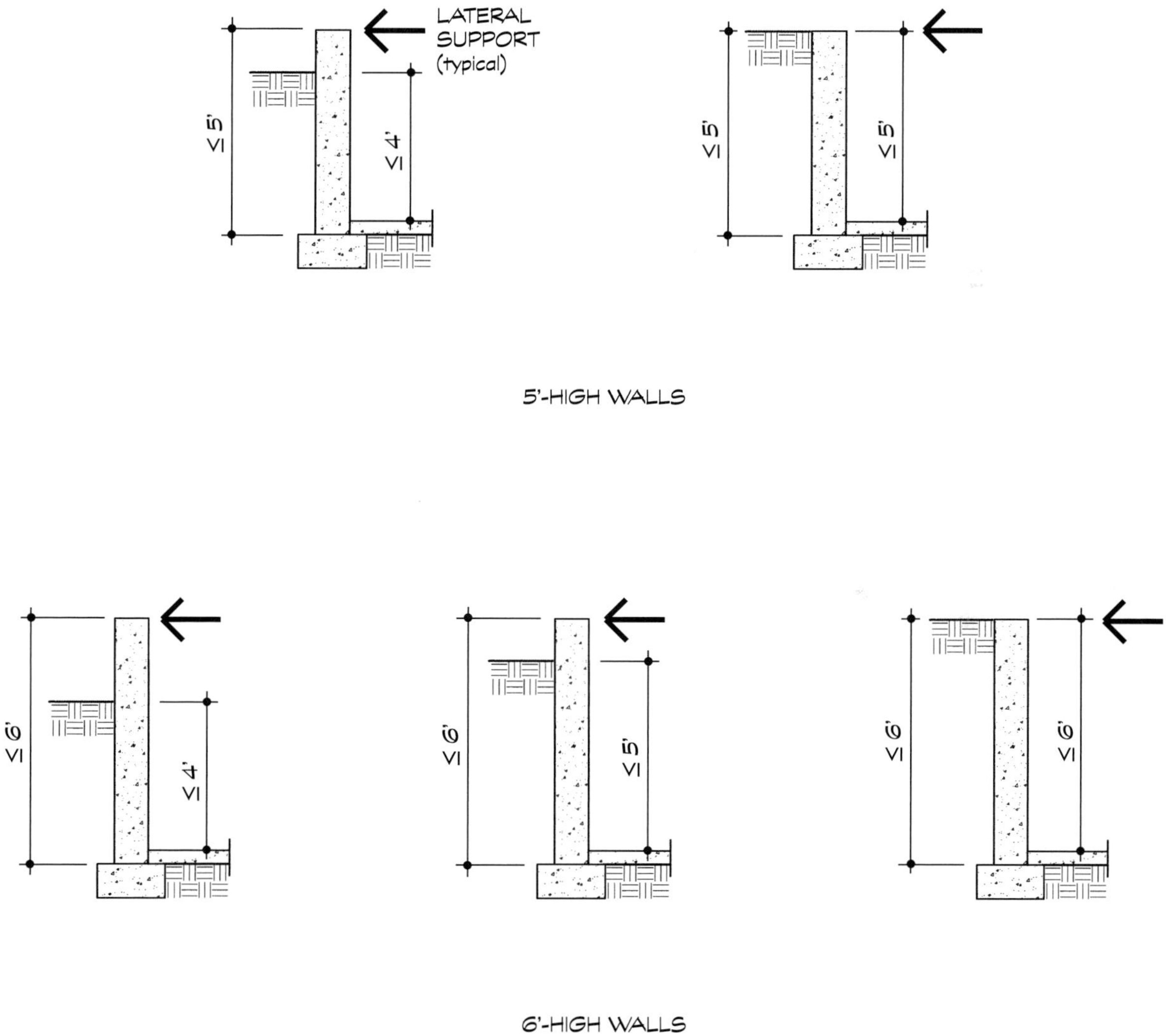

Fig. 1807.1.6.2-11. Concrete foundation wall (plain concrete where reinforcing is not shown), thickness ≥ 11 1/2" for a design lateral soil load of 30 psf/ft of depth. *[IBC Table 1807.1.6.2]*

1807 Foundation Walls, Retaining Walls and Embedded Posts and Poles

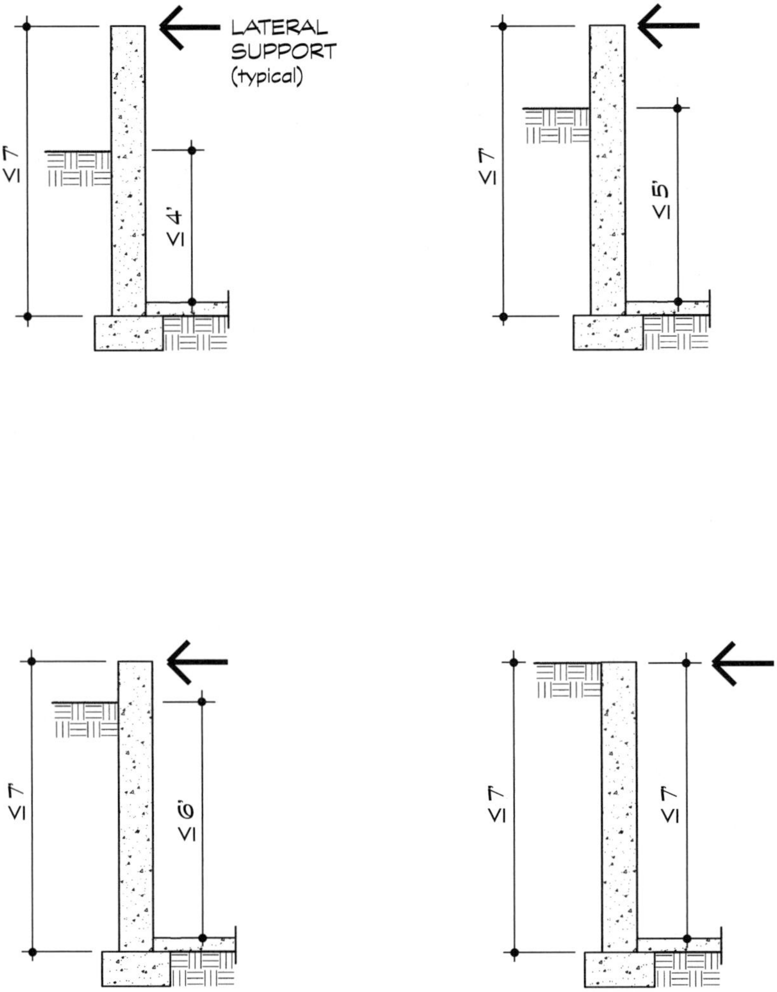

7'-HIGH WALLS

Fig. 1807.1.6.2-12. Concrete foundation wall (plain concrete where reinforcing is not shown), thickness ≥ 11 1/2" for a design lateral soil load of 30 psf/ft of depth. *[IBC Table 1807.1.6.2]*

1807 Foundation Walls, Retaining Walls and Embedded Posts and Poles

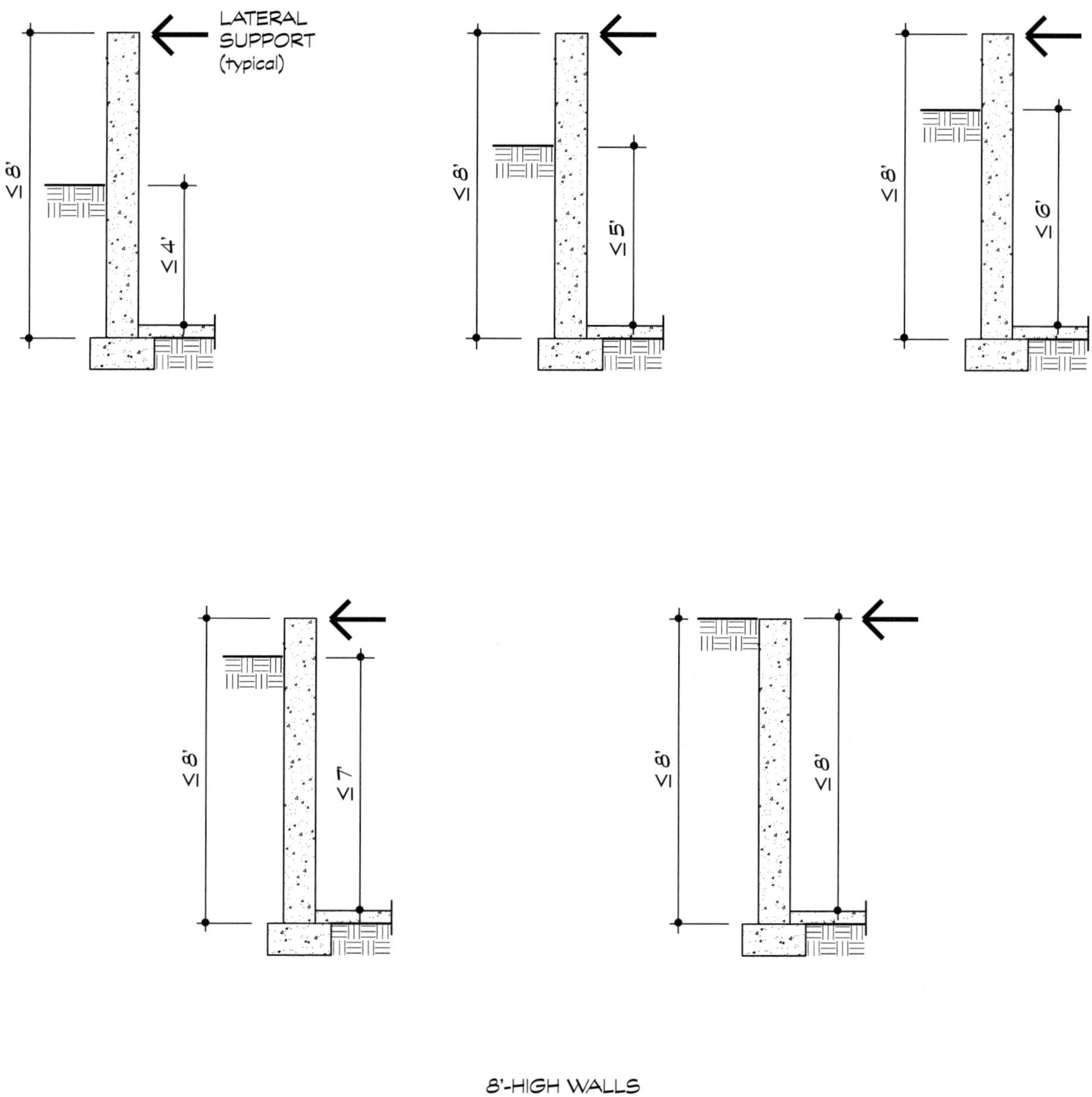

Fig. 1807.1.6.2-13. Plain concrete foundation wall, thickness ≥ 11 1/2" for a design lateral soil load of 30 psf/ft of depth. *[IBC Table 1807.1.6.2]*

1807 Foundation Walls, Retaining Walls and Embedded Posts and Poles

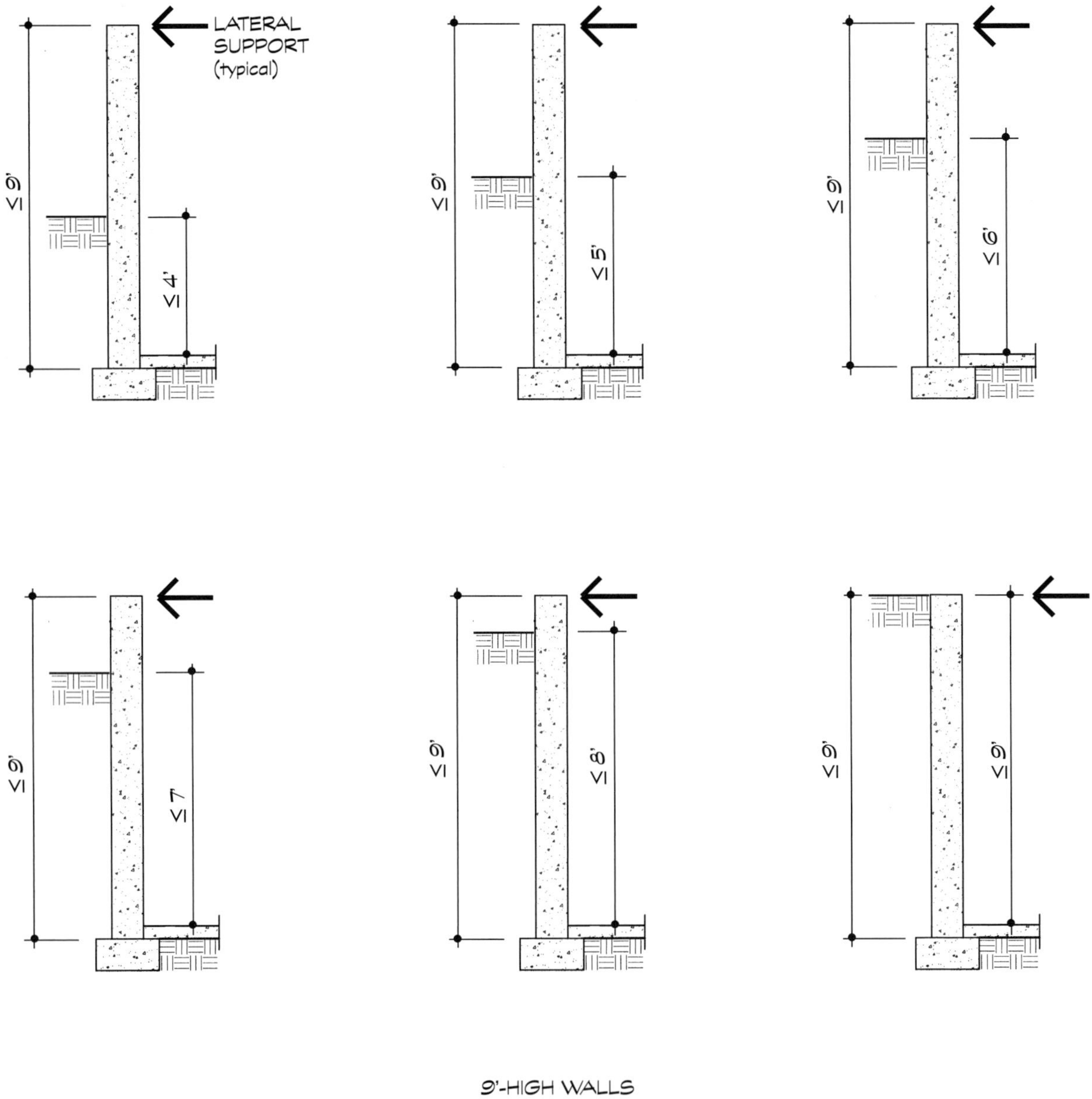

Fig. 1807.1.6.2-14. Plain concrete foundation wall, thickness ≥ 11 1/2" for a design lateral soil load of 30 psf/ft of depth. *[IBC Table 1807.1.6.2]*

1807 Foundation Walls, Retaining Walls and Embedded Posts and Poles

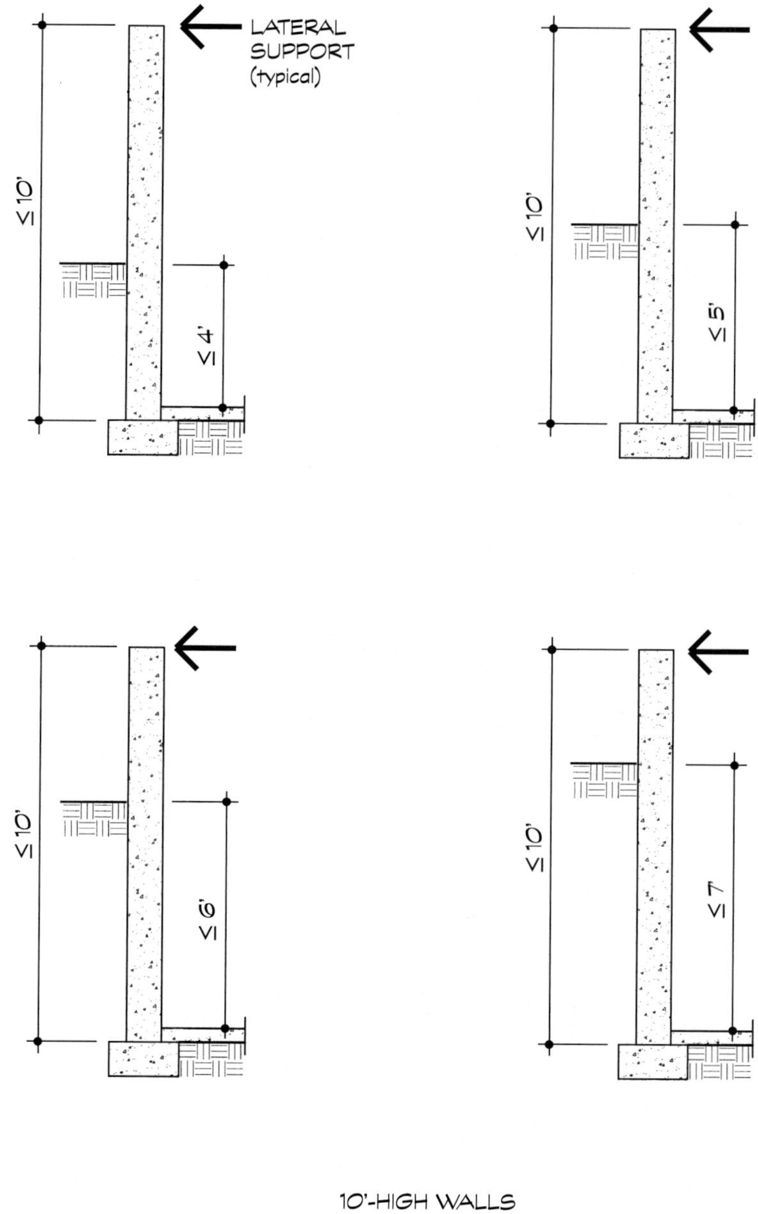

Fig. 1807.1.6.2-15a. Plain concrete foundation wall, thickness ≥ 11 1/2" for a design lateral soil load of 30 psf/ft of depth. *[IBC Table 1807.1.6.2]*

1807 Foundation Walls, Retaining Walls and Embedded Posts and Poles

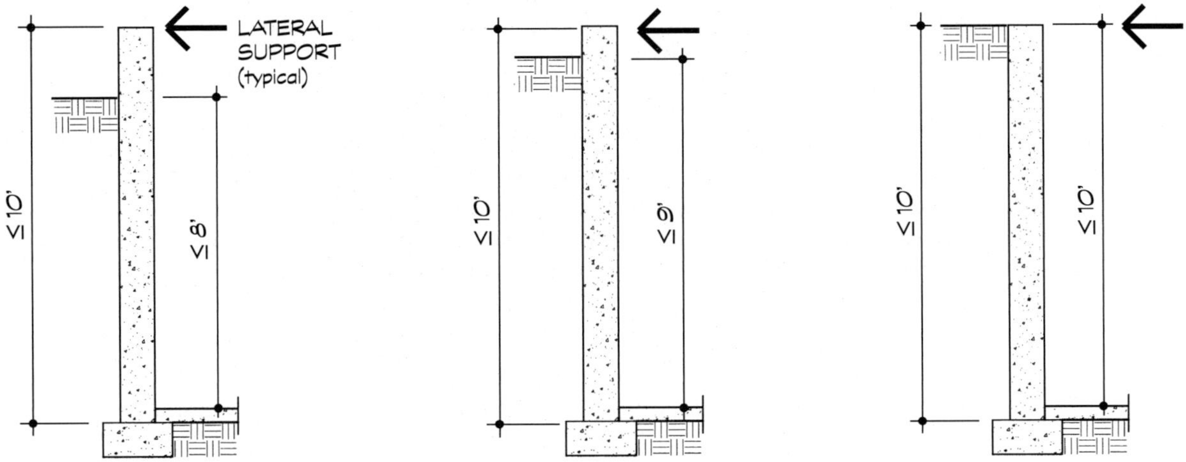

10'-HIGH WALLS

Fig. 1807.1.6.2-15b. Plain concrete foundation wall, thickness ≥ 11 1/2" for a design lateral soil load of 30 psf/ft of depth. *[IBC Table 1807.1.6.2]*

1807 Foundation Walls, Retaining Walls and Embedded Posts and Poles

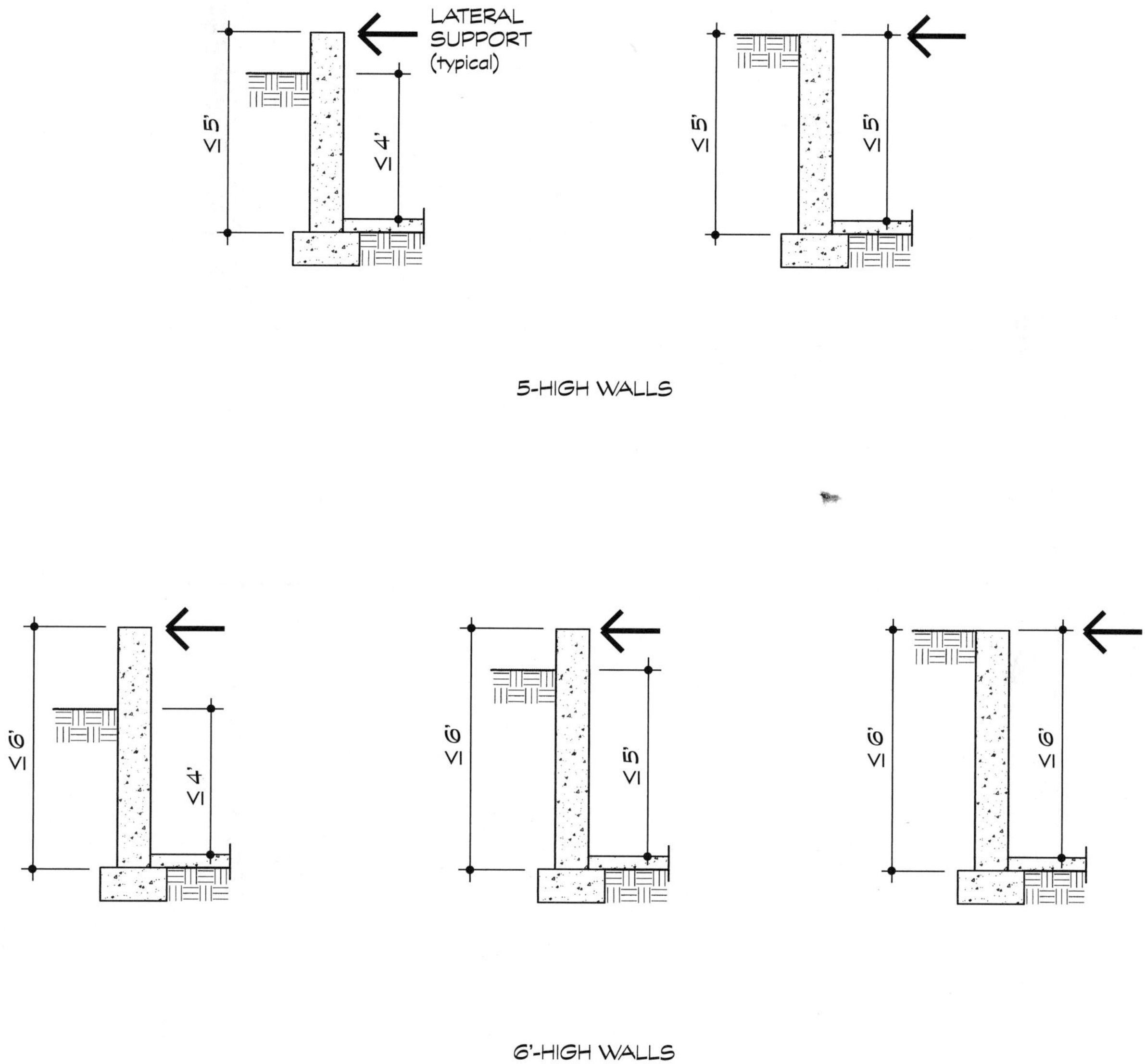

Fig. 1807.1.6.2-16. Concrete foundation wall (plain concrete where reinforcing is not shown), thickness ≥ 7 1/2" for a design lateral soil load of 45 psf/ft of depth. *[IBC Table 1807.1.6.2]*

1807 Foundation Walls, Retaining Walls and Embedded Posts and Poles

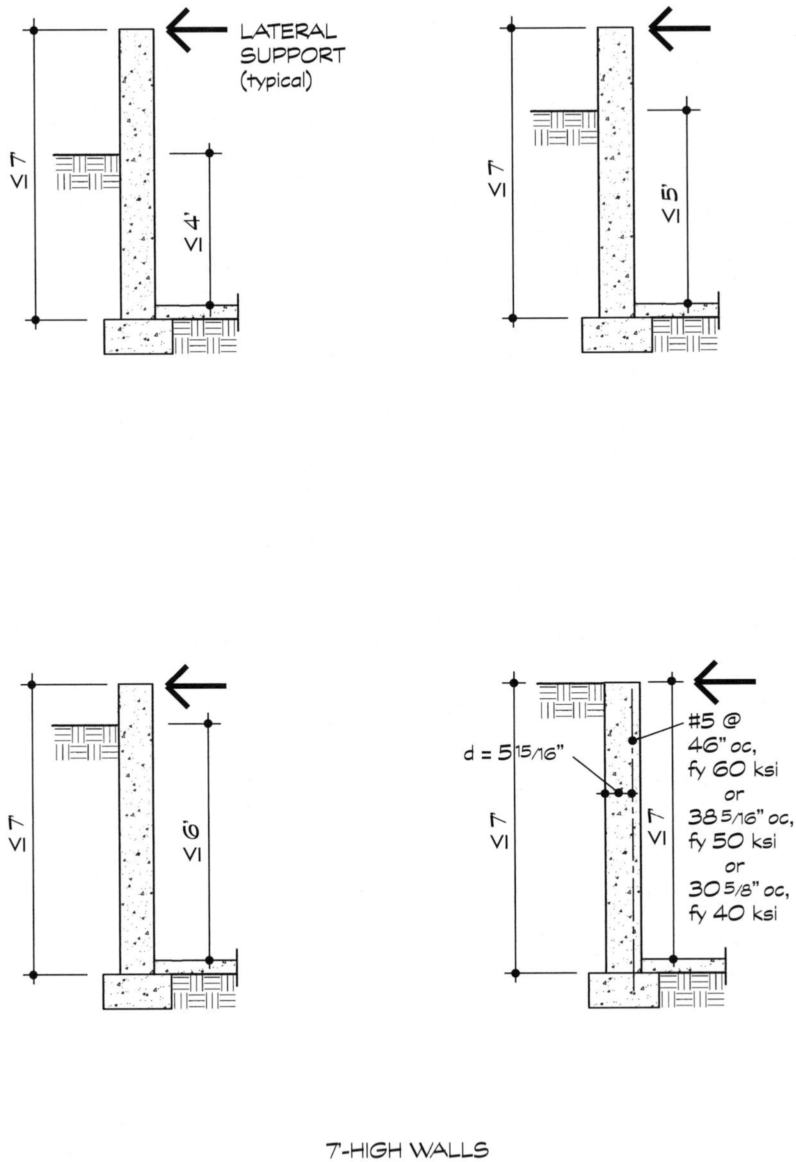

Fig. 1807.1.6.2-17. Concrete foundation wall (plain concrete where reinforcing is not shown), thickness ≥ 71/2" for a design lateral soil load of 45 psf/ft of depth. *[IBC Table 1807.1.6.2]*

1807 Foundation Walls, Retaining Walls and Embedded Posts and Poles

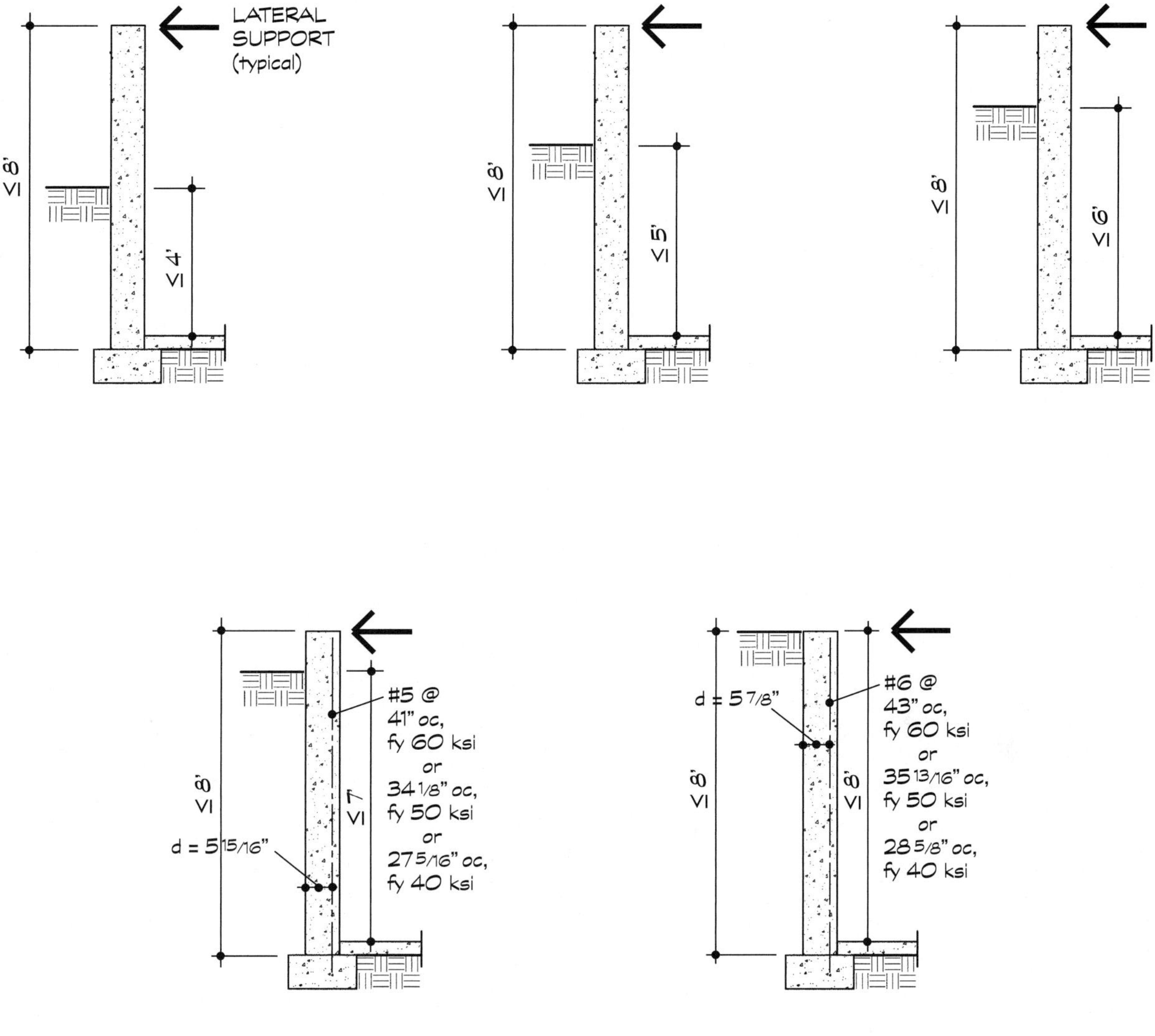

8'-HIGH WALLS

Fig. 1807.1.6.2-18. Concrete foundation wall (plain concrete where reinforcing is not shown), thickness ≥ 7 1/2" for a design lateral soil load of 45 psf/ft of depth. *[IBC Table 1807.1.6.2]*

1807 Foundation Walls, Retaining Walls and Embedded Posts and Poles

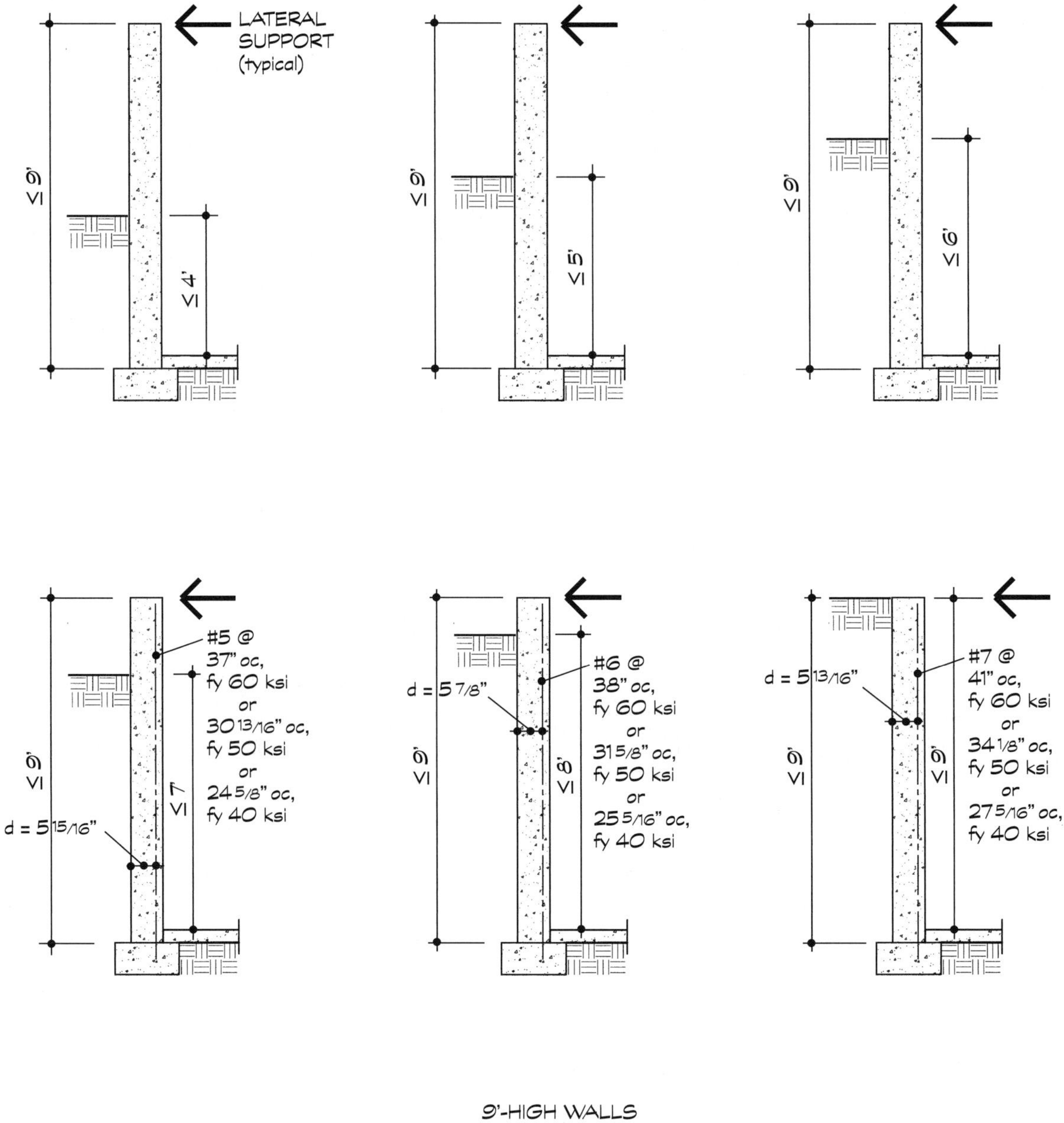

Fig. 1807.1.6.2-19. Concrete foundation wall (plain concrete where reinforcing is not shown), thickness ≥ 7 1/2" for a design lateral soil load of 45 psf/ft of depth. *[IBC Table 1807.1.6.2]*

1807 Foundation Walls, Retaining Walls and Embedded Posts and Poles

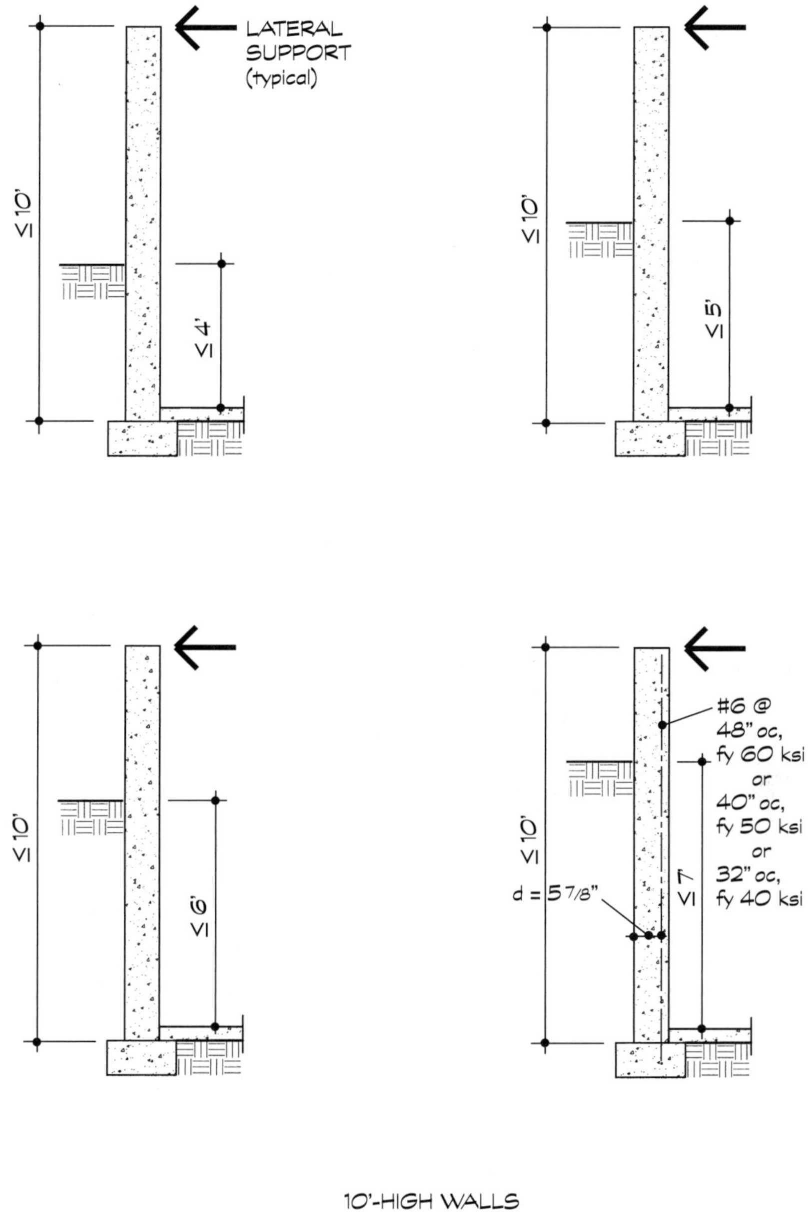

Fig. 1807.1.6.2-20a. Concrete foundation wall (plain concrete where reinforcing is not shown), thickness ≥ 71/2" for a design lateral soil load of 45 psf/ft of depth. *[IBC Table 1807.1.6.2]*

1807 Foundation Walls, Retaining Walls and Embedded Posts and Poles

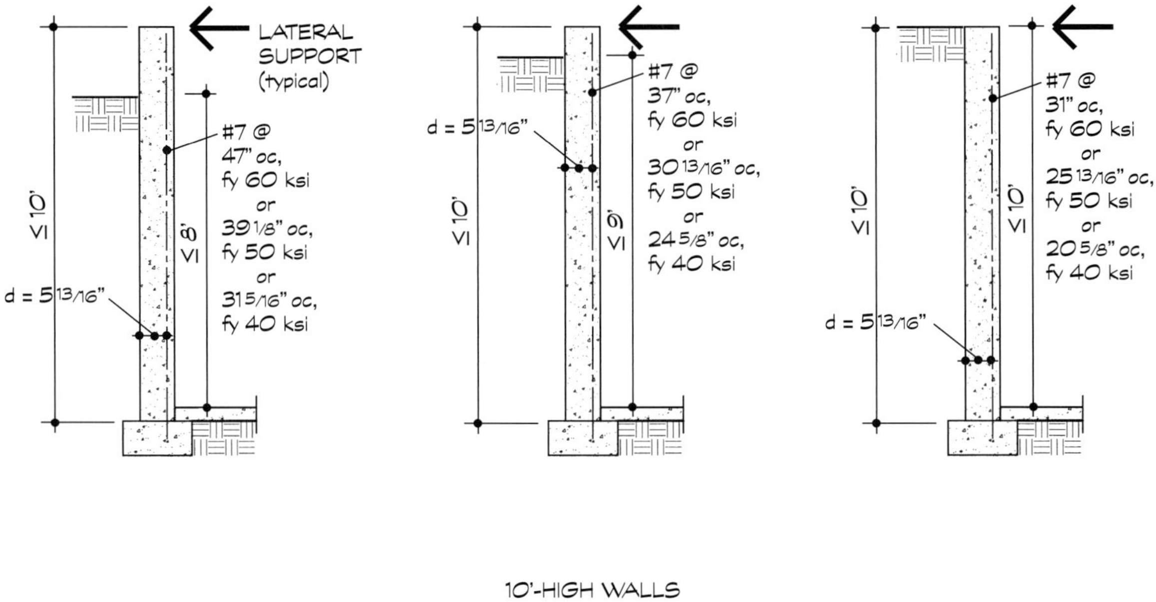

Fig. 1807.1.6.2-20b. Concrete foundation wall (plain concrete where reinforcing is not shown), thickness ≥ 7 1/2" for a design lateral soil load of 45 psf/ft of depth. *[IBC Table 1807.1.6.2]*

1807 Foundation Walls, Retaining Walls and Embedded Posts and Poles

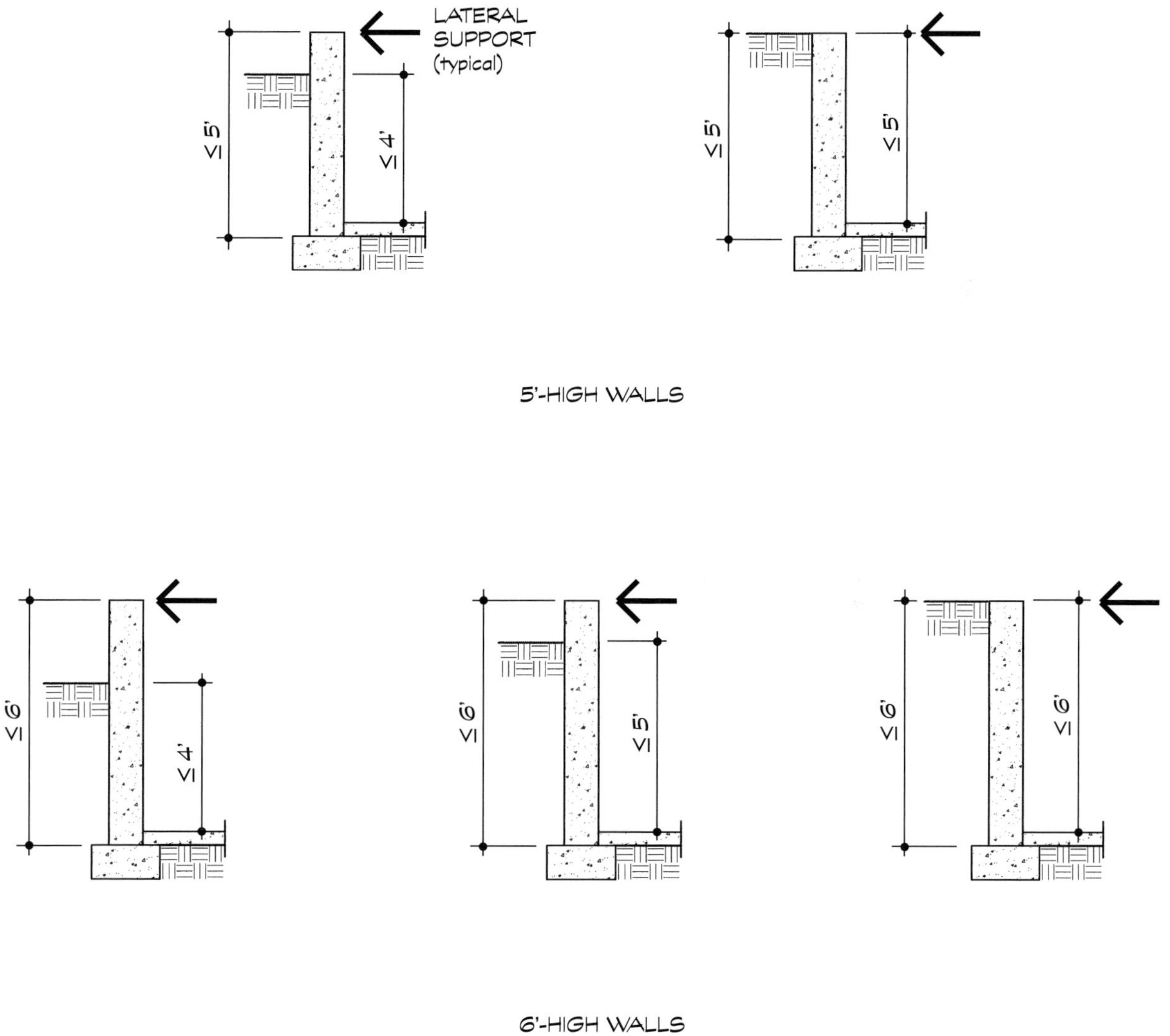

Fig. 1807.1.6.2-21. Plain concrete foundation wall, thickness ≥ 91/2" for a design lateral soil load of 45 psf/ft of depth. *[IBC Table 1807.1.6.2]*

1807 Foundation Walls, Retaining Walls and Embedded Posts and Poles

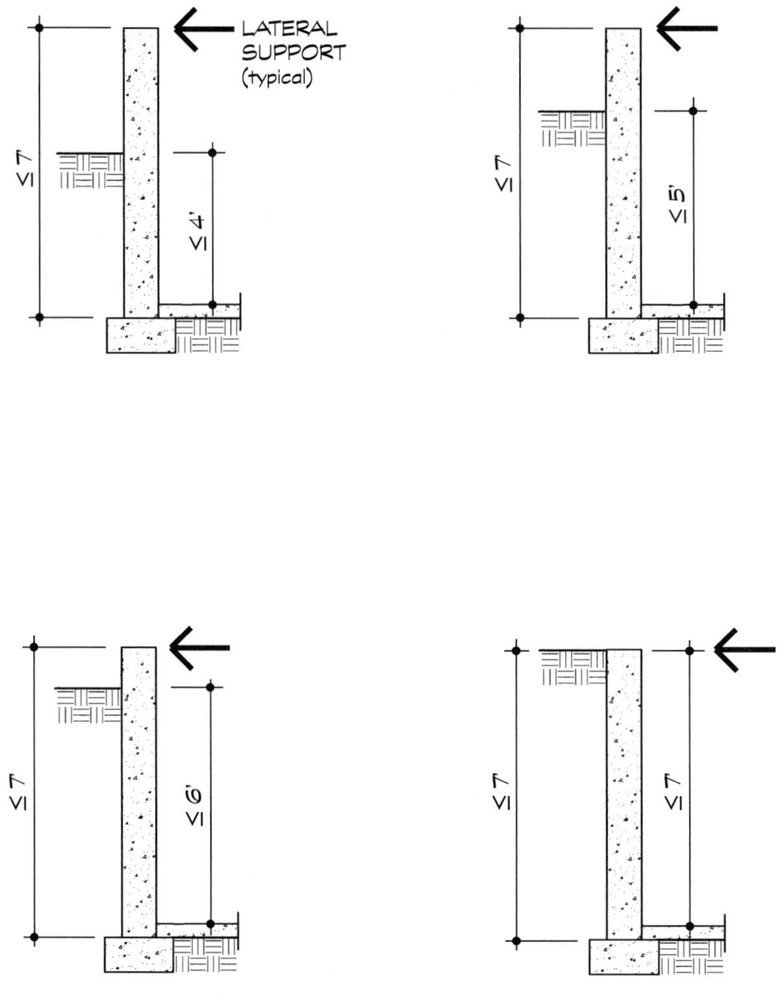

7'-HIGH WALLS

Fig. 1807.1.6.2-22. Plain concrete foundation wall, thickness ≥ 9 1/2" for a design lateral soil load of 45 psf/ft of depth. *[IBC Table 1807.1.6.2]*

1807 Foundation Walls, Retaining Walls and Embedded Posts and Poles

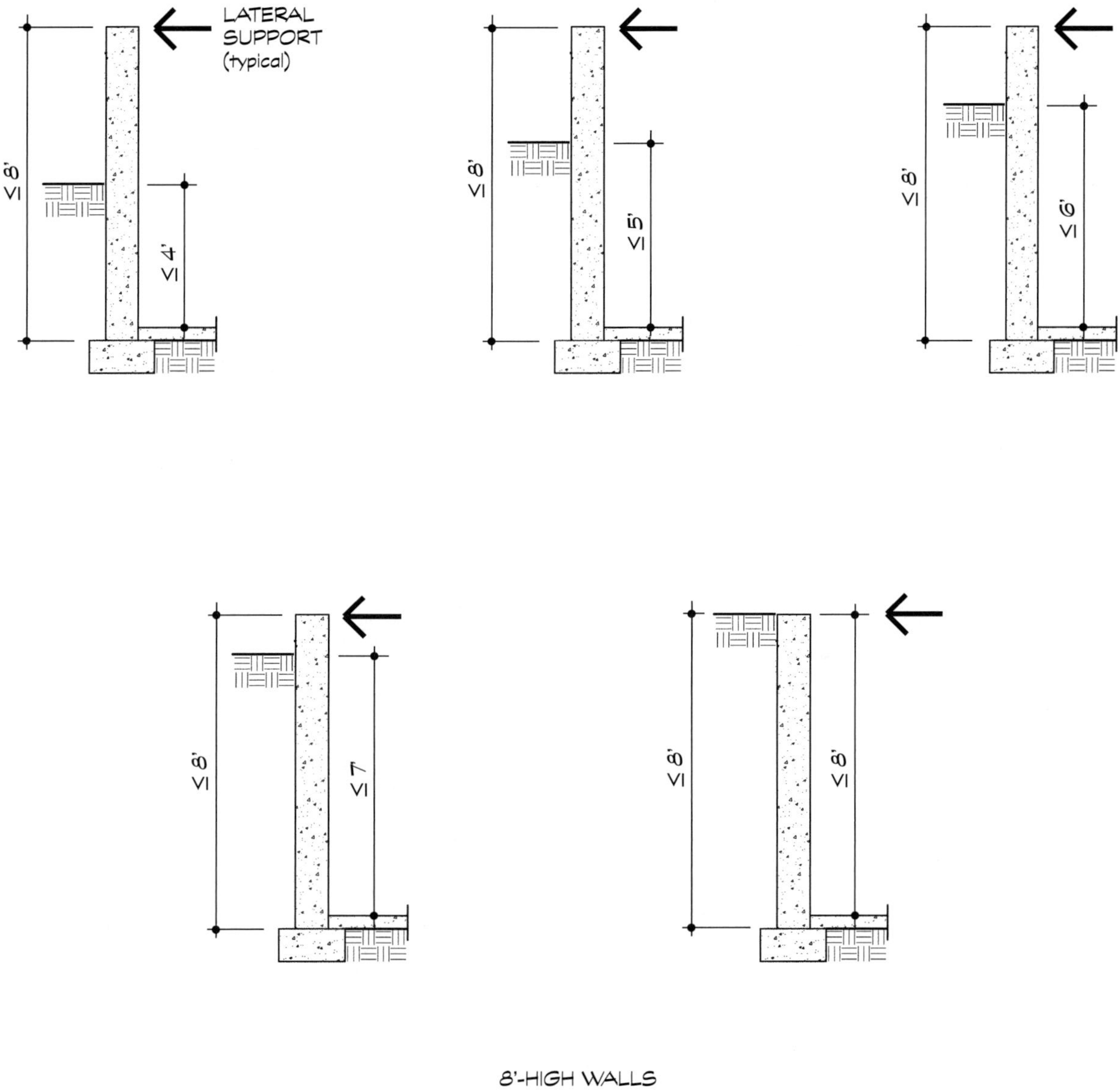

Fig. 1807.1.6.2-23. Plain concrete foundation wall, thickness ≥ 9 1/2" for a design lateral soil load of 45 psf/ft of depth. *[IBC Table 1807.1.6.2]*

1807 Foundation Walls, Retaining Walls and Embedded Posts and Poles

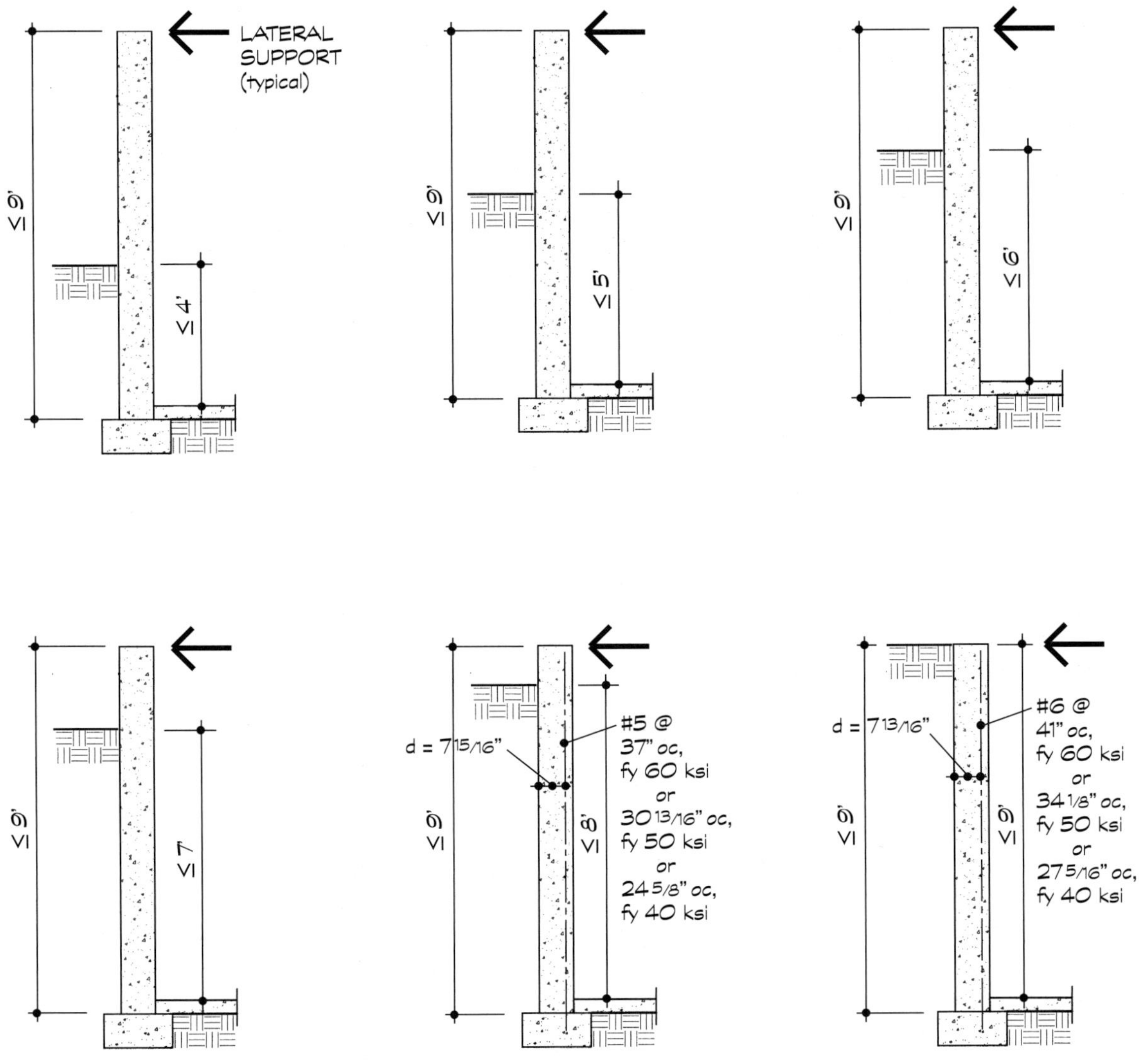

9'-HIGH WALLS

Fig. 1807.1.6.2-24. Plain concrete foundation wall, thickness ≥ 9 1/2" for a design lateral soil load of 45 psf/ft of depth. *[IBC Table 1807.1.6.2]*

1807 Foundation Walls, Retaining Walls and Embedded Posts and Poles

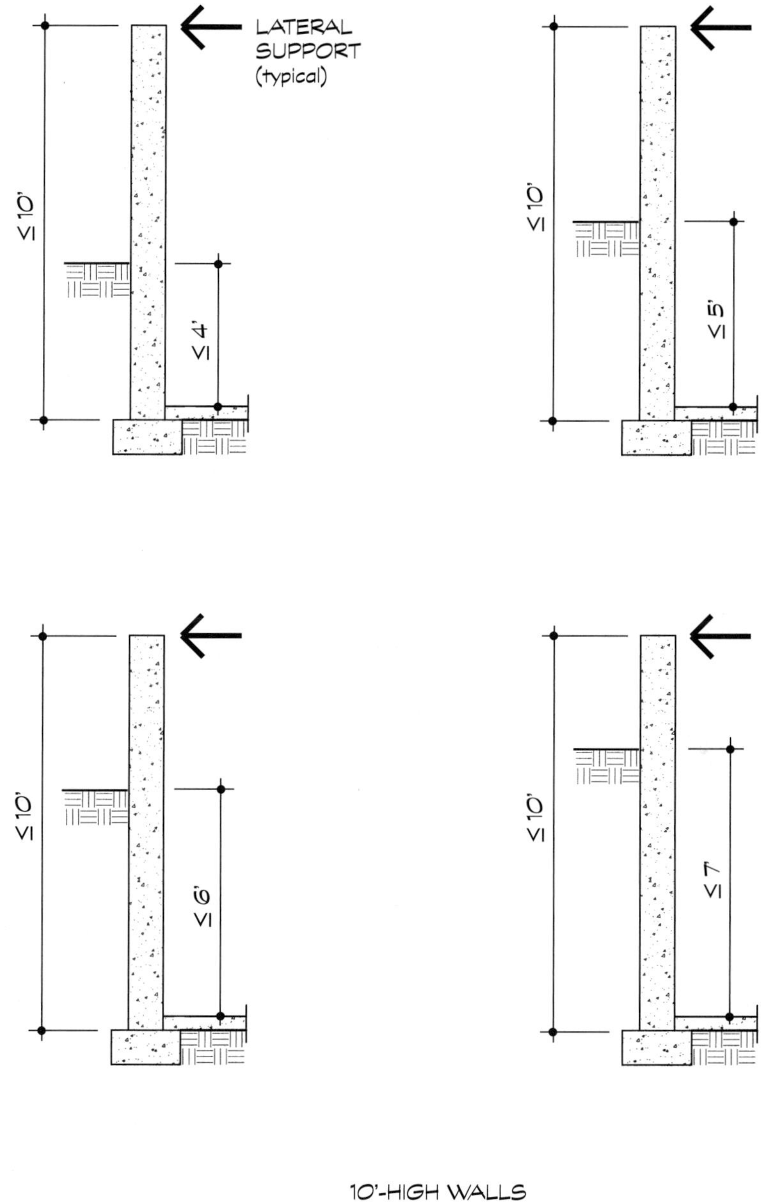

Fig. 1807.1.6.2-25a. Concrete foundation wall (plain concrete where reinforcing is not shown), thickness ≥ 9 1/2" for a design lateral soil load of 45 psf/ft of depth. *[IBC Table 1807.1.6.2]*

1807 Foundation Walls, Retaining Walls and Embedded Posts and Poles

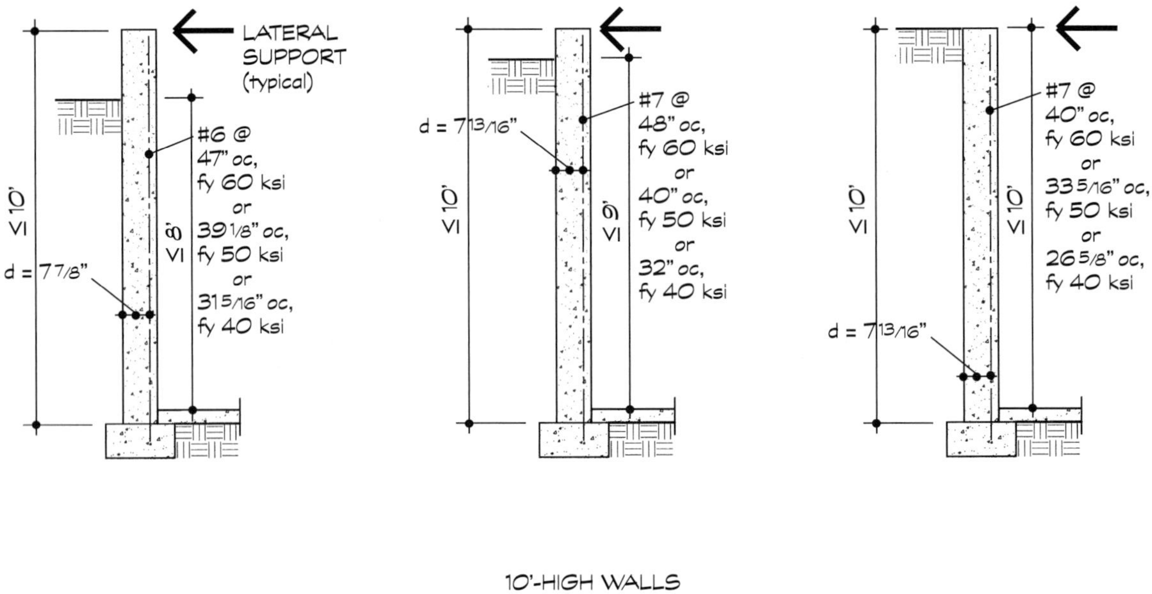

Fig. 1807.1.6.2-25b. Concrete foundation wall (plain concrete where reinforcing is not shown), thickness ≥ 91/2" for a design lateral soil load of 45 psf/ft of depth. *[IBC Table 1807.1.6.2]*

1807 Foundation Walls, Retaining Walls and Embedded Posts and Poles

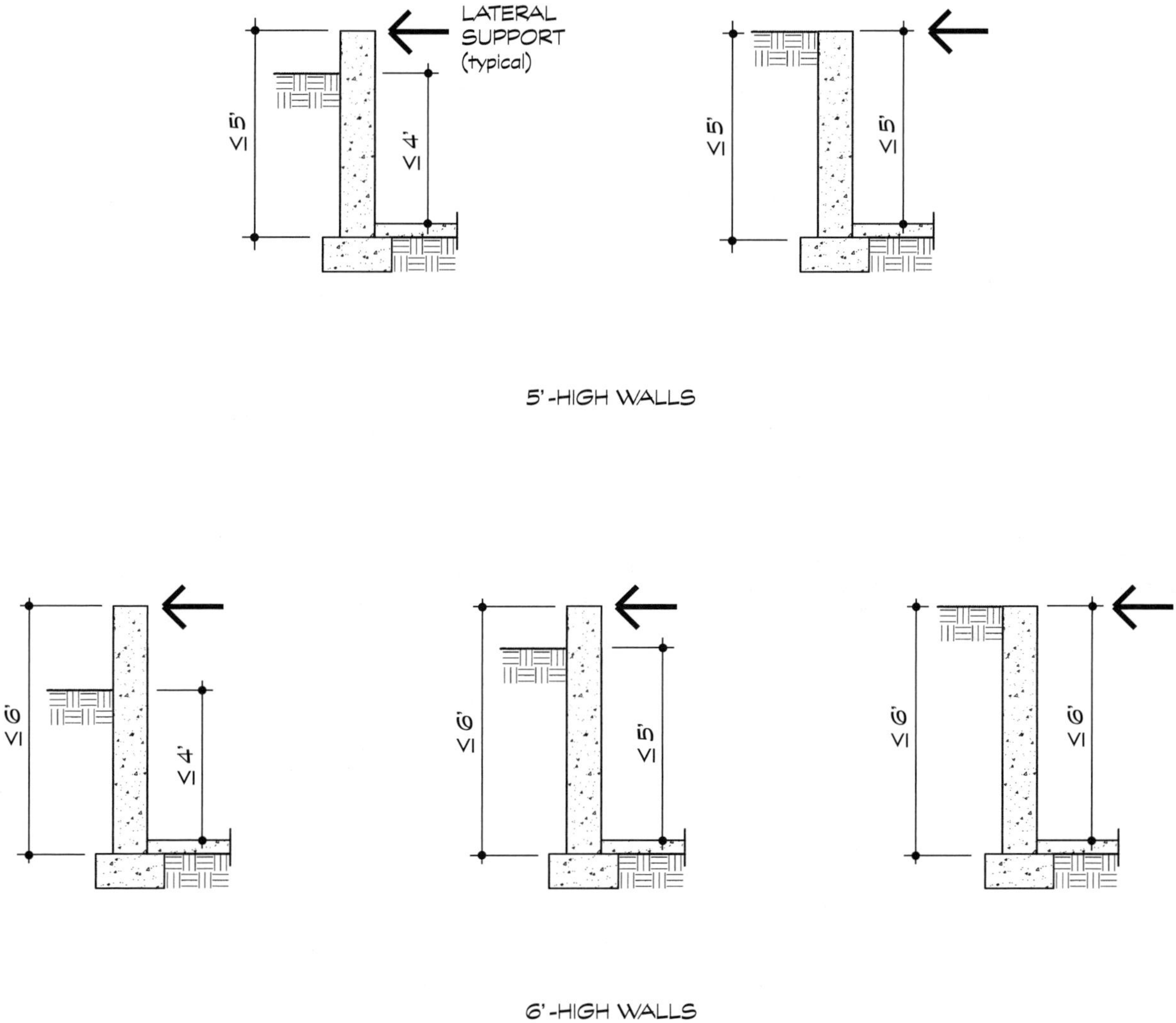

Fig. 1807.1.6.2-26. Plain concrete foundation wall, thickness ≥ 11 1/2" for a design lateral soil load of 45 psf/ft of depth. *[IBC Table 1807.1.6.2]*

1807 Foundation Walls, Retaining Walls and Embedded Posts and Poles

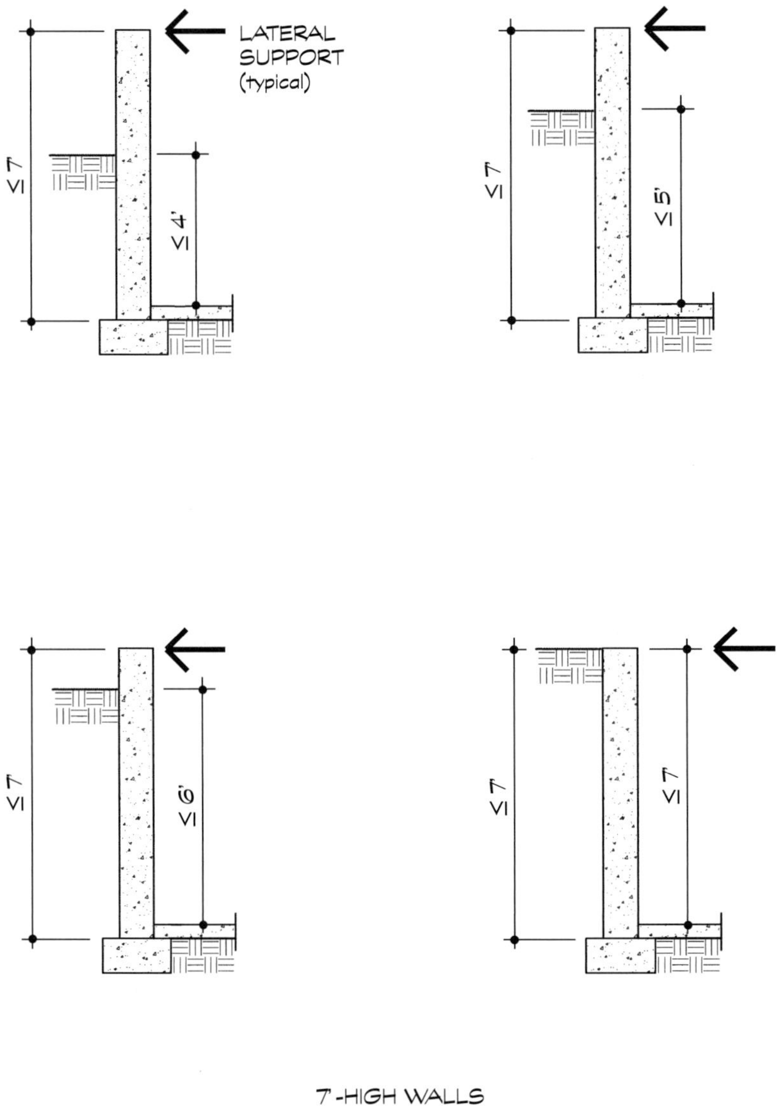

Fig. 1807.1.6.2-27. Plain concrete foundation wall, thickness ≥ 11 1/2" for a design lateral soil load of 45 psf/ft of depth. *[IBC Table 1807.1.6.2]*

1807 Foundation Walls, Retaining Walls and Embedded Posts and Poles

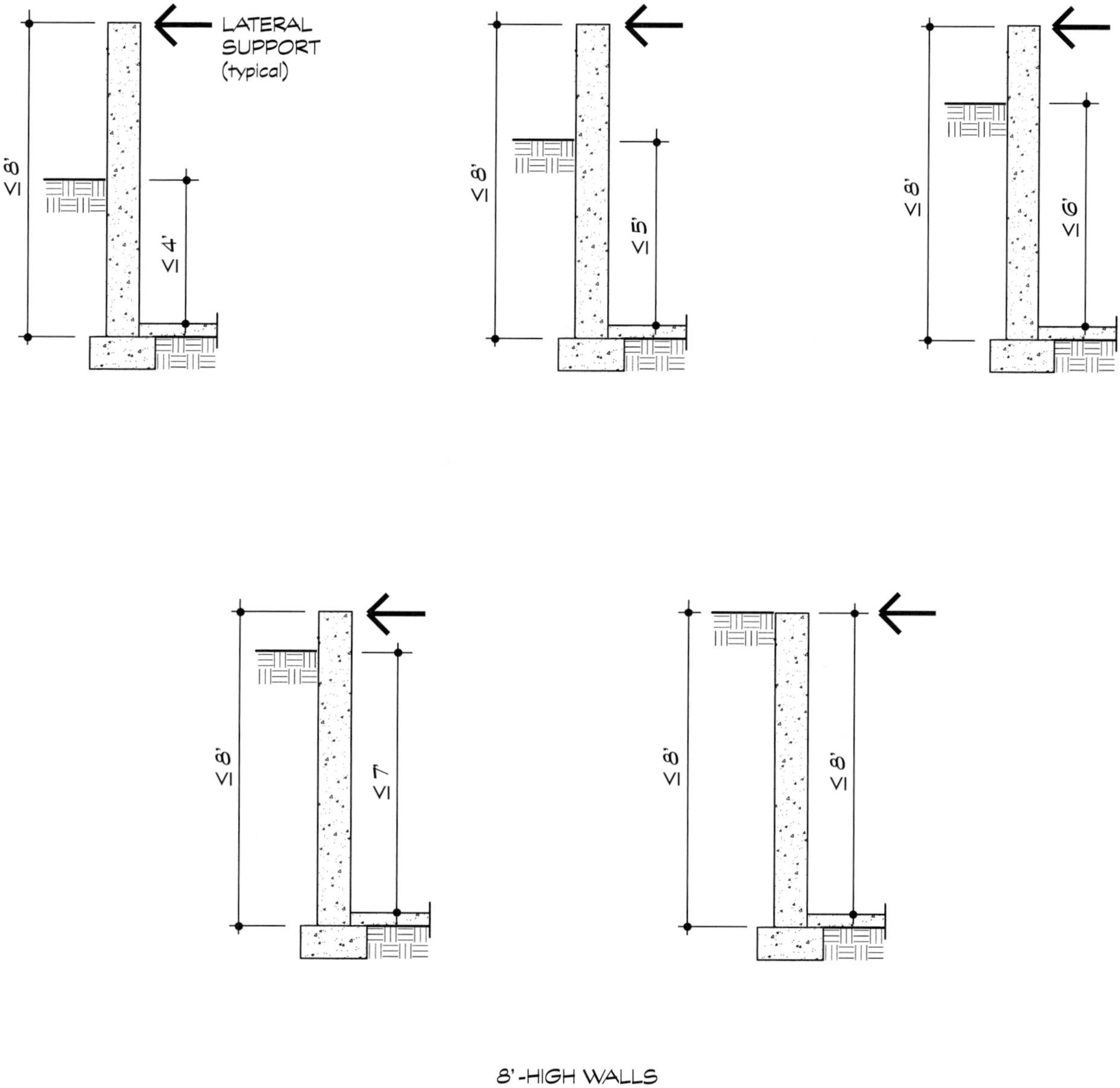

Fig. 1807.1.6.2-28. Plain concrete foundation wall, thickness ≥ 11 1/2" for a design lateral soil load of 45 psf/ft of depth. *[IBC Table 1807.1.6.2]*

1807 Foundation Walls, Retaining Walls and Embedded Posts and Poles

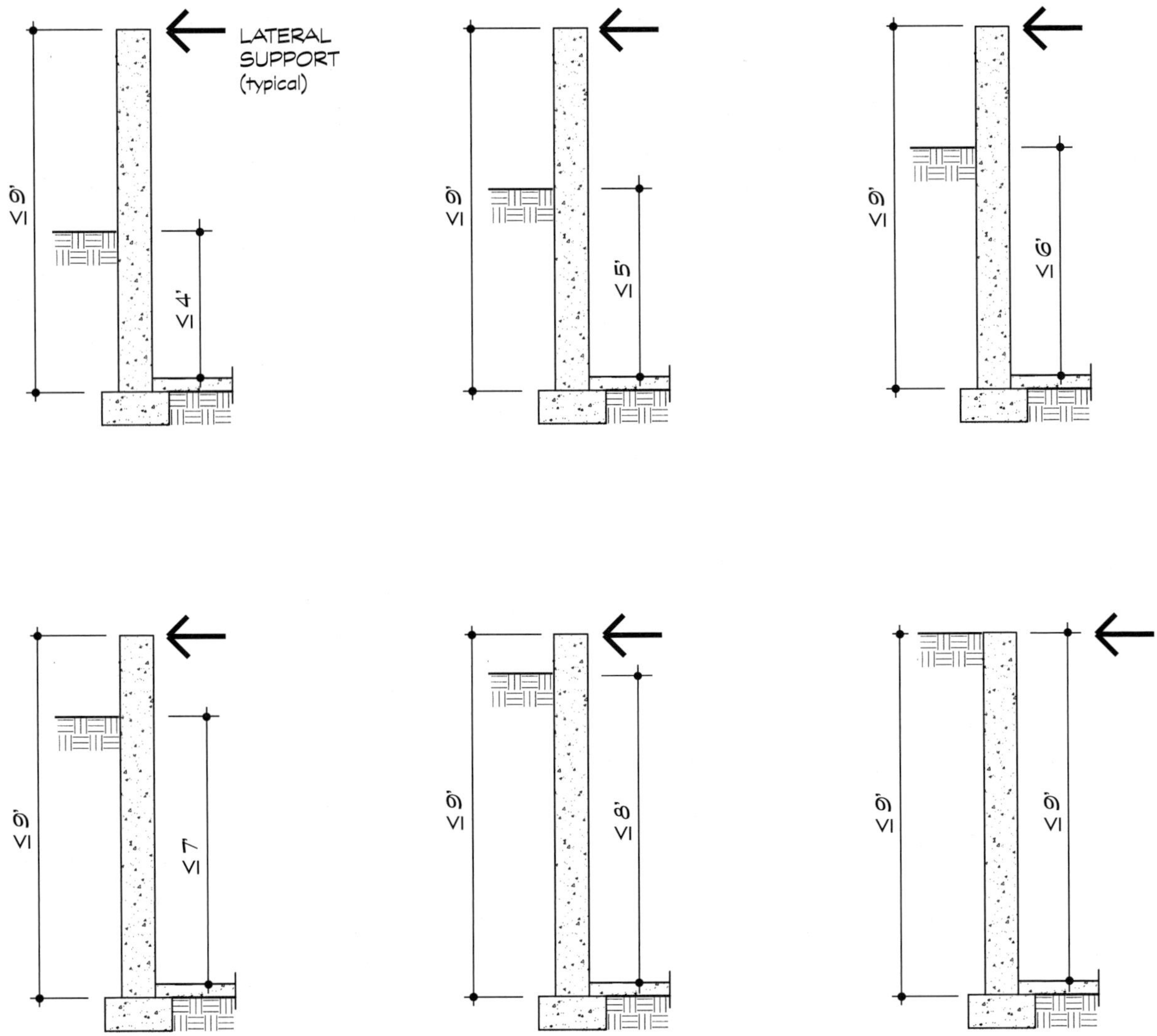

9'-HIGH WALLS

Fig. 1807.1.6.2-29. Plain concrete foundation wall, thickness ≥ 11 1/2" for a design lateral soil load of 45 psf/ft of depth. *[IBC Table 1807.1.6.2]*

1807 Foundation Walls, Retaining Walls and Embedded Posts and Poles

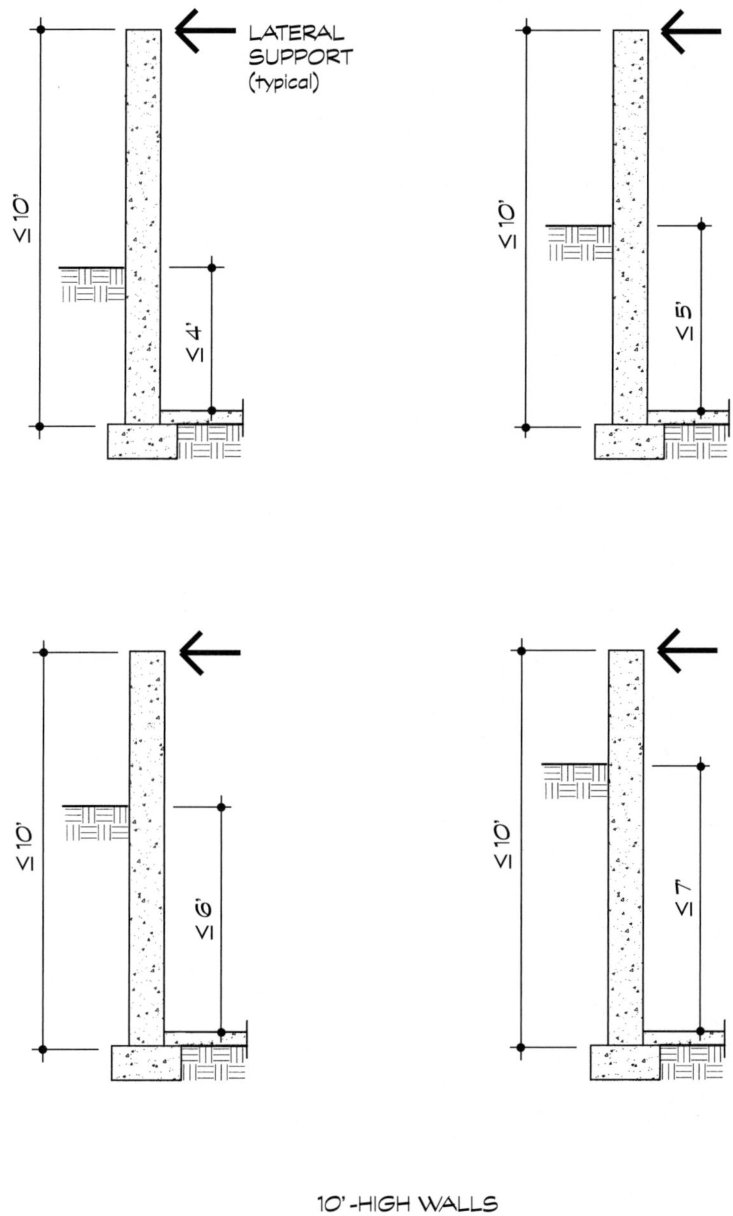

Fig. 1807.1.6.2-30a. Concrete foundation wall (plain concrete where reinforcing is not shown), thickness ≥ 11 1/2" for a design lateral soil load of 45 psf/ft of depth. *[IBC Table 1807.1.6.2]*

1807 Foundation Walls, Retaining Walls and Embedded Posts and Poles

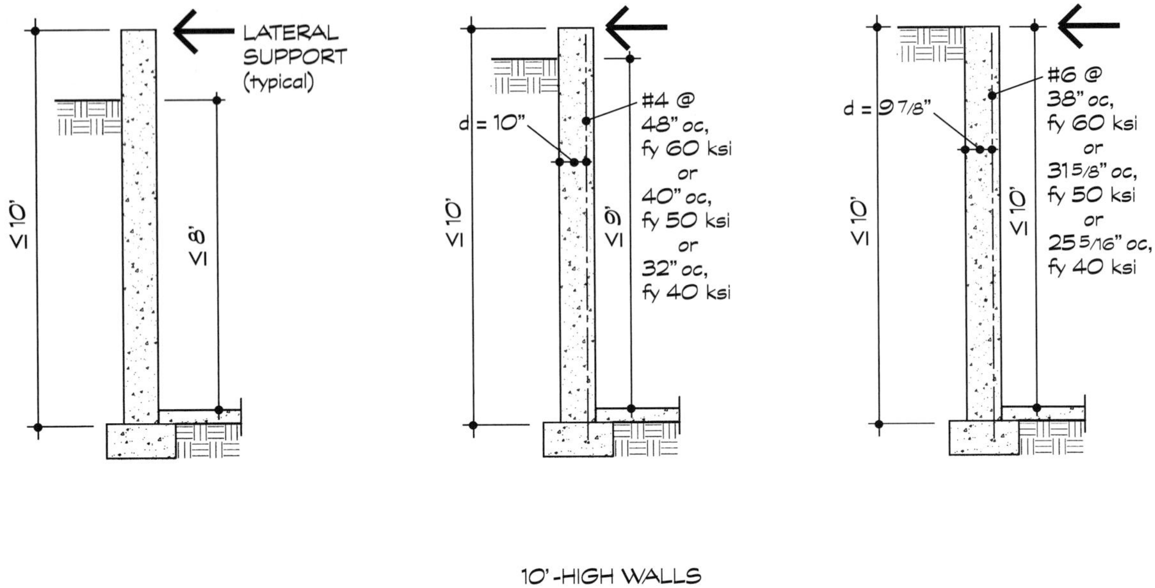

Fig. 1807.1.6.2-30b. Concrete foundation wall (plain concrete where reinforcing is not shown), thickness ≥ 11 1/2" for a design lateral soil load of 45 psf/ft of depth. *[IBC Table 1807.1.6.2]*

1807 Foundation Walls, Retaining Walls and Embedded Posts and Poles

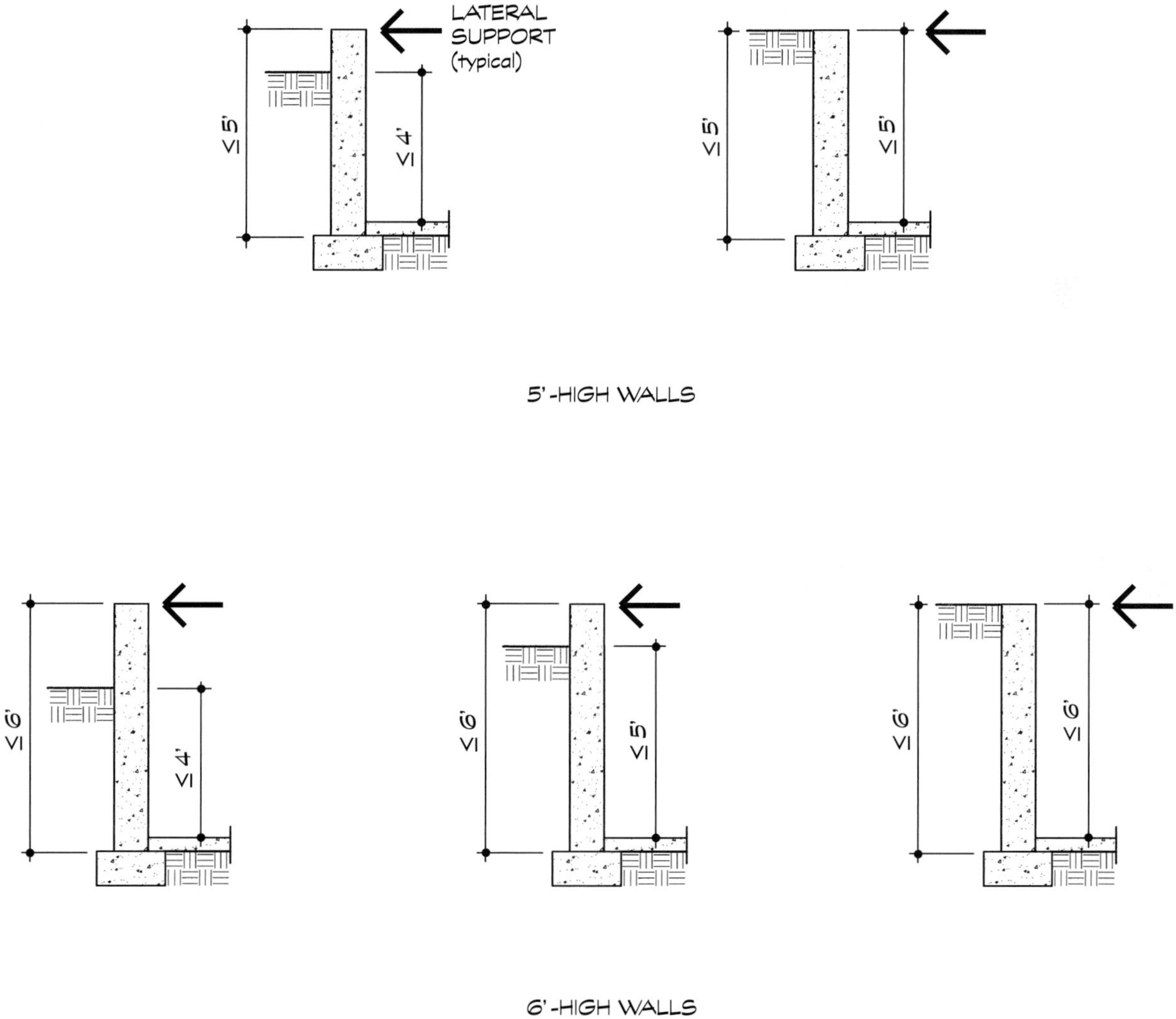

Fig. 1807.1.6.2-31. Concrete foundation wall (plain concrete where reinforcing is not shown), thickness ≥ 7 1/2" for a design lateral soil load of 60 psf/ft of depth. *[IBC Table 1807.1.6.2]*

1807 Foundation Walls, Retaining Walls and Embedded Posts and Poles

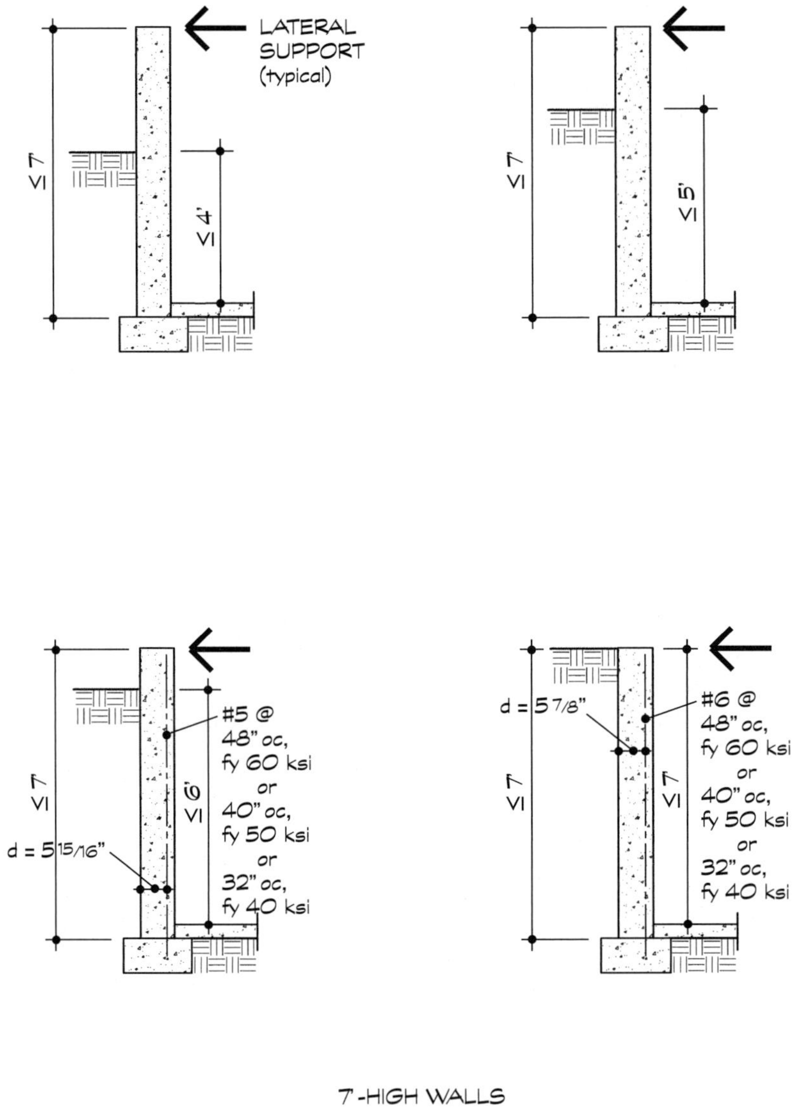

Fig. 1807.1.6.2-32. Concrete foundation wall (plain concrete where reinforcing is not shown), thickness ≥ 7 1/2" for a design lateral soil load of 60 psf/ft of depth. *[IBC Table 1807.1.6.2]*

1807 Foundation Walls, Retaining Walls and Embedded Posts and Poles

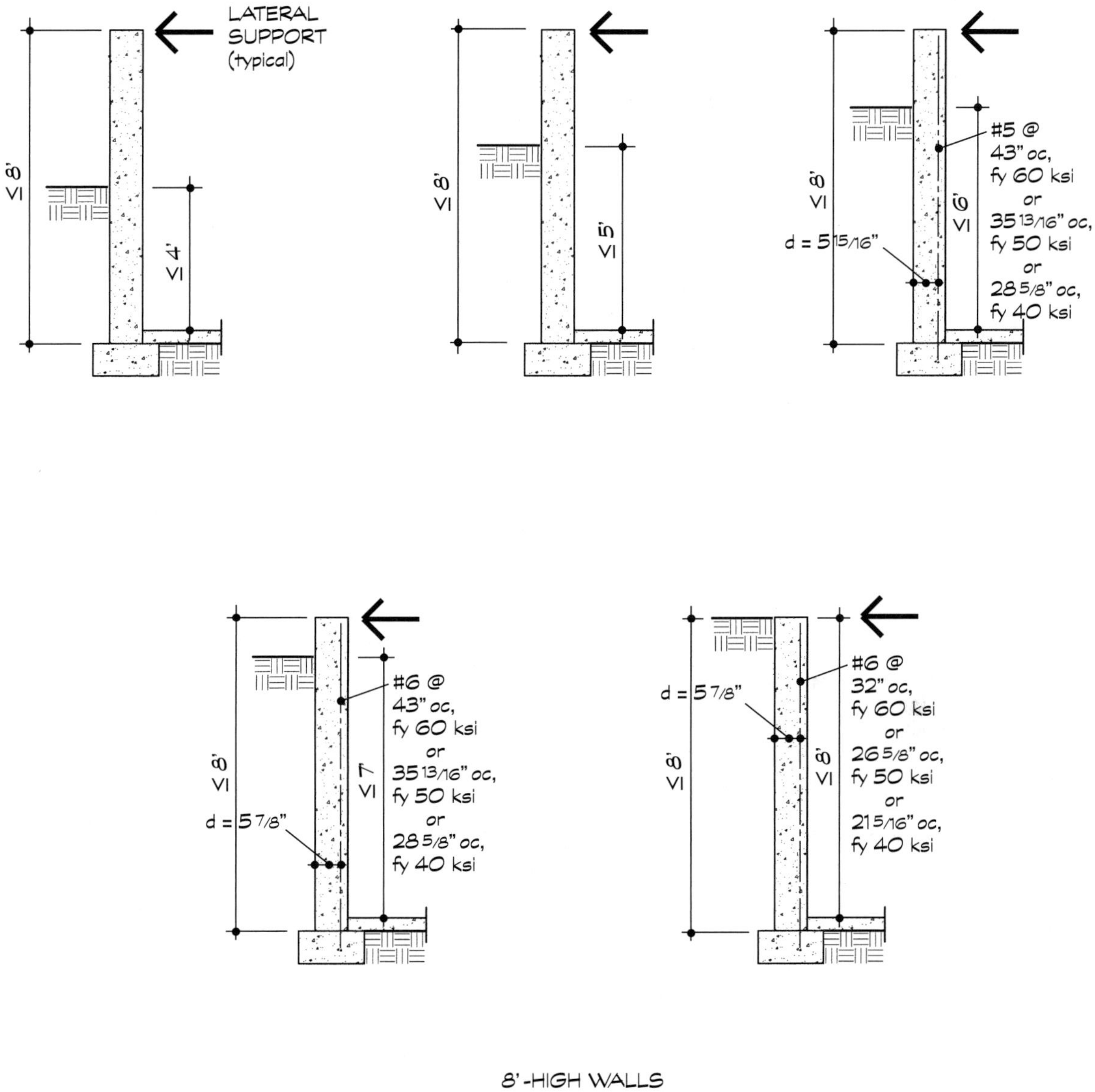

Fig. 1807.1.6.2-33. Concrete foundation wall (plain concrete where reinforcing is not shown), thickness ≥ 7 1/2" for a design lateral soil load of 60 psf/ft of depth. *[IBC Table 1807.1.6.2]*

1807 Foundation Walls, Retaining Walls and Embedded Posts and Poles

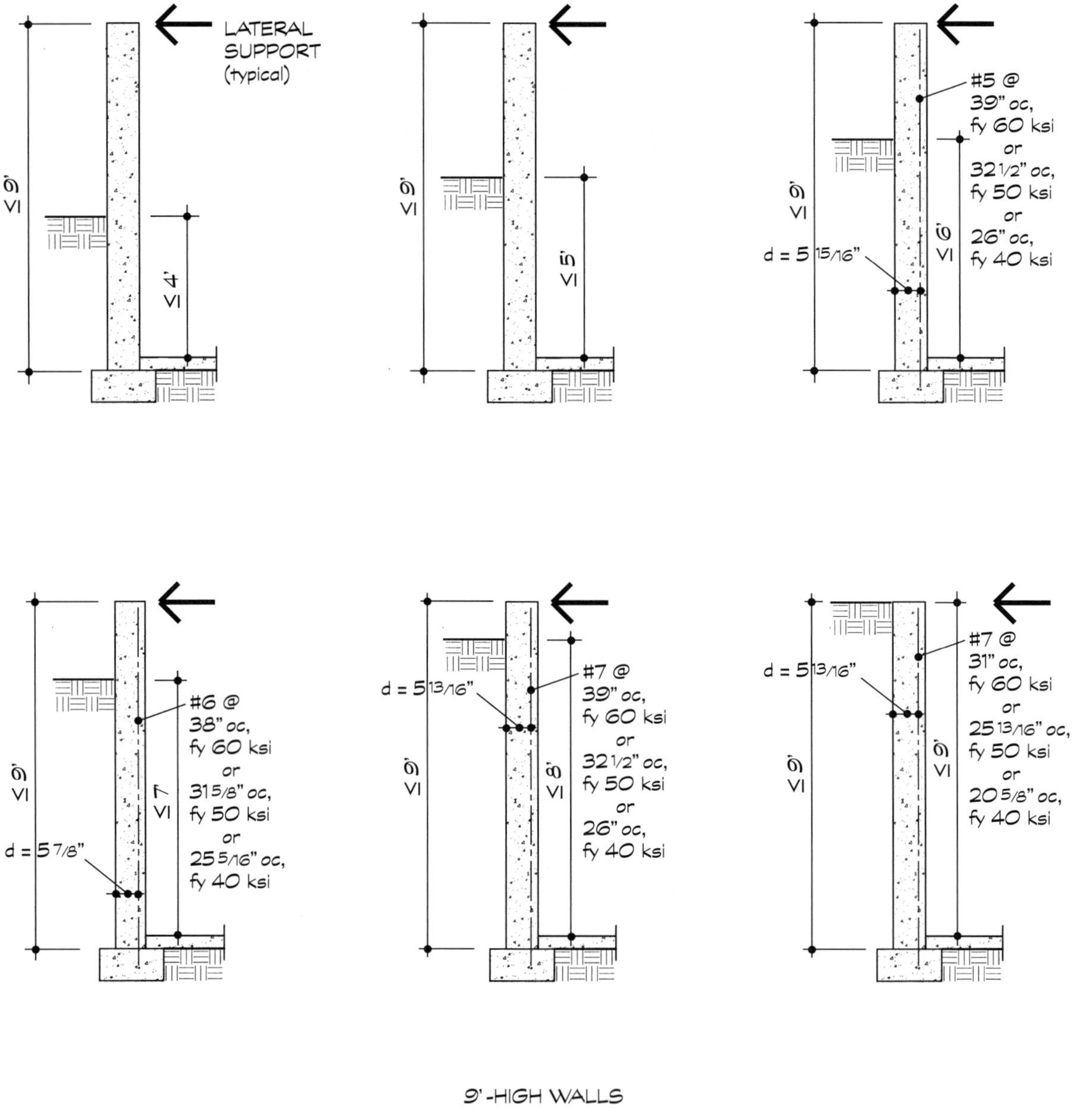

Fig. 1807.1.6.2-34. Concrete foundation wall (plain concrete where reinforcing is not shown), thickness ≥ 7 1/2" for a design lateral soil load of 60 psf/ft of depth. *[IBC Table 1807.1.6.2]*

1807 Foundation Walls, Retaining Walls and Embedded Posts and Poles

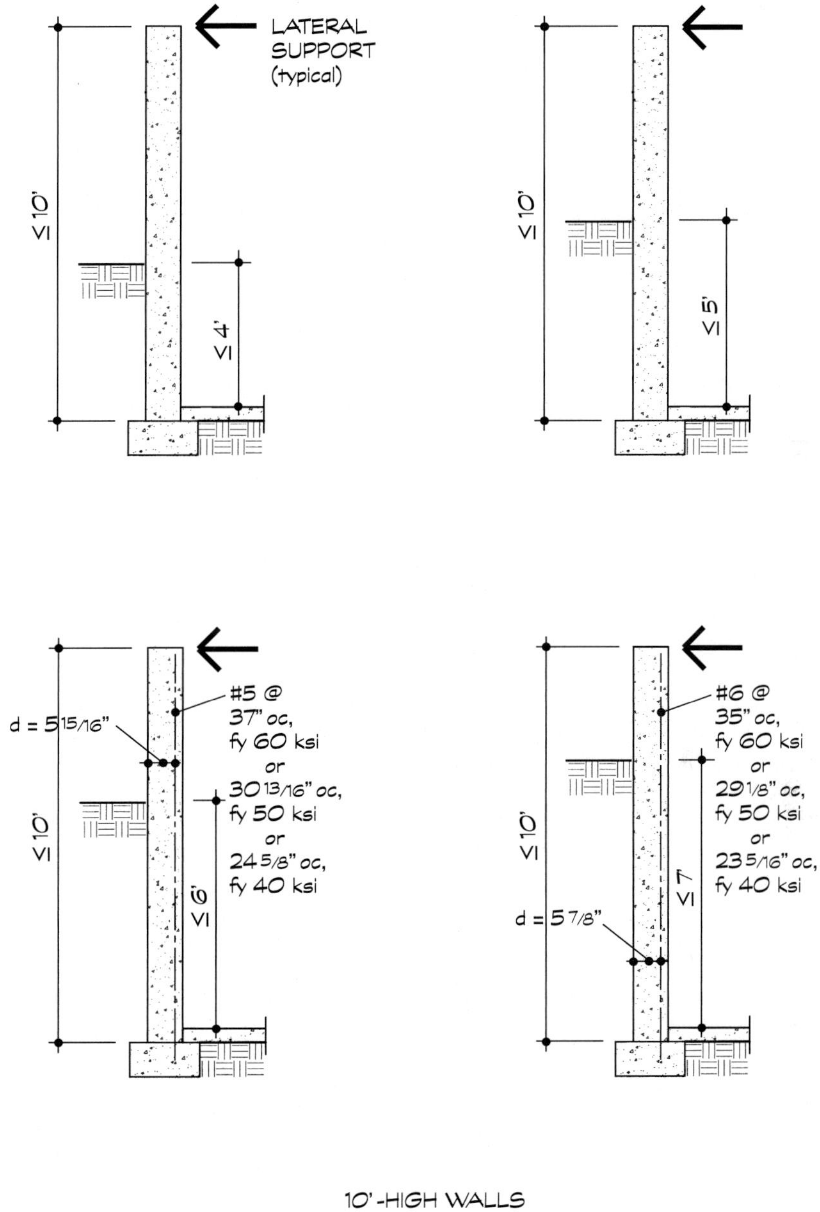

Fig. 1807.1.6.2-35a. Concrete foundation wall (plain concrete where reinforcing is not shown), thickness ≥ 7 1/2" for a design lateral soil load of 60 psf/ft of depth. *[IBC Table 1807.1.6.2]*

1807 Foundation Walls, Retaining Walls and Embedded Posts and Poles

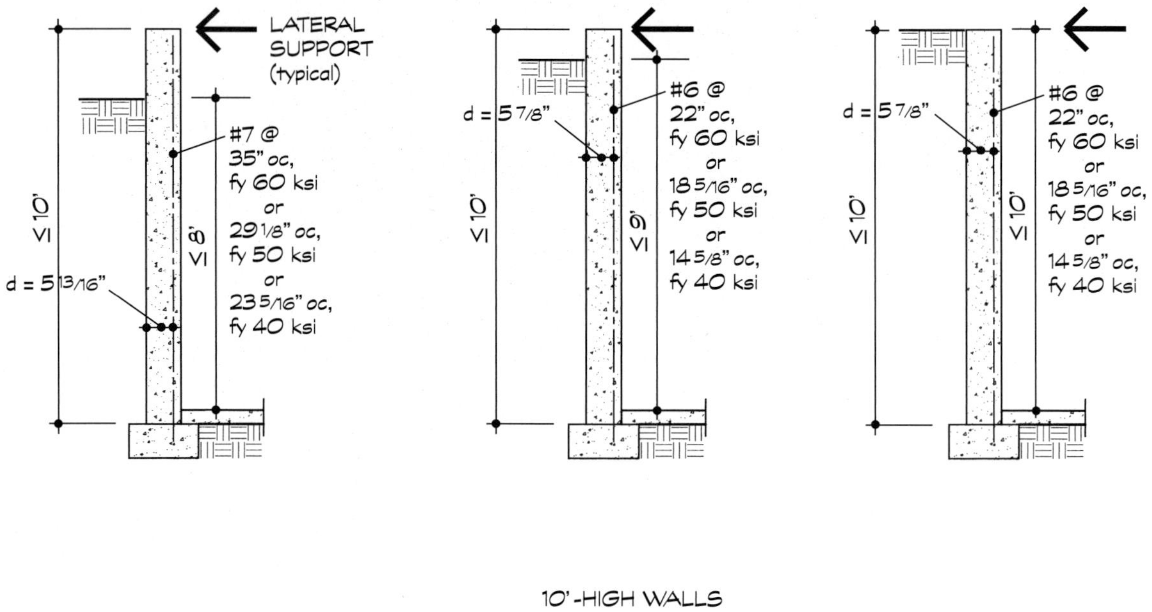

Fig. 1807.1.6.2-35b. Concrete foundation wall (plain concrete where reinforcing is not shown), thickness ≥ 7 1/2" for a design lateral soil load of 60 psf/ft of depth. *[IBC Table 1807.1.6.2]*

1807 Foundation Walls, Retaining Walls and Embedded Posts and Poles

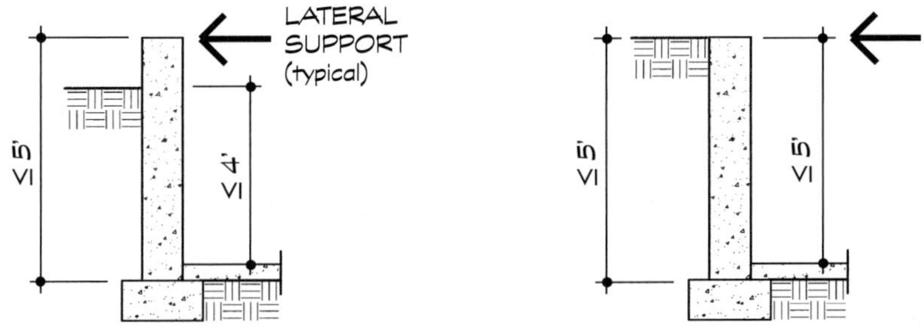

5'-HIGH WALLS

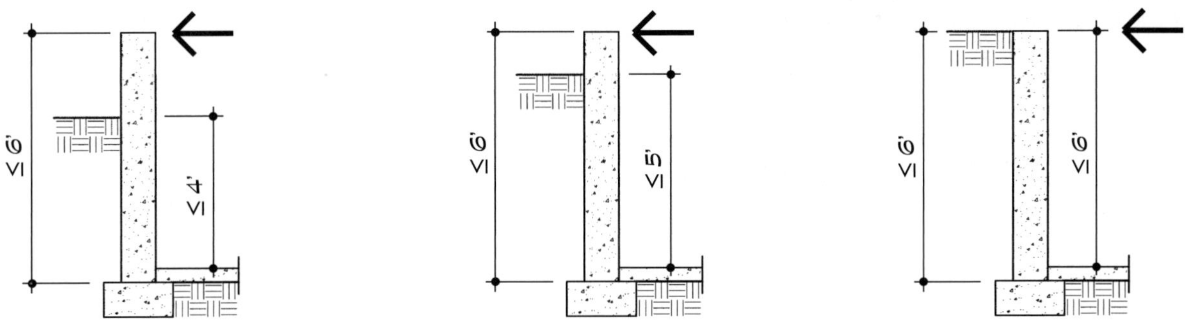

6'-HIGH WALLS

Fig. 1807.1.6.2-36. Plain concrete foundation wall, thickness ≥ 9 1/2" for a design lateral soil load of 60 psf/ft of depth. *[IBC Table 1807.1.6.2]*

1807 Foundation Walls, Retaining Walls and Embedded Posts and Poles

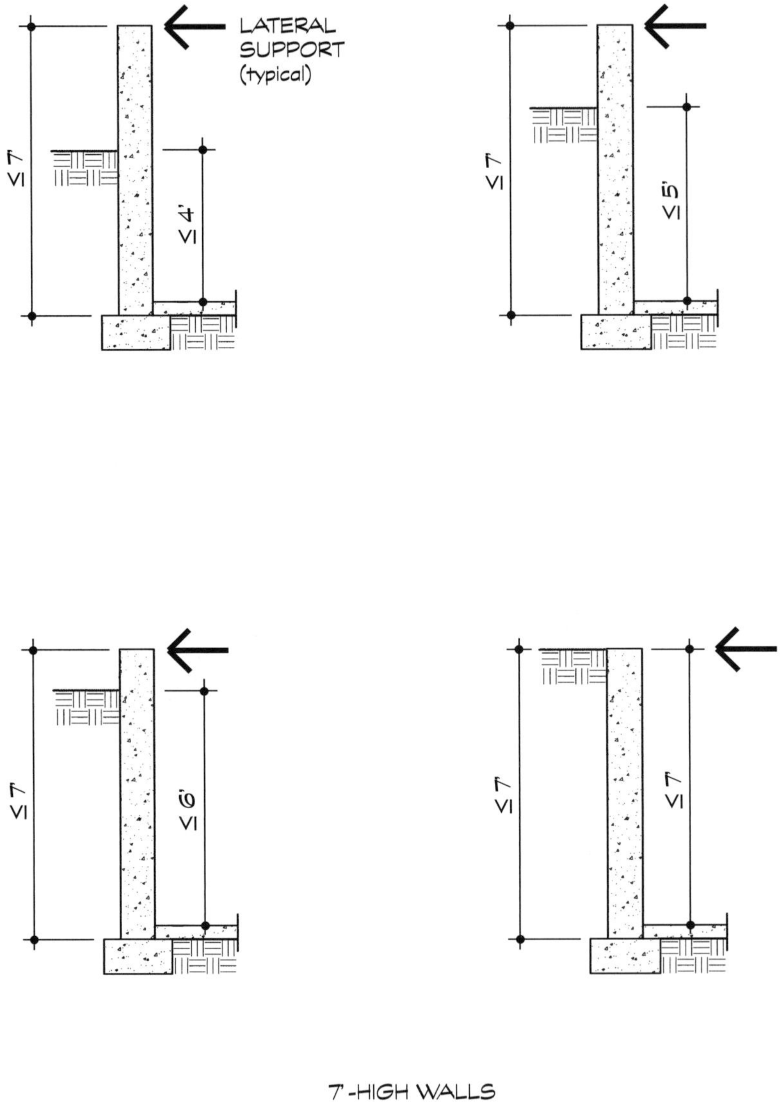

Fig. 1807.1.6.2-37. Plain concrete foundation wall, thickness ≥ 9 1/2" for a design lateral soil load of 60 psf/ft of depth. *[IBC Table 1807.1.6.2]*

1807 Foundation Walls, Retaining Walls and Embedded Posts and Poles

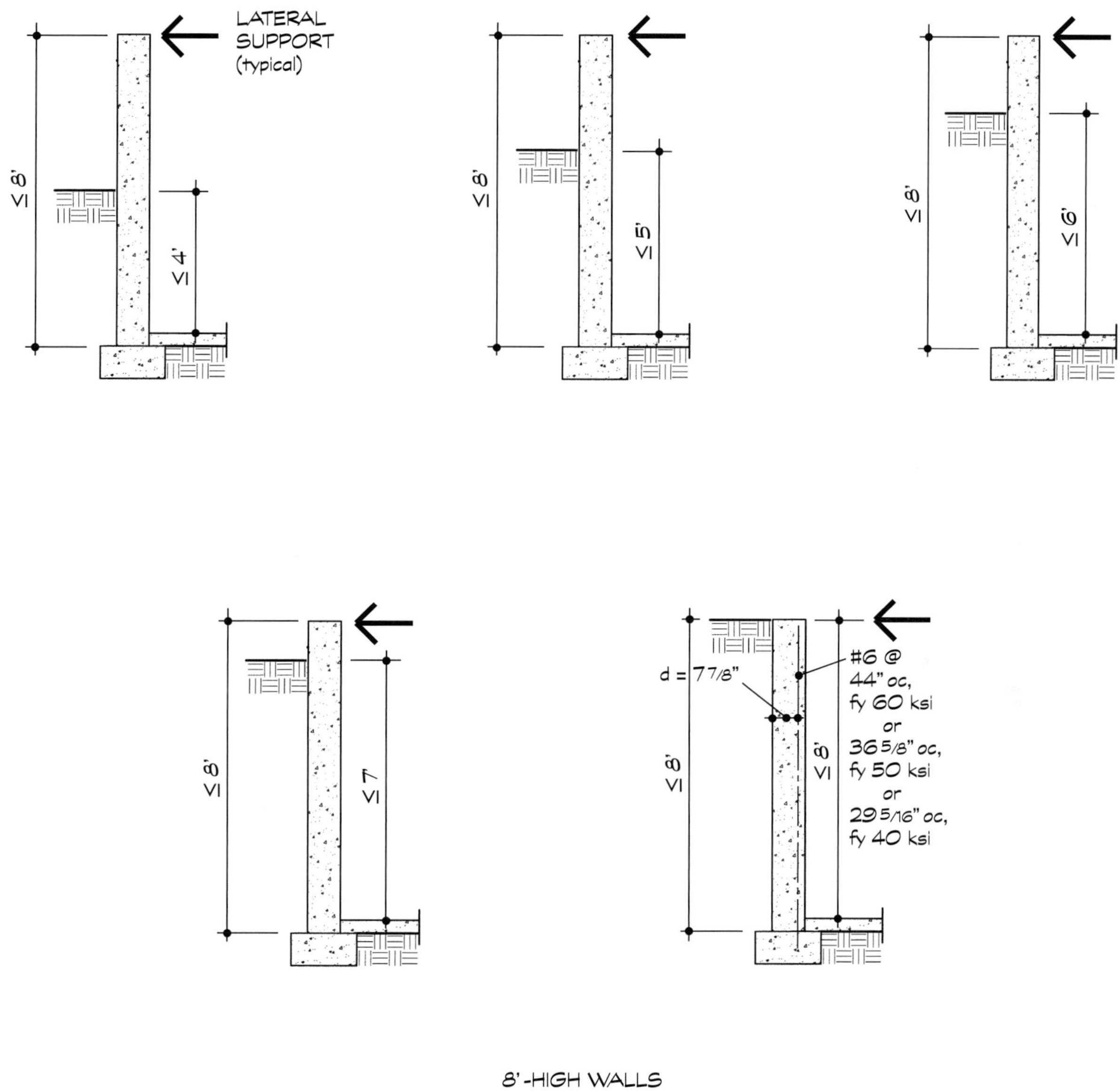

Fig. 1807.1.6.2-38. Concrete foundation wall (plain concrete where reinforcing is not shown), thickness ≥ 9 1/2" for a design lateral soil load of 60 psf/ft of depth. *[IBC Table 1807.1.6.2]*

1807 Foundation Walls, Retaining Walls and Embedded Posts and Poles

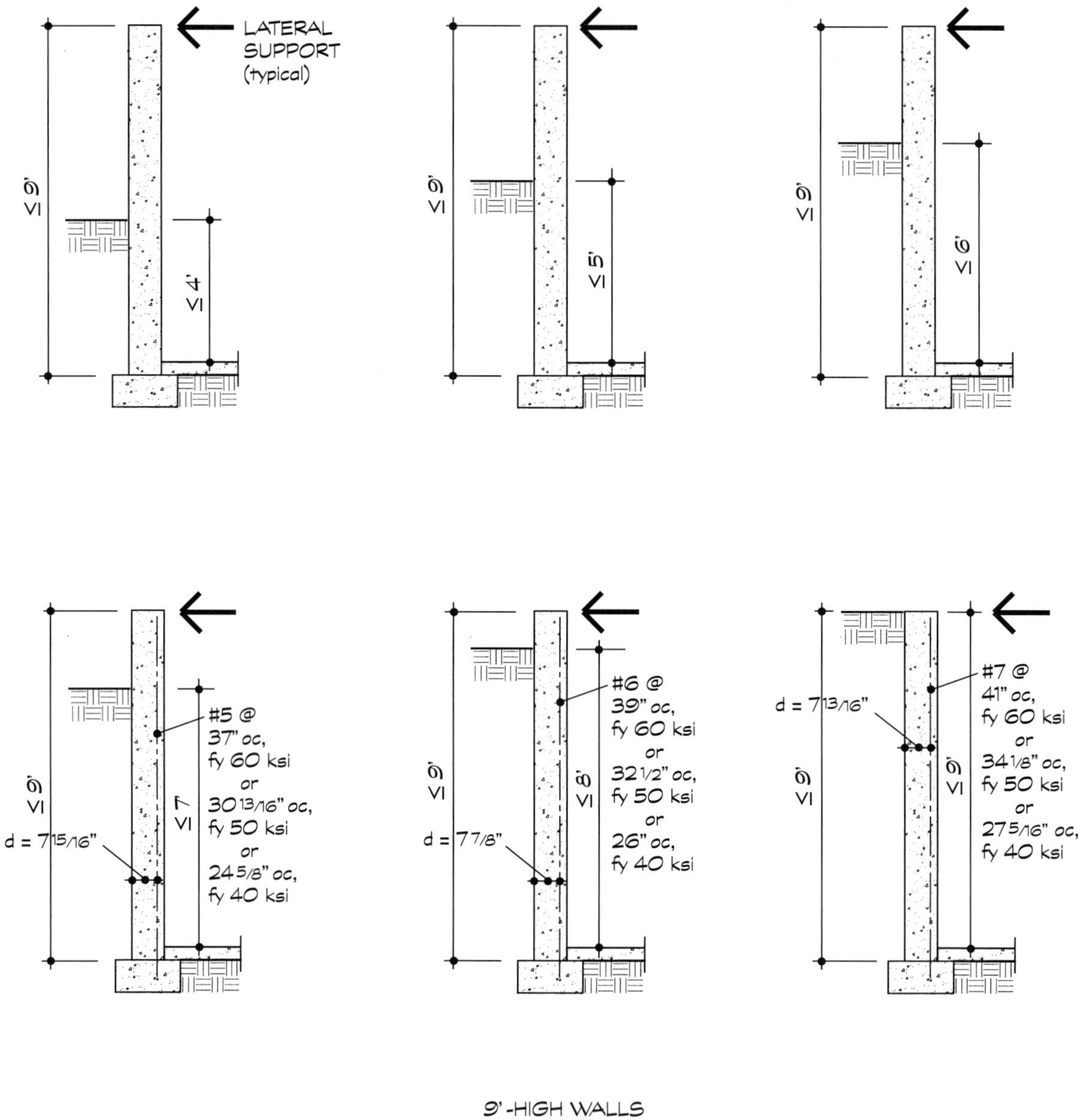

Fig. 1807.1.6.2-39. Concrete foundation wall (plain concrete where reinforcing is not shown), thickness ≥ 9 1/2" for a design lateral soil load of 60 psf/ft of depth. *[IBC Table 1807.1.6.2]*

1807 Foundation Walls, Retaining Walls and Embedded Posts and Poles

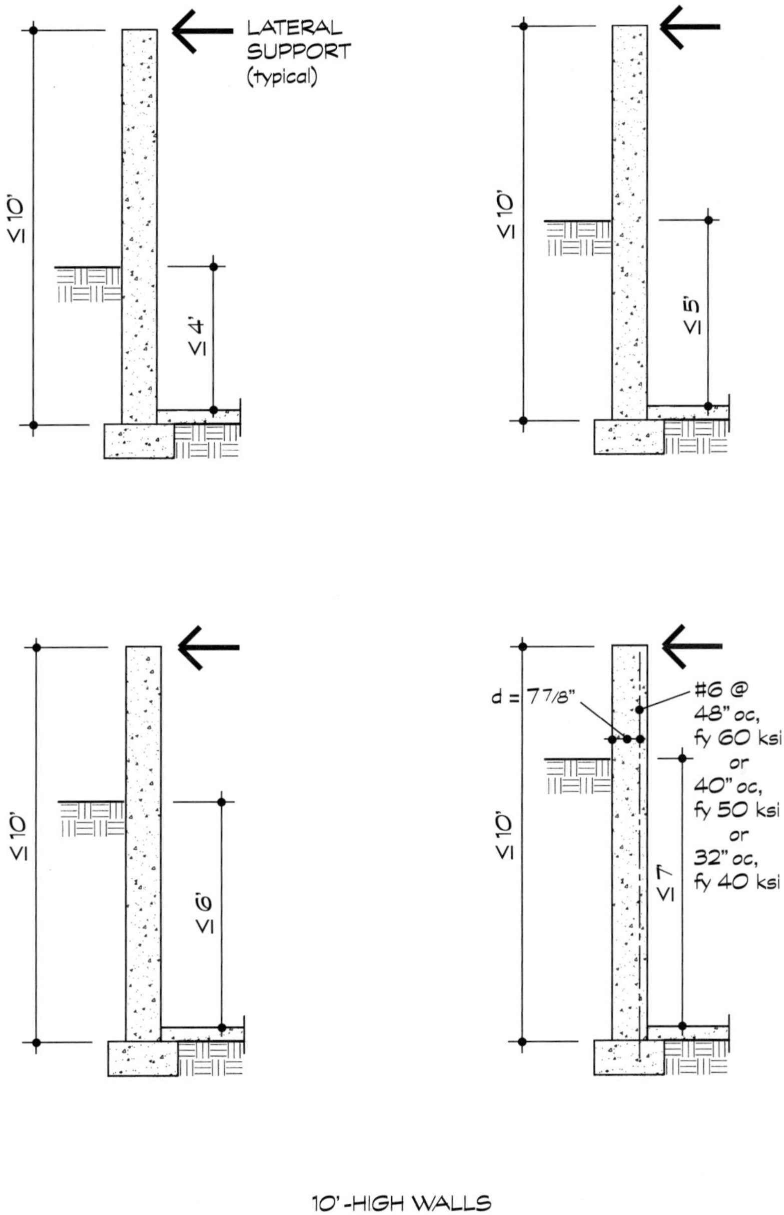

Fig. 1807.1.6.2-40a. Concrete foundation wall (plain concrete where reinforcing is not shown), thickness ≥ 9 1/2" for a design lateral soil load of 60 psf/ft of depth. *[IBC Table 1807.1.6.2]*

1807 Foundation Walls, Retaining Walls and Embedded Posts and Poles

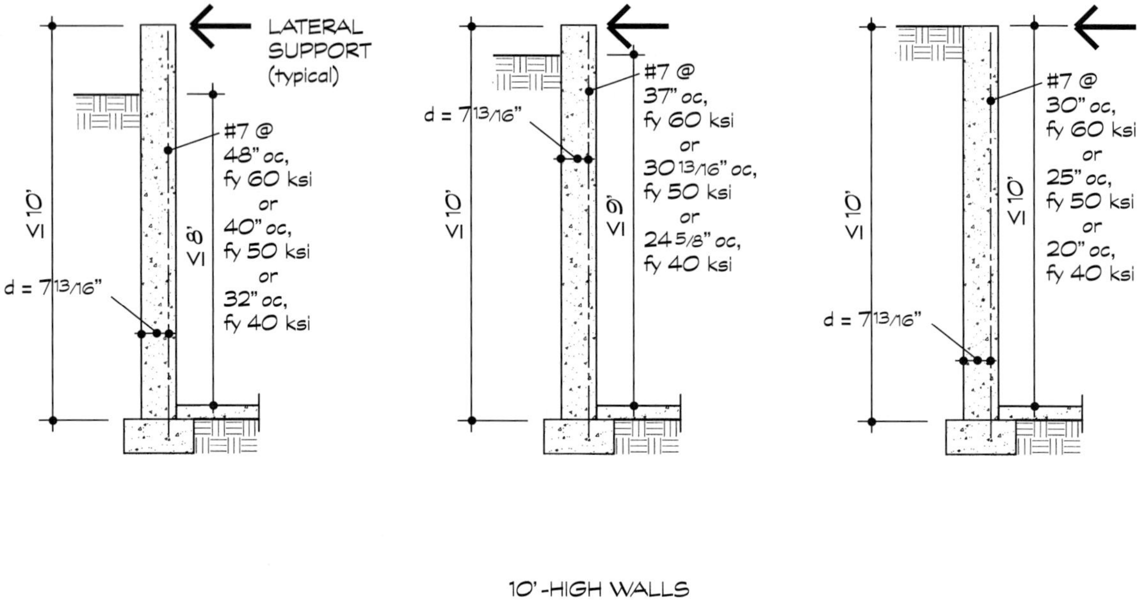

Fig. 1807.1.6.2-40b. Concrete foundation wall (plain concrete where reinforcing is not shown), thickness ≥ 9 1/2" for a design lateral soil load of 60 psf/ft of depth. *[IBC Table 1807.1.6.2]*

1807 Foundation Walls, Retaining Walls and Embedded Posts and Poles

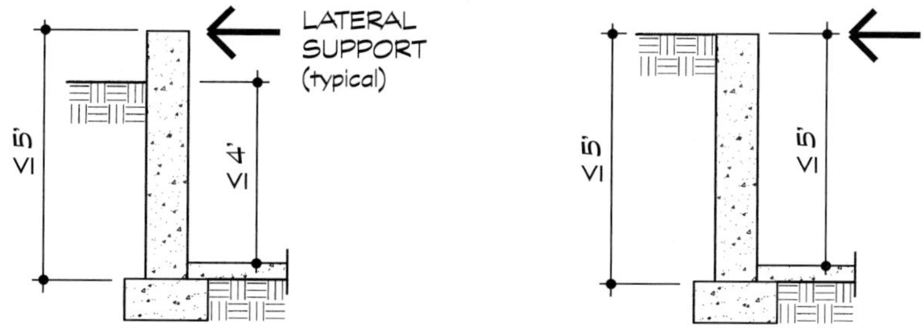

5'-HIGH WALLS

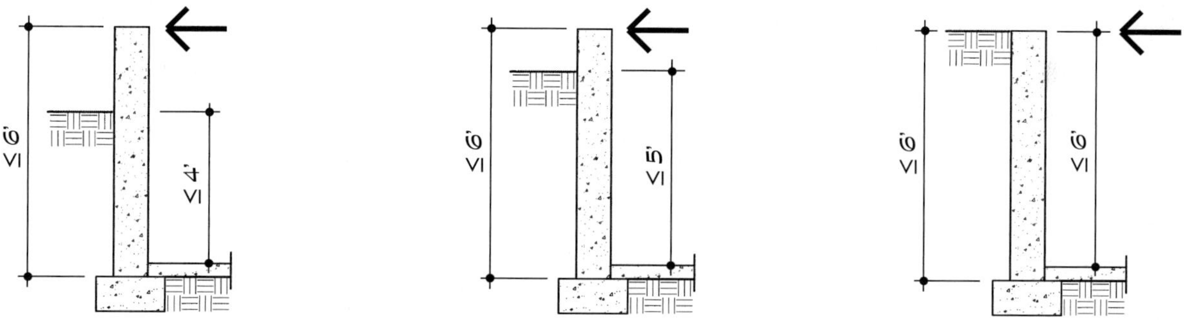

6'-HIGH WALLS

Fig. 1807.1.6.2-41. Plain concrete foundation wall, thickness ≥ 11 1/2" for a design lateral soil load of 60 psf/ft of depth. *[IBC Table 1807.1.6.2]*

1807 Foundation Walls, Retaining Walls and Embedded Posts and Poles

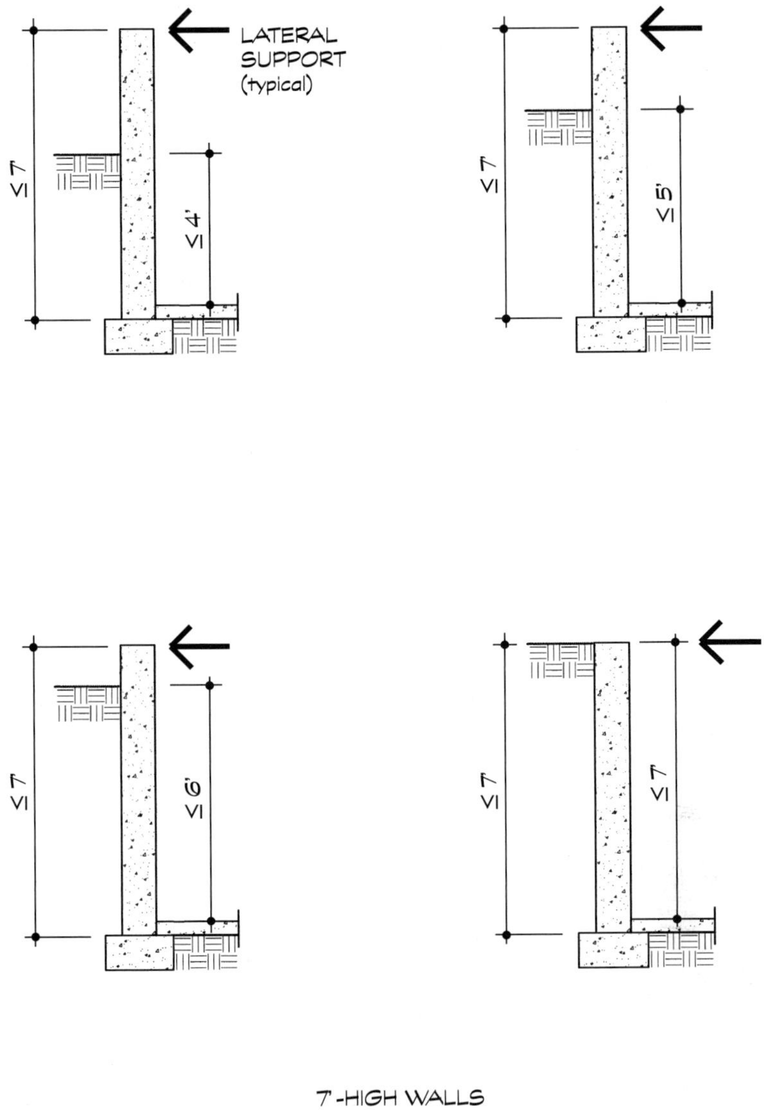

Fig. 1807.1.6.2-42. Plain concrete foundation wall, thickness ≥ 11 1/2" for a design lateral soil load of 60 psf/ft of depth. *[IBC Table 1807.1.6.2]*

1807 Foundation Walls, Retaining Walls and Embedded Posts and Poles

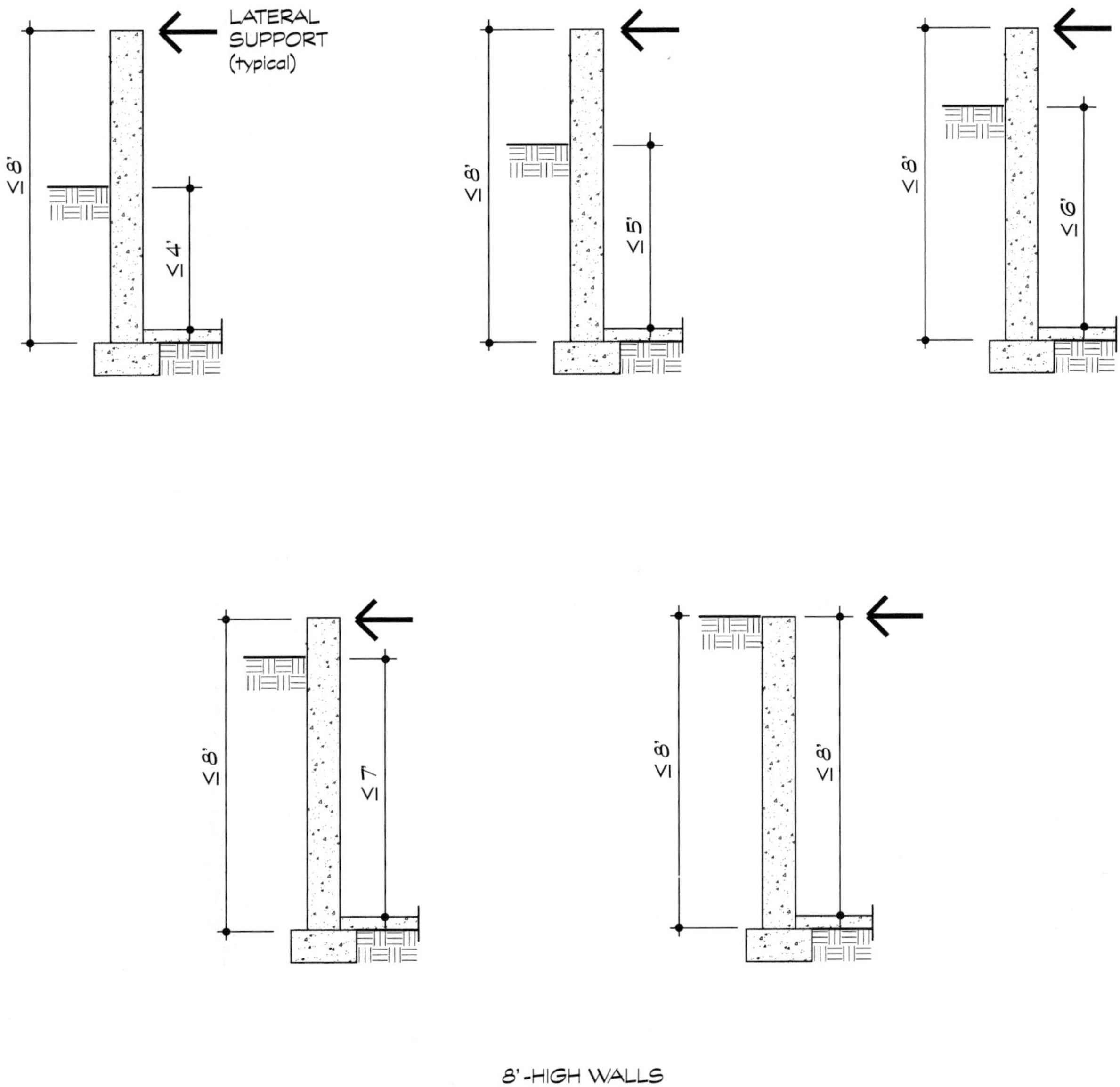

Fig. 1807.1.6.2-43. Plain concrete foundation wall, thickness ≥ 11 1/2" for a design lateral soil load of 60 psf/ft of depth. *[IBC Table 1807.1.6.2]*

1807 Foundation Walls, Retaining Walls and Embedded Posts and Poles

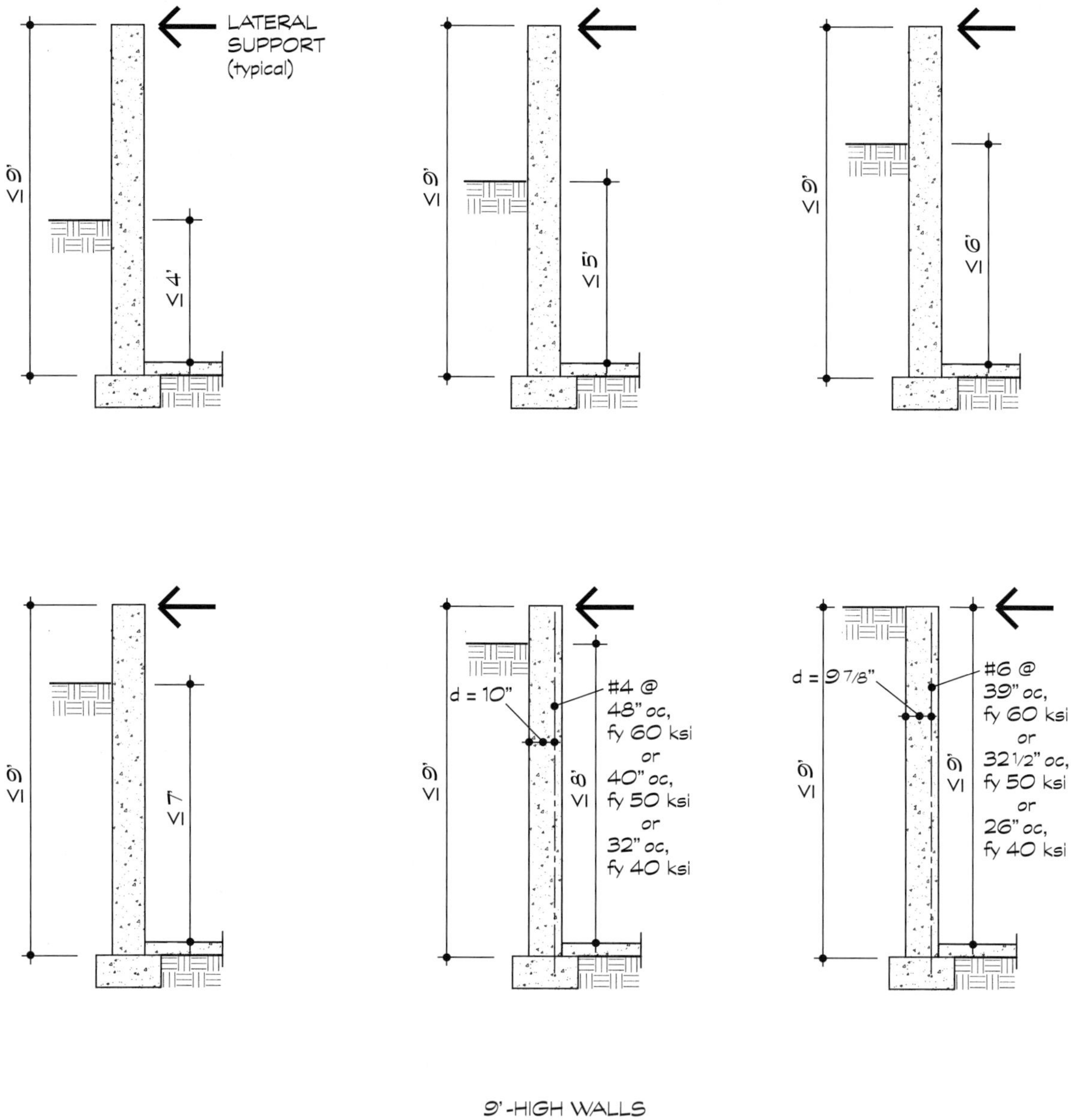

Fig. 1807.1.6.2-44. Plain concrete foundation wall, thickness ≥ 11 1/2" for a design lateral soil load of 60 psf/ft of depth. *[IBC Table 1807.1.6.2]*

1807 Foundation Walls, Retaining Walls and Embedded Posts and Poles

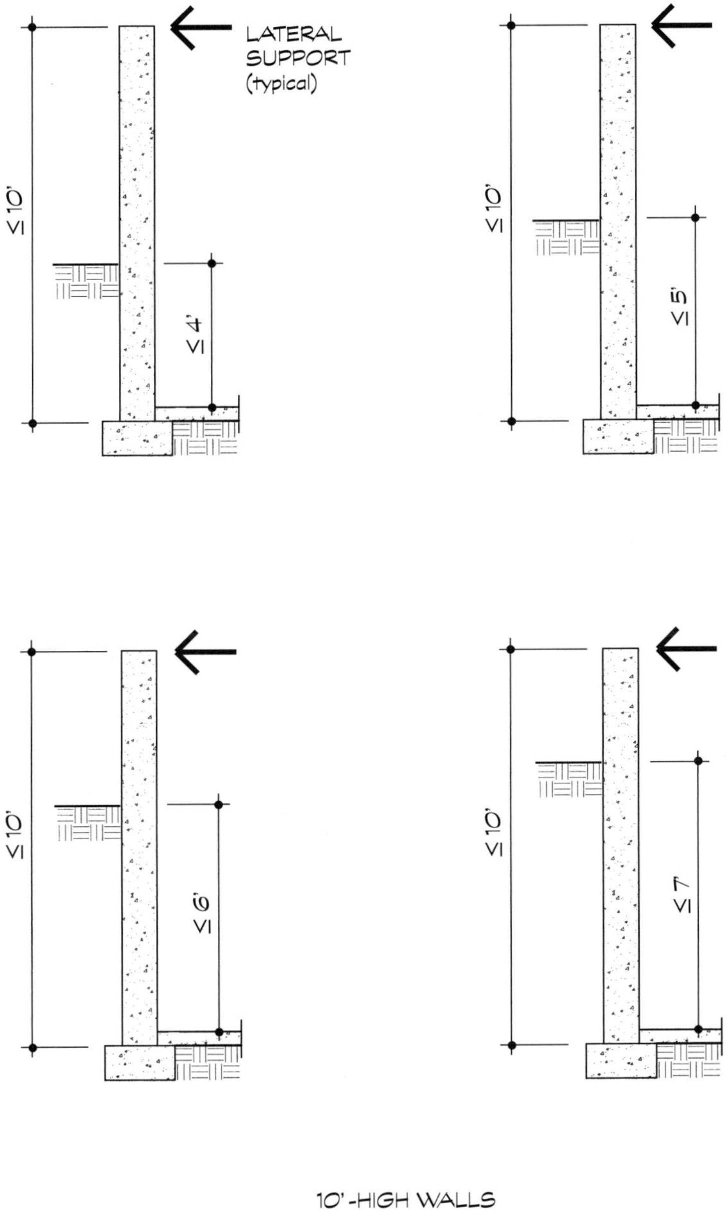

Fig. 1807.1.6.2-45a. Concrete foundation wall (plain concrete where reinforcing is not shown), thickness ≥ 11 1/2" for a design lateral soil load of 60 psf/ft of depth. *[IBC Table 1807.1.6.2]*

1807 Foundation Walls, Retaining Walls and Embedded Posts and Poles

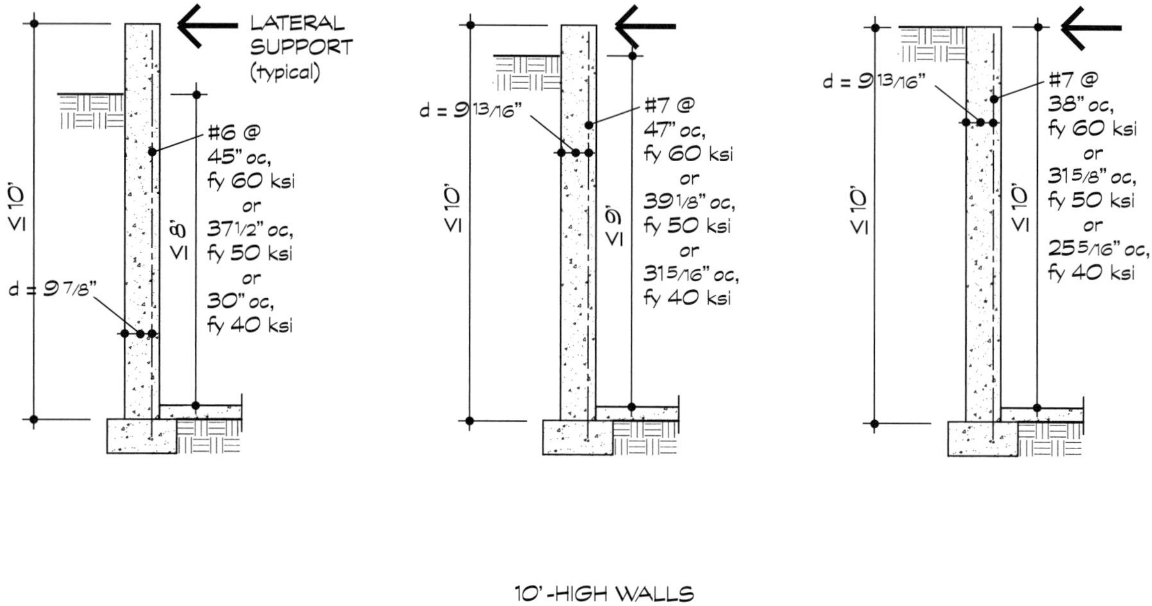

Fig. 1807.1.6.2-45b. Concrete foundation wall (plain concrete where reinforcing is not shown), thickness ≥ 11 1/2" for a design lateral soil load of 60 psf/ft of depth. *[IBC Table 1807.1.6.2]*

1807 Foundation Walls, Retaining Walls and Embedded Posts and Poles

1807.1.6.3 Masonry foundation walls *(part 1 of 2)*

- Masonry foundation walls are governed as follows:
  - The required thicknesses of masonry foundation walls are illustrated on the following pages.

Note: The following tables are cited as governing thicknesses of masonry foundation walls and are illustrated with reinforcing requirements on the following pages:
IBC Table 1807.1.6.3(1), "Plain Masonry Foundation Walls."
IBC Table 1807.1.6.3(2), "8-Inch Masonry Foundation Walls with Reinforcement Where d ≥ 5 Inches."
IBC Table 1807.1.6.3(3), "10-Inch Masonry Foundation Walls with Reinforcement Where d ≥ 6.75 Inches."
IBC Table 1807.1.6.3(4), "12-Inch Masonry Foundation Walls with Reinforcement Where d ≥ 8.75 Inches."

- Vertical reinforcing must have a yield strength ≥ 60 ksi.
- Vertical reinforcing must be placed as follows:
  At a distance ≥ the distance *d* specified in tabular data of this section as follows:
    Distance *d* is measured from the grade-side face of the wall to the center of the steel.
  Tolerances for placing vertical reinforcing is as follows:
    ± 1/2" ⊥ the face of the wall where *d* ≤ 8".
    ± 1" ⊥ the face of the wall where *d* > 8" and ≤ 24".
    ± 2" || to the face of the wall.

Note: TMS 602/ACI 530.1/ASCE 6, "Specification for Masonry Structures," Article 3.3.B.8, "Placement tolerances," is cited as governing tolerances for placement of vertical reinforcing steel. Key tolerances are summarized above.

- Masonry must be installed as follows:
  With Type M or S mortar.
  In running bond.

Note: 2103.8, "Mortar," is cited as governing mortar.

- The unfactored axial load per lineal foot of wall is limited to the following:
  ≤ 1.2 × wall thickness × compressive strength of masonry in lb/sq in.
- The following is required at the top of masonry-unit foundation walls at the bearing of girders:
  ≥ 4" of solid masonry.
- A corbeled 8" wall is governed as follows:
  The top corbel must have one of the following:
    A full course of headers ≥ 6" long.
    Be tied to the adjacent wythe with the following:
      With W2.8 ties as follows:
        Spaced ≤ 3' horizontally.
  The top of the corbel is limited to a height = to the bottom of the floor framing.
  The void behind the corbel must be filled with mortar or grout.

Note: 2104.2, "Corbeled masonry," is cited as governing corbeled masonry.

1807 Foundation Walls, Retaining Walls and Embedded Posts and Poles

1807.1.6.3 Masonry foundation walls *(part 2 of 2)*

Note: 2103.12, "Grout," is cited as governing grout where used in masonry walls.
ASTM C90, "Specification for Loadbearing Concrete Masonry Units," is cited as governing concrete masonry units.
ASTM C 652, "Specification for Hollow Brick (Hollow Masonry Units Made From Clay or Shale)," is cited as governing these units.
The following are cited as alternatives for governing solid masonry where used to comply with Table 1807.1.6.3(1), "Plain Masonry Foundation Walls":
ASTM C 62, "Specification for Building Brick (Solid Masonry Units Made from Clay or Shale)."
ASTM C 216, "Specification for Facing Brick (Solid Masonry Units Made from Clay or Shale)."
Section 1610, " Soil Lateral Loads," is cited as governing these loads.
IBC Table 1610.1, "Lateral Soil Load," is cited as the table that may not be used to identify lateral soil loads of 30 psf and 45 psf for walls with unbalanced backfill heights ≥ 8'.

- The height of unbalanced backfill is measured between the following two levels:
  - The finished exterior grade.
  - One of the following interior levels:
    - Top of concrete slab where provided.
    - The lower of the following two levels where there is no concrete slab:
      - Top of footing.
      - Interior grade.

Note: 1807.1.2, "Unbalanced backfill height," is cited as defining this height and is summarized above.

- Reinforcing where required is permitted to vary from that specified in the tabular data as follows:
  - Where spacing and bar sizes are adjusted so as to yield the following:
    - An average cross-sectional area of steel per foot of wall equal to that of the tabular requirements.

Note: 1807.1.6.3.1, "Alternative foundation wall reinforcement," is cited as governing variations permitted to tabular requirements for reinforcement size and spacing.

1807.1.6.3.1 Alternative foundation wall reinforcement

- Reinforcing where required is permitted to vary from that specified in the tabular data as follows:
  - Where spacing and bar sizes are adjusted so as to yield the following:
    - An average cross-sectional area of steel per foot of wall equal to that of the tabular requirements.
  - Bar spacing must not be > 6'.
  - Bar size must not be > #11.

Note: The following are cited as being applicable to this section:
IBC Table 1807.1.6.3(2), "8-Inch Masonry Foundation Walls with Reinforcement Where d ≥ 5 Inches."
IBC Table 1807.1.6.3(3), "10-Inch Masonry Foundation Walls with Reinforcement Where d ≥ 6.75 Inches."
IBC Table 1807.1.6.3(4), "12-Inch Masonry Foundation Walls with Reinforcement Where d ≥ 8.75 Inches."

1807 Foundation Walls, Retaining Walls and Embedded Posts and Poles

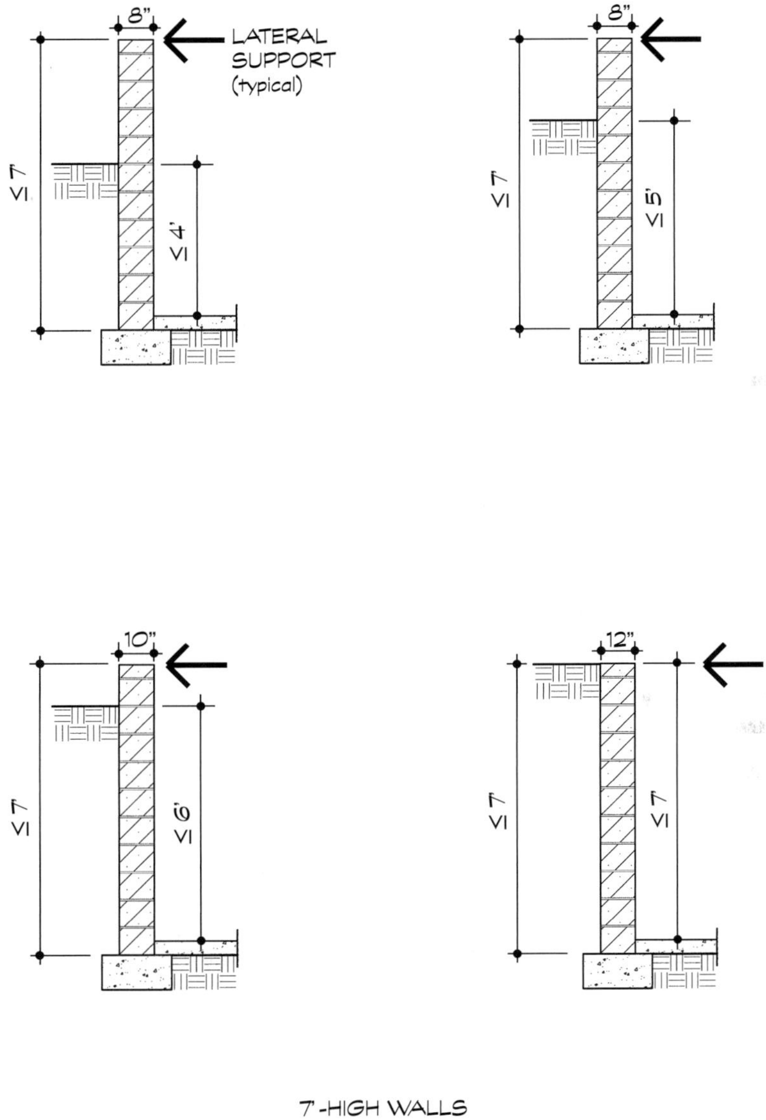

Fig. 1807.1.6.3(1)-1. Plain masonry foundation wall, minimum nominal thickness for 30 psf/ft of depth lateral soil load. *[IBC Table 1807.1.6.3(1)]*

1807 Foundation Walls, Retaining Walls and Embedded Posts and Poles

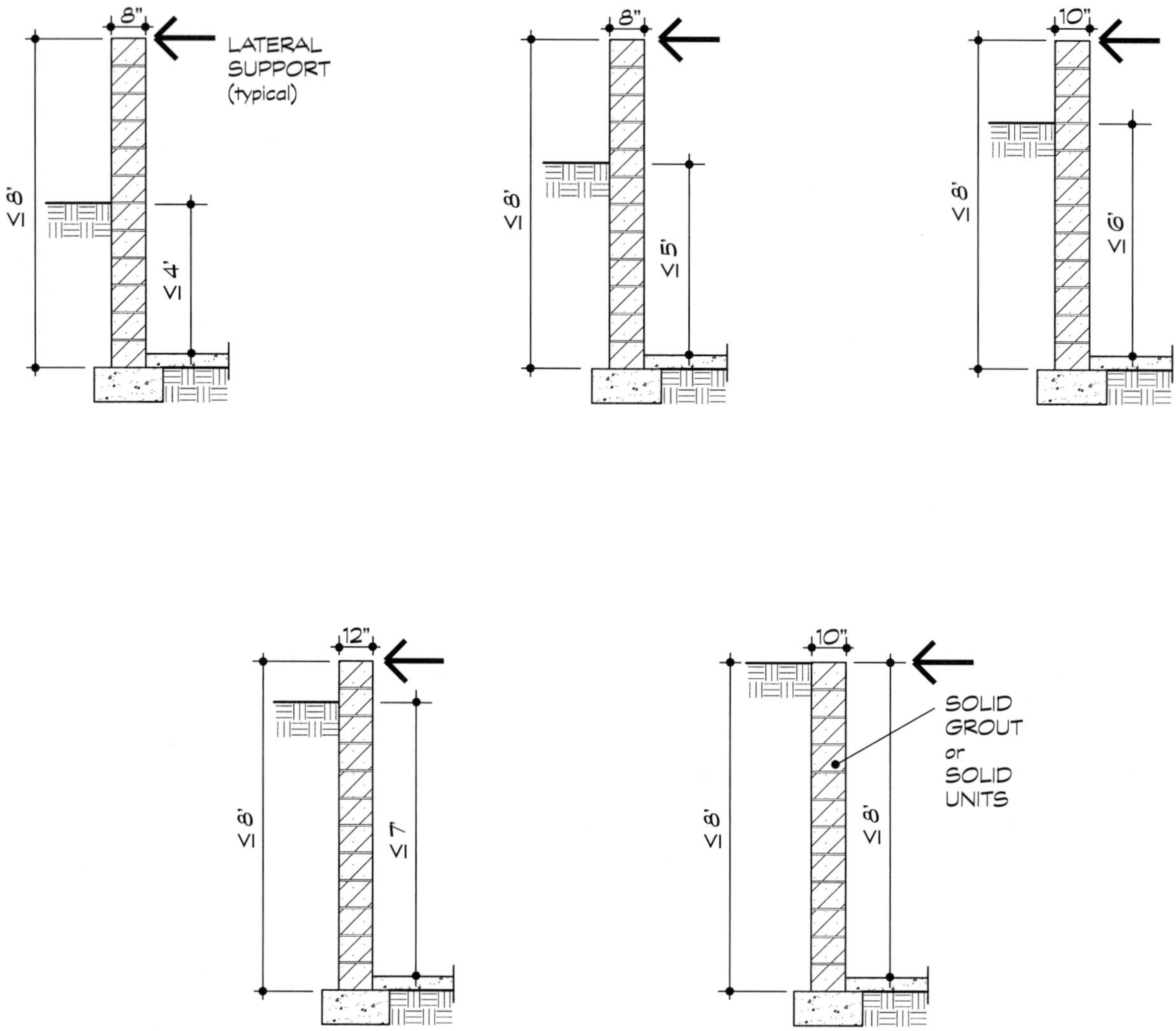

8'-HIGH WALLS

Fig. 1807.1.6.3(1)-2. Plain masonry foundation wall, minimum nominal thickness for 30 psf/ft of depth lateral soil load. *[IBC Table 1807.1.6.3(1)]*

1807 Foundation Walls, Retaining Walls and Embedded Posts and Poles

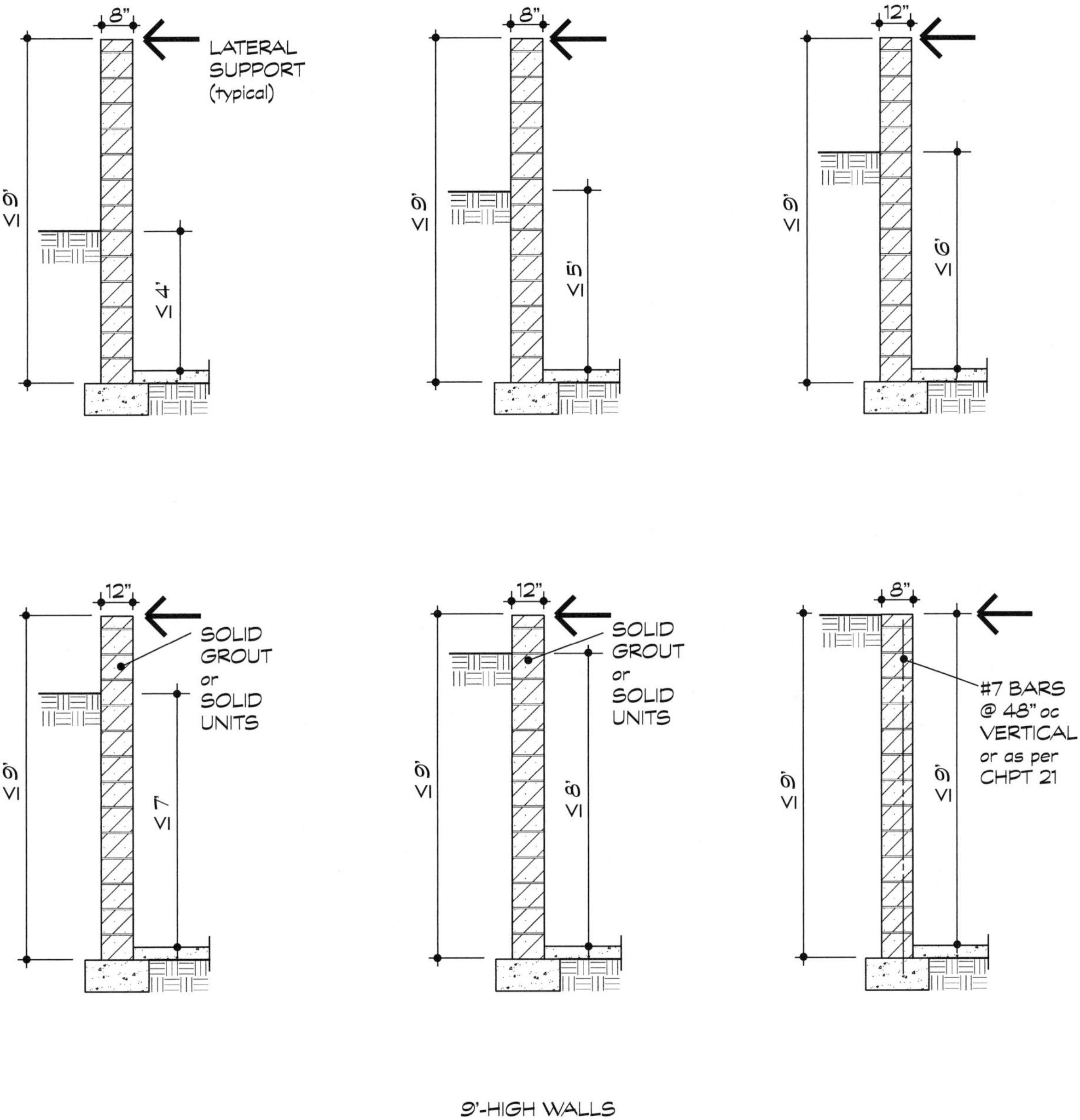

Fig. 1807.1.6.3(1)-3. Plain masonry foundation wall, minimum nominal thickness for 30 psf/ft of depth lateral soil load. *[IBC Table 1807.1.6.3(1)]*

1807 Foundation Walls, Retaining Walls and Embedded Posts and Poles

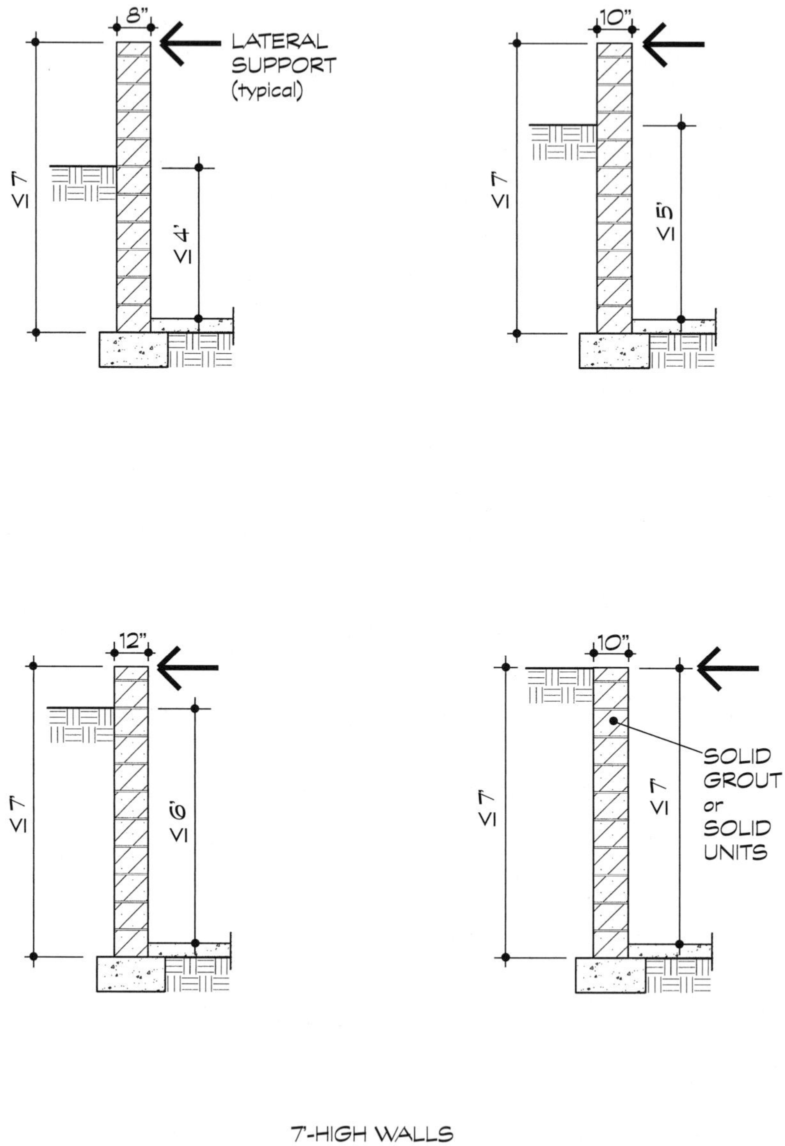

Fig. 1807.1.6.3(1)-4. Plain masonry foundation wall, minimum nominal thickness for 45 psf/ft of depth lateral soil load. *[IBC Table 1807.1.6.3(1)]*

1807 Foundation Walls, Retaining Walls and Embedded Posts and Poles

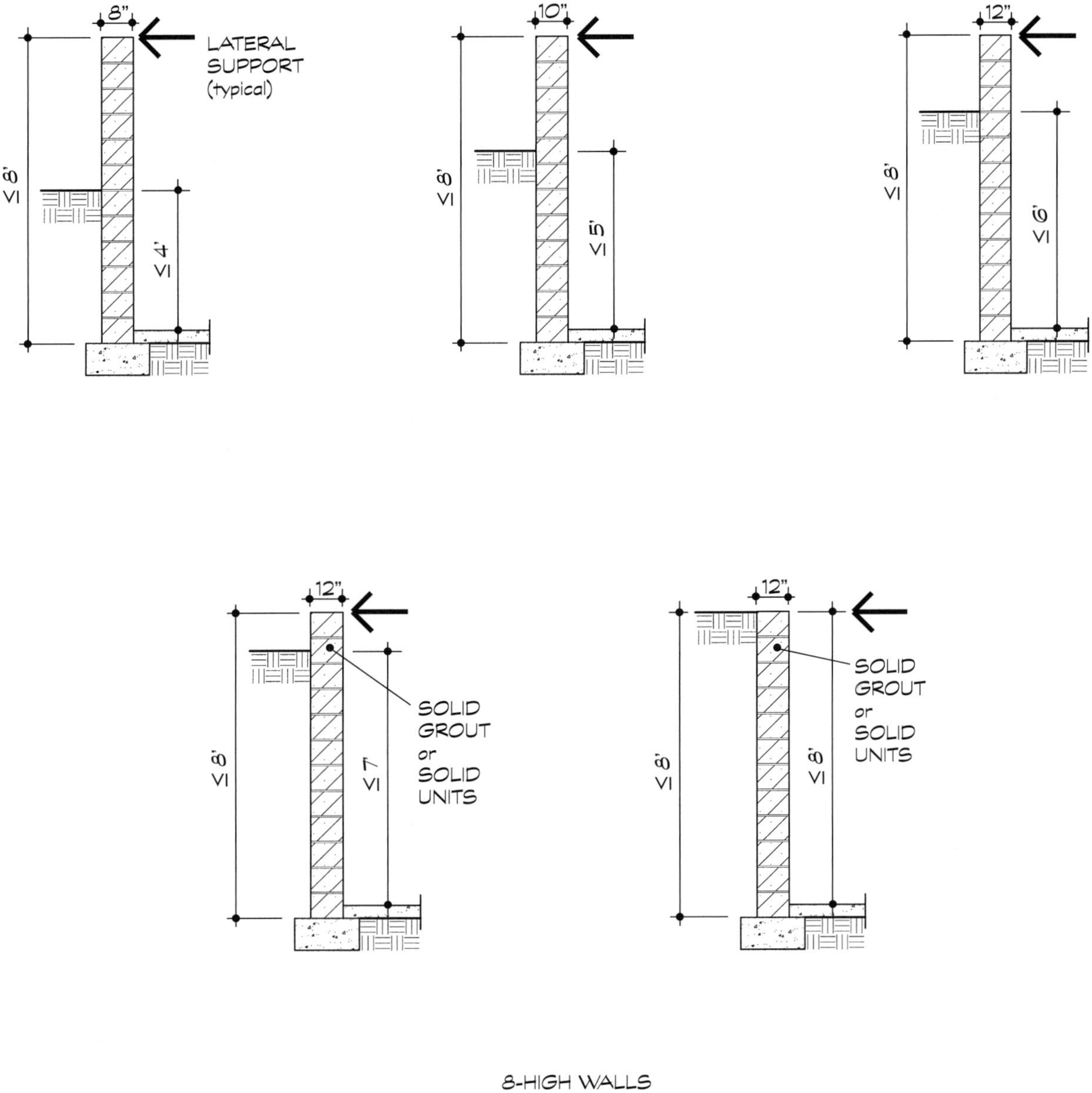

Fig. 1807.1.6.3(1)-5. Plain masonry foundation wall, minimum nominal thickness for 45 psf/ft of depth lateral soil load. *[IBC Table 1807.1.6.3(1)]*

1807 Foundation Walls, Retaining Walls and Embedded Posts and Poles

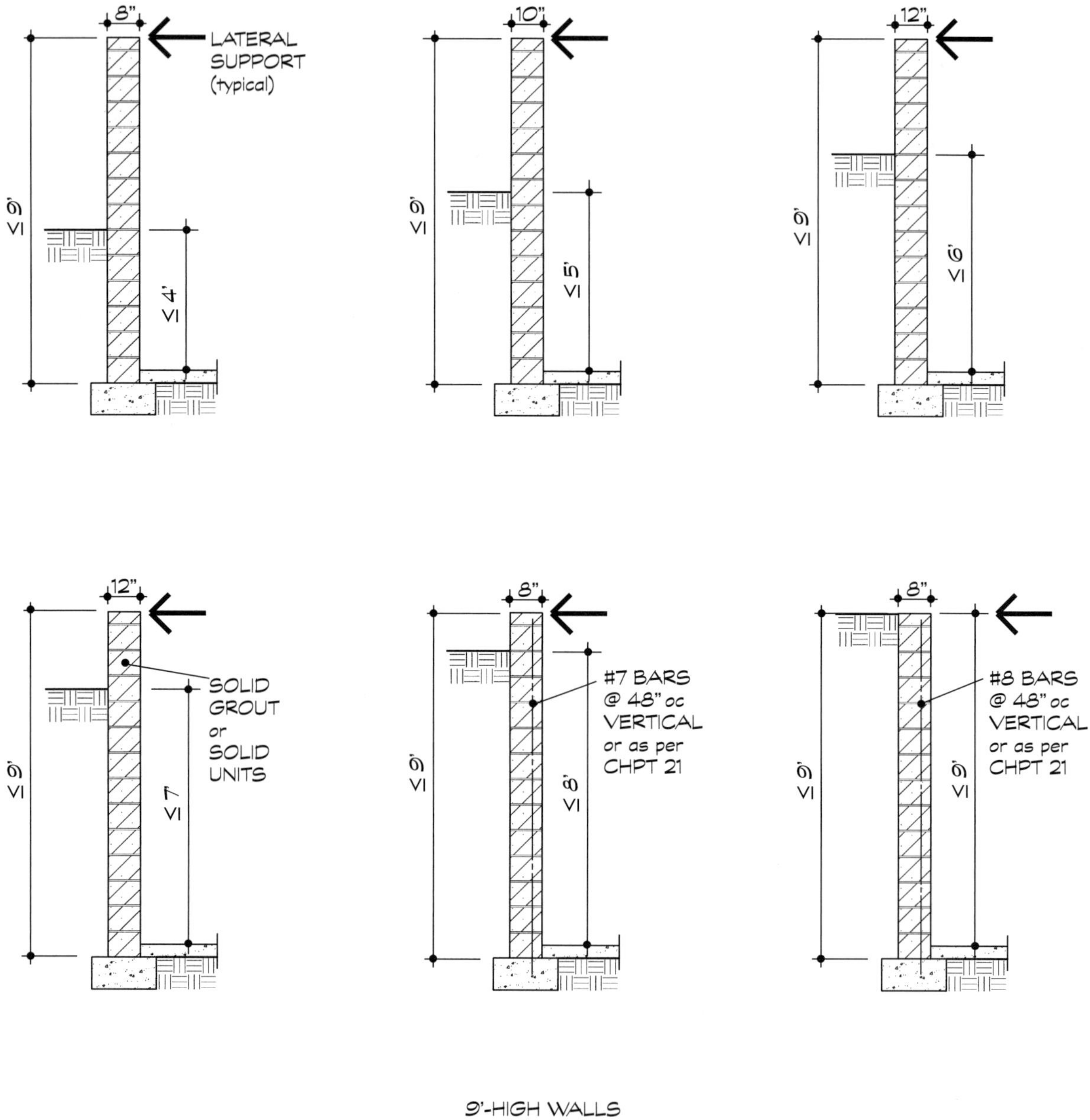

Fig. 1807.1.6.3(1)-6. Plain masonry foundation wall, minimum nominal thickness for 45 psf/ft of depth lateral soil load. *[IBC Table 1807.1.6.3(1)]*

1807 Foundation Walls, Retaining Walls and Embedded Posts and Poles

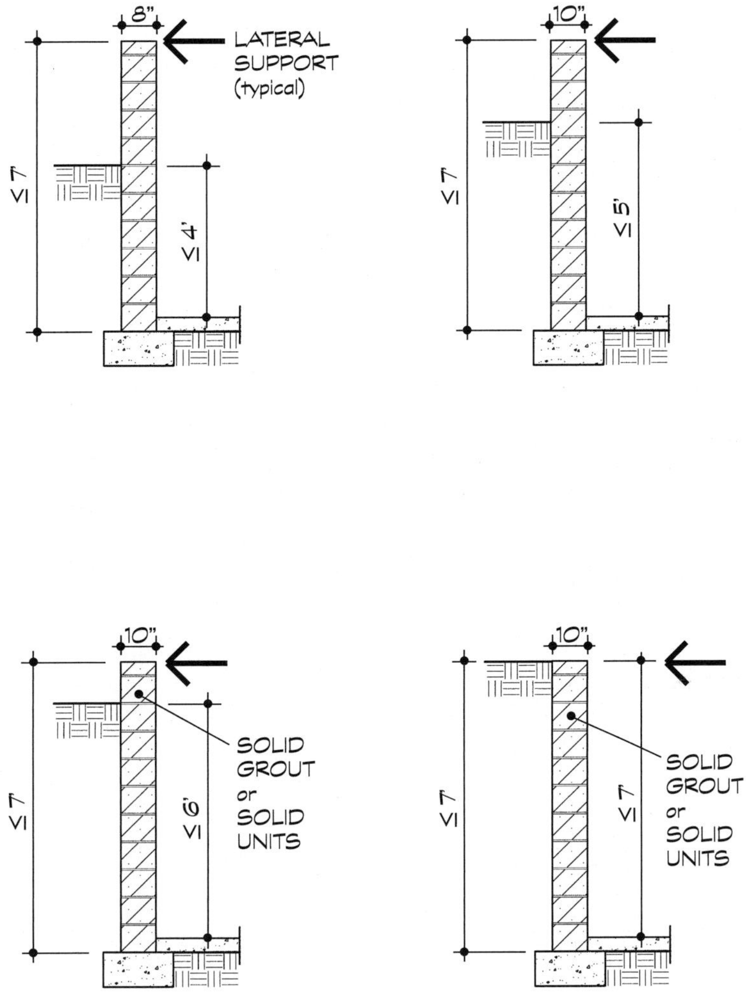

7-HIGH WALLS

Fig. 1807.1.6.3(1)-7. Plain masonry foundation wall, minimum nominal thickness for 60 psf/ft of depth lateral soil load. *[IBC Table 1807.1.6.3(1)]*

1807 Foundation Walls, Retaining Walls and Embedded Posts and Poles

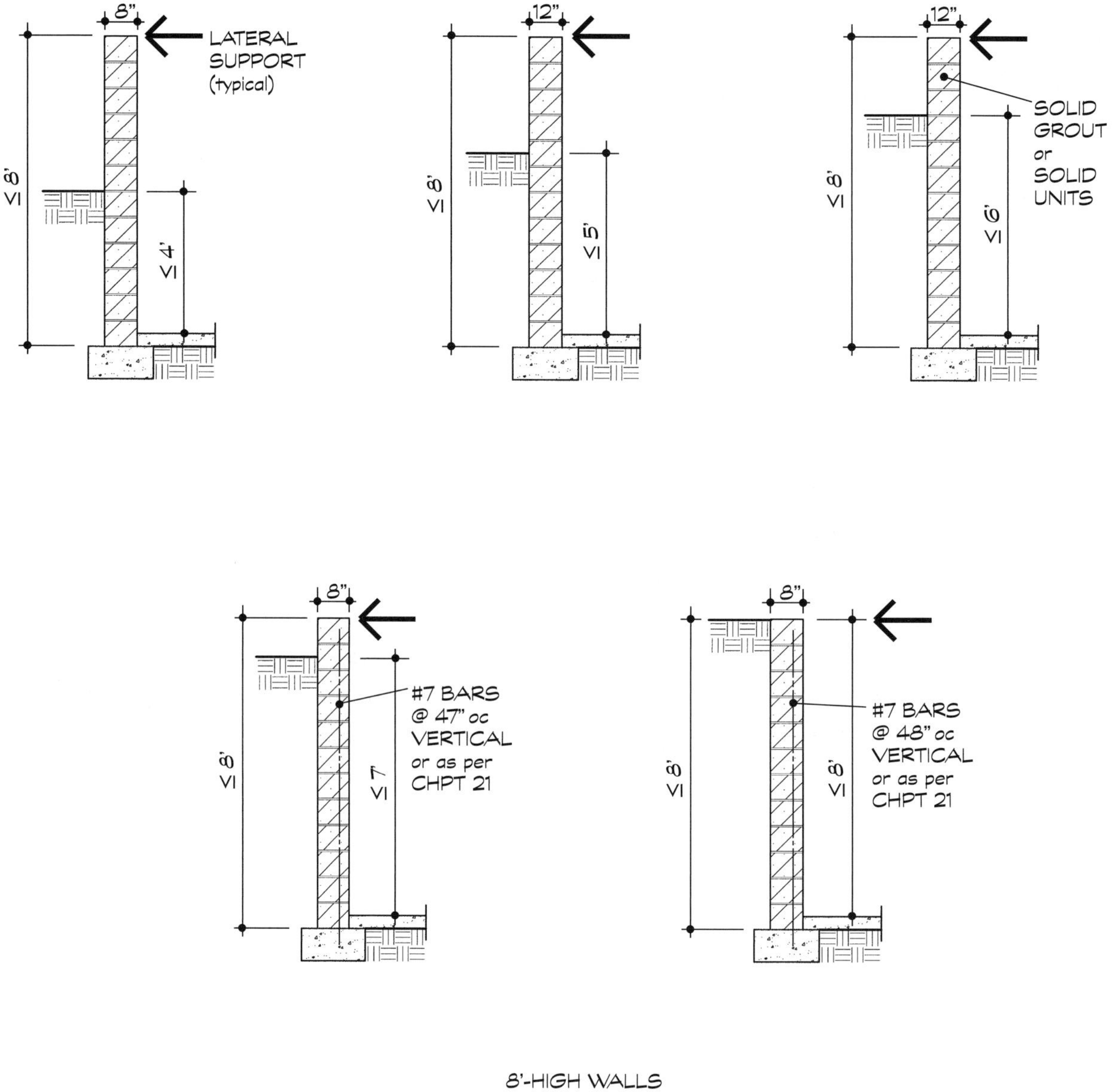

Fig. 1807.1.6.3(1)-8. Plain masonry foundation wall, minimum nominal thickness for 60 psf/ft of depth lateral soil load. *[IBC Table 1807.1.6.3(1)]*

1807 Foundation Walls, Retaining Walls and Embedded Posts and Poles

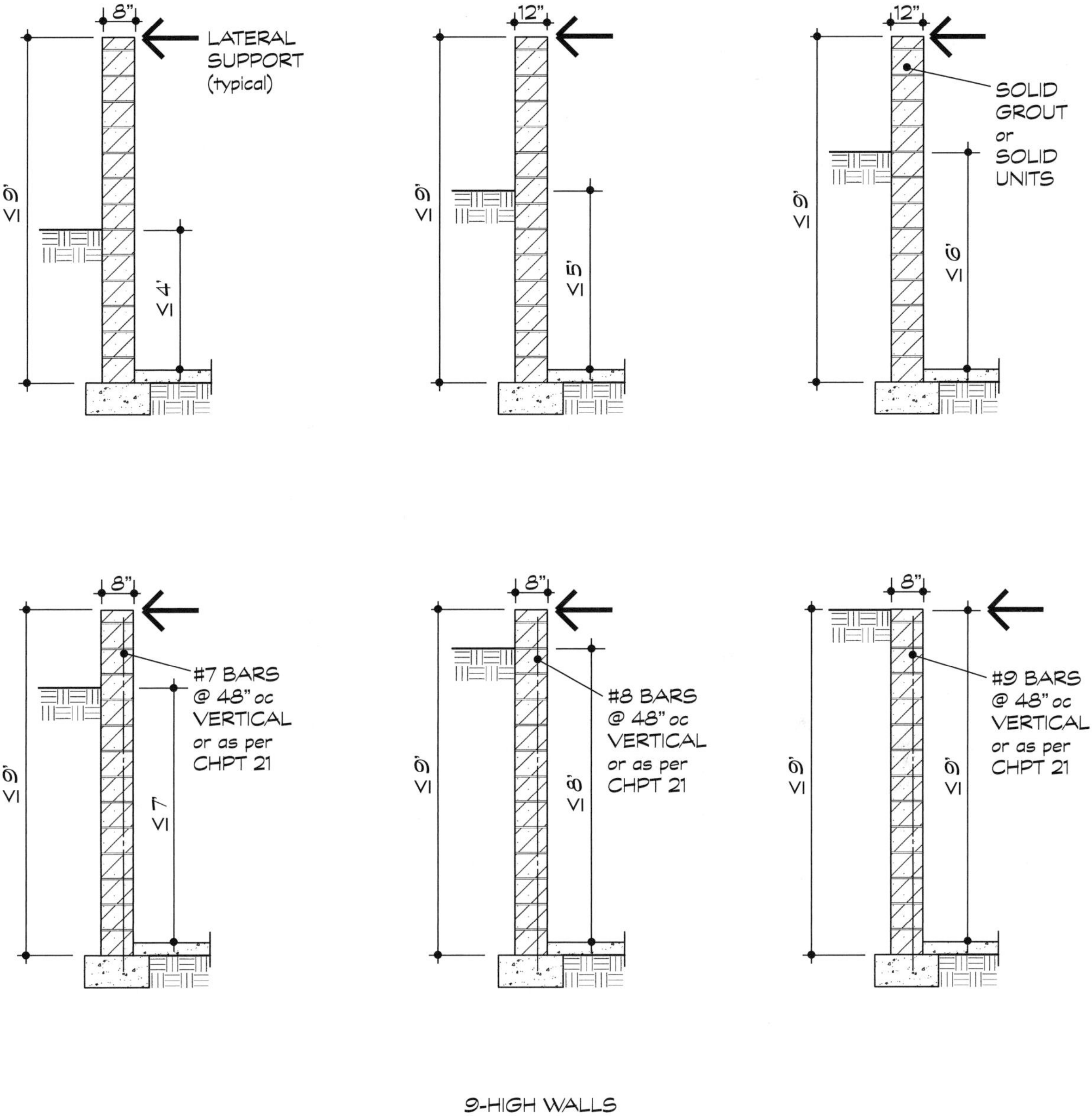

Fig. 1807.1.6.3(1)-9. Plain masonry foundation wall, minimum nominal thickness for 60 psf/ft of depth lateral soil load. *[IBC Table 1807.1.6.3(1)]*

1807 Foundation Walls, Retaining Walls and Embedded Posts and Poles

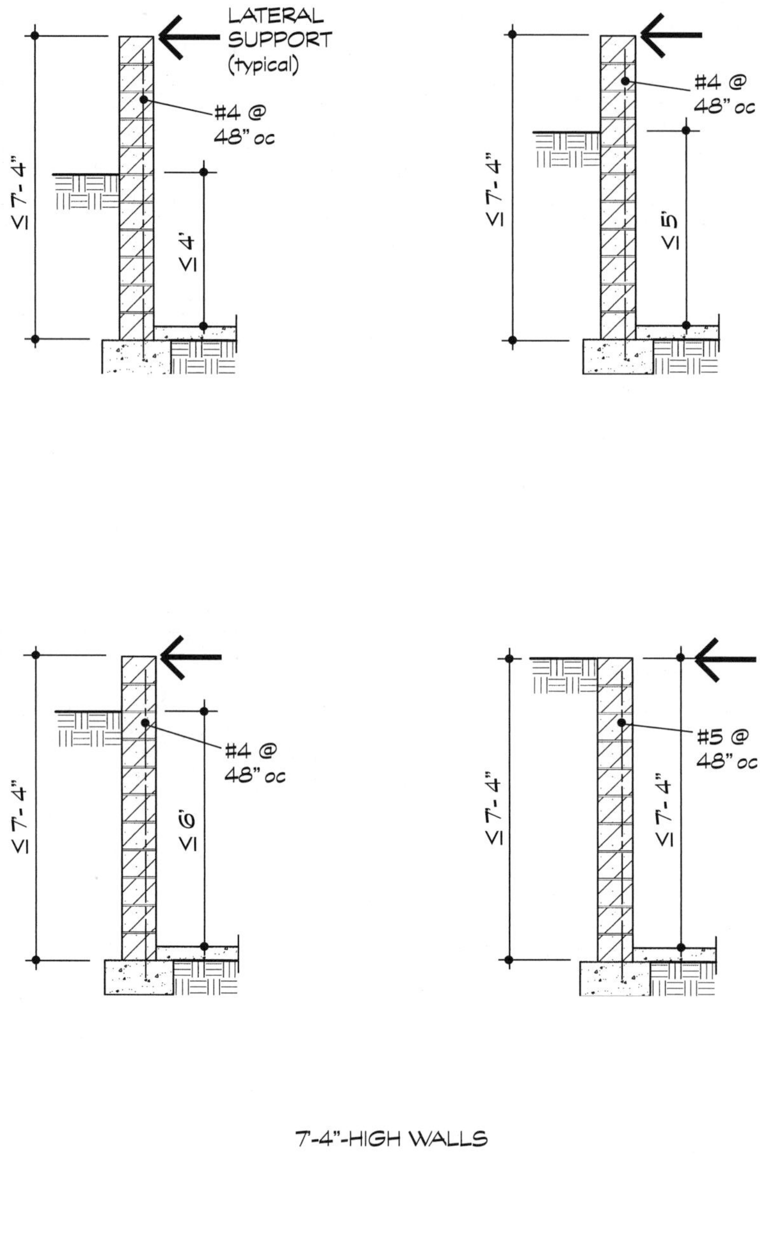

Fig. 1807.1.6.3(2)-1. Vertical reinforcement for 8" masonry foundation walls with $d \geq 5"$. Lateral soil load is 30 psf/ft of depth. *[IBC Table 1807.1.6.3(2)]*

1807 Foundation Walls, Retaining Walls and Embedded Posts and Poles

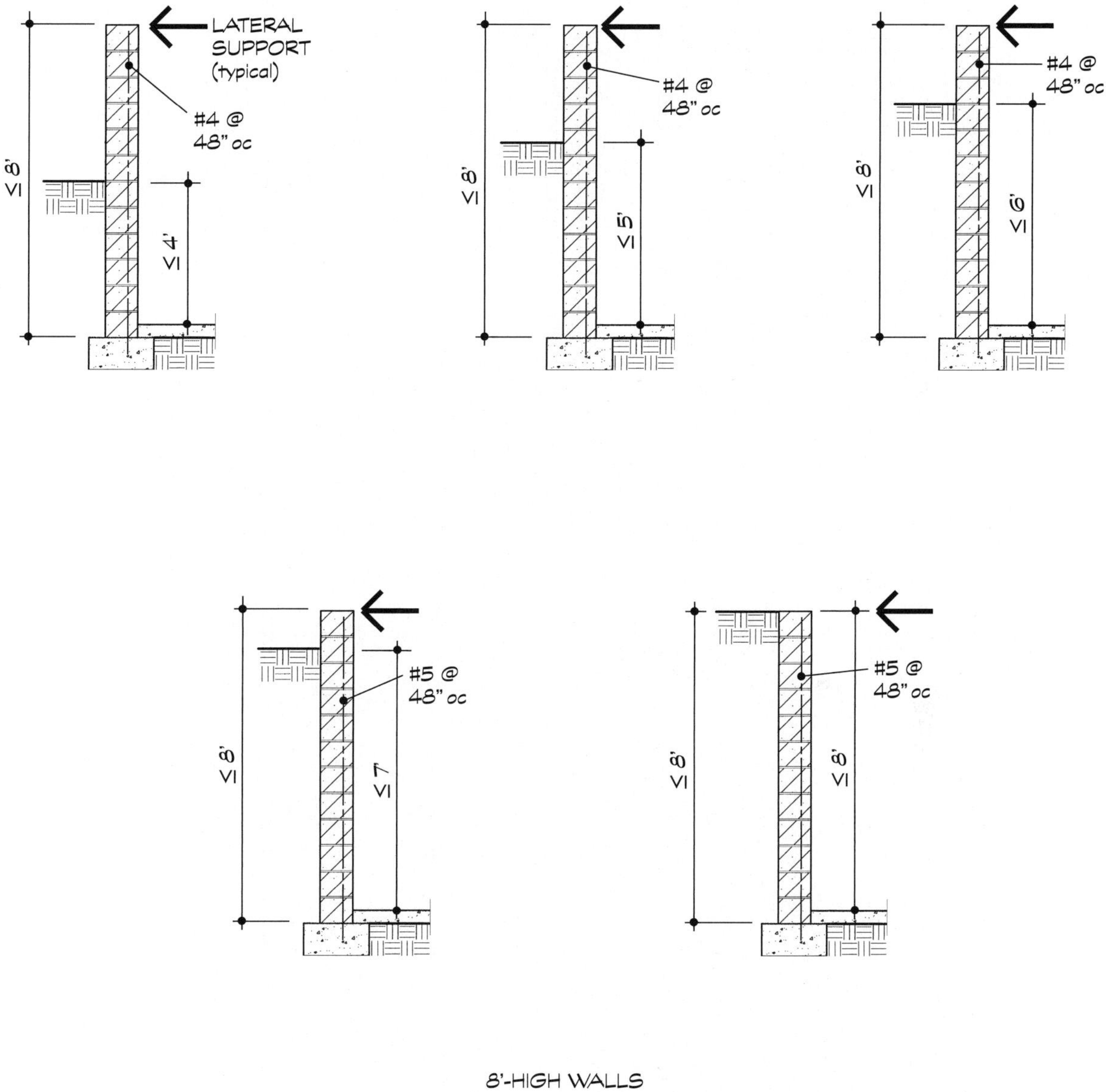

Fig. 1807.1.6.3(2)-2. Vertical reinforcement for 8" masonry foundation walls with $d \geq 5"$. Lateral soil load is 30 psf/ft of depth. *[IBC Table 1807.1.6.3(2)]*

1807 Foundation Walls, Retaining Walls and Embedded Posts and Poles

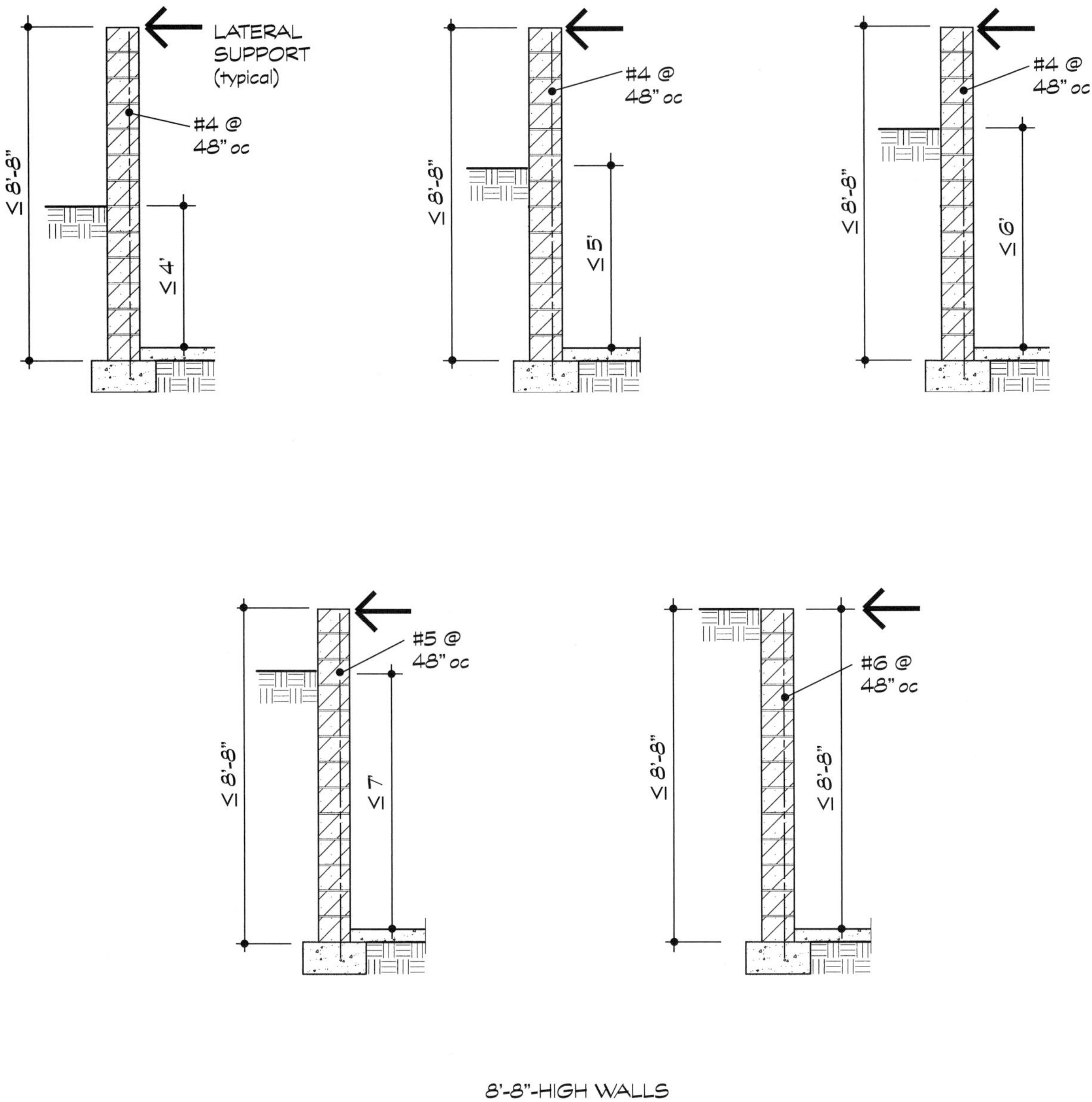

Fig. 1807.1.6.3(2)-3. Vertical reinforcement for 8" masonry foundation walls with $d \geq 5"$. Lateral soil load is 30 psf/ft of depth. *[IBC Table 1807.1.6.3(2)]*

1807 Foundation Walls, Retaining Walls and Embedded Posts and Poles

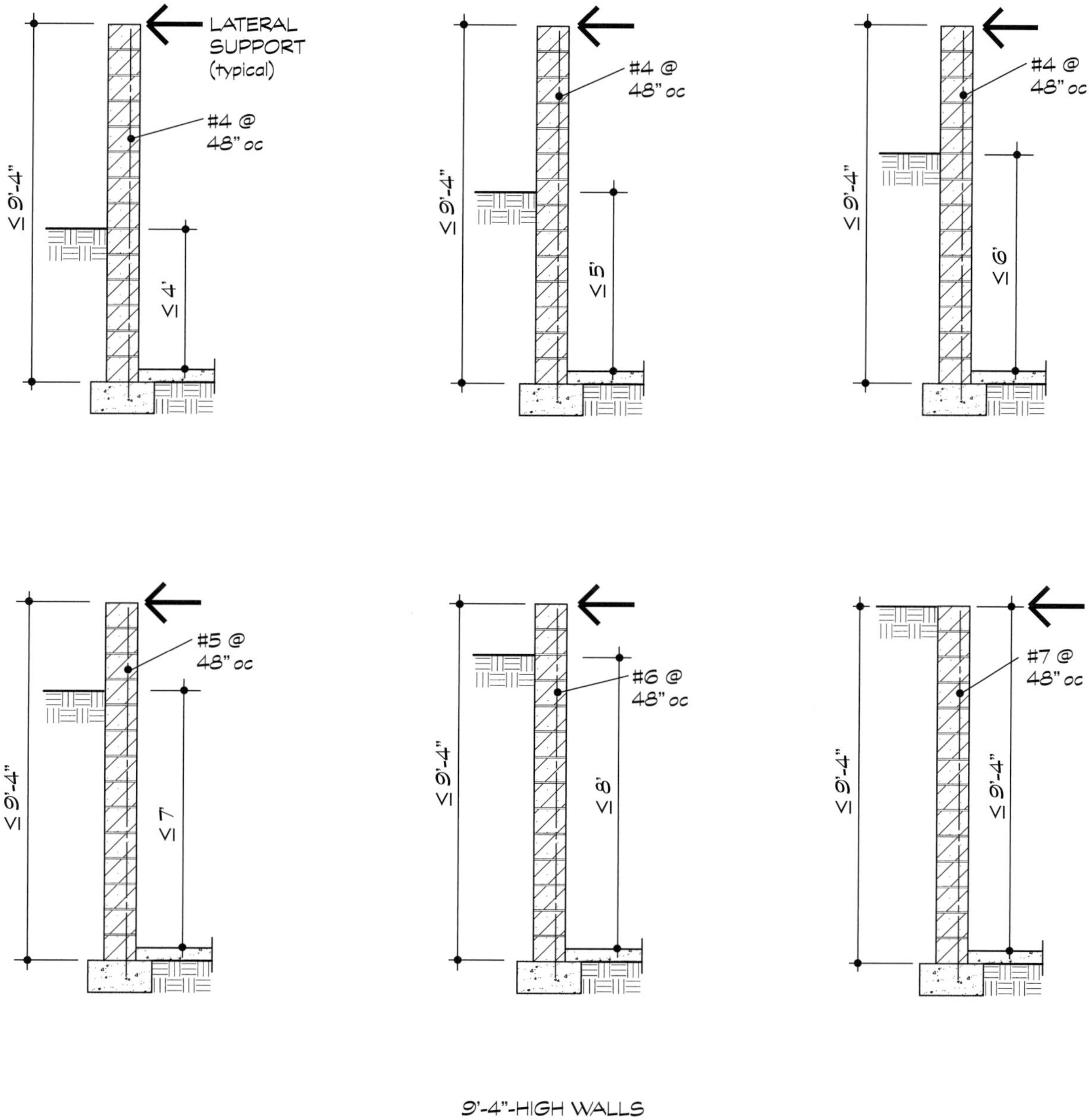

Fig. 1807.1.6.3(2)-4. Vertical reinforcement for 8" masonry foundation walls with $d \geq 5$". Lateral soil load is 30 psf/ft of depth. *[IBC Table 1807.1.6.3(2)]*

1807 Foundation Walls, Retaining Walls and Embedded Posts and Poles

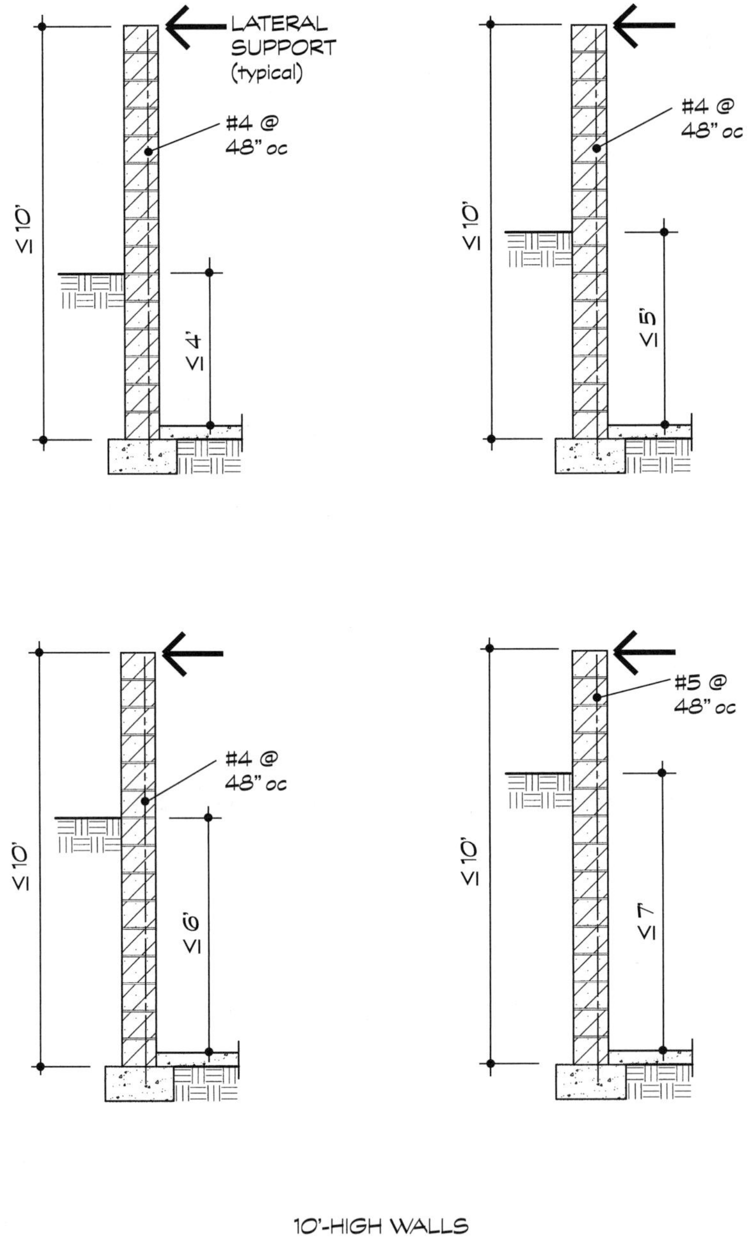

Fig. 1807.1.6.3(2)-5a. Vertical reinforcement for 8" masonry foundation walls with $d \geq 5"$. Lateral soil load is 30 psf/ft of depth. *[IBC Table 1807.1.6.3(2)]*

1807 Foundation Walls, Retaining Walls and Embedded Posts and Poles

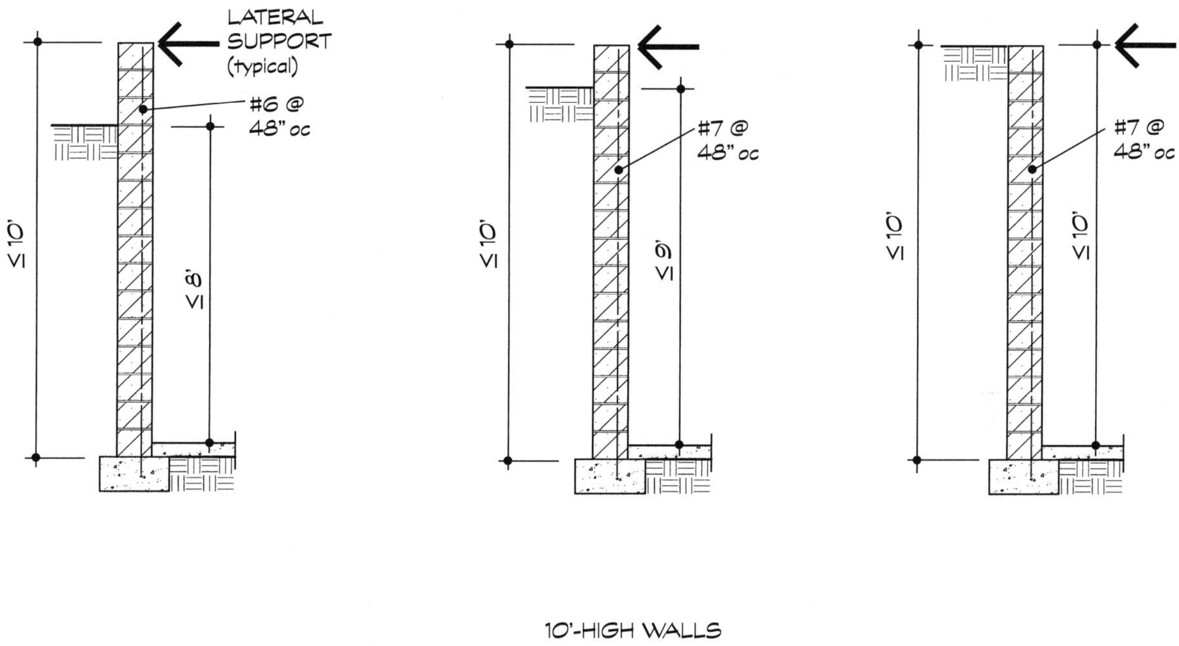

Fig. 1807.1.6.3(2)-5b. Vertical reinforcement for 8" masonry foundation walls with $d \geq 5$". Lateral soil load is 30 psf/ft of depth. *[IBC Table 1807.1.6.3(2)]*

1807 Foundation Walls, Retaining Walls and Embedded Posts and Poles

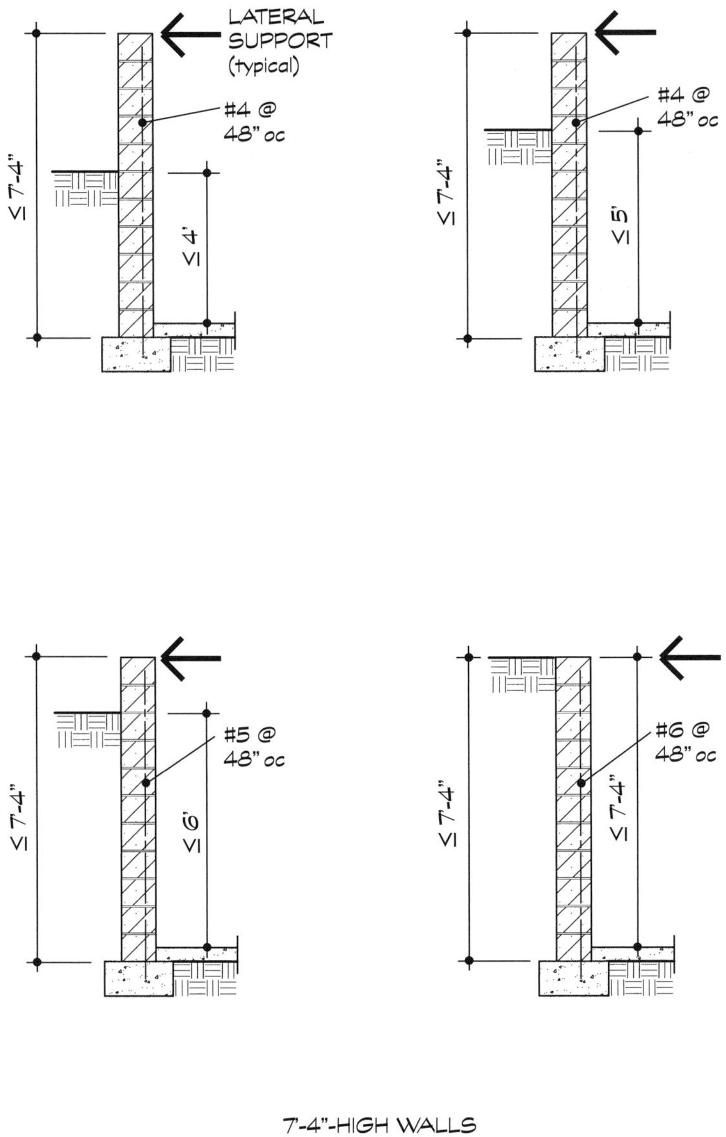

Fig. 1807.1.6.3(2)-6. Vertical reinforcing for 8" masonry foundation walls with $d \geq 5"$. Lateral soil load is 45 psf/ft of depth. *[IBC Table 1807.1.6.3(2)]*

1807 Foundation Walls, Retaining Walls and Embedded Posts and Poles

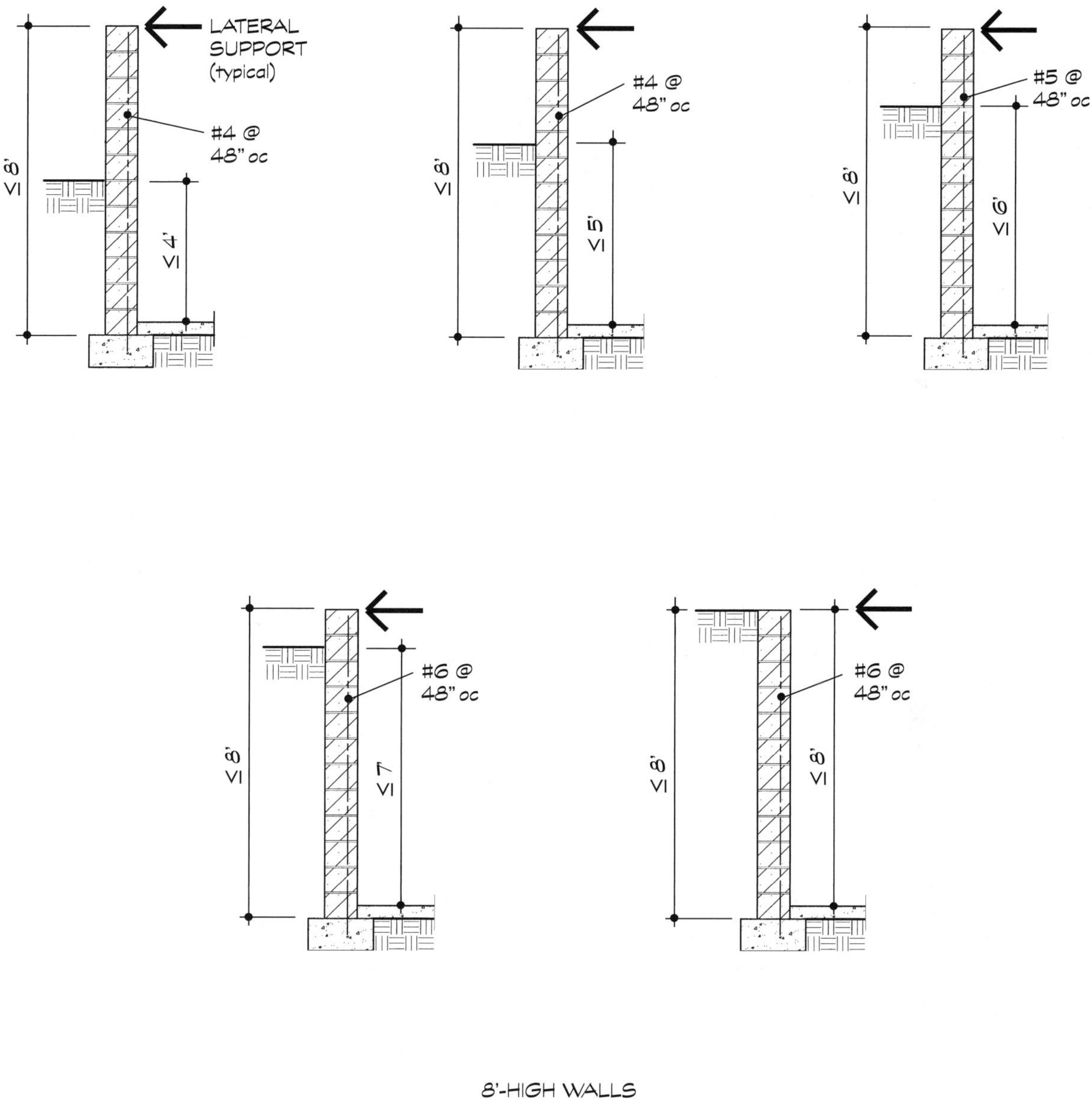

Fig. 1807.1.6.3(2)-7. Vertical reinforcing for 8" masonry foundation walls with $d \geq 5"$. Lateral soil load is 45 psf/ft of depth. *[IBC Table 1807.1.6.3(2)]*

1807 Foundation Walls, Retaining Walls and Embedded Posts and Poles

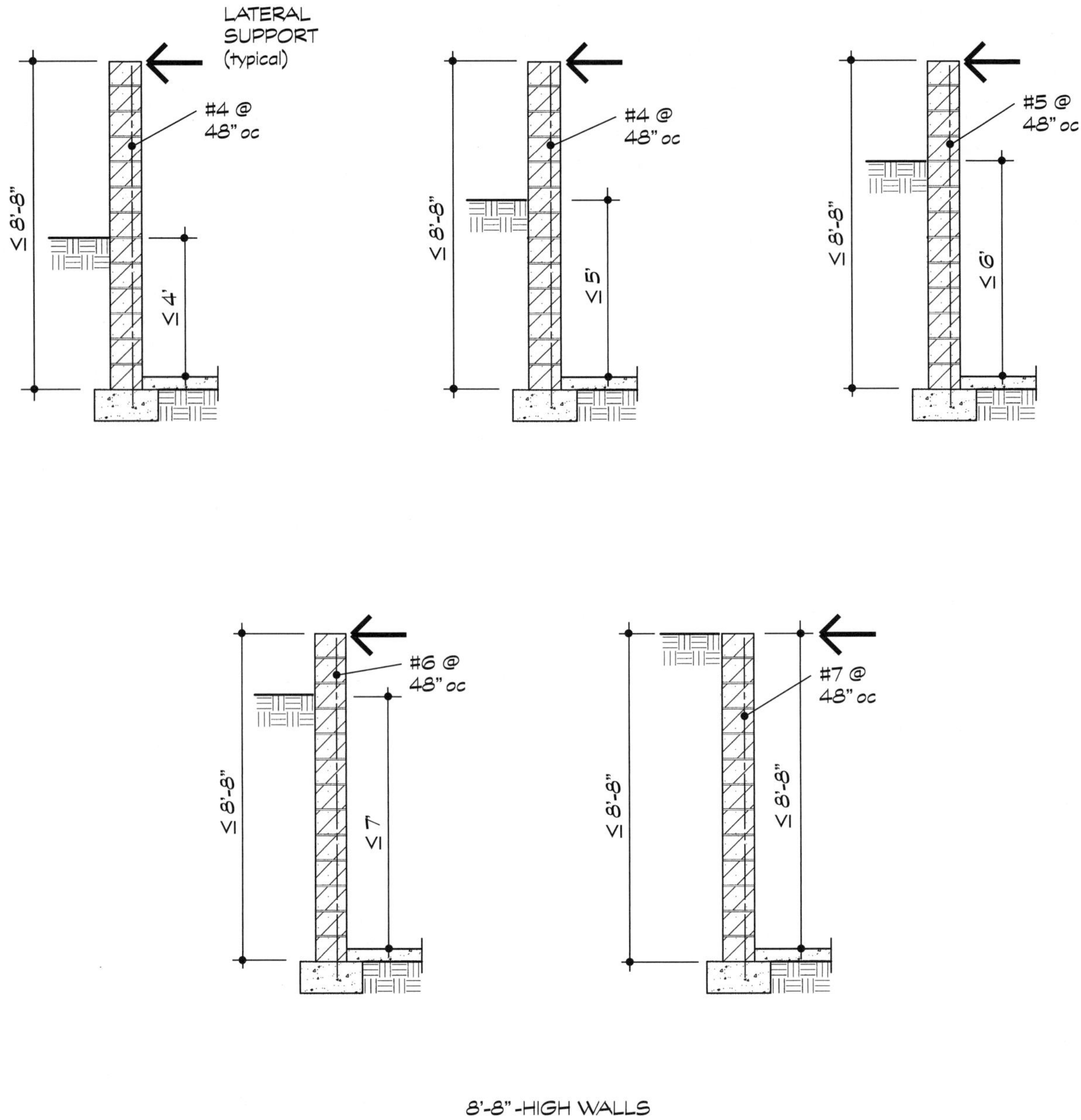

Fig. 1807.1.6.3(2)-8. Vertical reinforcing for 8" masonry foundation walls with $d \geq 5"$. Lateral soil load is 45 psf/ft of depth. *[IBC Table 1807.1.6.3(2)]*

1807 Foundation Walls, Retaining Walls and Embedded Posts and Poles

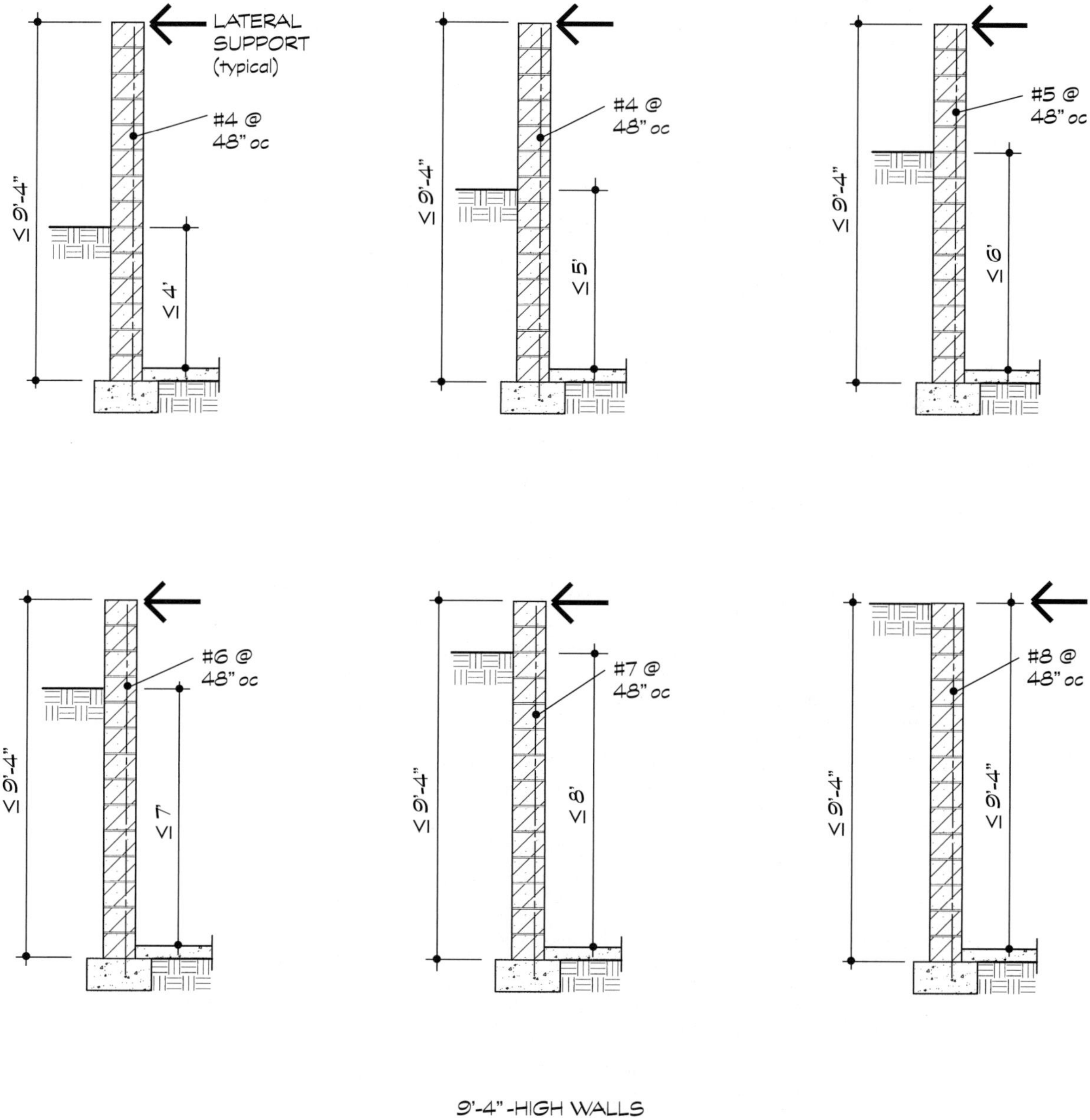

Fig. 1807.1.6.3(2)-9. Vertical reinforcing for 8" masonry foundation walls with $d \geq 5$". Lateral soil load is 45 psf/ft of depth. *[IBC Table 1807.1.6.3(2)]*

1807 Foundation Walls, Retaining Walls and Embedded Posts and Poles

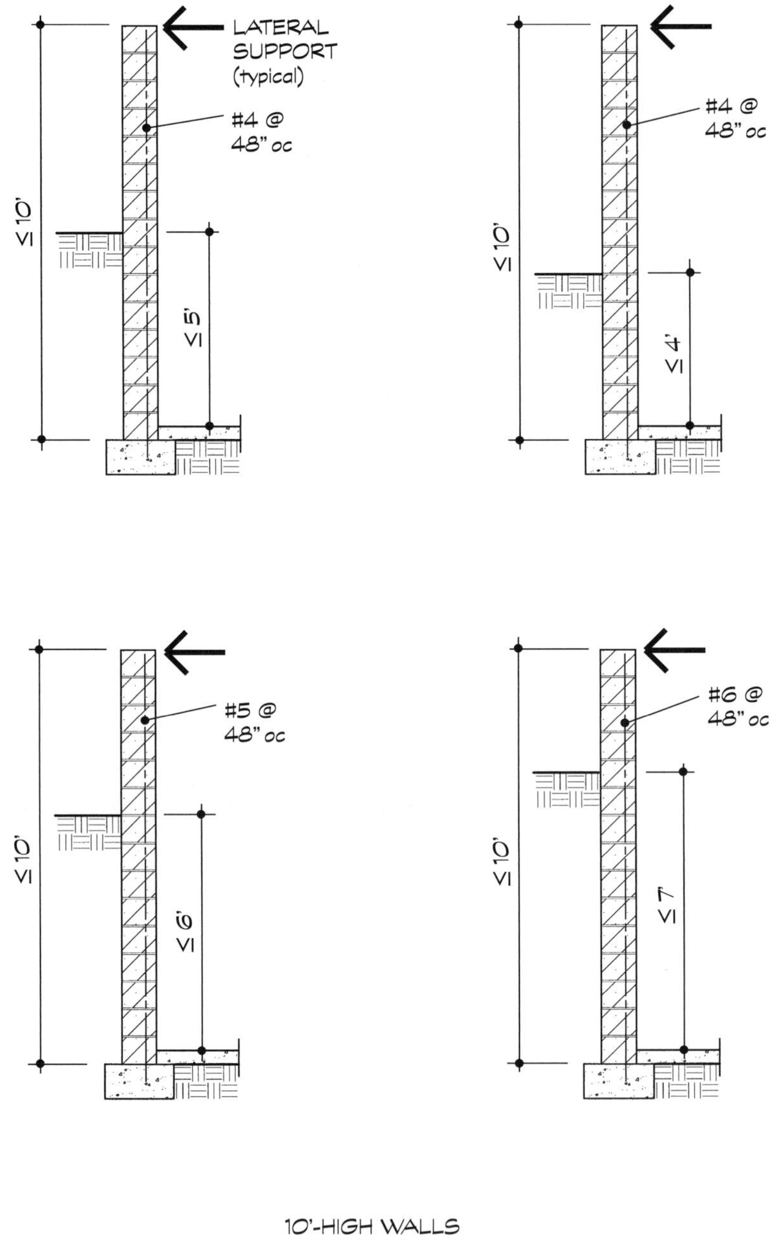

Fig. 1807.1.6.3(2)-10a. Vertical reinforcing for 8" masonry foundation walls with $d \geq 5$". Lateral soil load is 45 psf/ft of depth. *[IBC Table 1807.1.6.3(2)]*

1807 Foundation Walls, Retaining Walls and Embedded Posts and Poles

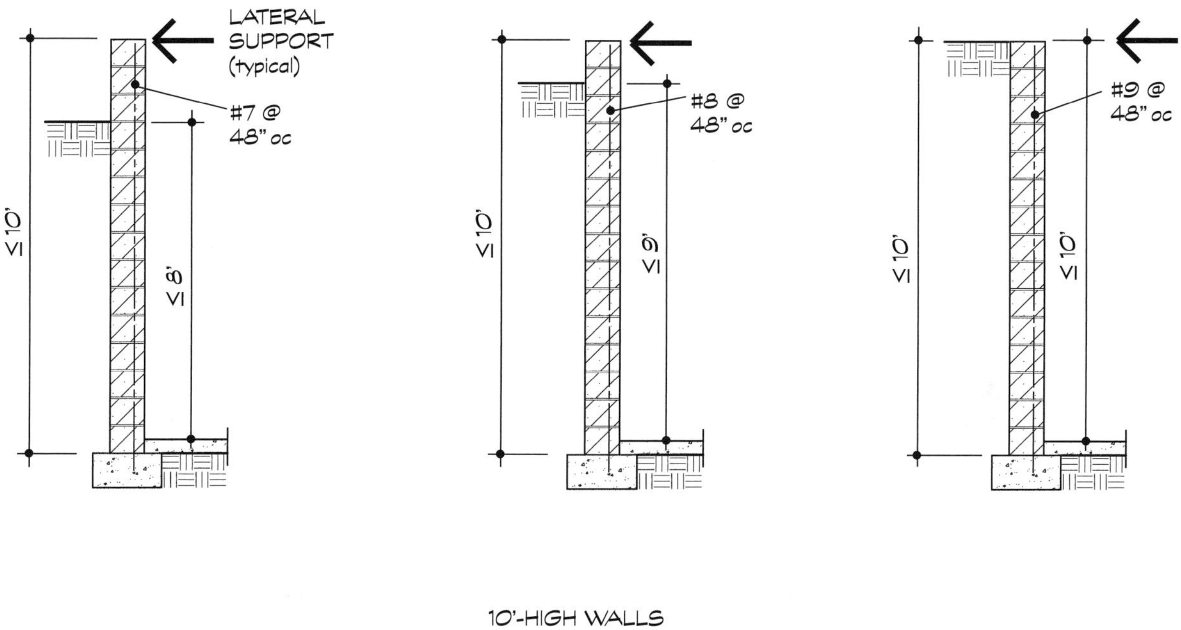

Fig. 1807.1.6.3(2)-10b. Vertical reinforcing for 8" masonry foundation walls with $d \geq 5"$. Lateral soil load is 45 psf/ft of depth. *[IBC Table 1807.1.6.3(2)]*

1807 Foundation Walls, Retaining Walls and Embedded Posts and Poles

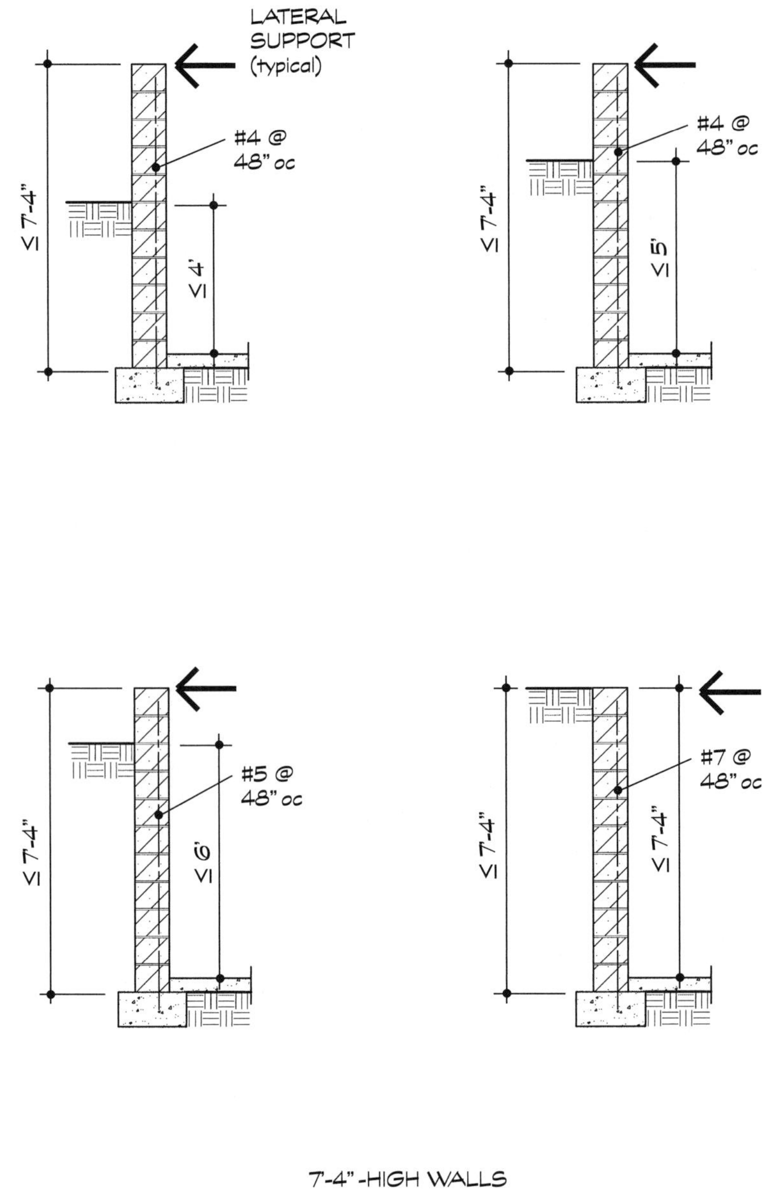

Fig. 1807.1.6.3(2)-11. Vertical reinforcing for 8" masonry foundation walls with $d \geq 5"$. Lateral soil load is 60 psf/ft of depth. *[IBC Table 1807.1.6.3(2)]*

1807 Foundation Walls, Retaining Walls and Embedded Posts and Poles

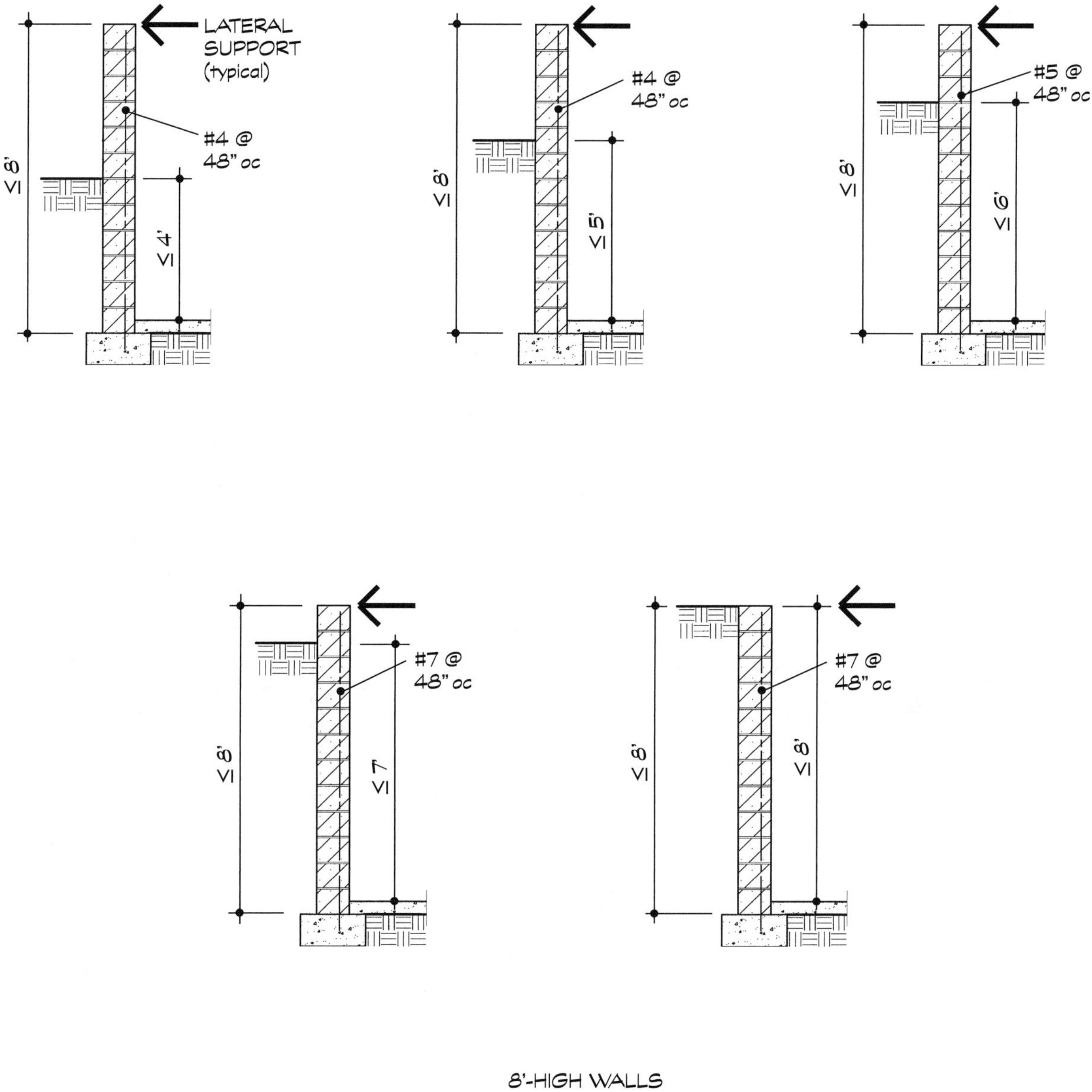

Fig. 1807.1.6.3(2)-12. Vertical reinforcing for 8" masonry foundation walls with $d \geq 5"$. Lateral soil load is 60 psf/ft of depth. *[IBC Table 1807.1.6.3(2)]*

1807 Foundation Walls, Retaining Walls and Embedded Posts and Poles

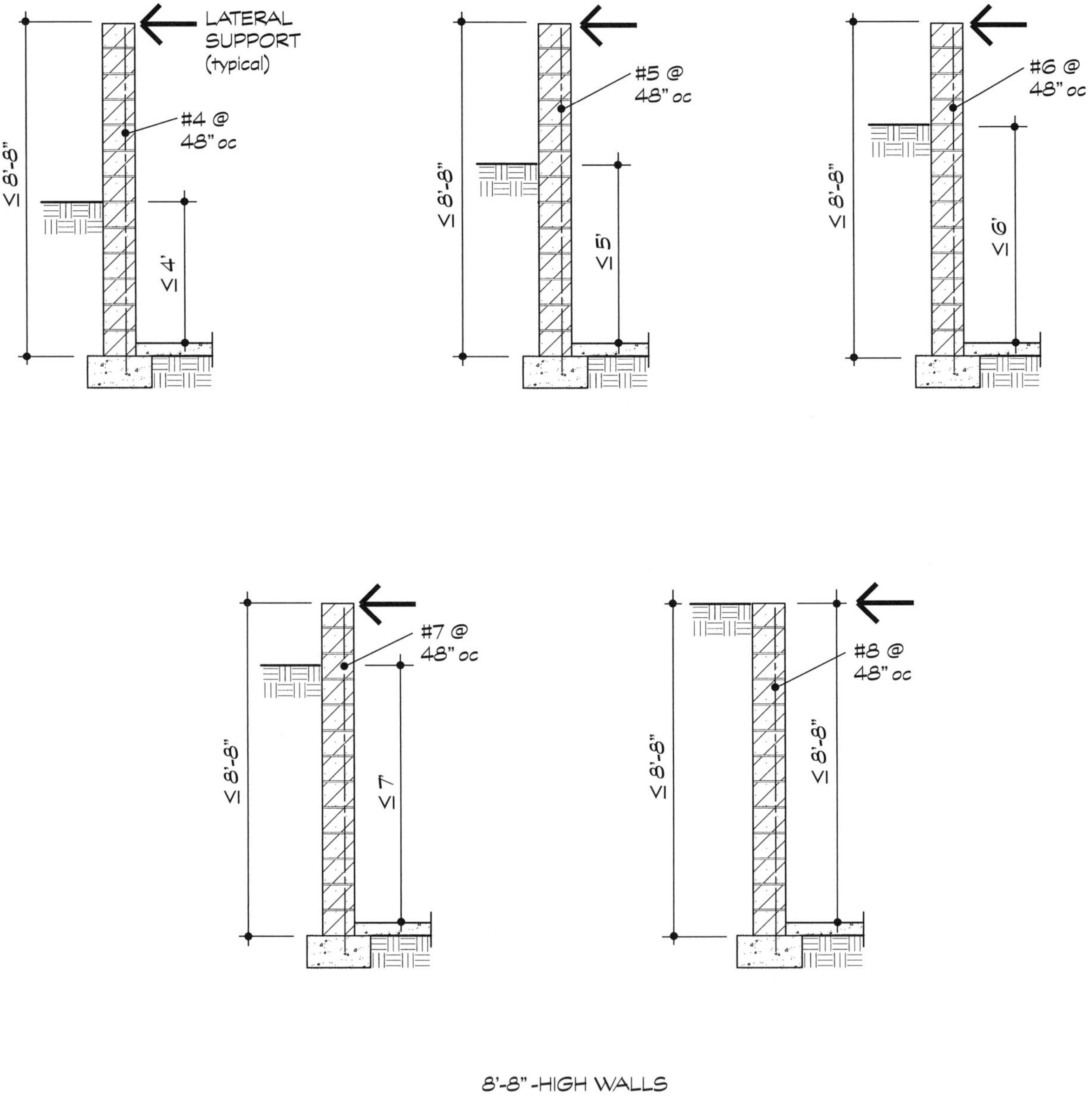

Fig. 1807.1.6.3(2)-13. Vertical reinforcing for 8" masonry foundation walls with $d \geq 5$". Lateral soil load is 60 psf/ft of depth. *[IBC Table 1807.1.6.3(2)]*

1807 Foundation Walls, Retaining Walls and Embedded Posts and Poles

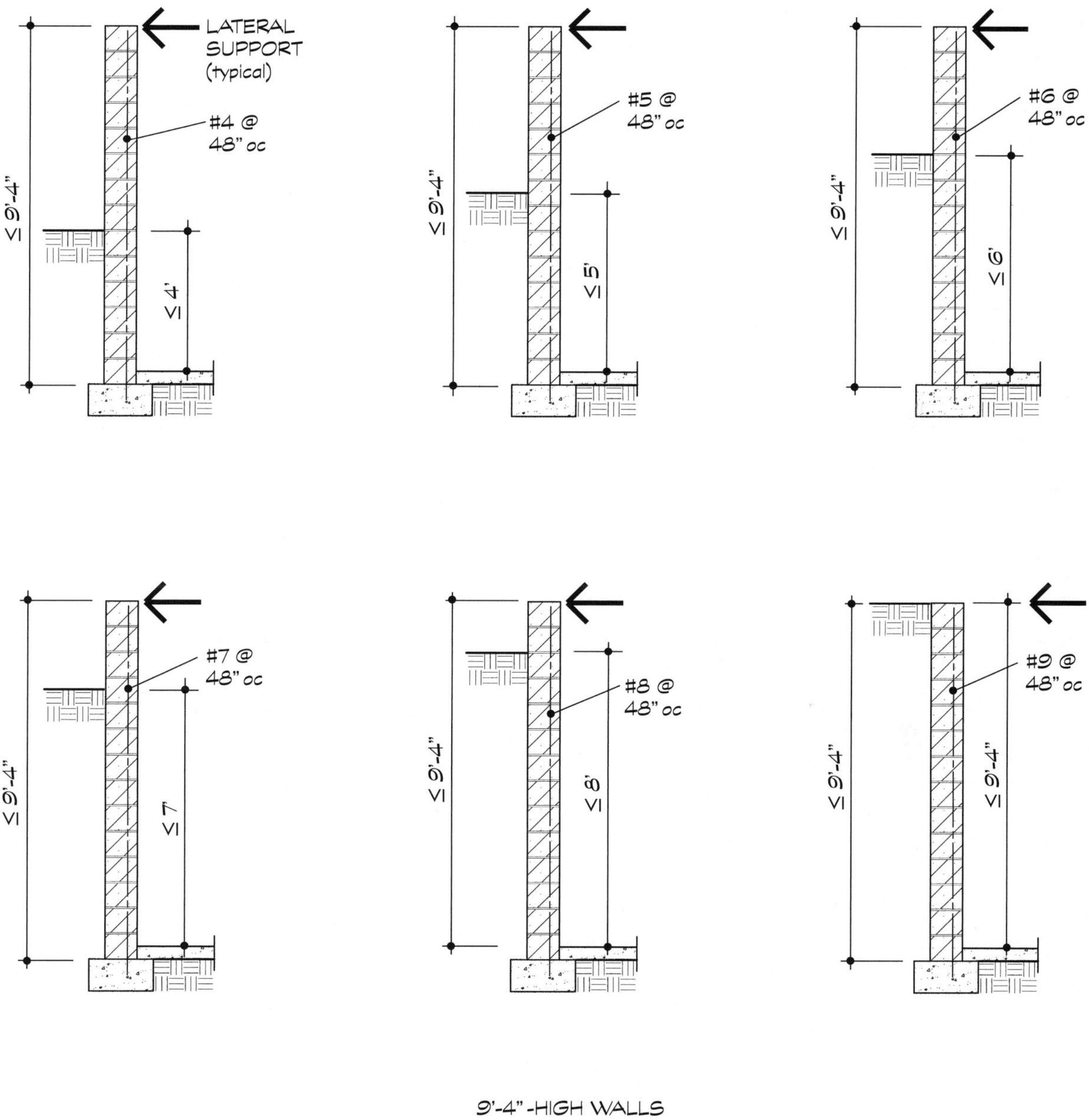

Fig. 1807.1.6.3(2)-14. Vertical reinforcing for 8" masonry foundation walls with $d \geq 5"$. Lateral soil load is 60 psf/ft of depth. *[IBC Table 1807.1.6.3(2)]*

1807 Foundation Walls, Retaining Walls and Embedded Posts and Poles

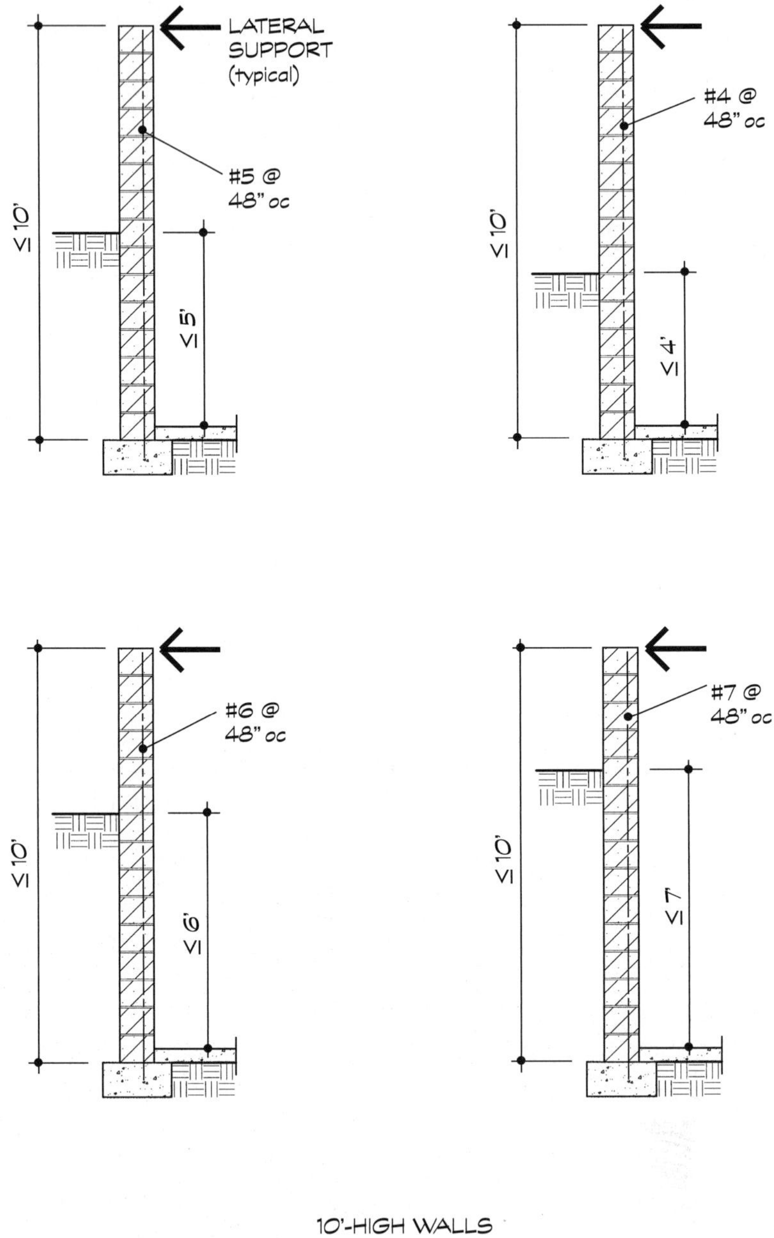

Fig. 1807.1.6.3(2)-15a. Vertical reinforcing for 8" masonry foundation walls with $d \geq 5"$. Lateral soil load is 60 psf/ft of depth. *[IBC Table 1807.1.6.3(2)]*

1807 Foundation Walls, Retaining Walls and Embedded Posts and Poles

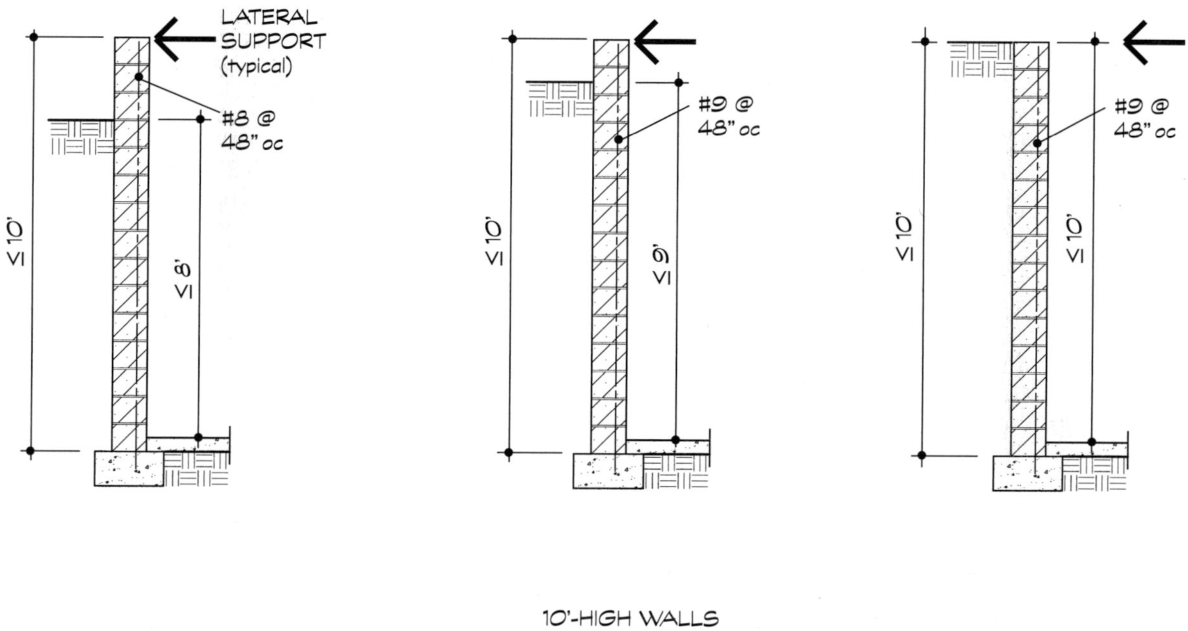

Fig. 1807.1.6.3(2)-15b. Vertical reinforcing for 8" masonry foundation walls with $d \geq 5$". Lateral soil load is 60 psf/ft of depth. *[IBC Table 1807.1.6.3(2)]*

1807 Foundation Walls, Retaining Walls and Embedded Posts and Poles

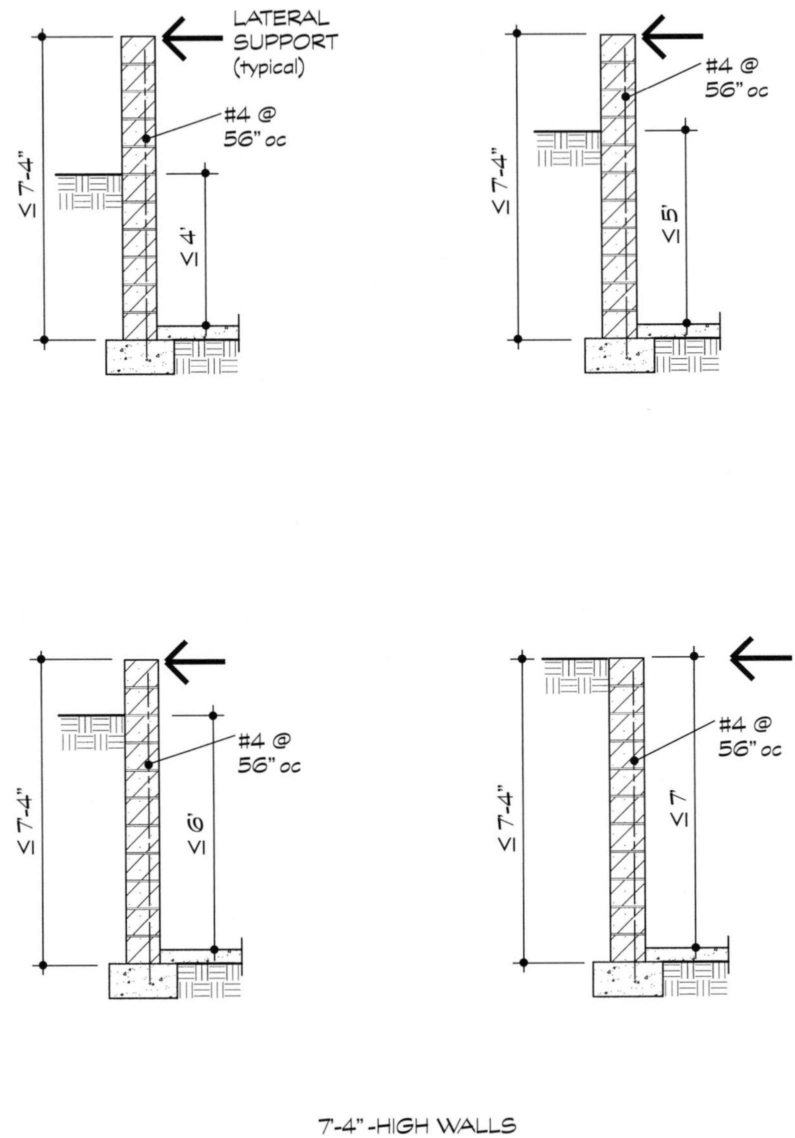

Fig. 1807.1.6.3(3)-1. Vertical reinforcing for 10" masonry foundation walls with $d \geq 6\,3/4$". Lateral soil load is 30 psf/ft of depth. *[IBC Table 1807.1.6.3(3)]*

1807 Foundation Walls, Retaining Walls and Embedded Posts and Poles

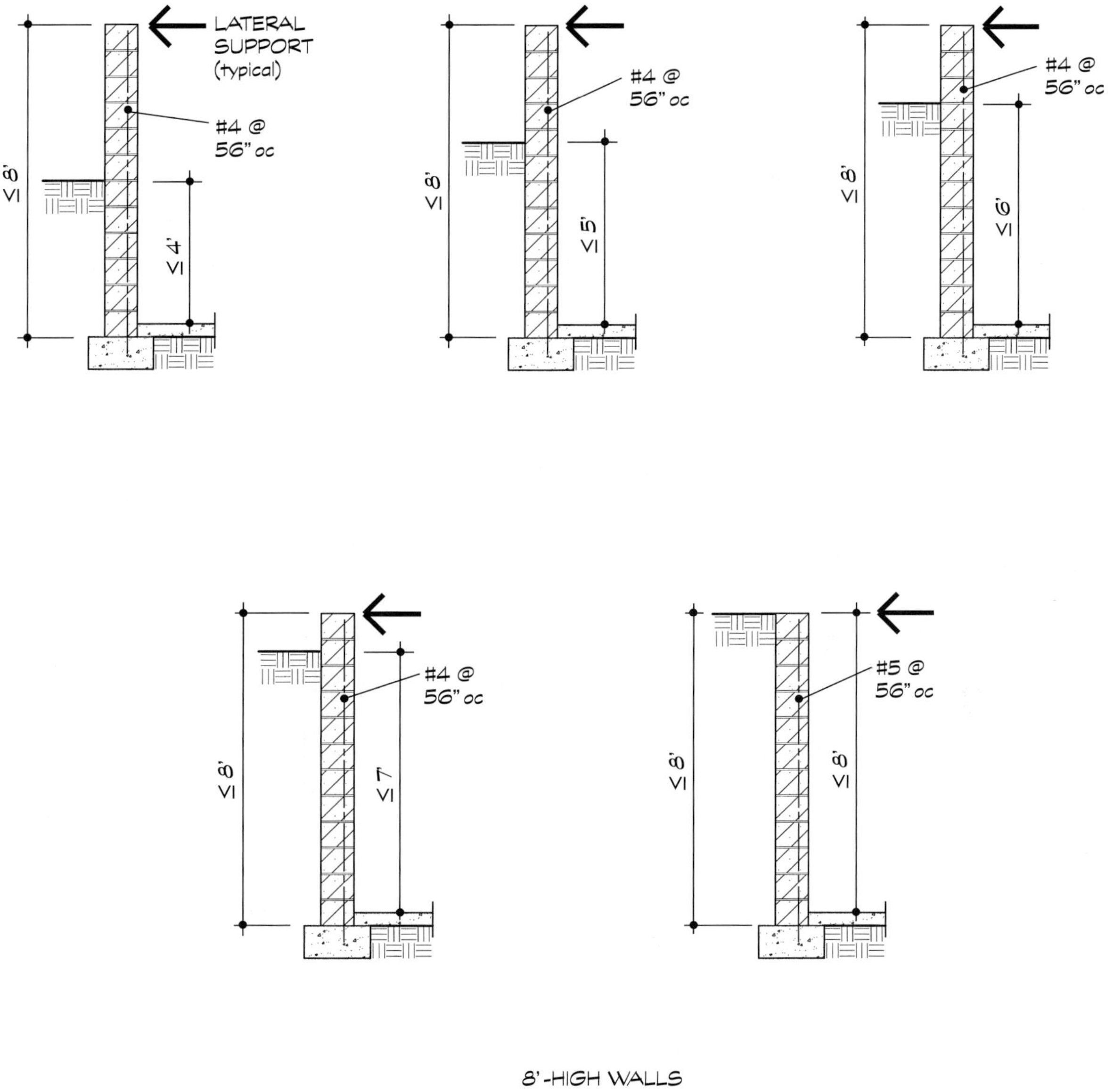

Fig. 1807.1.6.3(3)-2. Vertical reinforcing for 10" masonry foundation walls with $d \geq 6 3/4$". Lateral soil load is 30 psf/ft of depth. *[IBC Table 1807.1.6.3(3)]*

1807 Foundation Walls, Retaining Walls and Embedded Posts and Poles

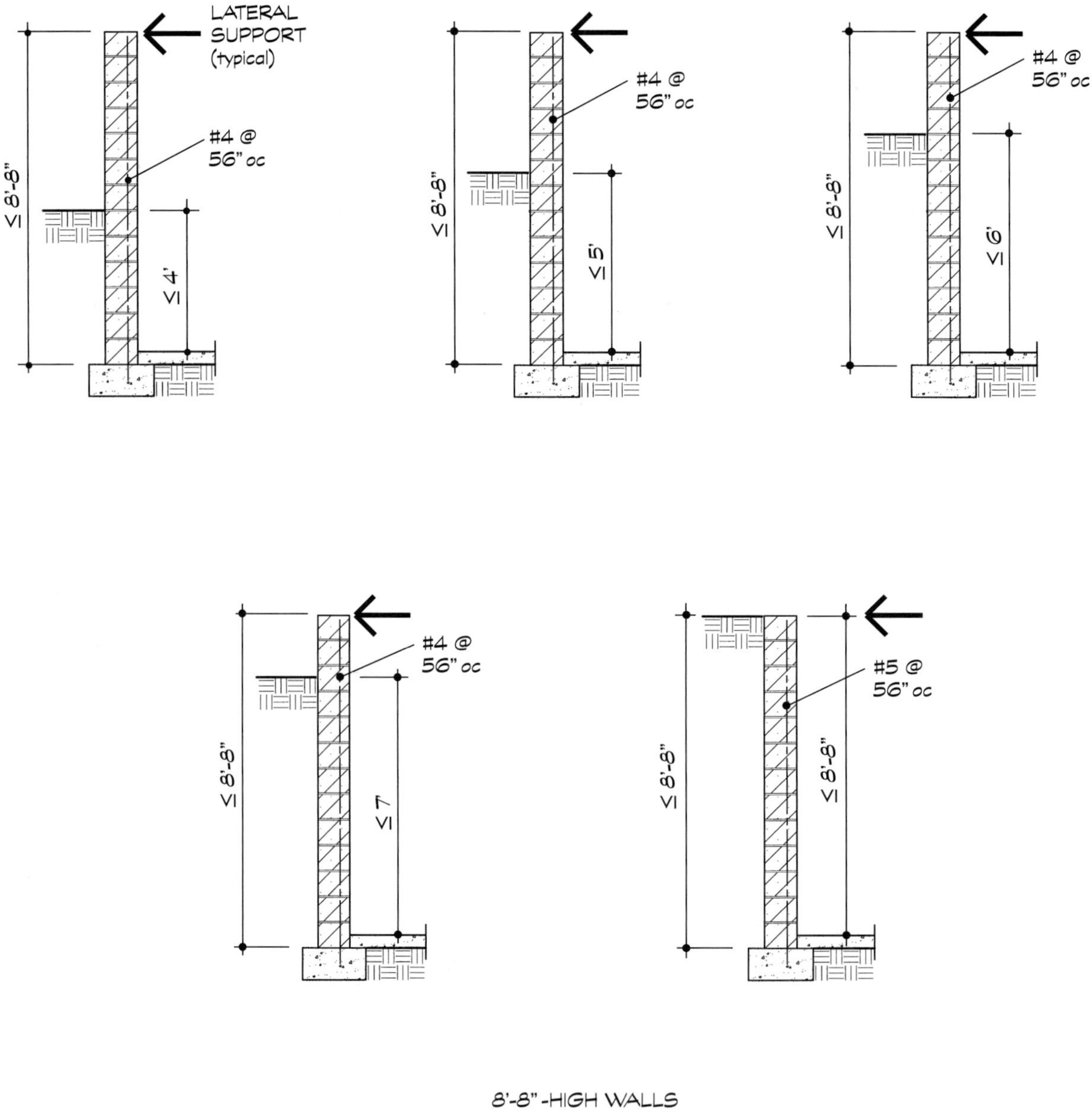

Fig. 1807.1.6.3(3)-3. Vertical reinforcing for 10" masonry foundation walls with $d \geq 6^3/_4$". Lateral soil load is 30 psf/ft of depth. *[IBC Table 1807.1.6.3(3)]*

1807 Foundation Walls, Retaining Walls and Embedded Posts and Poles

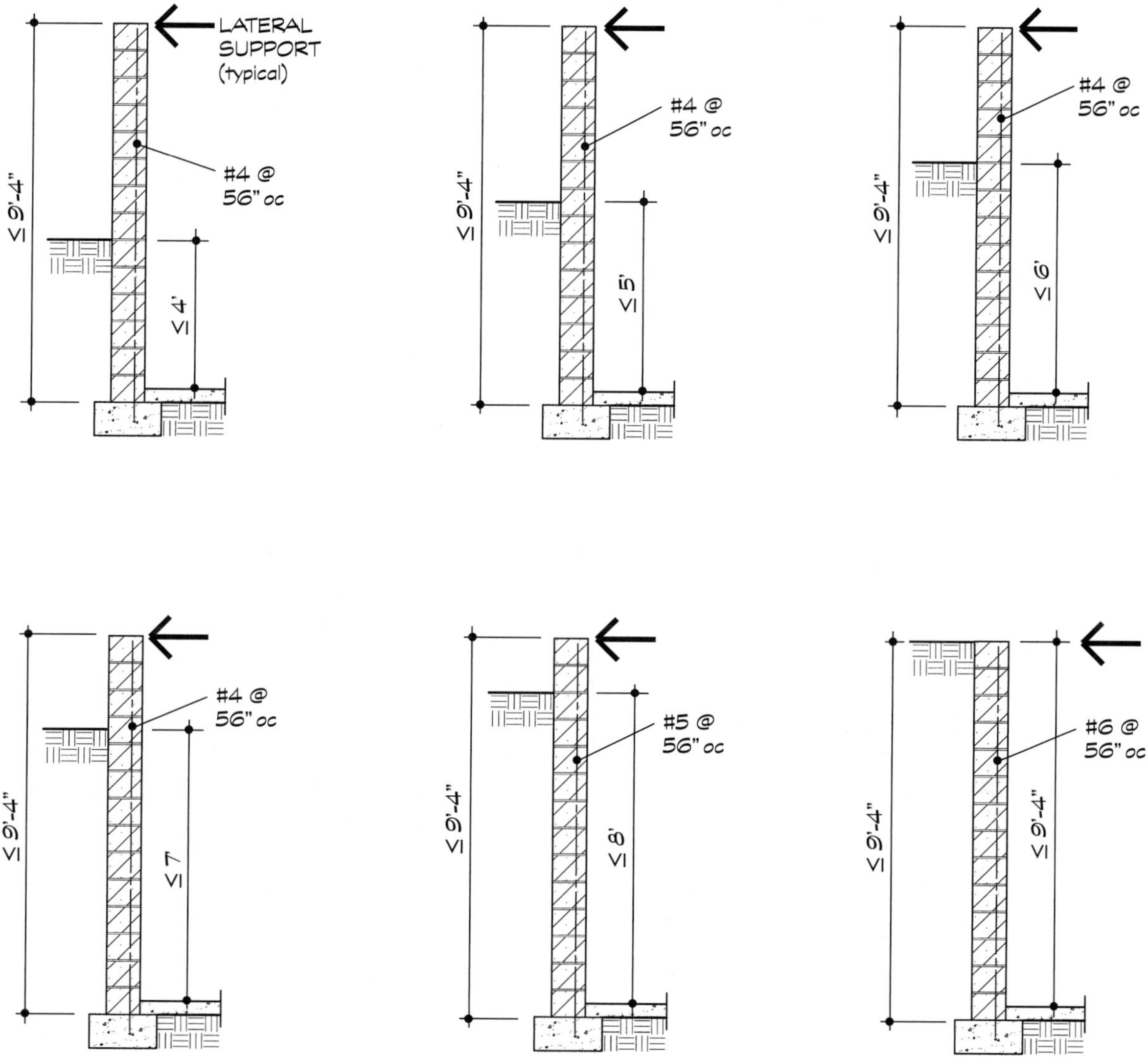

9'-4"-HIGH WALLS

Fig. 1807.1.6.3(3)-4. Vertical reinforcing for 10" masonry foundation walls with $d \geq 6\frac{3}{4}$". Lateral soil load is 30 psf/ft of depth. *[IBC Table 1807.1.6.3(3)]*

1807 Foundation Walls, Retaining Walls and Embedded Posts and Poles

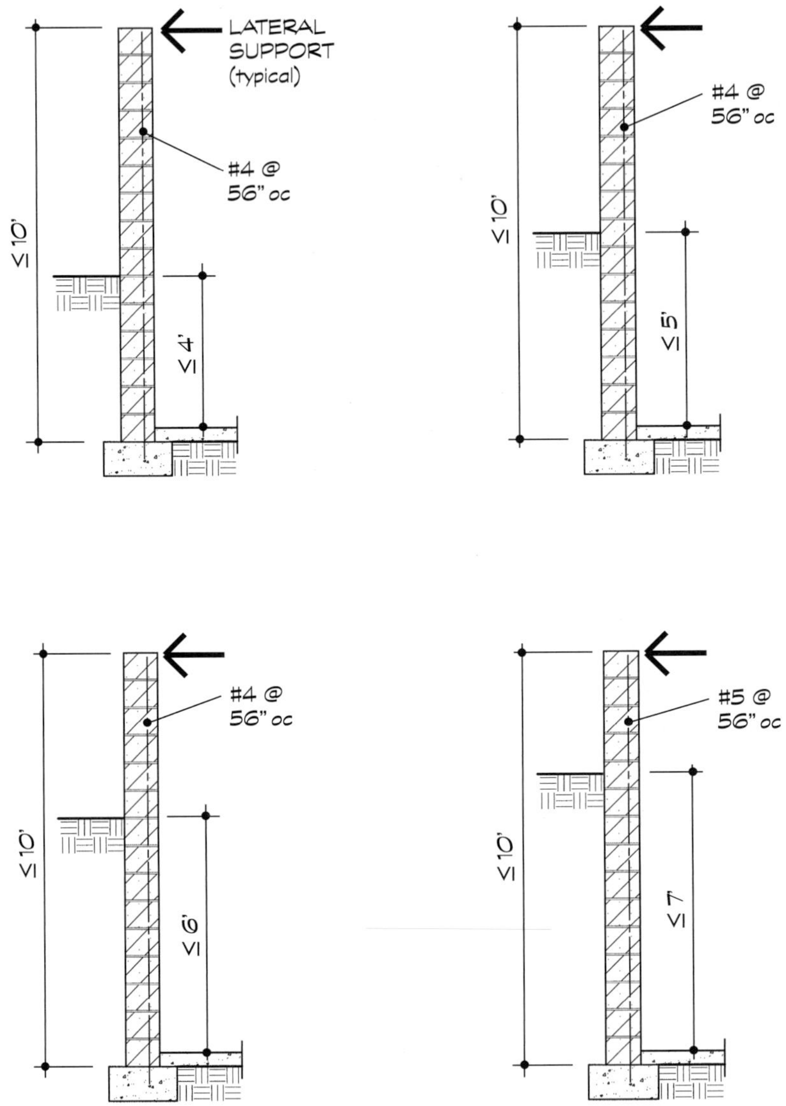

10'-HIGH WALLS

Fig. 1807.1.6.3(3)-5a. Vertical reinforcing for 10" masonry foundation walls with $d \geq 6^3/_4$". Lateral soil load is 30 psf/ft of depth. *[IBC Table 1807.1.6.3(3)]*

1807 Foundation Walls, Retaining Walls and Embedded Posts and Poles

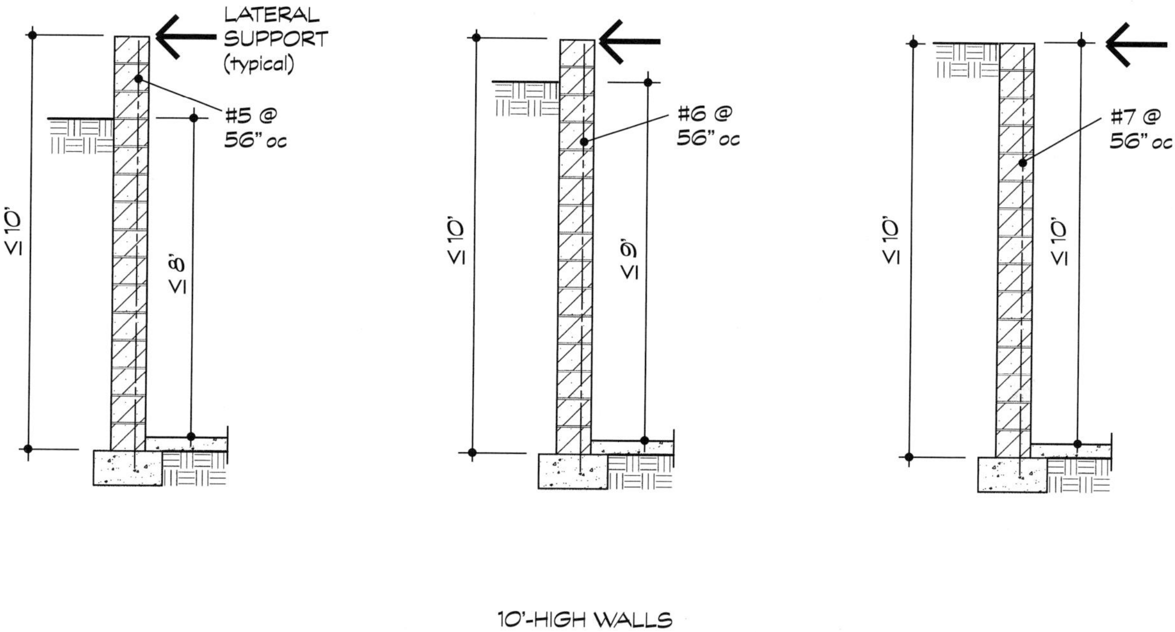

Fig. 1807.1.6.3(3)-5b. Vertical reinforcing for 10" masonry foundation walls with $d \geq 6\frac{3}{4}$". Lateral soil load is 30 psf/ft of depth. *[IBC Table 1807.1.6.3(3)]*

1807 Foundation Walls, Retaining Walls and Embedded Posts and Poles

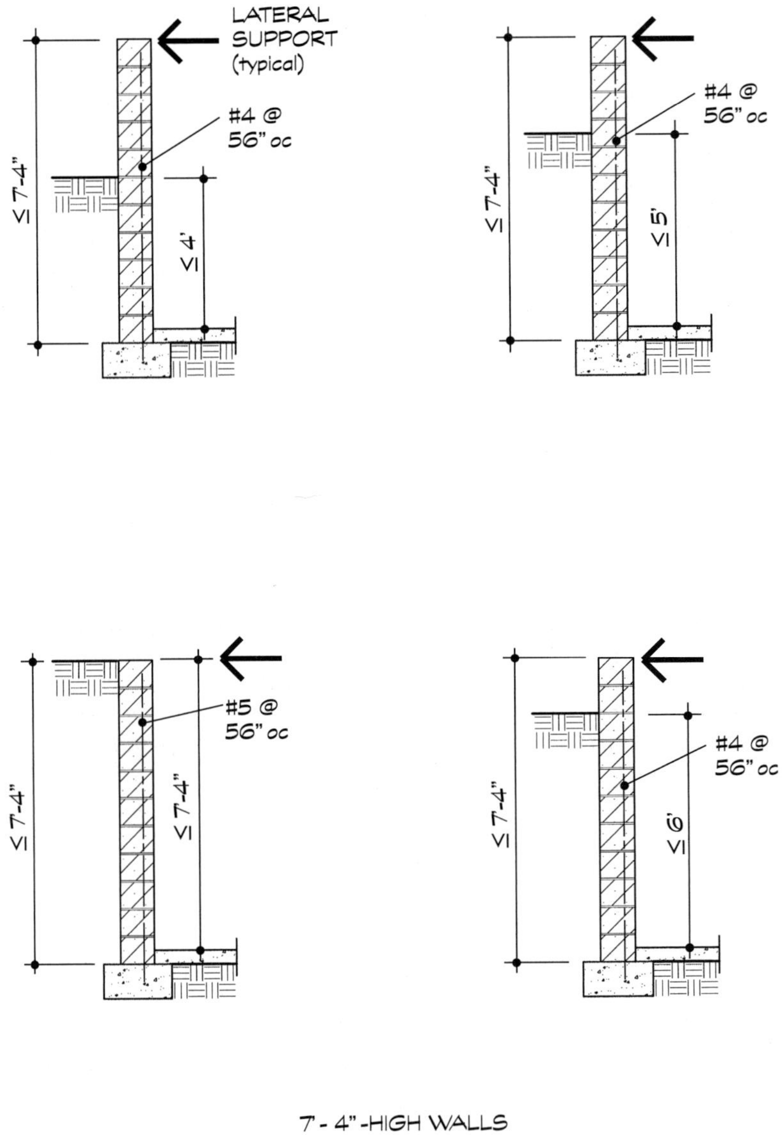

Fig. 1807.1.6.3(3)-6. Vertical reinforcing for 10" masonry foundation walls with $d \geq 6^3/_4$". Lateral soil load is 45 psf/ft of depth. [*IBC Table 1807.1.6.3(3)*]

1807 Foundation Walls, Retaining Walls and Embedded Posts and Poles

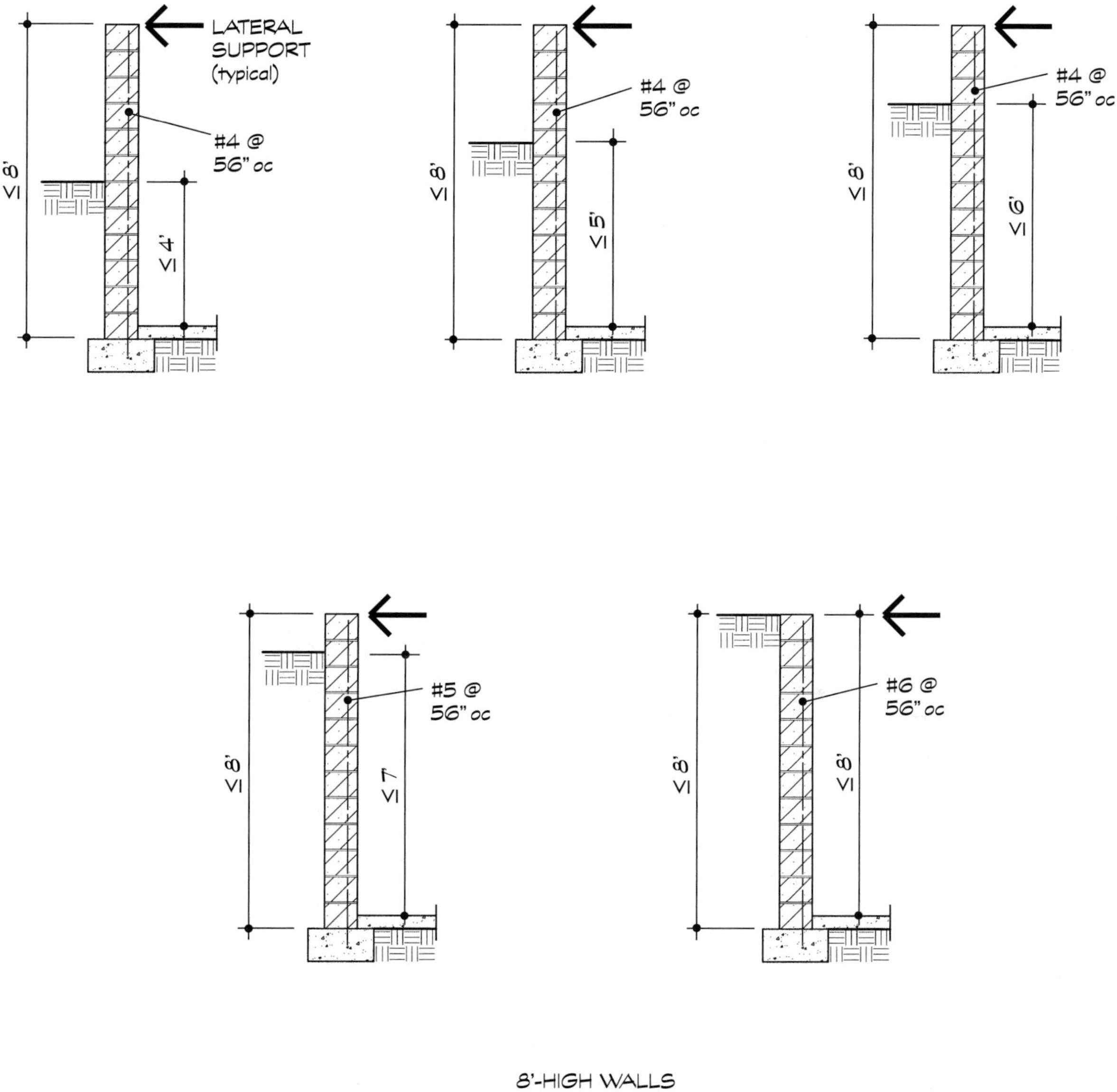

Fig. 1807.1.6.3(3)-7. Vertical reinforcing for 10" masonry foundation walls with $d \geq 6\frac{3}{4}$". Lateral soil load is 45 psf/ft of depth. *[IBC Table 1807.1.6.3(3)]*

1807 Foundation Walls, Retaining Walls and Embedded Posts and Poles

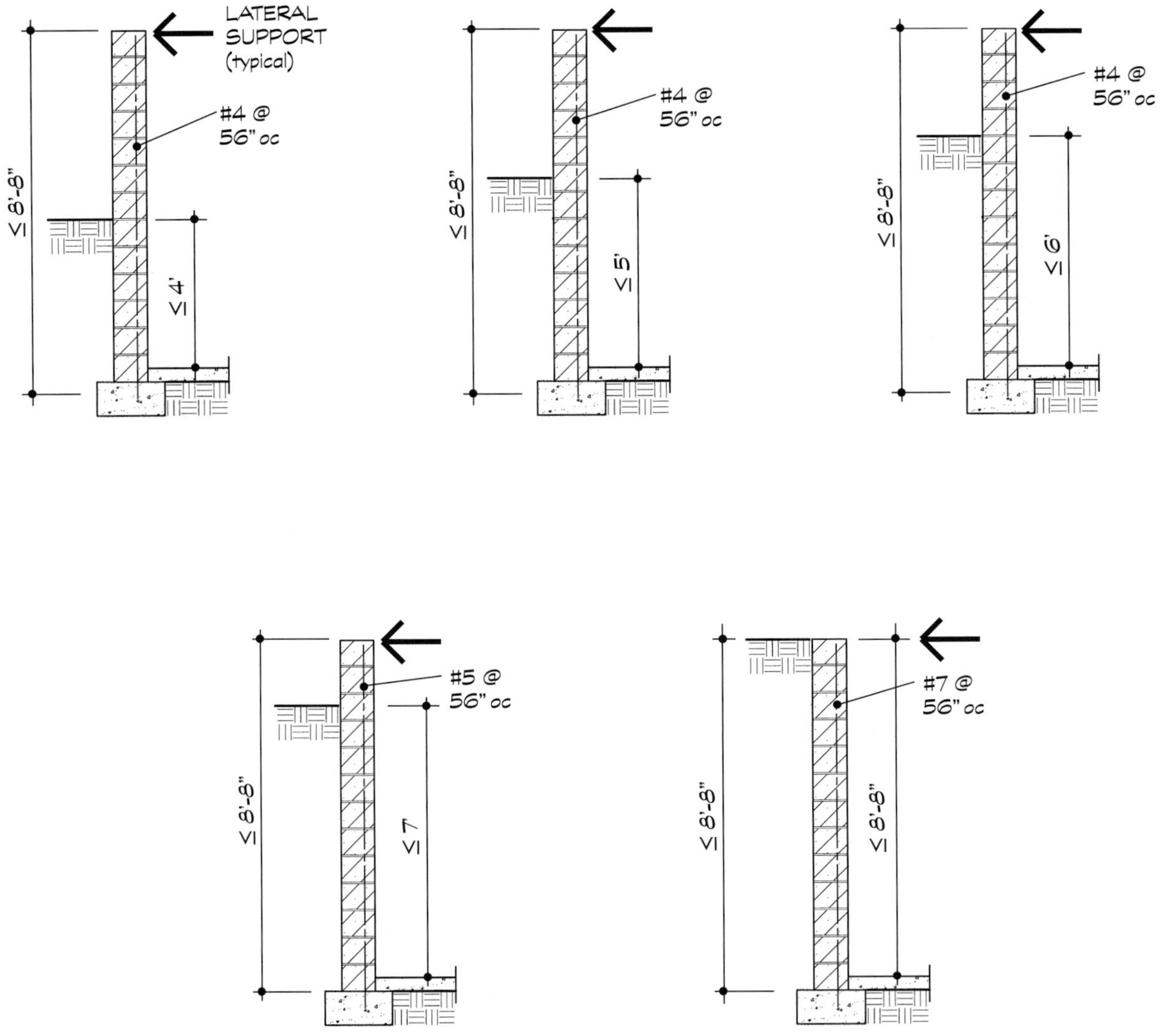

8' - 8" -HIGH WALLS

Fig. 1807.1.6.3(3)-8. Vertical reinforcing for 10" masonry foundation walls with $d \geq 6^3/_4$". Lateral soil load is 45 psf/ft of depth. [*IBC Table 1807.1.6.3(3)*]

1807 Foundation Walls, Retaining Walls and Embedded Posts and Poles

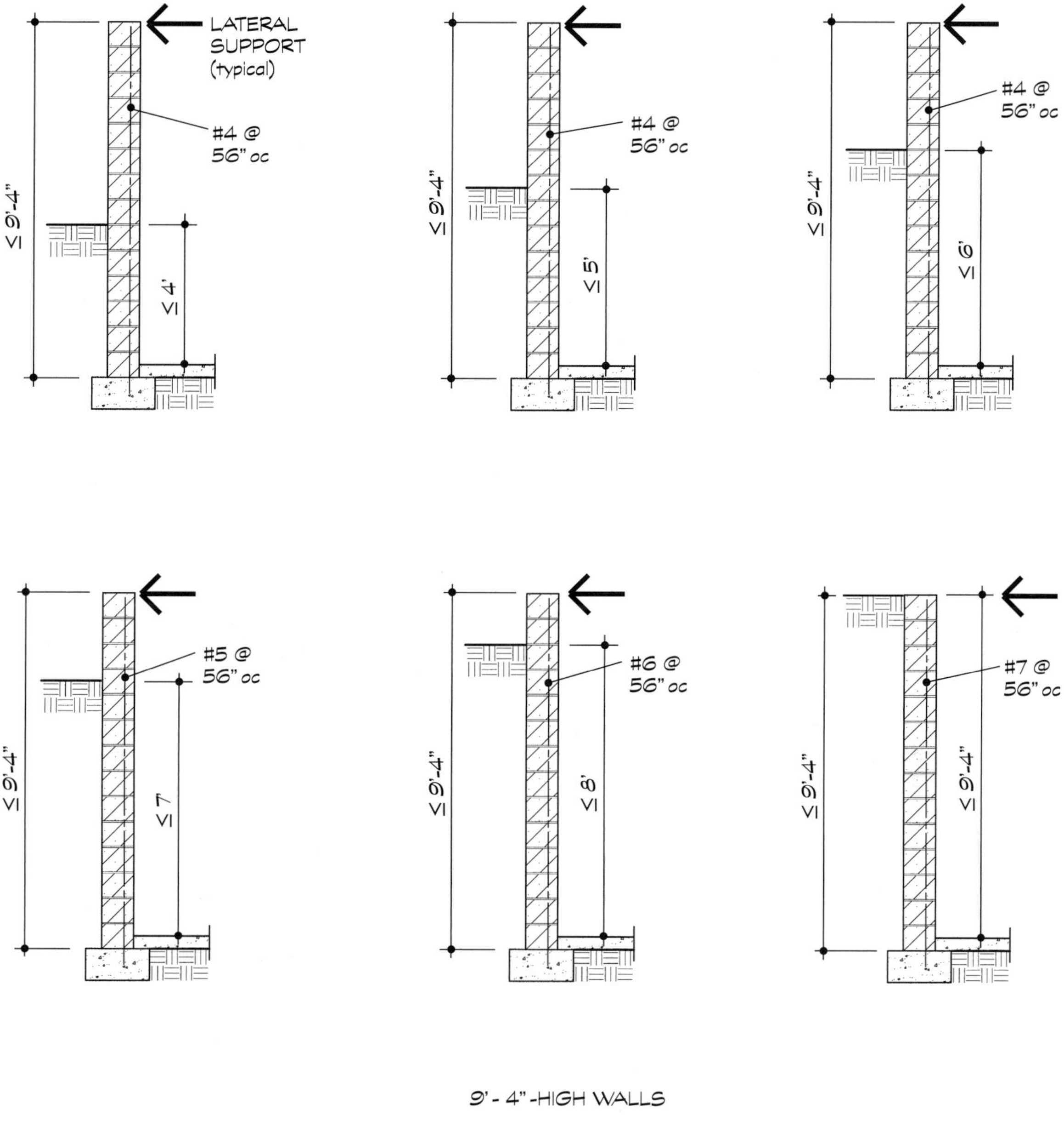

9'- 4"-HIGH WALLS

Fig. 1807.1.6.3(3)-9. Vertical reinforcing for 10" masonry foundation walls with $d \geq 6\frac{3}{4}$". Lateral soil load is 45 psf/ft of depth. *[IBC Table 1807.1.6.3(3)]*

1807 Foundation Walls, Retaining Walls and Embedded Posts and Poles

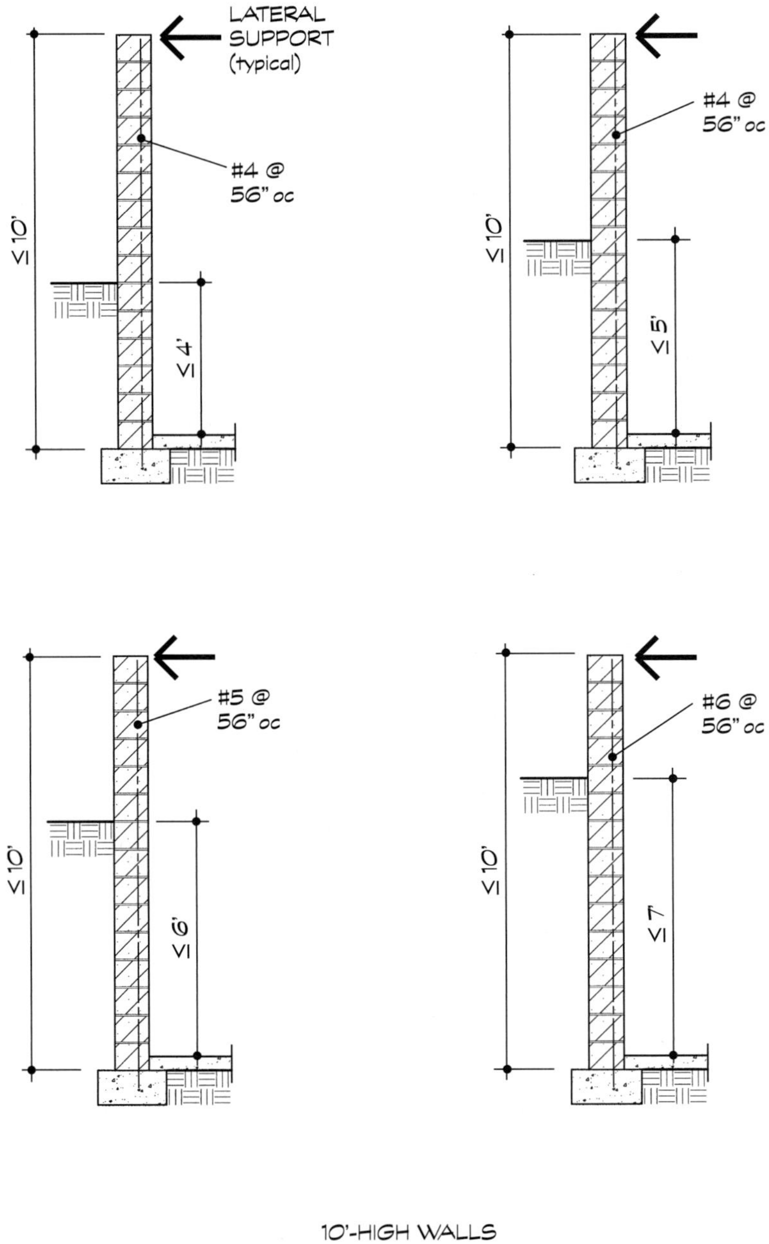

Fig. 1807.1.6.3(3)-10a. Vertical reinforcing for 10" masonry foundation walls with $d \geq 6^3/_4$". Lateral soil load is 45 psf/ft of depth. *[IBC Table 1807.1.6.3(3)]*

1807 Foundation Walls, Retaining Walls and Embedded Posts and Poles

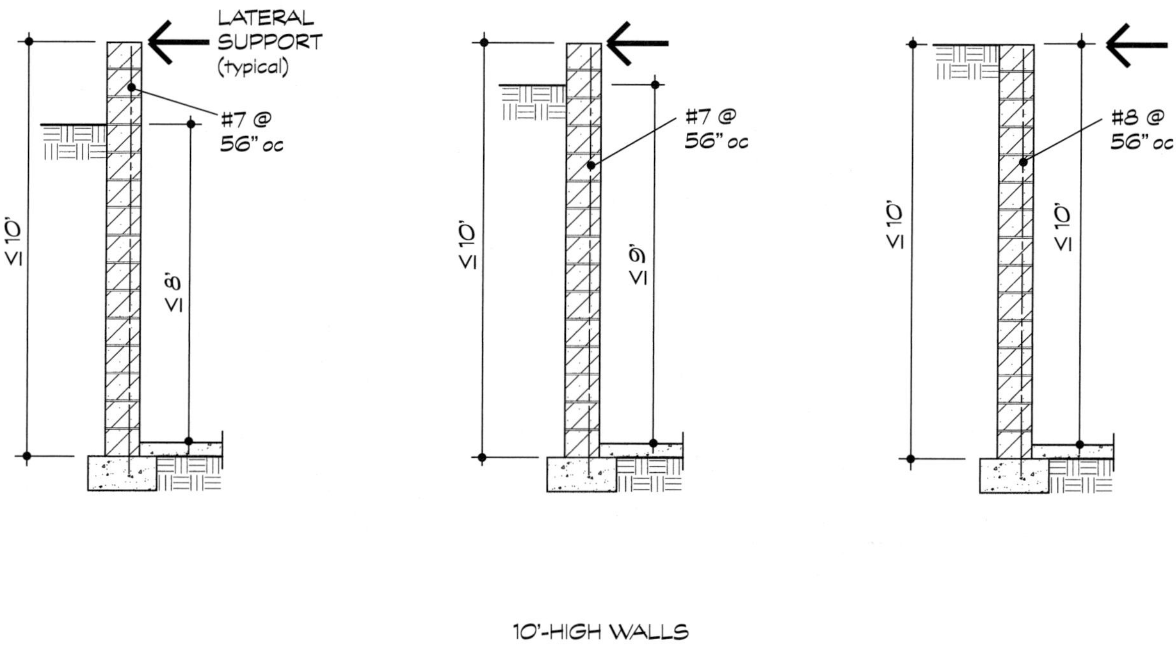

Fig. 1807.1.6.3(3)-10b. Vertical reinforcing for 10" masonry foundation walls with $d \geq 6^3/_4$". Lateral soil load is 45 psf/ft of depth. *[IBC Table 1807.1.6.3(3)]*

1807 Foundation Walls, Retaining Walls and Embedded Posts and Poles

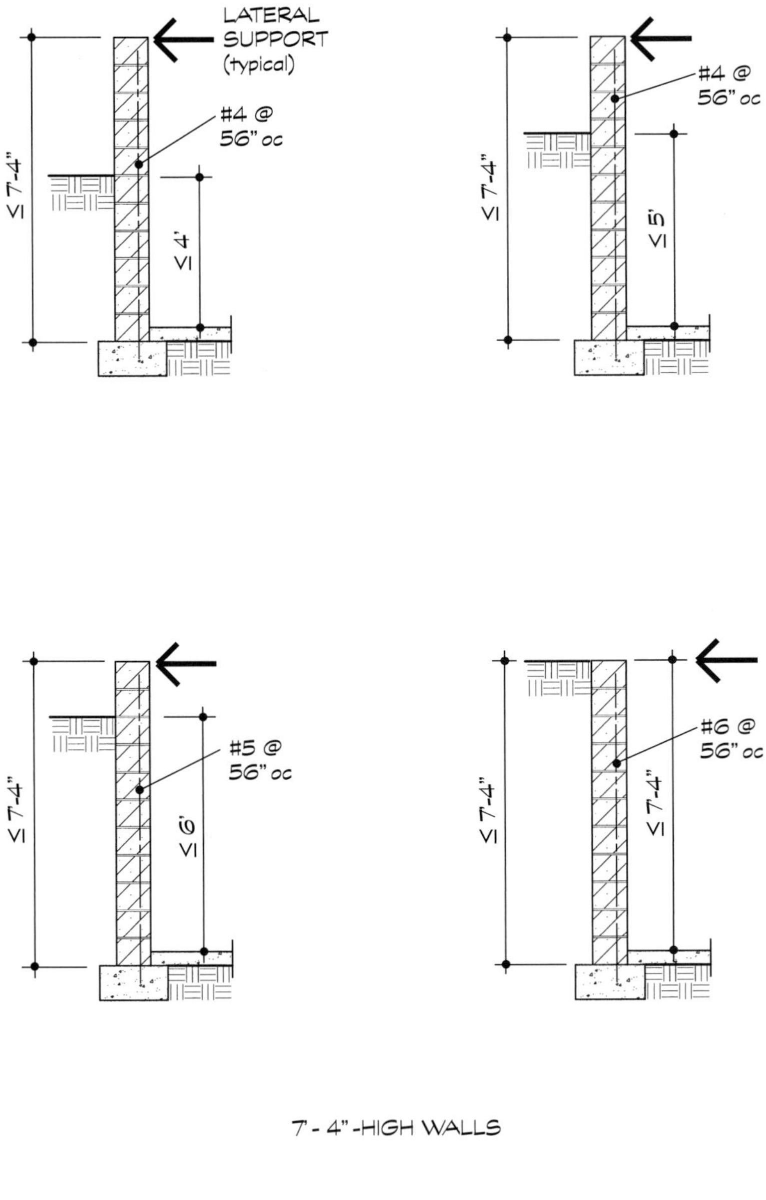

Fig. 1807.1.6.3(3)-11. Vertical reinforcing for 10" masonry foundation walls with $d \geq 6\frac{3}{4}$". Lateral soil load is 60 psf/ft of depth. *[IBC Table 1807.1.6.3(3)]*

1807 Foundation Walls, Retaining Walls and Embedded Posts and Poles

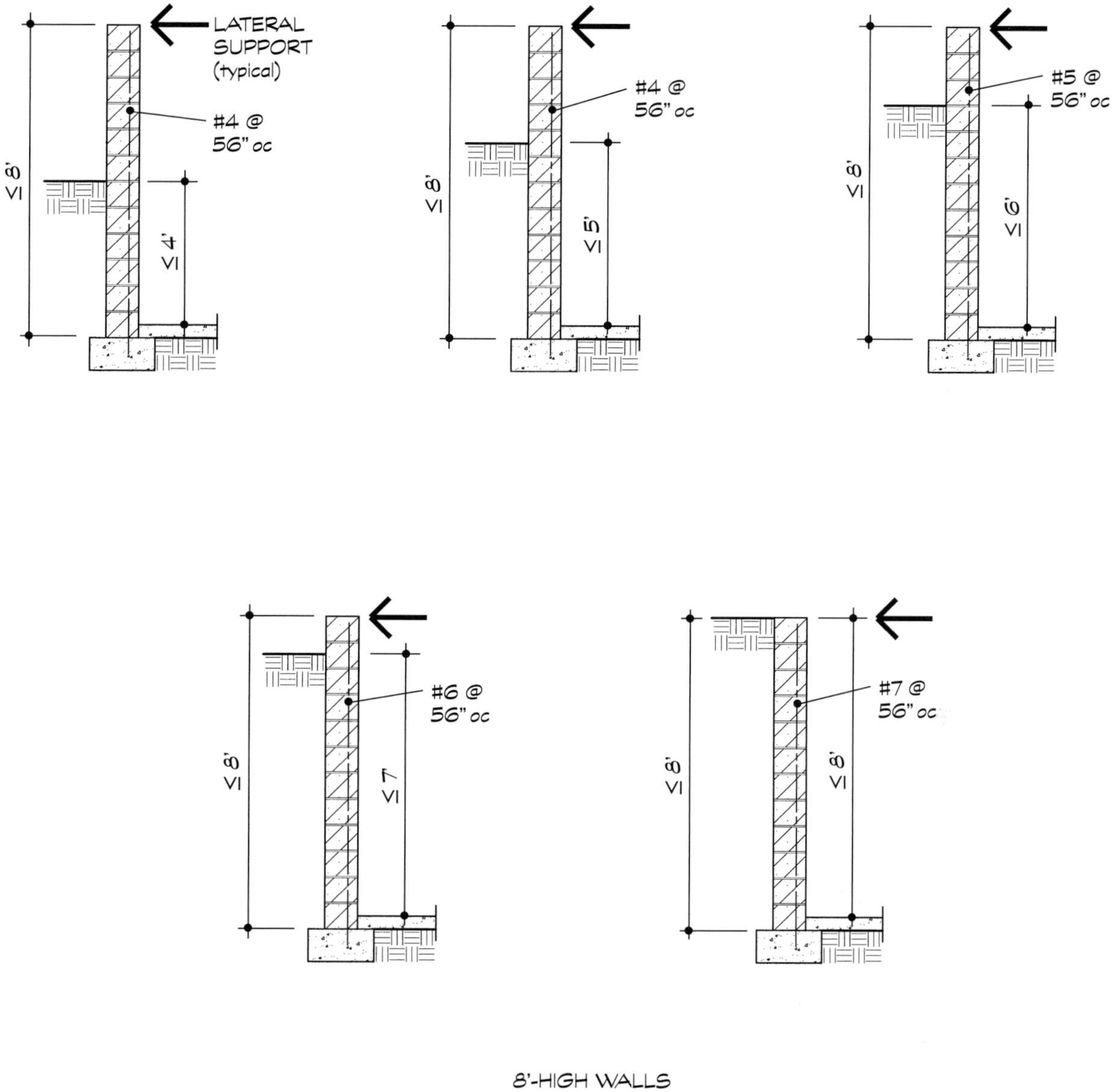

Fig. 1807.1.6.3(3)-12. Vertical reinforcing for 10" masonry foundation walls with $d \geq 6^3/_4$". Lateral soil load is 60 psf/ft of depth. *[IBC Table 1807.1.6.3(3)]*

1807 Foundation Walls, Retaining Walls and Embedded Posts and Poles

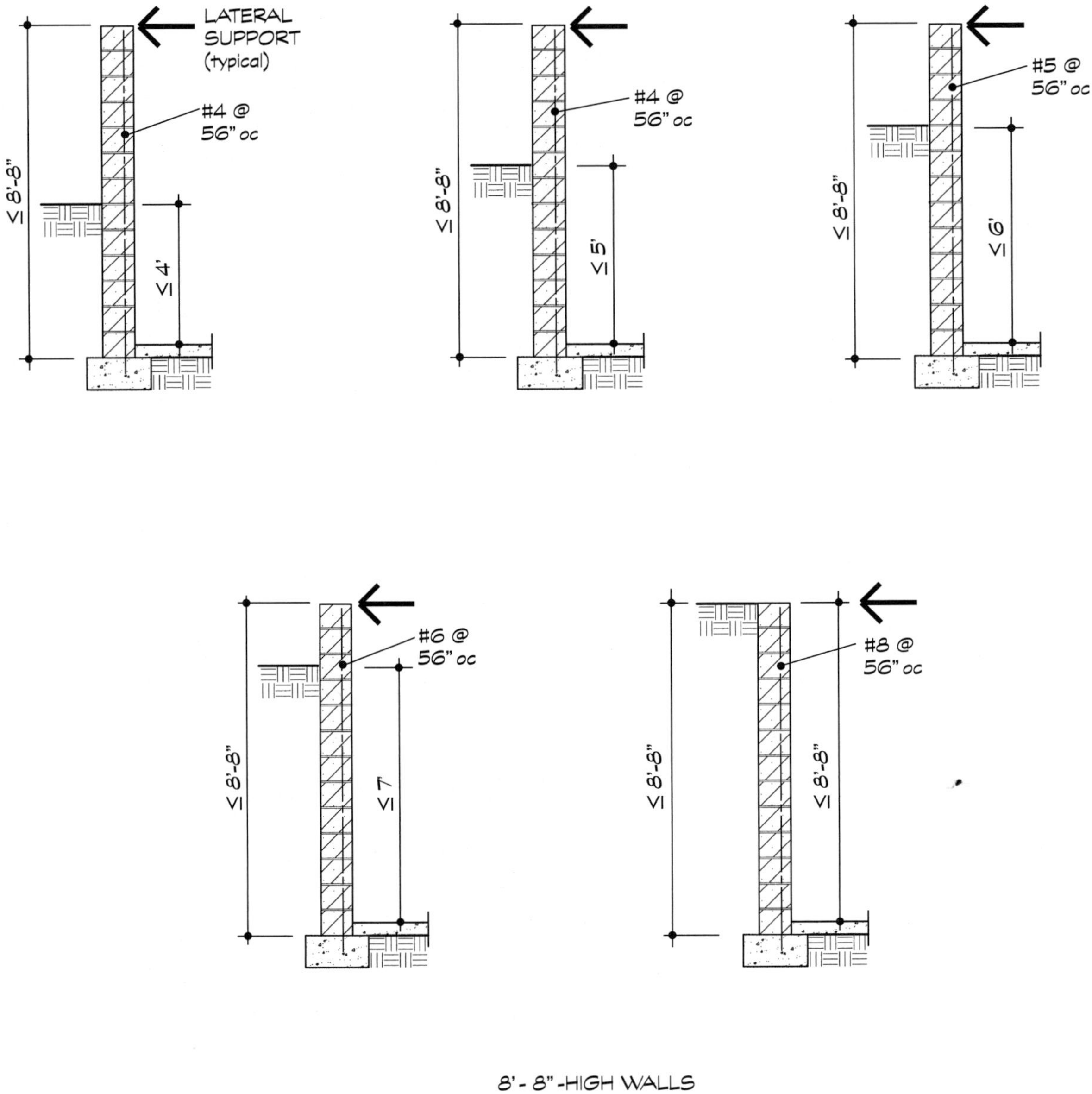

Fig. 1807.1.6.3(3)-13. Vertical reinforcing for 10" masonry foundation walls with $d \geq 6\frac{3}{4}$". Lateral soil load is 60 psf/ft of depth. *[IBC Table 1807.1.6.3(3)]*

1807 Foundation Walls, Retaining Walls and Embedded Posts and Poles

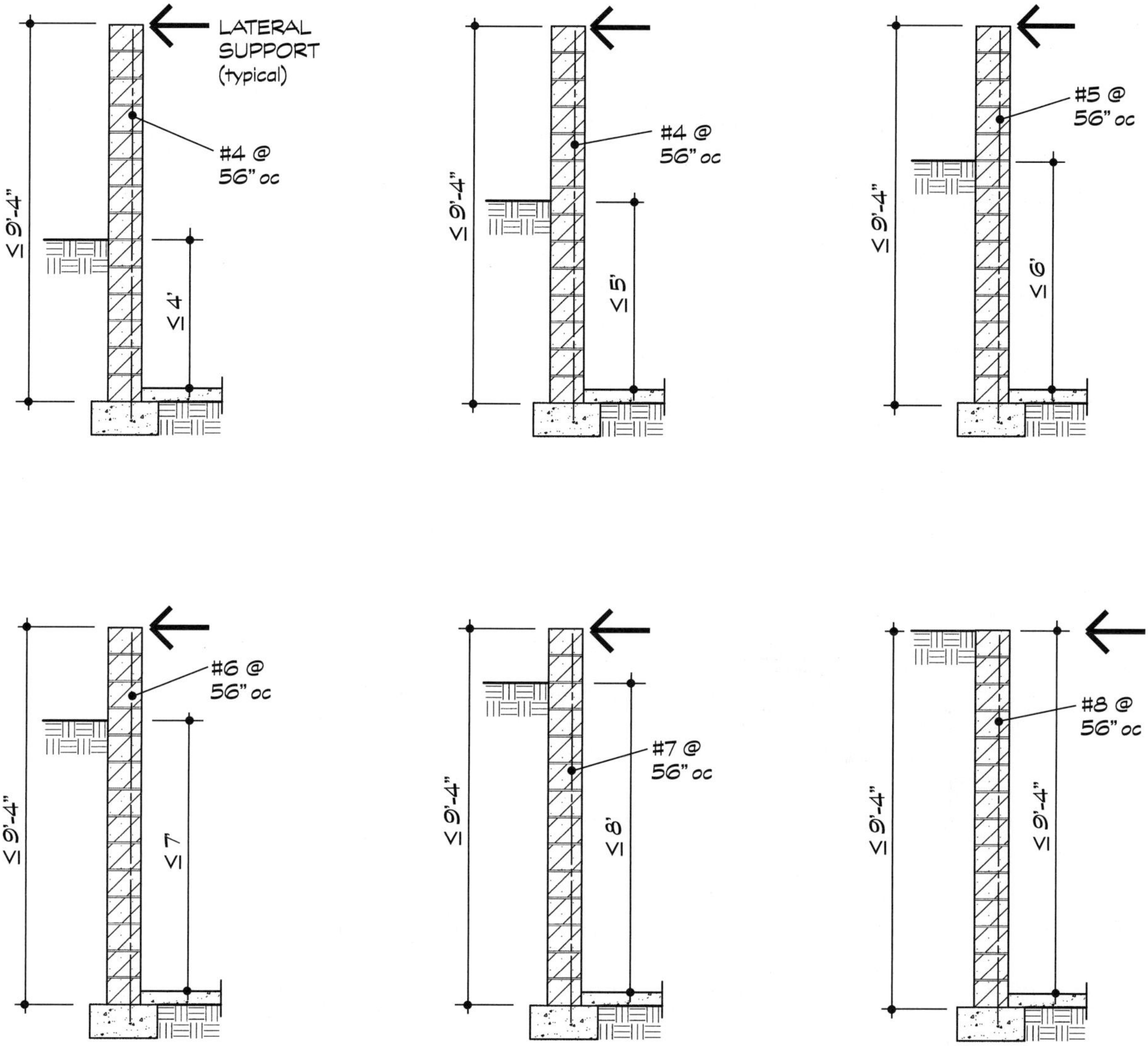

9'- 4"-HIGH WALLS

Fig. 1807.1.6.3(3)-14. Vertical reinforcing for 10" masonry foundation walls with $d \geq 6 3/4$". Lateral soil load is 60 psf/ft of depth. *[IBC Table 1807.1.6.3(3)]*

1807 Foundation Walls, Retaining Walls and Embedded Posts and Poles

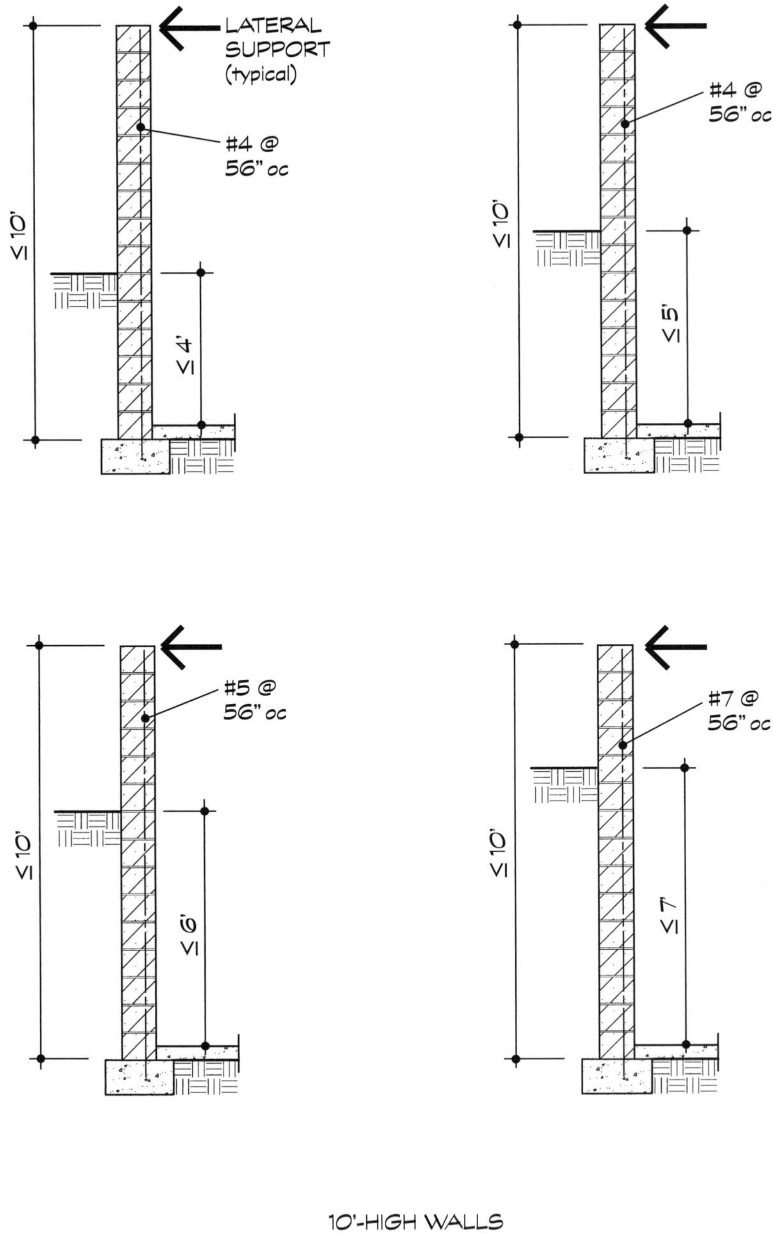

Fig. 1807.1.6.3(3)-15a. Vertical reinforcing for 10" masonry foundation walls with $d \geq 6\frac{3}{4}$". Lateral soil load is 60 psf/ft of depth. *[IBC Table 1807.1.6.3(3)]*

1807 Foundation Walls, Retaining Walls and Embedded Posts and Poles

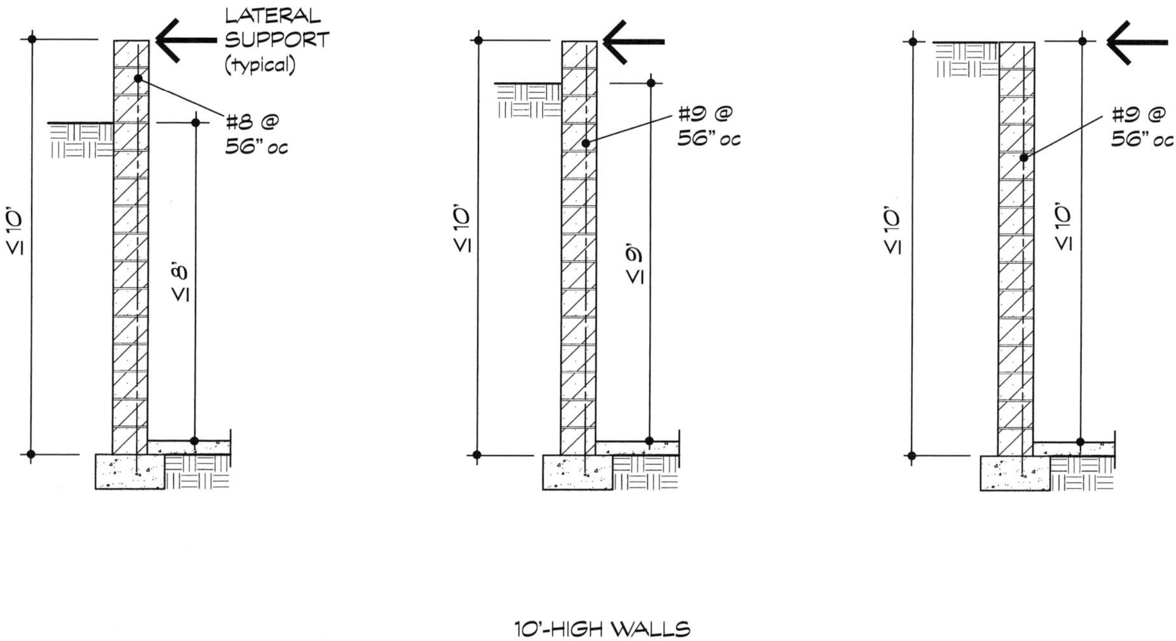

Fig. 1807.1.6.3(3)-15b. Vertical reinforcing for 10" masonry foundation walls with $d \geq 6^3/_4$". Lateral soil load is 60 psf/ft of depth. *[IBC Table 1807.1.6.3(3)]*

1807 Foundation Walls, Retaining Walls and Embedded Posts and Poles

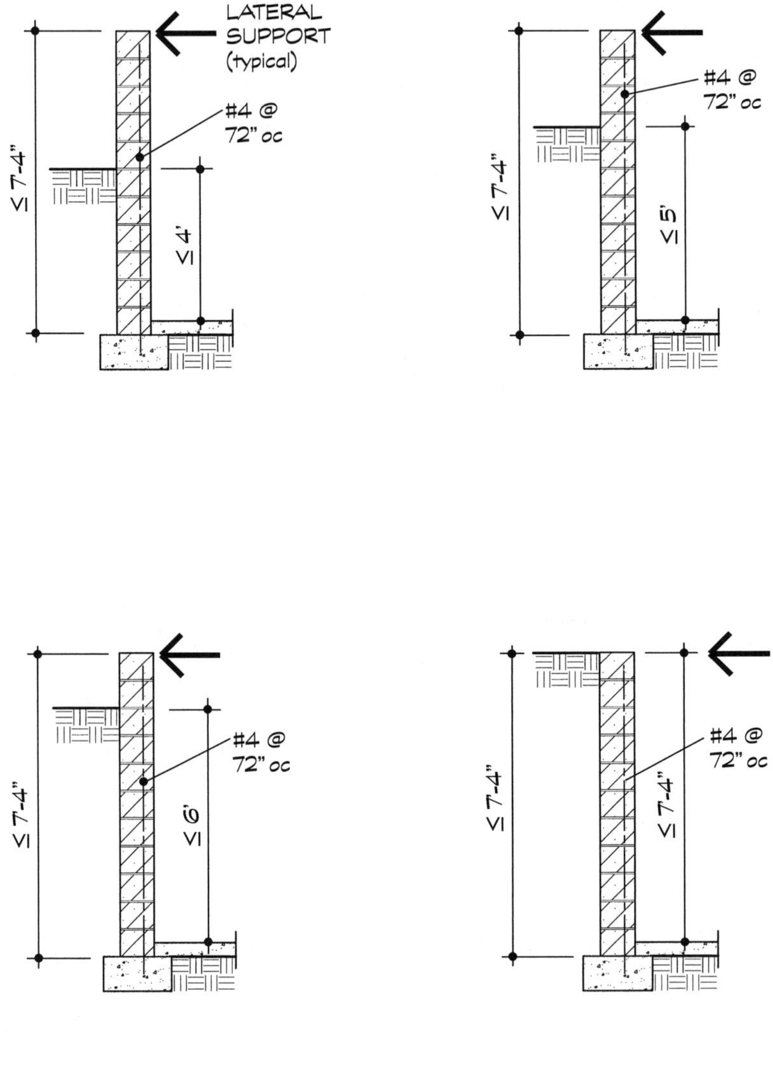

7'-4"-HIGH WALLS

Fig. 1807.1.6.3(4)-1. Vertical reinforcing for 12" masonry foundation walls with $d \geq 8^3/4$". Lateral soil load is 30 psf/ft of depth. *[IBC Table 1807.1.6.3(4)]*

1807 Foundation Walls, Retaining Walls and Embedded Posts and Poles

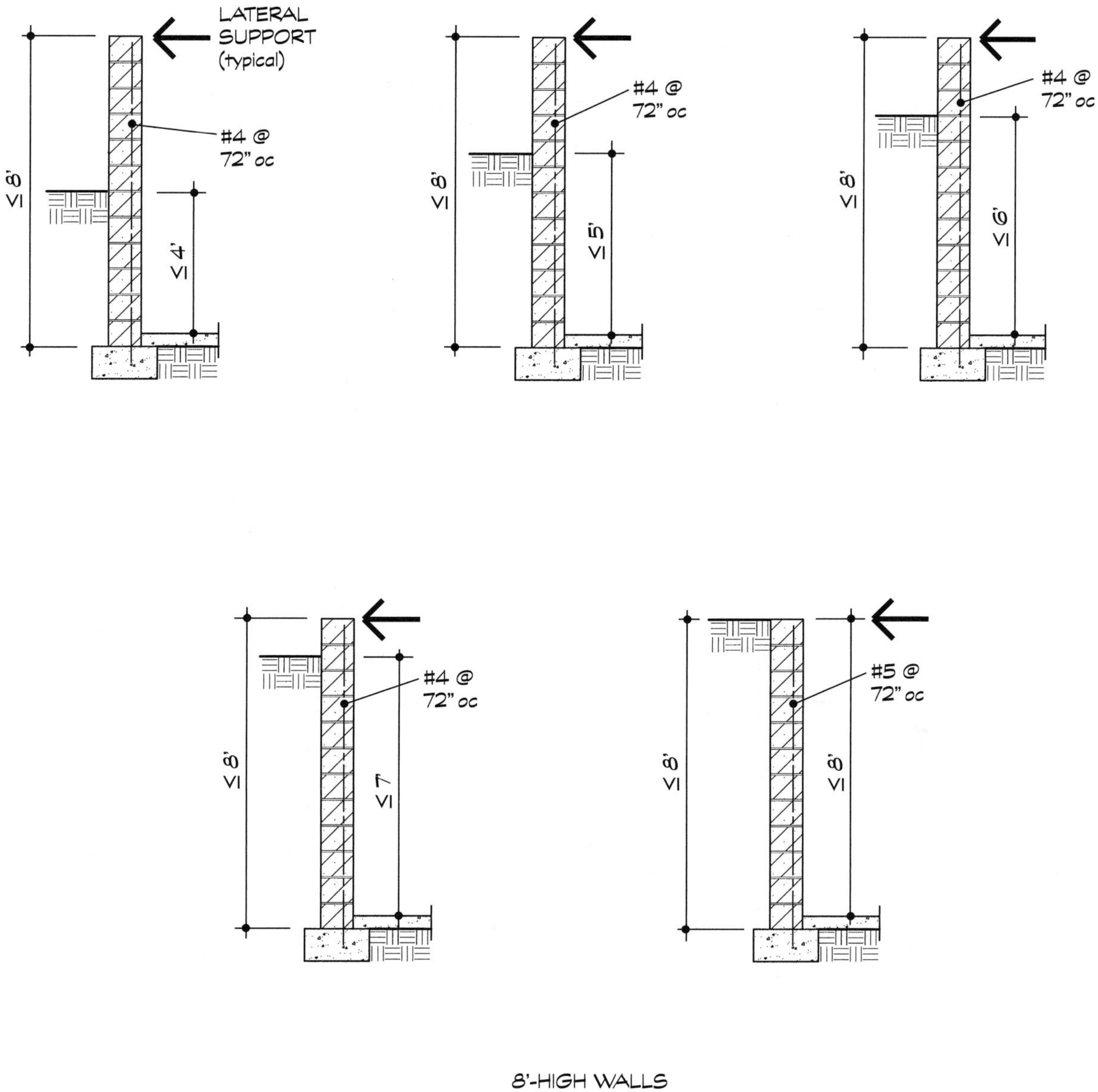

Fig. 1807.1.6.3(4)-2. Vertical reinforcing for 12" masonry foundation walls with $d \geq 8\frac{3}{4}$". Lateral soil load is 30 psf/ft of depth. *[IBC Table 1807.1.6.3(4)]*

1807 Foundation Walls, Retaining Walls and Embedded Posts and Poles

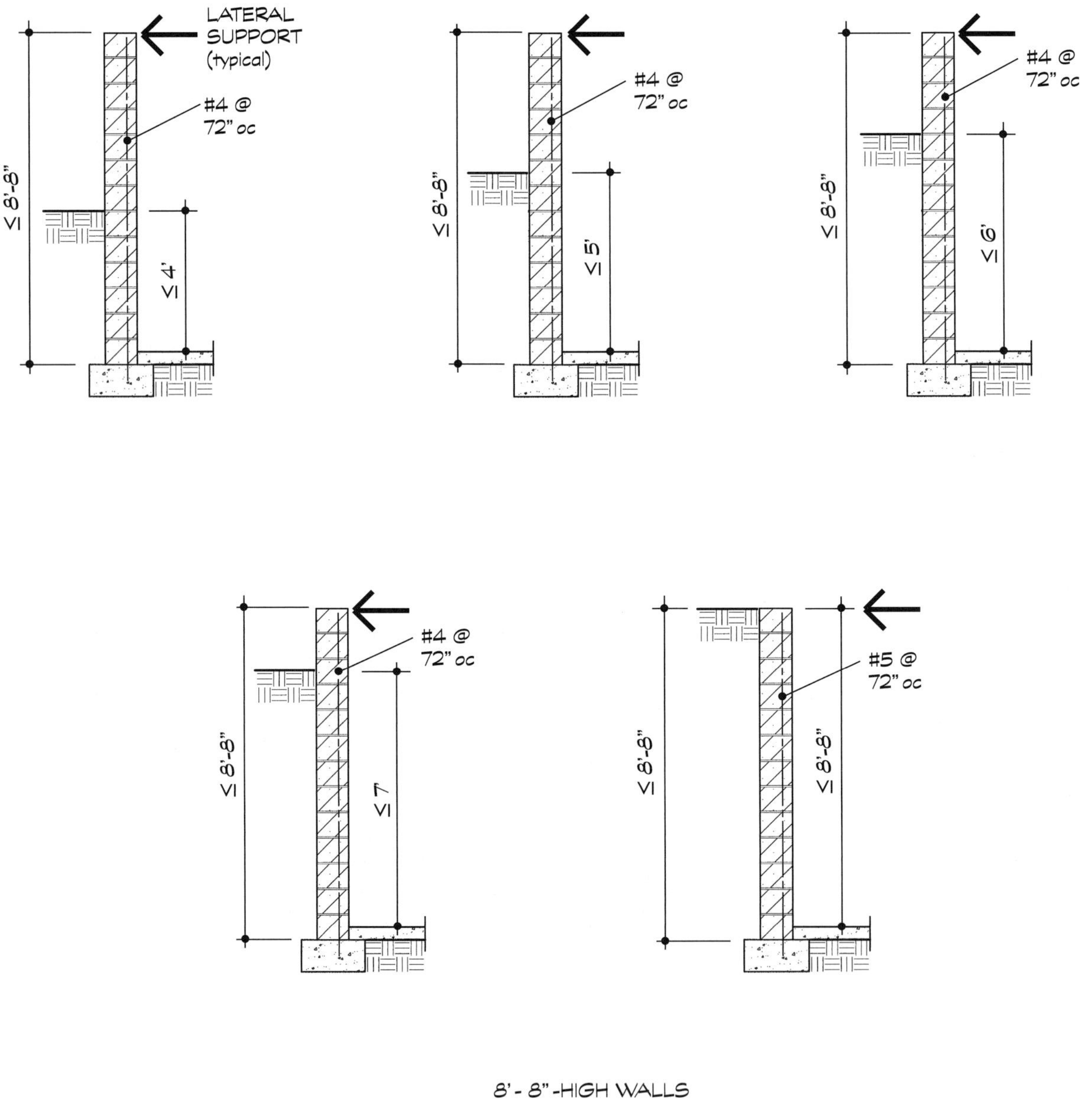

Fig. 1807.1.6.3(4)-3. Vertical reinforcing for 12" masonry foundation walls with $d \geq 8^3/_4$". Lateral soil load is 30 psf/ft of depth. *[IBC Table 1807.1.6.3(4)]*

1807 Foundation Walls, Retaining Walls and Embedded Posts and Poles

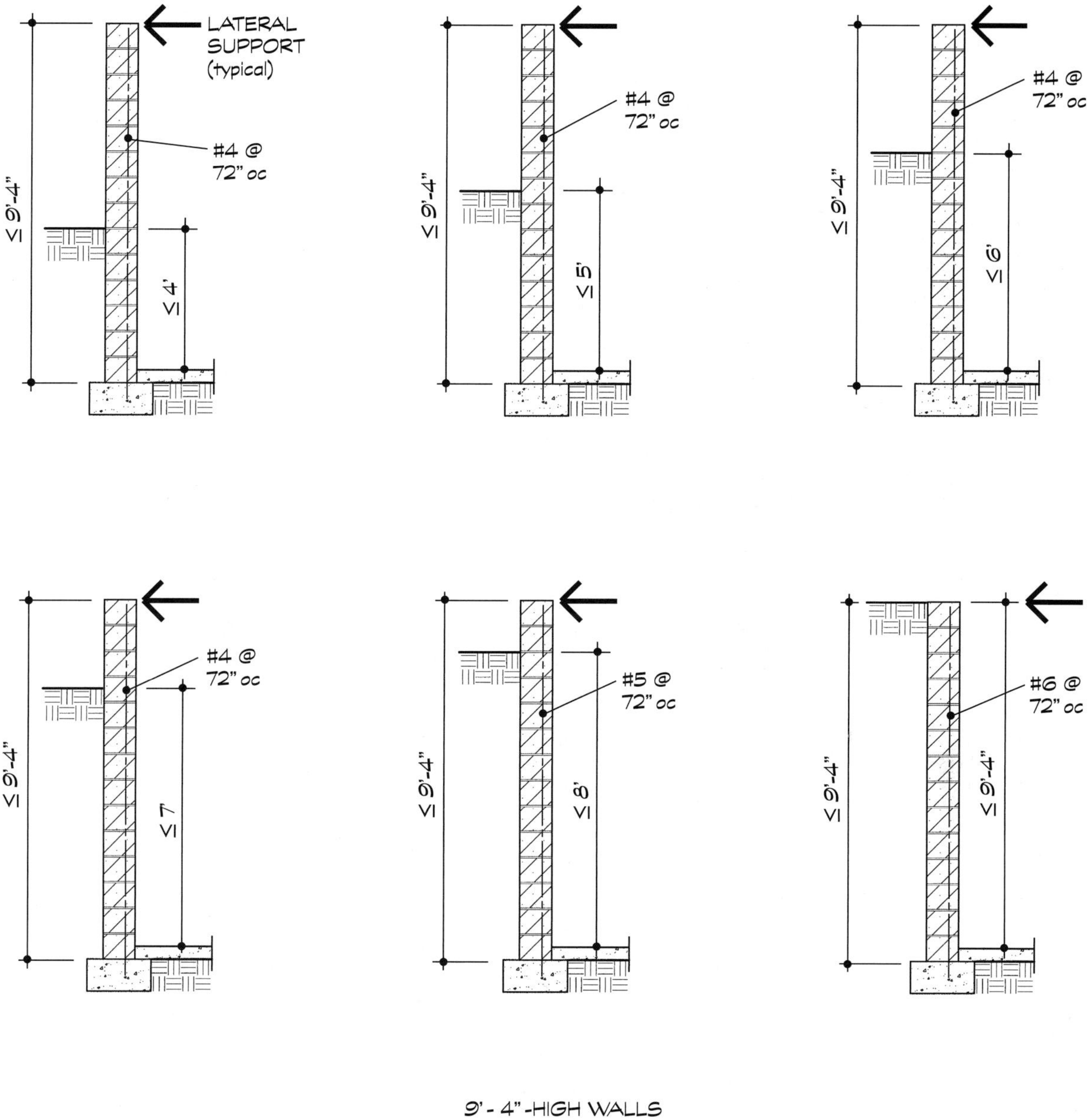

Fig. 1807.1.6.3(4)-4. Vertical reinforcing for 12" masonry foundation walls with $d \geq 8^3/_4$". Lateral soil load is 30 psf/ft of depth. *[IBC Table 1807.1.6.3(4)]*

1807 Foundation Walls, Retaining Walls and Embedded Posts and Poles

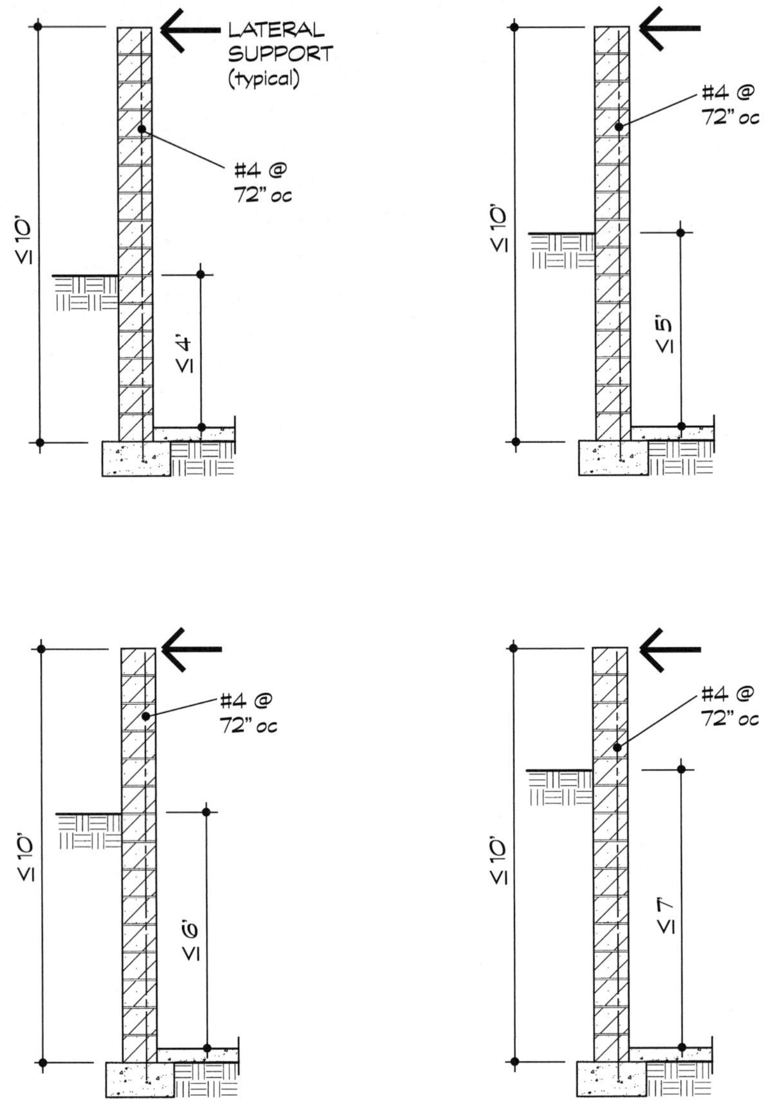

10'-HIGH WALLS

Fig. 1807.1.6.3(4)-5a. Vertical reinforcing for 12" masonry foundation walls with $d \geq 8\frac{3}{4}$". Lateral soil load is 30 psf/ft of depth. *[IBC Table 1807.1.6.3(4)]*

1807 Foundation Walls, Retaining Walls and Embedded Posts and Poles

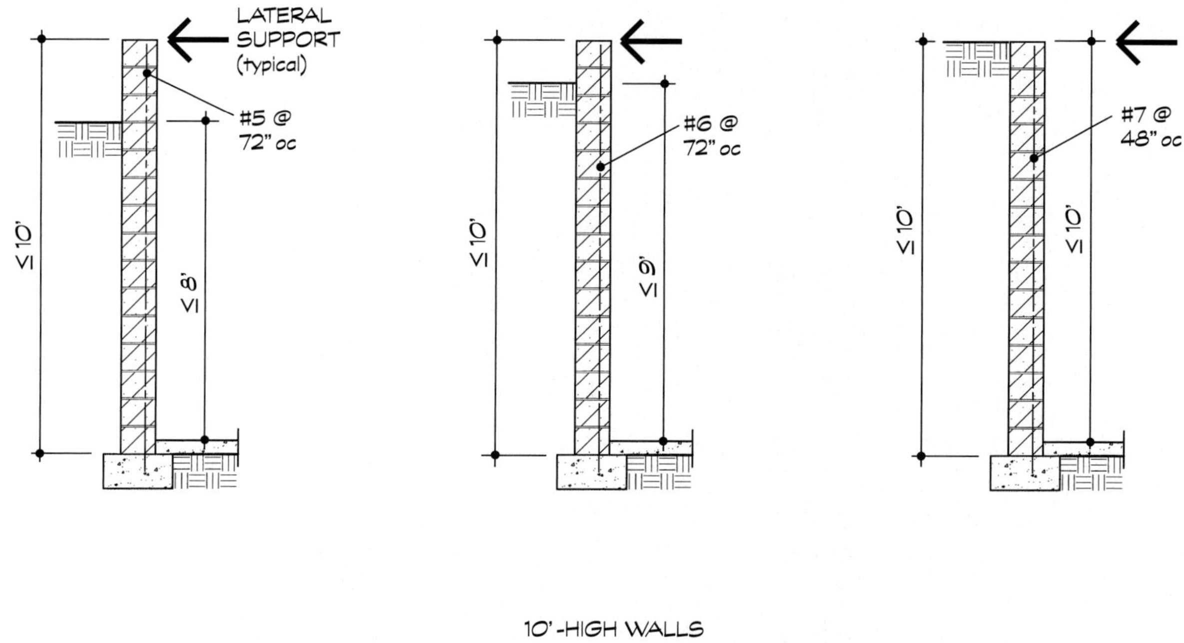

Fig. 1807.1.6.3(4)-5b. Vertical reinforcing for 12" masonry foundation walls with $d \geq 8\frac{3}{4}$". Lateral soil load is 30 psf/ft of depth. *[IBC Table 1807.1.6.3(4)]*

1807 Foundation Walls, Retaining Walls and Embedded Posts and Poles

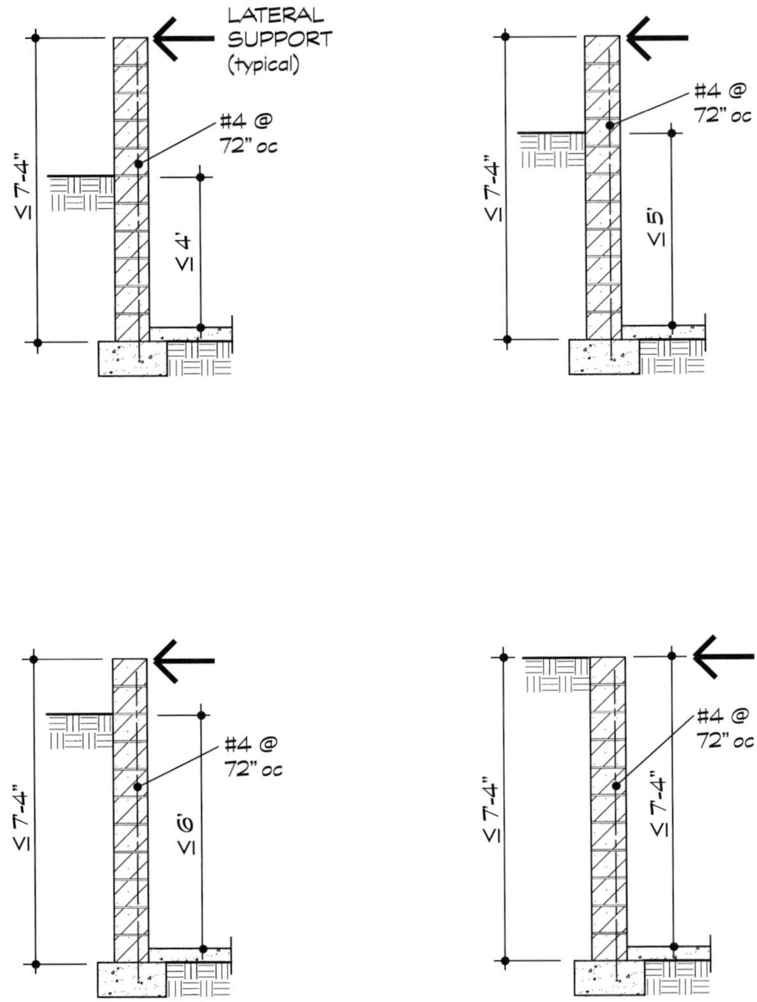

7' - 4" -HIGH WALLS

Fig. 1807.1.6.3(4)-6. Vertical reinforcing for 12" masonry foundation walls with $d \geq 8^{3/4}$". Lateral soil load is 45 psf/ft of depth. *[IBC Table 1807.1.6.3(4)].*

1807 Foundation Walls, Retaining Walls and Embedded Posts and Poles

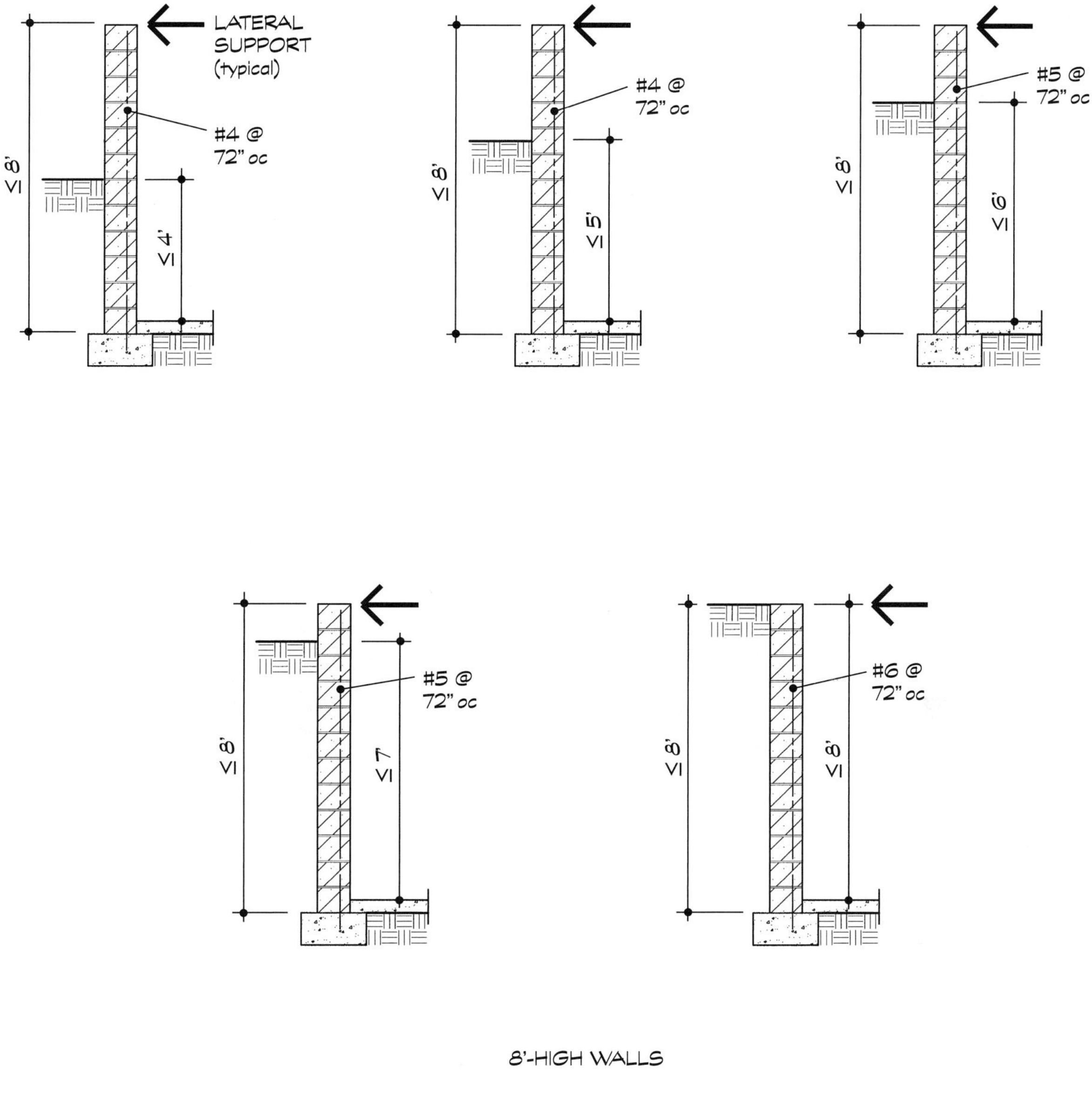

Fig. 1807.1.6.3(4)-7. Vertical reinforcing for 12" masonry foundation walls with $d \geq 8^3/_4$". Lateral soil load is 45 psf/ft of depth. *[IBC Table 1807.1.6.3(4)].*

1807 Foundation Walls, Retaining Walls and Embedded Posts and Poles

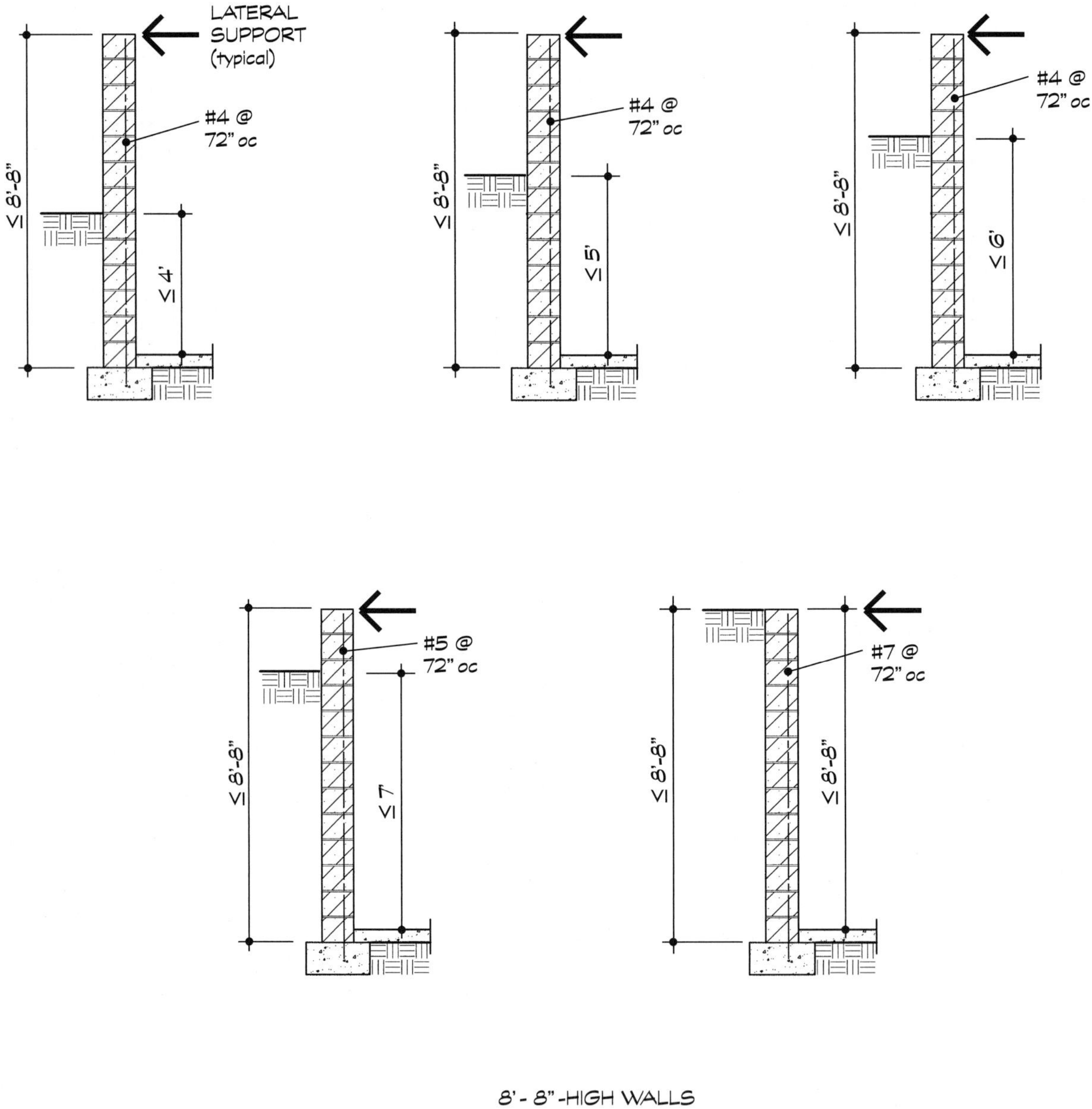

Fig. 1807.1.6.3(4)-8. Vertical reinforcing for 12" masonry foundation walls with $d \geq 8^3/_4$". Lateral soil load is 45 psf/ft of depth. *[IBC Table 1807.1.6.3(4)].*

1807 Foundation Walls, Retaining Walls and Embedded Posts and Poles

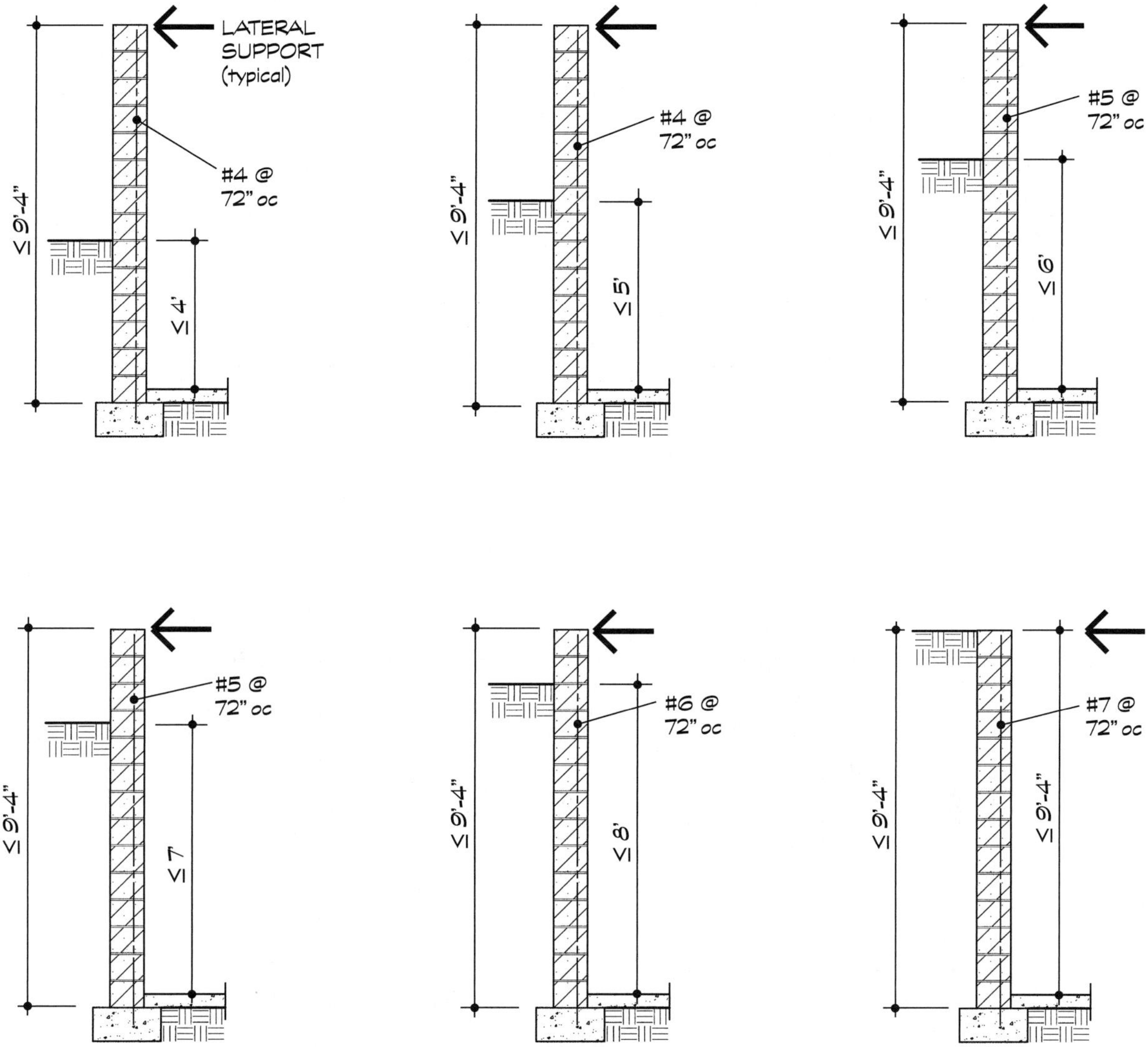

9'- 4"-HIGH WALLS

Fig. 1807.1.6.3(4)-9. Vertical reinforcing for 12" masonry foundation walls with $d \geq 8\frac{3}{4}$". Lateral soil load is 45 psf/ft of depth. *[IBC Table 1807.1.6.3(4)]*.

1807 Foundation Walls, Retaining Walls and Embedded Posts and Poles

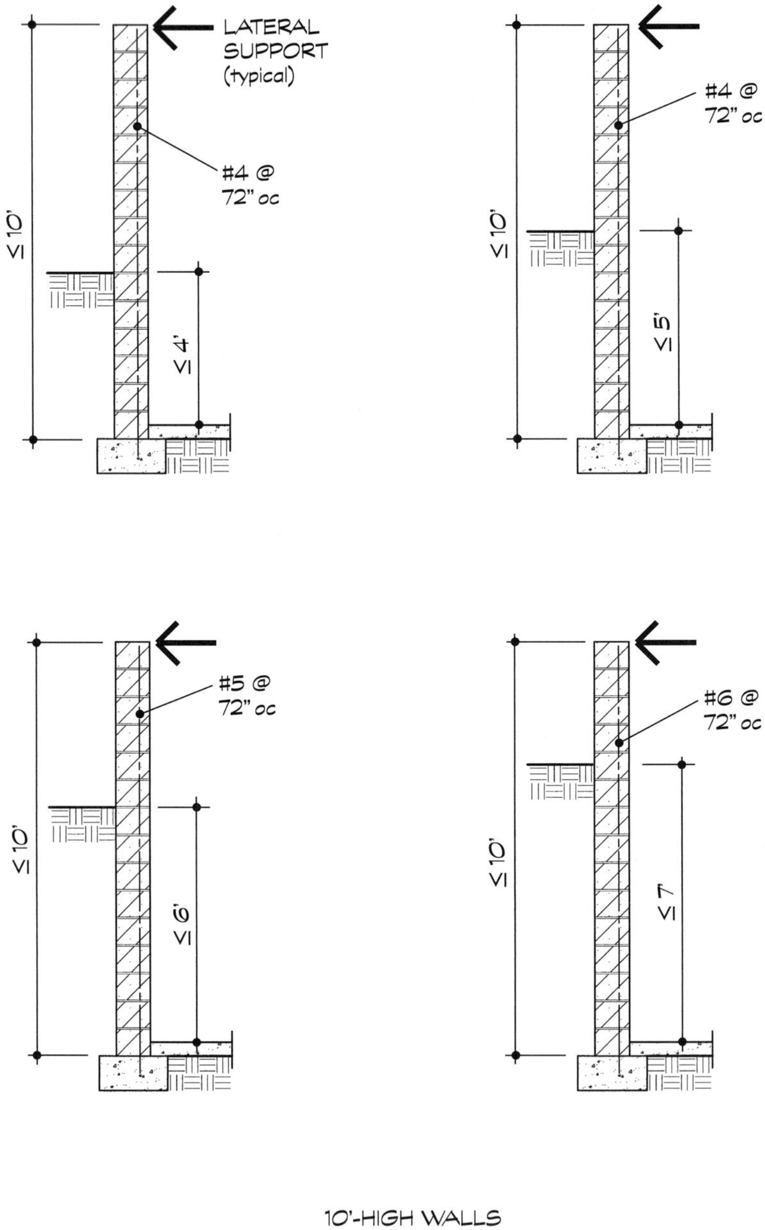

Fig. 1807.1.6.3(4)-10a. Vertical reinforcing for 12" masonry foundation walls with $d \geq 8\frac{3}{4}$". Lateral soil load is 45 psf/ft of depth. *[IBC Table 1807.1.6.3(4)].*

1807 Foundation Walls, Retaining Walls and Embedded Posts and Poles

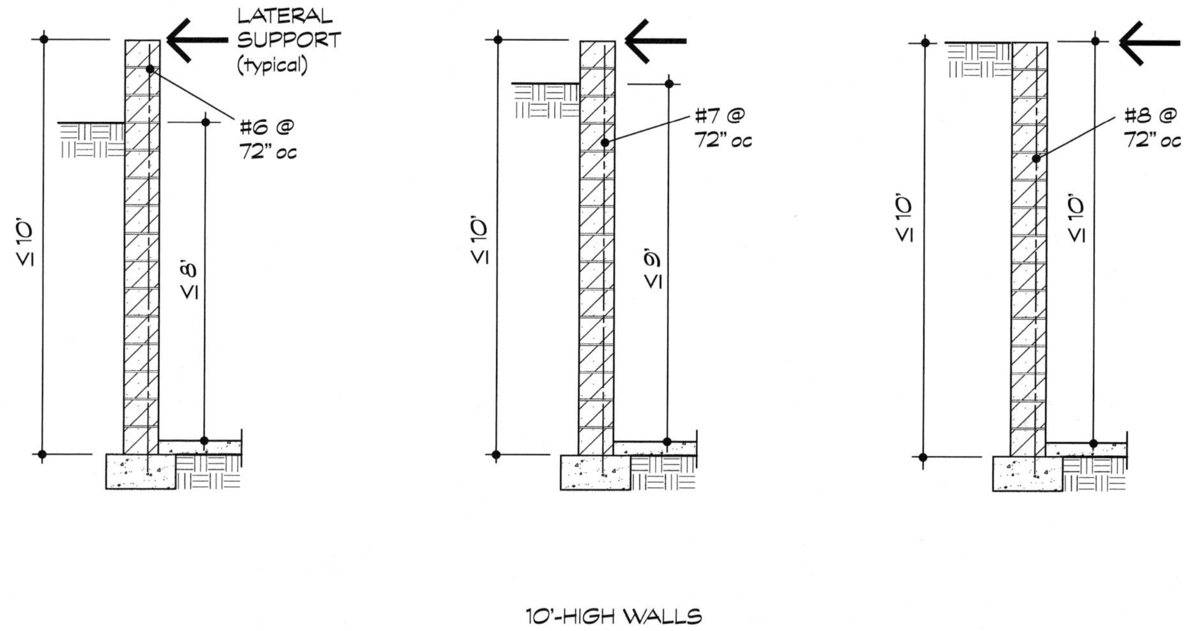

Fig. 1807.1.6.3(4)-10b. Vertical reinforcing for 12" masonry foundation walls with $d \geq 8\frac{3}{4}$". Lateral soil load is 45 psf/ft of depth. *[IBC Table 1807.1.6.3(4)].*

1807 Foundation Walls, Retaining Walls and Embedded Posts and Poles

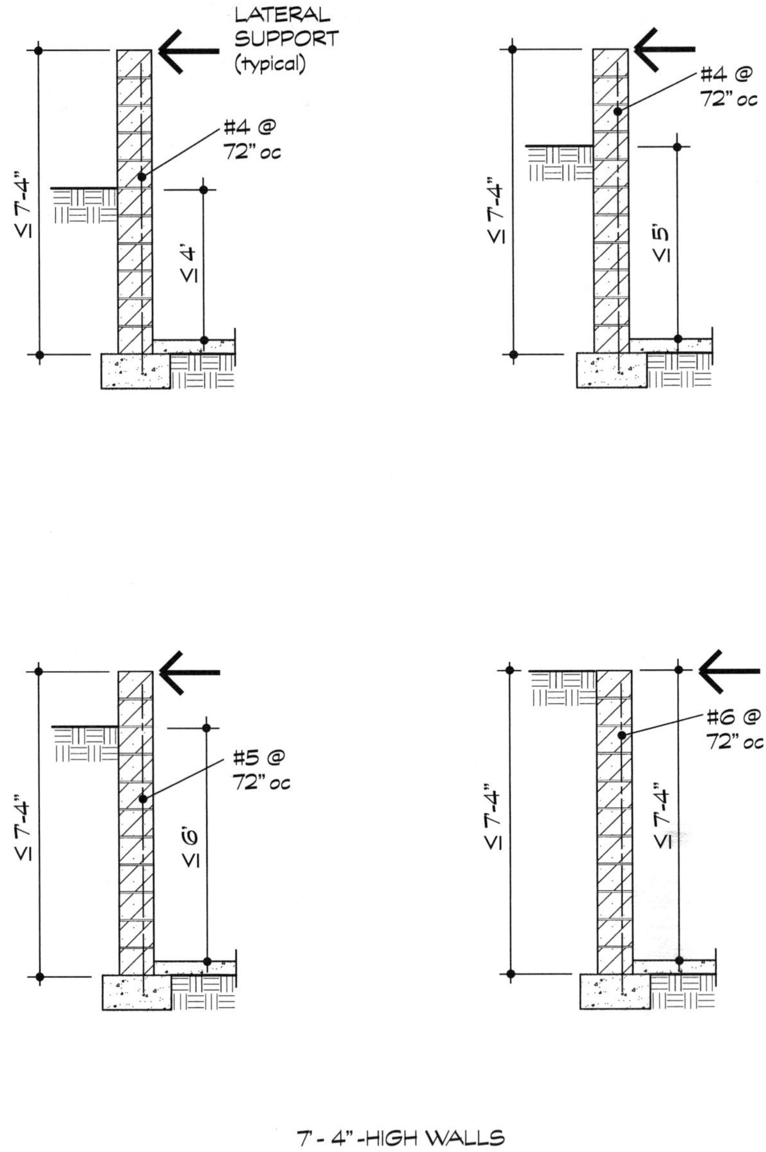

Fig. 1807.1.6.3(4)-11. Vertical reinforcing for 12" masonry foundation walls with $d \geq 8\frac{3}{4}$". Lateral soil load is 60 psf/ft of depth. *[IBC Table 1807.1.6.3(4)]*.

1807 Foundation Walls, Retaining Walls and Embedded Posts and Poles

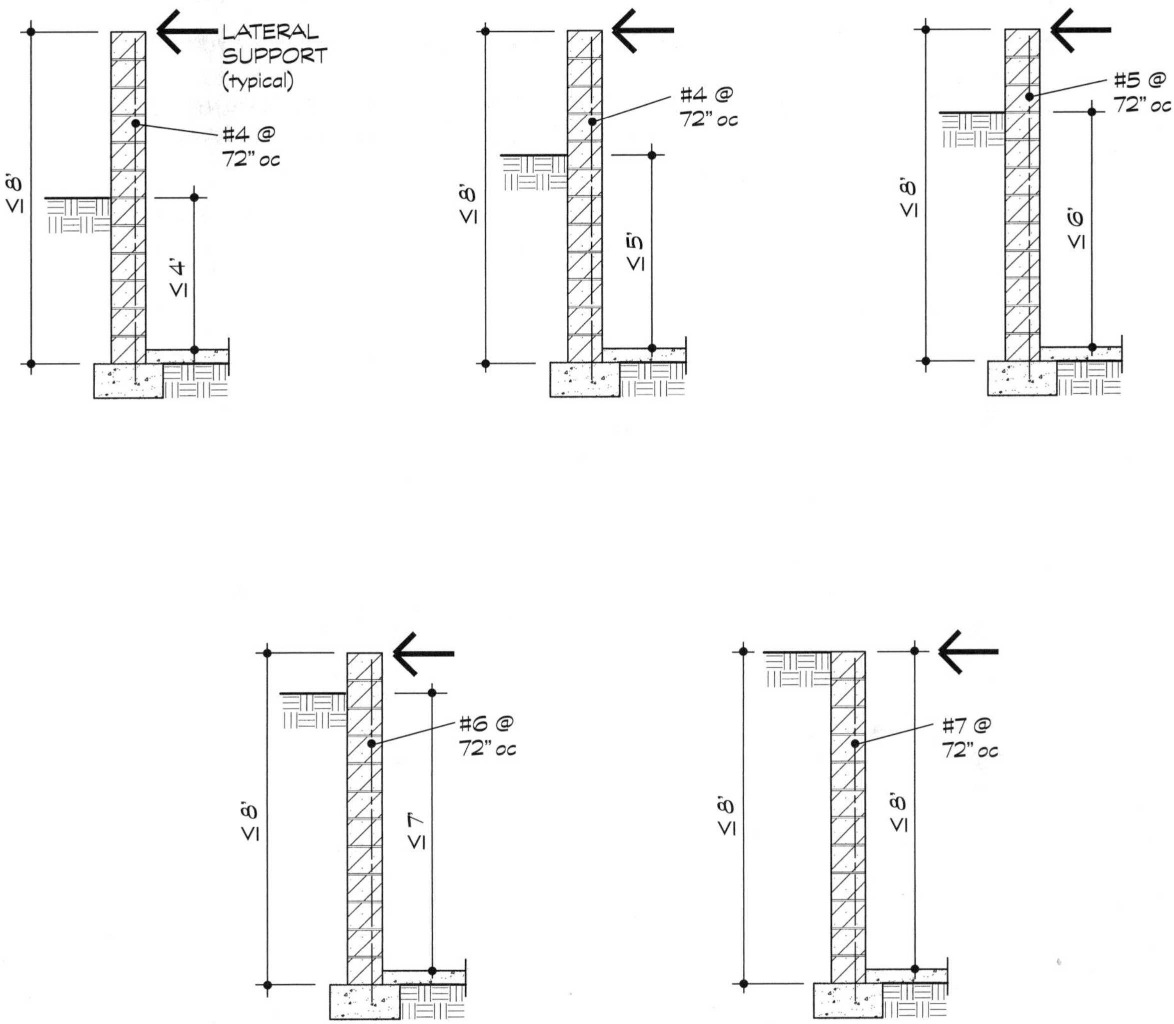

8'-HIGH WALLS

Fig. 1807.1.6.3(4)-12. Vertical reinforcing for 12" masonry foundation walls with $d \geq 8\frac{3}{4}$". Lateral soil load is 60 psf/ft of depth. *[IBC Table 1807.1.6.3(4)]*.

1807 Foundation Walls, Retaining Walls and Embedded Posts and Poles

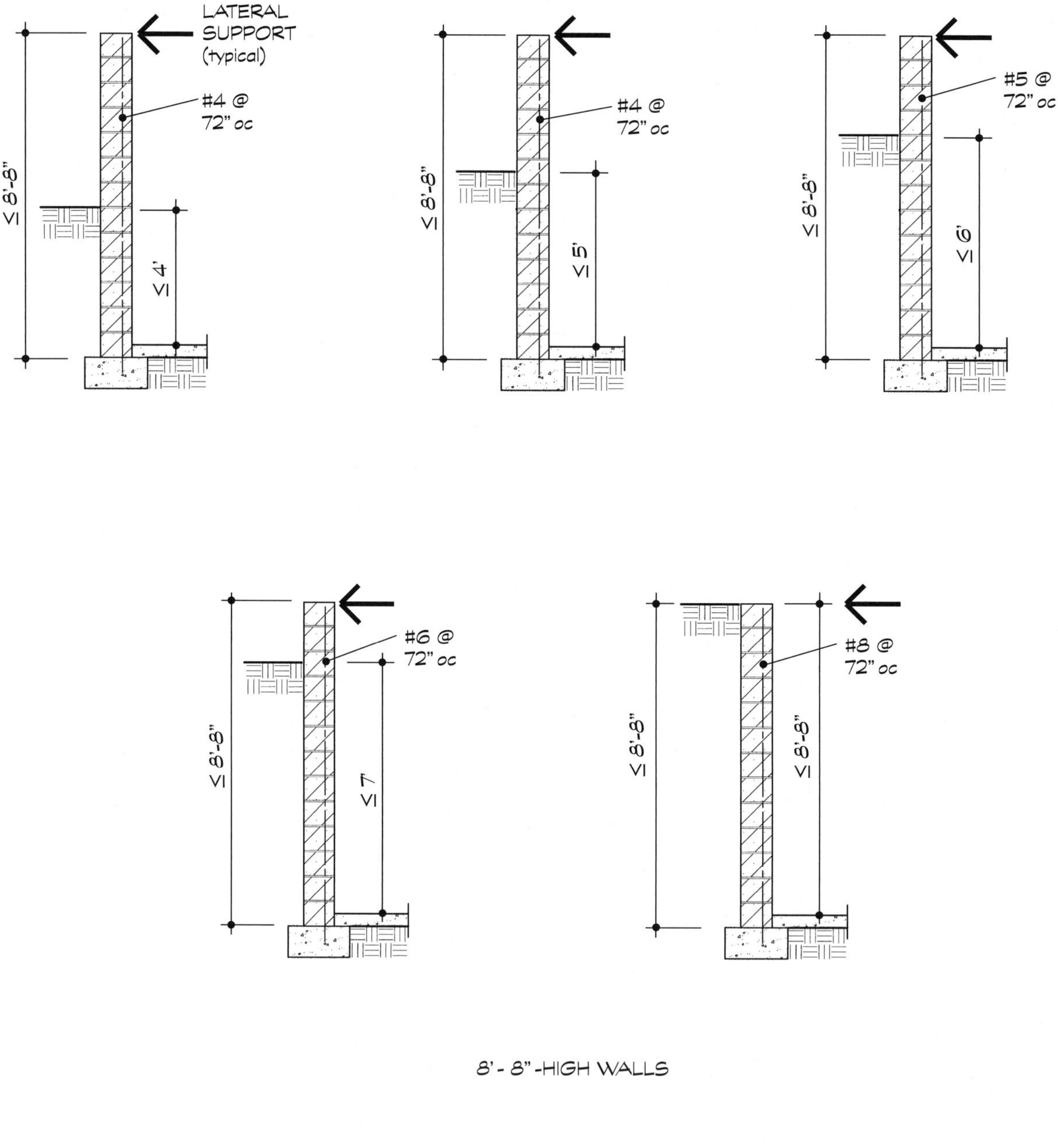

Fig. 1807.1.6.3(4)-13. Vertical reinforcing for 12" masonry foundation walls with $d \geq 8\frac{3}{4}$". Lateral soil load is 60 psf/ft of depth. *[IBC Table 1807.1.6.3(4)]*.

1807 Foundation Walls, Retaining Walls and Embedded Posts and Poles

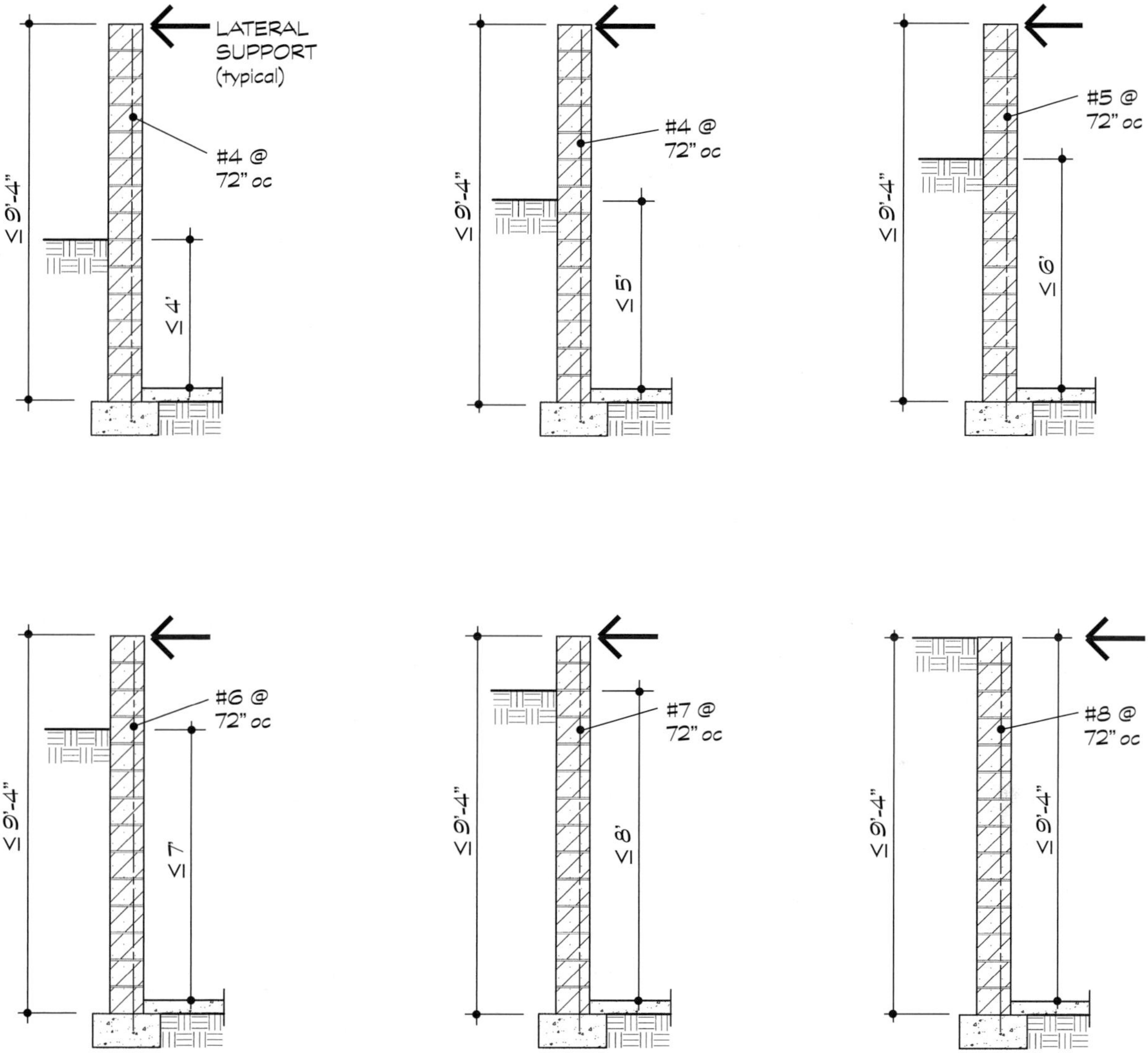

9'- 4"-HIGH WALLS

Fig. 1807.1.6.3(4)-14. Vertical reinforcing for 12" masonry foundation walls with $d \geq 8^3/_4$". Lateral soil load is 60 psf/ft of depth. *[IBC Table 1807.1.6.3(4)].*

1807 Foundation Walls, Retaining Walls and Embedded Posts and Poles

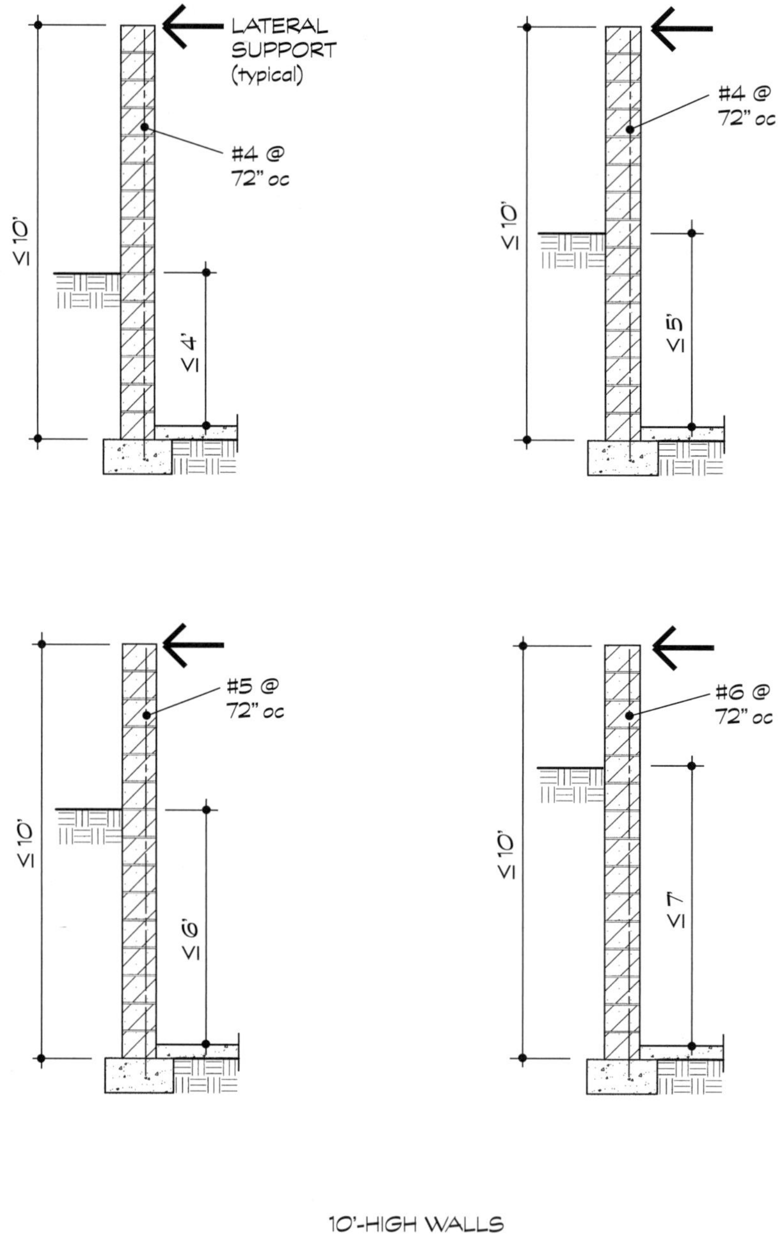

Fig. 1807.1.6.3(4)-15a. Vertical reinforcing for 12" masonry foundation walls with $d \geq 8^3/_4$". Lateral soil load is 60 psf/ft of depth. *[IBC Table 1807.1.6.3(4)]*.

1807 Foundation Walls, Retaining Walls and Embedded Posts and Poles

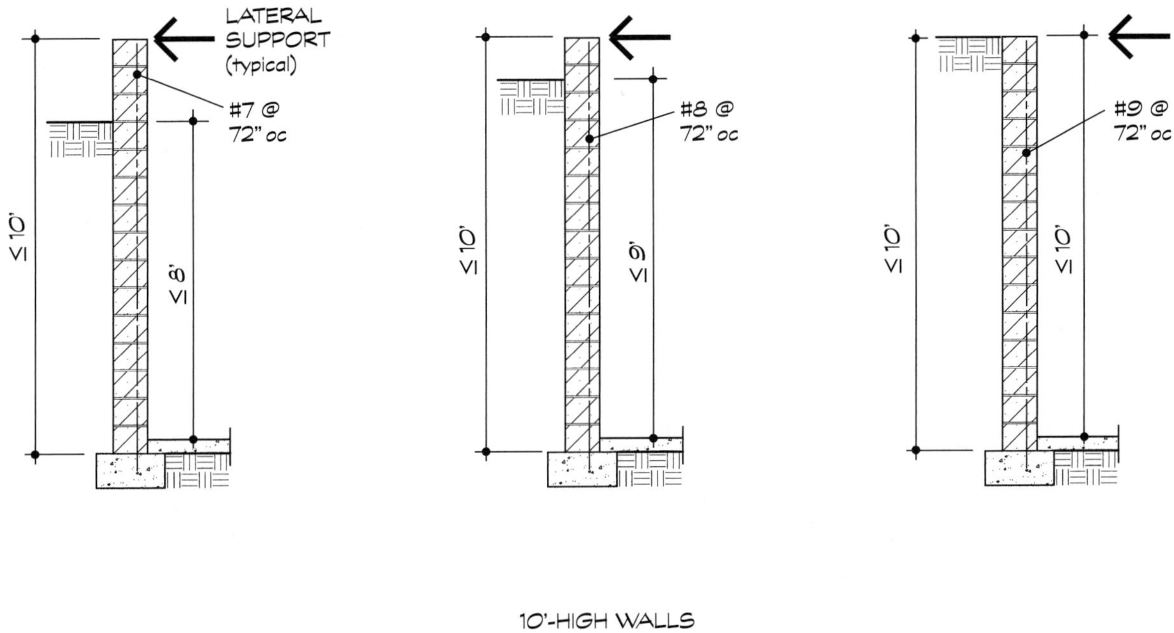

Fig. 1807.1.6.3(4)-15b. Vertical reinforcing for 12" masonry foundation walls with $d \geq 8\frac{3}{4}$". Lateral soil load is 60 psf/ft of depth. *[IBC Table 1807.1.6.3(4)].*

1808 Foundations

1808.5 Shifting or moving soils

- Where the soil is unstable, the following applies:
  - Footing depth must be adequate to provide stability.

1808.6.1 Foundations

- Foundations in active expansive soil must be designed as follows:
  - To resist differential volume changes.
  - To prevent structural damage to the structure.
  - To limit the following, which would interfere with the serviceability of the structure:
    - Deflection.
    - Racking.
- Foundations extending through expansive soils must be designed as follows:
  - To prevent uplift of the structure.
  - To accomplish one of the following:
    - To resist forces applied on the foundation due to the following:
      - Soil volume change.
    - To be isolated from the expansive soil.

1808.7.1 Building clearance from ascending slopes

- The distance required between building face and toe of such a slope is to be one of the following:
  - That verified to be adequate by engineering analysis.
  - ≥ the smaller of the following distances as defined below:
    - Distance:
      - 1/2 the rise of the slope.
      - 15'.
    - Definitions:
      - Where the slope is > 1:1, the toe is defined as follows:
        - The toe is located at the intersection of the following lines:
          - A 45° line tangent to the top of the slope.
          - A horizontal line at the top of the foundation.
      - Rise of slope is measured vertically between the following points:
        - Top of the slope.
        - Toe of slope if there is no retaining wall.
        - Top of any retaining wall at the toe of the slope.

Note: The following are cited as sources of pertinent requirements:

1808.7.5, "Alternate setback and clearance," addresses requirements pertaining to the engineering analysis as indicated above.

IBC Figure 1808.7.1, "Foundation clearances from slopes," includes setback requirements, a summary of which is provided above.

- Buildings near the bottom of a slope with a gradient > 1:3 are governed as follows:
  - Buildings must be located far enough from such a slope to be protected from the following:
    - Drainage.
    - Erosion.
    - Shallow failures.

1808 Foundations

1808.7.2 Foundation setback from descending slope surface

- This section applies to buildings on or near a slope with a gradient > 1:3.
- Foundations of such buildings are governed as follows:
  - They must be embedded in firm material.
  - They must be set back from the sloped surface as follows:
    - So as to avoid damaging settlement by the following:
      - Adequate vertical support.
      - Adequate lateral support.
  - They must be set back equal to one of the following distances:
    - That verified to be adequate by engineering analysis.
    - ≥ the smaller of the following distances measured as defined below:
      - Distance:
        - 1/3 the rise.
        - 40'.
      - Measurement:
        - The setback distance is measured from the near face of the footing as follows:
          - From the footing at bearing level on a horizontal line to one of the following:
            - Where the slope is ≤ 1:1:
              - To a point on the surface of the slope.
            - Where the slope is > 1:1:
              - To a point on a line as follows:
                - A 45° line connecting to the bottom of the slope.
          - Rise of slope is measured vertically between the following points:
            - Top of the slope.
            - Bottom of slope.

Note: The following are cited as sources of pertinent requirements:

1808.7.5, "Alternate setback and clearance," addresses requirements pertaining to the engineering analysis as indicated above.

IBC Figure 1808.7.1, "Foundation clearances from slopes," includes setback requirements, a summary of which is provided above.

1808.7.3 Pools

- This section addresses pools near a slope with a gradient > 1:3 as follows:
  - Pools regulated by the code.
- The required setback between pools and slopes is as follows:
  - 1/2 the setback required for building footings.
- Any part of a pool wall which is ≤ 7' from the top of a slope is governed as follows:
  - The wall must be able to support the water contained without the following:
    - Assistance from the soil.

1808 Foundations

1808.7.4 Foundation elevation

- On all sites, the top of the foundation wall must be at one of the following heights:
  - ≥ a height equal to the sum of the following dimensions:
    - 12".
    - 2% of the distance between the building and either of the following:
      - The surface of the street gutter as follows:
        - At the point where the site drains into it.
      - An approved drainage device as follows:
        - At the inlet where the site drains into it.
  - A height as follows:
    - Approved by the building official.
    - Where the following drainage is verified at all points on the site:
      - Site drainage is to the point of discharge.
      - Site drainage is away from the structure.

1808.7.5 Alternate setback and clearance

- Alternate setbacks and clearances are permitted in the following case:
  - Upon approval of the building official as follows:
    - The building official may require a geotechnical investigation.

Note: 1803.5.10, "Alternate setback and clearance," is cited as governing geotechnical investigations.

1809 Shallow Foundations

1809.2 Supporting soils

- Shallow foundations must be constructed on one of the soil types listed below:
  - Undisturbed.
  - Compacted fill.
  - Controlled low-strength material (CLSM).

*Note: 1804.5, "Compacted fill material," is cited as governing that material.
1804.6, "Controlled low-strength material," is cited as governing CLSM.*

1809.3 Stepped footings

- The top of a footing must be level.
- The bottom of a footing may slope ≤ 1:10.
- Footing surfaces must be stepped as follows:
  - At the top where required for to change level.
  - At the bottom where grade slopes > 1:10.

1809.4 Depth and width of footings

- Footing depth is required to be one or more of the following as applicable:
  - ≥ 12".
  - Detailed to be protected from frost.

Note: 1809.5, "Frost protection," is cited as governing this aspect of footings.

- Footing width is required to be ≥ 12".

1809.5 Frost protection

- Buildings that have all of the following characteristics are not governed by this section:
  - Freestanding.
  - Classified as Occupancy Category I.

Note: 1604.5, "Occupancy category," is cited as defining building types in Occupancy Category I.

  - ≤ 600 sf where the building is light-frame construction.
  - ≤ 400 sf where the building is not light-frame construction.
  - ≤ 10' eave height.
- This section governs foundations and permanent supports of buildings not otherwise protected from frost.
- One of the following frost protection methods is required for foundations governed by this section.
  - Setting the foundation at a level below the local frost line.
  - Setting foundations on solid rock.
  - By methods specified in the referenced standard.

Note: ASCE 32, "Design and Construction of Frost-Protected Shallow Foundations," is cited as an acceptable method of protection of shallow foundations from frost.

- The bearing of shallow foundations on frozen soil is governed as follows:
  - Permitted where the frozen condition is permanent.
  - Not permitted where the frozen condition is not permanent.

1809 Shallow Foundations

1809.6 Location of footings

- Footings on granular soil are governed as follows:
  - The slope of a line between the bottom edges of adjacent footings must comply with one of the following:

    ≤ 30° with the horizontal.

    As needed where the higher footing is laterally supported as follows:

    By an approved method.

    As permitted by engineering analysis of the soil.

1809.7 Prescriptive footings for light-frame construction

- Footings for light-frame construction are governed as follows:
  - Concrete footings may be designed according to details provided in this section on the following pages.

  Note: IBC Table 1809.7, "Prescriptive Footings Supporting Walls of Light-Frame Construction," is cited as listing design parameters for concrete or masonry unit footings supporting light-frame construction. The data of this table is summarized in the details provided in this section in Fig. 1809.7A–G on the following pages.

  1809.4, "Depth and width of footings," is cited as governing the depth of footings.

  Section 1908, "Modifications to ACI 318," is cited as listing additional requirements for concrete footings in Seismic Design Category C, D, E, or F.

  1807.1.6, "Prescriptive design of concrete and masonry foundation walls," is cited as governing thickness of foundation walls.

  - Grade under the floor may be excavated to a level at the top of the footing.

1809.8 Plain concrete footings

- An edge thickness ≥ 6" is required for plain concrete footings as follows:
  - In Occupancy R-3 where both of the following apply:

    Where the footing extends beyond the foundation wall as follows:

    A distance ≤ wall thickness.

    Where bearing on soil or rock.
- Plain concrete footings in the following locations have the requirement listed below:
  - Locations:

    Other than R-3.

    Other than in light-frame construction.
  - Requirement:

    Must have an edge thickness ≥ 8" where bearing on soil or rock.

1809 Shallow Foundations

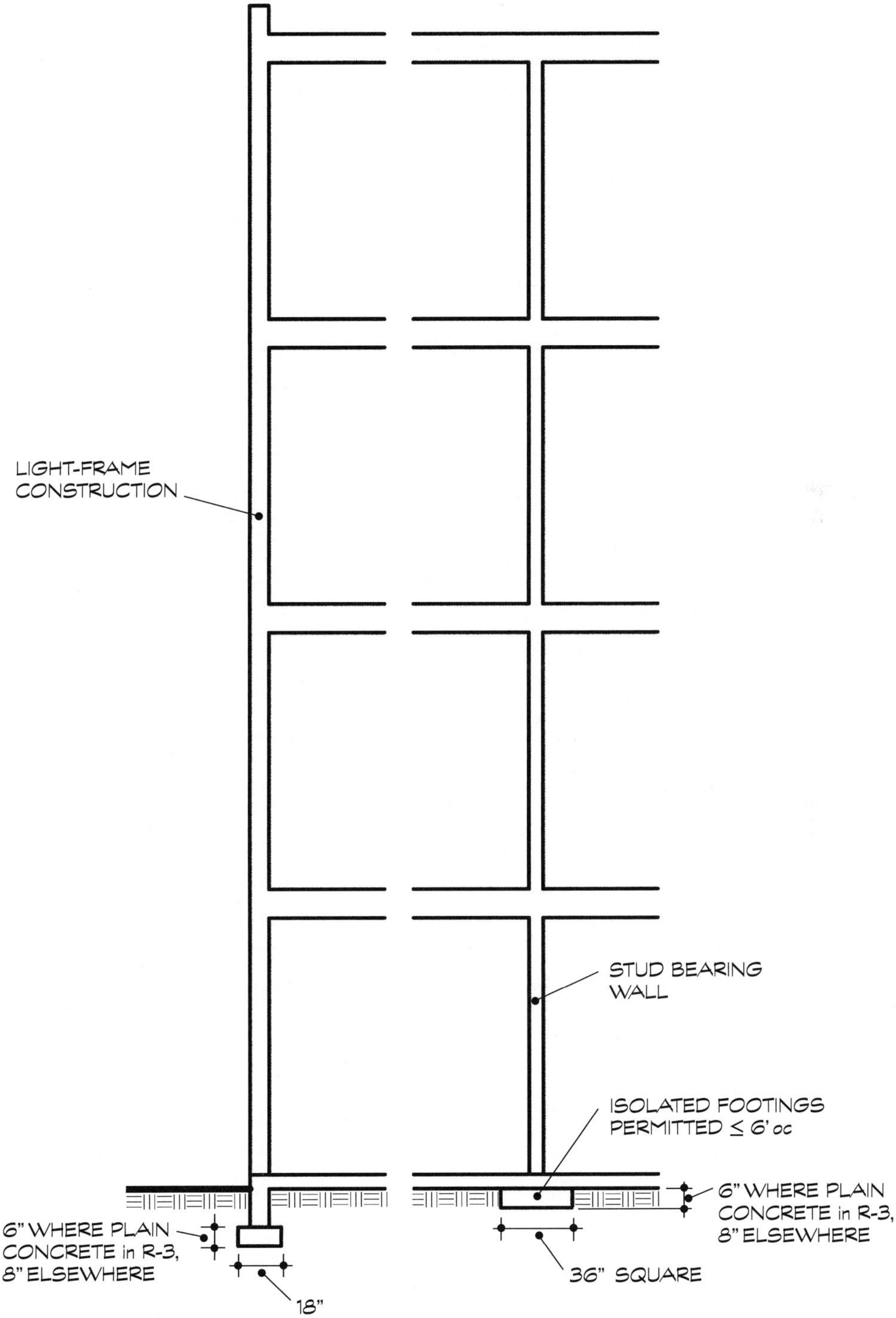

Fig. 1809.7A. Concrete or masonry unit foundations supporting 3 floors and a roof.

1809 Shallow Foundations

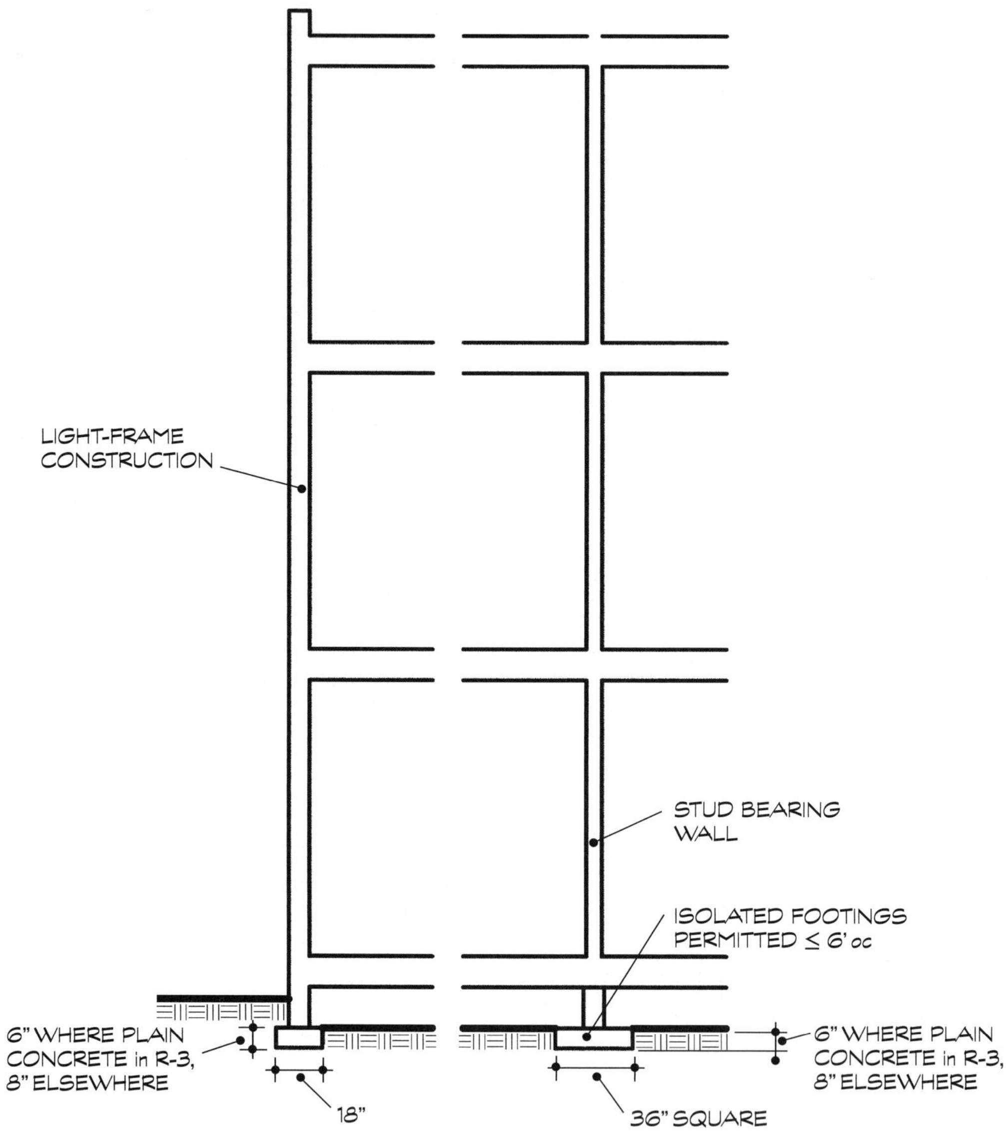

Fig. 1809.7B. Concrete or masonry unit foundations supporting 3 floors and a roof.

1809 Shallow Foundations

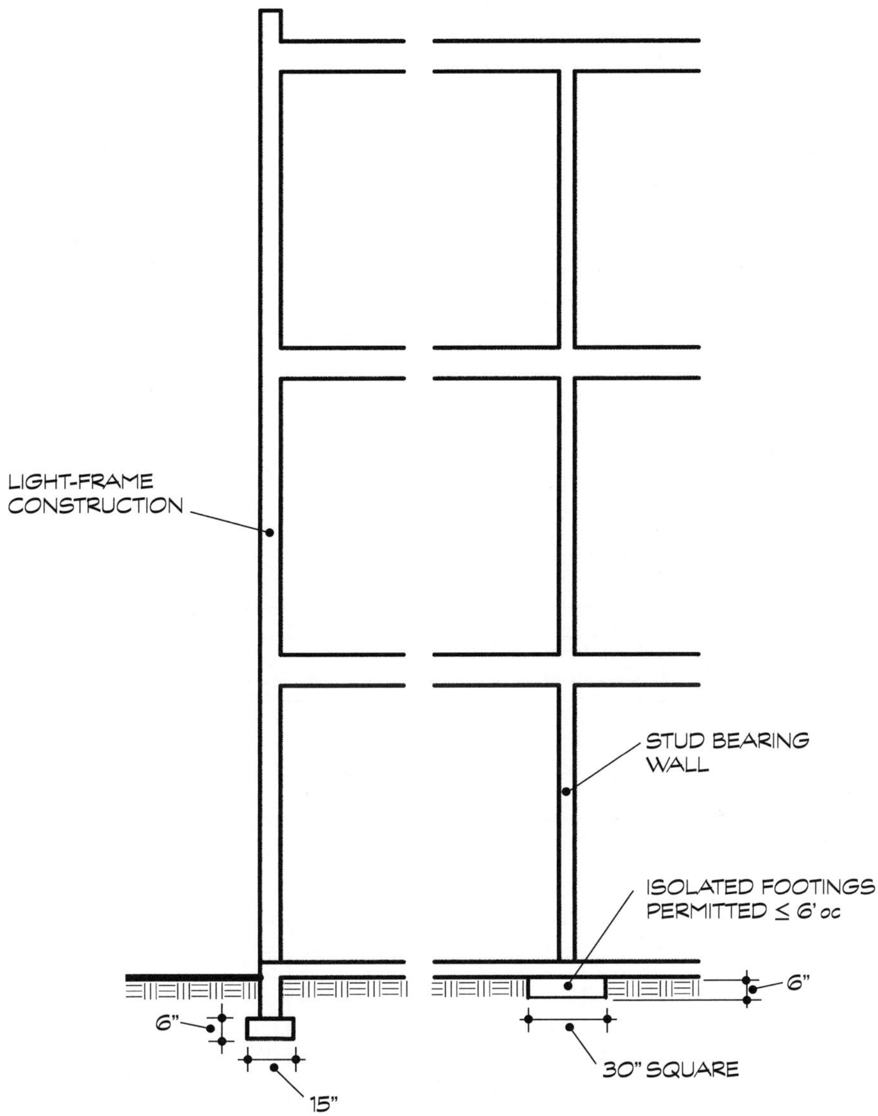

Fig. 1809.7C. Concrete or masonry unit foundations supporting 2 floors and a roof.

1809 Shallow Foundations

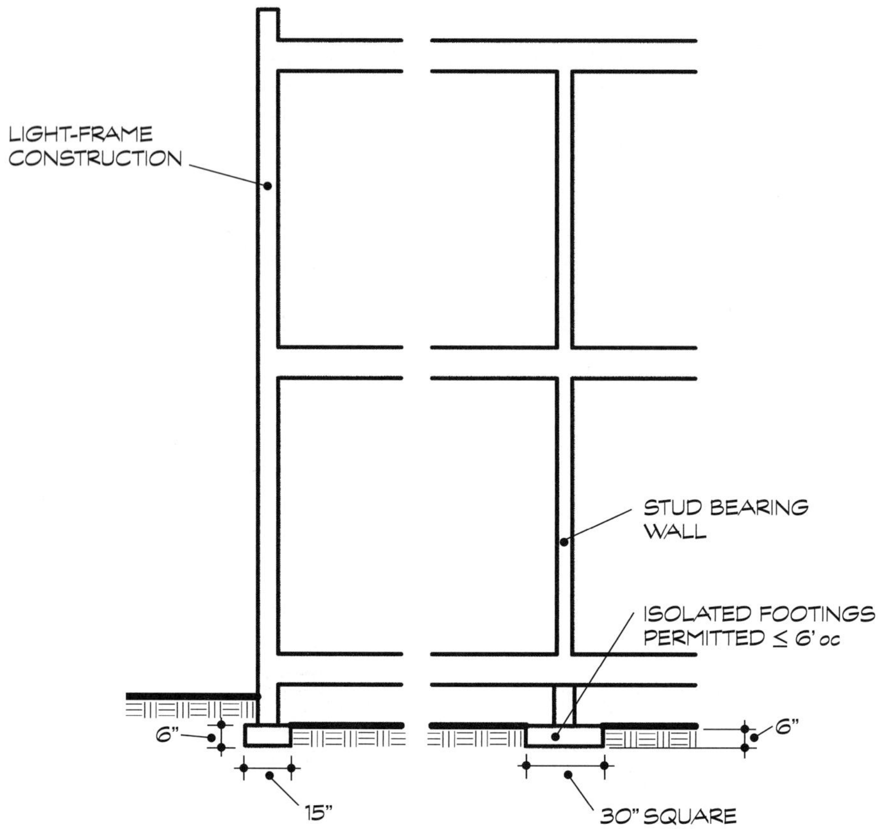

Fig. 1809.7D. Concrete or masonry unit foundations supporting 2 floors and a roof.

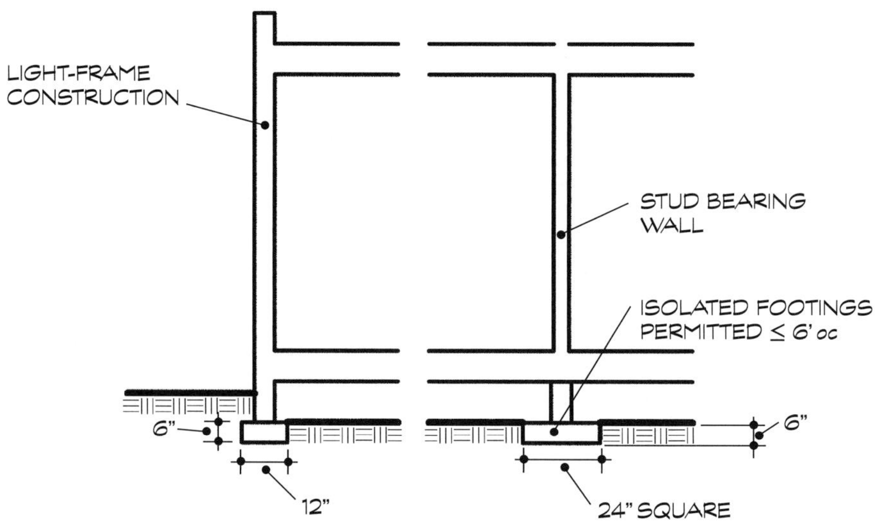

Fig. 1809.7E. Concrete or masonry unit foundations supporting 1 floor and a roof.

1809 Shallow Foundations

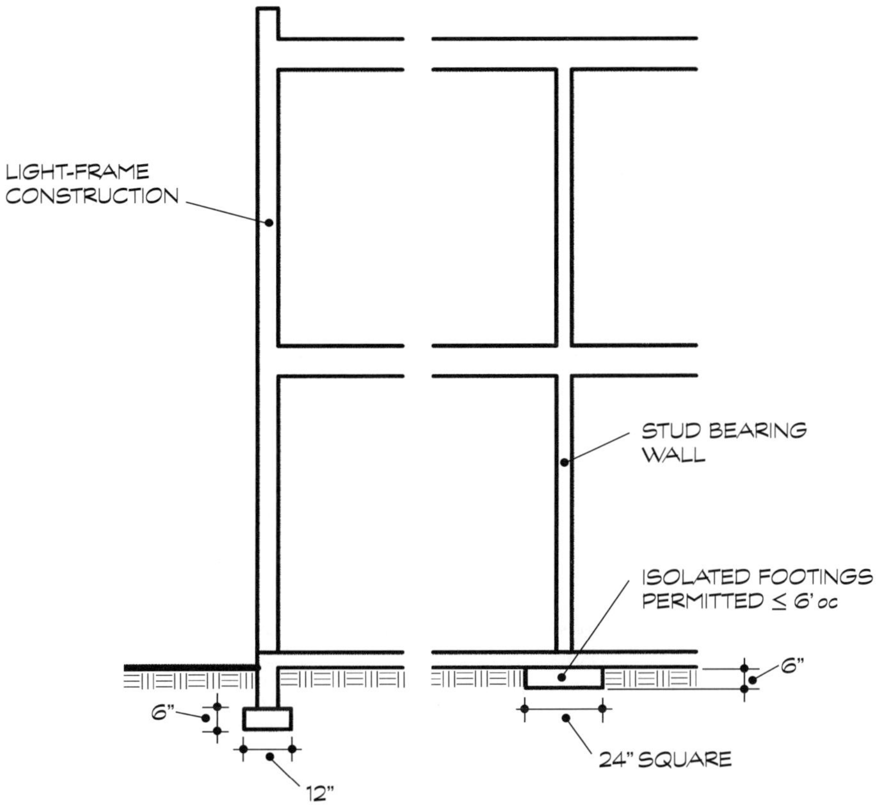

Fig. 1809.7F. Concrete or masonry unit foundations supporting 1 floor and a roof.

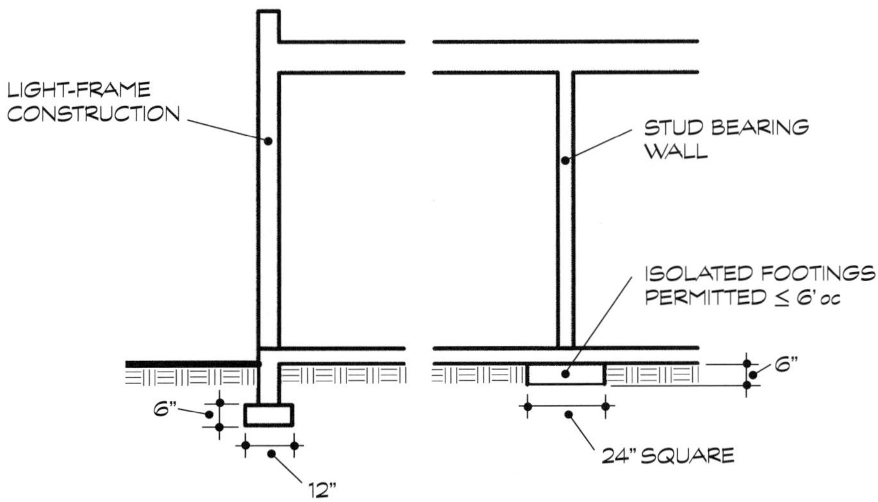

Fig. 1809.7G. Concrete or masonry unit foundations supporting a roof.

1809 Shallow Foundations

1809.9 Masonry-unit footings

- Light-frame construction is governed as follows:
  - Masonry-unit footings may be designed according to details provided on the previous pages.

  Note: *IBC Table 1809.7, "Prescriptive Footings Supporting Walls of Light-Frame Construction," is cited as listing design parameters for masonry unit footings supporting light-frame construction. The data of this table is summarized in the details provided in this section in Fig. 1809.7A–G on the previous pages.*

- For other construction, the following aspects of masonry-unit footings must comply with this section and other masonry design requirements of the code:
  - Design.
  - Materials.
  - Construction.

  Note: *The following are cited as governing masonry-unit footings:*
  1809.9.1, "Dimensions."
  1809.9.2, "Offsets."
  Chapter 21, "Masonry."

1809.9.1 Dimensions

- Masonry unit footings are governed as follows:
  - Footing depth (thickness) required is as follows:
    ≥ 2 × footing projection beyond the following:
    Wall.
    Pier.
    Column.
  - Footing width required is as follows:
    ≥ 8" wider than the foundation wall supported.
  - Masonry must be set in one of the following mortar types:
    M, S.

  Note: *2103.8, "Mortar," is cited as governing this material as indicated above.*

1809.9.2 Offsets

- This section governs brick foundations as follows:
  - Where successive brick courses step back from a wider footing to approach the thickness of the foundation wall.
- Where such foundations step back with each course, the following applies:
  - Courses must be set back ≤ 1 1/2" from the course below.
- Where each step back is a pair of courses flush with each other, the following applies:
  - Pairs of flush courses must be set back ≤ 3" from the pair of courses below.

1809 Shallow Foundations

1809.10 Pier and curtain wall foundations

- Pier and curtain wall foundations may be used as follows where in compliance with this section:
  - In locations other than the following Seismic Design Categories:
    D, E, F.
  - For light-frame construction as follows:
    ≤ 2 stories above grade plane.
- Piers must meet the following requirements:
  - Unsupported height of masonry piers is limited as follows:
    Height must be ≤ 10 × the least pier dimension.
  - The following units supporting beams and girders must meet requirements listed:
    Units:
      Structural clay tile.
      Hollow concrete masonry.
    Requirements:
      Where unsupported height of piers is > 4 × least dimension:
        Cellular spaces must be filled with the following:
          Type M or S mortar.
      Where unsupported height of piers is ≤ 4 × least dimension:
        Cellular spaces need not be filled.
      Hollow piers must be capped by one of the following methods:
        With ≥ 4" of solid masonry.
        With ≥ 4" of concrete.
        By filling the cells of the top course with concrete or grout.

Note: Chapter 21, "Masonry," is cited as governing masonry piers in addition to this section.

- Load-bearing walls must meet the following requirements:
  - They must bear on continuous footings with the following detail:
    Footings must tie into exterior wall footings.
  - Thickness must be one of the following:
    ≥ 4" nominal.
    ≥ 3 5/8" actual.
  - They must be integral with piers as follows:
    Spaced ≤ 6' on center.
- The height of a 4" load-bearing masonry wall is limited as follows:
  - To ≤ 4' where supporting the following:
    Wood-framed walls.
    Wood-framed floors.
- The height of unbalanced fill for 4" foundation walls is limited as follows:
  - To ≤ 2' for solid masonry.
  - To ≤ 1' for hollow masonry.

1809 Shallow Foundations

1809.11 Steel grillage footings

- Structural steel sections in grillage footings are governed as follows:
  - Components must be separated by approved steel spacers.
  - Components must be completely encased in concrete as follows:
    - ≥ 6" thick on the bottom.
    - ≥ 4" elsewhere.
  - Space between components must be filled with one of the following:
    - Concrete.
    - Cement grout.

1809.12 Timber footings

- Timber footings are permitted as follows:
  - In Type V construction.
  - As approved by the building official.
- Treatment of timber footings is governed as follows:
  - The timber is not required to be treated in the following locations:
    - Where permanently located completely below the water table.
    - Where used as caps for wood piling as follows:
      - Where above the water level in the following locations:
        - Submerged lands.
        - Marsh lands.
  - In other locations the timber must be treated.

Note: AWPA U1, "Use Category System: User Specification for Treated Wood," Commodity Specification A, Use Category 4B, is cited as governing the treatment of timber footings:

- Stress is limited in the following footings as indicated below:
  - Footings:
    - Untreated timber footings supported on treated piling.
  - Limitation:
    - Compressive stress ⊥ to the grain is limited as follows:
      - To ≤ 70% of the otherwise allowable stress.

Note: AF&PA NDS, "National Design Specifications for Wood Construction," is cited as governing the allowable stress of the timber.

19

Concrete

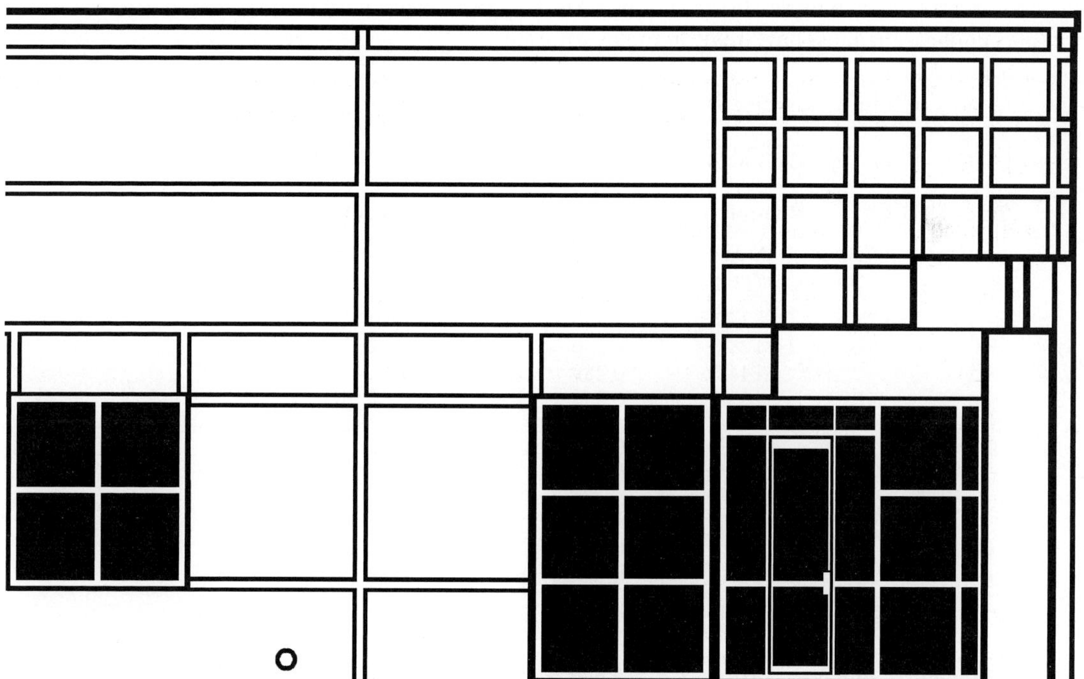

Garments to Go. Bastrop, Texas. *(partial elevation)*
Spencer Godfrey Architects. Round Rock, Texas.

1908 Modifications to ACI 318

1908.1.8 ACI 318, Section 22.10

- Delete Section 22.10 from the ACI *Building Code Requirements for Structural Concrete* in lieu of this section.
- This section specifies the only structural plain concrete permitted in the following locations:
  - In Seismic Design Categories C, D, E, or F.
- Structural plain concrete is permitted for the following walls for the conditions listed below:
  - Walls:
    - Basement walls.
    - Foundation walls.
  - Conditions:
    - In detached 1- and 2-family dwellings as follows:
      - ≤ 3 stories.
      - Constructed with stud bearing-walls.
      - Located in Seismic Design Category D or E.
    - Wall height is ≤ 8'.
    - Wall thickness is ≤ 71/2".
    - Any unbalanced fill is ≤ 4' in height.

Note: ACI, "Building Code Requirements for Structural Concrete" 22.6.6.5, is cited as governing reinforcing in these walls.

- In conditions other than those shown in Fig. 1908.1.15 B, C, and D (on the next page), plain concrete footings under walls must have reinforcing as follows:
  - ≥ 2 bars ≥ #4.
  - A total area of reinforcing bars is required as follows:

    ≥ 0.002 × gross cross-sectional area of the footing as shown in the partial list below:

Table 1908.1.8 Reinforcing Required in Plain Concrete Footings

Footing height (")	Footing width (")	Area of section (sq in)	Area of steel required (sq in)	Number of reinforcing bars required #4	#5	#6	#7
8.0	16	128.0	0.256	2	2	2	2
8.5	17	144.5	0.289	2	2	2	2
9.0	18	162.0	0.324	2	2	2	2
9.5	19	180.5	0.361	2	2	2	2
10.0	20	200.0	0.400	2	2	2	2
10.5	21	220.5	0.441	3	2	2	2
11.0	22	242.0	0.484	3	2	2	2
11.5	23	264.5	0.529	3	2	2	2
12.0	24	288.0	0.576	3	2	2	2
12.5	25	312.5	0.625	4	3	2	2
13.0	26	338.0	0.676	4	3	2	2
13.5	27	364.5	0.729	4	3	2	2
14.0	28	392.0	0.784	4	3	2	2
14.5	29	420.5	0.841	5	3	2	2
15.0	30	450.0	0.900	5	3	3	2

1908 Modifications to ACI 318

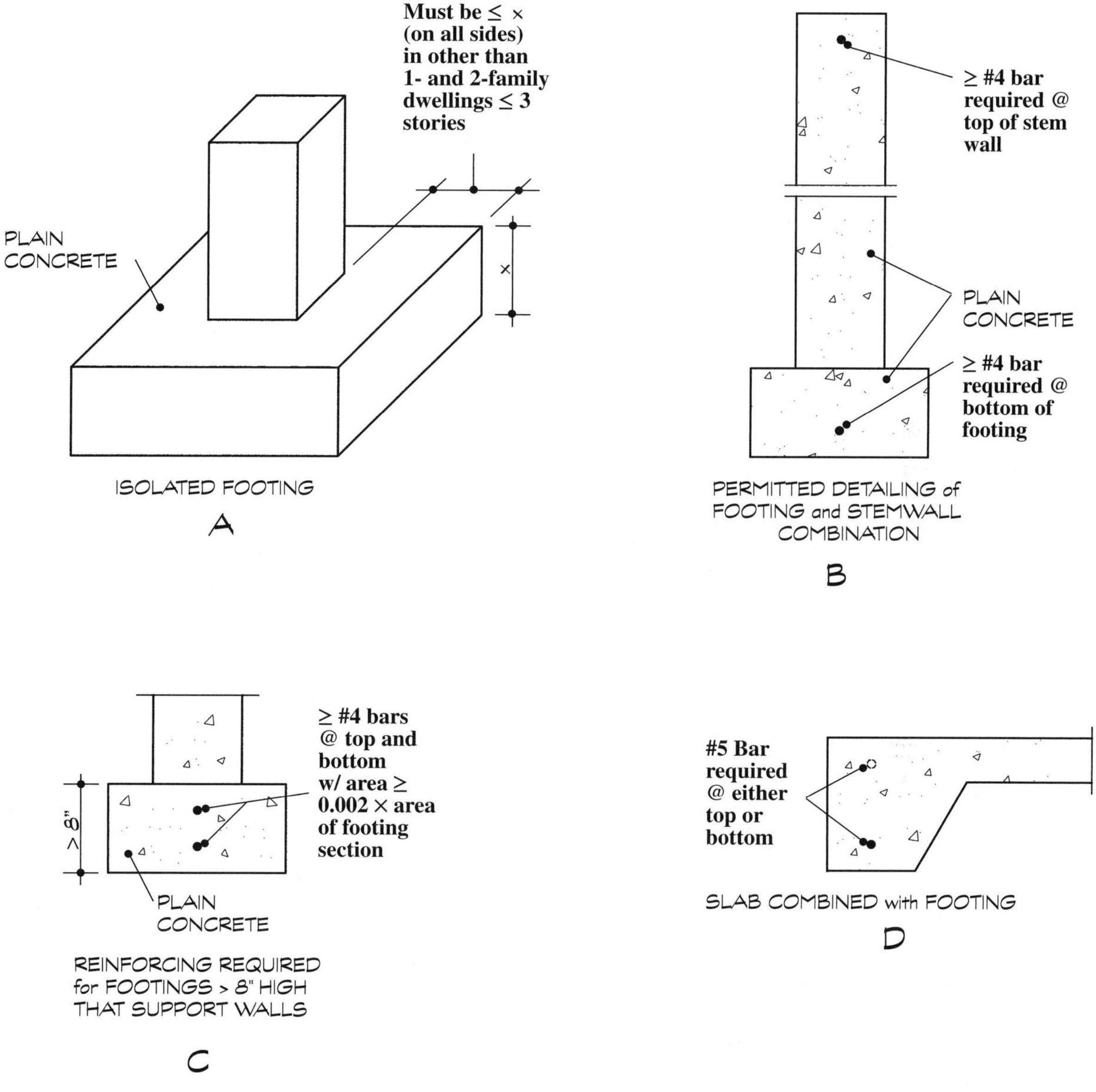

Fig. 1908.1.8. Plain concrete footings in Seismic Design Categories C, D, E, or F.

1909 Structural Plain Concrete

1909.6.1 Basement walls

- The detail provided in this section indicates the thickness required for the following:
  - Structural plain concrete walls as follows:
    Exterior basement walls.
    Foundation walls.

Note: Chapter 22 of ACI 318, "Building Code Requirements for Structural Concrete," is the source of requirements with which the walls above must comply in addition to this section.

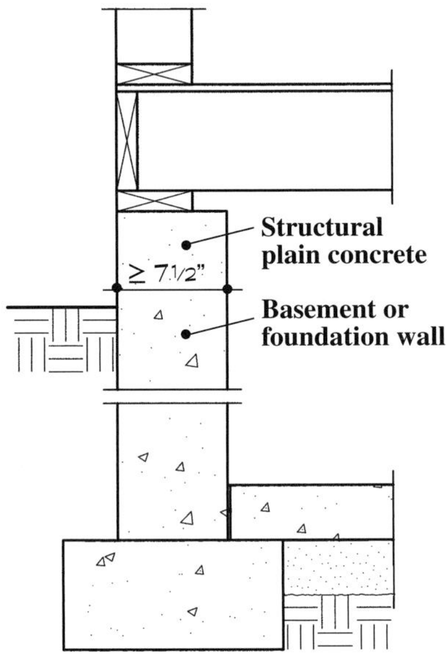

Fig. 1909.6.1. Minimum thickness of exterior basement or foundation wall of structural plain concrete.

1909 Structural Plain Concrete

1909.6.2 Other walls

- This section does not govern the following concrete bearing walls:
  - Basement walls.
  - Foundation walls.
- The thickness of other structural plain concrete bearing walls is governed as follows:
  - Required thickness is based on either unsupported dimensions as follows:
    Length.
    Height.
  - Required thickness of walls with an unsupported dimension ≤ 11' is ≥ 5 1/2".
  - Required thickness of walls with an unsupported dimension > 11' is as follows:
    Unsupported dimension ÷ 24.
- A partial list of minimum thickness for plain concrete bearing walls is provided below:

Table 1909.6.2 Minimum Thickness of Plain Concrete Bearing Walls

Dimension	Thickness	Dimension	Thickness	Dimension	Thickness
11'-0"	5 1/2"	15'-0"	7 1/2"	19'-0"	9 1/2"
11'-2"	5 9/16"	15'-2"	7 9/16"	19'-2"	9 9/16"
11'-4"	5 11/16"	15'-4"	7 11/16"	19'-4"	9 11/16"
11'-6"	5 3/4"	15'-6"	7 3/4"	19'-6"	9 3/4"
11'-8"	5 13/16"	15'-8"	7 13/16"	19'-8"	9 13/16"
11'-10"	5 15/16"	15'-10"	7 15/16"	19'-10"	9 15/16"
12'-0"	6"	16'-0"	8"	20'-0"	10"
12'-2"	6 1/16"	16'-2"	8 1/16"	20'-2"	10 1/16"
12'-4"	6 3/16"	16'-4"	8 3/16"	20'-4"	10 3/16"
12'-6"	6 1/4"	16'-6"	8 1/4"	20'-6"	10 1/4"
12'-8"	6 5/16"	16'-8"	8 5/16"	20'-8"	10 5/16"
12'-10"	6 7/16"	16'-10"	8 7/16"	20'-10"	10 7/16"
13'-0"	6 1/2"	17'-0"	8 1/2"	21'-0"	10 1/2"
13'-2"	6 9/16"	17'-2"	8 9/16"	21'-2"	10 9/16"
13'-4"	6 11/16"	17'-4"	8 11/16"	21'-4"	10 11/16"
13'-6"	6 3/4"	17'-6"	8 3/4"	21'-6"	10 3/4"
13'-8"	6 13/16"	17'-8"	8 13/16"	21'-8"	10 13/16"
13'-10"	6 15/16"	17'-10"	8 15/16"	21'-10"	10 15/16"
14'-0"	7"	18'-0"	9"	22'-0"	11"
14'-2"	7 1/16"	18'-2"	9 1/16"	22'-2"	11 1/16"
14'-4"	7 3/16"	18'-4"	9 3/16"	22'-4"	11 3/16"
14'-6"	7 1/4"	18'-6"	9 1/4"	22'-6"	11 1/4"
14'-8"	7 5/16"	18'-8"	9 5/16"	22'-8"	11 5/16"
14'-10"	7 7/16"	18'-10"	9 7/16"	22'-10"	11 7/16"

1909 Structural Plain Concrete

1909.6.3 Openings in walls

- This section addresses structural plain concrete walls.
- Reinforcing is required around openings as shown in the detail below:

Note: Chapter 22 of ACI 318, "Building Code Requirements for Structural Concrete," governs plain concrete walls as indicated above.

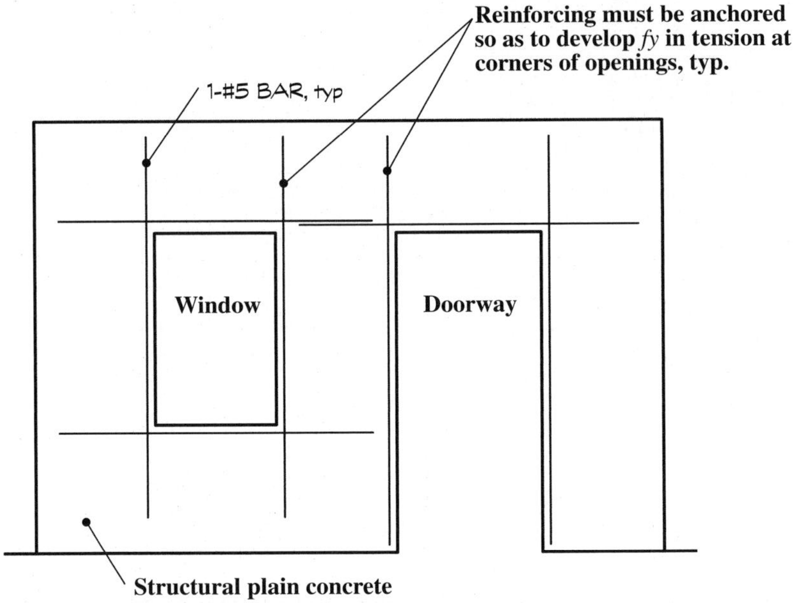

Fig. 1909.6.3. Reinforcing required around openings in structural plain concrete walls.

1910 Minimum Slab Provisions

1910.1 General

- Requirements for concrete floor slabs-on-grade are shown in the detail provided in this section.
- A vapor retarder is not required under a slab in the following locations:
  - In detached buildings which are accessory to Occupancy R-3 such as follows:
    Garages.
    Utility buildings.
    Other unheated facilities.
  - In unheated storage rooms as follows:
    < 70 sf in area.
  - In unheated carports attached to Occupancy R-3 buildings.
  - In buildings of occupancies other than R-3 as follows:
    Where moisture migrating to the surface of the slab will not be detrimental.
  - Where the following slabs will remain unenclosed:
    Driveways.
    Walks.
    Patios.
    Other similar slabs.
  - Where local site conditions permit as follows:
    Must be approved.

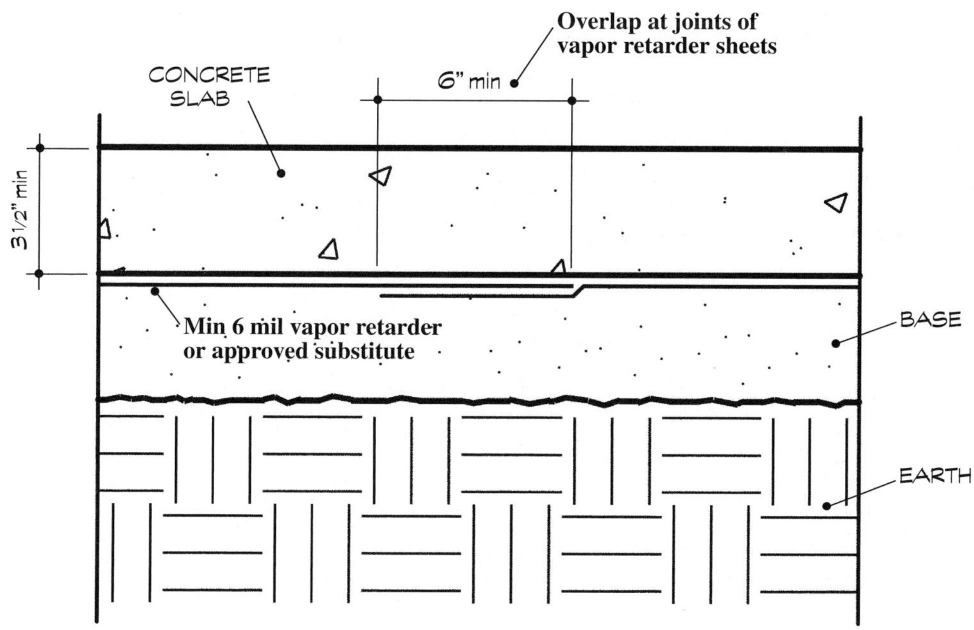

Fig. 1910.1A. Detail at concrete slab. The detail shows code requirements for concrete slabs on grade.

1910 Minimum Slab Provisions

Case study: Fig. 1910.1B. The concrete slab is 4" thick, which is > than the 31/2" required minimum. The 8 mil vapor retarder is thicker than the minimum 6 mils required. The concrete slab is in compliance with code requirements.

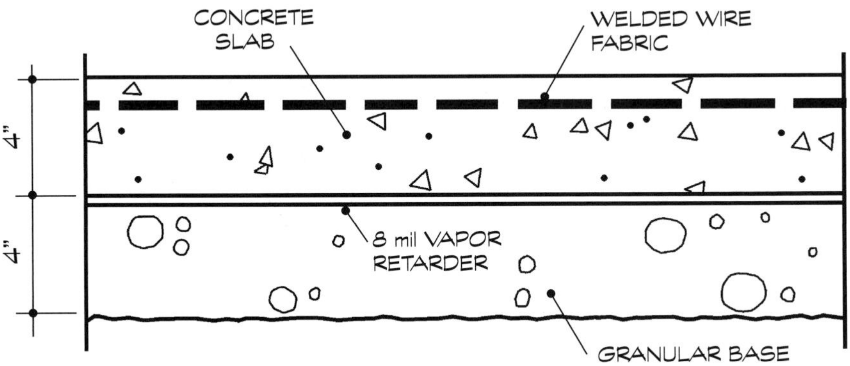

Fig. 1910.1B. Detail at concrete slab. Multipurpose Building Addition to Children's Home. Wilkes-Barre, Pennsylvania. C. Allen Mullins, Architect. Bear Creek, Pennsylvania.

1913 Shotcrete

1913.4.1 Size

- Required reinforcing for shotcrete construction is one of the following:
  - ≤ #5 bars.
  - Any size for which preconstruction tests verify that adequate encasement will be provided.

1913.4.2 Clearance

- Required clearances between reinforcing bars are shown in the details provided in this section as follows:
  - Required clearances may be reduced where both of the following apply:
    Where preconstruction tests verify that adequate encasement will be provided.
    Where approved by the building official.

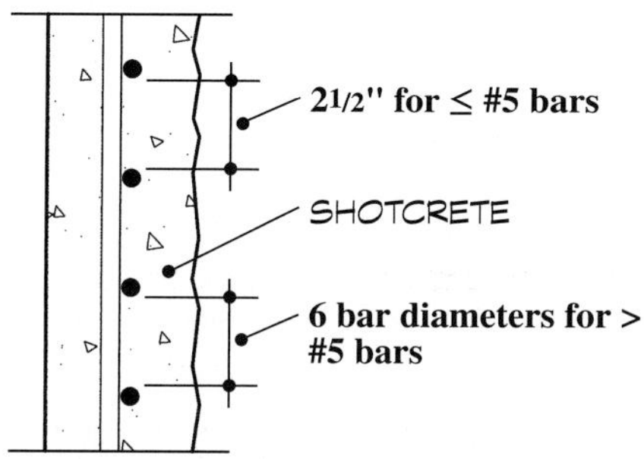

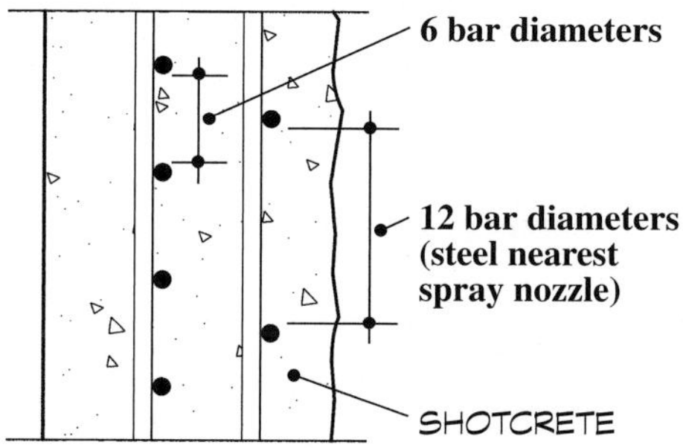

Fig. 1913.4.2. Minimum clearance required between reinforcing for shotcrete.

1914 Reinforced Gypsum Concrete

1914.2 Minimum thickness

- The thickness required for reinforced gypsum concrete is shown in the details provided in this section as follows:
  - Requirements for the reduced thickness include those indicated on the detail and the following:
    Diaphragm action must not be required of the assembly.
    The live load may not be > 40 psf.

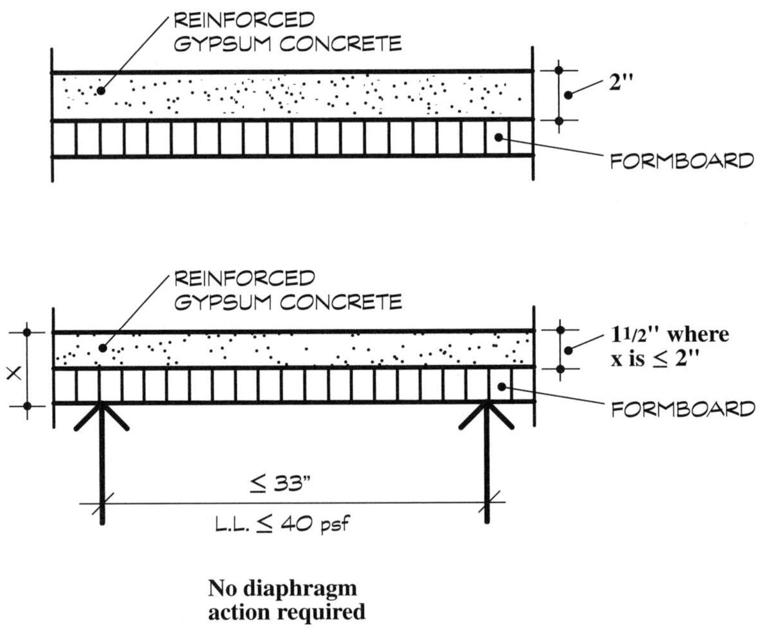

Fig. 1914.2. Minimum thickness of reinforced gypsum concrete.

1915 Concrete-Filled Pipe Columns

1915.4 Reinforcement

- Reinforcing in concrete-filled pipe columns is governed as follows:
  - Reinforcing is to be one of the following:
    Rods.
    Structural shapes.
    Pipe.
  - Structural shapes must be milled to provide bearing on the following:
    Cap plate.
    Baseplate.
  - Adequate clearance between pipe wall and reinforcing is required for composite action.
  - Minimum clearance between pipe wall and reinforcing is shown in the details provided.

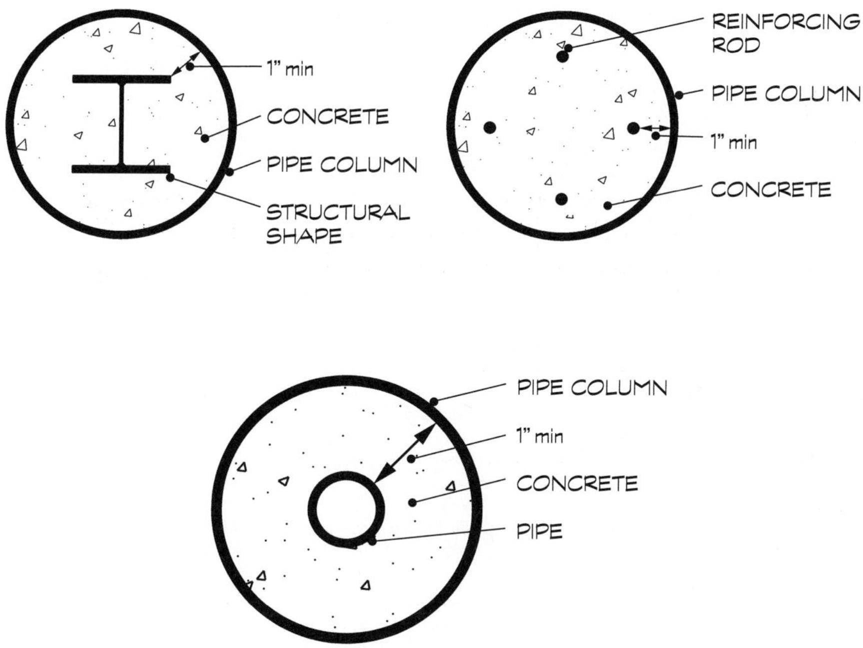

Fig. 1915.4. Clearance for reinforcement in concrete-filled pipe columns.

1915 Concrete-Filled Pipe Columns

1915.5 Fire-resistance-rating protection

- Pipe columns must have the fire-resistance rating required for the building type.

Note: IBC Table 601, "Fire-Resistance Rating Requirements for Building Elements," is cited as governing required fire-resistance ratings.

- Where a steel shell surrounds the fire protective covering, it may not be assumed to carry structural load.
- Required sizes are shown in the details provided below.

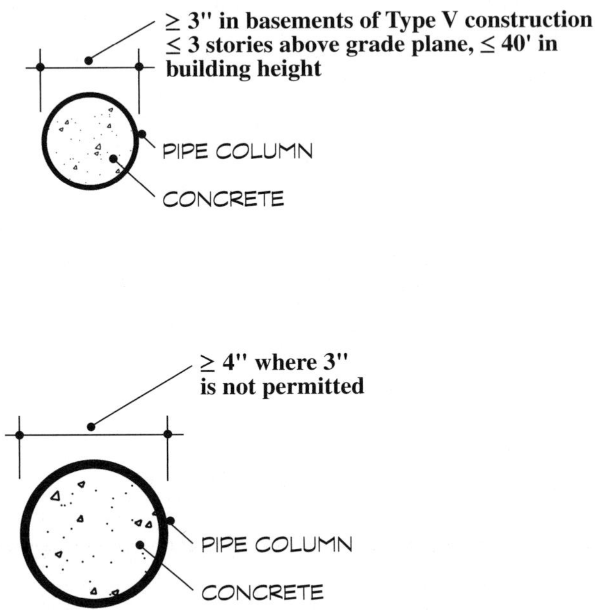

Fig. 1915.5. Minimum diameter of concrete-filled pipe columns.

20

Aluminum

Multipurpose Building Addition to Children's Home. Wilkes-Barre, Pennsylvania. *(partial elevation)*
C. Allen Mullins, Architects. Bear Creek, Pennsylvania.

2002 Materials

2002.1 General

- Design of aluminum components for structural application is governed as follows:
  - Design must comply with structural load requirements of the code.
  - Design must comply with industry standards.

Note: The following are cited as governing the design of aluminum components:
Chapter 16, "Structural Design."
AA ASM 35, "Aluminum Sheet Metal Work in Building Construction."
AA ADM 1, "Aluminum Design Manual."

21

Masonry

Country Club Park Building One. Wichita, Kansas. *(partial elevation)*
Gossen Livingston Associates, Inc., Architecture. Wichita, Kansas.

2103 Masonry Construction Materials

2103.6 Glass unit masonry

- Requirements for hollow glass blocks are as follows:
  - Reclaimed units may not be used.
  - Other requirements are shown on the detail provided below.

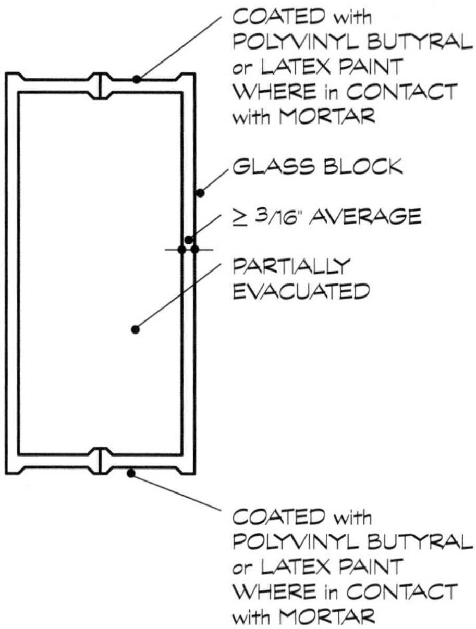

Fig. 2103.6. Section through hollow glass block.

2104 Construction

2104.1.2 Placing mortar and units *(part 1 of 2)*

- Masonry units must be placed in mortar with the following characteristics:
  - Soft.
  - Plastic.
- Where the bond of a unit is broken after it is set in place, the following is required:
  - The unit must be reset in new mortar.
- Detailing required for mortar and units is shown in Figures 2104.1.2A–C:

Note: TMS 602/ACI 530.1/ASCE 6, "Specifications for Masonry Structures," is cited as governing the placement of mortar and units in masonry construction. Requirements of 3.3 B, "Placing mortar and units," are partially summarized above and in Figures 2104.1.2A–C below and on the following pages.

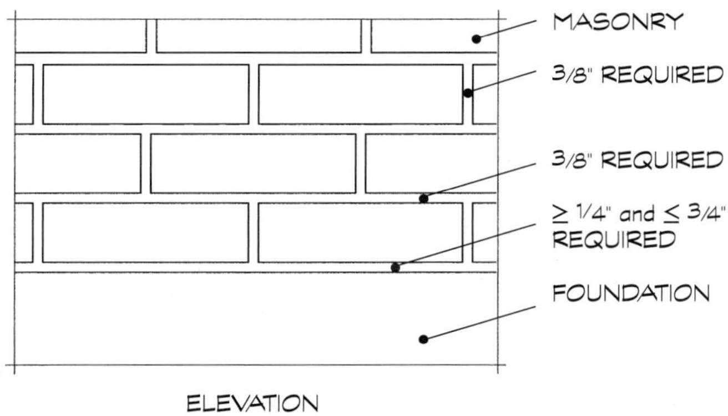

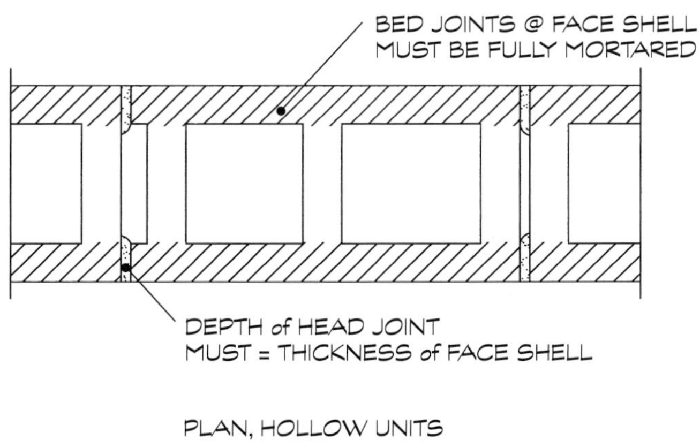

Fig. 2104.1.2A Detailing of mortar and unit placement.

2104 Construction

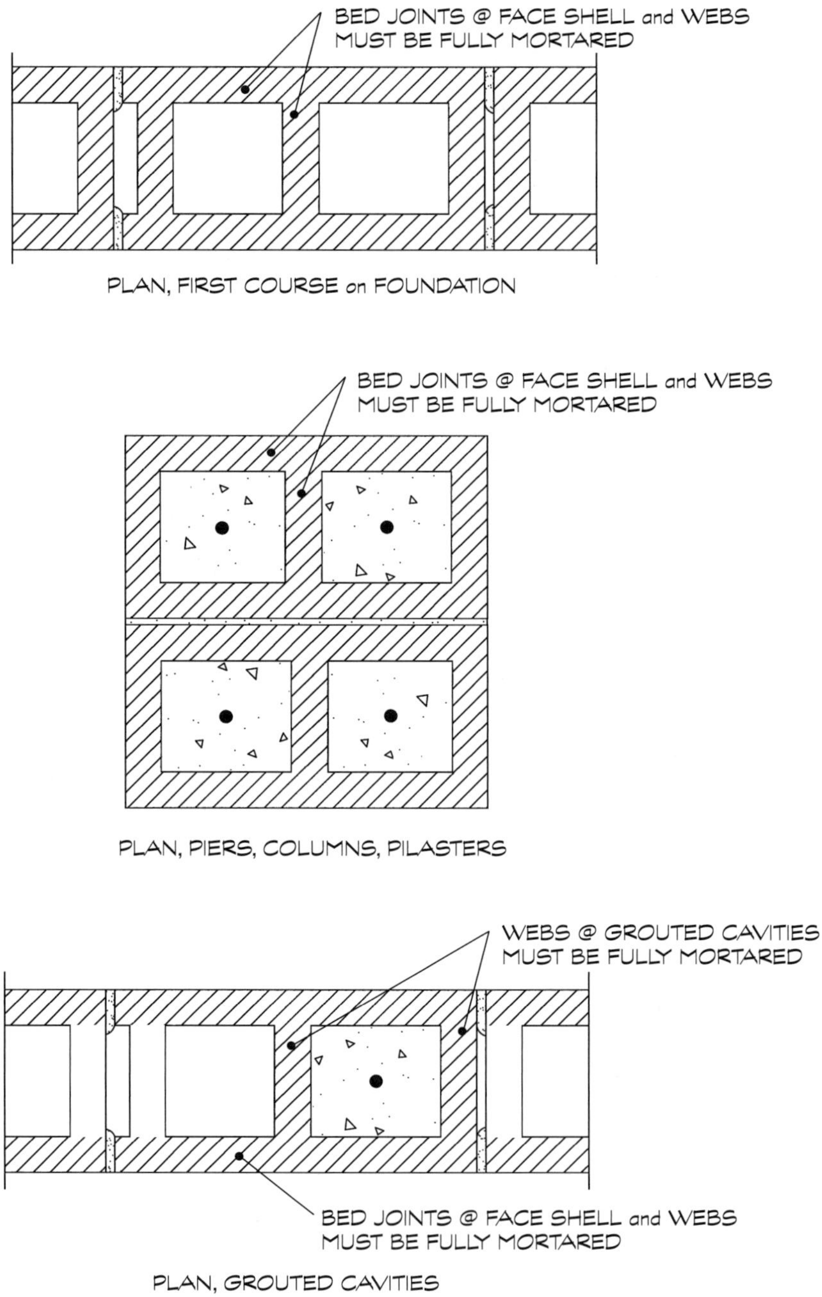

Fig. 2104.1.2B Detailing of mortar and unit placement.

2104 Construction

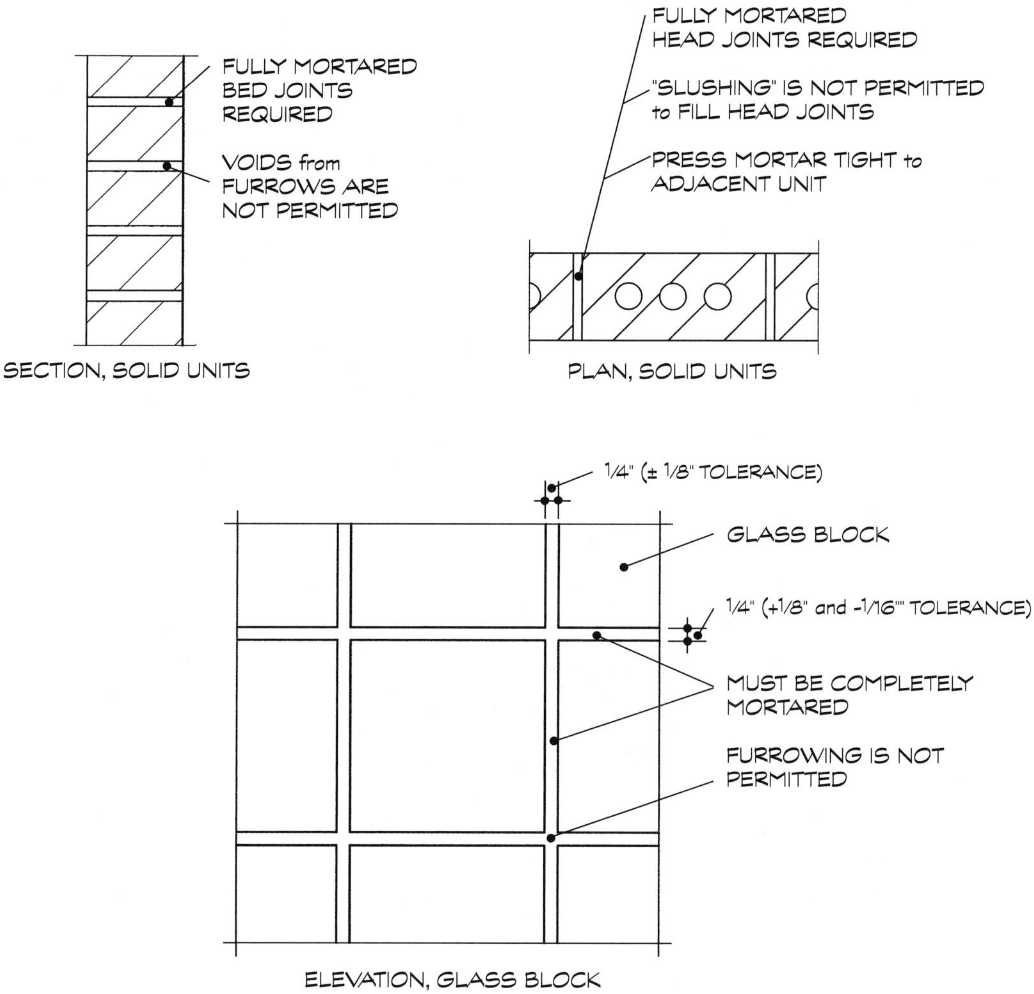

Fig. 2104.1.2C Detailing of mortar and unit placement.

2104.1.2 Placing mortar and units *(part 2 of 2)*

- A grout key is required between grout pours where the 1st lift sets prior to placing the 2nd lift as follows:
  - The top of a grout pour must stop $\geq 1\frac{1}{2}$" below a mortar joint where the grouting cavity is vertical.
  - The following applies at beams or lintels with closed bottoms:

    The grout must stop at the bottom with no grout key.

Note: TMS 602/ACI 530.1/ASCE 6, "Specifications for Masonry Structures," is cited as governing grout in masonry construction. Requirements of 3.5 F, "Grout key," are partially summarized above in Figure 2104.1.2C.

2104 Construction

2104.1.3 Installation of wall ties

- Requirements for embedding wall ties in masonry are as follows:
  - Wall ties must be embedded in the mortar bed.
  - Other requirements are shown in the details provided in Figures 2104.1.3A–B below and on the following page.

Note: TMS 602/ACI 530.1/ASCE 6, "Specifications for Masonry Structures," is cited as governing the installation of wall ties. Requirements of TMS 3.4 C, "Wall ties," are partially summarized above, and in Figures 2104.1.3A–B provided below and on the following page.

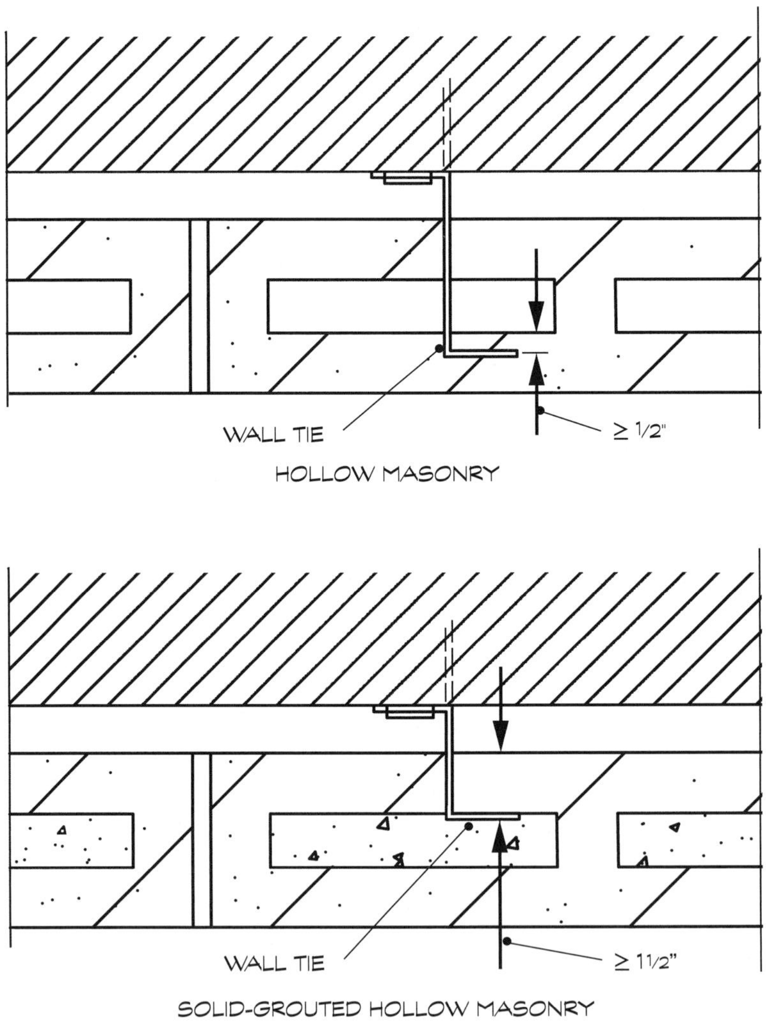

Fig. 2104.1.3A. Requirements for embedment of wall ties in masonry.

2104 Construction

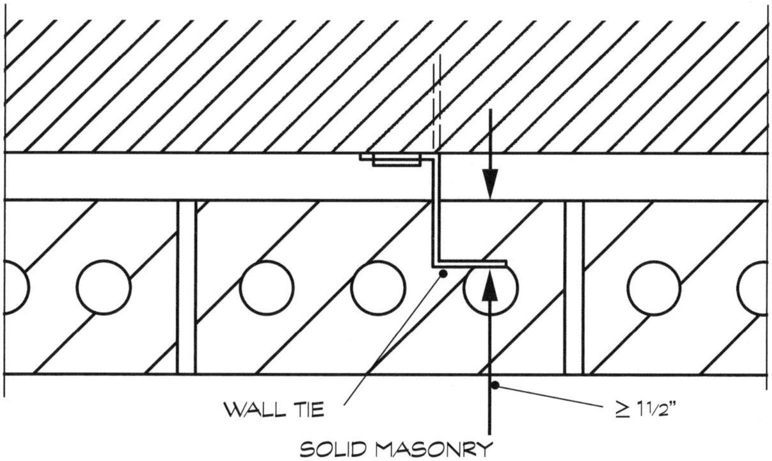

Fig. 2104.1.3B. Requirements for embedment of wall ties in masonry.

2104.1.4 Chases and recesses

- Where the following occur in masonry construction, they are to be constructed as masonry units are set:
  - Chases.
  - Recesses.
- A lintel is required for masonry above chases or recesses > 12" wide.

2104.1.5 Lintels

- Masonry lintels are to be designed by one of the following methods:
  - Allowable stress design.
  - Strength design.

Note: The following are cited as governing the design of masonry lintels:
Section 2107, "Allowable Stress Design."
Section 2108, "Strength Design of Masonry."

- The bearing requirement for lintels in masonry construction is shown in Figure 2104.1.5.

Note: TMS 402/ACI 530/ASCE 5, "Building Code Requirements for Masonry Structures," Chapter 2, "Allowable Stress Design in Masonry," is cited in Section 2107, "Allowable Stress Design," as governing allowable stress design. TMS Section 2.3.3.3, "Beams," specifies the length of bearing required for beams, which is shown in Figure 2104.1.5 on the following page.

2104.1.6 Support on wood

- Masonry may be supported on a wood structure only in certain cases.

Note: 2304.12, "Long-term loading," is cited as specifying conditions wherein a wood structure can support masonry.

2104 Construction

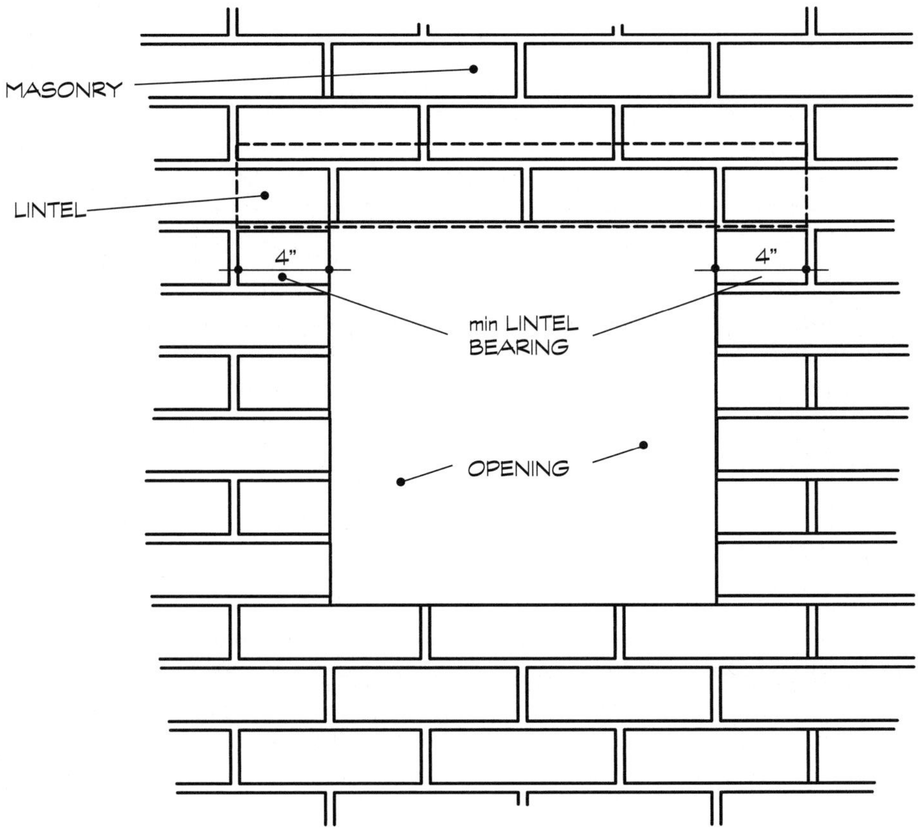

Fig. 2104.1.5. Minimum bearing length of lintels and beams for masonry.

2104 Construction

2104.2 Corbeled masonry

- Corbels are not governed by this section.

Note: TMS 402/ACI 530/ASCE 5, "Building Code Requirements for Masonry Structures," Section 1.12, "Corbels," is cited as governing corbels. Load-bearing corbels must be designed by engineering methods as specified in the reference. Nonload-bearing corbels may be designed by engineering methods or by prescriptive detailing as partially summarized in the details provided below:

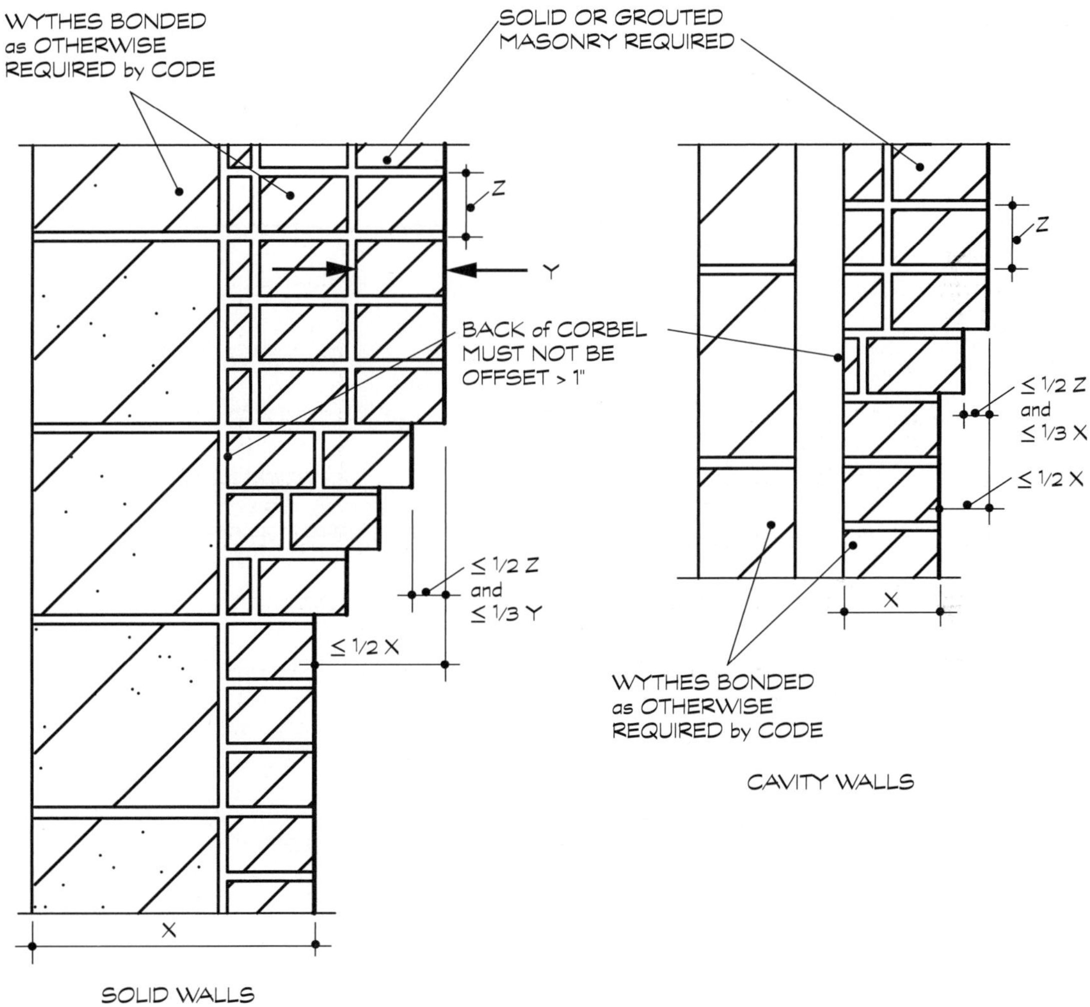

Fig. 2104.2. Projection limitations for corbeled masonry.

2106 Seismic Design

2106.1 Seismic design requirements for masonry

- This section does not address seismic design requirements for masonry.

Note: Section 1613 "Earthquake Loads," is cited as governing a structure's seismic design category which is used to determine which requirements of the governing TMS standard are applicable.

TMS 402/ACI 530/ASCE 5, "Building Code Requirements for Masonry Structures," Section 1.17, "Seismic design requirements," is cited as governing these requirements. TMS Section 17.3.1, "Nonparticipating elements," governs masonry elements that do not contribute to the seismic resisting structure. The requirements of this section are partially summarized in the detail provided below:

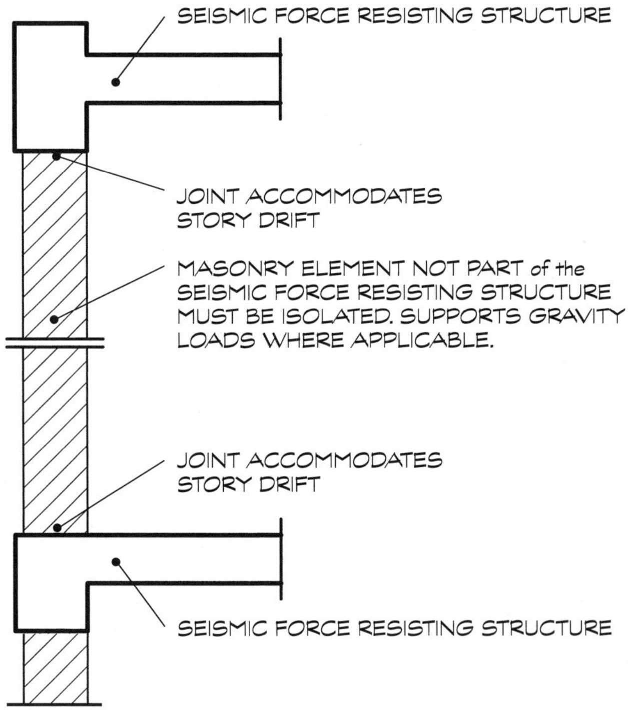

Fig. 2106.1. Isolation of masonry elements not part of the seismic force resisting structure.

2107 Allowable Stress Design

2107.1 General

- This section addresses masonry columns with all of the following characteristics:
  - Located in the following Seismic Design Categories:
    - Category A.
    - Category B.
    - Category C.
  - Supporting light-frame roofs of the following:
    - Carports.
    - Porches.
    - Sheds.
    - Similar structures.
  - Where subject to unfactored gravity loads ≤ 2000 lb as follows:
    - Acting within the perimeter of the column.
  - With a height ≤ 12'.
- Requirements are shown in the details provided below.

Note: The following are cited as governing allowable stress design for masonry structures:
Section 2106, "Seismic Design."
TMS 402/ACI 530/ASCE 5, "Building Code Requirements for Masonry Structures," Chapter 1, "General Design Requirements for Masonry," and Chapter 2, "Allowable Stress Design of Masonry," as modified by IBC sections 2107.2, "TMS 402/ACI 530/ASCE 5, Section 2.1.2, load combinations" through 2107.5, "TMS 402/ACI 530/ASCE 5, Section 2.1.9.7.1.1 lap splices." TMS 402/ACI 530/ASCE 5 Section 1.14.2, "Lightly loaded columns," is partially summarized above and in the details below.

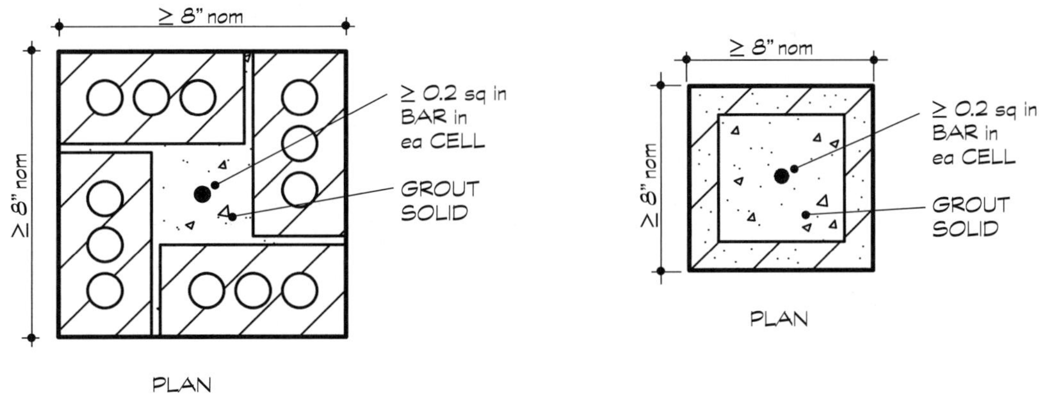

Fig. 2107.1. Masonry columns supporting light structures in Seismic Design Categories A, B, and C.

2109 Empirical Design of Masonry

2109.1 General *(part 1 of 17)*

- The length of shear walls where required is governed as follows:
  - Shear walls must be located parallel to lateral forces in ≥ 2 separate planes.
  - The sum of individual shear wall lengths does not include the following:
    A wall where length < 1/2 height.
    Openings.
  - The sum of individual shear wall lengths in any direction must total the following:
    ≥ 0.2 × the long dimension of the building, as illustrated in Figure 2109.1(1) on the next page.

Note: TMS 402/ACI 530/ASCE 5, "Building Code Requirements for Masonry Structures," Chapter 5, "Empirical Design of Masonry," is cited as governing this subject as modified by the IBC. Section 5.3.5, "Lateral stability," is partially summarized in the case study below and in Figure 2109.1(1).

Case study: Fig. 2109.1. The illustration shows a shear wall system in each direction. The minimum length in each direction where a shear wall is required, is indicated in the summation of the segments as shown below: dimension "x" indicates the long dimension of the building.

Where shear walls are required in the long dimension of the building:

B + D + F + H + J + K + L ≥ 0.2x.

Walls A, C, E, G, and I are not included because they have individual widths < 1/2 their height.

Where shear walls are required in the short direction of the building:

M + N + O + Q + R ≥ 0.2x.

Walls P and S are not included because they have individual widths < 1/2 their height.

2109 Empirical Design of Masonry

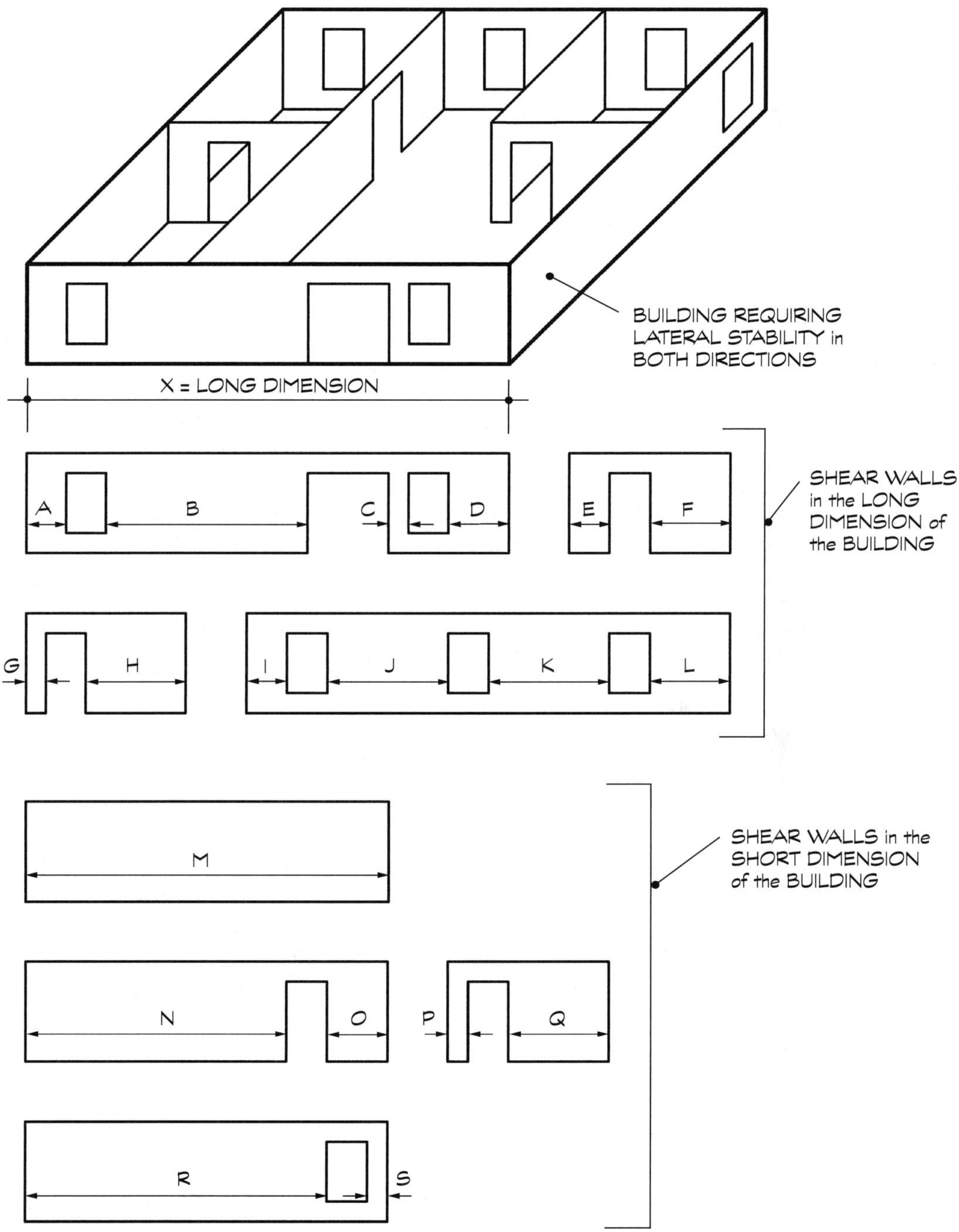

Fig. 2109.1(1). Cumulative length of shear walls.

2109 Empirical Design of Masonry

2109.1 General *(part 2 of 17)*

- Masonry shear walls are required in locations to create length-to-width ratios of diaphragms as follows:
  - As shown in the details below, for the following building elements:

    Roof diaphragms.

    Floor diaphragms.

Note: TMS 402/ACI 530/ASCE 5, "Building Code Requirements for Masonry Structures," Chapter 5, "Empirical Design of Masonry," is cited as governing this subject as modified by the IBC. TMS Section 5.3, "Lateral stability," is partially summarized in the illustration below, which is based on TMS Table 5.3.1, "Diaphragm Length-to-Width Ratios."

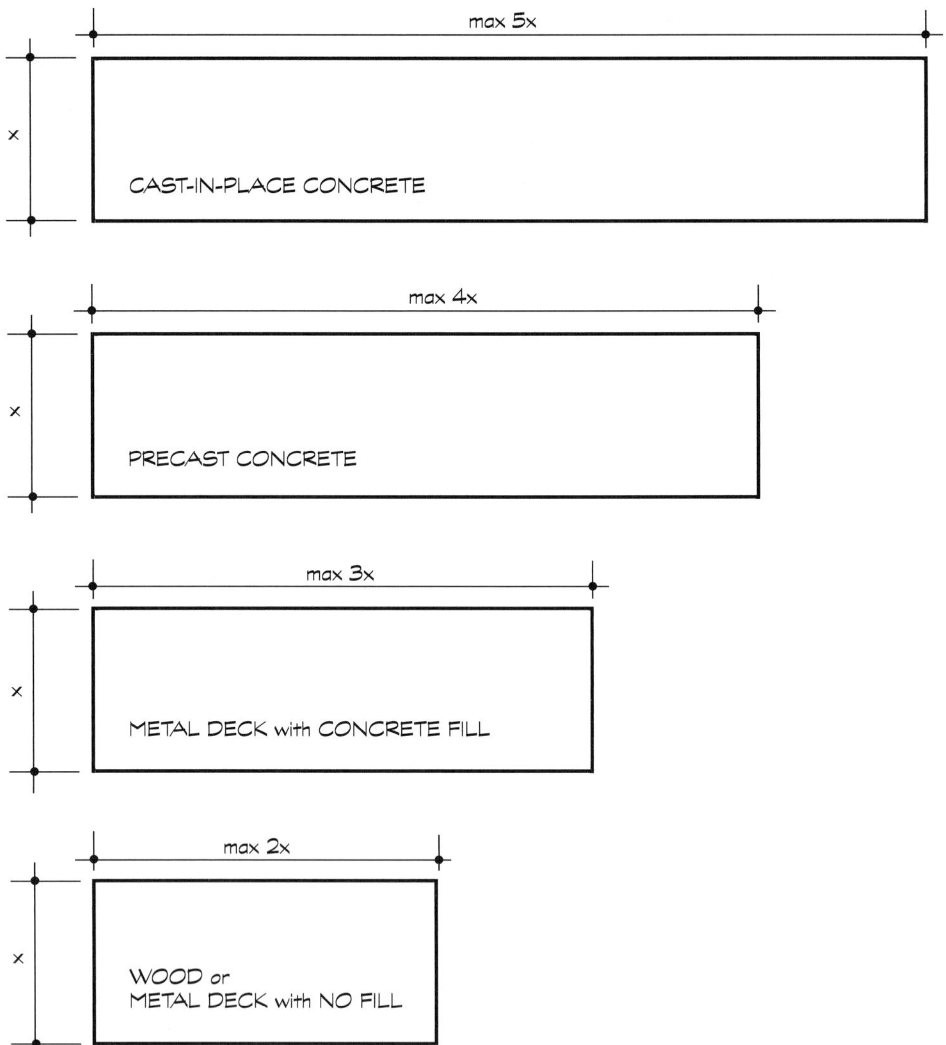

Fig. 2109.1(2). Plan views. Maximum diaphragm ratios of various floor and roof systems as defined by shear wall locations.

2109 Empirical Design of Masonry

2109.1 General *(part 3 of 17)*

- A partial list of maximum dimensions between lateral support is provided below for the following walls:
  - Load-bearing masonry walls with solid units or fully grouted and without openings as follows:

    Length must be ≤ 20 × thickness.

    Width must be ≤ 20 × thickness.

Note: TMS 402/ACI 530/ASCE 5, "Building Code Requirements for Masonry Structures," Chapter 5, "Empirical Design of Masonry," is cited as governing this subject as modified by the IBC. TMS Section 5.5, "Lateral support," is partially summarized in the Tables 2109.1a–d below and on the following page. The basis for these tables is TMS table 5.5.1, "Wall Lateral Support Requirements."

Maximum Distance between Lateral Supports for Masonry Walls

Table 2109.1a Load-bearing Walls with Solid Units or Fully Grouted without Openings

Wall thickness	Max. length Max. height	Wall thickness	Max. length Max. height	Wall thickness	Max. length Max. height
4"	6'-8"	12"	20'-0"	18"	30'-0"
6"	10'-0"	13"	21'-8"	19"	31'-8"
8"	13'-4"	14"	23'-4"	20"	33'-4"
9"	15'-0"	15"	25'-0"	21"	35'-0"
10"	16'-8"	16"	26'-8"	22"	36'-8"
11"	18'-4"	17"	28'-4"	23"	38'-4"

- A partial list of maximum dimensions between lateral support is provided below for the following walls:
  - Load-bearing walls other than solid units or fully grouted and without openings as follows:

    Length must be ≤ 18 × thickness.

    Width must be ≤ 18 × thickness.

Maximum Distance between Lateral Supports for Masonry Walls

Table 2109.1b Load-bearing Walls Other Than Solid Units or Fully Grouted without Openings

Wall thickness	Max. length Max. height	Wall thickness	Max. length Max. height	Wall thickness	Max. length Max. height
4"	6'-0"	12"	18'-0"	18"	27'-0"
6"	9'-0"	13"	19'-6"	19"	28'-6"
8"	12'-0"	14"	21'-0"	20"	30'-0"
9"	13'-6"	15"	22'-6"	21"	31'-6"
10"	15'-0"	16"	24'-0"	22"	33'-0"
11"	16'-6"	17"	25'-6"	23"	34'-6"

2109 Empirical Design of Masonry

2109.1 General *(part 4 of 17)*

- A partial list of maximum dimensions between lateral support is provided below for the following walls:
  - Exterior nonload-bearing walls:

    Length must be ≤ 18 × thickness.

    Width must be ≤ 18 × thickness.

Table 2109.1c Maximum Distance between Lateral Supports for Masonry Walls Exterior Nonload-bearing Walls without Openings

Wall thickness	Max. length Max. height	Wall thickness	Max. length Max. height	Wall thickness	Max. length Max. height
4"	6'-0"	12"	18'-0"	18"	27'-0"
6"	9'-0"	13"	19'-6"	19"	28'-6"
8"	12'-0"	14"	21'-0"	20"	30'-0"
9"	13'-6"	15"	22'-6"	21"	31'-6"
10"	15'-0"	16"	24'-0"	22"	33'-0"
11"	16'-6"	17"	25'-6"	23"	34'-6"

- A partial list of maximum dimensions between lateral support is provided below for the following walls:
  - Interior nonload-bearing walls:

    Length must be ≤ 36 × thickness.

    Width must be ≤ 36 × thickness.

Table 2109.1d Maximum Distance between Lateral Support for Masonry Walls Interior Nonload-bearing Walls without Openings

Wall thickness	Max. length Max. height	Wall thickness	Max. length Max. height	Wall thickness	Max. length Max. height
4"	12'-0"	12"	36'-0"	18"	54'-0"
6"	18'-0"	13"	39'-0"	19"	57'-0"
8"	24'-0"	14"	42'-0"	20"	60'-0"
9"	27'-0"	15"	45'-0"	21"	63'-0"
10"	30'-0"	16"	48'-0"	22"	66'-0"
11"	33'-0"	17"	51'-0"	23"	69'-0"

2109 Empirical Design of Masonry

2109.1 General *(part 5 of 17)*

- The ratios of height to thickness for cantilever walls are shown in the illustration below:

Note: TMS 402/ACI 530/ASCE 5, "Building Code Requirements for Masonry Structures," Chapter 5, "Empirical Design of Masonry," is cited as governing this subject as modified by the IBC. TMS Section 5.5.2, "Cantilever walls," governs height to thickness ratios for cantilever walls and is partially summarized in the illustration below. Thickness is nominal. Parapet walls are not included.

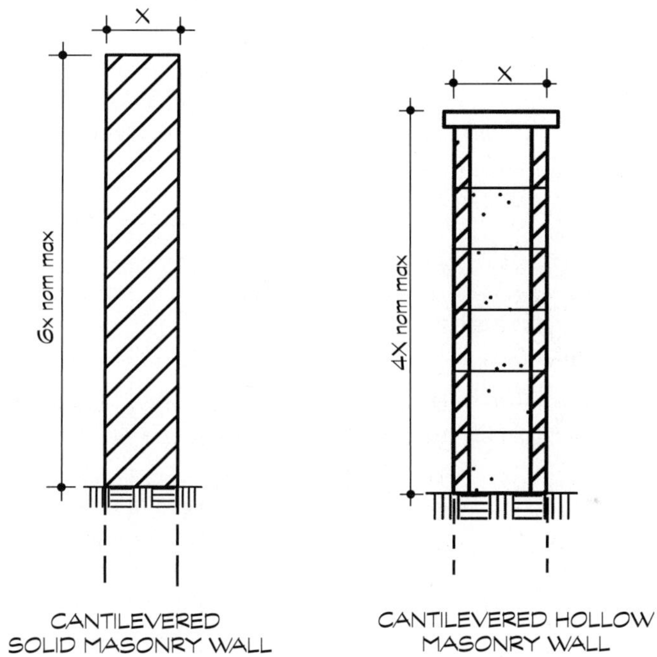

Fig. 2109.1(3). Ratio of height to thickness for cantilever masonry walls.

2109 Empirical Design of Masonry

2109.1 General *(part 6 of 17)*

- Lateral support must be provided to masonry walls as follows:
  - As listed in the details provided below.
  - Where the horizontal dimension is limited by thickness:

    Lateral support is to be provided by the following elements:

Buttresses	Pilasters
Cross walls	Structural frame members

  - Where the vertical dimension is limited by thickness:

    Lateral support is to be provided by the following elements:

    Floors acting as diaphragms.
    Roofs acting as diaphragms.
    Structural frame members.

Note: TMS 402/ACI 530/ASCE 5, "Building Code Requirements for Masonry Structures," Chapter 5, "Empirical Design of Masonry," is cited as governing this subject as modified by the IBC. TMS Section 5.5.3, "Support elements," governs these elements and is summarized in the illustration below.

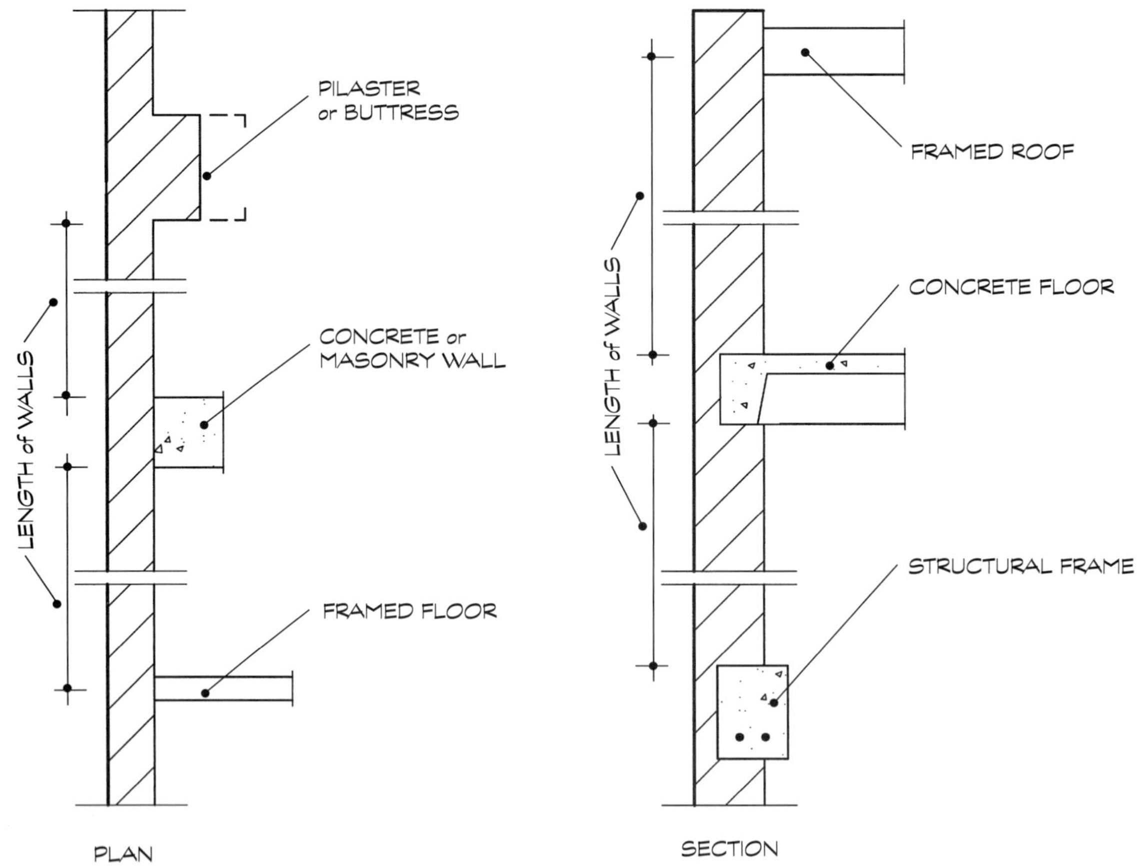

Fig. 2109.1(4). Lateral support for masonry walls.

2109 Empirical Design of Masonry

2109.1 General *(part 7 of 17)*

- The required thickness of rubble stone walls is ≥ 1'-4".
- The required thickness of masonry bearing walls is shown in the details provided below.

Note: TMS 402/ACI 530/ASCE 5, "Building Code Requirements for Masonry Structures," Chapter 5, "Empirical Design of Masonry," is cited as governing this subject as modified by the IBC. TMS Section 5.6, "Thickness of masonry," governs wall thickness.

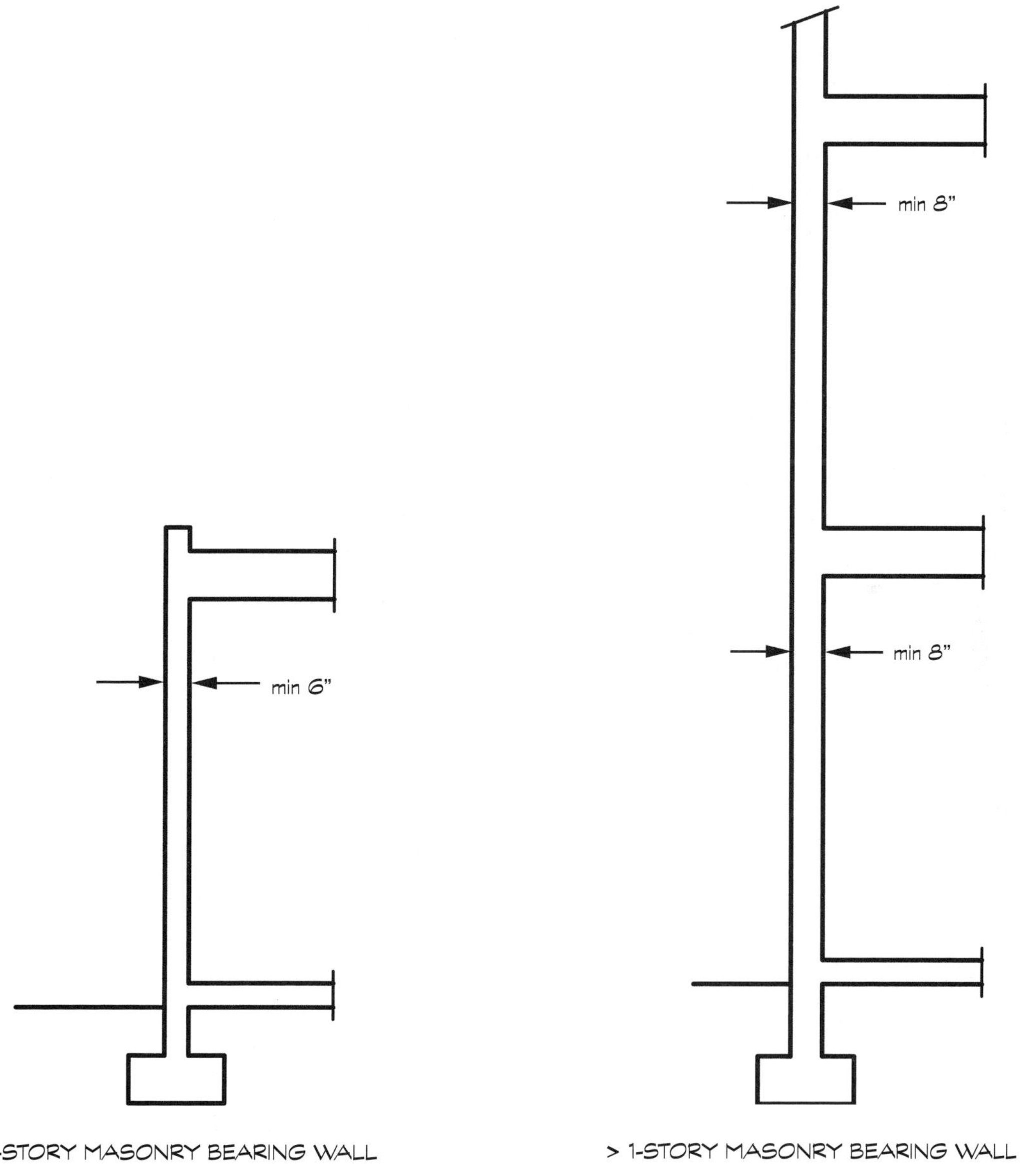

Fig. 2109.1(5). Minimum thickness of masonry bearing walls.

2109 Empirical Design of Masonry

2109.1 General *(part 8 of 17)*

- The thickness required for unreinforced masonry parapet walls is shown in details below.
- Where a higher part of a masonry wall constructed with the following units and is thinner than a lower part, the requirement listed below applies:
  - Walls:
    - Hollow units.
    - Bonded hollow walls.
  - Requirement:
    - One of the following is required at the transition of sizes to transfer loads:
      - One or more courses of solid masonry, as required.
      - Special units as shown in the details provided below.
      - Special construction.

Note: TMS 402/ACI 530/ASCE 5, "Building Code Requirements for Masonry Structures," Chapter 5, "Empirical Design of Masonry," is cited as governing this subject as modified by the IBC. TMS Section 5.6, "Thickness of masonry," governs parapet wall thickness and changes in masonry wall thickness.

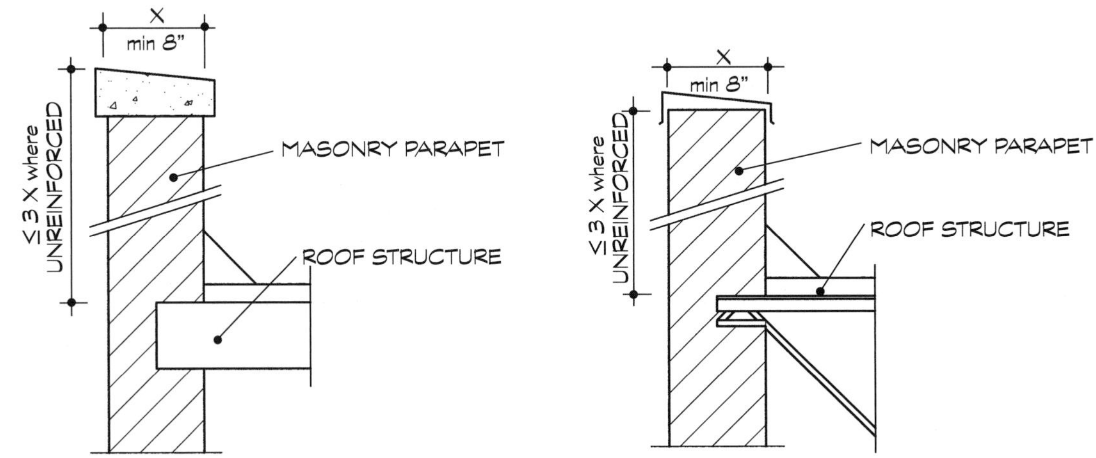

Fig. 2109.1(6). Minimum thickness required for unreinforced masonry parapet walls.

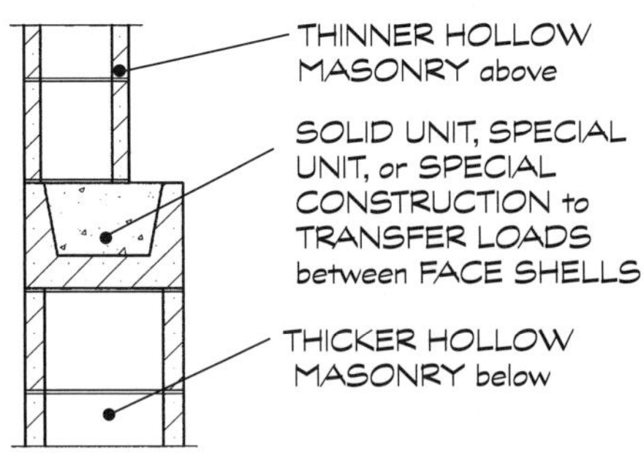

Fig. 2109.1(7). Change in masonry wall thickness at a grouted U-block.

2109 Empirical Design of Masonry

2109.1 General *(part 9 of 17)*

- Where adjacent wythes of solid units are bonded with masonry headers, the following is required:
  - Vertical and horizontal distance between headers must be ≤ 24".
  - ≥ 4% of the wall surface must be headers, examples of which are shown in the elevations provided.
  - Other requirements are shown in Figures 2109.1(8) – 2109.1(13) below and on the following pages.

Note: TMS 402/ACI 530/ASCE 5, "Building Code Requirements for Masonry Structures," Chapter 5, "Empirical Design of Masonry," is cited as governing this subject as modified by the IBC. TMS Section 5.7.2.1, "Solid units," governs bonding masonry walls with solid masonry units and is partially summarized in Figures 2109.1(8) – 2109.1(13) below and on the next pages.

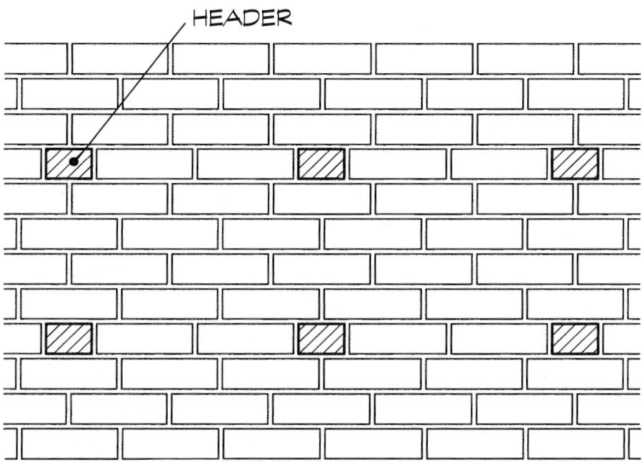

Fig. 2109.1(8). Bond pattern providing headers at the minimum 4% of wall surface.

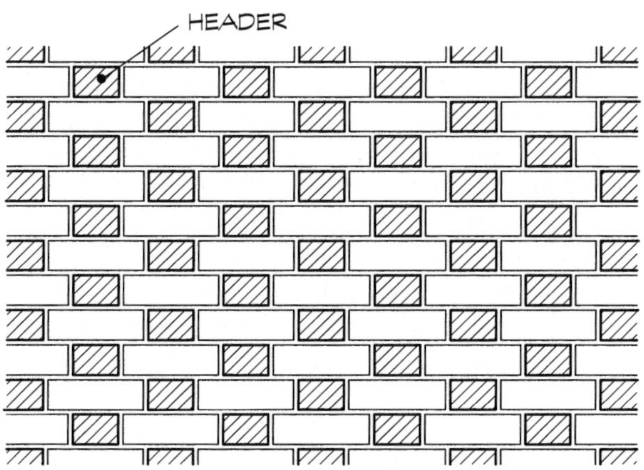

Fig. 2109.1(9). Flemish bond providing headers at 33% of the wall surface.

2109 Empirical Design of Masonry

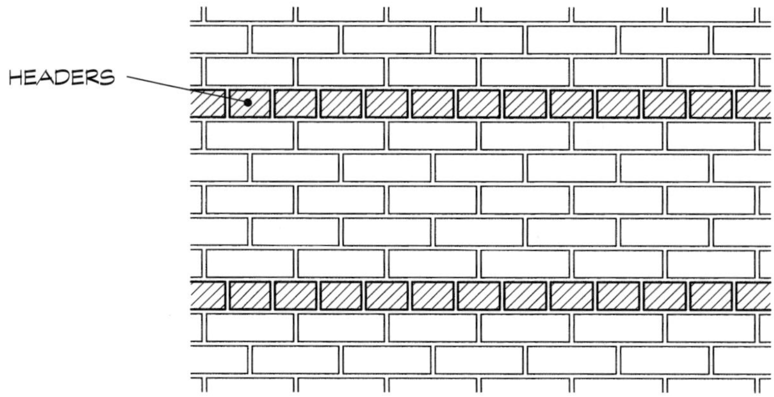

Fig. 2109.1(10). Common bond with 6th course headers providing headers at 16% of the wall surface.

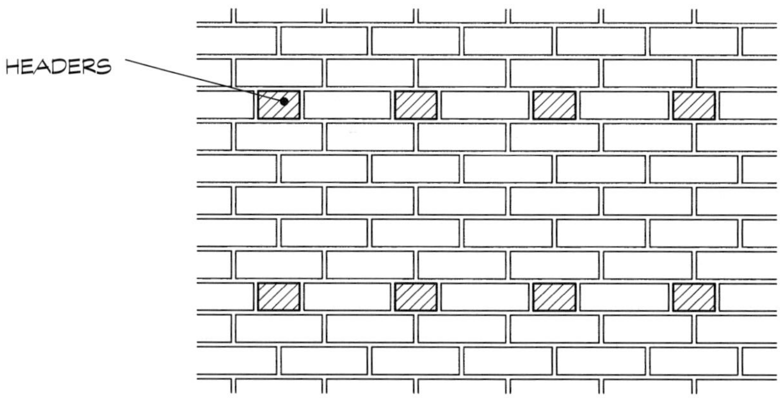

Fig. 2109.1(11). Common bond with 6th course Flemish headers providing headers @ 5% of the wall surface.

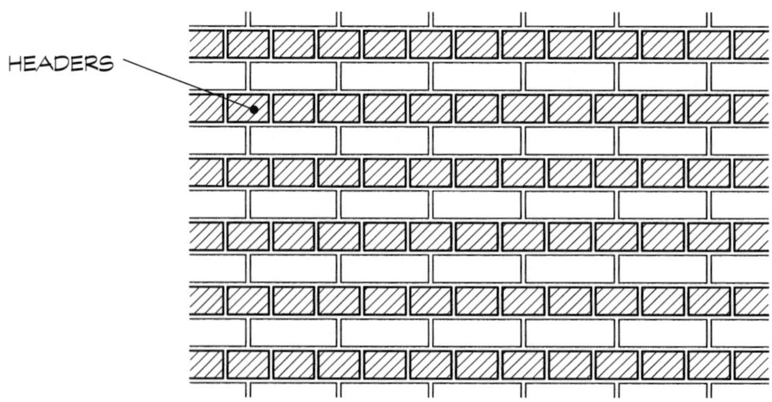

Fig. 2109.1(12). English bond providing headers at 50% of the wall surface.

2109 Empirical Design of Masonry

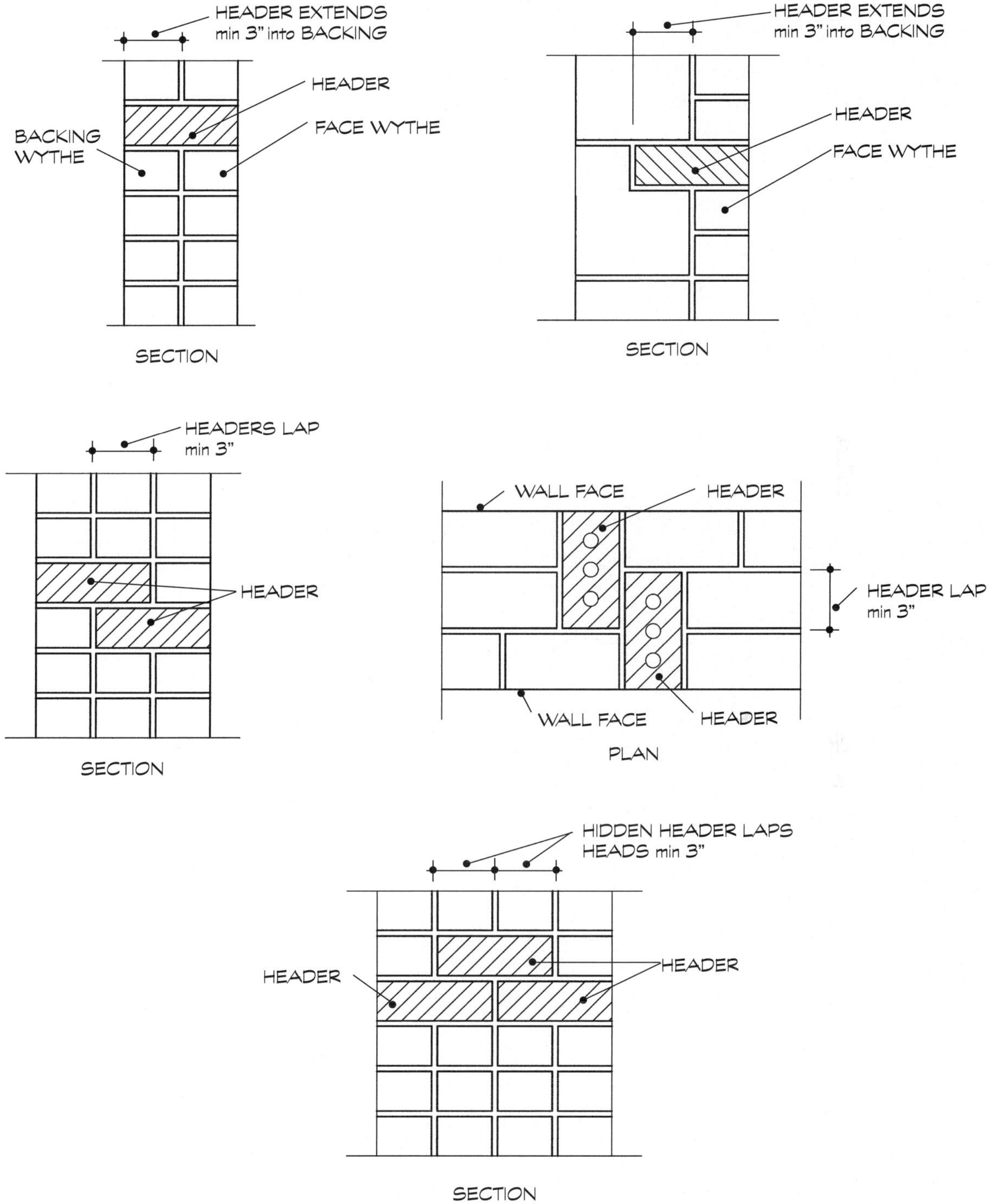

Fig. 2109.1(13). Overlap requirements for headers used to bond masonry walls.

2109 Empirical Design of Masonry

2109.1 General *(part 10 of 17)*

- Adjacent wythes of hollow masony bonded with masonry headers have requirements shown below:

Note: TMS 402/ACI 530/ASCE 5, "Building Code Requirements for Masonry Structures," Chapter 5, "Empirical Design of Masonry," is cited as governing this subject as modified by the IBC. TMS Section 5.7.2.2, "Hollow units," bonding masonry walls with hollow masonry units and is partially summarized in the illustration below.

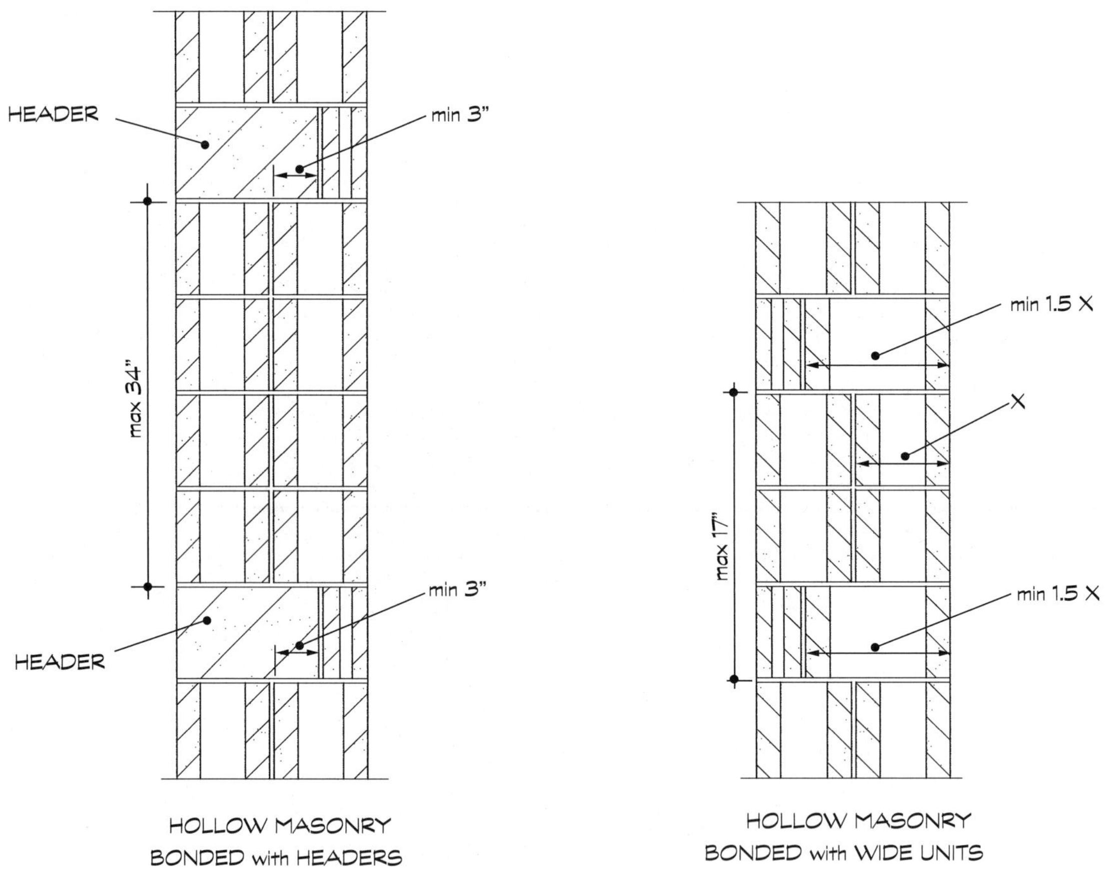

Fig. 2109.1(14). Requirements for bonding masonry walls with hollow units.

2109 Empirical Design of Masonry

2109.1 General *(part 11 of 17)*

- Where wall ties bond adjacent masonry wythes, the following applies:
  - Ties must be one of the following:
    $\geq$ W2.8 wire.
    Metal wire with stiffness $\geq$ that of W2.8 wire.
  - Other requirements are shown on the details provided in Figures 2109.1(15) – 2109.1(17) below and on the following pages.

Note: TMS 402/ACI 530/ASCE 5, "Building Code Requirements for Masonry Structures," Chapter 5, "Empirical Design of Masonry," is cited as governing this subject as modified by the IBC. TMS Section 5.7.2.2, "Hollow units," bonding masonry walls with hollow masonry units, and is partially summarized in Figures 2109.1(15) – 2109.1(17) below and on the following pages:

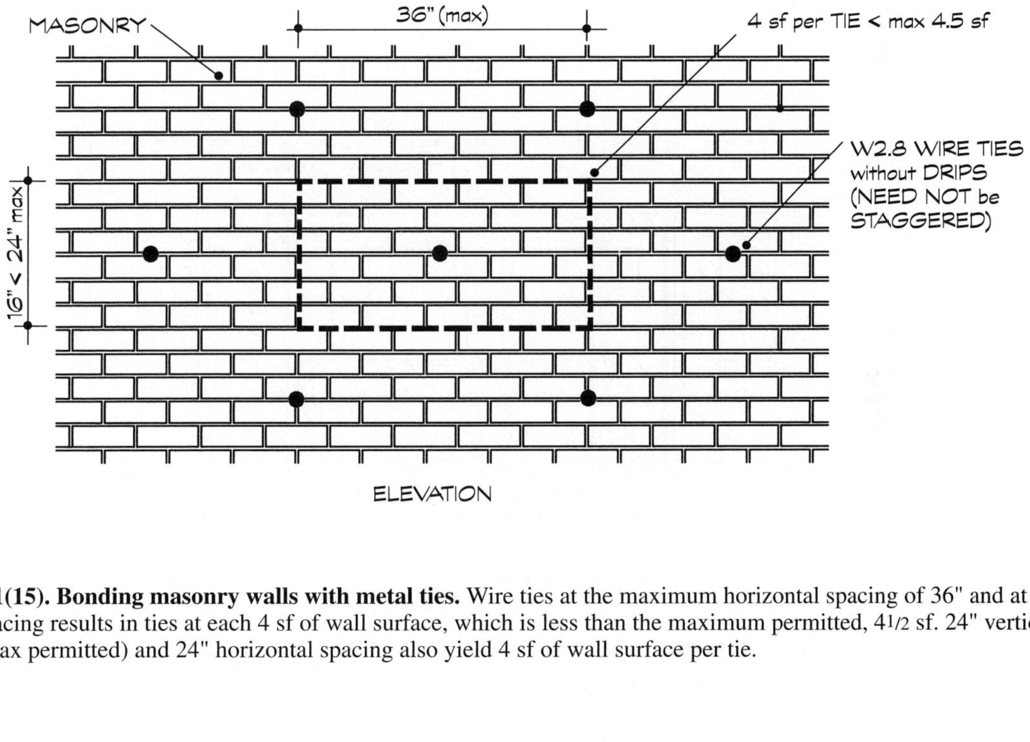

Fig. 2109.1(15). Bonding masonry walls with metal ties. Wire ties at the maximum horizontal spacing of 36" and at 16" vertical spacing results in ties at each 4 sf of wall surface, which is less than the maximum permitted, 4 1/2 sf. 24" vertical spacing (max permitted) and 24" horizontal spacing also yield 4 sf of wall surface per tie.

2109 Empirical Design of Masonry

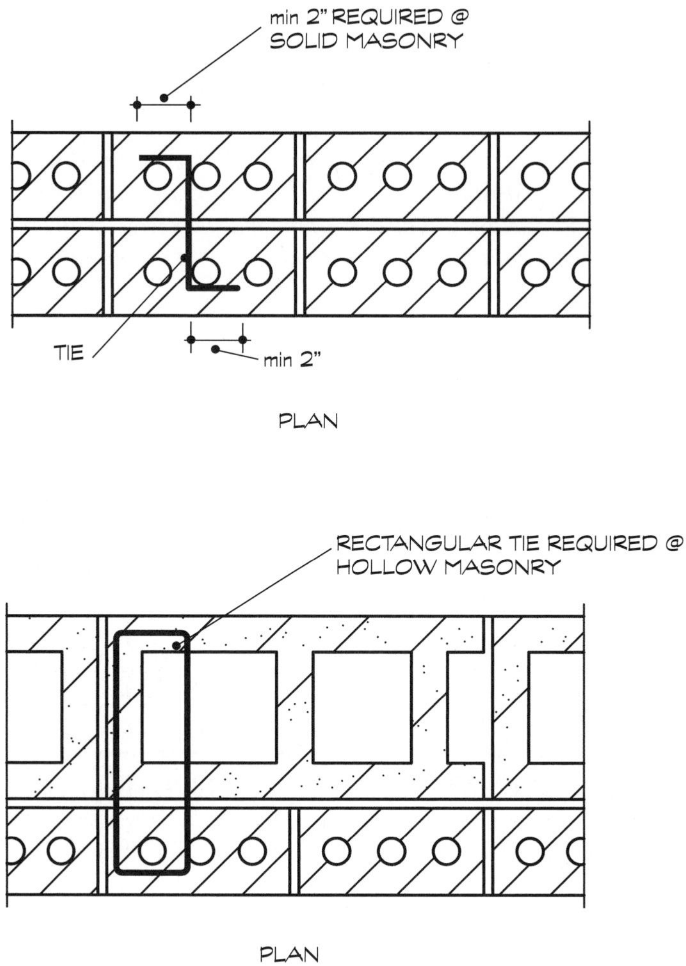

Fig. 2109.1(16). Bonding masonry walls with metal ties.

2109 Empirical Design of Masonry

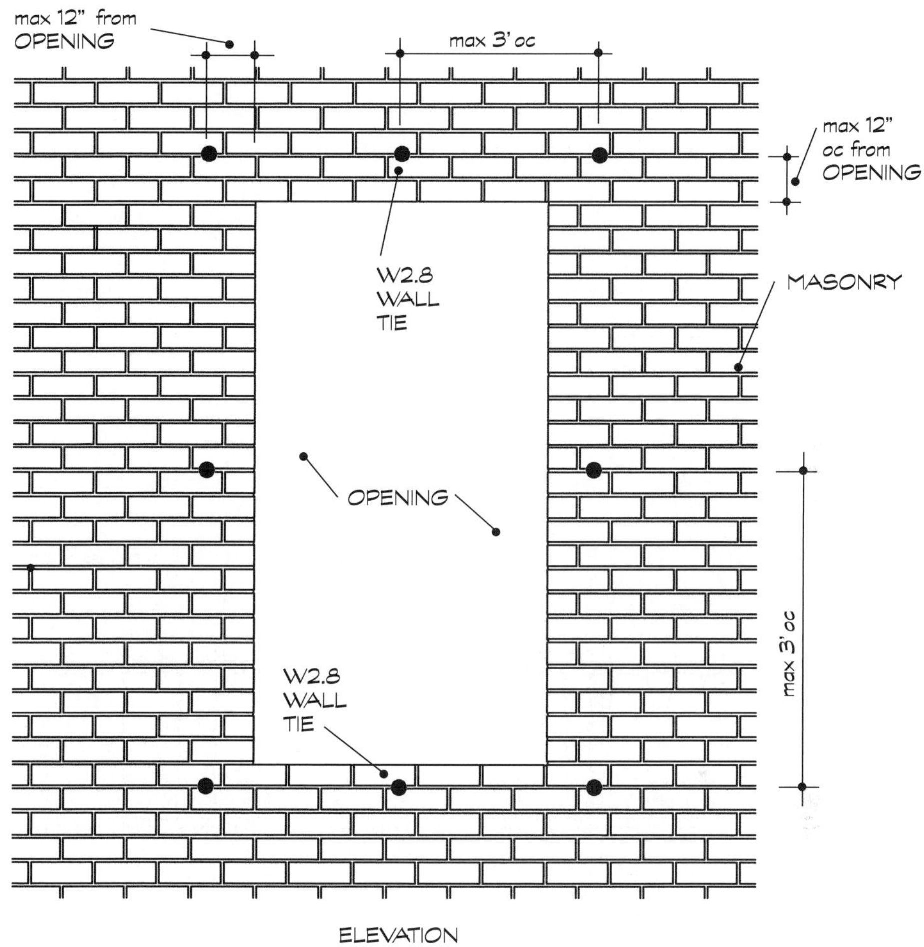

Fig. 2109.1(17). Bonding masonry walls with metal ties.

2109 Empirical Design of Masonry

2109.1 General *(part 12 of 17)*

- Requirements for masonry walls bonded with prefabricated joint reinforcement are shown in the following details:

  Note: TMS 402/ACI 530/ASCE 5, "Building Code Requirements for Masonry Structures," Chapter 5, "Empirical Design of Masonry," is cited as governing this subject as modified by the IBC. TMS Section 5.7.3, "Bonding with wall ties or joint reinforcement," governs joint reinforcement and is partially summarized in the illustrations below.

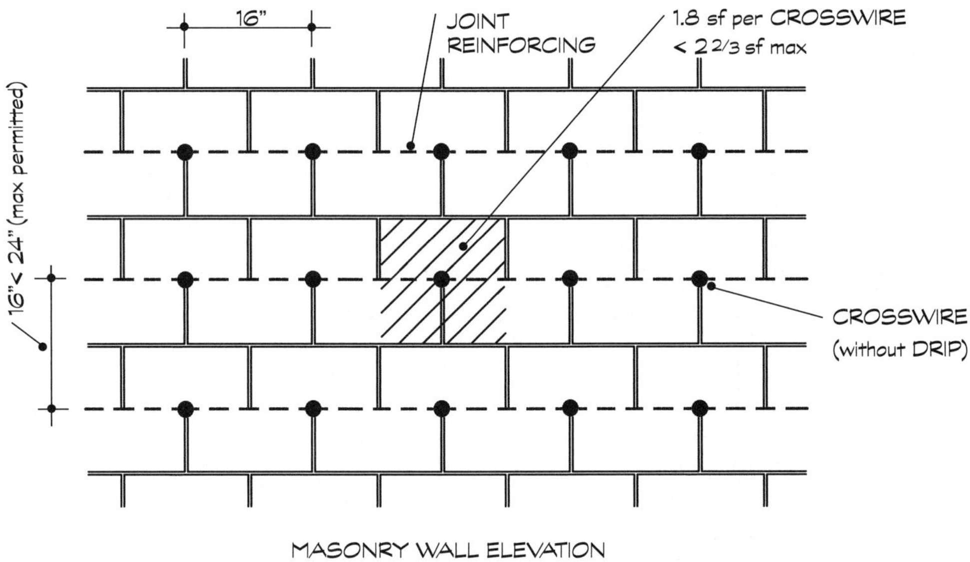

Fig. 2109.1(18). Bonding with prefabricated joint reinforcement. Cross wires @ 16" oc horizontally and joint reinforcement @ 16" oc vertically yield 1.8 sf of wall surface per cross wire. This is < 2 2/3 sf per cross wire, the maximum permitted.

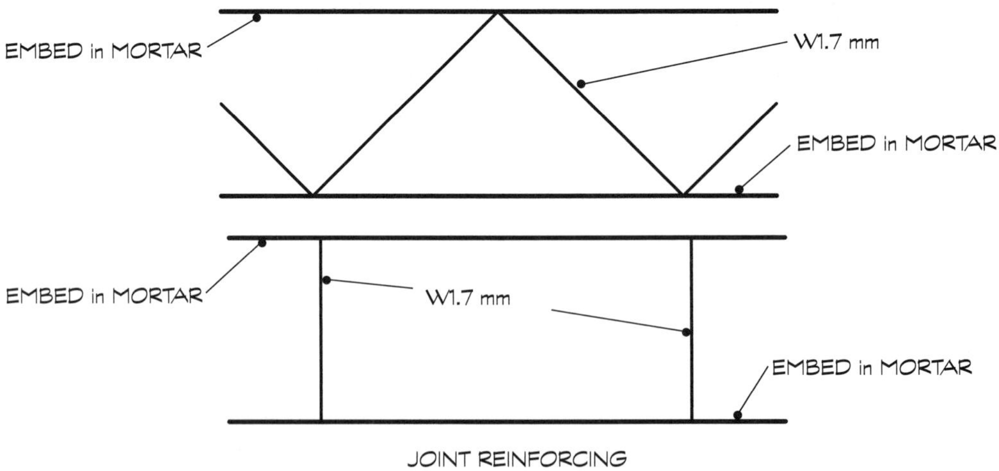

Fig. 2109.1(19). Minimum wire sizes for bonding with prefabricated joint reinforcement.

2109 Empirical Design of Masonry

2109.1 General *(part 13 of 17)*

- Requirements for bonding ashlar natural or cast stone masonry are shown in Figures 2109.1(20) and 2109.1(21) below and on the following page.

  Note: TMS 402/ACI 530/ASCE 5, "Building Code Requirements for Masonry Structures," Chapter 5, "Empirical Design of Masonry," is cited as governing this subject as modified by the IBC. TMS Section 5.7.4.1, "Ashlar masonry," governs bonding of ashlar masonry and is partially summarized in Figures 2109.1(20) and 2109.1(21) below and on the next page.

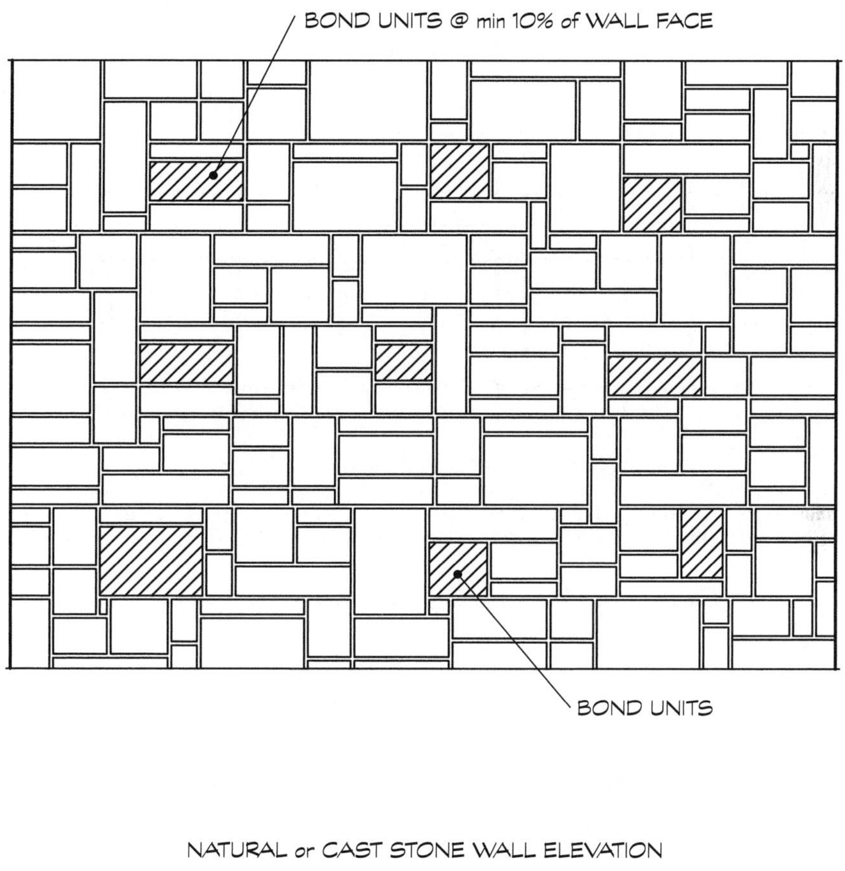

Fig. 2109.1(20). Bonding ashlar masonry with masonry units. Units used to bond ashlar masonry must = 10% minimum of the wall face and be uniformly distributed.

2109 Empirical Design of Masonry

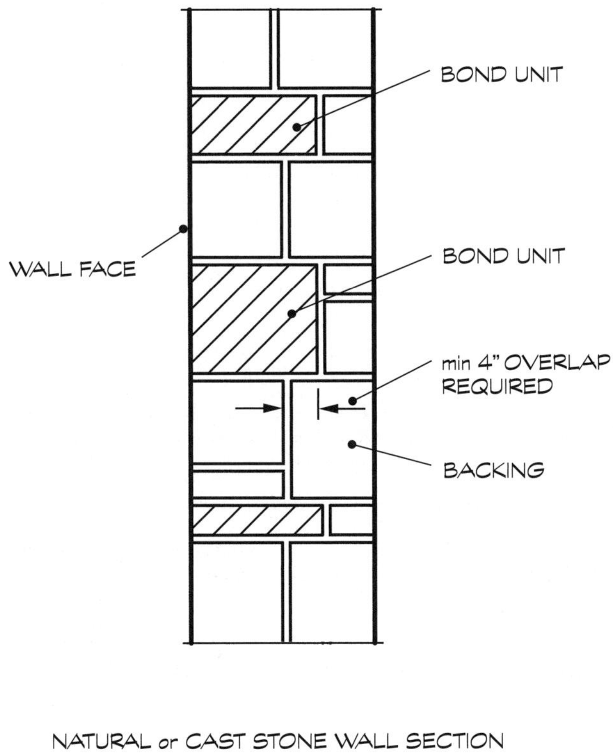

Fig. 2109.1(21). Required overlap for bonding ashlar masonry with masonry units.

2109 Empirical Design of Masonry

2109.1 General *(part 14 of 17)*

- Requirements for bonding with rubble stone masonry are shown in the details below.

Note: TMS 402/ACI 530/ASCE 5, "Building Code Requirements for Masonry Structures," Chapter 5, "Empirical Design of Masonry," is cited as governing this subject as modified by the IBC. TMS Section 5.7.4.2, "Rubble stone masonry," governs bonding of ashlar masonry and is partially summarized in the illustrations below:

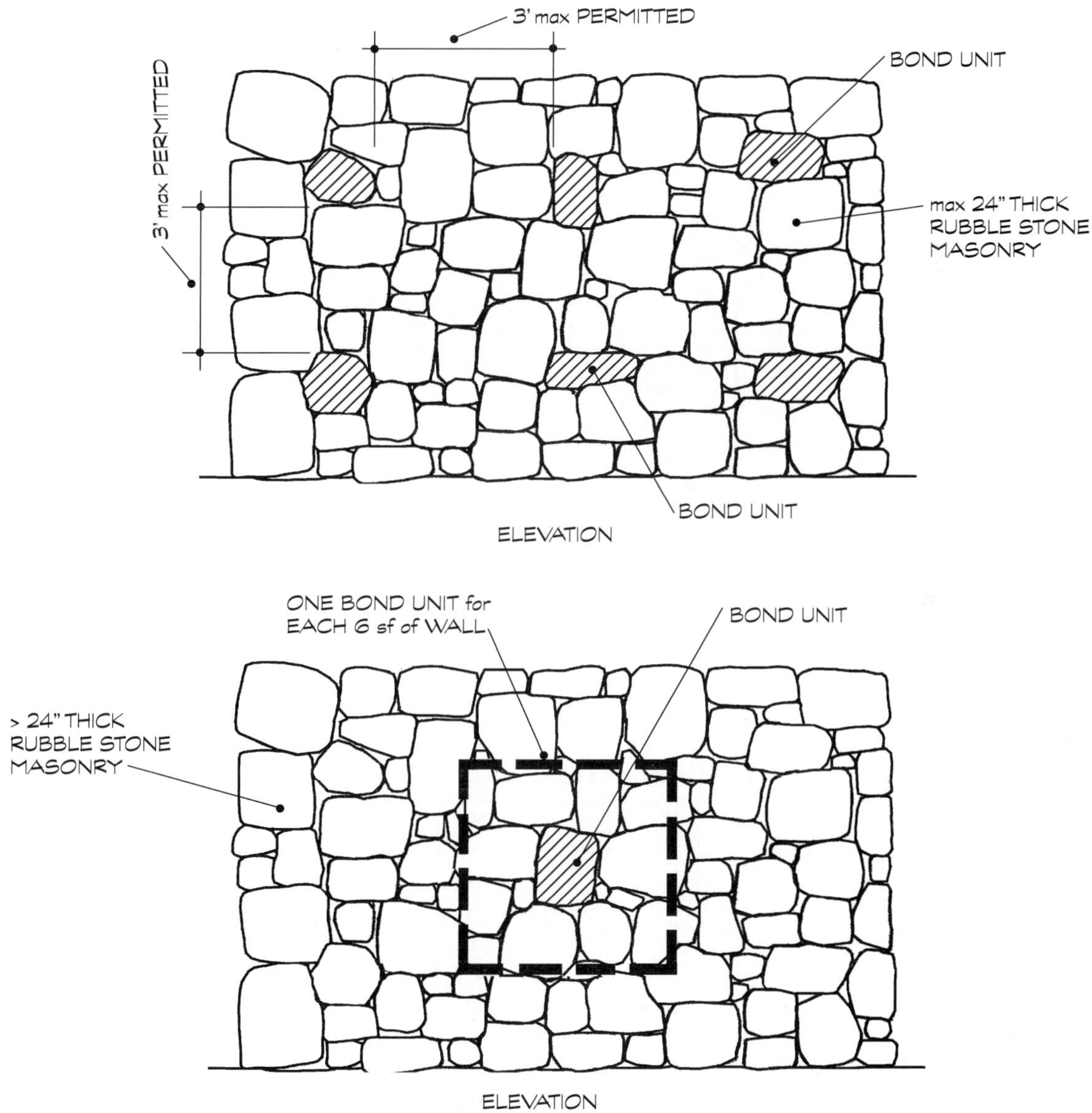

Fig. 2109.1(22). Bonding with rubble stone masonry.

2109 Empirical Design of Masonry

2109.1 General *(part 15 of 17)*

- The intersection of masonry walls that provide lateral support to each other are governed as follows:
  - Walls must be anchored to each other by one of the methods shown in Figures 2109.1(23) – 2109.1(26) below and on the following pages.

Note: TMS 402/ACI 530/ASCE 5, "Building Code Requirements for Masonry Structures," Chapter 5, "Empirical Design of Masonry," is cited as governing this subject as modified by the IBC. TMS Section 5.8.2, "Intersecting walls," governs these connections and is partially summarized in the Figures 2109.1(23) – 2109.1(26) below and on the following pages. Other anchor devices are permitted where an equivalent area of anchorage is provided as required by this section.

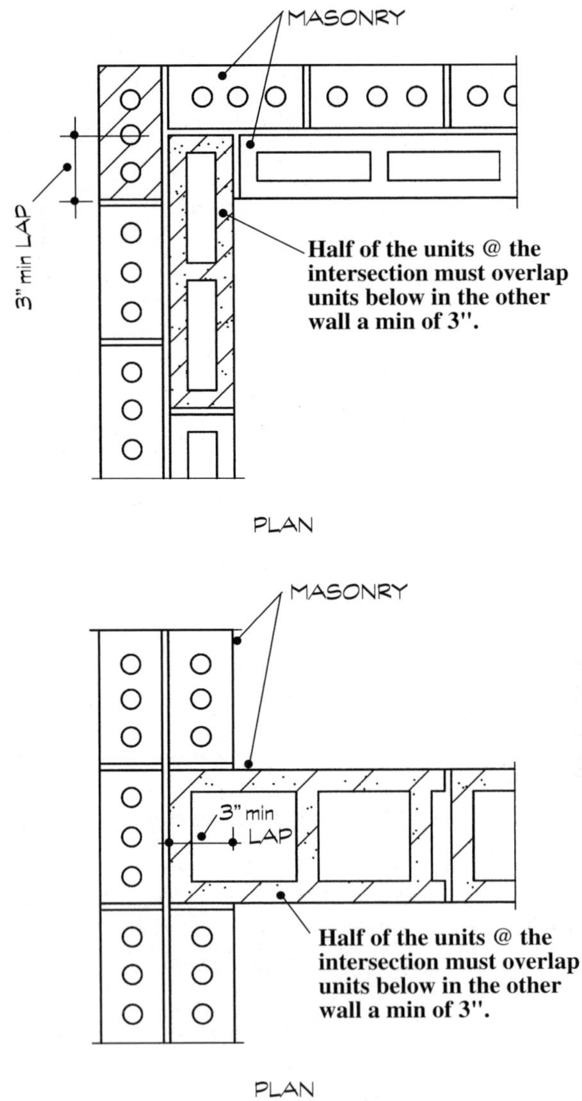

Fig. 2109.1(23). Bonding requirements at intersection of masonry walls.

2109 Empirical Design of Masonry

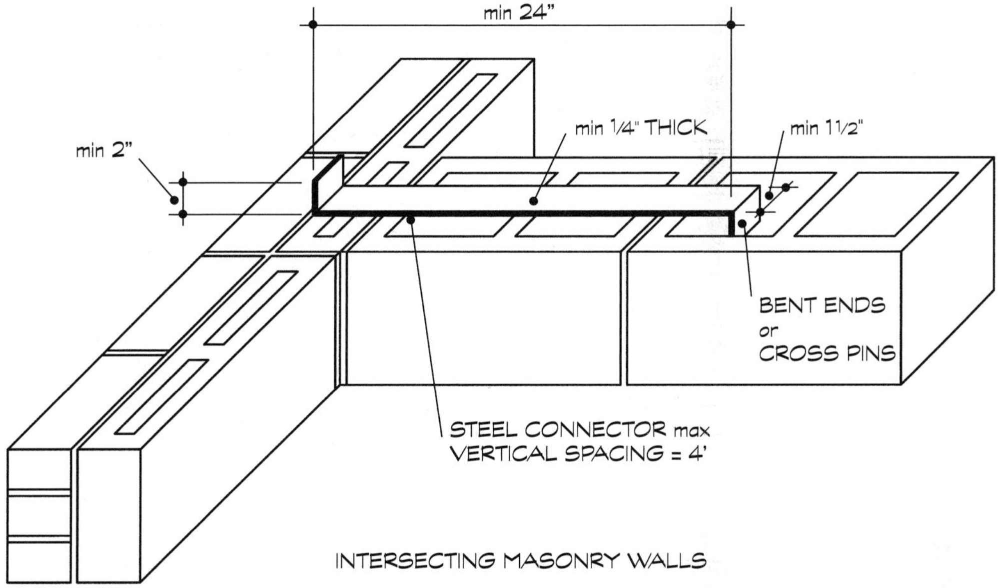

Fig. 2109.1(24). Steel connectors at intersections of masonry walls.

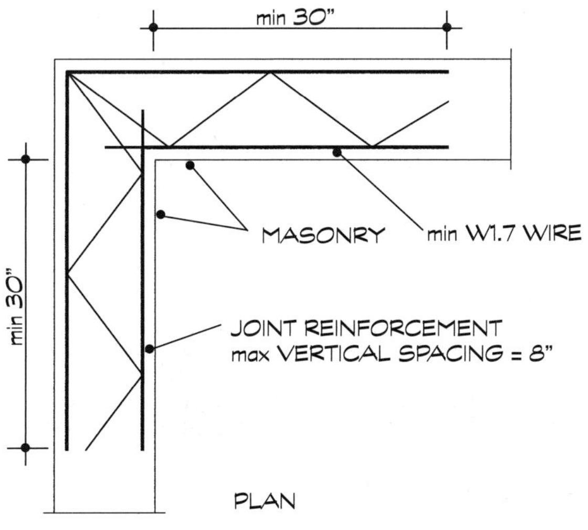

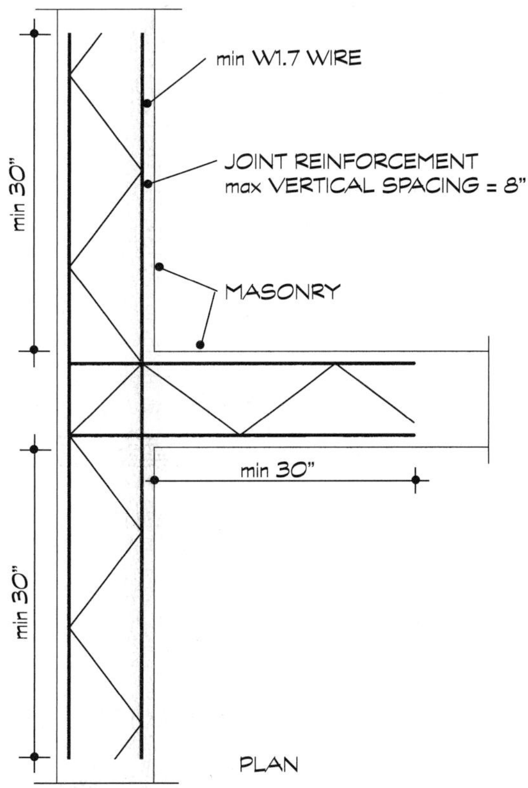

Fig. 2109.1(25). Joint reinforcement at intersection masonry walls.

2109 Empirical Design of Masonry

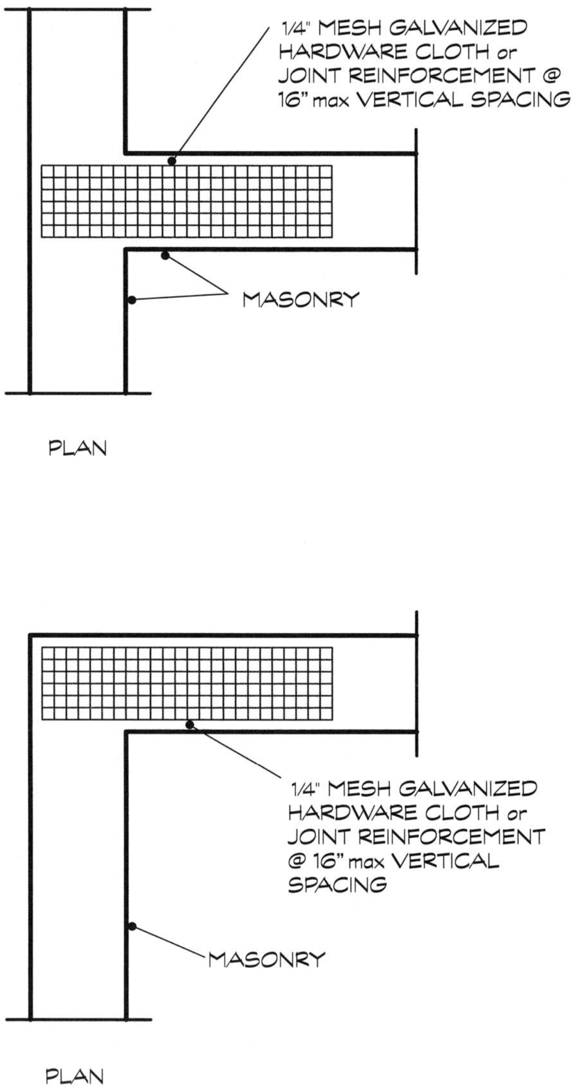

Fig. 2109.1(26). Anchorage at intersections of interior nonload-bearing walls.

2109 Empirical Design of Masonry

2109.1 General *(part 16 of 17)*

- Wood floor diaphragms providing lateral stability to masonry walls are governed as follows:
    - The floor joists must be anchored to the wall as shown in the details below.

Note: TMS 402/ACI 530/ASCE 5, "Building Code Requirements for Masonry Structures," Chapter 5, "Empirical Design of Masonry," is cited as governing this subject as modified by the IBC. TMS Section 5.8.3, "Floor and roof anchorage," governs the connections of floor and roof joists to masonry walls and is partially summarized in Figures 2109.1(27) – 2109.1(29) below and on the following page.

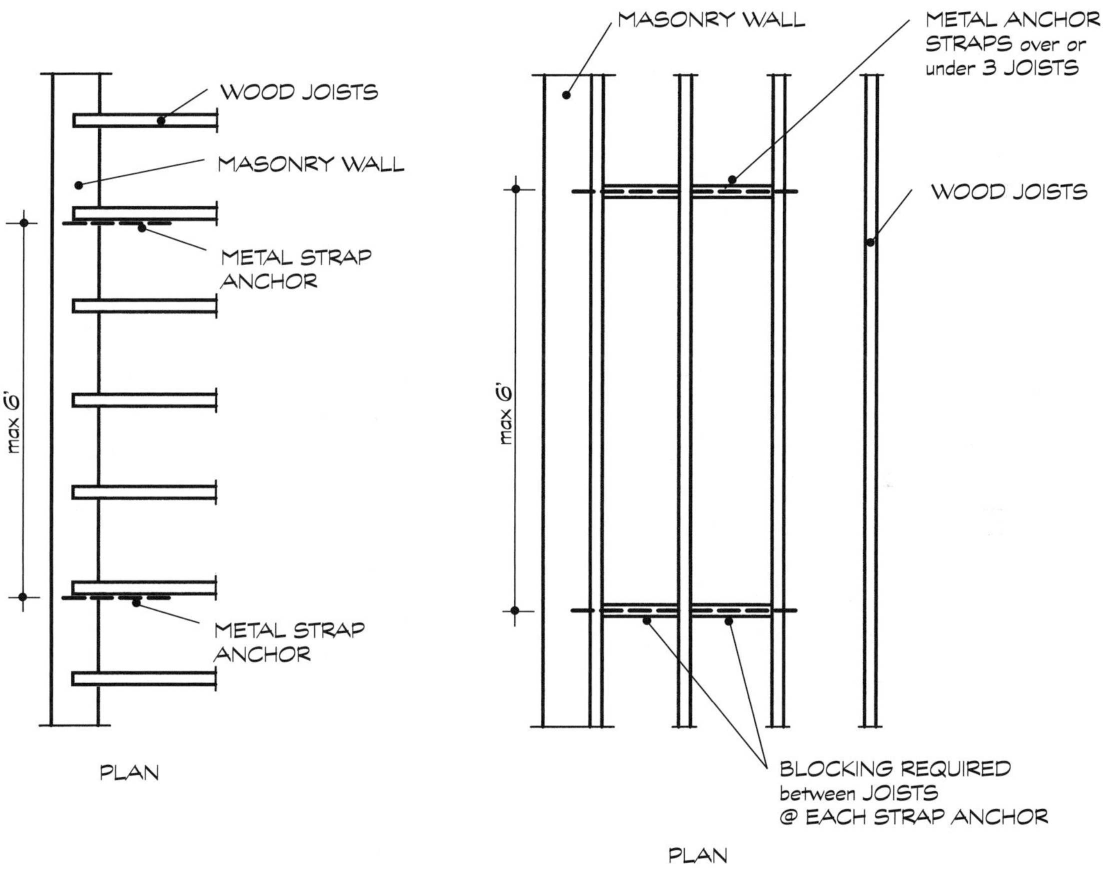

Fig. 2109.1(27). Anchorage of wood floor diaphragms providing lateral stability to masonry walls.

2109 Empirical Design of Masonry

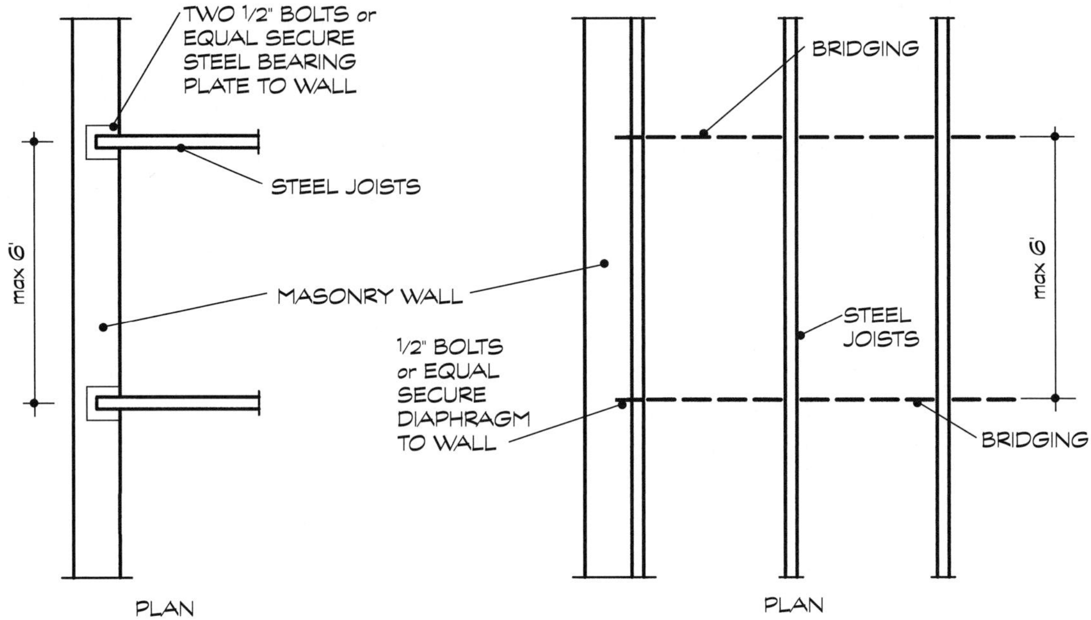

Fig. 2109.1(28). Anchorage of steel floor joist diaphragms providing lateral stability to masonry walls.

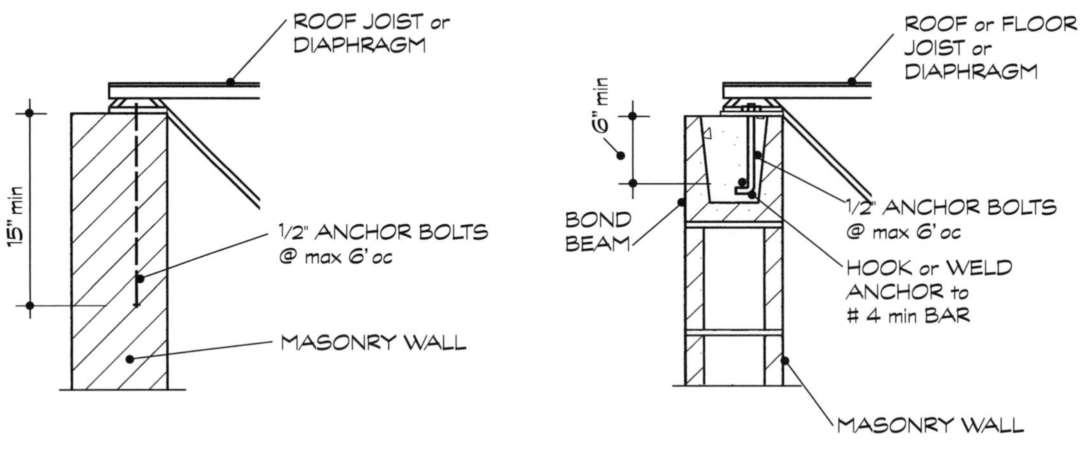

Fig. 2109.1(29). Anchorage of floor or roof diaphragms providing lateral stability to masonry walls.

2109 Empirical Design of Masonry

2109.1 General *(part 17 of 17)*

- Walls dependent on the structural frame for lateral support must be anchored to the frame as shown in the details below.

Note: TMS 402/ACI 530/ASCE 5, "Building Code Requirements for Masonry Structures," Chapter 5, "Empirical Design of Masonry," is cited as governing this subject as modified by the IBC. TMS Section 5.8.4, "Walls adjoining structural framing," governs the connections structural framing to masonry walls and is partially summarized in the illustrations below:

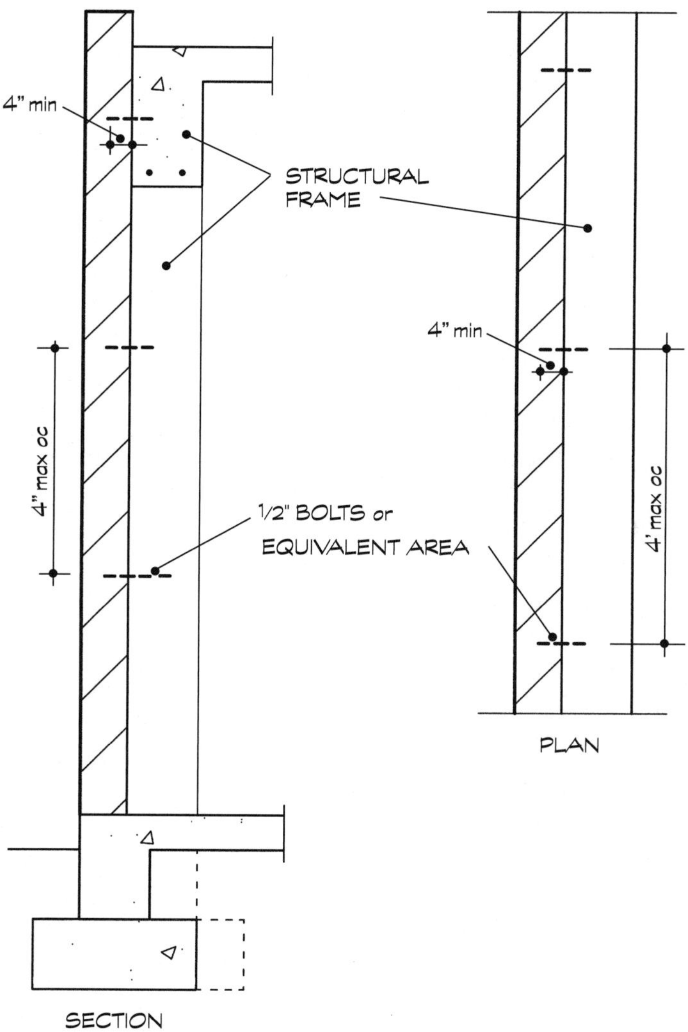

Fig. 2109.1(30). Anchorage of masonry walls adjoining structural framing.

2110 Glass Unit Masonry

2110.1 General *(part 1 of 10)*

- The thicknesses of glass masonry units are shown in the details below.

Note: TMS 402/ACI 530/ASCE 5, "Building Code Requirements for Masonry Structures," Chapter 7, "Glass Unit Masonry," is cited as governing this subject as supplemented by the IBC. TMS Section 7.1.3, "Units," governs glass masonry units and is partially summarized in the illustrations below.

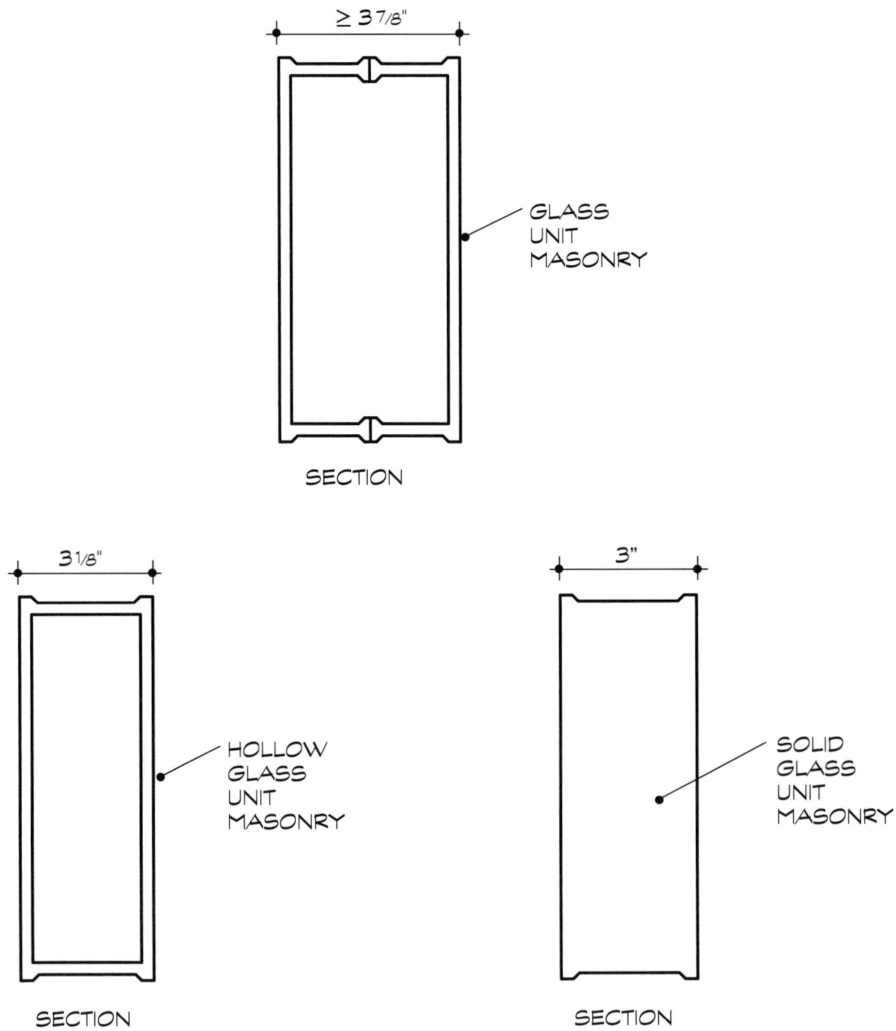

Fig. 2110.1(1). Minimum thickness of glass masonry units.

2110 Glass Unit Masonry

2110.1 General *(part 2 of 10)*

- This section addresses exterior glass block panels with standard units.
- Maximum panel height between structural supports is 20'.
- Maximum panel width between structural supports is 25'.
- Maximum panel area varies with wind load as follows:
  - Table 2110.1a provides a partial list of width and height combinations that yield an area within the maximum permitted for the wind pressure listed:
    Width and height are multiples of standard glass block dimensions.
    Either or both dimensions in any combination may be smaller than those listed below.

Note: TMS 402/ACI 530/ASCE 5, "Building Code Requirements for Masonry Structures," Chapter 7, "Glass Unit Masonry," is cited as governing this subject as supplemented by the IBC. TMS Section 7.2.1, "Exterior standard-unit panels," governs glass masonry panel size and is partially summarized in Table 2110.1a below and on the following pages:

Table 2110.1a Maximum Dimensions of Glass Block Panels

Width	Height	Width	Height	Width	Height	Width	Height
Widths and heights ≤ 57 sf as permitted by a wind pressure of 60 psf:							
24'-8"	2'-0"	18'-8"	2'-8"	12'-8"	4'-0"	7'-4"	7'-4"
24'-0"	2'-0"	18'-0"	2'-8"	12'-0"	4'-8"	6'-8"	8'-0"
23'-4"	2'-0"	17'-4"	2'-8"	11'-4"	4'-8"	6'-0"	9'-4"
22'-8"	2'-0"	16'-8"	3'-4"	10'-8"	5'-4"	5'-4"	10'-8"
22'-0"	2'-0"	16’-0”	3'-4"	10'-0"	5'-4"	4'-8"	12'-0"
21'-4"	2'-8"	15'-4"	3'-4"	9'-4"	6'-0"	4'-0"	14'-0"
20'-8"	2'-8"	14'-8"	3'-4"	8'-8"	6'-0"	3'-4"	16'-8"
20'-0"	2'-8"	14'-0"	4'-0"	8'-0"	6'-8"	2'-8"	20'-0"
19'-4"	2'-8"	13'-4"	4'-0"				
Widths and heights ≤ 62 sf as permitted by a wind pressure of 55 psf:							
24'-8"	2'-0"	18'-8"	2'-8"	12'-8"	4'-8"	7'-4"	8'-0"
24'-0"	2'-0"	18'-0"	3'-4"	12'-0"	4'-8"	6'-8"	8'-8"
23'-4"	2'-0"	17'-4"	3'-4"	11'-4"	5'-4"	6'-0"	10'-0"
22'-8"	2'-8"	16'-8"	3'-4"	10'-8"	5'-4"	5'-4"	11'-4"
22'-0"	2'-8"	16'-0"	3'-4"	10'-0"	6'-0"	4'-8"	12'-8"
21'-4"	2'-8"	15'-4"	4'-0"	9'-4"	6'-0"	4'-0"	15'-4"
20'-8"	2'-8"	14'-8"	4'-0"	8'-8"	6'-8"	3'-4"	18'-0"
20'-0"	2'-8"	14'-0"	4'-0"	8'-0"	7'-4"	2'-8"	20'-0"
19'-4"	2'-8"	13'-4"	4'-0"				

(Continued on next page)

2110 Glass Unit Masonry

2110.1 General *(part 3 of 10)*

Table 2110.1a—*Continued* **Maximum Dimensions of Glass Block Panels**

Width	Height	Width	Height	Width	Height	Width	Height
Widths and heights ≤ 70 sf as permitted by a wind pressure of 50 psf:							
24'-8"	2'-8"	18'-8"	3'-4"	13'-4"	4'-8"	8'-0"	8'-8"
24'-0"	2'-8"	18'-0"	3'-4"	12'-8"	5'-4"	7'-4"	9'-4"
23'-4"	2'-8"	17'-4"	4'-0"	12'-0"	5'-4"	6'-8"	10'-0"
22'-8"	2'-8"	16'-8"	4'-0"	11'-4"	6'-0"	6'-0"	11'-4"
22'-0"	2'-8"	16'-0"	4'-0"	10'-8"	6'-0"	5'-4"	12'-8"
21'-4"	2'-8"	15'-4"	4'-0”	10'-0"	6'-8"	4'-8"	14'-8"
20'-8"	3'-4"	14'-8"	4'-8"	9'-4"	7'-4"	4'-0"	17'-4"
20'-0"	3'-4"	14'-0"	4'-8"	8'-8"	8'-0"	3'-4"	20'-0"
19'-4"	3'-4"						
Widths and heights ≤ 77 sf as permitted by a wind pressure of 45 psf:							
24'-8"	2'-8"	18'-8"	4'-0"	13'-4"	5'-4"	8'-0"	9'-4"
24'-0"	2'-8"	18'-0"	4'-0"	12'-8"	6'-0"	7'-4"	10'-0"
23'-4"	2'-8"	17'-4"	4'-0"	12'-0"	6'-0"	6'-8"	11'-4"
22'-8"	3'-4"	16'-8"	4'-0"	11'-4"	6'-8"	6'-0"	12'-8"
22'-0"	3'-4"	16'-0"	4'-8"	10'-8"	6'-8"	5'-4"	14'-0"
21'-4"	3'-4"	15'-4"	4'-8"	10'-0"	7'-4"	4'-8"	16'-0"
20'-8"	3'-4"	14'-8"	4'-8"	9'-4"	8'-0"	4'-0"	18'-8"
20'-0"	3'-4"	14'-0"	5'-4"	8'-8"	8'-8"	3'-4"	20'-0"
19'-4"	3'-4"						
Widths and heights ≤ 84 sf as permitted by a wind pressure of 40 psf:							
24'-8"	3'-4"	19'-4"	4'-0"	14'-0"	6'-0"	8'-8"	9'- 4"
24'-0"	3'-4"	18'-8"	4'-0"	13'-4"	6'-0"	8'-0"	10'-0"
23'-4"	3'-4"	18'-0"	4'-8"	12'-8"	6'-0"	7'-4"	11'-4"
22'-8"	3'-4"	17'-4"	4'-8"	12'-0"	6'-8"	6'-8"	12'-0"
22'-0"	3'-4"	16'-8"	4'-8"	11'-4"	7'-4"	6'-0"	14'-0"
21'-4"	3'-4"	16'-0"	4'-8"	10'-8"	7'-4"	5'-4"	15'-4"
20'-8"	4'-0"	15'-4"	5'-4"	10'-0"	8'-0"	4'-8"	17'-4"
20'-0"	4'-0"	14'-8"	5'-4"	9'-4"	8'-8"	4'-0"	20'-0"

(Continued on next page)

2110 Glass Unit Masonry

2110.1 General *(part 4 of 10)*

Table 2110.1a—*Continued* **Maximum Dimensions of Glass Block Panels**

Width	Height	Width	Height	Width	Height	Width	Height
Widths and heights ≤ 94 sf as permitted by a wind pressure of 35 psf:							
24'-8"	3'-4"	19'-4"	4'-8"	14'-0"	6'-8"	8'-8"	10'-8"
24'-0"	3'-4"	18'-8"	4'-8"	13'-4"	6'-8"	8'-0"	11'-4"
23'-4"	4'-0"	18'-0"	4'-8"	12'-8"	7'-4"	7'-4"	12'-8"
22'-8"	4'-0"	17'-4"	5'-4"	12'-0"	7'-4"	6'-8"	14'-0"
22'-0"	4'-0"	16'-8"	5'-4"	11'-4"	8'-0"	6'-0"	15'-4"
21'-4"	4'-0"	16'-0"	5'-4"	10'-8"	8'-8"	5'-4"	17'-4"
20'-8"	4'-0"	15'-4"	6'-0"	10'-0"	9'-4"	4'-8"	20'-0"
20'-0"	4'-8"	14'-8"	6'-0"	9'-4"	10'-0"		
Widths and heights ≤ 104 sf as permitted by a wind pressure of 30 psf:							
24'-8"	4'-0"	19'-4"	5'-4"	14'-0"	7'-4"	8'-8"	11'-4"
24'-0"	4'-0"	18'-8"	5'-4"	13'-4"	7'-4"	8'-0"	12'-8"
23'-4"	4'-0"	18'-0"	5'-4"	12'-8"	8'-0"	7'-4"	14'-0"
22'-8"	4'-0"	17'-4"	6'-0"	12'-0"	8'-8"	6'-8"	15'-4"
22'-0"	4'-8"	16'-8"	6'-0"	11'-4"	8'-8"	6'-0"	17'-4"
21'-4"	4'-8"	16'-0"	6'-0"	10'-8"	9'-4"	5'-4"	19'-4"
20'-8"	4'-8"	15'-4"	6'-8"	10'-0"	10'-0"	4'-8"	20'-0"
20'-0"	4'-8"	14'-8"	6'-8"	9'-4"	10'-8"		
Widths and heights ≤ 119 sf as permitted by a wind pressure of 25 psf:							
24'-8"	4'-8"	18'-8"	6'-0"	12'-8"	9'-4"	8'-8"	13'-4"
24'-0"	4'-8"	18'-0"	6'-0"	12'-0"	9'-4"	8'-0"	14'-8"
23'-4"	4'-8"	17'-4"	6'-8"	11'-4"	10'-0"	7'-4"	16'-0"
22'-8"	4'-8"	16'-8"	6'-8"	10'-8"	10'-8"	6'-8"	17'-4"
22'-0"	5'-4"	16'-0"	7'-4"	10'-0"	11'-4"	6'-0"	19'-4"
21'-4"	5'-4"	15'-4"	7'-4"	9'-4"	12'-8"	5'-4"	20'-0"
20'-8"	5'-4"	14'-8"	8'-0"				
20'-0"	5'-4"	14'-0"	8'-0"				
19'-4"	6'-0"	13'-4"	8'-8"				

(Continued on next page)

2110 Glass Unit Masonry

2110.1 General *(part 5 of 10)*

Table 2110.1a—*Continued* **Maximum Dimensions of Glass Block Panels**

Width	Height	Width	Height	Width	Height	Width	Height
Widths and heights ≤ 144 sf as permitted by a wind pressure of 20 psf.							
24'-8"	5'-4"	20'-0"	6'-8"	15'-4"	9'-4"	10'-8"	13'-4"
24'-0"	6'-0"	19'-4"	7'-4"	14'-8"	9'-4"	10'-0"	14'-0"
23'-4"	6'-0"	18'-8"	7'-4"	14'-0"	10'-0"	9'-4"	15'-4"
22'-8"	6'-0"	18'-0"	8'-0"	13'-4"	10'-8"	8'-8"	16'-0"
22'-0"	6'-0"	17'-4"	8'-0"	12'-8"	11'-4"	8'-0"	18'-0"
21'-4"	6'-8"	16'-8"	8'-0"	12'-0"	12'-0"	7'-4"	19'-4"
20'-8"	6'-8"	16'-0"	8'-8"	11'-4"	12'-8"	6'-8"	20'-0"
Widths and heights ≤ 182 sf as permitted by a wind pressure of 15 psf:							
24'-8"	7'-4"	20'-0"	8'-8"	16'-0"	11'-4"	12'-0"	14'-8"
24'-0"	7'-4"	19'-4"	9'-4"	15'-4"	11'-4"	11'-4"	16'-0"
23'-4"	7'-4"	18'-8"	9'-4"	14'-8"	12'-0"	10'-8"	16'-8"
22'-8"	8'-0"	18'-0"	10'-0"	14'-0"	12'-8"	10'-0"	18'-0"
22'-0"	8'-0"	17'-4"	10'-0"	13'-4"	13'-4"	9'-4"	19'-4"
21'-4"	8'-0"	16'-8"	10'-8"	12'-8"	14'-0"	8'-8"	20'-0"
20'-8"	8'-8"						
Widths and heights ≤ 235 sf as permitted by a wind pressure of 10 psf:							
24'-8"	9'-4"	20'-8"	11'-4"	17'-4"	13'-4"	14'-0"	16'-8"
24'-0"	9'-4"	20'-0"	11'-4"	16'-8"	14'-0"	13'-4"	17'-4"
23'-4"	10'-0"	19'-4"	12'-0"	16'-0"	14'-8"	12'-8"	18'-0"
22'-8"	10'-0"	18'-8"	12'-0"	15'-4"	14'-8"	12'-0"	19'-4"
22'-0"	10'-8"	18'-0"	12'-8"	14'-8"	16'-0"	11'-4"	20'-0"
21'-4"	10'-8"						

2110 Glass Unit Masonry

2110.1 General *(part 6 of 10)*

- This section addresses exterior glass block panels with standard units.
- Panels may not be subjected to wind pressure > 20 psf.
- Maximum panel height between structural supports is 10'.
- Maximum panel width between structural supports is 15'.
- Maximum panel area between structural supports is 85 sf.
- The table below provides a partial list of widths and heights which yield an area within the maximum permitted as follows:
  - Width and height are multiples of standard glass block dimensions.
  - Either or both dimensions in any combination may be smaller than those listed.

Note: TMS 402/ACI 530/ASCE 5, "Building Code Requirements for Masonry Structures," Chapter 7, "Glass Unit Masonry," is cited as governing this subject as supplemented by the IBC.
TMS Section 7.2.2, "Exterior thin-unit panels," governs glass masonry panel size and is partially summarized in the table below.

Table 2110.1b Thin Glass Block Units: Widths × Heights ≤ 85 sf

Width	Height	Width	Height
14'-8"	5'-4"	11'-4"	7'-4"
14'-0"	6'-0"	10'-6"	7'-4"
13'-4"	6'-0"	10'-0"	8'-0"
12'-8"	6'-8"	9'-4"	8'-8"
12'-0"	6'-8"	8'-8"	9'-4"

2110 Glass Unit Masonry

2110.1 General *(part 7 of 10)*

- This section addresses interior glass block panels.
- Maximum panel height between structural supports is 20'.
- Maximum panel width between structural supports is 25'.
- Standard units are governed as follows:
  - Maximum panel area between structural supports is 250 sf.
- Thin units are governed as follows:
  - Maximum panel area between structural supports is 150 sf.
- The tables below provide a partial list of widths and heights which yield an area within the maximum permitted as follows:
  - Width and height are multiples of standard glass block dimensions.
  - Either or both dimensions in any combination may be smaller than those listed.
- Where wind pressure is > 10 psf, interior panels must meet requirements for exterior panels.

Note: TMS 402/ACI 530/ASCE 5, "Building Code Requirements for Masonry Structures," Chapter 7, "Glass Unit Masonry," is cited as governing this subject as supplemented by the IBC. TMS Section 7.2.3, "Interior panels," governs glass masonry panel size and is partially summarized in the tables below.

Table 2110.1c Standard Glass Block Units: Width × Height ≤ 250 sf

Width	Height	Width	Height	Width	Height	Width	Height
24'-8"	10'-0"	21'-4"	11'-4"	18'-0"	13'-4"	14'-8"	16'-8"
24'-0"	10'-0"	20'-8"	12'-0"	17'-4"	14'-0"	14'-0"	17'-4"
23'-4"	10'-8"	20'-0"	12'-0"	16'-8"	14'-8"	13'-4"	18'-8"
22'-8"	10'-8"	19'-4"	12'-8"	16'-0"	15'-4"	12'-8"	19'-4"
22'-0"	11'-4"	18'-8"	13'-4"	15'-4"	16'-0"	12'-0"	20'-0"

Table 2110.1d Thin Glass Block Units: Width × Height ≤ 150 sf

Width	Height	Width	Height	Width	Height
24'-8"	6'-0"	18'-8"	8'-0"	12'-8"	11'-4"
24'-0"	6'-0"	18'-0"	8'-0"	12'-0"	12'-0"
23'-4"	6'-0"	17'-4"	8'-0"	11'-4"	12'-8"
22'-8"	6'-0"	16'-8"	8'-8"	10'-8"	14'-0"
22'-0"	6'-8"	16'-0"	9'-4"	10'-0"	14'-8"
21'-4"	6'-8"	15'-4"	9'-4"	9'-4"	16'-0"
20'-8"	6'-8"	14'-8"	10'-0"	8'-8"	16'-8"
20'-0"	7'-4"	14'-0"	10'-8"	8'-0"	18'-8"
19'-4"	7'-4"	13'-4"	10'-8"	7'-4"	20'-0"

2110 Glass Unit Masonry

2110.1 General *(part 8 of 10)*

- The width of curved glass-block panels must comply with the same dimensional limitations that govern straight panels.
- Additional structural supports are required at the following locations as shown in the details below:
  - At the connection of a curved panel to a straight panel.
  - At inflection points of serpentine curves.

Note: TMS 402/ACI 530/ASCE 5, "Building Code Requirements for Masonry Structures," Chapter 7, "Glass Unit Masonry," is cited as governing this subject as supplemented by the IBC. TMS Section 7.2.4, "Curved panels," governs curved glass unit masonry panels and is partially summarized in the illustrations below.

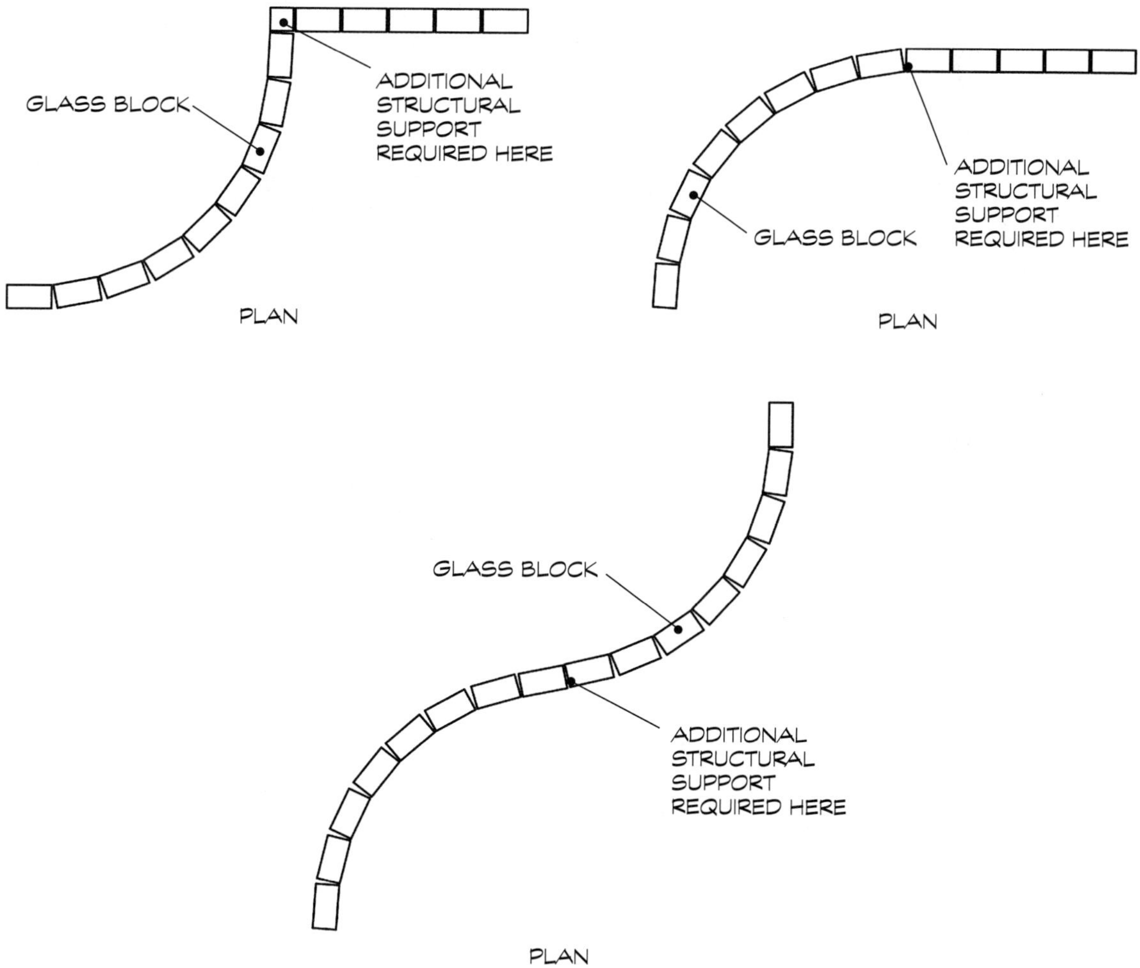

Fig. 2110.1(2). Additional structural support in curved glass block panels.

2110 Glass Unit Masonry

2110.1 General *(part 9 of 10)*

- Lateral support for glass block is required as follows:
  - At the top of a panel as follows:
    Where the panel is > one unit wide.
    Lateral support is not required where the panel is only 1 unit wide.
  - At the sides of a panel as follows:
    Where the panel is > one unit high.
    Lateral support is not required where the panel is only 1 unit high.
- Lateral supports must resist the greater of the following loads:
  - Actual loads applied.
  - 200 lbs/lineal foot.
- Dimensional requirements for lateral support are shown in the details below and on the facing page.

Note: TMS 402/ACI 530/ASCE 5, "Building Code Requirements for Masonry Structures," Chapter 7, "Glass Unit Masonry," is cited as governing this subject as supplemented by the IBC. TMS Section 7.3.3, "Lateral," governs glass unit masonry panels and is partially summarized in the Figures 2110.1(3) and 2110.1(4) below and on the next page.

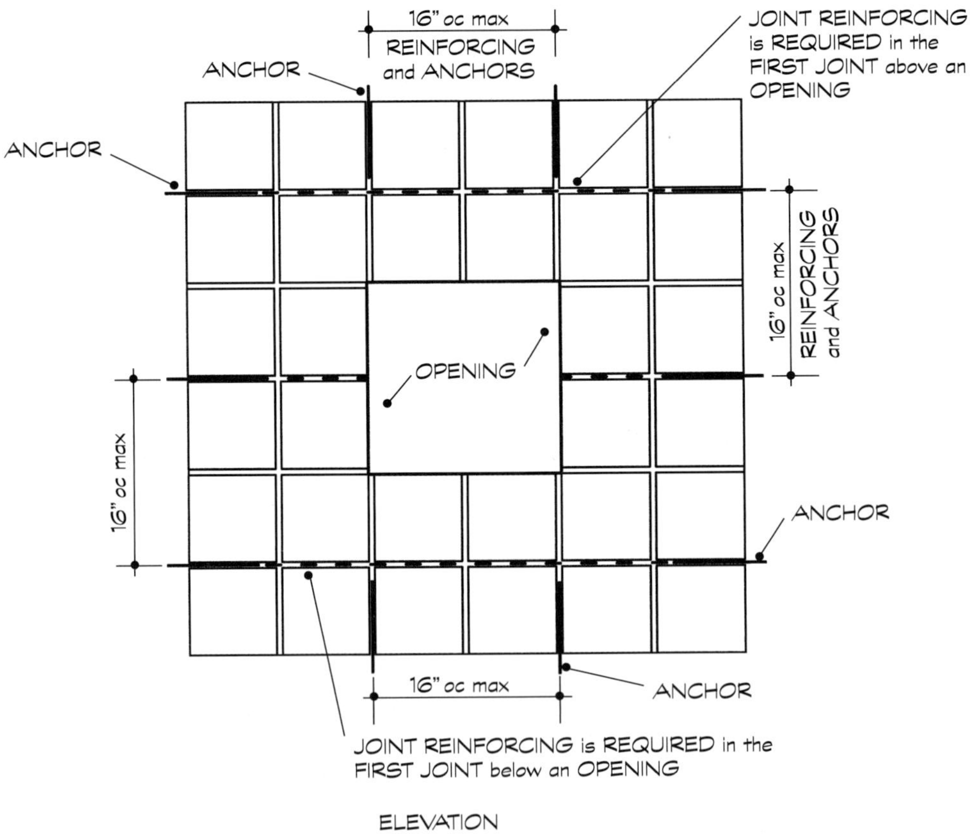

Fig. 2110.1(3). Lateral support for glass block and reinforcement required.

2110 Glass Unit Masonry

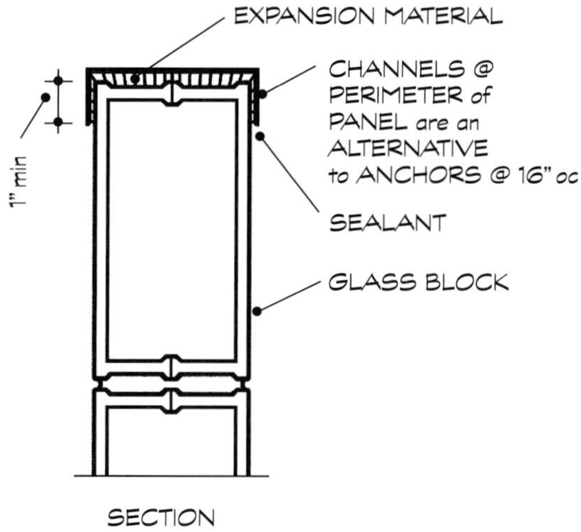

Fig. 2110.1(4). Lateral support for glass block.

2110.1 General *(part 10 of 10)*

- Glass block panels require horizontal joint reinforcing as follows:
  - Spaced vertically 1'-4' oc.
  - Placed in the mortar joint.
  - Extending the full length of the glass unit panel as follows:
    Reinforcing does not cross expansion joints.
  - Placed in mortar joints nearest openings as follows:
    Above openings.
    Below openings.
  - Other requirements are shown in the detail below.

Note: TMS 402/ACI 530/ASCE 5, "Building Code Requirements for Masonry Structures," Chapter 7, "Glass Unit Masonry," is cited as governing this subject as supplemented by the IBC. TMS Section 7.7, "Reinforcement," governs glass unit masonry reinforcing and is partially summarized in Figure 2110.1(3) and 2110.1(5).

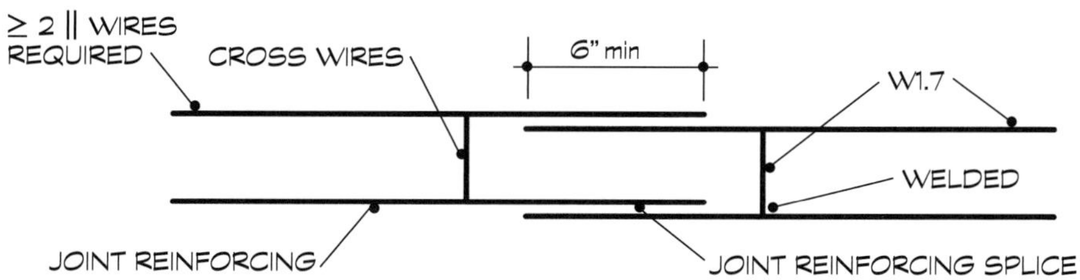

Fig. 2110.1(5). Size and splice requirements for joint reinforcement.

NOTES

22

Steel

Christus St. Michael Health Care Center. Texarkana, Texas. *(partial elevation)*
Watkins Hamilton Ross Architects, Inc. Houston, Texas.

2202 Definitions

2202.1 Definitions

- **Steel construction, cold-formed**
  - Steel construction with products, such as those below, produced from the following materials by cold-forming processes:
    - Materials:
      - Steel sheet.
      - Steel strip.
    - Products:
      - Floor joists.
      - Floor panels.
      - Roof deck.
      - Roof joists.
      - Studs.
      - Wall panels.
      - Other products.
- **Steel joist**
  - A steel structural member with characteristics as follows:
    - Formed by one of the following processes:
      - Cold-formed.
      - Hot-rolled.
    - With one of the following web conditions:
      - Closed.
      - Open.
    - Fabricated from any of the following components:
      - Riveted bars.
      - Sheet steel.
      - Slotted and expanded sections.
      - Strip steel.
      - Welded bars.
      - Other deformed rolled sections.
- **Steel member, structural**
  - Any rolled steel structural shape other than the following:
    - Cold-formed shapes.
    - Steel joists.

2206 Steel Joists

2206.2 Design

- This section addresses data that must be recorded on the construction documents as follows:
  - Data provided by a registered design professional.
  - Data regarding the following:
    - Steel joists.
    - Steel joist girders.
- The data must include the following from Steel Joist Institute specifications as applicable:
  - Steel joist designations.
  - Steel joist girder designations.

Note: 2206.1, "General," is cited as listing the necessary Steel Joist Institute specifications.

- The data must include the following requirements as applicable:
  - Anchorage.
  - Bearing connection design to resist the following:
    - Uplift.
    - Lateral loads.
  - Bridging termination connections.
  - Design.
  - End support.
  - Layout.
  - Non-SJI standard bridging.
- The data must identify the following special loads as applicable:
  - Axial loads.
  - Concentrated loads.
  - Connection forces.
  - End moments.
  - Net uplift loads.
  - Nonuniform loads.
- The data must address the following special considerations as applicable:
  - Nonstandard profiles.
  - Nonstandard web openings as follows:
    - Oversized web openings.
    - Other web openings.
  - Extended ends.

2206.3 Calculations *(part 1 of 2)*

- This section addresses documentation for the following:
  - Steel joist girders.
  - Steel joists.
- The joists and girders must be designed according to the following by the manufacturer:
  - Steel Joist Institute specifications.
  - Steel Joist Institute load tables as follows:
    - To support loads specified in the previous section.

Note: 2206.2, "Design," is cited as listing the necessary loads.

2206 Steel Joists

2206.3 Calculations *(part 2 of 2)*

- The project registered design professional is authorized to require the following from the manufacturer:
  - Design calculations for the joists and joist girders as follows:
    - As prepared by the registered design professional who designed the structural members.
    - Accompanied by a cover letter with:
      - The seal of the manufacturer's registered design professional.
      - The signature of the manufacturer's registered design professional.
    - Including the following:
      - Non-SJI standard bridging details such as:
        - For cantilevers.
        - For net uplift.
      - Connection details for the following:
        - Non-SJI standard connections such as:
          - Flush framed connections.
          - Framed connections.
        - Field splices.
        - Joist headers.

2206.4 Steel joist drawings

- Drawings are required as follows:
  - They must include the joists and joist girders as specified in the construction documents.
  - They must be used for the construction of the structure as required by the design.
  - They must list all applicable loads as required by the design.

Note: 2206.2, "Design," is cited as governing the loads as required by the design.

  - They must show the following as applicable:
    - The location of the joists and joist girders as follows:
      - Location drawings do not require:
        - The manufacturer's registered design professional's seal.
        - The manufacturer's registered design professional's signature.
    - Any profiles for nonstandard joist and joist girder configurations.

Note: The Steel Joist Institute catalog is cited as containing standard joist and joist girder configurations.

    - Connection requirements for the following as applicable:
      - Bridging attachments.
      - Field splices.
      - Joist girder supports.
      - Joists supports.
    - Deflection criteria for non-SJI standard joists as follows:
      - For live loads.
      - For total loads.
    - Joist headers.

23

Wood

Central Kitchen. Lompoc Unified School District.
Lompoc, California. *(partial elevation)*
Phillips Metsch Sweeney Moore Architects. Santa Barbara, California.

2304 General Construction Requirements

2304.6 Wall sheathing

- Wood framing for buildings must be sheathed by one of the following materials:
  - Weatherboarding.

Note: Section 1405, "Installation of Wall Coverings," is cited as governing weatherboarding.

  - Stucco construction.

Note: Section 2510, "Lathing and Furring for Cement Plaster (Stucco)," is cited as governing stucco.

  - Materials shown in Figures 2304.6A and 2304.6B below and on the next page.
  - Other material with both of the following characteristics:
    Approved.
    Of equal strength and durability.

Note: IBC Table 2304.6, "Minimum Thickness of Wall Sheathing," is cited as listing acceptable wall sheathing. This table is the basis of the details in Figures 2304.6A and 2304.6B below and on the next page.

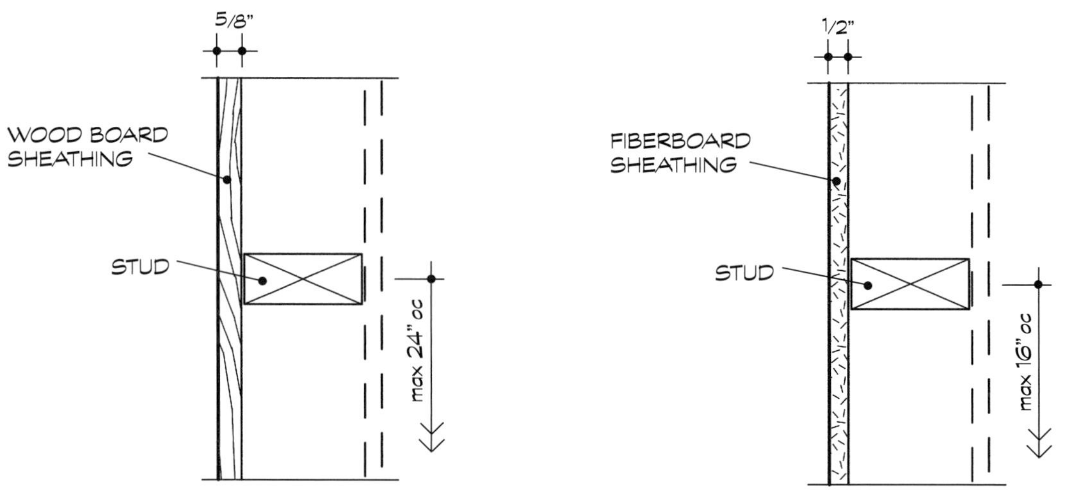

Fig. 2304.6A. Minimum thickness of wall sheathing. *[IBC Table 2304.6]*

2304 General Construction Requirements

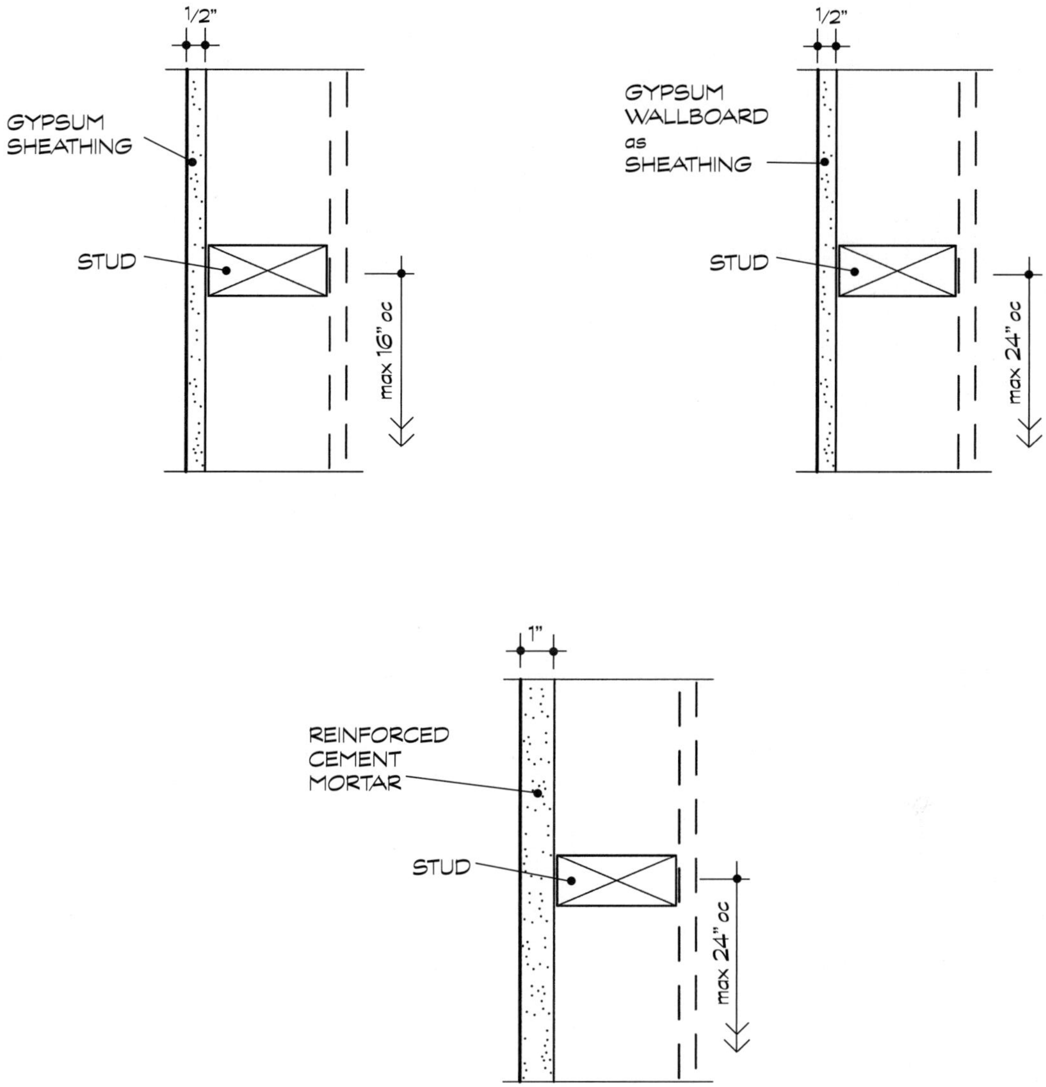

Fig. 2304.6B. Minimum thickness of wall sheathing. *[IBC Table 2304.6]*

2304 General Construction Requirements

2304.7.1 Structural floor sheathing

- Thicknesses required for structural floor sheathing are shown in the details provided in Figures 2304.7.1A–D below and on the following pages.

Note: IBC Table 2304.7(2), "Sheathing Lumber, Minimum Grade Requirements: Board Grade," is cited as governing floor sheathing. The table lists grading standards.
The following are cited as the sources of requirements for the structural floor sheathing indicated above and are the basis of the details provided in this section:
IBC Table 2304.7(1), "Allowable Spans for Lumber Floor and Roof Sheathing."
IBC Table 2304.7(3), "Allowable Spans and Loads for Wood Structural Panel Sheathing and Single-Floor Grades Continuous over Two or More Spans with Strength Axis Perpendicular to Supports."
IBC Table 2304.7(4), "Allowable Span for Wood Structural Panel Combination Subfloor-Underlayment (Single Floor)."

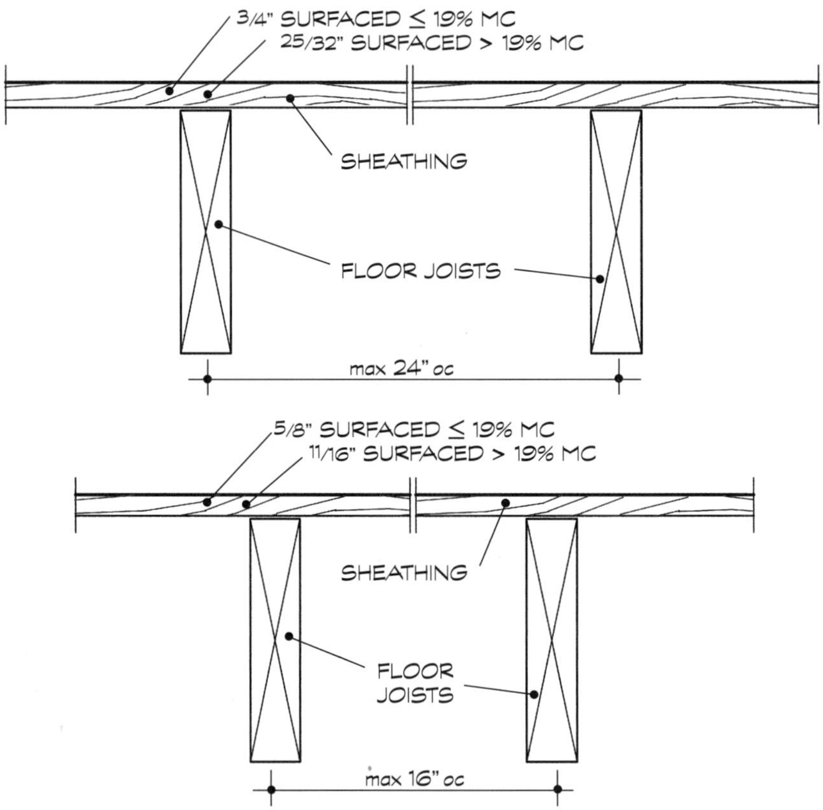

Note: 2304.6.1, "Wood structural panel sheathing," is cited as governing the detailing of the sheathing.

Fig. 2304.7.1A. Spans and minimum thicknesses for lumber floor sheathing perpendicular or diagonal to supports.
[IBC Table 2304.7(1)]

2304 General Construction Requirements

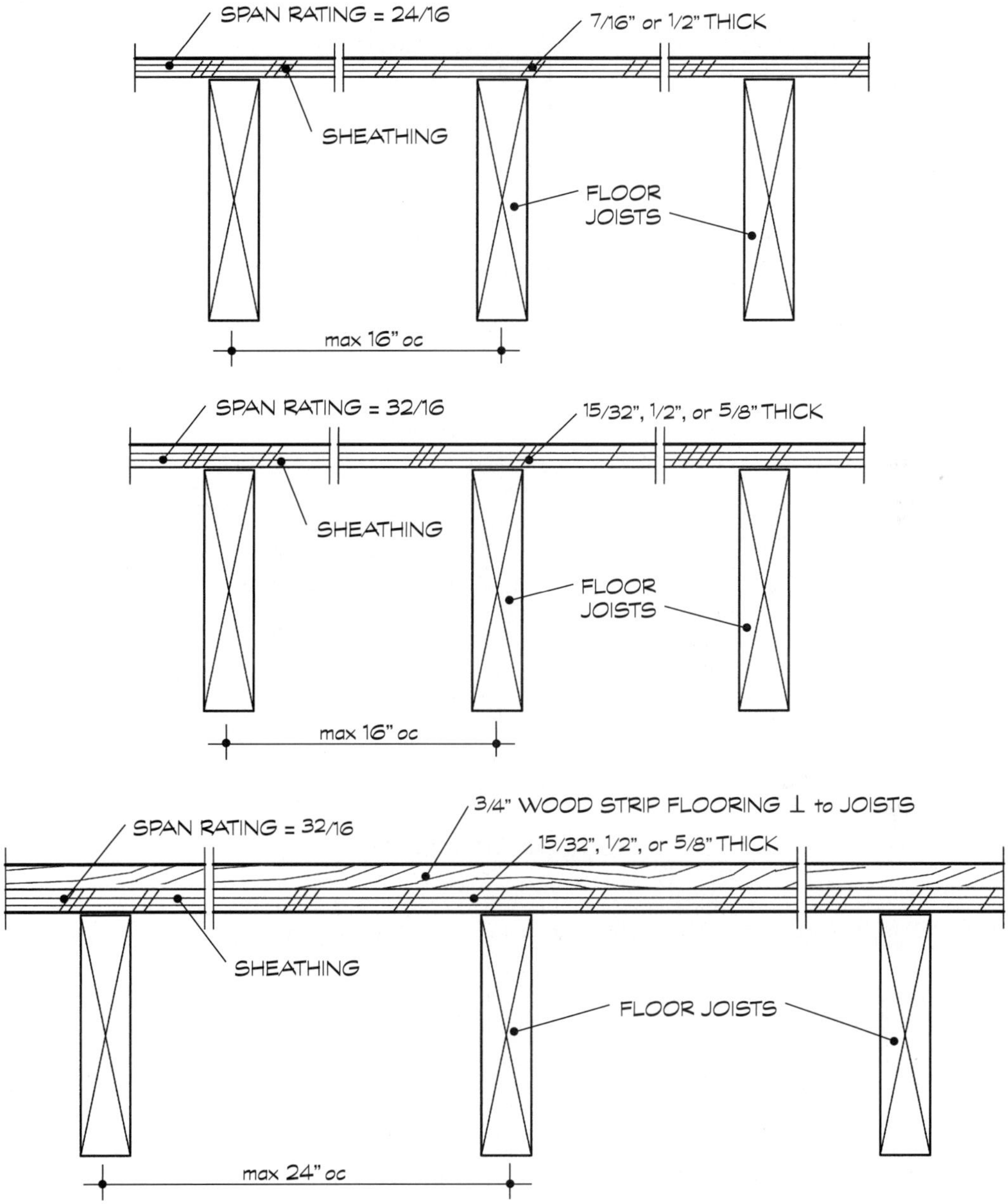

Note: 2304.7, "Floor and roof sheathing," is cited as governing the design criteria of the sheathing.
Uniform load is ≤ 100 psf.
Deflection is ≤ span ÷ 360.

Fig. 2304.7.1B(1). Spans, loads, and minimum thicknesses for wood structural panel floor sheathing ≥ 2' wide continuous over ≥ 2 spans, ⊥ to supports. *[IBC Table 2304.7(3)]*

2304 General Construction Requirements

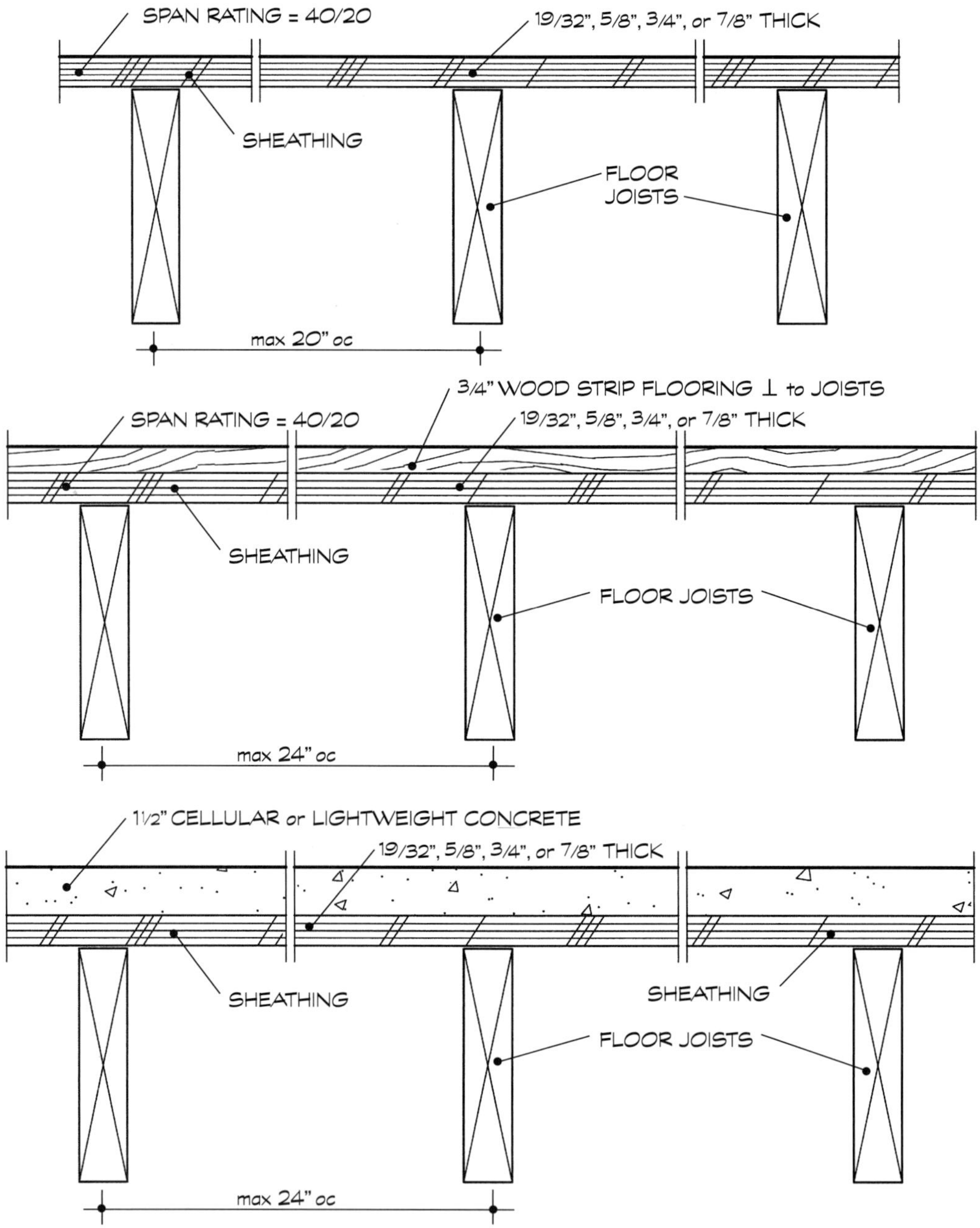

Note: 2304.7, "Floor and roof sheathing," is cited as governing the design criteria of the sheathing.
Uniform load is ≤ 100 psf.
Deflection is ≤ span ÷ 360.

Fig. 2304.7.1B(2). Spans, loads, and minimum thicknesses for wood structural panel floor sheathing ≥ 2' wide continuous over ≥ 2 spans, ⊥ to supports. *[IBC Table 2304.7(3)]*

2304 General Construction Requirements

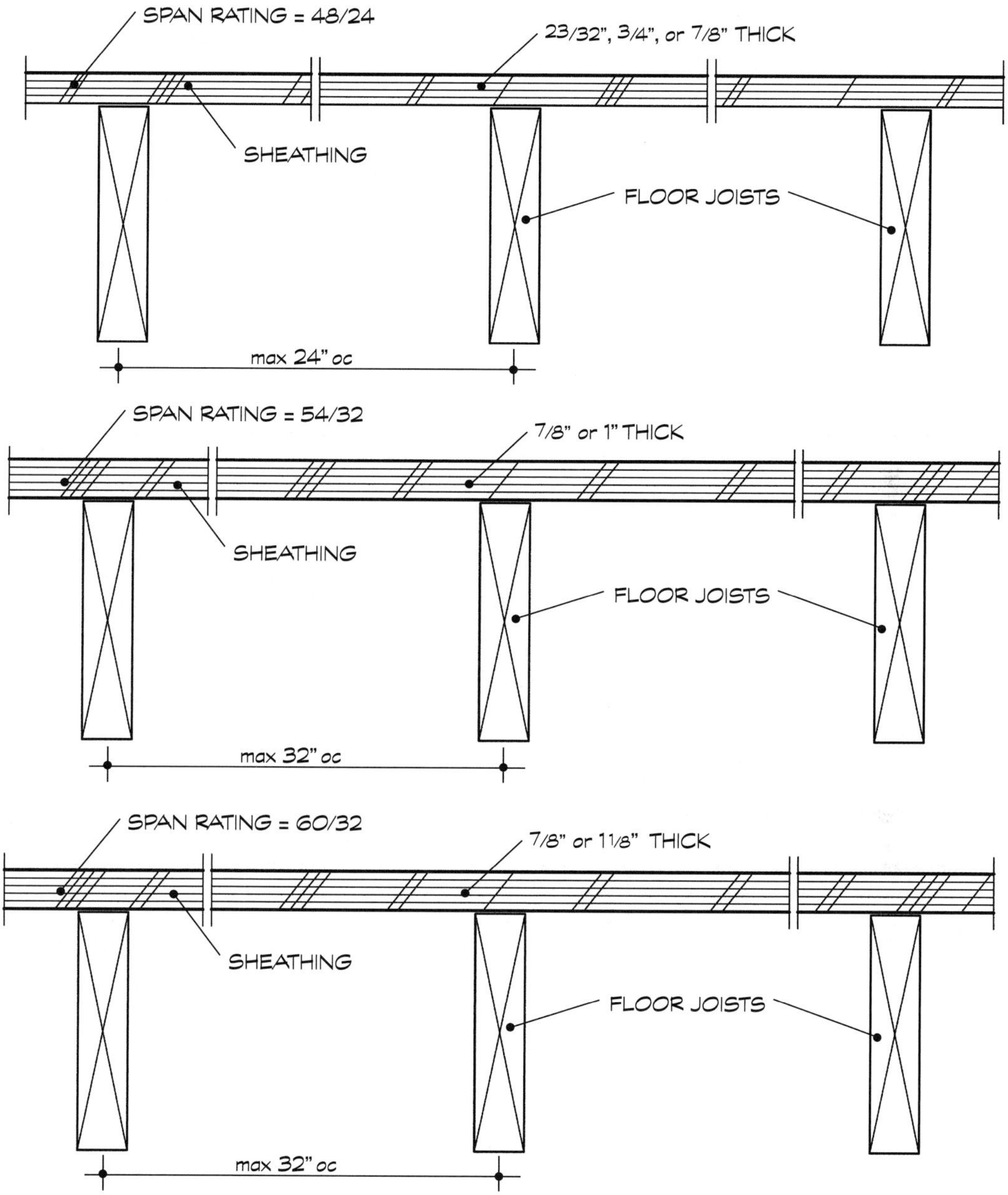

Note: 2304.7, "Floor and roof sheathing," is cited as governing the design criteria of the sheathing.
Uniform load is ≤ 100 psf.
Deflection is ≤ span ÷ 360.

Fig. 2304.7.1B(3). Spans, loads, and minimum thicknesses for wood structural panel floor sheathing ≥ 2' wide continuous over ≥ 2 spans, ⊥ to supports. *[IBC Table 2304.7(3)]*

2304 General Construction Requirements

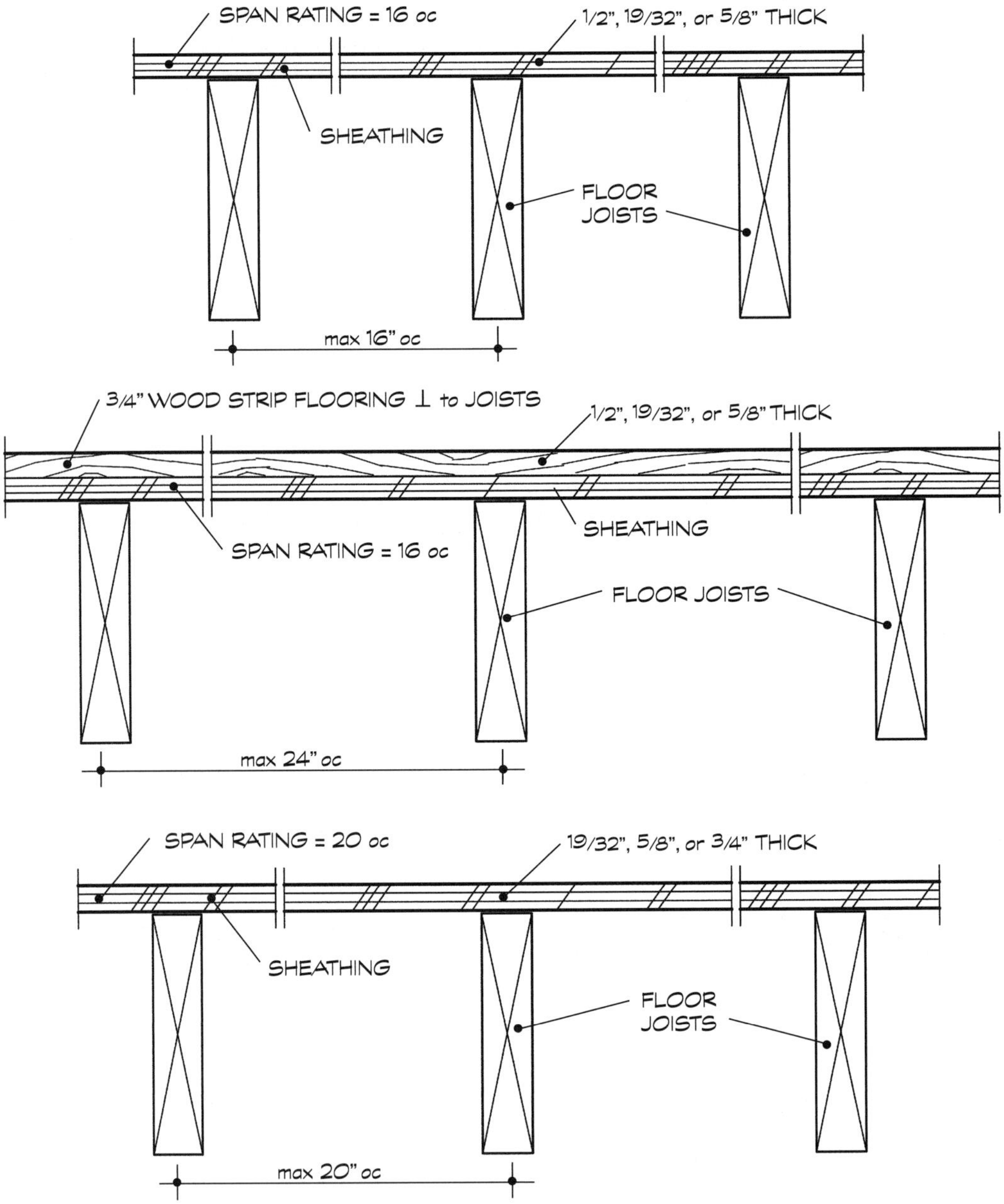

Note: 2304.7, "Floor and roof sheathing," is cited as governing the design criteria of the sheathing. Uniform load is ≤ 100 psf. Deflection is ≤ span ÷ 360.

Fig. 2304.7.1B(4). Spans, loads, and minimum thicknesses for wood structural panel floor sheathing (in grades designed to be used as a single layer) ≥ 2' wide continuous over ≥ 2 spans, ⊥ to supports. *[IBC Table 2304.7(3)]*

2304 General Construction Requirements

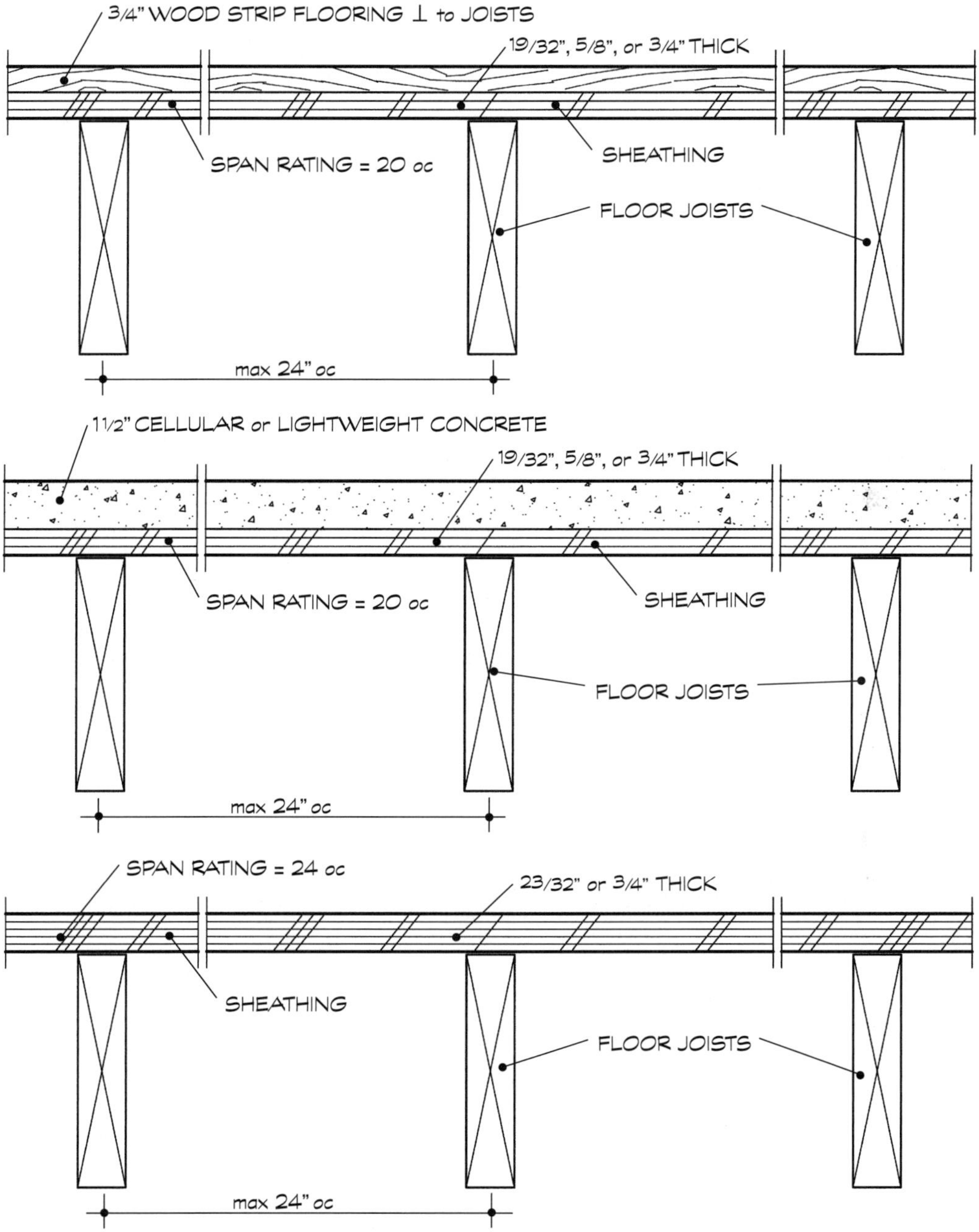

Note: 2304.7, "Floor and roof sheathing," is cited as governing the design criteria of the sheathing.
Uniform load is ≤ 100 psf.
Deflection is ≤ span ÷ 360.

Fig. 2304.7.1B(5). Spans, loads, and minimum thicknesses for wood structural panel floor sheathing (in grades designed to be used as a single layer) ≥ 2' wide continuous over ≥ 2 spans, ⊥ to supports. *[IBC Table 2304.7(3)]*

2304 General Construction Requirements

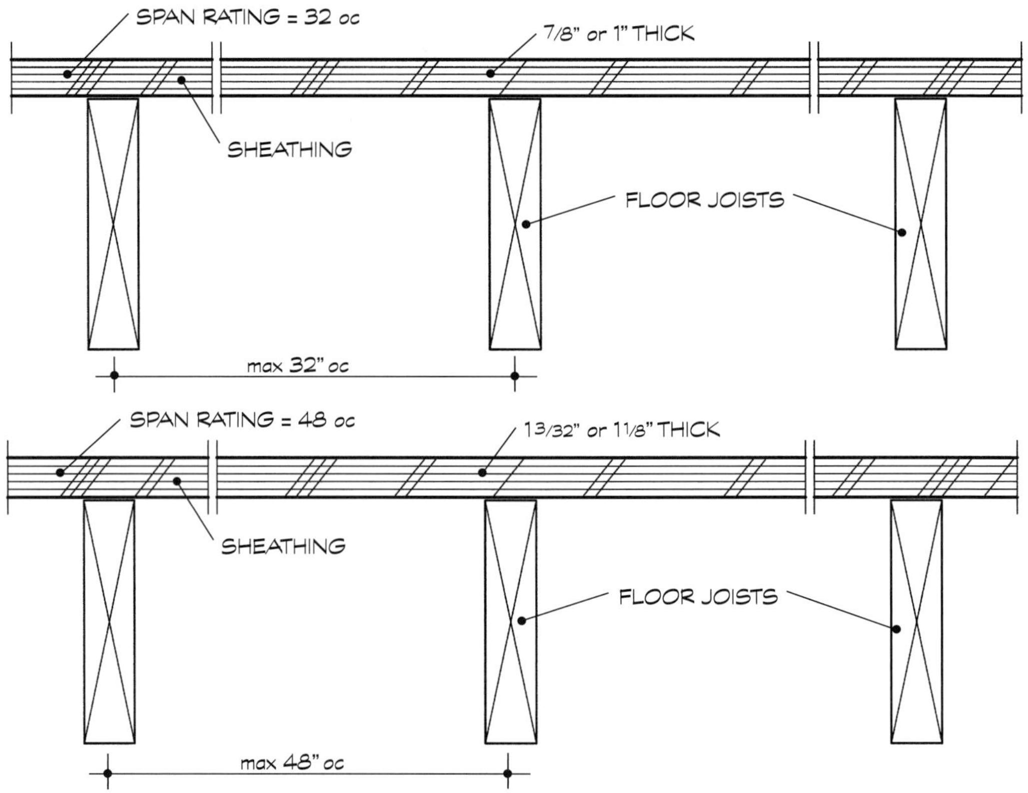

Note: 2304.7, "Floor and roof sheathing," is cited as governing the design criteria of the sheathing. Uniform load is ≤ 100 psf for 32" span and 65 psf for 48" span. Deflection is ≤ span ÷ 360.

Fig. 2304.7.1B(6). Spans, loads, and minimum thicknesses for wood structural panel floor sheathing (in grades designed to be used as a single layer) ≥ 2' wide continuous over ≥ 2 spans, ⊥ to supports. *[IBC Table 2304.7(3)]*

Fig. 2304.7.1C(1). Options for required edge support at subfloor panels. Subfloor panels in IBC Table 2304.7(4) require one of the edge support details shown in Figs. 2304.7.1C(1) and C(2).

2304 General Construction Requirements

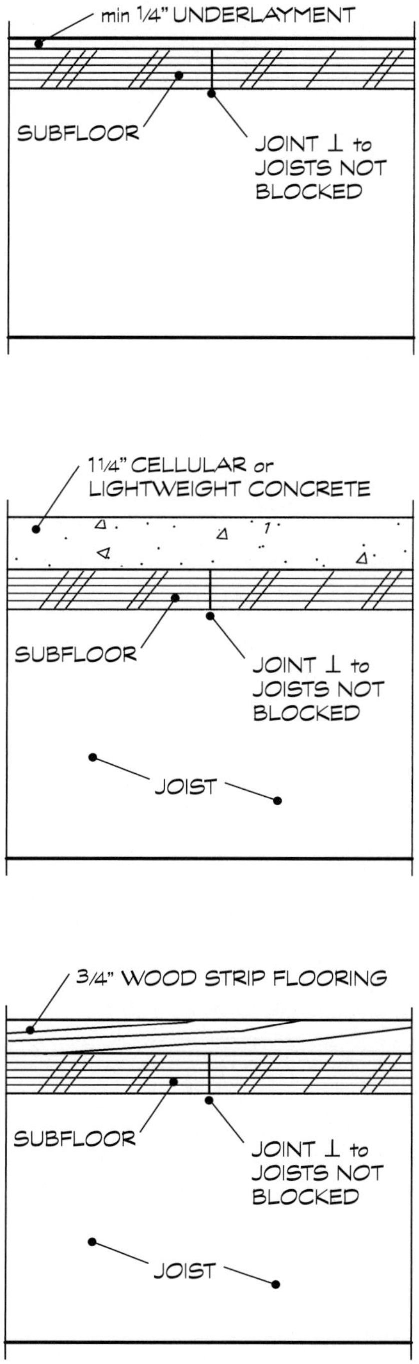

Fig. 2304.7.1C(2). Options for required edge support at subfloor panels. Subfloor panels in IBC Table 2304.7(4) require one of the edge support details shown in Figs. 2304.7.1C(1) and C(2).

2304 General Construction Requirements

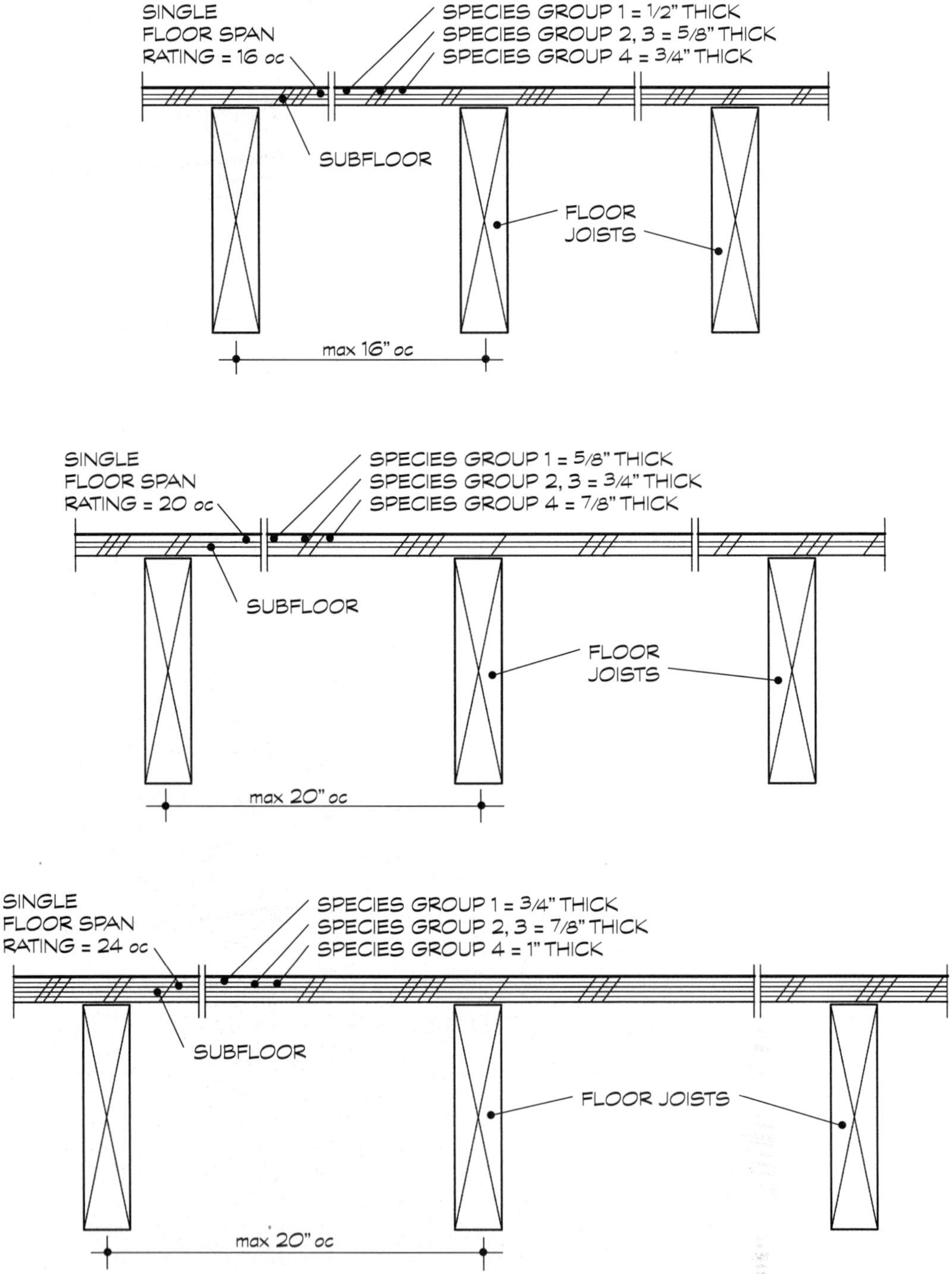

Fig. 2304.7.1D. Spans and minimum thicknesses of combination subfloor-underlayment (single floor) panels continuous over ≥ 2 spans, ⊥ to supports. *[IBC Table 2304.7(4)]*

2304 General Construction Requirements

2304.7.2 Structural roof sheathing

- Thicknesses required for structural roof sheathing are shown in the details in Figures 2304.7.2A – 2304.7.2C(4) provided below and on the following pages.
- Wood structural panel roof sheathing must be bonded as follows:
  - With exterior grade glue.

Note: IBC Table 2304.7(2), "Sheathing Lumber, Minimum Grade Requirements: Board Grade," is cited as governing roof sheathing. The table lists grading rules.

The following are cited as the sources of requirements for the structural roof sheathing indicated above and are the basis of the details provided:

IBC Table 2304.7(1), "Allowable Spans for Lumber Floor and Roof Sheathing."

IBC Table 2304.7(3), "Allowable Spans and Loads for Wood Structural Panel Sheathing and Single-Floor Grades Continuous over Two or More Spans with Strength Axis Perpendicular to Supports."

IBC Table 2304.7(5), "Allowable Load (psf) for Wood Structural Panel Roof Sheathing Continuous over Two or More Spans and Strength Axis Parallel to Supports."

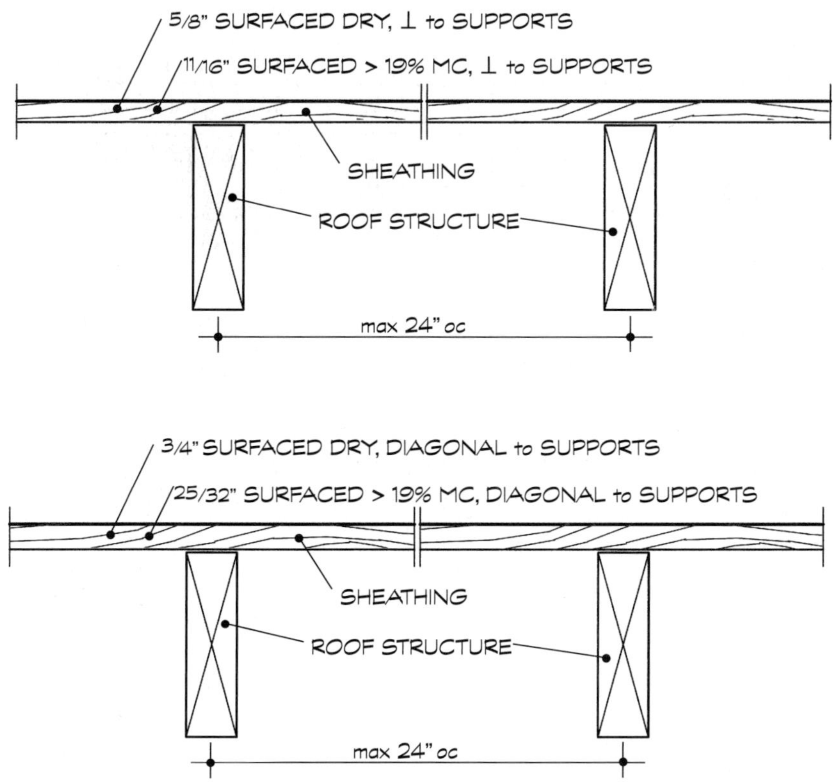

Fig. 2304.7.2A. Spans and minimum thicknesses for roof sheathing. *[IBC Table 2304.7(1)]*

2304 General Construction Requirements

Total load ≤ 40 psf, Live load ≤ 30 psf

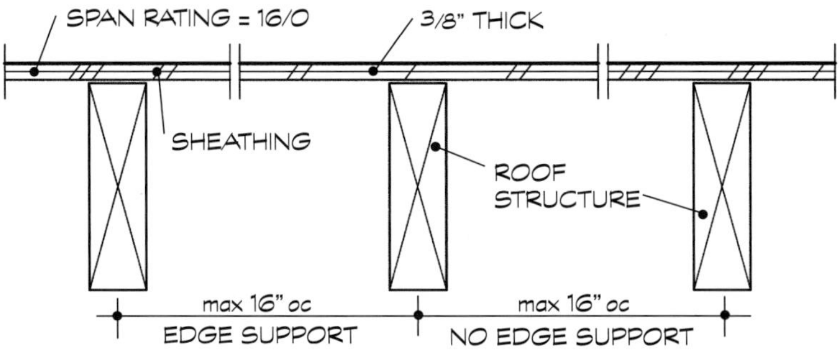

Total load ≤ 40 psf, Live load ≤ 30 psf

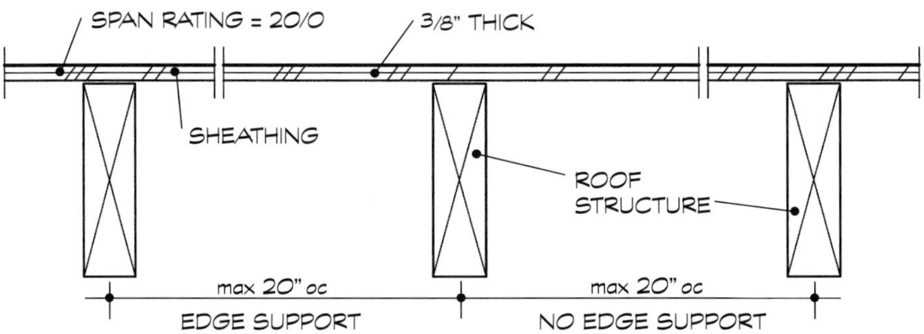

Total load ≤ 40 psf, Live load ≤ 30 psf

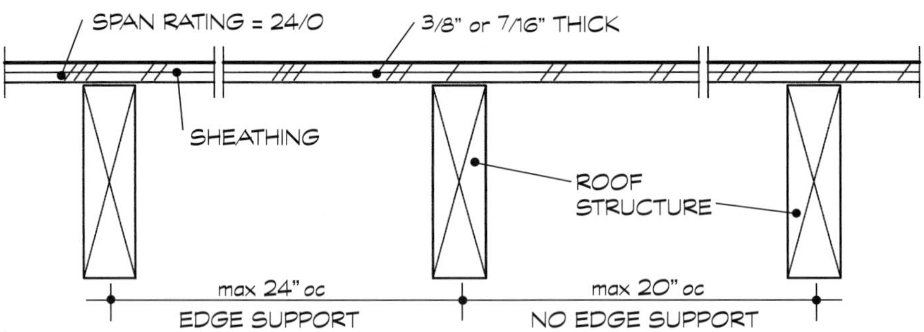

Fig. 2304.7.2B(1). Spans, loads, and minimum thicknesses for wood structural panel roof sheathing ≥ 2' wide continuous over ≥ 2 spans, ⊥ to supports. Uniform load deflection ≤ 1/180 × span (DL + LL), ≤ 1/240 × span (LL only). *[IBC Table 2304.7(3)]*

2304 General Construction Requirements

Total load ≤ 40 psf, Live load ≤ 30 psf

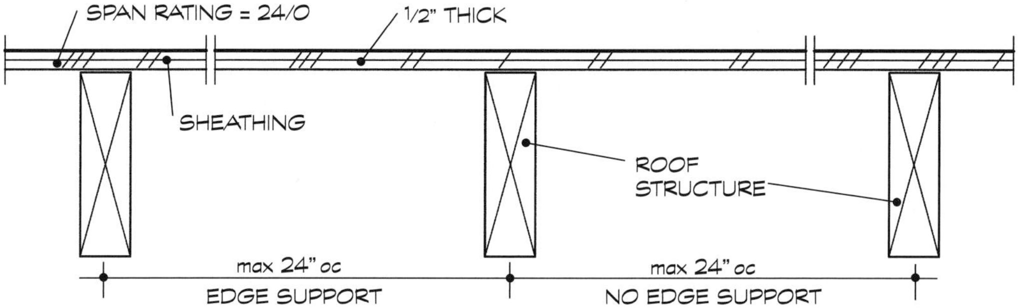

Total load ≤ 50 psf, Live load ≤ 40 psf

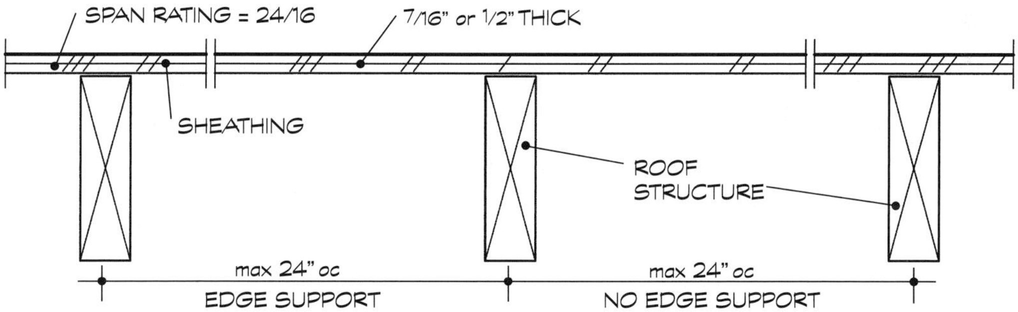

Total load ≤ 40 psf, Live load ≤ 30 psf

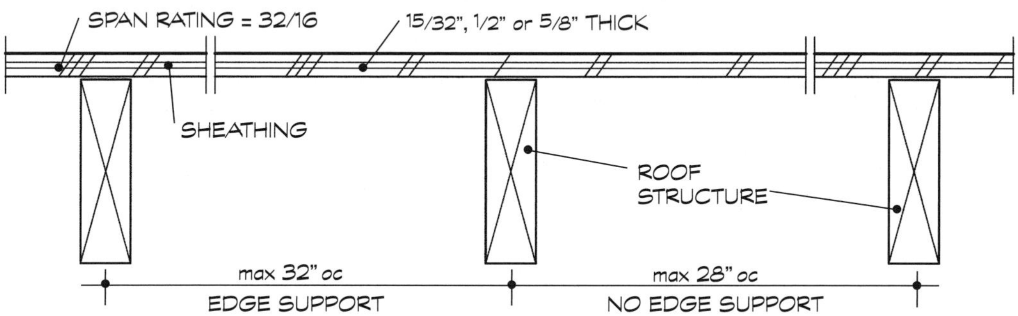

Fig. 2304.7.2B(2). Spans, loads, and minimum thicknesses for wood structural panel roof sheathing ≥ 2' wide continuous over ≥ 2 spans, ⊥ to supports. Uniform load deflection ≤ 1/180 × span (DL + LL), ≤ 1/240 × span (LL only). *[IBC Table 2304.7(3)]*

2304 General Construction Requirements

Total load ≤ 40 psf, Live load ≤ 30 psf

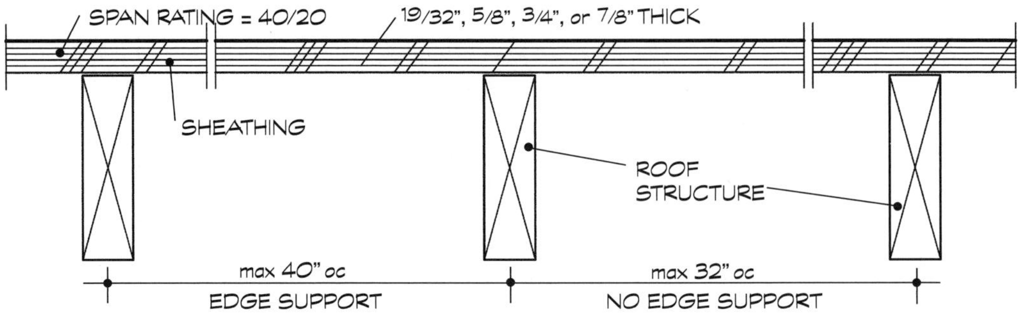

Total load ≤ 45 psf, Live load ≤ 35 psf

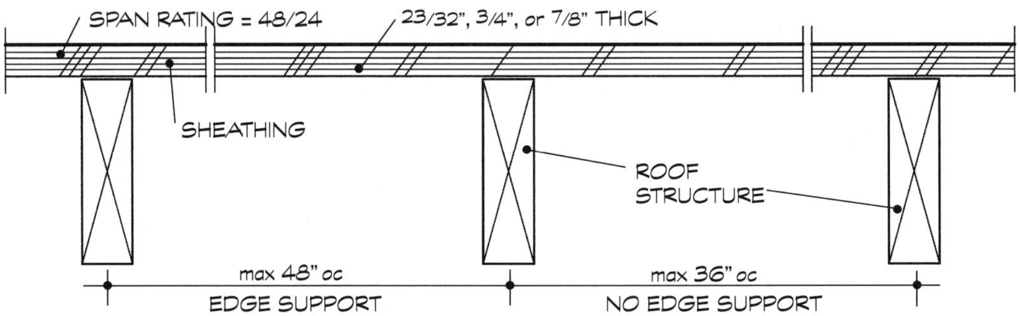

Total load ≤ 45 psf, Live load ≤ 35 psf

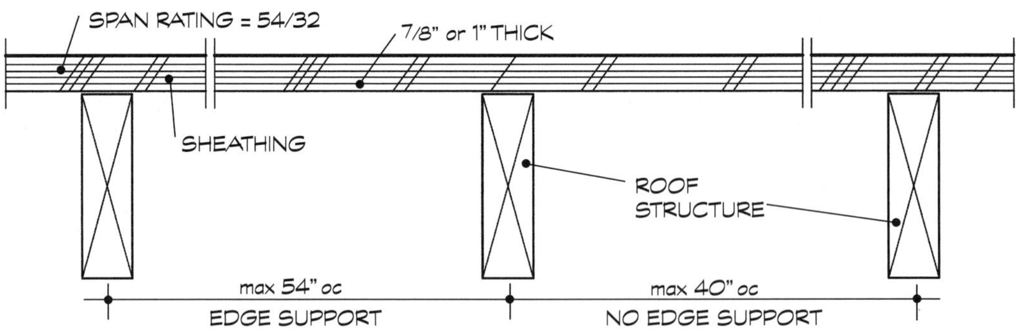

Fig. 2304.7.2B(3). Spans, loads, and minimum thicknesses for wood structural panel roof sheathing ≥ 2' wide continuous over ≥ 2 spans, ⊥ to supports. Uniform load deflection ≤ 1/180 × span (DL + LL), ≤ 1/240 × span (LL only). *[IBC Table 2304.7(3)]*

2304 General Construction Requirements

Total load ≤ 45 psf, Live load ≤ 35 psf

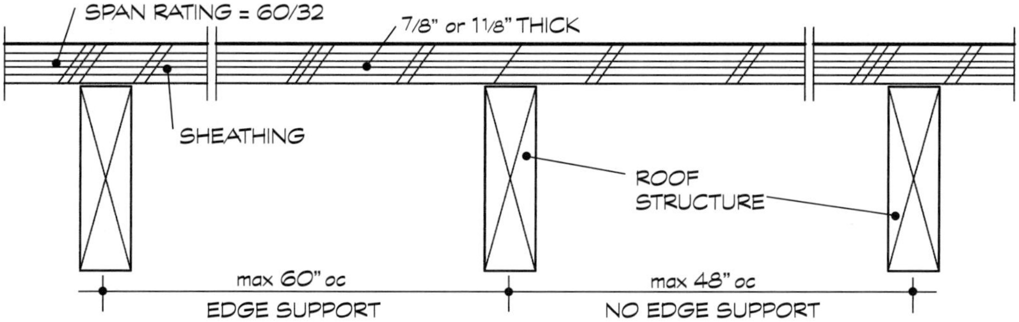

Total load ≤ 50 psf, Live load ≤ 40 psf

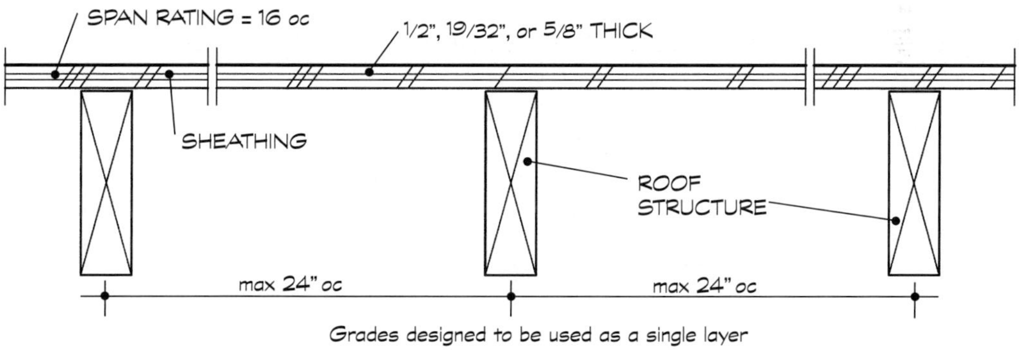

Total load ≤ 40 psf, Live load ≤ 30 psf

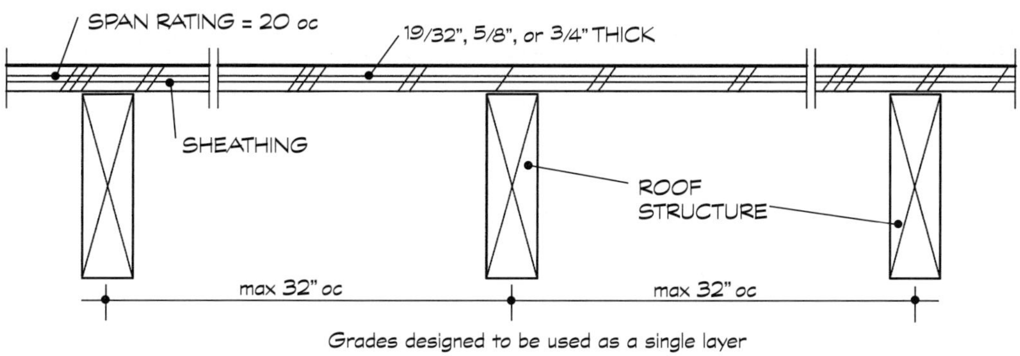

Fig. 2304.7.2B(4). Spans, loads, and minimum thicknesses for wood structural panel roof sheathing ≥ 2' wide continuous over ≥ 2 spans, ⊥ to supports. Uniform load deflection ≤ 1/180 × span (DL + LL), ≤ 1/240 × span (LL only). *[IBC Table 2304.7(3)]*

2304 General Construction Requirements

Total load ≤ 35 psf, Live load ≤ 25 psf

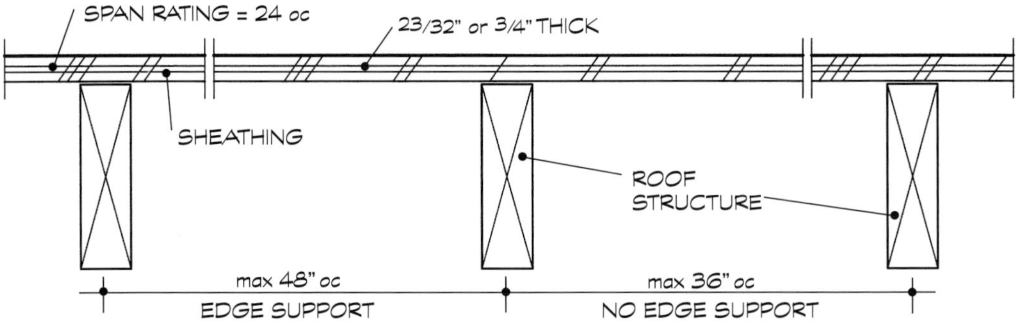

Total load ≤ 50 psf, Live load ≤ 40 psf

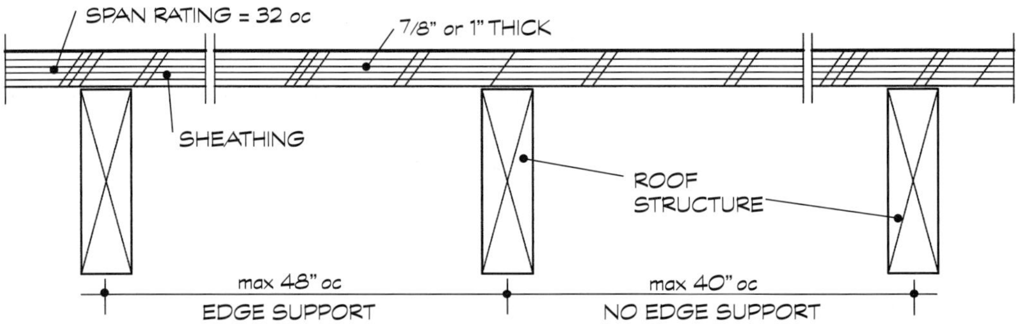

Total load ≤ 50 psf, Live load ≤ 40 psf

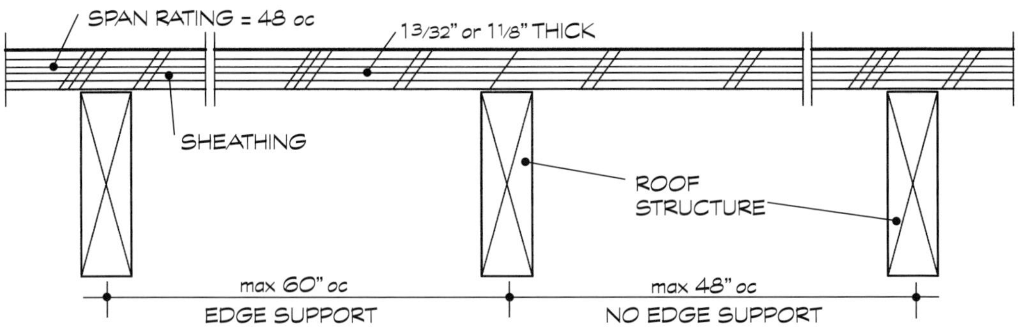

Fig. 2304.7.2B(5). Spans, loads, and minimum thicknesses for wood structural panel roof sheathing (in grades designed to be used as a single layer) ≥ 2' wide continuous over ≥ 2 spans, ⊥ to supports. Uniform load deflection ≤ 1/180 × span (DL + LL), ≤ 1/240 × span (LL only). *[IBC Table 2304.7(3)]*

2304 General Construction Requirements

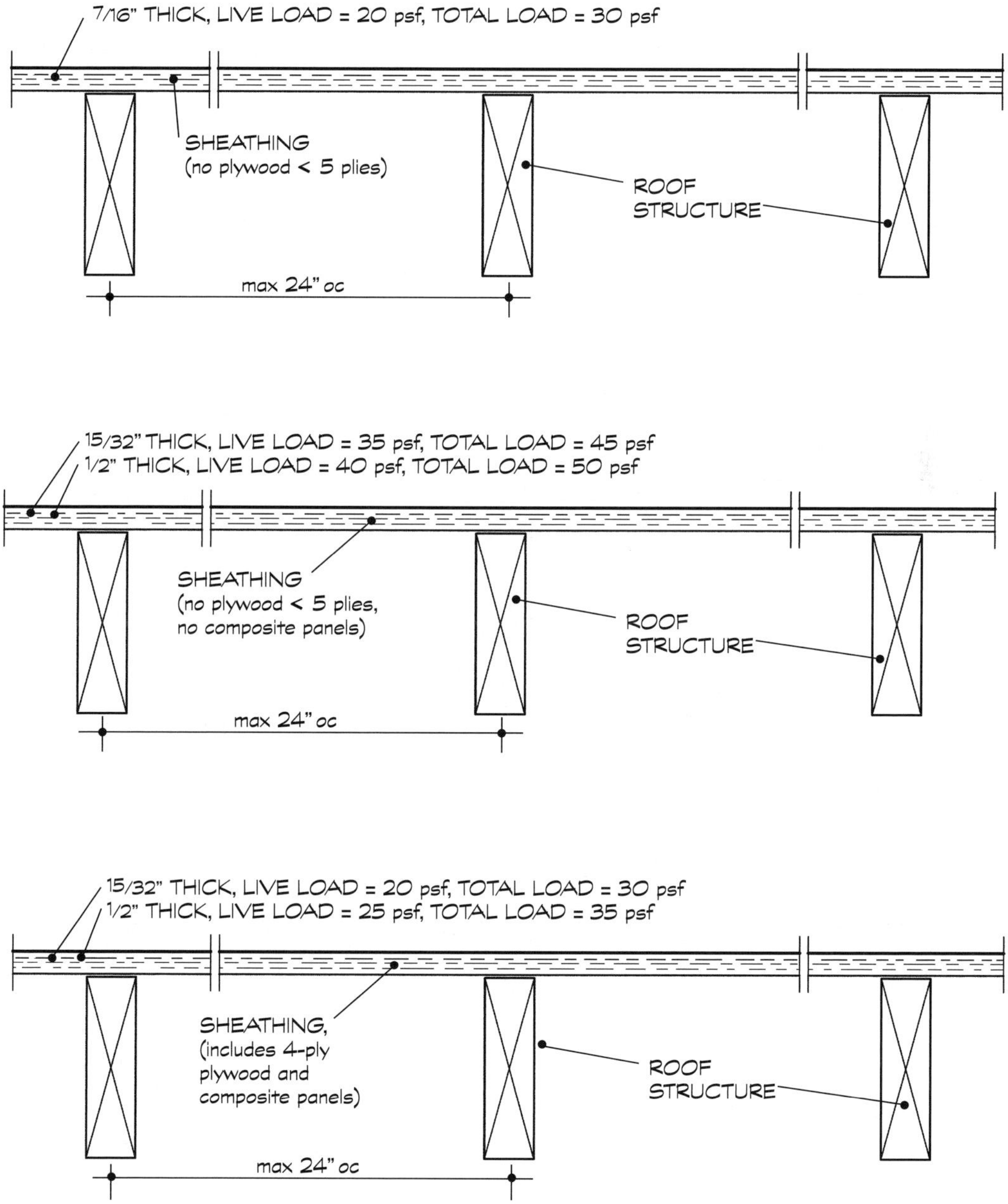

Fig. 2304.7.2C(1). Spans, loads, and minimum thicknesses for wood structural 1 roof sheathing panels continuous over ≥ 2 spans, ⊥ to supports. Uniform load deflection ≤ 1/180 × span (DL + LL), ≤ 1/240 × span (LL only). Edges supported with blocking. *[IBC Table 2304.7(5)]*

2304 General Construction Requirements

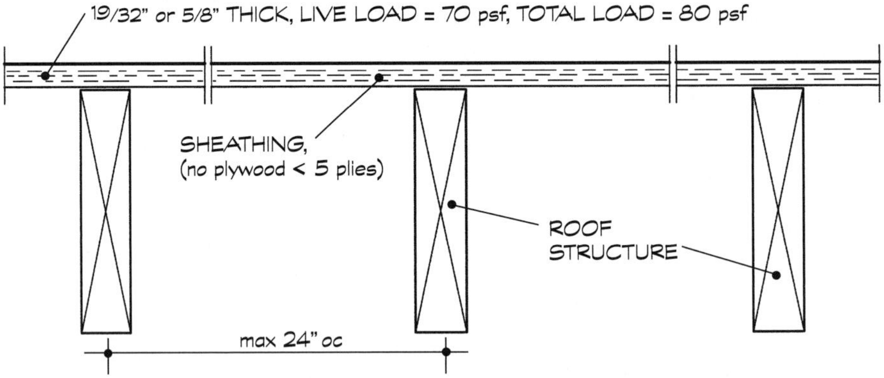

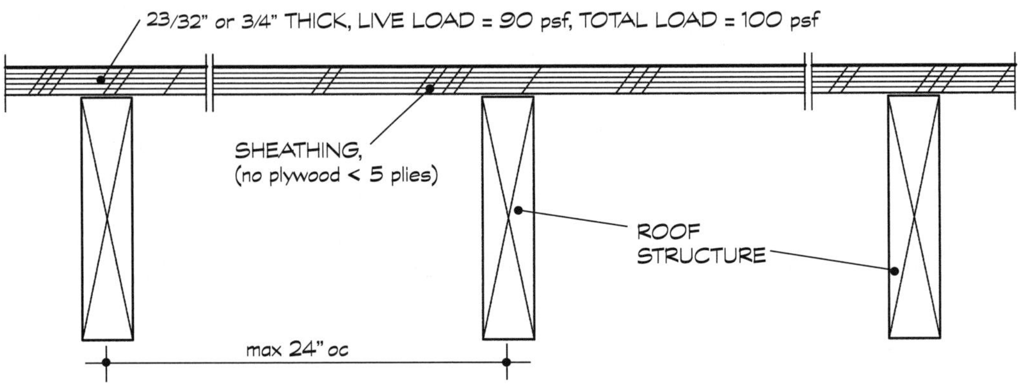

Fig. 2304.7.2C(2). Spans, loads, and minimum thicknesses for wood structural 1 roof sheathing panels continuous over ≥ 2 spans, ⊥ to supports. Uniform load deflection ≤ 1/180 × span (DL + LL), ≤ 1/240 × span (LL only). Edges supported with blocking. *[IBC Table 2304.7(5)]*

2304 General Construction Requirements

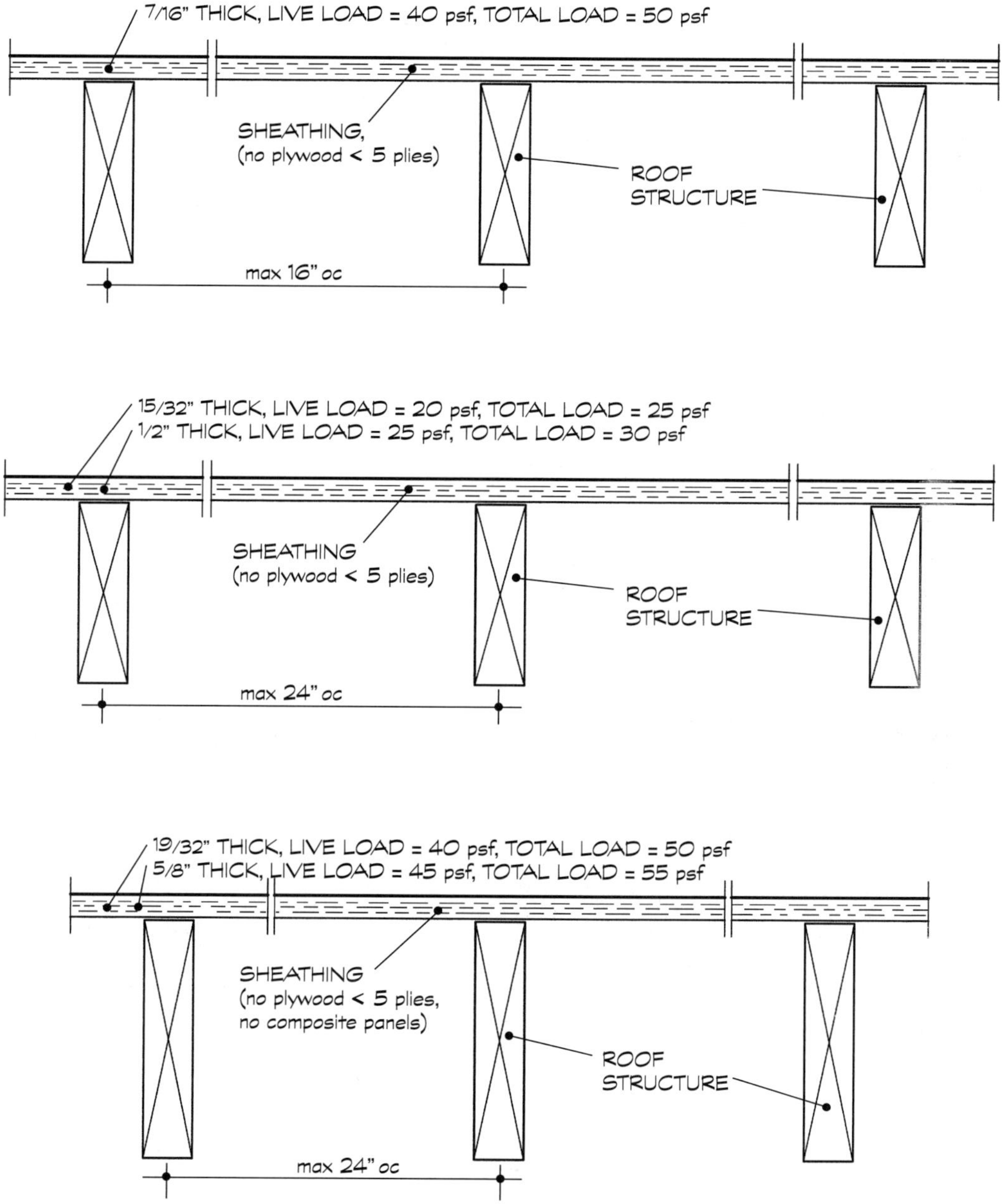

Fig. 2304.7.2C(3). Spans, loads, and minimum thicknesses for wood panel roof sheathing (as described in Department of Commerce documents PS 1, "Structural Sheathing," and PS 2 , "Performance Standard for Wood-based Structural-Use Panels," other than structural 1 sheathing) continuous over $\geq$ 2 spans, $\perp$ to supports. Uniform load deflection $\leq$ 1/180 $\times$ span (DL + LL), $\leq$ 1/240 $\times$ span (LL only). Edges supported with blocking. *[IBC Table 2304.7(5)]*

2304 General Construction Requirements

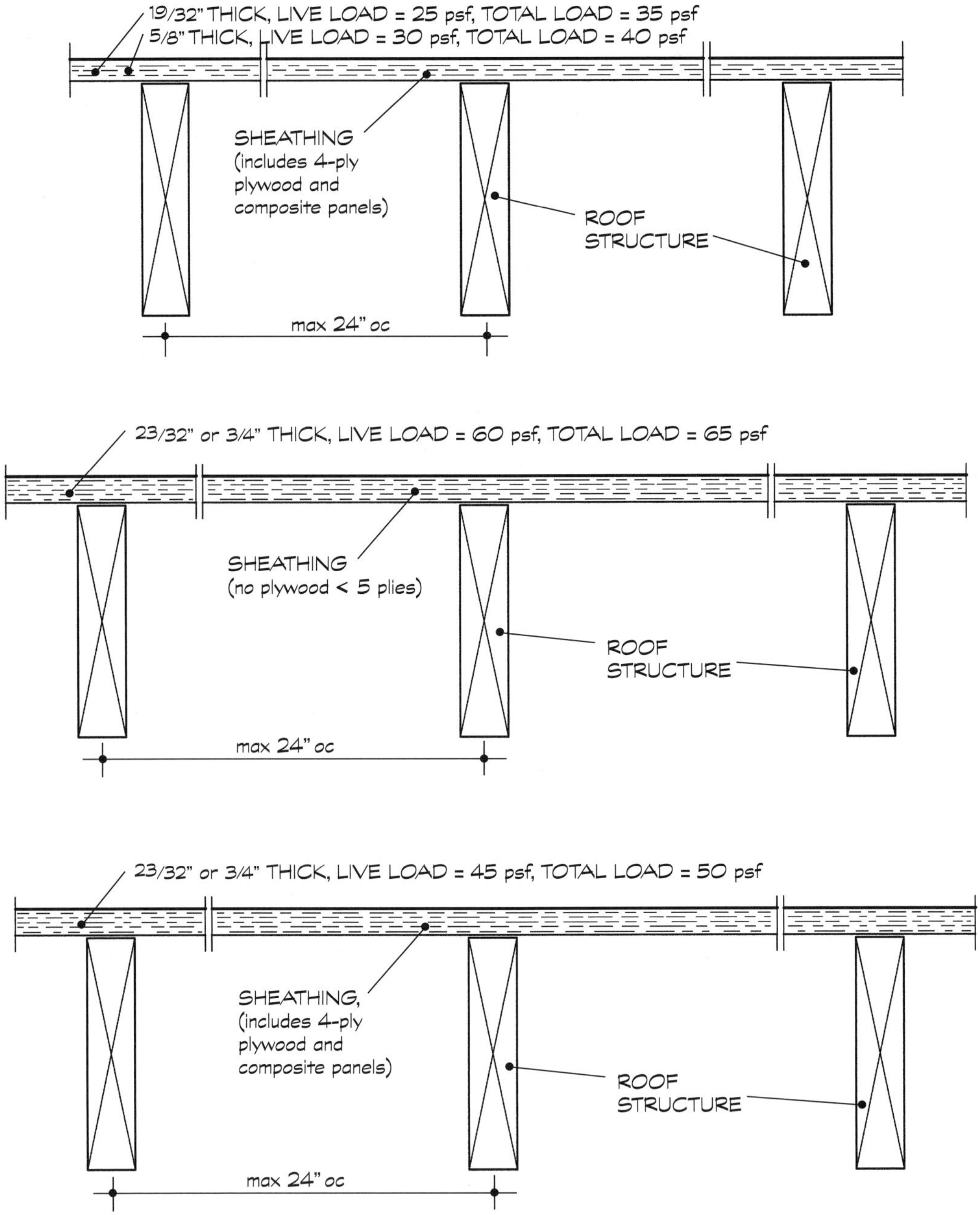

Fig. 2304.7.2C(4). Spans, loads, and minimum thicknesses for wood panel roof sheathing (as described in Department of Commerce documents PS 1, "Structural Sheathing," and PS 2 , "Performance Standard for Wood-based Structural-Use Panels," other than structural 1 sheathing) continuous over ≥ 2 spans, ⊥ to supports. Uniform load deflection ≤ 1/180 × span (DL + LL), ≤ 1/240 × span (LL only). Edges supported with blocking. *[IBC Table 2304.7(5)]*

2308 Conventional Light-Frame Construction

2308.9.1 Size, height and spacing

- The following studs are not governed by this section where they occur at certain openings:
  - Cripple studs.
  - Jack studs.
  - Trimmer studs.

Note: IBC Table 2308.9.5, "Header and Girder Spans for Exterior Bearing Walls," is cited as governing the openings where the abovelisted studs might occur and, therefore, be exempt from this section.

- Utility-grade studs are governed as follows:
  - Spacing must be ≤ 16" oc.
  - They may not support more than the following:
    A ceiling.
    A roof.
  - Length is limited to the following:
    ≤ 8' for exterior walls and load-bearing walls.
    ≤ 10' for interior walls and nonload-bearing walls.
- Studs must be continuous from the sole plate to the top plate as follows:
  - Studs must resist loads ⊥ to the wall.
  - Plates must be supported by one of the following as applicable:
    Ceiling diaphragm.
    Floor.
    Foundation.
    Roof diaphragm.
    An assembly designed by accepted engineering practice.
- Other studs are governed as shown in the details provided in Figures 2308.9.1A–2308.9.1E on the following pages.

Note: IBC Table 2308.9.1, "Size, Height and Spacing of Wood Studs," is cited as the source of requirements and is the basis for the details provided in this section.

2308 Conventional Light-Frame Construction

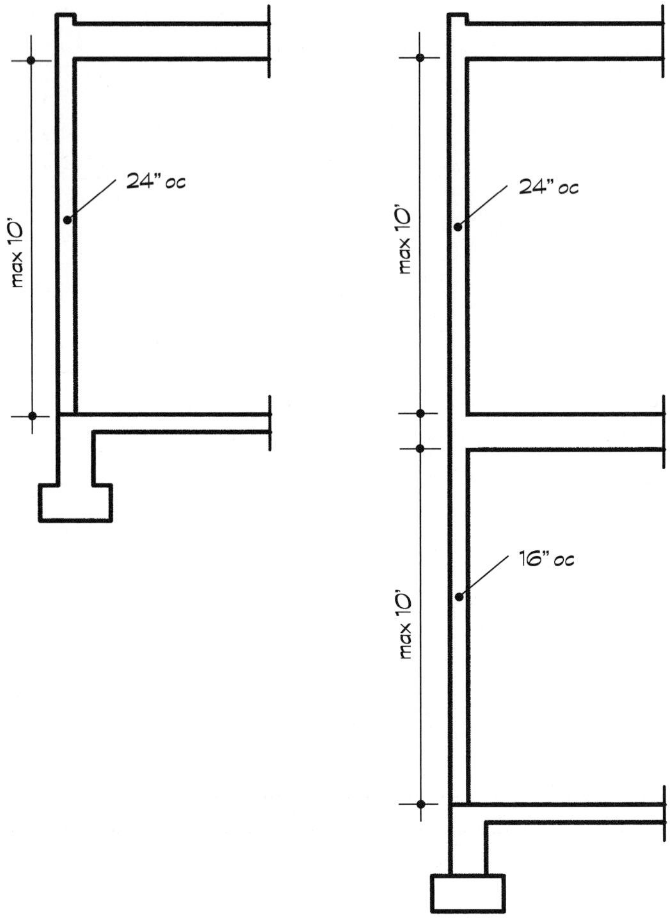

Fig. 2308.9.1A. Maximum height between lateral supports for 2" × 4" wood studs in load-bearing walls.

2308 Conventional Light-Frame Construction

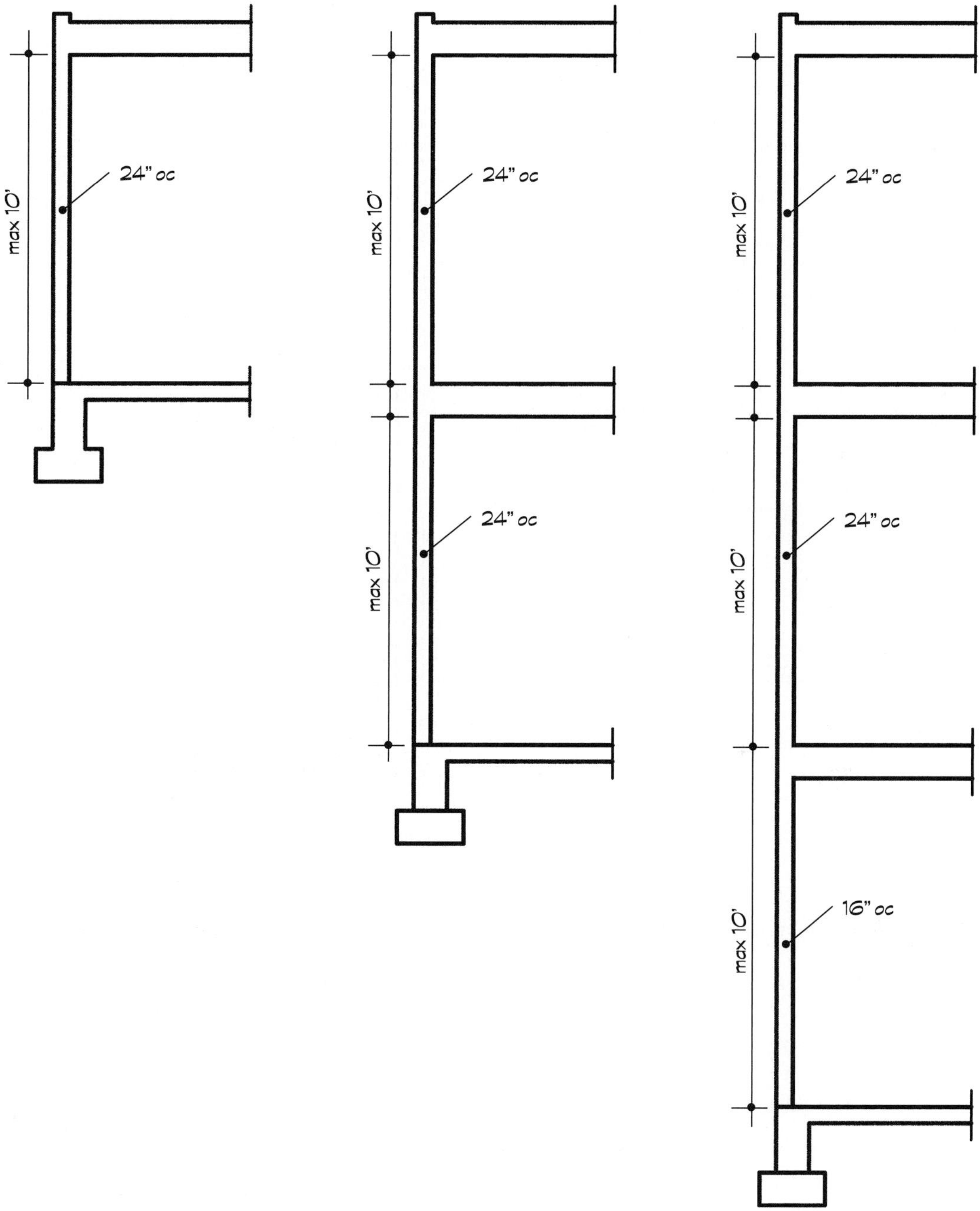

Fig. 2308.9.1B. Maximum height between lateral supports for 3"× 4" wood studs in load-bearing walls.

2308 Conventional Light-Frame Construction

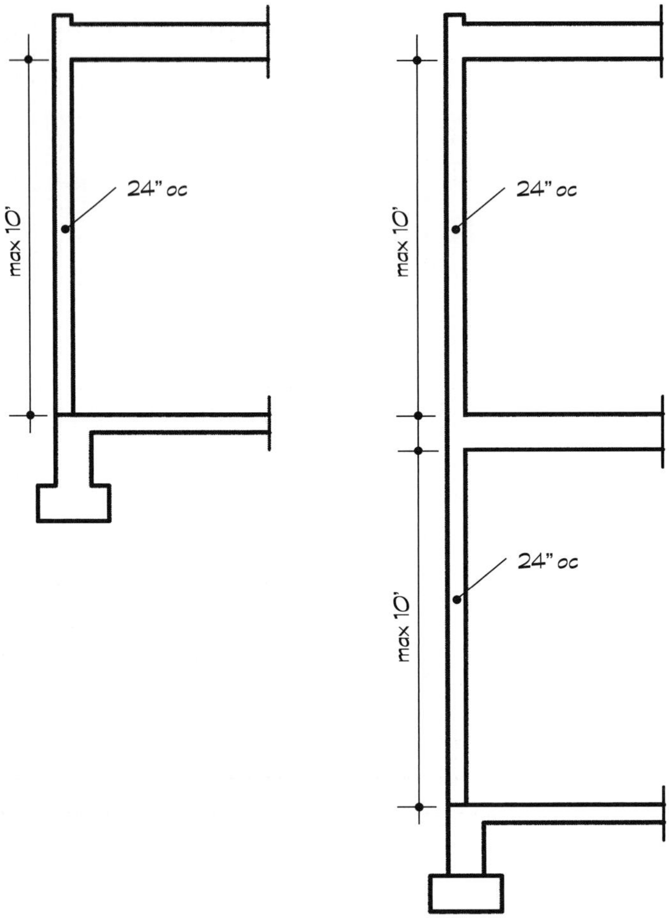

Fig. 2308.9.1C. Maximum height between lateral supports for 2"× 5" wood studs in load-bearing walls.

2308 Conventional Light-Frame Construction

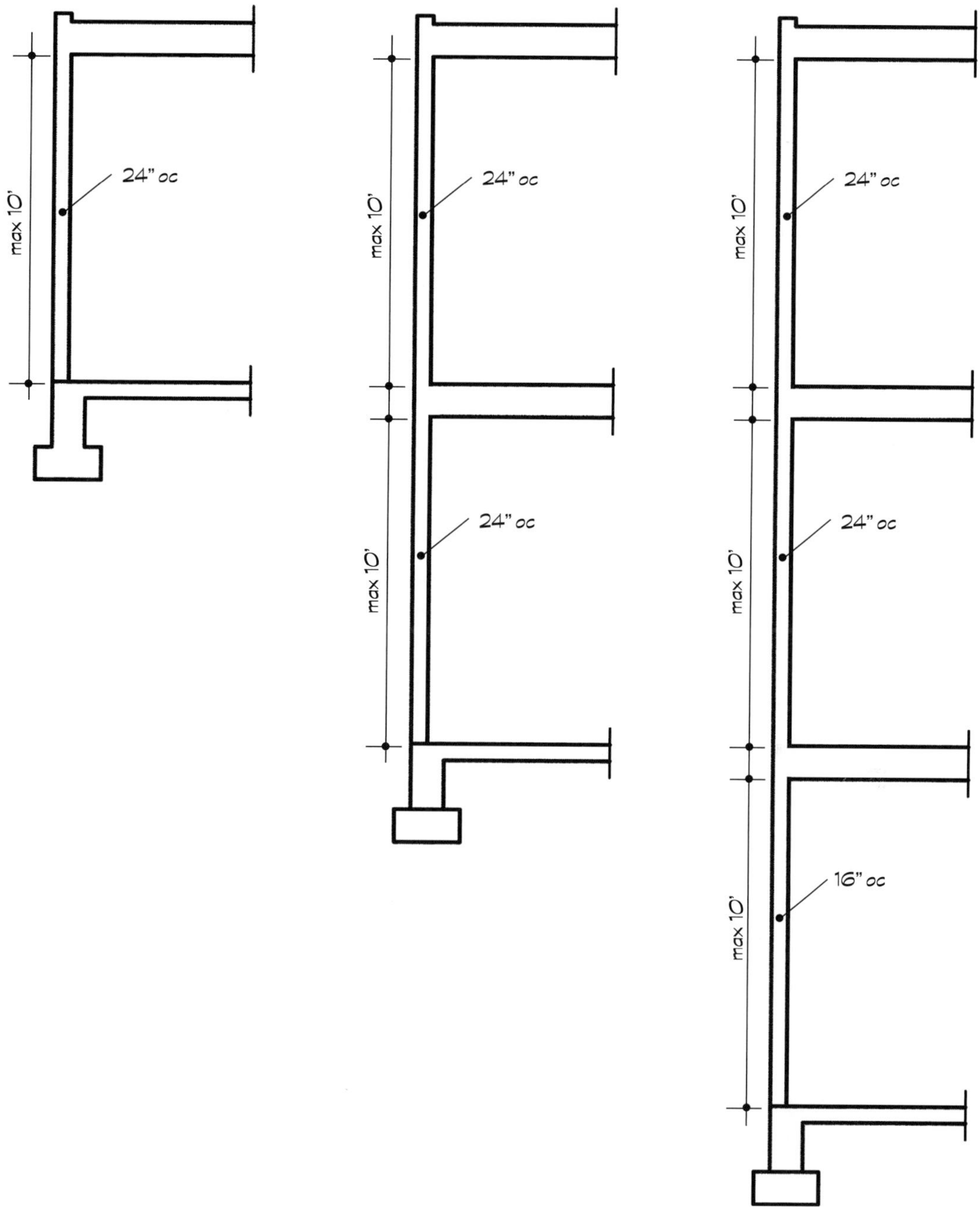

Fig. 2308.9.1D. Maximum height between lateral supports for 2" × 6" wood studs in load-bearing walls.

2308 Conventional Light-Frame Construction

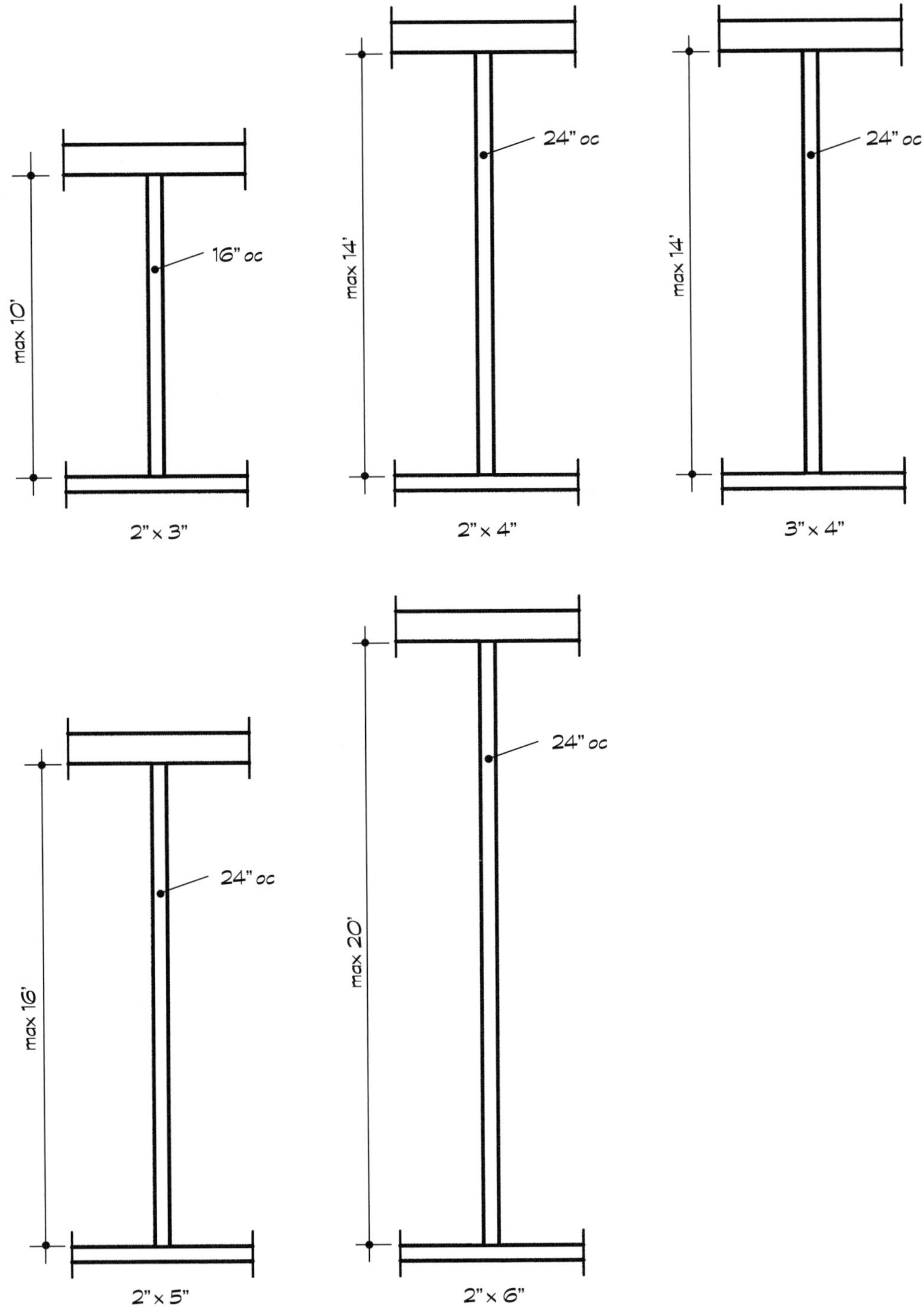

Fig. 2308.9.1E. Maximum height between lateral supports for wood studs in nonload-bearing walls.

2308 Conventional Light-Frame Construction

2308.9.3 Bracing *(part 1 of 3)*

- Braced wall lines must consist of the following:
  - Braced wall panels as follows:
    - Panels must meet the following requirements of this section:
      - Location within the wall.
      - Type of bracing.
      - Amount of bracing.

Note: The following are cited as sources of requirements for braced wall panels:
IBC Fig. 2308.9.3, "Basic components of the lateral bracing system."
IBC Table 2308.9.3(1), "Braced Wall Panels."

- Braced wall panels must be located as follows:
  - In one of the following positions:
    - Aligned with each other.
    - Offset ≤ 4'.
  - ≤ 12'- 6" from each end of a braced wall line.
- Braced wall panels must be identified on construction drawings.
- Braced wall panels must be detailed as follows:
  - Vertical joints of panel sheathing must fall on studs.
  - Adjacent panels must be fastened to the same framing member as follows:
    - Where edges meet.
  - Horizontal joints must fall on blocking or other framing as follows:
    - Blocking or other framing must be one of the following:
      - The same size as the studs.
      - The size specified by installation requirements for the sheathing material.
  - Blocking is required under braced wall lines in the following case:
    - Where the joists below are || to braced wall lines.
  - Sole plates must be nailed to floor framing.
  - Top plates must be nailed to the framing above.

Note: 2308.3.2, "Braced wall line connections," is cited as governing sole plate and top plate connections to adjacent framing.

- Cripple walls must meet the following bracing requirements:
  - Where stud height ≥ 1'-2":
    - Where located in Seismic Design Category A, B, or C:
      - Requirements are the same for full height walls.
    - Where located in Seismic Design Category D or E:
      - Requirements other than those for full height walls apply.

Note: 2308.9.4.1, "Bracing," is cited as the source of requirements for cripple walls.

2308 Conventional Light-Frame Construction

2308.9.3 Bracing *(part 2 of 3)*

- Braced wall panels must be constructed as one of the following systems:
  - Continuous diagonal lumber braces with the following characteristics:
    - Nominal 1" × 4".
    - Let into:
      - Bottom plates.
      - Intervening studs.
      - Top plates.
    - Installed at an angle as follows:
      - ≥ 45° and ≤ 60° from the horizontal.

Note: IBC Table 2304.9.1, "Fastening Schedule," is cited as governing the size, type, and location of fasteners required for the bracing system described above.

  - Wood boards with the following characteristics:
    - ≥ 5/8" net thickness.
    - Installed diagonally on studs:
      - Spaced ≤ 24" on center.
    - Length of panel (measured horizontally) must be ≥ 4'.
    - Where studs are spaced ≤ 16" on center:
      - Panel must span ≥ 3 stud spaces.
    - Where studs are spaced ≤ 24" on center:
      - Panel must span ≥ 2 stud spaces.
  - Wood structural panel sheathing as shown on the details provided in Figures 2308.9.3A(1) – 2308.9.3B on the following pages.

Note: The following are cited as governing the structural panels indicated above and are the basis of the details provided in this section:
IBC Table 2308.9.3(2), "Exposed Plywood Panel Siding."
IBC Table 2309.9.3(3), "Wood Structural Panel Wall Sheathing."

  - Fiberboard sheathing panels with the following characteristics:
    - ≥ 1/2" thickness.
    - Installed as follows on studs spaced ≤ 16" on center:
      - Vertically.
      - Horizontally.
    - Length of panel (measured horizontally) must be ≥ 4'.
    - Panel must span ≥ 3 stud spaces.

Note: The following are cited as governing the fastening of the fiberboard panels described above:
2306.6, "Fiberboard shear walls."
IBC Table 2306.6, "Allowable Shear Values (plf) for Wind or Seismic Loading on Shear Walls of Fiberboard Sheathing Board Construction for Type V Construction Only."

2308 Conventional Light-Frame Construction

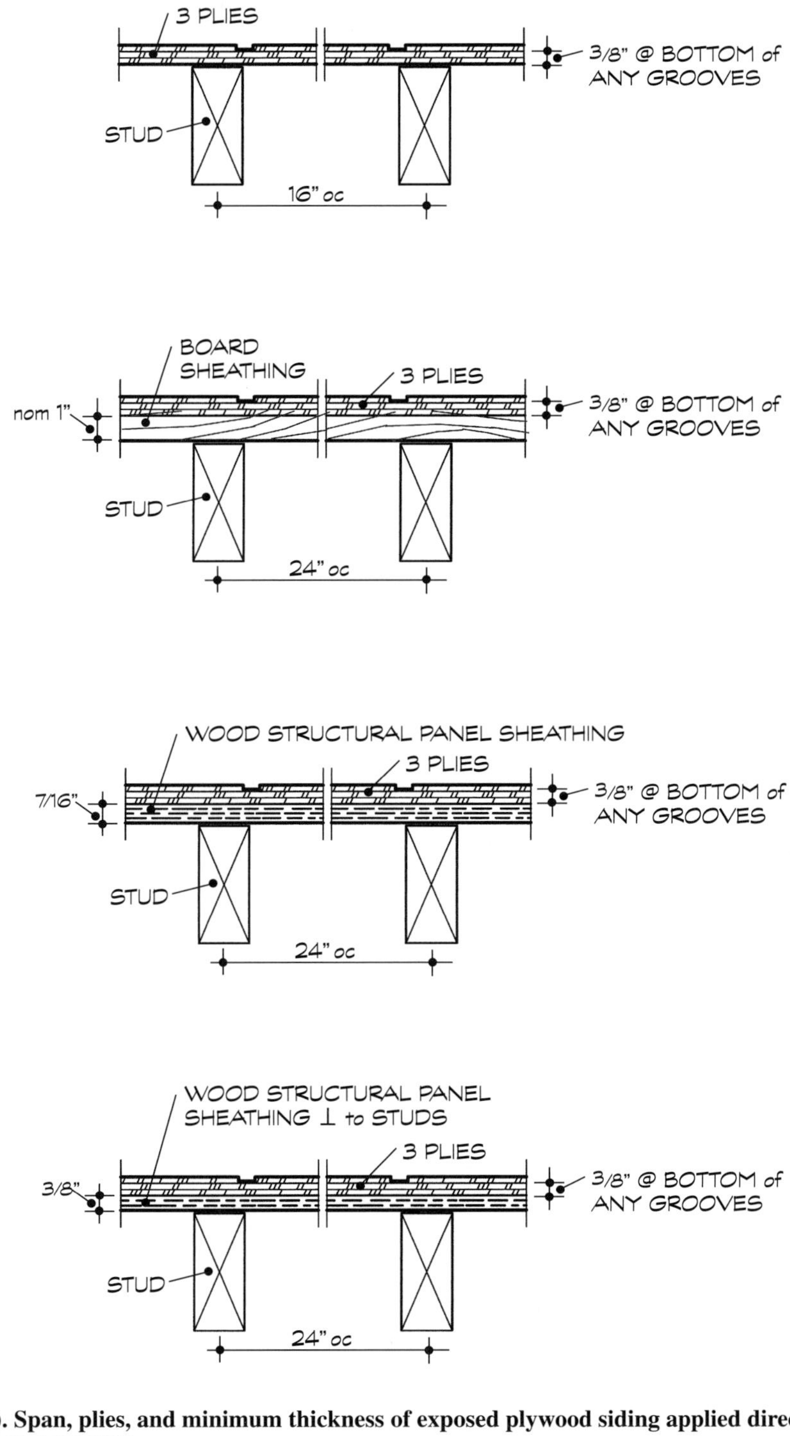

Fig. 2308.9.3A(1). Span, plies, and minimum thickness of exposed plywood siding applied directly to studs or over sheathing. *[IBC Table 2308.9(2)]*

2308 Conventional Light-Frame Construction

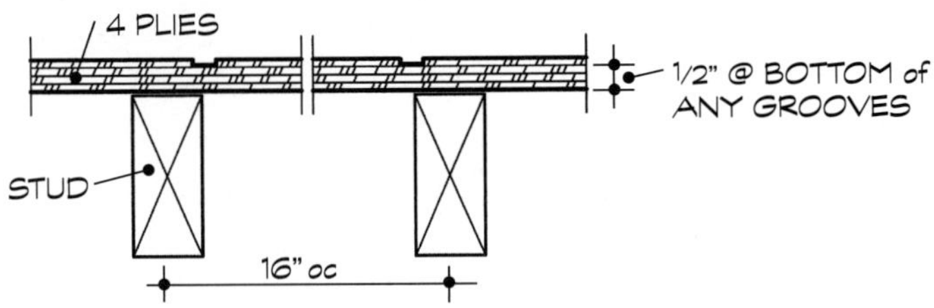

Fig. 2308.9.3A(2). Span, plies, and minimum thickness of exposed plywood siding applied directly to studs or over sheathing. *[IBC Table 2308.9(2)]*

2308 Conventional Light-Frame Construction

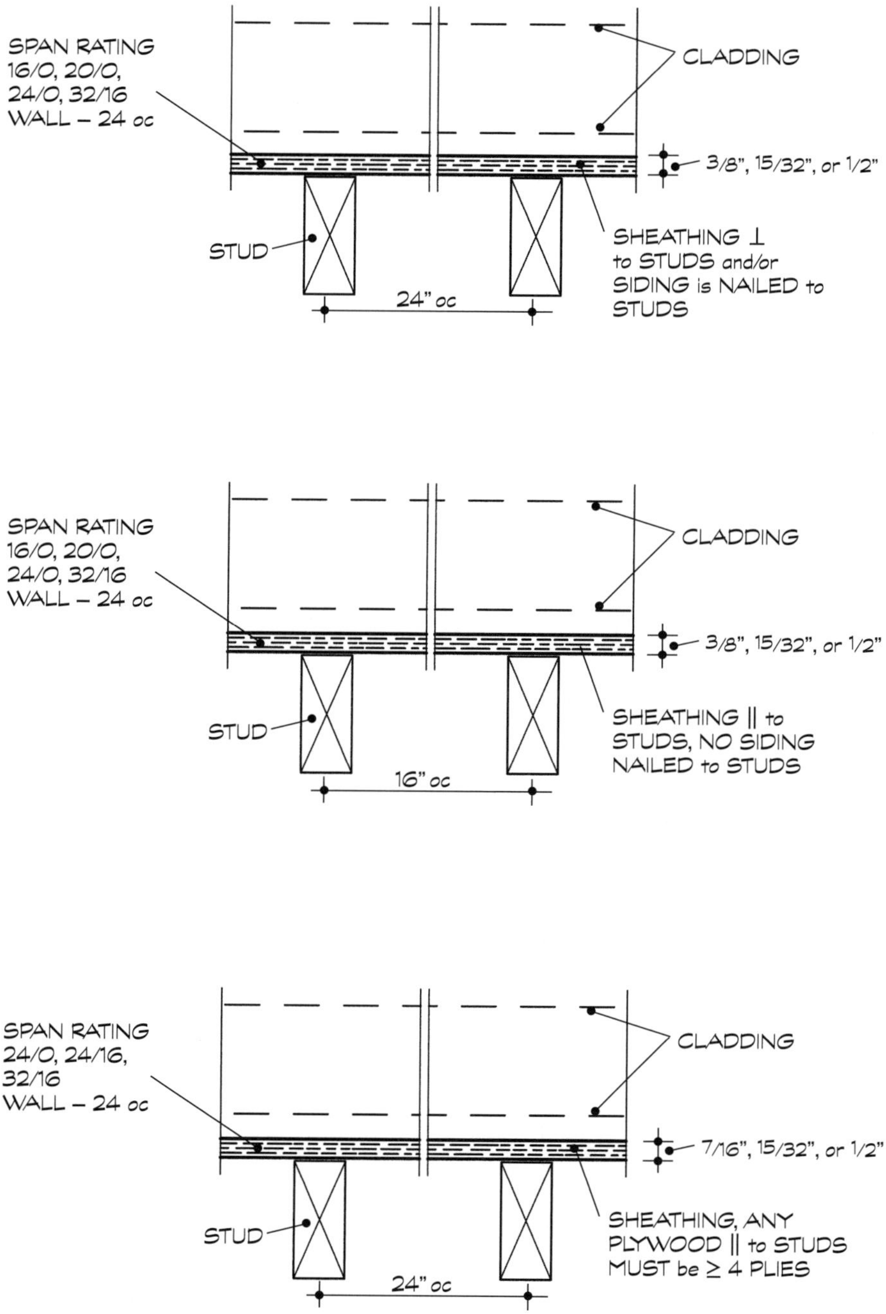

Fig. 2308.9.3B. Span and minimum thickness of wood structural wall sheathing not exposed to weather.
[IBC Table 2308.9(3)]

2308 Conventional Light-Frame Construction

2308.9.3 Bracing *(part 3 of 3)*

- Gypsum board as follows:
  - Types:
    - Sheathing.
    - Wallboard.
    - Veneer base.
  - Requirements:
    - 1/2" thick.
    - 4' wide.
    - Nailed 7" on center to studs ≤ 24" on center.
    - Where applied to one side of a bracing panel:
      - Length must be ≥ 8' (measured horizontally).
    - Where applied to both sides of a bracing panel:
      - Length must be ≥ 4' (measured horizontally).

Note: IBC Table 2306.7, "Allowable Shear for Wind or Seismic Forces for Shear Walls of Lath and Plaster or Gypsum Board Wood Framed Wall Assemblies," is cited as governing the fastening of gypsum board as described above.

- Particleboard wall sheathing panels with the following characteristics as shown in the details provided in this section:
  - Length of panel (measured horizontally) must be ≥ 4':
    - Where studs are spaced ≤ 16" on center:
      - Panel must span ≥ 3 stud spaces.
    - Where studs are spaced ≤ 24" on center:
      - Panel must span ≥ 2 stud spaces.

Note: IBC Table 2308.9.3(4), "Allowable Spans for Particleboard Wall Sheathing," is cited as governing particleboard wall sheathing and is the basis of the details provided in this section.

- Portland cement plaster as follows:
  - On studs ≤ 16" on center.
  - Length of panel (measured horizontally) must be ≥ 4'.
  - Panel must span ≥ 3 stud spaces.

Note: Section 2510, "Lathing and Furring for Cement Plaster (Stucco)," is cited as governing the application of portland cement plaster as described above.

- Hardboard panel siding as follows:
  - Length of panel (measured horizontally) must be ≥ 4'.
  - Where studs are spaced ≤ 16" on center:
    - Panel must span ≥ 3 stud spaces.
  - Where studs are spaced ≤ 24" on center:
    - Panel must span ≥ 2 stud spaces.

Note: 2303.1.6, "Hardboard," is cited as a source of requirements.
IBC Table 2308.9.3(5), "Hardboard Siding," is cited as a source of requirements.

2308 Conventional Light-Frame Construction

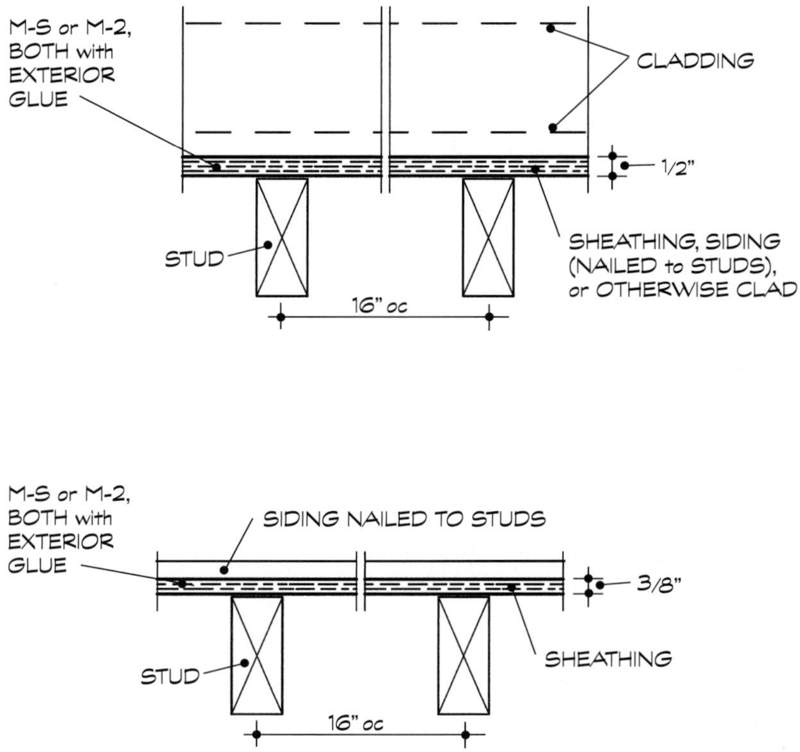

Fig. 2308.9.3C. Span and minimum thickness of particleboard wall sheathing not exposed to weather.
[IBC Table 2308.9(4)]

NOTES

24

Glass and Glazing

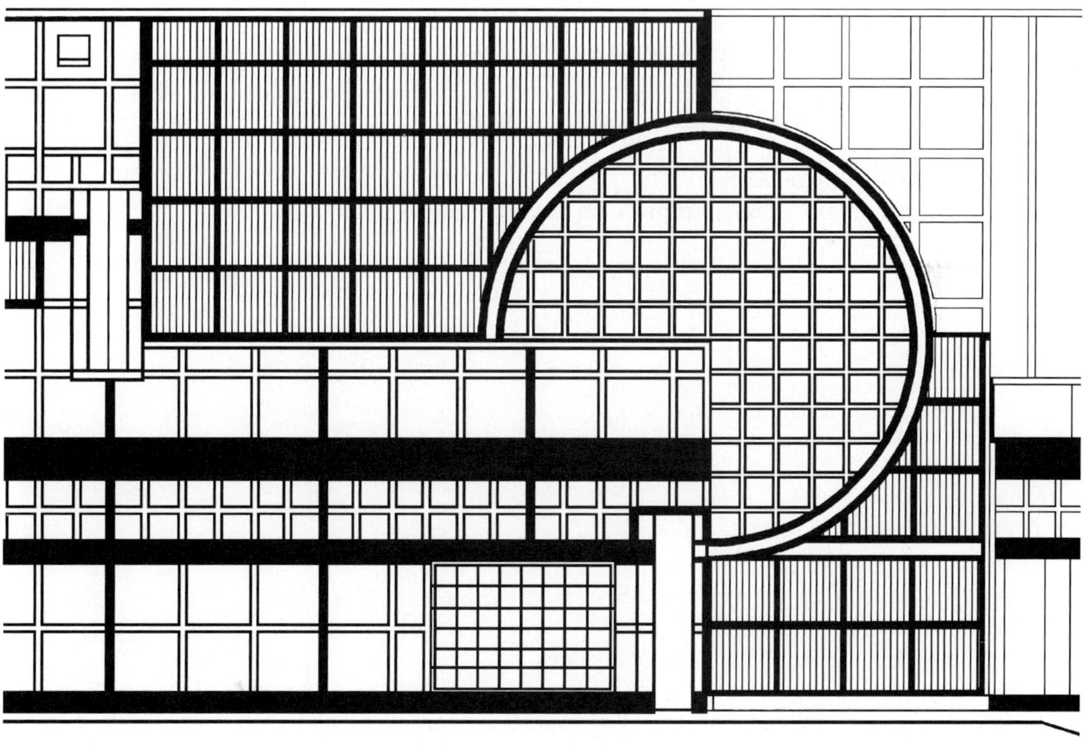

Lubrication Engineers, Inc. Wichita, Kansas. *(partial elevation)*
Gossen Livingston Associates, Inc. Architecture. Wichita, Kansas.

2403 General Requirements for Glass

2403.3 Framing

- A glass edge qualifies as having a firm support where deflection due to the loading listed below is limited as follows:
  - Deflection:
    - The edge deflection ⊥ to a pane of glass may not be > either of:
      - 1/175 the length of the glass edge.
      - 3/4".
  - Loading:
    - Where the larger of the following loads is applied where loads are combined:
      - The positive load.
      - The negative load.

Note: Section 1605, "Load Combinations."

2403.4 Interior glazed areas

- Deflection of interior glazing in the following location is limited as listed:
  - Location:
    - Adjacent to a walking surface.
  - Deflection:
    - The differential deflection of adjacent unsupported edges is governed as follows:
      - Deflection must be ≤ the thickness of the glass in the following case:
        - Where a force of 50 lbs/ft is applied:
          - Horizontally to 1 panel.
          - At any point ≤ 3'-6" above the walking surface.

2403.5 Louvered windows or jalousies

- In the following conditions, glass must meet the requirements listed:
  - Conditions:
    - Glass:
      - Float.
      - Wired.
      - Patterned.
    - Locations:
      - Louvered windows.
      - Jalousies.
  - Requirements:
    - Glass must be ≥ 3/16" thick.
    - Glass must be ≤ 4' long.
    - Exposed edges of glass must be smooth.
    - Wire glass may not be used:
      - Where wire is exposed on longitudinal edges.
    - Where other glass types are used:
      - Design must be provided to the building official for approval.

2405 Sloped Glazing and Skylights

2405.1 Scope

- This section applies to the following glazing where sloped as listed:
  - Glazing:
    - Glass.
    - Transparent materials.
    - Translucent materials.
    - Opaque glazing materials.
    - Glazing materials in the following:
      - Skylights.
      - Roofs.
      - Sloped walls.
  - Slope:
    - Where sloped > 15 degrees from vertical plane.

2405.2 Allowable glazing materials and limitations

- The following materials are permitted in sloped glazing:
  - Laminated glass with one of the following:
    - A polyvinyl butyral interlayer ≥ 30 mils thick.
    - An equivalent interlayer.
  - Wired glass.
  - Light-transmitting plastics.

Note: Section 2607, "Light-Transmitting Plastic Wall Panels," is cited as governing these plastics.

  - Heat-strengthened glass.
  - Fully tempered glass.
  - Annealed glass as follows:
    - Where there is no walking surface below.
    - Where any walking surface below is protected from falling glass.
    - In commercial or detached noncombustible greenhouses as follows:
      - Used only for growing plants.
      - Closed to the public.
      - Height of greenhouse at ridge is ≤ 30' above grade.

Note: 2405.3, "Screening," Exceptions 2 and 3, are cited as governing annealed glass and are summarized above.

Section 2610, "Light-Transmitting Plastic Skylight Glazing," is cited as the source of additional requirements for plastic skylights.

2101.2.5, "Glass unit masonry," is cited as governing glass block installations.

2405 Sloped Glazing and Skylights

2405.3 Screening *(part 1 of 2)*

- Screens are not required under fully tempered glass:
  - Where the glazing occurs between floors as follows:
    - Glazing is sloped ≤ 30 degrees from a vertical plane.
    - Highest point of glass is ≤ 10' above the walking surface.
- Screens are not required below the following glazing for the conditions listed:
  - Glazing:
    - Any glazing including annealed glass.
  - Conditions:
    - Where one of the following conditions applies:
      - Where there is no walking surface below.
      - Where any walking surface below is protected from falling glass.
- Screens are not required below the following glazing in the locations listed:
  - Glazing:
    - Any glazing including annealed glass.
  - Locations:
    - In commercial or detached noncombustible greenhouses as follows:
      - Used only for growing plants.
      - Closed to the public.
      - Height of greenhouse at ridge is ≤ 30' above grade.
- Screens are not required in the following locations for the conditions listed:
  - Locations:
    - In occupancies R-2, R-3, R-4.
  - Conditions:
    - Area of each pane of glass is ≤ 16 sf.
    - Highest point of glass is ≤ 12' above either:
      - A walking surface.
      - Any other area which may be accessed.
    - Where glazing is fully tempered glass:
      - Glass thickness is ≤ 3/16".
    - Where glazing is laminated glass:
      - One of the following interlayers must be provided:
        - Polyvinyl butyral ≥ 15 mils thick.
        - An equivalent interlayer.

2405 Sloped Glazing and Skylights

2405.3 Screening *(part 2 of 2)*

- Screens, as listed below, are required under the following sloped glazing:
  - Glazing:
    - Includes the following glass in the formats listed:
      - Glass:
        - Heat-strengthened glass.
        - Fully tempered glass.
      - Formats:
        - Glazing with a single layer of glass.
        - The bottom layer among multiple layers of glass.
  - Screens:
    - Must be able to support 2 × the weight of the glazing.
    - Must be securely fastened to framing.
    - Must be installed ≤ 4" of the glass.
    - Must be noncombustible.
    - Must be ≥ #12 B&S gage (0.0808") mesh ≤ 1" × 1".
    - Where located in a corrosive atmosphere:
      - Equivalent noncorrosive screening is required.

2405.4 Framing *(part 1 of 2)*

- The following must be noncombustible in Type I and II construction:
  - Frames for sloped glazing.
  - Frames for skylights.
- In environments with acid fumes that damage metals, the following components may be constructed of the materials listed:
  - Components:
    - Sash and frames of skylights.
    - Sash and frames of sloped glazing.
  - Materials:
    - Approved pressure-treated wood.
    - Other approved noncorrosive material.
- Curbs for skylights are governed as follows:
  - Occupancy R-3 has the following requirements regarding curbs:
    - On roofs sloping ≥ 3:12 curbs are not required.
    - On roofs sloping < 3:12 curbs are required as follows:
      - Skylights must be mounted ≥ 4" above the roof on a curb.
      - Curb must be constructed according to one of the following:
        - As required for the framing.
        - As per manufacturer's instructions.
  - In other locations, the following is required regarding curbs:
    - Where the roof slope is < 45°:
      - Curbs ≥ 4" high are required for skylights.
      - Curbs must be constructed as required for the framing.

2405 Sloped Glazing and Skylights

2405.4 Framing *(part 2 of 2)*

- Framing supporting skylights and sloped glazing must be designed as follows:
  - To resist tributary roof loads assigned by the code.

Note: Chapter 16, "Structural Design," is cited as the source of tributary roof loads.

Case study: Fig. 2405.4. The roof of the Occupancy B building slopes 4:12, thus, requiring a 4" curb at skylights. Such a curb is provided as indicated in the illustration below. Consequently, the skylight is in compliance with code requirements.

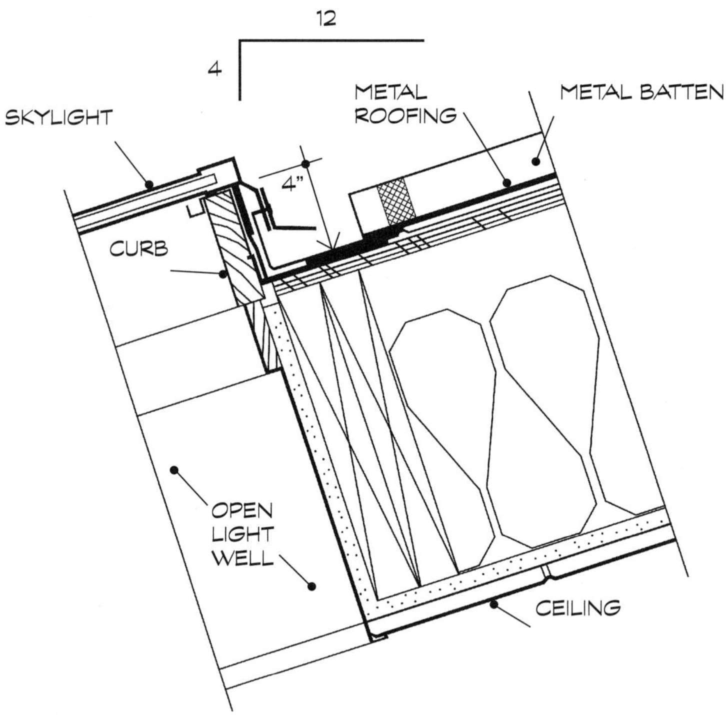

Fig. 2405.4. Detail at skylight. Central Kitchen. Lompoc Unified School District. Lompoc, California. Phillips Metsch Sweeney Moore Architects. Santa Barbara, California.

2406 Safety Glazing

2406.4 Hazardous locations *(part 1 of 3)*

- This section addresses the requirement for safety glazing where glazing is used.
- Safety glazing is not required in nonhazardous locations as noted in the next section.

Note: 2406.4.1, "Exceptions," lists locations not hazardous for glazing.

- The following locations require safety glazing:
  - Sliding door units as follows:
    - Sliding panels.
    - Fixed panels.
  - Sliding and bifold closet doors.
  - Storm doors.
  - Unframed swinging doors.
  - In the following elements at the bathing-type locations listed:
    - Elements:
      - Doors.
      - Enclosures.
      - Any building wall serving as an enclosure:
        - Where the lowest exposed glazing is < 5' above the standing surface.
    - Locations:
      - Bathtubs.
      - Hot tubs.
      - Saunas.
      - Showers.
      - Steam rooms.
      - Whirlpools.
- The following glazing near doors is governed as listed:
  - Glazing:
    - Fixed or operable.
    - Adjacent to a door with both the following characteristics:
      - Exposed glazing is ≤ 2' from the door as follows:
        - Measured on the shortest line to the nearest edge of the closed door.
      - Lowest exposed glazing is < 5' above the walking surface.
  - Requirements:
    - The following conditions do not require safety glazing:
      - Where the glazing is decorative glass.
      - Where there is a wall or barrier as follows:
        - Between the door and the glazing.
      - Where the door opens to one of the following spaces ≤ 3' deep:
        - Closet.
        - Storage.

Note: 2406.4, "Hazardous locations," item 7, is cited as governing this glazing.

      - Where the glazing is ⊥ to the closed door and both of the following apply:
        - In 1- and 2-family dwellings.
        - In Occupancy R-2.
    - Other conditions require safety glazing.

2406 Safety Glazing

Case study: Fig. 2406.4. Safety glazing is required on either side of the entry doors since it is within 2' of the door. The glazing above the door is not required to be safety glazing as it is above a height of 5'. Safety glazing is required in the swinging doors. Tempered glass is provided in the doors and on each side of the doors, thus, the entry is in compliance with the code.

Fig. 2406.4. Partial elevation at east entry. Hot Springs Police Department New Headquarters. Hot Springs National Park, Arkansas. Cromwell Architects Engineers. Little Rock, Arkansas.

2406 Safety Glazing

2406.4 Hazardous locations *(part 2 of 3)*

- Safety glazing in swinging doors is governed as follows:
  - It is not required in swinging jalousie doors.
  - It is not required where the glazing is decorative glass.
  - It is required in other swinging doors.

  Note: 2406.4.1, "Exceptions," is cited as the source of requirements for jalousies without safety glazing.

- The following glazing is governed as listed:
  - The following glazing is not required to be safety glazing:
    - That which is protected by a bar as follows:
      - Bar has a vertical dimension ≥ 1 1/2".
      - Bar is able to resist a 50-lb/ft load applied horizontally as follows:
        - Without deflecting to contact the glazing.
      - Bar is located as follows:
        - On the side of glazing to which there is access.
        - ≥ 2'-10" and ≤ 3'-2" above the walking surface.
    - The exterior pane of multiple layers of glazing as follows:
      - Where the lowest exposed glazing is ≥ 25' above:
        - Any of the following surfaces adjacent to the exterior of the glazing:
          - Grade.
          - Roof.
          - Walking surface.
          - Other horizontal or sloped surface.
  - The following glazing is governed elsewhere in this section:
    - That which is required to have safety glazing as follows:
      - Where glazing is < 5' above a standing surface in:
        - Bathing-type locations.
        - Near doors.
  - Glazing in locations other than those indicated above is governed as follows:
    - Decorative glass is not required to be safety glazing.
    - Otherwise, safety glazing is required where all of the following conditions apply:
      - Exposed surface has all the following characteristics:
        - Area of any pane is > 9 sf.
        - Bottom edge is < 1'-6" above the floor.
        - Top edge is > 3' above the floor.
      - Plane of glazing is ≤ 3' from a walking surface:
        - Measured horizontally.
- Safety glazing is required for the following components in the locations listed:
  - Components:
    - The following components with any area or height are included:
      - Structural baluster panels.
      - Nonstructural in-fill panels.
  - Locations:
    - Guards and railings.

2406 Safety Glazing

2406.4 Hazardous locations *(part 3 of 3)*

- Safety glazing is required in the following locations where the conditions listed apply:
  - Locations:
    - Walls and fences as follows:
      - Enclosing the following both indoors and outdoors:
        - Swimming pools.
        - Hot tubs and spas.
  - Conditions:
    - Where all of the following conditions apply:
      - Bottom edge of glazing is < 5' above the walking surface:
        - On the side where water is contained.
      - Glazing is ≤ 5' from the edge of the water:
        - Measured horizontally.
  - Glazing adjacent to the following elements is governed as listed:
    - Elements:
      - Stairways and ramps.
      - Landings.
    - Requirements:
      - Safety glazing is not required where the following condition applies:
        - The side of the element has:
          - A guard or handrail as follows:
            - With one of the components:
              - Balusters.
              - In-fill panels.
            - Located ≥ 1'-6" from the glazing.

Note: The following are cited as governing the guards and handrails:
Section 1013, "Guards."
1607.7, "Loads on handrails, guards, grab bars, seats and vehicle barrier systems."

- Safety glazing is required where the glazing meets both of the following conditions:
  - Glazing is ≤ 3' from a walking surface as follows:
    - Measured horizontally.
  - Bottom edge of glazing is < 5' above the adjacent walking surface.
- Safety glazing is required at stairways where the glazing meets both of the following conditions:
  - Glazing is ≤ 5' from the bottom stairway tread:
    - Measured horizontally in any direction.
  - Bottom edge of glazing is < 5' above the tread nosing.

2406 Safety Glazing

2406.4.1 Exceptions

- The following are not hazardous locations requiring safety glazing:
  - Openings in doors as follows:
    Able to pass a 3" sphere.
  - Decorative glass.

  Note: 2406.4, "Hazardous locations," item 1, 6, or 7 is cited as specifying locations and conditions wherein decorative glass need not be safety glazing.

  - Curved glazing as follows:
    In revolving doors.
  - Glazed doors as follows:
    In commercial refrigeration.
  - Glass block.

    Note: 2101.2.5, "Glass unit masonry," is cited as governing glass block.

  - Louvered glazing as follows:
    Windows.
    Jalousies.

    Note: 2403.5, "Louvered windows or jalousies," is cited as governing this glazing.

  - The following glazing located as listed:
    Glazing:
      Mirrors.
      Other glass panels.
    Location:
      On a surface providing the following:
        Support across the entire back of the glazing.

NOTES

25

Gypsum Board and Plaster

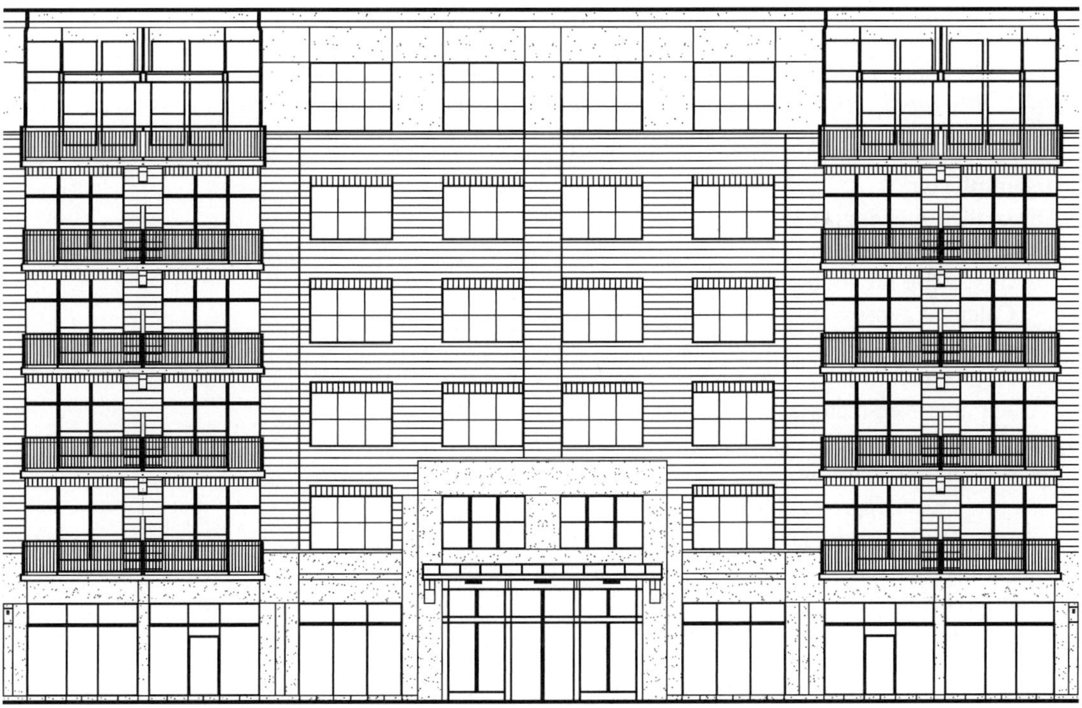

Hoyt Street Properties. Portland Oregon. *(partial elevation)*
Ankrom Moisan Associated Architects. Portland, Oregon.

2502 Definitions

2502.1 Definitions *(part 1 of 2)*

- **Cement plaster**
  - One of the following mixtures:
    - Blended cement, aggregate, or other approved materials.
    - Blended cement, hydrated lime, aggregate, other approved materials.
    - Masonry cement, aggregate, other approved materials.
    - Plastic cement, aggregate, other approved materials.
    - Portland cement, aggregate, other approved materials.
    - Portland cement, hydrated lime, aggregate, other approved materials.
  - Other approved materials as specified in the code.
- **Exterior surfaces**
  - Surfaces exposed to the weather.
- **Gypsum board**
  - Any of the following:
    - Exterior gypsum soffit board.
    - Gypsum sheathing.
    - Gypsum veneer plaster base.
    - Gypsum wallboard.
    - Predecorated gypsum board.
    - Water-resistant gypsum backing board.

Note: The following are cited as listing standards governing gypsum board:
IBC Table 2506.2, "Gypsum Board Materials and Accessories."
IBC Table 2507.2, "Lath, Plastering Materials and Accessories."
Chapter 35, "Referenced Standards."

- **Gypsum plaster**
  - Any of the following:
    - Calcined gypsum.
    - A mixture of the following:
      - Calcined gypsum.
      - Lime.
      - Aggregate.
    - Other approved materials.
- **Gypsum veneer plaster**
  - Gypsum plaster as follows:
    - Applied to an approved base.
    - Applied in 1 or more layers.
    - Usually ≤ 1/4" thick.
- **Interior surfaces**
  - Surfaces not exposed to the weather.

2502 Definitions

2502.1 Definitions *(part 2 of 2)*

- **Weather-exposed surfaces**
  - The following surfaces are not included:
    - Ceilings and roof soffit surfaces as follows:
      - Enclosed by the following components that extend ≥ 12" below such surfaces:
        - Beams.
        - Bulkheads.
        - Fascia.
        - Walls.
    - Walls or parts of walls under a roof as follows:
      - In an area that is not enclosed.
      - Set back from the roof edge as follows:
        - A distance ≥ 2 × the height of the open space under the roof edge.
    - Parts of ceilings and roof soffits located as follows:
      - A horizontal distance ≥ 10' from the outer edges of the following:
        - The ceiling or roof soffit.
  - Otherwise, the following surfaces where exposed to the weather are included:
    - Ceilings.
    - Floors.
    - Roofs.
    - Soffits.
    - Walls.
    - Similar surfaces.
  - Certain surfaces exposed to weather and not exposed to weather are shown in Fig. 2502.1.
- **Wire backing**
  - Horizontal strands of wire as follows:
    - Attached to surfaces of vertical supports.
    - Wire is taut.
    - Covered with building paper.
    - Serves as a backing for cement plaster.

2502 Definitions

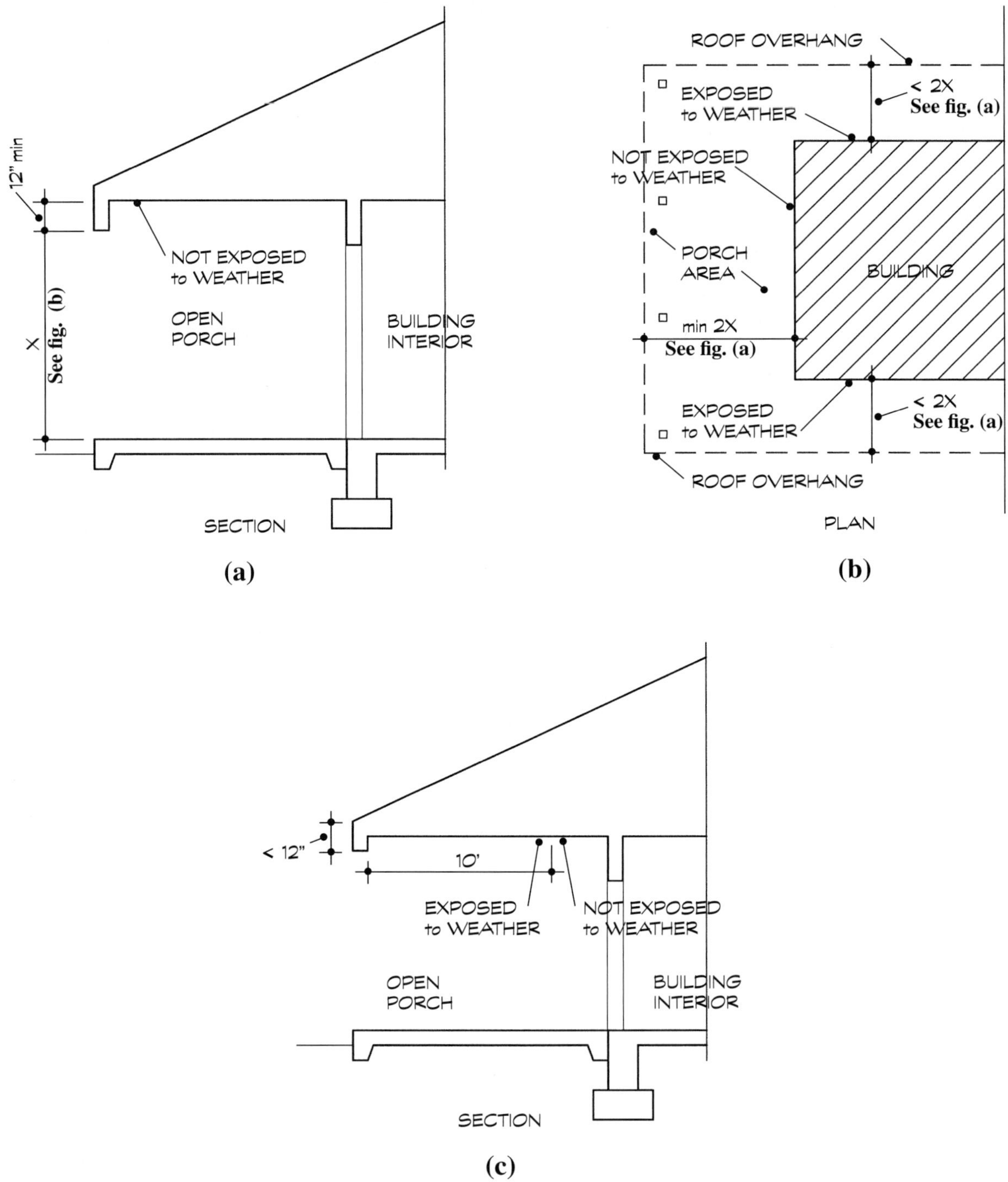

Fig. 2502.1. Conditions of exposure conditions for ceilings, soffits, and walls.

2504 Vertical and Horizontal Assemblies

2504.1.1 Wood framing

- The wood supports for the following must be ≥ 2" in any dimension:
  - Lath.
  - Gypsum board.
- Wood stripping or furring is governed as follows:
  - Where applied to solid backing, the following size is required:
    ≥ 1"× 2".
  - Otherwise, the following size is required:
    ≥ 2" in any dimension.

2504.1.2 Studless partitions

- The required thicknesses of the following vertical studless partitions are ≥ 2":
  - Solid plaster on the following:
    3/8" rib metal lath.
    3/4" rib metal lath.
    1/2" long-length gypsum lath.
  - Gypsum board partitions.

2508 Gypsum Construction

2508.2 Limitations

- The following materials may not be exposed directly to the weather:
  - Gypsum wallboard.
  - Gypsum plaster.
- Gypsum wallboard may not be used in the following locations:
  - With direct exposure to water.
  - With exposure to continuous high humidity.
- Gypsum sheathing may be installed in the following location:
  - On exterior surfaces.

Note: ASTM C 1280, "Specification for Application of Gypsum Sheathing," is cited as governing gypsum sheathing.

2508.2.1 Weather protection

- Weather protection must be provided prior to the installation of the following:
  - Gypsum wallboard.
  - Gypsum lath.
  - Gypsum plaster.

2508.3 Single-ply application

- The following parts of gypsum board must fall on framing members:
  - Edges || to framing members.
  - Ends || to framing members.
- The following parts of gypsum board are governed as listed:
  - Parts:
    - Edges.
    - Ends.
  - Requirements:
    - They need not touch adjacent panels in the following case:
      - In concealed spaces as follows:
        - Where there is no requirement for the following:
          - Fire-resistance-rated construction.
          - Shear resistance.
          - Diaphragm resistance.
      - In other locations they must be in contact with adjacent panels.

2508 Gypsum Construction

2508.3.1 Floating angles

- Where the following conditions apply, the fastening of gypsum board is governed as listed:
  - Conditions:
    - The assembly is not intended to resist shear.
    - The assembly does not require a fire-resistance rating.
  - Requirements:
    - Fasteners are not required at the following locations:
      - Top plates of walls.
      - Bottom plates of walls.
    - Fasteners are not required at the perimeters of the following:
      - Ceilings.
      - Soffits.
- Fastener heads may not break the following:
  - The face paper on the gypsum board.

2508.4 Joint treatment

- This section addresses the following details in fire-resistance-rated assemblies:
  - Joints between gypsum board panels.
  - Fasteners in gypsum board panels.
- The details need not be finished so as to seal them in any of the following cases:
  - Where the details will have a decorative finish such as follows that would seal them:
    - Acoustical finishes.
    - Battens.
    - Wood paneling.
    - Similar applications.
  - In one-layer assemblies as follows:
    - Where joints fall on wood framing.
  - Where any of the following gypsum board products are used:
    - Backing board.
    - Sheathing.
    - Square edge.
    - Tongue-and-groove (V-edge).
  - In multiple-layer assemblies as follows:
    - Where joints of adjacent layers are offset.
  - In assemblies which were tested for fire resistance as follows:
    - Without the details being sealed.
- In other cases, the details must be finished so as to seal them.

2508.5 Horizontal gypsum board diaphragm ceilings

- A horizontal diaphragm ceiling may be created by the following:
  - Gypsum board applied wood joists.

Note: IBC Table 2508.5, "Shear Capacity for Horizontal Wood Framed Gypsum Board Diaphragm Ceiling Assemblies," is cited as governing these diaphragms.

2508 Gypsum Construction

2508.5.1 Diaphragm proportions

- Horizontal diaphragm ceilings must have the following proportions:
  - ≤ 1 1/2:1 as defined at its edges by shear walls.
- The following conditions are not allowed for the diaphragm:
  - Rotation or cantilevers.

2508.5.2. Installation

- Installation of horizontal diaphragm ceilings is governed as follows:
  - Gypsum board must be applied ⊥ to the joists.
  - Joints between the ends of gypsum board panels have the following requirement:
    - Adjacent courses must have end joints on different joists.

2508.5.3 Blocking of perimeter edges

- Blocking of horizontal diaphragm ceilings are governed as follows:
  - The perimeter of the gypsum board ceiling must be blocked as follows:
    - With wood ≥ 2"× 6" as follows:
      - Installed with the wider dimension horizontal in the following location:
        - Over the wall top plate as follows:
          - To provide a nailing surface for the gypsum board ≥ 2" wide.

2508.5.4 Fasteners

- Fasteners in horizontal diaphragm ceilings have the following requirements:
  - Gypsum board must be secured in the following manner:
    - Fasteners must be spaced ≤ 7" oc as follows:
      - At all supports and perimeter blocking.
    - Fasteners must be located ≤ 3/8" from the following:
      - Edge of the gypsum board as applicable.
      - End of the gypsum board as applicable.

Note: IBC Table 2508.5, "Shear Capacity for Horizontal Wood Framed Gypsum Board Diaphragm Ceiling Assemblies," is cited as governing the fasteners.

2508.5.5 Lateral force restrictions

- A gypsum board diaphragm may not be used to resist forces from the following:
  - Masonry or concrete construction.

2509 Gypsum Board in Showers and Water Closets

2509.2 Base for tile

- One or more of the following materials must be installed according to manufacturer's recommendations in the locations listed below:
  - Materials:
    Glass mat water-resistant gypsum backing panels.
    Discrete nonasbestos fiber-cement interior substrate sheets.
    Nonasbestos fiber-cement reinforced cement substrate sheets.
  - Locations:
    Behind wall tile in tub and shower areas.
    Behind wall and ceiling panels in shower areas.

Note: The following are cited as governing the backing boards listed above:
ASTM C 1288, "Specification for Discrete Non-Asbestos Fiber-Cement Interior Substrate Sheets."
ASTM C 1325, "Specification for Non-Asbestos Fiber-Mat Reinforced Cementitious Backer Units."
ASTM C 1178, "Specification for Coated Glass Mat Water-Resistant Gypsum Backing Panel."
Manufacturer recommendations.

- Water-resistant gypsum backing board must be installed as follows:
  - Behind wall tile in WC compartments.
  - According to manufacturer's recommendations.
- Standard gypsum wallboard is allowed in the following locations:
  - Behind tile or panels on walls or ceilings.

Note: Either of the following is cited as governing the water-resistant gypsum backing board and the gypsum wallboard listed above:
GA-216, "Application and Finishing of Gypsum Panel Products."
ASTM C 840, "Specification for Application and Finishing of Gypsum Board."

2509.3 Limitations

- This section addresses water-resistant gypsum backing board.
- Such boards may not be used as follows:
  - Over a vapor retarder in the following areas:
    Bathtub compartments.
    Shower compartments.
  - In any of the following areas:
    Those directly exposed to water.
    Those exposed to continuous high humidity.
- Such boards 1/2" thick may not be used as follows:
  - On ceilings where supports are > 12" on center.
- Such boards 5/8" thick may not be used as follows:
  - On ceilings where supports are 16" on center.

2510 Lathing and Furring for Cement Plaster (Stucco)

2510.5.1 Support of lath

- Solid backing is required for lath and attachments as follows:
  - Where lath on a vertical surface extends between the following:
    - Rafters or similar projections.

2510.5.2.1 Use of gypsum board as a backing board

- Use of the following materials as backing board must meet the requirements listed below:
  - Materials:
    - Gypsum lath or wallboard.
  - Requirements:
    - Such materials may be used as a backing for cement plaster as follows:
      - Where the following materials are used in the locations listed:
        - Materials:
          - Water-resistive barrier between lath and sheathing.
          - One of the following types of lath:
            - Self-furred metal lath.
            - Self-furred wire fabric lath.
        - Locations:
          - On horizontal supports of ceilings or roof soffits.
          - On interior walls.
    - Such materials may not be used as backing for cement plaster in other cases.

2510.5.2.2 Use of gypsum sheathing backing

- Gypsum sheathing may be used as a backing for cement plaster where the following are used:
  - Water-resistive barrier between lath and sheathing.
  - Metal or wire fabric lath.

Note: 2510.6, "Water-resistive barriers," is cited as governing such barriers.

2510.5.3 Backing not required

- Wire backing is not required behind the following:
  - Expanded metal lath.
  - Paperbacked wire fabric lath.

2510.6 Water-resistive barriers

- This section does not govern weather-resistant barriers with all of the following characteristics:
  - Where applied over wood-based sheathing.
  - With a water resistance ≥ that of 60-minute Grade D paper.
  - Where separated from the stucco by one of the following:
    - A layer that does not absorb water.
    - A drainage space.
- Water-resistive barriers installed over wood-based sheathing require the following:
  - A weather-resistive vapor permeable barrier that performs as follows must be included:
    - Performance to be ≥ 2 layers of Grade D paper.
  - Water-resistive barriers are to be installed as specified elsewhere in the code.

Note: 1404.2, "Water-resistive barrier," is cited as governing such barriers.

2511 Interior Plaster

2511.2 Limitations

- Plaster may not be applied directly to the following:
  - Fiber insulation board.
- Cement plaster applied to the following materials is governed as indicated below:
  - Materials:
    Gypsum lath.
    Gypsum plaster.
  - Requirements:
    Such materials must be protected by a water-resistive barrier.
    Direct application to such materials is not permitted.

Note: The following are cited as sources of applicable requirements, a partial summary of which is provided above:
2510.5.1, "Support of lath."
2510.5.2, "Use of gypsum backing board."

2511.3 Grounds

- Where used, grounds must establish the required thickness for plaster.
- Plaster thickness is measured from the following:
  - Face of lath.
  - Face of other bases as applicable.

Note: The following are cited as governing the required thickness of plaster:
ASTM C 842, "Specification for Application of Interior Gypsum Plaster."
ASTM C 926, "Specification for Application of Portland Cement-Based Plaster."

2511.5 Wet areas

- Showers and public toilets require the following wall surfaces:
  - Smooth.
  - Nonabsorbent.

Note: The following are cited as sources of applicable requirements, a partial summary of which is provided above:
1210.2, "Walls and partitions."
1210.3, "Showers."

- Wood framing must be protected with an approved moisture barrier as follows:
  - Where the interior of walls and partitions have all of the following characteristics:
    Covered with one of the following materials:
      Cement plaster.
      Tile of similar material.
    Subject to splashed water.

2512 Exterior Plaster

2512.1 General

- Cement plaster is required to be applied as follows:
  - In ≥ 3 coats as follows:
    - Where applied over any of the following:
      - Metal lath.
      - Wire fabric lath.
      - Gypsum board.
  - In ≥ 2 coats as follows:
    - Where applied over any of the following:
      - Masonry.
      - Concrete.
    - Where both of the following conditions exist:
      - The plaster is of adequate thickness.
      - Where the plaster is covered by any of the following:
        - A veneer.
        - Other facing material.
        - A second wall.

Note: 2510.5, "Backing," is cited as governing the backing for cement plaster.
ASTM C 926, "Specification for Application of Portland Cement-Based Plaster," is cited as reporting the thickness of the plaster, which is adequate for the purposes noted above.

2512.1.1 On-grade floor slab

- On the following assemblies, exterior plaster is governed as listed:
  - Assemblies:
    - Wood framing on a concrete slab-on-grade.
    - Steel framing on a concrete slab-on-grade.
  - Requirements:
    - The plaster must cover the lath and paper.
    - The plaster may not extend below the lath and paper.
  - The installation of the following components must also comply with standards other than the code:
    - Drip screeds.
    - Flashing.
    - Lath.
    - Paper.

Note: ASTM C 1063, "Specification for Installation of Lathing and Furring to Receive Interior and Exterior Portland Cement-Based Plaster," is cited as governing the components listed above.

2512.1.2 Weep screeds

- Weep screeds are required for exterior cement plaster as follows:
  - Screeds must drain trapped water to the exterior.
  - Screed requirements are shown in the detail on the next page.

Note: ASTM C 926, "Specification for Application of Portland Cement-Based Plaster," is cited as governing weep screeds.

2512 Exterior Plaster

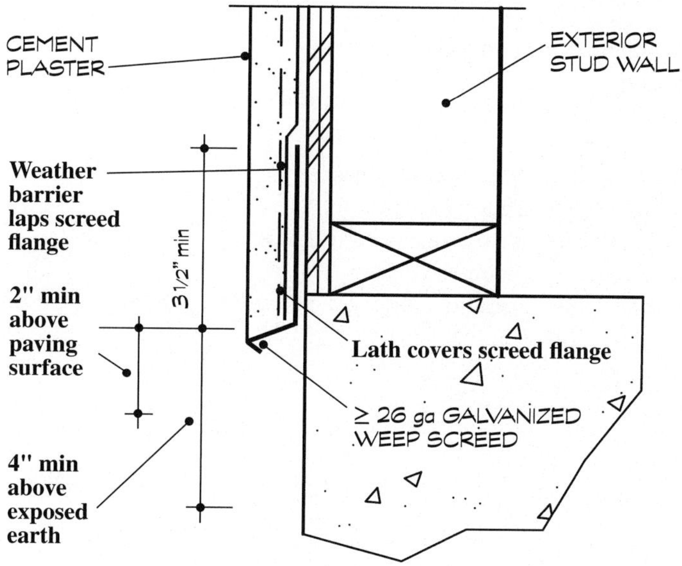

Fig. 2512.1.2. Cement plaster weep screed requirements.

2512.2 Plasticity agents

- The following substances may be added to the cements listed:
  - Substances:
    - Approved plasticity agents as follows:
      - In approved amounts
  - Cements:
    - Portland cement
    - Blended cement
- The following substances may not be added to the stucco mixtures listed:
  - Substances:
    - Lime.
    - Plasticizers.
  - Mixtures:
    - With plastic cement.
    - With masonry cement.
- The following plasticizers may be added to the stucco mixtures indicated below:
  - Plasticizers:
    - Hydrated lime.
    - Lime putty.
  - Mixtures:
    - Cement plaster.
    - Cement lime plaster.

Note: ASTM C 926, "Specification for Application of Portland Cement-Based Plaster," is cited as governing the amounts of the plasticizers that may be added.

2512 Exterior Plaster

2512.3 Limitations

- Gypsum plaster is not permitted as follows:
  - In exterior applications.

2512.5 Second-coat application

- The second coat of exterior cement plaster is governed as follows:
  - It must be applied to the required thickness.
  - It must be rodded and floated to a roughness as follows:

    Adequate for bonding with the finish coat.

    Roughness is limited to 1/4" variation under a 5' straight edge in any direction.

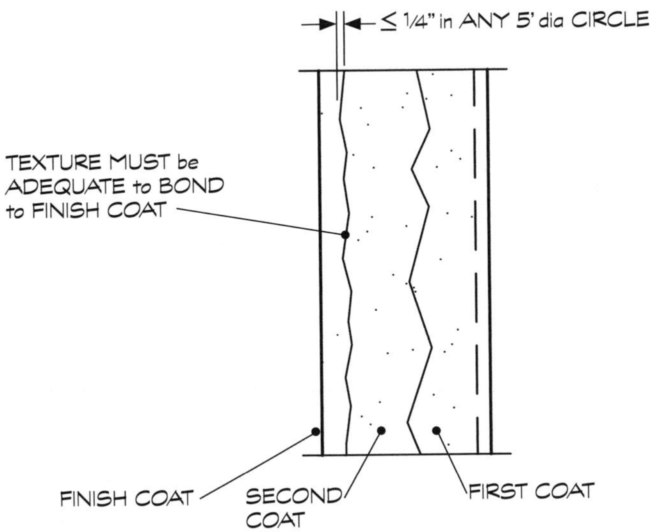

Fig. 2512.5. Three-coat exterior plaster installation.

26

Plastic

Alterations to 209 Main Street. Annapolis, Maryland.
Alt Breeding Schwarz, Architects, LLC. Annapolis, Maryland.

2603 Foam Plastic Insulation

2603.4 Thermal barrier

- Foam plastic insulation need not be separated from the interior of a building where specifically permitted by the code.

  Note: The following are cited as sources of conditions wherein such insulation need not be separated from the interior:
  2603.4.1, "Thermal barrier not required."
  2603.9, "Special approval."

- Otherwise, foam plastic insulation must be separated from the interior of a building as follows:
  - By the following thermal barrier:
    Barrier must limit temperature rise as follows:
    Average temperature rise on the unexposed surface is limited by the following:
    ≤ 250° F after 15 minutes of fire exposure.
    Barrier must remain in place for ≥ 15 minutes of fire exposure.
  - Other requirements are shown in the detail shown below.

  Note: The following are cited as alternatives for governing temperature rise as indicated above:
  ASTM E 119, "Test Methods for Fire Tests of Building Construction and Materials."
  UL 263, "Fire Tests of Building Construction and Materials."
  Section 717, "Concealed Spaces," is cited as governing such spaces.
  The following are cited as governing a barrier's performance in a fire:
  FM 4880, "Class 1 Fire Rating of Insulated Wall or Wall and Roof/Ceiling Panels, Interior Finish Materials or Coatings, and Exterior Wall Systems."
  UL 1040, "Fire Test of Insulated Wall Construction."
  UL 1715, "Fire Test of Interior Finish Material."
  NFPA 286, "Method of Fire Test for Evaluating Contribution of Wall and Ceiling Interior Finish to Room Fire Growth."

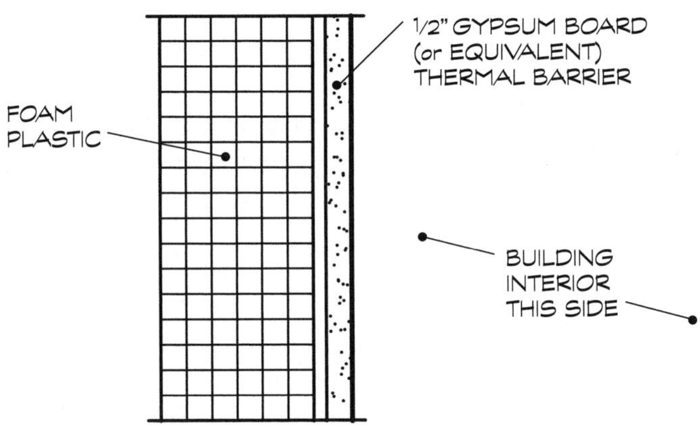

Fig. 2603.4. Thermal barrier.

2603 Foam Plastic Insulation

2603.4.1.1 Masonry or concrete construction

- This section addresses thermal barriers between foam plastic insulation and building interiors.
- Such a thermal barrier is not required in the following walls where conditions listed below apply:
  - Walls:
    - Concrete.
    - Masonry.
  - Conditions:
    - Where the insulation is covered on both sides by ≥ 1" of one of the following materials:
      - Concrete as shown in the detail below.
      - Masonry as shown in the detail below.

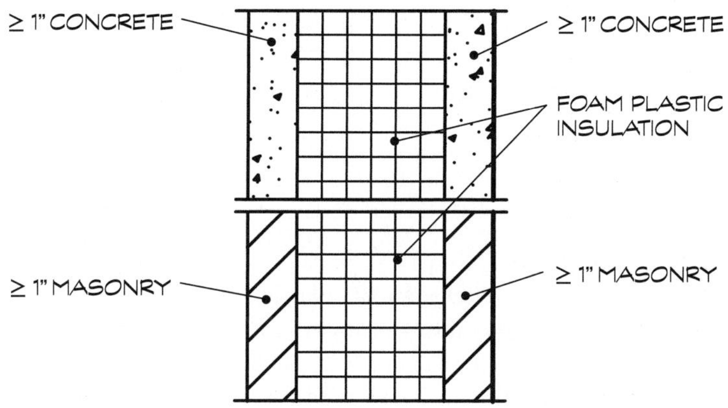

Fig. 2603.4.1.1. Alternative to thermal barrier.

2603.4.1.2 Cooler and freezer walls

- This section addresses thermal barriers between foam plastic insulation and building interiors.
- Such a thermal barrier is not required in walls of cooler and freezer walls where all of the following conditions apply:
  - The foam plastic insulation is tested in a thickness ≥ 4" to have the following properties:
    - Flame spread index ≤ 25.
    - Smoke-developed index ≤ 450.
  - The foam plastic insulation has the following temperature thresholds:
    - Flash ignition temperature is ≥ 600° F.
    - Self-ignition temperature is ≥ 800° F.
  - Cooler or freezer is sprinklered.
  - Where located in a building:
    - The portion of the building containing the cooler or freezer is sprinklered.

  Note: 903.3.1.1, "NFPA 13 sprinkler systems," is cited as governing the sprinklers.

  - Other requirements are shown in Figure 2603.4.1.2 on the following page.

2603 Foam Plastic Insulation

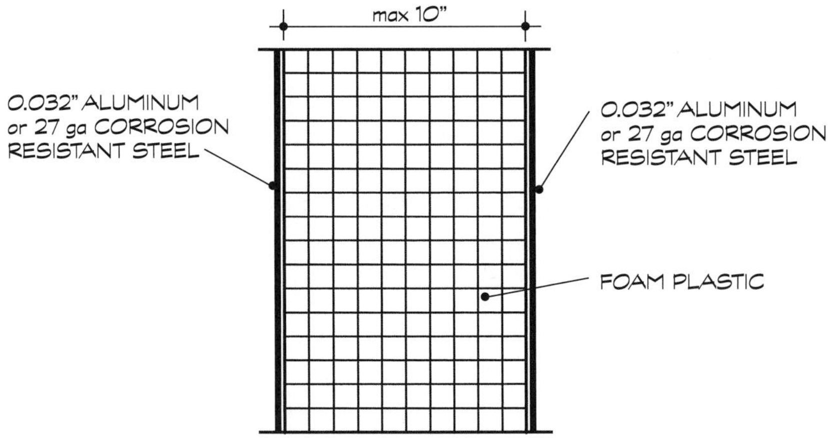

Fig. 2603.4.1.2. Alternative to thermal barriers in cooler and freezer walls.

2603.4.1.3 Walk-in coolers

- This section addresses thermal barriers between foam plastic insulation and building interiors.
- Where buildings are not sprinklered, the following applies to walk-in coolers and freezers:
  - A thermal barrier is not required where all of the following conditions are present:
    The sum of cooler and/or freezer floor areas is ≤ 400 sf.
    The foam plastic flame spread is ≤ 75.
    The foam plastic conforms to requirements shown in the detail provided.
  - Thicker foam plastic is permitted where meeting requirements shown in the detail below.

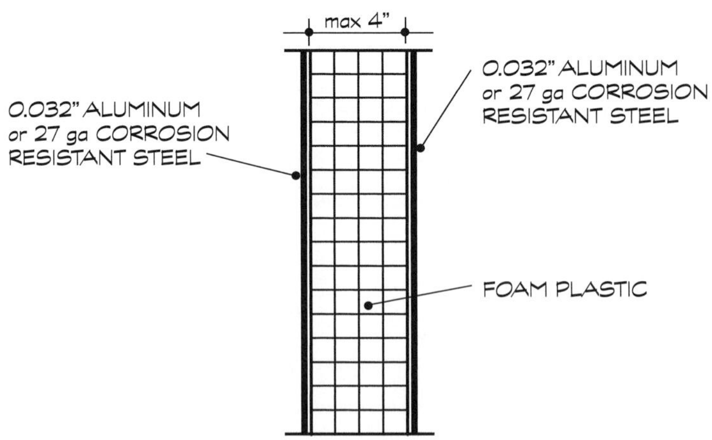

Fig. 2603.4.1.3. Alternative to thermal barrier in walk-in coolers.

2603 Foam Plastic Insulation

2603.4.1.4 Exterior walls — one-story buildings

- This section addresses thermal barriers between foam plastic insulation and building interiors.
- Such barriers are not required where all of the following conditions apply:
  - The foam plastic has the following properties:
    Flame spread index is ≤ 25.
    Smoke-developed index is ≤ 450.
  - Building is sprinklered.

Note: 903.3.1.1, "NFPA 13 sprinkler systems," is cited as governing the sprinklers.

  - Other requirements are met as shown in the detail below.

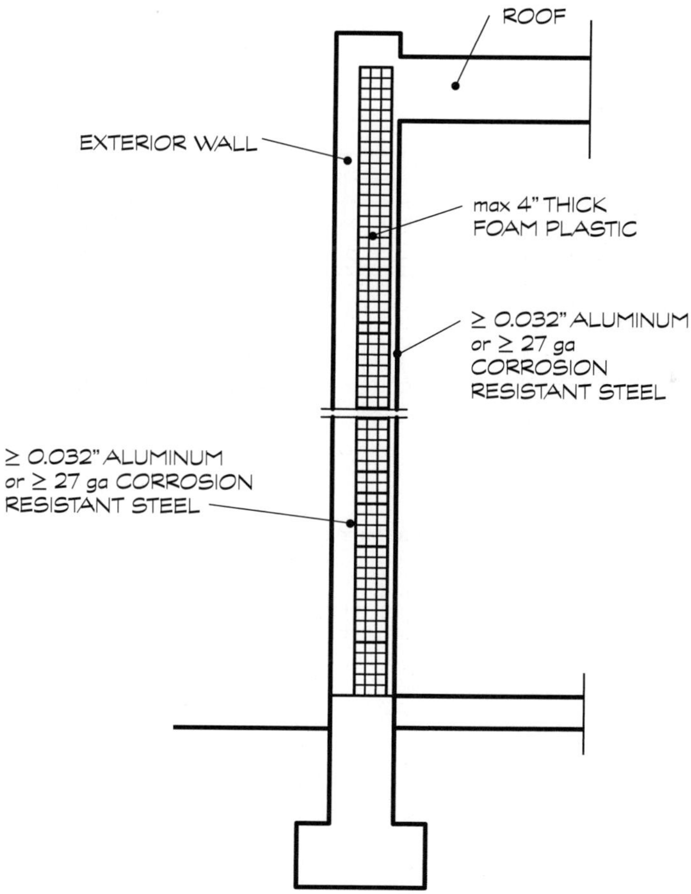

Fig. 2603.4.1.4. Alternative to a thermal barrier in exterior walls of 1-story buildings.

2603 Foam Plastic Insulation

2603.4.1.5 Roofing

- This subsection addresses thermal barriers between foam plastic insulation and building interiors.
- Such a barrier is not required where foam plastic insulation is used as follows:
  - Where the plastic is part of a roof assembly meeting all of the following conditions:
    - Assembly is one of the following:
      - Class A.
      - Class B.
      - Class C.
    - The assembly with the plastic insulation passes required tests.

Note: The following tests are cited as alternative standards of which the assembly indicated above must pass one:
FM 4450, "Approval Standard for Class 1 Insulated Steel Deck Roofs."
UL 1256, "Fire Test of Roof Deck Constructions."

- The insulation does not require a thermal barrier for conditions shown in the detail below.

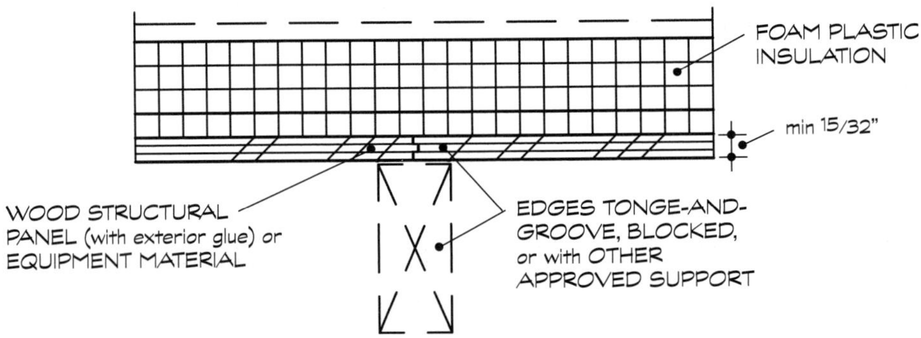

Fig. 2603.4.1.5. Roof sheathing not requiring a thermal barrier.

2603.4.1.6 Attics and crawl spaces

- This section addresses thermal barriers between foam plastic insulation and building interiors.
- Such a barrier is not required in the following spaces for conditions indicated below:
  - Spaces:
    - The following locations where accessed only for utility service:
      - Attics.
      - Crawl spaces.
  - Conditions:
    - Foam insulation is protected against ignition as shown in Figure 2603.4.1.6 on the next page.

2603 Foam Plastic Insulation

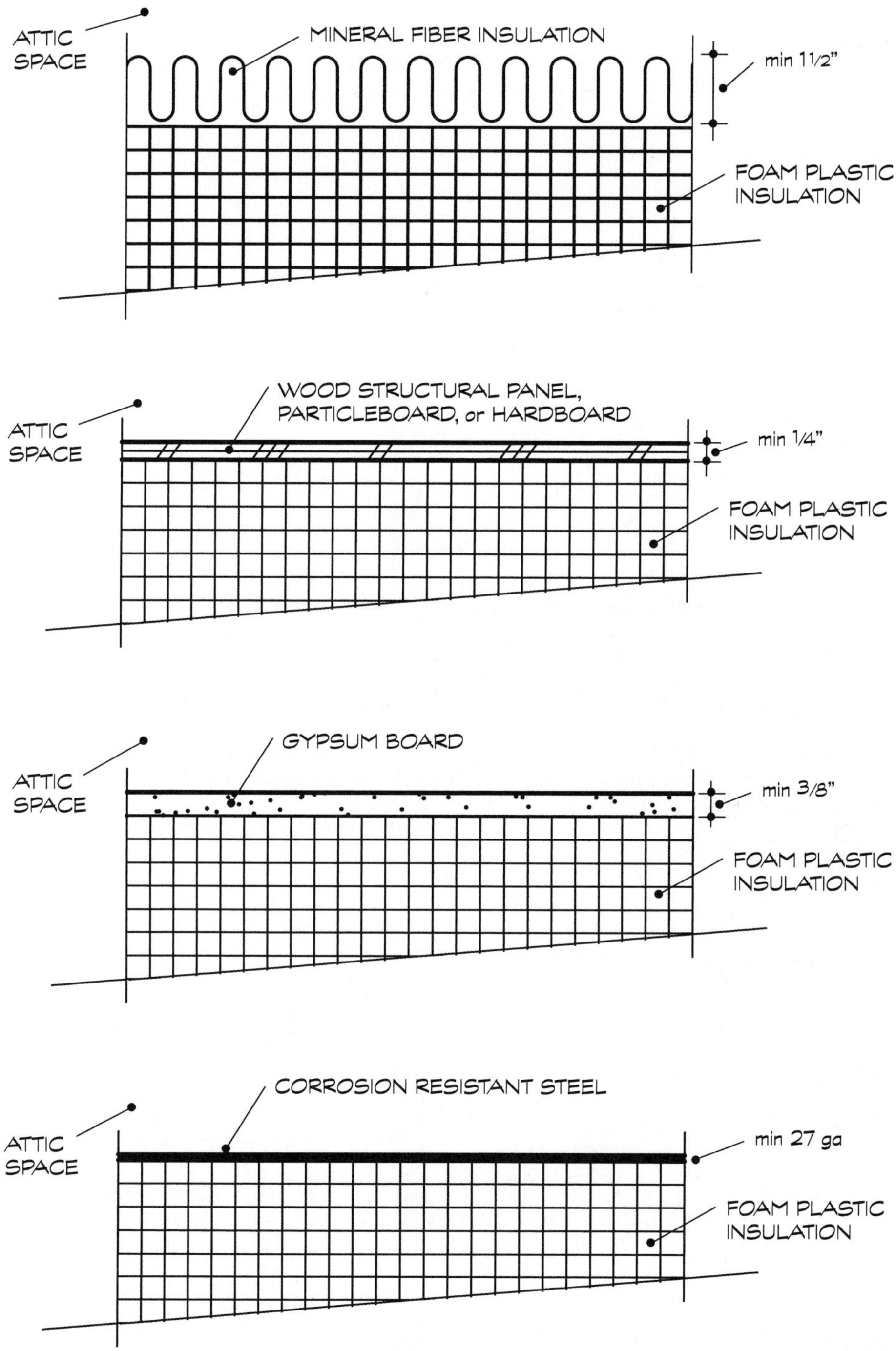

Fig. 2603.4.1.6. Alternatives to a thermal barrier at attics and crawl spaces. Orientation varies for crawl spaces and some attics.

2603 Foam Plastic Insulation

2603.4.1.7 Doors not required to have a fire protection rating

- This section addresses thermal barriers between foam plastic insulation and building interiors.
- Such a barrier is not required in doors with the following insulation and as shown in the detail below:
  - Foam plastic insulation must have the following properties:
    Flame spread index ≤ 75.
    Smoke-developed index ≤ 450.

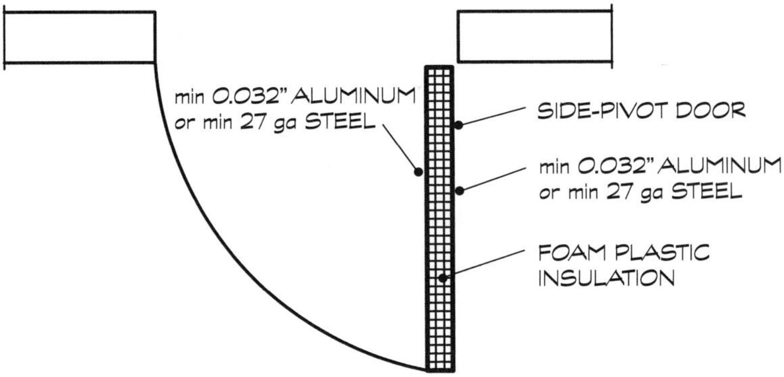

Fig. 2603.4.1.7. Door facing material where no fire protection rating is required.

2603.4.1.8 Exterior doors in buildings of Group R-2 or R-3

- This section addresses thermal barriers between foam plastic insulation and building interiors.
- Such barriers are not required in doors filled with foam insulation as follows:
  - Entrance doors as follows:
    In the following occupancies:
      R-2, R-3.
    To individual dwelling units.
    That do not require a fire-resistance rating.
    As shown in the detail below.

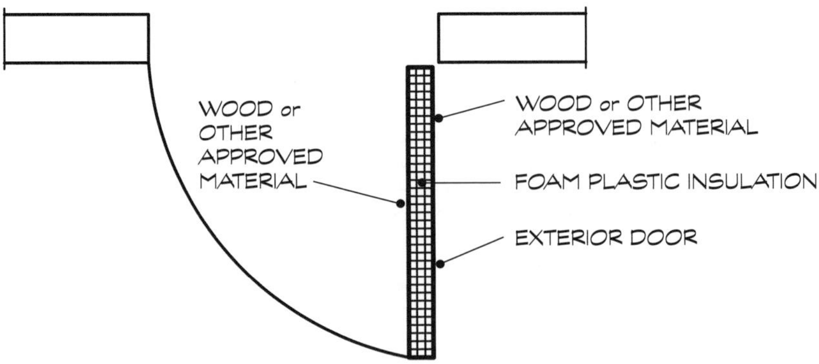

Fig. 2603.4.1.8. Exterior door facing at R-2 and R-3 where no fire-resistance rating is required.

2603 Foam Plastic Insulation

2603.4.1.9 Garage doors

- This section addresses thermal barriers between foam plastic insulation and building interiors.
- Such barriers are not required in garage doors filled with foam insulation where all of the following conditions are met:
  - Insulation has the following properties:
    - Flame spread index is ≤ 75.
    - Smoke developed index is ≤ 450.
  - Garage serves 1- or 2-family dwelling.
  - Garage may be either of the following:
    - Attached.
    - Detached.

Note: 2603.3, "Surface-burning characteristics," is cited as governing foam insulation, a partial summary of which is provided above.

- A thermal barrier is not required for the following garage doors filled with foam insulation:
  - Garage doors that do not require a fire-resistance rating as follows:
    - With facing materials shown in the detail below.
    - Other facing materials must meet required standards.

Note: DASMA 107, "Room Fire Test Standard for Garage Doors Using Foam Plastic Insulation," is cited as the standard that other facings must meet.

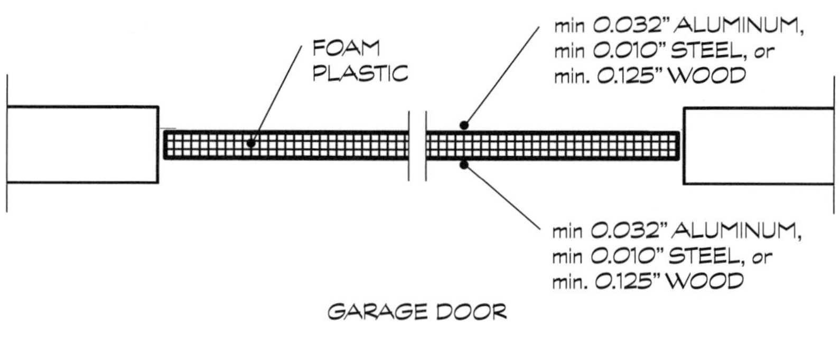

Fig. 2603.4.1.9. Garage door with foam insulation.

2603 Foam Plastic Insulation

2603.4.1.10 Siding backer board

- This section addresses thermal barriers between foam plastic insulation and building interiors.
- Such a barrier is not required for foam plastic insulation with all of the following characteristics:
  - Insulation of ≤ 2,000 Btu/sf.
  - Used as a backing for siding.
  - Meeting conditions shown in the detail below.

Note: NFPA 259, "Standard Test Method for Potential Heat of Building Materials," is cited as the test for determining Btu/sf.

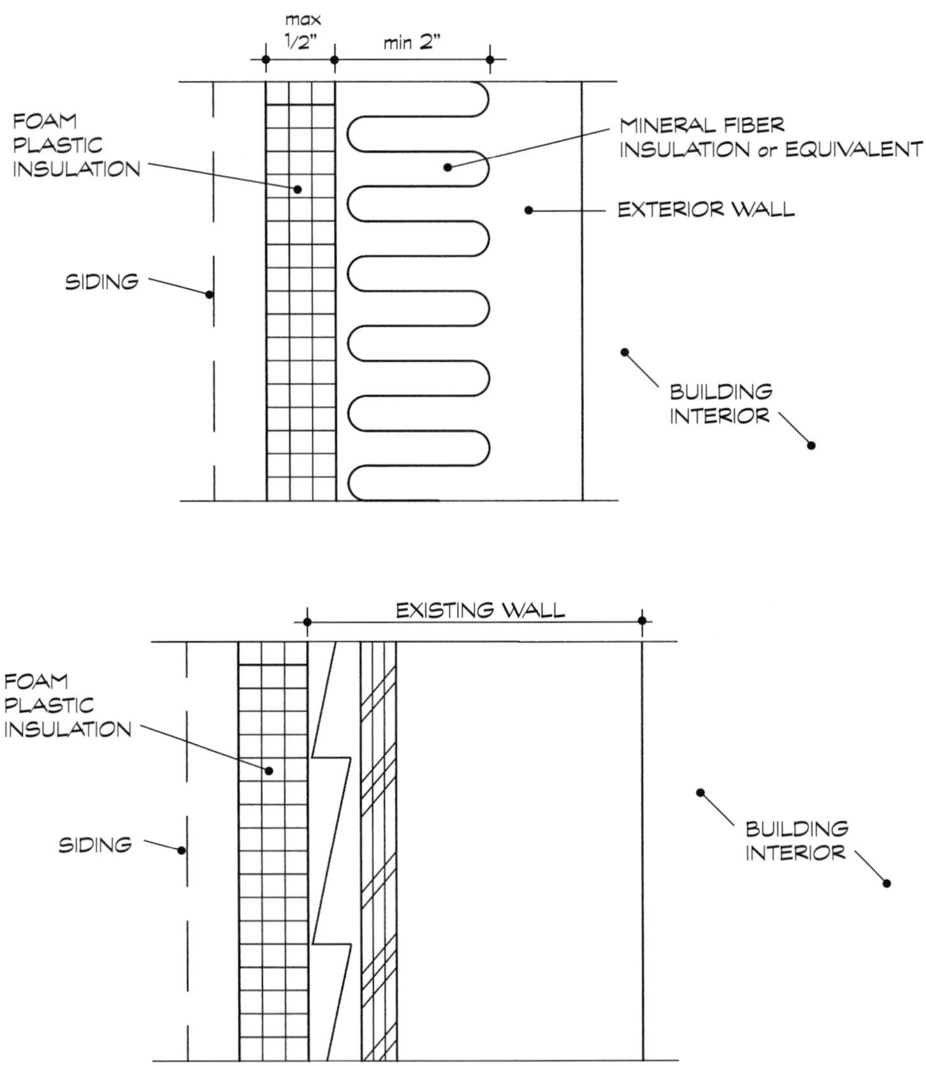

Fig. 2603.4.1.10. Permitted uses of foam plastic insulation as a backing for siding.

2604 Interior Finish and Trim

2604.2.2 Thickness

- Size limitations of interior trim of foam plastic are shown in the illustration below.

2604.2.3 Area limitation

- Limitations of surface coverage for trim of foam plastic are shown in the illustration below:

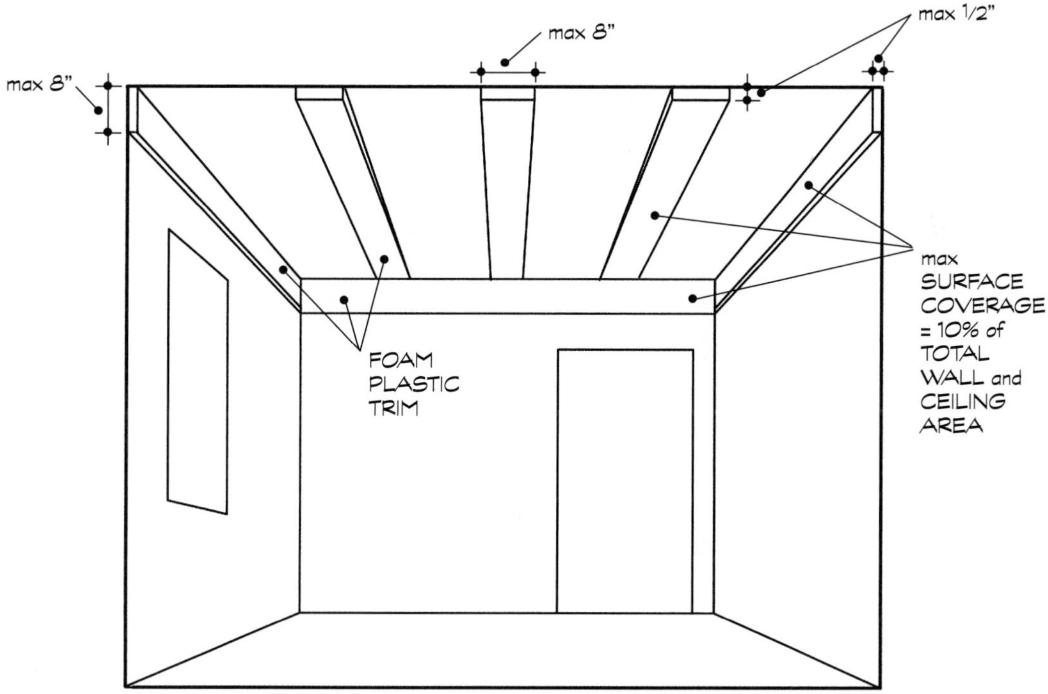

Fig. 2604.2.2. Foam plastic as interior trim.

2605 Plastic Veneer

2605.2 Exterior use

- Plastic siding is not governed by this section.
- Other plastic veneers on building exteriors must meet the following requirements:
  - Physical aspects are governed as follows:

    In Type VB construction:

    The following are not limited where the walls do not have a fire-resistance rating:

    Area of plastic panel.

    Minimum separation distance between panels.

    Smoke-density.

    Otherwise, plastic must meet the requirements shown in the illustration below.

Note: 2606.4, "Specifications," is cited as governing exterior plastic veneer.

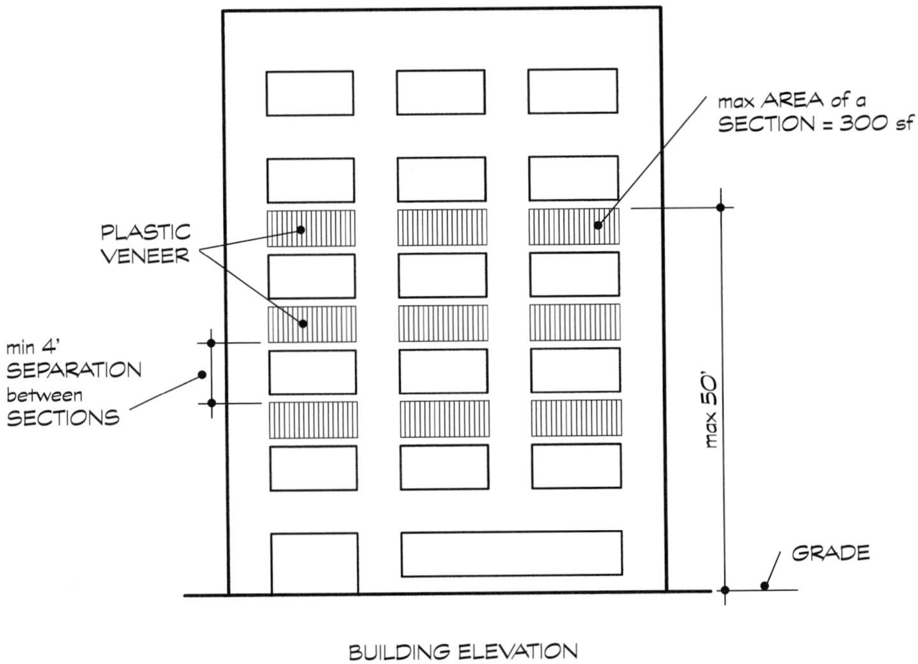

Fig. 2605.2. Building façade with plastic panels applied.

2605.3 Plastic siding

- Plastic siding must comply with requirements found elsewhere in this code.

Note: The following are cited as governing plastic siding:
Section 1404, "Materials."
Section 1405, "Installation of Wall Coverings."

2606 Light-Transmitting Plastics

2606.7.1 Support

- Light-transmitting plastic diffusers must be supported from overhead construction as follows:
  - By hangers as shown in the detail below:

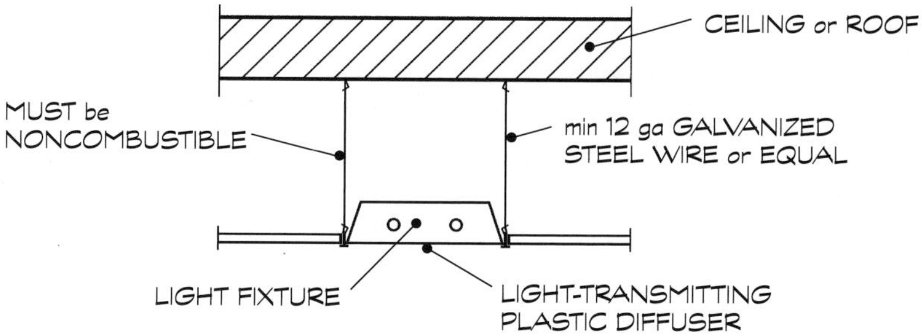

Fig. 2606.7.1. Supports for right-transmitting plastic diffusers.

2606.7.3 Size limitations

- Individual panels of light-transmitting plastics are limited in size as follows:
  - Length must be ≤ 10'.
  - Area must be ≤ 30 sf.
- A partial list of dimensions meeting size limitations is provided below:

Table 2606.7.3 Maximum Sizes of Light-Transmitting Plastic Panels

Width	Length	Width	Length	Width	Length
3'-0"	10'-0"	3'-10"	7'-9"	4'-8"	6'-5"
3'-1"	9'-8"	3'-11"	7'-7"	4'-9"	6'-3"
3'-2"	9'-5"	4'-0"	7'-6"	4'-10"	6'-2"
3'-3"	9'-2"	4'-1"	7'-4"	4'-11"	6'-1"
3'-4"	9'-0"	4'-2"	7'-2"	5'-0"	6'-0"
3'-5"	8'-9"	4'-3"	7'-0"	5'-1"	5'-10"
3'-6"	8'-6"	4'-4"	6'-11"	5'-2"	5'-9"
3'-7"	8'-4"	4'-5"	6'-9"	5'-3"	5'-8"
3'-8"	8'-2"	4'-6"	6'-8"	5'-4"	5'-7"
3'-9"	8'-0"	4'-7"	6'-6"	5'-5"	5'-6"

2606 Light-Transmitting Plastics

2606.7.5 Electrical luminaires

- Light-transmitting plastic diffusers in electrical luminaires located in the following areas have the requirements listed below:
  - Areas:
    - Required exits.
    - Required corridors.
  - Requirements:
    - Where the building is not sprinklered the following applies:
      - Area of plastic must be ≤ 30% of the ceiling area as follows:
        - A partial list of minimum on-center spacing of various fixtures is provided below.
        - Common acceptable layouts are shown below and on the facing page.

Note: The following are cited, one of which must govern the plastic panels above:
Chapter 8, "Interior Finishes."
2606.7.2, "Installation," which governs service temperatures.

Table 2606.7.5 Minimum Center-to-Center Spacing of Plastic Diffusers in Ceilings of Exits and Corridors

	Width of ceiling					
Diffuser size	3'	4'	5'	6'	7'	8'
1' × 4'	4'-5"	3'-4"	2'-8"	2'-3"	1'-11"	1'-8"
1' × 8'	8'-11"	6'-8"	5'-4"	4'-5"	3'-10"	3'-4"
2' × 2'	4'-5"	3'-4"	2'-8"	2'-3"	1'-11"	1'-8"
2' × 4'	8'-11"	6'-8"	5'-4"	4'-5"	3'-10"	3'-4"
2' × 8'	17'-9"	13'-4"	10'-8"	8'-11"	7'-7"	6'-8"
4' × 4'	17'-9"	13'-4"	10'-8"	8'-11"	7'-7"	6'-8"

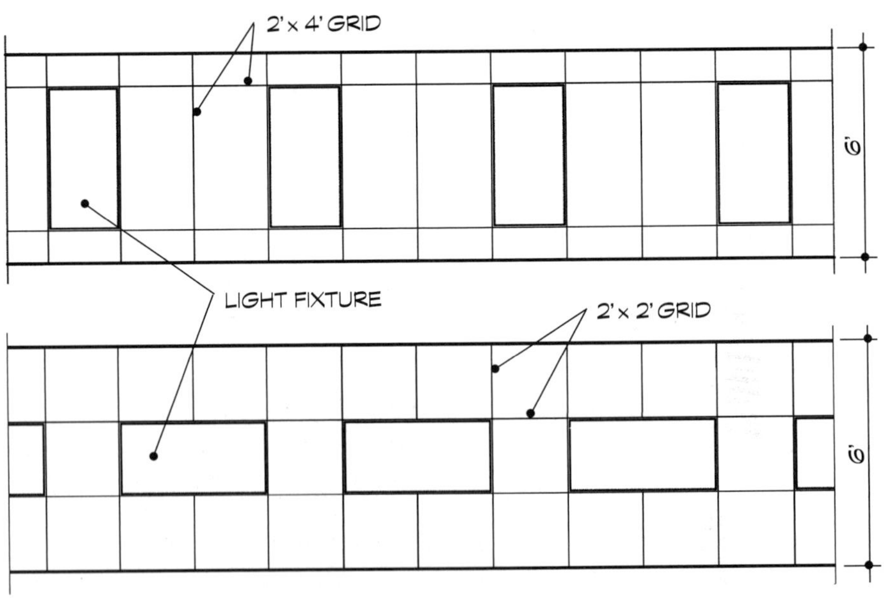

Fig. 2606.7.5A. Reflected ceiling plans with 22% fixture coverage.

2606 Light-Transmitting Plastics

Case study:
Fig. 2606.7.5A.
Plastic diffusers at the ceiling light fixtures are limited to 30% of the ceiling area. The corridor shown in this school is 8' wide. This means that the difussers cannot be closer than 3'- 4" as indicated in the handbook Table 2606.7.5. Since the actual spacing of the fixtures is 12' oc, they comply with the code requirement. The actual ceiling coverage of the fixtures is < 9%.

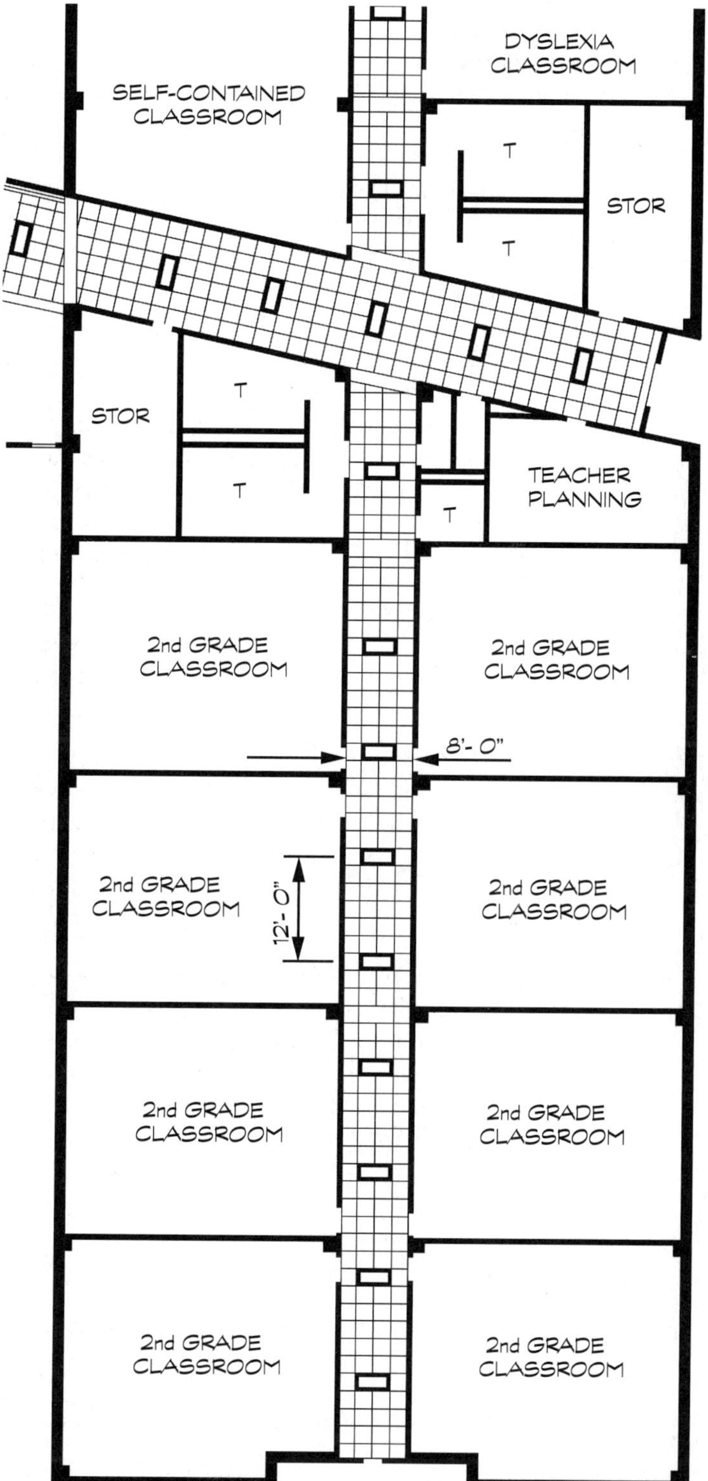

Fig. 2606.7.5B. Partial reflected ceiling plan. New Jasper Pre-K–2nd Grade School. Jasper, Texas. PBK Architects, Inc. Houston, Texas.

2606 Light-Transmitting Plastics

2606.12 Solar collectors

- This section addresses light-transmitting plastics as covers on solar collectors.
- Such plastic covers are limited in area as follows:
  - Where thickness is ≤ 0.010":
    The total area of any type plastic must be ≤ 1/3 the roof area.
  - Where thickness is > 0.010":
    The total area of type CC1 plastics as follows must be ≤ 1/3 the roof area:
      Such as polycarbonate.
    The total area of type CC2 plastics as follows must be ≤ 1/4 the roof area:
      Such as acrylic.
- Other requirements are shown on the illustration below.

Note: 2606.4, "Specifications," lists the properties and standards for CC1 and CC2 plastics, which are based on burning characteristics.

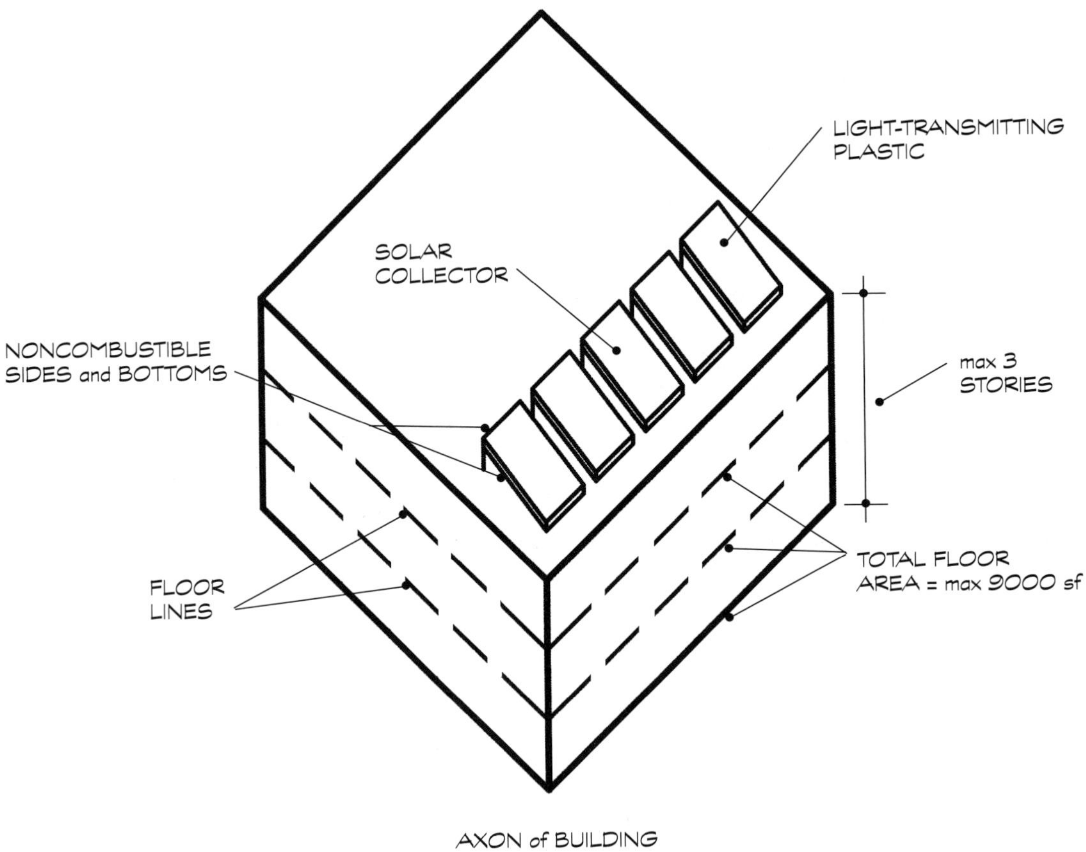

Fig. 2606.12. Building limits for roof-top solar collectors with light-transmitting plastic covers.

2607 Light-Transmitting Plastic Wall Panels

2607.3 Height limitation

- This section addresses light-transmitting plastic panels on exterior building walls.
- The height of such panels is not limited where all of the following apply:
  - Where building is sprinklered.
  - Where panel size is limited according to the following:
    Fire separation distance.
    Class of plastic.

Note: 2607.5, "Automatic sprinkler system," is cited as the source of requirements necessary for panels to have unlimited height as indicated above.

- Height requirement of other exterior wall panels is shown on the illustration below:

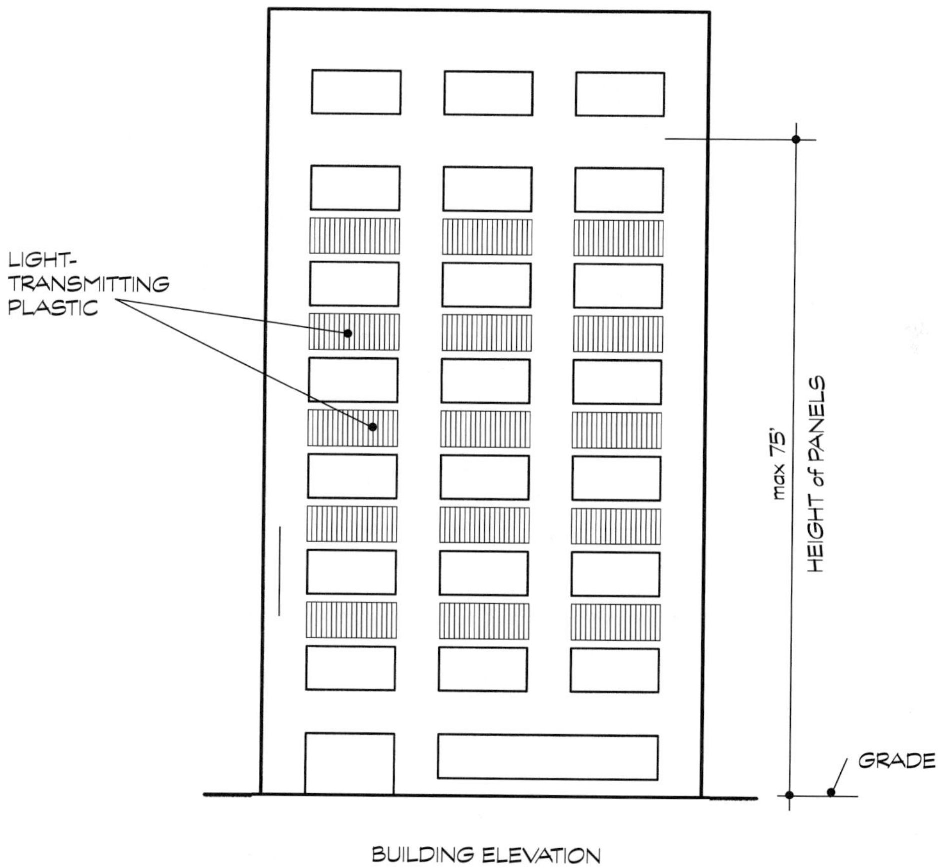

Fig. 2607.3. Maximum height for light-transmitting plastic on a building façade.

2607 Light-Transmitting Plastic Wall Panels

2607.4 Area limitation and separation

- This section addresses light-transmitting plastic panels on exterior building walls.
- This section does not apply to plastic veneers used as exterior siding in Type V construction.
- This section does not apply to plastic wall panels in greenhouses.

Note: Section 1406, "Combustible Materials on the Exterior Side of Exterior Walls," is cited as governing the plastic siding as indicated above.
705.8, "Openings," is cited as governing greenhouse unprotected openings.

- The following requirements are shown in Figures 2607.4A–2607.4F below and on the following pages:
  - Maximum area of an individual plastic panel on a building façade.
  - Minimum vertical and horizontal distances between panels as follows:
    Separation may be provided by either of the following:
    Distances shown in the illustrations provided with this section.
    A flame barrier as shown in the illustration provided with this section.
- The maximum % of wall area in any story that a plastic panel may cover is the smaller of the following:
  - The maximum % of unprotected openings permitted by the code.
  - The maximum % shown in the illustrations provided with this section.

Note: 705.8, "Openings," is cited as the source listing the maximum % of unprotected openings permitted in a building façade.
IBC Table 2607.4, "Area Limitation and Separation Requirements for Light-Transmitting Plastic Wall Panels," is cited as governing plastic panels in Figures 2607.4A –2607.4F below and on the following pages.

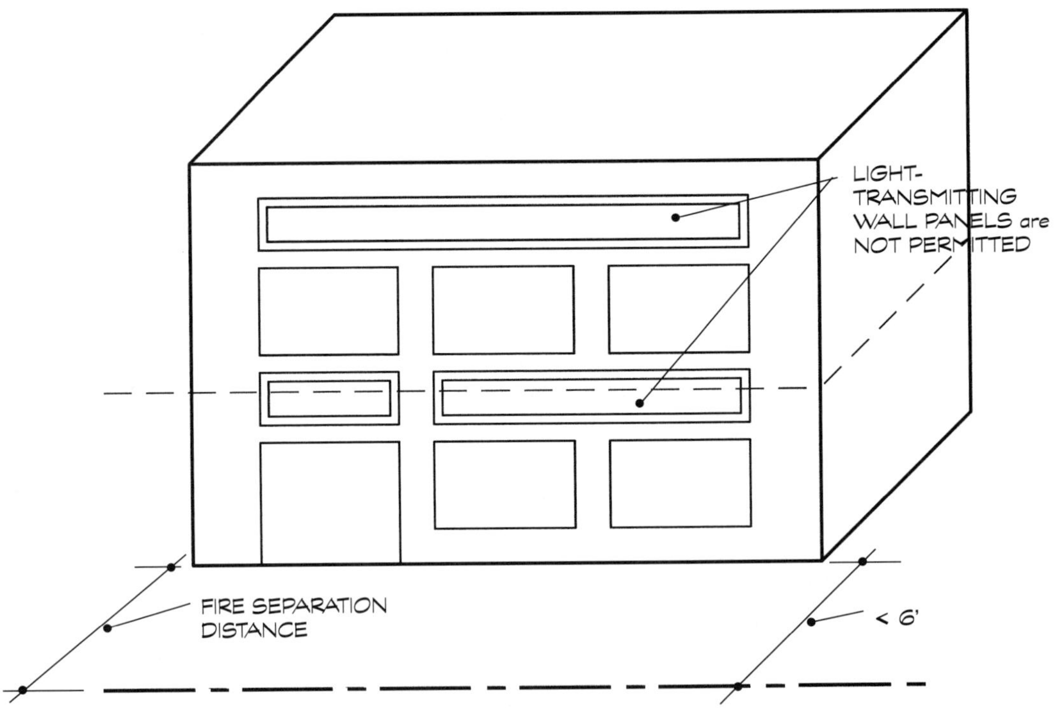

Fig. 2607.4A. Building with fire separation distance < 6'. *[IBC Table 2607.4]*

2607 Light-Transmitting Plastic Wall Panels

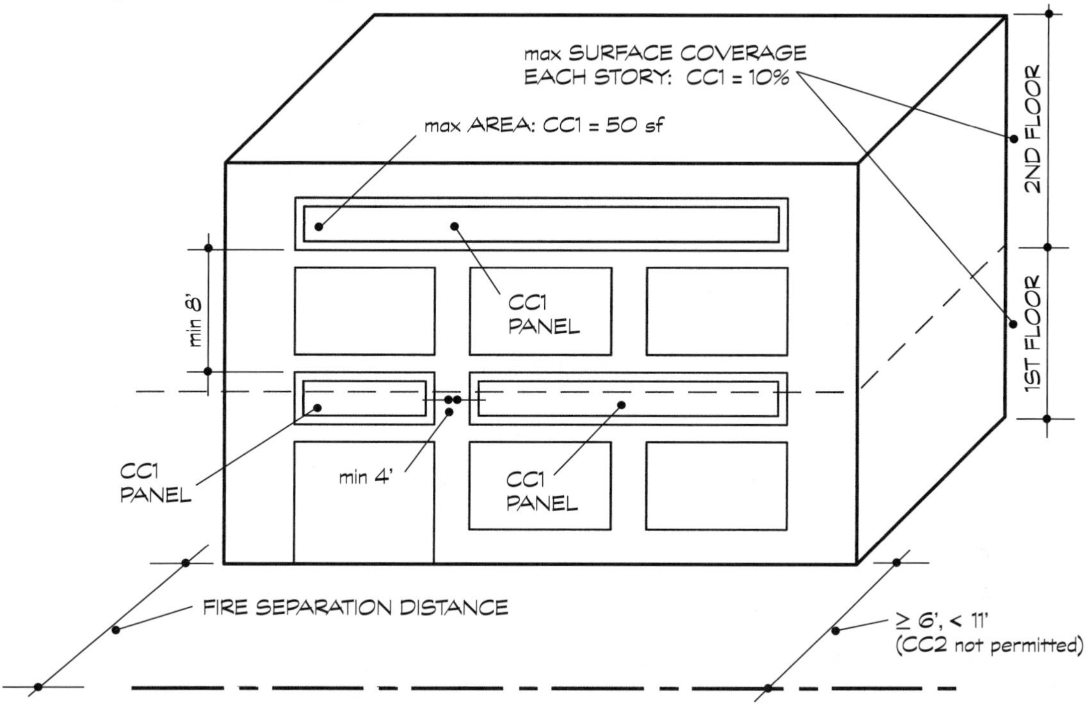

Fig. 2607.4B. Building with fire separation distance ≥ 6', < 11'. *[IBC Table 2607.4]*

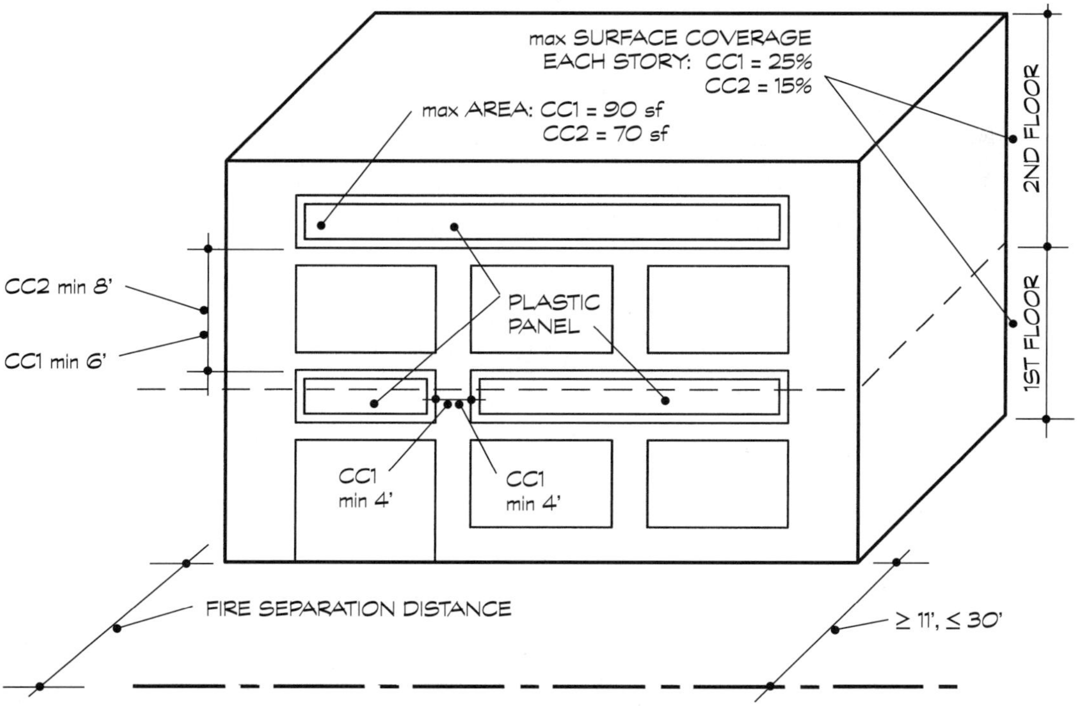

Fig. 2607.4C. Building with fire separation distance ≥ 11', ≤ 30'. *[IBC Table 2607.4]*

2607 Light-Transmitting Plastic Wall Panels

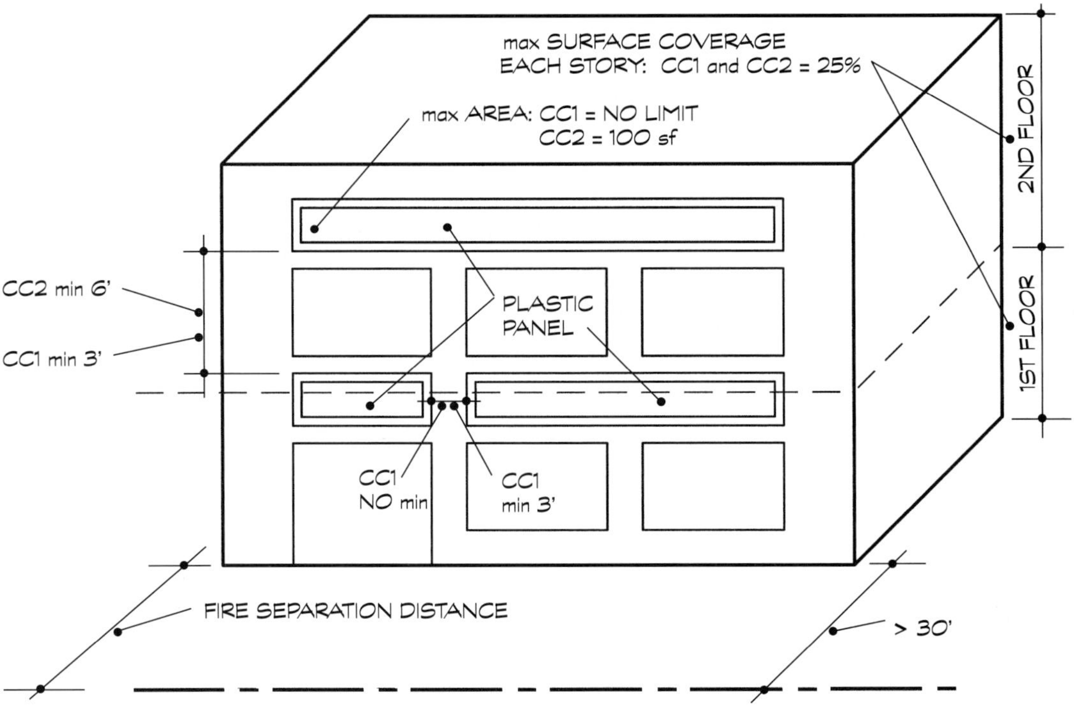

Fig. 2607.4D. Building with fire separation distance > 30'. *[IBC Table 2607.4]*

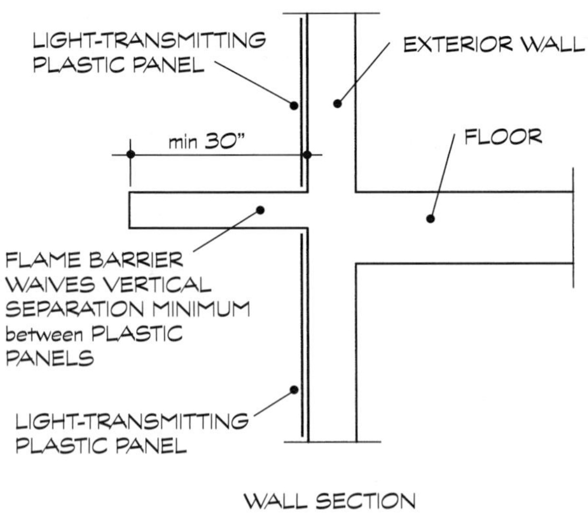

Fig. 2607.4E. Flame barrier at exterior wall.

2607 Light-Transmitting Plastic Wall Panels

Case study: Fig. 2607.4F. The CC1 light-transmitting plastic wall panels located above the lower roof are in a wall of high 1st-floor space. Neither the distance between panels nor the area of the panels is governed for this type plastic, as indicated by IBC Table 2607.4. The panel coverage is < 50% of the exterior wall area (which continues beyond the elevation shown), thus, meeting this code limitation.

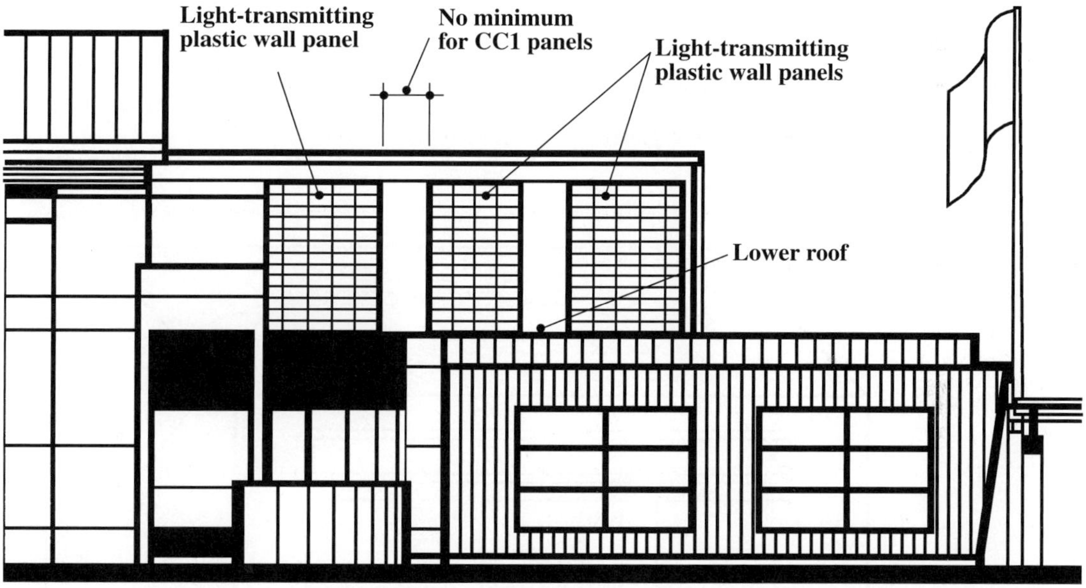

Fig. 2607.4F. Partial elevation. Wichita Transit Storage, Administration, and Maintenance Facility. Wichita, Kansas. Wilson Darnell Mann, P.A., Architects. Wichita, Kansas.

2607 Light-Transmitting Plastic Wall Panels

2607.5 Automatic sprinkler system

- This section addresses light-transmitting plastic panels on exterior building walls as follows:
  - At sprinklered buildings.
  - The following tabular limits set by the previous section are doubled and shown in Figures 2607.5A –2607.5C below and on the following pages:
    % of wall area covered.
    Area of a single panel.

Note: 903.3.1.1, "NFPA 13 sprinkler systems," is cited as governing the sprinklers.
IBC Table 2607.4, "Area Limitation and Separation Requirements for Light-Transmitting Plastic Wall Panels," is cited as listing the maximums that are doubled.

- The maximum % of wall area in any story that a plastic panel may cover is the smaller of the following three limits:
  - 50% of the wall area.
  - The maximum % shown in Figures 2607.5A –2607.5C are provided below and on the next page.
  - The maximum % of unprotected openings permitted elsewhere by the code.

Note: 705.8, "Openings," is cited as the other source listing the maximum % of unprotected openings permitted in a building façade.

  - There are no height limitations to the plastic panels meeting the above requirements.

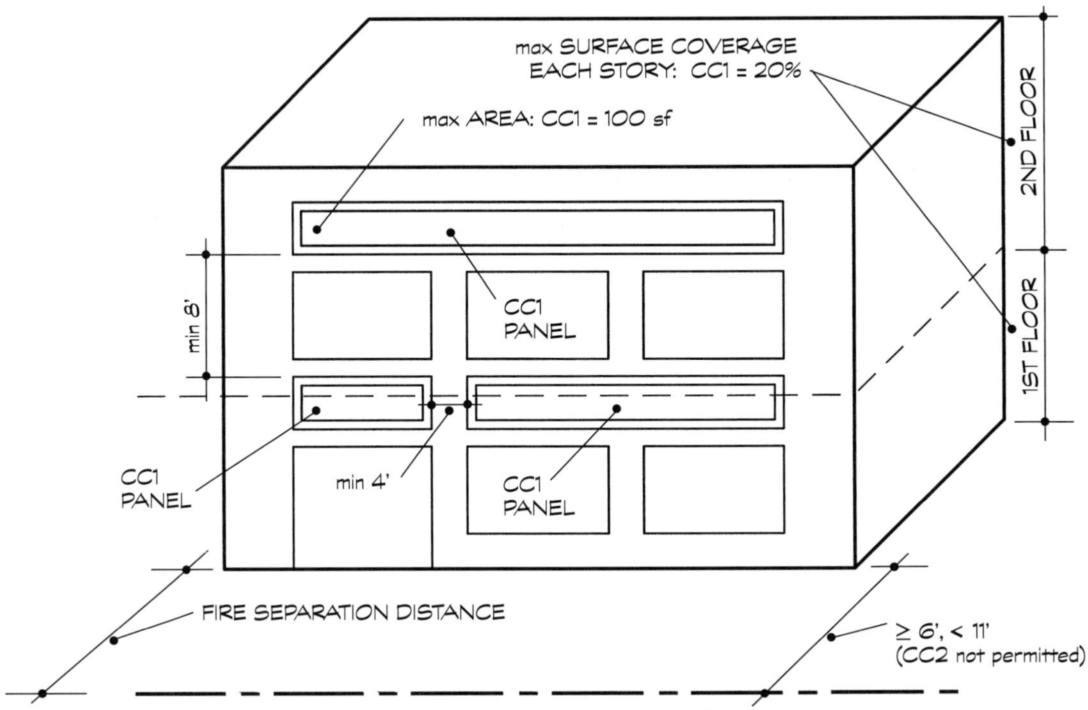

Fig. 2607.5A. Building with fire separation distance ≥ 6', < 11'; building sprinklered.

2607 Light-Transmitting Plastic Wall Panels

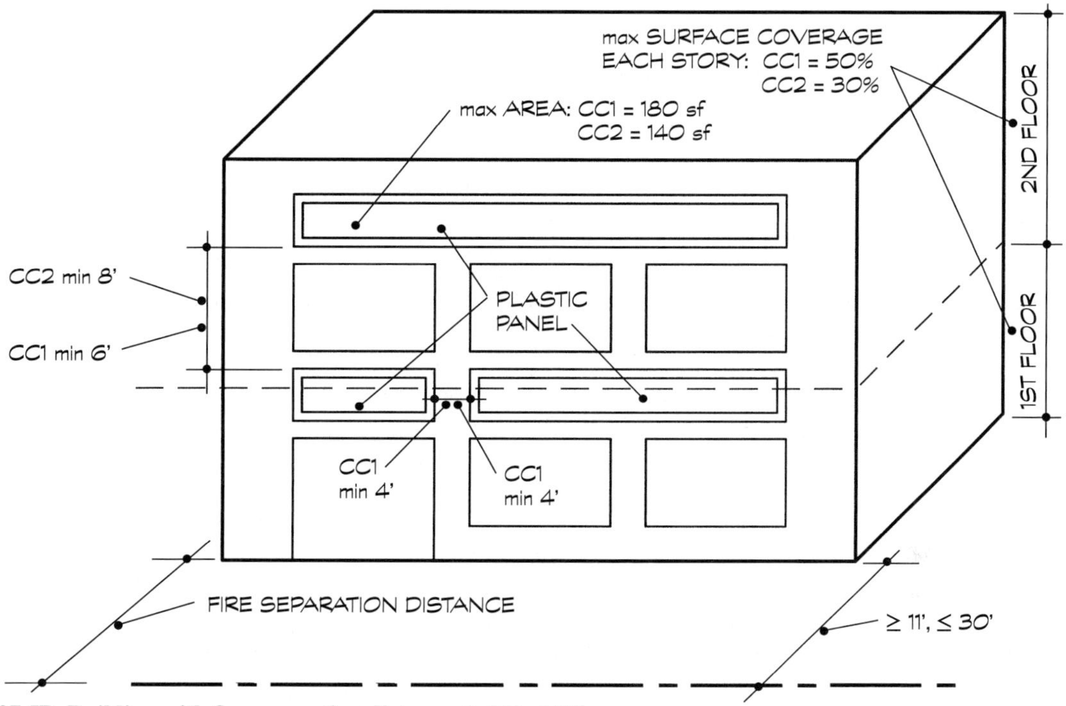

Fig. 2607.5B. Building with fire separation distance ≥ 11', ≤ 30'.

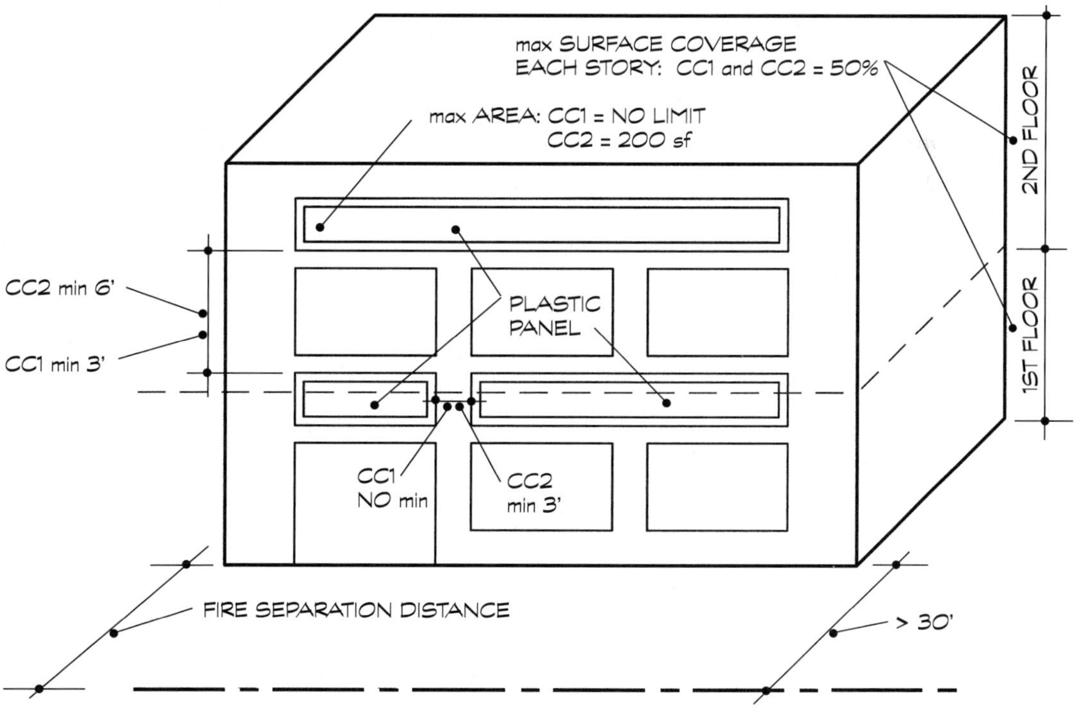

Fig. 2607.5C. Building with fire separation distance > 30'; building sprinklered.

2608 Light-Transmitting Plastic Glazing

2608.2 Buildings of other types of construction

- This section addresses light-transmitting plastic glazing in buildings other than Type VB construction.
- Where openings in exterior walls are not required to be protected, the following applies:
  - Requirements for light-transmitting plastic glazing are shown in Figures 2608.2A – 2608.2C below and on the next page as follows:

    Required vertical separation may be provided by either of the following:

    Distances shown in the illustrations provided with this section.

    A flame barrier as shown in the illustrations provided with this section.

Note: Section 705, "Exterior Walls," is cited as the source specifying conditions wherein exterior walls must be protected.

Section 2606, "Light-Transmitting Plastics," is cited as governing the light-transmitting glazing noted above.

903.3.1.1, "NFPA 13 sprinkler systems," is cited as governing the sprinklers required for reduced restrictions for plastic glazing.

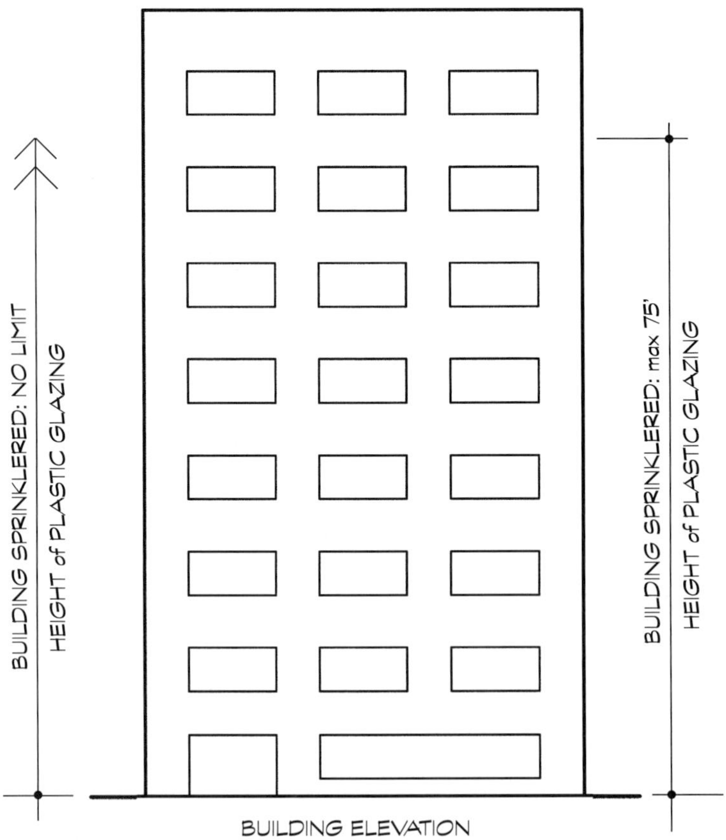

Fig. 2608.2A. Height limit for plastic glazing on building façades.

2608 Light-Transmitting Plastic Glazing

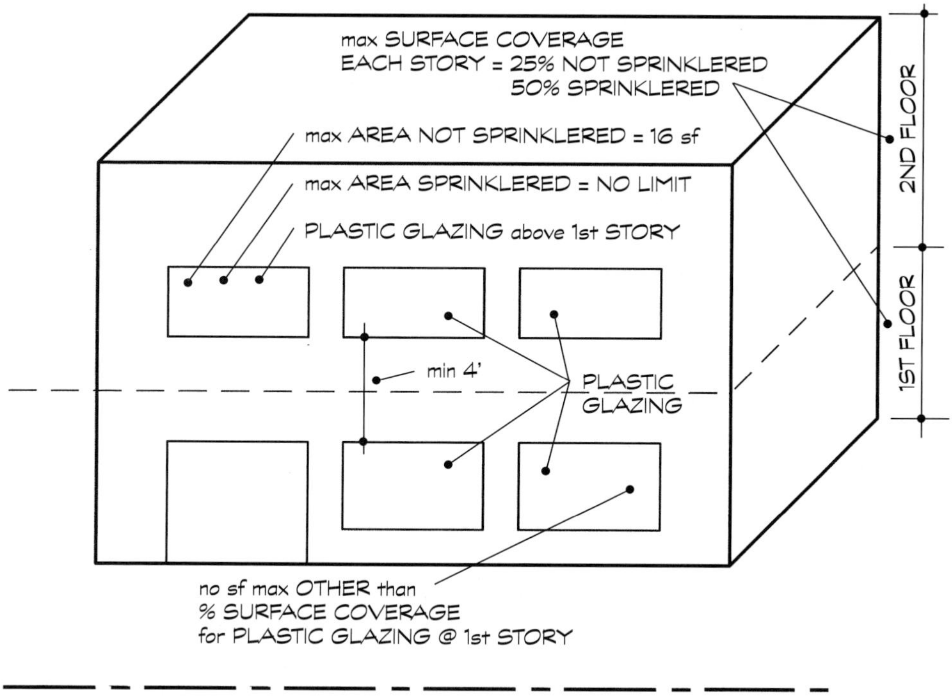

Fig. 2608.2B. Plastic glazing limits.

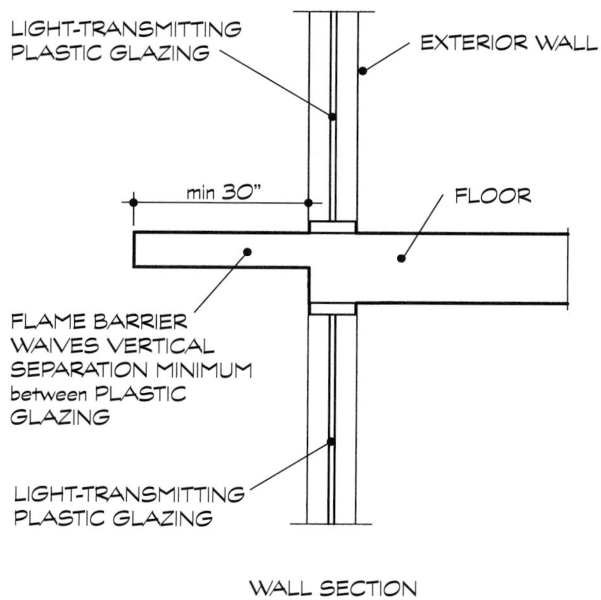

Fig. 2608.2C. Plastic glazing limits.

2609 Light-Transmitting Plastic Roof Panels

2609.2 Separation

- This section addresses light-transmitting plastic roof panels.
- The separation of plastic roof panels is not required in the following cases:
  - In low-hazard buildings such as the following:
    - The following swimming pool buildings:
      - ≤ 5,000 sf.
      - With a fire separation distance ≥ 10'.
    - The following greenhouses:
      - Used to grow plants for one of the following purposes:
        - On a production basis.
        - For research.
      - No public access.
      - With a fire separation distance ≥ 4'.
- Otherwise, separation requirements for plastic roof panels are as shown in Figures 2609.2A and 2609.2B below and on the next page.

Note: 2609.4, "Area limitations," Exception 2 or 3, is cited as the source waiving the requirements for roof panel separation for the cases indicated above.

903.3.1.1, "NFPA 13 sprinkler systems," is cited as governing the sprinklers required for waiving separation requirements for plastic roof panels.

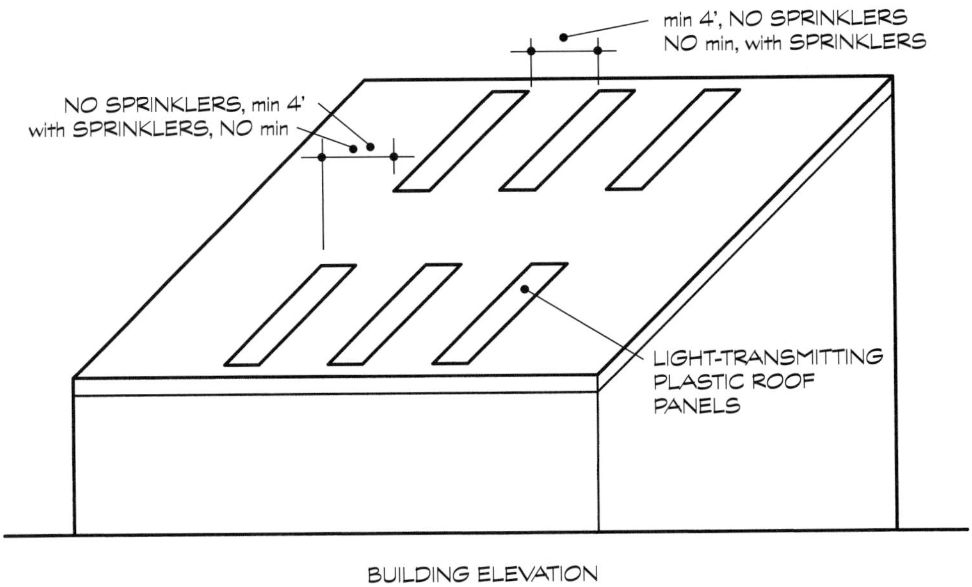

Fig. 2609.2A. Minimum separation of light-transmitting plaster roof panels measured in a horizontal plane.

2609 Light-Transmitting Plastic Roof Panels

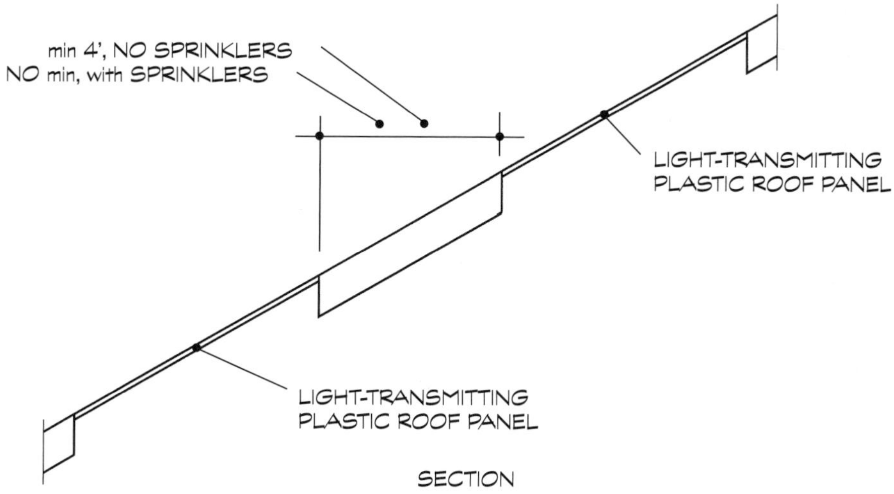

Fig. 2609.2B. Minimum separation of light-transmitting plastic roof panels measured in a horizontal plane.

2609.3 Location

- This section addresses light-transmitting plastic roof panels.
- The requirement for locating plastic roof panels is shown in the illustration below addressing the following case:
  - Where exterior wall openings are required to be protected.

Note: 705.8, "Openings," is cited as the source defining conditions wherein exterior wall openings must be protected.

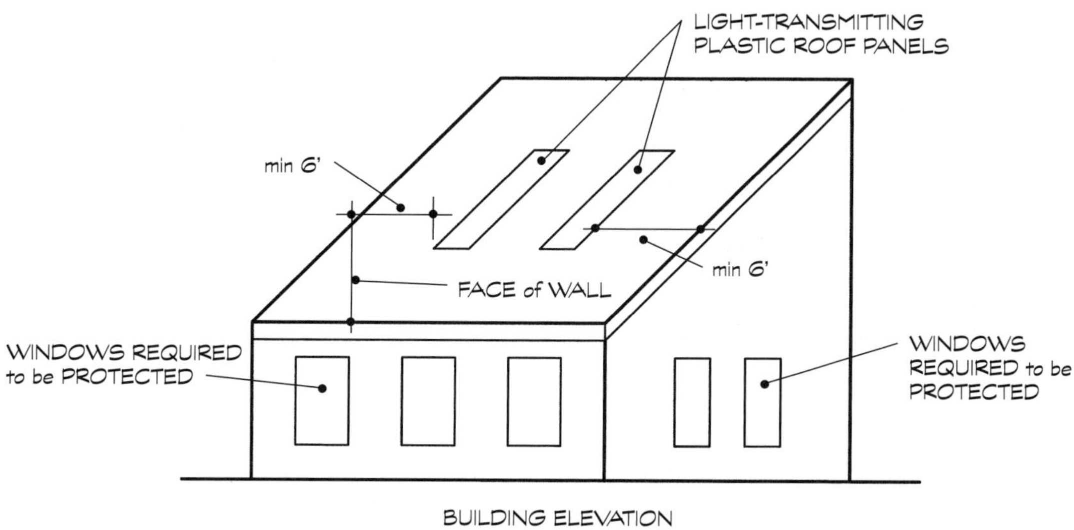

Fig. 2609.3. Plastic roof panels near walls where windows need to be protected.

2609 Light-Transmitting Plastic Roof Panels

2609.4 Area limitations *(part 1 of 5)*

- This subsection addresses light-transmitting plastic roof panels.
- This section does not apply to the following structures:
  - Low-hazard structures such as the following:
    - The following swimming pool buildings:
      - ≤ 5,000 sf.
      - With a fire separation distance ≥ 10'.
  - The following greenhouses:
    - Used to grow plants for one of the following purposes:
      - On a production basis.
      - For research.
    - No public access.
    - With a fire separation distance ≥ 4'.
  - Roof coverings over the following:
    - In Occupancy R-3:
      - Terraces.
      - Patios.
- Otherwise, areas of plastic roof panels are limited as shown in illustration provided below and in Tables 2609.4a–h on the following pages.

Note: IBC Table 2609.4, "Area Limitations for Light-Transmitting Plastic Roof Panels," is cited as the source of area limitations and is the basis for the tables provided with this section.
903.3.1.1, "NFPA 13 sprinkler systems," is cited as governing the sprinklers necessary for increasing area limitations for plastic roof panels.

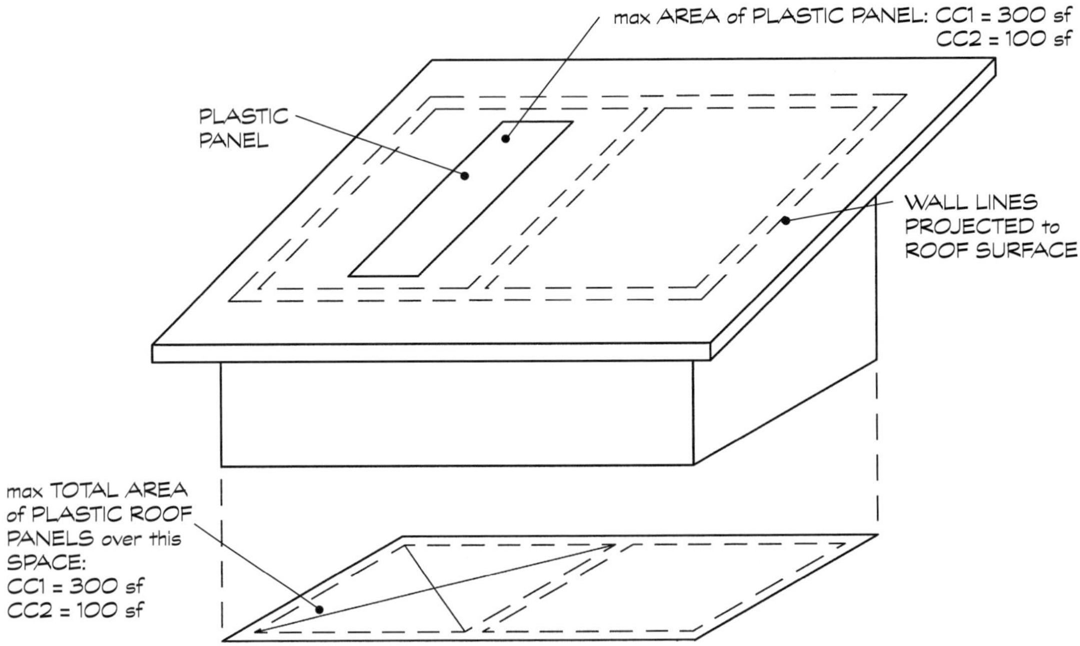

Fig. 2609.4. Maximum size of light-transmitting plastic roof panels.

2609 Light-Transmitting Plastic Roof Panels

2609.4 Area limitations *(part 2 of 5)*

- A partial list of maximum sizes for the following CC1-class plastic roof panels is shown below:
  - Panels must be ≤ 300 sf each for buildings with no sprinklers.

Table 2609.4a Maximum Roof Panel Sizes for CC1 Plastics with No Sprinklers

Given width	Max. length	Given width	Max. length	Given width	Max. length	Given width	Max. length
1'-0"	300'-0"	6'-0"	50'-0"	11'-0"	27'-3"	16'-0"	18'-9"
1'-4"	225'-6"	6'-4"	47'-4"	11'-4"	26'-5"	16'-4"	18'-4"
1'-8"	180'-0"	6'-8"	45'-0"	11'-8"	25'-8"	16'-8"	18'-0"
2'-0"	150'-0"	7'-0"	42'-10"	12'-0"	25'-0"	17'-0"	17'-7"
2'-4"	128'-6"	7'-4"	40'-11"	12'-4"	24'-3"	17'-4"	17'-3"
2'-8"	112'-6"	7'-8"	39'-1"	12'-8"	23'-8"	17'-8"	16'-11"
3'-0"	100'-0"	8'-0"	37'-6"	13'-0"	23'-0"	18'-0"	16'-8"
3'-4"	90'-0"	8'-4"	36'-0"	13'-4"	22'-6"	18'-4"	16'-4"
3'-8"	81'-9"	8'-8"	34'-7"	13'-8"	21'-11"	18'-8"	16'-0"
4'-0"	75'-0"	9'-0"	33'-4"	14'-0"	21'-5"	19'-0"	15'-9"
4'-4"	69'-3"	9'-4"	32'-1"	14'-4"	20'-11"	19'-4"	15'-6"
4'-8"	64'-3"	9'-8"	31'-0"	14'-8"	20'-5"	19'-8"	15'-3"
5'-0"	60'-0"	10'-0"	30'-0"	15'-0"	20'-0"	20'-0"	15'-0"
5'-4"	56'-3"	10'-4"	29'-0"	15'-4"	19'-6"	20'-4"	14'-9"
5'-8"	52'-11"	10'-8"	28'-1"	15'-8"	19'-1"	20'-8"	14'-6"

Source: IBC Table 2607.4.

- A partial list of maximum sizes for the following CC1-class plastic roof panels is shown below:
  - Panels must be ≤ 600 sf each for sprinklered buildings.

Table 2609.4b Maximum Roof Panel Sizes for CC1 Plastics on Sprinklered Buildings

Given width	Max. length	Given width	Max. length	Given width	Max. length	Given width	Max. length
1'-0"	600'-0"	7'-0"	85'-8"	13'-0"	46'-1"	19'-0"	31'-7"
1'-6"	400'-0"	7'-6"	80'-0"	13'-6"	44'-5"	19'-6"	30'-9"
2'-0"	300'-0"	8'-0"	75'-0"	14'-0"	42'-10"	20'-0"	30'-0"
2'-6"	240'-0"	8'-6"	70'-7"	14'-6"	41'-4"	20'-6"	29'-3"
3'-0"	200'-0"	9'-0"	66'-8"	15'-0"	40'-0"	21'-0"	28'-6"
3'-6"	171'-5"	9'-6"	63'-1"	15'-6"	38'-8"	21'-6"	27'-10"
4'-0"	150'-0"	10'-0"	60'-0"	16'-0"	37'-6"	22'-0"	27'-3"
4'-6"	133'-4"	10'-6"	57'-1"	16'-6"	36'-4"	22'-6"	26'-8"
5'-0"	120'-0"	11'-0"	54'-6"	17'-0"	35'-3"	23'-0"	26'-1"
5'-6"	109'-1"	11'-6"	52'-2"	17'-6"	34'-3"	23'-6"	25'-6"
6'-0"	100'-0"	12'-0"	50'-0"	18'-0"	33'-4"	24'-0"	25'-0"
6'-6"	92'-3"	12'-6"	48'-0"	18'-6"	32'-5"	24'-6"	24'-5"

Source: IBC Table 2607.4.

2609 Light-Transmitting Plastic Roof Panels

2609.4 Area limitations *(part 3 of 5)*

- A partial list of maximum areas for the following CC1-class plastic roof panels is shown below:
  - Sum of panel areas must be ≤ 30% of floor area served in buildings with no sprinklers.

Table 2609.4c Maximum Total Area Permitted for Roof Panels of CC1 Plastics with No Sprinklers

Floor area (sf)	Maximum panel area (sf)	Floor area (sf)	Maximum panel area (sf)	Floor area (sf)	Maximum panel area (sf)	Floor area (sf)	Maximum panel area (sf)
100	30	1,300	390	2,500	750	3,700	1110
200	60	1,400	420	2,600	780	3,800	1140
300	90	1,500	450	2,700	810	3,900	1170
400	120	1,600	480	2,800	840	4,000	1200
500	150	1,700	510	2,900	870	4,100	1230
600	180	1,800	540	3,000	900	4,200	1260
700	210	1,900	570	3,100	930	4,300	1290
800	240	2,000	600	3,200	960	4,400	1320
900	270	2,100	630	3,300	990	4,500	1350
1,000	300	2,200	660	3,400	1020	4,600	1380
1,100	330	2,300	690	3,500	1050	4,700	1410
1,200	360	2,400	720	3,600	1080	4,800	1440

Source: IBC Table 2607.4.

- A partial list of maximum areas for the following CC1-class plastic roof panels is shown below:
  - Sum of panel areas must be ≤ 60% of floor area served in sprinklered buildings.

Table 2609.4d Total Aggregate Area Permitted for Roof Panels of CC1 Plastics with Sprinklers

Floor area (sf)	Maximum panel area (sf)	Floor area (sf)	Maximum panel area (sf)	Floor area (sf)	Maximum panel area (sf)	Floor area (sf)	Maximum panel area (sf)
100	60	1,300	780	2,500	1,500	3,700	2,220
200	120	1,400	840	2,600	1,560	3,800	2,280
300	180	1,500	900	2,700	1,620	3,900	2,340
400	240	1,600	960	2,800	1,680	4,000	2,400
500	300	1,700	1,020	2,900	1,740	4,100	2,460
600	360	1,800	1,080	3,000	1,800	4,200	2,520
700	420	1,900	1,140	3,100	1,860	4,300	2,580
800	480	2,000	1,200	3,200	1,920	4,400	2,640
900	540	2,100	1,260	3,300	1,980	4,500	2,700
1,000	600	2,200	1,320	3,400	2,040	4,600	2,760
1,100	660	2,300	1,380	3,500	2,100	4,700	2,820
1,200	720	2,400	1,440	3,600	2,160	4,800	2,880

Source: IBC Table 2607.4.

2609 Light-Transmitting Plastic Roof Panels

2609.4 Area limitations *(part 4 of 5)*

- A partial list of maximum sizes of the following CC2-class plastic roof panels is shown below:
  - Panels must be ≥ 100 sf each for buildings with no sprinklers.

Table 2609.4e Maximum Roof Panel Sizes for CC2 Plastics with No Sprinklers

Given width	Max. length	Given width	Max. length	Given width	Max. length	Given width	Max. length
1'-0"	100'-0"	3'-6"	28'-6"	6'-0"	16'-8"	8'-6"	11'-9"
1'-2"	85'-8"	3'-8"	27'-3"	6'-2"	16'-2"	8'-8"	11'-6"
1'-4"	75'-0"	3'-10"	26'-1"	6'-4"	15'-9"	8'-10"	11'-3"
1'-6"	66'-8"	4'-0"	25'-0"	6'-6"	15'-4"	9'-0"	11'-1"
1'-8"	60'-0"	4'-2"	24'-0"	6'-8"	15'-0"	9'-2"	10'-10"
1'-10"	54'-6"	4'-4"	23'-1"	6'-10"	14'-7"	9'-4"	10'-8"
2'-0"	50'-0"	4'-6"	22'-2"	7'-0"	14'-3"	9'-6"	10'-6"
2'-2"	46'-1"	4'-8"	21'-5"	7'-2"	13'-11"	9'-8"	10'-4"
2'-4"	42'-10"	4'-10"	20'-8"	7'-4"	13'-7"	9'-10"	10'-2"
2'-6"	40'-0"	5'-0"	20'-0"	7'-6"	13'-4"	10'-0"	10'-0"
2'-8"	37'-6"	5'-2"	19'-4"	7'-8"	13'-0"	10'-2"	9'-10"
2'-10"	35'-3"	5'-4"	18'-9"	7'-10"	12'-9"	10'-4"	9'-8"
3'-0"	33'-4"	5'-6"	18'-2"	8'-0"	12'-6"	10'-6"	9'-6"
3'-2"	31'-7"	5'-8"	17'-7"	8'-2"	12'-2"	10'-8"	9'-4"
3'-4"	30'-0"	5'-10"	17'-1"	8'-4"	12'-0"	10'-10"	9'-2"

Source: IBC Table 2607.4.

- A partial list of maximum sizes for the following CC2-class plastic roof panels is shown below:
  - Panels must be ≤ 200 sf each for sprinklered buildings.

Table 2609.4f Maximum Roof Panel Sizes for CC2 Plastics on Sprinklered Buildings

Given width	Max. length	Given width	Max. length	Given width	Max. length	Given width	Max. length
1'-0"	200'-0"	5'-0"	40'-0"	9'-0"	22'-2"	13'-0"	15'-4"
1'-4"	150'-0"	5'-4"	37'-6"	9'-4"	21'-5"	13'-4"	15'-0"
1'-8"	120'-0"	5'-8"	35'-3"	9'-8"	20'-8"	13'-8"	14'-7"
2'-0"	100'-0"	6'-0"	33'-4"	10'-0"	20'-0"	14'-0"	14'-3"
2'-4"	85'-8"	6'-4"	31'-7"	10'-4"	19'-4"	14'-4"	13'-11"
2'-8"	75'-0"	6'-8"	30'-0"	10'-8"	18'-9"	14'-8"	13'-7"
3'-0"	66'-8"	7'-0"	28'-6"	11'-0"	18'-2"	15'-0"	13'-4"
3'-4"	60'-0"	7'-4"	27'-3"	11'-4"	17'-7"	15'-4"	13'-0"
3'-8"	54'-6"	7'-8"	26'-1"	11'-8"	17'-1"	15'-8"	12'-9"
4'-0"	50'-0"	8'-0"	25'-0"	12'-0"	16'-8"	16'-0"	12'-6"
4'-4"	46'-1"	8'-4"	24'-0"	12'-4"	16'-2"	16'-4"	12'-2"
4'-8"	42'-10"	8'-8"	23'-1"	12'-8"	15'-9"	16'-8"	12'-0"

Source: IBC Table 2607.4.

2609 Light-Transmitting Plastic Roof Panels

2609.4 Area limitations *(part 5 of 5)*

- A partial list of maximum areas for the following CC2-class plastic roof panels is shown below:
  - Sum of panel areas must be ≤ 25% of floor area served in buildings with no sprinklers.

Table 2609.4g Maximum Total Area Permitted for Roof Panels of CC2 Plastics with No Sprinklers

Floor area (sf)	Maximum panel area (sf)	Floor area (sf)	Maximum panel area (sf)	Floor area (sf)	Maximum panel area (sf)	Floor area (sf)	Maximum panel area (sf)
100	25	1,300	325	2,500	625	3,700	925
200	50	1,400	350	2,600	650	3,800	950
300	75	1,500	375	2,700	675	3,900	975
400	100	1,600	400	2,800	700	4,000	1,000
500	125	1,700	425	2,900	725	4,100	1,025
600	150	1,800	450	3,000	750	4,200	1,050
700	175	1,900	475	3,100	775	4,300	1,075
800	200	2,000	500	3,200	800	4,400	1,100
900	225	2,100	525	3,300	825	4,500	1,125
1,000	250	2,200	550	3,400	850	4,600	1,150
1,100	275	2,300	575	3,500	875	4,700	1,175
1,200	300	2,400	600	3,600	900	4,800	1,200

Source: IBC Table 2607.4.

- A partial list of maximum areas for the following CC2-class plastic roof panels is shown below:
  - Sum of panel areas must be ≤ 50% of floor area served in sprinklered buildings.

Table 2609.4h Total Aggregate Area Permitted for Roof Panels of CC2 Plastics with Sprinklers

Floor area (sf)	Maximum panel area (sf)	Floor area (sf)	Maximum panel area (sf)	Floor area (sf)	Maximum panel area (sf)	Floor area (sf)	Maximum panel area (sf)
100	50	1300	650	2500	1250	3700	1850
200	100	1400	700	2600	1300	3800	1900
300	150	1500	750	2700	1350	3900	1950
400	200	1600	800	2800	1400	4000	2000
500	250	1700	850	2900	1450	4100	2050
600	300	1800	900	3000	1500	4200	2100
700	350	1900	950	3100	1550	4300	2150
800	400	2000	1000	3200	1600	4400	2200
900	450	2100	1050	3300	1650	4500	2250
1000	500	2200	1100	3400	1700	4600	2300
1100	550	2300	1150	3500	1750	4700	2350
1200	600	2400	1200	3600	1800	4800	2400

Source: IBC Table 2607.4.

2610 Light-Transmitting Plastic Skylight Glazing

2610.2 Mounting

- This section addresses skylights with light-transmitting plastic glazing.
- Curbs are not required for skylights as follows:
  - Where the roof slopes ≥ 3:12 in either of the following locations:
    In Occupancy R-3.
    Where unclassified roof coverings are allowed.
- Neither of the following materials is required at the edge of a skylight where the condition indicated below applies:
  - Materials:
    Metal.
    Noncombustible material.
  - Condition:
    Where unclassified roof coverings are allowed.
- Otherwise, skylights are governed as follows:
  - Curbs are required for skylights as shown in the detail below.
  - Where edges are not constructed as per the detail provided, the following applies:
    Edge must be shown to resist combustion upon exposure to a standard test flame.

Note: The following are cited as governing exposure of the skylight edge to a test flame of which one be used: A Class B brand is specified for the test.
ASTM E 108, "Test Methods for Fire Tests of Roof Coverings."
UL 790, "Test Methods for Fire Tests of Roof Coverings."

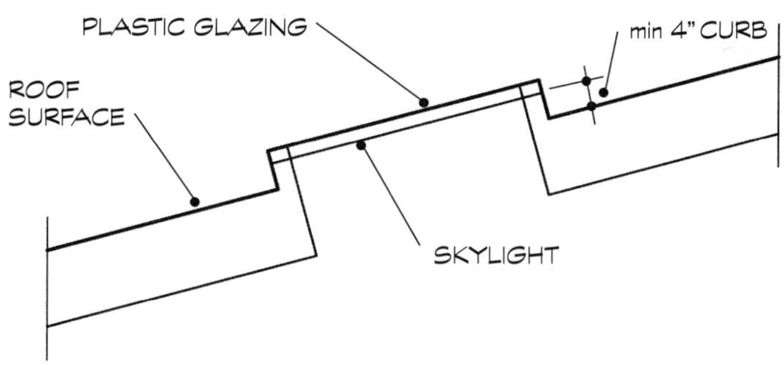

Fig. 2610.2. Required curb for light-transmitting plastic glazing in skylights.

2610 Light-Transmitting Plastic Skylight Glazing

2610.3 Slope

- This section addresses skylights with light-transmitting plastic glazing.
- This section does not apply to skylights that are shown to resist combustion upon exposure to a standard test flame.

Note: The following are cited as governing exposure of the skylight to a test flame. One of the standards must be used. A Class B brand is specified for the test.
ASTM E 108, "Test Methods for Fire Tests of Roof Coverings."
UL 790, "Test Methods for Fire Tests of Roof Coverings."

- Requirements for other skylights are shown on the details provided in Figures 2610.3A–B on the next page.
- A partial list of mounting heights required for domed skylights is provided with this section.
- Dome-shaped skylights must have a rise ≥ 10% of its longest span as indicated in the following partial list of rise requirements.

Table 2610.3 Required Rise of Domed Skylight vs. Span

Minimum		Minimum		Minimum		Minimum	
Span	rise	Span	rise	Span	rise	Span	rise
≤ 2'-6"	3"	3'-10"	4 5/8"	5'-2"	6 1/4"	6'-6"	7 13/16"
2'-7"	3 1/8"	3'-11"	4 3/4"	5'-3"	6 5/16"	6'-7"	7 15/16"
2'-8"	3 1/4"	4'-0"	4 13/16"	5'-4"	6 7/16"	6'-8"	8"
2'-9"	3 5/16"	4'-1"	4 15/16"	5'-5"	6 1/2"	6'-9"	8 1/8"
2'-10"	3 7/16"	4'-2"	5"	5'-6"	6 5/8"	6'-10"	8 1/4"
2'-11"	3 1/2"	4'-3"	5 1/8"	5'-7"	6 3/4"	6'-11"	8 5/16"
2'-0"	3 5/8"	4'-4"	5 1/4"	5'-8"	6 13/16"	7'-0"	8 7/16"
2'-1"	3 3/4"	4'-5"	5 5/16"	5'-9"	6 15/16"	7'-1"	8 1/2"
3'-2"	3 13/16"	4'-6"	5 7/16"	5'-10"	7"	7'-2"	8 5/8"
3'-3"	3 15/16"	4'-7"	5 1/2"	5'-11"	7 1/8"	7'-3"	8 3/4"
3'-4"	4"	4'-8"	5 5/8"	6'-0"	7 1/4"	7'-4"	8 13/16"
3'-5"	4 1/8"	4'-9"	5 3/4"	6'-1"	7 5/16"	7'-5"	8 15/16"
3'-6"	4 1/4"	4'-10"	5 13/16"	6'-2"	7 7/16"	7'-6"	9"
3'-7"	4 5/16"	4'-11"	5 15/16"	6'-3"	7 1/2"	7'-7"	9 1/8"
3'-8"	4 7/16"	5'-0"	6"	6'-4"	7 5/8"	7'-8"	9 1/4"
3'-9"	4 1/2"	5'-1"	6 1/8"	6'-5"	7 3/4"	7'-9"	9 5/16"

2610 Light-Transmitting Plastic Skylight Glazing

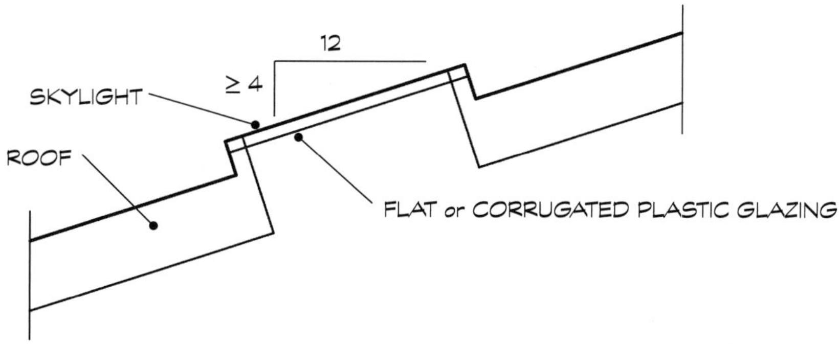

Fig. 2610.3A. Minimum slope of flat or corrugated light-transmitting plastic skylight glazing.

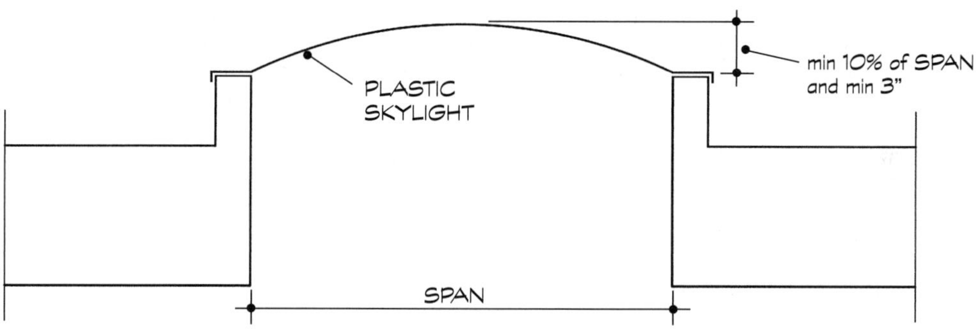

Fig. 2610.3B. Minimum rise of dome-shaped light-transmitting plastic skylight glazing.

2610 Light-Transmitting Plastic Skylight Glazing

2610.4 Maximum area of skylights

- This section addresses skylights with light-transmitting plastic glazing.
- Area limitations of this section do not apply to skylights in the following cases:
  - Where the building is sprinklered.
  - Where the building is equipped with the following:
    Smoke vents.
    Heat vents.

Note: 903.3.1.1, "NFPA 13 sprinkler systems," is cited as governing the sprinklers.
Section 910, "Smoke and Heat Vents," is cited as governing such devices as indicated above.

- Otherwise, the table below provides a partial list of maximum sizes permitted for skylights as follows:
  - Area inside the curb must ≤ 100 sf for an individual skylight.

Table 2610.4 Maximum Sizes Permitted for Skylights

Given width	Max. length	Given width	Max. length	Given width	Max. length	Given width	Max. length
1'-0"	100'-0"	3'-6"	28'-6"	6'-0"	16'-8"	8'-6"	11'-9"
1'-2"	85'-8"	3'-8"	27'-3"	6'-2"	16'-2"	8'-8"	11'-6"
1'-4"	75'-0"	3'-10"	26'-1"	6'-4"	15'-9"	8'-10"	11'-3"
1'-6"	66'-8"	4'-0"	25'-0"	6'-6"	15'-4"	9'-0"	11'-1"
1'-8"	60'-0"	4'-2"	24'-0"	6'-8"	15'-0"	9'-2"	10'-0"
1'-10"	54'-6"	4'-4"	23'-1"	6'-10"	14'-7"	9'-4"	10'-8"
2'-0"	50'-0"	4'-6"	22'-2"	7'-0"	14'-3"	9'-6"	10'-6"
2'-2"	46'-1"	4'-8"	21'-5"	7'-2"	13'-11"	9'-8"	10'-4"
2'-4"	42'-10"	4'-10"	20'-8"	7'-4"	13'-7"	9'-10"	10'-2"
2'-6"	40'-0"	5'-0"	20'-0"	7'-6"	13'-4"	10'-0"	10'-0"
2'-8"	37'-6"	5'-2"	19'-4"	7'-8"	13'-0"	10'-2"	9'-10"
2'-10"	35'-3"	5'-4"	18'-9"	7'-10"	12'-9"	10'-4"	9'-8"
3'-0"	33'-4"	5'-6"	18'-2"	8'-0"	12'-6"	10'-6"	9'-6"
3'-2"	31'-7"	5'-8"	17'-7"	8'-2"	12'-2"	10'-8"	9'-4"
3'-4"	30'-0"	5'-10"	17'-1"	8'-4"	12'-0"	10'-10"	9'-2"

2610 Light-Transmitting Plastic Skylight Glazing

2610.5 Aggregate area of skylights *(part 1 of 3)*

- This section addresses skylights with light-transmitting plastic glazing.
- The sum of the areas of skylights on a roof is limited based on the size of the floor area served, as shown in Figures 2610.5A–2610.5B provided below and on the next page.
- Higher limits are permitted where one of the following is present:
  - Sprinklers.
  - Smoke and heat vents.

Note: 903.3.1.1, "NFPA 13 sprinkler systems," is cited as governing the sprinklers. Section 910, "Smoke and Heat Vents," is cited as governing such devices as indicated above.

- Tables 2610.5a–d are provided with partial lists of skylight sizes complying with the limits of this section.

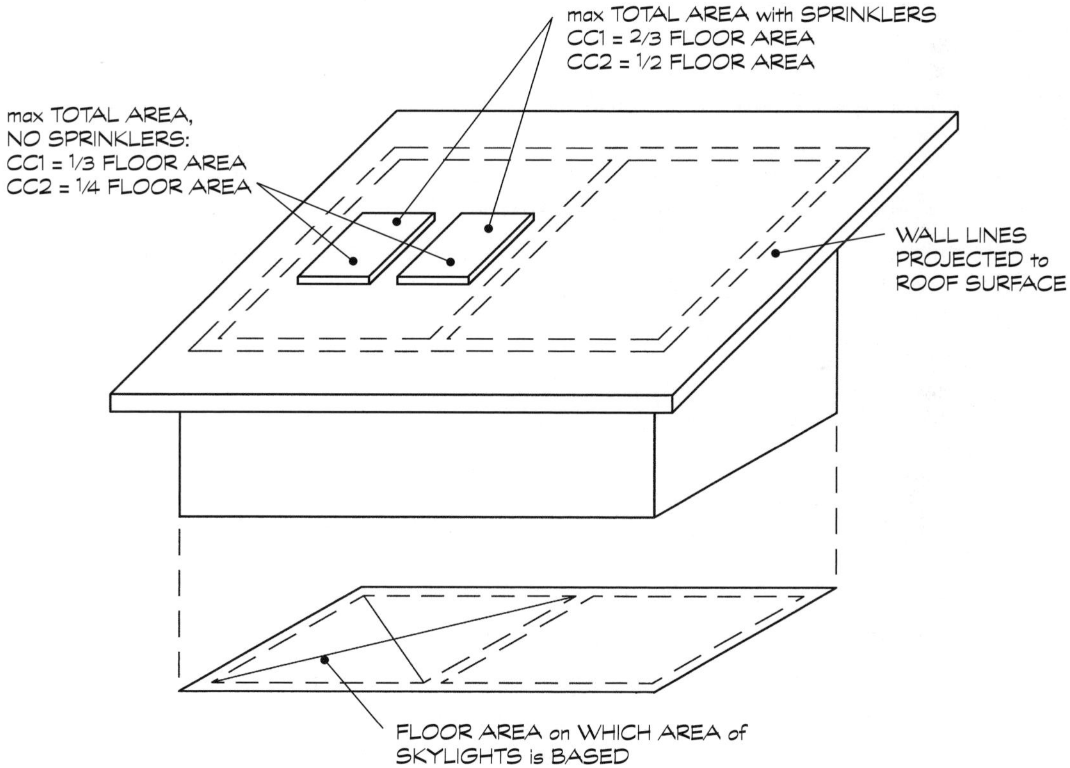

Fig. 2610.5A. Maximum area of skylights with light-transmitting plastic glazing.

2610 Light-Transmitting Plastic Skylight Glazing

Case study: Fig. 2610.5B. There are 4 light-transmitting polycarbonate (CC1) plastic skylights serving the atrium shown. The area of each 10' ×10' skylight is not limited due to the fact that the building is sprinklered. The aggregate area of the skylights is limited to 2/3 of the area they serve (twice the 1/3 limit for nonsprinklered buildings). In this case, the 400 sf total area of the 4 skylights is < the 1,410 sf limit. Separation of the skylights is not governed since the building is sprinklered. The skylights are in compliance with code requirements.

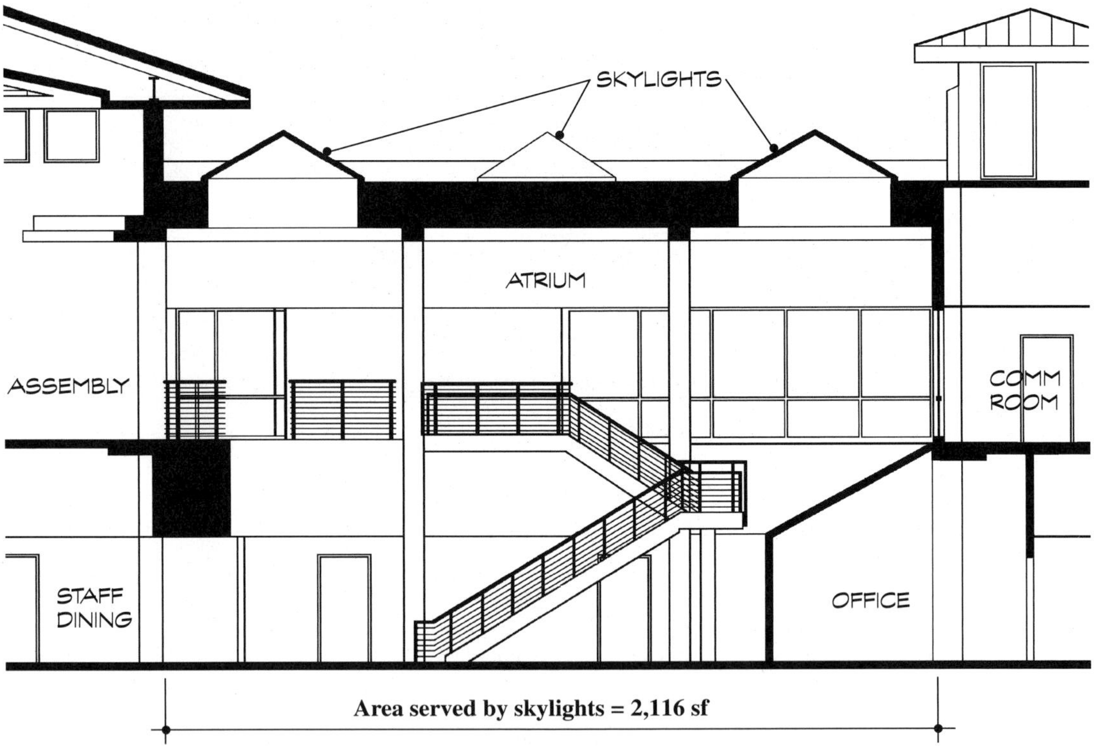

Fig. 2610.5B. Section at atrium. Lee's Summit Police and Court Facility. Lee's Summit, Missouri. The Hollis and Miller Group, Inc. Lee's Summit, Missouri.

2610 Light-Transmitting Plastic Skylight Glazing

2610.5 Aggregate area of skylights *(part 2 of 3)*

- The maximum permitted sum of skylight areas with CC1-class plastic is as follows:
  - Total area be ≤ 1/3 the floor area served where no sprinklers or vents are provided.

Table 2610.5a Maximum Total Skylight Area for CC1 Plastics with No Sprinklers or Vents

Floor area (sf)	Maximum panel area (sf)	Floor area (sf)	Maximum panel area (sf)	Floor area (sf)	Maximum panel area (sf)	Floor area (sf)	Maximum panel area (sf)
100	33.33	1,300	433.33	2,500	833.33	3,700	1,233.33
200	66.67	1,400	466.67	2,600	866.67	3,800	1,266.67
300	100.00	1,500	500.00	2,700	900.00	3,900	1,300.00
400	133.33	1,600	533.33	2,800	933.33	4,000	1,333.33
500	166.67	1,700	566.67	2,900	966.67	4,100	1,366.67
600	200.00	1,800	600.00	3,000	1,000.00	4,200	1,400.00
700	233.33	1,900	633.33	3,100	1,033.33	4,300	1,433.33
800	266.67	2,000	666.67	3,200	1,066.67	4,400	1,466.67
900	300.00	2,100	700.00	3,300	1,100.00	4,500	1,500.00
1,000	333.33	2,200	733.33	3,400	1,133.33	4,600	1,533.33
1,100	366.67	2,300	766.67	3,500	1,166.67	4,700	1,566.67
1,200	400.00	2,400	800.00	3,600	1,200.00	4,800	1,600.00

- The maximum permitted sum of skylight areas with CC1-class plastic is as follows:
  - Total area be ≤ 2/3 the floor area served with sprinklers or vents provided.

Table 2610.5b Maximum Total Skylight Area for CC1 Plastics with Sprinklers or Vents

Floor area (sf)	Maximum panel area (sf)	Floor area (sf)	Maximum panel area (sf)	Floor area (sf)	Maximum panel area (sf)	Floor area (sf)	Maximum panel area (sf)
100	66.67	1,300	866.67	2,500	1,666.67	3,700	2,466.67
200	133.33	1,400	933.33	2,600	1,733.33	3,800	2,533.33
300	200.00	1,500	1,000.00	2,700	1,800.00	3,900	2,600.00
400	266.67	1,600	1,066.67	2,800	1,866.67	4,000	2,666.67
500	333.33	1,700	1,133.33	2,900	1,933.33	4,100	2,733.33
600	400.00	1,800	1,200.00	3,000	2,000.00	4,200	2,800.00
700	466.67	1,900	1,266.67	3,100	2,066.67	4,300	2,866.67
800	533.33	2,000	1,333.33	3,200	2,133.33	4,400	2,933.33
900	600.00	2,100	1,400.00	3,300	2,200.00	4,500	3,000.00
1,000	666.67	2,200	1,466.67	3,400	2,266.67	4,600	3,066.67
1,100	733.33	2,300	1,533.33	3,500	2,333.33	4,700	3,133.33
1,200	800.00	2,400	1,600.00	3,600	2,400.00	4,800	3,200.00

2610 Light-Transmitting Plastic Skylight Glazing

2610.5 Aggregate area of skylights *(part 3 of 3)*

- The maximum permitted sum of skylight areas with CC2-class plastic is as follows:
  - Total area be ≤ 1/4 the floor area served where no sprinklers or vents are provided.

Table 2610.5c Maximum Total Skylight Area for CC2 Plastics with No Sprinklers or Vents

Floor area (sf)	Maximum panel area (sf)	Floor area (sf)	Maximum panel area (sf)	Floor area (sf)	Maximum panel area (sf)	Floor area (sf)	Maximum panel area (sf)
100	25	1,300	325	2,500	625	3,700	925
200	50	1,400	350	2,600	650	3,800	950
300	75	1,500	375	2,700	675	3,900	975
400	100	1,600	400	2,800	700	4,000	1,000
500	125	1,700	425	2,900	725	4,100	1,025
600	150	1,800	450	3,000	750	4,200	1,050
700	175	1,900	475	3,100	775	4,300	1,075
800	200	2,000	500	3,200	800	4,400	1,100
900	225	2,100	525	3,300	825	4,500	1,125
1,000	250	2,200	550	3,400	850	4,600	1,150
1,100	275	2,300	575	3,500	875	4,700	1,175
1,200	300	2,400	600	3,600	900	4,800	1,200

- The maximum permitted sum of skylight areas with CC2-class plastic is as follows:
  - Total area be ≤ 1/2 the floor area served with sprinklers or vents provided.

Table 2610.5d Maximum Total Skylight Area for CC2 Plastics with Sprinklers or Vents

Floor area (sf)	Maximum panel area (sf)	Floor area (sf)	Maximum panel area (sf)	Floor area (sf)	Maximum panel area (sf)	Floor area (sf)	Maximum panel area (sf)
100	50	1,300	650	2,500	1,250	3,700	1,850
200	100	1,400	700	2,600	1,300	3,800	1,900
300	150	1,500	750	2,700	1,350	3,900	1,950
400	200	1,600	800	2,800	1,400	4,000	2,000
500	250	1,700	850	2,900	1,450	4,100	2,050
600	300	1,800	900	3,000	1,500	4,200	2,100
700	350	1,900	950	3,100	1,550	4,300	2,150
800	400	2,000	1,000	3,200	1,600	4,400	2,200
900	450	2,100	1,050	3,300	1,650	4,500	2,250
1,000	500	2,200	1,100	3,400	1,700	4,600	2,300
1,100	550	2,300	1,150	3,500	1,750	4,700	2,350
1,200	600	2,400	1,200	3,600	1,800	4,800	2,400

2610 Light-Transmitting Plastic Skylight Glazing

2610.6 Separation

- This section addresses skylights with light-transmitting plastic glazing.
- This section does not address the following:
  - Sprinklered buildings.

  Note: 903.3.1.1, "NFPA 13 sprinkler systems," is cited as governing the sprinklers.

  - In Occupancy R-3 with both of the following characteristics:
    Multiple skylights above the same space.
    Sum of areas ≤ 100 sf.

  Note: 2610.4, "Maximum area of skylights," is cited as governing the multiple skylights in R-3.

- Otherwise, skylights must be separated as shown in the illustration below.

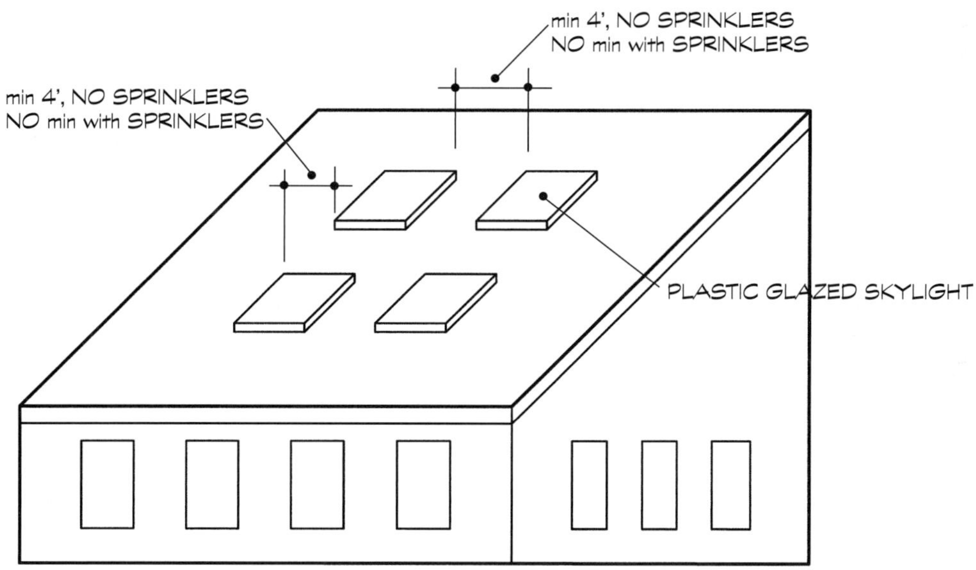

Fig. 2610.6. Minimum separation of skylights with light-transmitting plastic glazing measured in a horizontal plane.

2611 Light-Transmitting Plastic Interior Signs

2611.2 Aggregate area

- Wall signs of light-transmitting plastics in covered mall buildings may not be > 20% of the wall area as follows:
  - A partial list of signs within this limit is provided below:

Table 2611.2 Areas of Interior Signs Permitted Based on Wall Area

Wall area (sf)	Sign area (sf)	Wall area (sf)	Sign area (sf)	Wall area (sf)	Sign area (sf)	Wall area (sf)	Sign area (sf)
73	14.60	85	17.00	97	19.40	109	21.80
74	14.80	86	17.20	98	19.60	110	22.00
75	15.00	87	17.40	99	19.80	111	22.20
76	15.20	88	17.60	100	20.00	112	22.40
77	15.40	89	17.80	101	20.20	113	22.60
78	15.60	90	18.00	102	20.40	114	22.80
79	15.80	91	18.20	103	20.60	115	23.00
80	16.00	92	18.40	104	20.80	116	23.20
81	16.20	93	18.60	105	21.00	117	23.40
82	16.40	94	18.80	106	21.20	118	23.60
83	16.60	95	19.00	107	21.40	119	23.80
84	16.80	96	19.20	108	21.60	120	24.00

2611.3 Maximum area

- Wall signs of light-transmitting plastics in covered mall buildings may not be > 24 sf as follows:
  - A partial list of signs within this limit is provided below:

Table 2611.3 Dimensions of Interior Signs ≤ 24 sf in Area

Width	Length	Width	Length	Width	Length	Width	Length
1'-0"	24'-0"	2'-0"	12'-0"	3'-0"	8'-0"	4'-0"	6'-0"
1'-1"	22'-1"	2'-1"	11'-6"	3'-1"	7'-9"	4'-1"	5'-10"
1'-2"	20'-6"	2'-2"	11'-1"	3'-2"	7'-7"	4'-2"	5'-9"
1'-3"	19'-2"	2'-3"	10'-8"	3'-3"	7'-4"	4'-3"	5'-7"
1'-4"	18'-0"	2'-4"	10'-3"	3'-4"	7'-2"	4'-4"	5'-6"
1'-5"	16'-11"	2'-5"	9'-11"	3'-5"	7'-0"	4'-5"	5'-5"
1'-6"	16'-0"	2'-6"	9'-7"	3'-6"	6'-10"	4'-6"	5'-4"
1'-7"	15'-1"	2'-7"	9'-3"	3'-7"	6'-8"	4'-7"	5'-2"
1'-8"	14'-4"	2'-8"	9'-0"	3'-8"	6'-6"	4'-8"	5'-1"
1'-9"	13'-8"	2'-9"	8'-8"	3'-9"	6'-4"	4'-9"	5'-0"
1'-10"	13'-1"	2'-10"	8'-5"	3'-10"	6'-3"	4'-10"	4'-11"
1'-11"	12'-6"	2'-11"	8'-2"	3'-11"	6'-1"	4'-11"	4'-10"

2611 Light-Transmitting Plastic Interior Signs

2611.4 Encasement

- This section addresses interior signs of light-transmitting plastic wall signs in covered mall buildings.
- Edge requirements for such signs are shown in the illustration below:

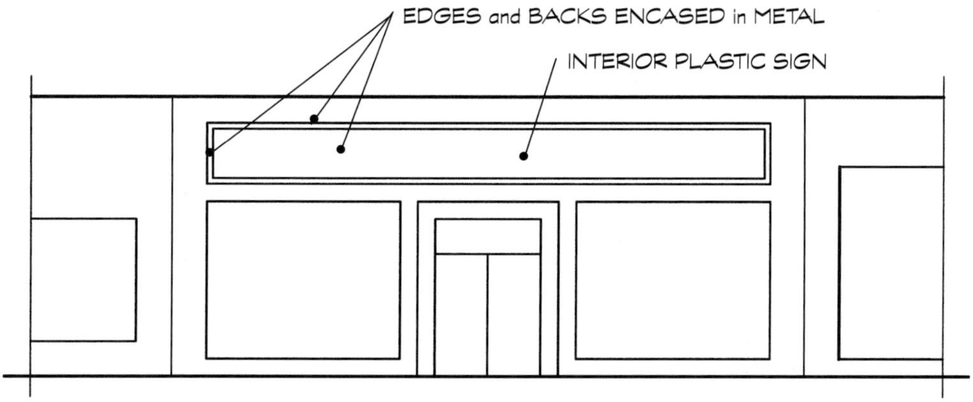

Fig. 2611.4. Light-transmitting plastic wall signs in covered mall buildings.

2612 Fiber Reinforced Polymer and Fiberglass Reinforced Polymer

2612.6 Exterior use *(part 1 of 2)*

- This section addresses the following reinforced polymers:
  - Fiberglass-reinforced polymers.
  - Other fiber-reinforced polymers.
- Reinforced polymers that meet all of the following conditions are not governed by this section:
  - Condition 1 is any one of the following:
    - Flame-spread index of the polymer is ≤ 25 in the following case:
      - Area of the polymer application is ≤ 20% of the wall area on which the application is made.
    - Flame-spread index of the polymer is ≤ 75 in the following case:
      - Area of the polymer application is ≤ 10% of the wall area on which the application is made.
    - Flame-spread index of the polymer is not governed in the following case:
      - A coating or paint is applied to the polymer in a thickness < 0.036".
  - Condition 2 is as follows:
    - Fireblocking is required.

  Note: 717.2.6, "Architectural trim," is cited as governing the fire blocking.

  - Condition 3 is either of the following:
    - The polymer is installed directly to a noncombustible substrate.
    - The polymer is separated from the exterior wall by one of the following:
      - Corrosion-resistant steel with a base metal thickness ≥ 0.016" throughout.
      - Aluminum with a thickness ≥ 0.019".
      - Other noncombustible material that is approved.
  - Condition 4 is as follows:
    - The polymer is designed for the following loads:
      - Uniform live loads.
      - Snow loads.
      - Wind loads.
      - Earthquake loads.

  Note: The following are cited as specifying the loads applicable to the polymer design.
  IBC Table 1607.1, "Minimum Uniformly Distributed Live Loads, L_o, and Minimum Concentrated Live Loads."
  Section 1608, "Snow Loads."
  Section 1609, "Wind Loads."
  Section 1613, "Earthquake Loads."

- Reinforced polymers installed on buildings ≤ 40' above grade plane have all of the following requirements:

  Note: 1406.2, "Combustible exterior wall coverings," is cited as governing these polymers.

  - One of the following applies:
    - Where fire separation distance is ≤ 5' the following applies:
      - Area of polymer is limited to 10% of the wall area.
    - Where the fire separation distance is > 5' the following applies:
      - Area of polymer is not limited.

2612 Fiber Reinforced Polymer and Fiberglass Reinforced Polymer

2612.6 Exterior use *(part 2 of 2)*

- One of the following applies:
  - A flame-spread index of ≤ 200 is required in the following case:
    - Where there is no coating or paint is applied to the polymer in a thickness < 0.036".
  - The flame-spread index of the polymer is not governed in the following case:
    - Where a coating or paint is applied to the polymer in a thickness < 0.036".
- Fireblocking is required.

Note: 717.2.6, "Architectural trim," is cited as governing the fireblocking.

- The polymer must be designed for the following loads:
  - Uniform live loads.
  - Snow loads.
  - Wind loads.
  - Earthquake loads.

Note: The following are cited as specifying the loads applicable to the polymer design:
IBC Table 1607.1, "Minimum Uniformly Distributed Live Loads, L_o, and Minimum Concentrated Live Loads."
Section 1608, "Snow Loads."
Section 1609, "Wind Loads."
Section 1613. "Earthquake Loads."

- Other reinforced polymers installed on exterior walls are governed as follows:

Note: 2603.5, "Exterior walls of buildings of any height," is cited as governing polymers.

- These polymers may be installed on any construction type.
- Fireblocking is required.

Note: Section 717, "Concealed Spaces," is cited as governing the fireblocking.

- The polymer must be designed for the following loads:
  - Uniform live loads.
  - Snow loads.
  - Wind loads.
  - Earthquake loads.

Note: The following are cited as specifying the loads applicable to the polymer design:
IBC Table 1607.1, "Minimum Uniformly Distributed Live Loads and Minimum Concentrated Live Loads."
Section 1608, "Snow Loads."
Section 1609, "Wind Loads."
Section 1613. "Earthquake Loads."

2613 Reflective Plastic Core Insulation

2613.1 General

- Reflective plastic core insulation is governed by the following:
  - Identification requirements of this section.

Note: 2613.2, "Identification," is cited as governing identification of this insulation.

  - One of the following sets of criteria:
    Flame-spread and smoke-developed indexes.
    Limitations on physical characteristics of the flame during fire testing.

Note: The following are cited as alternative sets of criteria for insulation performance:
2613.3, "Surface-burning characteristics."
2613.4, "Room corner test heat release."

2613.2 Identification

- The following information is required on reflective plastic core insulation packaging delivered to the job:
  - Manufacturer's or supplier's name.
  - Product identification.
  - Information verifying compliance with code requirements by the end use.

2613.3 Surface-burning characteristics

- Surface-burning criteria for reflective plastic core insulation are as follows:
  - Flame-spread index must be ≤ 25.
  - Smoke-developed index must be ≤ 450.

Note: One of the following tests is required to determine surface-burning characteristics:
ASTM E 84, "Test Method for Surface Burning Characteristics of Building Materials."
UL 723, "Test for Surface Burning Characteristics of Building Materials."
The insulation mounting method in one of the following is required for use in the cited tests:
2613.3.1, "Mounting of test specimen."
2613.3.2, "Specific testing."

2613.3.1 Mounting of test specimen

- The test specimen must be mounted on 2" high metal frames as follows:
  - So as to provide an air space between the following:
    The unexposed face of the insulation and the lid of the test device.

2613.3.2 Specific testing

- The following product preparation and mounting method is required for testing:
  - That designed specifically for reflective plastic core insulation.

1613.4 Room corner test heat release

- Reflective plastic core insulation must meet certain performance criteria when fire tested as follows:
  - Insulation to be tested in the format and maximum thickness intended for its building installation.

Note: 803.1.2.1, "Acceptance criteria for NFPA 286," is cited as listing required performance criteria.
NFPA, 286, "Methods of Fire Test for Evaluating Contribution of Wall and Ceiling Interior Finish to Room Fire Growth," is cited as an alternative test.
UL 1715, "Fire Test of Interior Finish Material," is cited as an alternative test.

27

Electrical

Methodist Community Health Center. Sugar Land, Texas. *(partial elevation)*
HKS, Inc., Architects, Engineers, Planners. Dallas, Texas.

2702 Emergency and Standby Power Systems

2702.2.1 Group A occupancies

- Emergency power is required for the following:
  - Voice communication systems where required in Occupancy A.

Note: 907.5.2.2.4, "Emergency power," is cited as governing these systems.

2702.2.2 Smoke control systems

- Standby power is required for smoke control systems.

Note: 909.11, "Power systems," is cited as governing these systems.

2702.2.3 Exit signs

- Emergency power is required for exit signs.

Note: 1011.5.3, "Power source," is cited as governing exit signs.

2702.2.4 Means of egress illumination

- Emergency power is required for means of egress illumination.

Note: 1006.3, "Illumination emergency power," is cited as governing means of egress illumination.

2702.2.5 Accessible means of egress elevators

- Standby power is required for the following elevators:
  - Where serving an accessible means of egress.

Note: 1007.4, "Elevators," is cited as governing these elevators.

2702.2.6 Accessible means of egress platform lifts

- Standby power is required for the following platform lifts:
  - Where serving an accessible means of egress.

Note: 1007.5, "Platform lifts," is cited as governing these lifts.
ASME A 18.1, "Safety Standard for Platform Lifts and Stairway Chairlifts," is cited as an alternative to this section to govern lifts.

2702.2.7 Horizontal sliding doors

- Standby power is required for horizontal sliding doors

Note: 1008.1.4.3, "Horizontal sliding doors," is cited as governing these doors.

2702.2.8 Semiconductor fabrication facilities

- Emergency power is required for semiconductor fabrication facilities.

Note: 415.8.10, "Emergency power system," is cited as governing emergency power for semiconductor fabrication facilities.

2702 Emergency and Standby Power Systems

2702.2.9 Membrane structures

- Standby power is required for the following:
  - Back-up inflation systems.
  - Exit signs in temporary tents and membrane structures.

Note: 3102.8.2, "Standby power," is cited as governing standby power for back-up inflation systems. The International Fire Code *is cited as governing exit signs for tents and membrane structures.*

2702.2.10 Hazardous materials

- One of the following types of power as applicable is required for locations with hazardous materials:
  - Standby power.
  - Emergency power.

Note: 414.5.4, "Standby or emergency power," is cited as governing this power.

2702.2.11 Highly toxic and toxic materials

- Emergency power is required for occupancies with the following materials:
  - Highly toxic materials.
  - Toxic materials.

Note: The International Fire Code *is cited as governing these occupancies.*

2702.2.12 Organic peroxides

- Standby power is required for occupancies where silane gas is located.

Note: The International Fire Code *is cited as governing these occupancies.*

2702.2.13 Pyrophoric materials

- Emergency power is required for occupancies where silane gas is located.

Note: The International Fire Code *is cited as governing these occupancies.*

2702.2.14 Covered mall buildings

- Standby power is required for voice/alarm communication as follows:
  - In certain covered malls.

Note: 402.14, "Standby power," is cited as governing such power for these voice/alarm systems.

2702.2.15 High-rise buildings

- The following types of power are required for certain applications in high-rise buildings:
  - Standby power.
  - Emergency power.

Note: 403.4.7, "Standby power," is cited as governing this type of power.
403.4.8, "Emergency power systems," is cited as governing this type of power.

2702 Emergency and Standby Power Systems

2702.2.16 Underground buildings

- The following types of power are required for certain applications in underground buildings:
  - Standby power.
  - Emergency power.

Note: 405.8, "Standby power," is cited as governing this type of power.
405.9, "Emergency power," is cited as governing this type of power.

2702.2.17 Group I-3 occupancies

- Emergency power is required for doors in Occupancy I-3.

Note: 408.4.2, "Power-operated doors and locks," is cited as governing these doors.

2702.2.18 Airport traffic control towers

- Standby power is required for airport traffic control towers.

Note: 412.3.5, "Standby power," is cited as governing these systems.

2702.2.19 Elevators

- Standby power is required for elevators.

Note: The following are cited as governing elevators:
3003.1, "Standby power."
3007.7, "Electrical power."
3008.15, "Electrical power."

2702.2.20 Smokeproof enclosures

- Standby power is required for smoke-proof enclosures.

Note: 909.20.6.2, "Standby power," is cited as governing these enclosures.

28

Mechanical Systems

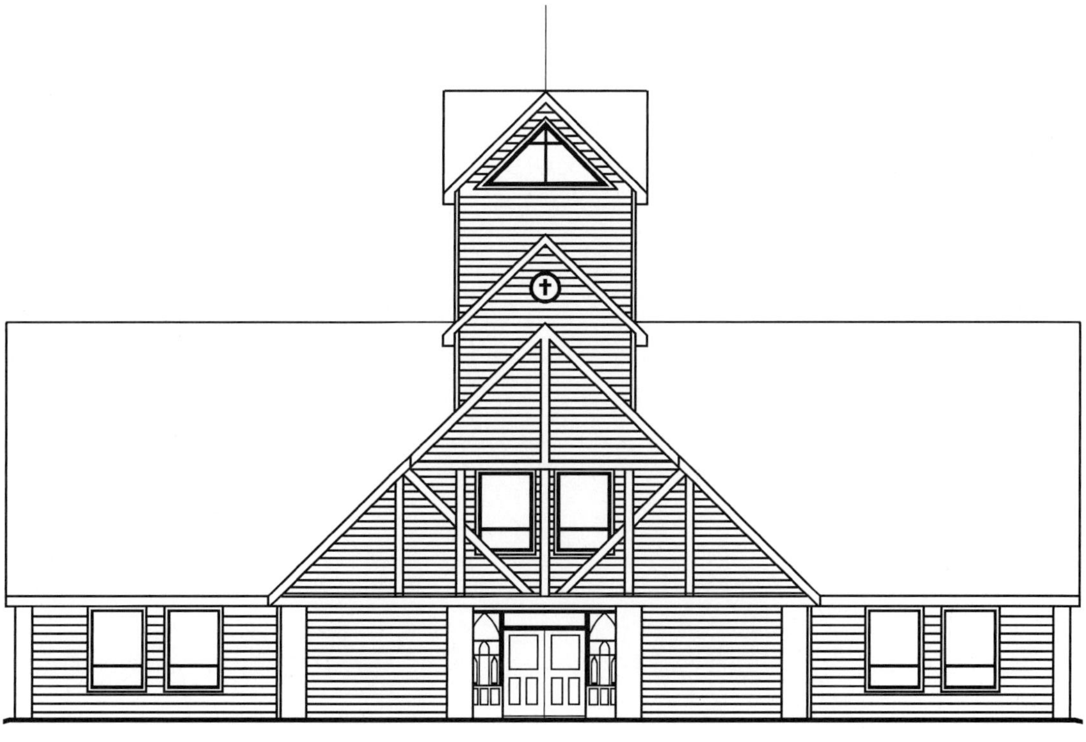

Glad Tidings Assembly of God Church. Naticoke, Pennsylvania.
Mullins and Weida, Architect and Associate. Bear Creek, Pennsylvania.

2801 General

2801.1 Scope

- The following components are governed by the codes listed below:
  - Components:
    - Mechanical appliances.
    - Mechanical equipment.
    - Mechanical systems.
  - Codes:
    - *International Fuel Gas Code.*
    - *International Mechanical Code.*
- The following components are governed by the codes listed below:
  - Components:
    - Fireplaces.
    - Barbecues.
    - Masonry chimneys.
  - Codes:
    - IBC Chapter 21, "Masonry."
    - *International Mechanical Code.*

29

Plumbing Systems

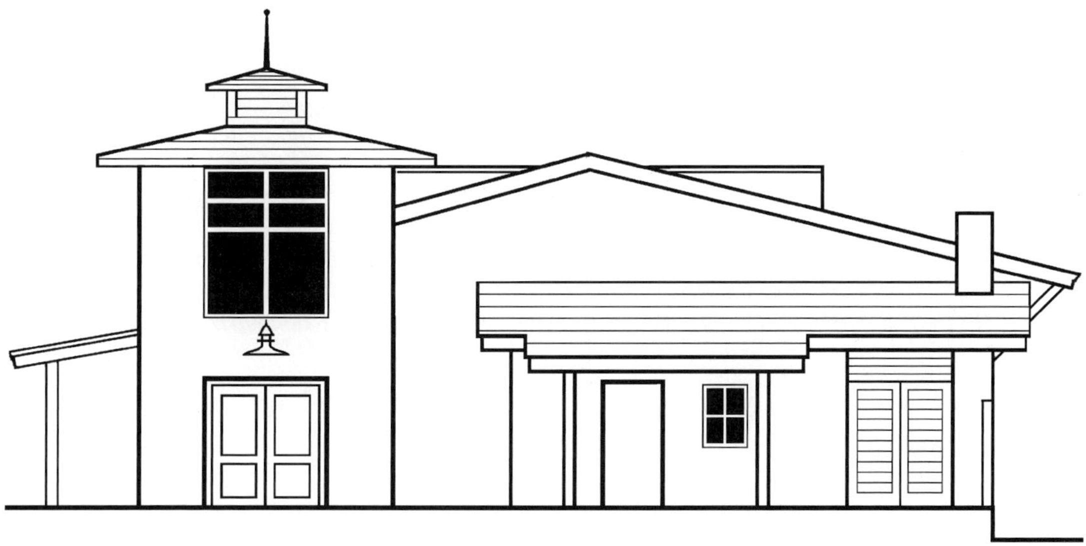

Creston Elementary Multipurpose Building. Creston, California.
Phillips Metsch Sweeney Moore Architects. Santa Barbara, California.

2902 Minimum Plumbing Facilities

2902.1 Minimum number of fixtures *(part 1 of 21)*

- Partial lists of plumbing fixtures required for each occupancy are provided in the tables on following pages.
- Minimum numbers of plumbing fixtures for occupancies not listed will be determined by the building official.
- Where not indicated otherwise, ≥ 1 service sink is required for all occupancies.
- In each toilet room, urinals may be installed in lieu of required water closets as follows:
  - For ≥ 67% of the required water closets in the following occupancies:
    Assembly.
    Education.
  - For ≥ 1/2 of the required water closets in all other occupancies.
- Drinking fountains are not required where restaurants serve water.
- The following may be substituted for ≤ 1/2 of required drinking fountains:
  - Water coolers.
  - Bottled-water dispensers.
- Where "1" is listed as the minimum number of drinking fountains, one of the following is required:
  - A drinking fountain that accommodates both seated (handicapped) persons and standing persons.
  - 2 drinking fountains as follows:
    1 for handicapped (seated) persons.
    1 for standing persons.
- Occupants in seasonal outdoor use areas of the following occupancies are added to the building occupant load in determining the number of plumbing fixtures required:
  - A-1, A-2, A-3.

Note: IBC Table 2902.1, "Minimum Number of Required Plumbing Fixtures," is cited as governing the number of plumbing fixtures required and is the basis of the tables provided in this section.
The following are cited as governing the number of drinking fountains:
IBC Table 2902.1, "Minimum Number of Required Plumbing Fixtures."
Chapter 11, "Accessibility."
Chapter 3, "Use and Occupancy Classification," is cited as governing the classification of occupancies.
The following sections of the International Plumbing Code *are cited as governing substitutions for certain plumbing fixtures:*
IPC 419.2, "Substitution for water closets," permits the substitution of urinals for water closets as summarized above.
IPC 410.1, "Approval," permits omitting drinking fountains as summarized above.

2902 Minimum Plumbing Facilities

2902.1 Minimum number of fixtures *(part 2 of 21)*

- The table below includes the following A-1 uses:
  - Legitimate theaters.
  - Movie theaters.
  - Other performing arts buildings.

Table 2902.1a Occupancies A-1: Minimum Plumbing Fixtures Required

Occupant count	WC men	WC women	Lavs	DF	Occupant count	WC men	WC women	Lavs	DF
1–65	1	1	1	1	751–780	7	12	4	2
66–125	1	2	1	1	781–800	7	13	4	2
126–130	2	2	1	1	801–845	7	13	5	2
131–195	2	3	1	1	846–875	7	14	5	2
196–200	2	4	1	1	876–910	8	14	5	2
201–250	2	4	2	1	911–975	8	15	5	2
251–260	3	4	2	1	976–1,000	8	16	5	2
261–325	3	5	2	1	1,001–1,040	9	16	6	3
326–375	3	6	2	1	1,041–1,105	9	17	6	3
376–390	4	6	2	1	1,106–1,125	9	18	6	3
391–400	4	7	2	1	1,126–1,170	10	18	6	3
401–455	4	7	3	1	1,171–1,200	10	19	6	3
456–500	4	8	3	1	1,201–1,235	10	19	7	3
501–520	5	8	3	2	1,236–1,250	10	20	7	3
521–585	5	9	3	2	1,251–1,300	11	20	7	3
586–600	5	10	3	2	1,301–1,365	11	21	7	3
601–625	5	10	4	2	1,366–1,375	11	22	7	3
626–650	6	10	4	2	1,376–1,400	12	22	7	3
651–715	6	11	4	2	1,401–1,430	12	22	8	3
716–750	6	12	4	2	1,431–1,495	12	23	8	3

Source: IBC Table 2902.1.

2902 Minimum Plumbing Facilities

2902.1 Minimum number of fixtures *(part 3 of 21)*

- The table below includes the following A-2 uses:
  - Bars.
  - Dance halls.
  - Nightclubs.
  - Similar functions.
  - Taverns.

Table 2902.1b Occupancy A-2 Nightclubs, Bars, etc.: Minimum Plumbing Fixtures Required

Occupant count	WC men	WC women	Lavs	DF
1–40	1	1	1	1
41–75	2	2	1	1
76–80	2	2	2	1
81–120	3	3	2	1
121–150	4	4	2	1
151–160	4	4	3	1
161–200	5	5	3	1
201–225	6	6	3	1

Occupant count	WC men	WC women	Lavs	DF
226–240	6	6	4	1
241–280	7	7	4	1
281–300	8	8	4	1
301–320	8	8	5	1
321–360	9	9	5	1
361–375	10	10	5	1
376–400	10	10	6	1
401–440	11	11	6	1

Source: IBC Table 2902.1.

- The table below includes the following A-2 uses:
  - Banquet halls.
  - Food courts.
  - Restaurants.

Table 2902.1c Occupancy A-2 Restaurants, etc.: Minimum Plumbing Fixtures Required

Occupant count	WC men	WC women	Lavs	DF
1–75	1	1	1	1
76–150	2	2	1	1
151–200	3	3	1	1
201–225	3	3	2	1
226–300	4	4	2	1
301–375	5	5	2	1

Occupant count	WC men	WC women	Lavs	DF
376–400	6	6	2	1
401–450	6	6	3	1
451–500	7	7	3	1
501–525	7	7	3	2
526–600	8	8	3	2
601–675	9	9	4	2

Source: IBC Table 2902.1.

2902 Minimum Plumbing Facilities

2902.1 Minimum number of fixtures *(part 4 of 21)*

- The table below includes the following A-3 uses:
  - Arcades.
  - Art galleries.
  - Auditoriums as follows:
    No permanent seating.
  - Exhibition halls.
  - Gymnasiums.
  - Lecture halls.
  - Libraries.
  - Museums.

Table 2902.1d Occupancy A-3 Auditoriums, Galleries, etc.: Minimum Plumbing Fixtures Required

Occupant count	WC men	WC women	Lavs	DF	Occupant count	WC men	WC women	Lavs	DF
1–65	1	1	1	1	751–780	7	12	4	2
66–125	1	2	1	1	781–800	7	13	4	2
126–130	2	2	1	1	801–845	7	13	5	2
131–195	2	3	1	1	846–875	7	14	5	2
196–200	2	4	1	1	876–910	8	14	5	2
201–250	2	4	2	1	911–975	8	15	5	2
251–260	3	4	2	1	976–1,000	8	16	5	2
261–325	3	5	2	1	1,001–1,040	9	16	6	3
326–375	3	6	2	1	1,041–1,105	9	17	6	3
376–390	4	6	2	1	1,106–1,125	9	18	6	3
391–400	4	7	2	1	1,126–1,170	10	18	6	3
401–455	4	7	3	1	1,171–1,200	10	19	6	3
456–500	4	8	3	1	1,201–1,235	10	19	7	3
501–520	5	8	3	2	1,236–1,250	10	20	7	3
521–585	5	9	3	2	1,251–1,300	11	20	7	3
586–600	5	10	3	2	1,301–1,365	11	21	7	3
601–625	5	10	4	2	1,366–1,375	11	22	7	3
626–650	6	10	4	2	1,376–1,400	12	22	7	3
651–715	6	11	4	2	1,401–1,430	12	22	8	3
716–750	6	12	4	2	1,431–1,495	12	23	8	3

Source: IBC Table 2902.1.

2902 Minimum Plumbing Facilities

2902.1 Minimum number of fixtures *(part 5 of 21)*

- The table below includes the following A-3 uses:
  - Passenger terminals.
  - Transportation facilities.

Table 2902.1e Occupancy A-3 Transportation Facilities: Minimum Fixtures Required

Occupant count	WC	Lavs	DF	Occupant count	WC	Lavs	DF
1–500	1	1	1	7,501–8,000	16	11	8
501–750	2	1	1	8,001–8,250	17	11	9
751–1,000	2	2	1	8,251–8,500	17	12	9
1,001–1,500	3	2	2	8,501–9,000	18	12	9
1,501–2,000	4	3	2	9,001–9,500	19	13	10
2,001–2,250	5	3	3	9,501–9,750	20	13	10
2,251–2,500	5	4	3	9,751–10,000	20	14	10
2,501–3,000	6	4	3	10,001–10,500	21	14	11
3,001–3,500	7	5	4	10,501–11,000	22	15	11
3,501–3,750	8	5	4	11,001–11,250	23	15	12
3,751–4,000	8	6	4	11,251–11,500	23	16	12
4,001–4,500	9	6	5	11,501–12,000	24	16	12
4,501–5,000	10	7	5	12,001–12,500	25	17	13
5,001–5,250	11	7	6	12,501–12,750	26	17	13
5,251–5,500	11	8	6	12,751–13,000	26	18	13
5,501–6,000	12	8	6	13,001–13,500	27	18	14
6,001–6,500	13	9	7	13,501–14,000	28	19	14
6,501–6,750	14	9	7	14,001–14,250	29	19	15
6,751–7,000	14	10	7	14,251–14,500	29	20	15
7,001–7,500	15	10	8	14,500–15,000	30	20	15

Source: IBC Table 2902.1.

2902 Minimum Plumbing Facilities

2902.1 Minimum number of fixtures *(part 6 of 21)*

- The table below includes the following A-3 uses:
  - Places of worship.
  - Other religious services.

Table 2902.1f Occupancy A-3 Places of Worship: Minimum Plumbing Fixtures Required

Occupant count	WC men	WC women	Lavs	DF	Occupant count	WC men	WC women	Lavs	DF
1–75	1	1	1	1	451–525	4	7	3	1
76–150	1	2	1	1	526–600	4	8	3	1
151–200	2	3	1	1	601–675	5	9	4	1
201–225	2	3	2	1	676–750	5	10	4	1
226–300	2	4	2	1	751–800	6	11	4	1
301–375	3	5	2	1	801–825	6	11	5	1
376–400	3	6	2	1	826–900	6	12	5	1
401–450	3	6	3	1	901–975	7	13	5	1

Source: IBC Table 2902.1.

- The table below includes A-4 and A-5 uses as follows:
  - A-4
    - Arenas.
    - Coliseums.
    - Skating rinks.
    - The following functions for indoor sporting events and activities:
      - Swimming pools.
      - Tennis courts.
  - A-5
    - Stadiums.
    - Amusement parks.
    - Grandstands for the following:
      - Outdoor sporting events.
      - Outdoor activities.

Table 2902.1g(1) Occupancy A-4 Coliseums, Arenas, etc., and Occupancy A-5 Stadiums, Amusement Parks, etc. 1–225 Seats: Minimum Plumbing Fixtures Required

Occupant count	WC men	WC women	Lavs men	Lavs women	DF	Occupant count	WC men	WC women	Lavs men	Lavs women	DF
1–40	1	1	1	1	1	121–150	2	4	1	1	1
41–75	1	2	1	1	1	151–160	3	4	1	2	1
76–80	2	2	1	1	1	161–200	3	5	1	2	1
81–120	2	3	1	1	1	201–225	3	6	2	2	1

Source: IBC Table 2902.1

2902 Minimum Plumbing Facilities

2902.1 Minimum number of fixtures *(part 7 of 21)*

- Occupancy A-4 Table 2902.1g—*continued*:

Table 2902.1g(2) Occupancy A-4 Coliseums, Arenas, etc., and Occupancy A-5 Stadiums, Amusement Parks, etc. 226–2360 Seats: Minimum Plumbing Fixtures Required

Occupant count	WC men	WC women	Lavs men	Lavs women	DF	Occupant count	WC men	WC women	Lavs men	Lavs women	DF
226–240	4	6	2	2	1	1,241–1,275	17	32	7	9	2
241–280	4	7	2	2	1	1,276–1,280	18	32	7	9	2
281–300	4	8	2	2	1	1,281–1,320	18	33	7	9	2
301–320	5	8	2	3	1	1,321–1,350	18	34	7	9	2
321–360	5	9	2	3	1	1,351–1,360	19	34	7	10	2
361–375	5	10	2	3	1	1,361–1,400	19	35	7	10	2
376–400	6	10	2	3	1	1,401–1,425	19	36	8	10	2
401–440	6	11	3	3	1	1,426–1,440	20	36	8	10	2
441–450	6	12	3	3	1	1,441–1,480	20	37	8	10	2
451–480	7	12	3	4	1	1,481–1,520	20	38	8	10	2
481–520	7	13	3	4	1	1,521–1,580	21	39	8	11	2
521–525	7	14	3	4	1	1,581–1,600	21	40	8	11	2
526–560	8	14	3	4	1	1,601–1,620	21	40	9	11	2
561–600	8	15	3	4	1	1,621–1,640	22	40	9	11	2
601–640	9	16	4	5	1	1,641–1,650	22	41	9	11	2
641–675	9	17	4	5	1	1,651–1,700	22	41	9	12	2
676–680	10	17	4	5	1	1,701–1,740	22	42	9	12	2
681–720	10	18	4	5	1	1,741–1,760	23	42	9	12	2
721–750	10	19	4	5	1	1,761–1,800	23	43	9	12	2
751–760	11	19	4	6	1	1,801–1,820	23	43	10	13	2
761–800	11	20	4	6	1	1,821–1,860	23	44	10	13	2
801–825	11	21	5	6	1	1,861–1,880	24	44	10	13	2
826–840	12	21	5	6	1	1,881–1,940	24	45	10	13	2
841–880	12	22	5	6	1	1,941–1,950	24	46	10	13	2
881–900	12	23	5	6	1	1,951–1,980	24	46	10	14	2
901–920	13	23	5	7	1	1,981–2,000	25	46	10	14	2
921–960	13	24	5	7	1	2,001–2,060	25	47	11	14	3
961–975	13	25	5	7	1	2,101–2,120	26	48	11	15	3
976–1,000	14	25	5	7	1	2,121–2,180	26	49	11	15	3
1,001–1,040	14	26	6	7	2	2,181–2,200	26	50	11	15	3
1,041–1,050	14	27	6	7	2	2,201–2,220	26	50	12	15	3
1,051–1,080	15	27	6	8	2	2,221–2,240	27	50	12	15	3
1,081–1,120	16	28	6	8	2	2,241–2,250	27	51	12	15	3
1,121–1,160	16	29	6	8	2	2,251–2,300	27	51	12	16	3
1,161–1,200	16	30	6	8	2	2,301–2,340	27	52	12	16	3
1,201–1,240	17	31	7	9	2	2,341–2,360	28	52	12	16	3

Source: IBC Table 2902.1.

2902 Minimum Plumbing Facilities

2902.1 Minimum number of fixtures *(part 8 of 21)*

- Occupancy A-4 Table 2902.1g—*continued*:

Table 2902.1g(3) Occupancy A-4 Coliseums, Arenas, etc., and Occupancy A-5 Stadiums, Amusement Parks, etc. 2,361–4,600 Seats: Minimum Plumbing Fixtures Required

Occupant count	WC men	WC women	Lavs men	Lavs women	DF
2,361–2,400	28	53	12	16	3
2,401–2,420	28	53	13	17	3
2,421–2,460	28	54	13	17	3
2,461–2,480	29	54	13	17	3
2,481–2,540	29	55	13	17	3
2,541–2,550	29	56	13	17	3
2,551–2,580	29	56	13	18	3
2,581–2,600	30	56	13	18	3
2,601–2,660	30	57	14	18	3
2,661–2,700	30	58	14	18	3
2,701–2,720	31	58	14	19	3
2,721–2,780	31	59	14	19	3
2,781–2,800	31	60	14	19	3
2,801–2,840	32	60	15	19	3
2,841–2,850	32	61	15	19	3
2,851–2,900	32	61	15	20	3
2,901–2,940	32	62	15	20	3
2,941–2,960	33	62	15	20	3
2,961–3,000	33	63	15	20	3
3,001–3,020	33	63	16	21	4
3,021–3,060	33	64	16	21	4
3,061–3,080	34	64	16	21	4
2,081–3,140	34	65	16	21	4
3,141–3,150	34	66	16	21	4
3,151–3,180	34	66	16	22	4
3,181–3,200	35	66	16	22	4
3,201–3,260	35	67	17	22	4
3,261–3,300	35	68	17	22	4
3,301–3,320	36	68	17	23	4
3,321–3,380	36	69	17	23	4
3,381–3,400	36	70	17	23	4
3,401–3,420	36	70	18	23	4
3,421–3,440	37	70	18	23	4
3,441–3,450	37	71	18	23	4
3,451–3,500	37	71	18	24	4
3,501–3,540	37	72	18	24	4
3,541–3,560	38	72	18	24	4
3,561–3,600	38	73	18	24	4
3,601–3,620	38	73	19	25	4
3,621–3,660	38	74	19	25	4
3,661–3,680	39	74	19	25	4
3,681–3,740	39	75	19	25	4
3,741–3,750	39	76	19	25	4
3,751–3,780	39	76	19	26	4
3,781–3,800	40	76	19	26	4
3,801–3,860	40	77	20	26	4
3,861–3,900	40	78	20	26	4
3,901–3,920	41	78	20	27	4
3,921–3,980	41	79	20	27	4
3,981–4,000	41	80	20	27	4
4,001–4,020	41	80	21	27	5
4,021–4,040	42	80	21	27	5
4,041–4,050	42	81	21	27	5
4,051–4,100	42	81	21	28	5
4,101–4,140	42	82	21	28	5
4,141–4,160	43	82	21	28	5
4,161–4,200	43	83	21	28	5
4,201–4,220	43	83	22	29	5
4,221–4,260	43	84	22	29	5
4,261–4,280	44	84	22	29	5
4,281–4,340	44	85	22	29	5
4,341–4,350	44	86	22	29	5
4,351–4,380	44	86	22	30	5
4,381–4,400	45	86	22	30	5
4,401–4,460	45	87	23	30	5
4,461–4,500	45	88	23	30	5
4,501–4,520	46	88	23	31	5
4,521–4,580	46	89	23	31	5
4,581–4,600	46	90	23	31	5

Source: IBC Table 2902.1.

2902 Minimum Plumbing Facilities

2902.1 Minimum number of fixtures *(part 9 of 21)*

- Occupancy B required plumbing fixtures are shown in the partial table below as follows:
  - Buildings for the following purposes are included:

    For transacting business.
    For providing professional services.
    For providing other services involving merchandise.
    Office building functions.
    Banking.
    Light industrial activity.

Table 2902.1h Occupancy B Business: Minimum Plumbing Fixtures Required

Occupant count	WC	Lavs	DF	Occupant count	WC	Lavs	DF
1–25	1	1	1	351–400	9	6	4
26–40	2	1	1	401–450	10	7	5
41–50	2	2	1	451–480	11	7	5
51–80	3	2	1	481–500	11	8	5
81–100	3	3	1	501–550	12	8	6
101–150	4	3	2	551–560	13	8	6
151–160	5	3	2	561–600	13	9	6
161–200	5	4	2	601–640	14	9	7
201–240	6	4	3	641–650	14	10	7
241–250	6	5	3	651–700	15	10	7
251–300	7	5	3	701–720	16	10	8
301–320	8	5	4	721–750	16	11	8
321–350	8	6	4	751–800	17	11	8

Source: IBC Table 2902.1.

2902 Minimum Plumbing Facilities

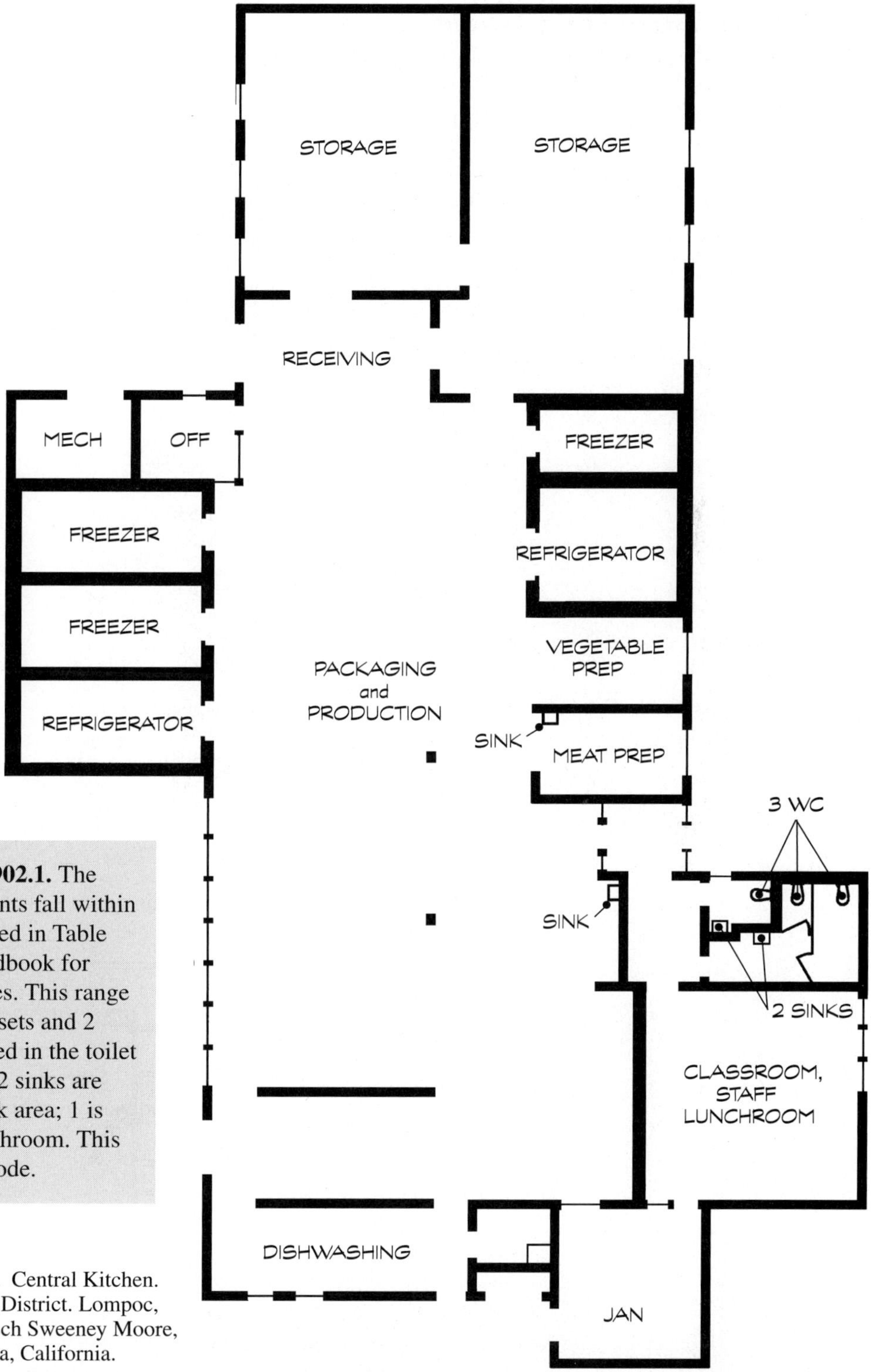

Case study: Fig. 2902.1. The kitchen's 56 occupants fall within the 51–80 range listed in Table 2902.1h of this handbook for business occupancies. This range requires 3 water closets and 2 lavatories as provided in the toilet rooms. In addition, 2 sinks are provided in the work area; 1 is provided in the lunchroom. This complies with the code.

Fig. 2902.1 Floor plan. Central Kitchen. Lompoc Unified School District. Lompoc, California. Phillips Metsch Sweeney Moore, Architects. Santa Barbara, California.

2902 Minimum Plumbing Facilities

2902.1 Minimum number of fixtures *(part 10 of 21)*

- Occupancy E required plumbing fixtures are shown in the table below:

Table 2902.1i Occupancy E Educational: Minimum Plumbing Fixtures Required

Occupant count	WC	Lavs	DF	Occupant count	WC	Lavs	DF
1–50	1	1	1	1,001–1,050	21	21	11
51–100	2	2	1	1,051–1,100	22	22	11
101–150	3	3	2	1,101–1,150	23	23	12
151–200	4	4	2	1,151–1,200	24	24	12
201–250	5	5	3	1,201–1,250	25	25	13
251–300	6	6	3	1,251–1,300	26	26	13
301–350	7	7	4	1,301–1,350	27	27	14
351–400	8	8	4	1,351–1,400	28	28	14
401–450	9	9	5	1,401–1,450	29	29	15
451–500	10	10	5	1,451–1,500	30	30	15
501–550	11	11	6	1,501–1,550	31	31	16
551–600	12	12	6	1,551–1,600	32	32	16
601–650	13	13	7	1,601–1,650	33	33	17
651–700	14	14	7	1,651–1,700	34	34	17
701–750	15	15	8	1,701–1,750	35	35	18
751–800	16	16	8	1,751–1,800	36	36	18
801–850	17	17	9	1,801–1,850	37	37	19
851–900	18	18	9	1,851–1,900	38	38	19
901–950	19	19	10	1,901–1,950	39	39	20
951–1,000	20	20	10	1,951–2,000	40	40	20

Source: IBC Table 2902.1.

2902 Minimum Plumbing Facilities

2902.1 Minimum number of fixtures *(part 11 of 21)*

- Factory and industrial occupancies require the following facilities, where required by the manufacturer, and other fixtures as shown in the partial table below:
  - Emergency eyewash stations.
  - Emergency showers.
- Functions included in the F-1 and F-2 tables are as follows:
  - Fabrication.
  - Assembly of products.
  - Assembly of materials.
  - Processing of products.
  - Processing of materials.

Table 2902.1j Occupancies F-1 and F-2 Factory and Industrial: Minimum Plumbing Fixtures Required

Occupant count	WC	Lavs	DF	Occupant count	WC	Lavs	DF
1–100	1	1	1	801–900	9	9	3
101–200	2	2	1	901–1,000	10	10	3
201–300	3	3	1	1,001–1,100	11	11	3
301–400	4	4	1	1,101–1,200	12	12	3
401–500	5	5	2	1,201–1,300	13	13	4
501–600	6	6	2	1,301–1,400	14	14	4
601–700	7	7	2	1,401–1,500	15	15	4
701–800	8	8	2	1,501–1,600	16	16	4

Source: IBC Table 2902.1.

- Plumbing fixtures for I-1 residential care are shown in the table below.

Table 2902.1k Occupancy I-1 Residential Care: Minimum Plumbing Fixtures Required

Occupant count	WC	Lavs	Bathtub/ shower	DF	Occupant count	WC	Lavs	Bathtub/ shower	DF
1–8	1	1	1	1	41–48	5	5	6	1
9–10	1	1	2	1	49–50	5	5	7	1
11–16	2	2	2	1	51–56	6	6	7	1
17–20	2	2	3	1	57–60	6	6	8	1
21–24	3	3	3	1	61–64	7	7	8	1
25–30	3	3	4	1	65–70	7	7	9	1
31–32	4	4	4	1	71–72	8	8	9	1
33–40	4	4	5	1	73–80	8	8	10	1

Source: IBC Table 2902.1.

2902 Minimum Plumbing Facilities

2902.1 Minimum number of fixtures *(part 12 of 21)*

- Fixtures for patients in occupancy I-2 hospitals and ambulatory nursing homes are governed as follows:
  - 1 toilet room may serve 2 adjacent rooms where all of the following conditions apply:
    Toilet has 1 lavatory and 1 water closet.
    Toilet is directly accessed from each room with provisions for privacy.
  - Otherwise, the following is required for each patient room:
    1 water closet and 1 lavatory.
  - 1 service sink is required for each floor.
  - Other fixtures are required as shown in the following table:

Table 2902.1l **Occupancy I-2 Hospitals, Ambulatory Nursing Homes: Minimum Fixtures Required for Patients**

Patient count	Bathtub/ shower	DF	Patient count	Bathtub/ shower	DF	Patient count	Bathtub/ shower	DF
1–15	1	1	166–180	12	2	331–345	23	4
16–30	2	1	181–195	13	2	346–360	24	4
31–45	3	1	196–200	14	2	361–375	25	4
46–60	4	1	201–210	14	3	376–390	26	4
61–75	5	1	211–225	15	3	391–400	27	4
76–90	6	1	226–240	16	3	401–405	27	5
91–100	7	1	241–255	17	3	406–420	28	5
101–105	7	2	256–270	18	3	421–435	29	5
106–120	8	2	271–285	19	3	436–450	30	5
121–135	9	2	286–300	20	3	451–465	31	5
136–150	10	2	301–315	21	4	466–480	32	5
151–165	11	2	316–330	22	4	481–495	33	5

Source: IBC Table 2902.1.

- Fixtures for institutional employees other than residential care are governed as follows:
  - Employee toilets are separate from patient toilets.
  - Fixture requirements are as shown in the partial table below:

Table 2902.1m **Occupancy I Institutional Employees Other Than Residential Care: Minimum Fixtures Required for Employees**

Employee count	WC	Lavs	DF	Employee count	WC	Lavs	DF
1–25	1	1	1	126–140	6	4	2
26–35	2	1	1	141–150	6	5	2
36–50	2	2	1	151–175	7	5	2
51–70	3	2	1	176–200	8	6	2
71–75	2	2	1	201–210	9	6	3
76–100	4	3	1	211–225	9	7	3
101–105	5	3	2	226–245	10	7	3
106–125	5	4	2	246–250	10	8	3

Source: IBC Table 2902.1.

2902 Minimum Plumbing Facilities

2902.1 Minimum number of fixtures *(part 13 of 21)*

- Fixtures for institutional visitors other than residential care are governed as follows:
  - A service sink is not required to serve visitors.
  - Bathing facilities are not required for visitors.
  - Otherwise, fixture requirements are as shown in the partial table below:

Table 2902.1n Occupancy I Institutional Visitors Other Than Residential Care: Minimum Plumbing Fixtures Required for Visitors

Visitor count	WC	Lavs	DF	Visitor count	WC	Lavs	DF
1–75	1	1	1	201–225	3	3	1
76–100	2	1	1	226–300	4	3	1
101–150	2	2	1	301–375	5	4	1
151–200	3	2	1	376–400	6	4	1

Source: IBC Table 2902.1.

- Fixtures for inmates of prisons are governed as follows:
  - Employee toilets are separate from inmate toilets and are not governed by guidelines for inmates or the table below.
  - 1 water closet for each cell.
  - 1 lavatory for each cell.
  - Other fixtures are required as shown in the partial table below:

Table 2902.1o Occupancy I-3 Prisons: Minimum Fixtures Required for Inmates

Inmate count	Bathtub/ shower	DF	Inmate count	Bathtub/ shower	DF	Inmate count	Bathtub/ shower	DF
1–15	1	1	211–225	15	3	436–450	30	5
16–30	2	1	226–240	16	3	451–465	31	5
31–45	3	1	241–255	17	3	466–480	32	5
46–60	4	1	256–270	18	3	481–495	33	5
61–75	5	1	271–285	19	3	496–500	34	5
76–90	6	1	286–300	20	3	501–510	34	6
91–100	7	1	301–315	21	4	511–525	35	6
101–105	7	2	316–330	22	4	526–540	36	6
106–120	8	2	331–345	23	4	541–555	37	6
121–135	9	2	346–360	24	4	556–570	38	6
136–150	10	2	361–375	25	4	571–585	39	6
151–165	11	2	376–390	26	4	586–600	40	6
166–180	12	2	391–400	27	4	601–615	41	7
181–195	13	2	401–405	27	5	616–630	42	7
196–200	14	2	406–420	28	5	631–645	43	7
201–210	14	3	421–435	29	5	646–660	44	7

Source: IBC Table 2902.1.

2902 Minimum Plumbing Facilities

2902.1 Minimum number of fixtures *(part 14 of 21)*

- Fixtures for inmates of the following I-3 occupancies the following are governed as indicated below and as listed in the following table:
  - Occupancies:
    Reformatories.
    Detention and correctional centers.
  - Employee toilets are separate from inmate toilets and are not governed here.
  - Otherwise, fixture requirements are as shown in the partial table below.
- The following I-3 table includes the following functions:
  - Reformatories.
  - Detention centers.
  - Correction centers.

Table 2902.1p **Occupancy I-3 Detention Centers, etc.: Minimum Plumbing Fixtures Required for Inmates**

Inmate count	WC	Lavs	Bathtub/ shower	DF	Inmate count	WC	Lavs	Bathtub/ shower	DF
1–15	1	1	1	1	106–120	8	8	8	2
16–30	2	2	2	1	121–135	9	9	9	2
31–45	3	3	3	1	136–150	10	10	10	2
46–60	4	4	4	1	151–165	11	11	11	2
61–75	5	5	5	1	166–180	12	12	12	2
76–90	6	6	6	1	181–195	13	13	13	2
91–100	7	7	7	1	196–200	14	14	14	2
101–105	7	7	7	2	201–210	14	14	14	3

Source: IBC Table 2902.1.

- Occupancy I-4 required plumbing fixtures are shown in the table below.

Table 2902.1q **Occupancy I-4 Adult Day Care and Child Care: Minimum Plumbing Fixtures Required for Patients**

Patient count	WC	Lavs	DF	Patient count	WC	Lavs	DF
1–15	1	1	1	106–120	8	8	2
16–30	2	2	1	121–135	9	9	2
31–45	3	3	1	136–150	10	10	2
46–60	4	4	1	151–165	11	11	2
61–75	5	5	1	166–180	12	12	2
76–90	6	6	1	181–195	13	13	2
91–100	7	7	1	196–200	14	14	2
101–105	7	7	2	201–210	14	14	3

Source: IBC Table 2902.1.

2902 Minimum Plumbing Facilities

2902.1 Minimum number of fixtures *(part 15 of 21)*

- Occupancy M required plumbing fixtures are shown in the table below:

Table 2902.1r Occupancy M Mercantile: Minimum Plumbing Fixtures Required

Occupant count	WC	Lavs	DF	Occupant count	WC	Lavs	DF
1–500	1	1	1	1,501–2,000	4	3	2
501–750	2	1	1	2,001–2,250	5	3	3
751–1,000	2	2	1	2,251–2,500	5	4	3
1,001–1,500	3	2	2	2,501–3,000	6	4	3

Source: IBC Table 2902.1.

- The following R-1 occupancies are governed as indicated below:
  - Occupancies:
    - Hotels.
    - Motels.
    - Transient boarding houses.
  - Requirements:
    - The following fixtures are required in each sleeping unit:
      - 1 water closet.
      - 1 lavatory.
      - 1 bathtub/shower.

2902 Minimum Plumbing Facilities

2902.1 Minimum number of fixtures *(part 16 of 21)*

- The beginning of Occupancy R-2 Table 2902.1s is shown below:

Table 2902.1s(1) Occupancy R-2 Dorms, Fraternities, Sororities, Nontransient Boarding Houses. 1–360 Occupants: Minimum Plumbing Fixtures Required

Occupant count	WC	Lavs	Bathtub/shower	DF
1–8	1	1	1	1
9–10	1	1	2	1
11–16	2	2	2	1
17–20	2	2	3	1
21–24	3	3	3	1
25–30	3	3	4	1
31–32	4	4	4	1
33–40	4	4	5	1
41–48	5	5	6	1
49–50	5	5	7	1
51–56	6	6	7	1
57–60	6	6	8	1
61–64	7	7	8	1
65–70	7	7	9	1
71–72	8	8	9	1
73–80	8	8	10	1
81–88	9	9	11	1
89–90	9	9	12	1
91–96	10	10	12	1
97–100	10	10	13	1
101–104	11	11	13	2
105–110	11	11	14	2
111–112	12	12	14	2
113–120	12	12	15	2
121–128	13	13	16	2
129–130	13	13	17	2
131–136	14	14	17	2
137–140	14	14	18	2
141–144	15	15	18	2
145–150	15	15	19	2
151–152	16	16	19	2
153–160	16	16	20	2
161–168	17	17	21	2
169–170	17	17	22	2
171–176	18	18	22	2
177–180	18	18	23	2
181–184	19	19	23	2
185–190	19	19	24	2
191–192	20	20	24	2
193–200	20	20	25	2
201–208	21	21	26	3
209–210	21	21	27	3
211–216	22	22	27	3
217–220	22	22	28	3
221–224	23	23	28	3
225–230	23	23	29	3
231–232	24	24	29	3
233–240	24	24	30	3
241–248	25	25	31	3
249–250	25	25	32	3
251–256	26	26	32	3
257–260	26	26	33	3
261–264	27	27	33	3
265–270	27	27	34	3
271–272	28	28	34	3
273–280	28	28	35	3
281–288	29	29	36	3
289–290	29	29	37	3
291–296	30	30	37	3
297–300	30	30	38	3
301–304	31	31	38	4
305–310	31	31	39	4
311–312	32	32	39	4
313–320	32	32	40	4
321–328	33	33	41	4
329–330	33	33	42	4
331–336	34	34	42	4
337–340	34	34	43	4
341–344	35	35	43	4
345–350	35	35	44	4
351–352	36	36	44	4
353–360	36	36	45	4

Source: IBC Table 2902.1.

2902 Minimum Plumbing Facilities

2902.1 Minimum number of fixtures *(part 17 of 21)*

- The conclusion of Occupancy R-2 Table 2902.1s is shown below:

Table 2902.1s(2) Occupancy R-2 Dorms, Fraternities, Sororities, Nontransient Boarding Houses. 361–720 Occupants: Minimum Plumbing Fixtures Required

Occupant count	WC	Lavs	Bathtub/ shower	DF	Occupant count	WC	Lavs	Bathtub/ shower	DF
361–368	37	37	46	4	541–544	55	55	68	6
369–370	37	37	47	4	545–550	55	55	69	6
371–376	38	38	47	4	551–552	56	56	69	6
377–380	38	38	48	4	553–560	56	56	70	6
381–384	39	39	48	4	561–568	57	57	71	6
385–390	39	39	49	4	569–570	57	7	72	6
391–392	40	40	49	4	571–576	58	58	72	6
393–400	40	40	50	4	577–580	58	58	73	6
401–408	41	41	51	5	581–584	59	59	73	6
409–410	41	41	52	5	585–590	59	59	74	6
411–416	42	42	52	5	591–592	60	60	74	6
417–420	42	42	53	5	593–600	60	60	75	6
421–424	43	43	53	5	601–608	61	61	76	7
425–430	43	43	54	5	609–610	61	61	77	7
431–432	44	44	54	5	611–616	62	62	77	7
433–440	44	44	55	5	617–620	62	62	78	7
441–448	45	45	56	5	621–624	63	63	78	7
449–450	45	45	57	5	625–630	63	63	79	7
451–456	46	46	57	5	631–632	64	64	79	7
457–460	46	46	58	5	633–640	64	64	80	7
461–464	47	47	58	5	641–648	65	65	81	7
465–470	47	47	59	5	649–650	65	65	82	7
471–472	48	48	59	5	651–656	66	66	82	7
473–480	48	84	60	5	657–660	66	66	83	7
481–488	49	49	61	5	661–664	67	67	83	7
489–490	49	49	62	5	665–670	67	67	84	7
491–496	50	50	62	5	671–672	68	68	84	7
497–500	50	50	63	5	673–680	68	68	85	7
501–504	51	51	63	6	681–688	69	69	86	7
505–510	51	51	64	6	689–690	69	69	87	7
511–512	52	52	64	6	691–696	70	70	87	7
513–520	52	52	65	6	697–700	70	70	88	7
521–528	53	53	66	6	701–704	71	71	88	8
529–530	53	53	67	6	705–710	71	71	89	8
531–536	54	54	67	6	711–712	72	72	89	8
537–540	54	54	68	6	713–720	72	72	90	8

Source: IBC Table 2902.1.

2902 Minimum Plumbing Facilities

2902.1 Minimum number of fixtures *(part 18 of 21)*

- Occupancy R-2 apartment houses require the following plumbing fixtures in each dwelling unit:
  - 1 water closet.
  - 1 lavatory.
  - 1 bathtub/shower.
  - 1 kitchen sink.
- Drinking fountains are not required.
- A service sink is not required.
- Automatic clothes washer connections as indicated in the following table:

Table 2902.1t Occupancy R-2 Apartment Houses: Minimum Automatic Clothes Washer Connections Required

Dwelling unit count	Connections required	Dwelling unit count	Connections required	Dwelling unit count	Connections required
20	1	220	9	320	17
30	2	230	10	330	18
40	3	240	11	340	19
50	4	250	12	350	20
60	5	260	13	360	21
80	6	280	14	380	22
90	7	290	15	390	23
100	8	300	16	400	24

Source: IBC Table 2902.1.

2902 Minimum Plumbing Facilities

2902.1 Minimum number of fixtures *(part 19 of 21)*

- The beginning of occupancies R-3/R-4 Table 2902.1u is shown below:

Table 2902.1u(1) Occupancy R-3 Congregate Living Facilities with 16 or Fewer Persons. Occupancy R-4 Residential Care/Assisted Living Facilities. 1–360 Occupants: Minimum Plumbing Fixtures Required

Occupant count	WC	Lavs	Bathtub/ shower	DF	Occupant count	WC	Lavs	Bathtub/ shower	DF
1–8	1	1	1	1	181–184	19	19	23	2
9–10	1	1	2	1	185–190	19	19	24	2
11–16	2	2	2	1	191–192	20	20	24	2
17–20	2	2	3	1	193–200	20	20	25	2
21–24	3	3	3	1	201–208	21	21	26	3
25–30	3	3	4	1	209–210	21	21	27	3
31–32	4	4	4	1	211–216	22	22	27	3
33–40	4	4	5	1	217–220	22	22	28	3
41–48	5	5	6	1	221–224	23	23	28	3
49–50	5	5	7	1	225–230	23	23	29	3
51–56	6	6	7	1	231–232	24	24	29	3
57–60	6	6	8	1	233–240	24	24	30	3
61–64	7	7	8	1	241–248	25	25	31	3
65–70	7	7	9	1	249–250	25	25	32	3
71–72	8	8	9	1	251–256	26	26	32	3
73–80	8	8	10	1	257–260	26	26	33	3
81–88	9	9	11	1	261–264	27	27	33	3
89–90	9	9	12	1	265–270	27	27	34	3
91–96	10	10	12	1	271–272	28	28	34	3
97–100	10	10	13	1	273–280	28	28	35	3
101–104	11	11	13	2	281–288	29	29	36	3
105–110	11	11	14	2	289–290	29	29	37	3
111–112	12	12	14	2	291–296	30	30	37	3
113–120	12	12	15	2	297–300	30	30	38	3
121–128	13	13	16	2	301–304	31	31	38	4
129–130	13	13	17	2	305–310	31	31	39	4
131–136	14	14	17	2	311–312	32	32	39	4
137–140	14	14	18	2	313–320	32	32	40	4
141–144	15	15	18	2	321–328	33	33	41	4
145–150	15	15	19	2	329–330	33	33	42	4
151–152	16	16	19	2	331–336	34	34	42	4
153–160	16	16	20	2	337–340	34	34	43	4
161–168	17	17	21	2	341–344	35	35	43	4
169–170	17	17	22	2	345–350	35	35	44	4
171–176	18	18	22	2	351–352	36	36	44	4
177–180	18	18	23	2	353–360	36	36	45	4

Source: IBC Table 2902.1.

2902 Minimum Plumbing Facilities

2902.1 Minimum number of fixtures *(part 20 of 21)*

- The conclusion of occupancies R-3/R-4 Table 2902.1u is shown below:

Table 2902.1u(2) Occupancy R-3 Congregate Living Facilities with 16 or Fewer Persons. Occupancy R-4 Residential Care/Assisted Living Facilities. 361–720 Occupants: Minimum Plumbing Fixtures Required

Occupant count	WC	Lavs	Bathtub/ shower	DF
361–368	37	37	46	4
369–370	37	37	47	4
371–376	38	38	47	4
377–380	38	38	48	4
381–384	39	39	48	4
385–390	39	39	49	4
391–392	40	40	49	4
393–400	40	40	50	4
401–408	41	41	51	5
409–410	41	41	52	5
411–416	42	42	52	5
417–420	42	42	53	5
421–424	43	43	53	5
425–430	43	43	54	5
431–432	44	44	54	5
433–440	44	44	55	5
441–448	45	45	56	5
449–450	45	45	57	5
451–456	46	46	57	5
457–460	46	46	58	5
461–464	47	47	58	5
465–470	47	47	59	5
471–472	48	48	59	5
473–480	48	84	60	5
481–488	49	49	61	5
489–490	49	49	62	5
491–496	50	50	62	5
497–500	50	50	63	5
501–504	51	51	63	6
505–510	51	51	64	6
511–512	52	52	64	6
513–520	52	52	65	6
521–528	53	53	66	6
529–530	53	53	67	6
531–536	54	54	67	6
537–540	54	54	68	6
541–544	55	55	68	6
545–550	55	55	69	6
551–552	56	56	69	6
553–560	56	56	70	6
561–568	57	57	71	6
569–570	57	7	72	6
571–576	58	58	72	6
577–580	58	58	73	6
581–584	59	59	73	6
585–590	59	59	74	6
591–592	60	60	74	6
593–600	60	60	75	6
601–608	61	61	76	7
609–610	61	61	77	7
611–616	62	62	77	7
617–620	62	62	78	7
621–624	63	63	78	7
625–630	63	63	79	7
631–632	64	64	79	7
633–640	64	64	80	7
641–648	65	65	81	7
649–650	65	65	82	7
651–656	66	66	82	7
657–660	66	66	83	7
661–664	67	67	83	7
665–670	67	67	84	7
671–672	68	68	84	7
673–680	68	68	85	7
681–688	69	69	86	7
689–690	69	69	87	7
691–696	70	70	87	7
697–700	70	70	88	7
701–704	71	71	88	8
705–710	71	71	89	8
711–712	72	72	89	8
713–720	72	72	90	8

Source: IBC Table 2902.1.

2902 Minimum Plumbing Facilities

2902.1 Minimum number of fixtures *(part 21 of 21)*

- The following S-1 and S-2 occupancies are governed as indicated below:
  - Occupancies:
    - Freight depots.
    - Moderate-hazard storage.
    - Storage of goods.
    - Storehouses.
    - Warehouses.
  - Requirements:
    - The following must be provided where required by the function of the occupancy:
      - Emergency eyewash stations.
      - Emergency showers.
    - S-1 and S-2 required plumbing fixtures are shown in the partial table below.

Note: The International Plumbing Code, *Section 411, "Emergency Showers and Eyewash Stations," is cited as governing plumbing fixtures for S occupancies.*

Table 2902.1v Occupancy S-1 and S-2 Storage: Minimum Plumbing Fixtures Required

Occupant count	WC	Lavs	DF	Occupant count	WC	Lavs	DF
1–100	1	1	1	301–400	4	4	1
101–200	2	2	1	401–500	5	5	1
201–300	3	3	1	501–600	6	6	1

Source: IBC Table 2902.1.

2902 Minimum Plumbing Facilities

2902.1.1 Fixture calculations

- The number of occupants of each gender is determined by one of the following methods:
  - According to approved statistical data.
  - By assigning 1/2 the total number of occupants to each gender.
- The number of plumbing fixtures required for each gender is determined as follows:
  - By applying IBC ratios for each fixture type to the occupant load of each gender as follows:
    - Where the application of IBC ratios result in fractions, do the following:
      - For single occupancies, round fractions up to the next higher whole number of fixtures.
      - For multiple occupancies do the following:
        - Sum the fractions for each fixture type for each occupancy.
        - Round up such sums to the next whole number of fixtures.

Note: *IBC Table 2902.1, "Minimum Number of Required Plumbing Fixtures," is cited as listing the required ratios for determining the minimum number of plumbing fixtures required.*

2902.1.2 Family or assisted use toilet and bath fixtures

- This section addresses fixtures required by the IBC in the following locations:
  - In family or assisted-use toilet and bathing rooms.
- Such fixtures may be counted in either of the fixture minimums required for the following:
  - For male occupants of the following occupancies:
    - Assembly.
    - Mercantile.
  - For female occupants of the following occupancies:
    - Assembly.
    - Mercantile.

Note: *1109.2.1, "Family or assisted-use toilet and bathing rooms," is cited as requiring this type of facility.*

2902.2 Separate facilities

- Separate sex toilet facilities are not required for the following cases:
  - Dwelling units.
  - Sleeping units.
  - For occupancies with ≤ 15 employees.
  - In buildings or tenant spaces where occupant load is ≤ 15 including both of the following:
    - Employees.
    - Customers.
  - In mercantile occupancies as follows:
    - With ≤ 50 occupants.
- Otherwise, where toilet fixtures are required, separate sex facilities must be provided.

2902 Minimum Plumbing Facilities

2902.3 Required public toilet facilities

- The following people must be provided public toilet facilities in the locations listed below:
  - People:
    - Customers.
    - Patrons.
    - Visitors.
  - Locations:
    - Public buildings.
    - Public tenant spaces.
- The accessible route to public toilet facilities must not pass through the following spaces:
  - Closets.
  - Kitchens.
  - Similar spaces.
  - Storage rooms.
- Employees must be provided toilet facilities as follows:
  - In all occupancies.
  - In one of the following arrangements:
    - For employees only.
    - Combined with public toilets.

2902.3.2 Location of toilet facilities in occupancies other than covered mall buildings

- This section does not govern toilet facilities in covered malls.
- Toilet facilities in factory and industrial occupancies are governed as follows:
  - Travel distance to toilets may exceed that otherwise permitted by this section in the following case:
    - Where approved.
- In other cases, public and employee toilet facilities must be located as follows:
  - ≤ 1 story above or below the location that they are required to serve.
  - Travel distance to the required toilet facilities must be ≤ 500'.

2902.3.3 Location of toilet facilities in covered mall buildings

- Covered mall buildings require public and employee toilet facilities as follows:
  - Toilet facilities must be located as follows:
    - ≤ 1 story above or below the location that they are required to serve.
    - Travel distance to required toilet facilities must be ≤ 300'.
  - Required toilet facilities must be provided in one of the following locations:
    - In each individual store.
    - In a central toilet area.
  - Travel distance to central toilet facilities is measured from the following:
    - The main entrance of each store or tenant space.
  - The following applies where employee toilets are not located in tenant spaces:
    - Travel distance to the toilets is measured from the employee's work space.

2902.3.4 Pay facilities

- Required toilet facilities must be free of charge.
- Any pay toilet facilities must be in excess of the required number of facilities.

2902 Minimum Plumbing Facilities

2902.4 Signage

- Each toilet room entrance requires the following signage:
  - The room's gender designation as follows:
    Must be visible.
    Must be legible.

Note: ICC A117.1, "Accessible and Usable Buildings and Facilities," is cited as governing these signs.

2902.4.1 Directional signage

- Signage indicating the route to public plumbing faciliites is required as follows:
  - It must be provided in the following locations for customers and visitors:
    In corridors or aisles.
    At the entrance to the facilities.
  - It must be provided according to other IBC requirements.

Note: Section 3107, "Signs," is cited as being required for compliance by such signage.

2903 Toilet Room Requirements

2903.1 Water closet compartment

- This section does not address toilet areas in Occupancy I-3 housing areas.
- A water closet compartment is not required in the following rooms:
  - Single-occupant toilet rooms with a door lock.
- Toilet rooms with all of the following characteristics may have one water closet without an enclosure:
  - In one of the following locations:
    - Day care facility.
    - Child care facility.
  - With ≥ 2 water closets.
- In other locations, toilet rooms used by the public or employees are governed as follows:
  - They must have a compartments for each water closet as follows:
    - With walls or partitions and a door.
    - Compartments must provide privacy.

2903.2 Urinal partitions

- A urinal partition is not required in the following toilet rooms with a door lock:
  - Single-occupant toilet rooms.
  - Family toilet rooms.
  - Assisted use toilet rooms.
- Toilet rooms with all of the following characteristics may have one urinal without a partition:
  - In one of the following locations:
    - Day care facility.
    - Child care facility.
  - With ≥ 2 urinals.
- In other locations, toilet rooms used by the public or employees are governed as follows:
  - They must have privacy walls or partitions at each urinal as follows:
    - Walls or partitions with the following height:
      - Bottom ≤ 1' from the floor.
      - Top ≥ 5' from the floor.
    - Walls or partitions to extend from the back wall to a point as follows:
      - ≥ 1'-6" from the wall.
      - ≥ 6" in front of the urinal lip.
    - Required dimensions are shown in Fig. 2903.2 on the next page.

2903 Toilet Room Requirements

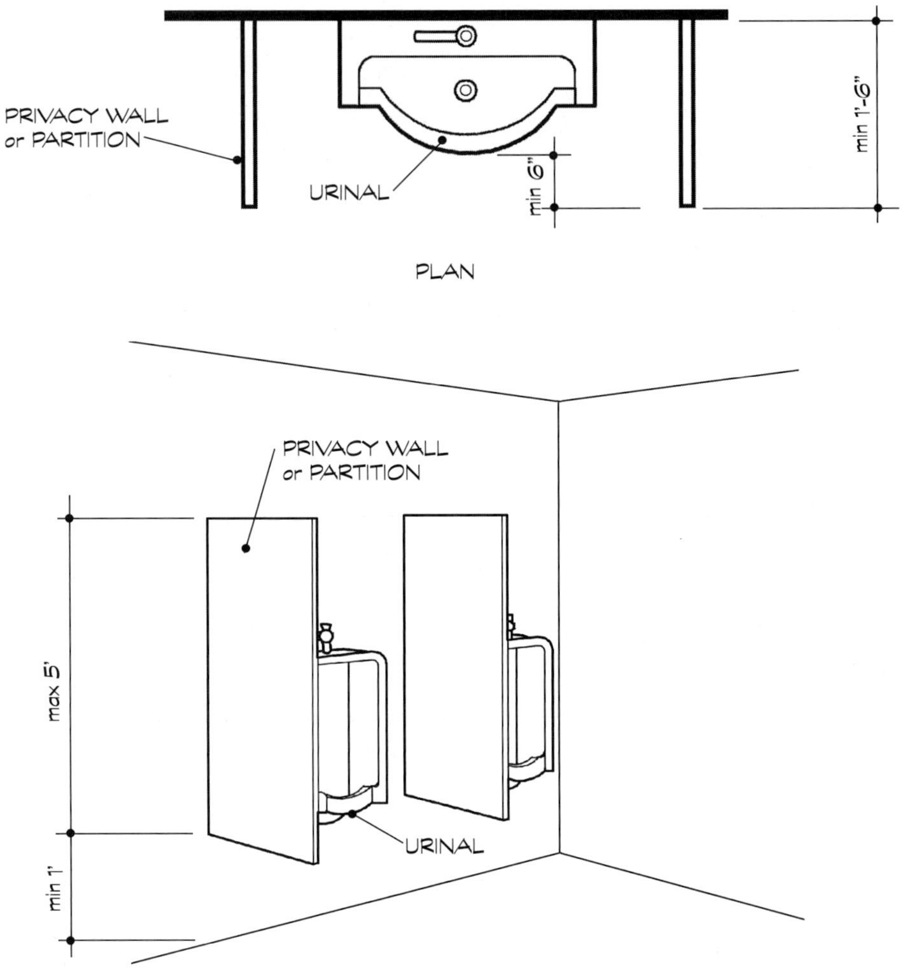

Fig. 2903.2 Urinal privacy panels

30

Elevators and Conveying Systems

Christus St. Michael Health Care Center. Texarkana, Texas. *(partial elevation)*
Watkins Hamilton Ross Architects, Inc. Houston, Texas.

3002 Hoistway Enclosures

3002.2 Number of elevator cars in a hoistway

- A hoistway is limited to ≤ 4 elevator cars as shown in the illustrations below.
- Where ≥ 4 elevators serve an area of a building the following applies:
  - ≤ 2 hoistways are required as shown in the illustration below.

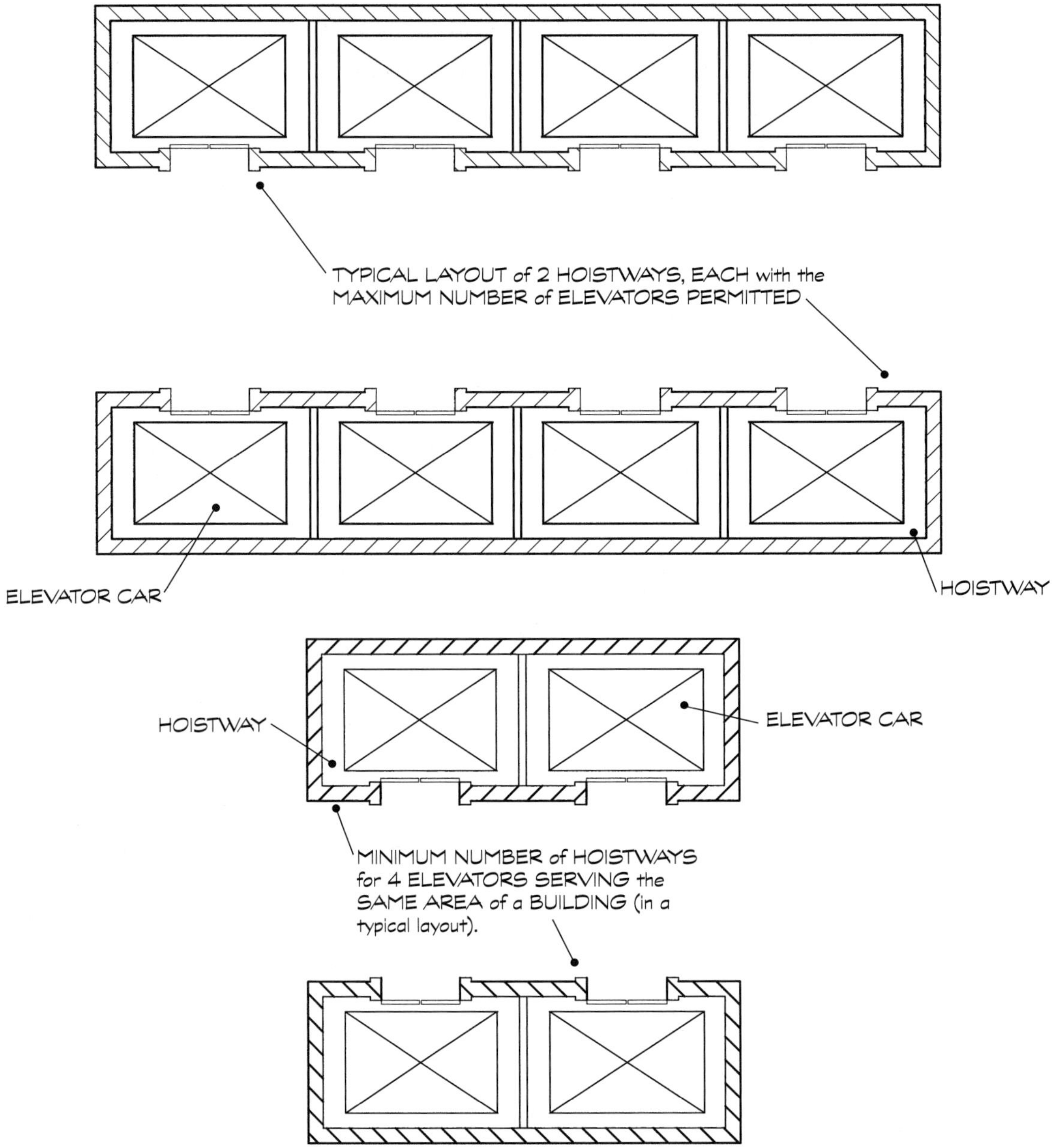

Fig. 3002.2. Maximum number of elevators in a hoistway and minimum number of hoistways.

3002 Hoistway Enclosures

3002.3 Emergency signs

- Signs at elevators prohibiting use of elevators during a fire are not required as follows:
  - Where elevators are part of an accessible means of egress.

  Note: 1007.4, "Elevators," is cited as governing elevators as indicated above.

  - Where elevators are used for self-evacuation by occupants.

  Note: Section 3008, "Occupant Evacuation Elevators," is cited as governing these elevators.

- Otherwise, a sign is required by each elevator call button as follows:
  - Sign must be approved.
  - Sign must be a standardized design.
  - Sign must state the message shown on the illustration below.

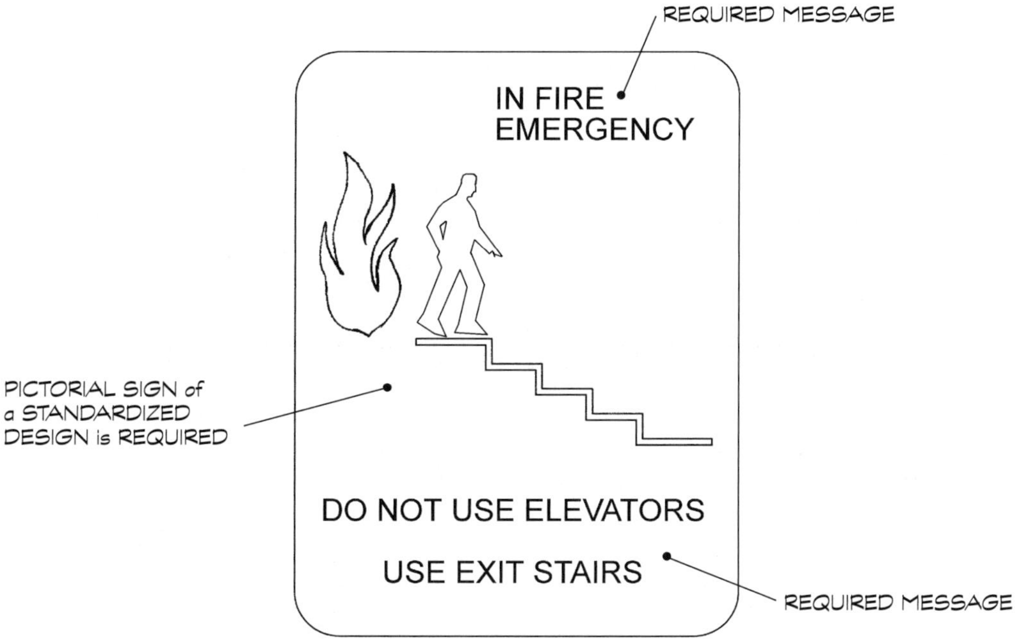

Fig. 3002.3. Acceptable design for a sign required to be posted at elevators prohibiting elevator use during fire.

3002.4 Elevator car to accommodate ambulance stretcher

- This section governs buildings with either or both of the following characteristics and with elevators:
  - ≥ 4 stories above the grade plane.
  - ≥ 4 stories below the grade plane.
- Buildings governed by this section must have ≥ 1 elevator as follows:
  - For fire department emergency access to all floors.
  - Size must accommodate an ambulance stretcher as follows:
    Required dimensions are shown in the detail provided with this section.
  - Must be identified with "star of life" signs as shown in the detail provided with this section.

3002 Hoistway Enclosures

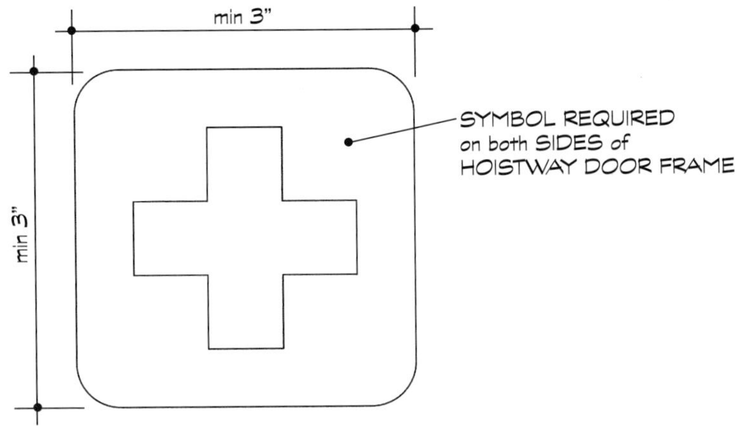

Fig. 3002.4A. "Star of life" symbol required at elevator designed for fire department use.

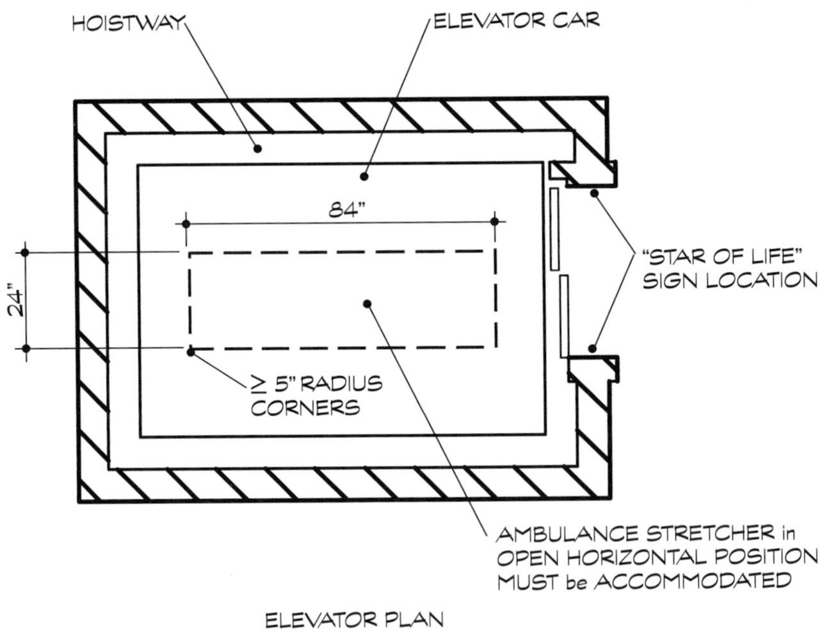

Fig. 3002.4B. Plan of elevator accommodating an ambulance stretcher.

3008 Occupant Evacuation Elevators

3008.1 General

- This section governs the following elevators:
  - Public passenger elevators as follows:
    Used for occupant self-evacuation in case of fire.
  - Other elevators used for occupant self-evacuation.

3008.2 Fire safety and evacuation plan

- Buildings affected by this section are required to have the following:
  - A fire safety and evacuation plan as follows:
    The plan must include instructions for use of evacuation elevators by occupants.

Note: International Fire Code, *Section 404, "Fire Safety and Evacuation Plans," is cited as governing these plans.*

3008.3 Operation

- Elevators governed by this section are to be used for occupant self-evacuation as follows:
  - Prior to Phase I Emergency Recall Operation.
  - In the normal elevator operating mode.
  - According to the building fire safety and evacuation plan.

Note: ASME A17.1/CSA B44, "Safety Code for Elevators and Escalators," is cited as applicable.

3008.4 Additional exit stairway

- This section applies to buildings > 420' in height as follows:
  - Where occupant evacuation elevators are used, the usual additional exit stairway is not required.

Note: 403.5.2, "Additional exit stairway," is cited as requiring an additional exit stairway for buildings > 420' in height.

3008.5 Emergency voice/alarm communication system

- Buildings affected by this section are required to have the following:
  - An emergency voice/alarm communication system as follows:
    Fire department must have access to the system.

Note: 907.5.2.2, "Emergency voice/alarm communication systems," is cited as governing.

3008.5.1 Notification appliances

- Lobbies serving elevators for self-evacuation by occupants require the following:
  - ≥ 1 audible notification appliance.
  - ≥ 1 visible notification appliance.

3008.6 Automatic sprinkler system

- Buildings affected by this section are required to have the following in certain cases:
  - A sprinkler system as follows:

Note: 903.3.1.1, "NFPA 13 Sprinkler systems," is cited as governing sprinklers.
903.3.1.1.1, "Exempt locations," is cited as identifying cases exempt from this section.
3008.6.1, "Prohibited locations," is cited as identifying locations prohibiting sprinklers.

3008 Occupant Evacuation Elevators

3008.6.1 Prohibited locations

- Sprinklers are not permitted in the following locations serving elevators for occupant evacuation:
  - Elevator machine rooms.
  - Elevator machine spaces.

3008.6.2 Sprinkler system monitoring

- The following is required for floors monitored by the fire alarm system:
  - A sprinkler control valve supervisory switch.
  - A water flow initiating device.

3008.7 High-hazard content areas

- Buildings affected by this section have the following requirement:
  - Quantities of high hazard may not exceed fire area limits.

Note: *414.2, "Control areas," is cited as defining quantity limits of hazardous materials.*

3008.8 Shunt trip

- This section applies to elevators used for occupant evacuation.
- The following means for shutting down elevator service is prohibited:
  - Where such means automatically disconnect power as follows:

    Prior to activation of sprinklers in the following locations:

      Machine room.

      Hoistway.

Note: *3006.5, "Shunt trip," is cited as governing the automatic disconnect system that is prohibited, a portion of which is incorporated into the wording above.*

3008.9 Hoistway enclosure protection

- Occupant evacuation elevators must be housed in a hoistway enclosure.

Note: *Section 708, "Shaft Enclosures," is cited as governing the hoistway shaft required above.*

3008.10 Water protection

- The elevator hoistway must be designed as follows:
  - Using an approved method to prevent the following:

    The entry of building sprinkler water into the hoistway.

3008.11 Occupant evacuation elevator lobby

- A lobby is required for elevators serving the evacuation of occupants.

Note: *The following are cited as governing such lobbies:*

3008.11.1, "Access."
3008.11.2, "Lobby enclosure."
3008.11.3, "Lobby doorways."
3008.11.3.1, "Vision panel."
3008.11.3.2, "Door closing."
3008.11.4, "Lobby size."

3008 Occupant Evacuation Elevators

3008.11.1 Access

- An elevator for evacuating occupants requires the following access:
  - Direct access to an exit enclosure.

3008.11.2 Lobby enclosure

- An enclosed lobby for elevators for occupant evacuation is not required in the following location:
  - At the level of exit discharge.
- In other locations, lobbies serving elevators for occupant evacuation require the following enclosure:
  - A smoke barrier as follows:
    With a fire-resistance rating ≥ 1 hr.
    With doors having a fire protection rating of 3/4 hr.

Note: 3008.11.3, "Lobby doorways," is cited as governing these doorways, a partial summary of which is provided above.

3008.11.3 Lobby doorways

- Lobbies serving elevators for occupant evacuation are required to have the following:
  - Doors having a fire protection rating of 3/4 hr.

Note: 715.4, "Fire door and shutter assemblies," is cited as governing these doors.

3008.11.3.1 Vision panel

- A vision panel is required in occupant evacuation elevator lobby fire door assemblies as follows:
  - Panel must have fire-protection-rated glazing.
  - Panel must afford a view of the lobby.

3008.11.3.2 Door closing

- Fire door assemblies serving occupant evacuation elevator lobbies must close as follows:
  - At a signal from the building emergency voice/alarm communication system.

3008.11.4 Lobby size

- Lobby size for multiple banks of occupant evacuation elevators is determined as follows:
  - According to the building fire safety and evacuation plan.
  - By approval.
- The size of other lobbies serving occupant evacuation elevators must comply with all of the following:
  - Floor area must have 3 sf/per person based on the following:
    1/4 the occupant load from the part of the floor served by the lobby.
  - Floor area must accommodate 1 wheelchair space for the following:
    For each 50 persons of the occupant load of the floor served as follows:
      Space is 30" x 48".
      Fractions of spaces are rounded up to the next whole number.

3008.11.5 Signage

- A sign is required at elevators used for occupant self-evacuation as follows:
  - The sign states that the elevators are suitable for self-evacuation.
  - A sign is posted on each floor adjacent to each call station.
  - The sign is approved.

3008 Occupant Evacuation Elevators

3008.12 Lobby status indicator

- A sign is required at each lobby serving occupant evacuation elevators as follows:
  - When elevators are operating and there is a fire alarm activated in the building, the following applies:
    - Sign has an illuminated green light.
    - Sign shows a message stating the following:
      - "Elevators available for occupant evacuation."
  - When elevators are in Phase 1 emergency recall status, the following applies:
    - Sign has an illuminated red light.
    - Sign shows a message stating the following:
      - "Elevators out of service, use exit stairs."
  - When elevators are in normal service, the following applies:
    - Sign has no illuminated light.
    - Sign has no message displayed.

3008.13 Two-way communication system

- Each occupant evacuation elevator lobby is required to have the following:
  - A two-way communication system as follows:
    - To communicate with one of the following:
      - The fire command center.
      - Other location approved by the fire department.

3008.13.1 Design and installation

- Two-way communication systems for occupant evacuation elevator lobbies must have the following:
  - Audible signals.
  - Visual signals.

Note: ICC A117.1, "Accessible and Usable Buildings and Facilities," is cited as governing the design and installation of the communications system.

3008.13.2 Instructions

- The following must be posted in occupant evacuation elevator lobbies:
  - Instructions for use of the two-way communication system as follows:
    - To be posted at the call station.
  - Identification of the call station location from which the call is being made as follows:
    - To be posted at the call station.

Note: ICC A117.1, "Accessible and Usable Buildings and Facilities," is cited as governing the signage.

3008 Occupant Evacuation Elevators

3008.14 Elevator system monitoring

- Occupant evacuation elevators are to be continuously monitored at one of the following locations:
  - The fire command center.
  - A central location approved by the fire department.
- The following is required to be displayed by the monitoring system:
  - The floor location of each elevator car.
  - Travel direction of each elevator car.
  - Indication of whether or not the elevator car is occupied.
  - The status of routine power to the following:
    Elevator equipment.
    Elevator controller cooling equipment.
    Elevator machine room ventilation equipment.
    Elevator machine room cooling equipment.
  - Status of standby or emergency power to the following:
    Elevator equipment.
    Elevator controller cooling equipment.
    Elevator machine room ventilation equipment.
    Elevator machine room cooling equipment.
  - Activation of any fire alarm initiating device in the following locations:
    In any elevator.
    In any elevator machine room.
    In any elevator machine space.
    In any elevator hoistway.

3008.14.1 Elevator recall

- This section addresses emergency recall of occupant evacuation elevators.
- Manual initiation of a Phase I Emergency Recall must be possible from one of the following locations:
  - The fire command center.
  - A location approved by the fire department.

Note: ASME A17.1/CSA B44, "Safety Code for Elevators and Escalators," is cited as governing such elevator recall.

3008.15 Electrical power

- Normal power and Type 60/Class 2/Level 1 standby power are required for the following:
  - Elevator equipment.
  - Elevator controller cooling equipment.
  - Elevator machine room ventilation equipment.
  - Elevator machine room cooling equipment.

3008 Occupant Evacuation Elevators

3008.15.1 Protection of wiring or cables

- This section addresses wiring or cables serving the following in fire service access elevators:
  - Normal power.
  - Standby power.
  - Control signals.
  - Communication with the car.
  - Lighting.
  - Heating.
  - Air conditioning.
  - Ventilation.
  - Fire-detection systems.
- Such wiring or cables must be one of the following:
  - Circuit integrity cable with a fire-resistance rating ≥ 1-hr.
  - Protected by ≥ 1-hr fire-resistance-rated construction.

31

Special Construction

AmberGlen Business Center. Hillsboro, Oregon. *(partial elevation)*
Ankrom Moisan Associated Architects. Portland, Oregon.

3104 Pedestrian Walkways and Tunnels

3104.5 Fire barriers between pedestrian walkways and buildings *(part 1 of 2)*

- This section applies to pedestrian walkways connecting buildings as follows:
  - At grade.
  - Above grade.
  - Below grade.
- This section does not require a fire-resistance rating for a wall (and glazing) separating a walkway from the building where the walkway is at one of the following heights and has any of the conditions below:
  - Heights:
    - ≤ 3 stories and ≤ 40' above grade where not sprinklered.
    - ≤ 5 stories and ≤ 55' above grade where sprinklered.
  - Condition 1:
    - Where connected buildings are > 10' apart and all of the following apply:
      - Walkway (to other than an open parking garage) and building are sprinklered.
      - The wall of the connected building has one of the following characteristics:
        - Resists the passage of smoke.
        - Is constructed of one of the following types of glass:
          - Tempered.
          - Wired.
          - Laminated.
      - The wall and/or glass of the connected building are protected by sprinklers as follows:
        - Sprinklers can wet the whole interior surface of the wall and/or glass.
        - Glass is mounted as follows:
          - In a gasketed frame.
          - So that the frame can deflect without breaking the glass as follows:
            - Prior to activation of the sprinklers.
        - No obstructions exist between sprinklers and wall and/or glass.
  - Condition 2:
    - Where connected buildings are > 10' apart and all of the following apply:
      - Both side walls of the walkway must be open as follows:
        - ≥ 50%.
        - Open area is uniformly distributed.
        - Openness prevents the accumulation of smoke and toxic gases.
  - Condition 3:
    - Where buildings are on the same lot.

  Note: 503.1.2, "Buildings on same lot," is cited as governing such buildings.

- Where exterior walls of the connected buildings are required to have a fire-resistance rating > 2 hrs, the following applies:
  - The walkway must be sprinklered.

  Note: Section 705, "Exterior Walls," is cited as governing the fire-resistance ratings of exterior walls.
  903.3.1.1, "NFPA 13 sprinkler systems," is cited as governing all sprinklers noted in this section.

3104 Pedestrian Walkways and Tunnels

3104.5 Fire barriers between pedestrian walkways and buildings *(part 2 of 2)*

- In other cases, walkways must be separated from the connected buildings as shown in the illustration below.

Note: Section 715, "Opening Protectives," is cited as governing the fire protection rating required for windows as noted in the illustration below.
The following are cited as governing fire barriers where required:
Section 707, "Fire Barriers."
Section 712, "Horizontal Assemblies."

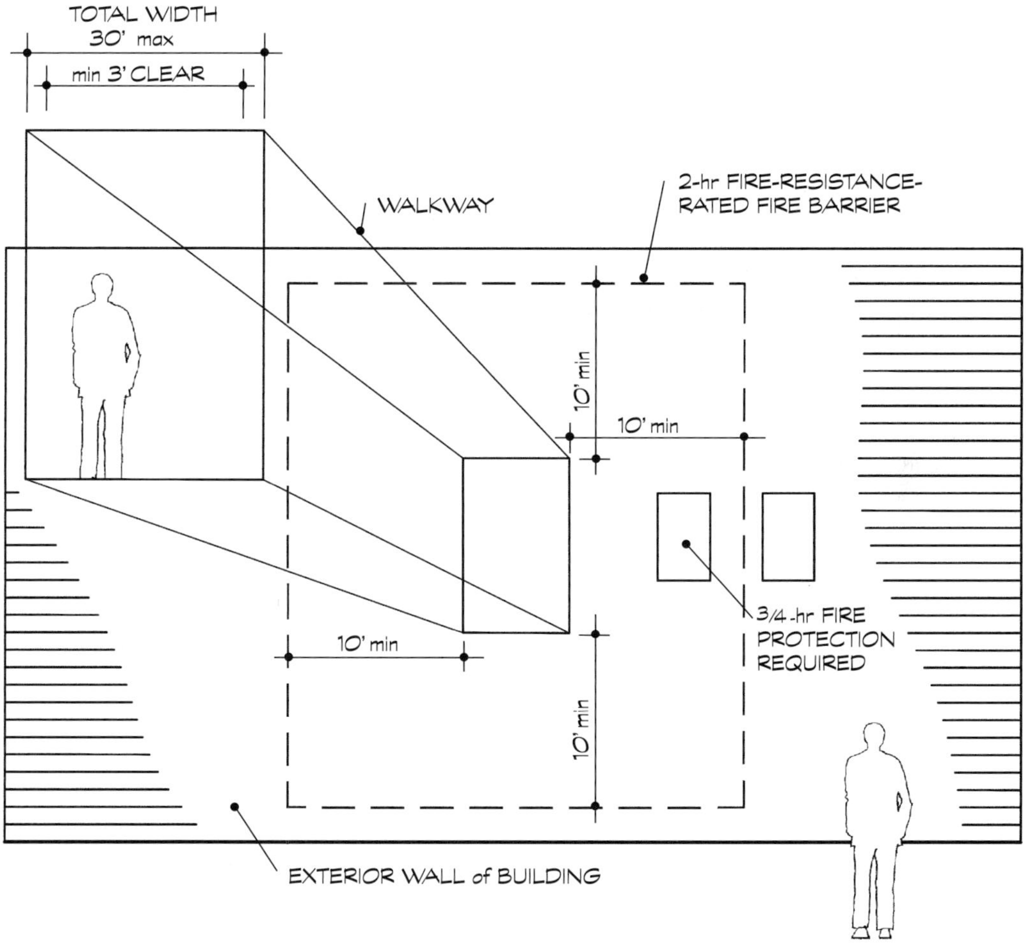

Fig. 3104.5. Pedestrian walkway, dimensions, and fire barrier at building.

3104 Pedestrian Walkways and Tunnels

3104.7 Egress

- Pedestrian walkways serving as a required exit are governed as follows:
  - They must be able to be accessed at all times.

3104.8 Width

- Width of pedestrian walkways must be as shown in Figure 3104.5 on the previous page.

3104.9 Exit access travel

- This section addresses pedestrian walkways and tunnels.
- Exit access travel distance within such elements is governed as follows:

Type of walkway	Sprinklered	Not sprinklered
Tunnel	≤ 200'	≤ 200'
Pedestrian walkway	≤ 250'	≤ 200'
Pedestrian walkway, both sides open ≥ 50%	≤ 400'	≤ 300'

Note: 903.3.1.1, "NFPA 13 sprinkler systems," is cited as governing sprinklers noted in this section.

3104.10 Tunneled walkway

- A separation is required between the following elements as listed:
  - Elements:

    Tunneled walkway.

    Building to which the tunneled walkway connects.
  - Separation:

    Must be ≥ 2 hr fire-resistance-rated construction.

    Openings require protection.

Note: IBC Table 715.4, "Fire Door and Fire Shutter Fire Protection Ratings," is cited as governing the openings in the separation construction.

3106 Marquees

3106.2 Thickness

- The vertical dimension of a marquee is as shown in the illustrations below:

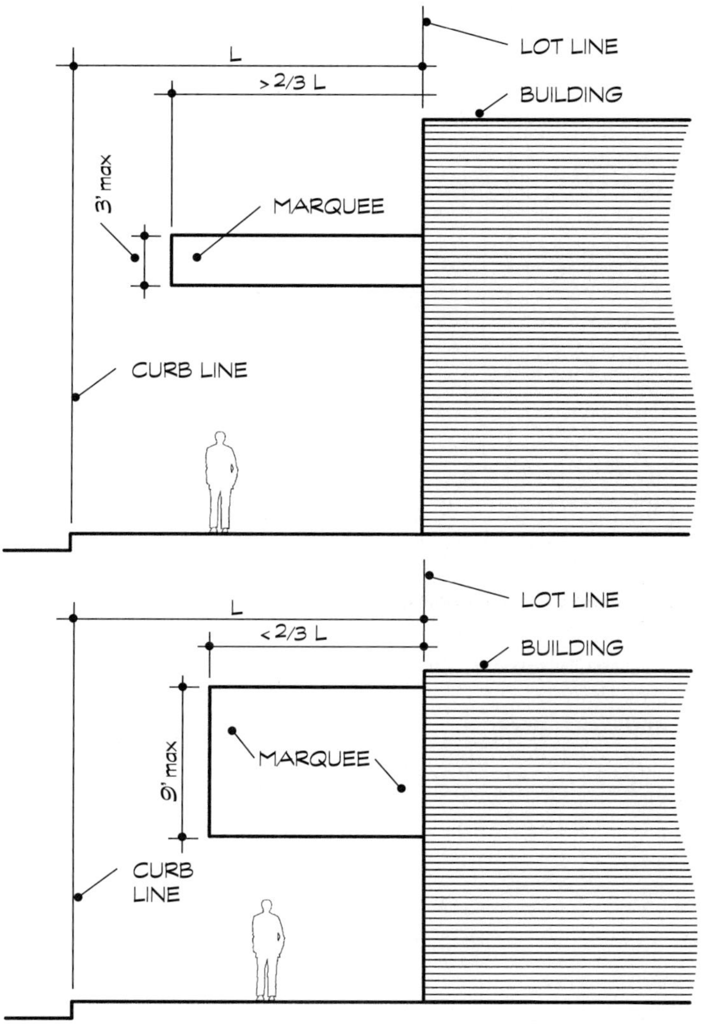

Fig. 3106.2. Maximum vertical dimension of marquees.

3106.5 Construction

- A marquee must be supported entirely from the building.
- A marquee must be constructed of noncombustible materials.
- A marquee's structure must be protected from deterioration.
- A marquee must meet structural requirements of the code.

Note: Chapter 16, "Structural Design," is cited as governing the structure of a marquee.

3109 Swimming Pool Enclosures and Safety Devices

3109.3 Public swimming pools

- Public swimming pools must be enclosed by a fence as shown in the illustration below:

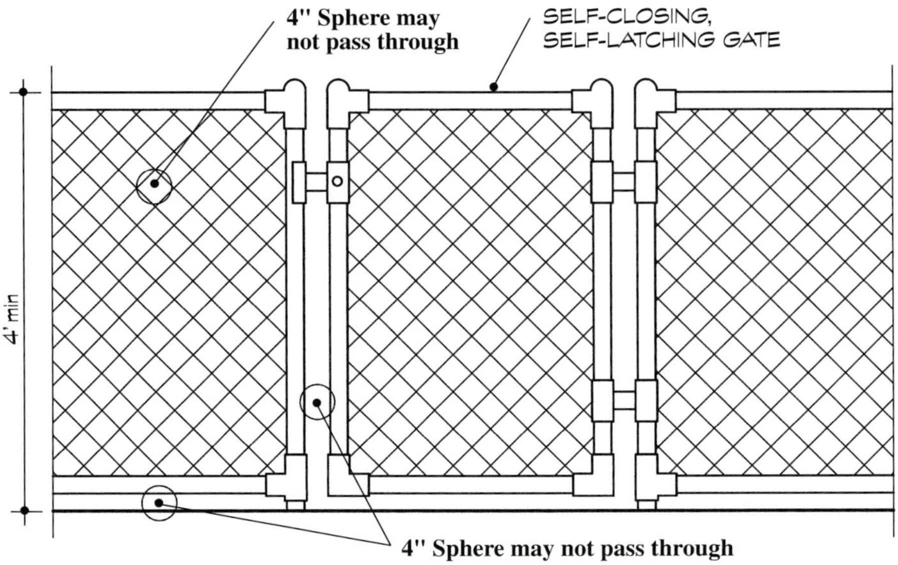

Fig. 3109.3. Barriers around public swimming pools.

3109.4.1 Barrier height and clearances

- This section addresses residential swimming pools.
- Such pools must be enclosed by a barrier as shown in the illustration below:

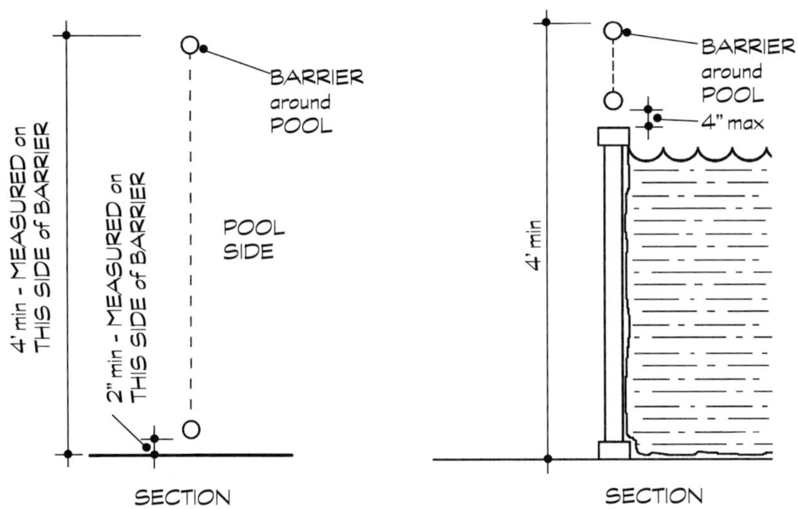

Fig. 3109.4.1. Residential swimming pool barrier height and clearances.

3109 Swimming Pool Enclosures and Safety Devices

3109.4 1.1 Openings

- This section addresses barriers enclosing residential swimming pools.
- Openings in the required barrier must be as shown in the detail below:

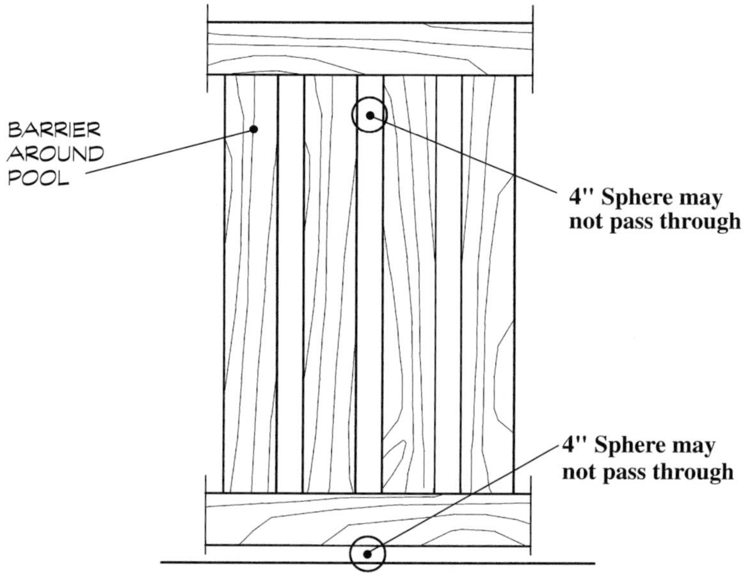

Fig. 3109.4 1.1. Openings in barriers around residential swimming pools.

3109.4.1.2 Solid barrier surfaces

- This section addresses barriers enclosing residential swimming pools.
- Where a required barrier is solid, the surface must be as shown in the illustration below:

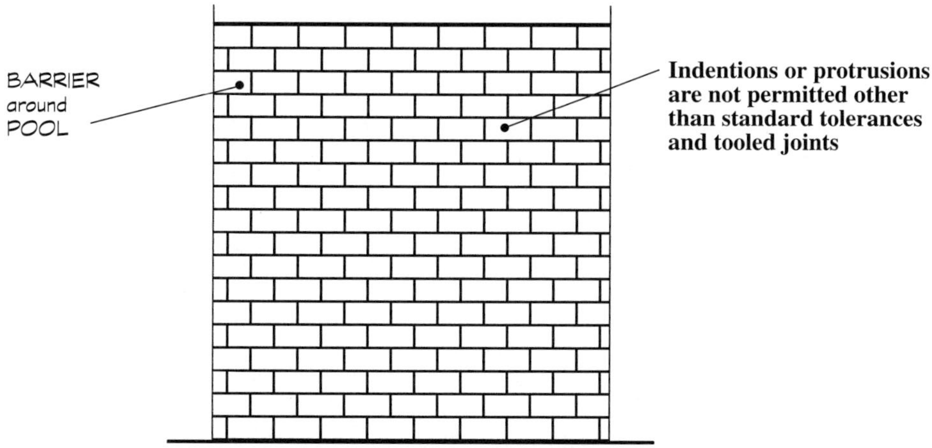

Fig. 3109.4.1.2. Solid barrier surfaces around residential swimming pools.

3109 Swimming Pool Enclosures and Safety Devices

3109.4.1.3 Closely spaced horizontal members

- This section addresses barriers enclosing residential swimming pools.
- Requirements for barriers constructed of the following are shown in the illustration below:
  - Horizontal members as follows:
    Vertical distance between tops of members is < 3'-9".
  - Vertical members.

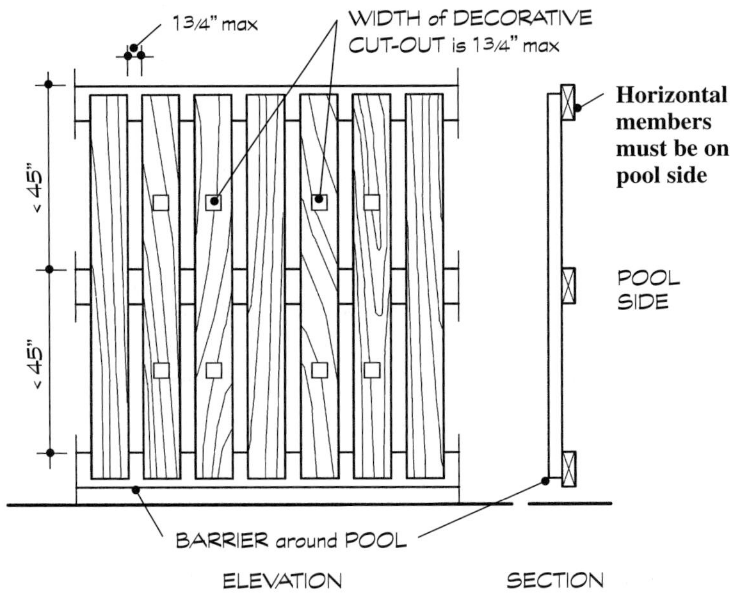

Fig. 3109.4.1.3. Closely spaced horizontal members in barriers around residential swimming pools.

3109.4.1.4 Widely spaced horizontal members

- This section addresses barriers enclosing residential swimming pools.
- Requirements for barriers constructed of the following are shown in Figure 3109.4.1.4 on the next page:
  - Horizontal members as follows:
    Vertical distance between tops of members is ≥ 3'-9".
  - Vertical members.

3109 Swimming Pool Enclosures and Safety Devices

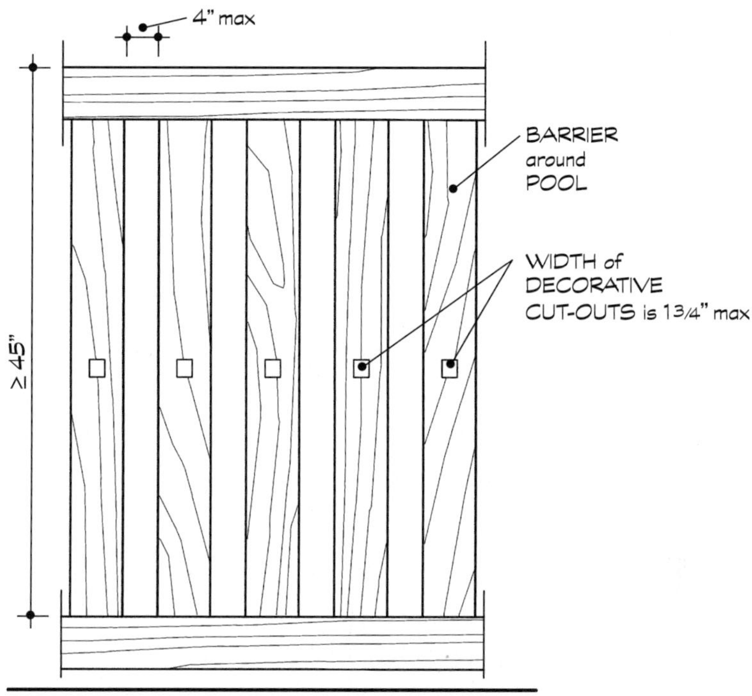

Fig. 3109.4.1.4. Widely spaced horizontal members in barriers around residential swimming pools.

3109.4.1.5 Chain link dimensions

- This section addresses barriers enclosing residential swimming pools.
- Requirements for chain link fencing are shown in the Figures 3109.4.1.5A–B below and on the next page.

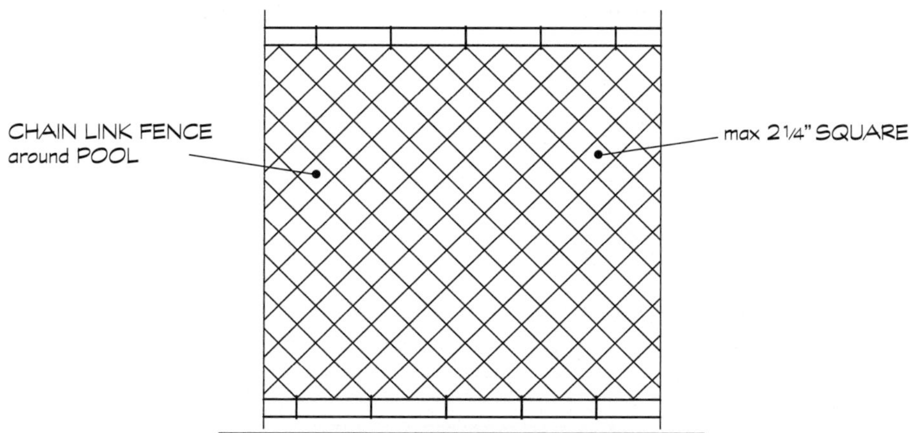

Fig. 3109.4.1.5A. Chain link dimensions in fences around residential swimming pools.

3109 Swimming Pool Enclosures and Safety Devices

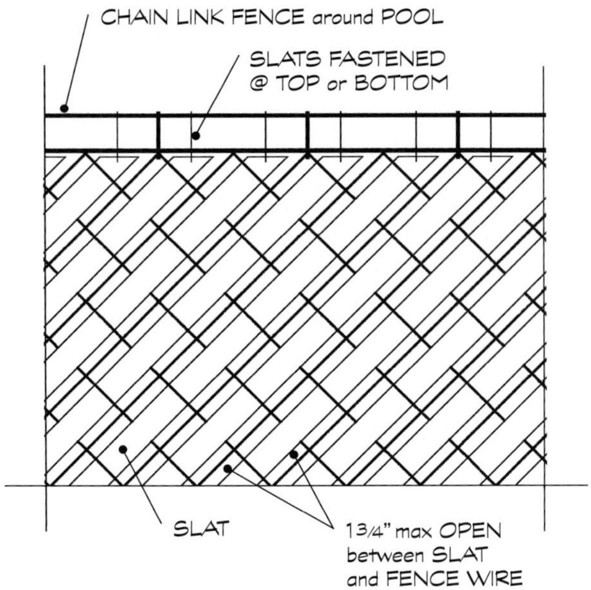

Fig. 3109.4.1.5B. Chain link dimensions at slats in fences around residential swimming pools.

3109.4.1.6 Diagonal members

- This section addresses barriers enclosing residential swimming pools.
- The limitation of openings for barriers with diagonal members is shown in the illustration below:

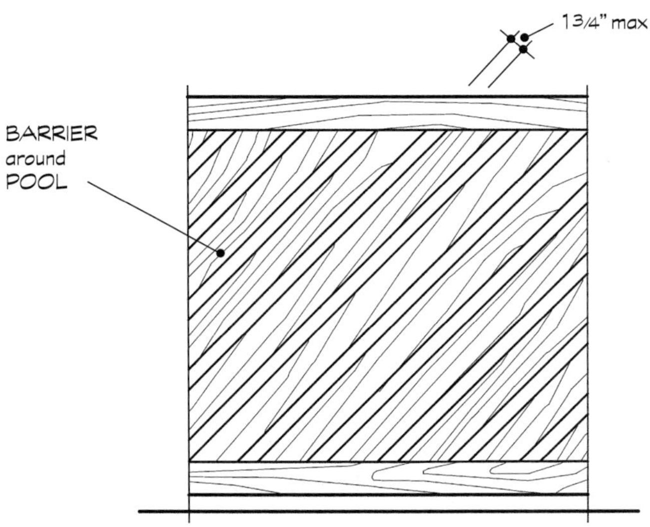

Fig. 3109.4.1.6. Diagonal members in barriers around residential swimming pools.

3109 Swimming Pool Enclosures and Safety Devices

3109.4.1.7 Gates

- Requirements for access doors or gates in barriers are shown in the illustration below.

Note: The following are cited as governing release mechanisms:
1008.1.9, "Door operations."
1109.12, "Controls, operating mechanisms and hardware."
The following are cited as governing access gates:
3109.4.1.1, "Openings."
3109.4.1.2, "Solid barrier surfaces."
3109.4.1.3, "Closely spaced horizontal members."
3109.4.1.4, "Widely spaced horizontal members."
3109.4.1.5, "Chain link dimensions."
3109.4.1.6, "Diagonal members."

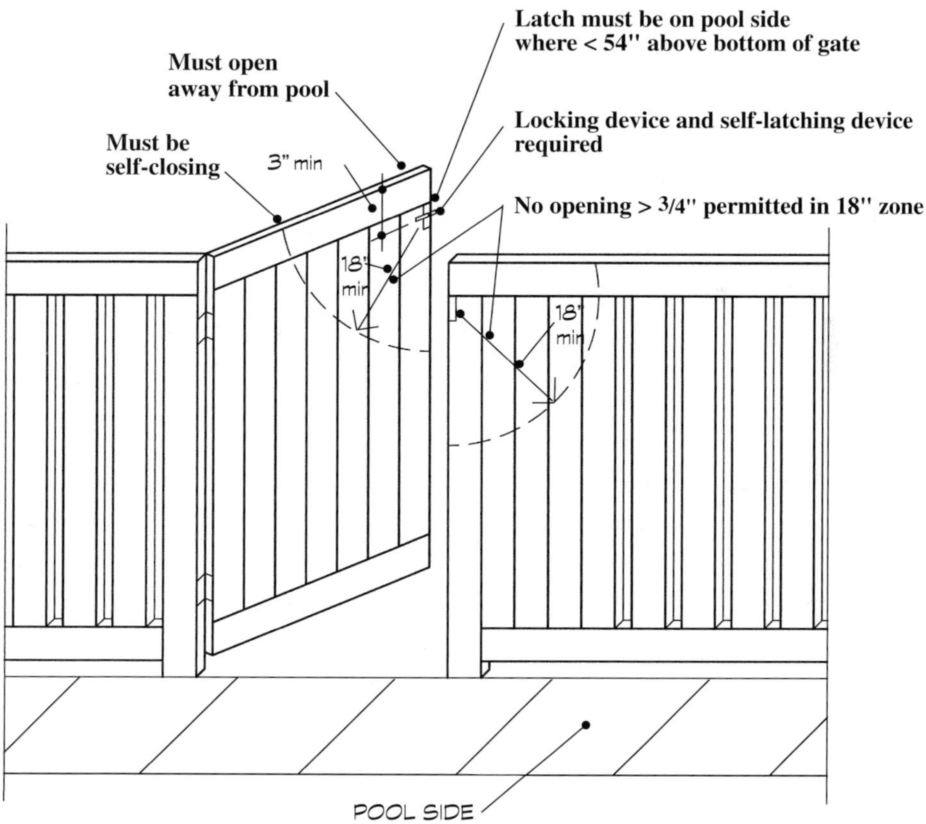

Fig. 3109.4.1.7. Gates in barriers around residential swimming pools.

NOTES

32

Encroachments into the Public Right-of-Way

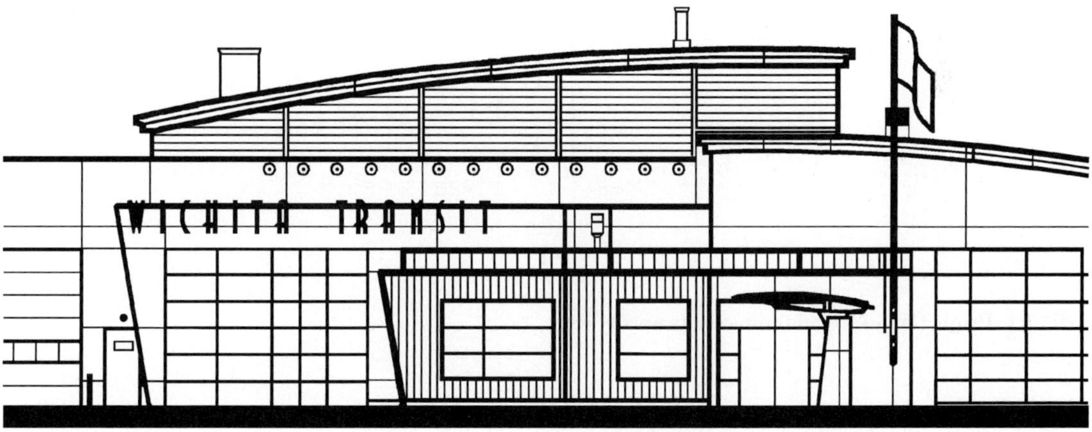

Wichita Transit Storage, Administration, and Maintenance Facility.
Wichita, Kansas. *(partial elevation)*
Wilson Darnell Mann, P.A., Architects. Wichita, Kansas.

3202 Encroachments

3202.1.1 Structural support

- The illustration below shows where a foundation may project beyond the property line.
- Otherwise, foundations may not project beyond the property line.

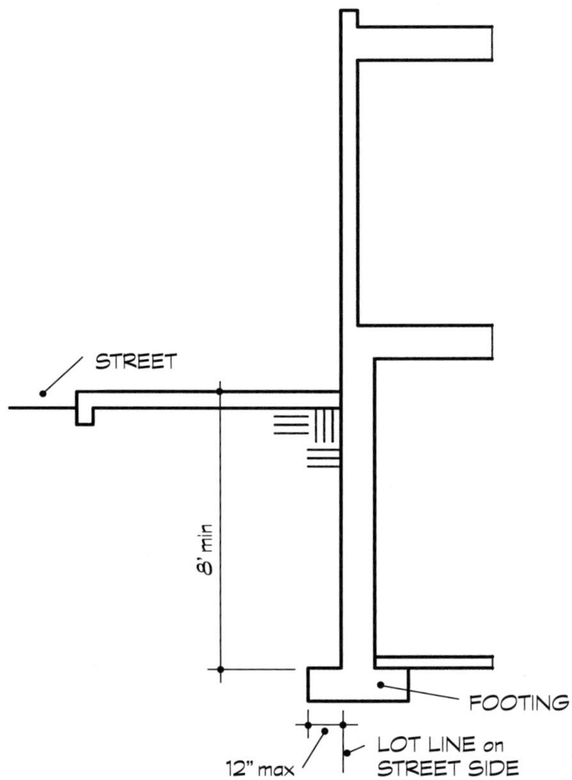

Fig. 3202.1.1. Projection of foundation over property line.

3202 Encroachments

3202.2 Encroachments above grade and below 8 feet in height

- Only the following elements may project into the public right-of-way within a height of 8':
  - Steps.
  - Architectural features.
  - Awnings.

Note: The following are cited as governing the elements that may project into the public right-of-way within a height of 8':
3202.2.1, "Steps."
3202.2.2, "Architectural features."
3202.2.3, "Awnings."

- Doors and window may not project into the public right-of-way within a height of 8'.

3202.2.1 Steps

- Requirements for steps that encroach on the public right-of-way are shown in the illustration below:

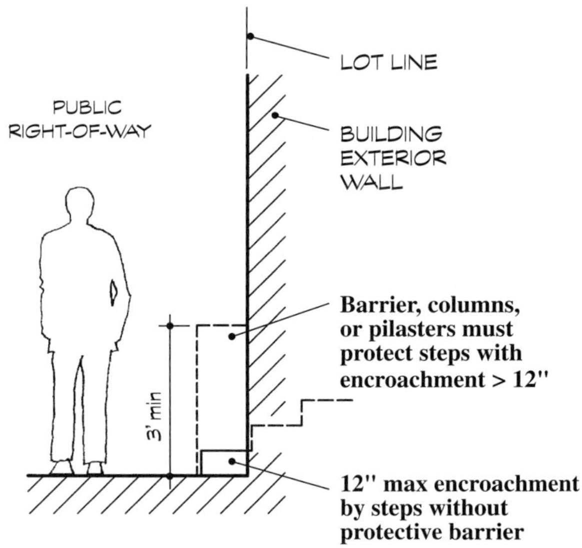

Fig. 3202.2.1. Projection of steps into the public right-of-way.

3202 Encroachments

3202.2.2 Architectural features

- Projection of the following features into the public right-of-way is limited as shown in the illustration below:
  - The following features are limited to a projection of ≤ 12":
    - Columns.
    - Pilasters.
  - The following details are limited to a projection of ≤ 4":
    - Belt courses.
    - Lintels.
    - Sills.
    - Architraves.
    - Pediments.

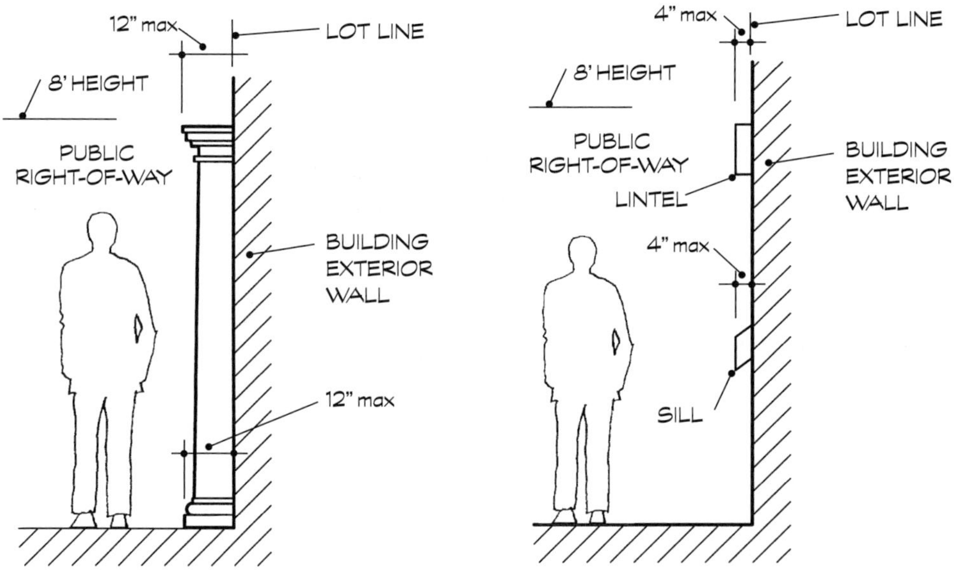

Fig. 3202.2.2. Projection of architectural features into the public right-of-way.

3202 Encroachments

3202.2.3 Awnings

- The required clearance for an awning projecting into the public right-of-way is shown in the illustration below:

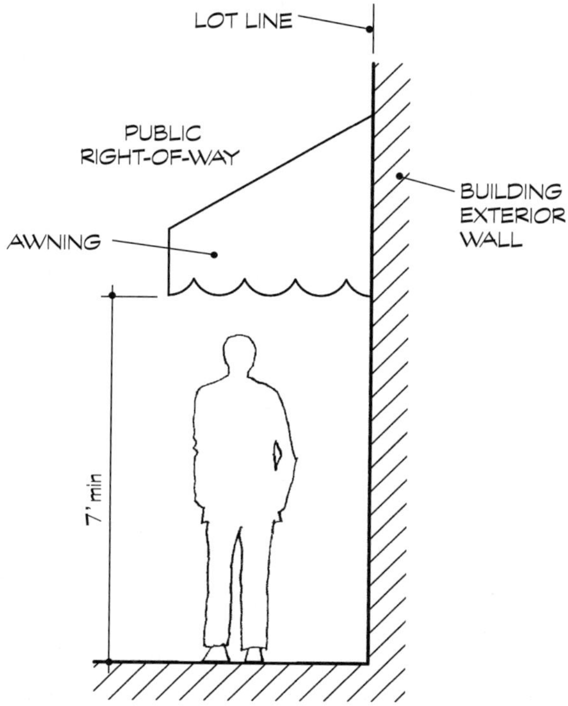

Fig. 3202.2.3. Projection of awnings into the public-right-of-way.

3202.3.1 Awnings, canopies, marquees and signs

- This section addresses encroachments into the public-right-of way above a height of 8'.
- The following elements are governed as indicated below:
  - Elements:
    - Awnings.
    - Canopies.
    - Marquees.
    - Signs.
  - Requirements:
    - Elements must meet code structural requirements.
    - Clearance requirements are shown in Figures 3203.3.1A–B on the next page.

Note: Chapter 16, "Structural Design," is cited as governing the structure of the elements listed above.

3202 Encroachments

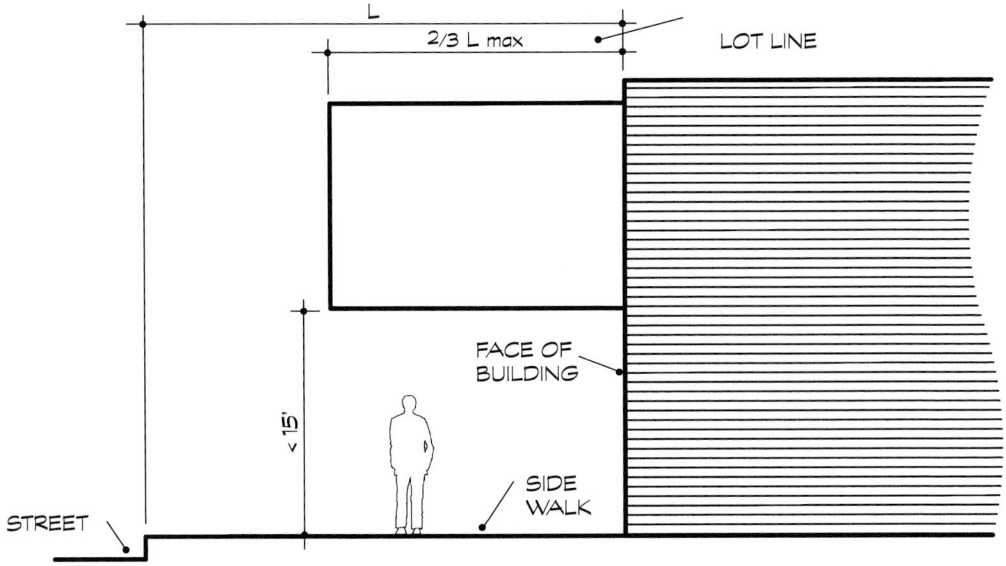

Fig. 3202.3.1A. Projection of awnings, canopies, marquees, and signs into the public right-of-way at a height ≥ 8' and < 15'.

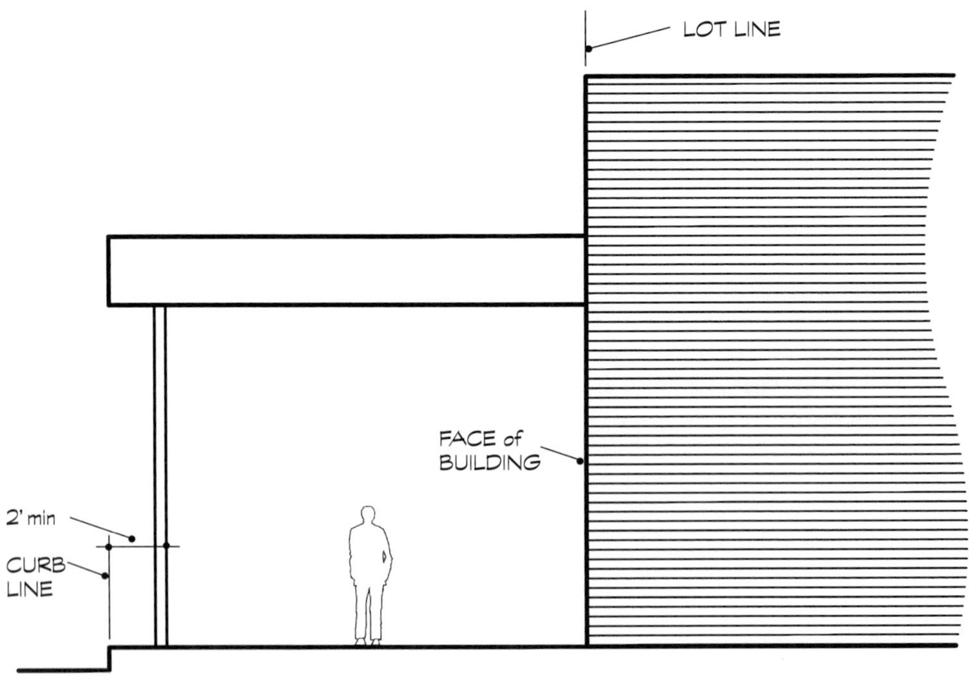

Fig. 3202.3.1B. Columns supporting awnings, canopies, marquees, and signs that project into the public right-of-way.

3202 Encroachments

3202.3.2 Windows, balconies, architectural features and mechanical equipment

- The following elements that project into the public right-of-way have the requirements listed below:
  - Elements:
    - Windows.
    - Balconies.
    - Architectural features.
    - Mechanical equipment.
  - Requirements:
    - Where located above 8', the limit of encroachment is indicated as follows:
      - In Figure 3203.3.2 on the next page.
      - In the partial table below as based on the following:
        - 1" of encroachment is permitted for each 1" of height above 8' as follows:
          - ≤ 4' of encroachment is permitted.

Table 3202.3.2 Encroachment into Public Right-of-Way Permitted: Windows, Balconies, Architectural Features, Mechanical Equipment

Vertical	Projection	Vertical	Projection	Vertical	Projection
8'-1"	0'-1"	9'-5"	1'-5"	10'-9"	2'-9"
8'-2"	0'-2"	9'-6"	1'-6"	10'-10"	2'-10"
8'-3"	0'-3"	9'-7"	1'-7"	10'-11"	2'-11"
8'-4"	0'-4"	9'-8"	1'-8"	11'-0"	3'-0"
8'-5"	0'-5"	9'-9"	1'-9"	11'-1"	3'-1"
8'-6"	0'-6"	9'-10"	1'-10"	11'-2"	3'-2"
8'-7"	0'-7"	9'-11"	1'-11"	11'-3"	3'-3"
8'-8"	0'-8"	10'-0"	2'-0"	11'-4"	3'-4"
8'-9"	0'-9"	10'-1"	2'-1"	11'-5"	3'-5"
8'-10"	0'-10"	10'-2"	2'-2"	11'-6"	3'-6"
8'-11"	0'-11"	10'-3"	2'-3"	11'-7"	3'-7"
9'-0"	1'-0"	10'-4"	2'-4"	11'-8"	3'-8"
9'-1"	1'-1"	10'-5"	2'-5"	11'-9"	3'-9"
9'-2"	1'-2"	10'-6"	2'-6"	11'-10"	3'-10"
9'-3"	1'-3"	10'-7"	2'-7"	11'-11"	3'-11"
9'-4"	1'-4"	10'-8"	2'-8"	12'-0"	4'-0"

3202 Encroachments

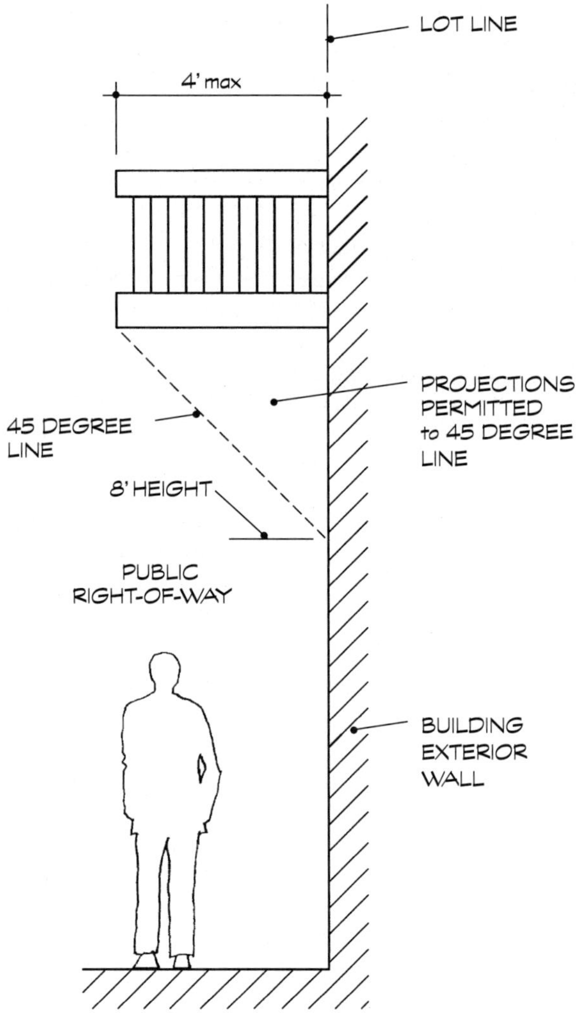

Fig. 3202.3.2. Projection of windows, balconies, architectural features, and mechanical equipment into the public right-of-way.

3202 Encroachments

3202.3.3 Encroachments 15 feet or more above grade

- Encroachments ≥ 15' above grade are governed as shown in the illustration below:

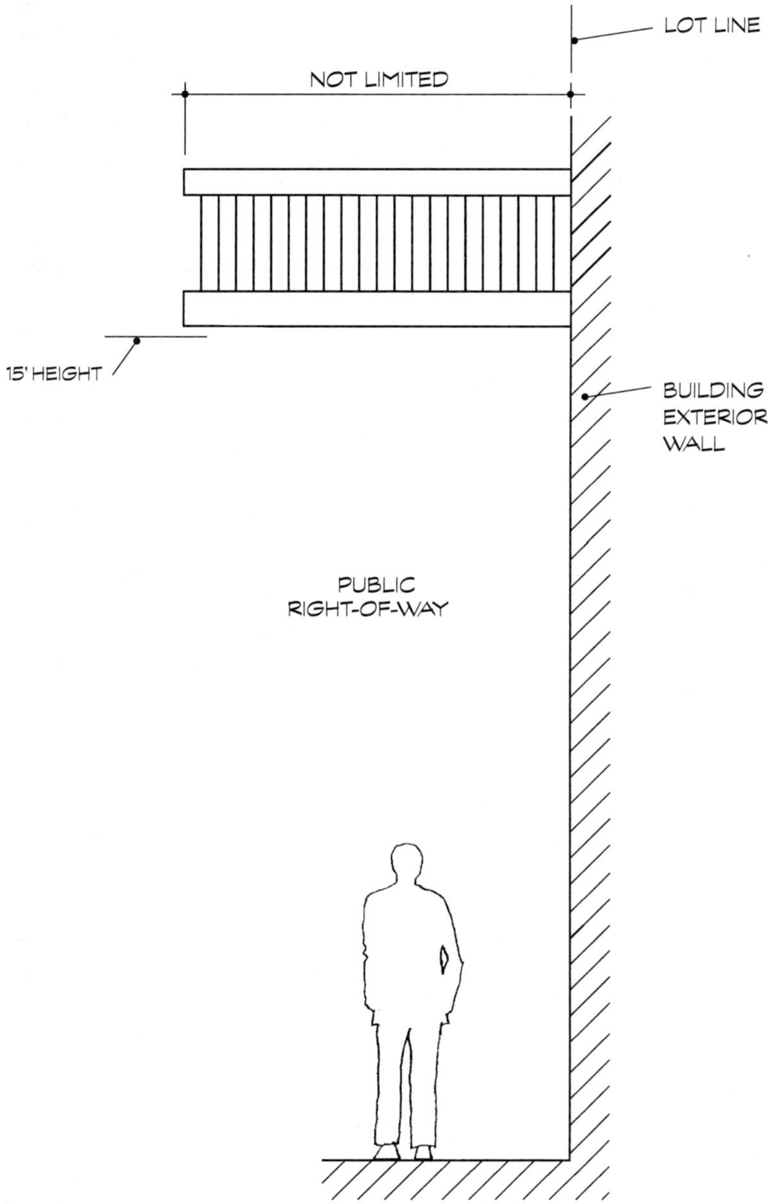

Fig. 3202.3.3. Projections into the public right-of-way ≥ 15' above grade.

3202 Encroachments

3202.3.4 Pedestrian walkways

- Pedestrian walkways over a public right-of-way are governed as follows:
  - Must be approved by the applicable governing authority.
  - Clearance required is shown in the illustration below:

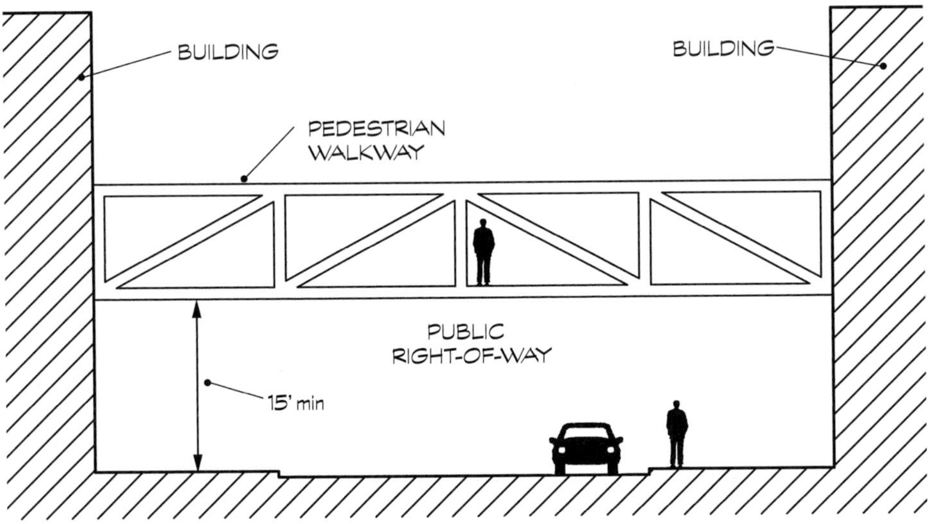

Fig. 3202.3.4 Pedestrian walkways over the public right-of-way.

3202.4 Temporary encroachments

- The following elements installed for ≤ 7 months in any calendar year have the requirements listed below:
  - Elements:
    - Vestibules.
    - Storm enclosures.
  - Requirements:
    - Must be approved by the applicable governing authority.
    - Encroachment permitted into a public right-of-way is shown as follows:
      - In the illustration provided with this section.
- Temporary entrance awnings are governed as shown in Figure 3202.4 on the next page.

3202 Encroachments

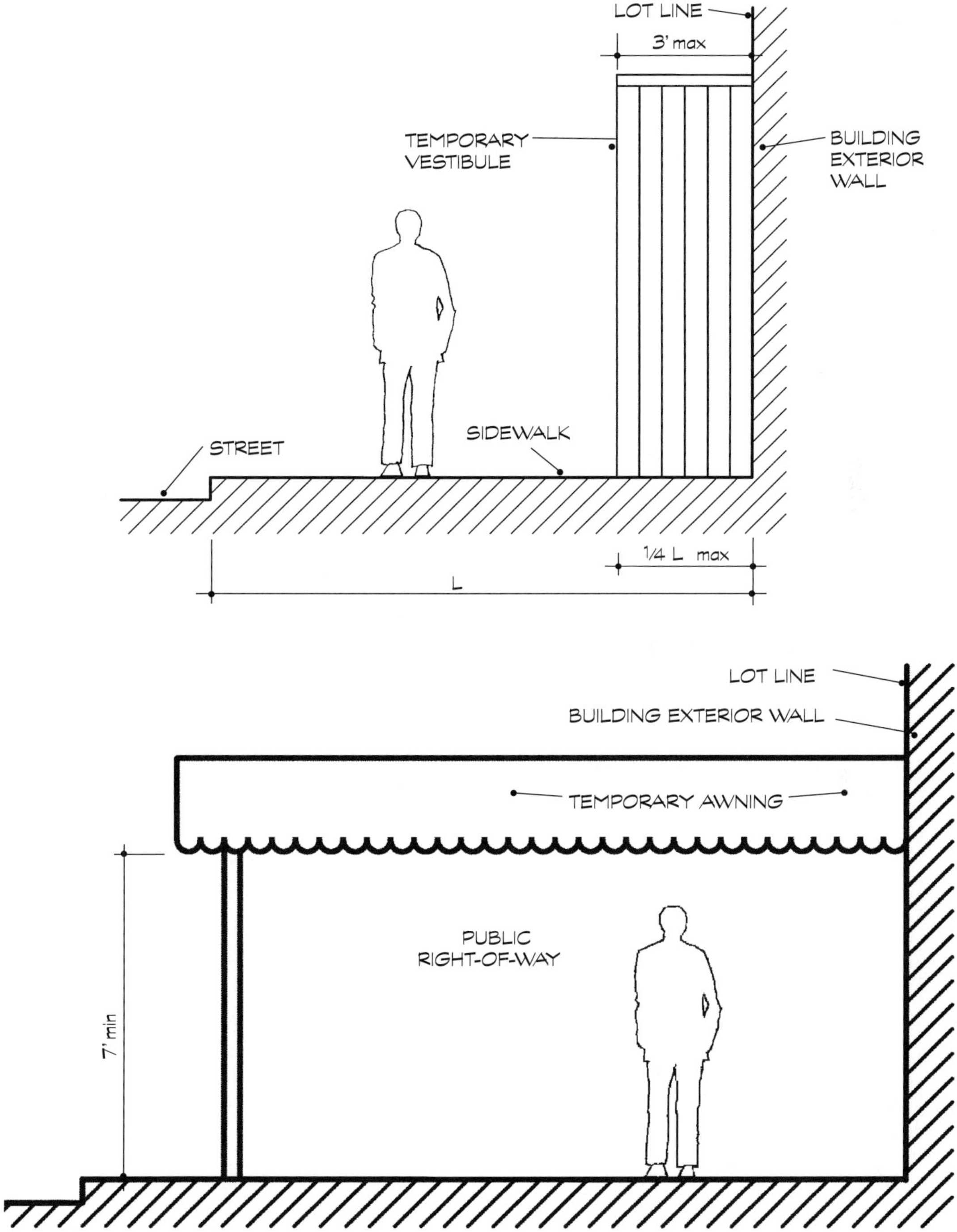

Fig. 3202.4. Temporary projections into the public right-of-way.

NOTES

33

Safeguards During Construction

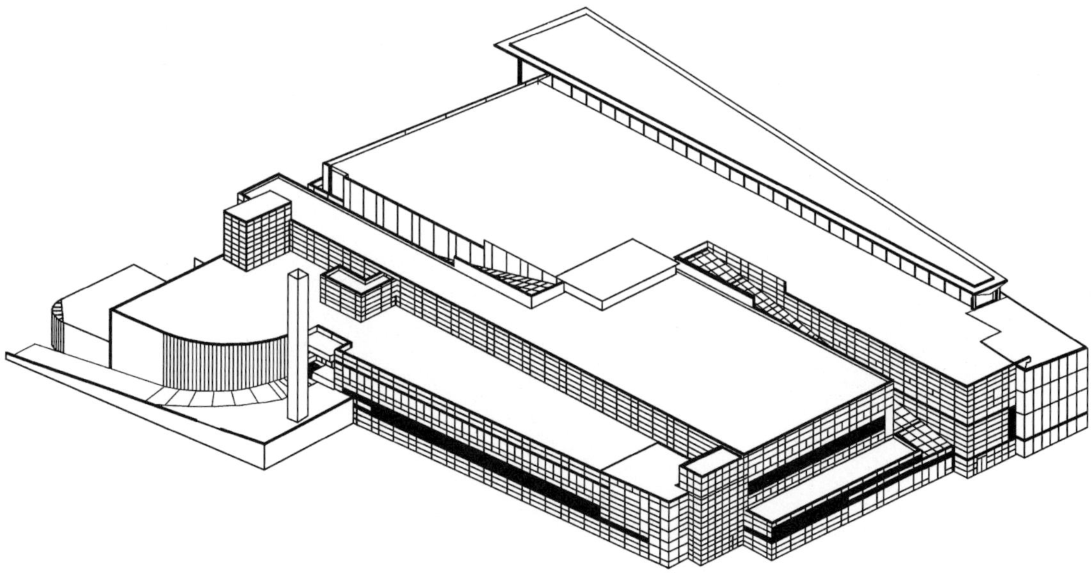

University of Connecticut New Downtown Campus at Stamford, Connecticut.
Perkins Eastman Architects, P.C. New York, New York.

3304 Site Work

3304.1 Excavation and fill

- Excavation and fill must be accommodated to protect the following:
  - Life safety.
  - Property.
- Stumps and roots must be removed as follows:
  - In the area to be occupied by the building:
    To a depth of ≥ 12" below grade.
- The following wood forms used for concrete must be removed before the building is used for any purpose:
  - Forms in the ground.
  - Forms between foundations sills and grade.
- Prior to completion of construction, the following must be removed:
  - Loose or miscellaneous wood as follows:
    Where in contact with grade under the building.

3304.1.1 Slope limits

- Permanent fill must slope as follows:
  - ≤ 1:2.
- Permanent slopes for excavated grade must comply with one of the following:
  - A gradient ≤ 1:2.
  - Other slopes are permitted where both of the following apply:
    Slope is justified by documentation from a soil investigation.
    Slope is approved by the building official.

3304.1.2 Surcharge

- Fill or other surcharge may be placed against a building only as follows:
  - Where the building is capable of resisting the additional loading.
- Where existing footings can be affected by excavation, the following applies:
  - Footings must be protected from present and future movement by one of the following:
    Underpinning.
    Otherwise protected against settlement.

34

Existing Structures

Lake Forest City Hall Renovation and Addition. Lake Forest, Illinois.
David Woodhouse Architects. Chicago, Illinois.

3401 General

3401.1 Scope

- This chapter does not govern the following elements where existing:
  - Bleachers.
  - Grandstands.
  - Folding seating.
  - Telescopic seating.

  Note: ICC 300-02, "Bleachers, Folding and Telescopic Seating, and Grandstands," is cited as governing these elements.

- This chapter governs the following for existing buildings:
  - Alterations.
  - Repairs.
  - Additions.
  - Changes of occupancy.

3401.4 Building materials

- This section governs building materials.

3401.4.1 Existing materials

- Existing materials in an existing building that comply with both of the following may remain in use:
  - They must comply with requirements at the time of their installation.
  - They must not be deemed unsafe or unhealthy by the code official.

3401.4.2 New and replacement materials

- The following materials may be used for new and replacement work:
  - Materials required or permitted by this code.
  - Materials like those being repaired or altered as follows:
    Where they cause no hazard to life, health, or property.
  - Materials that are not prohibited by the code for new construction due to being hazardous as follows:
    In a similar occupancy.
    For a similar purpose.
    In a similar location.

3404 Alterations

3404.1 General

- A stairway in an existing building being altered is governed as follows:
  - It is not required to meet requirements for stairways in new construction in the following cases:
    Where existing conditions do not permit a reduction in the following:
      Pitch.
      Slope.

Note: Section 1009, "Stairways," is cited as governing stairways in new construction.

  - Handrails otherwise required to meet construction rules are not required to comply with the following:
    Compliance with the full extention requirement is not required as follows:
      Where compliance would be hazardous due to the configuration of the plan.

Note: 1009.12, "Handrails," is cited as governing handrails in new construction.
1012.6, "Handrail extensions," is cited as governing handrail extensions in new construction.

- In certain cases, this section permits alterations not to comply with new construction requirements.

Note: 3401.4, "Building materials," is cited as another section permitting alterations not to comply with new construction requirements.

- Otherwise, alterations to existing buildings require compliance with requirements for new construction.
- After alterations, existing buildings must comply with the code as follows:
  - At least to the degree that it did before alterations.

Case study: Fig. 3404.1. The stairway is part of the existing portion of the renovation project. The winders, which would be prohibited in new construction by Section 1009, "Stairways," are permitted to remain in this stairway alteration. This is possible since the stairway space is restricted by the existing construction, which does not permit the additional steps that would be necessary to eliminate the winders.

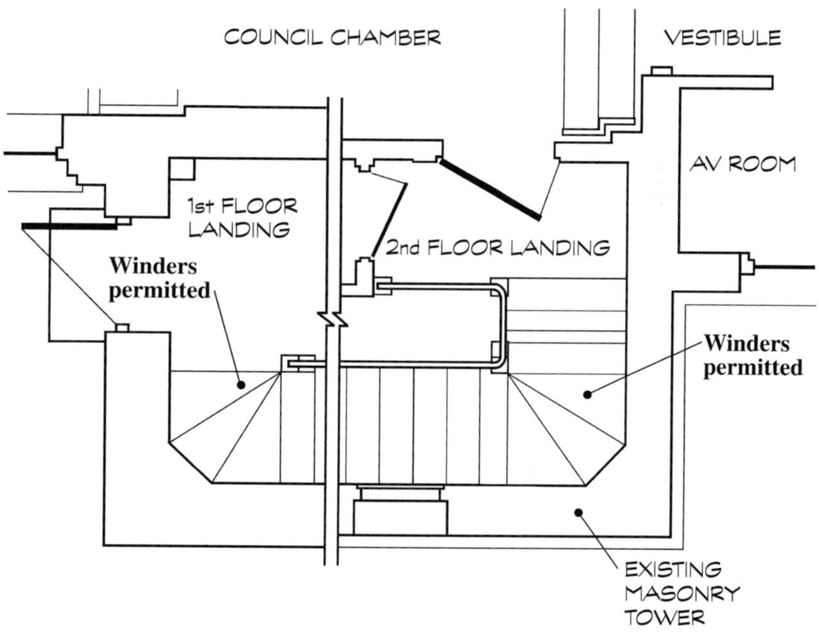

Fig. 3404.1. Partial plan at existing stairway. Lake Forest City Hall Renovation and Addition. Lake Forest, Illinois. David Woodhouse Architects. Chicago, Illinois.

3411 Accessibility for Existing Buildings

3411.8.5 Ramps

- This section addresses alterations to existing buildings.
- Where space does not permit compliance with standard ramp slopes, ramps are governed as shown in the illustrations below.

Note: 1010.2, "Slope," is cited as governing standard ramp slopes that may not be met due to space limitations.
IBC Table 3411.8.5, "Ramps," is cited as the source of requirements for ramps that cannot meet the standard slope requirements.

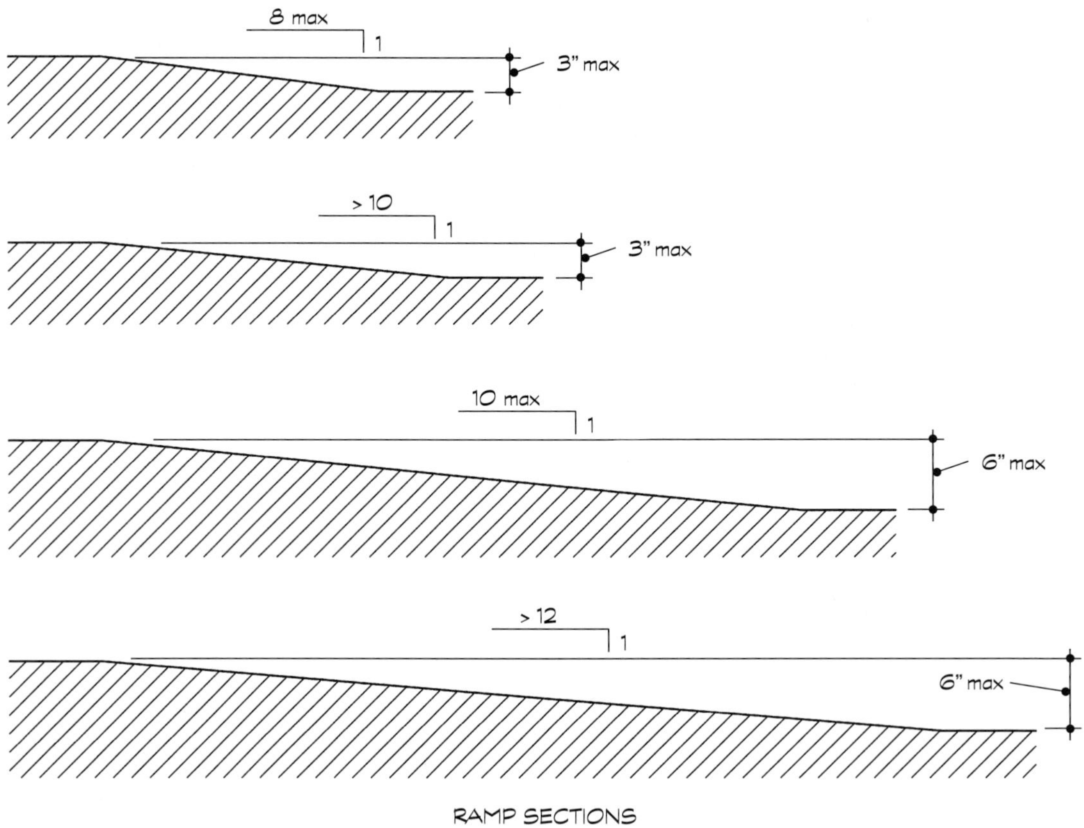

Fig. 3411.8.5. Ramps for accessibility in existing buildings.

3411 Accessibility for Existing Buildings

3411.9 Historic buildings

- This section applies to historic buildings as follows:
  - Where subject to either of the following:
    - Alterations.
    - Change of occupancy.
  - Where compliance with standard requirements for the following facilities would adversely affect historic value as indicated below:
    - Facilities:
      - Accessible routes.
      - Accessible ramps.
      - Accessible entrances.
      - Accessible toilet facilities.
    - Historic value:
      - Where approved by the authority having jurisdiction.
  - Where alternative access requirements are technically feasible.

Note: The following are cited as providing alternative requirements:
3411.9.1, "Site arrival points."
3411.9.2, "Multilevel buildings and facilities."
3411.9.3, "Entrances."
3411.9.4, "Toilet and bathing facilities."

3411.9.1 Site arrival points

- The following is an alternative access requirement for historic buildings:
  - ≥ 1 accessible route between the following locations is required:
    - An accessible site arrival point.
    - An accessible building entrance.

3411.9.2 Multilevel buildings and facilities

- The following is an alternative access requirement for historic buildings:
  - ≥ 1 accessible route between the following locations is required:
    - An accessible building entrance.
    - An accessible entrance to public spaces as follows:
      - On the same level as the accessible building entrance.

3411 Accessibility for Existing Buildings

3411.9.3 Entrances

- The following is an alternative access requirement for historic buildings:
  - ≥ 1 main entrance must be accessible as follows:
    - Where an accessible main entrance is not possible, one of the following applies:
      - A nonpublic entrance must be accessible as follows:
        - Entrance may not be locked while the building is occupied.
      - Another entrance must be accessible as follows:
        - Where the entrance is locked, one of the following is required:
          - A notification system.
          - A remote monitoring system.
  - Signs are required at the following entrances addressing accessibility.

Note: Section 1110, "Signage," is cited as governing the signs.

3411.9.4 Toilet and bathing facilities

- Where toilets are provided, the following is required:
  - ≥ 1 family or assisted-use toilet room must be accessible.

Note: 1109.2.1, "Family or assisted-use toilet and bathing rooms," is cited as governing this toilet room.

35

Referenced Standards

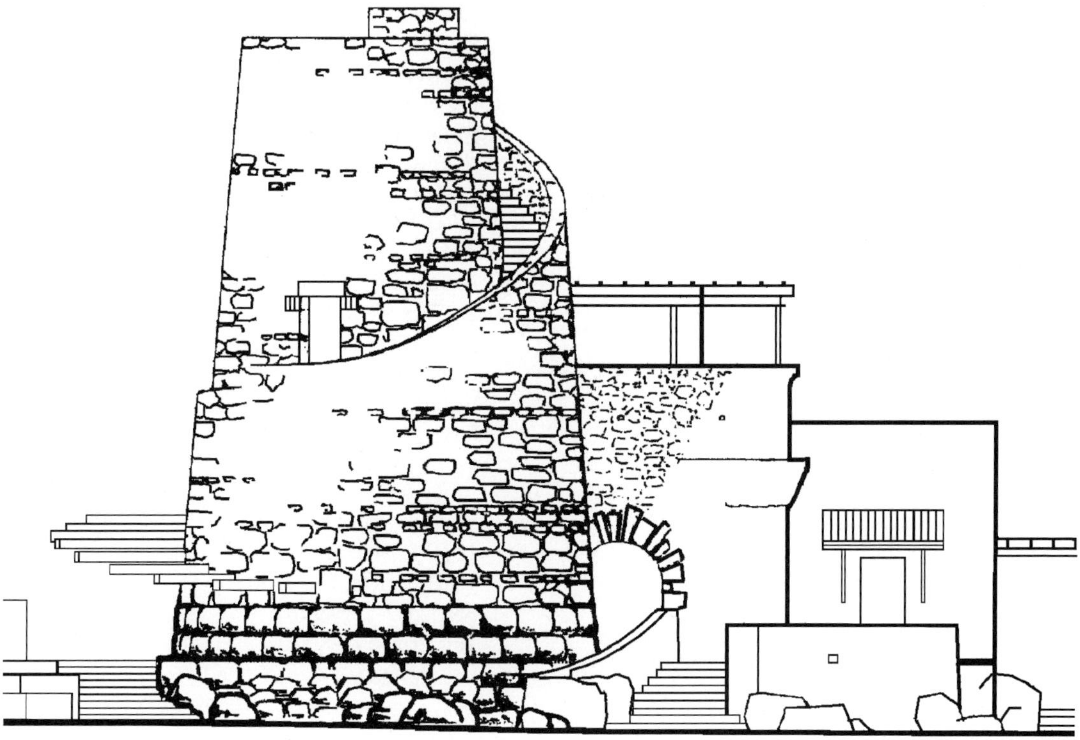

Lady Bird Johnson Wildflower Center. Austin, Texas. *(partial elevation)*
Overland Partners, Inc. San Antonio, Texas.

Referenced Standards

Listed below are agencies providing standards cited in the code including title abbreviations and contact information.

- AA
  Aluminum Association
  1525 Wilson Boulevard, Suite 600
  Arlington, VA 22209
  www.aluminum.org
  Email: slarkin@aluminum.org
  Tel: 703-358-2965

- AAMA
  American Architectural Manufacturers Association
  1827 Walden Office Square, Suite 550
  Schaumburg, IL 60173-4268
  www.aamanet.org
  Fax: 847-303-5774
  Tel: 847-303-5664

- ACI
  American Concrete Institute
  38800 Contry Club Drive
  Farmington Hills, MI 48331
  www.concrete.org
  Fax: 248-848-3701
  Tel: 248-848-3700

- AF&PA
  American Forest & Paper Association
  1111 19th St., NW, Suite 800
  Washington, DC 20036
  www.afandpa.org
  Email: info@afandpa.org
  Tel: 800-878-8878

- AISC
  American Institute of Steel Construction
  One East Wacker Drive, Suite 700
  Chicago, IL 60601-1802
  www.aisc.org
  Fax: 312-670-5403
  Tel: 312-670-2400

- AISI
  American Iron and Steel Institute
  1140 Connecticut Avenue, NW, Suite 705
  Washington, DC 20036
  www.steel.org
  Tel: 202-452-7100

- AITC
  American Institute of Timber Construction
  7012 S. Revere Parkway, Suite 140
  Centennial, CO 80112
  www.aitc-glulam.org
  Email: info@aitc-glulam.org
  Fax: 303-792-0669
  Tel: 303-792-9559

- ALI
  The Automotive Lift Institute
  P.O. Box 85
  Cortland, NY 13045
  www.autolift.org
  Email: info@autolift.org
  Fax: 607-756-0888
  Tel: 607-756-7775

- ANSI
  American National Standards Institute
  25 West 43rd Street, Fourth Floor
  New York, NY 10036
  www.ansi.org
  Email: info@ansi.org
  Fax: 212-398-0023
  Tel: 212-642-4900

  Washington DC Headquarters
  1819 L Street, NW, 6th Floor
  Washington, DC 20036
  Fax: 202-293-9287
  Tel: 202-293-8020

Referenced Standards

- APA
  APA - Engineered Wood Association
  7011 So. 19th St.
  Tacoma, WA 98466-5333
  www.apawood.org
  Fax: 253-565-7265
  Tel: 253-565-6600

- APSP
  The Association of Pool & Spa Professionals
  2111 Eisenhower Avenue, Suite 500
  Alexandria, VA 22314-4695
  www.apsp.org
  Email: memberservices@APSP.org
  Fax 703-549-0493
  Tel. 703-838-0083

- ASABE
  American Society of Agricultural
  and Biological Engineers
  2950 Niles Road
  St. Joseph, MI 49085
  www.asabe.org
  Email: hq@asabe.org
  Fax: 269-429-3852
  Tel: 269-429-0300

- ASCE/SEI
  American Society of Civil Engineers
  1801 Alexander Bell Drive
  Reston, VA 20191-4400
  www.asce.org
  Fax: 703-295-6333
  Tel: 800-548-2723

  Structural Engineering Institute
  www.seinstitute.org
  Email: sei@asce.org
  Tel: 800-548-2723

- ASME
  American Society of Mechanical Engineers
  Three Park Avenue
  New York, NY 10016-5990
  www.asme.org
  Email: infocentral@asme.org
  Fax: 973-882-1717
  Tel: 800-843-2763

- ASTM
  ASTM International
  100 Barr Harbor Drive
  P.O. Box C700
  West Conshohocken, PA 19428-2959
  www.astm.org
  Fax: 610-832-9555
  Tel: 610-832-9500

- AWCI
  Association of the Wall and Ceiling Industry
  513 West Broad Street, Suite 210
  Falls Church, VA 22046
  www.awci.org
  Fax: 703-534-8307
  Tel: 703-538-1600

- AWPA
  American Wood Protection Association
  P.O. Box 361784
  Birmingham, AL 35236-1784
  www.awpa.com
  Fax: 205-733-4075
  Tel: 205-733-4077

- AWS
  American Welding Society
  550 NW LeJeune Road
  Miami, FL 33126
  www.aws.org
  Email: info@aws.org
  Tel: 305-443-9353
  Tel: 800-443-9353

- BHMA
  Builders Hardware Manufacturers' Association
  355 Lexington Avenue, 15th Floor
  New York, NY 10017-6603
  www.buildershardware.com
  Fax: 212-370-9047
  Tel: 212-297-2122

Referenced Standards

- CGSB
  Canadian General Standards Board
  Place du Portage III, 6B1
  11 Laurier Street
  Gatineau, Quebec
  K1A 1G6
  Canada
  www.tpsgc-pwgsc.gc.ca
  Fax: 819-956-5740
  Tel: 800-665-2472

- CPA
  Composite Panel Association
  19465 Deerfield Avenue, Suite 306
  Leesburg, VA 20176
  www.pbmdf.com
  Fax: 703-724-1588
  Tel: 703-724-1128

- CPSC
  U.S. Consumer Product Safety Commission
  4330 East-West Highway
  Bethesda, MD 20814-4408
  www.cpsc.gov
  Fax: 301-504-0124
  Tel: 301-504-7923
  Tel: 800-638-2772

- CSA
  Canadian Standards Association
  5060 Spectrum Way, Suite 100
  Mississauga, Ontario
  L4W 5N6
  Canada
  www.csa.ca
  Fax: 416-747-2510
  Tel: 800-463-6727

- CSSB
  Cedar Shake and Shingle Bureau
  P.O. Box 1178
  Sumas, WA 98295-1178
  www.cedarbureau.org
  Email: info@cedarbureau.com
  Fax: 604-820-0266
  Tel: 604-820-7700

- DASMA
  Door and Access Systems Manufacturers
  Association International
  1300 Summer Avenue
  Cleveland, OH 44115-2851
  www.dasma.com
  Email: dasma@dasma.com
  Fax: 216-241-0105
  Tel: 216-241-7333

- DOC
  U.S. Department of Commerce
  National Institute of Standards
  and Technology
  100 Bureau Drive, Stop 1070
  Gaithersburg, MD 20899-1070
  www.commerce.gov
  www.nist.gov
  Email: inquiries@nist.gov
  Tel: 301-975-6478

- DOJ
  U.S. Department of Justice
  950 Pennsylvania Avenue, NW
  Civil Rights Division,
  Disability Rights Section -NYA
  Washington, DC 20530-0001
  www.usdoj.gov
  Email: askDOJ@USdoj.gov
  Tel: 202-307-2227

- DOL
  U.S. Department of Labor
  c/o Superintendent of Documents
  U.S. Government Printing Office
  Washington, DC 20402-9325
  www.gpo.gov
  Email: gpo@custhelp.com
  Fax: 202-512-2104
  Tel: 866-512-1800

- DOTn
  U.S. Department of Transportation
  1200 New Jersey Ave., SE
  Washington, DC 20590
  www.dot.gov
  Tel: 202-366-4000

Referenced Standards

- EN
  CEN
  European Committee for Standardization
  Avenue Marnix 17
  B-1000 Brussels
  www.cen.eu
  Email: infodesk@cen.eu
  Fax: + 32-2-550-08-19
  Tel: + 32-2-550-08-11

- FEMA
  Federal Emergency Management Agency
  Federal Center Plaza
  500 C Street, SW
  Washington, DC 20472
  www.fema.gov
  Tel: 202-646-2500

- FM
  Factory Mutual
  Standards Laboratories Department
  1301 Atwood Avenue
  P.O. Box 7500
  Johnston, RI 02919
  www.fmglobal.com
  Fax: 401-275-3029
  Tel: 401-275-3000

- GA
  Gypsum Association
  6525 Belcrest Road, Suite 480
  Hyattsville MD 20782
  www.gypsum.org
  Email: info@gypsum.org
  Fax: 302-277-8747
  Tel: 302-277-8686

- HPVA
  Hardwood Plywood and Veneer Association
  P.O. Box 2789
  1825 Michael Faraday Drive
  Reston, VA 20190
  www.hpva.org
  Email: hpva@hpva.org
  Fax: 703-435-2537
  Tel: 703-435-2900

- HUD
  U.S. Department of Housing
  and Urban Development
  451 7th Street S.W., Washington, DC 20410
  www.hud.gov
  Email: DC_Webmanager@hud.gov
  Fax: 202-275-6381
  Tel: 202-708-1112

- ICC
  International Code Council
  500 New Jersey Ave., NW, 6th Floor
  Washington, DC 20001-2070
  www.iccsafe.org
  Email: webmaster@iccsafe.org
  Fax: 202-783-2348
  Tel: 888-422-7233

- ISO
  International Organization for Standardization
  ISO Central Sectretariat
  1, ch. de la Voie-Creuse,
  Case postale 56
  CH-1211 Geneva 20, Switzerland
  www.iso.org
  Fax: + 41 22 733 34 30
  Tel: + 41 22 749 01 11

- NAAMM
  National Association of Architectural
  Metal Manufacturers
  800 Roosevelt Rd.
  Bldg. C, Suite 312
  Glen Ellyn, IL 60137
  Fax: 630-790-3095
  Tel: 630-942-6591

- NCMA
  National Concrete Masonry Association
  13750 Sunrise Valley Drive
  Herndon, VA 20171-4662
  www.ncma.org
  Fax: 703-713-1910
  Tel: 703-713-1900

Referenced Standards

- NFPA
  National Fire Protection Association
  1 Batterymarch Park
  Quincy, MA 02169-7471
  www.nfpa.org
  Fax: 617-770-0700
  Tel: 617-770-3000

- PCI
  Precast/Prestressed Concrete Institute
  209 W. Jackson Blvd. #500
  Chicago, IL 60606
  www.pci.org
  Tel: 312-786-0300

- PTI
  Post-Tensioning Institute
  38800 Country Club Drive
  Farmington Hills, MI 48331
  www.post-tensioning.org
  Email: miroslav.vejvoda@post-tensioning.org
  Fax: 248-848-3181
  Tel: 248-848-3180

- RMI
  Rack Manufacturers Institute
  8720 Red Oak Boulevard, Suite 201
  Charlotte, NC 28217-3992
  www.mhia.org
  Email: jnofsinger@mhia.org
  Fax: 704-676-1199
  Tel: 704-676-1190

- SDI
  Steel Deck Institute
  P.O. Box 25
  Fox River Grove, IL 60021
  www.sdi.org
  Email: steve@sdi.org
  Fax: 847-458-4648
  Tel: 847-458-4647

- SJI
  Steel Joist Institute
  196 Stonebridge Drive, Unit 1
  Myrtle Beach, SC 29588
  www.steeljoist.org
  Email: sji@steeljoist.org
  Fax: 843-293-7500
  Tel: 843-293-1995

- SPRI
  Single-Ply Roofing Institute
  411 Waverley Oaks Road, Suite 331B
  Waltham, MA 02452
  www.spri.org
  Email: info@spri.org
  Fax: 781-647-7222
  Tel: 781-647-7026

- TIA
  Telecommunications Industry Association
  2500 Wilson Boulevard
  Arlington, VA 22201-3834
  www.tiaonline.org
  Fax: 703-907-7727
  Tel: 703-907-7700

- TMS
  The Masonry Society
  3970 Broadway, Unit 201-D
  Boulder, CO 80304-1135
  www.masonrysociety.org
  Email: info@masonrysociety.org
  Fax: 303-541-9215
  Tel: 303-939-9700

- TPI
  Truss Plate Institute
  218 North Lee Street, Suite 312
  Alexandria, VA 22314
  www.tpinst.org
  Email: info@tpinst.org
  Fax: 866-501-4012
  Tel: 703-683-1010

Referenced Standards

- UL

  Underwriters Laboratories, Inc.
  2600 N.W. Lake Rd.
  Camas, WA 98607-8542
  www.ul.com
  Email: cec.us@us.ul.com
  Fax: 360-817-6278
  Tel: 877-854-3577

- ULC

  Underwriters Laboratories of Canada
  7 Underwriters Road
  Toronto ON M1R 3B4
  www.ulc.ca
  Email: customerservice@ulc.ca
  Fax: 416-757-8727
  Tel: 866-937-3852

- USC

  United States Code
  c/o Superintendent of Documents
  U.S. Government Printing Office
  Washington, DC 20402-9325
  www.gpo.gov
  Email: gpo@custhelp.com
  Fax: 202-512-2104
  Tel: 866-512-1800

- WDMA

  Window and Door Manufacturers Association
  401 N. Michigan Avenue, Suite 2200
  Chicago, Illinois 60611
  www.wdma.com
  Email: admin@wdma.com
  Fax: 312-673-6922
  Tel: 800-223-2301

- WRI

  Wire Reinforcement Institute, Inc.
  942 Main Street, Suite 300
  Hartford, CT 06103
  www.wirereinforcementinstitute.org
  Email:wwrinfo@wirereinforcementinstitute.org
  Fax: 860-808-3009
  Tel: 800-552-4974

NOTES

Appendices

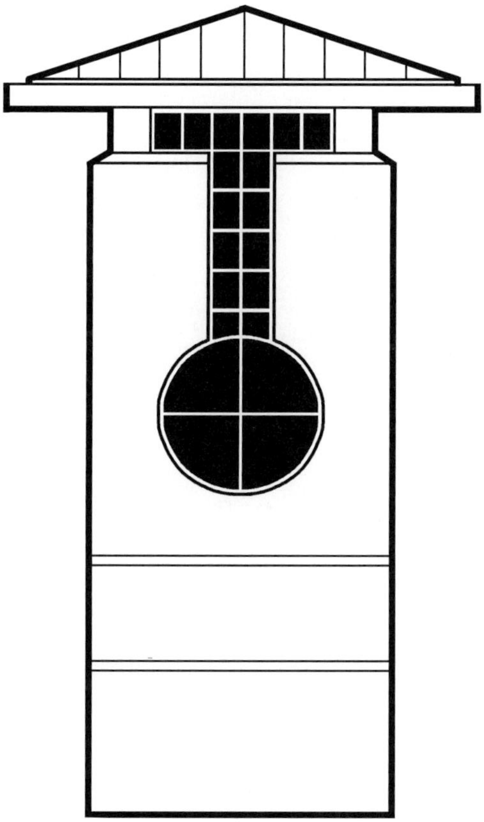

Lee's Summit Police and Court Facility. Lee's Summit, Missouri. *(partial elevation)*
The Hollis and Miller Group, Inc. Lee's Summit, Missouri.

Appendix A: Abbreviations

These abbreviated terms may appear in both upper- and lowercase forms in the text, tables, and drawings.

act	actual
admin	administration
all	allowable
bd	board
bldg	building
bm	beam
Btu	British thermal unit
cmu	concrete masonry unit
col	column
conc	concrete
conf	conference
const	construction
cu ft	cubic foot; cubic feet
cu in	cubic inch; cubic inches
df	drinking fountain
dia	diameter
dim	dimension
dr	door
ea	each
elec	electrical
elev	elevation
eq	equal
equip	equipment
exist	existing
ext	exterior
fdn	foundation
fe	fire extinguisher
fec	fire extinguisher cabinet
fin	finish
flr	floor
ft	foot; feet
ga	gage
galv	galvanized
gyp	gypsum
horiz	horizontal
ht	height
IBC	*International Building Code*
insul	insulation
int	interior
j	janitor
jan	janitor
lav	lavatory
lb	pound
lbs/ft	pounds per foot; pounds per feet
m	men
max	maximum
mech	mechanical
mezz	mezzanine
mgr	manager
min	minimum
na	not applicable
no.	number

nom	nominal
np	not permitted
oc	on center
occ	occupancy; occupants
off	office
opng	opening
psf	pounds per square foot
psi	pounds per square inch
r	radius
rec	reception
recpt	reception
ref	refrigerator
reinf	reinforcing
req'd	required
rm	room
sched	schedule
sec	secretary
sf	square foot; square feet
sht	sheet
sq cm	square centimeters
sq in	square inch; square inches
sq yd	square yard
stl	steel
sto	storage
stor	storage
struct	structural
t	toilet
typ	typical
ul	unlimited
vert	vertical
vest	vestibule
w	women
wc	water closet
wd	wood
wt	weight
wwf	welded wire fabric
wwm	welded wire mesh

Appendix B: Symbols

@ at

< less than

> greater than

≥ greater than or equal to; maximum

≤ less than or equal to; minimum

⊥ perpendicular to

|| parallel to

° degrees, temperature

° degrees, radial

÷ divide by

× multiply by

= equal to

% percent

' foot; feet

" inch; inches

Index

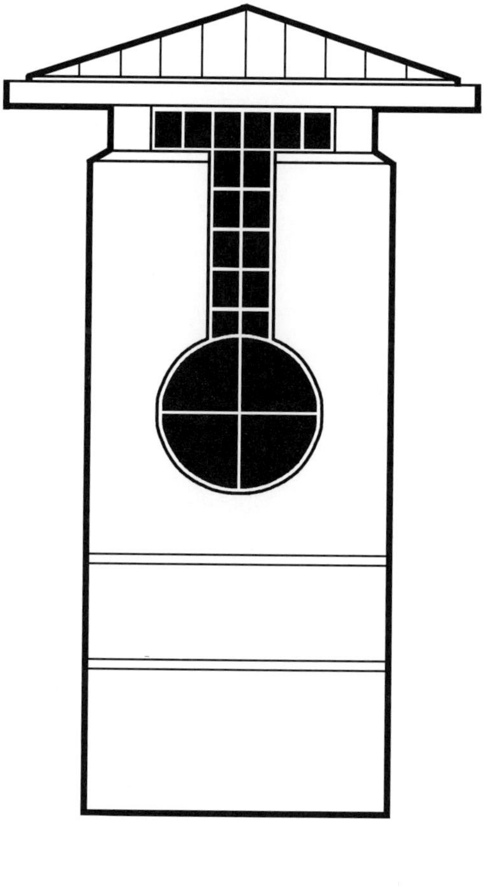

Lee's Summit Police and Court Facility. Lee's Summit, Missouri. *(partial elevation)*
The Hollis and Miller Group, Inc. Lee's Summit, Missouri.

Index

References to illustrations are printed in italic type.
References to tables are printed in bold type.